SOUTH-WESTERN

MATHMATTERS

BOOK 1

An Integrated Approach

TEACHER'S ANNOTATED EDITION

CHICHA LYNCH
Capuchino High School
San Bruno, California

EUGENE OLMSTEAD
Elmira Free Academy
Elmira, New York

SOUTH-WESTERN PUBLISHING CO.

Teacher's Edition **Contents**

[1]Complete contents of the student text appears on pages v–xii.

WHEN
MATHMATTERS

Everyone Can Learn

And math *does* matter— to us, to our students, to our country.

As teachers of mathematics, we are entrusted with the often daunting responsibility of structuring learning environments and encouraging learning, of meeting students' needs, of reaching our schools' overall educational goals, of implementing the National Council of Teachers of Mathematics <u>Curriculum and Evaluation Standards,</u> and of developing ourselves professionally. We are shaping not only the mathematical skills but also the accompanying attitudes of our nation's future leaders of business, government, academia, the media, medicine, social sciences, natural and physical sciences, and technology. Those empowered with mathematical know-how and confidence today will be the leaders of tomorrow. "Yet, for lack of mathematical power," as <u>Everybody Counts</u> says, "many of today's students are not prepared for tomorrow's jobs."[1]

Today's Students Will Need More Than "Get-By" Skills

"For the many students in secondary mathematics who are not specially talented in mathematics and not headed for careers in science or technology, current programs are a source of discouragement, anxiety, and repetition in a dull 'basic skills' program which serves them poorly."[2]

These are the students for whom **Math Matters** has been designed.

Rather then teaching elementary "get-by" skills or following the traditional—and sometimes intimidating—algebra-geometry sequence, **Math Matters** combines all of the important mathematical topics in an integrated program. The advantages of **Math Matters** are many.

Students See Mathematics As A Whole

Traditional algebra-geometry programs "tend to separate the branches of mathematics, so that students feel they are studying two distinct subjects with little connection between them..." [3]

Math Matters is organized around topics, not subjects. Each book expands the central ideas of number sense, algebra, geometry, statistics, and logic. Students see and use mathematics as a tool with which to investigate phenomena and to explore new math concepts and generalizations.

Students Develop And Use Reasoning Skills

Since the themes of mathematics are stressed throughout an integrated mathematics program, "students should be better able to reason both sequentially and holistically and be more appreciative of the interrelationships among the different parts of mathematics." [4] Every chapter in Math Matters opens with opportunities for students to use data and to work with others. Every lesson begins with an Explore opportunity. In both cases, the activities are designed to tap students' reasoning powers and engage in discovery for themselves.

Applied Mathematics And The Integrated Approach: Promoting Lifelong Math Skills

"The teaching of mathematics is shifting from emphasis on tools for future courses to greater emphasis on topics that are relevant to students' present and future needs," states *Everybody Counts*.[5] "Most mathematics should be presented in the context of its uses, with appreciation of mathematics as a deductive logical system built up slowly through the rising levels of education." [6]

Chambers in *The Agenda in Action* recommends a similar emphasis, broken down into goals for arithmetic, algebra, geometry, statistics and probability, and problem solving in an applied mathematics program.[7] Participation in an integrated program designed to follow these goals will help prepare your students for life and careers in the 21st century. Following, you'll find a synopsis of the content, goals, and student benefits of **Math Matters**.

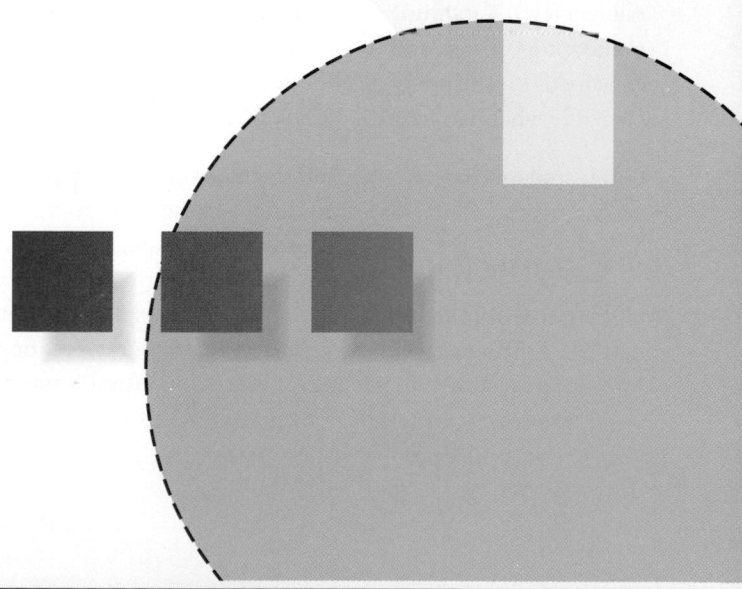

Each Branch Of Mathematics Contributes Its Own Goals

Number Sense and Technology

Thoughtful participation in **Math Matters** helps students gain understanding of numbers and their representations and learn how to use calculators to solve numerical problems. In addition, they'll see how estimation and approximation skills are essential life skills and learn to estimate the results of all computations. Indeed, the power to estimate and approximate will enable your students to judge the quality of answers found by use of calculators, computers, or pencil and paper. Ultimately, students will learn how to determine not only the best method (mental calculation, calculator, computer, or paper-and-pencil algorithm) for coming up with a precise answer, but also whether a precise answer is necessary in a given situation.

Algebra

With **Math Matters**, students will find out for themselves how algebra fits into their lives, overcoming the attitude that "I'll never use it again." Students progress from intuitive notions about arithmetic to generalizations involving variables and finally to more formal algebraic manipulations. Algebraic topics spiral throughout, so students continually develop and review their knowledge using applications in geometry, probability, and statistics (just to name a few). Graphing skills are developed, as well. Applications involving formulas, proportions, and statistics are highlighted. As students acquire stronger skills, they focus on the use of algebra in problem solving, developing their ability to think critically whatever the situation confronting them.

Geometry

Often students can't get beyond the traditional axiomatic method and proof of geometry to gain real understanding, leading them to frustration and discouragement. **Math Matters** builds your students' confidence by de-emphasizing proofs and emphasizing intuitive development of properties and relationships in real-world applications. Hands-on methods let your students discover for themselves how geometry really works, creating vivid, memorable impressions. By seeing, feeling, and experimenting, your students can achieve understanding of geometric concepts without frustration.

Statistics and Probability

We are all bombarded by statistics—statistics that can be manipulated and twisted to support just about any case the user wishes to prove. Studying statistics and probability helps students discern for themselves the value of the statistics bandied about each day in the headlines, on TV, in conversation—even in the classroom and in your textbooks! Students learn about statistics and probability in a problem-solving mode, collecting and organizing data and communicating results. Ideas about randomness, bias, and the variability of samples are developed and discussed. Indeed, the study of statistics and probability in **Math Matters** lends itself to active research and lively classroom discussion.

Problem Solving

In the real world, little calculation is done by hand anymore. Computers and calculators provide quick, reliable answers (when backed by a mental estimate and double-checked). Thus, calculators become an invaluable tool in the problem-solving setting, freeing students to focus on the process rather than on the computation. Learning how to attack problems critically and come up with the best method to solve them is a vital, everyday, on-the-job and at-home skill.

Throughout **Math Matters** students will learn to take the offensive using such skills and strategies as guess and check, make a chart and table, find a simpler problem, write a mathematical sentence, and find relevant data. While used throughout, these problem-solving skills are introduced in specific contexts. Then, the problem-solving strategies reappear in a variety of settings and in applications of mathematical content previously introduced. As students work with the material, they'll gain confidence in their abilities and see a future where math is a critical, everyday part of their lives.

MATHMATTERS

The Right Approach For Students Of The '90s

Surely, these are the mathematical accomplishments every student needs to assure his or her lifelong success. While old skills may give way to new technologies, the ability to reason, the eagerness to explore and discover, the freedom to try without fear of failure are what prepare and empower students to take charge of their futures, whatever they may hold. **Math Matters** endeavors to help your students achieve this success.

When presented with purpose and meaning, math will matter to all your students and *every one* can learn. **Math Matters** helps you lead your students to discover the purpose and power of mathematics in their lives.

Chicha Lynch   *Eugene Olmstead*

[1] National Research Council, *Everybody Counts: A Report to the Nation on the Future of Mathematics Education* (Washington, D.C., National Academy Press, 1989), 1.

[2] Donald L. Chambers and Henry S. Kepner, Jr., "Applied Mathematics: A Three-Year Program for Non-College-Bound Students," in *The Secondary School Mathematics Curriculum*, 1985 Yearbook of the NCTM, 18:211.

[3] Warren B. Manhard II, "Let's Teach Mathematics: A Case for Integrated Mathematics Programs," in *The Secondary School Mathematics Curriculum*, 1985 Yearbook of the NCTM, 16:189-90.

[4] Manhard, 190.

[5] Manhard, 190.

[6] National Research Council, 83.

[7] Donald L. Chambers, "New Directions for General Mathematics," in *The Agenda in Action*, 1983 Yearbook of the NCTM, 200-211.

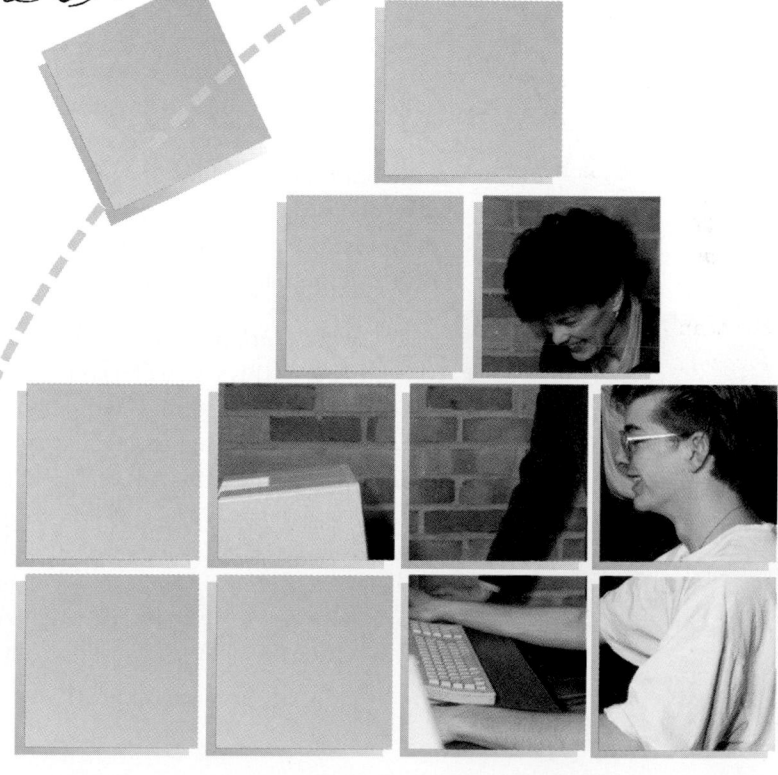

MATHMATTERS
IN YOUR CLASSROOM

Leader. Manager. Facilitator. Structurer. Resource. Coach. As mathematics teachers today, we are all of these. Our goal? To empower students, helping them to develop confidence, skills, concepts, and understanding to use math in their personal lives and in their work.

The National Council of Teachers of Mathematics has portrayed a vision for us—"a vision of mathematics power for all in a technological society; mathematics as something one does—solve problems, communicate, reason; a curriculum for all that includes a broad range of content, a variety of contexts, and deliberate connections; the learning of mathematics as an active, constructive process; instruction based on problems; evaluation as a means of improving instruction, learning, and programs."[1]

As we wrote **Math Matters**, we thought continuously of this vision of a core curriculum and research-based teaching methods. **Math Matters** offers opportunities for your students to explore mathematics through each other's eyes, to infer mathematical concepts through the use of manipulatives, to speak and to write about mathematics with confidence, to relate mathematical concepts to everyday life, and to critically examine problems and arrive at solutions.

Cooperative Learning

Teamwork: people with different backgrounds, strengths, and skills working together to solve common problems. It's great in the workplace—even better in the classroom. In a small group, students can benefit from their peers' explanations and evaluate the worth of their own ideas as well as those of others. Every chapter and every lesson of **Math Matters** offers opportunities for exactly this kind of interaction.

As you help students learn how to interact with each other for learning, we suggest a few ground rules. First and foremost, all should understand and practice respect for each other and all ideas. Every member in the group should be encouraged and expected to participate in some way. Importantly, your students should understand your goals and expectations for each group activity and know how you are going to evaluate their individual and group contributions.

Using Manipulatives

Hands-on experiments with concrete materials before attacking concepts on an abstract level allow students to get a feel for the concept or generalization, to develop some intuitive or common sense notion for what should be correct. Geometry, perhaps, first comes to mind as a natural for use of such concrete tools. Topics in algebra, probability, number sense, and measurement also lend themselves so well to letting students discover through actual experience.

Among the items you may find most helpful as you use *Math Matters* are

- Containers
- Maps and atlases
- Balances and weights
- Algebra tiles
- Spinners
- Thermometer
- Grid paper
- Games
- Reference materials
- Tangrams
- Number cubes
- Marbles

Communication, Connections, And Reasoning

It's almost impossible to discuss these concepts apart from each other. By talking and writing about mathematical activities, students reason through problems and make connections that otherwise might escape them.

In *Math Matters*, several features promote communication and reasoning. Each chapter opener includes a "Working Together" project that takes students beyond the textbook to find and use data. Features such as "Reading Math," "Check Understanding," "Talk It Over," and "Writing About Math" enhance daily lessons. And, students frequently are asked to discuss, explain, and justify their thinking in the "Extend" and "Think Critically" portions of lesson exercise sets.

Two additional features of *Math Matters*, "Math: Who, Where, When" and "Connections" help students understand the development of mathematics and its tools and relate mathematics to its uses.

Get set for a truly rewarding—and eye-opening—experience for you and your students. *Math Matters* is a remarkably innovative program that has been designed to bring out the best in your students. It helps you meet the recommendation of the *NCTM Standards* while you help your students prepare to meet the demands of mathematical literacy in the fast-paced, technology-based world of their future.

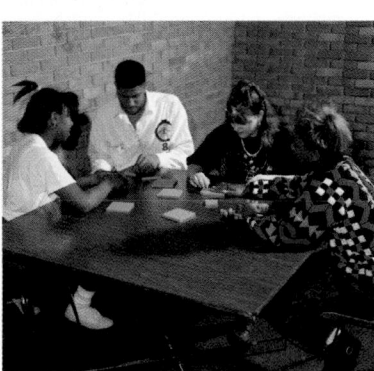

[1] National Council of Teachers of Mathematics, *Curriculum Standards for School Mathematics* (Reston, Va., 1989), 255.

Order
MATHMATTERS
Today!

Math Matters is the innovative, integrated approach to mathematics that ensures that your students will

Integrated Mathematics, *Math Matters*, And You

have the skills they need for the 21st century. They will learn number sense, geometry, algebra as well as problem solving in a realistic manner that makes sense to them. The complete package includes:

Pupil's Edition

Teacher's Annotated Edition

Student Supplement

Spanish Student Supplement

Overhead Transparencies

Assessment Options

Tests

Answer Key

Teacher's Resources

- Teacher's Annotated
- Reteaching Activities
- Enrichment Activities
- Transparency Masters
- Assessment Options
- Answer Key

MicroExam II formats for

   Apple   Macintosh

   IBM (3.5" and 5.25")

Call South-Western Publishing Co. today to order this exciting, new integrated approach to math.

1-800-543-7972

South-Western knows what works in mathematics!

Authors

Chicha Lynch	Eugene Olmstead
Capuchino High School	Elmira Free Academy
San Bruno, California	Elmira, New York

SOUTH WESTERN
PUBLISHING CO.

Strategies For Teaching

Teaching today is a very complex and challenging job. No longer can the teacher simply stand and lecture on mathematical concepts; instead, he or she needs to collaborate with students in the discovery process. In so doing, the teacher makes the classroom into a center of inquiry where students are active learners who construct their own meaning from the world around them. In this section you will find suggestions for new approaches and methods for using *Math Matters* and working with your students.

COOPERATIVE LEARNING   In real life, when a problem arises it is seldom solved by one person, whether that person is a community leader or fast-food store manager. For this reason, students need to be given opportunities to learn to work cooperatively. However, working in cooperative learning groups, like any other task, requires training and motivation.

Procedures   1. Explain to the entire class the goals of working in cooperative learning groups. For example:
   - To help all students learn and enjoy mathematics;
   - To learn to work cooperatively within a group setting;
   - To learn and expand abilities in taking responsibility, being part of a team, and being a leader.
2. Assign students to groups. This can be done randomly, alphabetically, or by some other method.
   - So that students learn to work with persons different from themselves, create heterogeneous groups where ability levels, genders, and ethnicity are mixed.
   - Consider rotating group members frequently with the goal of giving every student the chance to work with every other student in the class.
3. Establish rules, such as these:
   - Use soft voices.
   - Respect all members of the group *and* their opinions. No contributions, ideas, or opinions are to be disregarded or made fun of.
   - Each member of the group is to contribute to the activity.

Allow students to establish other rules that they feel are important.

Expectations   Have positive expectations for all groups. Set a model for students by using positive and encouraging language. Emphasize that each group is responsible for its own behavior and output. By having positive expectations, you convey to students that when they work in cooperative learning groups, the experience will be productive and enjoyable.

Assessment   Whether students are working in cooperative learning groups or individually, you should clearly state the desired outcome. If the activity is exploratory or discovery in nature, explain what students are to do, but do not give away the conclusion to be reached. Students need to understand how assessment for cooperative learning groups will be completed. Here you must decide issues such as the following:

- Will all students in the group receive the same grade?
- How will participation by *all* students be encouraged? Consider ways to set group goals or rewards that motivate students to care about helping one another learn.
- How will leadership be assessed?
- How will the work in cooperative learning groups be integrated with individual work, for assessment purposes?

Whenever possible, it is wise to allow students to participate in assessment of their work in cooperative learning groups. This can be accomplished orally or in writing. Encourage students to be honest and open when completing such assessments. If a problem is uncovered during the assessment process, you may want to have a class discussion to decide how to remedy the problem.

Helpful Suggestions
- Students have been taught to compete throughout most of their academic lives. It may take time before students function successfully in cooperative learning groups.
- Students need structure. They also need to understand all rules that they are to follow, as well as what is expected of them.
- Do not intervene unless the group asks. Try to allow the group to settle its own conflicts. If a group asks for help, try to ask probing or thought-provoking questions rather than giving outright answers.
- Encourage active participation by all members of a group. Often a student can understand a peer's explanation better than a teacher's. Encourage the group to make the best use of individual skills and abilities.
- Use cooperative learning groups for brief activities (5–15 minutes) from one to three times a week. Use cooperative learning groups for longer activities (30–60 minutes) at least once a month. In the *Math Matters* student text, each chapter opener and many Explore activities provide specific opportunities for cooperative learning. The Teacher Edition contains additional suggestions for section activities, as well as four special Integrated Units that describe long-term group projects.
- Set aside time for groups to decide how successfully students performed. Also set aside time for class discussion of advantages and disadvantages of cooperative learning groups.

HELPING STUDENTS

As a teacher committed to engaging in the discovery process with students, you often will function more as a facilitator than as an information source. You may find these hints helpful:

Act as a Model
- Talk through the steps you use to solve any exercise.
- Talk about and show alternative ways to solve exercises.
- Be sure to check the reasonableness of your answers to *all* exercises.

Question Wisely

- Do not ask questions that can be answered with a simple *yes* or *no.* The Check Understanding feature, denoted in the student text by the icon at the left, provides questions designed to evaluate students' grasp of key concepts and skills. Strive to formulate other, similar questions.
- Ask students to explain their answers and tell how they arrived at those answers.
- When asking a question, do not give away the answer.
- Encourage a variety of answers. Do not disregard or belittle any answer.
- Try to let the class decide on the correct answer. If necessary, ask another question to get the class on the right track.

Other Suggestions
- Encourage participation. Let students know that there is no such thing as a "bad question."
- Encourage organized approaches to all exercises. Point out that neatness is a helpful organizational tool.
- Encourage the use of correct mathematical language.
- Encourage alternative approaches. Have appropriate manipulatives, transparencies, and measuring instruments available.

DEVELOPING THINKING SKILLS

All students need to be encouraged to develop and apply higher-level critical thinking skills. In doing so, they must learn to question, to take risks, to verbalize ideas, to listen to others' ideas, and to critique their own and others' ideas. These critical thinking skills must be practiced regularly for growth:

Mental Reasoning
- Recall similar exercises and relate that information to new situations.
- Analyze components of a problem and apply useful information.
- Use logical reasoning and inference.
- Look for similarities, differences, and patterns.
- Synthesize ideas to create something new.
- Each four-page section of the *Math Matters* student text includes several Think Critically problems.

Spatial Reasoning
- Develop spacial perception.
- Visualize objects from different perspectives.

COMMUNICATING In order to succeed in the study of mathematics, students need to talk, listen, read, and write about mathematics.

Verbalizing

- When working with the entire class, encourage verbalizing through individual reports and explanations, group brainstorming, and discussions among all students.
- The icon at the left denotes a Talk It Over feature in the student text. It indicates that the text provides questions designed to motivate lively discussion among all students.
- For individual assignments, discuss the student's solution and thinking process with that student.
- When students work in cooperative learning groups, encourage all students to participate in discussions.
- When possible, have students follow these steps:
  Explain or restate in their own words what is known and what is to be done.
  Question how it can be done.
  Clarify what is known, what is to be done, and how it is to be done.
  Solve the exercise or problem.
  Check that the answer is reasonable.

Writing

Students learn best by being involved. To develop student proficiency at writing, create assignments that require more than just numerical answers.

- Have students keep a notebook or journal. Students are encouraged to do this in the Introduction to *Math Matters* in the student text.
- The icon shown at the left indicates that students are being asked to express and apply mathematical ideas through some written form. All students can benefit from these writing opportunities.
- You may prefer to have students use five or ten minutes per week to write an explanation of how they solved an exercise or problem that was causing them trouble.

Reading

- Students need practice in reading mathematics—both words and symbols, orally and silently. Allow time for such reading and also for discussion after the reading.
- Have all students use the Reading Mathematics feature, indicated by the icon at the left.
- Encourage use of correct vocabulary. Emphasize special mathematical meaning of ordinary words.
- Provide time for students to read and interpret charts, tables, graphs, and diagrams.
- Have students read completed vocabulary exercises and summary statements in the Chapter Review.

MAKING CONNECTIONS

Because no subject can be taught in isolation, students need to learn more than basic mathematical concepts—they need to extend their understanding to include how mathematics connects to other disciplines, to problems in society and the environment, and to diverse people from around the world. *Math Matters* has been planned to help students discover the relationship between mathematics and the real world.

Chapter Themes

High-interest chapter themes quickly involve students in the material and help to sustain their motivation. Moreover, the themes build students' awareness of the relevance of mathematics to almost every aspect of their lives. The Chapter Opener presents an extended activity relating to the theme , and the chapter theme is echoed in problems throughout the chapter. The chapter themes are as follows:

1. Health and Fitness
2. Money Around the World
3. Investigating, Solving Mysteries, Detective Work
4. Environment
5. Architecture Around the World
6. Earth Science
7. Recreation
8. Hobbies
9. Shopping
10. Road Signs and Other Signs
11. Sports
12. Astronomy
13. Forecasting and Choices
14. Patterns

Special Features

Special features (denoted by the icons at the left) in *Math Matters* can help you broaden the focus of the course:

• The Connections feature helps students link their mathematics learning to other content areas and to real-world situations.
• Math In the Workplace exposes students to the interesting and wide range of applications of mathematics in business and industry and provides examples of career choices available for those with a knowledge of mathematics.
• Math: Who, Where, When focuses on ideas from different cultures around the world and the contributions that individuals have made to mathematics.

Other Suggestions

• You may also wish to encourage students to bring items clipped from newspapers or magazines, items that illustrate interesting mathematical connections.
• Encourage students to write in their journals about mathematical ideas from their ethnic and family background.
• The Making Connections boxes in the Teacher's Annotated Edition provide additional interesting activities for students.

LEARNING STYLES

Learning styles identify how students learn and perceive what is going on around them. Students may prefer to use different perceptual modalities.
• Some students may be visual, others auditory, and others kinesthetic.
• Some students work best individually, some competitively, and others cooperatively.
• Some students prefer bright light, others dim light.
• Some students prefer warmth, others coolness.

To accommodate different styles of learning, students need to have available a variety of room settings and materials. In reality, the room setting is difficult to change. However, students who prefer bright light could be seated near windows. Students who prefer cooler temperatures could be seated farthest from the heat source. Materials can be varied by allowing students to work with reteaching and enrichment worksheets, calculators and computers, and other resource materials.

Some students can listen to a lecture and understand exactly what is being taught. Others need to read the text and do examples. Still others may need to work with manipulatives or with other students to develop understanding of concepts. For most students, direct, actual experience with opportunities for cooperation and collaboration are often best. Consulting with a friend or working in groups should be encouraged. To accommodate these different styles, teaching methods should be varied and flexible.

Students also need to recognize that they learn in different ways. One effective way to help students do this is through informal discussions. The Explore activities in the text often provide chances for discussion. Allow students to brainstorm to decide how to solve such problems. These informal discussions will allow students to discover additional strategies that they can add to their problem-solving repertoires.

Hands-on experience with concrete materials before attacking concepts on an abstract level allows students to get a feel for the concept and develop some intuitive or common sense feeling for "what should be correct." Many of the Explore Activities provide a setting where concrete materials can be used.

By using manipulatives, students are also introduced to a concept in a non-threatening way. Encourage students to describe, orally or in writing, what they did and what they learned. By using this procedure, students improve their own communication and logical reasoning skills.

Below is a suggested list of manipulative materials needed. Some you may have to purchase, but some can be handmade either by you or by the students themselves. Others you may be able to obtain from different departments—especially the science department, for measuring instruments.

Mathematical Tools	• Protractor • Compass
Measurement Tools	• Ruler—inch and centimeter • Yardstick or tape measure • Meterstick • Thermometer—preferably showing both degrees Celsius and degrees Fahrenheit • Beakers marked in cubic centimeters • Containers—from a teaspoon to a gallon • Balance scale with weights

Other Materials

The following materials can be handmade or commercially purchased. Specific manipulatives are mentioned throughout the Teacher's Annotated Edition where they are used.

- algebra tiles
- calculators, standard
- calculators, graphing
- calculators, scientific
- centimeter cubes
- clock with second hand
- coins
- containers, 3-qt and 5-qt
- crayons or markers
- dot paper
- envelopes
- erasers
- geoboards and bands
- graph paper
- grid paper
- index cards
- integer chips
   and two-color counters

- marbles
- masking tape
- mirrors
- number cubes
- paper bags
- paper clips
- paper cups
- paper, construction and plain
- posterboard
- scissors
- spinners
- straightedge
- string
- strips of paper
- straws
- tape
- tiles, one-inch (24)
- tracing paper
- wood blocks

TECHNOLOGY

When students use either calculators or computers, they must understand that the information going in must be accurate for the output to be valid. (This fact can be made vivid to students by introducing them to the expression GIGO, signifying "garbage in, garbage out.") For this reason, proficiency in estimation is essential when utilizing technology.

Calculators

Calculators allow students to concentrate on the concept or problem rather than on the computation. Students should have calculators available most of the time. As the year progresses, students should be encouraged to experiment with more advanced calculators, including graphing calculators.

When a calculator is especially helpful in learning a concept or in solving an exercise, the icon shown at the left appears in the text. The same icon also indicates features that offer tips to help students work successfully with calculators.

Computers

Students need to learn the capabilities and limitations of computers by using them. Throughout the year, students will be introduced to simple BASIC programs that should run on most computers. When possible, allow students to enter and run such programs.

As with calculators, an icon (shown at the left) is used to indicate when the use of a computer is especially appropriate.

Try to integrate different computer spreadsheet programs and simulations into your instruction.

PROBLEM SOLVING

The central focus of *Math Matters* is problem solving. The importance of being able to solve problems correctly and efficiently is emphasized throughout the text.

The 5-Step Plan

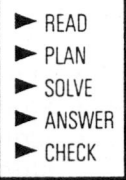

► READ
► PLAN
► SOLVE
► ANSWER
► CHECK

A key element in the coverage of problem solving is a basic strategy referred to as the 5-Step Problem Solving Plan. By utilizing this plan, you can encourage students to follow an organized approach to problem solving. Emphasize that this plan provides a framework for solving all kinds of problems.

• The steps of the plan and the identifying logo are introduced and discussed in detail as the first Problem Solving Strategy in Chapter 1, and again in Chapter 2.

• In subsequent Problem Solving sections, the steps are not highlighted individually. However, you may wish to have students identify each step as they work through the solution.

Strategies and Skills

Strategies

• The 5-Step Plan
• Find a Pattern
• Using Logical Reasoning
• Solve a Simpler Problem
• Draw a Picture
• Choose a Strategy
• Work Backward
• Make an Organized List
• Act It Out
• Guess and Check
• Simulate the Experiment
• Make a Model

Skills

• Choose an Appropriate Scale
• Estimation
• Choose the Operation
• Using Venn Diagrams
• Estimating with Fractions and Decimals
• Use a Formula
• Choose the Operation
• Writing an Equation
• Using Proportions to Solve Problems
• Choose a Computation Method
• Make a Conjecture
• Use a Formula for Perimeter Area
• Chunking
• Using Samples to Make Predictions
• Using a Grid to Solve Distance Problems

Special features in the text offer Problem Solving, Estimation, and Mental Math Tips for a variety of problems.

Decision Making

When solving a problem, students must decide which strategy will work best to solve the problem. In the student text, students are presented with several opportunities to select from among strategies they have learned.

- As any problem is discussed, emphasize the need to *think fast.* Then allow discussion of which strategy would work best. Sometimes only one strategy will work, other times any one of several strategies will lead to a solution. Sometimes, several strategies must be used together to solve a problem.
- Point out that experience in applying the strategies will help students decide which one to choose for solving a particular problem.
- Some students may continue to choose the same strategy even when the strategy is inappropriate. Encourage students to take risks and try different approaches.

Procedures

- Encourage cooperative learning groups to brainstorm to solve problems, especially those found in the Explore activities.
- Have students keep notes on how they solved various problems.
- Allow the use of calculators and computers, especially where the issue is the problem, not the computation.
- Accept alternative solutions or methods of solving problems. Allow for creativity.
- Discuss various solutions to problems. By analyzing such solutions, students can learn about, adapt, and use alternative methods.

Discussions

- Individual discussions might involve your asking probing or leading questions to help a student around a block or an obstacle. You should also discuss individual student strategies used to solve different types of problems.
- Discussions with cooperative learning groups should be similar to the individual discussions above. However, you should also use the group discussion to be sure all students are active in the process.
- Class discussions are essential. When discussing word problems in class, encourage students to share "how" they solved a problem. The Extend or Think Critically problems often can be solved in different ways and provide ideal opportunities for such discussions.
- Intuitive math students often think they "just know" how to find an answer without really realizing how they got their answer. However, it is important to follow up by asking probing questions so that these students understand their own thinking processes.

Helpful Suggestions	•	"Think out loud" to let students understand the process you use to solve problems.
	•	Many National Council of Teachers of Mathematics convention sessions focus on the use of humor in the classroom. Humor is important—use it.
	•	Encourage estimation *before* solving a problem as well as after to check for reasonableness of answers.
	•	Include regular practice in choosing strategies and skills.
	•	Use problems relevant to students. You can make up your own problems or change names, places, and data in additional examples and questions found in the Teacher's Annotated Edition side margins.
	•	Encourage students to bring other problems to class—whether from other school subjects, other people, or other sources. Allow time for solution of such problems.
	•	Have students write their own problems. Then have other students edit and critique the problems. By writing and editing problems, students become more aware of what is needed to solve problems.

Data File

Using data to solve problems and make decisions is a unifying theme emphasized throughout *Math Matters*. In the back of the student text, you will find a mini-book called the Data File. The purpose of the Data File is to provide students with information they will need to solve problems that appear throughout the student text under the heading Using Data. In addition, students must often make their own selection of appropriate data.

ERROR ANALYSIS

Errors in Reading

When reading mathematics material, a student may be unable to follow and understand what is being read.
• Encourage students to concentrate when reading. If possible, reading should be done in a quiet location.
• Encourage students to take notes as they read. Students should restate the main idea in their own words.
• Encourage students to work examples on paper as they read them in the textbook.
• Remind students to review their notes before starting the assignment.

Errors in Writing

Students often err in the use of symbols and labels.
• Encourage students to take time to think before they start writing.
• Encourage students to review their work after it is written, including carefully checking all labels and the use of all variables.
• Encourage neatness.
Additional suggestions for error analysis and remediation appear in the 5-Minute Clinics found throughout the Teacher's Annotated Edition.

	Errors in Calculation	Calculation errors are especially common in solving word problems.

- Encourage the use of estimation to check that answers are reasonable.
- Allow students to use a calculator when completing word problems or other exercises where practice of computation skills is not the main goal.
- Again, encourage neatness.

INDIVIDUAL OPTIONS

Reteaching

Teachers must adjust their instructional approaches to what students are learning. Some students will need a second or third try at learning a particular concept. You can use several different methods to accommodate different learning styles in an effort to reteach students.

- When possible, allow students to use manipulative materials. You will find suggestions for such concrete experiences throughout the Teacher's Annotated Edition.
- Use Reteaching worksheets, which often present the topic in a different way.
- Use transparencies. Allow students to work along with you, step-by-step, as you complete an exercise or solve a word problem.
- Encourage students to explain what they do or do not understand to you or to members of a cooperative learning group.
- Use peer tutoring. Often another student can communicate more effectively than a teacher when students have trouble understanding a concept.
- Have students use their math journals. (See the Introduction to the student text.) Encourage students to put an * or a ? in the margins for any concept or topic that they do not understand. If possible, have students write out questions or concerns. Then, when discussing the student's work, you can quickly see remediation needs.

Enrichment

Some students do not need additional help in understanding the lesson. These students often complete the section more quickly, as well as demonstrate above average ability, strong learning commitment, and creativity. To maintain their interest and further develop their abilities, enrich and challenge them in a variety of ways.

- Use Enrichment worksheets, which often extend the topic or introduce new concepts.
- Discuss special interests with students and help them decide on independent projects they might enjoy.
- Have students investigate math-related careers.
- Encourage students to use the journals to write down their ideas and observations about concepts, relationships, patterns, and generalizations.

- Have students start with problems in the text and extend them with "what if" questions. Pairs of students can exchange work and solve each other's problems.

Assessment   In order to know whether students are achieving the goals of mathematics education, teachers need a variety of complementary assessment options. Opportunities for assessment are found in the Pupil's and Teacher's editions as well as in Assessment Options.

- Assessment Options provides the following in blackline master form: Section Quizzes, Chapter Tests A and B, Final Exam A and B, Cumulative Tests, Journal prompts, Integrated Units, Attitude Assessment scales, and Chapter Evaluations. In addition, rubrics for evaluating open-ended assessments and guidance for building portfolios is offered for teachers.

- In the *Math Matters* program, formal assessment is readily available in the student text through Skills Previews, Mixed Reviews, Chapter Reviews, Chapter Tests, Cumulative Reviews, and Cumulative Tests. Additional Section Quizzes are offered in the Teacher's Annotated Edition. Also available in the Teacher Resource package are Chapter Tests (Forms A and B) and a MicroExam.

- Suggestions for alternative or informal, performance-based assessment activities are offered in the Teacher's Annotated Edition side columns for Chapter Reviews and Chapter Tests. Oral questioning, written reports, and manipulative activities are among the possibilities given.

- The discussions encouraged in the Chapter Openers, Explore activities, and Talk It Over features provide opportunities for the teacher to observe and assess student performance. Make notes about student attitudes, participation, reasoning ability, and understanding of mathematical concepts.

- Students' journals and responses to Writing About Math features offer another source for performance-based assessment, particularly the ability to communicate mathematical ideas and intellectual processes.

- An informal interview is another means for judging how comfortable students are with mathematics and how useful they feel their learning is in the real world.

Any situation in which the student uses mathematics is an opportunity for assessment. By employing a multiplicity of means, the teacher will obtain a more complete and accurate picture of student achievement.

Bibliography

Articles in Periodicals

Blume, Glendon W. "Enriching and Redirecting the Secondary Mathematics Curriculum with Computer-Based Materials." *The Mathematics Curriculum: Issues and Perspectives* (1987), pp. 21–24.

Brutag, Dan. "Making Your Own Rules." *Mathematics Teacher,* Vol 83 (November 1990), pp. 608–611

Cornish, E. "Educating Children for the 21st Century." *Curriculum Review,* Vol. 25 (1986), pp. 12–17.

Crowell, Sam. "A New Way of Thinking: The Challenge of the Future." *Educational Leadership,* Vol. 47 (September 1989), pp. 60–63.

Dede, Christopher. "The Evolution of Information Technology: Implications for Curriculum." *Educational Leadership,* Vol. 47 (September 1989), pp. 23–26.

Dehan, Harriet Stone. "A Mathematical Magic Show." *Mathematics Teacher,* Vol. 83 (October 1990), pp. 516–521.

Dion, Gloria. "The Graphics Calculator: A Tool for Critical Thinking." *Mathematics Teacher,* Vol. 83 (October 1990), pp. 564–571.

Dossey, John A. "Transforming Mathematics Education." *Educational Leadership,* Vol. 47 (November 1989), pp. 22–24.

Ford, Margaret I. "The Writing Process: A Strategy for Problem Solvers." *Arithmetic Teacher,* Vol. 38 (November 1990), pp. 35–38.

Frye, Shirley M. "President's Report: Maintaining Momentum As We Reach New Heights." *Journal for Research in Mathematics Education,* Vol. 21 (November 1990), pp. 427–436.

Glasser, William. "The Quality School." *Phi Delta Kappan,* Vol. 71 (February 1990), pp. 425–435.

Graves, Ted. "Non-Academic Benefits of Cooperative Learning." *Cooperative Learning,* Vol. 10 (March 1990), pp. 16–17.

Hoeffner, Karl. "Teaching Mathematics with Technology: Problem Solving with a Spreadsheet." *Arithmetic Teacher,* Vol. 38 (November 1990), pp. 52–56.

Levinson, Eliot. "Will Technology Transform Education or Will the Schools Co-opt Technology?" *Phi Delta Kappan,* Vol. 72 (October 1990), pp. 121–126.

Long, Clavin T. "On Pigeons and Problems." *Mathematics Teacher,* Vol. 81 (January 1988), pp. 28–30, 64.

McCarthy, Bernice. "Using the 4MAT System to Bring Learning Styles to Schools." *Educational Leadership,* Vol. 48 (October 1990), pp. 31–37.

Mecklenberger, James A. "Educational Technology Is Not Enough." *Phi Delta Kappan,* Vol. 72 (October 1990), pp. 105–108.

Schillow, Ned W. "Writing in Mathematics—Long Overdue." *The Mathematics Curriculum: Issues and Perspectives* (1987), pp. 1–4.

Sowell, Evelyn J. "Effects of Manipulative Materials in Mathematics Instruction." *Journal for Mathematics Education,* Vol. 20 (November 1989), pp. 498–505.

Steen, Lynn Arthur. "Teaching Mathematics for Tomorrow's World." *Educational Leadership,* Vol. 47 (September 1989), pp. 18–21.

Walberg, Herbert J. "Productive Teaching and Instruction: Assessing the Knowledge Base." *Phi Delta Kappan,* Vol. 71 (February 1990), pp. 470–478.

Books for Teachers

Ascher, Marcia. *Ethnomathematics: A Multicultural View of Mathematical Ideas.* Pacific Grove, CA: Brooks/Cole Publishing Co., 1991.

Beasley, John D. *The Mathematics of Games.* New York: Oxford University Press, 1989.

Bendick, Jeanne. *Mathematics Illustrated Dictionary: Facts, Figures and People.* Danbury, CT: Franklin Watts, 1989.

Botermans, Jack. *Paper Capers.* New York: Henry Holt & Co., 1986.

Burton, Leone, ed. *Gender and Mathematics: An International Perspective.* New York: Cassell Education, 1991.

Carson, Joan C. and Ruby N. Bostick. *Math Instruction Using Media and Modality Strengths: How . . . It Figures.* Springfield, IL: Charles C. Thomas, 1988.

Connolly, P. and T. Vilardi. *Writing to Learn Mathematics and Science.* New York: Teachers College Press, 1989.

Davidson, Neil, ed. *Cooperative Learning in Mathematics: A Handbook for Teachers.* Menlo Park, CA: Addison-Wesley Publishing Company, 1989.

Dixon, Robert. *Mathographics.* Mineola, NY: Dover Publications, 1987.

Dunham, William. *Journey through Genius.* New York: Penguin Books, 1990.

Ernest, Paul. *The Philosophy of Mathematics Education.* Bristol, PA: Falmer Press, 1991.

Fennema, E. and G.C. Leder, eds. *Mathematics and Gender.* New York: Teachers College Press, 1990.

Gay, David. *Solving Problems Using Elementary Mathematics.* New York: Dellen/Macmillan Publishing Company, 1992.

Hall, James W. *Beginning Algebra.* Boston: PWS-Kent Publishing Co., 1992

Hyde, Arthur A. and Pamela R. Hyde. *Mathwise: Teaching Mathematical Thinking and Problem Solving.* Portsmouth, NH: Heinemann Educational Books, 1991.

Kanigel, Robert. *The Man Who Knew Infinity: A Life of the Genius Ramanujan.* New York: Macmillan Publishing Co., 1991.

McKnight, C. C. et al. *The Underachieving Curriculum: Assessing U.S. School Mathematics from an International Perspective.* Champaign, IL: Stipes Publishing Co., 1987.

National Research Council. *Everybody Counts: A Report to the Nation on the Future of Mathematics Education.* Washington, DC: National Academy Press, 1989.

National Council of Teachers of Mathematics. *An Agenda for Action: Recommendations for School Mathematics of the 1980s.* Reston, VA: The Council, 1980.

———. *Curriculum and Evaluation Standards for School Mathematics.* Reston, VA: NCTM, 1989.

———. *Research Agenda for Mathematics Education.* Reston, VA: Lawrence Erlbaum Associates, 1988–89.

———. *Teaching and Learning Mathematics in the 1990s.* Reston, VA: NCTM, 1990.

———. *The Agenda in Action, 1983 Yearbook.* Reston, VA: NCTM, 1983.

National Council of Teachers of Mathematics, Commission on Standards for School Mathematics. *Curriculum and Evaluation Standards for School Mathematics.* Reston, VA: NCTM, 1989.

———. *Guidelines for the Use of Calculators in Competitions.* Reston, VA: NCTM Task Force, 1989.

Ore, Oystein. *Graphs, and Their Uses.* Washington, DC: The Mathematical Association of America, 1990.

Pappas, Theoni. *What Do You See? A Slide Show on Optical Illusions.* San Carlos, CA: Math Aids/Math Products Plus, 1985.

Paulos, J. A. *Innumeracy: Mathematical Illiteracy and Its Consequences.* New York: Hill and Wang, 1988.

Primm, David. *Speaking Mathematically: Communication in Mathematics Classrooms.* New York: Routledge, 1987.

Resnick, L. B. *Education and Learning to Think.* Washington, DC: National Academy Press, 1987.

Rucker, Rudy. *Mind Tools: The Five Levels of Mathematical Reality.* Boston: Houghton Mifflin Co., 1987.

Shane, H. *Teaching and Learning in a Microelectronic Age.* Bloomington, IN: Phi Delta Kappa, 1987.

Slavin, Robert E. *Cooperative Learning: Theory, Research, and Practice.* Englewood Cliffs, NJ: Prentice-Hall, 1990.

Stacey, Kaye with Susie Groves. *Strategies for Problem Solving: Lesson Plans for Developing Mathematical Thinking.* Victoria, Australia: Latitude Publications, 1990.

Sterrett, Andrew, ed. *Using Writing to Teach Mathematics.* Washington, DC: The Mathematical Association of America, 1990.

Books For Students

Bell, E. T. *The Last Problem.* Washington, DC: The Mathematical Association of America, 1990.

———. *Mathematics: Queen and Servant of Science.* Washington, DC: The Mathematical Association of America, 1987.

Cajori, Florian. *A History of Mathematics.* New York: Chelsea Publishing Co., 1991.

Devlin, Kevin. *Mathematics: The New Golden Age.* Washington, DC: The Mathematical Association of America, 1990.

Eves, Howard. *Great Moments in Mathematics Before 1650.* Washington, DC: The Mathematical Association of America, 1982.

———. *Great Moments in Mathematics After 1650.* Washington, DC: The Mathematical Association of America, 1982.

Gardner, Martin. *Mathematical Magic Show.* Washington, DC: The Mathematical Association of America, 1990.

———. *Riddles of the Sphinx and Other Mathematical Puzzles and Tales.* Washington, DC: The Mathematical Association of America, 1990, 1987.

Halmos, Paul R. *I Want to Be A Mathematician.* Washington, DC: The Mathematical Association of America, 1987.

Hollingdale, Stuart. *Makers of Mathematics.* Washington, DC: The Mathematical Association of America, 1989.

Honsberger, Ross. *Mathematical Morsels.* Washington, DC: The Mathematical Association of America, 1979.

———. *More Mathematical Morsels.* Washington, DC: The Mathematical Association of America, 1990.

Townsend, Charles Barry. *World's Toughest Puzzles.* New York: Sterling Publishing Co., 1991.

Software

Davidson & Associates, Torrance, CA:
Alge-Blaster Plus.

First Byte Software, Santa Ana, CA:
MathTalk Fractions.

Great Wave Software, Scotts Valley, CA:
NumberMaze Decimals & Fractions.

Key Curriculum Press, Berkeley, CA:
The Geometer's Sketchpad.

Milliken Publishing Co., St. Louis, MO:
Discovery! Experiences with Scientific Reasoning.

Minnesota Educational Computing Consortium, St. Paul, MN:
Number Munchers.

National Council of Teachers of Mathematics, Reston, VA:
Data Analysis.

Paracomp, San Francisco, CA:
Milo.

True BASIC, West Lebanon, NH:
Probability Theory.

Scholastic Inc., New York:
A.I., an Experience with Artificial Intelligence,
Algebra Shop.

Sunburst Communications, Pleasantville, NY:
Creature Cube,
Data Insights and Statistics Workshop,
Elastic Lines: The Electronic Geoboard,
Explorer Metros,
The Factory,
Gears,
The Geometric Supposer: Triangles,
Safari Search,
Semcalc,
Teasers by Tobbs with Integers,
Tobbs Learns Algebra,
What Do You Do with a Broken Calculator?

Unicorn Software, Las Vegas, NV:
Decimal Dungeon.

Ventura Educational Systems, Newbury Park, CA:
Algebra Concepts,
Geometry Concepts.

William K. Bradford Publishing Co., Acton, MA:
Algebra Xpresser,
Graph Wiz,
3D Images.

Pacing Chart

This pacing chart is intended as a guide only. Adjustments that need to be made are left to the discretion of the teacher. Most lessons can be completed in one class period. Lessons for which two days are suggested often include student involvement in cooperative learning groups, using manipulatives, or obtaining and organizing data. It is important to allow students to have such hands-on or group experiences to develop mathematical skills.

Student Text Content	Number of Class Periods	Teacher's Resource Guide			
		Reteaching	Enrichment	Technology	Transparencies
CHAPTER 1 EXPLORING DATA					
Chapter 1 Skills Preview	1				
1–1 Collecting Data	2	1–1	1–1		
1–2 Stem-and-Leaf Plots	1	1–2	1–2		TM1
1–3 Pictographs	1	1–3	1–3		TM2
1–4 Problem Solving Strategies: Introducing the 5-Step Plan	1	1–4	1–4		TM3
1–5 Problem Solving Skills: Choose an Appropriate Scale	1	1–5	1–5		
1–6 Bar Graphs	1	1–6	1–6		TM4
1–7 Problem Solving Skills: Estimation	1	1–7	1–7		
1–8 Line Graphs	1	1–8	1–8		TM5
1–9 Mean, Median, Mode, Range	1	1–9	1–9	1–9	TM6
Chapter 1 Review	1				
Chapter 1 Test	1				
Chapter 1 Cumulative Review/Test	1	Chapter 1 Tests			
CHAPTER 2 EXPLORING WHOLE NUMBERS AND DECIMALS					
Chapter 2 Skills Preview	1				
2–1 Whole Numbers	1	2–1	2–1		TM7, 8
2–2 Variables and Expressions	1	2–2	2–2		
2–3 Problem Solving Skills: Choose the Operation	1	2–3	2–3		
2–4 Adding and Subtracting Decimals	1	2–4	2–4		TM9
2–5 Multiplying and Dividing Decimals	1	2–5	2–5	TM10	
2–6 Exponents and Scientific Notation	1	2–6	2–6	2–6	
2–7 Laws of Exponents	1	2–7	2–7		
2–8 Order of Operations	1	2–8	2–8		TM11
2–9 Properties of Operations	1	2–9	2–9		TM12
2–10 Distributive Property	1	2–10	2–10		
2–11 Problem Solving Strategies: Find a Pattern	1	2–11	2–11	2–11	
Chapter 2 Review	1				
Chapter 2 Test	1				
Chapter 2 Cumulative Review/Test	1	Chapter 2 Tests			

Student Text Content	Number of Class Periods	Teacher's Resource Guide			
		Reteaching	Enrichment	Technology	Transparencies
CHAPTER 3 REASONING LOGICALLY					
Chapter 3 Skills Preview	1				
3–1 Optical Illusions	1	3–1	3–1		
3–2 Inductive Reasoning	1	3–2	3–2		
3–3 Deductive Reasoning	1	3–3	3–3		
3–4 Problem Solving Skills: Using Venn Diagrams	1	3–4	3–4		TM13
3–5 Problem Solving Strategies: Using Logical Reasoning	1	3–5	3–5		TM14
3–6 Brainteasers	2	3–6	3–6	3–6	
Chapter 3 Review	1				
Chapter 3 Test	1				
Chapter 3 Cumulative Review/Test	1	Chapter 3 Tests			
CHAPTER 4 EXPLORING FRACTIONS					
Chapter 4 Skills Preview	1				
4–1 Factors and Multiples	1	4–1	4–1		TM15
4–2 Equivalent Fractions	1	4–2	4–2		TM16
4–3 Fractions and Decimals	1	4–3	4–3		
4–4 Zero and Negative Exponents	1	4–4	4–4		
4–5 Multiplying and Dividing Fractions	1	4–5	4–5		TM17, 18, 19
4–6 Adding and Subtracting Fractions	2	4–6	4–6	4–6	
4–7 Problem Solving Skills: Estimating with Fractions and Decimals	1	4–7	4–7		
4–8 Problem Solving Strategies: Solve a Simpler Problem	1	4–8	4–8		
Chapter 4 Review	1				
Chapter 4 Test	1				
Chapter 4 Cumulative Review/Test	1	Chapter 4 Tests			
CHAPTER 5 MEASUREMENT AND GEOMETRY					
Chapter 5 Skills Preview	1				
5–1 Customary Units of Measurement	1	5–1	5–1	TM20	
5–2 Metric Units of Measurement	1	5–2	5–2	TM21	
5–3 Working with Measurements	1	5–3	5–3		
5–4 Perimeter	1	5–4	5–4	5–4	
5–5 Problem Solving Skills: Use a Formula	1	5–5	5–5		
5–6 Circumference	1	5–6	5–6	5–6	TM22
5–7 Problem Solving Strategies: Draw a Picture	1	5–7	5–7		
Chapter 5 Review	1				
Chapter 5 Test	1				
Chapter 5 Cumulative Review/Test	1	Chapter 5 Tests			

Student Text Content	Number of Class Periods	Teacher's Resource Guide			
		Reteaching	Enrichment	Technology	Transparencies
CHAPTER 6 EXPLORING INTEGERS					
Chapter 6 Skills Preview	1				
6–1 Exploring Integers	1	6–1	6–1		TM23–25
6–2 Adding Integers	1	6–2	6–2	6–2	TM26
6–3 Subtracting Integers	1	6–3	6–3		TM27
6–4 Multiplying Integers	1	6–4	6–4		TM28
6–5 Dividing Integers	1	6–5	6–5	6–5	TM29
6–6 Problem Solving Skills: Choose the Operation	1	6–6	6–6		TM23
6–7 Rational Numbers	1	6–7	6–7		
6–8 Problem Solving/Decision Making: Choose a Strategy	1	6–8	6–8		
Chapter 6 Review	1				
Chapter 6 Test	1				
Chapter 6 Cumulative Review/Test	1	Chapter 6 Tests			
CHAPTER 7 EQUATIONS AND INEQUALITIES					
Chapter 7 Skills Preview	1				
7–1 Introduction to Equations	1	7–1	7–1		
7–2 Using Addition or Subtraction to Solve an Equation	1	7–2	7–2		
7–3 Problem Solving Skills: Writing an Equation	1	7–3	7–3		
7–4 Using Multiplication or Division to Solve an Equation	1	7–4	7–4		
7–5 Using More Than One Operation to Solve an Equation	1	7–5	7–5		TM30
7–6 Using a Reciprocal to Solve an Equation	1	7–6	7–6	7–6	
7–7 Working with Formulas	1	7–7	7–7		
7–8 Problem Solving Strategies: Work Backward	1	7–8	7–8		
7–9 Combining Like Terms	1	7–9	7–9		
7–10 Graphing Open Sentences	1	7–10	7–10	7–10	TM31, 32
7–11 Solving Inequalities	2	7–11	7–11		
Chapter 7 Review	1				
Chapter 7 Test	1				
Chapter 7 Cumulative Review/Test	1	Chapter 7 Tests			
CHAPTER 8 EXPLORING RATIO AND PROPORTION					
Chapter 8 Skills Preview	1				
8–1 Ratios	2	8–1	8–1		TM33
8–2 Rates	1	8–2	8–2		
8–3 Equivalent Ratios	1	8–3	8–3		
8–4 Solving Proportions	1	8–4	8–4		

Student Text Content	Number of Class Periods	Teacher's Resource Guide			
		Reteaching	Enrichment	Technology	Transparencies
8–5 Problem Solving Skills: Using Proportions to Solve Problems	1	8–5	8–5	8–5	
8–6 Scale Drawings	1	8–6	8–6		
8–7 Problem Solving Strategies: Make an Organized List	1	8–7	8–7		
Chapter 8 Review	1				
Chapter 8 Test	1				
Chapter 8 Cumulative Review/Test	1	Chapter 8 Tests			
CHAPTER 9 EXPLORING PERCENT					
Chapter 9 Skills Preview	1				TM34
9–1 Exploring Percent	1	9–1	9–1		
9–2 Finding the Percent of a Number	1	9–2	9–2		
9–3 Finding What Percent One Number Is of Another	1	9–3	9–3		
9–4 Percent of Increase and Decrease	1	9–4	9–4		
9–5 Finding a Number When a Percent of It Is Known	1	9-5	9–5		
9–6 Problem Solving Skills: Choose a Computation Method	1	9–6	9–6		
9–7 Discount	1	9–7	9–7		
9–8 Simple Interest	1	9–8	9–8	9–8	
9–9 Problem Solving/Decision Making: Choose a Strategy	1	9–9	9–9		
Chapter 9 Review	1				
Chapter 9 Test	1				
Chapter 9 Cumulative Review/Test	1	Chapter 9 Tests			
CHAPTER 10 GEOMETRY OF SIZE AND SHAPE					
Chapter 10 Skills Preview	1				
10–1 Points, Lines, and Planes	1	10–1	10–1		TM35
10–2 Angles and Angle Measures	1	10–2	10–2		TM36
10–3 Parallel and Perpendicular Lines	1	10–3	10–3		TM37, 38
10–4 Triangles	1	10–4	10–4		TM39
10–5 Polygons	1	10–5	10–5	10–5	TM 40–45
10–6 Three-Dimensional Figures	1	10–6	10–6		TM46–53
10–7 Problem Solving Skills: Make a Conjecture	1	10–7	10–7		
10–8 Problem Solving Strategies: Act It Out	1	10–8	10–8		
10–9 Constructing and Bisecting Angles	1	10–9	10–9		TM54, 55
10–10 Constructing Line Segments and Perpendiculars	1	10–10	10–10		
Chapter 10 Review	1				
Chapter 10 Test	1				
Chapter 10 Cumulative Review/Test	1	Chapter 10 Tests			

Student Text Content	Number of Class Periods	Teacher's Resource Guide			
		Reteaching	Enrichment	Technology	Transparencies
CHAPTER 11 AREA AND VOLUME					
Chapter 11 Skills Preview	1				
11–1 Area and Volume	1	11–1	11–1		
11–2 Area of Rectangles	1	11–2	11–2		
Problem Solving Applications: Baseball Strike Zone					
11–3 Area of Parallelograms and Triangles	1	11–3	11–3		
11–4 Problem Solving Skills: Use a Formula for Perimeter or Area	1	11–4	11–4		
11–5 Squares and Square Roots	2	11–5	11–5		
11–6 The Pythagorean Theorem	1	11–6	11–6	11–6	TM56
11–7 Volume of Prisms	1	11–7	11–7		TM57
11–8 Area of Circles	1	11–8	11–8		
11–9 Volume of Cylinders	1	11–9	11–9		
Problem Solving Applications: Thunderstorms and Tornados					
11–10 Volume of Pyramids, Cones, and Spheres	1	11–10	11–10		
11–11 Surface Area	1	11–11	11–11	11–11	
11–12 Problem Solving Strategies: Guess and Check	1	11–12	11–12		
Chapter 11 Review	1				
Chapter 11 Test	1				
Chapter 11 Cumulative Review/Test	1	Chapter 11 Tests			
CHAPTER 12 POLYNOMIALS					
Chapter 12 Skills Preview	1				
12–1 Polynomials	1	12–1	12–1	TM30	
12–2 Adding and Subtracting Polynomials	1	12–2	12–2		
12–3 Multiplying Monomials	1	12–3	12–3		
12–4 Multiplying a Polynomial by a Monomial	1	12–4	12–4		
12–5 Problem Solving/Decision Making: Choose a Strategy	1	12–5	12–5		
12–6 Factoring	1	12–6	12–6	12–6	
Problem Solving Applications: Using Polynomials in Geometry					
12–7 Problem Solving Skills: Chunking	1	12–7	12–7		
12–8 Dividing by a Monomial	1	12–8	12–8	12–8	
Chapter 12 Review	1				
Chapter 12 Test	1				
Chapter 12 Cumulative Review/Test	1	Chapter 12 Tests			

Student Text Content	Number of Class Periods	Teacher's Resource Guide			
		Reteaching	Enrichment	Technology	Transparencies
CHAPTER 13 PROBABILITY					
Chapter 13 Skills Preview	1				
13–1 Probability	1	13–1	13–1		TM58
13–2 Sample Spaces and Tree Diagrams	1	13–2	13–2	13–2	
13–3 The Counting Principle	1	13–3	13–3	13–3	
Problem Solving Applications: Decision Making in Television Programming					
13–4 Independent and Dependent Events	1	13–4	13–4		
13–5 Experimental Probability	1	13–5	13–5		
13–6 Problem Solving Skills: Using Samples to Make Predictions	1	13–6	13–6		
13–7 Problem Solving Strategies: Simulate the Experiment	1	13–7	13–7		
Chapter 13 Review	1				
Chapter 13 Test	1				
Chapter 13 Cumulative Review/Test	1	Chapter 13 Tests			
CHAPTER 14 GEOMETRY OF POSITION					
Chapter 14 Skills Preview	1				
14–1 The Coordinate Plane	2	14–1	14–1		TM59
14–2 Problem Solving Skills: Using a Grid to Solve Distance Problems	1	14–2	14–2		
14–3 Graphing Linear Equations	1	14–3	14–3		TM60
14–4 Working with Slope	1	14–4	14–4		
Problem Solving Applications: Inclined Planes					
14–5 Problem Solving Strategies: Make a Model	1	14–5	14–5		
14–6 Translations	1	14–6	14–6		TM61
14–7 Reflections	1	14–7	14–7		TM62
Problem Solving Applications: Reflections in Billiards					
14–8 Rotations	1	14–8	14–8	14–8	TM63, 64
14–9 Working with Symmetry	1	14–9	14–9		
Chapter 14 Review	1				
Chapter 14 Test	1				
Chapter 14 Cumulative Review/Test	1	Chapter 14 Tests			

More About Alternative Assessment Methods

MATH MATTERS AND ASSESSMENT OPTIONS

Math Matters provides teachers with a full set of traditional and alternative assessment options. Alternative assessment options include informal and performance-based assessment activities in the Teacher's Annotated Edition, as well as journal prompts and alternative assessment worksheets in the Assessment Options package. Worksheets are provided for writing projects, attitude assessment scales, and chapter evaluations.

Using alternative, performance-based assessments as well as traditional "test and score" methods of assessments may be especially important to *Math Matters* teachers. Especially important because *Math Matters* has been designed for students who do not feel they are talented or gifted in mathematics. Indeed, many of these students may feel they cannot be successful in mathematics.

Math Matters' integrated approach, emphasis on cooperative group learning, problem solving, reasoning, real-world application of mathematical concepts, and use of data make it especially well-suited to alternative assessment methods that allow students an opportunity to communicate their understanding of mathematical thinking, ideas, and techniques.

ASSESSMENT ALTERNATIVES

Assessment alternatives include the following:

- **Interviews:** Interviews can be used frequently and consist simply of talking with students in an effort to assess their thinking—a two-way communication effort.

- **Oral Activities:** Oral questions can help you assess more fairly students who have difficulty reading.

- **Writing Samples:** Encouraging students to record their ideas can help them organize their thinking.

- **Observations:** Observing students individually and in groups furnishes valuable insights into student thinking and learning. The language used and behavior exhibited indicate understanding of relationships, levels of interest, diversity of thinking, and evolving strengths and weaknesses. Observations may be recorded in a variety of ways, including narrative descriptions, in teacher logs, or on rating sheets.

- **Student Self-Assessment:** It is important to involve students in the process of making informed assessments about their own work so that they can plan ways of improving their performance in the future. Self-assessment skills develop as students are offered opportunities to reflect on their own progress and challenged to accept responsibility for their own learning.

- **Classroom Discussion:** Teachers can use in-class discussion to assess student thinking and understanding of the "big idea."

- **Performance Tasks:** Performance tasks involve presenting students with a mathematical task or project. Teachers can observe students as they complete the task or review students' solutions to assess what students have learned and how they apply this knowledge to a variety of situations.

- **Portfolios:** A student portfolio provides a record of student progress in mathematics, allows recognition of different learning styles, and provides evidence of student development. Determining the contents of a portfolio should be a collaborative effort between the teacher and the student. Typical portfolio contents include reports or projects, drawings or constructions, analyses of problem situations, homework, computations, observations, and student self-assessments.

A portfolio should contain samples of the student's best work and of typical work. Including draft and final work will allow a student to assess her or his efforts at various stages of completion and to reflect on improvement. Portfolios allow students to learn to evaluate progress, compare past and current work, and see their development over a period of time.

WHAT TO CONSIDER WHEN SELECTING ASSESSMENT METHODS

Your choice of assessment methods should be determined by what you want to learn about student performance. For example, traditional testing may work very well for testing computation skills. However, a more open-ended assessment, such as performance task or writing assignment, can help you distinguish a student who understands mathematical concepts but has careless computation skills from a student who does not understand mathematical concepts but has good computation skills.

Other variables that may affect your choice of assessment method include:

- **Open-Ended:** Some problems may allow several answers to a problem; others may allow multiple approaches to reach a single answer.

- **Duration:** Assessment duration may vary in length according to the time needed to complete the task successfully.

- **Individual or Group Work:** Some types of assessment lend themselves more readily to individual, partner, or group work.

- **On-Demand Work, Draft Work, and Perfected Work:** Each type of work provides a different view of the student's performance. On-demand work provides limited information because it is simply a student's first-draft work. Perfected work, on the other hand, is the student's final product after being given the opportunity to revise, access multiple resources, and attain feedback.

- **Presentation and Response Modes:** Different types of responses—oral, written, demonstration, projects—should be used whenever appropriate.

- **Cross-Content and Mathematics Specific:** Often an assessment can embed mathematics in other disciplines, such as science and literature.

- **Choice:** You may wish to have students choose samples of their best work to showcase their accomplishments in a portfolio. In doing so, students demonstrate their ability to assess their work and provide you with additional information about their work.

SETTING PERFORMANCE STANDARDS

Performance-based standards—one or more statements about intended levels of student performance in mathematics—should be developed so that you can evaluate your students' mathematical work.

Performance-based standards should include desired concepts and skills needed to perform the task in such a way that students' work can be judged on many levels of performance. This, in effect, allows students to show what they know. *Performance benchmark* is a term used to describe an indicator of progress toward a performance standard.

NCTM recommends following these six steps to create performance standards:

1. Identify important "big ideas" in mathematics.
2. Identify the expected knowledge of concepts and skills, know-how, effort, and context related to the "big ideas" of mathematics.
3. Collect or create several examples of problem situations and questions about those problems that could be used to assess students' capabilities on this standard.
4. Develop new scoring procedures to judge student performance.
5. Report the results of student performance.
6. Conduct an equity review of the entire process and its consequences.

For performance-based and open-ended assessments, you can use scoring criteria, a rubric, along with some previous scored examples known as "anchors." A sample scoring rubric is shown below. Ideally each problem should have its own rubric to take into account that problem's specific characteristics. You may wish to discuss the rubric with your students and adjust it or create your own for specific situations.

**GENERIC
RUBRIC**

- **4 or 4+   Accomplishes or Exceeds the Task (Excellent):** This score indicates that the response by the student accomplishes the intended purpose. The response is clear, complete, and logically presented. It may show an accurate and understandable diagram, graph, or chart. Computations are correct. The student has used strategies and skills that will solve the problem. A 4+ indicates a perfect response. A 4 contains only minor defects that students can easily fix without written feedback or additional teaching. Answers reflect an in-depth understanding of mathematical concepts and methods.

- **3   Ready for Needed Revision (Good):** Response shows an understanding of mathematical concepts and methods, and presents appropriate explanations, examples, and arguments. Only written feedback from the teacher is needed for the student to revise the task.

- **2   Partial Success (Acceptable):** Only part of the task is completed or the response is adequate but not entirely clear or complete. Students will probably need additional teaching and discussion to revise their work.

- **1   Little Success (Incomplete):** The response is not complete, does not reflect understanding of the problem or the strategies needed to solve the problem. Lacks or has unclear explanations, diagrams, charts, or graphs.

- **0   No Response or Off Task:** The student shows no understanding of the problem or did not complete the task.

Chapter 1 Objectives

Section	
Opener	• To explore how data can be used as a basis for making real-life decisions
1–1	• Explore the nature of data and types of samples
1–2	• Read and create stem-and-leaf plots
1–3	• Read and create pictographs
1–4	• Use the 5-step plan to solve problems
1–5	• Analyze, then simplify numbers to develop the problem solving skill of choosing an appropriate scale for a graph
1–6	• Read and create bar graphs
1–7	• Use the problem solving skill, estimation, to interpret data on graphs
1–8	• Read and create line graphs
1–9	• Identify and calculate mean, median, mode, and range

Chapter 2 Objectives

Section	
Opener	• To explore how whole numbers and decimal numbers affect our understanding of world economy
2–1	• Round whole numbers to evaluate reasonableness of answers • Solve problems involving whole numbers
2–2	• Write and evaluate variable expressions
2–3	• Identify and use the operations needed to solve problems
2–4	• Find decimal sums and differences • Solve problems involving addition and subtraction of decimals
2–5	• Find decimal products and quotients • Solve problems involving multiplication and division of decimals
2–6	• Express numbers in exponential form and scientific notation • Simplify expressions with exponential form and scientific notation
2–7	• Apply the laws of exponents in multiplication, division, and raising an exponential number to a power
2–8	• Use order of operations in evaluating expressions and solving word problems
2–9	• Apply properties of operations in evaluating expressions and solving word problems
2–10	• Apply the distributive property of multiplication over addition and subtraction in evaluating expressions and solving word problems
2–11	• Use the strategy of finding a pattern to solve problems

Chapter 3 Objectives

Section	
Opener	• To explore how data can be used as a basis for making real-life decisions
3-1	• Use optical illusions to make statements and determine their truth or falsity
3-2	• Use inductive reasoning to make and test conjectures
3-3	• Write conditional statements • Identify conditional statements as true or false • Identify valid and invalid deductive arguments
3-4	• Draw Venn diagrams to solve nonroutine problems
3-5	• Make a table to solve nonroutine problems involving logic
3-6	• Solve nonroutine problems involving multiple-steps and various reasoning processes

Chapter 4 Objectives

Section	
Opener	• To explore the meaning of fractions and use data in solving real-life problems
4–1	• Find the GCF and LCM of a pair of numbers
4–2	• Find equivalent fractions and write fractions in simplest form
4–3	• Write fractions as decimals and decimals as fractions • Compare fractions and decimals
4–4	• Use zero and negative exponents and scientific notation
4–5	• Find the product or quotient of two fractions
4–6	• Find the sum of, or the difference between, two fractions
4–7	• Estimate with fractions and mixed numbers
4–8	• Use the strategy of solving a simpler problem

Chapter 5 Objectives

Section	
Opener	• To explore how data can be used as a basis for making real-life decisions
5–1	• Use customary units of measure to solve problems • Change to larger/smaller customary units
5–2	• Use metric units of measure to solve problems • Change to larger/smaller metric units
5–3	• Write customary and metric units of measure • Change units of measure when performing operations with measurements • Compare units of measure in solving problems
5–4	• Find perimeter of polygons in customary and metric units • Solve problems involving perimeter
5–5	• Solve problems using a formula for perimeter
5–6	• Find the circumference of circles • Solve problems involving circumference
5–7	• Draw pictures to solve problems

Chapter 6 Objectives

Section	
Opener	• To explore how data involving integers can be used as a basis for making real-life decisions
6–1	• Use integers to represent points on a number line • Give the absolute value of an integer • Compare and order integers
6–2	• Use a number line to find the sum of two integers • Add to find the sum of two integers
6–3	• Subtract integers • Solve problems involving the subtraction of integers
6–4	• Multiply integers • Solve problems involving the multiplication of integers
6–5	• Divide integers • Solve problems involving division of integers
6–6	• Choose the operation needed to solve a problem
6–7	• Write rational numbers as the ratio of two integers • Use a number line to compare and order rational numbers
6–8	• Choose a strategy to solve problems

Chapter 7 Objectives

Section	
Opener	• To explore how data can be used as a basis for making real-life decisions
7–1	• Determine if an equation is true, false, or open
7–2	• Use addition or subtraction to solve an equation
7–3	• Write an equation needed to solve a problem
7–4	• Use multiplication or division to solve an equation
7–5	• Solve an equation using more than one operation
7–6	• Use the reciprocal of a number to solve an equation
7–7	• Solve a formula to find the value of a specified variable
7–8	• Use the problem solving strategy of working backward to solve a problem
7–9	• Use the distributive property to combine like terms
7–10	• Graph the solution of an equation or inequality
7–11	• Solve inequalities with one variable

Chapter 8 Objectives

Section	
Opener	• To explore how rates and ratios can be used to determine real-life statistics
8–1	• To read and write ratios
8–2	• To read and write rates • To simplify rates
8–3	• To find a ratio or a group of ratios equivalent to a given ratio • To write a ratio in lowest terms
8–4	• To write and solve proportions
8–5	• To solve problems by writing and solving proportions
8–6	• To use and interpret scale drawings
8–7	• To solve problems by making an organized list

Chapter 9 Objectives

Section	
Opener	• To explore how data involving percents can be used as a basis for making real-life decisions
9–1	• Write a percent as a decimal or fraction • Write a decimal or fraction as a percent
9–2	• Find the percent of a number using an equation or proportion • Solve problems involving finding the percent of a number
9–3	• Find the percent one number is of another number
9–4	• Find the percent of increase or decrease • Solve problems involving percent of increase or decrease
9–5	• Find the whole amount when given a percent of it • Solve problems involving finding the whole amount when given a percent of it
9–6	• Choose a computation method to solve a problem
9–7	• Find the discount and sale price of items • Solve problems involving discount and sale price
9–8	• Solve problems involving simple interest
9–9	• Choose a strategy to solve problems

Chapter 10 Objectives

Section	
Opener	• To explore how geometry can be used as a basis for acquiring information needed in real-life decisions
10–1	• To identify basic geometric figures
10–2	• Use a protractor to measure and draw angles • Classify angles
10–3	• Identify parallel lines and perpendicular lines • Recognize vertical lines • Classify the angles formed by a transversal
10–4	• Classify triangles by the measures of the angles or by the lengths of their sides
10–5	• To find the sum of the angles of a polygon • To identify types of quadrilaterals
10–6	• To define and identify types of polyhedra and other three-dimensional figures
10–7	• Use the strategy of making a conjecture to solve problems
10–8	• Use the strategy of acting it out to solve problems
10–9	• Construct a copy of a given angle • Construct an angle bisector
10–10	• Construct a copy of a given line segment • Construct the perpendicular bisector of a line segment

Chapter 11 Objectives

Section	
Opener	• To explore how geometry can be used as a basis for acquiring information needed in real-life situations
11-1	• Use patterns to find the number of square units or cubic units in a figure • Choose appropriate units to measure area or volume
11-2	• Use the area formula to find the areas of rectangular figures and irregular figures
11-3	• Find the area of a parallelogram • Find the area of a triangle
11-4	• Use the strategy of choosing the appropriate formula to solve problems involving perimeter or area
11-5	• Calculate squares and square roots
11-6	• Apply the Pythagorean Theorem
11-7	• Find the volume of a prism
11-8	• Find the area of a circle
11-9	• Find the volume of a cylinder
11-10	• Find the volume of a pyramid, cone, and a sphere
11-11	• Find the surface area of a prism, a cylinder, a cone, and a sphere
11-12	• Use the strategy of guess and check to solve problems

Chapter 12 Objectives

Section	
Opener	• To explore how data can be used as a basis for making real-life decisions
12–1	• Write polynomials in standard form • Simplify polynomials
12–2	• Add and subtract polynomials
12–3	• Multiply monomials
12–4	• Multiply a polynomial by a monomial
12–5	• Choose a strategy to solve problems
12–6	• Factor polynomials using the greatest common monomial factor
12–7	• Use chunking to factor polynomials and to solve problems
12–8	• Divide a monomial by a polynomial • Divide a polynomial by a monomial

Chapter 13 Objectives

Section	
Opener	• To explore how data can be used as a basis for making real-life decisions
13–1	• Find the probability of an event
13–2	• Use a tree diagram to find all the possible outcomes in a sample space
13–3	• Use the counting principle to find the number of outcomes in a sample space
13–4	• Find the probabilities of independent and dependent events
13–5	• Find experimental probability
13–6	• Use samples to make predictions
13–7	• Use simulations to estimate probabilities

Chapter 14 Objectives

Section	
Opener	• To explore how patterns can be used to represent numbers
14–1	• Identify points and graph ordered pairs on coordinate planes
14–2	• To solve distance problems using a grid
14–3	• Find and use ordered pairs to graph linear equations
14–4	• Find the slopes of the lines • Use slopes to write equations
14–5	• To solve problems by making a model
14–6	• Identify and draw translations
14–7	• Identify and draw reflections • Explore an application of reflections
14–8	• Identify and draw rotations
14–9	• Identify line symmetry and rotational

SOUTH-WESTERN

MATH MATTERS

BOOK 1

An Integrated Approach

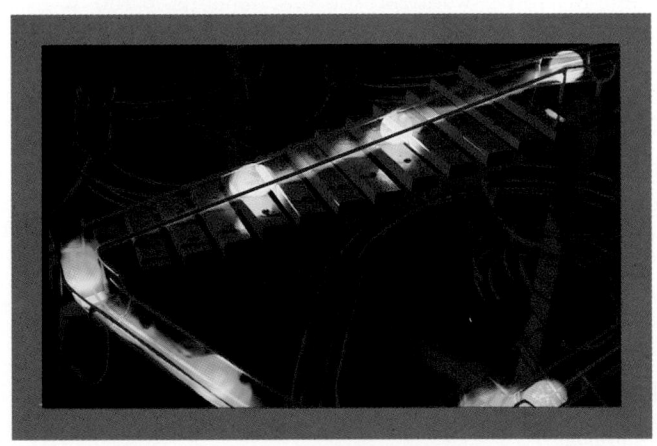

CHICHA LYNCH
Capuchino High School
San Bruno, California

EUGENE OLMSTEAD
Elmira Free Academy
Elmira, New York

SOUTH-WESTERN PUBLISHING CO.

Managing Editor: Eve Lewis
Mathematics Consultant: Carol Ann Dana
Production Editor: Thomas N. Lewis
Editorial Production Manager: Carol Sturzenberger
Cover Photo Photographer: Wayne Sorce © June 1988
Audio Kinetic Sculptures by George Rhoads

About the Math Matters Cover
Artist George Rhoads' "audio kinetic sculptures" appear in shopping centers, malls, terminals, and museums throughout North America. The mechanisms in the sculptures are visible and easy to understand, but the mathematical design provides infinite variety. Crowds gather to watch the acrobatic activity and hear the clanging noises triggered by randomly propelled balls traveling down twisting skeletal ramps.

Copyright © 1993

by SOUTH-WESTERN PUBLISHING CO.

Cincinnati, Ohio

Authorized adaptation of a work first published by Nelson Canada,
A Division of Thomson Canada Limited,
1120 Birchmount Road,
Scarborough, Ontario
M1K 5G4

ISBN: 0-538-61100-6

Library of Congress Catalog Card Number: 91-62724

3 4 5 6   Ki   98 97 96 95 94 93

Printed in the United States of America

About the **Math Matters** Authors

Chicha Lynch

Ms. Lynch is mathematics department head at Capuchino High School, San Bruno, California. As a specialist with the California Mathematics Project, Ms. Lynch helped her school district redesign courses to meet the state framework for Math A. Since 1988 she has served the state of California as a Mentor Teacher for Math A/B. Ms. Lynch is a graduate of the University of Florida and received the LaBoskey Award in 1988 from Stanford University for her contribution to its teacher education program. Most recently Ms. Lynch has been a research associate at the Far West Educational Laboratory, San Francisco, developing assessment for secondary mathematics teachers for certification by the National Board of Professional Teaching Standards. She was a state finalist in 1988 for the Presidential Award for Excellence in Mathematics Teaching.

Eugene Olmstead

Mr. Olmstead is a mathematics teacher at Elmira Free Academy, Elmira, New York. He has had 20 years of public school experience teaching courses from general mathematics through calculus. Mr. Olmstead earned his B.S. in mathematics at State University College at Geneseo, New York, and received his M.S. in mathematics education at Elmira College, Elmira, New York. Mr. Olmstead has worked as a counselor at the Calculator and Computer Precalculus Institutes at The Ohio State University and frequently gives workshops on teaching mathematics using technology. He was a state finalist in 1991 for the Presidential Award for Excellence in Mathematics Teaching.

Program Consultants

Bert K. Waits
Professor of Mathematics
The Ohio State University
Columbus, Ohio

Henry S. Kepner, Jr.
Professor of Mathematics
and Computer Education
University of Wisconsin-Milwaukee
Milwaukee, Wisconsin

Tommy Eads
Mathematics Teacher
North Lamar High School
Paris, Texas

Ann M. Farrell
Assistant Professor of
Mathematics and Statistics.
Wright State University
Dayton, Ohio

Reviewers

Lucy Duffley
Mathematics Department Chair
Canyon Springs High School
Moreno Valley, California

David D. Molina
Assistant Professor
Department of Education
Trinity University
San Antonio, Texas

Valmore Guernon
Mathematics Teacher
Lincoln Junior/Senior High School
Lincoln, Rhode Island

David M. Otte
Principal
St. Henry High School
Erlanger, Kentucky

Gerald A. Haber
Assistant Principal
Winthrop Intermediate School
Brooklyn, New York

Janice Udovich
Mathematics Teacher
Washington High School
Milwaukee, Wisconsin

Sonja Hubbard
Mathematics Teacher
Milton High School
Alpharetta, Georgia

Mary Weber
Mathematics Teacher
Washington High School
Milwaukee, Wisconsin

Joan C. Lamborne
Supervisor of Mathematics
and Computers
Egg Harbor Township High School
Pleasantville, New Jersey

Peter Westergard
Mathematics Teacher
Nicolet High School
Glendale, Wisconsin

Contents

ix

Credits

TEXT ACKNOWLEDGMENTS

The authors and editors have made every effort to trace the ownership of all copyrighted selections found in this book and to make full acknowledgments of their use. Grateful acknowledgment is made to the following authors, publishers, agents, and individuals for their permission to reprint copyrighted materials.

Marshall Cavendish Limited: "Cases of Worldwide Smallpox" and "Distribution of Blood in the Human Body" in *The Marshall Cavendish Illustrated Encyclopedia of Family Health.* Copyright © 1988.
D. C. Heath and Company: "How Many Calories Are Used In Ten Minutes?" in *Houghton Mifflin Health* by Bud Getchell, Rusty Pippin, Jill Varnes. Copyright © 1987. Reprinted by permission of D. C. Heath and Company.
St. Martin's Press, Inc.: "Top Speeds of Athletes in Various Sports." Copyright © 1980 by The Diagram Group. From the book *Comparisons* and reprinted with permission from St. Martin's Press, Inc.
U.S. News & World Report: "On a Roll," Copyright © 1989. August 21, 1989 issue.
USA TODAY: "Le Tour de France." Copyright © 1991, USA TODAY. Reprinted with permission.

PHOTO ACKNOWLEDGMENTS

CONTENTS

p. v: Robert E. Daemmrich/Tony Stone Worldwide; p. vii: Sheila Turner/Monkmeyer Press Photo Service, Inc. (top); Viesti Associates, Inc. (bottom); p. viii: Joe Viesti/Viesti Associates, Inc.; p. ix: Joe Viesti/Viesti Associates, Inc.; p. x: Diane Johnson/Tony Stone Worldwide; p. xi: Boris Berranger/Viesti Associates, Inc.; p. xii: Luis Villota/The Stock Market

CHAPTER 1

p. 3: Bill Stanton/International Stock Photography, Ltd. (top); Agence Vandysladt, Paris, France/Allsport Photography, Inc. (center); p. 4: Don and Pat Valenti/Tony Stone Worldwide (left); Peter Frank/Tony Stone Worldwide (middle); David Brownell/The Image Bank (right); p. 5: Courtesy of University of California at Berkeley Public Information Office; p. 7: Gerhard Gscheidle/Peter Arnold, Inc.; p. 8: Obremski/The Image Bank; p. 9: Manny Millan/Sports Illustrated (right); Wayne Sproul/International Stock Photography, Ltd. (left); p. 10: Wayne Sproul/International Stock Photography, Ltd.; p. 11: David R. Frazier/Photo Researchers, Inc. (top); Jan Halaska/Photo Researchers, Inc. (bottom); p. 12: Chip Henderson/Woodfin Camp & Associates; p. 13: Gabe Palmer/The Stock Market; p. 16: Jim Pickerell/Tony Stone Worldwide; p. 17: Robert Isear/Photo Researchers, Inc. (top); Salmoiraghi/The Stock Market (bottom); p. 18: Sports Illustrated; p. 19: Bill Stanton/International Stock Photography, Ltd. (top); Bob Firth/International Stock Photography, Ltd. (middle); Ken Levinson/International Stock Photography, Ltd. (bottom); p. 22: Michael Salas/The Image Bank; p. 24: Tom McCarthy/The Stock Market; p. 26: Courtesy of Barbara and Steven Griffel (both); p. 28: Mimi Cotter/International Stock Photography, Ltd.; p. 29: Jose Azel/Woodfin Camp & Associates; p. 30: Elyse Lewin/The Image Bank (left); Dick Luria/FPG International Corp. (right); p. 32: Sports Illustrated; p. 33: Paul Solomon/Woodfin Camp & Associates

CHAPTER 2

p. 40: Barry Seidman/The Stock Market (top); Ed Bock/The Stock Market (center); Pete Salantos/The Stock Market (bottom); p. 41: Gabe Palmer/ Mugshots/The Stock Market; p. 43: Jerry Cooke/Photo Researchers, Inc.; p. 44: Ray Shaw/The Stock Market; p. 45: Joanne Savio/Falletta Associates; p. 47: Joanne Savio/Falletta Associates (top); Craig Hammell/The Stock Market (bottom); p. 48: Roy Morsch/The Stock Market; p. 50: John G. Zimmerman/Sports Illustrated; p. 52: Joanne Savio/Falletta Associates; p. 53: Michael Simpson/FPG International Corp. (top); William D. Adams/FPG International Corp. (bottom); p. 54: Stephen Johnson/Tony Stone Worldwide; p. 55: Guido Alberto Rossi/The Image Bank; p. 59: Chip Henderson/Tony Stone Worldwide; p. 60: Robert H. McNaught/SPL/Photo Researchers, Inc.; p. 63: Obremski/The Image Bank; p. 64: Dr. William C. Keel/SPL/Photo Researchers, Inc.; p. 66: NASA; p. 67: Art Wilkinson/Woodfin Camp & Associates; p. 68: Don Johnson/The Stock Market; p. 69: Richard Steedman/The Stock Market; p. 70: Joanne Savio/Falletta Associates; p. 71: Carroll Seghers/Photo Researchers, Inc.; p. 72: Jose Fuste Raga/The Stock Market; p. 73: Stephanie Stokes/The Stock Market; p. 75: Joanne Savio/Falletta Associates; p. 77: Eric Curry/H. Armstrong Roberts, Inc. (top); Sonya Jacobs/The Stock Market (bottom); p. 79: Barbara Filet/Tony Stone Worldwide; p. 80: Martin Chaffer/Tony Stone Worldwide; p. 81: John P. Kelly/The Image Bank (top); Snowdon/Hoyer/Woodfin Camp & Associates (bottom)

CHAPTER 3

p. 88: Angus McBean Estate (top left); EMI Film Distributors Ltd./Movie Still Archives (top center); Courtesy of Paramount Pictures Release (top right); Snark International/Art Resource (bottom left); Weber-Sipa Press/Art Resource (bottom right); p. 89: Culver Pictures (left); Wide World Photos (right); p. 94: Joanne Savio/Falletta Associates (top); CBS (bottom); p. 98: Travelpix/FPG International Corp.; p. 100: M. Lyons/The Image Bank; p. 102: Bob Rashid/Monkmeyer Press Photo Service, Inc.; p. 103: Michael Chua/The Image Bank; p. 104: Roy Morsch/The Stock Market (left); Focus on Sports (right); p. 105: Larry Busacca/Retna Ltd.; p. 107: Sam Griffith/Tony Stone Worldwide; p. 109: David Walberg/Sports Illustrated

CHAPTER 4

p. 116: Jon Riley/Tony Stone Worldwide (top); Gabe Palmer/Mugshots/The Stock Market (right); Joe Sohm/Chromosohm/Nawrocki Stock Photo, Inc. (bottom); p. 117: Will McIntyre/Photo Researchers, Inc. (top); Joe Sohm/Chromosohm/Nawrocki Stock Photo, Inc. (bottom); p. 120: David Woods/The Stock Market; p. 121: Don Smetzer/Tony Stone Worldwide; p. 124: Photri, Inc.; p. 125: Photri, Inc. (left); Photri, Inc. (center); Chris Jones Photo/The Stock Market (right); p. 126: Joanne Savio/Falletta Associates (both); p. 129: Suzanne Murphy/Tony Stone Worldwide; p. 133: Manfred Kage/Peter Arnold, Inc. (both); p. 135: Joe Sohm/Chromosohm/Nawrocki Stock Photo, Inc.; p. 136: Joseph Nettes/Photo Researchers, Inc.; p. 137: Roy Morsch/The Stock Market; p. 139: Russ Kinne/Photo Researchers, Inc.; p. 141: Lawrence Migdale Photography; p. 142: Charles Gupton/The Stock Market; p. 143: Don Mason/The Stock Market; p. 145: Comnet/Nawrocki Stock Photo, Inc. (top); Michael A. Keller/The Stock Market (bottom)

CHAPTER 5

p. 152: Melanie Carr/Nawrocki Stock Photo, Inc. (top); Candee Prod./Nawrocki Stock Photo, Inc. (middle); Jose Fernandez/Woodfin Camp & Associates (bottom); p. 153: Jose Fernandez/Woodfin Camp & Associates (right top); Nick Nicholson/The Image Bank (left); N. Ferguson/Photri, Inc. (middle); Jeffrey Apoian/Nawrocki Stock Photo, Inc. (bottom); p. 154: Bachman/Photri, Inc.; p. 157: Joanne Savio/Falletta Associates (top); Al Clayton, International Stock Photography, Ltd. (bottom); p. 158: H. Armstrong Roberts, Inc. (top left); Sepp Seitz/Woodfin Camp & Associates (top center); Joanne Savio/Falletta Associates (all others); p. 161: Joanne Savio/Falletta Associates; p. 162: Robert E. Daemmrich/Tony Stone Worldwide; p. 165: Bettye Lane/Photo Researchers, Inc.; p. 166: Joanne Savio/Falletta Associates; p. 167: Peter Russell Clemens/International Stock Photography, Ltd.; p. 170: Hugh Sitton/Tony Stone Worldwide; p. 172: from *The Man Who Knew Infinity* by Robert Kanigel, published by Simon & Schuster; p. 173: Superstock; p. 175: Courtesy of Oldsmobile; p. 176: Chuck Keeler/Tony Stone Worldwide (top left and right);) Don Smetzer/Tony Stone Worldwide (bottom left); Brent Jones/Tony Stone Worldwide (bottom right)

CHAPTER 6

p. 184: Steve Vidler/Nawrocki Stock Photo, Inc. (top); David Weintraub/Photo Researchers, Inc. (bottom); Ed Zirkle/Woodfin Camp & Associates (left); p. 185: Del Mulkey/Photo Researchers, Inc. (left); Rob Goldman/FPG Inernational, Corp. (right); p. 186: Photri, Inc. (left); Andrew Conway/ACE/Nawrocki Stock Photo, Inc. (right); p. 190: Melanie Carr/Nawrocki Stock Photo, Inc.; p. 193: Will/Deni McIntyre/Photo Researchers, Inc.; p. 196: Photo Library International, Ltd./Nawrocki Stock Photo, Inc. (left); Steve Vidler/Nawrocki Stock Photo, Inc. (right); p. 203: Vito Palmisano/Nawrocki Stock Photo, Inc.; p. 204: Courtesy of Ruth Gonzales (top); Robert J. Bennett/Photri, Inc. (bottom); p. 205: C. P. George/H. Armstrong Roberts, Inc.; p. 207: Peter Russell Clemens/ International Stock Photography, Ltd.; p. 212: Richard Shock/Nawrocki Stock Photo, Inc. (left); David Lissy/Nawrocki Stock Photo, Inc. (right); p. 213: Photri, Inc.

CHAPTER 7

p. 220: Ken Lax/Photo Researchers, Inc.; p. 221: Ken Lax/Photo Researchers, Inc.; p. 222: David Vance/The Image Bank; p. 227: Photri, Inc.; p. 229: John D. Luke/Index Stock Photography, Inc.; p. 230: Photri, Inc. (top); Richard Hutchings/Photo Researchers, Inc. (bottom); p. 235: Sepp Seitz/Woodfin Camp & Associates; p. 238: H. Abernathy/H. Armstrong Roberts, Inc.; p. 239: Courtesy of Chevrolet; p. 240: Addison Geary/Nawrocki Stock Photo, Inc.; p. 242: Chris Hackett/The Image Bank; p. 243: FPG International Corp.(left); Timothy Eagan/Woodfin Camp & Associates (right); p. 244: Lawrence Migdale Photography; p. 247: Mark Scott/FPG International Corp.; p. 248: Photri, Inc.; p. 249: Peter B. Kaplan/Photo Researchers, Inc.; p. 251: Photri, Inc.; p. 252: Photri, Inc.; p. 257: Tony Stone Worldwide; p. 259: Guy Gillette/Photo Researchers, Inc.; p. 261: Jose Fernandez/Woodfin Camp & Associates

CHAPTER 13

p. 472: Ken Biggs/Photo Researchers, Inc. (top); Ray Ellis/Photo Researchers, Inc. (bottom); p. 473: Fulvio Roiter/The Image Bank; (left) Bill C. Kenny/The Image Bank (right); p. 474: Joanne Savio/Falletta Associates; p. 476: Joanne Savio/Falletta Associates; p. 477: Ted Kawalerski/The Image Bank; p. 478: Blair Seitz/Photo Researchers, Inc. (top); Joanne Savio/Falletta Associates (bottom); p. 481: Joanne Savio/Falletta Associates; p. 482: Joanne Savio/Falletta Associates; p. 483: Palmer/Kane, Inc./The Stock Market; p. 484: Courtesy of Oldsmobile; p. 485: Steve Dunwell/The Image Bank (top); Mel Digiacomo/The Image Bank (bottom); p. 486: Photri, Inc.; p. 487: Brock May/Photo Researchers, Inc.; p. 489: Jeffrey W. Myers/FPG International Corp.; p. 490: Joanne Savio/Falletta Associates; p. 491: Joanne Savio/Falletta Associates; p. 492: Joanne Savio/Falletta Associates; p. 494: Grant LeDuc/Monkmeyer Press Photo Service, Inc. (left); Michael Heron/Woodfin Camp & Associates (right); p. 495: Roger A. Clark, Jr./Photo Researchers, Inc.; p. 497: Roger Dollarhide/Monkmeyer Press Photo Service, Inc.

CHAPTER 14

p. 504: Viviane Moos/The Stock Market (left); Masahiro Sano/The Stock Market (right); p. 505: Paul Jablonka/International Stock Photography Ltd. (top left); Masahiro Sano/The Stock Market (left center); Viviane Moos/The Stock Market(right center and bottom right); p. 509: George Hall/Woodfin Camp & Associates; p. 511: ABY/Photri, Inc.; p. 515: Dennis Hallinan/FPG International Corp.; p. 516: David Lissy/Nawrocki Stock Photo, Inc.; p. 518: Steven Burr Williams/The Image Bank (left); Jürgen Vogt/The Image Bank (right); p. 521: Renate Hiller/Monkmeyer Press Photo Service, Inc. (top); Alexas Urba/The Stock Market (bottom); p. 522: Joanne Savio/Falletta Associates; p. 526: Sybil Shackman/Monkmeyer Press Photo Service, Inc.; p. 528: Photri, Inc.; p. 529: Ch. Petit/Photo Researchers, Inc.; p. 530: James F. Palka/Nawrocki Stock Photo, Inc.; p. 531: Courtesy of Rosemary Chang; p. 533: Hungerfond/Photri, Inc.; p. 534: Joanne Savio/Falletta Associates

DATA FILES

p. 548: G. Galicz/Photo Researchers, Inc.; p. 549: Bob Firth/International Stock Photography, Inc. (top); Art Montes de Oca/FPG International Corp. (bottom); p. 550: Guy Gillette/Photo Researchers, Inc. (top); David J. Maenzo/The Image Bank (bottom); p. 551: Camerique/H. Armstrong Roberts, Inc. (top); Cyril Toker/Photo Researchers, Inc. (center); p. 552: Van Bucher/Photo Researchers, Inc. (top); David R. Frazier/Photo Researchers, Inc. (bottom); p. 553: Ron Sherman/Tony Stone Worldwide (top); William Caram/Index Stock Photography, Inc. (bottom left); Menschenfreund/The Stock Market (bottom right); p. 554: Rev. Ronald Royer/SPL/Photo Researchers, Inc. (top); Hale Observatories/Photo Researchers, Inc. (bottom); p. 555: Bill W. Marsh/Photo Researchers, Inc. (top); Hale Observatories/Photo Researchers, Inc. (left); p. 556: Photri, Inc. (top); William Strode/Woodfin Camp & Associates (bottom); p. 557: Jeff Friedman Photography (top); Fotex/Drewa/Nawrocki Stock Photo, Inc. (bottom); p. 558: J. Barry O'Rourke/The Stock Market (top); Richard L. Carlton/Photo Researchers, Inc. (right bottom); p. 559: David J. Cross/ Peter Arnold, Inc. (top left); John V.A. F. Neal/International Stock Photography, Ltd. (top right); Simon Fraser/SPL/Photo Researchers, Inc. (center); p. 560: Verna Brainard/Photri, Inc. (top); R. Kord/H. Armstrong Roberts, Inc. (bottom); p. 561: Joanne Savio/Falletta Associates (top); Index Stock Photography, Inc. (center); p. 562: Dr. C. W. Biedel/Photri, Inc.; p. 563: Suzanne Szasz/Photo Researchers, Inc. (top); Jon Riley/Tony Stone Worldwide (bottom left); Alfred Gescheidt/The Image Bank (bottom right); p. 564: Walter Iooss, Jr./Sports Illustrated (top); Lawrence Migdale Photography (bottom); p. 565: Wayne Sproul/International Stock Photography; p. 566: Tom McHugh/Photo Researchers, Inc.; p. 568: Gregory Sams/SPL/Photo Researchers, Inc. (top left); Character House/Photo Researchers, Inc. (bottom right); p. 571: Gregory Sams/SPL/ Photo Researchers, Inc. (bottom left); p. 570: Gregory Sams/SPL/Photo Researchers, Inc. (top); Paul Jablonica/International Stock Photorgraphy, Ltd. (bottom); p. 571: Gregory Sams/SPL/Photo Researchers, Inc. (top); Dale Boyer/USCF/Photo Researchers, Inc. (center); John Wells/SPL/Photo Researchers, Inc. (bottom)

All statistical information that appears in the Data Files is in the public domain unless otherwise noted. We gratefully acknowledge the following sources:

Playing with Fire — Hazardous Waste Incineration, © 1990 Greenpeace U.S.A., Washington, DC, "Quantities of Hazardous Waste Burned in U.S.," p. 559.

Acknowledgments are continued on page 586.

Preface

Welcome to *Math Matters*. You may find this textbook different from other math books you have used. Unlike more traditional textbooks which focus on one specific subject, *Math Matters* combines mathematic topics in an integrated program. Number sense, algebra, geometry, statistics, and logic are presented as tools for investigating phenomena and exploring new math concepts.

As you work through *Math Matters*, you will find new ways to use estimation and approximation skills to use computers and calculators more effectively. You will discover that algebraic concepts can enhance your critical thinking skills. You will see how geometry relates to reasoning and problem solving. You will find that you can evaluate the meaning of statistics that are presented on TV and in newspapers. In short, you will learn how to use mathematics to your advantage in school, at home, and at work.

To help you achieve success with this program, we suggest you take time now to learn how to use this textbook. Here are a few ideas you may find worthwhile.

GETTING STARTED WITH NOTEBOOKS AND LEARNING JOURNALS

Before you begin using *Math Matters,* you will find that keeping an organized notebook or learning journal is essential to successful learning throughout the school year. It is important that the organization of your notebook or learning journal reflect your particular needs and learning style. Begin by thinking about the kinds of information you will want to record and how you might easily get access to this information. Below are some ideas to help you organize your notebook or journal. Be sure, however, to listen carefully to what your teacher requires of you, since he or she may wish to add or change the kinds of information you record.

Vocabulary List

For each unfamiliar word, definition, or property, write the correct definition as given in the text or Glossary. Also include an example if that will help. Then rewrite the definition and example in your own words. Review your vocabulary list periodically to make sure that you maintain a clear idea of what each term means. A few minutes of vocabulary review each week will go a long way to improving your opportunities for success.

Notes and Questions

The contents of this part of your notebook or learning journal will depend somewhat on your teacher's method of instruction. This section should include any important problems, exercises, and solutions that your teacher puts on the chalkboard for class discussion. It also might include highlights of key points that

your teacher makes during a class or lecture. You should write any questions, examples, or exercises that need further explanation. You may want to mark your questions in some way—perhaps, with an asterisk (*) or a question mark in the margin. Then, when you get a satisfactory explanation or you are able to answer your own question, you can cross out or circle the symbol you used to indicate that you no longer need to be concerned about that problem.

Summaries and Conclusions

In this section, you can reflect in your own words on what you have learned, which topics interest you the most, and which topics have been difficult. Summarizing and drawing conclusions is often the last step in mastering what you have learned. These skills are also helpful in clarifying areas for further study.

Formulas, Properties, and Drawings

You may wish to put this section at the back of your notebook or learning journal so that you can quickly find what you need. Here you can record commonly used formulas, properties, and drawings that explain or summarize key information.

Homework and Assignments

Have a special place in your notebook or learning journal where you can write the complete assignment and make a regular habit of recording all the assignments you receive. Doing these things will prevent last-minute "panic" phone calls to classmates, asking, "Do you know what the math homework is?" Be sure that you have some special way of indicating long-term projects so that the due day doesn't sneak up on you unexpectedly. Your teacher will tell you whether you should keep your homework in your notebook or turn it in each day.

LEARN HOW TO USE THIS BOOK

Math Matters has a consistent organization and many recurring features. It is wise to get to know how the book is organized and what the purpose of each feature is. Knowing what to expect will allow you to focus efficiently on what you need to learn. Take time now to become familiar with these features of your text, and you will save valuable time later on.

Skills Preview

Each chapter begins with a Skills Preview that will help you to discover which topics of the chapter you may already know and which will require careful study. Don't be concerned by what you get wrong on the Skills Preview. By identifying what you already know, you will be able to spend more time on those topics that are new to you and require your full attention.

Chapter Opener

The Chapter Opener helps to set the stage for the topics you are about to study. It tells you about the subject matter and theme of the chapter and presents some

fascinating data and visual information for you to examine. Each opener requires that you use the data presented to make decisions and that you complete a group project or exploration tied to the mathematical content and theme of the chapter.

Every chapter is divided into a group of numbered sections, each of which is either four pages or two pages long. Every four-page section touches on a group of related math concepts or skills and follows a consistent organization.

Explore and Explore/Working Together

Each four-page section begins with an activity that provides a stimulating introduction to the mathematical concepts or skills to be studied. These activities often involve hands-on investigations that lead to meaningful mathematical insights. Many of the Explore activities require you to work cooperatively with other students. Look over each Explore activity before you begin working on it. In this way you will understand what is really expected of you, and you can gather together what you will need to complete the activity.

Skills Development

This section explains the concepts and skills you are expected to master. Here you will find helpful discussions, along with essential definitions. The Skills Development presents the key points of the lesson through model Examples and worked-out Solutions for you to study and apply later on.

Try These

These are questions and problems that you will likely discuss in your classroom. The items in Try These are similar to the ones presented in the Examples and Solutions of the Skills Development, so they are an excellent way for you to determine if you really have absorbed the key points of each lesson. If you have trouble with any Try These items, it is worth your while to take a second look at the relevant model Examples and Solutions in the Skills Development.

Exercises

The Exercises provide numerous opportunities to practice and apply the mathematical concepts and skills that you have studied and to make connections with other areas of mathematics and with the use of mathematics in real-world situations and in other subject areas. Each section contains three levels of Exercises, which provide increasingly challenging practice and problem solving. These items appear under the headings Practice/Solve Problems, Extend/Solve Problems, and Think Critically/Solve Problems.

From time to time in each chapter a four-page section ends with a special full-page feature called Problem Solving Applications. These pages make connections between the mathematics presented in the section and the practical application of those mathematical skills and concepts within a particular career or professional field.

Problem Solving Strategies

Each two-page Problem Solving Strategies section focuses on one important strategy. Each Problem Solving Strategies section has a brief introduction, model Problems and Solutions, and practice Problems. Problem Solving Strategies, introduced and applied throughout the book, include, for example, Using Logical Reasoning, Finding a Pattern, Working Backward, and Making an Organized List. From time to time throughout the book there will be a special Choose a Strategy lesson that presents a variety of problems involving any of the strategies that have been presented up to that point in the book.

Problem Solving Skills

These are special two-page sections that introduce a skill essential to successful problem solving. Each Problem Solving Skills section has a brief introduction, model Problems and Solutions, and practice Problems to help you apply the problem solving skill being taught. Problem Solving Skills, introduced and applied throughout the book, include Choosing an Operation, Choosing an Appropriate Scale, Using Venn Diagrams, and Making a Conjecture.

Within the sections of each chapter you will encounter a variety of special features, each of which is identified by a special visual symbol, or icon. The following is a description of the various features and their corresponding icons.

Check Understanding

This feature helps you to evaluate your understanding of key concepts and skills by posing questions about model Examples or Exercises.

Talk It Over

This feature provides thought-provoking questions that will spark interesting group discussions between you and your classmates.

Reading Mathematics

This feature provides an opportunity to become familiar with the vocabulary of mathematics and with the special meanings that ordinary words sometimes have within the discipline.

Writing About Math

In this feature you will be asked to demonstrate your knowledge of mathematical concepts and skills by writing about them in a variety of meaningful ways, including notes, advertisements, and brief essays.

Connections

This feature helps you to make connections between the mathematics you are learning and other areas of mathematics, other disciplines, and real-world problem solving situations.

Math in the Workplace
Here you will learn how the math you are studying is applied in the everyday work world.

Math: Who, Where, When
In this feature you will learn about mathematical ideas from different cultures around the world and the contributions that people have made to the development of mathematics.

Computer and Computer Tips
These features provide useful computer programs for solving math problems, along with helpful tips to make them work effectively and to adapt them to other uses.

Calculator and Calculator Tips
These features explain how to solve problems using a calculator, as well as tips to improve your efficiency and accuracy when using a calculator.

Problem Solving, Estimation, and Mental Math Tips
These features will help you to use a variety of strategies to solve problems that are presented in the text.

Mixed Review
This feature presents a quick way to refresh and maintain your mastery of previously learned mathematical skills and concepts.

At the end of each chapter you will find several features to help you maintain your mastery of concepts and skills and to help you prepare for tests.

Chapter Review
Each chapter ends with a two-page Chapter Review that highlights all the major rules and definitions in the chapter and provides additional examples and practice items. It is very helpful to look at the Chapter Review even before beginning a new chapter since the Review will help to focus your attention on the chapter's key concepts and skills.

Chapter Test
The one-page Chapter Test is very much like the Skills Preview at the beginning of the chapter. Try comparing your answers on both these features to learn which concepts and skills you have mastered and which you should review more carefully before taking any out-of-book examinations.

Cumulative Review

The Cumulative Review feature provides on-going opportunities to maintain math concepts and skills learned earlier in the course of study.

Cumulative Test

The Cumulative Test feature provides additional opportunities to practice test-taking skills for concepts and skills taught throughout the year. The test items are in multiple-choice format similar to the type encountered on standardized tests.

Your teacher may ask you to do other things to evaluate how well you are learning and using the math ideas that have been presented. You may be asked to build a portfolio of your work or to work with other students to complete extended projects that pull together the math ideas with which you have worked. You also may be asked to reflect on your experiences in mathematics class and your attitudes toward mathematics throughout the school year. In any case, remember that you are responsible for your own learning and for producing your best work.

Data Files

This section provides fascinating data organized around major themes, such as Health and Fitness, Economics, Ecology, Useful Math References, and so on. The tables, charts, graphs, and other visual and verbal data provided in this section are needed to answer questions that appear from time to time under the heading Using Data.

Glossary

This is an alphabetical listing of all the key terms introduced and defined in the text. Use the Glossary entries as you develop your vocabulary lists. Note that after each term there is a reference indicating the page of the chapter section where the term is first introduced and defined.

Selected Answers

This section provides answers to selected odd-number items of lessons and features. Using the Selected Answers on a regular basis is a good way to assess your own learning progress in an on-going way.

CHAPTER 1 SKILLS PREVIEW

A survey was taken in a town to find out how people felt about the new water-conservation program. The interviewer polled people as they boarded the 7:05 train one morning.

1. What kind of sampling does this represent? **convenience sampling**

2. What is an advantage of this kind of sampling? **The population is readily available.**

The pictograph shows the number of books read last month by the students in four grades.

3. The students in which grade read the most books? **Grade 11**

4. How many books did the juniors read? **150 books**

5. How many books were read in all? **460 books**

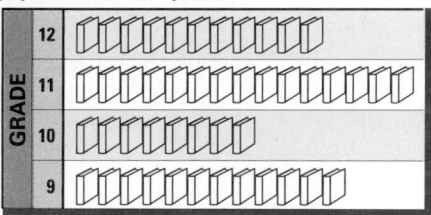

Key: 📖 = 10 books

At a large family gathering, the heights of the family members were recorded in inches. The data are shown in this stem-and-leaf plot:

4|7 represents 47 in.

3	9
4	6 7 7 8
5	0 0 2 4 5 5 8
6	1 1 7 8 8 8
7	0 2 2 4 5

The heights of how many people fall within each interval?

6. 30−39 in. **1**   **7.** 60−69 in. **6**   **8.** 70 in. or more **5**

How many people are there of each height?

9. 47 in. **2**

10. 54 in. **1**

11. 68 in. **3**

The bar graph gives the number of aluminum cans brought to a recycling center one week.

On which days were more cans brought to the recycling center?

12. Tuesday or Wednesday   **13.** Monday or Thursday

14. Friday or Saturday

The number of books read during the summer by eight students were: 3, 6, 4, 4, 5, 2, 8, 9.

15. Find the mean, median, mode, and range of the data. **5.125, 4.5, 4, 7**

Introduction The purpose of this Skills Preview is to assess students' knowledge of all the major objectives of Chapter 1. Test results may be used
- to determine those topics which may need only to be reviewed and those topics which need to be more carefully developed;
- for class placement;
- in prescribing for individual differences.

If you prefer, you may use the Skills Preview as an alternative form of the Chapter Test (page 36) to evaluate mastery of chapter objectives. The items here and on the Chapter Test correspond in content and level of difficulty.

CHAPTER 1

EXPLORING DATA

OVERVIEW

In this chapter, students explore methods of collecting, displaying, and analyzing data. They learn how to determine scales for graphs, how to use estimation to interpret data on graphs, and how to use a 5-step plan for solving problems. They construct frequency tables, line plots, stem-and-leaf plots, pictographs, bar graphs, and line graphs. Students also learn how to describe and analyze data using the three measures of central tendency — mean, median, mode — and range.

SPECIAL CONCERNS

Although the concepts introduced in this chapter are not difficult, students may confuse the new vocabulary introduced in Sections 1–1, 1–2, and 1–9. You may wish to suggest that students use individual index cards to list each new term, its definition, and an example which illustrates the term. Encourage students to carry these cards with them and to briefly review them whenever they have a few minutes to spare.

VOCABULARY

back-to-back stem-and-leaf plot	gap	pictograph
	leaf	poll
bar graph	line graph	population
cluster sampling	line plot	random sampling
cluster	mean	range
convenience sampling	median	sample
data	mode	stem-and-leaf plot
frequency table	outlier	systematic sampling

MATERIALS

clock with second hand	almanac and/or encyclopedia
graph paper	rulers
calculators	

BULLETIN BOARD

Display news, sports, and business articles and classified ads on the bulletin board. Have students choose an article or ad, write a word problem for it, and then display the problem below the article. Other students can then solve the word problems for extra credit.

INTEGRATED UNIT 1

The skills and concepts involved in Chapters 1–4 are included within the special Integrated Unit 1 entitled "Voting and Citizenship." This unit is in the Teacher's Edition beginning on page 150H. Worksheets for this integrated unit appear in the Enrichment Activities booklet, pages 47–49.

TECHNOLOGY CONNECTIONS
- Calculator Worksheets, 19–20
- Computer Worksheets, 21–22
- MicroExam, Apple Version
- MicroExam, IBM Version

- *Data Analysis*, National Council of Teachers of Mathematics
- *Data Insights* and *Statistics Workshop*, Sunburst Communications

TECHNOLOGY NOTES

Computer statistics programs, such as those listed in Technology Connections, allow the students to input their own data. The computer can find the mean, median, mode, and range of the data and display them in a table, plot, or graph of the student's choosing. Such graphic presentations allow the students to focus on interpreting and analyzing the data.

EXPLORING DATA

PLANNING GUIDE

SECTIONS	TEXT PAGES	ASSIGNMENTS			
		BASIC	AVERAGE	ENRICHED	
Chapter Opener/Decision Making	2–3				
1–1 Collecting Data; Problem Solving Applications	4–7	1–3, 4–5, 8–9, PSA 1–3	4–7, 8–11; PSA 1–5	4–12; PSA 4–6	
1–2 Stem-and-Leaf Plots; Problem Solving Applications	8–11	1–4, 5, PSA 1–4	1–4, 5–6, 7, PSA 1–5	5–10; PSA 1–7	
1–3 Pictographs	12–15	1–5. 6–10, 14–16	6–13, 14–19	6–13, 17–22	
1–4 Problem Solving Strategies: Introducing the 5-Step Plan	16–17	1–4	1–5	1–6	
1–5 Problem Solving Skills: Choose an Appropriate Scale	18–19	1, 2, 5	1, 3–5, 7	3–7	
1–6 Bar Graphs	20–23	1–6, 8–11	1–6, 8–11, 12–14	7–11, 12–15	
1–7 Problem Solving Skills: Estimation	24–25	1–4	1–6	1–6	
1–8 Line Graphs	26–29	1–3, 5, 7–9, 15	1–5, 7–10, 11–13, 18–19	6, 11–14, 15–20	
1–9 Mean, Median, Mode, Range	30–33	1–6, 11–17	3–10, 11–18, 26–28	7–10, 19–25, 26–31	
Technology	31	✓	✓	✓	

ASSESSMENT				
Skills Preview	1	All	All	All
Chapter Review	34–35	All	All	All
Chapter Test	36	All	All	All
Cumulative Review	37	All	All	All
Cumulative Test	38	All	All	All

	ADDITIONAL RESOURCES			
	RETEACHING	**ENRICHMENT**	**TECHNOLOGY**	**TRANSPARENCY**
	1–1	1–1		
	1–2	1–2		TM1
	1–3	1–3		TM2
	1–4	1–4		TM3
	1–5	1–5		
	1–6	1–6		TM4
	1–7	1–7		
	1–8	1–8		TM5
	1–9	1–9	1–9	TM6

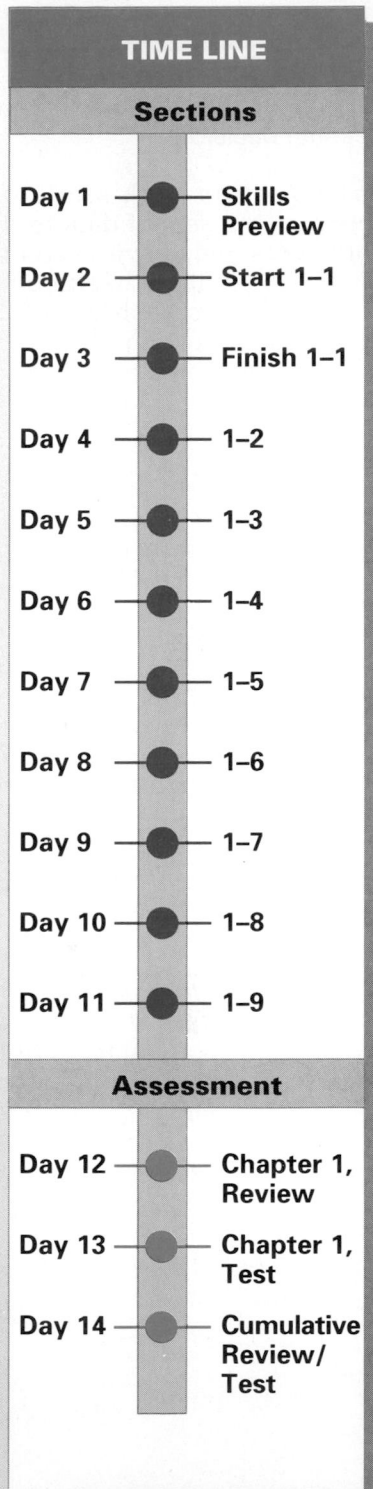

TIME LINE

Sections

Day 1 — Skills Preview
Day 2 — Start 1–1
Day 3 — Finish 1–1
Day 4 — 1–2
Day 5 — 1–3
Day 6 — 1–4
Day 7 — 1–5
Day 8 — 1–6
Day 9 — 1–7
Day 10 — 1–8
Day 11 — 1–9

Assessment

Day 12 — Chapter 1, Review
Day 13 — Chapter 1, Test
Day 14 — Cumulative Review/ Test

ASSESSMENT OPTIONS

Chapter 1, Test Forms A and B	
Chapter 1, Test	Text, 36
Alternative Assessment	TAE, 36
Chapter 1, MicroExam	

CHAPTER OPENER

Objective To explore how data can be used as a basis for making real-life decisions

Introduction Ask students to examine the sets of data to determine what they have in common. (They all relate to health and fitness.) Discuss each set of data, asking the following.
- If you were researching this topic for your own personal interest, how might you use these data?
- What questions might these data answer for you?
- What decisions might you make after analyzing the data?

Decision Making

Using Data Tell students that the air-pollution data was compiled by the American Lung Association and the vitamin-C data comes from the U. S. Department of Agriculture. Then ask, *Why is it important to know the source of each set of data?* **to ascertain the credibility of the data**

Working Together Have students share the data they collect. Discuss why someone might have need of that data and how the data would help that person make a decision.

CHAPTER 1

EXPLORING DATA

THEME   Health and Fitness

Pieces of information are called **data**. **Statistics** is the branch of mathematics that involves collecting data and organizing them in such a way that they can be used as a basis for making decisions. In this chapter you will have a chance to collect and organize data and display it in stem-and-leaf plots, pictographs, bar graphs, and line graphs. You will also work with statistical measures that help you describe a set of data.

DECISION MAKING

Using Data

1. How do you suppose the data were collected for the sports table? Do you think each of the people represented in the table was polled individually? **A poll; Answers will vary.**

2. If you wanted to be sure to drink enough juice to take in 200 mg of vitamin C today, what would you drink? **fresh orange juice**

3. If you could have just two days to videotape Le Tour de France, which dates would you choose? In which cities would you be taping? **Answers will vary.**

AMERICANS WHO REGULARLY PARTICIPATE IN SPORTS	
Sport	**Number of Participants**
Swimming	105,400,000
Bicycling	69,800,000
Bowling	43,300,000
Jogging/running	35,700,000
Tennis	32,300,000
Softball	28,500,000
Roller skating	25,400,000
Basketball	24,000,000
Ice skating	18,900,000
Golf	15,900,000
Skiing	15,400,000
Football	14,300,000
Racquetball	10,700,000
Soccer	6,500,000
Archery	5,500,000
Ice hockey	1,700,000

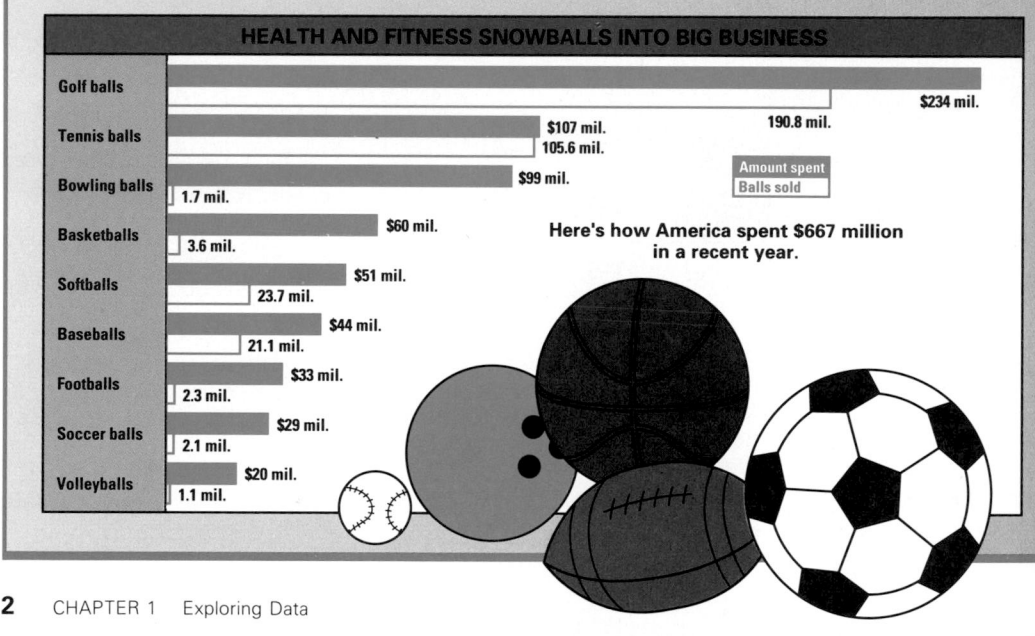

HEALTH AND FITNESS SNOWBALLS INTO BIG BUSINESS

Golf balls — $234 mil. / 190.8 mil.
Tennis balls — $107 mil. / 105.6 mil.
Bowling balls — $99 mil. / 1.7 mil.
Basketballs — $60 mil. / 3.6 mil.
Softballs — $51 mil. / 23.7 mil.
Baseballs — $44 mil. / 21.1 mil.
Footballs — $33 mil. / 2.3 mil.
Soccer balls — $29 mil. / 2.1 mil.
Volleyballs — $20 mil. / 1.1 mil.

Amount spent / Balls sold

Here's how America spent $667 million in a recent year.

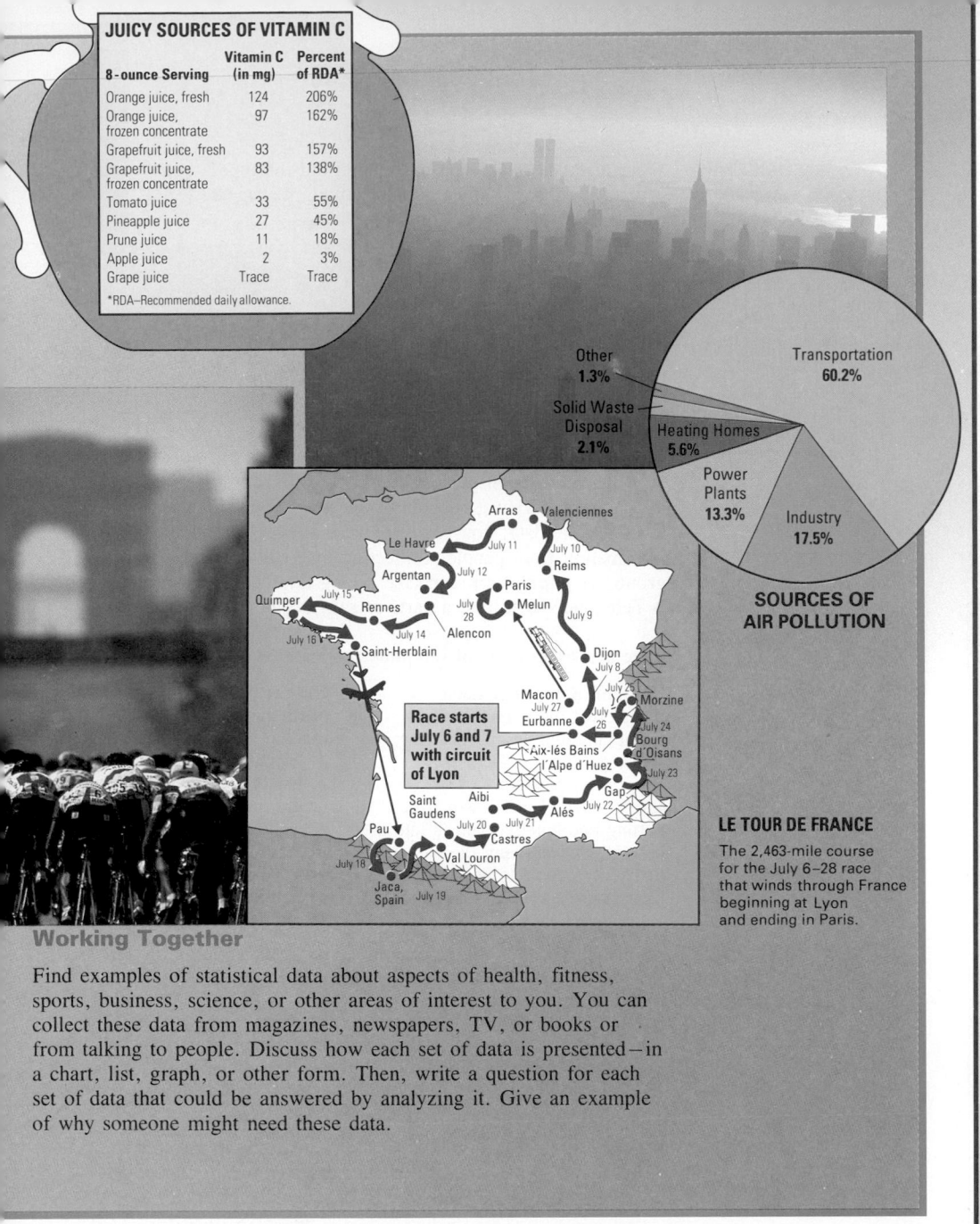

JUICY SOURCES OF VITAMIN C

8-ounce Serving	Vitamin C (in mg)	Percent of RDA*
Orange juice, fresh	124	206%
Orange juice, frozen concentrate	97	162%
Grapefruit juice, fresh	93	157%
Grapefruit juice, frozen concentrate	83	138%
Tomato juice	33	55%
Pineapple juice	27	45%
Prune juice	11	18%
Apple juice	2	3%
Grape juice	Trace	Trace

*RDA—Recommended daily allowance.

SOURCES OF AIR POLLUTION

Transportation 60.2%
Industry 17.5%
Power Plants 13.3%
Heating Homes 5.6%
Solid Waste Disposal 2.1%
Other 1.3%

LE TOUR DE FRANCE

The 2,463-mile course for the July 6–28 race that winds through France beginning at Lyon and ending in Paris.

Working Together

Find examples of statistical data about aspects of health, fitness, sports, business, science, or other areas of interest to you. You can collect these data from magazines, newspapers, TV, or books or from talking to people. Discuss how each set of data is presented—in a chart, list, graph, or other form. Then, write a question for each set of data that could be answered by analyzing it. Give an example of why someone might need these data.

Exploring Data   **3**

SPOTLIGHT

OBJECTIVE
• Explore the nature of data and types of samples

VOCABULARY
cluster sampling, convenience sampling, data, frequency table, line plot, poll, population, random sampling, sample, survey, systematic sampling

WARM-UP

Write the next three numbers.
1. 27, 62, 97, 132, ■, ■, ■
   167, 202, 237
2. 15, 30, 26, 41, 37, ■, ■, ■
   52, 48, 63

1 MOTIVATE

Explore Have students discuss the question in small groups. List each group's method on the board and discuss which methods are best and why.

2 TEACH

Use the Pages/Skills Development Have students read this part of the section and then discuss the examples. Elicit that *records of events* might refer to the records of athletes' performances in each game in a series. A coach might use such records to determine which athletes to use in the next game. Have students share any experiences they have had with questionnaires. Discuss the reason for sampling, and the various methods of sampling populations. Emphasize that, to be valid, the sampling method must be appropriate for the entire population.

1-1 Collecting Data

EXPLORE What methods might you use to find out the types of exercise that are most popular among teenagers?

SKILLS DEVELOPMENT

READING MATH

The word *data* is the plural form of the Latin word *datum.*

Whenever you say or write the word *data*, be sure to follow it with the plural form of the verb *to be.* Say, for example,

The data *are* correct.
These data *were* collected.

Data are information from which decisions can be made. Data can be gathered through personal or telephone interviews, records of events, or questionnaires. In order to obtain the data, you need to take a **survey**, or **poll**, to collect information from people. Collecting information from an entire group, or **population**, can take a long time and can be very costly. A better way is to poll a part, or **sample**, of the population.

Here are some ways of sampling populations.

Random Sampling: Each member of the population is given an equal chance of being selected. The members are chosen independently of one another. An example would be putting one hundred names into a hat and then drawing the names of twenty people to ask about their favorite rock groups.

Convenience Sampling: The population is chosen only because it is readily available. An example would be polling the ten students who happen to be sitting near you in the cafeteria to find out teenagers' favorite styles of jeans.

Cluster Sampling: The members of the population are chosen at random from a particular part of the population and are then polled in clusters. An example would be choosing parts of town at random, visiting the pet shops in these areas, and then asking *all* the pet owners you meet to name their pets' favorite brands of pet food.

Systematic Sampling: After a population has been ordered in some way, its members are chosen according to a pattern. An example would be pulling every tenth item off a production line for quality-control testing.

In order to be able to make decisions about an entire population based on a sample alone, you must first be sure to choose an appropriate sampling method.

4 CHAPTER 1 Exploring Data

TEACHING TIP

To help students understand that time and cost limitations prevent the surveying of entire populations, discuss surveys and polls that may be familiar to them, such as the Nielsen Ratings (of TV programming), Gallup Polls (taken before and after major elections), and the U.S. Census (taken every 10 years).

Example 1

Which of these reflects the *random sampling* method that a health club owner might use to identify the most popular exercise machine in the club?

a. Ask the first twenty members who enter the club one morning.
b. Ask the members whose phone numbers end with the digit 7.
c. Ask the members who live on the six busiest streets in town.

Solution

Choice **b**. If phone numbers are assigned at random, those members whose phone numbers end in 7 represent a random sample. ◄

Example 2

Every twentieth rowing machine in a run of 500 was quality tested as it came off the assembly line. Two were found to be defective.
a. What kind of sampling does this situation represent?
b. What might be an advantage of this kind of sampling?
c. What might be a disadvantage of this kind of sampling?

Solution

a. This example represents *systematic* sampling.
b. An advantage is that the sample comes from the whole population.
c. If there were additional runs, then a sample taken from the last run might differ in quality from a sample taken from the first run. ◄

**MATH:
WHO, WHERE, WHEN**

David Blackwell (1919-), a professor emeritus of statistics at the University of California, Berkeley, originally planned to become an elementary school teacher. Often portrayed today as a theoretician who presents his ideas in beautiful and elegant ways, Professor Blackwell has made contributions to Bayesian probability, game theory, set theory, dynamic programming, and information theory. At the outset of his teaching career, he taught at Howard University in Washington, D.C.

TRY THESE

1. Which of these reflects the *cluster sampling* method that a committee of high school students might use to find out how many adults would attend a school fair?

   a. Ask all the adults who leave the local supermarket on Tuesday between 5:00 and 6:00 p.m.
   b. Ask all the members of the men's and women's softball leagues.
   c. Ask all the adults who live in houses chosen at random on some of the streets surrounding the school.
   Choice "c" represents cluster sampling.

In a telephone survey of the first fifty people who said that they earned their living by working in their home offices, it was found that nine out of ten people eat breakfast at least some of the time.

2. What kind of sampling does this situation represent? **convenience sampling**

3. What might be an advantage of this kind of sampling? **ease of getting responses**

4. What might be a disadvantage of this kind of sampling? **It does not include people who work outside the home. (It may be that people who work at home eat breakfast with greater frequency than do those who work outside the home.)**

1–1   Collecting Data   **5**

Example 1: Have students identify the sampling methods used in choices **a** and **c**. **convenience, cluster** Then discuss why choice **a** might be most beneficial to the health club owner and why. **These people might be representative of members who attend most often.**

Example 2: Have students predict the number of defective machines in groups of 1,000 and 2,000. **4, 8**

Additional Questions/Examples

1. A class of 25 students must decide where to go on a field trip. Should they poll their entire population or poll a sample? Explain. **Poll the entire population since it is not too large.**

2. You want to find out how many students will vote for your friend as students council representative. Which sampling method would you use and why? **Answers will vary.**

3. What are some school or community activities or events for which you could collect data? **Answers will vary.**

4. If you were doing a survey of heavy-metal rock groups, what group of people would you

poll? **teenagers who are knowledgeable about heavy-metal music** Which sampling method would you use? **cluster sampling**

Guided Practice/Try These Have students complete these exercises in small groups. For Exercise 1 have students give advantages and disadvantages of each sampling method given.

3 SUMMARIZE

Write About Math In their math journals, have students explain and give an example of each of the sampling methods.

4 PRACTICE

You may wish to have students work through this lesson in pairs or in small groups.

Practice/Solve Problems Have students identify the sampling method described in each example. **systematic, random, convenience**

Extend/Solve Problems Have students determine a sampling method to be used for a class poll to find out students' favorite fast-food restaurants.

Think Critically/Solve Problems This polling activity may take several days. When students are ready, allow time for them to share the results of their surveys. Challenge them to draw conclusions based on the data they gathered. Have them save their data for use in Section 1–3.

EXERCISES

**PRACTICE/
SOLVE PROBLEMS**

Suppose you need to determine the most popular video arcade game. Name the advantages and disadvantages of choosing a sample in each of the following ways. **See Additional Answers.**

1. Ask every twelfth person who enters the local video arcade.

2. Ask a randomly chosen sample of twelve students in your school.

3. Ask the first twelve people you meet in a video-rental store.

**EXTEND/
SOLVE PROBLEMS**

Use the following information for Exercises 4–7. The owners of a shopping mall have to fill two vacant stalls in their food court. They want to find out what kinds of fast food shoppers like best. Suppose they station interviewers outside the food court between 12:00 noon and 1:00 p.m. and have them ask the first 100 people who pass about their food preferences.

4. What kind of sampling is represented by this situation? **convenience**

5. How could this method be changed to produce a systematic sample? **See Additional Answers.**

6. How could the survey be conducted to produce a random sample? **Answers will vary.**

7. Which kind of sample do you think would be of most use to the owners of the mall? Explain. **Answers will vary.**

**THINK CRITICALLY/
SOLVE PROBLEMS**

Conduct your own poll! Work alone or in a group. Ask forty students about a subject of your choice. **Answers will vary.**

8. Decide on a subject whose popularity or frequency you would like to know more about. Then write a question about the subject that can be answered "yes" or "no."

9. Decide on the population from which you will take your sample.

10. Give a pretest! Ask your question of a few students in your class to see if others understand the question as you mean it to be understood. (If necessary, change the wording of your question.)

11. Choose forty students for your sample. Decide on a way to record your data. Then take the poll.

12. Write an article for the school or community newspaper in which you describe your poll. Tell why you decided on the question, tell why you chose the particular population to poll, and describe the data that resulted.

6 CHAPTER 1 Exploring Data

CHALLENGE

Have students work in small groups to prepare a questionnaire that reflects the interests of the students in their school. Possible topics include favorite subjects, sports, music groups, or books. Suggest that each group have forty students respond to their questionnaire, record the results, and present the results to the class.

Problem Solving Applications:
FREQUENCY TABLES AND LINE PLOTS

Brad interviewed people about their favorite type of TV program. He used this code to record his data.

S	Sports	Q	Quiz shows
M	Mysteries	V	Music videos
So	Soap operas	A	Adventure
N	News	C	Comedies

These are the data Brad collected.

V	C	So	N	Q	A	M	S	M	So	A	So
Q	C	C	N	A	C	N	M	Q	S		So
A	S	Q	A	C	S	N	Q	A	Q		S
A	A	M	S	So	V	N	C	A	C		S

Brad displayed the data in a frequency table and in a line plot.

In a **frequency table**, a tally mark is used to record each response. The total number of tally marks for a given response is the frequency of that response. In a **line plot**, an X is made to record each response. The X's may be stacked one on top of another until all the data are recorded. Read the scale along the side, then follow it across to the X at the top of each stack to find the frequency of each response.

TYPES OF TV PROGRAMS

Program	Tally	Frequency
Sports	JHT II	7
Mysteries	IIII	4
Soap operas	JHT	5
News	JHT	5
Quiz shows	JHT I	6
Music videos	II	2
Adventures	JHT IIII	9
Comedies	JHT II	7

TYPES OF TV PROGRAMS

Frequency	Sports	Mysteries	Soap operas	News	Quiz shows	Music videos	Adventures	Comedies
9							X	
8							X	
7	X						X	X
6	X				X		X	X
5	X		X	X	X		X	X
4	X	X	X	X	X		X	X
3	X	X	X	X	X		X	X
2	X	X	X	X	X	X	X	X
1	X	X	X	X	X	X	X	X

Use Brad's frequency table or line plot for Exercises 1–3.

1. How many more people chose sports programs than chose news? **2**

2. How many fewer people chose music videos than chose comedies? **5**

3. As many people chose adventures as chose which two possible pairs of programs together? **sports and music videos; mysteries and soap operas; mysteries and news; music videos and comedies**

Gather your own data about popular types of exercise among teenagers. **Answers will vary.**

4. Decide on how to word a survey question about exercising. Then decide on a method of sampling and the number of people that will make up your sample.

5. Ask your question, record the data, then display them in a frequency table or in a line plot.

6. Use your data display in a chart or poster that could serve as an ad for fitness and good health.

MAKING CONNECTIONS

Have students find examples of sampling of populations in different areas such as politics, business, entertainment, and education. Discuss their findings, particularly the sampling procedures and predictions, to determine how representative the sampling is.

5 FOLLOW-UP

Extra Practice What is the disadvantage of using each of the samples described below? Suggest a better way of choosing each sample.

1. Fifty people who live near some railroad tracks were asked whether or not they wanted railroad service discontinued in the town. **They would probably want to have the service discontinued because of the noise; better to take a random sampling of people living in the town.**

2. As they exited the most expensive restaurant in town, thirty people were asked to name the most popular restaurant in town. **People may express preferences, but may not know which is the most popular; better to poll all the restaurants about their numbers of patrons.**

Extension Encourage students to graph the results of their surveys in Exercises 8–11. Display the graphs on a bulletin board.

Section Quiz
1. Define the following in your own words: data, poll, population, sample, random sampling, convenience sampling, cluster sampling, systematic sampling.
2. Give an example to illustrate the meaning of each word.

Get Ready clock with second hand, almanac or encyclopedia

Additional Answers
See page 572.

Additional answers for odd-numbered exercises are found in the Selected Answers portion of the page.

1-2 Stem-and-Leaf Plots

You can find your heart rate by checking your pulse. Find the beat at the side of your neck or at your wrist. Have a partner watch the second hand of a clock for exactly 1 minute and tell you when to start counting the number of beats and when to stop. Record the pulse rates for the class.

a. Which pulse rate was the fastest?
b. Which pulse rate occurred most frequently?
c. How did you record the data?
d. Suggest other ways to record the data so that they are easy to analyze.

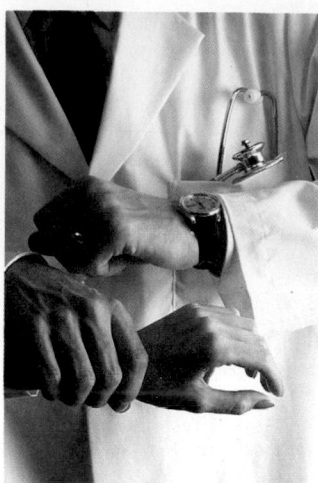

A **stem-and-leaf plot** can be used to organize data. In the example below, the digits in the tens places make up the **stems.** The digits in the ones places make up the **leaves.**

Example 1
Organize these pulse rates into a stem-and-leaf plot.

51	73	79	90	86	66	71	86	64
67	81	67	53	59	91	73	87	86
61	66	68	58	93	76	73	82	76

Solution
To start making a stem-and-leaf plot, first identify the least and greatest values in the set of data. The digit in the tens place in each of these values forms the least and greatest extremes of the stems. Write the stems in a column, from least to greatest. Draw a vertical line to the right of the stems. Write the ones digits of the values, in the order in which the values appear, to the right of their stems. Write an explanation of the plot to the left of the stems.

Stems	Leaves	
5	1 3 9 8	Least pulse rate is 51.
6	6 4 7 7 1 6 8	
7	3 9 1 3 6 3 6	
8	6 6 1 7 6 2	
9	0 1 3	Greatest pulse rate is 93. ◄

7|3 represents a pulse rate of 73

8   CHAPTER 1   Exploring Data

8

Analyze the data in a stem-and-leaf plot by looking for the greatest and least values and outliers, clusters, and gaps. **Outliers** are data values that are much greater than or much less than most of the other values. **Clusters** are isolated groups of values. **Gaps** are large spaces between values.

Example 2

This stem-and-leaf plot shows the heights, in inches, of twenty-two students. Write a description of the data. Be sure to note any outliers, clusters, and gaps.

```
              4 | 7  8  9  9
5|1 represents  5 | 1  5  7  7  8  9  9
51 in. tall     6 | 2  2  2  2  3  3  5
              7 | 1  1  2  7
```

Solution

This set of data indicates that the shortest of the students is 47 in. tall and the tallest is 77 in. tall. The value 77 is an outlier, since it is considerably separated from the other values. There are several clusters of values—those in the high 40s, those in the high 50s, and those in the low 60s. The greatest gaps in values fall between 65 and 71 and between 72 and 77. ◄

TRY THESE

1. Organize these basketball scores into a stem-and-leaf plot.

```
61   70   36   52   49   77   62
37   55   32   44   82   96   73
55   61   43   38   37   56   59
62   32   46   58   63   72   84
```
See Additional Answers.

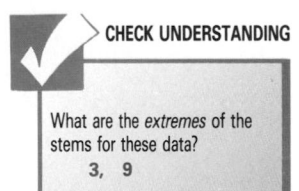

2. This stem-and-leaf plot shows the number of points scored by a high school basketball team during a recent season. Write a description of the data, noting any outliers, clusters, and gaps.

```
              3 | 4
              4 | 0  2
              5 | 1  3  7  8  9
5|1 represents  6 | 0  0  3  5  5  6  6  6  7  7  8
51 points       7 | 0  3  4  4  5  5  6  7  8
              8 | 7  7  8
              9 | 2  2  3  9
```

See Additional Answers.

✓ **CHECK UNDERSTANDING**

What are the *extremes* of the stems for these data?
3, 9

ASSIGNMENTS

BASIC
1–4, 5, PSA 1–4

AVERAGE
1–4; 5-6, 7, PSA 1–5

ENRICHED
5–10; PSA 1–7

ADDITIONAL RESOURCES
Reteaching 1–2
Enrichment 1–2
Transparency Master 1

Example 2: Stress that a description of a plot should include the greatest and least values and any outliers, clusters, and gaps.

Additional Questions/Examples
1. What other kinds of data could you present in a stem-and-leaf plot? **Answers will vary. Possible answers: weight of students in a class, basketball season scores.**
2. What are the advantages of organizing data in a stem-and-leaf plot? **It is quick; it makes it easy to analyze and draw conclusions about the data.**

Guided   Practice/Try   These
Observe students and provide guidance where necessary as they complete these exercises independently.

9

Key Questions
1. What is a stem-and-leaf plot?
2. What kind of data is best represented on a stem-and-leaf plot?
3. How do you read a stem-and-leaf plot?

4 PRACTICE

Extend/Solve Problems Make sure students understand how to plot decimal numbers. Have students share the paragraphs they wrote for Exercise 6.

Think Critically/Solve Problems Allow time for students to ask and answer each other's questions for Exercise 9.

Problem Solving Applications Have students compare the back-to-back stem-and-leaf plot with the data in the table. Ask them to explain how to organize the data using such a plot. For Exercise 6, discuss and list on the board several possible sets of data that can be compared.

5 FOLLOW-UP

Extra Practice The number of people using the school library each hour during one week was as follows:

```
27  18  32  25   9  12  19  31
35  22  15  21  27  17  20  28
23  16  24  10   7  13  24  14
```

1. Construct a stem-and leaf plot for the data.
2. Make a second plot, arranging the leaves from least to greatest.
3. For how many hours did more than 30 people use the library? **3**

EXERCISES

PRACTICE/ SOLVE PROBLEMS These data represent the numbers of baskets sunk by basketball players, each taking 50 shots.

```
27  34  41  35  48  36
12  32  42  50  39  38
42  17   8  42  28  37
 3  15  23   6  25  43
 9   4  14  41  21  29
11  17   9  26  31  34
```

1. How many players are represented? **36**

2. Organize the data into a stem-and-leaf plot. **See Additional Answers.**

3. How many players sank exactly 17 baskets? 21 baskets? **2; 1** 46 baskets? **0**

4. Write a description of the data. Note the greatest and least numbers of baskets sunk and any outliers, clusters, and gaps. **See Additional Answers.**

EXTEND/ SOLVE PROBLEMS **USING DATA** Use the Data Index on page 546 to find the statistics for the numbers of miles walked by workers in various jobs during the course of their workdays. Use these data for Exercises 5 and 6.

5. Make a stem-and-leaf plot for the data. (Hint: Think of the vertical line as replacing the decimal points.) **See Additional Answers.**

6. Write a paragraph telling which of the jobs might be of interest to you and whether or not the amount of walking involved is important to your decision. **Answers will vary.**

THINK CRITICALLY/ SOLVE PROBLEMS These data represent the number of voters at forty polling stations.

```
300  580  390  520  460  290  270  440  200  290
230  320  240  620  350  400  420  600  340  620
670  380  290  290  690  410  300  410  280  530
530  520  350  490  410  500  230  150  290  150
```

7. Make a stem-and-leaf plot for these data. Use the explanation that 2|3 represents 230 voters. **See Additional Answers.**

8. Write two questions based on the data. Then answer them. **Answers will vary.**

9. Ask your questions of others. Answer their questions. **Answers will vary.**

10. **USING DATA** Use the Data Index on page 546 to find the number of calories burned during various kind of exercise by a person weighing 110 lb. Round the values to the nearest ten. Then make a stem-and-leaf plot for the data. **See Additional Answers.**

Problem Solving Applications:

BACK-TO-BACK STEM-AND-LEAF PLOTS

A **back-to-back stem-and-leaf plot** can be used to organize two sets of data so that they can be compared. The stems form the "backbone" of this kind of plot. This back-to-back stem-and-leaf plot shows age data for the first three U.S. presidents.

Age at Inauguration		Age at Death
9	0	
8	3	
7		
1	6	7
7 7	5	
	4	

1. Copy and complete the back-to-back stem-and-leaf plot to include the data for all forty-one presidents. **See Additional Answers.**

2. How many presidents were inaugurated when they were in their fifties? **24**

3. Who was the youngest president to be inaugurated? How old was he? Who was the oldest president to be inaugurated? How old was he?
youngest—T. Roosevelt, 42; oldest—Reagan, 69

4. Which president died the youngest? How old was he?
Kennedy, 46

5. Describe any outliers, clusters, and gaps in the data. **Answers will vary.**

6. Research two sets of data that can be compared. You might use, for example, the numbers of games won by the teams in two leagues or the amounts of fat and cholesterol found in various foods. Organize the data into a back-to-back stem-and-leaf plot.
Answers will vary.

7. Exchange the plot you made for Exercise 6 for a classmate's plot. Write a description of the data in your classmate's plot.
Answers will vary.

AGES OF UNITED STATES PRESIDENTS AT INAUGURATION AND AT DEATH

Name	Term	Age at Inaug.	Age at Death
1. Washington	1789–1797	57	67
2. J. Adams	1797–1801	61	90
3. Jefferson	1801–1809	57	83
4. Madison	1809–1817	57	85
5. Monroe	1817–1825	58	73
6. J. Q. Adams	1825–1829	57	80
7. Jackson	1829–1837	61	78
8. Van Buren	1837–1841	54	79
9. W. H. Harrison	1841	68	68
10. Tyler	1841–1845	51	71
11. Polk	1845–1849	49	53
12. Taylor	1849–1850	64	65
13. Fillmore	1850–1853	50	74
14. Pierce	1853–1857	48	64
15. Buchanan	1857–1861	65	77
16. Lincoln	1861–1865	52	56
17. A. Johnson	1865–1869	56	66
18. Grant	1869–1877	46	63
19. Hayes	1877–1881	54	70
20. Garfield	1881	49	49
21. Arthur	1881–1885	50	56
22. Cleveland	1885–1889	47	71
23. R. Harrison	1889–1893	55	67
24. Cleveland	1893–1897	55	—
25. McKinley	1897–1901	54	58
26. T. Roosevelt	1901–1909	42	60
27. Taft	1909–1913	51	72
28. Wilson	1913–1921	56	67
29. Harding	1921–1923	55	57
30. Coolidge	1923–1929	51	60
31. Hoover	1929–1933	54	90
32. F. D. Roosevelt	1933–1945	51	63
33. Truman	1945–1953	60	88
34. Eisenhower	1953–1961	62	78
35. Kennedy	1961–1963	43	46
36. L. B. Johnson	1963–1969	55	64
37. Nixon	1969–1974	56	81
38. Ford	1974–1977	61	—
39. Carter	1977–1981	52	—
40. Reagan	1981–1989	69	—
41. Bush	1989–	64	—

 **CHECK UNDERSTANDING**

Why are there dashes in the Age-at-Death column for the last four presidents?

Why is there a dash in the Age-at-Death column for the twenty-fourth president, Cleveland?

Because they are still living; Cleveland was also the twenty-second president, so his age at death is listed under entry number 22.

4. For how many hours did fewer than 10 people use the library?
2

Extension Have pairs of students measure each other's height. List each student's name and height on the chalkboard. Have students make stem-and-leaf plots of the data, noting the greatest and least heights and any outliers, clusters, and gaps. Then have them make a back-to-back stem-and-leaf plot using this data and the list of heart rates developed for the Explore/Working Together activity. Have them see if they can find a correlation between a person's height and heart rate.

Section Quiz Use the data in Exercise 1 of Try These to answer these questions.

1. Suppose that each score had been 8 points higher. Construct a stem-and-leaf plot showing the new scores.

2. Make a second plot, based on your new plot, arranging the leaves from least to greatest.

Additional Answers
See page 572. Additional answers for odd-numbered exercises are found in the Selected Answers portion of the page.

MAKING CONNECTIONS

Have students predict the three most commonly used letters of the alphabet. Then have them check their predictions by tallying the frequency of occurrence of each letter of the alphabet in a literary passage of their choosing. Have them organize their data in a stem-and-leaf plot. **The letters e, t, and o are the most commonly used.**

1-3 Pictographs

EXPLORE

This graph shows the results of a survey taken of joggers one day in a big city park.

Use this clue to complete the key: *One hundred twenty-five joggers ran a distance of from 2 to 4 km on the day of the survey.*

a. Did more joggers run 0–2 km or over 10 km? **0–2 km**
b. The fewest joggers ran how many kilometers? **10 or more**
c. How many joggers ran 4–6 km? **125**
d. How many joggers were surveyed? **525**
e. Write a title for this graph.
   Distances Run by Joggers

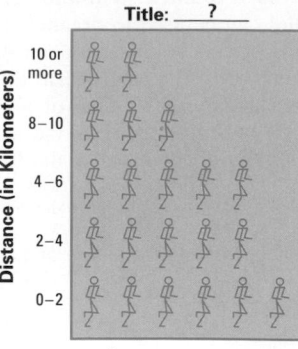

Title: ___?___

Distance (in Kilometers): 10 or more, 8–10, 4–6, 2–4, 0–2

Key: 🏃 represents __?__ joggers

SKILLS DEVELOPMENT

A picture graph, or **pictograph**, is a means of displaying data by using graphic symbols. The **key** identifies the number of data items represented by each symbol. The symbols often represent rounded, rather than exact, amounts. To read a pictograph, start by interpreting the key. Then multiply the value of one symbol by the number of symbols in a row to find the value for the row.

Example 1

Use this pictograph.

Key:
🐟 represents 10 lb of fish

FISH CONSUMPTION AROUND THE WORLD (in Pounds Eaten, Per Person, Per Year)
U.S.A. 🐟🐟🐟
France 🐟🐟🐟🐟
Japan 🐟🐟🐟🐟🐟🐟🐟🐟🐟🐟🐟🐟🐟🐟
India 🐟
Brazil 🐟🐟

a. In which country is the most fish eaten? About how many pounds of fish does the average person in that country eat in a year?
b. How much fish does the average person in the U.S. eat per year?
c. The average Norwegian eats about 110 lb of fish per year. How many fish symbols would be needed to represent this information on the pictograph?

12 CHAPTER 1   Exploring Data

Solution

a. The average person in Japan eats about 190 lb of fish per year.

b. The average person in the U.S. eats about 35 lb of fish per year.

c. Eleven fish symbols would be needed to represent the amount of fish eaten by the average Norwegian. ◀

To construct a pictograph, choose a symbol for the key, then determine a value for the symbol. Draw the symbols to represent the data. Write a title for the graph.

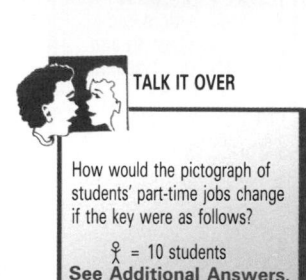

Example 2

Construct a pictograph for these data.

Number of Students Holding Part-Time Jobs:

Convenience-store worker	25	Newspaper deliverer	10
Supermarket cashier	15	Waiter	30
Short-order cook	25	Recycling-center worker	15
Babysitter	40		

Solution

NUMBER OF STUDENTS HOLDING PART-TIME JOBS

Convenience-store worker	☻ ☻ ☻ ☻ ☻
Supermarket cashier	☻ ☻ ☻
Short-order cook	☻ ☻ ☻ ☻ ☻
Babysitter	☻ ☻ ☻ ☻ ☻ ☻ ☻ ☻
Newspaper deliverer	☻ ☻
Waiter	☻ ☻ ☻ ☻ ☻ ☻
Recycling-center worker	☻ ☻ ☻

Key: ☻ represents 5 students   ◀

TALK IT OVER

How would the pictograph of students' part-time jobs change if the key were as follows?

☻ = 10 students
See Additional Answers.

TRY THESE

🧩 Use this pictograph.

AMOUNT OF SODIUM IN SOME FOODS

Pizza (1 slice)	🧂 🧂 🧂 🧂
Tomato juice (1 cup)	🧂 🧂 🧂 🧂 🧂 🧂 🧂 🧂
Pita bread (1)	🧂 🧂
Potato salad (1 cup)	🧂 🧂 🧂 🧂 🧂 🧂 🧂 🧂 🧂 🧂 🧂 🧂 🧂
Carrot cake (1 slice)	🧂 🧂 🧂 🧂

Key: 🧂 represents 100 mg of sodium

1. Which food contains the most sodium? About how much sodium does it contain?
potato salad; 1,300 mg

2. About how much sodium is there in two slices of pizza?
800 mg

3. There are 1,075 mg of sodium in 1 cup of cream of mushroom soup. How many saltshaker symbols would be needed to represent this information on the pictograph? $10\frac{3}{4}$ **symbols**

4. Construct a pictograph for these data. **See Additional Answers.**

Number of Students in After-School Activities:

Writer's Workshop	13	Softball Team	10	Math Team	18
Soccer Team	11	Crafts Club	8	Runner's Club	30

1–3   Pictographs   **13**

ASSIGNMENTS

BASIC
1–5, 6–10, 14–16

AVERAGE
6–13, 14–19

ENRICHED
6–13, 17–22

ADDITIONAL RESOURCES
Reteaching 1–3
Enrichment 1–3
Transparency Master 2

Additional Questions/Examples

1. What symbols and values would you use to show this data on a pictograph? **Answers may vary.**

Math Test Results

Score	Number of Students
90–100	6
80–89	8
70–79	4
below 70	3

2. Suppose 15 additional students took the math test. Seven scored 80–89, 6 scored 70–79, and 2 scored below 70. What changes would you make in the key for your pictograph to reflect this data? **Answers may vary.**

Guided Practice/Try These You may want students to work in pairs for Exercise 5.

5-MINUTE CLINIC

Exercise	Student's Error	Error Diagnosis
How many students like vanilla ice cream? How many like chocolate? **vanilla** ☻ ☻ ☻ ☻ **chocolate** ☻ ☻ ☻ Key: ☻ = 10 students	4, 10	• Student incorrectly calculates the value of several symbols. • Student incorrectly evaluates partial symbols.

SUMMARIZE

3

Talk it Over Discuss how the information in the graph in Try These would look as a table. Ask, *What advantage does a pictograph have over a table?* **It is easier to make comparisons from a pictograph**

PRACTICE

4

Practice/Solve Problems Discuss students' responses to Exercise 5. You may want students to work in pairs to construct the pictographs in Exercises 6 and 11. Have students share the reasoning they used in constructing their graphs.

Think Critically/Solve Problems Have students share the paragraphs they wrote for Exercise 22.

FOLLOW-UP

5

Extra Practice Have students construct pictographs for these data:

Touchdowns Scored

Turnbull	11	Lofthouse	6
Rausse	3	Jarvis	8
Lane	10	Ramirez	4

Have students answer these questions using their pictographs

1. Who scored the most touchdowns? **Turnbull**
2. How many more touchdowns did Ramirez score than Rausse? **1**
3. Which player scored twice as many touchdowns as Ramirez? **Jarvis**

14

PRACTICE/ SOLVE PROBLEMS

MIXED REVIEW

Add or subtract.

1. 4,685 + 1,709 **6,394**
2. 17,505 − 8,936 **8,569**
3. 30,710 − 19,666 **11,044**
4. 23,006 + 10,084 **33,090**

Find the fractional part.

5. $\frac{1}{3}$ of 360 **120**
6. $\frac{1}{4}$ of 360 **90**
7. $\frac{1}{9}$ of 360 **40**
8. $\frac{3}{4}$ of 360 **270**

Solve.

9. A dump truck contains 1,440 lb of sand that must be moved by wheelbarrow to a construction site. If the wheelbarrow can move 90 lb in one trip from the dump truck to the site, how many trips must be made to move all the sand? **16 trips**

Use this pictograph for Exercises 1–5.

STUDENT PREFERENCES— KINDS OF MUSIC

Rock, Jazz, Country & Western, Classical, Blues, Other

Key: ♪ represents 50 students

1. Which type of music is most popular? **rock**
2. Which type of music is least popular? **classical**
3. How many students prefer country and western? **200**
4. How many more students prefer blues than prefer classical music? **150**
5. What kinds of music could be represented by "other"? How many students might prefer each of the "others"?

Answers will vary. Students should suggest different kinds of music and the number of students (totaling 150) that prefer each kind.

Roberto has scored the most points on the basketball team. Here are his scores for a recent season's game.

Game	1	2	3	4	5	6	7	8
Number of points	12	20	16	12	20	4	16	24

6. Construct a pictograph for the above data. For your key use a basketball symbol to represent 4 points. **See Additional Answers.**
7. In which game did Roberto score the greatest number of points? **8**
8. In which game did he score the least number of points? **6**
9. What was Roberto's total number of points in the first half of the season? **60**
10. What was Roberto's total in the second half of the season? **64**

The number of points scored by the hockey players on one team are shown in the chart.

Player	Kelly	Green	Tookey	Currie	Smith	Charron
Points	60	74	45	35	40	48

11. Construct a pictograph to show the numbers of points scored. For your key use a hockey puck, ⬬, to represent 10 points and ◖ to represent 5 points. **See Additional Answers.**
12. How many more points did Green score than Charron? **26**
13. Kelly scored half again as many points as which player? **Smith**

14 CHAPTER 1 Exploring Data

AT-RISK STUDENTS

Have students gather data on the number of students in 5 different classes. Have them use grid–paper strips of ten to represent 10 students. Help them to construct a pictograph using the strips, and parts of strips when necessary.

This pictograph shows the number of situps done by the students in one school. Use this pictograph for Exercises 14–19.

NUMBER OF SITUPS	
0 to 35 (poor)	∿∿∿∿
36 to 45 (satisfactory)	∿∿∿∿∿∿
46 to 60 (good)	∿∿∿∿∿∿∿
61 or more (excellent)	∿∿∿

Key: ∿ represents 25 students

14. How many students are in the excellent category? **75**

15. There are 75 more students in the satisfactory category than in which other category? **excellent**

16. How many students do these data represent? **500**

In the second half of the school year, 20 more students enrolled in school. They were all tested. Each did between 36 and 45 situps.

17. How many students are now in the satisfactory category? **170**

18. How many students could now be represented by the pictograph? **520**

19. How would you change the graph to include the new students?
Answers will vary.

Here are the greatest numbers of home runs hit by the top American and National League players for three recent years. Use these data to help you make decisions for Exercises 20–22.

THINK CRITICALLY/
SOLVE PROBLEMS

American League players: 42, 36, 51
National League players: 39, 47, 40

This graph shows the number of home runs hit by the top players on two high school baseball teams for five recent years.

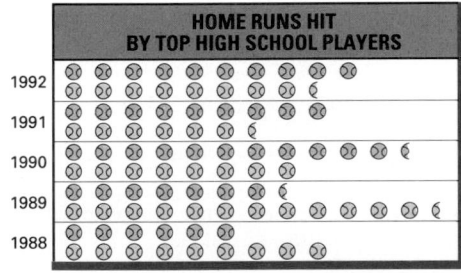

HOME RUNS HIT
BY TOP HIGH SCHOOL PLAYERS

1992
1991
1990
1989
1988

Key: ⊗ represents ?
Crain H.S.
Home Runs
⊗ represents ?
St. Clair H.S.
Home Runs

WRITING ABOUT MATH

Write about the following in your journal or notebook.

What are some similarities between stem-and-leaf plots and pictographs? When would you display data in a stem-and-leaf plot? When would you display data in a pictograph?
Answers will vary.

20. Which of these is a reasonable number with which to complete the key for the high school baseball data—2, 3, 5, or 10? **2 or 3**

21. The players on which team have the better five-year record? According to the number you determined for the key, by how many home runs is this team ahead of the other team? **St. Clair; Answers will vary.**

22. Write a paragraph explaining your answer to Exercise 20.
Answers will vary, but should show that the major league players' scores reflect greater ability and a longer playing season.

EXTEND/
SOLVE PROBLEMS

Extension Have students cut out pictographs from newspapers or magazines. Ask them to write five questions about their pictographs. Display the graphs and questions on a bulletin board. Have students answer the questions for extra credit.

Section Quiz Use the pictograph for Exercises 1–5 in Practice/Solve Problems to answer these questions.

1. How many more students like rock than jazz? **150**
2. How would you show that 125 students like rap music? **with 2 1/2 note symbols**
3. Construct a new pictograph that shows the same information but uses a different key. **Pictographs will vary.**

Additional Answers
See page 572.

CHALLENGE

Have students construct a pictograph using the data they obtained from their survey in Section 1–1, Think Critically/Solve Problems. Pair students who conducted similar surveys, challenging them to prepare a double pictograph with which they can compare both sets of data.

WARM-UP

If $9 \times 12 = 108$
       $9 \times 123 = 1{,}107$
       $9 \times 1{,}234 = 11{,}106$
       $9 \times 12{,}345 = 111{,}105$
Then $9 \times 123{,}456 = \mathbf{1{,}111{,}104}$

1 MOTIVATE

Introduction Discuss the types of problems computers can solve and how they differ from problems that require logical reasoning.

2 TEACH

Use the Pages/Problem Have students read this part of the section and then discuss the problem.

Use the Pages/Solution Discuss the solution as presented in the text. Stress steps 4 and 5.

3 SUMMARIZE

Write About Math Ask students to pretend they are writing a letter to a friend in which they teach their friend the 5-step plan for solving problems.

1-4   Problem Solving Strategies:
INTRODUCING THE 5-STEP PLAN

You know that people use computers as powerful problem solving tools. Computers can store data and then organize and compute them at breakneck speeds. However, computers still cannot solve problems the way people can—by thinking, or reasoning!

PROBLEM

A computer magazine published a survey for home-computer users. Of those who responded to the survey, 5,410 users said they used only daisy-wheel printers; 11,068 said they used only dot-matrix printers; and 3,709 said they used only laser printers. How many users responded to the survey?

SOLUTION

5-STEP PROBLEM SOLVING PLAN

► READ
► PLAN
► SOLVE
► ANSWER
► CHECK

1. READ . . . Read the entire problem slowly and carefully. Then go back and read the question again.

How many users responded to the survey?

2. PLAN . . . Decide what you can do to solve the problem.

Add to find the number of users that responded.

3. SOLVE . . . Put your plan into action.

$5{,}410 + 11{,}068 + 3{,}709 = 20{,}187$

4. ANSWER . . . State your answer in terms of the original question.

A total of 20,187 users responded to the survey.

5. CHECK . . . Determine whether or not your answer is *reasonable*.

Round each addend to the nearest thousand. Then add.
$5{,}000 + 11{,}000 + 4{,}000 = 20{,}000$
So, the answer is reasonable.

TEACHING TIP

Encourage students to create a mnemonic device to help them remember the 5-step plan. They might suggest, for example, Ruth Picks Sunflowers And Carnations.

PROBLEMS

Use the 5-step plan to solve these problems.

This stem-and-leaf plot shows the number of students in the computer classes in a school system. Use it for Exercises 1 and 2.

```
        0 | 7 9 9
        1 | 0 3 5 5 7 8 9
0/7 represents    2 | 1 4 5 8 9
7 students        3 | 3 6 7 7 8
        4 | 0 1 1 2
```

1. If the classes of 40 or more students were disbanded, with the students redistributed among the other classes, how many computer classes would remain? **20 classes**

2. What is the total student enrollment in the computer classes? **604 students**

For the cost of an airline ticket, a passenger may check two suitcases. Checking a third suitcase costs an additional $35. Checking the fourth to the sixth suitcases costs $45 each. Checking a seventh suitcase or more costs $55 each.

3. How much money would the airline collect for additional suitcases if nine passengers each checked a third suitcase? **$315**

4. How much would a passenger pay for checking a total of five suitcases? **$125**

5. If a family of three moving to another state paid $289 each for their airline tickets and checked eight suitcases, what would be their total cost? **$937**

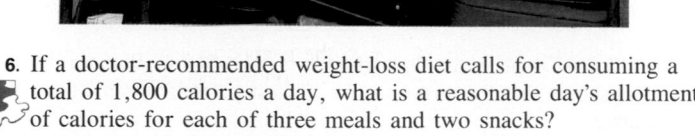

6. If a doctor-recommended weight-loss diet calls for consuming a total of 1,800 calories a day, what is a reasonable day's allotment of calories for each of three meals and two snacks?
Answers will vary. A possible solution follows: breakfast—350 calories; lunch—550 calories; dinner—800 calories; snacks—100 calories each

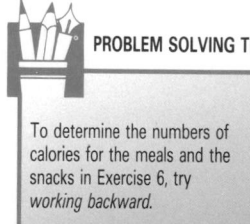

PROBLEM SOLVING TIP

To determine the numbers of calories for the meals and the snacks in Exercise 6, try *working backward.*

1–4   Problem Solving Strategies: Introducing the 5-Step Plan   **17**

WARM-UP

If 5,000,000 can be written as 5.0 million and 28,000,000 can be written as 2.8 million, how would you write 7,000,000?
7.0 million

1 MOTIVATE

Introduction This lesson demonstrates how to simplify data in order to present it graphically. Discuss the differences between showing data on a chart and on a graph.

2 TEACH

Use the Pages/Problem Have students read this part of the section and then discuss the problem. You may wish to discuss the kind of graph to use to present these data.

Use the Pages/Solution Emphasize that students should always identify the high and low numbers before choosing a scale. Also, point out that when data is simplified, it is not changed.

1-5

► READ
► PLAN
► SOLVE
► ANSWER
► CHECK

Problem Solving Skills:
CHOOSE AN APPROPRIATE SCALE

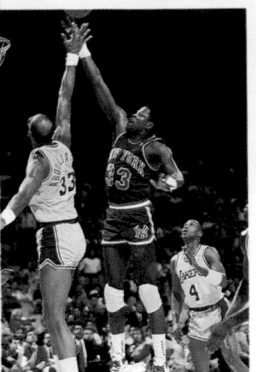

In the lessons that follow you will be constructing various types of graphs. For each type of graph, you will need to determine an appropriate scale. When determining the scale, you will find that it helps to first write the data in a simpler form.

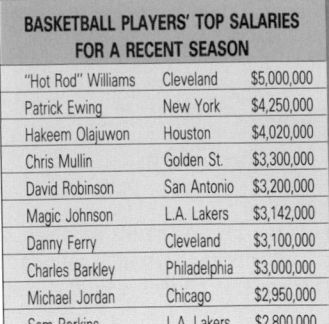

BASKETBALL PLAYERS' TOP SALARIES FOR A RECENT SEASON		
"Hot Rod" Williams	Cleveland	$5,000,000
Patrick Ewing	New York	$4,250,000
Hakeem Olajuwon	Houston	$4,020,000
Chris Mullin	Golden St.	$3,300,000
David Robinson	San Antonio	$3,200,000
Magic Johnson	L.A. Lakers	$3,142,000
Danny Ferry	Cleveland	$3,100,000
Charles Barkley	Philadelphia	$3,000,000
Michael Jordan	Chicago	$2,950,000
Sam Perkins	L.A. Lakers	$2,800,000

PROBLEM

Assume you have to graph the data in this table. What size interval would you use for the scale of the graph?

SOLUTION

The data are given in millions of dollars. Simplify the data. Express the greatest and least salaries in simpler forms.

You can write $5,000,000 as 5.0 million dollars. You can write $2,800,000 as 2.8 million dollars. So, all the data lie between 5.0 and 2.8 million dollars.

Now, determine the size of the intervals for your scale. If you use intervals of 0.2, you would need 25 intervals. If you use intervals of 0.5, you would need 10 intervals. If you use intervals of 1.0, you would need 5 intervals.

You might decide that 9 intervals of 0.6 would best reflect the data.

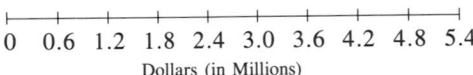

0   0.6 1.2 1.8 2.4 3.0 3.6 4.2 4.8 5.4
Dollars (in Millions)

PROBLEMS

Determine the intervals of the scale you would use to graph each set of data. **See Additional Answers**.

1. *USING DATA* Use the Data Index on page 546 to find statistics on the shows that have run for the longest periods of time on Broadway.

AT-RISK STUDENTS

Have students actually work through the various scales suggested for the sample problem (Basketball Players' Top Salaries for a Recent Season). Have them work together to complete scales using intervals of 0.2 million, 0.5 million, and 1.0 million. Discuss which is the best scale to present the data and why.

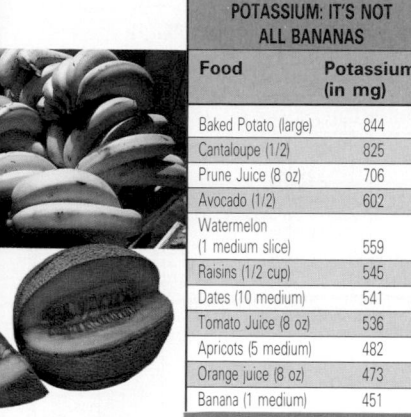

2. While bananas are often considered to be the best source of potassium, many other foods actually have more.

POTASSIUM: IT'S NOT ALL BANANAS	
Food	Potassium (in mg)
Baked Potato (large)	844
Cantaloupe (1/2)	825
Prune Juice (8 oz)	706
Avocado (1/2)	602
Watermelon (1 medium slice)	559
Raisins (1/2 cup)	545
Dates (10 medium)	541
Tomato Juice (8 oz)	536
Apricots (5 medium)	482
Orange juice (8 oz)	473
Banana (1 medium)	451

3.

Most Popular Programs at Summer Camp

Program	Camps Offering Program
Swimming	3,426
Art	1,986
Cycling	1,679
Horseback riding	1,573
Canoeing	1,540
Tumbling, gymnastics	1,266
Archery	1,256
Drama	807
Tennis	766
Sailing	733
Hockey	693
Physical fitness	668
Softball	655
Basketball	619
Rowing	554
Backpacking	483
Music	442
Nutrition	434
Hiking	379

4. A recent survey showed the time, to the nearest 15 minutes, spent watching TV each week by an average household in selected locations.

Market	Time (h:min)
Dallas–Fort Worth	56:45
Detroit	56:45
Houston	55:30
Atlanta	54:15
Philadelphia	53:00
Boston	50:30
Chicago	50:30
Sacramento–Stockton	50:30
Hartford–New Haven	49:15
Los Angeles	49:15
Memphis	49:15
New Orleans	49:15
New York	49:15
Seattle–Tacoma	49:15
Washington, D.C.	49:15
Cleveland–Akron	48:00
Denver	48:00
Miami–Ft. Lauderdale	48:00
Minneapolis–St. Paul	46:30
Nashville	46:30
Pittsburgh	46:30
Raleigh–Durham	46:30
Cincinnati	45:15
Oklahoma City	45:15
San Francisco–Oakland–SanJose	45:15
St. Louis	45:15
Portland, Oregon	44:30
Buffalo, N.Y.	44:00
Charlotte	44:00
Indianapolis	44:00
Orlando–Daytona Beach–Melbourne	44:00
Tampa–St. Petersburg–Sarasota	44:00
Baltimore	42:45
Columbus, Ohio	42:45
Kansas City, Mo.	42:45
Greenville–Spartanburg, S.C.–Asheville	42:45
Milwaukee	41:30
Grand Rapids–Kalamazoo–Battle Creek	41:30
Phoenix	40:15
San Diego	37:45

5.

Money Spent on Concessions

Event	Money Spent on Concessions
Amusement park	$5.25
Tractor pull	$5.00
Boxing	$4.75
NFL football game	$4.00
Major-league baseball game	$3.90
Wrestling match	$3.75
Ice-hockey game	$3.00
Jai alai match	$3.00
NBA basketball game	$3.00

6.

Major League Lifetime Records	
Player	Average
Ty Cobb	.366
Rogers Hornsby	.358
Joe Jackson	.356
Pete Browning	.347
Ed Delahanty	.346
Willie Keeler	.345
Billy Hamilton	.344
Ted Williams	.344
Tris Speaker	.344
Dan Brouthers	.342
Jesse Burkett	.342
Babe Ruth	.342
Harry Heilmann	.342
Bill Terry	.341
George Sisler	.340
Lou Gehrig	.340

7.

Some of the World's Largest Dams		
Largest[1]	Location	yd³
New Cornelia	Arizona	274,026,000
Tarbela	Pakistan	186,000,000
Fort Peck	Montana	125,600,000
Oahe	S. Dakota	92,000,000
Mangla	Pakistan	85,870,000
Gardiner	Canada	85,740,000
Oroville	California	78,000,000

[1]All earthfill except Tarbela (earth and rockfill).

BASIC
1, 2, 5

AVERAGE
1, 3–5, 7

ENRICHED
3–7

ADDITIONAL RESOURCES
Reteaching 1–5
Enrichment 1–5

3 SUMMARIZE

Talk It Over Have students brainstorm other examples of data they would need to simplify to present graphically. **Possible answers: distance of planets from the sun, population figures**

4 PRACTICE

Problems For Exercises 3, point out the format in which time is written in hours and minutes. Discuss students answers for all exercises, allowing them to give opinions about which are the best scales and why.

5 FOLLOW-UP

Extra Practice Have students use an almanac to research the 10 most populous states in the United States. Have them list the data in a table and then simplify the data and show the scale they would use to present it graphically.

Get Ready graph paper, rulers

Additional Answers
See page 572.

1-6   Bar Graphs

**EXPLORE/
WORKING
TOGETHER**

The data in the table show the number of injuries for which emergency-room treatment was needed in one recent year.

CHILDREN'S INJURIES CAUSED BY TOYS

Toy	Number of Injuries
Marbles	1,845
Balloons	1,913
Toy guns	2,581
Blocks and pull toys	2,799
Nonwheeled riding toys	4,022
Flying toys	4,263
Wheeled riding toys	6,553
Children's wagons	7,935

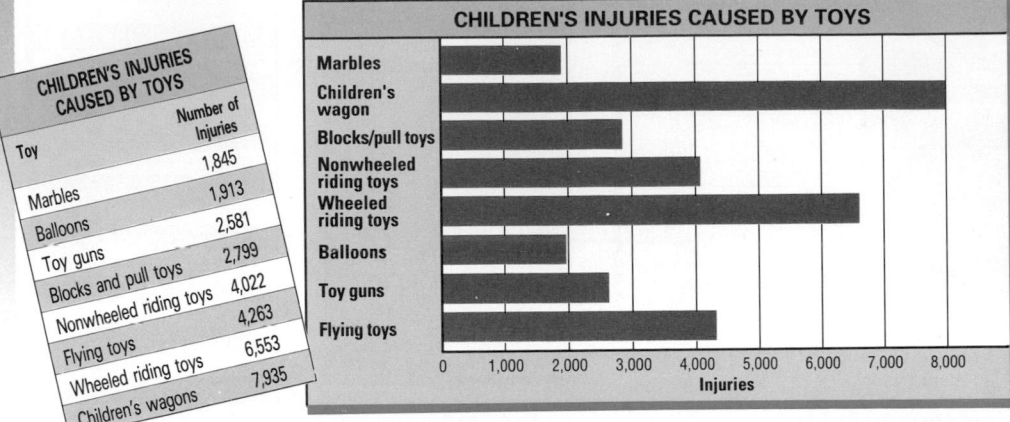

Use the bar graph of the data in the chart to answer these questions.

a. The scale at the bottom of the graph is marked in intervals of what number? **1,000**

b. The left edges of all the bars represent what number on the scale? **0**

c. Find the bar for injuries caused by flying toys. Imagine a vertical line drawn from the right edge of this bar downward to meet the scale. Between which two numbers on the scale would the vertical line fall? **between 4,000 and 5,000**

d. What does the shortest bar tell you? **that marbles were the cause of almost 2,000 injuries**

e. What does the longest bar tell you? **that children's wagons were the cause of about 8,000 injuries**

**SKILLS
DEVELOPMENT**

In a **bar graph**, either horizontal or vertical bars are used to display data. A scale is used to show number values.

To read a horizontal bar graph, look at the right edge of each bar. Match that edge with the number on the scale at the bottom of the graph to find the value for that bar. To read a vertical bar graph, match the top edge of each bar to the scale at the side of the graph to find the value of the bar.

Example 1

Use the bar graph at the right.

a. Which bar represents the longest river? About how long is that river?

b. About how much longer is the Nile than the Yangtze?

c. Which rivers are over 5,400 km long?

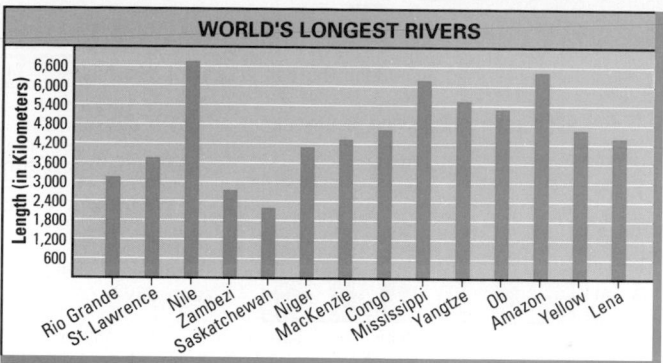

WORLD'S LONGEST RIVERS

Solution

a. The longest bar represents the Nile River. The top of the bar is very close to the 6,600-km mark on the scale. The Nile is about 6,600 km long.

b. The Nile is about 1,200 km longer than the Yangtze.

c. The Nile, the Mississippi, the Yangtze, and the Amazon are each over 5,400 km long. ◄

When you construct a bar graph, you may have to estimate the lengths of some of the bars.

Example 2

Construct a bar graph for these data.

NATIONAL HOCKEY LEAGUE STANDINGS			
Team	**Points**	**Team**	**Points**
N.Y. Islanders	65	Philadelphia	59
Detroit	44	N.Y. Rangers	53
Montreal	56	Vancouver	26
Toronto	31	Washington	14
Boston	47	Edmonton	14

Solution

To construct a bar graph, choose an appropriate scale, then label it. Draw and label the bars. Write a title for the graph.

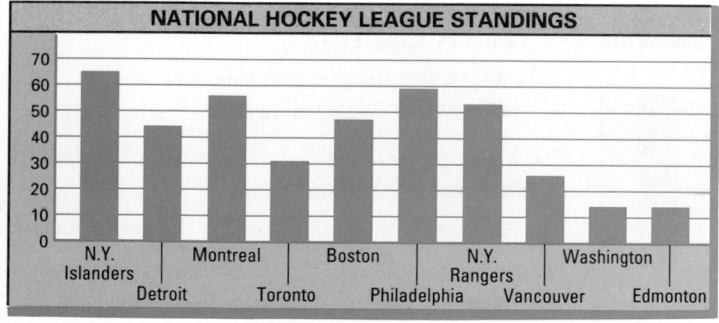

NATIONAL HOCKEY LEAGUE STANDINGS

BASIC
1–6, 8–11

AVERAGE
1–6, 8–11, 12–14

ENRICHED
7–11, 12–15

ADDITIONAL RESOURCES
Reteaching 1–6
Enrichment 1–6
Transparency Master 4

Example 2: Discuss how to choose an appropriate scale to display the data. Have students suggest another scale that would work other than the one shown in the solution. **Possible answer: intervals of 5's**

Additional Questions/Examples
1. Explain how to choose a scale when constructing a bar graph. **Identify the greatest and least data, then choose increments that best show all the data.**

2. What are some other examples of data you could display on a bar graph? **Answers will vary.**

Guided Practice/Try These You may wish to have students answer Exercises 1–11 orally. Observe students as they complete Exercise 12, providing guidance as necessary.

CHALLENGE

Have students survey two different classes of students about their hobbies. Then have them construct a double bar graph which compares the results of the survey.

Write About Math Have students write in their math journals the steps to follow when constructing a bar graph. **Choose an appropriate scale and label it, estimate and draw the bars, label the bars, then write a title.**

4 PRACTICE

Extend/Solve Problems For Exercise 7, have students include at least five buildings on their graph. Explain that the bars on their graphs can be either horizontal or vertical. Have students include the data given in Exercises 9 and 10 on the graphs they constructed for Exercise 8.

Think Critically/Solve Problems Point out that, to answer Exercises 12 and 13, students must compare the heights of the bars since there are no scales on these graphs.

5 FOLLOW-UP

Extra Practice Which of the following questions could you answer using the bar graph in the Explore activity? Answer them. **2 and 3**

1. How many children were injured by swallowing puzzle pieces?
2. About how many more children were injured using wheeled riding toys than nonwheeled riding toys? **about 2,500 children**
3. Which toys caused more than 4,000 injuries? **wagons, wheeled riding toys, flying toys, non-wheeled riding toys**

TRY THESE

Use the bar graph for Exercises 1–11.

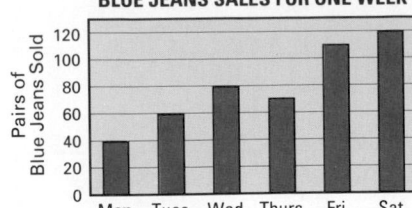

BLUE JEANS SALES FOR ONE WEEK

Which day had the lower number of sales?

1. <u>Monday</u> or Thursday   2. <u>Thursday</u> or Friday
3. Wednesday or <u>Saturday</u>   4. Saturday or <u>Tuesday</u>

About how many pairs of blue jeans were sold on each day?

5. Thursday **70**   6. Friday **110**   7. Saturday **120**

8. On which days were more than 60 pairs of blue jeans sold? **Wed., Thurs., Fri., Sat.**
9. On which days were more than 80 pairs of blue jeans sold? **Fri., Sat.**
10. About how many more pairs of blue jeans were sold on Thursday than on Monday? **about 30**
11. About how many pairs of blue jeans were sold on Wednesday and Friday together? **about 190**

12. Construct a bar graph for the video-rental data.

    See Additional Answers.

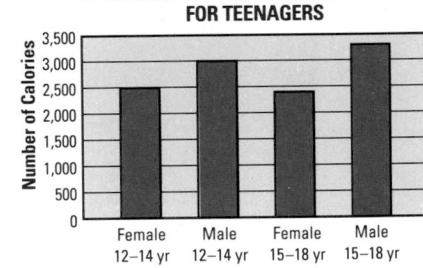

Student Preferences for Video Rentals

Types of Videos	Number of Videos Rented
Westerns	32
Musicals	17
Comedies	41
Action/Adventure	53
Science Fiction	70
Horror	50
Other	20

EXERCISES

PRACTICE/ SOLVE PROBLEMS

Use the bar graph below for Exercises 1–6.

AVERAGE DAILY CALORIE REQUIREMENTS FOR TEENAGERS

1. Which age group requires the greatest number of calories? **male, 15–18**
2. Which age group requires the least number of calories? **female, 15–18**

Carla is 14 years old and so far today has consumed 1,700 calories.

3. How many more calories does she need for the day?
4. How many fewer calories does she need for the day than does a boy of her age? **500**

22   CHAPTER 1   Exploring Data

Michael is 16 years old. He consumed 1,100 calories at breakfast.

5. How many more calories does he need for the day? **2,200**

6. How many more calories does he need for the day than does a girl of his age? **900**

7. **USING DATA**   Use the Data Index on page 546 to find the heights of the tallest buildings in the world. Construct a bar graph for these data. Show either the numbers of stories or the heights of the buildings, in meters or in feet. **Check students' graphs.**

Use the table below.

DEPTHS OF LARGE BODIES OF WATER			
Body of Water	Average Depth (m)	Body of Water	Average Depth (m)
Hudson Bay	90	Atlantic Ocean	3,740
Mediterranean Sea	1,500	Pacific Ocean	4,190
Arctic Ocean	1,330	Black Sea	1,190

8. Construct a horizontal bar graph for the data. **See Additional Answers.**

9. The average depth of the Caribbean Sea is 2,575 m. Which of the bodies of water listed here are shallower? **Hudson, Mediterranean, Arctic, Black**

10. The average depth of the Red Sea is 540 m. Which of the bodies of water listed here are deeper than the Red Sea? **all but Hudson Bay**

11. Create a question about the bar graph. Then answer it. **Answers will vary.**

A double-bar graph allows you to compare two sets of data.

USING DATA   Use the Data Index on page 546 to find the data showing the links between vegetable consumption and lung cancer. For Exercises 12–14, compare the data within each double-bar graph. Then compare the two graphs.

12. About how much greater is the risk of lung cancer in women who consume few tomatoes than in women who consume many tomatoes? **about $3\frac{1}{2}$ times greater**

13. About how much lower is the risk of lung cancer in men who consume many cruciferous vegetables than in men who consume few of these vegetables? **about $2\frac{1}{2}$ times lower**

14. For whom, men or women, do all vegetables have the greater effect on the risk of lung cancer? **women**

15. **USING DATA**   Use the Data Index on page 546 to find the sleep-patterns data. About how much sleep does a 1-year-old child get daily? **about $14\frac{1}{2}$ h**

EXTEND/ SOLVE PROBLEMS

THINK CRITICALLY/ SOLVE PROBLEMS

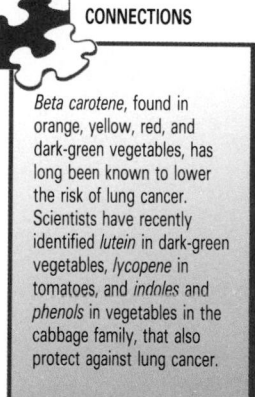

CONNECTIONS

Beta carotene, found in orange, yellow, red, and dark-green vegetables, has long been known to lower the risk of lung cancer. Scientists have recently identified *lutein* in dark-green vegetables, *lycopene* in tomatoes, and *indoles* and *phenols* in vegetables in the cabbage family, that also protect against lung cancer.

1–6   Bar Graphs   **23**

WARM-UP

Evaluate.
1. Is 27 closer to 25 or 30? **25**
2. Is 170,000 closer to 150,000 or 200,000? **150,000**

1 MOTIVATE

Introduction Point out that the scale on the bar graph uses rounded numbers in increments of 50,000.

2 TEACH

Use the Pages/Problem Have students read this part of the section and then discuss the problem. Emphasize that students should identify the number on the scale closest to the top of each bar before estimating.

Use the Pages/Solution Point out that the estimated solution is not exact. Discuss other answers that might be reasonable.

1-7

► READ
► PLAN
► SOLVE
► ANSWER
► CHECK

Problem Solving Skills:
ESTIMATION

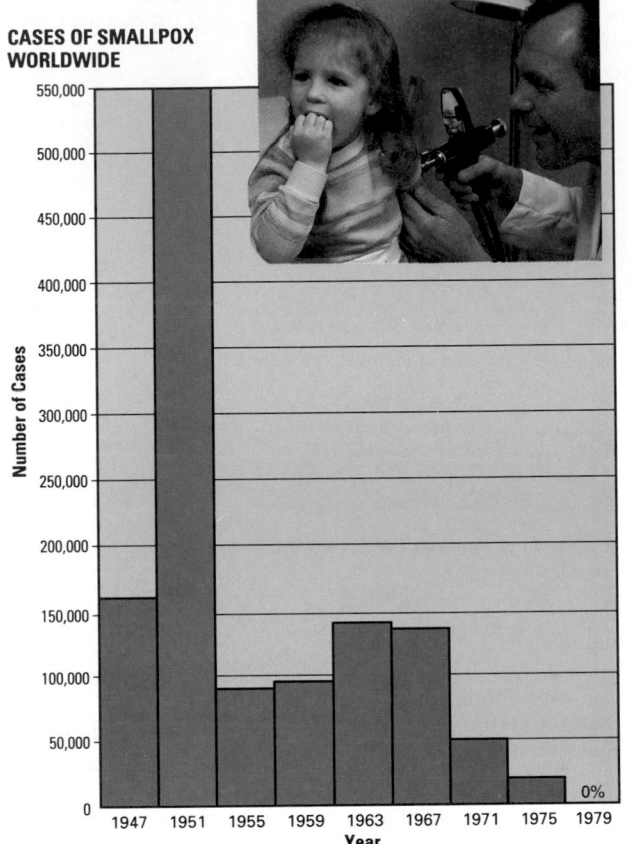

CASES OF SMALLPOX WORLDWIDE

Some graphs display greater numbers in rounded terms, instead of in exact terms. You have to make estimates in order to interpret such graphs.

PROBLEM

Use the graph to estimate about how many times as many smallpox cases there were in 1951 as compared to the number of cases in 1971.

SOLUTION

In 1951, there were about 550,000 cases of smallpox. So, the top of the bar for 1951 is at the 550,000 mark. There were about 50,000 cases in 1971. So, the top of the bar for 1971 is at the 50,000 mark. So there were about 11 times as many smallpox cases in 1951 as there were in 1971.

PROBLEMS

Use the bar graph above for Exercises 1–2.

1. Estimate in which years the number of smallpox cases was about twice the number that occurred in 1971. **1955; 1959**

2. Estimate in which year the number of smallpox cases was about one-third the number of cases that occurred in 1947. **1971**

MAKING CONNECTIONS

Have students bring in graphs from newspapers and magazines. Tell them to write five questions about their graphs. Have students exchange graphs and answer the questions.

Use these graphs for Exercises 3–4. Estimate each answer.

DISTRIBUTION OF BLOOD IN THE HUMAN BODY

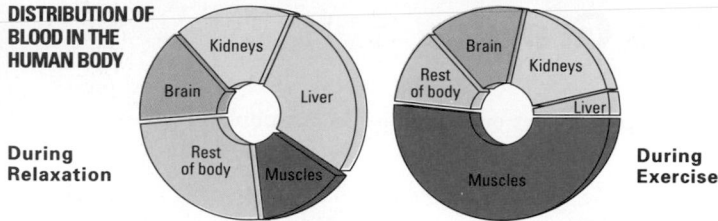

During Relaxation   During Exercise

3. How many times as much blood is in the muscles during exercise as during relaxation? **about 4 times**

4. How many times as much blood is in the liver during relaxation as during exercise? **about 10 times**

Use this bar graph for Exercises 5–6.

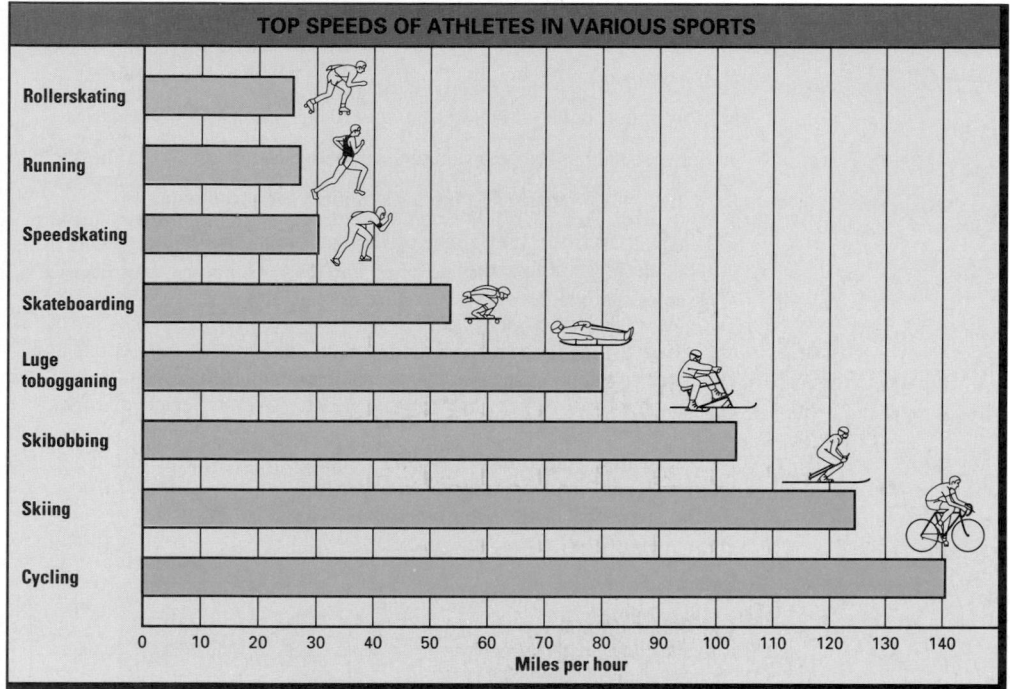

TOP SPEEDS OF ATHLETES IN VARIOUS SPORTS

Rollerskating
Running
Speedskating
Skateboarding
Luge tobogganing
Skibobbing
Skiing
Cycling

0 10 20 30 40 50 60 70 80 90 100 110 120 130 140
Miles per hour

5. The record speed for cycling is about twice as fast as the record speed for which sport? **tobogganing**

6. The record speed for roller skating is about one-third the record speed for which sport? **tobogganing**

TEACHING TIP

Some students may have difficulty working with greater numbers. When working with the smallpox graph, suggest that students drop the last three zeros and work with the simpler numbers to answer the questions. They can then reinstate the three zeros to make their estimates.

ASSIGNMENTS

BASIC
1–4

AVERAGE
1–6

ENRICHED
1–6

ADDITIONAL RESOURCES
Reteaching 1–7
Enrichment 1–7

3 SUMMARIZE

Write About Math Have students explain how to estimate the number of smallpox cases in the world in 1959. **A possible estimate would be 90,000.**

4 PRACTICE

Problems No numerical data is given in the graphs for Exercises 3–4. Point out that students must compare like segments on the two graphs to derive their estimates.

5 FOLLOW-UP

Extra Practice Have students use the circular graphs from Exercises 3 and 4 to answer these questions.
1. In which parts of the body is there little or no change in the distribution of blood during relaxation and during exercise? **kidneys and brain**
2. Does more change take place in the liver or in the muscles? **in the liver**

Get Ready graph paper, rulers

WARM-UP

Are the data increasing or decreasing?
1. 36, 47, 54, 87 **increasing**
2. 903, 901, 890, 889
   decreasing

1 MOTIVATE

Explore/Working Together Have students work in pairs. Discuss their answers to question **e**.

2 TEACH

Use the Pages/Skills Development
Have students read this part of the section and then discuss the example. Have students give examples of kinds of data which would best be represented with line graphs.

Example 1: Discuss how to use estimation to help read the line graph. Ask, *What does the trend of the data indicate when you consider that the population of the United States is constantly increasing?* **Possible answer: Today's farms must be much larger than those in the past to accommodate the increasing population.**

26

1-8   Line Graphs

EXPLORE/ WORKING TOGETHER

This line graph shows one child's height from birth to age 10.

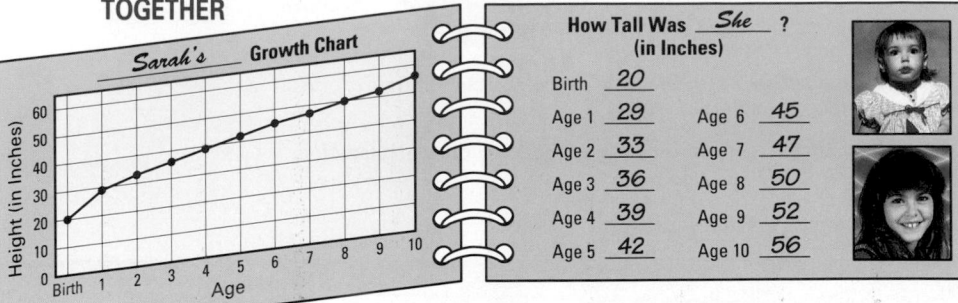

a. What was Sarah's height at age 4? How tall was she at age 8?
   39 in., 50 in.
b. How much taller was Sarah at age 10 than at age 5? **14 in. taller**
c. Between which two consecutive years did Sarah grow the most?
   between birth and age 1
d. Work with a partner. Make a bar graph for these data.
   See Additional Answers.
e. Which graph, the line graph or the bar graph, better represents these data? Explain. **The line graph better represents the data, because it shows that the data are continuous, or ongoing.**

SKILLS DEVELOPMENT

On a **line graph**, points representing data are plotted, then connected with line segments. Because the points are connected in sequence, a line graph shows trends, or changes in data, over a period of time.

To read a line graph, locate the first data point. Relate it to the corresponding labeled points on the vertical and horizontal scales. Do the same for each of the other data points.

Example 1

Use the line graph at the right.
a. About how many farms were there in the United States in 1940? About how many were there in 1990?
b. About how many fewer farms were there in 1960 than in 1950?
c. Describe the trend of the data as *increasing* or *decreasing*.

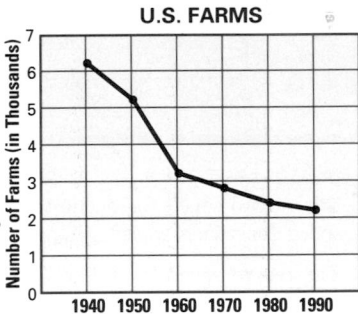

TEACHING TIP

Remind students that in order to choose appropriate scales for their graphs, they should begin by identifying the greatest and least values of the data.

Solution

a. In 1940, there were about 6,000 farms. In 1990, there were about 2,000 farms.

b. There were about 2,000 fewer farms in 1960 than in 1950.

c. The trend is decreasing. ◀

Example 2

These data show the prices paid to farmers for a bushel of corn. Construct a line graph for the data.

PRICE OF CORN (per bushel)							
Year	1930	1940	1950	1960	1970	1980	1990
Price	$0.60	$0.62	$1.52	$13.00	$1.33	$3.11	$2.25

Solution

To construct a line graph, draw and label the horizontal and vertical axes. Locate and mark the data points. Draw straight lines to connect the points. Write a title for the graph.

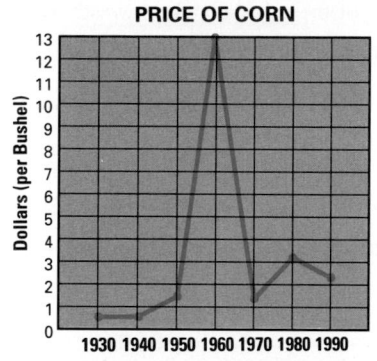

PRICE OF CORN

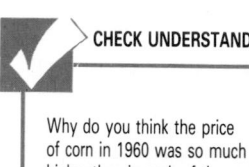

CHECK UNDERSTANDING

Why do you think the price of corn in 1960 was so much higher than in each of the other years?

The high price probably reflects drought conditions. ◀

TRY THESE

Use the line graph below for Exercises 1–3.

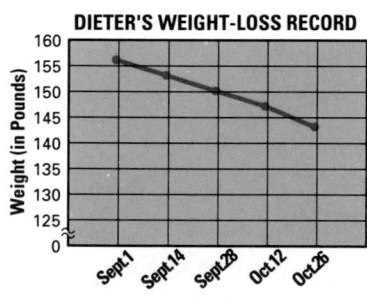

DIETER'S WEIGHT-LOSS RECORD

1. About how much did the dieter weigh on Sept. 1? On Oct. 26?
   about 156 lb; about 143 lb

2. About how many pounds were lost over the eight-week period? about 13 lb

3. Describe the trend of the data as *increasing* or *decreasing*. decreasing

4. These data show the prices paid by consumers for a pound of fish. Construct a line graph for the data. Check students' graphs.

PRICE OF FISH (per pound)							
Year	1960	1965	1970	1975	1980	1985	1990
Price	$0.85	$0.95	$1.29	$1.89	$3.59	$4.99	$6.99

$1.19 (3 quarters, 4 dimes, 4 pennies)

JUST FOR FUN

What is the greatest amount of money you can have in coins and still not have change for a dollar?

1–8   Line Graphs   **27**

ASSIGNMENTS

BASIC
1–3, 5, 7–9, 15

AVERAGE
1–5, 7–10, 11–13, 18–19

ENRICHED
6, 11–14, 15–20

ADDITIONAL RESOURCES
Reteaching 1–8
Enrichment 1–8
Transparency Master 5

Example 2: Copy the data on the chalkboard and guide students in constructing the line graph. Have them compare their graphs with the one presented in the text.

Additional Questions/Examples
Have students tell whether they would use a line graph or a bar graph to display the following data.

1. change in car prices over the last ten years **line graph**

2. senior citizens' favorite television programs **bar graph**

3. number of households that have been using computers over the last five years **line graph**

Guided Practice/Try These Have students complete these exercises independently. For Exercise 4, have students describe the trend of the data as *increasing* or *decreasing*. **increasing**

MAKING CONNECTIONS

Have students research why the price of corn jumped so high in 1960. They should be able to link the high price of corn to existing drought conditions during the previous growing season.

PRACTICE/ SOLVE PROBLEMS

The line graph shows the changes in the price of the best-selling hand-held calculator from 1965 to 1990.

1. What was the price of the calculator in 1965? **$125**

2. Between which two years did the price of a calculator drop the most? **1975 and 1980**

3. By which year was the price of the calculator the lowest? **1985**

4. **Answers will vary, but students should mention that improved technological findings often lead to lower prices. (The upturn in 1990, however, might reflect additional factors.)**

4. Write a paragraph describing the trend in the data. Suggest a reason for this trend.

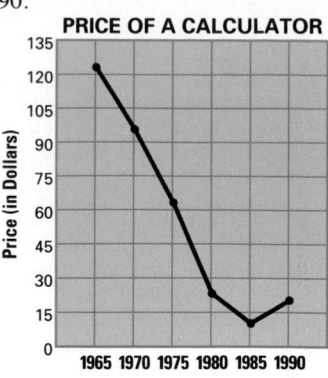

PRICE OF A CALCULATOR

Draw a line graph to show the temperature data for each city.

5.

Average Monthly Temperature in Seattle, Washington

Month	J	F	M	A	M	J	J	A	S	O	N	D
Temp. (°C)	4	4	6	9	13	15	17	17	14	10	7	4

6.

Average Monthly Temperature in Miami, Florida

Month	J	F	M	A	M	J	J	A	S	O	N	D
Temp. (°C)	22	23	24.5	26.5	27.5	28	29.5	30.5	29	27.5	23.5	22

EXTEND/ SOLVE PROBLEMS

Use this line graph for Exercises 7–10.

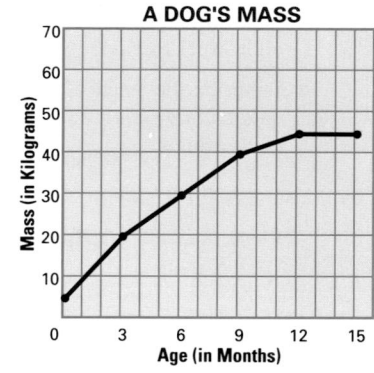

A DOG'S MASS

7. What was the dog's mass at 6 months? What was its mass at 12 months? **30 kg; 45 kg**

8. Until what age was the dog's mass 30 kg or less? **6 months**

9. For which 3-month interval did the dog's mass increase the most? By how much did it increase? **from birth to 3 mo; 15 kg**

10. Predict the dog's mass at 18 months. Explain your prediction. **Answers will vary.**

28 CHAPTER 1 Exploring Data

CHALLENGE

Divide the class into small groups. Have each group choose a state, research its population figures for each ten-year period for the last 50 years, and construct a line graph to show the changes. Have groups share and discuss their results.

11. Beginning at age 8, Dana kept track of how much she weighed on each birthday. Construct a line graph for her weight data.

Dana's Weight   Check students' graphs.

Age	8	9	10	11	12	13	14
Weight (lb)	56	62	69	77	86	98	107

12. By how much did Dana's weight increase between her eighth and fourteenth birthdays? **51 lb**

13. Between which two birthdays did Dana's weight increase the most? **between her twelfth and thirteenth birthdays**

14. Can you predict what Dana's weight will be on her fifteenth birthday? Can you predict her weight on her eigtheenth birthday? Explain your answers. **Answers will vary.**

Use this multiple-line graph.

15. If you weighed 120 lb, about how many more calories would you use swimming for 10 min than walking for 10 min? **about 25 calories**

16. About how many calories would a person weighing 140 lb use by walking for 10 min and then playing tennis for 10 min? **about 120 calories**

17. Of what use could these data be to a diet-conscious person? **The data could be used in helping to determine the proper amount of exercise needed to gain, lose, or maintain weight.**

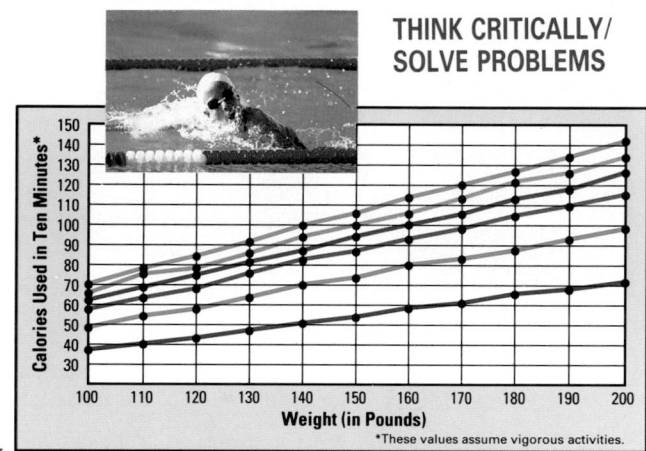

THINK CRITICALLY/ SOLVE PROBLEMS

HOW MANY CALORIES ARE USED IN TEN MINUTES?

— Bicycling   — Swimming
— Running   — Tennis
— Basketball   — Walking

These values assume vigorous activities.

USING DATA Use the Data File on page 546 to find life-expectancy information.

18. What was the projected life expectancy of a baby boy born in 1960? **66.8 years**

19. What is the approximate projected life expectancy of a baby girl born today? **79 years**

20. Predict what the life expectancies will be for boys and for girls born in the year 2010. Give reasons for your predictions.

20. Answers will vary, but should reflect a generally upward trend from the year 2000.

1–8   Line Graphs   **29**

2. How much did the price drop between 1980 and 1985? **about $15**

3. Predict the price of the calculator in 1995. Explain your answer. **Answers will vary.**

Extension Have students use the data in the line graph for Exercises 15–17 to plan a 1-hour-long exercise program for themselves in which they would use between 250 and 500 calories.

Get Ready calculators

Additional Answers
See page 573.

WARM-UP

Evaluate.
Write 7 three-digit whole numbers. Order the numbers from least to greatest. Find their sum. Divide by 7. Compare answers.

1 MOTIVATE

Explore/Working Together Have students count to find the number of books they have with them. List the numbers on the board. Then have students work in pairs to answer questions **a–d**. For question **d**, have students calculate the average to verify their guesses.

2 TEACH

Use the Pages/Skills Development Have students read this part of the section and then discuss the example.

Example: To illustrate how extreme data can affect measures of central tendency, have students add the numbers 1 and 12 to the numbers of books. Ask, *Which*

30

1-9 Mean, Median, Mode, Range

EXPLORE/ WORKING TOGETHER

What do you think is the average number of books that a high school student carries to school each day?

Count the number of books you brought with you today. (Remember to include this math book in your count.) Record the number of books for each student in the class.
Answers will vary.

a. Is there one number in your data that occurs more frequently than the others?

b. If you arrange the data in numerical order, what is the middle number? (If the number of data is even, there are two middle numbers.)

c. What is the difference between the greatest and the least numbers of books per student?

d. Guess the average number of books for the students in your class.

SKILLS DEVELOPMENT

Four statistical measures that help you describe a set of data are the mean, the median, the mode, and the range.

▶ The **mean**, or arithmetic average, is found by dividing the sum of the data by the number of data.

▶ The **median** is the middle number when the data are arranged in numerical order. (If the number of data is even, the median is the average of the *two* middle numbers.)

▶ The **mode** is the number that occurs most frequently in the set of data. Some sets of data have no mode; some have more than one mode.

▶ The **range** is the difference between the greatest and the least values in a set of data.

The mean, median, mode, and range enable you to analyze data and to determine how the data are grouped. Because the mean, median, and mode help you to locate the center of a set of data, these terms are sometimes called **measures of central tendency**.

ESL STUDENTS

Students may confuse the terms *mean, median, mode,* and *range.* Have them brainstorm techniques which will help them remember the definition of each.

When you need to analyze a set of data, it usually helps to begin by arranging the data in order, from least to greatest.

 **COMPUTER**

Example

The seven students at one table in the cafeteria had these numbers of books with them: 4 3 6 5 5 7 5

a. Find the mean.   **b.** Find the median.

c. Find the mode.   **d.** Find the range.

Solution

a. The mean is 5 because the sum of the data is 35 and 35 divided by 7 is 5.

b. The median is 5 because when the data are arranged in order— 3, 4, 5, 5, 5, 6, 7—the middle number is 5.

c. The mode is 5 because it is the number that occurs most frequently.

d. The range is 4, because the difference between 7 and 3 is 4. ◄

Calculating the mean, especially for a large set of data, can be tedious. In this program when you enter the number of data items and the data, the computer will calculate the sum and the mean for you.

```
10 INPUT "HOW MANY
   DATA ITEMS? ";N
20 LET S = 0
30 FOR K = 1 TO N
40 PRINT : INPUT "DATA "; D
50 LET S = S + D
60 NEXT K
70 LET M = S / N
80 PRINT : PRINT "SUM = ";
   S; "MEAN = ";M
90 PRINT : INPUT "RUN
   AGAIN? Y OR N";X$:
   PRINT : PRINT
100 IF X$ = "Y" GOTO 10
```

TRY THESE

Sports Shoes Unlimited had 11 pairs of men's running shoes on sale in these sizes: 10 7 7 9 10 8 11 12 10 8 10

1. Find the mean to the nearest tenth. **9.3**   **2.** Find the median. **10**

3. Find the mode. **10**   **4.** Find the range. **5**

EXERCISES

Here are one cyclist's competition times in seconds for eight tries in a season: 9.1 8.7 9.2 9.0 8.7 8.9 15.0 9.2

PRACTICE/ SOLVE PROBLEMS

1. Find the mean, median, and mode. **9.7; 9.05; 8.7 and 9.2**

2. Which measure of central tendency is the best indication of the cyclist's ability? **median**

Tell which measure of central tendency would best represent each of the following.

3. The average mass of an adult polar bear. **mean**

4. The most requested song at a radio station. **mode**

5. The usual number of goals scored by a hockey team in one game. **mean or median**

6. The average height of a female student in your grade. **mean**

1-9 Mean, Median, Mode, Range **31**

MAKING CONNECTIONS

Have students find the measures of central tendency of the data in the table of U.S. Presidents on page 11.

ASSIGNMENTS

BASIC
1–6, 11–17

AVERAGE
3–10, 11–18, 26–28

ENRICHED
7–10, 19–25, 26–31

ADDITIONAL RESOURCES
Reteaching 1–9
Enrichment 1–9
Transparency Master 6

measures of central tendency do you think will be affected by this change? **mean and range** Have students recalculate the mean to the nearest tenth and then the range to verify. **Answers will vary.**

Additional Questions/Examples

1. Which measure of central tendency is most representative of the data? **mean**

2. When would the median be more representative than the mean? **When there are extreme highs and/or lows in the data.**

Guided Practice/Try These Have students complete these exercises independently.

3 SUMMARIZE

Write About Math Have students write in their math journals the definitions and procedures for finding each measure of central tendency. Tell them to create a set of data with ten numbers and show an example of each measure.

4 PRACTICE

Practice/Solve Problems If students have difficulty finding the median in Exercise 1, point out that since there there are an even number of data items, the two middle numbers (9.1 and 9.0) must be averaged to find the median (9.05). Have students explain their answers to Exercises 2, 8, and 10.

Extend/Solve Problems For Exercise 25, discuss possible methods of solution. **The sum of the given data is 759, thus (759 + x) ÷ 6 = 153, so x =159.**

Think Critically/Solve Problems Have students work in small groups to complete these exercises.

5 FOLLOW-UP

Extra Practice Here are the speeds of fifteen cars in a 40-mi/h zone: 35, 45, 42, 43, 37, 50, 41, 38, 40, 45, 43, 39, 43, 44, 30

1. Find the mean, median, mode, and range of the data. **41, 42, 43, 20**
2. What conclusion can you draw from the measures of central tendency? **Possible answer: The drivers go slightly above the speed limit.**

Extension Have students survey their classmates to find the number of children in each student's family. Tell them to record the results in a frequency table or line plot. Then have them find the mean, median, and mode and see if any of the following occur.

1. more than one mode
2. mean or median not being a whole number

Position	Salary
President	$125,000
Vice President	$ 86,000
Plant Manager	$ 71,500
Accounting Manager	$ 53,000
Personnel Manager	$ 41,250
Research Manager	$ 40,975

7. Find the mean, median, and mode of the salaries. **$69,620.83; $62,250; none**

8. Which measure of central tendency best describes the average executive salary? **median**

During a gymnastics competition, the judges awarded these scores:

7.0 7.0 9.6 9.4 10.0 8.8 9.2 9.8 9.0 8.5

9. Find the mean, median, and mode of the scores. **8.8; 9.1; 7.0**

10. Which of the measures of central tendency do you think best describes the data? **median**

EXTEND/ SOLVE PROBLEMS

The stores at a mini-mall recorded the number of customers they had each day for a week.

Find each of the following for Jeans City:

11. the mean number of customers per day **191**

12. the median number of customers per day **180**

13. How many customers should Jeans City expect in one 30-day month? **5,730**

	Mon.	Tues.	Wed.	Thurs.	Fri.	Sat.
Keon's Electronics	206	184	212	253	267	184
Computer Store	212	241	208	279	296	197
Jeans City	198	147	164	196	281	162
Record Shop	146	162	180	234	294	184
A and M Dept. Store	553	607	692	756	914	833

14. On which three days of the week might Jeans City have to hire extra part-time help? **Mon., Thurs., Fri.**

15. Find the mean number of customers for the five stores on Thursday. **344**

16. Which store would you go to if you were looking for a part-time job on Thursdays? **A and M Dept. Store**

17. Find the total number of customers in all five stores for the entire week. **9,442**

18. Find the mean number of customers for the week for all the stores. **1,888**

CHALLENGE

Have students research the daily selling prices of a single stock for one month. Have them calculate the mean, median, mode, and range of the prices and use the information to predict the selling price for the next day. Have students check their predictions and discuss their accuracy.

19. Create a problem involving measures of central tendency based on the information in the table. Then, write a solution for your problem. **Answers will vary.**

Person	Time (in min)
Ravi	80
Agnes	45
Rosalie	75
Liz	30
Cathy	50
Ian	80
George	25
Phil	40

The number of minutes that a sample of customers spent in a record store is shown in the table at the right. Use the table for Exercises 20–21.

20. Find the mean, median, and mode for these data. **53.1 min; 47.5 min; 80 min**

21. How would your answers to Exercise 20 be affected if each person had spent 5 minutes more in the store? **Each would be 5 min longer.**

Eight crates were loaded onto a delivery truck. The mass of each crate was recorded in kilograms as follows:

40.2 50.0 39.8 46.1 34.8 39.8 40.2 44.7

22. Find the mean, medain, and mode for the data. Which measure best represents the data? **41.95, 40.2, 39.8 and 40.2; mean**

23. If four larger crates with masses 80.4 kg, 91.6 kg, 86.2 kg, and 90.5 kg were added to the load in the truck, which measures of central tendency would change for these loads? **mean; median**

24. Which measure best represents the data after the four crates were loaded? **mean**

25. Lorimer High School's wrestling team has six wrestlers with a mean weight of 153 lb. If the weights of five of the wrestlers are 158 lb, 143 lb, 140 lb, 162 lb, and 156 lb, what is the weight of the sixth wrestler? **159 lb**

26. Find the mean of the first one hundred counting numbers. **50.5**

 27. Calculate the mean and median of any number of consecutive whole numbers. What conclusion can you draw? **The mean and median are the same.**

28. In 1990, the minimum wage was raised from $3.80 to $4.25 per hour. How do you think this affected the median income? **It did not affect it at all.**

Create sample data to illustrate your answers to Exercises 29–31. **Sample data will vary.**

29. If the median of a set of data is 15, must one of the data be 15? **no**

30. If the mean of a set of data is 15, must one of the data be 15? **no**

31. If the mode of a set of data is 15, must one of the data be 15? **yes**

MATH:
WHO, WHERE, WHEN

As a child, the German mathematician Karl Friedrich Gauss (1777–1855) astounded his teacher with his almost instantaneous response to a challenging problem.

The teacher had asked the class to find the sum of the numbers from 1 to 100. Karl found the sum mentally. How did he do this? What was Karl's trick?
See Additional Answers.

THINK CRITICALLY/ SOLVE PROBLEMS

3. one or two large families causing the mean to be high

Section Quiz For Exercises 1–3, write *mean, median, mode,* or *range.*

1. The age most students graduate from high school. **mean**

2. The usual number of baskets scored by a basketball team. **mean or median**

3. The score most often obtained on a math quiz. **mode**

4. Find the mean, median, mode, and range for this set of data: 18, 25, 16, 22, 19, 22, 27, 20 **21.1, 21, 22, 11**

5. The median for a set of data is 50. Does 50 need to be one of the pieces of the data? Give an example to illustrate your answer. **no; Examples will vary.**

6. Find the mode for this set of data: 0, 1, 3, 4, 6, 7, 135, 135. Why is the mode not a useful piece of information for the data? **135; The mode is not representative of the data.**

Additional Answers
See page 573.

1 CHAPTER REVIEW

Introduction The Chapter Review emphasizes the major concepts, skills, and vocabulary presented in this chapter and can be used for diagnosing students' strengths and weaknesses. Page references direct students back to appropriate sections for additional review and reteaching.

Using Pages 34–35 Allow students to quickly scan the Chapter Review and ask questions about any section they find confusing. Exercises 1–8 review key vocabulary. For Exercise 12 make sure students include a key statement explaining the stem-and-leaf plot. **They might write, for example, 8|1 represents a grade of 81.** Exercise 13 prompts students to focus on the advantages of each type of graph. Observe students as they complete Exercise 14. Rounding each amount to the nearest hundred should make the scale interval apparent.

Informal Evaluation Have students explain how they arrived at any incorrect answers. They will probably find their own mistakes and give you some clues as to the nature of their errors. Make sure students understand this material before administering the Chapter Test.

1. If each member of a population being polled has an equal chance of being selected, the sample is ___?___ . **f**

2. Information from which decisions can be made is called ___?___ . **e**

3. A random sample chosen from a particular part of the population is a ___?___ sample. **a**

4. The number that occurs most often in a set of data is the ___?___ . **b**

5. If a population has been ordered and the members are sampled according to a pattern, the sample is ___?___ . **d**

6. A ___?___ sample is data obtained from a readily available population. **c**

7. Adding the data, then dividing the sum by the number of data, gives the ___?___ of the data. **h**

8. When data are arranged in numerical order, the middle number is the ___?___ . **g**

a. cluster
b. mode
c. convenience
d. systematic
e. data
f. random
g. median
h. mean

SECTION 1–1 COLLECTING DATA (pages 4–7)

► **Data,** or information from which decisions can be made, can be collected by sampling a population. Some ways of sampling are *random, convenience, cluster,* and *systematic.*
► Data may be organized in a *frequency table* or a *line plot.*

A population of homeowners is to be surveyed to determine the benefits of various kinds of home-heating systems. Name the kind of sampling that is represented.

9. All the homeowners on various streets in four neighborhoods are surveyed. **cluster sampling**

10. A telephone survey is conducted of names chosen from the phone book. **random sampling**

11. Which of the samples outlined in Exercises 9 and 10 would be the more valuable? Explain. **Cluster sample; This kind of sample polls homeowners, whereas the telephone sample may include renters and apartment dwellers.**

SECTION 1–2 STEM-AND-LEAF PLOTS (pages 8–11)

► Numerical data may be organized in a **stem-and-leaf plot.**
► In a stem-and-leaf plot for two-digit numbers, the digits in the tens places make up the *stems.* The digits in the ones places make up the *leaves.*

12. These data represent one class's grades on an English exam. Make a stem-and-leaf plot for the data.

53 68 82 61 91 69 72 67 72 67
66 55 79 87 67 90 84 62 98 67
55 81 75 79 55 79 82 67

See Additional Answers.

SECTIONS 1–3, 1–6, 1–8 GRAPHS (pages 12–15, 20–23, 26–29)

► **Pictographs** and **bar graphs** are used to compare data.
► **Line graphs** are used to show trends, or changes over time.

13. Decide which kind of graph—pictograph, bar graph, or line graph—would best represent the exam-grade data in Exercise 12. Explain. **A bar graph would be best since these grades do not lend themselves to picturing and do not represent changes over time.**

SECTIONS 1–4, 1–5, 1–7 PROBLEM SOLVING (pages 16–19, 24–25)

► To graph a set of data involving greater numbers, it helps to first simplify the numbers in order to determine the interval for the scale.
► Some graphs present data in rounded, instead of in exact, terms. To read these graphs, you have to make estimates.

14. What intervals would you use on a graph of these annual salaries for five supermarket employees? **$500 intervals**

$23,900 $24,000 $19,570 $22,550 $21,050

About how many books are there in each library system?

NUMBER OF BOOKS IN LIBRARY SYSTEM	
Chicago	📖 📖 📖 📖 📖 📖 📖 📖
Dallas	📖 📖
Detroit	📖 📖 📖
San Francisco	📖 📖 Key: 📖 = 1 million books

15. Chicago **7.5 million**

16. Dallas **1.75 million**

17. Detroit **2.5 million**

18. San Francisco **1.5 million**

SECTION 1–9 MEAN, MEDIAN, MODE, AND RANGE (pages 30–33)

► The **mean** of a set of data is found by dividing the sum of the data by the number of data.
► The **median** is the middle number in a set of data. If there are two middle numbers, the mean of the middle numbers is the median.
► The **mode** is the most frequently occurring number in a set of data. Some sets of data have no mode. Some have more than one mode.
► The **range** is the difference between the greatest and least values in a set of data.

USING DATA For Exercises 19–22, use the pulse-rate data on page 8.

19. What is the mean? **73.4**   20. What is the median? **73**

21. What is the mode? **73**   22. What is the range? **42**

23. **USING DATA** Use the data on page 2 to find out how much more money was spent in a year on bowling balls than on baseballs. **$55,000,000**

Follow-Up Have students find the populations of the ten most populous states, round these data to the nearest 500,000, and display the data in a graph. (Either a pictograph or a bar graph would be appropriate.) Then have them find the mean, median, mode, and range for the set of data.

Study Skills Tip Have students write each of the following terms on index cards, one to a card: *bar graph, cluster sampling, convenience sampling, frequency table, line graph, line plot, mean, median, mode, pictograph, random sampling, range,* and *systematic sampling.* Then have them make an additional card for each of these terms: *stem-and-leaf plot, gaps, outliers, clusters,* and *extremes.* For each term, have them write the definition and give an example. Encourage students to carry these cards with them and to review the terms whenever they have time.

Additional Answers
See page 573.

CHAPTER TEST

Introduction The Chapter Test uses a variety of questioning techniques to assess students' mastery of the major objectives of Chapter 1. If you prefer, you may use the Skills Preview (page 1) as an alternate form of the Chapter Test. The items on this test and the Skills Preview correspond in content and level of difficulty.

Alternative Assessment
Write About Math Choose a subject you are curious about. Write an explanation of how you would conduct a survey on that subject, how you would organize the information, and how you would display it. Explain the reasons for your choices.

To test the quality of a shipment of light bulbs, a sample of 500 was chosen randomly from different lots, then tested. In the sample, 14 bulbs were defective. **Answers may vary.**

1. What kind of sampling does this represent? **systematic sampling**

2. What is an advantage of this kind of sampling?
   The sampling is representative of the entire population.

The pictograph below shows how students travel to Laurier High School.

Walk	🧍 🧍 🧍 🧍 🧍 🧍 🧍 🧍
Bicycle	🧍 🧍 🧍 🧍 🧍
Bus	🧍 🧍 🧍 🧍 🧍 🧍 🧍
Car	🧍 🧍 🧍 🧍
Other	🧍

Key: 🧍 represents 30 students

3. How many students were surveyed? **780**

4. Which method of travel is used by the greatest number of students? **walking**

5. How many students walk to school? **240**

Linda took a survey of people's ages and recorded the data in this stem-and-leaf plot.

```
0 | 7 9 9
1 | 1 4 5 5 6 6 6 8 9
2 | 2 2 3 4 5 5 6 7 7 8
3 | 1 1 1 2 3 3 5 8
4 | 1 2 2 5 5 6 7 9 9
5 | 3 5 6 6 7
6 | 0 2 4 6 6 6
7 | 4 5 8
8 | 6
```

4|1 represents an age of 41

How many people were in each age group?

6. 0–9 **3**   7. 10–19 **9**   8. 30–39 **8**

How many people were there of each age?

9. 66 **3**   10. 15 **2**   11. 57 **1**

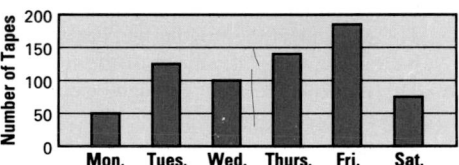

The bar graph gives the number of tapes sold in one store in a week. On which day were more tapes sold?

12. <u>Tuesday</u> or Wednesday   13. Monday or <u>Thursday</u>   14. <u>Friday</u> or Saturday

The number of repairs needed for 20 TVs during the first five years were as follows:

```
3  2  0  0  4  5  7  3  2  1
1  3  6  4  8  3  10  4  0  2
```

15. Find the mean, median, mode, and range of the data. **3.4, 3, 3, 10**

The following data show the numbers of pairs of blue jeans sold during one week by various clothing stores.

```
        0 | 8 9
        1 | 5 6 8 8
        2 | 2 2 4 5 5 5
6|0 represents 60
pairs of blue jeans  3 | 4 4 6 6 9 9 9
        4 | 1 1 2 3 3 5 5 6 7 7
        5 | 2 3 3 4
        6 | 0 0 1
        7 | 1
```

How many stores sold the following numbers of pairs of jeans?

1. 43 **2**   2. 25 **3**   3. 67 **none**

4. How many stores sold 46 or fewer pairs of jeans during the week? **27**

5. How many stores sold more than 50 pairs of jeans during the week? **8**

Students were asked the following question: "How far is it from your home to your school?"

DISTANCE FROM HOME TO SCHOOL (mi)						
1.7	4.3	5.6	0.5	2.5	3.1	3.7
1.9	5.0	6.8	3.4	3.7	2.1	4.3
6.2	0.8	2.4	3.7	2.5	0.9	1.9
5.6	7.4	4.3	3.6	13.6	8.1	5.0

6. Make a stem-and-leaf graph for these data. **See Additional Answers.**

7. How many students were interviewed? **28**

8. How many students travel 10 mi or more? **1**

9. How many students travel less than 2.6 mi? **10**

10. What is the average number of miles traveled from home to school? **4.09 mi**

11. Make a bar graph to show the data.
See Additional Answers.

Lake	Depth (in feet)
Erie	210
Huron	751
Michigan	922
Ontario	777
St. Clair	33
Superior	1,332

Check students' graphs.

Use the line graph for Exercises 12–15.

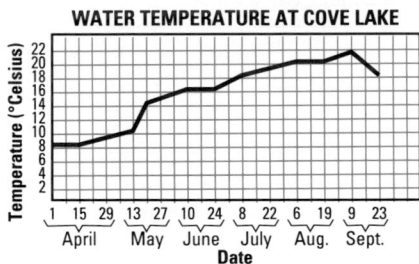

WATER TEMPERATURE AT COVE LAKE

12. Approximately when was the water the warmest? What was the temperature?
Sept. 9; 21°C

13. At what time intervals was the lake temperature measured? **14 days**

14. During which two intervals did the temperature increase the most? **May 13–27**

15. Estimate the temperature of the lake water in October. **12°–14°**

The following numbers of boats going through a strait were recorded on ten days.

36 41 58 45 36 39 52 40 43 47

16. Find the mean, median, and mode for the data. **43.7; 42; 36**

17. Which measure of central tendency in Exercise 16 best represents the data? **mean**

18. If on the last day the number was 94 instead of 47, which measure of central tendency would change? Which measure would then best represent the data?
mean; median

---

1 CUMULATIVE REVIEW

Introduction The purpose of this Cumulative Review is to maintain previously taught skills and concepts and to apply them to the material presented in this chapter. At least one major objective of each chapter is included in the review.

Item Analysis The table below correlates the Cumulative Review items with the chapter and section that are being reviewed.

Section	Items
1-2	1–10
1-6	11
1-8	12–15
1-9	16–18

Additional Answers
See page 573.

CUMULATIVE TEST

Introduction The Cumulative Test uses a standardized-test format of multiple-choice questions to assess retention of previously-learned concepts and test-taking skills. Test results may be used to diagnose students' strengths and weaknesses.

Item Analysis The table below correlates the Cumulative Test items with the chapter and section that are being tested.

Section	Items
1-1	1
1-2	2
1-3	3
1-6	4–7
1-9	8

1. A radio newscaster asked every fifth caller to answer the question "Do you think the President is doing a good job?" Which term describes the sample population?
   A. random
   B. clustered
   C. systematic
   D. convenience

2. These data represent the number of hamburgers sold each day for one month in a school cafeteria.

98	86	112	94	133	89	138	111
102	124	130	126	94	105	93	107
114	126	83	104	137	122	100	
135	132	101	116	121	136	108	

   What numbers would you use for the stems of a stem-and-leaf plot of this data?
   A. 6–4
   B. 1–9
   C. 8–13
   D. 4–14

3. Use this pictograph. What are the *most common* and the *least common* shoe sizes in the group sampled?

   FEMALE STUDENTS' SHOE SIZES

   Size $5\frac{1}{2}$ or smaller
   Size 6
   Size $6\frac{1}{2}$
   Size 7
   Size $7\frac{1}{2}$
   Size 8
   Size $8\frac{1}{2}$ or larger

   Key: represents 40 students

   A. $6\frac{1}{2}$ and 8
   B. 7 and $8\frac{1}{2}$
   C. $5\frac{1}{2}$ and $8\frac{1}{2}$
   D. 7 and 8

   The bar graph shows the number of animals in some zoos, according to species. Use the graph for Exercises 4–7.

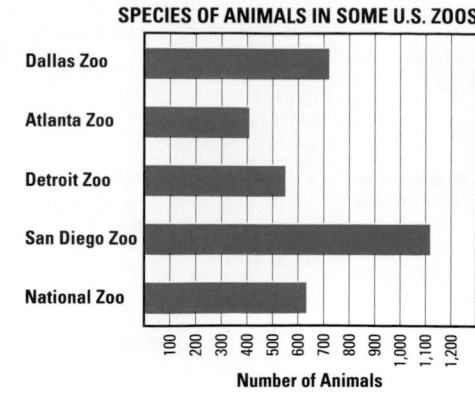

SPECIES OF ANIMALS IN SOME U.S. ZOOS

Dallas Zoo
Atlanta Zoo
Detroit Zoo
San Diego Zoo
National Zoo

Number of Animals

4. The zoo with the greatest number of species is
   A. San Diego
   B. Detroit
   C. Dallas
   D. Atlanta

5. Which zoo has about two times as many species as Detroit?
   A. Atlanta
   B. Detroit
   C. San Diego
   D. Dallas

6. The zoo with about 700 species of animals is
   A. Atlanta
   B. Detroit
   C. San Diego
   D. Dallas

7. The zoo with the least number of species is
   A. Atlanta
   B. National
   C. San Diego
   D. Detroit

8. The weekly wages of 10 students working part time are as follows:
   $35 $40 $34 $75 $36
   $39 $41 $38 $42 $37

   Which of the following best represents the typical wage?
   A. mean
   B. median
   C. mode
   D. range

Name _____ Date _____

Collecting Data

Taking a sample is a means of collecting information, or **data,** from a large group of people. Some ways of sampling an entire group, or population, are **random sampling, cluster sampling, systematic sampling,** and **convenience sampling.**

► **Example**

To determine public opinion concerning an Arizona state senator, every tenth person visiting the Grand Canyon in Arizona was interviewed. What kind of sampling does this represent? Name a disadvantage of this kind of sampling for this situation.

Solution

convenience sampling—Many visitors to the Grand Canyon in Arizona are not Arizona residents. It would be better to interview Arizona residents whose names were chosen from voter lists in several Arizona cities.

EXERCISES

Is the sample appropriate for each situation? If not, what is its disadvantage? Suggest a better way of choosing the sample.

1. To determine whether people like a new brand of bran muffin, ask every person who tastes a free sample at a health-food exposition.
No, sample probably biased in favor of people who like health foods. Try surveying consumers who have bought the new product.

2. To determine the quality of ball-point pens delivered to a school-supply store, refer to company records showing 4 out of every 500 pens were found defective.
Yes, systematic sample.

3. To determine the average amount spent at a store in a day, poll everyone who leaves the store between 5:00 and 7:00 p.m.
No. Not random because you poll only the dinner time crowd. Sample a few customers each hour, throughout the day.

4. To determine the proportion of students in your school who own cars, ask everyone in your math class.
No. The sample is not representative of the whole school. Take a poll of every 10th student over age 15 leaving school on a given day.

5. To determine the "hottest" fall fashion color, count the number of garments of each color on racks in department stores in different parts of the city.
Yes.

6. To determine whether town residents think there is too much traffic on local Route 44, ask all the homeowners who live along the route.
No. Homeowners along the route are likely to object to the traffic. They are not representative of population. Use a telephone sample of all town residents.

Reteaching • SECTION 1-1   1

Name _____ Date _____

Telephone Book Sampling

Here is part of a column of listings taken from a local telephone book that covers twenty-four towns. Think about whether or not this sample could be used for various purposes.

Noyes M Durham
Noyes Q Poland
Noyes Tire Co. Lewiston
Nu Look Hair Fashions Lewiston
Nugene A Greene
Nugent R Lewiston
Nulty S Lewiston
Nunon A Lewiston
Nussinow B R Lewiston
Nutri System Weight Loss Center Auburn
Nutta G Lewiston
Nutting G Auburn
Nuzzato B Auburn
Nuzzo G Wales
Nuzzo K Auburn
Nyberg L S Auburn
Nye C Lewiston
Nyen N Lewiston
Nygard P Lewiston
Nyulaszy A Lewiston
Oak Hill High School Wales
Oak Hill Park, Inc. Sabattus
Oakdale Poultry Farm Auburn
Oakes D North Raymond
Oakes F G Auburn
Oakes L Lewiston
Oasis Restaurant Auburn
Oates R Auburn
Oates Z Lewiston
Obermeyer L Lewiston
Obertautsch P Lisbon
Obie H Auburn
Obie H L Lewiston
Obie K Lewiston
Obie N Auburn
Obie S T Lewiston
Obie's Shoe Repair Lewiston
O'Bounder Shoe Co. Lewiston
Obrey T Sabattus
O'Brien Consolidated Industries, Inc Lewiston

► **Example**

Could you use this sample to determine the approximate number of listings for each town?

Solution

No. The sample is too small. Only nine towns are represented in this sample. A larger sample would be needed.

EXERCISES

1. Could this sample be used to estimate the number of businesses in the area as compared to the number of residences? Explain.
Yes, but a larger sample of listings from several pages chosen at random would be better.

2. Would this be a good sample to use to determine the kind of businesses that exist in the area?
No. It is not representative of the entire business community. It would be better to use the "yellow pages" for such a purpose.

3. How could the sample be used to determine whether people like to shop in the stores in the area?
Conduct a telephone survey of everyone listed.

4. Does this represent a random sample of all the residents of the twenty-four towns?
No. There may be residents without telephones or with unlisted telephone numbers.

5. How could you use the telephone book to devise a systematic sample?
Answers will vary. Two possibilities are to choose every 200th name listed in the book or to choose every 100th name on a page.

2   Enrichment • SECTION 1-1

Name _____ Date _____

Stem-and-Leaf Plots

A **stem-and-leaf plot** can help you organize data so it can be easily analyzed.

MONDAY'S TEMPERATURES				
	High	Low	High	Low
Anchorage	70	54	Miami Beach	88	73
Boston	79	72	New York	89	73
Chicago	99	74	Richmond	97	77
Denver	84	60	St Ste Marie	75	56
Honolulu	87	75	Seattle	82	59
Houston	89	74	Washington	97	79
Los Angeles	76	62			

► **Example**

Make a stem-and-leaf plot for the high temperatures listed in the chart. Then write a description of the data.

Solution

Step 1: Form the *stems.*
The high temperatures range from 70 to 99. Use the digits in the tens place, 7, 8, and 9, as the stems. Write them in a column. Draw a vertical line to the right.

Step 2: Form the *leaves.*
Show the first high temperature, 70° for Anchorage, by writing a "leaf," 0, next to the "stem," 7. To show 79° for Boston, write 9 next to the 0. Enter the rest of the leaves in the same way.

```
                    7 | 0 9 6 5
9 | 9 represents 99°   Stems  8 | 4 7 9 8 9 2   Leaves
                    9 | 9 7 7
```

Step 3: The 70 is an *outlier* (extremely high or low value). *Clusters* (groups of values close to one another) appear in the high 80s and high 90s. There is a large *gap* (space between values) between 89 and 97.

EXERCISES

1. On a separate sheet of paper, make a stem-and-leaf plot for the low temperatures. **Check students' graphs.**

2. Write a description of the data noting any outliers, clusters, or gaps.
There are no outliers. There is a cluster in the low to mid 70s. There is a gap between 63 and 73.

3. Complete the stem-and-leaf plot for advertised monthly rents for two-bedroom apartments. Use the numbers in the hundreds as the stem.

Rents for 2-Bedroom Apartments			
$400	$400	$425	$225
$295	$360	$300	$280
$435	$360	$350	$400
$350	$280	$550	$375
$400	$415	$325	$395
$320	$300	$380	$470

```
2 | 95 80 25 80
3 | 50 20 60 60 00 00 50 25 80 75 95
4 | 00 35 00 00 15 25 00 70
5 | 50
```

4. On a separate sheet of paper, describe the data to a family who is looking for a two-bedroom apartment. **The lowest rent is an outlier, 225 is far from others. The highest 550 is also an outlier. Rents cluster in the high 200s, and from 350 to 435.**

Reteaching • SECTION 1-2   3

Name _____ Date _____

Data Duo

A **back-to-back stem-and-leaf plot** displays two sets of data at once.

FRIDAY'S TEMPERATURES IN U.S. CITIES	
Albany, NY	80/60
Albuquerque	86/61
Anchorage	67/52
Atlanta	87/69
Atlantic City	86/69
Austin	93/74
Bismarck	82/56
Boston	80/68
Buffalo	76/56
Burlington, VT	78/58
Charleston, SC	93/75
Chicago	77/54
Cleveland	76/56
Daytona Beach	92/74
Denver	80/57
Detroit	78/54
Fargo	81/57
Green Bay	74/49
Honolulu	87/64
Kansas City	81/64
Las Vegas	100/73
Los Angeles	79/62
Memphis	88/72
Miami	90/78
New York City	83/68
Philadelphia	84/70
Pittsburgh	80/58
San Francisco	66/55
Seattle	75/57
Washington, DC	86/73

► **Example 1**

Make a back-to-back stem-and-leaf plot of the high and low temperatures one Friday in the cities listed.

Solution

Identify the extremes of the data. Write the stems in a column to form the "backbone" of the plot. The highs and lows for the first four cities are plotted.

```
LOWS          HIGHS
         4
      2  5
  9 1 0  6  7
         7
         8  0 6 7
         9
        10
```

Plot the high and low temperatures on a scattergram to get a different picture of the same data. A **scattergram** shows a pattern in the relationship of high and low temperatures.

► **Example 2**

Make a scattergram to show the relationship of high/low temperatures.

Solution

For Albany: Low 60°, High 80°
Look across the bottom scale. Find the 60 point. Move up the scale from here to the 80 point. Mark the point. Each pair of high and low temperatures is plotted as a point.

HIGH/LOW TEMPERATURES IN U.S. CITIES

EXERCISES

1. Copy, then complete the back-to-back stem-and-leaf plot. **Check students' stem-and-leaf plots.**

2. Copy, then complete the scattergram for the temperature data. *Do not connect the points.* Look for a pattern. **Check students' graphs.**

3. What are the outliers on the stem-and-leaf plot?
A high of 100° at Las Vegas is an outlier compared to next highest of 93°.

4. What are the outliers on the scattergram?
The point for Las Vegas is not in line with the other points.

5. Where are the clusters on the stem-and-leaf plot?
There are lows from 54 to 58 and 68–69 and highs from 76 to 81 and 86–87.

6. How do clusters look on the scattergram?
Points clustered together as in square between 55 and 60 lows and 75 to 80 highs.

4   Enrichment • SECTION 1-2

Name _____ Date _____

Pictographs

A **pictograph** displays data using symbols and a key. The key tells the number of data items represented by each symbol.

▶ **Example**

Use the pictograph to answer the questions.
a. What does the symbol $ represent?
b. What chore do the greatest number of students get paid to do? How many students is that?
c. How many more students get paid for yardwork than for housework?

NUMBER OF STUDENTS PAID TO DO CHORES

Housework	$
Set and clear table	$ $
Care for pets	$ $ $ $
Take out trash	$ $
Yard work	$ $ $

Key: $ represents 10 students

Solution
a. The symbol represents 10 students.
b. Forty students get paid to take care of pets.
c. yardwork, 25; housework, 15; $25 - 15 = 10$. Ten more students get paid for yardwork.

EXERCISES

Use the pictograph to answer the questions.

1. How many pairs of size-10 shoes are in stock?
   18 pairs

2. How many pairs of size-8½ shoes are in stock?
   72 pairs

3. Based on this store's stock, what do you think is the most popular shoe size?
   Size 8, because there are more of them than any other size.

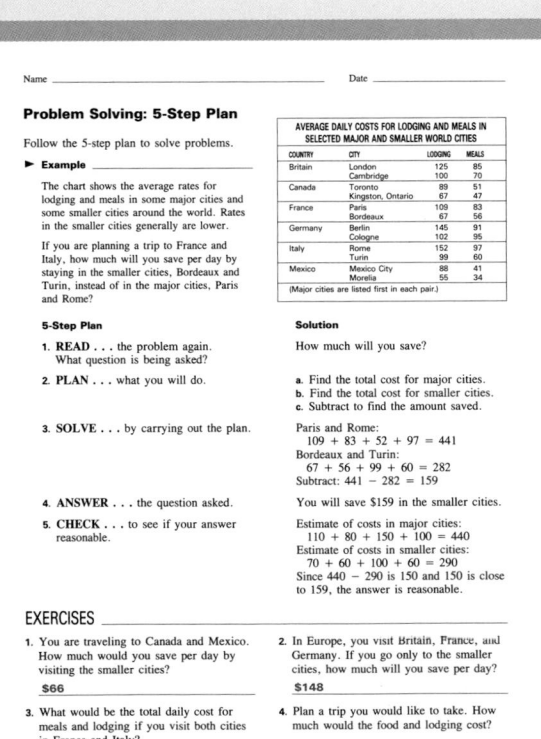

PAIRS OF WOMEN'S SHOES IN STOCK

Key: represents 1 dozen pairs

4. If the store sells 17 pairs of size-6½ shoes, how many pairs of shoes will be left in that size?
   43 pairs

5. If the store sells 16 pairs of size-8 shoes, which shoe size will it then have the most of in stock?
   size 7½

6. This chart shows the number of minutes for each device to use 1 kilowatt of electricity. On another piece of paper, construct a pictograph for these data. **Check students' graphs.**

clothes dryer	shower	iron	dishwasher	vacuum cleaner
30	15	50	20	90

Name _____ Date _____

Picture This

Sometimes graphs shown as pictures can be misleading. Some graphs with pictures are not pictographs. Sometimes a pictograph is not the best choice for displaying data.

▶ **Example 1**

Is the graph to the right a pictograph? Explain.

Solution
No. The graph has pictures but no key. The car symbols are not the same size and shape. The information is not accurately shown because the size of the cars is not in proportion to the number of owners.

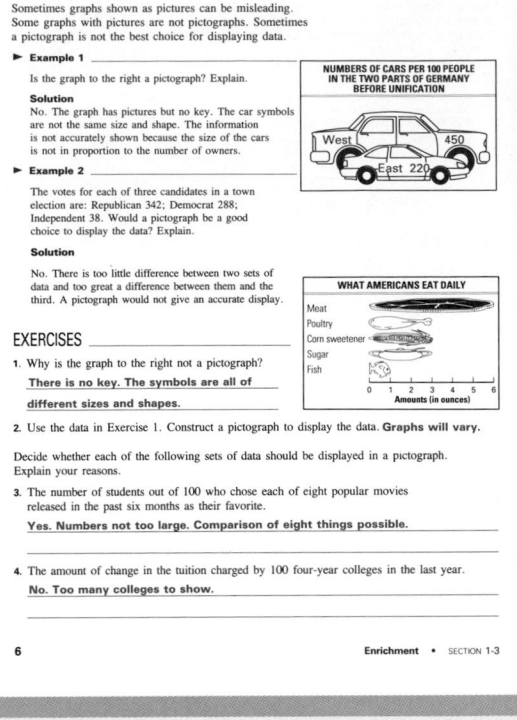

NUMBERS OF CARS PER 100 PEOPLE IN THE TWO PARTS OF GERMANY BEFORE UNIFICATION

West 450
East 220

▶ **Example 2**

The votes for each of three candidates in a town election are: Republican 342; Democrat 288; Independent 38. Would a pictograph be a good choice to display the data? Explain.

Solution

No. There is too little difference between two sets of data and too great a difference between them and the third. A pictograph would not give an accurate display.

EXERCISES

1. Why is the graph to the right not a pictograph?
   There is no key. The symbols are all of different sizes and shapes.

WHAT AMERICANS EAT DAILY

Meat
Poultry
Corn sweetener
Sugar
Fish

Amounts (in ounces)

2. Use the data in Exercise 1. Construct a pictograph to display the data. **Graphs will vary.**

Decide whether each of the following sets of data should be displayed in a pictograph. Explain your reasons.

3. The number of students out of 100 who chose each of eight popular movies released in the past six months as their favorite.
   Yes. Numbers not too large. Comparison of eight things possible.

4. The amount of change in the tuition charged by 100 four-year colleges in the last year.
   No. Too many colleges to show.

Name _____ Date _____

Problem Solving: 5-Step Plan

Follow the 5-step plan to solve problems.

▶ **Example**

The chart shows the average rates for lodging and meals in some major cities and some smaller cities around the world. Rates in the smaller cities generally are lower.

If you are planning a trip to France and Italy, how much will you save per day by staying in the smaller cities, Bordeaux and Turin, instead of in the major cities, Paris and Rome?

AVERAGE DAILY COSTS FOR LODGING AND MEALS IN SELECTED MAJOR AND SMALLER WORLD CITIES

COUNTRY	CITY	LODGING	MEALS
Britain	London	125	85
	Cambridge	100	70
Canada	Toronto	89	51
	Kingston, Ontario	67	47
France	Paris	109	83
	Bordeaux	67	56
Germany	Berlin	145	91
	Cologne	102	95
Italy	Rome	152	97
	Turin	99	60
Mexico	Mexico City	88	41
	Morelia	55	34

(Major cities are listed first in each pair.)

5-Step Plan

1. **READ** . . . the problem again. What question is being asked?

2. **PLAN** . . . what you will do.

3. **SOLVE** . . . by carrying out the plan.

4. **ANSWER** . . . the question asked.

5. **CHECK** . . . to see if your answer is reasonable.

Solution

How much will you save?

a. Find the total cost for major cities.
b. Find the total cost for smaller cities.
c. Subtract to find the amount saved.

Paris and Rome:
$109 + 83 + 52 + 97 = 441$
Bordeaux and Turin:
$67 + 56 + 99 + 60 = 282$
Subtract: $441 - 282 = 159$

You will save $159 in the smaller cities.

Estimate of costs in major cities:
$110 + 80 + 150 + 100 = 440$
Estimate of costs in smaller cities:
$70 + 60 + 100 + 60 = 290$
Since $440 - 290$ is 150 and 150 is close to 159, the answer is reasonable.

EXERCISES

1. You are traveling to Canada and Mexico. How much would you save per day by visiting the smaller cities?
   $66

2. In Europe, you visit Britain, France, and Germany. If you go only to the smaller cities, how much would you save per day?
   $148

3. What would be the total daily cost for meals and lodging if you visit both cities in France and Italy?
   $723

4. Plan a trip you would like to take. How much would the food and lodging cost?
   Answers will vary.

Name _____ Date _____

Problem Solving for Standardized Tests

College-entrance tests often have problems that require more than one operation to solve.

▶ **Example**

Some members of the Student Council at Community High School washed cars to raise money. Each person washed 4 cars at $2.50 per car. They collected a total of $163.00, including $13.00 in tips. How many people washed cars?

Solution

a. Subtract earnings from tips from total earnings to get earnings from car washes alone.
b. Multiply cost per wash by number of washes per person to get amount earned per person.
c. Divide total earned from washes by amount earned per person to get the number of people who washed cars.

$163 - 13 = 150$
$4 \times 2.50 = 10.00$
$150 \div 10 = 15$

There were 15 people who washed cars.

EXERCISES

1. In a sophomore class of 320 students, 5 students were running for class president. If each student received at least 10 votes, according to the votes shown in the table, what is the maximum number of votes that David could receive?

Candidate	Votes
Ashley	?
Beau	107
Celeste	59
David	?
Emily	84

   a. 64   **b. 60**   c. 35   d. none of these

2.

CROSS-COUNTRY RELAY (400 m)	
Runner	Seconds
Alonso	52.0
Black	51.2
Cucio	52.8
Denton	49.3

The table shows the time for each runner in a 400-meter relay race. The total running times exceeded 3 minutes by how many seconds?

   a. 3   b. 180   c. 205.3   d. none of these

3. At the school store, some students are stamping notepads with the school logo. Others are stacking the notepads on shelves. One student can stamp 50 pads per minute. Another student can stack 1 pad per second. How many student stampers are needed to keep up with 5 student stackers?

   a. 5   **b. 6**   c. 10   d. none of these

4. At Evergreen High School, the photography club has 15 members and the French club has 17 members. Sixteen students belong to just one of the two clubs. How many students belong to both clubs?

   a. 7   **b. 8**   c. 9   d. 32

38B

Problem Solving Skills: Choose an Appropriate Scale

In order to graph data, you may have to simplify it. Then you can decide on an appropriate scale.

► **Example**

What scale intervals would you use to graph the data at the right?

Solution

For the first half of one recent year, the earnings for the top players in the LPGA (Ladies Professional Golf Association) ranged from just over $496,000 to almost $103,000.

Simplify the data:
Think of 496,000 as 496 thousands.
Think of 103,000 as 103 thousands.

Consider different intervals:
• If you use intervals of 100, you would need 5 scale intervals.
• If you use intervals of 50, you would need 10 scale intervals.

Assume you choose 10 intervals. Your scale might be written horizontally or vertically.

Be sure to include a label for the scale.

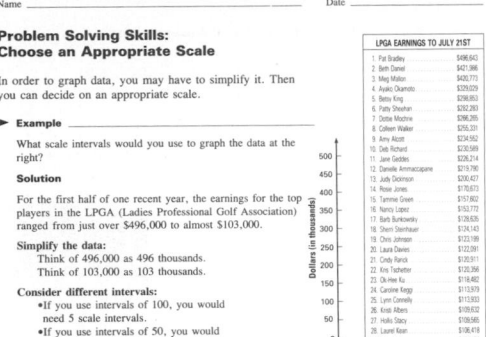

LPGA EARNINGS TO JULY 21ST	
1. Pat Bradley	$496,643
2. Beth Daniel	$421,996
3. Meg Mallon	$420,773
4. Ayako Okamoto	$329,029
5. Betsy King	$298,853
6. Patty Sheehan	$282,283
7. Dottie Mochrie	$266,285
8. Colleen Walker	$255,331
9. Amy Alcott	$234,552
10. Deb Richard	$230,589
11. Jane Geddes	$226,214
12. Danielle Ammaccapane	$219,780
13. Judy Dickinson	$200,427
14. Rosie Jones	$170,673
15. Tammie Green	$157,902
16. Nancy Lopez	$152,772
17. Barb Bunkowsky	$128,635
18. Sherri Steinhauer	$124,143
19. Chris Johnson	$123,189
20. Laura Davies	$122,091
21. Cindy Rarick	$120,911
22. Kris Tschetter	$120,266
23. Ok-Hee Ku	$118,482
24. Caroline Keggi	$113,979
25. Lynn Connelly	$113,933
26. Kristi Albers	$109,632
27. Hollis Stacy	$108,585
28. Laurel Kean	$108,619
29. Sally Little	$103,136
30. Michelle McGann	$102,690

EXERCISES

Describe the scale you would use to graph each set of data. **Answers will vary. Suggested intervals are given.**

1.

ESTIMATED ENERGY REQUIREMENTS OF SOME PHYSICAL ACTIVITIES	
Activity	Calories/h
Basketball	560
Bowling	215
Calisthenics	200
Cleaning	185
Cycling (easy pace)	450
Racquet sports	870
Running (7-min. mile)	950
Studying/Reading	105
Swimming (steady pace)	500
Walking (normal pace)	180
Walking up stairs	300

10 intervals of 100 calories each,

0 to 1,000; label calories per hour

2.

MAINE—APPALACHIAN TRAIL	
Mountain Peaks	Height (ft)
Saddleback	4,116
The Horn	4,023
Saddleback Junior	3,640
Crocker	4,168
Sugarloaf	4,237
Bigelow	4,150
Little Bigelow	3,001
Stewart	2,671
Pierce Pond	2,516
Pleasant Pond	2,480
Bald	2,630
Boarstone	1,947

10 intervals of 500 feet each, 0 to

5,000; label scale height (in feet)

Making Decisions About Graph Scales

You may need to make several decisions before choosing an appropriate scale for a graph.

• Will the scale display the data accurately?
• Can the data be simplified without losing important information?

EXERCISES

Determine the appropriate scale and intervals for each set of data. Explain your decision.

1.

STATE LABOR DEPARTMENT REPORT	
Average Work Week (in hours)	
August '90	41.5
September '90	42.1
October '90	41.9
November '90	42.3
December '90	42.4
January '91	41.5
February '91	42.0
March '91	41.9
April '91	41.5
May '91	41.2
June '91	41.5
July '91	41.2
August '91	41.4

Use intervals of 0.5, from 40.5 to 42.5. Label hours. Average work week always around 40 hours, so not distorted by using scale without 0.

2.

LENGTH OF SOME AMERICAN BIRDS	
Bird	Length (in.)
Barn Swallow	5.75–7.75
Bobwhite	9.25–10.75
Common Meadowlark	9–11
Common Nighthawk	8.25–10
Crow	17–21
Downy Woodpecker	6.25–7.25
Horned Lark	6.75–8
Horned Owl	18–25
House Wren	4.25–5.25
Song Sparrow	6–6.75
Yellow Warbler	4.75–5.25

Intervals of 2, from 0 to 26. Label length in inches. May graph shortest, longest, or average length of each bird.

3.

AUCTION PRICES OF CLASSIC COLLECTIBLE CARS	
'74 Porsche 911	$11,000
'64 Studebaker Avanti	$13,000
'31 Chevrolet cabriolet	$18,900
'70 Plymouth Hemi 'Cuda	$25,100
'67 Chevrolet Camaro RS convertible	$14,200
'57 Ford Thunderbird Roadster	$29,500
'57 Chevrolet Bel Air convertible	$52,000
'64 Mercedes-Benz 220 SEb cabriolet	$15,000
'36 Ford Deluxe street rod	$17,300
'65 Chevrolet Impala Super Sport convertible	$18,000

Use intervals of 10 thousands, 0 to 60 thousands. Label thousand dollars. Range from 11 to 52 thousands. Need to show 0 on scale. Possibly round each to number of thousands.

4.

NBA ATTENDANCE	
(Regular Season Home Games)	
1990–91	16,876,125
1989–90	17,368,659
1988–89	15,464,994
1987–88	12,654,374
1986–87	12,065,351
1985–86	11,214,888
1984–85	10,506,355

Use intervals of 2 million, from 0 to 16 million. Label number of millions. Round to millions as 16,876,125 → 16.9 million. Just need to compare growth over the years.

Bar Graphs

A bar graph can be used to display data that you find in a list, see in a report, or read in a newspaper or book.

► **Example**

SPORTS EQUIPMENT SOLD DAILY IN U.S.	
Hockey sticks	3,014
Soccer balls	6,153
Skateboards	7,611
Basketballs	8,493
Golf clubs (sets)	33,973

Construct a bar graph to illustrate these facts about what sports equipment Americans buy.

Solution

• Choose an appropriate scale.
  Use intervals of 3,000 from 0 to 36,000.
• Get your data ready.
  Estimate the number of thousands for each choice. For example:
    3,014 is about 3 thousands
    7,611 is about 7.5 thousands
• Label your scale.
• Draw the bars, leaving equal spaces between them. Be sure to label each bar.
• Write a title for your graph.

EXERCISES

Use the bar graph for Exercises 1–4.

1. Which three vehicles had about the same amount of sales?
   Ford Ranger, Explorer, Honda Civic

2. How much money came from sales of the Honda Accord?
   $380,000

3. How many more dollars in sales came from sales of the Ford pick-up than from the Ford Ranger?
   $178,000 to $182,000

4. How much higher were the total sales for Chevrolets than for Hondas?
   about $80,000 higher

SALES LEADERS FOR 1991 MODELS

(Bar graph: Ford F-Series pickup, Chevrolet C/K pickup, Honda Accord, Ford Taurus, Chevrolet Cavalier, Toyota Camry, Ford Escort, Ford Ranger, Ford Explorer, Honda Civic; Dollars (in thousands) 0–400)

Construct a bar graph for the following data.

5. On an average day, Americans enroll in the following foreign language courses:
   46 in Chinese; 64 in Japanese; 93 in Russian
   Check students' graphs.

Fancy Facts

Double- or triple-bar graphs are used to display data about more than one group, or about the same group at different periods of time. This double-bar graph shows that there were fewer students in elementary school in 1986 (about 27,000,000) than in 1970 (about 34,000,000).

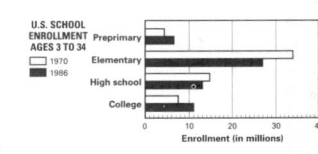

Sometimes the bars in a graph are stacked, one on top of another. This graph shows that in 1990 there were about 30,000,000 elderly people. The stacked bars show about 27,000,000, aged 65 to 84, and about 3,000,000 at age 85 and over.

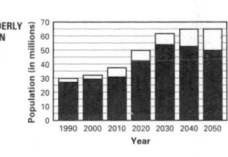

EXERCISES

Use the bar graphs above to answer Exercises 1–4.

1. About how many more people attended college in 1986 than in 1970?
   about $3\frac{1}{2}$ million

2. At what other level on the School Enrollment Graph do you see an increase in the attendance between 1970 and 1986? What do you think caused the increase?
   preprimary; answers will vary; more working mothers

3. Describe the projected trend in the 65- to 84-year-old population from 1990 to 2050.
   increases until 2030 then decreases slowly

4. By what year will the 85-and-over population be the greatest?
   2050

5. Use the table at the right to construct a double-bar graph.
   Check students' graphs.

FOOTBALL PLAYERS' DIETS HAVE IMPROVED		
(Daily Servings for About 100 Players)		
	5 Years Ago	Today
Prime rib	250	200
Chicken	100	150
Fish	30	100
Vegetables	30	65
Fruit	50	125
Pasta	100	200

38C

Name _____ Date _____

Problem Solving Skills: Estimation

To estimate figures from a graph, first study the scale.

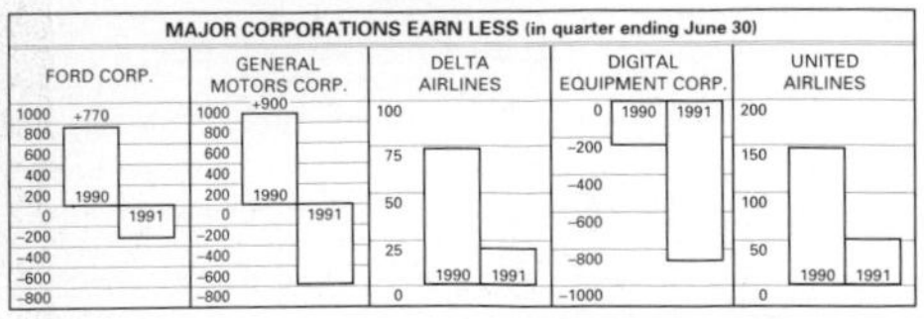

MAJOR CORPORATIONS EARN LESS (in quarter ending June 30)

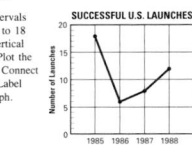

(Graph showing Ford Corp., General Motors Corp., Delta Airlines, Digital Equipment Corp., United Airlines — Dollars in millions for 1990 and 1991)

▶ **Example** _____

Estimate the 1991 earnings or loss for the Ford Corporation.

Solution

The scale on the graph for Ford is for millions of dollars, marked in intervals of 200. The scale runs above 0, from 0 to 1,000, to show earnings. The scale runs below 0, from 0 to −800, to represent losses. For 1991, the bar extends downward from 0 to a point between −200 and −400, or −300. So, Ford had a loss of about $300 million.

EXERCISES

Estimate the earnings or losses from the graphs above. **Answers will vary. Sample answers are given.**

1. General Motors in 1991
   Loss of $780 or $790 million

2. Delta in 1990
   Earnings of $73 or $74 million

3. Digital in 1991
   Loss of $870 to $890 million

4. United Airlines in 1990
   Earnings of $148 or $149 million

5. Use the Bradley Airport graph to estimate the difference in profits between 1990 and 1991.
   About $4\frac{1}{2}$ million dollars' difference

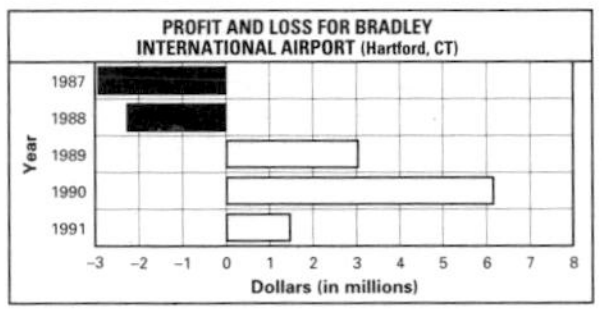

PROFIT AND LOSS FOR BRADLEY INTERNATIONAL AIRPORT (Hartford, CT)

(Horizontal bar graph showing Years 1987–1991, Dollars in millions from −3 to 8)

6. On a separate piece of paper, write a question based on the Bradley Airport graph. Ask a classmate to answer it.

Reteaching • SECTION 1-7   13

Name _____ Date _____

Problem Solving Skills: Estimates and Assumptions

Demographers are statisticians who study population. They make estimates and assumptions about population in order to determine what it will be like in the future.

EXERCISES

Make estimates and assumptions to answer Exercises 1–2.

1. Use the stacked bar graph at the right to determine about how many high school seniors there will be in the U.S. the year you graduate from high school.

   Assumption (year of graduation)
   Assuming 1997

   Estimate (number of high school students in U.S. in year of graduation):
   About 44,000,000 less about 26,000,000 is about 18,000,000
   Assumption (approximate number of students in each of grades 9–12): **There will be an equal number of students in each of the four grades.**

   Calculation: **18,000,000 ÷ 4 = 4,500,000**

   Conclusion: **There will be about 4,500,000 seniors in 1997.**

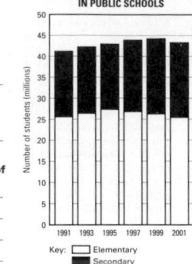

PROJECTED ENROLLMENT IN PUBLIC SCHOOLS

(Stacked bar graph, Number of students (millions) 0–50, years 1991–2001)
Key: ☐ Elementary ■ Secondary

2. Use the bar graph below to determine about how many 15-year-old girls will be living in the Chicago metropolitan area in 1993.

   Hint: The population growth in Chicago from 1980 to 1987 was 2.5% (read *two point five percent*). Assume the same amount of growth from 1987 to 1993. Use a calculator to multiply the number of people in Chicago in 1987 by 2.5% to find the additional number of people in Chicago for the 1987–1993 period.

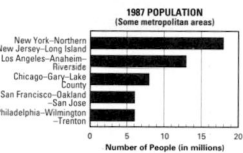

1987 POPULATION (Some metropolitan areas)

(Horizontal bar graph: New York–Northern New Jersey–Long Island; Los Angeles–Anaheim–Riverside; Chicago–Gary–Lake County; San Francisco–Oakland–San Jose; Philadelphia–Wilmington–Trenton — Number of People (in millions) 0 to 20)

Population estimate for 1993 is 8.2 million or 8,200,000. Assume population from 0 to 80 is evenly distributed. 8,200,000 ÷ 80 = 102,500 (in each age group). So, there are 102,500 15-year-old boys and girls. 102,500 ÷ 2 = 51,250. So there are 51,250 15-year-old girls.

14   Enrichment • SECTION 1-7

Name _____ Date _____

Line Graphs

To display data on a **line graph** points are plotted and connected in order.

▶ **Example 1** _____

Use the line graph to answer the questions.

a. In which year were the most cars sold?
b. About how many were sold in 1983?
c. Describe the trend in the data.

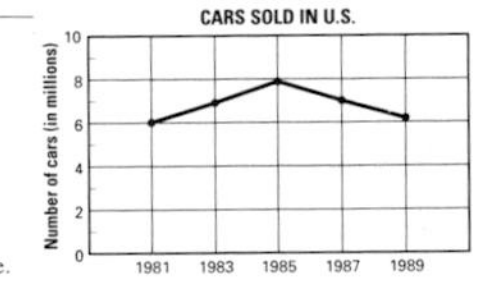

CARS SOLD IN U.S.

(Line graph, Number of cars in millions 0–10, years 1981–1989)

Solution

a. 1985
b. about 6.7 million
c. Sales increase to a peak in 1985, then decrease.

▶ **Example 2** _____

The chart shows the number of successful U.S. space launches for four years. Construct a line graph from the data.

Year	Successful Launches
1985	17
1986	6
1987	8
1988	12

Solution

Use intervals from 0 to 18 on a vertical scale. Plot the points. Connect them. Label the graph.

SUCCESSFUL U.S. LAUNCHES

(Line graph, Number of Launches 0–20, years 1985–1988)

EXERCISES

Use the weather graph for July to answer Exercises 1–5.

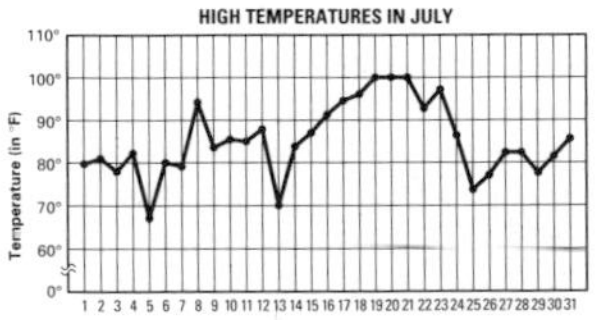

1. What was the lowest high temperature of the month?
   66° or 67°

2. On how many days was the high temperature above 90°?
   9 times

3. What is the difference between the highest and the lowest high temperature?
   34° or 35°

4. On another piece of paper, construct a line graph. Use the data in the chart at the right. **Check students' graphs.**

HIGH TEMPERATURES IN JULY

(Line graph, Temperature (in °F) 0–110°, days 1–31)

Billions of Dollars Spent in the U.S. on Shoes	
1985	22.9
1986	24.3
1987	25.9
1988	27.5
1989	29.8

Reteaching • SECTION 1-8   15

Name _____ Date _____

Multiple-Line Graphs

Sometimes you can show two or more sets of data by plotting two or more lines on the same graph.
• Use the same scales for all the sets of data.
• Show differences in the lines by using different colors or patterns.

▶ **Example** _____

a. Draw a double-line graph to display the numbers of U.S. Army and Navy officers from 1950 through 1980.

b. Use the double-line graph showing numbers of officers to compare the numbers of Army to Navy officers over the 30-year period.

NUMBERS OF OFFICERS		
Year	U.S. Navy	U.S. Army
1950	42,678	67,784
1960	67,456	91,056
1970	78,488	143,704
1980	60,100	80,339

Solution

a. Combined data vary from about 43,000 to about 144,000.
   • Round data values to thousands.
   • Use intervals of 10 thousands from 40 to 150 thousands.
   • Use ●——●——● for Army and x——x——x for Navy.

b. There were more Army than Navy officers in each year, but increases and decreases were similar.

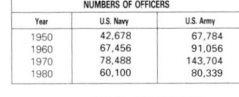

NUMBER OF U.S. OFFICERS

(Double-line graph, Thousands 0–140, years 1950–1980)
●—— Army   x—— Navy

EXERCISES

Use the double-line graph shown above to answer Exercises 1–2.

1. Would the graph look different if data for every 5-year-period were plotted?
   Yes, more ups and downs, more variability.

2. What do you think caused the peak in the numbers of officers in 1970?
   Vietnam War, wartime would show peaks.

3. The *consumer price index* measures the change in average prices over time as compared to a particular standard. The standard in this chart is 100. The chart represents the average prices for the years 1982–1984. Construct a multiple-line graph for the data in the chart.

CONSUMER PRICE INDEX				
Fruits and vegetables	106.4	109.4	119.1	128.1	138.0
Footwear	102.3	101.9	105.1	109.9	114.4
Gasoline	98.6	77.0	80.1	80.8	88.5
	1985	1986	1987	1988	1989

16   Enrichment • SECTION 1-8

38D

Name _____ Date _____

Mean, Median, Mode, Range

Baseball statistics are in the newspaper every day during the baseball season.

▶ **Example**

Find the mean, median, mode, and range for the number of lifetime stolen bases for these top 28 players.

Solution

Mode is the most frequent score. The mode is 244.

Median is the middle number. Write the numbers in order from highest to lowest or lowest to highest. Then find the middle score.

For 28 scores, the median is halfway between the 14th and 15th numbers. The median is 293.

Arithmetic **mean** is the sum of the data divided by the number of data. The sum of the stolen bases, 9,821, divided by the number of leaders, 28, is 350.75, or about 351. The mean is 351.

Subtract the lowest number from the highest to get the **range**. 966 − 207 = 759. The range is 759.

28 ACTIVE LEADERS LIFETIME STOLEN BASES	
Rickey Henderson	966
Tim Raines	660
Willie Wilson	623
Vince Coleman	582
Ozzie Smith	484
Steve Sax	390
Brett Butler	377
Paul Molitor	369
Lonnie Smith	352
Juan Samuel	330
Gary Pettis	326
Mookie Wilson	321
Andre Dawson	304
Gary Redus	298
Willie McGee	288
Ryne Sandberg	285
Lloyd Moseby	276
Willie Randolph	268
Eric Davis	244
Robin Yount	244
Kirk Gibson	242
Von Hayes	240
Garry Templeton	240
Otis Nixon	234
Carney Lansford	217
Dave Winfield	210
Alan Trammell	207

EXERCISES

Find the mean, median, mode, and range for each set of baseball data.

1.

ACTIVE LEADERS LIFETIME HITS	
Robin Yount	2,826
George Brett	2,760
Dave Parker	2,659
Dave Winfield	2,629
Eddie Murray	2,423
Dwight Evans	2,411
Andre Dawson	2,284
Carlton Fisk	2,255
Keith Hernandez	2,182
Ken Griffey, Sr.	2,143

Number of Hits

mean: **2,457.2**

median: **2,417**

mode: **no mode**

range: **683**

2.

AMERICAN LEAGUE LEADERS						
	G	AB	R	H	Pct.	
CRipken Blt.	91	368	63	122	.332	
Tartabull KC	80	311	56	103	.331	
Palmeiro Tex	88	366	66	120	.328	
Puckett Min	90	363	57	119	.328	
Molitor Mil	87	364	68	119	.327	
Boggs Bsn	87	329	54	106	.322	
Franco Tex	84	344	60	109	.317	
Baines Oak	82	290	47	91	.314	
Joyner Cal	87	338	52	105	.311	
Greenwell Bsn	89	334	46	104	.311	

Batting Percentages

mean: **.322**

median: **.324 or .325**

mode: **.328 and .311**

range: **.021**

3.

TOP 10 YANKEES Through Thursday's Game							
Batters	AVG	AB	H	HR	RBI	SB	
Hall	.311	280	87	14	58	9	
Mattingly	.297	367	109	6	41	2	
Sax	.291	188	113	6	36	17	
Nokes	.284	278	79	20	58	0	
Williams	.270	74	20	2	12	4	
Espnoza	.266	282	75	3	19	3	
R. Kelly	.263	297	78	10	36	16	
Sheridan	.238	63	15	3	5	0	
P. Kelly	.237	194	46	3	18	8	
Barfield	.225	284	64	17	48	1	

Batting Averages

mean: **.268**

median: **.268**

mode: **no mode**

range: **.086**

Name _____ Date _____

Using Statistics for Decision Making

The chart below lists facts about a college in each of ten states.

FACTS ON COLLEGES A–J

	State	Tuition (excluding room and board)	Population of Town or City	Size of Campus (in acres)	Volumes in Library	Number of Students
A	Virginia	$13,100	6,000	332	337,189	1,600
B	Vermont	$20,000	9,000	350	600,000	1,900
C	Rhode Island	$16,375	150,000	40	2,015,469	5,585
D	Ohio	$14,000	600,000	128	1,500,000	2,530
E	New York	$15,600	70,000	1,100	655,361	2,270
F	Massachusetts	$15,975	50,000	140	689,000	4,792
G	Maine	$15,430	42,000	125	487,238	1,500
H	Pennsylvania	$15,725	9,000	300	500,000	3,297
I	New Jersey	$16,390	27,000	600	4,015,599	4,497
J	Connecticut	$14,300	60,000	200	208,570	2,937

EXERCISES

Find the mean, median, mode, and range for each set of college data.

		mean	median	mode for rounded values	range
1.	Tuition	$15,689.50	$15,662	$16,000	$6,900
2.	Population	102,300	46,000	9,000	594,000
3.	Acres	331.5	250	for rounded values 300	about 1,060
4.	Library	1,100,843	627,681	none	about 3.8 million
5.	Students	3,091	2,733 or 2,734	none	4,085

6. Ryan wants a lower than average tuition, a larger than average library, and a smaller than average campus. Which college would he choose?

College D in Ohio

7. Desiree likes a large student population in a large city that has an exceptionally large library. Which college would she choose?

College C in Rhode Island

Name _____ Date _____

Calculator Activity: Finding the Mean from a Bar Graph

You can use the memory keys on a calculator to find the mean. Check the manual for your calculator to see exactly how it works.

▶ **Example**

This bar graph shows the scores that twenty-one students received on one math quiz. Find the mean of the set of scores to the nearest whole number.

MATH QUIZ SCORES

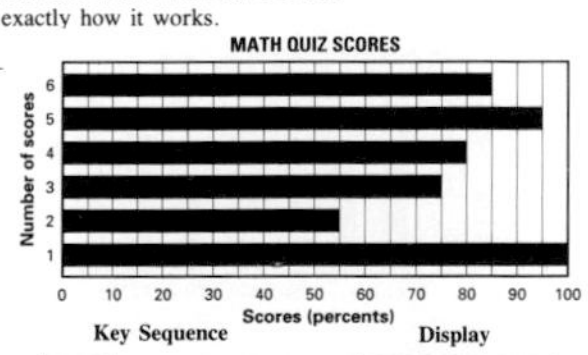

Solution

Here is a typical key sequence.

Step	Key Sequence	Display
1. Clear the calculator. If necessary, clear the memory.	ON/C	0.
2. Enter data and store in the memory.	100 M+ 2 × 55 = M+ etc.	M 100. M 110. etc.
3. Recall data from the memory.	MRC	M 1740.
4. Divide by number of scores.	÷ 21 =	M 82.857142
5. Round to the nearest whole number, 83.		

The mean score is about 83.

EXERCISES

1. This graph shows some teenagers' hourly babysitting rates. Find the mean rate to the nearest cent.

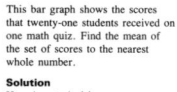

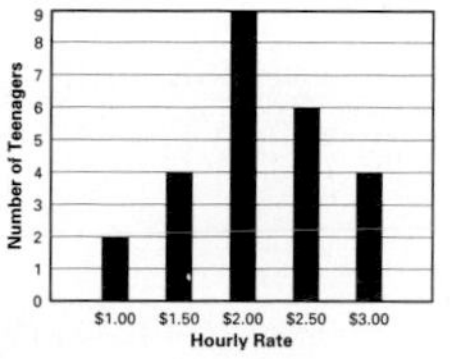

mean rate: **$2.12**

2. This graph shows the distances that people jog each week. Find the mean distance to the nearest hundredth.

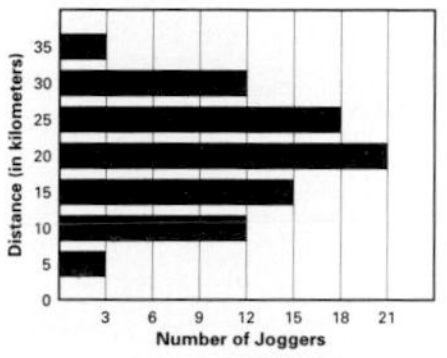

mean distance: **20.18 km**

Name _____ Date _____

Finding the Mean from a Bar Graph (continued)

3. This graph shows the amount of money spent by various shoppers on holiday gifts. Find the mean amount to the nearest dollar.

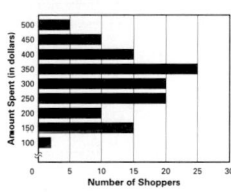

mean amount spent: **$305**

4. This graph shows the heights of the students in one high school math class. Find the mean to the nearest hundredth.

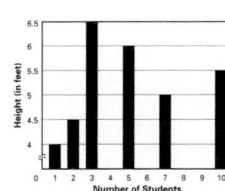

mean height: **5.45 ft**

5. This graph shows how much money was raised by various students at Eastside High School's Rally Day. Find the mean amount to the nearest cent.

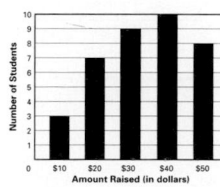

mean amount raised: **$33.51**

6. This graph shows the heights of the dogs brought to a veterinarian's office one day. Find the mean height to the nearest hundredth.

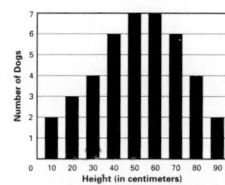

mean height: **51.95 cm**

Name _____ Date _____

Computer Activity: Statistics With a Spreadsheet

A **spreadsheet** is a collection of data arranged in **rows** and **columns**. A row is named by a number and runs horizontally. A column is named by a letter and runs vertically. A **cell** is the location formed by the intersection of a column and a row. A cell is identified by the column and row that form it, for example, B3. The data in a cell can be either a **value** or a **label**. A value can be used in calculations. A label cannot be used in calculations. A sample spreadsheet is shown below.

	A	B	C	D	E	F	G
1:	STUDENT		Quiz 1	Quiz 2	Quiz 3	Average	
2:			2/6	2/7	2/8		
3:							
4:	Descartes, R.		86	97	91	91.33	
5:	Euler, L.		92	89	94	91.67	
6:	Gauss, C.		98	96	100	98.00	

F6: (value, Layout–F2) @AVG(C6…E6)

– Rows 1 and 2 contain column headings (Labels)
– Rows 4–6 contain student names and grades (Labels and Values)
– Column A contains the student names (Labels)
– Cells C4 through E6 contain the grades (Values)

The lower left hand corner gives cursor location, type of entry, special instructions, and indicates the actual contents of the cell or how the contents of the cell were obtained.

F6: (Value, Layout—F2) @AVG(C6…E6)
└─ Cursor location └─ The 98.00 in cell F6 is the average of the data in cells C6, D6, and E6.
└─ Type of entry in that cell. The entry has a special layout with 2 fixed decimal places.

EXERCISES

1. Name the contents of each cell in the sample spreadsheet.
   a. D5 **89**   b. A6 **Gauss, C.**   c. E2 **2/8**   d. C4 **86**

2. Name the cell that contains each of the following:
   a. 94 **E5**   b. 2/6 **C2**   c. 98.00 **F6**   d. 91.67 **F5**

3. In what cell is the cursor located? **F6**

4. How was the value in F6 obtained? **It is the average, or mean, of the values in cells C6; D6; and E6.**

Name _____ Date _____

Statistics With a Spreadsheet (continued)

The spreadsheet below shows the number of situps done by each of five students.

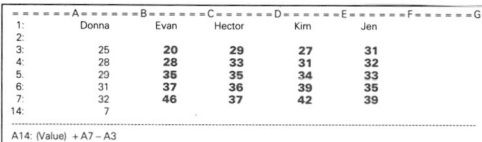

	A	B	C	D	E	F	G
1:	Donna	Evan	Hector	Kim	Jan		
2:							
3:	25	35	33	27	32		
4:	31	46	37	34	35		
5:	32	20	35	31	31		
6:	29	37	36	39	33		
7:	28	28	29	42	39		
8:							
9:	145	166	170	173	170		
10:	5	5	5	5	5		
11:	29	33.2	34	34.6	34		
12:	29	33.2	34	34.6	34		

B12: (Value) @AVG(B3…B7)

Use the lower-left-hand-corner information to tell how values of each row were obtained.

5. B9:   (Value) @SUM(B3…B7)   Row 9: **Sum of values B3 through B7**

6. B10: (Value) @COUNT(B3…B7) Row 10: **Number of entries in B3 through B7**

7. B12: (Value) @AVG(B3…B7)   Row 12: **Mean, or average, of entries B3 through B7**

8. How were B9 and B10 used to get B11? **B9 was divided by B10.**

A spreadsheet will "arrange" data in order. Donna's scores were arranged. Arrange the other scores.

	A	B	C	D	E	F	G
1:	Donna	Evan	Hector	Kim	Jen		
2:							
3:	25	20	29	27	31		
4:	28	28	33	31	32		
5:	29	35	35	34	33		
6:	31	37	36	39	35		
7:	32	46	37	42	39		
14:	7						

A14: (Value) + A7 – A3

9. Use the arranged data to find the median for each person.
   a. Donna **29**   b. Evan **35**   c. Hector **35**   d. Kim **34**   e. Jen **33**

10. What is represented in A14? **Range**

11. Find the corresponding value for the following cells.
    a. B14 **26**   b. C14 **8**   c. D14 **15**   d. E14 **8**

ACHIEVEMENT TEST
Exploring Data
CHAPTER 1 FORM A

MATH MATTERS BOOK 1
Chicha Lynch
Eugene Olmstead

SOUTH-WESTERN PUBLISHING CO.

Name _____
Date _____

SCORING RECORD	
Possible	Earned
15	

A survey was taken to determine the popularity of radio stations in the Boston area. The interviewers polled all the patrons at restaurants and shops throughout the area.

1. What kind of sampling does this represent?
   cluster sampling

2. What is a disadvantage of this kind of sampling?
   members of the population are not chosen independently of one another

The pictograph at the right shows the number of students at Riverview High School who own dogs.

3. In which grade are there the fewest dog owners? **grade 9**

4. How many dog owners are there in the eleventh grade? **95**

5. How many Riverview High School students own dogs? **305**

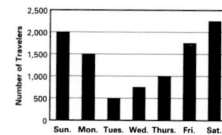

Key: 🦴 represents 10 dog owners

The weights of all the boys in a summer camp are recorded in this stem-and-leaf plot.

```
       5 | 8 8 9
       6 | 1 7 7 8 8 9 9 9
7|0 represents 70 lbs  7 | 0 0 2 5 7 8
       8 | 3 3 4 6 6 6 8 8 9
       9 | 3 4 4
```

The weights of how many boys fall within each interval?

6. 90 – 99 lb **3**   7. 70 – 79 lb **6**   8. 80 – 89 lb **9**

How many boys are there of each weight?

9. 70 lb **2**   10. 69 lb **3**   11. 86 lb **3**

Name _____ Date _____

The bar graph gives the number of travelers on one airline's cross-country flight from New York to San Francisco in a recent week. Use the bar graph to answer Exercises 12–14.

Number of Travelers bar graph: y-axis labeled Number of Travelers from 0 to 2,500; x-axis days Sun. Mon. Tues. Wed. Thurs. Fri. Sat.

On which day were there more travelers?

12. Saturday or Sunday   **Saturday**

13. Monday or Friday   **Friday**

14. Wednesday or Thursday   **Thursday**

The number of home computers sold at a Compusale store over a recent 7-day period was as follows: 3, 3, 5, 1, 3, 6, 4

15. Find the mean, median, mode, and range of the data.
    3.57, 3, 3, 5

38F

ACHIEVEMENT TEST

Exploring Data

CHAPTER 1 FORM B

MATH MATTERS BOOK 1

Chicha Lynch
Eugene Olmstead

SOUTH-WESTERN PUBLISHING CO.

Name _____

Date _____

SCORING RECORD	
Possible	Earned
15	

In a survey to determine the best candidate for mayor of Los Angeles, the passengers on all flights that arrived one morning in Los Angeles were interviewed.

1. What kind of sampling does this represent?   convenience sampling

2. What is a disadvantage of this kind of sampling?
Those people interviewed might not be from Los Angeles and so might not know about the candidates.

The pictograph shows the number of guests at a ski resort over a 6-month period.

3. In which month were there the greatest number of skiers?   January

4. How many skiers were there in December?   475

5. How many skiers were at the resort for the 6-month period?   2,075

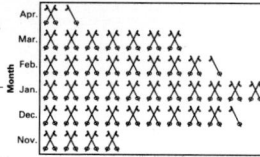

Key: X represents 50 guests

The ages of the members of a cousins' club are shown in this stem-and-leaf plot.

```
        0 | 5 6 9
        1 | 0 4 4 4 6 5
        2 | 3 6 7
2|2 represents   3 | 5 7 7 7 9
an age of 22     4 | 0 0 3 4 4 6 7 8
        5 | 5 5 8
        6 | 1 3 3 8 9
        7 | 2
```

How many people are in each age group?

6. 20 – 29   3   **7.** 40 – 49   8   **8.** 60 – 69   5

How many people are there of each age?

9. 63   2   **10.** 14   3   **11.** 45   0

Name _____   Date _____

The bar graph gives the number of people who banked at First Federal Savings Bank in a recent week. Use the bar graph to answer Exercises 12–14.

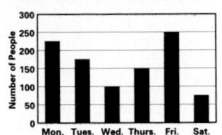

On which day did more people bank?

12. Monday or Friday   Friday

13. Tuesday or Thursday   Tuesday

14. Wednesday or Saturday   Wednesday

The number of paperback books sold each day at the Readon Bookstore for a recent month were as follows:

```
13 10  8 12 16  9 22 17 16 15
25 22 13 18 18 17  7 18 28  9
```

15. Find the mean, median, mode, and range of the data.
15.65, 16, 18, 21

Teacher's Notes

CHAPTER 2 SKILLS PREVIEW

Find each answer. Estimate and compare to see if your answer is reasonable.

1. $3,876 - 1,624$ **2,252**   2. 68×49 **3,332**   3. $98 \times 4,152$ **406,896**

4. $7,332 \div 78$ **94**   5. $6.38 + 11.3 + 0.367$ **18.047**   6. $19.3 - 0.782$ **18.518**

7. 0.75×0.27 **0.2025**   8. 8×8.6 **68.8**   9. $6,930 \div 35$ **198**

Evaluate each expression. Let $a = 2$ and $b = 8$.

10. $a + 15$ **17**   11. $b \div 4$ **2**   12. $b \div 4a$ **1**   13. $12a \div b$ **3**

Write in exponential form.

14. $4 \times 4 \times 4$ **4^3**   15. $9 \times 9 \times 9 \times 9 \times 9$ **9^5**   16. $7 \times 7 \times 7 \times 7$ **7^4**

Write in standard form.

17. 8^3 **$8 \times 8 \times 8 = 512$**   18. 2^7 **$2 \times 2 \times 2 \times 2 \times 2 \times 2 \times 2 = 128$**   19. 3^3 **$3 \times 3 \times 3 = 27$**   20. 4^2 **$4 \times 4 = 16$**

Write in scientific notation.

21. $5,483$ **5.483×10^3**   22. 228 **2.28×10^2**   23. $549,761$ **5.49761×10^5**   24. $1,475,265$ **1.475265×10^6**

Write each answer in exponential form.

25. $7^4 \times 7^3$ **7^7**   26. $9^{10} \div 9^5$ **9^5**   27. $x^9 \div x^2$ **x^7**   28. $y^3 \times y^7$ **y^{10}**

Simplify.

29. $5 \times 6 - 27 \div 3$ **21**   30. $2 \times (8 - 4) + 3$ **11**   31. $(5 + 3)^2 \times 2$ **128**

Complete.

32. $7(\blacksquare - 3) = (7 \times 2) - (7 \times 3)$ **2**   33. $\blacksquare(4 + 8) = (3 \times 4) + (3 \times 8)$ **3**

34. $9.6 + (\blacksquare + 2.4) = (9.6 + 5) + 2.4$ **5**   35. $\blacksquare \times 2,158 = 2,158$ **1**

36. $4.89 \times \blacksquare = 0$ **0**   37. $2.1 \times 3.92 = 3.92 \times \blacksquare$ **2.1**

Simplify.

38. $3(4 - a)$ **$12 - 3a$**   39. $2(b + 3)$ **$2b + 6$**   40. $6(8 - x)$ **$48 - 6x$**   41. $5(n + 9)$ **$5n + 45$**

Solve.

42. Keisha put $15 in her savings account in January, $22 in February, $29 in March, and $36 in April. Following this pattern, find how much money she put into the account in October. **$78**

J	F	M	A	M	J	J	A	S	O
$15	$22	$29	$36	$43	$50	$57	$64	$71	$78

Skills Preview **39**

2 SKILLS PREVIEW

Introduction The purpose of this Skills Preview is to ascertain students' abilities on all the major objectives of Chapter 2. Test results may be used
- to determine those topics which may need only to be reviewed and those topics which need to be more carefully developed;
- for class placement;
- in prescribing for individual differences.

If you prefer, you may use the Skills Preview as an alternate form of the Chapter Test (page 84) to evaluate mastery of chapter objectives. The items on the Skills Preview and the Chapter Test correspond in content and level of difficulty.

2

WHOLE NUMBERS AND DECIMALS

OVERVIEW

In this chapter, students discover various applications of the uses of whole numbers and decimals. They learn how to round and estimate whole numbers and decimals and apply these skills to problem solving situations. They evaluate variable expressions, apply the order of operations, and discover the relationship between standard number form and scientific notation. Exponential form and the laws of exponents are explored, along with the properties of number operations.

SPECIAL CONCERNS

Many basic concepts are introduced and practiced in this chapter. Students may forget some of the processes introduced in the first five sections (rounding, estimating, operations with decimals) by the time they have completed the subsequent sections. After each section is finished, you may wish to display a poster with the overall concepts and procedures taught. Display posters around the room for students to refer to when necessary.

VOCABULARY

associative property
base
commutative property
compatible numbers
distributive property
dividend
divisor
estimation

evaluate
exponent
exponential form
front-end estimation
identity properties
  of addition and
  multiplication
law of exponents

order of operations
properties of zero
  for addition
  and multiplication
rounding
scientific notation
variable

MATERIALS

calculators
grid paper
almanac or encyclopedia

BULLETIN BOARD

Display news, sports, and business articles and classified ads on the bulletin board. Have students choose an article or ad, write a word problem for it, and then display the problem below the article. Other students can then solve the word problems for extra credit.

INTEGRATED UNIT 1

The skills and concepts involved in Chapters 1–4 are included within the special Integrated Unit 1 entitled "Voting and Citizenship." This unit is in the Teacher's Edition beginning on page 150H. Worksheets for this integrated unit appear in the Enrichment Activities booklet, pages 47–49.

TECHNOLOGY CONNECTIONS

- Calculator Worksheet, 45
- Computer Worksheet, 46
- MicroExam, Apple Version
- MicroExam, IBM Version

- *Number Munchers,* MECC
- *What Do You Do with a Broken Calculator?* Sunburst Communications

TECHNOLOGY NOTES

Computer numeration programs, such as those listed in Technology Connections, provide critical-thinking activities that challenge students' understanding of whole numbers and decimals. Some programs may build quick-recall ability through engaging game formats. Others may reinforce the concept that there is more than one way to reach a solution.

2 WHOLE NUMBERS AND DECIMALS

PLANNING GUIDE

SECTIONS	TEXT PAGES	ASSIGNMENTS			
		BASIC	AVERAGE	ENRICHED	
Chapter Opener/Decision Making	40–41				
2–1 Whole Numbers	42–45	1–18, 21–25, 30, 33	1–20, 21–28, 31, 32–34	7–18, 20, 23–31, 32–34	
2–2 Variables and Expressions	46–49	1–38, 40–48, 52, 56	7–39, 52–54, 55–56	39, 53–54, 55–57	
2–3 Problem Solving Skills: Choose the Operation	50–51	1–12	3–13, 15–16	6–14, 16–17	
2–4 Adding and Subtracting Decimals	52–55	1–10, 11–13, 15–17, PSA 1, 5, 6, 8	5–10, 11–13, 18–21, PSA 1–8	8–10, 12–14, 19–20, PSA 4–8	
2–5 Multiplying and Dividing Decimals	56–59	1–20, 29, 38	5–13, 18–36, 41–42	21–42	
2–6 Exponents and Scientific Notation	60–63	1–17, 18–21, 31–32, PSA 1–6	3–17, 22–30, 31–34, PSA 1–7	9–17, 22–30, 33–35, PSA 4–12	
2–7 Laws of Exponents	64–67	1–24, 27–38, 49–51	5–8, 13–16, 31–47, 49–56	25–26, 31–48, 52–58	
2–8 Order of Operations	68–71	1–12, 13–16, 19–20, 31–32	1–12, 13–29, 31–36	11–12, 22–30, 31–46	
2–9 Properties of Operations	72–75	1–23, 24–26, 31, 33	13–23, 24–32, 33–34	22–23, 28–32, 33–37	
2–10 Distributive Property	76–79	1–20, 21–26, 39, 42	1–20, 21–41, 45, 46	8–13, 20, 30–41, 42–47	
2–11 Problem Solving Strategies: Find a Pattern	80–81	1–11	3–13	5–15	
Technology	43, 45, 61, 63, 69, 80	✔	✔	✔	

ASSESSMENT					
Skills Preview	39	All	All	All	
Chapter Review	82–83	All	All	All	
Chapter Test	84	All	All	All	
Cumulative Review	85	All	All	All	
Cumulative Test	86	All	All	All	

ADDITIONAL RESOURCES			
RETEACHING	**ENRICHMENT**	**TECHNOLOGY**	**TRANSPARENCY**
2–1	2–1		TM 7, 8
2–2	2–2		
2–3	2–3		
2–4	2–4		TM 9
2–5	2–5		TM 10
2–6	2–6	2–6	
2–7	2–7		
2–8	2–8		TM 11
2–9	2–9		TM 12
2–10	2–10		
2–11	2–11	2–11	

ASSESSMENT OPTIONS

Chapter 2, Test Forms A and B	
Chapter 2, Test	Text, 84
Alternative Assessment	TAE, 84
Chapter 2 MicroExam	

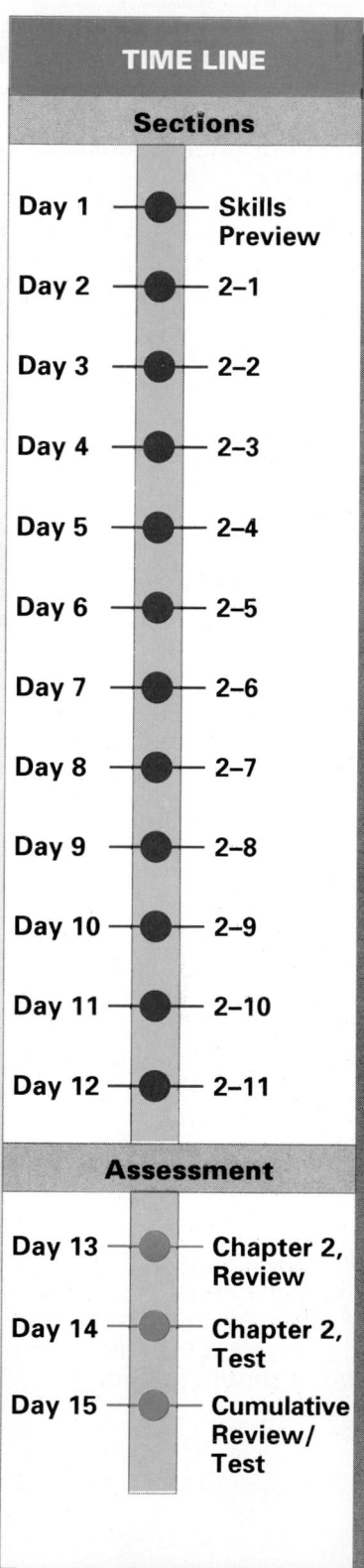

TIME LINE

Sections

Day 1	Skills Preview
Day 2	2–1
Day 3	2–2
Day 4	2–3
Day 5	2–4
Day 6	2–5
Day 7	2–6
Day 8	2–7
Day 9	2–8
Day 10	2–9
Day 11	2–10
Day 12	2–11

Assessment

Day 13	Chapter 2, Review
Day 14	Chapter 2, Test
Day 15	Cumulative Review/ Test

Objective To explore how whole numbers and decimal numbers affect our understanding of world economy

Introduction Ask students to determine the relationships between the values of the various currencies. Discuss different relationships by asking the following:
- How do the exchange rates of the British pound and the Canadian dollar compare in terms of U.S. dollars? **The British exchange rate is about twice that of the Canadian dollar.**
- Based on the table, which two currencies' exchanges are almost the same? **Swiss, British**
- What kinds of decisions would the data in this table help you to make if you were traveling to the countries listed? **Answers will vary.**

Decision Making
Using Data After discussing Questions 1–4, ask students which is a better time to convert U.S. currency into foreign currency: when the exchange rate is high or low for the foreign currency? **high** When is it better to exchange foreign currency for U.S. currency? **when exchange rate is low** Have them explain their answers.

Working Together Based on their tables, have groups decide which of the countries they chose had the best exchange rate for the twelve-month period.

Numbers are an important part of everyday life. We use numbers daily as we check the time, run a specific distance, or pay for lunch. In many countries, as in the United States, the monetary system is a decimal system, in which the basic monetary unit is equivalent to 100 smaller units. For example, in the United States the dollar is the basic monetary unit, and one dollar is equivalent to 100 cents. In this chapter, you will use whole numbers and decimals to solve problems involving money. You will also work with numbers expressed in exponential form.

Most countries of the world issue their own currency in coins and paper notes. The table shows the values of the currencies of five countries as compared with the currency of the United States. The column headed "Late Wed" shows the exchange rate for one Wednesday in a recent year. Note that the table gives the equivalent U.S. dollar values for one British pound and for one Canadian dollar, but gives the value of one U.S. dollar in Swiss francs, in Japanese yen, and in German marks.

CURRENCY	Late Wed	Day's High	Day's Low	12-Month High	12-Month Low
British pound (in U.S. dollars)	1.6215	1.6255	1.6015	2.0040	1.6010
Canadian dollar (in U.S. dollars)	0.8707	0.8720	0.8697	0.8859	0.8504
Swiss franc (per U.S. dollar)	1.5715	1.5690	1.5955	1.2325	1.5903
Japanese yen (per U.S. dollar)	138.62	138.10	138.75	124.33	150.85
German mark (per U.S. dollar)	1.8135	1.8080	1.8370	1.4475	1.8356

DECISION MAKING

Using Data

Use the information in the table above to answer each question. Use the exchange rates in the "Late Wed" column for Exercises 1 and 2.

1. What U.S. dollar amount was equivalent to one Canadian dollar?
   $0.8707 or $0.87

2. Would you have received more German marks or more Swiss francs for one U.S. dollar? Explain.
   more German marks, since 1.8135 > 1.5715

3. What was the difference in the exchange rates for the 12-month high and the 12-month low for the British pound? $0.403

4. Why was the 12-month high a lesser amount than the 12-month low for the Swiss franc, the Japanese yen, and the German mark?

Answers to Exercise 4 will vary. A sample answer follows. The table shows the foreign currency received for one U.S. dollar. The high shows that you receive less foreign currency for each U.S. dollar and indicates that it takes *more* U.S. dollars to "buy" the currency of these countries.

Working Together

Your group's task is to research the currency exchange rates for five countries during the past twelve months. Choose a date, such as the third of the month, and find the high and low exchange rates on that date for each of the previous twelve months. Find these data for five countries. You may refer to newspapers, business magazines, or other sources of business information for your data. Assemble the results of your research in a table. Compare your group's table with those made by other groups. Did you choose the same or different countries? Which country had the greatest fluctuation in the exchange rate? Which country had the least fluctuation in the rate? What are some reasons for changes in the exchange rate? Answers will vary.

Exploring Whole Numbers and Decimals   **41**

WARM-UP

Use mental math to complete.
1. 2,400 + 3,500 = **5,900**
2. 126,000 - 99,000 = **27,000**
3. 1,200 × 60 = **72,000**
4. 7,000 ÷ 50 = **140**

1 MOTIVATE

Explore Have students make paper models of Napier's rods and model the solution to the first problem before trying additional problems. Then have them work in pairs with their models. Have one partner make up a problem for the other to solve using the rods. Have the second partner demonstrate the solution with the rods, then check the answer with a calculator. Have partners switch roles and repeat the activity.

2 TEACH

Use the Pages/Skills Development Have students read this part of the section and then discuss the examples.

Example 1: Before discussing the example, you may wish to review place value.

2-1 Whole Numbers

EXPLORE Napier's rods can be used to multiply and divide. Use them to multiply 3 × 425.

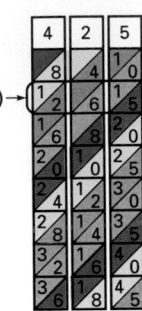

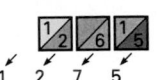

Add the numbers inside each colored diagonal section.

3 × 425 = 1,275

a. Position the rods so that the digits in the top row form the factor 425.

b. From the top of the rods, count down 3 boxes.

c. Add the numbers inside each colored diagonal section as shown in the diagram. The product is 1,275.

Explain how you could use the rods to find the product 7 × 254. **1,778**

SKILLS DEVELOPMENT

MATH: WHO, WHERE, WHEN

John Napier, born in the sixteenth century in Scotland, is credited with many discoveries and inventions. While some of his discoveries had military uses, many others had mathematical applications.

Napier is best known for devising the logarithmic tables. But Napier's rods were his most unusual invention. (They are sometimes called Napier's bones, since animal bones were first used for this purpose.)

Answers to computations can be checked for reasonableness by **estimating**. One technique to help you estimate is **rounding**. To round a number follow these steps.
► Locate the digit in the place to which you are rounding.
► Increase this digit by 1 if the next digit to the right is 5 or greater.
► Leave the digit unchanged if the next digit to the right is less than 5.

Example 1

Round 1,538 to the indicated place-value position.
a. the nearest ten   b. the nearest hundred   c. the nearest thousand

Solution

a. 1,5ǐ38 → 1,540 The digit to the right of 3 is greater than 5. So, increase 3 to 4.

b. 1,5ǐ38 → 1,500 The digit to the right of 5 is less than 5. So, leave 5 unchanged.

c. ǐ1,538 → 2,000 The digit to the right of 1 is 5. So, increase 1 to 2. ◄

Example 2

Find the actual answer. Use rounding to estimate the answer and compare to see if your actual answer is reasonable.
a. 3,234 + 2,578   b. 796 − 308   c. 93 × 198   d. 836 ÷ 22

42   CHAPTER 2   Exploring Whole Numbers and Decimals

TEACHING TIP

Review with students rules for determining the number of zeros in a product or quotient. Use these examples:
How many zeros are in each product?
230 × 100 **3**   2,000 × 20 **4**
How many zeros are in each quotient?
3,600 ÷ 120 **1**   40,000 ÷ 500 **2**

Solution

a.
Actual		Estimate
3,234	⟶	3,000
+2,578	⟶	+3,000
5,812		6,000

b.
Actual		Estimate
796	⟶	800
−308	⟶	−300
488		500

c.
Actual		Estimate
198	⟶	200
× 93	⟶	× 90
18,414		18,000

d.
Actual	Estimate
38	40
22)836	20)800

Since each estimate is close to the actual answer, the answers are all reasonable. ◀

Compatible numbers, or numbers with which you can compute mentally, are often used in estimating quotients.

Example 3

Find the quotient 2,535 ÷ 13. Then use compatible numbers to estimate the answer. Compare the estimate with the actual answer.

Solution
Actual   Estimate

$$\begin{array}{r} 195 \\ 13)\overline{2,535} \end{array}$$ 13 is close to 12 and 2,535 is close to 2,400. $$\begin{array}{r} 200 \\ 12)\overline{2,400} \end{array}$$

12 and 24 are compatible numbers.

So, mentally divide 24 hundred by 12.

Since 195 is close to 200, the answer is reasonable. ◀

Example 4

Friday's attendance at a football game was 21,904. Saturday's attendance was 38,053.

a. Estimate the total attendance to the nearest ten thousand.

b. Estimate how many more people attended on Saturday than on Friday.

Solution

a.
21,904	⟶	20,000
+38,053	⟶	+40,000
		60,000

The total attendance was about 60,000 people.

b.
38,053	⟶	40,000
− 21,904	⟶	−20,000
		20,000

About 20,000 more people attended on Saturday. ◀

TRY THESE

Round each of the following.

1. 68 to the nearest ten **70**

2. 21 to the nearest ten **20**

3. 358 to the nearest hundred **400**

4. 844 to the nearest hundred **800**

5. 7,333 to the nearest thousand **7,000**

6. 2,724 to the nearest thousand **3,000**

COMPUTER

In BASIC, the function INT finds the greatest INTeger in a number. All digits to the right of the decimal are dropped. Only the whole number is given. For example:

INT(8.32) = 8
INT(8.95) = 8

You can round in BASIC by adding 0.5 to the number and then finding the INT of the sum.

```
10 INPUT "ENTER THE
   NUMBER: ";N
20 R = INT (N + 0.5)
30 PRINT : PRINT
   "ROUNDED TO THE
   NEAREST INTEGER: ";R:
   PRINT
40 INPUT "ENTER T TO
   TRY AGAIN OR Q TO
   QUIT ";X$
50 PRINT : PRINT : IF X$ =
   "T" GOTO 10
```

ASSIGNMENTS

BASIC
1–18, 21–25, 30, 33

AVERAGE
1–20, 21–28, 31, 33–34

ENRICHED
7–18, 20, 23–31, 32–34

ADDITIONAL RESOURCES
Reteaching 2–1
Enrichment 2–1
Transparency Masters 7, 8

Example 2: Point out that it is important to check the reasonableness of an answer when solving word problems. You may wish to have students make up a word problem for each problem in this example.

Example 3: Point out that if two numbers are compatible, then one is a multiple of the other.

Example 4: Students can use calculators to check their estimates with the actual answers.

Additional Questions/Examples
Which is a better method for estimating quotients: rounding or using compatible numbers? Why? **Answers will vary. Possible answer: compatible numbers, since this method guarantees you will be able to use mental math to estimate.**

Guided Practice/Try These After students have completed these problems, have volunteers discuss their criteria for determining if an estimate was reasonable or not.

43

Actual	Estimate
7. 1,298	1,300
8. 9,100	9,000
9. 1,143	1,000
10. 298	300
11. 76	80
12. 35	40

③ SUMMARIZE

Write About Math Give each students two index cards. Have them write an explanation of how to estimate, either by rounding or by using compatible numbers, on one side of each card. On the other side of each card, they can write an example of the corresponding method of estimation.

④ PRACTICE

Practice/Solve Problems For Exercises 7–18, you may wish to have students use calculators to check their answers.

Extend/Solve Problems For Exercise 31 review how to find the average of a set of numbers.

Think Critically/Solve Problems For Exercises 32 and 33, point out that students can use a guess-and-check strategy.

⑤ FOLLOW-UP

Extra Practice

1. Find the actual answer. Use estimation by rounding to determine the reasonableness of your answer.
   a. $6,806 \div 82 =$ **83**
   b. $92,876 - 87,231 =$ **5,645**
   c. $3,100 \times 29 =$ **89,900**
2. Estimate the least amount of money you would need in order to pay the following dinner bill:
   Beverages $24.67
   Dinner $52.65
   Dessert $13.14
   about $90

Find the actual answer. Use rounding to estimate the answer. Then compare to see if your actual answer is reasonable.

7. $345 + 953$ 8. 28×325 9. $3,127 - 1,984$

Find the actual answer. Use compatible numbers to estimate the answer. Compare your estimate with your actual answer.

10. $9,834 \div 33$ 11. $2,432 \div 32$ 12. $1,820 \div 52$

Solve.

13. On Friday, there were 5,146 students at the Ice Palace. On Saturday, 3,821 students were there.
    a. Estimate the total attendance to the nearest thousand. **9,000 students**
    b. Estimate how many more students were at the Ice Palace on Friday than on Saturday. **1,000 students**

PRACTICE/ SOLVE PROBLEMS

EXERCISES

Round each of the following.

1. 34 to the nearest ten **30**   2. 12 to the nearest ten **10**

3. 684 to the nearest hundred   4. 8,237 to the nearest hundred **8,200**
   700

5. 1,255 to the nearest thousand   6. 4,328 to the nearest thousand **4,000**
   1,000

Find the actual answer. Use rounding to estimate the answer. Then compare to see if your actual answer is reasonable.

7. $2,168 + 3,804$ 8. $8,796 - 4,521$ 9. 24×352

10. $1,932 \div 21$ 11. $8,142 - 3,490$ 12. $5,942 + 1,298$

Actual	Estimate
7. 5,972	6,000
8. 4,275	4,000
9. 8,448	8,000
10. 92	100
11. 4,652	5,000
12. 7,240	7,000
13. 29	30
14. 84	80
15. 81	80
16. 32	30
17. 61	60
18. 193	200

Find the actual answer. Use compatible numbers to estimate the answer. Compare your estimate with your actual answer.

13. $2,175 \div 75$ 14. $5,712 \div 68$ 15. $7,452 \div 92$

16. $1,888 \div 59$ 17. $4,392 \div 72$ 18. $2,895 \div 15$

Solve.

19. There were 6,758 people at a local theater production on Friday and 9,056 people on Saturday.
    a. Estimate the total attendance to the nearest thousand. **16,000 people**
    b. Estimate how many more people were at the theater on Saturday than on Friday. **2,000 people**

20. XYZ Corporation sold 64,918 cassette tapes in November and 75,985 cassette tapes in December.
  a. Estimate, to the nearest thousand, the total number of tapes sold in November and December. **141,000 tapes**
  b. Estimate how many more tapes were sold in December than in November. **11,000 more tapes**

Estimate. Then choose the letter of the actual answer.

	A	B	C
21. $27 \times 32 \times 49$	423,360	4,233	42,336
22. $83,978 \div 398$	21	211	441
23. $1,594 + 375 + 8,946$	10,915	9,695	18,904
24. $96,350 + 235 + 2,145$	100,845	98,730	141,300
25. $21,914 - 13,450$	45,816	18,917	8,464

USING DATA Use the Data Index on page 546 to find the seigniorage chart. Round the seigniorage of coin and silver to the nearest million dollars for the given year.

26. 1974 **$321,000,000**   **27.** 1984 **$498,000,000**   **28.** 1988 **$470,000,000**

29. A student gave the sum of the seigniorage for 1970, 1980, and 1986 as 1,629,478,350.10. Is this a reasonable answer? Explain. **See Additional Answers.**

Solve.

30. There are 12 rows of seats in an auditorium with 48 seats in each row. It was reported that 850 students were seated in the auditorium on Monday. Is this a reasonable answer? Explain. **See Additional Answers.**

31. A student had these five test scores: 92, 75, 84, 89, 75.
  a. Estimate to see if a mean score of 83 is reasonable.
  b. Find the median and the mode. **See Additional Answers.**

Round to estimate.

32. Three numbers are given as the missing factor for the following: $56 \times \blacksquare = 4,536$. Which one is reasonable?
  a. 903   **b.** 81   **c.** 93 **b.**

33. Three numbers are given as a divisor for $1,862 \div \blacksquare = 38$. Which one seems reasonable?
  a. 4   **b.** 49   **c.** 490 **b.**

34. It was reported that about 80,000 people attended a football game. If this number was rounded to the nearest ten thousand:
  a. What is the least number of people that could have attended? **75,000 people**
  b. What is the greatest number of people that could have attended? **84,999 people**

EXTEND/SOLVE PROBLEMS

Estimate
21. 45,000
22. 210
23. 11,000
24. 98,500
25. 9,000

COMPUTER TIP

Data base software is a powerful application tool of the computer. It allows you to organize, store, and retrieve information in a variety of ways. A computer that is equipped with a CD ROM can search very large data bases, for example, all the volumes of an encyclopedia. It enables you to retrieve information, complete with pictures, very quickly.

THINK CRITICALLY/SOLVE PROBLEMS

3. Choose the most reasonable answer for each.
  a. $85 \times \blacksquare = 5,865$
  1. 51   2. 69   3. 700 **2**
  b. $30,218 \div \blacksquare = 521$
  1. 58   2. 601   3. 512 **1**

Extension Have students work in small groups. Have each group make a list of situations, for some of which they would need an actual or exact answer (for example, the total number of employees to be paid) and for some of which they would need only an estimate (for example, the number of people at a concert). Have groups exchange lists and identify which situations need exact answers and which do not.

Section Quiz Give an estimate or exact answer for each.

1. $32 \times 12 \times 48$ **18,432**
2. $8,960 \div 32$ **280**
3. $76,532 + 614$ **77,146**
4. $87 \times 31 - 710$ **1,987**
5. Estimate the average of the following 5 student heights: 64 inches, 72 inches, 74 inches, 67 inches, and 68 inches.
  70 inches = estimate; actual = 69
6. A national computer company has sold 3,542 computers so far this year. If their goal is to sell 6,000 computers per year, about how many more must they sell? **about 2,500**

Additional Answers
See page 573.

Additional answers for odd-numbered exercises are found in the Selected Answers portion of the page.

AT-RISK STUDENTS

Some students may still have difficulty understanding place value when rounding. Provide additional place-value practice by having them organize the following numbers on a place-value chart and then round each to the nearest thousand, hundred, ten:

12,456   124,367   23,890

SPOTLIGHT

OBJECTIVE
- Write and evaluate variable expressions

VOCABULARY
evaluate, value, variable, variable expression

WARM-UP

Write numerical expressions for each of the following.
5 students, each with 2 hands
5 × 2   36 eggs, 12 of which break **36 − 12**

1 MOTIVATE

Explore Have students discuss whether this activity would work if they used decimal numbers or fractions instead of whole numbers. Have them try it several times to see if they are correct.

2 TEACH

Use the Pages/Skills Development
Have students read this part of the section and then discuss the examples.
   Example 1: Explain that to write an expression, the students must pay attention to key words and phrases, such as *more than, less than,* and *times.*

2-2 Variables and Expressions

EXPLORE

Try the following number trick four times, each time starting with a different number. Suppose you choose 3 and then add 5. Adding 5 to 3 leads to the statement $3 + 5 = 8$. Write a statement for each step from b–g.

> **Predict the Number!**
> a. Choose any number.   e. Divide by 2.
> b. Add 5.   f. Subtract the original number.
> c. Multiply by 2.   g. Record your answer.
> d. Subtract 4.

What conclusions can you draw?

SKILLS DEVELOPMENT

Often it is necessary to use a **variable,** a letter such as a, n, x, or y, to represent an unknown number. Expressions with at least one variable, such as $n + 5$ and $x \div 9$, are **variable expressions.** You can write variable expressions to describe word phrases.

Example 1

Write a variable expression. Let n represent "a number."
a. 8 more than a number   **b.** 9 less than a number
c. a number divided by 3   **d.** 3 times a number

Solution
a. $n + 8$ **b.** $n - 9$

c. $n \div 3$ or $\frac{n}{3}$ A bar indicates division.   **d.** $3 \times n$ or $3n$ 3n means multiply n by 3. ◄

When you **evaluate** a variable expression, you replace the variable with a number. (The number is called a **value** of the variable.) Then you perform any indicated operations, keeping in mind these important rules.

► First do all multiplications and divisions in order from left to right.
► Then do any additions and subtractions in order from left to right.

Example 2

Evaluate each expression.
a. $9 + m$, if $m = 12$ **b.** $a - 7$, if $a = 15$
c. $3y - 4$, if $y = 9$ **d.** $m \div 7 + 1$, if $m = 21$

MIXED REVIEW

Find the mean, median, and mode for each set of data.

1. 178, 235, 679, 384, 384
   372, 384, 384
2. 13, 98, 45, 67, 67, 98, 17, 28, 98 **59, 67, 98**
3. 31, 56, 94, 79, 85, 62, 31, 92, 39, 31 **60, 59, 31**
4. Give the difference between the value of the mean and the value of the median of these numbers: 7, 9, 2, 3, 8, 9, 5, 1, 9, 4, 3, 6, 8, 0, 1 **0**

Solution

a. $9 + m$
$9 + 12 = 21$

b. $a - 7$
$15 - 7 = 8$

c. $3y - 4$
$3 \times 9 - 4$ Multiply first.
$27 - 4 = 23$ Then subtract.

d. $m \div 7 + 1$
$21 \div 7 + 1$ Divide first.
$3 + 1 = 4$ Then add. ◄

Some expressions have more than one variable.

Example 3

Evaluate the expression $2x + 5y$, if $x = 2$ and $y = 3$.

Solution

$2x + 5y$
$2 \times 2 + 5 \times 3$ Multiply first.
$4 + 15 = 19$ Then add. ◄

You can write and evaluate expressions to describe real-world situations.

Example 4

Write an expression for each real-world situation.
a. c pens shared equally by 4 boys   **b.** s dollars of savings decreased by $27
c. d dollars more than $7
d. 5 times p points
e. t movie tickets at $7 each

Solution
a. $c \div 4$ or $\frac{c}{4}$ **b.** $s - 27$ **c.** $7 + d$
d. $5 \times p$ or $5p$ **e.** $7 \times t$ or $7t$ ◄

TRY THESE

Write a variable expression. Let n represent "a number."

1. 7 greater than a number $n + 7$ **2.** 5 less than a number $n - 5$

3. the product of 3 and a number $3n$ **4.** a number divided by 3 $n \div 3$ or $\frac{n}{3}$

Evaluate each expression. Let $a = 8$.

5. $13 - a$ 5   **6.** $a + 12$ 20   **7.** $a \div 2$ 4   **8.** $5a$ 40

Evaluate each expression. Let $a = 3$ and $b = 6$.

9. $a + b$ 9   **10.** $b \div a$ 2   **11.** $b - 2a$ 0   **12.** $2b \div a$ 4

Write an expression to describe each situation.

13. b baseballs less 7 $b - 7$ **14.** s dollars of savings equally deposited in 5 accounts $s \div 5$ or $\frac{s}{5}$

15. s dollars increased by $12 $s + 12$ **16.** three times d dollars $3d$

ASSIGNMENTS

BASIC
1-38, 40–48, 52, 56

AVERAGE
7–39, 52–54, 55–56

AVERAGE
33–39, 53–54, 55–57

ADDITIONAL RESOURCES
Reteaching 2–2
Enrichment 2–2

Example 2: Discuss the result of changing the value of the variable in an expression. **The value of the expression often changes as well.**
Example 3: Have students interpret the expression in words. **two times one number plus five times another number**
Example 4: Remind students of the words used to suggest number operations in verbal expressions. **shared equally, decreased, more than, times**

Additional Questions/Examples
Create three real-life situations involving numbers and describe each using an expression. **Answers will vary.**

Guided Practice/Try These
For additional practice you may wish to have students assign values to the variables in Exercises 1–4 and 13–16 and then evaluate the expressions.

5-MINUTE CLINIC

Exercise	Student's Error	Error Diagnosis
Evaluate $2a + b$. Let $a = 2$ and $b = 4$.	$2 \times 2 + 4 = 2 \times 6 = 12$	• Student does not realize that multiplication must be performed before addition: $2 \times 2 + 4 = (2 \times 2) + 4 = 4 + 4 = 8$

17. A carton holds c cans of food. The cans were shared equally by 8 people.
    a. Write an expression to describe the situation. $c \div 8$ or $\frac{c}{8}$
    b. Evaluate the expression. Let $c = 48$. **6**

EXERCISES

Write a variable expression. Let n represent "a number."

1. half of a number $\frac{n}{2}$
2. 4 less than a number $n - 4$
3. six times a number $6n$
4. a number increased by 9 $n + 9$
5. 7 greater than a number $n + 7$
6. a number divided by 7 $n \div 7$ or $\frac{n}{7}$
7. the product of 10 and a number $10n$
8. 16 decreased by a number $16 - n$

Evaluate each expression. Let $a = 9$.

9. $a - 2$ **7**
10. $2a$ **18**
11. $27 \div a$ **3**
12. $21 + a$ **30**
13. $a \div 3$ **3**
14. $7a$ **63**
15. $12 - a$ **3**
16. $a + 13$ **22**
17. $3a$ **27**
18. $3a - 2$ **25**
19. $3a + 4$ **31**
20. $\frac{4a}{9}$ **4**

Evaluate each expression. Let $a = 2$ and $b = 8$.

21. $a + b$ **10**
22. $b - a$ **6**
23. $b \div a$ **4**
24. $9a - b$ **10**
25. $b - 4a$ **0**
26. $3b \div a$ **12**
27. $7a \times 4b$ **448**
28. $9a - 2b$ **2**
29. $12a \div b$ **3**
30. $b \div 2a$ **2**
31. $b - 3a$ **2**
32. $5a + 6b$ **58**

Write an expression to describe each situation.

33. 9 more than w wallets $w + 9$
34. d dollars shared equally by 3 people $d \div 3$ or $\frac{d}{3}$
35. p party favors at 25 cents each $25p$
36. e erasers decreased by 2 $e - 2$

Solve.

37. A box holds c pencils. The pencils were shared equally by 8 students.
    a. Write an expression to describe the situation. $c \div 8$ or $\frac{c}{8}$
    b. Evaluate the expression if c is 24. **3**

38. Jenny has c cassette tapes. She gives 5 tapes away.
    a. Write an expression to describe the situation. $c - 5$
    b. Evaluate the expression if c is 38. **33**

3 SUMMARIZE

Key Questions
What is a variable? **A variable is a letter used to represent an unknown number.**
What is a variable expression? **an expression that contains at least one variable**
How can you evaluate a variable expression? **Replace the variable with a number and perform the indicated operations.**

4 PRACTICE

Practice/Solve Problems For Exercises 37–39, have students identify key words and phrases that will help them write accurate expressions. **shared equally, gives away, rose**

Extend/Solve Problems For Exercise 51 remind students to divide before they subtract.

Think Critically/Solve Problems For Exercise 57 give students a hint that they should work backwards to create their own trick.

5 FOLLOW-UP

Extra Practice
Write an expression to describe each situation:
1. the cost of n books at $4 each
2. 15 more than two times n students
3. 25 miles per hour faster than the speed limit of n miles per hour
   1. **4n** 2. **2n + 15** 3. **n + 25**

39. The temperature in the morning was 65°F. It rose d degrees before noon.
  a. Write an expression to describe the situation. **65 + *d***
  b. Evaluate the expression if d is 14. **79**

Evaluate each expression. Let $a = 2$, $b = 3$, and $c = 6$.

40. $a + c - b$ **5**   **41.** $c - 2a + 1$ **3**   **42.** $a + \dfrac{12}{b} - 1$ **5**

43. $a + b + c$ **11**   **44.** $c + a - b$ **5**   **45.** $\dfrac{c}{a} + b$ **6**

46. $b - a + c$ **7**   **47.** $c - b + a$ **5**   **48.** $b + c - 2a$ **5**

49. $3c + 2a + b$ **25**   **50.** $2c - 4a + b$ **7**   **51.** $12b \div 6 - 1$ **5**

Solve.

52. Ian shares d dollars equally with 5 friends.
  a. Write an expression to describe the situation. ***d* ÷ 6 or $\dfrac{d}{6}$**
  b. Evaluate the expression if d is 30. **5**

53. In May, Chang had d dollars in his bank account. Three months later, he had 4 more than 3 times that amount.
  a. Write an expression to describe the situation. **3*d* + 4**
  b. Evaluate the expression if d is 25. **79**

54. Ramona's test scores — 96, 88, x, 88, and 85 — are given in order from greatest to least.
  a. The median is represented by x. Find its value. **88**
  b. Find the mean score. **89**

55. Let x represent an odd number. Then write an expression that describes each of the following.
  a. next odd number ***x* + 2**
  b. preceding odd number ***x* − 2**
  c. next whole number ***x* + 1**
  d. preceding whole number ***x* − 1**
  e. sum of the odd number and the next whole number ***x* + *x* + 1**
  f. sum of the odd number and the next odd number ***x* + *x* + 2**

56. Try this number trick. Pick a number. Add 20, multiply by 6, divide by 3, subtract 40, then divide by 2.
  a. Write a conclusion about your result.
  b. Change this number trick so that your result is twice the number you started with.

57. Write a number trick in which the result is your age. Try it out with a classmate. **Answers will vary.**

CHALLENGE

Display the following trick to guess someone's age: Write the year of birth. Double it. Add 5. Multiply by 50. Add the age. Add x. Subtract y. The display will show the person's age preceded by the year of birth. Have students use a guess-and-check strategy to find the value of the variables x and y.
Answers will vary. Sample answers:
***x* = 365, *y* = 615; *x* = 1, *y* = 251**

► READ
► PLAN
► SOLVE
► ANSWER
► CHECK

Remember to use the 5-step problem solving plan to help you organize your work when you solve any problem.

PROBLEM

A professional athlete earned the same amount of money for each game played. The athlete earned a total of $67,536 for 36 games. How much did the athlete earn for each game?

SOLUTION

These steps can help you solve the problem.

READ   *What information do you know?*
         Amount earned: $67,536
         Number of games played: 36
         Equal amounts of money earned for each game.

         What information are you asked to find?
         How much was earned for each game?

PLAN   *Decide what operation(s) you need to use.*
         Because the same amount of money was earned for each game, divide the total amount by the number of games to find how much was earned for each game.
$$67,536 \div 36 = n$$

SOLVE   *Carry out the plan.*
$$67,536 \div 36 = 1,876$$

ANSWER   *Give the answer in a sentence.*
         The athlete earned $1,876 per game.

CHECK   *Check to see if the answer is reasonable.*
         Use compatible numbers to estimate.
$$70,000 \div 35 = 2,000$$

         Since 1,876 is close to 2,000, the answer is reasonable.

PROBLEMS

Choose the operation. Do not solve.

1. Willie Mays of the Giants hit 660 career home runs. Joe DiMaggio hit 361 home runs. How many more home runs did Mays hit than DiMaggio? **subtract**

2. A Ford V8 needs spark plugs at a cost of $2 each. What is the total cost of the spark plugs? **multiply**

Choose the operation(s). Then solve.

3. There are 174 people who work at a tourist attraction. During the summer, 37 more people are hired. How many people in all are working at the attraction during the summer? **add; 211 people**

4. Twelve couples bought a load of 288 lb of potatoes. How many pounds will each of the couples receive if they share equally? **divide; 24 lb**

5. Ron bought 14 stickers at 3 cents each. What was the total cost? **multiply; 42 cents**

6. Three brothers shared a birthday gift of $135 equally. How much money did each brother receive? **divide; $45**

7. Mandy raises worms for bait. One year she raised 23,167 worms. The next year she raised 15,154 more worms than the year before. How many worms did she raise the second year? **add; 38,321 worms**

8. Kevin bought 8 boxes of erasers. Each box contained 44 erasers. How many erasers did he buy in all? **multiply; 352 erasers**

9. There were 175 benches in the auditorium, each holding the same number of people. There were 1,400 people in all. How many people were seated on each bench? **divide; 8 people**

10. The combined height of two acrobats, one standing on the other's head, was 128 inches. The height of one acrobat was 68 inches. What was the height of the other acrobat? **subtract; 60 inches**

11. How many cartons of 12 eggs could be filled with 144 eggs? **divide; 12 cartons**

12. On election night, it was reported that out of 1,545 polling stations that were closed, 698 had completed counting their votes and 150 stations were still counting. How many stations had not yet begun counting? **add, subtract; 697 stations**

13. A sleeping person breathes an average of 15 times per minute. How many breaths would someone take during 8 h of sleep? **multiply, multiply; 7,200 breaths**

14. Kesper Auto Limited carries spark plugs in boxes of 10. They have 24 boxes of plugs in stock. How many cars requiring 6 spark plugs each could they service with this amount? **multiply, divide; 40 cars**

USING DATA Use the Data Index on page 546 to find the listing of the early monetary system used in the Pacific Islands.

15. How many coconuts would have to have been gathered to equal the monetary value of 1 dog's tooth? **100 coconuts**

16. What is the value in dog's teeth of 500 coconuts? **5 dog's teeth**

17. What is the value in porpoise teeth of 100 strings of white teeth? **50 porpoise teeth**

 **READING MATH**

When you read a problem, how do you decide which operation is involved in the solution? There are no hard-and-fast rules, but certain words in the problem may suggest which operation(s) you should consider. Consider these words and word phrases, for example. Then see if you can think of others to add to the list.

Addition
plus
sum
total
altogether
combined
in all

Subtraction
minus
difference
how much more
how much less

Multiplication
times
product
doubled, tripled, and so on
total
altogether
in all

Division
quotient
shared equally
equal parts

ASSIGNMENTS

BASIC
1–12, 15

AVERAGE
3–13, 15–16

ENRICHED
6–14, 16–17

ADDITIONAL RESOURCES
Reteaching 2–3
Enrichment 2–3

3 SUMMARIZE

Write About Math Have students write the operation they would use to solve the following problem along with an explanation of why they chose that operation: There are 15 windows on each of 12 houses. How many windows are there in all? **multiplication; 180 windows in all**

4 PRACTICE

For Problems 12–14 point out that more than one operation is needed.

5 FOLLOW-UP

Have students work in pairs. Have each pair choose three of the word problems they solved in this section and write a new problem for each, using the answer and one of the given numbers as the information in their new problem.

Get Ready calculators

OBJECTIVES
- Find decimal sums and differences
- Solve problems involving addition and subtraction of decimals

MATERIALS NEEDED
calculators

VOCABULARY
front-end estimation

WARM-UP

Estimate to the nearest thousand.
1. 2,345 + 1,256 = **3,000**
2. 12,123 + 2,658 = **15,000**
3. 16,789 − 4,987 = **12,000**

1 MOTIVATE

Explore After students discuss similarities between addition of whole numbers and addition of decimals, have them draw conclusions about similarities between subtraction of whole numbers and subtraction of decimals.

2 TEACH

Use the Pages/Skills Development
Have students read this part of the section and then discuss the examples.

Example 1: Point out that although it is not necessary to annex zeros for addition of decimals, annexing zeros is usually necessary when subtracting decimals with different numbers of decimal places as in **b** of this example.

2-4 Adding and Subtracting Decimals

EXPLORE/ WORKING TOGETHER

Choose a sum of money between 25 cents and one dollar. Take turns in your group showing various combinations of coins that are equivalent in value to this sum. For example, to show 47 cents, someone might choose the following coin combination:

Number of Coins	Coins	Number of Cents	Decimal Form
1	Quarter	25	$0.25
2	Dimes	20	$0.20
2	Pennies	2	$0.02
Totals 5		47	$0.47

How many combinations of coins did your group find for the sum you chose? Repeat for four more sums, determining for each sum all the possible combinations of coins that are equivalent in value. Make a table like the one above to record all the combinations.

For each sum, compare the totals in the last two columns.

SKILLS DEVELOPMENT

What you learn in one part of mathematics often can be used to guide you as you learn other skills. For example, you can use what you know about adding and subtracting whole numbers to help you add and subtract decimals.

▶ To add or subtract whole numbers, line up the digits in the ones, tens, and hundreds columns and within any other columns that represent values greater than 1. Then add or subtract in each column.

▶ To add or subtract decimals, line up the digits in the tenths, hundredths, and thousandths columns and within any other columns that represent values less than 1. Aligning the decimal points helps you align the digits correctly. Then add or subtract in each column. Place the decimal point in the answer directly below the other decimal points.

Example 1

Find each answer. Use rounding to estimate the answer and compare to see if the answer is reasonable.
a. 8.6 + 3.89 + 6.2   b. 86.2 − 19.76

52 CHAPTER 2 Exploring Whole Numbers and Decimals

MAKING CONNECTIONS

Have students choose two different years in which the Summer Olympics were played. Have them research the winners of the Men's and Women's 100-, 200-, and 400-meter dashes in Track and Field. Tell students to write a brief report comparing the winning runners' times.

Solution

Align the decimal points. If necessary, annex zeros so all numbers have the same number of decimal places. Then add or subtract.

<table>
<tr><td colspan="2">a. Actual   Estimate</td><td colspan="2">b. Actual   Estimate</td></tr>
</table>

a. Actual	Estimate
8.60 →	9
3.89 →	4
+ 6.20 →	+ 6
18.69	19

b. Actual	Estimate
86.20 →	86
−19.76 →	−20
66.44	66

Since 18.69 is close to 19, the answer is reasonable.

Since 66.44 is close to 66, the answer is reasonable. ◄

You also can use **front-end estimation.**

Example 2

Find the total cost of these purchases, including tax: tape, $6.99; CD, $15.99; earphones, $19.99; sales tax, $2.58. Use front-end estimation and compare your estimate with the actual answer.

Solution

Actual	Add front-end digits.	Adjust remaining digits.
$ 6.99	6	0.99 → 1
15.99	10	5.99 → 6
19.99	10	9.99 → 10
+ 2.58	+ 2	0.58 → 1
$45.55	28 +	18 = 46

The total, $45.55, is close to 46, so the answer is reasonable. ◄

TRY THESE

Compute. Round to estimate. Compare to check your answer.

1. $6.29 + $16.29 + $4.28 **$26.86**   **2.** 294.8 − 124.7 **170.1**

Compute. Use front-end estimation to see if the answer is reasonable.

3. 6.94 + 41.77 + 5.2 + 8.4 **62.31**   **4.** $2,456.69 − $1,297.70 **$1,158.99**

Solve. Estimate and compare to see if the answer is reasonable.

5. Maria works part time. One week, her gross pay was $34.55. She paid $8.75 in taxes. What was her take-home pay that week? **$25.80**

EXERCISES

Compute. Round to estimate. Compare to check your answer.

PRACTICE/ SOLVE PROBLEMS

1. $16.52 + $5.03 + $166.29   See Additional Answers.   **2.** $3.23 + $17.58 + $100.01   See Additional Answers.

3. 1,568.33 − 422.88 **1,145.45**   **4.** 24,367.1 − 14,368.9 **9,998.2**

ASSIGNMENTS

BASIC
1–10, 13, 15–17, PSA 1, 5, 6, 8

AVERAGE
5–10, 11–13, 18–21, PSA 1–8

ENRICHED
8–10, 12–14, 19–20, PSA 4–8

ADDITIONAL RESOURCES
Reteaching 2-4
Enrichment 2-4
Transparency Master 9

Example 2: Discuss the steps involved in front-end estimation. Have students suggest another method of estimation that could have been used. **rounding to the nearest dollar**

Additional Questions/Examples
1. Why should you align the decimal points when you add or subtract decimals? **Answers may vary, but students should understand that they must add or subtract numbers that are in like place-value positions.**

Use estimation to choose the most sensible answer for each.
2. 3.45 + 125.007
   a. 128.457   **b.** 460.007   **c.** 42.457   **a**
3. 563.978 − 2.007
   a. 363.971   **b.** 561.971   **c.** 3.6307   **b**

Guided Practice/Try These Allow students to double-check their answers with calculators. For Exercise 5 you may wish to define *gross pay* and *take-home pay*.

5-MINUTE CLINIC

Exercise	Student's Error	Error Diagnosis

45.8 − 2.87	45.8
	− 2.87
	1.71
		• Student aligned digits rather than decimal points.
		Correct solution: 45.80
		   − 2.87
		   42.93

53

3 SUMMARIZE

Talk It Over Have students discuss the similarities and differences between adding and subtracting whole numbers and adding and subtracting decimals. Encourage students to give examples of real-life situations in which they would need to perform these operations.

4 PRACTICE

Practice/Solve Problems For Exercise 10 point out that, by writing a check, Anwar is withdrawing money from his account.

Extend/Solve Problems Point out that Exercise 12 is based on the answers to Exercise 11.

Think Critically/Solve Problems For Exercises 15–20 students may use the guess-and-check strategy.

Problem Solving Applications For Exercise 2 you may wish to review division of decimals and whole numbers.

5 FOLLOW-UP

Extra Practice Solve, then estimate to check the reasonableness of each answer.
1. $72.65 + $321.02 + $4.98 = **$398.65**
2. 645.03 – 5.124 = **639.906**
3. 2,345.1 – 245.78 = **2,099.32**
Use the following data to answer Examples 4–6.

Cost of Rita's New Clothes
Shoes $34.75
Suit $125.60
Coat $99.50
Dress $102.04

4. Rita gave the clerk $400 for all the clothes. What was her

---

Compute. Use front-end estimation to see if the answer is reasonable.

5. 356.5 + 3,496.80 + 524.6 **4,377.9**
6. $525.49 − $137.84 **$387.65**
7. 6.66 + 66.66 + 666.66 + 66.666 **806.646**
8. $1,489.65 − $1,211.15 **$278.5**

Solve. Estimate and compare to see if your answer is reasonable.

9. Bill earned $15.60, $21.75, and $16.80 for baby-sitting one week. He also earned $12.50 doing yard work. How much money did he earn in all that week? **$66.65**

10. Anwar had a balance of $157.92 in his checking account. He wrote a check for $62.39. What was his new balance? **$95.53**

EXTEND/ SOLVE PROBLEMS

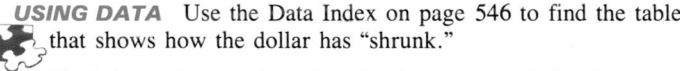

USING DATA Use the Data Index on page 546 to find the table that shows how the dollar has "shrunk."

11. Find the total cost of the four foods for each of the six years.
See Additional Answers.

12. Find the change in the total cost of the four foods between the following pairs of years.
$0.10 a. 1890 and 1910   b. 1910 and 1930 **$0.56**   c. 1930 and 1950 **$0.94**
d. 1950 and 1970 **$1.02**   e. 1970 and 1975 **$1.19**   f. 1890 and 1975 **$3.81**

Solve. Estimate and compare to see if your answer is reasonable.

13. Allison purchased a sweater for $72.21 and a jacket for $102.99. She gave the clerk $200. How much change did she receive? **$24.80**

14. Lynne and Rico each loaned Damon $5.00. Damon used the money to buy a bouquet of flowers for his mother. How much money did Damon have left after spending $8.37 for the flowers? **$1.63**

THINK CRITICALLY/ SOLVE PROBLEMS

Complete. Estimate and compare to see if your answers are reasonable.

15. 43.25■ **9**
+ 0.352
4■.■11 **3, 6**

16. 5.004
−1.9■■ **9, 9**
■.■05 **3, 0**

17. 9.■16 **8**
−3.7■9 **6**
■.04■ **6, 7**

18. 2.16
3.■■ **1, 3**
+■.05 **4**
9.34

19. 4.■7 **2**
−■.3■ **2, 5**
1.92

20. 2.1■5 **2**
6.■94 **3**
+■.058 **0**
8.57■ **7**

21. Enrico had some money. He earned $15 for baby-sitting. Then he bought food for $18.74 and gasoline for $21.93. He has $35.84 left. How much money did he start out with? **$61.51**

Problem Solving Applications:

TRAVEL EXPENSES

After graduation, three students took a 7-day trip across part of the United States. They recorded their expenses.

Day	Food	Motel	Gas
Sunday	$65.00	$35.00	$45.29
Monday	72.16	48.00	56.16
Tuesday	34.91	27.00	18.92
Wednesday	52.38	54.00	14.38
Thursday	43.27	38.00	10.21
Friday	63.64	29.00	54.98
Saturday	38.72		21.37

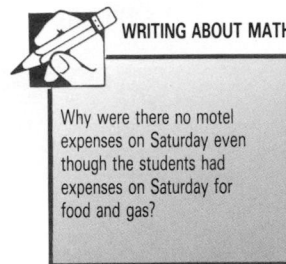

WRITING ABOUT MATH

Why were there no motel expenses on Saturday even though the students had expenses on Saturday for food and gas?

There were no motel expenses because they arrived home on Saturday.

Use the expense record to solve these problems.

1. Find the total spent during the week for each.
   a. food **$370.08**   b. motels **$231.00**   c. gas **$221.31**

2. The students shared the costs equally. Find out each student's cost for each item.
   a. food **$123.36**   b. motels **$77.00**   c. gas **$73.77**

3. What was each student's total expense? **$274.13**

4. Each student started the trip with $400. How much did each have left to spend on other items after paying for food, motels, and gas? **$125.87**

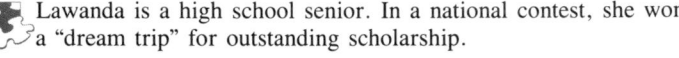 Lawanda is a high school senior. In a national contest, she won a "dream trip" for outstanding scholarship.

5. In Mexico, Lawanda bought a pair of maracas for 2,500 pesos. She paid with a 5,000-peso note. How many pesos did she receive in change? **2,500 pesos**

6. In India, she bought a bag for 1,200 rupees, a sari for 2,300 rupees, and a brass dish for 750 rupees. How many rupees did she spend? **4,250 rupees**

7. In Canada, she purchased a coat for $123.45 and a scarf for $11.98. She handed the clerk $150.00. How much change did she receive? **$14.57**

8. While in Canada, Lawanda attended two games in a hockey exhibition. Each ticket cost $25.75. She bought food totaling $12.35 at the snack bar. How much did she spend? **$63.85**

2-4   Adding and Subtracting Decimals   **55**

---

change? **$38.11**

5. How much more did the dress and shoes cost than the suit? **$11.19**

6. How much less was the cost of the coat than the suit? **$26.10**

Extension Have students work in pairs to decide if and where a decimal should be placed in each whole number to make sentences true:

1.   267 + 2146 + 321 = 51.37
     26.7 + 21.46 + 3.21 = 51.37

2.   5604 – 12987 = 43.093
     56.04 – 12.987 = 43.053

Suggest students use estimation to check.

Section Quiz Solve. Estimate and compare to see if your answer is reasonable.

1. 456.1 + 3.567 + 21.43 =   **481.097**
2. 567.2 – 23.409 =   **543.791**
3. Dave worked for two days and earned $25.50 per day. He spent $2.50 on travel each day and $4.75 on lunch each day. How much of his salary did he have left?   **$36.50**

Complete and estimate to check the reasonableness of your answers

4.   52.5 _
     –1_.98
     _____
     38.52
     52.50 – 13.98 = 38.52

5.   4,_12._ 8
        23._ 0
     + 45.6 _ _
     _____
     4,581.689
     4,512.98 + 23.1 + 45.609 =
     4,581.689

Get Ready calculators, grid paper

Additional Answers
See page 573.

56

2-5 Multiplying and Dividing Decimals

EXPLORE Mark off a 10 by 10 grid on graph paper.

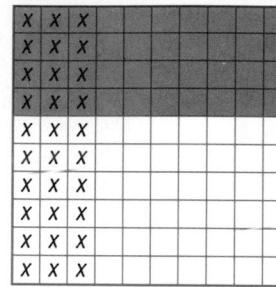

0.4 × 0.3 = 0.12

a. Shade all the squares that make up four rows. Write the decimal that describes the shaded part.
b. Mark an X in all the squares that make up three columns. Next to your first decimal, write the decimal for the squares with X's.
c. Count the squares that are both shaded and marked with an X. Write a decimal for these squares next to your second decimal.
d. Put a multiplication sign between the first two decimals and an equals sign between the last two.

You have just multiplied two decimals!
Experiment with other numbers.

SKILLS DEVELOPMENT

Multiply and divide decimals just as you multiply and divide whole numbers. Then place a decimal point in the answer.

Study and analyze these examples. See if you can find the pattern that was used to place the decimal points in the products.

64	64	64	6.4	6.4	6.4	0.64	0.64
× 2	×0.2	×0.02	× 2	×0.2	×0.02	× 2	×0.02
128	12.8	1.28	12.8	1.28	0.128	1.28	0.0128

► To find the number of decimal places to put in a product, find the sum of the decimal places in the factors. Count off that number of places in the product, counting from right to left.

Example 1

Find the product. Estimate and compare to see if it is reasonable.
a. 4×8.7 b. 6.3×9.2 c. 0.98×16.4

Solution

a. Actual Estimate
$$\begin{array}{rr} 8.7 & \to \quad 9 \\ \times\ 4 & \to \quad \times 4 \\ \hline 34.8 & 36 \end{array}$$

b. Actual Estimate
$$\begin{array}{rr} 9.2 & \to \quad 9 \\ \times 6.3 & \to \quad \times 6 \\ \hline 276 & 54 \\ 5520 & \\ \hline 57.96 & \end{array}$$

c. Actual Estimate
$$\begin{array}{rr} 16.4 & \to \quad 16 \\ \times 0.98 & \to \quad \times\ 1 \\ \hline 1312 & 16 \\ 14760 & \\ \hline 16.072 & \end{array}$$

Each estimate is close to the actual answer, so each product is reasonable. ◄

Study and analyze these examples. See if you can find the pattern that was used to place the decimal points in the quotients.

$$\frac{12}{3\overline{)36}} \qquad \frac{1.2}{3\overline{)3.6}} \qquad \frac{0.12}{3\overline{)0.36}} \qquad \frac{0.0012}{3\overline{)0.0036}} \qquad \frac{0.07}{7\overline{)0.49}}$$

$$\frac{125}{0.2\overline{)25.0}} \qquad \frac{12.5}{0.2\overline{)2.50}} \qquad \frac{1.25}{0.2\overline{)0.250}} \qquad \frac{8}{0.08\overline{)0.64}} \qquad \frac{0.08}{0.08\overline{)0.0064}}$$

► To divide a decimal by a whole number, place the decimal point in the quotient above the decimal point in the dividend.

► To divide a decimal by a decimal, first move the decimal point in the divisor to the right to get a whole number. Then move the decimal point in the dividend the same number of places to the right. Annex zeros if necessary.

Example 2

Find the quotient. Estimate and compare to see if it is reasonable.

a. $23.76 \div 24$ b. $11.178 \div 2.3$

Solution

a.   Actual   Estimate   b.   Actual   Estimate

```
      0.99                    1                        4.86                  5
  24)23.76              24)24                    2.3)11.1 78            2)10
     21 6                                            9 2
      2 16                                           1 9 7
      2 16                                           1 8 4
         0                                             1 38
                                                       1 38
                                                          0
```

Each estimate is close to the actual answer, so each quotient is reasonable. ◄

Try These

Write the letter for the product.

	A	**B**	**C**
1. 4.6×3	138	<u>13.8</u>	1.38
2. 42×5.8	24.36	<u>243.6</u>	2.436
3. 1.9×2.8	<u>5.32</u>	0.532	53.2

Find each product. Estimate and compare to see if it is reasonable.

4. 10×1.3 **13**   5. 100×0.31 **31**   6. $1,000 \times 5.63$ **5,630**

7. 4.8×4 **19.2**   8. 92.3×0.35 **32.305**   9. 0.75×0.92 **0.69**

2–5   Multiplying and Dividing Decimals   **57**

5-MINUTE CLINIC

Exercise	Student's Error	Error Diagnosis
$0.07\overline{)0.0049}$	$\underset{\frown\frown\,\,\,\frown\frown}{0.07\overline{)0.0049}}^{\,.7}$	• Student forgets to include a zero as a place-holder. Correct solution: $\underset{\frown\frown\,\,\,\frown\frown}{0.07\overline{)0.0049}}^{\,.07}$

ASSIGNMENTS

BASIC
1–20, 29, 38

AVERAGE
5–13, 18–36, 41–42

ENRICHED
21–42

ADDITIONAL RESOURCES
Reteaching 2–5
Enrichment 2–5
Transparency Master 10

Example 1: Remind students to find the actual products before they estimate to check each one.

Example 2: Before discussing the solutions, review rounding to the nearest whole number.

Additional Questions/Examples
Write *true* or *false*. If *false*, tell why.
1. The product of a whole number and a decimal less than 1 is always a number greater than the whole number. **false, the product is a number less than the whole number.**
2. When you divide a whole number by a decimal number less than 1, the quotient is always less than the whole number. **false, the quotient is greater than the whole number.**

Guided Practice/Try These Point out that for Exercises 1–3 and 10–12 students need not calculate the products or quotients. For Exercises 13–18 have students explain the methods of estimation they used.

3 SUMMARIZE

Write About Math Have students write the rules for placing decimal points in the answers when multiplying and dividing decimals. Have them give an example of each.

4 PRACTICE

Practice/Solve Problems Suggest students use mental math for Exercises 5–8 and 18–20. For Exercises 27 and 28, have students who have difficulty with problem solving identify key words and phrases and the operation they suggest. *shared equally* suggests "division"; *each* suggests "multiplication"

Extend/Solve Problems Point out that Exercises 34–36 are all two-step problems. Have students choose the correct operations in each before they do their calculations. Suggest they follow the problem solving plan discussed in section 2–3.

Think Critically/Solve Problems For Exercises 41 and 42, have students provide examples to support their answers and explanations.

5 FOLLOW-UP

Extra Practice Solve each problem. Estimate and compare to see if the answer is reasonable.
1. $23.4 \times 9.2 =$ **215.28**
2. $12 \times 0.078 =$ **0.936**
3. 1.05×0.0098 **0.01029**
4. $45.6 \div 2.5$ **18.24**
5. $\$1,652.40 \div \$45.90 =$ **36**

Write the letter for the quotient.

		A	B	C
10.	$17.28 \div 4.8$	36	<u>3.6</u>	0.36
11.	$39.2 \div 14$	28	<u>2.8</u>	0.28
12.	$56.7 \div 6.3$	0.09	<u>9</u>	0.9

Find each quotient. Estimate and compare to see if it is reasonable.

13. $10\overline{)12.3}$ **1.23**   14. $100\overline{)18.96}$ **0.1896**   15. $1,000\overline{)489}$ **0.489**
16. $6\overline{)94.56}$ **15.76**   17. $1.8\overline{)36.36}$ **20.2**   18. $0.05\overline{)16.75}$ **335**

EXERCISES

PRACTICE/ SOLVE PROBLEMS

Write the letter of the product.

		A	B	C
1.	4×3.68	1.472	<u>14.72</u>	147.2
2.	5.1×38	19.38	<u>193.8</u>	1,938
3.	4.6×3.8	1.748	<u>17.48</u>	174.8
4.	3.36×4.2	<u>14.112</u>	141.12	1,411.2

Find each product. Estimate and compare to see if it is reasonable.

5. 10×2.6 **26**   6. 10×0.45 **4.5**   7. 100×0.86 **86**
8. $1,000 \times 4.62$ **4,620**   9. 4×7.2 **28.8**   10. 3.2×5 **16**
11. 3×6.9 **20.7**   12. 13.5×1.8 **24.3**   13. 90.5×0.26 **23.53**

Write the letter of the quotient.

		A	B	C
14.	$29.16 \div 12$	243.0	24.3	<u>2.43</u>
15.	$54.30 \div 15$	36.2	<u>3.62</u>	0.362
16.	$63.18 \div 1.3$	<u>48.6</u>	4.86	0.486
17.	$10.07 \div 2.65$	0.38	<u>3.8</u>	38

Find each quotient. Estimate and compare to see if it is reasonable.

18. $10\overline{)1.36}$ **0.136**   19. $100\overline{)12.4}$ **0.124**   20. $1,000\overline{)148.36}$ **0.14836**
21. $14\overline{)1.694}$ **0.121**   22. $22\overline{)16.94}$ **0.77**   23. $12\overline{)1.44}$ **0.12**
24. $116\overline{)40.6}$ **0.35**   25. $2.5\overline{)3.75}$ **1.5**   26. $0.02\overline{)0.354}$ **17.7**

Solve. Estimate and compare to see if the answer is reasonable.

27. A class of 50 students went to a baseball game together. The total cost of admission was $346.00. If the students shared the costs equally, how much did each student pay? **$6.92**

28. Tickets for a school play cost $4.25 each. The attendance at the play was 1,006 people. How much ticket money was collected? **$4,275.50**

Find each answer rounded to the indicated place-value position.

29. 52.7 × 6.2 to the nearest tenth **326.7**

30. 484.27 ÷ 8.14 to the nearest hundredth **59.49**

31. 4.799 ÷ 2.658 to the nearest thousandth **1.805**

32. 174.8 × 16.3 to the nearest hundredth **2,849.24**

33. 86.42 ÷ 26.17 to the nearest hundredth **3.30**

Solve.

34. A class of 40 students went to a museum. The cost of renting a bus was $156.40. Admission was $3.45 per student. If they shared the costs equally, how much did each student pay? **$7.36**

35. Tickets to a charity dance cost $45.50 each. The attendance at the dance was 1,262. The cost of refreshments averaged $6.21 per person. What was the profit for the charity? **$49,583.98**

36. A pet-store owner bought 6 mollies at $0.79 each, 7 guppies at $0.59 each, 12 swordtails at $1.89 per pair, and 14 platies at $1.69 per pair. What was the total cost of the fish? **$32.04**

 USING DATA Use the Data Index on page 546 to find the currency exchange rates. All exchanges were made on Thursday.

37. Find the equivalent number of Finnish markkas you would receive for 30 U.S. dollars. **132.42 markkas**

38. Find the equivalent number of Portuguese escudos you would receive for 75 U.S. dollars. **11,738.25 escudos**

39. Find the U.S. dollar equivalent for 4,687 Irish punts. **$6,833.36**

40. Find the U.S. dollar equivalent for 467 Pakistani rupees. **$19.15**

Tell whether the statement is *true* or *false*. Explain.

41. When a decimal is divided by a whole number the quotient always has the same number of decimal places as the dividend. **True**

42. The product of two decimals cannot be a whole number.
False. The product of two decimals can be a whole number as, for example, 5.2 × 2.5 = 13.

**EXTEND/
SOLVE PROBLEMS**

**THINK CRITICALLY/
SOLVE PROBLEMS**

6. In a 15-mile race a runner averaged 7.2 minutes per mile. In hours and minutes, how long did it take the runner to finish? **1 h 48 min**

7. Ryan and George own a lawn–mowing business. They earned $75 on Monday, and $50.75 each of the remaining weekdays. How much did each get paid if they share the earnings equally? **$139**

Extension Have students work in pairs to identify the errors in each of the following. Have them classify each error as a miscalculation, a misplaced decimal point, or both. Then have students find the correct answer.

1. 23.456 × 1.7 = 3.98752 **misplaced decimal point, 39.8752**

2. 387.684 ÷ 12.1 = 330.4 **both, 32.04**

3. 0.0708 × 145 = 10.376 **miscalculation, 10.266**

4. 0.0567 ÷ 0.045 = 12.6 **misplaced decimal point, 1.26**

Section Quiz Solve. Estimate and compare to see if your answer is reasonable.

1. $456.17 × 32 = **$14,597.44**

2. 359.94 ÷ 0.42 = **857**

3. 0.0032 × 1.09 = **0.003488**

4. 0.01564 ÷ 0.068 = **0.23**

5. Round your answers for Exercises 1–2 to the nearest whole number. Round your answers to Exercises 3–4 to the nearest tenth. **$14,597; 857; 0; 0.2**

Get Ready calculators

WARM-UP

Write the product of 3.42 and the number given.

1. 10	2. 100	3. 0.10
34.2	**342**	**0.342**

1 MOTIVATE

Explore Explain that students must make up a multiplication problem in which the factors are all 10s and the product equals the number listed in the column. Have students start at 100.

2 TEACH

Use the Pages/Skills Development Have students read this part of the section and then discuss the examples.
    Example 1: Point out that decimals are written the same way as whole numbers in exponential form.
    Example 2: Remind students that the exponent tells how many factors there are.
    Example 3: Point out that a power of 10 means 10 written as a base with an exponent.

2-6 Exponents and Scientific Notation

EXPLORE

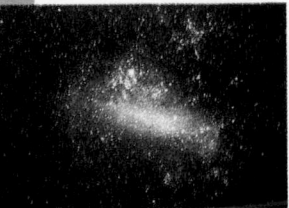

Compare these numbers. The top number, 10, can be written as the product 1×10. Write each of the other numbers as a product, using only 10 as a factor. Then write the number of factors beside each number. What do you notice? How many factors of 10 would there be for the number 100,000,000,000? Explain how to use mental math to determine your answer.

10
100
1,000
10,000
100,000
1,000,000
10,000,000
100,000,000
1,000,000,000

SKILLS DEVELOPMENT

You can make large multiples of ten easier to use and to read by writing them in **exponential form**. A number written in exponential form has a base and an exponent.

$$10 \times 10 \times 10 \times 10 = 10^4 \quad \leftarrow \text{exponent}$$
$$\underset{\text{base}}{}$$

The **base** tells what factor is being multiplied. The **exponent** tells how many equal factors there are. The expression above, 10^4, is read as "10 to the fourth power."

The exponents zero and one are special.

Any number raised to the first power is that number itself.
$$2^1 = 2, \ 3^1 = 3$$

Any number, except 0, raised to the zero power is equal to one.
$$2^0 = 1, \ 3^0 = 1$$

A number multiplied by itself is said to be **squared**. Read 10^2 as "10 squared." You will learn to express the area of a plane figure in square units. A number multiplied by itself two times is said to be that number **cubed**. Read 10^3 as "10 cubed." You will learn to express the volume of a solid figure in cubic units.

Example 1

Write in exponential form.
a. $4 \times 4 \times 4$ **b.** $7 \times 7 \times 7 \times 7 \times 7$ **c.** 0.4×0.4

Solution
a. 4^3 **b.** 7^5 **c.** $(0.4)^2$ ◄

ESL STUDENTS

To help students remember the new terms introduced, have them make flashcards with basic information such as:

Exponent ⌐
     7³ ◄
⌐► **7³** ◄
└ Base

Exponential form: **7³**

Standard form: $7 \times 7 \times 7 = 343$

Example 2

Write in standard form.

a. 3^4 **b.** 5^2 **c.** 7^3

Solution

a. $3^4 = 3 \times 3 \times 3 \times 3$ **b.** $5^2 = 5 \times 5$ **c.** $7^3 = 7 \times 7 \times 7$
$\quad\quad = 81$ $= 25$ $= 343$ ◄

Frequently, very large or very small numbers are written in **scientific notation**. A number written in scientific notation has two factors. The first factor is a number greater than or equal to 1 and less than 10. The second factor is a power of 10.

These numbers are written in scientific notation:

$$4 \times 10^3 \quad\quad\quad 9.9 \times 10^8 \quad\quad\quad 6.001 \times 10^{20}$$

Example 3

Write in scientific notation.
a. 50,000   **b.** 4,183,000

Solution

a. $50,000 = 5 \times 10,000$ *Ask yourself, What number times 5 equals 50,000?*
$\quad\quad\quad\quad = 5 \times 10^4$

b. $4,183,000 = 4.183 \times 1,000,000$
$\quad\quad\quad\quad\quad = 4.183 \times 10^6$ ◄

Example 4

Write in standard form.
a. 2×10^3 **b.** 6.47×10^8

Solution

a. $2 \times 10^3 = 2 \times 1,000$ **b.** $6.47 \times 10^8 = 6.47 \times 100,000,000$
$\quad\quad\quad\quad = 2,000$ $= 647,000,000$ ◄

TRY THESE

Write in exponential form.

1. $5 \times 5 \times 5$ 5^3 **2.** $3 \times 3 \times 3 \times 3$ 3^4

Write in standard form.

3. 2^2 4   **4.** 3^2 9   **5.** 3^3 27   **6.** 4^2 16

Write in scientific notation.

7. 7,302   **8.** 452,968   **9.** 209   **10.** 6,000,432
   7.302×10^3 4.52968×10^5 2.09×10^2 6.000432×10^6

Write in standard form.

11. 5.629×10^3 5,629   **12.** 9.04×10^5 904,000

CHECK UNDERSTANDING

Explain why these numbers are *not* written in scientific notation.

0.621×10^2
1.23×5^4
$4.1 \times 1.6 \times 10^3$
21×10^5
$1.25 + 10^3$

Each is not the product of a number greater than or equal to 1, but less than 10, and a power of 10.

 COMPUTER TIP

The computer memory can store numbers with up to nine digits and a decimal point. For numbers with more digits, the computer uses scientific notation. Here are some examples. Can you figure out what the E stands for?

]PRINT 98000000*520000
5.096E+13

]PRINT 82000*560000
4.592E+10

Answers will vary.
The E indicates the exponent for the power of 10 that one would write if one were not using a computer.

Example 4: Point out that mental math involves moving the decimal point the number of places indicated by the exponent.

Additional Questions/Examples
Which can always be calculated by using mental math and why?
a. any number with an exponent less than 4
b. any number expressed in scientific notation
c. any two factors expressed in standard form
   b. because a power of 10 can always be calculated mentally

Guided Practice/Try These For Exercises 1 and 2, have students calculate the value of each. **125, 81** Explain that for Exercises 3–6 the standard form represents the value of the expression.

3 SUMMARIZE

Key Questions
1. How do you change a number from standard form to scientific notation? **Rewrite the number as a product of two factors; the first is a number ≥ 1 and < 10; the second is a power of 10.**

2. How do you write a product of like factors in exponential form? **write the factor once and the number of factors as the exponent**

3. How do you change numbers written in scientific notation to standard form? **calculate the product of the factors**

4 PRACTICE

Practice/Solve Problems For Exercises 1–4, you may wish to have students calculate each number and write it in scientific notation: **2.7 × 10¹; 1.28 × 10²; 1.296 × 10³; 1.0 × 10⁵**

Extend/Solve Problems For Exercise 30 point out that the numbers must be in the same form before they can be compared.

Think Critically/Solve Problems You may wish to show Exercises 31–34 in the following formats. Have students find the value of x in each.
31. $2^x = 32$
32. $4^x = 1,024$
33. $x^3 = 343$
34. $x^6 = 1.0$

Problem Solving Applications Point out that, in the photo, the display shows an exponent as a full-sized number at the right of the display. Other scientific calculators show an exponent as a small-sized superior number at the right of the display.

5 FOLLOW-UP

Extra Practice Write in exponential form and scientific notation. Use a calculator if needed.
1. $12 \times 12 \times 12 \times 12 \times 12$ **12^5; 2.48832 × 10⁵**

62

PRACTICE/ SOLVE PROBLEMS

MENTAL MATH TIP

When you write a number in scientific notation, the exponent of 10 is the same as the number of places that you move the decimal point.

$700,000 = 700,000$
  5 places
$= 7 \times 10^5$

$9.04 \times 10^3 = 9.040$
  3 places
$= 9,040$

Write in exponential form.
1. $3 \times 3 \times 3$ **3^3**   **2.** $2 \times 2 \times 2 \times 2 \times 2 \times 2 \times 2$ **3^2**
3. $6 \times 6 \times 6 \times 6$ **6^4**   **4.** $10 \times 10 \times 10 \times 10 \times 10$ **10^5**

Write in standard form.
5. 2^4 **16**   **6.** 0.5^2 **0.25**   **7.** 6^2 **36**   **8.** 1^{10} **1**

Write in scientific notation.
9. $625,984$ **10.** $1,961,048$ **11.** 892 **12.** $54,203$
  6.25984 × 10⁵   **1.961048 × 10⁶** **8.92 × 10²** **5.4203 × 10⁴**

Write in standard form.
13. 6.39×10^4 **63,900**   **14.** 8.7231×10^9 **8,723,100,000**
15. 3×10^2 **300**   **16.** 4.124×10^3 **4,124**

Solve. Explain your answer.
17. A student said that 3^2 equals 2^3. Is this correct? **No. 9 ≠ 8**

EXTEND/ SOLVE PROBLEMS

Evaluate. Let $x = 2$, $y = 3$, and $z = 4$.
18. $x^5 - 15$ **17** **19.** $z^3 + 21$ **85** **20.** $x^2 + y^2$ **13** **21.** $x^3 + y^4$ **89**
22. $y^2 - x^2$ **5** **23.** $z^4 - x^3$ **248** **24.** $x^2 + y^3 + z^2$ **47** **25.** $z^3 - y^2 + x^5$ **87**

Write in standard form. Use a calculator for the computation.
26. 29^5 **20,511,149** **27.** 13^7 **62,748,517** **28.** 2^{12} **4,096** **29.** $1,452^1$ **1,452**

Solve. Explain your answer.
30. One light year equals 9,460,000,000,000 km. The mean distance between Earth and the sun is 1.496×10^8 km. Is the distance between Earth and the sun greater than, less than, or equal to one light year? **less than one light year because 9,460,000,000,000 = 9.46 × 10¹² and 9.46 × 10¹² > 1.496 × 10⁸**

THINK CRITICALLY/ SOLVE PROBLEMS

Write each number with the given base or exponent.
31. 32 with base of 2 **2^5**   **32.** 1,024 with base of 4 **4^5**
33. 343 with exponent of 3 **7^3**   **34.** 1.0 with exponent of 6 **1.0^6 or 1^6**

35. A fund-raising drive began with $3.00. Every hour the amount in the fund doubled. The goal was met in 8 hours. How long was it before the goal was half met? **7 hours**

Problem Solving Applications:

USING A SCIENTIFIC CALCULATOR

All calculators are not created equal. A scientific calculator has more keys than most standard calculators and uses a different program to perform calculations.

A scientific calculator may have a key labeled x^y or y^x to compute numbers in exponential form. How you use a calculator to write a number in standard form depends on the type of calculator.

Scientific Calculator Key Sequence:	Standard Calculator Key Sequence:
$2^4 \rightarrow$ $\boxed{2}$ $\boxed{x^y}$ $\boxed{4}$ $\boxed{=}$	$\boxed{2}$ $\boxed{\times}$ $\boxed{=}$ $\boxed{=}$ $\boxed{=}$

Press $\boxed{=}$ one time less than the number indicated by the exponent.

CALCULATOR TIP

If your calculator has an $\boxed{EE}$ or an $\boxed{EXP}$ key, you can use it as a quick way to enter a number that is in scientific notation. For example, to enter 8×10^5, use this key sequence:

$\boxed{8}$ $\boxed{EXP}$ $\boxed{5}$

Sometimes a number is too large to fit in the display of a calculator. Depending on the calculator, the display either will indicate an error or it will display the number in scientific notation.

You may see the complete answer or one of the following displays if you use a calculator to find $600 \times 962,431 = 577,458,600$.

Error Message: ERROR   or   E5.7745860

Scientific Notation: 5.774586^8 or 5.774586^{08}

Choose the letter of the key sequence to be used on a scientific calculator to find the number given in exponential form.

1. 6^{15}
   a. 15 $\boxed{x^y}$ 6
   b. 6 $\boxed{x^y}$ 15
   c. $\boxed{x^y}$ 6 15

2. 15^6
   a. 15 $\boxed{x^y}$ 6
   b. 6 $\boxed{x^y}$ 15
   c. $\boxed{x^y}$ 15 6

3. 9^8
   a. $\boxed{x^y}$ 8 9
   b. 8 $\boxed{x^y}$ 9
   c. 9 $\boxed{x^y}$ 8

What is the number for which the answer is given in the calculator display? Write it in scientific notation and then in standard form.

4. $\boxed{9.6578^{10}}$

5. $\boxed{1.39^{11}}$

6. $\boxed{2.5555^{15}}$

Write each number in scientific notation. Then tell which of the numbers has the least value and which has the greatest value.

7. 27^{14}

8. 30^{10}

9. 25^{19}

10. $\boxed{2.6359^{16}}$

11. $\boxed{2.39245^{20}}$

12. $\boxed{1.962^{25}}$

4. 9.6578×10^{10}
   96,578,000,000
5. 1.39×10^{11}
   139,000,000,000
6. 2.5555×10^{15}
   2,555,500,000,000,000
7. 1.0941899×10^{20}
8. 5.9049×10^{14}
9. 3.6379788×10^{26}
10. 2.6359×10^{16}
11. 2.39245×10^{20}
12. 1.962×10^{25}
least value: 30^{10}; greatest value: 25^{19}

2. $32 \times 32 \times 32$ **32^3; 3.2768×10^4**
3. $1.5 \times 1.5 \times 1.5 \times 1.5$ **1.5^4; 5.0625×10^0**

Tell which number is greater.
4. 328,009 or 3.28×10^5 **328,009**
5. 14 or 1.4×10^3 **1.4×10^3**
6. Twelve tickets to a concert were sold during the first hour tickets were on sale. If the number of tickets sold tripled every hour, how many tickets were sold during the sixth hour? **2,916**
7. Write the answer to Exercise 6 in exponential form. **12×3^5 or $2^2 \times 3^6$**

Extension Explain that very small numbers often occur in science. They are represented with negative exponents. Give the following example: $.00004 = 4 \times 10^{-5}$ in scientific notation. Have students write the following in scientific notation:
1. .0007 **7×10^{-4}**
2. .000000089 **8.9×10^{-8}**

Have students change the following to standard form:
3. 2.7×10^{-4} **.00027**
4. 8.92×10^{-6} **.00000892**

Section Quiz Evaluate each expression. Let $a = 2$, $b = 3$, $c = 5$.
1. $a^3 + b^3 - c^2$ **10**
2. $3^b + 4^a + 2^c$ **75**
3. $c^b \times 4^b$ **8,000**

Replace ● with <, >, or = .
4. 23^5 ● 6.436343×10^7 **<**
5. 31.2×10^3 ● 31,200 **=**
6. 3,408,900 ● 3.408×10^6 **>**

Get Ready calculators

MAKING CONNECTIONS

Have students use an almanac or encyclopedia to find the following information and indicate each in both standard form and scientific notation: The diameter of the sun **864,900 mi; 8.649×10^5 mi** The mass of the sun **2,000,000,000,000, 000,000,000,000,000,000 kg; 2×10^{30} kg**

SPOTLIGHT

OBJECTIVE
- Apply the laws of exponents in multiplication, division, and raising an exponential number to a power

MATERIALS NEEDED
calculators

VOCABULARY
laws of exponent, product rule, quotient rule

WARM-UP

Write in standard form.
7^3 $7 \times 7 \times 7 = 343$
3^5 $3 \times 3 \times 3 \times 3 \times 3 = 243$
9^3 $9 \times 9 \times 9 = 729$

1 MOTIVATE

Explore Have students try to formulate a generalization about multiplying and dividing like factors in exponential form before they read the rules below. Have them make up simple examples to prove their generalizations. For example: $2^2 \times 2^3 = 4 \times 8$
$\qquad\qquad = 32$
$\qquad\qquad = 2^5$

2 TEACH

Use the Pages/Skills Development Have students read this part of the section and then discuss the examples.
   Example 1: Have students substitute several different values for *a* in Example 1b to demonstrate the rule in specific cases. For example, $a^2 \times a^2$ for *a* =2: $2^2 \times 2^2 =$ $4 \times 4 = 16$ and $2^2 \times 2^2 = 2^4 = 16$.

2-7   Laws of Exponents

EXPLORE   Find the product by writing each factor in expanded form. Then rewrite it as a power of ten.

$10^2 \times 10^3 = 10 \times 10 \times 10 \times 10 \times 10 = 10^? \;\; \mathbf{5}$
$10^4 \times 10^5 = \underline{\;\;?\;\;} = 10^? \;\; \mathbf{9}$
$10^{10} \times 10^2 = \underline{\;\;?\;\;} = 10^? \;\; \mathbf{12}$
$10^1 \times 10^2 = \underline{\;\;?\;\;} = 10^? \;\; \mathbf{3}$
$10^0 \times 10^3 = \underline{\;\;?\;\;} = 10^? \;\; \mathbf{3}$

How are the exponents in the factors related to the exponent in the product?

Find the quotient by first writing each number in standard form, then dividing. Then rewrite the quotient as a power of ten.

$10^6 \div 10^2 = 1,000,000 \div 100 = 10,000 = 10^? \;\; \mathbf{4}$
$10^4 \div 10^3 = \underline{\;\;?\;\;} = 10^? \;\; \mathbf{1}$
$10^3 \div 10^0 = \underline{\;\;?\;\;} = 10^? \;\; \mathbf{3}$
$10^5 \div 10^1 = \underline{\;\;?\;\;} = 10^? \;\; \mathbf{4}$
$10^2 \div 10^1 = \underline{\;\;?\;\;} = 10^? \;\; \mathbf{1}$

How are the exponents in the dividend and divisor related to the exponent in the quotient?

SKILLS DEVELOPMENT   A number can be easier to work with if it is written in exponential form than if it is written in standard form. For example, 10^{10} is easier to work with than 10,000,000,000. Several rules govern operations with numbers written in exponential form. These rules are called the **laws of exponents.**

To find the product $2^2 \times 2^3$, you could write out the factors for each term and then write the product using exponents.

$$2^2 \times 2^3 = 2 \times 2 \times 2 \times 2 \times 2 = 2^5$$

A faster method is to use the **product rule** which states:
▶   To multiply numbers with the same base, add the exponents.
$$a^m \times a^n = a^{m+n}, \text{ so } 2^2 \times 2^3 = 2^{2+3} = 2^5$$

64   CHAPTER 2   Exploring Whole Numbers and Decimals

TEACHING TIP

Problems involving numbers in exponential form may be difficult for students to understand. Suggest that they use the problem solving strategy of creating a simpler or easier problem when this difficulty arises.

Example 1

Use the product rule to multiply.

a. $10^5 \times 10^2$ **b.** $a^2 \times a^2$

Solution

a. $10^5 \times 10^2$ **b.** $a^2 \times a^2$
     $= 10^{5 + 2}$ $= a^{2 + 2}$
     $= 10^7$ $= a^4$ ◄

To find the quotient $3^4 \div 3^3$, you could write out the factors for each term and then write the quotient using exponents.

$$3^4 \div 3^3 \text{ can be written as } \frac{3^4}{3^3} = \frac{3 \times 3 \times 3 \times 3}{3 \times 3 \times 3} = 3^1$$

Another method is to use the **quotient rule** which states:

► To divide numbers with the same base, subtract the exponents.

$$a^m \div a^n = a^{m - n}, \text{ so } 3^4 \div 3^3 = 3^{4 - 3} = 3^1$$

Example 2

Use the quotient rule to divide.

a. $10^3 \div 10^1$ **b.** $b^7 \div b^6$

Solution

a. $10^3 \div 10^1$ **b.** $b^7 \div b^6$
     $= 10^{3 - 1}$ $= b^{7 - 6}$
     $= 10^2$ $= b^1$
                         $= b$ ◄

A number written in exponential form can be raised to a power. You could first write out the factors and then write the product in exponential form. For example:

$$\begin{aligned}(3^3)^2 &= 3^3 \times 3^3 \\ &= 3 \times 3 \times 3 \times 3 \times 3 \times 3 \\ &= 3^6\end{aligned}$$

Compare the exponents in the original number to the exponents in the final number to determine their relationship. This relationship is described by the **power rule** which states:

► To raise an exponential number to a power, multiply the exponents.

$$(a^m)^n = a^{mn}, \text{ so } (3^3)^2 = 3^{3 \times 2} = 3^6$$

Example 3

Use the power rule.

a. $(4^3)^6$ **b.** $(y^9)^8$

Solution

a. $(4^3)^6 = 4^{3 \times 6} = 4^{18}$ **b.** $(y^9)^8 = y^{9 \times 8} = y^{72}$ ◄

BASIC
1–24, 27–38, 49–51

AVERAGE
5–8, 13–26, 31–47, 49–56

ENRICHED
25–26, 31–48, 52–58

ADDITIONAL RESOURCES
Reteaching 2–7
Enrichment 2–7

Example 2: Have students substitute different values for b in Example 2b to demonstrate the rule in specific cases. For example, $b^7 \div b^6$ for $b = 2$: $2^7 \div 2^6 = 128 \div 64 = 2$ and $2^7 \div 2^6 = 2^1 = 2$.

Example 3: Have students solve the following example to demonstrate the rule: $(2^3)^2 = 2^3 = 8$; $8^2 = 64$ and $(2^3)^2 = 2^6 = 64$.

Example 4: Point out that the answer gives the number of years it takes for one crater to form.

Additional Questions/Examples
The answers to three of the following five exercises are wrong. Find the errors and correct them.

1. $3^4 \times 3^3 = 6^7$ **3^7**

2. $(13^2)^7 = 13^{14}$ **correct**

3. $10^9 \div 10^2 = 10^7$ **correct**

4. How many times greater is 9^{10} than 9^4? Explain how you arrived at your answer.   **9^6; divided**

5. Prove that your answer to Exercise 4 is correct by using multiplication.   **$9^6 \times 9^4 = 9^{10}$**

Guided Practice/Try These   For Exercise 13 suggest to students that they express 1,000 as 10^3.

3 SUMMARIZE

3 SUMMARIZE

Write About Math Have students write out each of the following rules and provide an example that demonstrates it.
   product rule
   quotient rule
   power rule

4 PRACTICE

Practice/Solve Problems For Exercises 25 and 26, suggest students use the problem solving strategies of writing a simpler problem and choosing the correct operation, respectively.

Extend/Solve Problems For Exercises 27–38 you may wish to have students also specify the law they used to solve each problem.

Think Critically/Solve Problems For Exercise 48 remind students that they should express both numbers in exponential form before calculating the answer.

5 FOLLOW-UP

Extra Practice Use the laws of exponents to find each answer.
1. $32^{12} \times 32^2$ **32^{14}**
2. $(1.4^5)^3$ **1.4^{15}**
3. $21^{14} \div 21^7$ **21^7**
4. $2^a \times 2^b$ **2^{a+b}**
5. $6^x \div 6^y$ **6^{x-y}**
6. $(12^n)^m$ **$12^{n \times m}$**
Which number is greater and why?
7. a. $12^{14} \times 12^{12}$ or b. $(12^{12})^2$
   a; $12^{26} > 12^{24}$
8. a. $2^8 \div 2^4$ or b. $2^2 \times 2^3$
   b; $2^4 < 2^5$
9. a. $(10^5)^6$ or b. $10^{20} \times 10^8$
   a; $10^{30} > 10^{28}$

The laws of exponents can be used when solving some word problems.

Example 4

It is believed that 10^4 craters form on the moon every 10^9 years. On average, about how many years are there between the formation of one crater and the next?

Solution

Divide the entire span of 10^9 years by 10^4, the number of craters.
$10^9 \div 10^4 = 10^{9-4} = 10^5 = 100,000$

A crater forms about once every 100,000 years. ◀

TRY THESE

Use the product rule to multiply.

1. $10^4 \times 10^2$ **10^6** **2.** $5^8 \times 5^5$ **5^{13}** **3.** $m^{21} \times m^7$ **m^{28}** **4.** $d^6 \times d^3$ **d^9**

Use the quotient rule to divide.

5. $9^8 \div 9^2$ **9^6** **6.** $4^{11} \div 4^0$ **4^{11}** **7.** $p^{20} \div p^{10}$ **p^{10}** **8.** $a^9 \div a^3$ **a^6**

Use the power rule.

9. $(3^6)^8$ **3^{48}** **10.** $(7^2)^{10}$ **7^{20}** **11.** $(x^4)^4$ **x^{16}** **12.** $(a^8)^1$ **a^8**

Solve.
13. The total number of bacteria on a surface was estimated to be $(10^{10})^{50}$. If the same number of bacteria was on each of 1,000 of these surfaces, how many bacteria would there be? **10^{503} bacteria**

EXERCISES

PRACTICE/ SOLVE PROBLEMS

Use the product rule to multiply.

1. $10^{21} \times 10^8$ **10^{29}** **2.** $10^1 \times 10^9$ **10^{10}** **3.** $5^6 \times 5^2$ **5^8** **4.** $6^2 \times 6^3$ **6^5**
5. $n^3 \times n^3$ **n^6** **6.** $d^{11} \times d^0$ **d^{11}** **7.** $m^9 \times m^3$ **m^{12}** **8.** $a^6 \times a^9$ **a^{15}**

Use the quotient rule to divide.

9. $10^4 \div 10^2$ **10^2** **10.** $5^8 \div 5^5$ **5^3** **11.** $2^8 \div 2^5$ **2^3** **12.** $6^{16} \div 6^{15}$ **6^1**
13. $b^9 \div b^1$ **b^8** **14.** $x^{21} \div x^7$ **x^{14}** **15.** $y^7 \div y^5$ **y^2** **16.** $n^6 \div n^3$ **n^3**

Use the power rule.

17. $(45^6)^{10}$ **45^{60}** **18.** $(9^8)^8$ **9^{64}** **19.** $(5^7)^9$ **5^{63}** **20.** $(11^4)^5$ **11^{20}**
21. $(a^2)^{15}$ **a^{30}** **22.** $(x^{20})^5$ **x^{100}** **23.** $(n^7)^7$ **n^{49}** **24.** $(p^3)^3$ **p^9**

Solve.

25. There are 10^2 centimeters in a meter and 10^3 meters in a kilometer. How many centimeters are there in a kilometer? **10^5 cm**

26. A googol is 10^{100}. It has been estimated that there are 10^{79} electrons in the universe. About how many times greater is the googol than the number of electrons? **about 10^{21} times greater**

Use the laws of exponents.

27. $(4^{10})^{15}$ **4^{150}**   28. $1^{16} \div 1^8$ **1^8**   29. $8^{24} \times 8^3$ **8^{27}**   30. $11^8 \div 11^4$ **11^4**

31. $7^{15} \div 7^3$ **7^{12}**   32. $6^{20} \times 6^5$ **6^{25}**   33. $(12^5)^5$ **12^{25}**   34. $2^{90} \times 2^9$ **2^{99}**

35. $8^{25} \times 8^5$ **8^{30}**   36. $(103^4)^6$ **103^{24}**   37. $10^7 \div 10^1$ **10^6**   38. $(5^7)^{10}$ **5^{70}**

Replace ● with $<$, $>$, or $=$.

39. $5^8 \div 5^6$ ● 5^4 **$<$**   40. $3^6 \times 3^5$ ● 3^{30} **$<$**   41. $(7^6)^2$ ● $(7^3)^4$ **$=$**

42. $2^3 \times 2^6$ ● 2^8 **$>$**   43. $(9^{24})^8$ ● $(9^8)^{24}$ **$=$**   44. $8^{12} \div 8^6$ ● 8^3 **$>$**

USING DATA   Use the Data Index on page 546 to find the Richter Scale. Find the difference in the ground movement between an earthquake measuring 2.5 and one with each magnitude below.

45. 3.5 **3.5 shows 10 times more ground movement than 2.5.**

46. 4.5 **4.5 shows 100 times more ground movement than 2.5.**

47. 5.5 **5.5 shows 1,000 times more ground movement than 2.5.**

Solve.

48. The number of atoms of oxygen in an average thimble is 1,000,000,000,000,000,000,000,000,000. Since a googol is equal to 10^{100}, about how many times greater is a googol than the number of atoms of oxygen in an average thimble? **$10^{100} \div 10^{27} = 10^{73}$, about 10^{73} times greater**

Find the value of each variable.

49. $6^3 \times 6^a = 6^{18}$ **$a = 15$** 50. $7^4 \div 7^b = 7^2$ **$b = 2$** 51. $(6^c)^3 = 6^{12}$ **$c = 4$**

52. $5^d \div 5^8 = 5^5$ **$d = 13$** 53. $4^e \times 4^2 = 4^{24}$ **$e = 22$** 54. $(9^7)^f = 9^{49}$ **$f = 7$**

55. $(2^8)^2 = 2^g$ **$g = 1$**   56. $3^5 \times 3^h = 3^{45}$ **$h = 40$** 57. $8^i \div 8^{15} = 8^{15}$ **$i = 30$**

Solve.

58. Use the laws of exponents to write two factors that would have each of these products. **Answers will vary. Sample answers are given.**

   a. a googol (10^{100}) **$10^{35} \times 10^{65}$**   b. a googolplex (10^{googol}) **$10^0 \times 10^{\text{googol}}$**

2–7   Laws of Exponents   **67**

**EXTEND/
SOLVE PROBLEMS**

**THINK CRITICALLY/
SOLVE PROBLEMS**

Extension A googol is a quantity expressed as 10^{100}. DNA molecules, the hereditary material in living cells, can be measured in units of length called *angstroms*. There are 10 billion angstroms in a meter.

Which is greater, a googol or the number of angstroms in a meter, and by how much? **a googol; $10^{100} \div 10^9 = 10^{91}$**
How many angstroms are in 1,000 meters? **$10^{16} \div 10^9 = 10^7$**

Section Quiz Use the laws of exponents to find each answer.
1. $a^3 \times a^6$ **a^9**
2. $x^7 \div x^4$ **x^3**
3. $(n^6)^7$ **n^{42}**
Replace ● with $<$, $>$, or $=$.
4. $2^3 \times 2^6$ ● $2^{12} \div 2^{10}$ **$>$**
5. $(12^4)^5$ ● $12^4 \times 12^5$ **$>$**
6. $14^{12} \div 14^8$ ● $14^2 \times 14^2$ **$=$**
Find the value of each variable.
7. $3^a \times 3^6 = 3^{12}$ **6**
8. $4^6 \div 4^x = 4^4$ **2**
9. $(12^n)^4 = 12^{32}$ **8**
10. The length of a running course is 10^4 cm. How many meters long is the course? **$10^4 \div 10^2 = 10^2$ m or 100 m**

Get Ready calculators

CHALLENGE

Display the following equation:

$$x = a - b^{a-b}$$

Challenge students to find x when $a = -1$ and $b = -2$. Explain that students must simplify the exponent first. **$x = 1$**

OBJECTIVE
• Use order of operations in evaluating expressions and solving word problems

MATERIALS NEEDED
calculators

VOCABULARY
order of operations

WARM-UP

Evaluate.

32×23	**736**	12^4	**20,736**
$450 \div 9$	**50**	$6^2 + 24$	**60**
$32 - 3^3$	**5**		

1 MOTIVATE

Explore To demonstrate how the order of operations can affect an answer, have some students use standard calculators, others use scientific calculators, and still others use pencil and paper to evaluate the following expressions. Then discuss how any differences in answers reflect the different methods used to find them.

$2 + 4 \times 2 + 6 \qquad 3 \times 2 + 5 \times 3$

2 TEACH

Use the Pages/Skills Development
Have students read this part of the section and then discuss the examples.

Example 1: Point out that after multiplication and division are performed from left to right, addition and subtraction are then performed from left to right as well.

2-8 Order of Operations

EXPLORE/ WORKING TOGETHER

List the steps needed to perform a simple activity, such as getting lunch. Cut your list apart, with each step of the activity on a separate strip of paper. Mix up the strips and place them face down on your desk. Have a partner choose one strip at a time, turning it face up and placing each succeeding strip below the previous one. Then read the steps in this new order. Explain why the order of the steps is important.

SKILLS DEVELOPMENT

To simplify an expression is to perform as many of the indicated operations as possible. Sometimes, to achieve a goal, you must take steps in a specified order. For instance, to start a car, you first have to put the key in the ignition. To simplify mathematical expressions, certain steps must be followed in a certain order. These rules are called the **order of operations.**

► First, perform all calculations within parentheses and brackets.
► Then, do all calculations involving exponents.
► Next, multiply or divide in order, from left to right.
► Finally, add or subtract in order, from left to right.

Sometimes there are no grouping symbols or exponents.

Example 1

Simplify.

a. $7 - 3 \times 2 + 9$ **b.** $8 \div 4 + 2 \times 3$

Solution

a.
$$\begin{aligned} 7 &- 3 \times 2 + 9 \\ = 7 &- \quad 6 \quad + 9 \\ = 1 &\qquad\quad + 9 \\ = &\quad 10 \end{aligned}$$

b.
$$\begin{aligned} 8 &\div 4 + 2 \times 3 \\ = &\quad 2 \quad + \quad 6 \\ = &\qquad 8 \quad \blacktriangleleft \end{aligned}$$

Example 2

Simplify.

a. $6 - (4 + 8) \div 2$ **b.** $(3 \times 4)^2 + 6$

Solution

a.
$$6 - (4 + 8) \div 2$$
$$= 6 - 12 \div 2$$
$$= 6 - 6$$
$$= 0$$

b.
$$(3 \times 4)^2 + 6$$
$$= (12)^2 + 6$$
$$= 144 + 6$$
$$= 150 \quad \blacktriangleleft$$

The way that you enter an expression on your calculator will depend on the type of calculator you have. Some calculators compute according to the order of operations. On calculators you compute through the use of memory keys.

Example 3

Use a calculator to simplify $25 - 2 \times 7 + 1$.

Solution 1

$2 \boxed{\times} 7 \boxed{=} \boxed{M+}$ Calculate 2 × 7 and enter it into memory.

$25 \boxed{-} \boxed{MR} \boxed{+} 1 \boxed{=}$

Solution 2

Use this key sequence on a scientific calculator.

$25 \boxed{-} 2 \boxed{\times} 7 \boxed{+} 1 \boxed{=}$

With either calculator, you will find that $25 - 2 \times 7 + 1 = 12$.   ◀

You can write and evaluate expressions to solve some word problems.

Example 4

Ed took 100,000 pesos on his trip to Mexico. He bought 3 rings for 20,000 pesos each and 2 sombreros for 10,000 pesos each. Then he exchanged U.S. money for 150,000 pesos. How many pesos did he have then?

Solution

$$100{,}000 - 3 \times 20{,}000 - 2 \times 10{,}000 + 150{,}000$$
$$= 100{,}000 - 60{,}000 - 20{,}000 + 150{,}000$$
$$= 20{,}000 + 150{,}000$$
$$= 170{,}000$$

Ed had 170,000 pesos.   ◀

TRY THESE

Simplify.

1. $8 - 4 \div 2 \times 3$ **2**
2. $8 \div 4 - 2 + 3$ **3**
3. $16 + 4 \times 3 \times 5$ **76**
4. $16 \times 4 + 3 \div 3$ **65**

Simplify. Remember to work within parentheses first.

5. $3 \times (8 + 2) \div 6$ **5**
6. $(12 - 2) \times 4 - 8$ **32**
7. $45 \div (3 + 9) + 5$ **10**
8. $6^2 \div 4 \div 2$ **4.5**
9. $6^2 \div (4 \div 2)$ **18**
10. $(3 + 3)^2 \times 5$ **180**

COMPUTER

The computer's computations reflect the order of operations. Here is a list of how a computer "understands" mathematical operations.

Expression	Computer
3 + 2	3 + 2
3 − 2	3 − 2
3 × 2	3 * 2
3 ÷ 2	3/2
3^2	3^2 or 3↑2

Use the PRINT command with the expression you want to evaluate. For example:

```
]PRINT 3*4^2+6
54

]PRINT 24/4−2*3+5
5
```

Example 2: You may wish to show students how the wrong order of operations would result in incorrect answers for both examples.

Example 3: If possible, have students input the expression in the order in which it is written, using a standard calculator, to see the difference in results.
$25 - 2 \times 7 + 1 = 23 \times 7 + 1 = 162$

Example 4: Discuss other possible ways to express the problem.
$100{,}000 - (3 \times 20{,}000 + 2 \times 10{,}000) + 150{,}000 = 170{,}000$

Additional Questions/Examples
Explain the difference in procedure for evaluating the following expressions:
1a. $3 + 4 \times 7$ **1b.** $(3 + 4) \times 7$
In 1a multiply first, then add; in 1b add first, then multiply.
2a. $5^2 + 2^2$ **2b.** $(5 + 2)^2$
In 2a first square, then add; in 2b first add, then square.

Guided Practice/Try These For Exercises 1–10 you may wish to have students list the order of operations to be performed.

5-MINUTE CLINIC

Exercise	Student's Error	Error Diagnosis
$4 + 7 \times 3 - 4$	$4 + 7 \times 3 - 4 =$ $11 \times 3 - 4 =$ $33 - 4 = 29$	• Student added before multiplying. Correct solution: $4 + 7 \times 3 - 4 = 4 + 21 - 4$ $= 25 - 4$ $= 21$

3 SUMMARIZE

Write About Math Have students write out the order of operations and create an expression for each operation in which that operation occurs first. For example:

Parentheses $(2 \times 2) + 5$
Exponents $2^2 + 6 \times 7$
Multiply, Divide $3 \times 4 \div 2$
Add, Subtract $(3 + 4 - 5) \div 2$

4 PRACTICE

Practice/Solve Problems For Exercise 11 remind students to include Maya when dividing the money equally. For Exercises 11 and 12, suggest students use parentheses in their expressions.

Extend/Solve Problems You may wish to have students evaluate the expressions they write for Exercises 19–21.

Think Critically/Solve Problems For Exercises 39–44 point out that the black boxes are only for the symbols, not for parentheses. For Exercise 46 have students use a calculator to solve the problem after they determine the correct sequence.

5 FOLLOW-UP

Extra Practice
Evaluate.
1. $(3 + 4)^2 \times 2$ **98**
2. $32 - (4 \times 6) \div 2$ **20**
3. $(3 + 2)^2 + 4 + 2^2$ **33**
Write a numerical expression for each. Then evaluate.
4. seventeen less than the quotient of four hundred twenty-seven divided by seven
$427 \div 7 - 17 = 44$

Answers may vary. Sample key sequences are given.

11. First key sequence:
$18 \boxed{\div} 6 \boxed{\times} 25 - 5 \boxed{=}$
Second key sequence:
$25 \boxed{\times} 18 \boxed{\div} 6 - 5 \boxed{=}$

12. First key sequence:
$3 \boxed{\times} 3 \boxed{\times} 3 \boxed{=} \boxed{MR}$
$2 \boxed{\times} 16 - \boxed{MR} \boxed{=}$
Second key sequence:
$2 \boxed{\times} 16 \boxed{=} 3 \boxed{x^y} 3 \boxed{=}$

PRACTICE/ SOLVE PROBLEMS

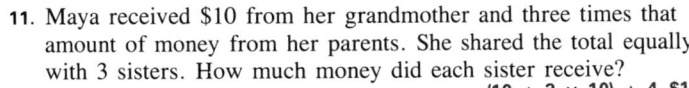

EXTEND/ SOLVE PROBLEMS

Simplify using a calculator. Show two different key sequences.

11. $25 \times (18 \div 6) - 5$ 12. $(2 \times 16) - 3^3$

Write an expression and solve.

13. Colin ran 6 miles last week. He ran twice that distance this week. How far did he run over the two-week period?
$6 + 2 \times 6$, 18 miles

EXERCISES

Simplify. Use the order of operations.

1. $6 \times 4 - 4 + 1$ **21**   2. $14 - 4 \times 3 - 2$ **0**

3. $72 \div (3 + 5) \times 9$ **81**   4. $3 + 2^4 \times 4$ **67**

5. $49 - 8 \times (42 \div 7)$ **1**   6. $(18 + 7) \times 4 \div 2$ **50**

7. $4^3 - (6 \div 3)$ **62**   8. $(3 + 2)^4 \times 4$ **2,500**

Simplify using a calculator. Show two different key sequences.
See Additional Answers.
9. $4^2 + 2 \times 6$ 10. $9 \times 8 - 6 + 3$

Write an expression and solve.

11. Maya received $10 from her grandmother and three times that amount of money from her parents. She shared the total equally with 3 sisters. How much money did each sister receive?
$(10 + 3 \times 10) \div 4$, $10

12. Kirk had $35.00. He bought 5 tubes of paint for $2.50 each and 3 canvases for $5.50 each. How much money did he have left?
$35 - (5 \times 2.5 + 3 \times 5.5)$, $6

Simplify.

13. $8 + 6 - 2 \times 2 - 3^2$ **1**   14. $4 \times 42 \div (56 \div 8 \times 3)$ **8**

15. $27 \div (7 + 2) \times (2 + 3)$ **15**   16. $63 \div (7 + 2) - 12 \div 3$ **3**

17. $19 + 2 \times (6 - 4)^3$ **35**   18. $(65 - 7^2 + 4) \times 9$ **180**

Write a numerical expression for each of the following.

19. twenty-five less the product of six and four **$25 - (6 \times 4)$**

20. the sum of eight and nine, multiplied by the difference of ten and two **$(8 + 9) \times (10 - 2)$**

21. five squared multiplied by the difference of thirty-two and twenty-eight, divided by twenty **$5^2 \times (32 - 28) \div 20$**

CHALLENGE

Challenge students to use parentheses, exponents, and the order of operations to write expressions equal to 1 – 10 using only one number, e.g., 2. (Allow them to use the number to the 0 power.) Show them the following expressions using 2. Have them try writing expressions using 3, 4, 5, 8, and 9.

$(2 \times 2) + 2^0 - 2^2 = 1$ $(2 \times 2) + 2 = 6$ $(2 \times 2) + 2 - 2^2 = 2$
$2^2 + 2^2 = 8$ $2^2 = 4$ $(2 \times 2) + 2^0 = 5$

Use only one of the symbols $+$, $-$, $\times$, or $\div$, to make each number sentence true.

22. $8 \blacksquare 4 \blacksquare 2 = 1$ $\div$

23. $12 \blacksquare 9 \blacksquare 3 = 0$ $-$

24. $6 \blacksquare 2 \blacksquare 4 = 12$ $+$

25. $7 \blacksquare 3 \blacksquare 1 = 11$ $+$

26. $8 \blacksquare 3 \blacksquare 6 = 144$ $\times$

27. $12 \blacksquare 3 \blacksquare 4 = 1$ $\div$

Solve. Write an expression if one is not given.

28. Explain why the key sequences for the expression $14 - 6 \times 2 \div 3$ might be different on two different calculators. **See Additional Answers.**

29. Jerome ran 4 miles farther than the mean distance of the other runners. The others ran 18 miles, 22 miles, 10 miles, and 26 miles. How far did Jerome run? **See Additional Answers.**

30. Lyn has $60.00. She wants to purchase a hat for $32.50, two scarves for $15.75 each, and one hair ribbon priced at 4 for $1.00. How much more money does she need?

$$\$32.50 + (2 \times \$15.75) + (\$1.00 \div 4) - 60 = \$4.25$$

Write *true* or *false* for each. For any that are false, insert parentheses to make them true.

31. $4 - 3 \times 7 = 7$
$(4 - 3) \times 7 = 7$

32. $7 \times 0 + 8 = 56$
$7 \times (0 + 8) = 56$

33. $5 \times 24 - 16 = 40$
$5 \times (24 - 16) = 40$

34. $15 \div 3 - 2 = 3$ **True**

35. $16 + 2^2 \div 5 \times 3 = 12$
$(16 + 2^2) \div 5 \times 3 = 12$

36. $4^2 \div 0 + 8 = 2$
$4^2 \div (0 + 8) = 2$

37. $3^2 - 6 \div 3 = 1$
$(3^2 - 6) \div 3 = 1$

38. $8^2 + 5 \times 4 \div 2^2 = 69$
$(8^2 + 5) \times (4 \div 2^2) = 69$

Use parentheses and the symbols $+$, $-$, $\times$, and $\div$ to make each number sentence true. **Answers may vary. Samples are given.**

39. $7 \blacksquare 3 \blacksquare 1 \blacksquare 1 = 9$
$7 + 3 - 1 \times 1 = 9$

40. $8 \blacksquare 3 \blacksquare 6 \blacksquare 5 = 4$
$(8 - 3) - (6 - 5) = 4$

41. $10 \blacksquare 5 \blacksquare 5 \blacksquare 2 = 10$
$(10 + 5 + 5) \div 2 = 10$

42. $12 \blacksquare 0 \blacksquare 4 \blacksquare 3 = 1$
$12 \times 0 + 4 - 3 = 1$

43. $30 \blacksquare 5 \blacksquare 5 \blacksquare 23 = 30$
$(30 + 5) \div 5 + 23 = 30$

44. $12 \blacksquare 3 \blacksquare 4 \blacksquare 5 = 16$
$(12 \times 3) - (4 \times 5) = 16$

Solve.

45. You can change the value of $3 + 5 \times 4 - 1$ by using parentheses. Show four possible ways to do this. Then find the value of each.

46. Show two different key sequences you could use on two different kinds of calculators to simplify the following expression: $(5 \times 6^3 \div 10) \div 81$

THINK CRITICALLY/ SOLVE PROBLEMS

45. $3 + 5 \times 4 - 1 = 22$
or $3 + (5 \times 4) - 1 = 22$
$(3 + 5) \times 4 - 1 = 31$
$(3 + 5) \times (4 - 1) = 24$
$3 + 5 \times (4 - 1) = 18$

46. Answers may vary. Sample key sequences are given.
First key sequence:
6 $\boxed{\times}$ 6 $\boxed{\times}$ 6 $\boxed{\times}$ 5 $\boxed{\div}$
10 $\boxed{\div}$ 81 $\boxed{=}$ 1.33
Second key sequence:
5 $\boxed{\times}$ 6 $\boxed{x^y}$ 3 $\boxed{\div}$ 10
$\boxed{=}$ $\boxed{\div}$ 81 $\boxed{=}$ 1.33

5. four cubed multiplied by the difference of sixteen and twelve divided by two
$$4^3 \times (16 - 12) \div 2 = 128$$

Extension Have students write an expression for the following problem. Have them solve and express the answer in scientific notation. Jupiter is about 300 thousand more than 778 million km from the sun. Earth is about 5.2 times closer to the sun than is Jupiter. About how many kilometers is Earth from the sun?
$$(300{,}000 + 778{,}000{,}000) \div 5.2 =$$
$$149{,}673{,}080 =$$
$$1.4967308 \times 10^8 \text{ km}$$

Section Quiz
Evaluate.

1. $(7 + 6)^2 \div 13$ **13**

2. $(3 + 4) \times (9 - 6)^2$ **63**

3. $8 + 2 - 3 \times 6 \div 9$ **8**

4. There are three art exhibits going on simultaneously at the art museum. Each exhibit has 5 paintings and each painting shows two men and three women. How many people are there in all?
$$3 \times (5 \times 2 + 5 \times 3) = 75$$

Use the symbols $+$, $-$, $\times$, or $\div$ to make each number sentence true.

5. $(8 \bullet 4) \bullet 20 = 52$ $\times$, $+$

6. $4 \bullet 4 \bullet 4 = 20$ $\times$, $+$

7. $25 \bullet 5 \bullet 2 \bullet 2 \bullet 1 = 0$
$\div$, $-$, $-$, $-$

Get Ready calculators

Additional Answers
See page 573.

WARM-UP

Evaluate, then replace ● with <, >, or = .
41×52 ● 52×41 **2,132; =**
$542 + 32$ ● $34 + 542$ **574 < 576**
$(12 + 6) + 2$ ● $12 + (6 + 1)$ **20 > 19**

1 MOTIVATE

Explore For those sentences in which the meaning is affected by the order, have students explain how the meaning is affected.

2 TEACH

Use the Pages/Skills Development
Have students read this part of the section and then discuss the examples.
   Example 1: Point out that the numbers on one side of the equals sign must be the same as the numbers on the other side.
   Example 2: Point out that although the groupings on each side of the equals signs are different, the numbers are the same. Be sure students understand that the associative property is true for addition and multiplication individually, not together. For example: $3 \times 4 + 2$ does not equal $3 \times 2 + 4$.

2-9 Properties of Operations

EXPLORE

Describe how the sentences in each pair differ. Then, tell if the order affects the meaning.

A. One peso was exchanged for 0.0003305 dollars.
One dollar was exchanged for 0.0003305 pesos.

B. Jared had one 1,000-peso note and two 500-peso coins.
Jared had two 500-peso coins and one 1,000-peso note.

C. Bo and Jin bought souvenirs from Paco at the Pyramid of the Sun.
Bo bought souvenirs from Jin and Paco at the Pyramid of the Sun.

D. Peso notes and peso coins along with centavos are Mexican monies.
Peso notes along with peso coins and centavos are Mexican monies.

SKILLS DEVELOPMENT

You know that the order of operations always affects the answer to a mathematics exercise. However, the order in which addends or factors are arranged does not affect the answer. Certain patterns, stated as **properties,** can help you understand this.

The **commutative property** states that the order in which numbers are added or multiplied does not affect the sum or product:

$$a + b = b + a, \text{ so } 15 + 12 = 12 + 15$$

$$x \times y = y \times x, \text{ so } 3 \times 2 = 2 \times 3$$

Example 1

Complete.
a. $12 + 7 = \blacksquare + 12$ **b.** $6 \times 12 = 12 \times \blacksquare$

Solution
Use the commutative property.

a. $12 + 7 = 7 + 12$
$19 = 19$
The unknown number is 7.

b. $6 \times 12 = 12 \times 6$
$72 = 72$ ◄
The unknown number is 6.

The **associative property** states that the grouping of factors does not affect the sum or product:

$$(a + b) + c = a + (b + c), \text{ so } (3 + 5) + 6 = 3 + (5 + 6)$$

$$(x \times y) \times z = x \times (y \times z), \text{ so } (4 \times 8) \times 2 = 4 \times (8 \times 2)$$

72   CHAPTER 2   Exploring Whole Numbers and Decimals

TEACHING TIP

Tell students that when using mental math, they should always look for numbers that will result in a sum or product that is a multiple of 10, then group these together first. For example: $32 + 31 + 18 + 9$ should be regrouped as $(32 + 18) + (31 + 9)$. Students should readily see that the numbers are easier to add mentally by regrouping in this way.

Example 2

Complete.

a. $(10 + 4) + 6 = \blacksquare + (4 + 6)$ **b.** $(2 \times \blacksquare) \times 3 = 2 \times (5 \times 3)$

Solution

Use the associative property.

a. $(10 + 4) + 6 = 10 + (4 + 6)$ **b.** $(2 \times 5) \times 3 = 2 \times (5 \times 3)$
$\qquad\qquad 14 + 6 = 10 + 10 \qquad\qquad\qquad 10 \times 3 = 2 \times 15$
$\qquad\qquad\quad 20 = 20 \qquad\qquad\qquad\qquad\quad 30 = 30$ ◄
The unknown number is 10.   The unknown number is 5.

The **identity property of addition** states that the sum of zero and a number is always that number:

$$a + 0 = a, \text{ so } 236 + 0 = 236$$

The **identity property of multiplication** states that the product of 1 and any other factor is always that factor:

$$x \times 1 = x, \text{ so } 863 \times 1 = 863$$

The **property of zero for multiplication** states that any number multiplied by zero has a product of zero:

$$x \times 0 = 0, \text{ so } 213 \times 0 = 0$$

Example 3

Complete. Tell which property you used.

a. $17 + \blacksquare = 17$ **b.** $1 \times \blacksquare = 36$ **c.** $45 \times \blacksquare = 0$

Solution

a. $17 + 0 = 17$ **b.** $1 \times 36 = 36$ **c.** $45 \times 0 = 0$
identity property of addition   identity property of multiplication   property of zero for multiplication ◄

The unknown number is 0.   The unknown number is 36.   The unknown number is 0.

Example 4

Lina bought a ceramic bowl for 3,325 pesos, a tea set for 2,425 pesos, and mugs for 4,675 pesos. How much did she spend in all?

Solution

$3,325 + 2,425 + 4,675$
$= 2,425 + (3,325 + 4,675)$ **Use the Commutative and Associative Properties.**
$= 2,425 + 8,000 = 10,425$

Lina spent 10,425 pesos. ◄

5-MINUTE CLINIC

Exercise	Student's Error	Error Diagnosis
$28 \times 0 \times 1$	$28 \times 0 \times 1 = 28$	• The property of zero for multiplication was not recognized. Correct solution: $28 \times 0 \times 1 = 0$

ASSIGNMENTS

BASIC
1–23, 24–26, 31, 33

AVERAGE
13–23, 24–32, 33–34

ENRICHED
22–23, 28–32, 33–37

ADDITIONAL RESOURCES
Reteaching 2–9
Enrichment 2–9
Transparency Master 12

Example 3: For Example 3c point out that 45 can be replaced with any number and the answer will still be zero.

Example 4: Have students group the numbers another way to solve the problem. **Possible answer: 2,425 + (4,675 + 3,325) = 10,425**

Additional Questions/Examples
Do the associative and commutative properties work for addition and multiplication of decimals or fractions? Write examples to prove your answer. **yes; 1.2 + 6 = 6 + 1.2; 1.4 × (1/2 × 6) = (1.4 × 1/2) × 6**

Guided Practice/Try These   For Exercises 1–8 you may wish to have students use calculators to check that their answers are correct.

3 SUMMARIZE

Key Questions
1. What are the commutative and associative properties of addition and multiplication?
2. What is the identity property of addition?
3. What is the identity property of multiplication?

4. What is the property of zero for multiplication?
Answers may vary, but should be similar to definitions given in the text.

Practice/Solve Problems Point out that students should be able to use mental math for Exercises 1–8 and 9–12 as well as for 13–21.

Extend/Solve Problems Have students explain how they used mental math in solving Exercises 30–32.

Think Critically/Solve Problems For each of Exercises 35–37 have students make up an example to see if the statement is *true* or *false*.

5 FOLLOW-UP

Extra Practice
Complete.
1. $45.6 \times 12 = \blacksquare \times 45.6$ **12**
2. $(3 \times 6) \times 3.2 = 3 \times (6 \times \blacksquare)$ **3.2**
3. $\blacksquare \times 89 = 0$ **0**
4. $\blacksquare \times 1 = 3.456$ **3.456**
5. $2,498 + \blacksquare = 2,498$ **0**
Evaluate using mental math.
6. $0.08 \times 9 \times 100$ **72**
7. $12 \times 0 \times 18$ **0**
8. $13 + 26 + 7$ **46**
9. $2 \times 876 \times 5$ **8,760**
10. $4 \times 57 \times 25$ **5,700**
Solve.
11. The supermarket clerk charged $21.20 for two steaks that were $7.90 each and a salad that was $6.20. Is $21.20 the correct total? Why or why not? **No, the total should have been $7.90 + $7.90 + $6.20 = $22.**

Complete.
1. $6 \times \blacksquare = 7 \times 6$ **7**
2. $1,588 + \blacksquare = 72 + 1,588$ **72**
3. $6 \times \blacksquare = 0$ **0**
4. $\blacksquare + 963 = 963$ **0**
5. $2.3 + 6.8 = 6.8 + \blacksquare$ **2.3**
6. $23 \times 48 = 48 \times \blacksquare$ **23**
7. $\blacksquare \times 1 = 2,586.25$ **2,586.25**
8. $248 \times \blacksquare = 248$ **1**

Complete.
9. $(6 \times 8) \times 5 = 6 \times (8 \times \blacksquare)$ **5**
10. $(9 + 6) + 8 = 9 + (6 + \blacksquare)$ **8**
11. $4 + (4 + 6) = (\blacksquare + 4) + 6$ **4**
12. $(3 \times \blacksquare) \times 9 = 3 \times (8 \times 9)$ **8**

Simplify. Name the properties you used. **See Additional Answers.**
13. $8 \times 9 \times 0$
14. 85×1
15. $0.35 + 0$
16. $8 + 6 + 4$
17. $1.2 + 0.32 + 0.8$
18. $5 \times 16 \times 2$
19. $0.05 \times 1.32 \times 200$
20. $8 + 19 + 2$
21. $9 \times 3 \times 3$

Solve.
22. Wanda spent 36,400 pesos on Wednesday, 106,925 pesos on Thursday, and 65,000 pesos on Friday. How many pesos did she spend on the three days? **208,325 pesos**

EXERCISES

PRACTICE/ SOLVE PROBLEMS

Complete.
1. $8.5 \times 5.8 = 5.8 \times \blacksquare$ **8.5**
2. $\blacksquare \times 72 = 72$ **1**
3. $\blacksquare + 0 = 963$ **963**
4. $\blacksquare + 768 = 768 + 432$ **432**
5. $9.36 + 2.58 = \blacksquare + 9.36$ **2.58**
6. $2.5 = \blacksquare \times 1$ **2.5**
7. $713 \times \blacksquare = 0$ **0**
8. $19 \times \blacksquare = 25 \times 19$ **25**

Complete.
9. $\blacksquare \times (3 \times 9) = (2 \times 3) \times 9$ **2**
10. $(7 + 8) + 2 = \blacksquare + (8 + 2)$ **7**
11. $6 \times (8 \times 5) = (\blacksquare \times 8) \times 5$ **6**
12. $1 + (\blacksquare + 2) = (1 + 7) + 2$ **7**

Simplify using mental math. Name the properties you used.
See Additional Answers.

13. $30 + 0 + 30$ **14.** $0.25 + 72 + 0.75$ **15.** $13 + 84 + 16$

16. $5 \times 6 \times 2$ **17.** $0.09 \times 9 \times 100$ **18.** $4 \times 1 \times 6$

19. $3 \times 0 \times 8$ **20.** $1 \times 11 \times 5$ **21.** $5 \times 82 \times 2$

Solve.

22. A server totaled a customer's lunch check. The customer had a salad for $4.75 and iced tea for $0.59. Tax on the meal was $0.25. What was the total? **$5.59**

23. Carla bought 5 packages of socks. Each package had 4 rows of socks with 2 socks in each row. How many socks did she buy?
40 socks

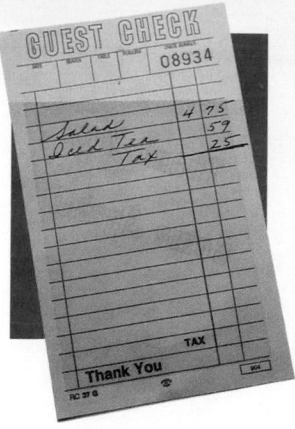

Simplify using mental math.

24. $6 + 9 + 4 + 1$ **20** **25.** $2 \times 8 \times 5 \times 9$ **720**

26. $19 + (2 \times 2) + 1$ **24** **27.** $14 + 8 + 206 + 92$ **320**

28. $0.68 + 0.7 + 0.32 + 1.3$ **3.0** **29.** $0.029 \times 100 \times 0 \times 8$ **0**

Solve. Use mental math.

30. A student had saved $4. She earned four times that amount on Friday and six times the original savings on Saturday. How much did she have then? **$44**

31. A customer was checking her restaurant bill. She had charges of $6.72, $9.01, $2.28, $0.75, $1.25, and $1.98. The written total, $22.00, was incorrect. What was the correct total? **$21.99**

32. Janelle rolled coins to take to the bank. She had 6 rolls of 20 dimes each. How much paper money did she receive from the bank in exchange? **$12.00**

 Solve. **Explanations may vary.**

33. To simplify $4 \times 16 + 6 + 5 + 4$, can you use the commutative property with $16 + 6$? Explain.

34. To simplify $24 \div 4 \times (3 + 2) \div 6$, can you use the commutative property with 4×3? Explain.

 Decide whether each statement is *true* or *false*. Explain.

35. The commutative property cannot be used when dividing.

36. The associative property cannot be used when subtracting.

37. A property can be used only once to evaluate each expression.

**EXTEND/
SOLVE PROBLEMS**

**THINK CRITICALLY/
SOLVE PROBLEMS**

35 and 36. True. It can be used only to add and multiply.

37. False. You can use a property as often as necessary to evaluate an expression since the use of the properties does not affect the value of an expression.

2–9   Properties of Operations   **75**

Extension Have students work in pairs to draw four models that show the associative and commutative properties of multiplication and of addition. You may need to show the following model of the commutative property of multiplication as an example:

```
* * * *              * * *
* * * *     =        * * *
* * * *              * * *
                     * * *

3 × 4        =       4 × 3
```

Section Quiz
Complete.
1. $327 \times 33 = \blacksquare \times 327$ **33**
2. $(2 \times 4) \times 7 \times 8 = 2 \times (4 \times \blacksquare) \times 8$ **7**
3. $0 \times \blacksquare = 0$ **any number**
4. $\blacksquare \times 1 = 657$ **657**
5. $\blacksquare + 0 = 34{,}567$ **34,567**
Evaluate using mental math.
6. $2 \times 4.678 \times 50$ **467.8**
7. $1 \times 456 \times 0 \times 3$ **0**
8. $25 + 98 + 75$ **198**
9. Can you use the commutative property in the expression $3 \times 18 + 7 + 8 + 6$? Explain your answer. **No, you must follow the order of operations and multiply 3×18 first.**

Get Ready calculators

Additional Answers
See page 574.

76

2-10 Distributive Property

EXPLORE

Study each diagram and the expression written below it. Then simplify each expression.

Compare the rows, columns, and the total number of squares in the two diagrams. Then compare the two expressions. How are they alike? How are they different?

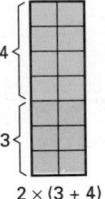
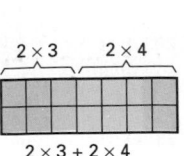

$2 \times (3 + 4)$ $2 \times 3 + 2 \times 4$

Make similar diagrams for numbers of your choice. Then, write two expressions for your diagrams.

SKILLS DEVELOPMENT

There may be more than one expression that will solve a problem. Examine these expressions. Both could be used to answer the same problem and both have the same answer.

$7(12 - 3)$	$7 \times 12 - 7 \times 3$
$= 7 \times 9$	$= 84 - 21$
$= 63$	$= 63$

The **distributive property** states that a factor outside parentheses can be used to multiply *each* term within the parentheses:
$a(b + c) = (a \times b) + (a \times c)$, so $5(3 + 2) = (5 \times 3) + (5 \times 2)$
$x(y - z) = (x \times y) - (x \times z)$, so $6(4 - 1) = (6 \times 4) - (6 \times 1)$

TALK IT OVER

Explain why the term *distributive property* is an appropriate name for this property.

Example 1

Complete.
a. $2(3 + 4) = (2 \times 3) + (\blacksquare \times 4)$
b. $\blacksquare(6 - 1) = (3 \times 6) - (3 \times 1)$

Solution
Use the distributive property.
a. $2(3 + 4) = (2 \times 3) + (2 \times 4)$
   $2(7) = 6 + 8$
   $14 = 14$ The unknown number is 2.
b. $3(6 - 1) = (3 \times 6) - (3 \times 1)$
   $3(5) = 18 - 3$
   $15 = 15$ The unknown number is 3. ◄

The distributive property can also be applied in reverse.

$a \times b + a \times c = a(b + c)$, so $8 \times 6 + 8 \times 5 = 8(6 + 5)$
$x \times y - x \times z = x(y - z)$, so $7 \times 5 - 7 \times 3 = 7 \times (5 - 3)$

Example 2

Use the distributive property to rewrite each expression.

a. $6 \times 45 + 6 \times 3$ 　　　　　　**b.** $9 \times 5 - 9 \times 4$

Solution

a. 　$6 \times 45 + 6 \times 3$ 　　　　**b.** 　$9 \times 5 - 9 \times 4$
　$= 6(45 + 3)$ 　　　　　　　　$= 9(5 - 4)$ 　◀

You can use the distributive property to break up a number by thinking of it as the sum or difference of two numbers that are easy to work with mentally.

Example 3

Simplify using mental math.

a. 20×130 　　　　　　　**b.** 48×40

Solution

a. 　20×130 　　　　　　**b.** 　48×40
　$= 20 \times (100 + 30)$ 　　　　　$= (50 - 2) \times 40$
　$= (20 \times 100) + (20 \times 30)$ 　　$= (50 \times 40) - (2 \times 40)$
　$= 2{,}000 + 600$ 　　　　　　　$= 2{,}000 - 80$
　$= 2{,}600$ 　　　　　　　　　　$= 1{,}920$ 　◀

Use the distributive property to simplify variable expressions.

Example 4

Simplify.

a. $6(4 + x)$ 　　　　　　　　**b.** $2(y - 7)$

Solution

Use the distributive property to simplify.

a. 　$6(4 + x)$ 　　　　　　**b.** 　$2(y - 7)$
　$= 6 \times 4 + 6 \times x$ 　　　　　$= 2 \times y - 2 \times 7$
　$= 24 + 6x$ 　　　　　　　　　$= 2y - 14$ 　◀

Example 5

Alicia and her friends crocheted 21 afghans for the homeless shelter. Each afghan was made up of 64 squares. How many squares did Alicia and her friends crochet?

Solution

To find the total number of squares, multiply 21 and 64.

　21×64 　　　　　Break up one factor to use the
　$= 20 \times 64 + 1 \times 64$ 　distributive property.
　$= 1{,}280 + 64$ 　　　　Compute, using the order of operations.
　$= 1{,}344$

Alicia and her friends crocheted 1,344 squares. 　◀

BASIC
1–20, 21–26, 39, 42

AVERAGE
1–20, 21–41, 45, 46

ENRICHED
8–13, 20, 30–41, 42–47

ADDITIONAL RESOURCES
Reteaching 2–10
Enrichment 2–10

multiply the 4. Ask students how many times 2 would be used in applying the distributive property to $2(3 + 4 + 5)$ **3 times**

Example 2: Point out that the distributive property can only be applied to a pair of expressions with common factors. Be sure students know when to use addition and when to use subtraction, as shown in the examples. Stress the use of parentheses. Recall with students the rules for the order of operations. Then have students evaluate the expression.

Example 3: Point out that when breaking apart one of the factors, students should always look for a factor that is, or can be, expressed as a multiple of 10.

Example 4: You may wish to assign a value to x in Example 4a and to y in Example 4b and have students evaluate the expressions before and after they are simplified.

Example 5: Be sure students understand that the factor 21 was broken apart and used as $20 + 1$ in this example.

5-MINUTE CLINIC

Exercise	Student's Error	Error Diagnosis
Use the distributive property to solve: 25×18	$(25 + 10) \times (25 + 8)$ 35×33 $1{,}155$	• Student confuses the positions of $\times$ and $+$. Correct solution: $(25 \times 10) + (25 \times 8)$ $250 + 200$ 450

78

Additional Questions/Examples

1. Why would you use the distributive property to multiply factors such as 298 and 15? **Answers will vary. Possible answer: less chance of making an error when multiplying multiples of 10.**

2. How do you decide how to break apart a factor in order to use the distributive property? **Answers will vary.**

Guided Practice/Try These For Exercises 7–12 have students tell how they broke apart the factors in the expressions.

3 SUMMARIZE

Write About Math Students can have a discussion about whether or not they think the distributive property is helpful to them. Be sure they support their opinions with facts about the property.

4 PRACTICE

Practice/Solve Problems For Exercises 8–13 have students explain how they used the distributive property to simplify mentally.

Extend/Solve Problems Select some exercises from 21–38 and have volunteers explain how they used the distributive property to simplify them.

Think Critically/Solve Problems For Exercises 42–44 have students change the signs, parentheses, and solutions so that the expressions are examples of the distributive property. **Possible answers: 42. (6 × 5) + (6 × 4) = 54; 43. (8 × 1) + (8 × 1) = 16; 44. (17 × 1) + (17 × 1) = 34**

MIXED REVIEW

Use the Data Index on page 546 for Exercises 1 and 2.

1. Decide whether a line graph is an appropriate way to express the data in the exchange-rate table. Explain.

2. Explain how you could use a pictograph to show the relative values of U.S. coins to one another.

Find the mean, the median, and the mode for each set of data.

3. 5.4, 2.9, 3.785, 6.04, 2.9

4. 7.3, 8.72, 7.3, 9.085, 4.7

See Additional Answers.

TRY THESE

Complete.

1. $5(4 + 2) = (5 \times 4) + (\blacksquare \times 2)$ **5**

2. $\blacksquare(3 - 1) = (2 \times 3) - (2 \times 1)$ **2**

Rewrite each expression using the distributive property.

3. $9 \times 3 + 9 \times 2$ **9(3 + 2)** **4.** $7 \times 3 + 7 \times 6$ **7(3 + 6)**

5. $8 \times 8 - 8 \times 4$ **8(8 – 4)** **6.** $4 \times 9 - 4 \times 3$ **4(9 – 3)**

Simplify using mental math.

7. 9×62 **558** **8.** 360×2 **720** **9.** 7×39 **273**

10. 82×4 **328** **11.** 105×3 **315** **12.** 28×4 **112**

Simplify.

13. $2(x - 3)$ **2x – 6** **14.** $18(2 + y)$ **36 + 18y** **15.** $21(z + 1)$ **21z + 21**

16. $9(8 - a)$ **72 – 9a** **17.** $11(4 + b)$ **44 + 11b** **18.** $8(c - 6)$ **8c – 48**

Write an expression using the distributive property and solve.

19. The student body filled 11 buses to travel to a football game. Each bus held 36 students. How many students went to the game? **Expressions may vary. Sample: 11 × 36 = 11 × (40 – 4) = 396 students**

EXERCISES

PRACTICE/ SOLVE PROBLEMS

Complete.

1. $6(3 + 1) = (\blacksquare \times 3) + (6 \times 1)$ **6**

2. $\blacksquare(9 + 5) = (4 \times 9) + (4 \times 5)$ **4**

3. $9(8 - 5) = (9 \times \blacksquare) - (9 \times 5)$ **8**

Rewrite each expression using the distributive property.

4. $7 \times 4 + 7 \times 8$ **7(4 + 8)** **5.** $9 \times 7 - 9 \times 3$ **9(7 – 3)**

6. $2 \times 18 - 2 \times 11$ **2(18 – 11)** **7.** $14 \times 12 + 14 \times 16$ **14(12 + 16)**

Simplify using mental math.

8. 6×28 **168** **9.** 5×26 **130** **10.** 16×9 **144**

11. 108×6 **648** **12.** 7×18 **126** **13.** 3×198 **594**

Simplify.

14. $4(7 - a)$ $^{28 - 4a}$ **15.** $6(4 + b)$ $^{24 + 6b}$ **16.** $10(c + 10)$ $^{10c + 100}$

17. $7(d + 1)$ $^{7d + 7}$ **18.** $5(e - 8)$ $^{5e - 40}$ **19.** $3(f - 2)$ $^{3f - 6}$

Write an expression using the distributive property and solve.

20. There were 17 workers on an assmbly line. In one hour, each worker was expected to put together 9 kits. How many kits in all should be assembled each hour? **Expressions may vary.**
Sample: 9 × 17 = 9(20 − 3) = 180 − 27 = 153 kits

Find each answer using mental math.

21. 90×62 ^{5,580} **22.** 360×20 ^{7,200} **23.** 70×39 ^{2,730}

24. 82×40 ^{3,280} **25.** 105×30 ^{3,150} **26.** 25×12 ^{300}

27. 108×60 ^{6,480} **28.** 70×18 ^{1,260} **29.** 30×198 ^{5,940}

Simplify using the distributive property.

30. 108×210 ^{22,680} **31.** 310×498 ^{154,380} **32.** 961×50 ^{48,050}

33. $2,005 \times 11$ ^{22,055} **34.** 603×306 ^{184,518} **35.** 492×201 ^{98,892}

36. $7,924 \times 80$ ^{633,920} **37.** $2,900 \times 109$ ^{316,100} **38.** 815×501 ^{408,315}

Write an expression using the distributive property and solve.

39. A warehouse contained 198 cases of product. Each case held 25 packages. How many packages were in the warehouse?
25 × 198 = 25(200 − 2) = 5,000 − 50 = 4,950 packages

40. A gas pipeline pumps 402,000 gallons of crude oil per day. At this rate, how many gallons are pumped in one week? **See Additional Answers.**

41. A school homecoming button costs $1.90. How much would it cost if every student in your math class bought a button? **Answers will vary.**

Write *yes* if the expression is an example of the distributive property. Write *no* if it is not and explain why. **Explanations may vary.**

42. $(6 \times 5) \times (6 \times 4) = 30 \times 24$ **No. It is a multiplication sentence.**

43. $8 \times 1 + 8 \times 3 = 8(1 + 3)$ **Yes**

44. $17 \times 1 \div 17 \times 1 = 17 \div 17$ **No. Division is used.**

Decide whether each is *true* or *false*. Explain.

45. The distributive property cannot be used with division.

46. The distributive property cannot be used when any other properties are used to solve an expression.

47. $a(b - c) = (b - c)a$

**EXTEND/
SOLVE PROBLEMS**

**THINK CRITICALLY/
SOLVE PROBLEMS**

45. True. It is used only with multiplication over addition or subtraction.

46. False. Any number of properties can be used to solve expressions.

47. True. $ab - ac = ba - ca$. The order of the variables as factors in a product does not affect the value of the product.

2–10   Distributive Property   **79**

5 **FOLLOW-UP**

Extra Practice Complete, using the distributive property.
1. $(7 \times 4) + (7 \times 6) = \blacksquare (4 + 6)$ **7**
2. $4(6 + 12) = (\blacksquare \times 6) + (4 \times \blacksquare)$ **4, 12**
3. $8(13 - 4) = (8 \times 13) - (\blacksquare \times \blacksquare)$ **8, 4**
Simplify, using the distributive property and mental math.
4. 60×28
60(30 − 2) = (60 × 30) − (60 × 2) = 1,800 − 120 = 1,680
5. 312×30
(300 × 30) + (12 × 30) = 9,000 + 360 = 9,360
Write an expression using the distributive property and solve.
6. There are 403 pages in each of 12 volumes of an encyclopedia. How many pages are there in all? **12 × 403 = (12 × 400) + (12 × 3) = 4,836**

Extension Explain that the distributive property also works with multiplication of fractions. Have students solve these problems mentally using the distributive property and explain their solutions.
1. $8 \times 3\ 3/4$ **(8 × 3) + (8 × 3/4) = 24 + 6 = 30**
2. $9\ 1/3 \times 6$ **(6 × 9) + (6 × 1/3) = 54 + 2 = 56**

Section Quiz
1. Which are examples of the distributive property? **a and c**
a. $(7 \times 4) + (7 \times 290) = 7(4 + 290)$
b. $(31 \times 3) \times 3(31 \times 2) = 31(2 \times 3) \times 31$
c. $7(3 - 1) = 7 \times 2$
Write an expression using the distributive property and solve.
2. There are 198 seats in each of 30 theaters. How many seats are there in all? **198 × 30 = (200 × 30) − (2 × 30) = 5,940**

Get Ready calculators

Additional Answers
See page 574.

79

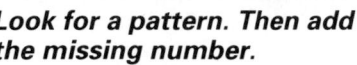

WARM-UP

Look for a pattern. Then add the missing number.
1. 3, 6, 9, 12, ■ **18**
2. 4.50, 4.25, 4.00, ■ , 3.50 **3.75**
3. 1, 3, 2, 4, 3, 5, ■ , 6 **4**

1 MOTIVATE

Introduction Have students discuss what they think a pattern means in mathematics. Encourage them to discuss the difference between number patterns and geometric patterns.

2 TEACH

Use the Pages/Problem Have students read this part of the section and then discuss the problem. Point out that sometimes more than one strategy can be used to solve a problem. In this case, making a list or table will help students to find the pattern.

2-11 Problem Solving Strategies:
FIND A PATTERN

► READ
► PLAN
► SOLVE
► ANSWER
► CHECK

Sometimes a problem is easier to understand and solve if you can find a pattern in the problem.

PROBLEM

Danny's parents put $1 into his savings account on his first birthday, $2 on his second birthday, $4 on his third birthday, $8 on his fourth birthday, and so on. How much money will Danny's parents put into his account on his tenth birthday?

SOLUTION

READ	Known:	Danny received $1 for his first birthday, $2 for his second, $4 for his third, and $8 for his fourth.
PLAN	Find:	How much he will receive on his tenth birthday
SOLVE		Make a table to organize the information. Then look for a pattern.

Birthday	1	2	3	4	5	6	7	8	9	10
Money	$1	$2	$4	$8	$16	$32	$64	$128	$256	$512

$\times 2 \quad \times 2 \quad \times 2$

Look at the first four entries in the table. Each amount of money is multiplied by 2 to get the next amount. The pattern continues through Danny's tenth birthday.

ANSWER Danny's parents will put $512 into his account on his tenth birthday.

CHECK Use this key sequence on your calculator to check:

$2 \;\boxed{\times}\; 1 \;\boxed{=}\;\boxed{=}\;\boxed{=}\;\boxed{=}\;\boxed{=}\;\boxed{=}\;\boxed{=}\;\boxed{=}\;\boxed{=}$

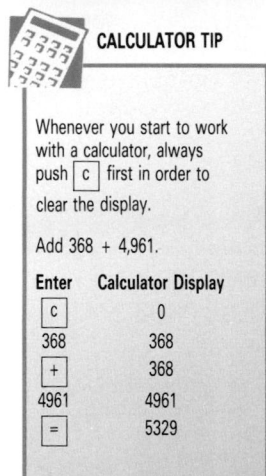

CALCULATOR TIP

Whenever you start to work with a calculator, always push ⌷c⌷ first in order to clear the display.

Add 368 + 4,961.

Enter	Calculator Display
⌷c⌷	0
368	368
⌷+⌷	368
4961	4961
⌷=⌷	5329

PROBLEMS

Which rule describes the pattern? Copy and complete.

1. 0.08, 0.16, 0.24, ■, ■
   a. Add 0.8.   **0.32, 0.40**
   b. Subtract 0.8.
   c. Add 0.08.
   d. Subtract 0.08.

2. 2.45, 2.05, ■, 1.25, ■
   a. Subtract 0.5.   **1.65, 0.85**
   b. Subtract 0.4.
   c. Subtract 0.05.
   d. Subtract 0.04.

AT-RISK STUDENTS

Students may need practice in recognizing simple number patterns. Begin by giving simple examples such as 1, 3, 5, 7, . . . ; 2, 4, 6, 8, . . . ; and 1, 1, 2, 3, 3, 4, 5, 5, 6, 7, 7, ... When student are comfortable analyzing such simple sequences, have them create number patterns of their own to share with other students.

3. 7.432, 8.543, 9.654, ■, ■

2.5, 2.0

a. Add 1.   **10.765, 11.876**
b. Add 1.111.
c. Add 1.11.
d. Add 1.001.

4. 1.5, 1.0, 2.0, 1.5, ■, ■

a. Add 0.5, subtract 1.0.
b. Add 2.0, subtract 0.5.
c. Subtract 0.5, add 0.5.
d. Subtract 0.5, add 1.0.

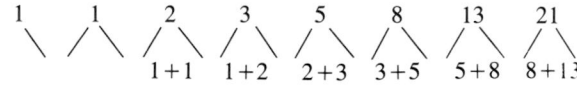 Write the rule. Then copy and fill in the missing numbers.

5. 1.5, 1.3, 1.1, ■, ■, 0.5 **Subtract 0.2; 0.9, 0.7**

6. 0.3, 0.6, 0.9, ■, ■, 1.8 **Add 0.3; 1.2, 1.5**

7. 0.9, 4.9, 4.8, 8.8, ■, ■ **Add 4, subtract 0.1; 8.7, 12.7**

8. 5.2, 3.2, 5.2, 3.2, ■ **Subtract 2, then add 2; 5.2**

9. 9.34, 8.34, 8.54, 7.54, ■, ■ **Subtract 1, add 0.20; 7.74, 6.74**

10. 1.005, 1.001, 1.003, ■, 1.001, ■ **Subtract 0.004, add 0.002, 0.999, 0.997**

Solve. **See Additional Answers for Exercises 11–14.**

11. Sondra plans to put $5 in her savings account in January, $8 in February, $11 in March, $14 in April, and so on.

a. How much money will she put into the account in December?
b. How much money will she put into the account for the entire year.

12. A movie theater plans to give away 2 theater tickets on Monday, 8 tickets on Tuesday, 32 tickets on Wednesday, and so on for Thursday and Friday.

a. How many tickets will they give away on Friday?
b. How many tickets will they give away for the five days?

13. Leroy went on a cross-country bike trip. He traveled 9 miles the first day, 15 miles the second day, 21 miles the third day, and so on. At this rate, how many miles will he have traveled by the end of the eighth day?

14. Melinda jogs 1 lap around the track the first week, 3 laps the second week, 6 laps the third week, and 10 laps the fourth week. If she continues this pattern, how many laps will she jog the eighth week?

15. The occurrence of the Fibonacci sequence of numbers appears frequently in nature. The arrangement of the parts of many plants reflects this sequence.

1   1   2   3   5   8   13   21

1+1   1+2   2+3   3+5   5+8   8+13

a. Continue the pattern to find the next three numbers. **34, 55, 89**
b. Find at least three other instances in which the Fibonacci sequence occurs in nature. **Answers will vary.**

Use the Pages/Solution Point out that although there are many problem solving strategies to choose from, the same procedure—Read, Plan, Solve, Answer, and Check—should be followed.

3 SUMMARIZE

Talk About Math Discuss students' strategies for finding patterns in problems. Do they make tables or lists? Why or why not?

4 PRACTICE

Allow students to use calculators to help identify and complete the patterns in Exercises 1–10. For Exercises 11–15 have students make tables to find patterns.

5 FOLLOW-UP

Extra Practice
1. Write the rule. Then give the missing numbers.
0.8, 3.8, 3.6, 6.6, ■ , ■   **Add 3, subtract 0.2; 6.4, 9.4**

Additional Answers
See page 574.

2 CHAPTER REVIEW

Introduction The Chapter Review emphasizes the major concepts, skills, and vocabulary presented in this chapter and can be used for diagnosing students' strengths and weaknesses. Page references direct students back to appropriate sections for additional review and reteaching.

Using Pages 82–83 Before students do the exercises in this review, allow them to scan the information and discuss any concepts or skills they may still find confusing. Exercises 1–5 present a brief review of key vocabulary in this chapter. For Exercises 18–19 be sure students name the operation they must use before they solve the problems. For Exercise 58 suggest students make a table to find the pattern.

Informal Evaluation Have students identify any sections in which they had half or more incorrect answers. Discuss the nature of their errors and clear up any confusion in concepts or skills.

Follow-Up Have students find the distances (in kilometers) between their town or city and two other cities they would want to visit. Have them use this information to create word problems involving addition, subtraction, division, and multiplication of decimals, rounding and estimation. Then have students solve the problems and identify any instances in which the associate, commutative, or distributive properties were used.

1. A letter representing an unknown number is a(n) __?__ . **e**
2. 3^4 is written in __?__ **c** .
3. Frequently very large or very small numbers are written in __?__ . **d**
4. Rules that must be followed when evaluating expressions are called the __?__ . **b**
5. According to the __?__ property, the order in which two numbers are added or multiplied does not affect the answer. **a**

a. commutative
b. order of operations
c. exponential form
d. scientific notation
e. variable

SECTION 2–1 WHOLE NUMBERS (pages 42–45)

► To round numbers, locate the digit in the place to which you are rounding.
  a. Increase this digit by 1 if the next digit to the right is 5 or greater.
  b. Leave the digit the same if the next digit to the right is less than 5.

Find each answer. Round and compare to see if your answer is reasonable.

6. $456 + 938$ **1,394**   7. $1,092 - 657$ **435**   8. 89×31 **2,759**   9. $1,189 \div 29$ **41**

SECTION 2–2 VARIABLES AND EXPRESSIONS (pages 46–49)

► When writing an expression to describe a real-world situation, you can use a variable to represent an unknown number.

Evaluate each expression. Let $a = 2$, $b = 4$, and $c = 6$.

10. $a + b$ **6**   11. $c - a$ **4**   12. $a + c$ **8**   13. $b - 2a$ **0**
14. $5c \div 3a$ **5**   15. $2c \div b$ **3**   16. $\dfrac{3b}{2a}$ **3**   17. $6a \times 2c$ **144**

SECTIONS 2–3 and 2–11 PROBLEM SOLVING (pages 50–51, 80–81)

► Before you can solve a problem, you need to decide on a strategy and choose the operation needed to solve the problem.

USING DATA Use the table on page 41 to help you answer the following.

18. What is the difference in the exchange rate for the day's low and the 12-month low for the German mark? **0.0014**

19. Use the day's high to find how many U.S. dollars are equal to 15 British pounds. **24.3825**

► When adding or subtracting decimals, align the decimal points.
► When multiplying or dividing decimals, compute just the way you do
   with whole numbers. In multiplication, the answer has as many
   decimal places as the sum of the decimal places in the factors.

Find each answer. Estimate and compare to see if your answer is reasonable.

20. $8.679 + 8.543$ **17.222**   21. $9.768 - 4.389$ **5.379**   22. $4.6 - 2.037$ **2.563**

23. 6×3.954 **23.724**   24. 0.32×1.4 **0.448**   25. $2.65 \div 0.5$ **5.3**

SECTIONS 2–6 and 2–7 EXPONENTS AND SCIENTIFIC NOTATION (pages 60–67)

► It is easier to work with large numbers when they are written in
   scientific notation.

Write in scientific notation.

26. $8,409$ 27. $732,037$ 28. 388 29. $7,000,849$
    8.409×10^3   **7.32037×10^5**   **3.88×10^2**   **7.000849×10^6**

Write each answer in exponential form.

30. $5^8 \times 5^3$ **5^{11}**   31. $9^8 \div 9^4$ **9^4**   32. $x^8 \times x^3$ **x^{11}**   33. $y^5 \div y^2$ **y^3**

SECTION 2–8 ORDER OF OPERATIONS (pages 68–71)

► Follow the **order of operations** to simplify mathematical expressions.

Simplify.

34. $8 \div 4 + 9 \times 3$ **29**   35. $6 + (2 \times 3) - 7$ **5**   36. $9 \times 2^2 + 1$ **37**

37. $7 \times (3 - 1) + 2$ **16**   38. $4 \times 5 + 12 \div 4$ **23**   39. $3^2 \div 9 + 8$ **9**

SECTIONS 2–9 and 2–10 NUMBER PROPERTIES (pages 72–79)

► Some of the ways in which we perform operations on numbers are
   governed by the **commutative, associative, distributive,** and **identity**
   properties and by the **property of zero for multiplication.**

Complete. Identify the property.

40. $(8 + 3) + 5 = 8 + (\blacksquare + 5)$
    3; associative property
41. $3,492 \times \blacksquare = 0$
    0; property of zero for multiplication
42. $16 + 9 + \blacksquare = 9 + 43 + 16$
    43; commutative property
43. $(a + b) + \blacksquare = (a + b)$
    0; identity property of addition

Review **83**

Study Skills Tip Have students
list the following terms in a column on a sheet of paper: *round,*
estimate, variable expression,
exponential form, standard form,
scientific notation, order of opera-
tions, associative property,
commutative property, and *dis-*
tributive property.

Next to each have them give an
example. Encourage them to use
this paper as a study guide when
they need to review the concepts
in this chapter.

2 CHAPTER TEST

Introduction The Chapter Test uses a variety of questioning techniques to assess students' mastery of the major objectives of Chapter 2. If you prefer, you may use the Skills Preview (page 39) as an alternative form of the Chapter Test. The items on this test and the Skills Preview correspond in content and level of difficulty.

Alternative Assessment
Making Connections in Science
Choose a planet to report on. Research facts about the planet and the distance between it, its nearest planet, the Sun, and Earth. Include in your report the size of the planet's diameter. Use scientific notation wherever possible to relate information. Include several sentences comparing the various distances.

Find each answer. Estimate and compare to see if your answer is reasonable.

1. $2{,}154 + 3{,}895$ **6,049**　　2. 38×49 **1,862**　　　　3. $97 \times 3{,}125$ **303,125**

4. $5{,}551 \div 61$ **91**　　5. $3.45 + 12.3 + 0.967$ **16.717**　　6. $48.2 - 0.345$ **47.855**

7. 0.34×0.34 **0.1156**　　8. 5×2.4 **12**　　　　9. $36.54 \div 18$ **2.03**

Evaluate each expression. Let $a = 2$ and $b = 6$.

10. $a + 19$ **21**　　11. $b \div 2$ **3**　　12. $2b \div 2a$ **3**　　13. $6a \div b$ **2**

Write in exponential form.

14. $8 \times 8 \times 8 \times 8$ **8^4**　　15. $5 \times 5 \times 5$ **5^3**　　16. $6 \times 6 \times 6 \times 6 \times 6 \times 6$ **6^6**

Write in standard form.

17. 7^3 **$7 \times 7 \times 7 = 343$**　　18. 2^6 **$2 \times 2 \times 2 \times 2 \times 2 \times 2 = 64$**　　19. 3^4 **$3 \times 3 \times 3 \times 3 = 81$**　　20. 4^3 **$4 \times 4 \times 4 = 64$**

Write in scientific notation.

21. $6{,}893$ **6.893×10^3**　　22. 349 **3.49×10^2**　　23. $435{,}982$ **4.35982×10^5**　　24. $4{,}358{,}231$ **4.358231×10^6**

Write each answer in exponential form.

25. $8^2 \times 8^9$ **8^{11}**　　26. $7^{12} \div 7^3$ **7^9**　　27. $x^8 \div x^3$ **x^5**　　28. $y^5 \times y^4$ **y^9**

Simplify.

29. $7 \times 8 - 36 \div 4$ **47**　　30. $4 \times 2^3 + 10$ **42**　　31. $(2 + 2)^2 \times 3$ **48**

Complete.

32. $(7 \times 9) \times 4 = \blacksquare \times (9 \times 4)$ **7**　　33. $\blacksquare \times 3{,}458 = 3{,}458$ **1**

34. $3.75 \times \blacksquare = 0$ **0**　　35. $7.6 \times 8.4 = 8.4 \times \blacksquare$ **7.6**

36. $873 + \blacksquare = 954 + 873$ **954**　　37. $87.3 + \blacksquare = 87.3$ **0**

Simplify.

38. $5(8 - x)$ **$40 - 5x$**　　39. $4(a - 3)$ **$4a - 12$**　　40. $7(5 - y)$ **$35 - 7y$**　　41. $6(n + 5)$ **$6n + 30$**

Solve.

42. Joan put \$25 in her savings account in January, \$28 in February, \$31 in March, \$34 in April, and so on. Following this pattern, find how much money she put into the account in December. **\$58**

J	F	M	A	M	J	J	A	S	O	N	D
\$25	\$28	\$31	\$34	\$37	\$40	\$43	\$46	\$49	\$52	\$55	\$58

1. A car dealer hired students to determine the ages of the cars owned by the people in one neighborhood. The students recorded the following ages (in years).

3 5 1 2 6 8 2 4 5 5 6 2 1 2 3 4 4 5
4 2 3 7 7 1 3 3 1 6 7 7 6 8 2 2 3 1

Make a pictograph for these data.
Check students' graphs.

Find the mean, median, mode, and range for each of the following sets of data. Round your answers to the nearest tenth.

2. 43 49 56 61 45 52 74   **54.3, 52, none, 31**

3. 46 48 50 50 52 53 49   **49.7, 50, 50, 7**

4. 6.6 5.9 6.8 9.6 5.2 6.9 9.9 8.8 9.6
7.7, 6.9, 9.6, 4.7

5. 30 30 45 28 31 31 45 30
33.8, 30.5, 30, 17

6. What number must be included in the following data for the median to be 14?

18 11 22 11 10   **17**

7. What number must be included in the following data for the mean to be 6.44?

4.25 3.14 8.72 9.19   **6.9**

Use the following data for Exercises 8–12. The number of basketball free throws sunk by 40 students each taking 50 free throws was recorded as follows:

17	8	29	32	15	31	22	4	35	11
23	39	38	6	42	17	50	34	48	25
3	37	27	43	28	36	41	9	41	29
26	12	42	34	14	9	27	37	21	35

8. Construct a stem-and-leaf plot for these data.   **See Additional Answers.**

How many students had the following numbers of successful throws?

9. 16–24   **5**   10. 40–48   **6**

11. less than 16   **10**   12. 32 or more   **17**

Use the bar graph for Exercises 13–15.

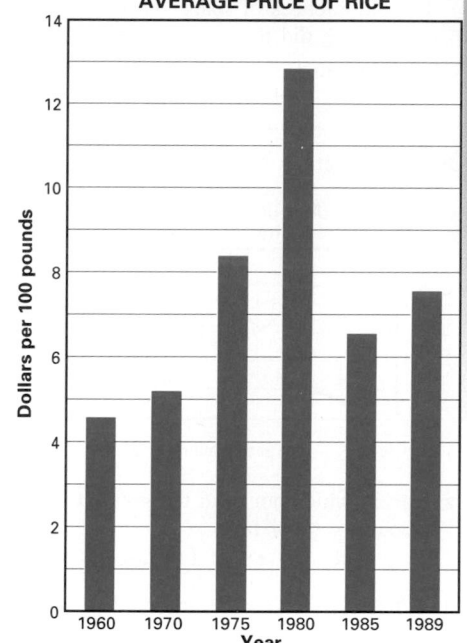

AVERAGE PRICE OF RICE

13. In which year was the price of rice the greatest?   **1980**

14. In which years was the price under $6?
1960, 1970

15. About how much of a price difference was there between 1970 and 1975?
about $3

16. In which year was the price about one-half of what it was in 1980?   **1985**

Find each answer. Estimate to see if the answer is reasonable.

17. 8.349 + 13.83   **22.179**   18. 18.2 − 0.375   **17.825**

19. 4.8 × 0.31 **1.488**   20. 4.941 ÷ 6.1 **0.81**

21. 24.96 ÷ 1.3 **19.2**   22. 0.02 × 0.08   **0.0016**

23. 83.5 − 9.678 **73.822**   24. 1.836 ÷ 1.2 **1.53**

---

2 CUMULATIVE REVIEW

Introduction   The purpose of this Cumulative Review is to maintain previously taught skills and concepts and to apply them to the material presented in this chapter. At least one major objective of each chapter is included in the review.

Item Analysis   The table below correlates the Cumulative Review items with the chapter and section that are being reviewed.

Section	Items
1–2	8–12
1–3	1
1–6	13–16
1–9	2–7
2–4	17–18, 23
2–5	19–22, 24

Additional Answers
See page 574.

2 CUMULATIVE TEST

Introduction The Cumulative Test uses a standardized-test format of multiple-choice questions to assess retention of previously learned concepts and test-taking skills. Test results may be used to diagnose students' strengths and weaknesses.

Item Analysis The table below correlates the Cumulative Test items with the chapter and section that are being tested.

Section	Items
1–1	1
1–2	5
1–8	2–3
1–9	4, 6–10
2–4	11
2–5	12, 13

1. A soap manufacturer sampled people who crossed at a busy intersection about their preferences of soap. What type of sampling did the manufacturer use?
   A. systematic   B. convenience
   C. random   D. cluster

Use the line graph for Exercises 2–3.

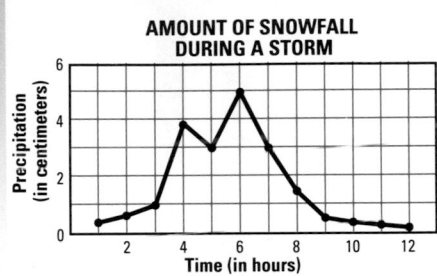

AMOUNT OF SNOWFALL DURING A STORM

2. During which hour did the greatest amount of snow fall?
   A. 4   B. 5   C. 8   D. none of these

3. What was the greatest amount of snow that fell in a single hour?
   A. 3 cm   B. 4 cm   C. 5 cm   D. 6 cm

4. If a car cost $15,000 and the expected life of the car is 10 years, what is the average cost per year?
   A. $15,000   B. $1,500
   C. $1,250   D. $150

5. Here is the amount of rainfall for a year recorded in centimeters.

J	F	M	A	M	J	J	A	S	O	N	D
63	67	56	43	19	0	0	21	45	57	59	60

If a stem-and-leaf plot were made for these data, what would be the extremes of the stems?
   A. 6 and 6   B. 1 and 6
   C. 0 and 6   D. 0 and 9

6. A statistic in the newspaper stated that in 1987 the average family size was 3.19 people. What measure of central tendency does 3.19 represent?
   A. mean   B. median
   C. mode   D. range

7. The median age in the United States in 1987 was 32.1. What can you conclude from this?
   A. Children made up about half of the population.
   B. Half of the population was older and half was younger than 32.1 years.
   C. The mean age of the population was 32.1 years.
   D. More people were 32.1 years old than any other age.

Use this data for Exercises 8–10.

| 8 | 19 | 34 | 5 | 39 | 41 | 58 | 63 | 12 |

8. What is the mean of the data?
   A. 41   B. 39   C. 31   D. none of these

9. What is the range of the data?
   A. 34   B. 31   C. 58   D. none of these

10. What is the median of the data?
    A. 34   B. 63   C. 31   D. none of these

11. Find 16.3 − 0.361.
    A. 16.061   B. 15.939
    C. 16.661   D. 15.949

12. Find 2.7 × 0.49.
    A. 0.1323   B. 13.23
    C. 1.323   D. 0.01323

13. Find 2.91 ÷ 1.5
    A. 1.94   B. 194
    C. 19.4   D. 1,940

Name _____ Date _____

Whole Numbers

When computing with whole numbers, estimate to see that your answer is reasonable. To estimate a sum or difference, round each number to the greatest place of the number with least value. To estimate a product or quotient, round each number to its greatest place.

► Example 1

Find the actual sum: 847 + 279. Then estimate to see if your answer is reasonable.

Solution

Actual	Estimate
847 →	800 Round each number to hundreds.
+279 →	+300
1,126	1,100

1,126 is close to 1,100, so the answer is reasonable.

► Example 2

Find the actual product: 267 × 1,105. Then estimate to see if your answer is reasonable.

Solution

Actual	Estimate
1,105 →	1,000 Round each number to its greatest place.
× 267 →	× 300
295,035	300,000

295,035 is close to 300,000, so the answer is reasonable.

You can also use compatible numbers to estimate.

► Example 3

Find 7,632 ÷ 48. Then use compatible numbers to estimate the answer and compare to see if your answer is reasonable.

Solution

Actual	Estimate
159 50 is close to 48; 8,000 is close	160
48)7,632 to 7,632. You can find 8,000 ÷ 50	50)8,000
mentally. (To divide by 50,	
multiply by 2 and divide by 100.)	

EXERCISES

Round each number to the greatest place value.

1. 78 **80**   2. 335 **300**   3. 899 **900**   4. 56,421 **60,000**

Find each answer. Then estimate using rounding to see if the answer is reasonable.

5. 703,261 + 385,592 **1,088,853**   6. 4,361 − 1,822 **2,539**

7. 5,694 ÷ 73 **78**   8. 37 × 308 **11,396**

Choose a pair of compatible numbers you would use to solve each problem.

9. 763 ÷ 29 **750; 30**   10. 2,786 ÷ 84 **2,700; 90**   11. 58,201 ÷ 792 **56,000; 800**

Find each answer. Then estimate using compatible numbers to see if the answer is reasonable.

12. 2,542 ÷ 82 **31**   13. 5,185 ÷ 61 **85**   14. 3,619 ÷ 47 **77**

Reteaching • SECTION 2-1   23

Name _____ Date _____

The Lattice Method

The lattice method of multiplying, based on Napier's rods, was used in the schools of England over 400 years ago. To multiply two whole numbers such as 645 × 732, you make a lattice such as the one below.

Step 1: Write two factors, one above the lattice and one to the right.

Step 2: Multiply 5 × 2 and write the result in the box common to the column for 5 and the row for 2. Fill in the rest of the boxes in a like manner, writing the tens digits on the top and the ones digits below.

Step 3: Add the figures diagonally, beginning with the lower right corner.

Step 4: Read down on the outside left and continuing across the bottom to find the product.
645 × 732 = 472,140

5 + 1 + 8 =14; write 4 and carry 1

1 + 5 + 1 + 2 + 2 =11; write 1 and carry 1

EXERCISES

Complete each lattice and give the product. Estimate to see if your answer is reasonable.

1. **1,176,768**   2. **2,065,878**

Find each answer using the lattice method. Estimate to see if your answer is correct.

3. 215 × 318 **68,370**   4. 27 × 913 **24,651**

5. 2,356 × 812 **1,913,072**   6. 4,376 × 5,297 **23,179,672**

24   Enrichment • SECTION 2-1

Name _____ Date _____

Variables and Expressions

A **variable** is a letter that is used to represent an unknown number. To **evaluate** an expression means to find the value of the expression for a particular value of the variable or variables.

Variables: a, b

Variable Expressions: $3x + 2$, $n - 5$ ← variables

► Example 1

Evaluate $14 - n$, if $n = 3$.

Solution

$14 - n =$
$14 - 3 = 11$ ← Write 3 in place of n.

Example 2

Evaluate $2n + 4$, if $n = 8$.

Solution

$2n + 4 =$
$2 \times 8 + 4 = 16 + 4 = 20$ ← Write 8 in place of n.

EXERCISES

Match each variable expression with its meaning.

1. $\frac{n}{2}$ **e**   a. 2 more than a number

2. $2 + n$ **a**   b. 2 divided by a number

3. $2n$ **d**   c. 2 less than a number

4. $n - 2$ **c**   d. twice a number

5. $\frac{2}{n}$ **b**   e. a number divided by 2

Write an expression for each situation.

6. 12 cookies divided among n people **12 ÷ n**

7. 8 more than n dollars **n + 8**

8. 7 less than d compact discs **d − 7**

9. 8 times x people **8x**

Evaluate each expression. Let $n = 18$.

10. $n - 5$ **13**   11. $8n$ **144**   12. $n \div 9$ **2**   13. $14 + n$ **32**

Complete each table.

14.
n	$6n$
0	**0**
1	**6**
2	**12**
3	**18**

15.
a	$12 - a$
12	**0**
11	**1**
10	**2**
6	**6**

16.
t	$t \div 6$
0	**0**
6	**1**
12	**2**
180	**30**

Reteaching • SECTION 2-2   25

Name _____ Date _____

Variable Magic

You can use what you know about writing expressions with variables to analyze number tricks. Examine the following table, which gives the steps in one number trick. The variable expressions help you to understand how the trick works.

Number Operations	Example	Your Number	Variable Expression
1. Pick any whole number.	3		x
2. Add 7.	10		$x + 7$
3. Now double the result.	20		$2x + 14$
4. Subtract 4.	16		$2x + 14 - 4$
5. Take one half of the result.	8		$\frac{2x}{2} + \frac{10}{2}$ or $x + 5$
6. Subtract your original number.	5		$x + 5 - x$ or 5

EXERCISES

1. Write variable expressions to complete the table.

Number Operations	Variable Expression
1. Pick any whole number.	x
2. Double the number.	$2x$
3. Multiply by 4.	$8x$
4. Add 40.	$8x + 40$
5. Divide by 8.	$\frac{8x}{8} + \frac{40}{8}$ or $x + 5$
6. Subtract your original number.	$x + 5 - x$, or 5

2. Write variable expressions to complete the table.

Number Operations	Variable Expression
1. Pick any whole number.	x
2. Triple the number.	$3x$
3. Add 12.	$3x + 12$
4. Double the result.	$6x + 24$
5. Divide by 6.	$\frac{6x}{6} + \frac{24}{6}$ or $x + 4$
6. Subtract your original number.	$x + 4 - x$, or 4

26   Enrichment • SECTION 2-2

86A

Name _____ Date _____

Problem Solving Skills: Choose the Operation

When you read a word problem, there are often words in the problem that give you a clue about which operation to use.

▶ **Example**

What operation would you use to solve?
A golfer earned $20,000 in one tournament, $8,000 in the next, and $75,000 in the third. How much did the golfer earn in all?

Solution
The words *in all* suggest addition. You would add to solve.

EXERCISES

Choose the operation. Do not solve. What words did you use to help you decide?

1. In 1991 Kevin's water bill averaged $28 per month. What was Kevin's total expense during the year for water?

   ×; total

2. Four people split the cost of dinner equally. How much did each spend if the dinner was $104?

   ÷; split equally

3. The top scorer for the Bullets scored 53 points and the top scorer for the Blazers earned 39 points. How many more points did the Bullets' scorer make?

   −; how many more

The table shows "great circle," or shortest air distances between major cities in the world. Tell what operation you would use to solve each problem and then find the solution.

	Berlin	London	Moscow	New York	Paris
Berlin	—	574	996	3,961	542
London	574	—	1,549	3,459	213
Moscow	996	1,549	—	4,662	1,541
New York	3,961	3,459	4,662	—	3,622
Paris	542	213	1,541	3,622	—

4. How much farther is it from London to Moscow than from London to Berlin?

   subtract; 975 mi

5. If you fly from New York to London and then on to Moscow, how many air miles will you have flown?

   add; 5,008 mi

6. About how many times as far is it from Paris to Berlin as from Paris to New York?

   divide; about 7 times as far

Reteaching • SECTION 2-3   27

Name _____ Date _____

Calories

We obtain our energy from the food we eat. The energy content of food is measured in calories. The numbers of calories in common foods are given in the table. Whether you sleep, think, or exercise, you burn calories. If you burn as many calories as you take in, you do not gain or lose weight. The numbers of calories used per minute in some common activities are given in the table.

Food	Calories	Food	Calories
3 oz tuna	170	carrot	20
3 oz chicken	115	4 oz spaghetti	65
1 c skim milk	90	1 c cornflakes	210
1 oz cottage cheese	24	1 slice bread	56
1 banana	100	4 oz peanuts	660
		1 apple	70

If you eat a banana, you take in 100 calories. If you swim for 10 min, you burn 100 calories.

Activity	Calories used (per min)	Activity	Calories used (per min)
sleeping	1	swimming	10
walking	5	bicycling	11
dancing	8	running	14

EXERCISES

1. How many calories do you burn if you run for 15 min? **210**

2. Swimming for 1 min burns how many times as many calories as walking for 1 min?

   2 times as many calories

3. Ellen ate an apple, 1 oz of cottage cheese, 1 slice of bread, and 3 oz of tuna. How many calories did she consume in all? **320**

4. Rob ate a meal in which he consumed 840 calories. How long would he have to run to burn off those calories? **60 min**

5. Theresa walks 30 min a day. How many calories does she burn in a week from walking? **1,050**

6. Write a problem that you can solve using addition. **Answers may vary.**

7. Write a problem that you can solve using subtraction. **Answers may vary.**

8. Write a problem that you can solve using multiplication. **Answers may vary.**

9. Write a problem that you can solve using division. **Answers may vary.**

28   Enrichment • SECTION 2-3

Name _____ Date _____

Adding and Subtracting Decimals

To add or subtract decimals, follow these steps.
• Line up the numbers according to place value.
• Annex zeros so that all the numbers have the same number of place values.
• Add or subtract.

▶ **Example**

Find each answer. Use rounding to estimate the answer and compare to see if the answer is reasonable.

a. 32.02 + 421.6 + 9.07   b. 800 − 273.6

Solution

a.
Align the decimal points.	Annex zeros.	Add.
32.02	32.02	32.02
421.6	421.60	421.60
+ 9.07	+ 9.07	+ 9.07
		462.69

b.
Align the decimal points.	Annex zeros.	Subtract.
800	800.0	800.0
−273.6	−273.6	−273.6
		526.4

Estimate

30 + 400 + 10 = 440   800 − 300 = 500

Since 462.60 is close to 440, the answer is reasonable.

Since 526.4 is close to 500, the answer is reasonable.

EXERCISES

Find each answer. Use rounding to estimate and compare to see if the actual answer is reasonable.

1. $29.50 + $18.75 + $8.50 **$56.75**   2. 27,861.8 − 18,401.91 **9,459.89**

3. 7.3 + 8.403 + 6.09 **21.793**   4. $562.10 − $429.53 **$132.57**

5. 593.88 + 726.153 + 201.5 **1,521.533**   6. 2,361 − 851.9 **1,509.1**

Find each answer. Use front-end estimation to see if the actual answer is reasonable.

7. 674.2 + 4,251.99 + 100.7 **5,026.89**   8. $679.02 − $388.71 **$290.31**

9. $2,388.09 − $2,125.78 **$262.31**   10. 5.23 + 55.64 + 8.799 **69.669**

11. 424.98 + 3,297.1 + 788.023 **4,510.103**   12. $56,982.23 − $32,918.56 **$24,063.67**

Solve. Estimate and compare to see if your answer is reasonable.

13. Elias had $462.88 in his savings account. How much did he have left after he spent $288.56 on a bicycle and $38.95 on a helmet? **$135.37**

Reteaching • SECTION 2-4   29

Name _____ Date _____

Magic Squares

A magic square is a square in which the sum of the numbers in each row, column, and diagonal is the same. The sum in this magic square is 3.3.

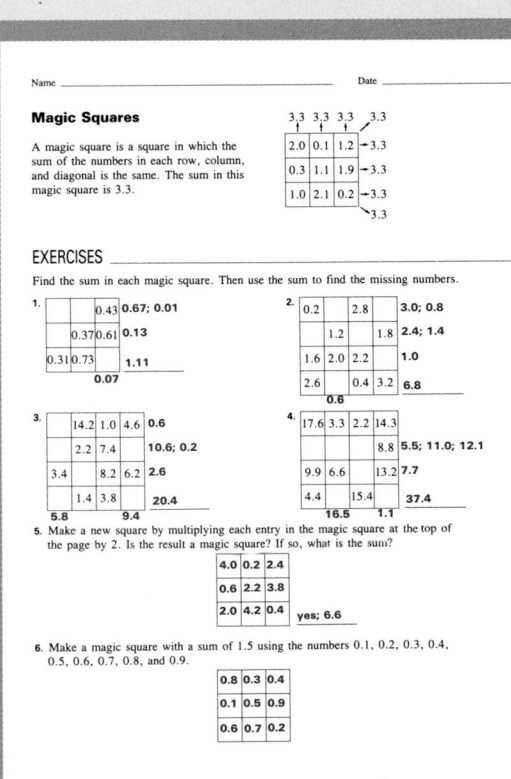

EXERCISES

Find the sum in each magic square. Then use the sum to find the missing numbers.

1.
	0.43	**0.67; 0.01**
0.37	0.61	**0.13**
0.31	0.73	**1.11**

0.07

2.
0.2		2.8	**3.0; 0.8**
	1.2		1.8	**2.4; 1.4**
1.6	2.0	2.2	**1.0**
2.6		0.4	3.2	**6.8**

0.6

3.
14.2	1.0	4.6	**0.6**
2.2	7.4		**10.6; 0.2**
3.4	8.2	6.2	**2.6**
1.4	3.8		**20.4**

5.8   9.4

4.
17.6	3.3	2.2	14.3	
			8.8	**5.5; 11.0; 12.1**
9.9	6.6		13.2	**7.7**
4.4		15.4		**37.4**

16.5   1.1

5. Make a new square by multiplying each entry in the magic square at the top of the page by 2. Is the result a magic square? If so, what is the sum?

4.0	0.2	2.4
0.6	2.2	3.8
2.0	4.2	0.4

yes; 6.6

6. Make a magic square with a sum of 1.5 using the numbers 0.1, 0.2, 0.3, 0.4, 0.5, 0.6, 0.7, 0.8, and 0.9.

0.8	0.3	0.4
0.1	0.5	0.9
0.6	0.7	0.2

30   Enrichment • SECTION 2-4

86B

Name _____ Date _____

Multiplying and Dividing Decimals

To multiply decimals, multiply as with whole numbers. Then find the sum of the number of decimal places in the factors. This is the number of decimal places in the product.

To divide decimals, move the decimal point in the divisor to the right to get a whole number. Then move the decimal point in the dividend the same number of places. Annex zeros if necessary. Place a decimal point above the decimal point in the dividend and then divide as with whole numbers.

► **Example** _____

Find the product 7.38×2.1.

Solution

Estimate and compare to see if the answer is reasonable.

Actual		Estimate
7.38	← 2 places	7
× 2.1	← 1 place	× 2
738		14
1476		
15.498	← 3 places	

Since 15.498 is close to 14, the answer is reasonable.

► **Example** _____

Find the quotient $7.98 \div 2.1$.

Solution

Estimate and compare to see if the answer is reasonable.

Actual		Estimate
	3.8	4
2.1)7.9 8	← Move the decimal point 1 place.	2)8
6 3		8
1 6 8		
1 6 8		

Since 3.8 is close to 4, the answer is reasonable.

EXERCISES

Estimate to place the decimal point correctly.

1. $3.6 \times 2 = 72$ **7.2**   2. $20.3 \times 4.2 = 8526$ **85.26**   3. $56.1 \div 17 = 33$ **3.3**

Find each product. Estimate and compare to see if it is reasonable.

4. 3.4×12.8 **43.52**   5. 0.09×0.3 **0.027**   6. 121.6×0.02 **2.432**

7. 5.25×4.4 **23.1**   8. $1,000 \times 465.9$ **465,900**   9. 100×3.852 **385.2**

Find each quotient. Estimate and compare to see if it is reasonable.

10. $100.7 \div 26.5$ **3.8**   11. $660.8 \div 16$ **41.3**   12. $1.7028 \div 85.14$ **0.02**

13. $25.23 \div 0.52$ **45.25**   14. $45.07 \div 100$ **0.4507**   15. $75.1 \div 1,000$ **0.0751**

Reteaching • SECTION 2-5

31

Name _____ Date _____

Expressions and Decimals

A variable expression can be evaluated for decimal values of the variable.

► **Example** _____

Evaluate $2n + 7$ if $n = 0.6$.

Solution

$2n + 7$
$2 \times 0.6 + 7 = 1.2 + 7 = 8.2$

EXERCISES

Complete each table.

1.
n	$0.4 + n$
0.2	0.6
0.4	0.8
1.55	1.95
2.38	2.78

2.
x	$x - 3.8$
4.5	0.7
5.2	1.4
15.02	11.22
27	23.2

3.
a	$0.8a$
1.4	1.12
2.6	2.08
6.08	4.864
10.2	8.16

4.
w	$w \div 0.5$
5.8	11.6
19.3	38.6
3.078	6.156
0.02	0.04

5.
t	$\frac{t}{5} - 0.8$
15.2	6.8
20	9.2
100.55	49.475
500.2	249.3

6.
x	$4x + 5.1$
0.5	7.1
12.6	55.5
34.27	142.18
50.6	207.5

A delivery company charges $0.60 for the first pound of shipment weight plus $0.20 for each additional pound.

7. Write an expression to calculate the delivery cost of a shipment. **$0.6 + 0.2(n - 1)$**

8. How much will the company charge for a 20-lb shipment? **$4.40**

9. How much will the company charge for a 45-lb shipment? **$9.40**

10. The charge for one delivery was $37. How many pounds was the delivery? **183**

32

Enrichment • SECTION 2-5

Name _____ Date _____

Exponents and Scientific Notation

Multiples of a number can be written in **exponential form**. The base tells what factor is being multiplied. The exponent tells how many equal factors there are.

$6 \times 6 \times 6 = 6^3$ (Read "six to the third power.") ← exponent
← base

Very large and very small numbers can be written as the product of a number greater than or equal to 1 but less than 10 and a power of 10. A number expressed in this form is in **scientific notation.**

► **Example 1**

a. Write $3 \times 3 \times 3 \times 3$ in exponential form.

b. Write 7^5 in standard form.

Solution

a. $3 \times 3 \times 3 \times 3 = 3^4$ 4 factors

b. $7^5 = 7 \times 7 \times 7 \times 7 \times 7 = 16,807$ 5 factors

► **Example**

a. Write 5,600,000 in scientific notation.
b. Write 3.8×10^2 in scientific form.

Solution

a. Move the decimal point to the right to get a number greater than or equal to 1 and less than 10.

5,600,000. ← 6 places

Write 5,600,000 as the product of 5.6 and a power of 10 equal to the number of decimal places you moved the decimal point.

$5,600,000 = 5.6 \times 1,000,000 = 5.6 \times 10^6$

b. $3.8 \times 10^2 = 3.8 \times 100 = 380$

EXERCISES

Write in exponential form.

1. $5 \times 5 \times 5 \times 5 \times 5 \times 5$ **5^6**   2. $10 \times 10 \times 10 \times 10$ **10^4**

Write in standard form.

3. 7^3 **343**   4. 8^0 **1**   5. 2^6 **64**   6. 30^2 **900**

Write in scientific notation.

7. 3,700 **3.7×10^3**   8. 24,000,000 **2.4×10^7**

Write in standard form.

9. 1.6×10^4 **16,000**   10. 3.088×10^8 **308,800,000**

Reteaching • SECTION 2-6

33

Name _____ Date _____

The Solar System

1. The diameter of the Sun is about 865,500 mi. Write the diameter in scientific notation. **8.655×10^5**

2. The diameter of the largest planet, Jupiter, is 8.87×10^4 mi. Write the diameter in standard form. **88,700**

3. At a steady rate of 6 mi/h, how many days would it take a jogger to run around Jupiter at its equator if its circumference is about 2.786×10^5 mi? **about 1,935 d**

4. The diameter of the smallest planet, Mercury, is about 0.1 that of Neptune. If the diameter of Mercury is 3.032×10^3 mi, about what is that of Neptune? **3.032×10^4 or 30,320 mi**

5. The estimated diameter of the largest star, IRS_5, is 9,200,000,000 mi. Write the diameter in scientific notation. **9.2×10^9**

6. About how many times greater is the diameter of IRS_5 than that of the Sun? Write your estimate as a power of 10. **10^4**

Use the table at the left to answer the following questions.

DISTANCE OF PLANETS FROM SUN	
Planet	Average Distance (mi)
Neptune	2.794×10^9
Mercury	3.6×10^7
Jupiter	4.84×10^8
Venus	6.72×10^7
Saturn	8.87×10^8
Earth	9.3×10^7
Uranus	1.783×10^9
Mars	1.42×10^8
Pluto	3.674×10^9

7. Write the distance from Pluto to the Sun in standard form. **3,674,000,000**

8. Which planet is farther from the Sun, Saturn or Neptune? **Neptune**

9. How much farther is Jupiter from the Sun than Venus? **416,800,000 m or 4.168×10^8 m**

10. Write the distances in order from least to greatest.
3.6×10^7
6.72×10^7
9.3×10^7
1.42×10^8
4.84×10^8
8.87×10^8
1.783×10^9
2.794×10^9
3.674×10^9

34

Enrichment • SECTION 2-6

86C

Laws of Exponents

You can use the rules of exponents in simplifying expressions containing numbers in exponential form.

Product Rule: To multiply numbers with the same base, add the exponents:
$$a^m \times a^n = a^{m+n}$$

Quotient Rule: To divide numbers with the same base, subtract the exponents:
$$a^m \div a^n = a^{m-n}$$

Power Rule: To raise an exponential number to a power, multiply the exponents:
$$(a^m)^n = a^{m \times n}$$

▶ **Example 1**

Simplify.

a. $3^3 \times 3^2$ b. $5^4 \div 5^2$

Solution

a. $3^3 \times 3^2 = 3^{3+2} = 3^5$ b. $5^4 \div 5^2 = 5^{4-2} = 5^2$
$3^3 \times 3^2 = (3 \times 3 \times 3) \times (3 \times 3) = 3^5$ $5^4 \div 5^2 = \frac{5 \times 5 \times 5 \times 5}{5 \times 5} = 5^2$

▶ **Example 2**

Simplify $(4^2)^3$.

Solution

$(4^2)^3 = 4^{2 \times 3} = 4^6$
$(4^2)^3 = 4^2 \times 4^2 \times 4^2 = (4 \times 4) \times (4 \times 4) \times (4 \times 4) = 4^6$

EXERCISES

Use the product rule to simplify.

1. $2^6 \times 2^8$ **2^{14}**   2. $5^3 \times 5^6$ **5^9**
3. $10^4 \times 10^7$ **10^{11}**   4. $4^5 \times 4$ **4^6**

Use the quotient rule to simplify.

5. $7^8 \div 7$ **7^7**   6. $2^3 \div 2^2$ **2**   7. $4^6 \div 4^6$ **1**   8. $10^9 \div 10^6$ **10^3**

Use the power rule.

9. $(2^3)^4$ **2^{12}**   10. $(5^2)^0$ **1**   11. $(7^2)^{10}$ **7^{20}**   12. $(10^3)^6$ **10^{18}**

13. Which of the following are equal to 2^{10}?   14. Which of the following are equal to twice 2^{16}?
a. $2^2 \times 2^5$ b. $(2^5)^5$ c. $2^{10} + 2^0$ d. $2^{20} \div 2^2$ **b, c**   a. 4^{16} b. 2^{17} c. 131,072   d. 2^{32} **b, c**

Using Scientific Notation

Scientific notation and the rules of exponents can be used to make computation with large numbers easier than it would be if you had to multiply or divide the numbers using long division.

▶ **Example 1**

Find $3,744,000 \times 18,350$. Write the product in scientific notation and then in standard form.

Solution
First, write each number in scientific notation.

$3,744,000 = 3.744 \times 10^6$ $18,350 = 1.835 \times 10^4$

Next, find the product of the decimals and the powers of 10.

$(3.744 \times 10^6) \times (1.835 \times 10^4) = (3.744 \times 1.835) \times (10^6 \times 10^4)$
$= 6.87024 \times 10^{10} = 68,702,400,000$

▶ **Example 2**

Divide 175,421,500 by 259,124.

Solution
$175,421,500 = 1.754215 \times 10^8$ $259,124 = 2.59124 \times 10^5$

$\frac{1.754215 \times 10^8}{2.59124 \times 10^5} = \frac{1.754215}{2.59124} \times \frac{10^8}{10^5} = 0.6769789 \times 10^3$

The quotient is not in scientific notation. Multiply by $\frac{10}{10}$.

$0.6769789 = 6.769789 \times 10^3 \times \frac{10}{10} = \frac{(6.769789 \times 10^3)}{10}$
$= 6.769789 \times 10^2$

EXERCISES

Use you calculator to find each product or quotient. Write your answer in scientific notation.

1. $(4.7 \times 10^3) \times (2.05 \times 10^2)$ **9.635×10^5**   2. $\frac{5.85 \times 10^{12}}{9 \times 10^7}$ **6.5×10^4**
3. $(3.2 \times 10^8)(4.07 \times 10^2)$ **13.024×10^{10}**   4. $(1.112 \times 10)(8 \times 10^5)$ **8.896×10^6**
5. $\frac{8.76 \times 10^4}{1.2 \times 10^2}$ **7.3×10^2**   6. $\frac{5.29 \times 10^{10}}{2.3}$ **2.3×10^{10}**
7. $(9.7 \times 10^6)(3.1 \times 10^5)$ **3.007×10^{12}**   8. $\frac{8.442 \times 10^6}{2.1 \times 10^2}$ **4.02×10^4**

Write each number in scientific notation. Then find the product or quotient. Use your calculator. Then write your answer in standard form.

9. $76,800,000 \times 324,000$ **2.48832×10^{13}; 24,883,200,000,000**
10. $4,343,000,000 \div 43,000$ **1.01×10^5; 101,000**
11. $\frac{71,500 \times 3,864,000}{1,100,000}$ **2.5116×10^5; 251,160**

Order of Operations

It would be very confusing if people got different answers when evaluating an expression. To avoid that problem, operations are always done in a specified order.

1. **Parentheses**—perform all calculations within grouping symbols first.
2. **Exponents**—do all calculations with exponents.
3. **Multiplication and division**—do all multiplications and divisions in order from left to right.
4. **Addition and subtraction**—add or subtract in order from left to right.

▶ **Example 1**

Simplify $21 - 7 \times 2 + 4$.

Solution

    Multiply.   Subtract.
$21 - 7 \times 2 + 4 = 21 - 14 + 4$
              $= 7 + 4$ ← Add.
              $= 11$

▶ **Example 2**

Simplify $3^2 \times (7 + 1) - 8 \times 2 \div 4$.

Solution

     Parentheses
$3^2 \times (7 + 1) - 8 \times 2 \div 4 = 3^2 \times 8 - 8 \times 2 \div 4$ ← Exponents.
                       $= 9 \times 8 - 8 \times 2 \div 4$ ← Multiply.
                       $= 72 - 16 \div 4$ ← Divide.
                       $= 72 - 4$ ← Subtract.
                       $= 68$

EXERCISES

What operation should you do first?

1. $7 \times 3 - 5 + 2$ **multiply, 7×3**   2. $74 - (2 \times 3)^2 + 1$ **multiply, 2×3**
3. $56 \div (4 + 4) \times 3$ **add, $4 + 4$**   4. $2^3 + 5 \times 3$ **simplify, 2^3**
5. $210 - 7 \times 8 \div 4$ **multiply, 7×8**   6. $25 - 16 + 2 - 8$ **subtract, $25 - 16$**

Evaluate each expression.

7. $7 \times 3 - 5 + 2$ **18**   8. $74 - (2 \times 3)^2 + 1$ **39**
9. $56 \div (4 + 4) \times 3$ **21**   10. $2^3 + 5 \times 3$ **23**
11. $2 \times 2 + 0 \times (18 + 56)$ **4**   12. $(25 - 8) + 6^2 - 15$ **38**

One on One

1. What is the value of $(1 + 1 + 1)^{1 \times 1 + 1}$?   **9**

2. Can you use six 1's in another way to get 9?   **$(1 + 1 + 1)(1 + 1 + 1)$, for example**

3. Use six 1's to write an expression whose value is 1.   **$1 + 1 - 1 + 1 - 1 \div 1$, for example**

4. Use six 1's to write an expression whose value is 2.   **$1 + 1 + 1 - 1 + 1 - 1$, for example**

5. Use six 1's to write an expression whose value is 3.   **$(1 + 1 + 1) \times 1 \times 1 \times 1$, for example**

6. Use six 1's to write an expression whose value is 4.   **$(1 + 1)(1 + 1)^{1 \times 1}$, for example**

7. Use six 1's to write an expression with each given value. **Sample answers are given.**

   5   **$(1 + 1 + 1 + 1)^1$**   6   **$1 + 1 + 1 + 1 + 1 + 1$**
   7   **$(1 + 1)^{1 + 1 + 1} - 1$**   8   **$(1 + 1 + 1)^{1 + 1} - 1$**
   9   **$(1 + 1 + 1)(1 + 1 + 1)$**   10   **$(1 + 1 + 1)^{1 + 1} + 1$**

8. Is there more than one correct expression for each number in Ex. 7?   **yes**

9. What other numbers can you write using six 1's?

   for ex, $(1 + 1 + 1)^{1 + 1 + 1} = 27$; $(1 + 1)^{1 + 1 + 1 + 1} = 16$

10. Use five 2's to write each of the following numbers.

   7   **$2 + 2 + 2 + 2 - 2$**   8   **$2 \times 2 \times 2 + 2 - 2$**
   9   **$2 \times 2 \times 2 + 2 \div 2$**   10   **$(2 + 2 + 2 \times 2) \times 2$**

11. Express 100 as a product of two factors involving the numbers 1–9.

   $(1 + 2 - 3 + 4)(9 + 8 + 7 + 6 - 5)$

Use the numbers 1, 2, 4, and 8 to complete each of the following.

12. $\blacksquare + (\blacksquare \times \blacksquare)(\blacksquare + \blacksquare) = 113$ **$1 + (2 \times 4)(6 + 8)$**
13. $\blacksquare^{\blacksquare} - \blacksquare + (\blacksquare \times \blacksquare) = 63$ **$4^2 - 1 + (6 \times 8)$**
14. $\blacksquare \times (\blacksquare + \blacksquare) - \blacksquare + \blacksquare = 43$ **$8 \times (2 + 4) - 6 + 1$**
15. $\blacksquare(\blacksquare^{\blacksquare} \div \blacksquare - \blacksquare) = 64$ **$8(6^2 + 4 - 1)$**

86D

Name _____ Date _____

Properties of Operations

Commutative Property	The order of addends or factors does not affect the answer.	$a + b = b + a$ $a \times b = b \times a$	$3 + 2 = 5; 2 + 3 = 5$ $7 \times 4 = 28; 4 \times 7 = 28$
Associative Property	The grouping of addends or factors does not affect the answer.	$a + (b + c) = (a + b) + c$ $a \times (b \times c) = (a \times b) \times c$	$3 + (4 + 5) = (3 + 4) + 5$ $4 \times (2 \times 3) = (4 \times 2) \times 3$
Identity Property of Addition	The sum of a number and 0 is the number.	$a + 0 = a$	$156 + 0 = 156$
Identity Property of Multiplication	The product of a number and 1 is the number.	$a \times 1 = a$	$375 \times 1 = 375$
Property of 0 for Multiplication	The product of a number and 0 is 0.	$a \times 0 = 0$	$210 \times 0 = 0$

EXERCISES

Match each equation with the property illustrated.

1. $5 \times (2 \times 12) = 10 \times 12$ **d**
2. $2 \times 7 = 7 \times 2$ **b**
3. $0 \times 746 = 0$ **g**
4. $0 + 582 = 582$ **e**
5. $1 \times 312 = 312$ **f**
6. $15 + (21 + 7) = 15 + (7 + 21)$ **a**
7. $(8 + 3) + 17 = 8 + (3 + 17)$ **c**

a. commutative property of addition
b. commutative property of multiplication
c. associative property of addition
d. associative property of multiplication
e. identity property of addition
f. identity property of multiplication
g. property of 0 for multiplication

Complete.

8. **10** $\times 73 = 73 \times 10$
9. $(5.6 + 8.2) + $ **1.8** $ = 5.6 + (8.2 + 1.8)$
10. $837 \times $ **0** $ = 0$
11. $6 \times (5 \times 12) = (6 \times $ **5** $) \times 12$

Evaluate, using mental math. Name the property or properties you used.

12. $56 \times 21 \times 0 \times 17$ **0; prop of 0**
13. $0.36 + 0$ **0.36, iden prop of +**
14. $7 \times 6 \times 5$ **210; assoc prop**
15. $20 \times 17 \times 5$ **1,700; comm & assoc**

Reteaching • SECTION 2-9   39

Name _____ Date _____

Adding and Multiplying on a Clock

You can have a number system with fewer than 10 digits. The dial at the right shows a system with 4 digits, 0, 1, 2, 3. You can use the dial to perform clock addition and multiplication in a 4-digit system.

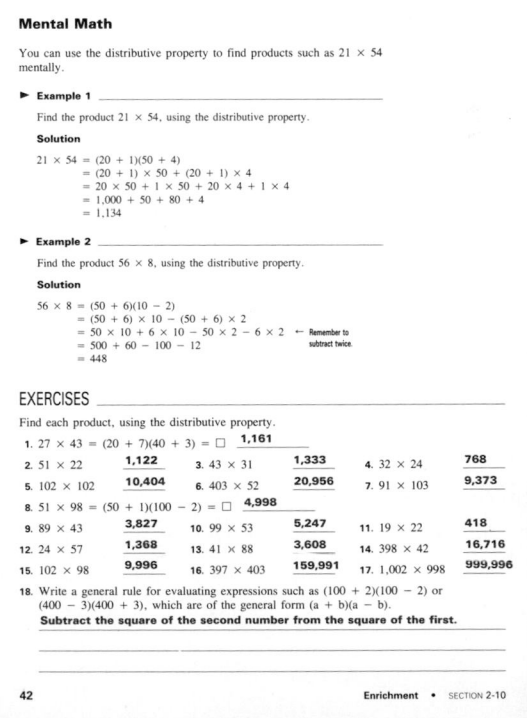

▶ **Example 1**

Find $3 \oplus 2$.

Solution

Begin at 3 and move 2 units around the clock. You arrive at 1. So, in the 4-digit number system, $3 \oplus 2 = 1$.

▶ **Example 2**

Find $2 \otimes 3$.

Solution

Begin at 2. Move 2 units and then move 2 more units around the clock. You arrive at 2. So, $2 \otimes 3 = 2$.

EXERCISES

1. Use the clock above to complete this table for clock addition in the 4-digit system.

$\oplus$	0	1	2	3
0	0	1	2	3
1	1	2	**3**	**0**
2	2	**3**	**0**	**1**
3	**3**	**0**	**1**	**2**

2. Use the clock above to complete this table for clock multiplication in the 4-digit system.

$\otimes$	0	1	2	3
0	0	0	0	0
1	0	1	**2**	**3**
2	0	**2**	**0**	**2**
3	**0**	**3**	**2**	**1**

Find each of the following sums.

3. $(2 \oplus 3) \oplus 2$ **3**
4. $2 \oplus (3 \oplus 2)$ **3**
5. $(2 \oplus 2) \oplus 2$ **2**
6. $2 \oplus (2 \oplus 2)$ **2**
7. $(1 \oplus 0) \oplus 3$ **0**
8. $1 \oplus (0 \oplus 3)$ **0**
9. Does $\otimes$ appear to be associative? **Yes**

Find each of the following products.

10. $(2 \otimes 3) \otimes 2$ **0**
11. $2 \otimes (3 \otimes 2)$ **0**
12. $(1 \otimes 2) \otimes 3$ **2**
13. $1 \otimes (2 \otimes 3)$ **2**
14. $(3 \otimes 2) \otimes 3$ **2**
15. $3 \otimes (2 \otimes 3)$ **2**
16. Does $\otimes$ appear to be associative? **Yes**

Answer these questions for the 5-digit system.

17. Does $\oplus$ appear to be commutative? Give an example. **Yes; 2 + 3 = 1 and 3 + 2 = 1**
18. Does $\otimes$ appear to be commutative? Give an example. **Yes; 3 × 2 = 2 and 2 × 3 = 2**

40   Enrichment • SECTION 2-9

Name _____ Date _____

Distributive Property

If you find the value of the expressions $6 \times (2 + 3)$ and $6 \times 2 + 6 \times 3$, you will find they are the same.

$6 \times (2 + 3) = 6 \times 5 = 30$
$(6 \times 2) + (6 \times 3) = 12 + 18 = 30$

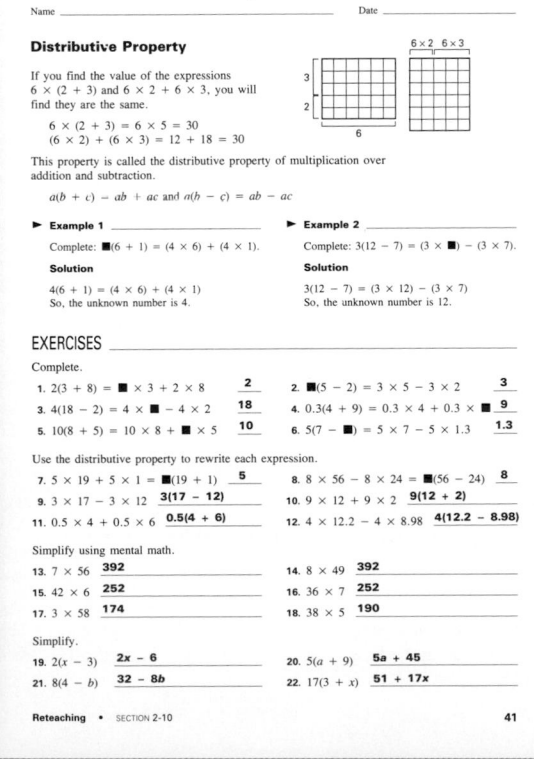

This property is called the distributive property of multiplication over addition and subtraction.

$a(b + c) = ab + ac$ and $a(b - c) = ab - ac$

▶ **Example 1**

Complete: $\blacksquare(6 + 1) = (4 \times 6) + (4 \times 1)$.

Solution

$4(6 + 1) = (4 \times 6) + (4 \times 1)$
So, the unknown number is 4.

▶ **Example 2**

Complete: $3(12 - 7) = (3 \times \blacksquare) - (3 \times 7)$.

Solution

$3(12 - 7) = (3 \times 12) - (3 \times 7)$
So, the unknown number is 12.

EXERCISES

Complete.

1. $2(3 + 8) = \blacksquare \times 3 + 2 \times 8$ **2**
2. $\blacksquare(5 - 2) = 3 \times 5 - 3 \times 2$ **3**
3. $4(18 - 2) = 4 \times \blacksquare - 4 \times 2$ **18**
4. $0.3(4 + 9) = 0.3 \times 4 + 0.3 \times \blacksquare$ **9**
5. $10(8 + 5) = 10 \times 8 + \blacksquare \times 5$ **10**
6. $5(7 - \blacksquare) = 5 \times 7 - 5 \times 1.3$ **1.3**

Use the distributive property to rewrite each expression.

7. $5 \times 19 + 5 \times 1 = \blacksquare(19 + 1)$ **5**
8. $8 \times 56 - 8 \times 24 = \blacksquare(56 - 24)$ **8**
9. $3 \times 17 - 3 \times 12$ **3(17 − 12)**
10. $9 \times 12 + 9 \times 2$ **9(12 + 2)**
11. $0.5 \times 4 + 0.5 \times 6$ **0.5(4 + 6)**
12. $4 \times 12.2 - 4 \times 8.98$ **4(12.2 − 8.98)**

Simplify using mental math.

13. 7×56 **392**
14. 8×49 **392**
15. 42×6 **252**
16. 36×7 **252**
17. 3×58 **174**
18. 38×5 **190**

Simplify.

19. $2(x - 3)$ **2x − 6**
20. $5(a + 9)$ **5a + 45**
21. $8(4 - b)$ **32 − 8b**
22. $17(3 + x)$ **51 + 17x**

Reteaching • SECTION 2-10   41

Name _____ Date _____

Mental Math

You can use the distributive property to find products such as 21×54 mentally.

▶ **Example 1**

Find the product 21×54, using the distributive property.

Solution

$21 \times 54 = (20 + 1)(50 + 4)$
$= (20 + 1) \times 50 + (20 + 1) \times 4$
$= 20 \times 50 + 1 \times 50 + 20 \times 4 + 1 \times 4$
$= 1,000 + 50 + 80 + 4$
$= 1,134$

▶ **Example 2**

Find the product 56×8, using the distributive property.

Solution

$56 \times 8 = (50 + 6)(10 - 2)$
$= (50 + 6) \times 10 - (50 + 6) \times 2$
$= 50 \times 10 + 6 \times 10 - 50 \times 2 - 6 \times 2$ ← Remember to subtract twice.
$= 500 + 60 - 100 - 12$
$= 448$

EXERCISES

Find each product, using the distributive property.

1. $27 \times 43 = (20 + 7)(40 + 3) = \square$ **1,161**
2. 51×22 **1,122**
3. 43×31 **1,333**
4. 32×24 **768**
5. 102×102 **10,404**
6. 403×52 **20,956**
7. 91×103 **9,373**
8. $51 \times 98 = (50 + 1)(100 - 2) = \square$ **4,998**
9. 89×43 **3,827**
10. 99×53 **5,247**
11. 19×22 **418**
12. 24×57 **1,368**
13. 41×88 **3,608**
14. 398×42 **16,716**
15. 102×98 **9,996**
16. 397×403 **159,991**
17. $1,002 \times 998$ **999,996**
18. Write a general rule for evaluating expressions such as $(100 + 2)(100 - 2)$ or $(400 - 3)(400 + 3)$, which are of the general form $(a + b)(a - b)$.
   Subtract the square of the second number from the square of the first.

42   Enrichment • SECTION 2-10

86E

Name _____ Date _____

Problem Solving Strategies: Find a Pattern

In solving some problems, sometimes you can organize the information in a table that will show a pattern.

► **Example**

Every day, Maria saves twice as many pennies as she did the day before. If Maria begins the year by putting away one penny, how many pennies will she have by January 10?

Solution

Make a table.

Date	1	2	3	4	5	6	7	8	9	10
Number	1	2	4	8	16	32	64	128	256	512

Add the total number of pennies saved.

$1 + 2 + 4 + 8 + 16 + 32 + 64 + 128 + 256 + 512 = 1,023$

Maria will have saved 1,023 pennies.

EXERCISES

Look for a pattern and choose the rule from those given. Then write the unknown numbers.

1. 0.24, 0.48, 0.72, ■, ■ **Rule b; 0.96, 1.2**
   a. Double the previous number.
   b. Add 0.24 to the previous number.
   c. Add 0.024 to the previous number.

2. 12.6, 12.4, ■, 12.0, ■ **Rule b; 12.2, 11.8**
   a. Add 0.2 to the previous number.
   b. Subtract 0.2 from the previous number.
   c. Subtract 0.02 from the previous number.

3. Find each product and look for a pattern.

   $10 \times 10 =$ **100** $15 \times 15 =$ **225**

   $9 \times 11 =$ **99** $14 \times 16 =$ **224**

4. If 45×45 is 2,025, what is 44×46? **2,024**

5. If 33×35 is 1,155, what is 34×34? **1,156**

6. Nikia put $10 in her savings account in January, $13 in February, $16 in March, and so on. If the pattern continues, how much money will she put into her account in December? How much will she have put into the account for the entire year?

   $43; $318

Name _____ Date _____

More Number Patterns

Use the patterns illustrated in Exercises 1 and 2 to predict your answers. Then check your predictions.

EXERCISES

1. $11^2 = 121$
   $111^2 = 12,321$
   $1,111^2 = 1,234,321$

   Predict:

   $11,111^2 =$ **123,454,321**

   $111,111^2 =$ **12,345,654,321**

   $1,111,111^2 =$ **1,234,567,654,321**

2. $12,345,679 \times 9 = 111,111,111$
   $12,345,679 \times 18 = 222,222,222$
   $12,345,670 \times 27 = 333,333,333$

   Predict

   $12,345,679 \times 81 =$ **999,999,999**

   $12,345,679 \times 243 =$ **2,999,999,997**

Look at the following pattern of powers of 11. Write the missing powers.

3. $11^0 = 1$ $11^1 = 11$ $11^2 =$ **121** $11^3 =$ **1,331**

What would you predict will be

4. the first two digits of 11^4?   5. the first two digits of 11^5?
   14   **16**

6. the first two digits of 11^6?   7. the first two digits of 11^7?
   17   **19**

8. the first two digits of 11^8?   9. the first two digits of 11^9?
   21   **25**

10. Check your predictions. Does the pattern continue? Explain.

    No; the pattern changes with 11^5; $11^5 = 161,051$, $11^6 = 1,771,561$,

    $11^7 = 19,487,171$, $11^8 = 214,358,881$, and $11^9 = 2,357,947,691$.

Name _____ Date _____

Calculator Activity: Evaluating Expressions

The $\boxed{x^2}$ and $\boxed{x^y}$ or $\boxed{y^x}$ keys on a scientific calculator can help you evaluate expressions.

► **Example 1**

Use a calculator to evaluate $x^3 + yz$ if $x = 11$, $y = 14$, and $z = 25$.

Solution

First, write the expression using the numbers in place of the variables. Also, write in any multiplication signs.

$$x^3 + yz = 11^3 + 14 \times 25$$

Then use a calculator.

$11 \boxed{x^y} 3 \boxed{+} 14 \boxed{\times} 25 \boxed{=} 1681$

When the result is too large to fit the display, the calculator automatically displays it using scientific notation.

► **Example 2**

Use a calculator to evaluate $a^{12}b^{15}$ if $a = 5$ and $b = 4$.

Solution

First rewrite the expression.

$$a^{12}b^{15} = 5^{12} \times 4^{15}$$

Then, use a calculator.

$5 \boxed{x^y} 12 \boxed{\times} 4 \boxed{x^y} 15 \boxed{=} 2.62144^{17}$

The display shows that $5^{12} \times 4^{15} = 2.62144 \times 10^{17}$.

EXERCISES

Use your calculator to evaluate each expression. Write your answer to the nearest hundredth.

1. $cd + d^2$ if $c = 23$ and $d = 18$ 2. $3.14g - 2.7h$ if $g = 2.5$ and $h = 5$
   738   **−5.65**

3. $3c^4 \times 8d$ if $c = 12$ and $d = 18$ 4. $s^4 + t^5$ if $s = 37$ and $t = 29$
   8,957,952   **22,385,310**

5. $\frac{m^2 n^3}{p^4}$ if $m = 4$, $n = 12$, and $p = 8$ 6. $5w^6 x^8 + 5x^5$ if $w = 1.5$ and $x = 6$
   972   **95,698,260**

7. $\frac{w^2 z^5}{x^3 y^3}$ if $x = -3$, $y = 5$, $w = -2$,   8. $\frac{a^1 b^3 - a^2 b}{a}$ if $a = 15$ and $b = 8$
   and $z = 10$
   −118.52   **7,372,680**

Name _____ Date _____

Computer Activity: Find a Pattern

If you can find a pattern in the numbers in a problem, you can use the rule of that pattern in a computer program. The computer can then quickly and accurately generate all or some of the steps in the pattern.

For example, in Exercise 11 on page 81, the pattern is to increase each month's amount by 3. This program will find the amount deposited in a specific month (December this time) and the total amount deposited.

```
10  INPUT "WHAT IS THE STARTING          Asks for the first value
    NUMBER? ";N: PRINT                    in the pattern
20  M = N
30  FOR X = 1 TO 11
40  M = M + 3                             Adds 3 to the value
50  T = T + M
60  NEXT X                                Keeps a running total
70  PRINT "DECEMBER AMOUNT",M: PRINT
80  PRINT "TOTAL AMOUNT",T + 5
```

EXERCISES

1. Why does X go from 1 to 11 instead of 1 to 12?

   The first value is entered as the beginning amount so you need one fewer
   loops than months.

2. Why is 5 added to T in line 70?

   The loops total months February through December so you have to add the

   first value for January's amount.

3. What would you change to find the amounts for June? Will they be half of those for December? Why or why not? Make the changes and RUN the program to check your answer.

   30 FOR X = 1 to 5

   70 ? "JUNE",M: ?   No because the previous month's total is used as the
   first addend.

4. What changes would you need to make to answer the questions in Exercise 12? Make them and RUN to check.

   30 FOR X = 1 to 4

   40 M = 4 * M

   70 PRINT "FRIDAY", M:PRINT

   80 ? "TOTAL",T + 2

ACHIEVEMENT TEST
Name _____

Exploring Whole Numbers and Decimals Date _____

CHAPTER 2 FORM A

MATH MATTERS BOOK 1
Chicha Lynch
Eugene Olmstead

SOUTH-WESTERN PUBLISHING CO.

SCORING RECORD	
Possible	Earned
42	

Find each answer. Estimate and compare to see if your answer is reasonable.

1. $4,793 - 1,749$ **3,044**
2. 84×37 **3,108**
3. $74 \times 5,183$ **383,542**
4. $39,846 \div 87$ **458**
5. $7.49 + 12.6 + 0.736$ **20.826**
6. $21.7 - 0.674$ **21.026**
7. 0.84×0.29 **0.2436**
8. 9×7.8 **70.2**
9. $1,305.72 \div 27$ **48.36**

Evaluate each expression. Let $a = 3$ and $b = 5$.

10. $a + 21$ **24**
11. $2b - 3$ **7**
12. $b + 3a$ **14**
13. $14a - 2b$ **32**

Write in exponential form.

14. $7 \times 7 \times 7 \times 7$ **7^4**
15. $9 \times 9 \times 9$ **9^3**
16. $4 \times 4 \times 4 \times 4 \times 4$ **4^5**

Write in standard form.

17. 3^3 **27**
18. 2^5 **32**
19. 5^4 **625**
20. 4^4 **256**

Write in scientific notation.

21. $8,784$ **8.784×10^3**
22. 726 **7.26×10^2**
23. $524,800$ **5.248×10^5**
24. $14,781,253$ **1.4781253×10^7**

Write each answer in exponential form.

25. $8^4 \times 8^3$ **8^7**
26. $7^{14} \div 7^7$ **7^7**
27. $x^8 \div x^5$ **x^3**
28. $y^3 \times y^8$ **y^{11}**

Simplify.

29. $5 \times 8 - 21 \div 3$ **33**
30. $2 \times (10 - 6) + 9$ **17**
31. $(3 + 8)^2 \times 4$ **484**

2A-1

Name _____   Date _____

Complete.

32. $8(\blacksquare - 2) = (8 \times 9) - (8 \times 2)$ **9**
33. $\blacksquare(3 + 9) = (5 \times 3) + (5 \times 9)$ **5**
34. $8.4 + (\blacksquare + 3.5) = (8.4 + 3) + 3.5$ **3**
35. $\blacksquare \times 3.459 = 3.459$ **1**
36. $9.82 \times \blacksquare = 0$ **0**
37. $4.56 \times 8.41 = 8.41 \times \blacksquare$ **4.56**

Simplify.

38. $4(5 + a)$ **$20 + 4a$**
39. $5(b - 2)$ **$5b - 10$**
40. $4(9 - x)$ **$36 - 4x$**
41. $6(y + 5)$ **$6y + 30$**

Solve.

42. Mara put \$8 into her savings account in June, \$12 in July, and \$16 in August. If Mara continues this pattern, how much money will she put into her account in December?

\$32

2A-2

ACHIEVEMENT TEST
Name _____

Exploring Whole Numbers and Decimals Date _____

CHAPTER 2 FORM B

MATH MATTERS BOOK 1
Chicha Lynch
Eugene Olmstead

SOUTH-WESTERN PUBLISHING CO.

SCORING RECORD	
Possible	Earned
42	

Find each answer. Estimate and compare to see if your answer is reasonable.

1. $5,843 - 2,944$ **2,899**
2. 49×52 **2,548**
3. $67 \times 6,214$ **416,338**
4. $17,088 \div 48$ **356**
5. $82.4 + 12.2 + 0.467$ **95.067**
6. $16.07 - 0.674$ **15.396**
7. 0.76×0.74 **0.5624**
8. 7×8.63 **60.41**
9. $1,000.16 \div 19$ **52.64**

Evaluate each expression. Let $a = 2$ and $b = 6$.

10. $a + 9$ **11**
11. $2b - 5$ **7**
12. $b + 3a$ **12**
13. $15a - 2b$ **18**

Write in exponential form.

14. $6 \times 6 \times 6$ **6^3**
15. $3 \times 3 \times 3 \times 3$ **3^4**
16. $8 \times 8 \times 8 \times 8 \times 8 \times 8 \times 8$ **8^7**

Write in standard form.

17. 4^4 **256**
18. 2^7 **128**
19. 3^5 **243**
20. 6^3 **216**

Write in scientific notation.

21. $30,756$ **3.0756×10^4**
22. $8,240$ **8.24×10^3**
23. $735,624$ **7.35624×10^5**
24. $23,438,600$ **2.34386×10^7**

Write each answer in exponential form.

25. $7^6 \times 7^4$ **7^{10}**
26. $3^{12} \div 3^4$ **3^8**
27. $x^{17} \div x^2$ **x^{15}**
28. $(y^4)^3$ **y^{12}**

Simplify.

29. $5 + 8 \div 2 - 3$ **6**
30. $3 \times (14 - 8) + 9$ **27**
31. $(5 + 2)^2 \times 4$ **196**

2B-1

Name _____   Date _____

Complete.

32. $7(\blacksquare - 3) = (7 \times 4) - (7 \times 3)$ **4**
33. $\blacksquare(5 + 7) = (4 \times 5) + (4 \times 7)$ **4**
34. $6.9 + (\blacksquare + 5.4) = (6.9 + 4) + 5.4$ **4**
35. $\blacksquare \times 6.542 = 6.542$ **1**
36. $76.4 \times \blacksquare = 0$ **0**
37. $9.83 \times 3.47 = 3.47 \times \blacksquare$ **9.83**

Simplify.

38. $3(6 + a)$ **$18 + 3a$**
39. $7(b - 3)$ **$7b - 21$**
40. $9(3 - x)$ **$27 - 9x$**
41. $4(y + 8)$ **$4y + 32$**

Solve.

42. Juan put \$15 into his savings account in January, \$18 in February, \$21 in March, and \$24 in April. If he continues this pattern, how much money will Juan put into his savings account in July?

\$33

2B-2

86G

Teacher's Notes

CHAPTER 3 SKILLS PREVIEW

1. Refer to the figure below. Make a statement about line segments *HJ* and *JK*. Then check to see whether your statement is true or false.

Segments *HJ* and *JK* have the same length.

2. Choose three single-digit numbers. Multiply the first by 2. Add 1. Multiply by 5. Add the second number. Multiply by 10. Add the third number and subtract 50. What number do you get? Repeat this process several times with different sets of digits. Make a conjecture about the result of the process. **The result is a three-digit number consisting of the three original numbers from left to right in the order chosen.**

3. Give an example to show why the following conjecture is false:

   The product of two one-digit whole numbers is a two-digit whole number. **Answers may vary. Possible answers: 1 × 1 = 1; 2 × 3 = 6**

4. Write two *if–then* statements, using the following two sentences:

   The temperature of the water is 212°F.
   The water will boil. **If the temperature of the water is 212°F, then the water will boil.**
   If the water will boil, then the temperature of the water is 212°F.

For each of the following statements, write *true* or *false*. If false, give a counterexample.

5. If a number is divisible by 6, then it is divisible by 12. **false: 18**

6. If a woman lives in Dallas, then she lives in Texas. **true**

Determine whether the following arguments are *valid* or *invalid*.

7. If two numbers are odd,
   then their sum is even.
   The sum of *a* and *b* is even.
   _____
   Therefore, *a* and *b* are odd. **invalid**

8. If a plant is a cactus, then
   it has thorns.
   This plant is a cactus.
   _____
   Therefore, this plant has thorns. **valid**

9. In a survey of 150 students, it was found that 83 participated in sports, 62 participated in chorus, and 70 participated in band. There were 56 who participated in chorus and band, 12 who participated in sports and chorus, and 15 who participated in sports and band. There were 7 who participated in all three. How many students did *not* participate in sports or chorus or band? **11 students**

10. Max, Bert, and Laura each play one sport. One plays football, one plays basketball, and one plays baseball. Bert is a poor student. Laura is an only child and a cheerleader for the football team. Each basketball player has a brother or sister, and all of them are good students. What sport does each person play? **Max: basketball; Bert: football; Laura: baseball**

3 SKILLS PREVIEW

Introduction The purpose of this Skills Preview is to assess students' abilities on all the major objectives of Chapter 3. Test results may be used
- to determine those topics which may need only to be reviewed and those topics which need to be more carefully developed;
- for class placement;
- in prescribing for individual differences.

OVERVIEW

In this chapter, students examine optical illusions, solve problems, and draw conclusions using inductive reasoning, deductive reasoning, and other logical thinking skills. Students make and test the truth or falsity of statements, make and test inductive conjectures, write and test the truth or falsity of conditional statements, identify valid and invalid forms of deductive reasoning, make Venn diagrams and tables to solve logic problems, and solve nonroutine problems using a variety of logical approaches.

SPECIAL CONCERNS

Every day, and throughout their lives, students will be confronted with situations in which they must make decisions and solve problems. Thus, students need to be provided with a variety of problem solving strategies and instilled with a feeling of confidence in their ability to handle problems. This requires that the teacher have patience, communicate a confidence that *all* students can achieve a significant degree of competence in problem solving, and have skill in asking questions, rather than in giving specific instructions to be emulated.

VOCABULARY

conclusion   hypothesis
conditional statement   inductive reasoning
conjecture   invalid argument
counterexample   statement
deductive reasoning   Venn diagram

MATERIALS

calculators   rulers
masking tape   tracing paper
scissors

Optional:
3-qt and 5-qt containers   cubes

BULLETIN BOARD

Have students post their solutions to the *Problem of the Day* on the bulletin board. Each problem should be nonroutine, requiring students to use either inductive reasoning, deductive reasoning, or both. In the *Clues and Hints* envelope, include questions and ideas which may help students solve the problem. Write each one on its own index card, explaining that the students may refer to only one card at a time. In the *Possible Strategies* envelope, include strategies which may be used to solve the problem.

INTEGRATED UNIT 1

The skills and concepts involved in Chapters 1–4 are included within the special Integrated Unit 1 entitled "Voting and Citizenship." This unit is in the Teacher's Edition beginning on page 150H. Worksheets for this integrated unit appear in the Enrichment Activities booklet, pages 47–49.

TECHNOLOGY CONNECTIONS
- Computer Worksheets, 59–60
- Calculator Worksheet, 61–62
- MicroExam, Apple Version
- MicroExam, IBM Version

- *Creature Cube,* Sunburst Communications
- *Discovery! Experiences with Scientific Reasoning,* Milliken Publishing Co.

TECHNOLOGY NOTES

Computer logic programs, such as those listed in Technology Connections, challenge students' problem solving and logical-reasoning abilities. Some programs require the use of inductive and deductive reasoning through search-and-find activities, such as mysteries and mazes. Other programs build students' inference skills by involving them in the development and testing of hypotheses.

PLANNING GUIDE

SECTIONS	TEXT PAGES	ASSIGNMENTS			
		BASIC	AVERAGE	ENRICHED	
Chapter Opener/Decision Making	88–89				
3–1 Optical Illusions	90–93	1–2, 3–5, PSA 1–5	1–2, 3–7, PSA 1–6	1–2, 5–7, 8, PSA 1–7	
3–2 Inductive Reasoning	94–97	1–7, 9–10, 12	3–8, 10–15	7–8, 9–11, 12–15	
3–3 Deductive Reasoning	98–101	1–4, 7–9, 13, 15, 17	1–6, 7–11, 13–15, 17	2–6, 7–16, 17–18	
3–4 Problem Solving Skills	102–103	1–3, 5–6	1–8	3–10	
3–5 Problem Solving Strategies	104–105	1–4	1–5	2–5	
3-6 Brainteasers	106–109	1–3, 5–6, 8–11, 15	1–4, 5–15, 16	3–4, 5–15, 16–17	
Technology	95, 96	✔	✔	✔	

ASSESSMENT				
Skills Preview	87	All	All	All
Chapter Review	110–111	All	All	All
Chapter Test	112	All	All	All
Cumulative Review	113	All	All	All
Cumulative Test	114	All	All	All

ADDITIONAL RESOURCES			
RETEACHING	**ENRICHMENT**	**TECHNOLOGY**	**TRANSPARENCY**
3–1	3–1		
3–2	3–2		
3–3	3–3		
3–4	3–4		TM 13
3–5	3–5		TM 14
3–6	3–6	3–6	

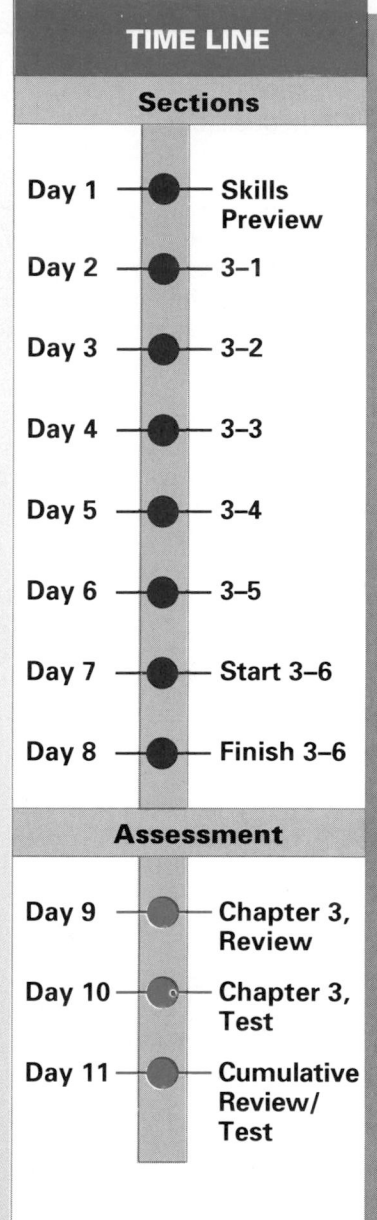

TIME LINE

Sections

Day 1 — Skills Preview

Day 2 — 3–1

Day 3 — 3–2

Day 4 — 3–3

Day 5 — 3–4

Day 6 — 3–5

Day 7 — Start 3–6

Day 8 — Finish 3–6

Assessment

Day 9 — Chapter 3, Review

Day 10 — Chapter 3, Test

Day 11 — Cumulative Review/ Test

ASSESSMENT OPTIONS

Chapter 3, Test Forms A and B	
Chapter 3, Test	Text, 112
Alternative Assessment	TAE, 112
Chapter 3, MicroExam	

Objective To explore how data can be used as a basis for making real-life decisions

Introduction After reading and discussing the introductory material and the fictional-detectives list, have students share any experiences they may have had with the characters and authors listed.

Decision Making
Using Data You may wish to discuss these statements orally. Have students justify their responses.

Working Together Brainstorm possible questions students might ask before making a decision about which writer and story to select. Discuss ways in which the groups might read the story they have chosen. For example, each group member might read a chapter, or a certain number of pages, and report on the contents to the group; members might take turns reading the story aloud. Give each group an opportunity to present its paragraph and its evaluation of the detective's logic to the rest of the class.

REASONING LOGICALLY

THEME Investigating, Solving Mysteries, Detective Work

Mystery and detective novels are among the best-selling books of all time. The most popular American mystery writer is Erle Stanley Gardner (1889–1970). So far more than 320,000,000 copies of his books, printed in 37 languages, have been sold.

The top-selling female mystery writer is British author Agatha Christie (1890–1976). Her 78 mystery novels have been printed in over 183 languages. To date, more than 2 billion copies of Christie's books have been sold.

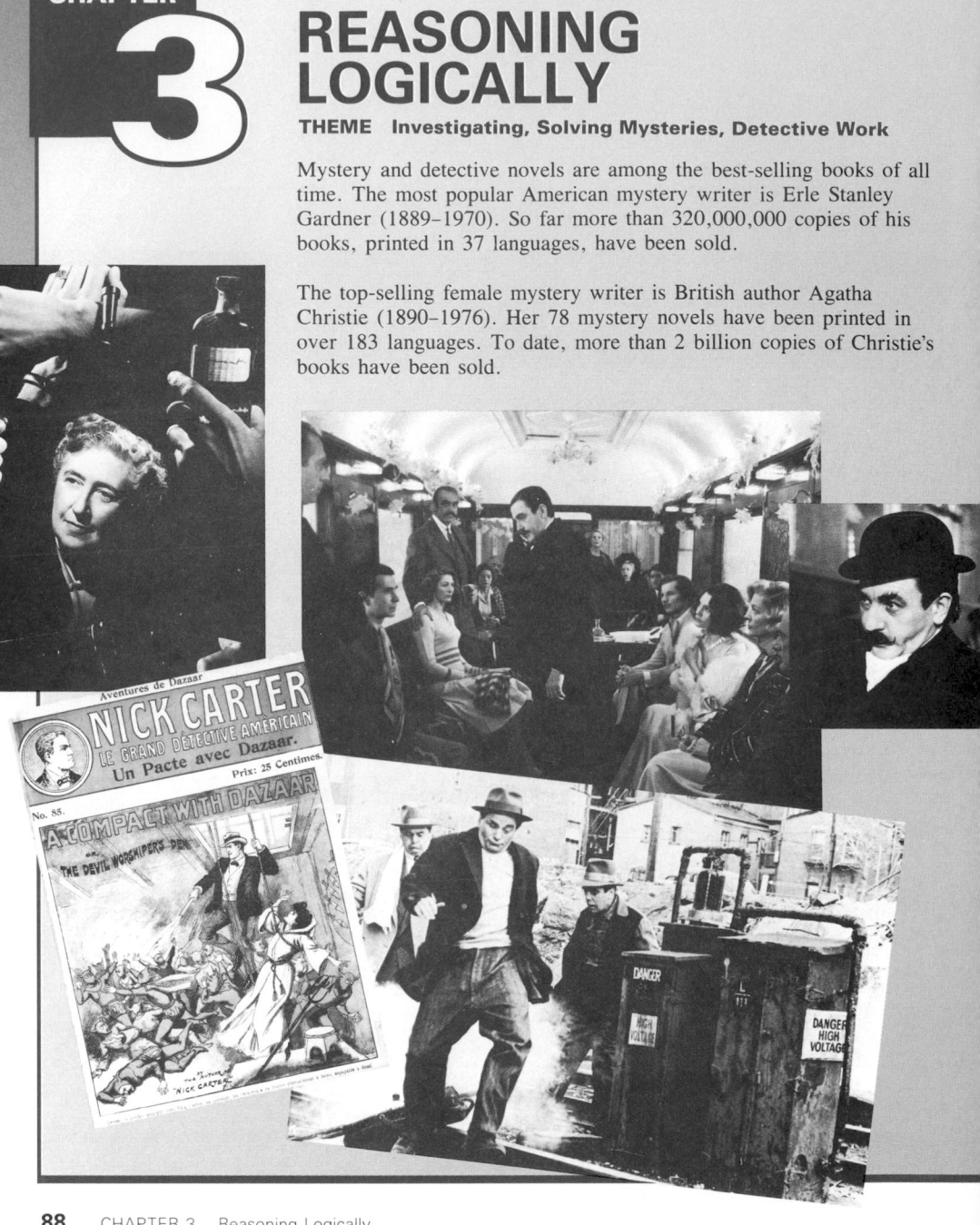

Two mystery writers named Frederic Dannay and Manfred B. Lee wrote as a team. The two men were cousins, and they published their work under the name Ellery Queen. The following list of fictional detectives and their creators was prepared by Ellery Queen for *The Book of Lists*.

ELLERY QUEEN'S 17 GREATEST FICTIONAL DETECTIVES OF ALL TIME (in alphabetical order)	
Fictional Detectives	**Authors Who Created Them**
1. Uncle Abner	Melville Post
2. Lew Archer	Ross Macdonald
3. Father Brown	G. K. Chesterton
4. Albert Campion	Margery Allingham
5. Charlie Chan	Earl Derr Biggers
6. C. Auguste Dupin	Edgar Allan Poe
7. Dr. Gideon Fell	John Dickson Carr
8. Sherlock Holmes	A. Conan Doyle
9. Inspector Maigret	Georges Simenon
10. Philip Marlowe	Raymond Chandler
11. Miss Marple	Agatha Christie
12. Perry Mason	Erle Stanley Gardner
13. Hercule Poirot	Agatha Christie
14. Ellery Queen	Ellery Queen (Frederic Dannay and Manfred B. Lee)
15. Sam Spade	Dashiell Hammett
16. Lord Peter Wimsey	Dorothy L. Sayers
17. Nero Wolfe	Rex Stout

DECISION MAKING

Using Data

Read each statement. Then determine from the information above whether or not the statement is true. Write *true, false,* or *cannot tell.*

1. Agatha Christie created more than one fictional detective. true
2. Sherlock Holmes is the best of the detectives listed. cannot tell
3. Lord Peter Wimsey was created by G. K. Chesterton. false
4. Ellery Queen was a real person. false
5. There are only three female authors on the list. cannot tell

Working Together

Your group's task is to read a story by one of the writers on the list above. You might choose A. Conan Doyle's "The Red-headed League," Dorothy L. Sayers's "The Inspiration of Mr. Budd," or Agatha Christie's *Nemesis*. Write a paragraph describing the reasoning the detective in the story uses to solve the mystery. Discuss whether or not you think the detective's conclusions were drawn logically. Then suggest other ways the mystery could have been solved.

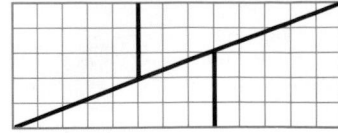

SPOTLIGHT

OBJECTIVE
• Use optical illusions to make statements and determine their truth or falsity.

MATERIALS NEEDED
graph paper, rulers

VOCABULARY
false, optical illusion, statement, true

WARM-UP

Discuss.
Can you think of an example of how our eyes sometimes deceive us? **Answers will vary; flatness of the earth**

1 MOTIVATE

Explore Before students copy the figure, have them carefully examine it. Ask them to count the number of squares horizontally and vertically and to notice where the interior lines of the figure begin and end.

2 TEACH

Use the Pages/Skills Development Have students read this part of the section and then discuss the examples. Then ask students what kinds of statements are either true or false (*statements of fact*) and what kinds are neither true nor false (*opinions, value judgments*).
    Example 1: Have students discuss what in the figure makes it seem as if the lines are not equal in length.

EXPLORE   A 5-unit by 13-unit rectangle is shown below.

a. Copy the rectangle as shown.

b. Cut it into pieces along the heavy lines shown and rearrange the pieces to form a square. Tape the pieces together.

c. What do you think is the relationship between the area of the rectangle and the area of the figure you made? **The areas of the two figures appear to be equal.**

The area of the rectangle is **d.** Count square units to compute the area of the original rectangle. 65 square units. The area of the square is 64 square units.
Now count square units to compute the area of the figure that you formed. What do you discover?

e. Discuss the answers to parts c and d. What explains your conclusion? **The figure fitted together from the pieces of the rectangle appears to be square, but the pieces do not make a perfect square.**

SKILLS DEVELOPMENT   Look carefully at the set of railroad tracks shown at the right. The two rails appear to get closer and closer together, but they do not actually do so. The distances between the railroad ties seem to get shorter and shorter, but they do not really do so. The trees along the sides of the railroad tracks appear to become smaller and smaller, yet they are all actually about the same height. The picture you see is an **optical illusion.** In an optical illusion, the human eye perceives, or pictures, something to be true that is actually not true.

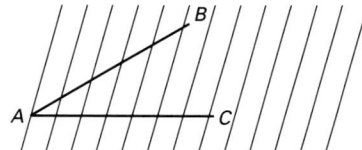

You can make a statement about what you believe to be true of something. A **statement** is a kind of sentence that is either true or false. You can find out if a statement is true or false by testing it.

Example 1

Refer to the figure at the right. Make a statement about the lengths of line segments *AB* and *AC*. Then check to see if your statement is true or false.

TEACHING TIP

Encourage students to make more than one testable statement whenever possible. This will allow them to focus on the different ways in which a figure can appear to be what it is not.

Solution

Line segment *AC* crosses more slant lines than does line segment *AB*.

Statement: *Line segment* AC *is longer than line segment* AB.

Measure both segments with a ruler. You will find that they have the same length. The statement is false. ◄

Example 2

Refer to the figure below. Make statements about the length and direction of *DE* and *FG*. Then check to see if each statement is true or false.

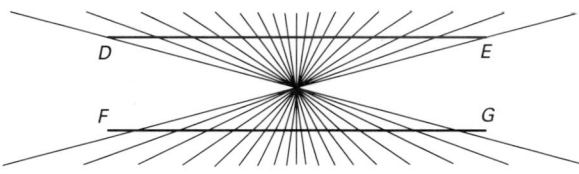

Solution

FG crosses the same number of lines as does *DE*. *FG* extends farther out past the lines, however. *FG* and *DE* seem to bow away from each other in the middle.

Statement 1: FG *is longer than* DE.
Statement 2: FG *and* DE *are not parallel to each other.*

Place a ruler next to *DE*, then next to *FG*. You will find that both are straight line segments, and they have the same length. Statement 1 is false. Now measure the distance between the left end of each segment, the right end of each segment, and the middle of each segment. The distance is the same. The line segments are parallel. Statement 2 is false. ◄

TRY THESE

Answers may vary.

1. Refer to the figure at the right. Write a statement about the lengths of line segments *EF* and *GH*. Then check to see if your statement is true or false.
The segments are equal in length.

2. Refer to the figure at the right. Write a statement about *AB*, *AC*, and *BC*. Then check to see if your statement is true or false.
The lines connecting points *A, B,* and *C* are straight lines.

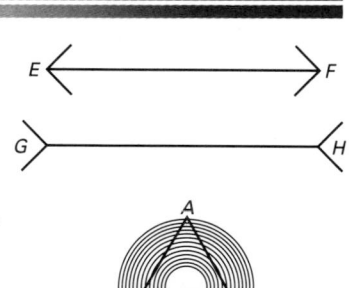

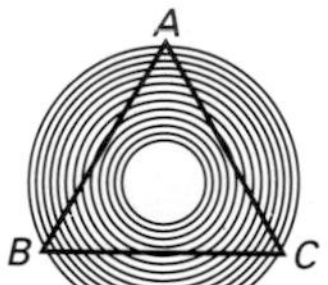

MATH IN THE WORKPLACE

Architects and interior designers use their knowledge of visual perception when they design buildings or interior spaces. For example, to make a building look taller, an architect might use tall, narrow windows. To make a room look wider, an interior designer might use horizontal stripes.

A fashion designer would seldom put horizontally striped clothing on a short person. Why?

The horizontal stripes would make the person look shorter and broader.

ASSIGNMENTS

BASIC
1–2, 3–5, PSA 1– 5

AVERAGE
1–2, 3–7, PSA 1–6

ENRICHED
1–2, 5–7, 8, PSA 1–7

ADDITIONAL RESOURCES
Reteaching 3–1
Enrichment 3–1

Example 2: Have students discuss what in the figure makes it seem as if the lines are not equal in length.

Additional Questions/Examples Write a statement about the figure surrounding the square. Test your statement.

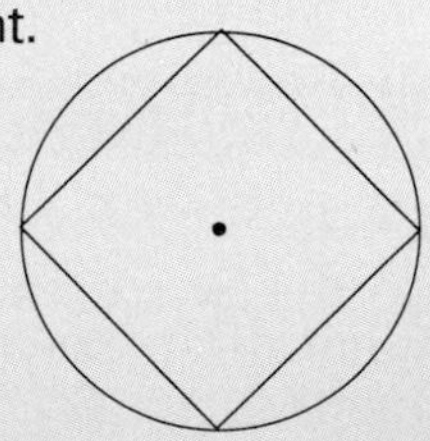

The figure is a circle, although it may not appear to be one.

Guided Practice/Try These Have students work in pairs or in small groups to complete the exercises.

3 SUMMARIZE

Talk It Over Have students explain what an optical illusion is. **In an optical illusion, the eye perceives something to be true that is not true.**

ESL STUDENTS

Students for whom English is a second language may benefit from describing the illusions in their native language first. Once they have identified what they wish to say, have them try to express it in English.

4 PRACTICE

Practice/Solve Problems Students should observe that the illusion in each exercise is such that they must use a ruler or straightedge to test their statement.

Extend/Solve Problems Call on volunteers to read their descriptions for Exercises 5 and 6 to the class. Have students read their descriptions for Exercise 7 to other members of the class and discuss.

Think Critically/Solve Problems Ask students to read their descriptions for Exercise 8 aloud.

Problem Solving Applications Ask for volunteers to explain whether or not it would be possible to make a model of each of the drawings in Exercises 1–4.

5 FOLLOW-UP

Extra Practice

1. Refer to the figure and write a statement about the length of segments *AB* and *CD.* Then test your statement.

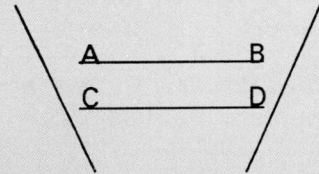

The segments are the same length.

**PRACTICE/
SOLVE PROBLEMS**

1. Write a statement about *l* and *m* that is suggested by the figure. Check to see if it is true.

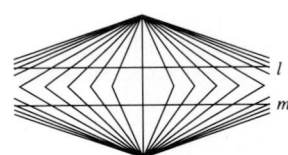

Lines *l* and *m*
are parallel.

2. Write a statement about line segments *AB* and *BC* that is suggested by the picture.

Segments *AB*
and *BC* have
the same length.

**EXTEND/
SOLVE PROBLEMS**

3. Write a statement about line segments *DE* and *FG* that is suggested by the figure.

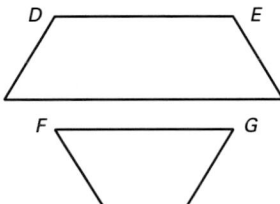

Line segments *DE* and *FG*
have the same length.

4. Write a statement about the dots in the two figures.

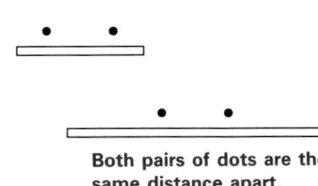

Both pairs of dots are the
same distance apart.

 Each drawing below can be viewed in two ways. Write a paragraph that describes what you see in each figure. Justify your statement.

5.

Viewed from one
perspective, the face of an
old woman is seen. From
another perspective, the
side of the head of a young
woman is seen.

6.

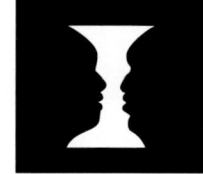

From one perspective, a
vase is seen. From another,
two people facing each
other can be seen.

7. ***USING DATA*** Use the Data Index on page 546 to find a reference to two drawings that show optical illusions. Describe what you first see in each drawing. Then look at the figures until you see something different. Describe what you see this time.

**THINK CRITICALLY/
SOLVE PROBLEMS**

 8. Choose one of the optical illusions shown in the lesson. Write a paragraph in which you describe the illusion and explain what it is in the picture that tricks the eye. **Answers will vary.**

Problem Solving Applications:
IMPOSSIBLE FIGURES IN ART

► READ
► PLAN
► SOLVE
► ANSWER
► CHECK

Examine the two-dimensional figures shown. If you were to attempt to make a three-dimensional model using straws, sticks, or other objects, could you do it? Justify your answer.   **Only 5 is possible.**

1.

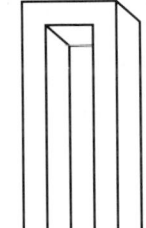

2.

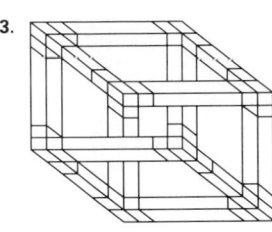

3.

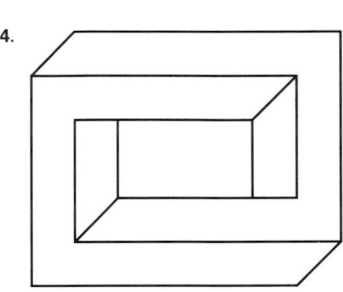

4.

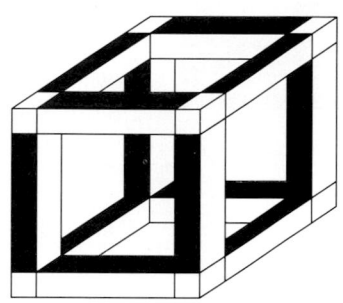

5.

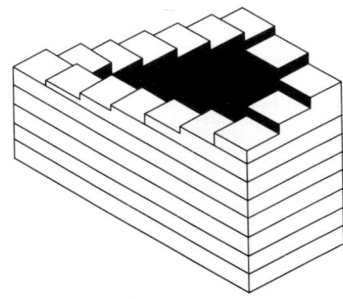

Study the two drawings below carefully. You will notice some strange things. Describe at least one unusual thing in each picture.

6.

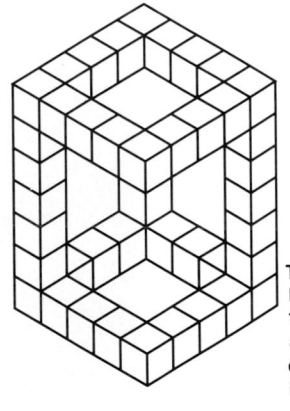

No matter whether you go up or down the stairs, you keep going around and passing the same spot.

7.

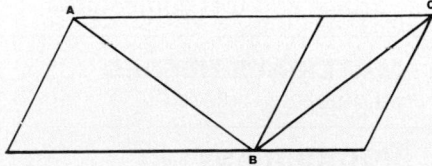

The connection between the top front cube and the bottom cube at back is impossible.

---

2. Refer to the figure and write a statement about the length of segments *AB* and *BC*. Then test your statement.

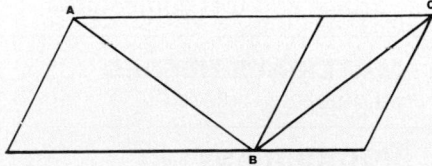

Segments *AB* and *BC* are equal in length.

Extension Have each student find an example of an optical illusion and share it with a classmate. Tell students to write a description or an explanation of what makes the illusion.

Section Quiz
1. What is an optical illusion?   **A perception by the eye of something that seems to be true but is actually not so.**
2. What is a statement?   **A kind of sentence that is either true or false.**

Get Ready calculators, rulers

MAKING CONNECTIONS

Have students research the work of Dutch artist M.C. Escher. Have each student select one work and describe the optical illusions or impossible figures that appear in the work of art.

WARM-UP

Write the next three numbers.
1. 1, 5, 9 **13, 17, 21**
2. 2, 6, 10 **14, 18, 22**
3. 3, 7, 11 **15, 19, 23**
4. 4, 8, 12 **16, 20, 24**

1 MOTIVATE

Explore Have students complete the activity in small groups. Remind them that all group members must come to an agreement about each answer. Discuss any problems, disagreements, or questions students have.

2 TEACH

Use the Pages/Skills Development Have students read this section of the lesson and then discuss the examples.

Example 1: Have students find the ones digit for 3^{14}, 3^{25}, and 3^{28}. Ask them to explain their reasoning. **9, 3, 1**

Example 2: Read the example with students. Have them test Pete's conjecture on their own before analyzing the solution.

3-2 Inductive Reasoning

EXPLORE/ WORKING TOGETHER

Agatha Applegate is a private investigator. She is gathering clues about four separate burglaries committed in Centertown. She discovers a white glove on the floor at the scene of each burglary.

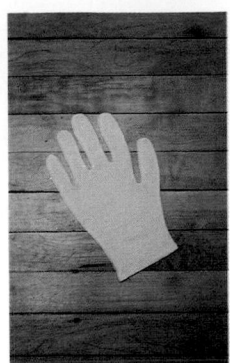

a. Discuss with your classmates whether each of the following conclusions is *warranted* or *not warranted*, given the evidence.
- The thief, or someone who is helping the thief, is the person who leaves the white glove on the floor. **warranted**
- The thief always wears white gloves while committing burglaries. **not warranted**
- It is possible that the thief wears a white glove while committing burglaries. **warranted**
- Everyone who wears white gloves is a thief. **not warranted**
- A white glove will be found at the scene of the next burglary committed in Centertown. **not warranted**

b. Suppose that Agatha investigates ten more burglaries and finds a white glove at the scene of each one. Do these added instances prove that a white glove will be found at the scene of the next burglary? **no**

c. With your classmates, determine if there are any other conclusions that might be warranted, given the facts. **Answers will vary.**

SKILLS DEVELOPMENT

People in many walks of life—detectives, scientists, advertisers, politicians, mathematicians, and others—pay special attention to things that happen repeatedly. They hope that by examining some events, they might discover a pattern or rule that applies to all such events. That is, they hope to arrive at a conclusion by using *inductive reasoning*. **Inductive reasoning** is the process of reaching a conclusion based on a set of specific examples. The conclusion is called a **conjecture.**

TEACHING TIP

As students work through the examples and exercises, encourage them to use calculators whenever possible. The use of calculators allows students to focus more on number patterns and relationships than on the computations.

Example 1

Use inductive reasoning to make a conjecture about the ones digit of 3^{16}.

COMPUTER

Solution
First, use a calculator to find several successive powers of 3 and look for a pattern in the ones digits.

$3^1 = 3$ $3^2 = 9$ $3^3 = 27$ $3^4 = 81$
$3^5 = 243$ $3^6 = 729$ $3^7 = 2{,}187$ $3^8 = 6{,}561$
$3^9 = 19{,}683$ $3^{10} = 59{,}049$ $3^{11} = 177{,}147$ $3^{12} = 531{,}441$

In these examples, the ones digits repeat in a pattern:

3 9 7 1   3 9 7 1   3 9 7 1

Next, make a conjecture based on the pattern. For example,

"When the exponent of 3 is a multiple of 4, the ones digit is 1."

Finally, make a specific prediction based on the conjecture and test it. For example,

"Since 16 is a multiple of 4, the ones digit for 3^{16} is 1."

In fact, $3^{16} = 43{,}046{,}721$, a number whose ones digit is 1. ◄

> The computer program below lets you print the successive powers of 3 whose exponents are multiples of 4, to check your conjecture in Example 1. How would you change line 10 to find the answers that end in 3, 9, or 7?
>
> ```
> 10 FOR X = 4 TO 16 STEP 4
> 20 V = INT ( 3 ↑ X + 0.5)
> 30 PRINT : PRINT "3↑";X;
>        = ";V
> 40 NEXT X
> ```
>
> **See Additional Answers.**

A conjecture may be true, but it is not *necessarily* true. To prove that a conjecture is false, you need to find just one case where the conjecture does not hold true.

Example 2

Pete observed that he could express fractions with two-digit numerators and denominators in lowest terms by this method.

$$\frac{1\cancel{6}}{\cancel{6}4} = \frac{1}{4} \qquad \frac{1\cancel{9}}{\cancel{9}5} = \frac{1}{5} \qquad \frac{2\cancel{6}}{\cancel{6}5} = \frac{2}{5}$$

On the basis of the three examples, he made this conjecture: "Whenever the ones digit in the numerator is the same as the tens digit in the denominator, you can cancel the ones digit in the numerator and the tens digit in the denominator to get an equivalent fraction."

Is Pete's conjecture always true?

 CHECK UNDERSTANDING

Solution
If possible, find a case in which Pete's conjecture is false.

Use Pete's method   Use division to write
for $\frac{12}{24}$. $\frac{12}{24}$ in lowest terms.

$$\frac{1\cancel{2}}{\cancel{2}4} \rightarrow \frac{1}{4} \qquad\qquad \frac{12}{24} = \frac{12 \div 12}{24 \div 12} = \frac{1}{2}$$

Since $\frac{1}{4} \neq \frac{1}{2}$, Pete's conjecture is false. ◄

> From what you have observed about inductive reasoning, why can't you state that if a rule or pattern holds for several examples, then it will hold for all examples?
>
> **The pattern might not hold for examples that have not been checked.**

ASSIGNMENTS

BASIC
1–7, 9–10, 12

AVERAGE
3–8, 10–15

ENRICHED
7–8, 9–11, 12–15

ADDITIONAL RESOURCES
Reteaching 3–2
Enrichment 3–2

After working through the two examples, ask a volunteer to make a conjecture and have the class use inductive reasoning to support or disprove it.

Additional Questions/Examples
1. At a certain instant it is 3:00. Use inductive reasoning to find what time it is 11,999,999,996 hours later. **12 billion hours later, it would be 3:00. Four hours before that would be 11:00.**
2. Use inductive reasoning to find the number of 1s in the product of $9 \times 12{,}345{,}678$.
   $9 \times 1 = 9$
   $9 \times 12 = 108$
   $9 \times 123 = 1{,}107$
   $9 \times 1234 = 11{,}106$
   When the factor has 2 digits, there is one 1; 3 digits, two 1s; 4 digits, three 1s. With 8 digits, there should be seven 1s.
3. Use inductive reasoning to find the missing number.
   1, 3, ■, 15, 31   **7; Double the preceding number and add 1.**

95

Guided Practice/Try These Have students work in pairs to complete these exercises.

3 SUMMARIZE

Talk It Over Have students explain inductive reasoning and how it works. **Examine several examples and look for a pattern; make a conjecture; test the conjecture, trying to find a counterexample.**

4 PRACTICE

Practice/Solve Problems Students may not immediately see the pattern in Exercise 8. If necessary, tell them that the pattern involves addition and subtraction.

Extend/Solve Problems Have students use calculators to work through Exercises 9–11. You may wish to challenge students to explain the reasoning behind Exercise 11. **The product of 7, 11, and 13 is 1,001 which, when multiplied by any three-digit number (abc), will always give a product with repeating digits (abc, abc).**

5 FOLLOW-UP

Extra Practice
1. Use inductive reasoning to find the ones digit of 7^{1000}. **The pattern of the ones digit for powers of 7 is 7, 9, 3, 1 and repeats in cycles of fours, so 7^{1000} ends in 1.**
2. Draw two intersecting lines. Measure the opposite angles. Repeat with other intersecting lines. Make a conjecture. **Opposite angles formed by two intersecting lines are equal.**

COMPUTER TIP

You can adapt the program on page 95 to test your conjecture for Try These Exercises 1 and 2. Change line 10 so X = 1 to 5. What change do you need to make to line 20? to line 30? How could you change line 10 to check your answer?

See Additional Answers.

TRY THESE

1. Use inductive reasoning to find the ones digit for 4^6. **6**

2. Use inductive reasoning to find the ones digit for 4^7. **4**

3. Merola found that the sum for each of three pairs of numbers was a two-digit number whose digits are the same:

$$16 + 61 = 77 \qquad 17 + 71 = 88 \qquad 18 + 81 = 99$$

She made this conjecture: Take any two-digit number and reverse the digits. The sum of that number and the original number is a two-digit number whose digits are the same. Is Merola's conjecture always true? **3. No. The rule holds true only for pairs of numbers whose sum is less than 100. The conjecture is false for these examples: 91 + 19 = 110; 55 + 55 = 110; 67 + 76 = 143**

EXERCISES

PRACTICE/
SOLVE PROBLEMS

1. Use inductive reasoning to find the ones digit of any odd-numbered power of 4. **4**

2. Use inductive reasoning to find the ones digit of any even-numbered power of 4. **6**

Use the given examples to complete each conjecture.

3. $3 \times 5 = 15 \quad 7 \times 9 = 63 \quad 13 \times 5 = 65 \quad 23 \times 27 = 621$
   The product of two odd numbers is an __?__ number. **odd**

4. $8 \times 12 = 96 \quad 12 \times 16 = 192 \quad 36 \times 74 = 2,664$
   The product of two even numbers is an __?__ number. **even**

5. $2 + 8 = 10 \quad 14 + 16 = 30 \quad 6 + 62 = 68 \quad 46 + 76 = 122$
   The sum of two even numbers is an __?__ number. **even**

6. Choose an even number. Add 20. Multiply the sum by 2. Divide the product by 4. Subtract 10 from the quotient. Multiply the difference by 2. What number results? Repeat this experiment with a different starting number. What number results? Make a conjecture. **The result is always the number you started with.**

7. Choose any number. Add 12 to it. Multiply the result by 2. Subtract 4. Divide by 4. Finally, subtract half the number you started with. What number results? Repeat this experiment with a different starting number. What number results? Make a conjecture. **The result is always 5.**

8. Examine the following sequence of numbers. Give a pattern for the sequence and give the next two numbers. **7, 5; To get the second number, add 3 to the first number; then subtract 2 from the result; repeat the process.**

$$1, 4, 2, 5, 3, 6, 4, \ldots$$

CHALLENGE

Have students work in groups to prove that Goldbach's conjecture is true for the first fifty natural numbers. Assign ten numbers to each group for examination.

9. Multiply any number by 9. Take the sum of the digits of this product. Keep taking the sum of the digits until the sum has only a single digit. Make a conjecture about the digit that results from this process. **The digit will always be 9.**

10. Add 5 to your age. Double that sum and then multiply by 50. Add the date of the month on which you were born. Double that sum. Subtract 1,000. Divide by 2. Make a conjecture about the number that results. Test your conjecture with several other people.

11. Pick a three-digit number. Multiply it by 7, then multiply that product by 11, and then multiply that product by 13. Test this process several times. Make a conjecture. **If the original number is *abc*, the final number is *abc,abc*.**

? × 7 = ?

? × 11 = ?

? × 13 = ?

For each of the following exercises, test the conjecture by finding at least five examples that support it *or* at least one example that disproves it. All the conjectures will be true for at least some cases. Try to explain what is wrong with each false conjecture and modify it if possible.

12. The difference between any two whole numbers is always positive.

13. The sum of consecutive odd whole numbers starting with 1 is always a perfect square. For example, $1 + 3 + 5 = 9$, and 9 is a perfect square. **Answers will vary. This conjecture is always true.**

14. The sum of n consecutive positive whole numbers starting with 1 is always equal to $\frac{n(n + 1)}{2}$. For example, the sum of
$$1 + 2 + 3 + 4 + \ldots + 99 + 100 = \frac{100 (100 + 1)}{2} = 5,050.$$
Answers will vary. This conjecture is always true.

15. *USING DATA* Use the Data Index on page 546 to find the list of prime numbers. A *prime number* is a number greater than 1 having only itself and 1 as factors. Use the formula $n^2 + n + 11$, and replace n with the whole numbers from 1 to 9. Is each resulting number prime? Replace n with 10. Is the result prime? **yes; no**

EXTEND/ SOLVE PROBLEMS

10. The result is a four-digit number. Its first two digits will be your age. Its last two digits are those of the date of your birth.

12. False; true only when the number being subtracted from is larger than the number being subtracted.

THINK CRITICALLY/ SOLVE PROBLEMS

MATH: WHO, WHERE, WHEN

A mathematician named Christian Goldbach made the following conjecture in 1742.

Every even whole number greater than 2 can be written as the sum of two prime numbers.

Examples: 8 = 3 + 5
20 = 7 + 13
100 = 11 + 89

This conjecture has never been proved, but no one has ever found an example that disproves it.

3. Use inductive reasoning to find the product $9,999,999 \times 9,999,999$.
$9 \times 9 = 81$
$99 \times 99 = 9,801$
$999 \times 999 = 998,001$
Therefore the product is 99,999,980,000,001.

Extension Have each student make a conjecture and exchange with a classmate. Tell students to test the conjecture by finding at least five examples that support it or one counterexample that disproves it.

Section Quiz
1. What is a conjecture? **An assumption based on the examination of many examples.**
2. What is a counterexample? **An example that disproves a conjecture.**
3. Use inductive reasoning to find the ones digit of 2^{40}. **6**
Test these hypotheses to see if they are always true.
4. The square of a number ending in 5 always ends in 25. **always true**
5. A number is divisible by 6 if the sum of the digits is divisible by 3 and by 2. **always true**
6. Pick any 2-digit number. Multiply it by 111. Multiply the product by 91. The product is a 6-digit number which repeats the 2-digit number three times. **always true**

Additional Answers
See page 575.

Additional answers for odd-numbered exercises are found in the Selected Answers portion of the page.

MAKING CONNECTIONS

Have students research Pascal's triangle, along with the Fibonacci sequence, to find other mathematical patterns and relationships.

3-3 Deductive Reasoning

EXPLORE/ WORKING TOGETHER

Study the map below and read these four statements.

A. If you live in Chicago, then you live in Illinois.
B. If you live in Illinois, then you live in Chicago.
C. If you do not live in Chicago, then you do not live in Illinois.
D. If you do not live in Illinois, then you do not live in Chicago.

Which of the statements are true? **A, D** Which of the statements are false? **B, C**

SKILLS DEVELOPMENT

In daily life, we often make statements that involve *if* and *then*. Such statements are called **if-then statements,** or **conditional statements.** A conditional statement has two parts, a *hypothesis* and a *conclusion.*

$$\underset{\downarrow}{\text{hypothesis}} \qquad \underset{\downarrow}{\text{conclusion}}$$

If **it rains,** then **it is cloudy.**

Example 1

Write two conditional statements that can be made from *The temperature is 0°C* and *Water freezes*.

Solution
Make one statement the hypothesis and the other the conclusion.

If the temperature is 0°C, then water freezes.
If water freezes, then the temperature is 0°C. ◄

Some conditional statements are true and others are false. A **counterexample** shows that a conditional sentence is false by satisfying the hypothesis but not the conclusion.

98 CHAPTER 3 Reasoning Logically

Example 2

Is the conditional statement *true* or *false*? If false, give a counterexample.
a. If a number is divisible by 10, then it is divisible by 5.
b. If a number is divisible by 5, then it is divisible by 10.

Solution
a. The statement is true, because any number ending in 0 is divisible by 5.
b. The number 25 is a counterexample, because 25 is divisible by 5 but not by 10. So, the statement is false. ◄

When you use a conditional statement along with given information to draw a new conclusion, you use **deductive reasoning,** or make a **logical argument.** The following is an example of deductive reasoning.

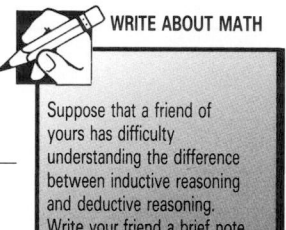
WRITE ABOUT MATH

Suppose that a friend of yours has difficulty understanding the difference between inductive reasoning and deductive reasoning. Write your friend a brief note explaining the difference between the two kinds of thinking. Give an example of each type of thought process.

Answers will vary.

If I do my homework by 6:00 p.m., then I may go to the movies. I *did* my homework by 6:00 p.m.

Therefore, I may go to the movies.

In a **valid** argument, a new statement is obtained from given statements by deductive reasoning. If there is an error in the reasoning, the argument is called **invalid.**

Example 3

Is this argument *valid* or *invalid*? Use a picture to help you decide.

If a runner wears Speedy Shoes, then the runner will win the race.
Constance, a contestant in the race, wears Speedy Shoes.

Therefore, Constance will win the race.

Solution
Constance is in the set of people who run and wear Speedy Shoes. That group is in the set of people who win. So, Constance is in the set of winners of the race. The argument is valid. ◄

Example 4

Is this argument *valid* or *invalid*? Use a picture to help you decide.

If an animal is a cat, then it loves tuna.
Fido loves tuna.

Therefore, Fido is a cat.

Solution
Draw a shaded region for cats inside the region for tuna lovers. The argument is not valid, because Fido might be a dog that loves tuna. ◄

MAKING CONNECTIONS

People use deductive reasoning in many different situations, for example, when playing games and when grocery shopping. Have students bring in pictures which depict or imply situations in which the use of deductive reasoning may occur. Ask students to provide an example of deductive reasoning for each situation.

ASSIGNMENTS

BASIC
1–4, 7–9, 13, 15, 17

AVERAGE
1–6, 7–11, 13–15, 17

ENRICHED
2–6, 7–16, 17–18

ADDITIONAL RESOURCES
Reteaching 3–3
Enrichment 3–3

Example 2: Be sure the students understand what is meant by a counterexample.

Examples 3 and 4: Point out that in deductive reasoning, the conclusion necessarily follows from the given statements. Emphasize the importance of using a diagram to help determine whether an argument is valid or invalid.

Additional Questions/Examples
1. Write two conditional statements using these two sentences:
The number is odd. The square of the number is odd. **If a number is odd, then the square of the number is odd. If the square of a number is odd, then the number is odd.**
Is the conditional statement *true* or *false*? If *false*, give a counterexample.
2. If a number is divisible by 4, then it is divisible by 2.   **true**
3. If a figure is a square, then it has four equal sides. **true**

Guided Practice/Try These Have students work in pairs or small groups to complete these exercises.

Write About Math Have students define the following terms and give examples of each in their math journals: *conditional statement, hypothesis, conclusion.*

Practice/Solve Problems In Exercise 5, point out that even though the first conditional statement is false (i.e., birds, such as penguins and ostriches, cannot fly), the argument is still valid, as a diagram will show. In Exercise 6, point out that the second statement denies the *hypothesis* of the conditional statement. Thus, you cannot know whether the fruit is or is not a citrus fruit.

Extend/Solve Problems Point out that any *All*-statement can be expressed as a conditional statement, and vice-versa. Exercise 11 is an example of another kind of valid deductive argument. The second statement *negates*, or denies, the conclusion of the hypothetical statement. The only conclusion possible is that figure *ABCD* is not a rectangle.

Think Critically/Solve Problems Point out that in Exercises 17 and 18, it helps to rewrite the *All-* and *No*-statements as *if-then*, conditional statements.

---

1. If there is no gasoline in the tank, then the car will not run.
   If the car will not run, then there is no gasoline in the tank.

PRACTICE/SOLVE PROBLEMS

1. If the number is divisible by 9, then the number is divisible by 3.
   If the number is divisible by 3, then the number is divisible by 9.

TRY THESE

1. Write two conditional statements that can be made from *There is no gasoline in the tank* and *The car will not run.*

2. Is the conditional statement *true* or *false*? If false, give a counterexample.
   a. If a woman lives in Atlanta, then she lives in Georgia. **true**
   b. If two numbers are even, then their sum is an odd number. **false: 2 + 2 = 4**

3. Is this argument *valid* or *invalid*? Use a picture to help you decide.
   If a flower is red, then it is pretty.
   This rose is red.

   Therefore, this rose is pretty. **valid**

4. Is this argument valid or invalid? Use a picture to help you decide.
   If the battery is dead, then the car will not start.
   The car did not start.

   Therefore, the battery is dead. **invalid**

EXERCISES

1. Write two conditional statements that can be made from *The number is divisible by 9* and *The number is divisible by 3.*

2. Is the conditional statement *true* or *false*? If false, give a counterexample.
   a. If two numbers are odd, then their product is odd. **true**
   b. If a number is divisible by 6, then it is divisible by 12. **false, 18**

Is the argument *valid* or *invalid*? Use a picture to help you decide.

3. If today is Monday, then tomorrow is Tuesday.
   Today is Monday.

   Therefore, tomorrow is Tuesday. **valid**

4. If two numbers are even, then their sum is even.
   The sum of 7 and 9 is even.

   Therefore, 7 and 9 are even. **invalid**

5. If an animal is a bird, then it can fly.
   An ostrich is a bird.

   Therefore, an ostrich can fly. **valid**

6. If a fruit is an orange, then it is a citrus fruit.
   This fruit is not an orange.

   Therefore, this fruit is not a citrus fruit. **invalid**

Write each statement as an *if-then* statement.

7. All even numbers are divisible by 2.

8. All squares are quadrilaterals.

9. All whales are mammals.

Is the argument *valid* or *invalid*? Use a picture to help you decide.

10. All giraffes have long necks.
    Titus has a long neck.
    Therefore, Titus is a giraffe. **invalid**

11. All rectangles have four right angles.
    Figure *ABCD* does not have four right angles.
    Therefore, figure *ABCD* is not a rectangle. **valid**

12. All multiples of 100 are multiples of 10.
    500 is a multiple of 100.
    Therefore, 500 is a multiple of 10. **valid**

Sometimes conditional statements can be strung together. Write an *if-then* statement of your own from the given conditional statements.

13. If a number is a whole number, then it is an integer.
    If a number is an integer, then it is a rational number.
    If a number is a whole number, then it is a rational number.

14. If people live in Mexico City, then they live in Mexico.
    If people live in Mexico, then they live in North America.
    If people live in Mexico City, then they live in North America.

    See Additional Answers.

Write a valid logical argument for each picture.

15.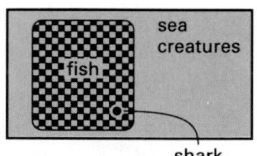
    sea creatures
    fish
    shark

16.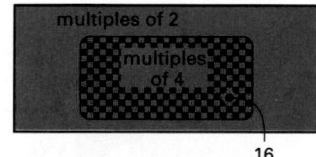
    multiples of 2
    multiples of 4
    16

 Lewis Carroll was the nineteenth-century English writer who wrote *Alice in Wonderland*. However, he was also a mathematician, and he wrote a book called *Symbolic Logic*. For the given statements from his book, write a valid argument. (You will need to draw your own conclusion.) **See Additional Answers.**

17. All well-fed canaries sing loud.
    No canary is melancholy if it sings loud.

18. All puddings are nice.
    No nice things are wholesome.
    This dish is a pudding.

3-3 Deductive Reasoning **101**

EXTEND/SOLVE PROBLEMS

7. **If a number is even, then it is divisible by 2.**

8. **If a figure is a square, then it is a quadrilateral.**

9. **If a creature is a whale, then it is a mammal.**

CONNECTIONS

Advertisements are sometimes misleading. An advertisement might show a young person being invited to a party. A voice simply says "Drink Sizzling Soda." The advertiser wants you to think "If I drink Sizzling Soda, then I will be popular". Find examples of misleading conditional statements like this and discuss them with your classmates.

Answers will vary.

THINK CRITICALLY/ SOLVE PROBLEMS

SPOTLIGHT

OBJECTIVE
- Draw Venn diagrams to solve nonroutine problems

VOCABULARY
universal set, Venn diagram

WARM-UP

Is the statement true, false, *or possibly true?* *Explain.*
The sum of 5 two-digit numbers is less than 100. Each number is less than 20. **false**

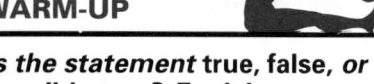

1 MOTIVATE

Introduction Read and discuss the first paragraph with the students.

2 TEACH

Use the Pages/Problem Have students read this part of the section and then discuss the problem. Explain that a Venn diagram will easily show the relationships in this example.

Use the Pages/Solution Work through the solution on the board as it is presented in the text. Continue with additional examples as needed.

3 SUMMARIZE

Talk it Over How does a Venn diagram show relationships? **shows the relationship between sets**

► READ
► PLAN
► SOLVE
► ANSWER
► CHECK

3-4 Problem Solving Skills:
USING VENN DIAGRAMS

Some problems can be solved with a diagram. In a **Venn diagram,** a collection is represented by a circular region inside a rectangle. The rectangle is called the **universal set.** For instance, the diagram at the right shows the relationships between basketball players and baseball players in a certain school.

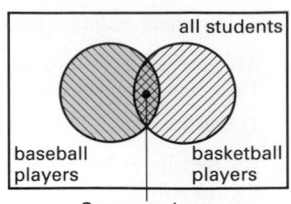

Some students play both sports.

PROBLEM

In a poll of 100 students, the following information was learned.

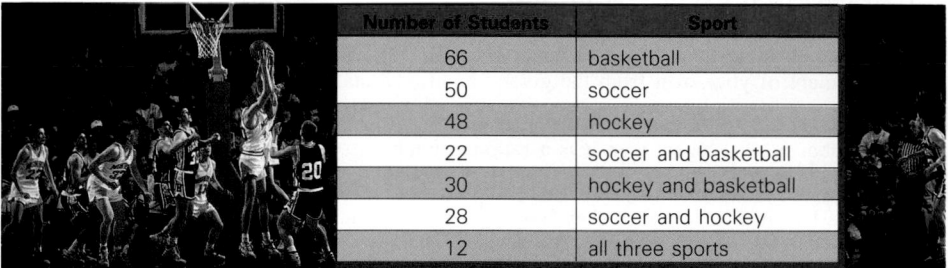

Number of Students	Sport
66	basketball
50	soccer
48	hockey
22	soccer and basketball
30	hockey and basketball
28	soccer and hockey
12	all three sports

How many play soccer or basketball? How many play no sport?

SOLUTION

Make a Venn diagram like the one at the left below. Put 12 in the region marked *A*, because there are 12 students who play all three sports.

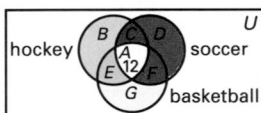 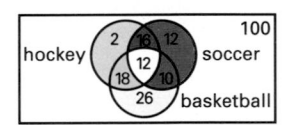

Since 28 students play hockey and soccer (regions *A* and *C* together) and region *A* has 12 people, region *C* has 28 − 12 = 16 people. Use addition and subtraction to find the value for each other letter. The completed diagram is at the right above.

The number of students who play soccer or basketball is
$$16 + 12 + 12 + 10 + 18 + 26 = 94.$$

The number of students who play no sport is
$$100 - (2 + 16 + 12 + 12 + 18 + 10 + 26) = 4.$$

TEACHING TIP

You may wish to introduce the idea of Venn diagrams by having students work with large overlapping circles made from lengths of yarn and a collection of buttons, seashells, or rocks. Students can position the items in the collection on the Venn diagram according to size, shape, color, or other characteristics.

PROBLEMS

Refer to the problem on the preceding page.

1. Tell how you know that 18 students play hockey and basketball but not soccer (region *E*).

2. Tell how you know that 2 students play hockey but not soccer or basketball (region *B*). (You will need the answer to Exercise 1.)

Use a Venn diagram to solve each problem.

3. Dan surveyed 120 people at a shopping mall one day. Of the people surveyed, 62 said they would like to see a bookstore included in the mall; 48 would like to see a music store; and 23 would like to see both included. How many did not want either type of store to be included in the mall? **33**

4. In a class of 30 students, 20 were going on family vacations during the summer, 15 were planning on working, and 3 were doing neither. How many were planning on both working and going on a family vacation? **8**

At a school sports banquet, a survey was conducted concerning what sports each person played. The following facts were recorded: 18 students played a fall sport; 16 played a winter sport; 21 played a spring sport. It was also noted that 5 played a fall and a winter sport, 8 played a fall and a spring sport, and 7 played a winter and a spring sport. Only 3 students played a fall, a winter, and a spring sport.

5. How many students played only in fall? in winter? in spring?
   fall, 8; winter, 7; spring, 9
6. How many students were surveyed at the banquet? **38**

A survey of 200 commuters produced the following data: 118 used the expressway; 98 wore glasses; 84 listened to the radio; 40 listened to the radio and wore glasses; 58 listened to the radio and used the expressway; 62 used the expressway and wore glasses; 24 did all three.

7. How many wore glasses but did not use the expressway or listen to the radio? **20**

8. How many only used the expressway? **22**

9. How many listened to the radio or used the expressway? **144**

10. How many did none of these things? **36**

1. **30 students play hockey and basketball.**
   30 – 12 = 18
2. **48 students play hockey. So, 48 – (16 + 12 + 18) = 2**

ASSIGNMENTS

BASIC
1–3, 5–6

AVERAGE
1–8

ENRICHED
3–10

ADDITIONAL RESOURCES
Reteaching 3–4
Enrichment 3–4
Transparency Master 13

---

4 PRACTICE

Problems For Problem 4 students can also draw two intersecting circles and one circle which does not intersect the others. Since they are looking for the number that belongs in the intersection of the two circles, they can label that section x, those going on family vacations $20 - x$, and those planning on working $15 - x$. The equation $20 - x + 15 - x + 3 + x = 30$ will give them the solution.

5 FOLLOW-UP

Extra Practice A survey on reading material showed that 75 people preferred adventure; 82 preferred mysteries; 65 preferred biographies; 40 read adventure and mysteries; 46 read mysteries and biographies; 34 read adventure and biographies; 28 read all three.

1. How many people read only mysteries? **24**
2. How many read mysteries or adventure? **117**
3. How many people were surveyed? **130**

CHALLENGE

Have students solve this problem by drawing a Venn diagram. A space ship is filled with CLOPS, CLIPS, AND CLEEPS. Every CLOP is a CLIP. Half of all CLEEPS are CLIPS. Half of all CLIPS are CLOPS. There are 30 CLEEPS and 20 CLOPS. No CLEEP is a CLOP. How many CLIPS are neither CLOPS nor CLEEPS? **5   See margin for Venn diagram.**

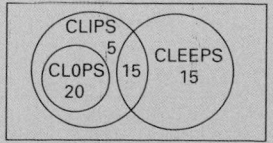

► READ
► PLAN
► SOLVE
► ANSWER
► CHECK

3-5

Problem Solving Strategies:
USING LOGICAL REASONING

In order to solve many logic problems, it helps to make a table. Using the *process of elimination*, you can complete the table to answer questions.

PROBLEM

Dexter, Maria, and Phil each play baseball, badminton, and tennis, but not necessarily in that order. Each person plays only one sport. Dexter does not play a sport that requires a ball. Phil does not play baseball.

Which sport does each play?

SOLUTION

Step 1: Make a table listing each person and each sport.

	baseball	badminton	tennis
Dexter			
Maria			
Phil			

Step 2: Since Dexter does not play a sport that requires a ball, place an **X** in the boxes for baseball and tennis. Place a ✔ in the box for badminton. Since Phil does not play baseball, place an **X** in the box for baseball.

	baseball	badminton	tennis
Dexter	X	✔	X
Maria			
Phil	X		

Step 3: Since each person plays only one sport, place an **X** in the badminton boxes for Maria and Phil.

	baseball	badminton	tennis
Dexter	X	✔	X
Maria		X	
Phil	X	X	

Step 4: There is only one choice of sport for Phil. Place a ✔ in the box for tennis for Phil.

	baseball	badminton	tennis
Dexter	X	✔	X
Maria		X	
Phil	X	X	✔

Step 5: By the process of elimination, Maria must be the baseball player. So, Dexter plays badminton, Maria plays baseball, and Phil plays tennis.

TEACHING TIP

Point out to students that each time they place a ✔ in a table, they should also be able to place some *x*'s.

PROBLEMS

Make a table to help solve each problem.

1. Three dogs are named Tillie, Spot, and Tippie. Two are long-haired dogs, a collie and a Pekingese. One is a short-haired dog, a basset hound. Tippie does not have short hair. The collie lives next door to Tillie. Tippie lives next door to the Pekingese. Tillie is not a basset hound. Who is the Pekingese? **Tillie**

2. Three singers appear at a concert. One is named Jones, one is named Salazar, and one is named Friedman. One sings only jazz, one sings only opera, and one sings only folk music. Jones dislikes the opera singer, but the folk singer is a friend of Jones. Salazar performs before the opera singer in the concert. Who is the folk singer? **Salazar**

3. **Sue married Tom; Betty married Harry; DeeDee married Bill.**

3. My three cousins, Sue, DeeDee, and Betty, just got married, but I have forgotten each husband's name. I remember that my cousins married Tom, Bill, and Harry. I heard Betty say that she hoped Tom didn't argue with her husband. I also remember that DeeDee married Tom's brother and that Harry married DeeDee's sister. Who married whom?

4. Hector Hollingsworth is a private investigator called to the scene of a crime. There are four suspects—the butler, the cook, the chauffeur, and the maid. All four suspects are wearing identical uniforms and refuse to divulge their occupations. Through his investigation, however, Hector is sure that the butler committed the crime. The four suspects are named Alexandra, Beatrice, Cecilia, and Delphine. Hector discovers that Delphine cannot drive. Unknown to Cecilia, her uniform is spattered with kitchen grease. Neither Alexandra nor Delphine can correctly tell Hector where the mops are kept. Who was the butler? **Delphine**

5. Armand, Belinda, Colette, and Dimitri are four artists. One is a potter, one is a painter, one is a violinist, and one is a writer. Armand and Colette saw the violinist perform. Belinda and Colette have modeled for the painter. The writer wrote a story about Dimitri and plans to write a story about Armand. Armand does not know the painter. Who is the writer?

TALK IT OVER

Discuss a mystery or detective story you have seen on TV or in the movies. Were you able to solve the mystery before the story ended or were you surprised at the outcome?

Answers will vary.

Armand is the potter. Belinda is the violinist. Colette is the writer. Dimitri is the painter.

3-5   Problem Solving Strategies: Using Logical Reasoning   **105**

ASSIGNMENTS

BASIC
1–4

AVERAGE
1–5

ENRICHED
2–5

ADDITIONAL RESOURCES
Reteaching 3–5
Enrichment 3–5
Transparency Master 14

4 PRACTICE

Problems Encourage all students to try all the problems. In Problem 5, students should deduce that Armand is not the painter, since Collette, the writer, modeled for the painter and Armand does not know the painter.

5 FOLLOW-UP

Extra Practice Jared, Danielle, Ralph and Colleen are playing hide-and-seek. One of the children is *it*. The others are hiding behind the garage, the house, and the tree. From where he is, Jared can see the girl behind the house. Ralph can't see the garage from his position. From where she's hiding, Colleen can see the girl who is *it*. Who is *it*, and where is each child hiding? **Jared: garage; Danielle: it; Ralph: tree; Colleen: house**

Get Ready masking tape, scissors, tracing paper

WARM-UP

A father and son were injured. The doctor in charge looked at the boy and said, "I cannot treat this boy. He is my son." Explain. **The doctor is his mother.**

1 MOTIVATE

Explore/Working Together Arrange students in groups of four. Have each group choose one person to work with the tape. After their first attempt, have them repeat the activity, choosing a different person to work with the tape. When they seem to understand the logic, discuss the answer to the question.

2 TEACH

Use the Pages/Skills Development Have students read this part of the section and then discuss the examples.

Example 1: Have students work in small groups. If possible, allow them to use 3- and 5-quart containers and act out the problem. Compare their solutions with the one presented in the text.

106

3-6 Brainteasers

EXPLORE/ WORKING TOGETHER

If one person has an X, the other two people should know immediately that their tapes are blank. If no one has an X, after a short time they should all realize that they have blank tapes. Otherwise, someone would have seen an X and reported that his or her tape was blank.

Work in groups of four students each.
1. Three students sit in a triangle facing each other.
2. The fourth student prepares three pieces of tape, marking one or none of them with an **X**. This student then places one piece of tape on the forehead of each student in the triangle.
3. By looking only at the other students, each student decides whether the tape on his or her own forehead is marked with an **X**.

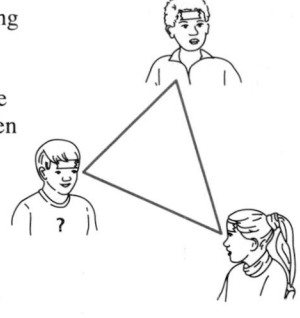

How can each player determine whether his or her tape is blank or marked with an **X**?

SKILLS DEVELOPMENT

Seemingly difficult problems can often be solved if you try a different point of view. In addition to using a different point of view, you will find experimentation, or *trial and error*, helpful.

Example 1

Suppose that you need exactly four quarts of water for a recipe. You have the water available, but you only have 3-quart and 5-quart containers. How can you measure exactly four quarts of liquid by using only the two containers you have on hand?

Solution

It might help to use pictures to help you think of the actions you could take. There are many possible solutions. Here is one of them. In this solution, container A holds 3 quarts and container B holds 5 quarts.

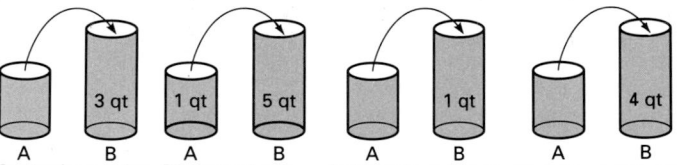

A	B	A	B	A	B	A	B
	3 qt	1 qt	5 qt		1 qt		4 qt

Pour 3 qt from container A into container B.

Fill container A again. From it, fill container B. Now container A has 1 qt and container B has 5 qt.

Empty container B. Place the 1 qt from container A into B.

Fill container A. Empty the 3 qt from container A into container B. Now container B has 4 qt. ◄

Let's see what happens to container B if we use numbers to symbolize the actions pictured above.

$$3 + 2 - 5 + 1 + 3 = 4$$

TEACHING TIP

Some students will struggle with the problems in this lesson. Refrain from solving the problems for them. Encourage students to work together as a group, and not give up. Grading may be difficult. Have students explain their answers, even though it may be time-consuming. Give partial credit for effort and logical thinking.

Example 2

Divide the figure shown into two pieces having the same size and shape.

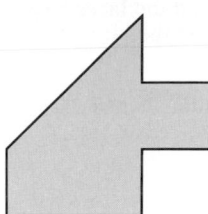

Solution

Suppose that you think about making one cut. Try it. The two figures that result probably won't have the same shape. See the figure at the left below.

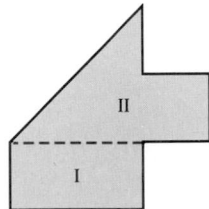
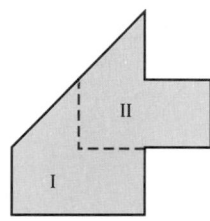

Think about making two cuts. The figure at the right above shows that two cuts will give two figures that have the same size and shape. ◄

TRY THESE

1. A girl put a caterpillar in a glass jar at 8:00 p.m. when she went to bed. Every hour the caterpillar climbed up the wall of the jar 2 inches and slid back 1 inch. If the jar was 6 inches high, at what time did the caterpillar reach the top of the jar?

2. If it takes one minute to make a cut, how long will it take to cut a 10-foot log into ten equal parts? **9 minutes**

3. What is the greatest number of pieces into which you could cut a pizza with four straight cuts of a knife?

1. **1:00 a.m.; on the last move, the caterpillar could leave the jar without sliding back.**

3. **11 pieces; each line should intersect with all the others, but no more than two lines should intersect at any point.**

5-MINUTE CLINIC

Exercise	Student's Error	Error Diagnosis
Is the conditional statement *true* or *false*? *If a number is divisible by 3, then it is an odd number.*	true	• Student fails to think of a counterexample, such as 12 and 30, both of which are even numbers divisible by 3.

ASSIGNMENTS

BASIC
1–3, 5–6, 8–11, 15

AVERAGE
1–4, 5–15, 16

ENRICHED
3–4, 5–15, 16–17

ADDITIONAL RESOURCES
Reteaching 3–6
Enrichment 3–6

Additional Questions/Examples

1. How many squares are in this figure?

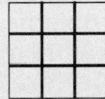

There are 9 one-unit squares, 4 two-unit squares, and 1 three-unit square for a total of 14 squares.

2. Yori has exactly $1.19. He does not have a half dollar. He cannot give exact change for one dollar. How many pennies, nickels, dimes, and quarters could Yori have? **One possible solution is 3 quarters, 4 dimes, and 4 pennies.**

Guided Practice/Try These Have students work in small groups. For Problem 1 students might draw a diagram. For Problem 2 students must realize there will only be 9 cuts. For Problem 3 students might start with a simpler problem and look for a pattern. They should realize that each cut must intersect each of the other cuts to get the maximum number of pieces.

108

3 SUMMARIZE

Talk It Over Discuss this question: *What does this section teach you about solving problems?*

4 PRACTICE

Have students work in small groups over a period of several days. You may wish to give them time to work on two or three problems each day for six days. Encourage them to try each problem and not give up. Explain there are no wrong ideas and that brainstorming is important. Have students explain their answers.

Practice/Solve Problems Students could use the guess-and-check strategy for Problems 1 and 4. For Problem 2 students might want to trace the diagram several times and shade in the triangles in different ways. A diagram would be helpful for Problem 3.

Extend/Solve Problems It would be helpful for students to model Problems 8–14 using cubes.

Think Critically/Solve Problems Students can act out Problem 17 in groups of four.

5 FOLLOW-UP

Extra Practice

1. Move exactly three circles in diagram A to make it look like diagram B.

A. 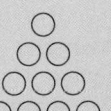   B.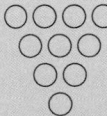

PRACTICE/ SOLVE PROBLEMS

1. A pair of sneakers and the laces cost $15.00 and the sneakers cost $10.00 more than the laces. How much do the laces cost? **$2.50**

2. How many triangles are there in the figure? **35 triangles**

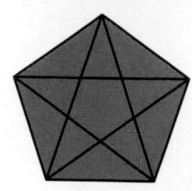

3. A property owner wants to build a fence around a square lot that is 50 feet long on each side. If the fencing comes in 10-foot sections, then how many posts will the owner need? **20 posts**

4. How can you score 100 points with 6 shots if all of them hit the target? **Two 16s and four 17s**

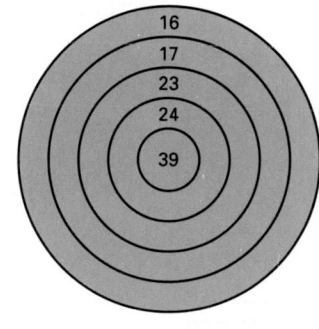

EXTEND/ SOLVE PROBLEMS

5. Suppose you have a 5-gallon, an 11-gallon, and a 13-gallon container. How could you take a 24-gallon container of water and divide it equally into three parts? **See Additional Answers.**

6. There are eight marbles exactly alike, except that one is heavier than the others. Using a balance scale, how can you find the heavier marble in two weighings? **See Additional Answers.**

7. Trace the four shapes below. Cut them out and arrange the pieces to form an equilateral triangle. Then rearrange them to form a square. **See Additional Answers.**

108 CHAPTER 3 Reasoning Logically

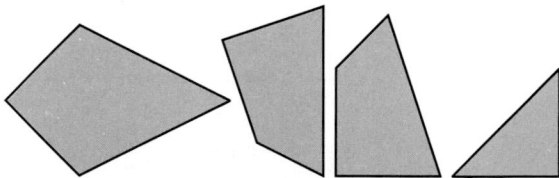

Suppose you have a 3-inch cube painted black.

8. How many cuts would it take to cut the cube into 1-inch cubes? **6 cuts**

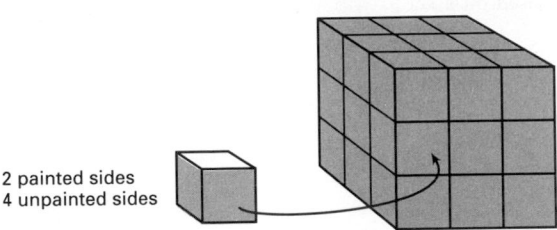

2 painted sides
4 unpainted sides

9. How many cubes would you have? **27 cubes**

10. How many cubes would have no black sides? **1 cube**

11. How many cubes would have exactly one black side? **6 cubes**

12. How many cubes would have exactly two black sides? **12 cubes**

13. How many cubes would have exactly three black sides? **8 cubes**

14. How many cubes would have exactly four black sides? **0 cubes**

15. Add three straight lines to get nine triangles, not counting those that overlap. **See Additional Answers.**

16. Two mothers and three daughters went to see a ballgame. There were only three tickets, yet each one had a ticket. How could this be? **They were a grandmother, mother, and a daughter.**

17. A man has to take a wolf, a goat, and some cabbage across a river. His rowboat has enough room for the man plus either the wolf or the goat or the cabbage. If he takes the cabbage with him, the wolf will eat the goat. If he takes the wolf, the goat will eat the cabbage. Only when the man is present is the goat safe from the wolf and the cabbage safe from the goat. How does he get them across? *Hint:* He makes the round trip three times before making his final crossing.

Find the median, mean, mode, and range for these scores.

1. 82, 87, 90, 77, 65, 81, 82, 88
82; 81.5; 82; 25

Add or subtract.

2. 2,159 + 2,607 **4,766**

3. 5,942 − 4,998 **944**

4. 6.94 + 128.526 **135.466**

5. 307 − 291.75 **15.25**

Multiply or divide.

6. 1.8
×1.8
3.24

7. 0.008
× 0.3
0.0024

8. 1.96 ÷ 0.7 **2.8**

9. 36 ÷ 1.8 **20**

Solve.

10. How much change would you receive if you purchased a sweater that cost $39.99 with a hundred dollar bill? **$60.01**

THINK CRITICALLY/ SOLVE PROBLEMS

17. **He first takes the goat across and returns alone. Next, he takes the cabbage across and returns with the goat. Then he leaves the goat and takes the wolf across and returns alone to get the goat. At no time are the potential enemies alone.**

3–6   Brainteasers   **109**

Move the top circle down below the middle of the last row and move one circle from each end of the last row up to what is now the top row.

2. Place parentheses to make this equation true:
(36 ÷ 4) + (5 +1) + (2 × 3) − (12 ÷ 4) = 6

3. How many different ways is it possible to make change for 50¢? **49**

Extension Ask students to bring in similar problems and puzzles to share with the class.

Section Quiz

1. How many squares are in this figure?

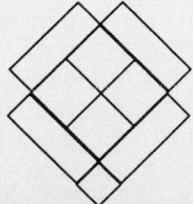

11

2. What system was used to arrange the ten digits in this order? 8, 5, 4, 9, 1, 7, 6, 3, 2, 0 **alphabetical order**

3. Remove two segments so exactly two squares remain.

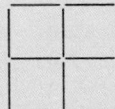

Remove any 2 adjacent interior segments.

Additional Answers
See page 575.

3 CHAPTER REVIEW

Introduction The Chapter Review emphasizes the major concepts, skills, and vocabulary presented in this chapter and can be used for diagnosing students' strengths and weaknesses. Page references direct students back to appropriate sections for additional review and reteaching.

Using Pages 110–111 Allow students to quickly scan the Chapter Review and ask questions about any section they find confusing. Exercises 1–9 review key vocabulary. Students may find Exercise 16 challenging. The only figures that fit the category described are those people who preferred only white-wall tires (8) or who preferred white walls and bucket seats but not front-wheel drive (25). For Exercise 18, students might use the guess-and-check strategy, or they might use the equation $4x + 20 = 2(x + 20)$ where $x =$ Li's age 20 years ago.

Informal Evaluation Have students explain how they arrived at any incorrect answers. They will probably find their own mistakes and give you some clues as to the nature of their errors. Make sure students understand this material before administering the Chapter Test.

Follow-Up Have students use the three views below to determine which sides are opposite each other on the cube. You may wish to suggest that students label a cube to help them.

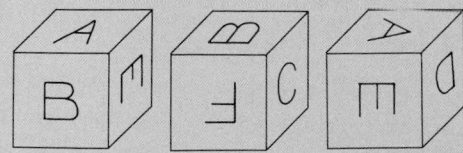

C is opposite A, F is opposite E, and D is opposite B.

CHAPTER 3 ● REVIEW

Write the letter for the word at the right that matches each description.

1. Process of reaching a conclusion based on a set of specific examples **f**

2. Trial generalization **e**

3. Shows that a conditional statement is false **b**

4. *If* part of a conditional statement **a**

5. *Then* part of a conditional statement **g**

6. Kind of argument in which there is an error in reasoning **d**

7. Reasoning from a conditional statement and other information, to draw a conclusion **h**

8. A picture that the eye sees as true that is actually not true **i**

9. Shows the relationships between two or three different classes of things **c**

a. hypothesis
b. counterexample
c. Venn diagram
d. invalid
e. conjecture
f. inductive reasoning
g. conclusion
h. deductive reasoning
i. optical illusion

SECTION 3–1   OPTICAL ILLUSIONS   (pages 90–93)

► An optical illusion is a picture in which the human eye perceives something to be true that is actually not true.
► A statement is a kind of sentence that is either true or false. You can often make a test to find out whether a statement is true or false.

Refer to the figures below. Make a statement about each figure. Then check to see whether your statement is true or false.

10.
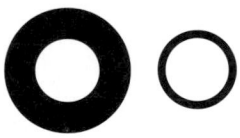

11.

The interior circles are the same size.   **The segments running from left to right are parallel.**

SECTION 3–2   INDUCTIVE REASONING   (pages 94–97)

► You can use a large number of examples to support a conjecture, or trial generalization, but the examples do not necessarily prove that the conjecture is always true. Yet you can prove a conjecture false with just one example that contradicts the generalization.

12. Pick a number. Double it. Add 6. Add the original number. Divide by 3. Add 4. Subtract the original number. What do you get? Repeat the process several times using a different original number. Then make a conjecture about the result of the process.   **The result is always 6.**

SECTION 3-3   DEDUCTIVE REASONING   (pages 98–101)

▶ An *if-then* statement is called a conditional statement. The statement in the *if* part is called the hypothesis. The statement in the *then* part is called the conclusion. A conditional statement is false if a counterexample satisfies the hypothesis but not the conclusion.

▶ In a valid argument, a new statement can be derived from given statements by deductive reasoning. If there is an error in the reasoning, the argument is invalid.

13. Write two conditional statements that can be made from these two sentences:
    If it has been snowing, then the streets are slippery.
    It has been snowing. The streets are slippery.
    If the streets are slippery, then it has been snowing.

Determine whether the following arguments are *valid* or *invalid*.

14. If today is Monday, then I will play tennis.
    Today is Monday.
    Therefore, I will play tennis.   **valid**

15. If it rains on me, then I shall get wet.
    It did not rain on me.
    Therefore, I did not get wet.   **invalid**

SECTIONS 3-4 AND 3-5   PROBLEM SOLVING   (pages 102–105)

▶ Some problems involving relationships between classes of things can be solved using a Venn diagram.
▶ Many problems involving logical relationships are easier to solve if you use a table to keep track of the clues.

16. In a survey of 250 people, 156 said they preferred front-wheel-drive cars, 183 said they preferred bucket seats, and 147 said they preferred white-wall tires. There were 126 who preferred front-wheel drive and bucket seats, 102 who preferred bucket seats and white-wall tires, and 114 who preferred front-wheel drive and white-wall tires. In all, 87 people preferred all three. How many preferred white-wall tires, but not front-wheel drive? **33**

17. Three students named Alfred, Betty, and Charlie receive three different grades on a test. One received an A, one received a B, and one received a C. No student received a grade matching the first letter of the student's name. Alfred did not receive a B. What grade did each receive?**Alfred: C; Betty: A; Charlie: B**

SECTION 3-6   BRAINTEASERS   (pages 106–109)

▶ Solving problems such as brainteasers often requires you to change your point of view toward the problem.

18. Suppose that, 20 years ago, Rhoda was four times as old as Li. Rhoda is only twice as old as Li now. How old is each person now? **Rhoda: 60; Li: 30**

3 CHAPTER TEST

Introduction The Chapter Test uses a variety of questioning techniques to assess students' mastery of the major objectives of Chapter 3. If you prefer, you may use the Skills Preview (page 87) as an alternative form of the Chapter Test. The items on this test and the Skills Preview correspond in content and level of difficulty.

Alternative Assessment
Critical Thinking Draw the missing shape.

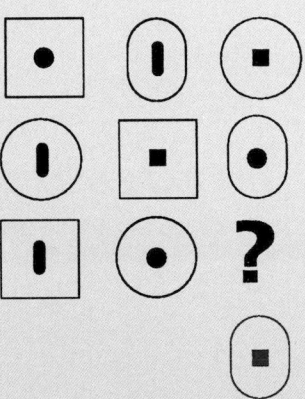

1. Refer to the figure below. Make a statement about line segments *MN* and *ST*. Then check to see whether your statement is true or false.

M ———————— N

S ———————— T

Line segments *MN* and *ST* have the same length.

2. Choose any single-digit number. Multiply it by 9. Then multiply by 12,345,679. What number do you get? Repeat this process with several numbers. Then make a conjecture about the result of the process. **For any digit *n*, the result will be *nnn,nnn,nnn*.**

3. Experiment with several different sets of numbers and make a conjecture about the product of an even number, an odd number, and an even number. **The product will always be even.**

4. Write two conditional statements, using the following two sentences:
   Something is made of wood. It will burn.
   If something is made of wood, then it will burn. If something will burn, then it is made of wood.

Is each of the following statements *true* or *false*? If false, give a counterexample.

5. If a man is a bachelor, then he is unmarried. **true**

6. If a number is divisible by 7, then it is divisible by 14. **false; 21**

Determine whether the following arguments are *valid* or *invalid*.

7. If today is Friday, then I will
      go to school.
   I went to school today.
   Therefore, today is Friday. **invalid**

8. If a horse wins the race, then
      it runs fast.
   Lightning won the race.
   Therefore, Lightning ran fast. **valid**

9. In a survey of employees in an office, it was found that 42 wanted peanuts to be placed in the vending machine, 56 wanted raisins, and 30 wanted granola bars. The survey showed that 18 wanted both peanuts and granola bars, 31 wanted peanuts and raisins, and 12 wanted raisins and granola bars. There were 9 who wanted all three. How many employees wanted either raisins or granola bars in the vending machine? **There were 74 who wanted raisins or granola bars.**

10. Terri, Jonathan, and Kathy each have different jobs. One is a writer, one is a bus driver, and one is a chef. Terri cannot drive. Kathy cannot cook. Jonathan does not know the chef. Kathy is the writer's sister. What job does each person have? **Terri: chef; Jonathan: writer; Kathy: bus driver**

The following data show the number of pies eaten by each person in a pie-eating contest.

$$8 \quad 4 \quad 12 \quad 10 \quad 3 \quad 5 \quad 7 \quad 9 \quad 11$$
$$14 \quad 6 \quad 11 \quad 2 \quad 1 \quad 6 \quad 14 \quad 8 \quad 4$$
$$10 \quad 13 \quad 12 \quad 5 \quad 7 \quad 15 \quad 6$$

1. Construct a stem-and-leaf plot for the data. **See Additional Answers**.

2. How many pies did the winner eat? **15**

3. How many contestants ate 10 or more pies? **9**

4. How many contestants ate between 8 and 12 pies? **9**

The table shows the bowling scores for four games played by Jennifer and Ricardo.

Name	Game 1	Game 2	Game 3	Game 4
Jennifer	144	151	136	157
Ricardo	153	146	132	161

5. Find the mean score for Jennifer and for Ricardo. **147; 148**

6. Who had the higher average? **Ricardo**

The annual salaries of six people follow:

$18,600 $19,400 $24,000
$20,000 $21,000 $45,000

7. Find the mean salary and the median salary. **$24,667; $20,500**

8. Is the mean or the median the more representative of the data? Give reasons for your answers. **median; 45,000 is extreme**

Use the pictograph for Exercises 9–11.

MEMBERS OF THE BIG SERVE TENNIS CLUB

Novice	🎾 🎾 🎾 🎾
Intermediate	🎾 🎾 🎾 🎾 🎾 🎾
Advanced-Intermediate	🎾 🎾 🎾 🎾
Advanced	🎾

Key: 🎾 represents 8 players

9. Which league has the most players? How many players are there? **intermediate, 48**

10. How many more novice players are there than advanced-intermediate? **4**

11. Which group has four times more players than the advanced group? **intermediate**

Evaluate.

12. $6 + 9 \div 3 - 2$ **7**

13. $8^2 \div 4 - 10$ **6**

14. $3 \times 9 + 12 \div 3$ **31**

15. Determine whether the following argument is *valid* or *invalid*.

   If the sun is shining, then
   I will go swimming.
   The sun is shining.
   _____
   Therefore, I will go swimming. **valid**

16. Three girls named Jo, Myra, and Tawanda have different hobbies. One collects stamps, one collects shells, and one collects bottles. Jo's collection will not fit in a book. Myra does not collect shells or stamps. What does each collect? **Jo: shells; Myra: bottles; Tawanda: stamps**

Introduction The purpose of this Cumulative Review is to maintain previously taught skills and concepts and to apply them to the material presented in this chapter. At least one major objective of each chapter is included in the review.

Item Analysis The table below correlates the Cumulative Review items with the chapter and section that are being reviewed.

Section	Items
1–2	1–4
1–3	9–11
1–9	5–8
2–8	12–14
3–3	15
3–5	16

Additional Answers
See page 576.

3 CUMULATIVE TEST

Introduction The Cumulative Test uses a standardized-test format of multiple-choice questions to assess retention of previously learned concepts and test-taking skills. Test results may be used to diagnose students' strengths and weaknesses.

Item Analysis The table below correlates the Cumulative Test items with the chapter and section that are being tested.

Section	Items
1–1	1, 2
1–6	11–13
1–9	3–6
3–3	14, 15

1. A forester selected one tree out of every 25 along the roadside to test for the effects of pollution. What type of sampling is this?
   A. random   B. systematic
   C. convenience   D. cluster

2. A stereo manufacturer tested every twentieth stereo coming off the assembly line. What type of sampling is this?
   A. systematic   B. convenience
   C. cluster   D. none of these

Use these data for Exercises 3–6.

| 7 | 14 | 14 | 8 | 9 | 11 | 8 | 10 | 7 | 6 | 8 |

3. What is the mean of the data?
   A. 8   B. 9.3   C. 11.3   D. 7

4. What is the median of the data?
   A. 8   B. 9   C. 10   D. 9.3

5. What is the mode of the data?
   A. 14   B. 8   C. 7   D. 9.3

6. What is the range of the data?
   A. 9.3   B. 8   C. 14   D. 7

7. Evaluate. $8 - (4 \div 2) + 2$
   A. 8   B. 4   C. 2   D. 0

8. Evaluate. $12 + 3 \div 3 - 4$
   A. 1   B. 9   C. 15   D. 14

9. Evaluate. $3 + 6 \div (3 - 1)$
   A. 2   B. 4   C. 5   D. 6

10. Evaluate. $6^2 \div 2 - 4$
    A. 0   B. 5   C. 14   D. 2

Use the bar graph for Exercises 11–13.

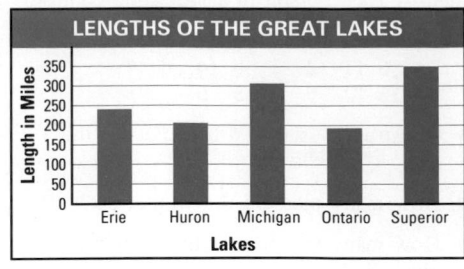

LENGTHS OF THE GREAT LAKES

11. Which is the longest lake?
    A. Michigan   B. Superior
    C. Huron   D. Ontario

12. Which lakes are over 250 miles long?
    A. Superior, Huron
    B. Superior, Erie
    C. Michigan, Erie
    D. Michigan, Superior

13. About how long is Lake Michigan?
    A. 350 miles   B. 300 miles
    C. 250 miles   D. 275 miles

14. Which is a counterexample for the following statement?

    If a number is prime, then it is not divisible by 7.
    A. 11   B. 8   C. 13   D. 7

15. Which conclusions can be drawn from these statements?

    If the vase is made of glass, then it is fragile.
    The vase is made of glass.
    A. The vase is fragile.
    B. The vase is not fragile.
    C. The vase is made of glass.
    D. none of these

Name _____ Date _____

Optical Illusions

Sometimes you can look at a picture and see something that you think is there, but in reality it is not there. Such a picture, which deceives your eyes, is called an **optical illusion.**

▶ **Example 1** _____

Refer to the figure at the right. Make a statement about AB, BC, CD, and DA in the figure. Then check to see if your statement is true or false.

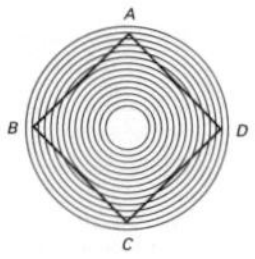

Solution

In the figure AB, BC, CD, and DA appear to bend in toward the circle in the middle.

Make a statement: AB, BC, CD, *and* DA *are not straight lines.*

Test the statement. Lay a ruler down along each line segment. AB, BC, CD, and DA are all straight. All have the same length.

So, the statement is false.

EXERCISES _____

Look carefully at the figure. Make a statement about each figure. Then check to see whether the statement is true or false.

1.

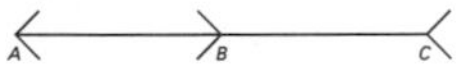

Both lines have the same length.

2.

Both bars are the same size and shape.

3.

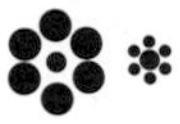

The circles in the center of each figure are both the same size.

Name _____ Date _____

Optical Illusions

Optical illusions are a special and very interesting kind of art in which lines or curves seem to "get in the way," or prevent the eye from seeing things truly. Lines, circles, and rectangles, for example, whose dimensions are actually equal often appear to have dimensions that are not equal.

▶ **Example** _____

Is this figure a spiral?

Solution

Trace each curve around the circle and measure its distance from the center of the figure.

The figure is a series of concentric circles.

EXERCISES _____

1. Examine Figure A at the right. Make a statement about the size of the light and dark regions.
   The two regions are
   equal in size.

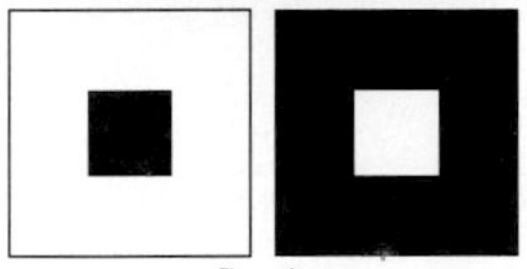

Figure A

2. Examine Figure B at the right. Make a statement about the line connecting 8 and 1.
   Though the line appears
   to be curved, it is actually
   a series of straight lines.

3. Create your own optical illusion and try it out on friends and classmates.

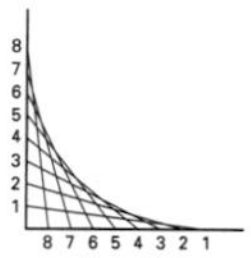

Figure B

Name _____ Date _____

Inductive Reasoning

You are using **inductive reasoning** when you:
- examine several examples;
- find a pattern or rule that explains those examples;
- make a trial generalization or **conjecture** stating the rules that explain those examples.

You might find examples that support your conjecture, but you cannot prove it is true by examples. In fact, *just one example* that does not support your conjecture, called a **counterexample,** proves that your conjecture is not true.

▶ **Example** _____

Joann notices the following:
   $12 + 21 = 33$ $34 + 43 = 77$ $62 + 26 = 88$
She makes the conjecture:
   To a 2-digit number, adding the same 2-digit number with the digits reversed gives a sum with repeating digits.
Test this conjecture. Can you find a counterexample?

Solution

Try other examples. Do they support the conjecture?
   $72 + 27 = 99$ (yes)
   $78 + 87 = 165$ (no)
You found a counterexample. Joann's conjecture is not true.

EXERCISES _____

Make a conjecture. Give at least 3 examples to support it. **Answers will vary.**

1. $2 \times 4 = 8$
   $3 \times 10 = 30$
   $4 \times 6 = 24$
   $5 \times 12 = 60$
   Any multiple of an even number is
   even

2. Take your age and double it. Add 10 and double the result. Subtract 20 and divide the result by 4. Repeat with someone else's age.
   The result is the age you started with.

Here are some conjectures. Test them. Can you find a counterexample?
                                   Counterexamples may vary.

3. Any multiple of an odd number is odd.
   Counterexample: $2 \times 3 = 6$

4. The sum of any number and an odd number is an even number.
   Counterexample: $0 + 3 = 3$; $2 + 3 = 5$

5. When a number that ends in 2 zeros is multiplied by a number that ends in 3 zeros, their product always ends in exactly 5 zeros.
   Counterexample: $500 \times 2000 = 1,000,000$

Name _____ Date _____

Palindromes

A **palindrome** is a group of digits or letters that read the same from left to right or right to left. Some examples are:

   Hannah   Mom   1221   1467641

▶ **Example** _____

Study this conjecture:

   Begin with any number. Reverse its digits to form a second number. Find the sum of the numbers. If the result is not a palindrome, use the sum as the first number, and repeat the process. The result will eventually be a palindrome.

Solution

Try several examples to test the conjecture. Can you find a counterexample?

Try 47. $\begin{array}{r} 47 \\ +74 \\ \hline 121 \end{array}$ Try 85. $\begin{array}{r} 85 \\ +58 \\ \hline 143 \\ +341 \\ \hline 484 \end{array}$ Try 113. $\begin{array}{r} 113 \\ +311 \\ \hline 424 \end{array}$

In each example, the result was a palindrome.

EXERCISES _____

Make and test conjectures. Give 3 or more examples to support a conjecture or find one counterexample to disprove it.

1. Conjecture: You can form palindromes using decimal numbers and the process above. Use a decimal number with any number of decimal places to test it.
   Conjecture is true for any example. Examples will vary.

2. Start with any 2-digit number. Reverse the digits. Subtract the smaller number from the larger. Make a conjecture. Test it.
   The result is always divisible by 9.

3. Try the conjecture you made in Exercise 2 for 3-digit numbers. Is the same conjecture true for 3-digit numbers?
   Yes; the conjecture is true for 3-digit numbers.

4. Examine the table at the right. Write a conjecture. Test it.
   Conjecture: Palindromes are divisible by 11.
   Not true: counterexample: 15,451 is not divisible by 11.

   $121 \div 11 = 11$
   $1331 \div 11 = 121$
   $14641 \div 11 = 1331$
   $12221 \div 11 = 1111$

Name _____ Date _____

Deductive Reasoning

A **conditional statement** has two parts, a hypothesis and a conclusion.

      hypothesis   conclusion

   If a person lives in Salt Lake City, then that person lives in Utah.

A conditional statement is false if you can find a counterexample that satisfies the hypothesis but does not satisfy the conclusion.

▶ **Example 1** _____

Write two conditional statements, using these two sentences:
A figure is a triangle. A figure has three angles.

Solution

If a figure is a triangle, then that figure has exactly three angles.
If a figure has exactly three angles, then that figure is a triangle.

▶ **Example 2** _____

Is the statement true or false? If false, give a counterexample.
If the number is divisible by 3, then it is divisible by 6.

Solution

The number 9 is a counterexample: 9 is divisible by 3 (satisfies the hypothesis) but is not divisible by 6 (does not satisfy the conclusion). So, the statement is false.

EXERCISES _____

1. Write two conditional statements, using these pairs of sentences: *The pond has ice a foot thick. We can skate on the pond.*

   If the pond has ice a foot thick, then we can skate on the pond.

   If we can skate on the pond, then the pond has ice a foot thick.

Is each conditional true or false? If false, give a counterexample.

2. If a number is divisible by 9, then the number is divisible by 3.

   true

3. If a number is divisible by 12, then it is divisible by 18.

   false; 24 is a counterexample; 24 is divisible by 12 but not by 18.

Name _____ Date _____

Yes and No

The truth or falsity of a statement is its **truth value.**

If a statement is true, then its opposite, or **negation,** is false.
If the statement is false, then its negation is true.

▶ **Example** _____

Write the negation of the statement. Give the truth value of the statement and its negation.

  a. The sum of 5 and 7 is an even number.
  b. The product of two odd numbers is an even number.

Solution

  a. Negation: The sum of 5 and 7 is not an even number.
     The statement is true. Its negation is false.

  b. Negation: The product of two odd numbers is not an even number.
     The statement is false. Its negation is true.

EXERCISES _____

Write the negation of the statement and give the truth value of the statement and its negation.

1. The product of 12 and 12 is 144.   **(true)**

   Negation: The product of 12 and 12 is not 144. (false)

2. Shakespeare was not a playwright.   **(false)**

   Negation: Shakespeare was a playwright. (true)

3. The product of a number multiplied by itself is the square of the number.   **(true)**

   Negation: The product of a number multiplied by itself is not the square of the number. (false)

4. The sun is not a star.   **(false)**

   Negation: The sun is a star. (true)

5. What statement will you get if you write the negation of both the hypothesis and the conclusion of this conditional statement: *If you live in San Diego, then you live in California*?

   If you do not live in San Diego, then you do not live in California.

6. Which statement is true and which false?

   The statement is true; its negation is false.

Name _____ Date _____

Problem Solving Skills: Using Venn Diagrams

Venn diagrams show relationships of sets. In a Venn diagram there are two or three intersecting. Each circle represents one set. The circles are in a rectangle representing the universe, the total number of things being considered. Different parts of the intersecting circles represent those parts of the sets that share one or more characteristics.

▶ **Example** _____

Here are the results of a survey of the reading habits of 100 people: 18, mystery; 20, romance; 14, nonfiction; 22, mystery and romance; 18, romance and nonfiction; 25, mystery and nonfiction; 10, all three types.

How many readers in all read mystery books?

Solution

Draw a Venn diagram. Label each circle for one kind of book. Write the total number of items in the universe.

In the center section, enter the number who read all three types (10). In the parts of the circles not intersected, enter the numbers who read only mysteries (18), only romance (20), and only nonfiction (14).

From each group who read only two kinds of books, subtract 10, the number who read all three. Enter those numbers in the sectors representing each group.

Add all the numbers *inside* the circle representing mystery books.

There are 55 people in all who read mystery books.

EXERCISES _____

Use a Venn diagram to answer the questions.

A survey of 178 drivers asked which of these features drivers want in a car: a sunroof, 4-wheel drive, or stereo tape deck. The survey showed that 48 drivers want all three; 8 want only the sunroof; 8 want 4-wheel drive only; 10 want only the stereo tape deck. In all, 58 drivers want a sunroof and 4-wheel drive; 118 want 4-wheel drive and stereo tape deck; and 70 want sunroof and stereo tape deck.

1. How many in all want a stereo tape deck?   **140**

2. How many in all look for a sunroof?   **88**

3. How many in all look for 4-wheel drive?   **136**

4. How many want 4-wheel drive and tape deck but not a sunroof?   **70**

5. Out of those surveyed, how many want none of these features?   **2**

Name _____ Date _____

Problem Solving, Sets, and Venn Diagrams

The circles in Venn diagrams can represent sets of numbers.

▶ **Example** _____

Laney needs to make a doctor's appointment for sometime in March. Laney is free any Monday, her day off. Her doctor alternates office hours with a partner. Her doctor is there only on odd-numbered days. Assuming March begins on a Monday, what dates are possibilities for Laney's doctor appointment?

Solution

Laney is available only on Mondays. Since March 1st is a Monday, Laney can make an appointment on the 1st, and every 7 days thereafter.

Her doctor is available on odd-numbered days only. March has 31 days.

The Venn diagram shows only three days when both Laney and her doctor are available. The intersection of the two circles contains the numbers 1, 15, and 29. Laney must make her appointment for March 1st, 15th or 29th.

EXERCISES _____

Pete, Rene, and Santos are three old friends who are trying to find a time to get together. Pete can meet only on even-numbered days. Rene is available on the third and every third day thereafter, and Santos is out of town for the first two weeks in every month. Draw a Venn diagram to solve.

1. When can Pete, Rene, and Santos meet?   **18th, 24th, and 30th**

2. When can Pete and Rene, but not Santos meet?   **6th and 12th**

3. When can Santos and Pete, but not Rene meet?   **16th, 20th, 22nd, 26th, and 28th**

4. When can Rene and Santos, but not Pete meet?   **15th, 21st, and 27th**

5. Make up a question that can be answered from your Venn diagram.

   Answers will vary.

114B

Name _____ Date _____

Problem Solving Strategies: Using Logical Reasoning

For logic problems, it is useful to make a table for keeping track of clues.

PROBLEM

Lou, Sue, and Drew ran in the same race. Drew finished second. Lou congratulated Sue on doing better than he himself did. Who finished third?

SOLUTION

Make a table as shown. Put ✓ in a box to indicate that a fact is true. Put X in any box for a fact that is not true.

Since Drew finished second, put ✓ in the box where the row labeled "Drew" and the column labeled "2nd" meet. Put X's in the other two boxes in the "Drew" row. Put X's in the other boxes in the "2nd" column, since neither Lou nor Sue came in second.

	1st	2nd	3rd
Lou		X	
Sue		X	
Drew	X	✓	X

Since Sue did better than Lou, she must have come in first. Put ✓ in the box where the row labeled "Sue" and the column labeled "1st" meet. Put X's in the empty boxes in Sue's row and in the "1st" column. That leaves only one empty box: Lou must have finished third.

	1st	2nd	3rd
Lou	X	X	
Sue	✓	X	X
Drew	X	✓	X

EXERCISES

Solve each problem. Complete the table to help solve each problem.

1. The Smiths have three children, Loretta, Tom, and Steve. The middle child is a girl. Tom is not the oldest. Who is the youngest child?

   Tom

	Oldest	Middle	Youngest
Loretta	X	✓	X
Tom	X	X	✓
Steve	✓	X	X

2. Amy, Bud, and José have different pets. One has a bird, one a snake, and one has a kitten. José's pet has fur. Bud's pet does not have wings. Who has the snake?

   Bud

	Bird	Snake	Kitten
Amy	✓	X	X
Bud	X	✓	X
José	X	X	✓

Name _____ Date _____

More Logical Reasoning for Problem Solving

In order to solve challenging logical problems,
- make a table to organize your thoughts.
- use *if-then* (conditional) thinking. Assume something is true. Then look for counterexamples to eliminate that possibility.

▶ **Example**

Three friends ate lunch with a fourth friend. Afterwards, the three gave conflicting stories about what the fourth friend ate:

Amy: She had fruit cocktail, then ordered roast turkey and dessert. She finished with a cup of tea.

Bonnie: She skipped the fruit cocktail, ordered steak, and ended with chocolate cake.

Chris: She had fruit cocktail, ate a steak, had strawberry shortcake for dessert, then ordered tea.

One of the friends had no facts correct; a second reported only one item incorrectly; the third friend told every detail accurately. Use logical reasoning to determine what the fourth friend had for dessert.

Solution

- Make a table.
- *Think:* If Amy is accurate, Bonnie is inaccurate. Amy's account has two things in common with Chris's. Therefore, Amy's account is correct and Bonnie's is wrong. Chris was wrong about the steak, so must be correct about the strawberry shortcake.

	fruit cocktail	meat	dessert	tea
Amy	yes	turkey	dessert	yes
Bonnie	no	steak	chocolate cake	no
Chris	yes	steak	strawberry shortcake	yes

The fourth friend had strawberry shortcake for dessert.

Use a table and logical reasoning to solve the following problem.

Three schoolmates, Tom, Bill, and Bo, went shopping with a fourth friend, Doug. Afterwards, the three gave conflicting stories about what the fourth friend bought.

Tom: Doug first bought a sweater. Then he bought a pair of walking shoes and next, rainwear. His last purchase was a birthday card.

Bill: Doug looked at a sweater but didn't buy it. He did buy a pair of basketball shoes, and last, a get-well card.

Bo: Doug bought a sweater. Then he bought a pair of basketball shoes. After that, he bought an umbrella. His last purchase was a birthday card.

One person got no facts right; another reported one purchase incorrectly; the third remembered every purchase accurately. What was Doug's third purchase?

Doug's third purchase was an umbrella.

Name _____ Date _____

Brainteasers: Different Points of View

Mysteries, problems, and brainteasers can often be solved by changing your approach—looking at the problem in a new way.

Study the picture at the right. Do you see the stairs from above (side B nearer to you) or from below (side A nearer to you)? Keep trying! You can shift your focus.

▶ **Example**

Study the arrangement of toothpicks at the right. How can you remove 2 toothpicks so that exactly 2 squares remain?

Solution

A possible solution is shown. Notice that the 2 squares that remain are not the same size, and one is within the other. Can you find another solution?

EXERCISES

Use the toothpick arrangement in the Example for Exercises 1–3.

1. Remove 4 toothpicks to make 1 square.

2. Remove 4 toothpicks to make 2 squares.
   Answers will vary.

3. Move 3 toothpicks and rearrange them so that there are 3 small squares, all the same size and shape.
   Answers will vary.

4. Move two dots so that the arrangement of dots is inverted (points up rather than down).

5. Study the arrangement of toothpicks at the right. How can you remove 2 toothpicks to make exactly 2 triangles?
   Answers will vary.

6. Study the arrangement of 15 toothpicks at the right. Remove 3 toothpicks so that exactly 3 squares are left.

7. Using the arrangement of 15 toothpicks from Exercise 6, remove 2 toothpicks to make 3 squares.

Name _____ Date _____

More Brainteasers

Sometimes you have to change your perspective to avoid getting stuck on a problem.

▶ **Example**

Study the pattern of 9 dots. Without lifting your pencil, draw exactly 4 straight lines that cross all 9 dots.

Solution

You might get stuck on this problem because you assume that you must restrict your lines to the boundaries set by the dots. Changing your perspective allows you to draw lines beyond the boundary and solve the puzzle.

EXERCISES

Divide each figure below into 2 identical smaller pieces that have the same shape as the large figure.

1.

2.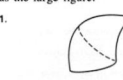

3. Show how the figure at the right can be divided into 4 figures having the same size and shape.

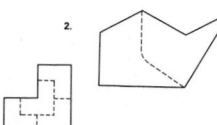

4. Pictured are 13 identical lengths of fence that a farmer used to make 6 animal enclosures. If one of the fences is broken, how can the farmer arrange the remaining 12 lengths to make 6 enclosures of the same size and shape?

5. Place 4 points on a flat surface. Connect all the points with straight lines so that the lines are of only two different lengths. One solution is shown. Find the other five solutions.

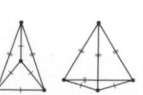
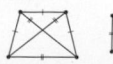
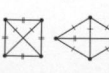

114C

Name _____ Date _____

Computer Activity: Testing Conjectures

When making a conjecture, you can use the computer to help you generate examples. By examining these examples, you can test the validity of your conjecture. For example, this program gives results for Exercise 6 on page 96.

```
10  INPUT "ENTER AN EVEN NUMBER:     Asks for an even number
    ";E: PRINT
20  IF E / 2 < > INT (E / 2) THEN    Checks that it was even
    GOTO 10
30  S = E + 20                        Adds 20 to the number
40  P = S • 2                         Multiplies the sum by 2
50  Q = P / 4                         Divides the product by 4
60  D = Q – 10                        Subtracts 10 from quotient
70  R = D • 2                         Multiplies difference by 2
80  PRINT "RESULT IS ";R              Prints the result
```

Each time you want another result, RUN the program.

EXERCISES

1. What happens when you change lines 10, 20, and 80 and add line 90 as shown below?

```
10 ? "NUMBER", "RESULT"        80 ? E,R
20 FOR E = 2 TO 10 STEP 2      90 NEXT E
```

A running list of the even numbers from 2 to 10 is printed in one column and the results in another column.

2. What does STEP 2 do?

Adds an increment of 2 to each value of E

3. Adjust the program to check if the conjecture works for odd numbers also.

20 for E = 1 to 11 STEP 2   It does work.

Name _____ Date _____

Testing Conjectures (continued)

4. Write a program that allows you to test your conjecture for Exercise 7. If you use a FOR-NEXT loop, would you need a STEP? Why or why not?

```
10  INPUT "ENTER A NUMBER:
    ";N: PRINT
20  S = N + 12
30  P = S • 2
40  D = P – 4
50  Q = D / 4
60  R = Q – 1 / 2 • N
70  PRINT "RESULT IS ";R
```

5. Supply the missing steps for this program that will test your conjecture for Exercise 10 on page 97. Have several of your friends RUN the program to check your conjecture.

```
10 INPUT "ENTER YOUR AGE: ";A:?
20 INPUT "ENTER THE DAY OF THE MONTH
   YOU WERE BORN: ";D: ?
30 S = A = 5
40 T = 2 • S • 50
50 U = T + D
60 F = 2 • U – 1000
70 G = F/2
80 ?? A; " ";
90 IF D < 10 THEN ? "0";
100 ? D;" ";G: ?
110 IF D > = 10 THEN ? D; " ";G: ?
```

6. Would it be helpful to put a FOR-NEXT loop in the program in Exercise 4? Why or why not?

No because the information is not INPUT in consecutive order. Also there are 2 INPUTS.

CHALLENGE

Write a program that will give results to let you check your conjecture for Exercise 14.

```
10  INPUT "HOW MANY NUMBERS? ";N: PRINT
20  FOR X = 1 TO N
30  S1 = S1 + X
40  NEXT X
50  S2 = N • (N + 1) / 2
60  PRINT S1,S2
```

Name _____ Date _____

Calculator Activity: Patterns

A calculator can help you discover number patterns.

► **Example**

Use a calculator to find the first four multiples of 99. Look for a pattern and predict the fifth multiple. Then use a calculator to check your answer.

Solution

With some calculators, press the operation key (+, –, ×, or ÷) twice to store the displayed number as a constant. The letter K appears on the display. Check your manual to find out how to enter a constant on your calculator.

```
99 ×  × 1 =  99      Notice that the hundreds digits increase by 1.
       2 = 198       The tens digits remain the same, 9. The ones
       3 = 297       digits decrease by 1. So the fifth multiple of 99
       4 = 396       must be 495. Use a calculator to check this
                     answer. 5 × 99 = 495
```

EXERCISES

1. What do you notice about the sum of the hundreds digit and the ones digit of each multiple of 99?

The sum is always 9.

2. Keep finding multiples of 99. What happens to the pattern when you pass 10 × 99?

The products become four-digit numbers. The hundreds digits continue increasing by 1; the tens digits decrease to 8. The ones digits decrease by ones.

Name _____ Date _____

Patterns (continued)

Use a calculator to find the first four products in each exercise. Look for a pattern in the products. Then write the fifth product without using a calculator.

3.
$4 \times 3,478 =$ **13,912**
$4 \times 34,978 =$ **139,912**
$4 \times 349,978 =$ **1,399,912**
$4 \times 3,499,978 =$ **13,999,912**
$4 \times 34,999,978 =$ **139,999,912**

4.
$909 \times 1 =$ **909**
$909 \times 2 =$ **1,818**
$909 \times 3 =$ **2,727**
$909 \times 4 =$ **3,636**
$909 \times 5 =$ **4,545**

5. Write the next six products for Exercise 4 without using a calculator.

909 × 6 = 5,454
909 × 7 = 6,363
909 × 8 = 7,272
909 × 9 = 8,181
909 × 10 = 9,090
909 × 11 = 9,999

6.
$6 \times 7 =$ **42**
$66 \times 67 =$ **4,422**
$666 \times 667 =$ **444,222**
$6,666 \times 6,667 =$ **44,442,222**
$66,666 \times 66,667 =$ **4,444,422,222**

7.
$9 \times 9 + 7 =$ **88**
$9 \times 98 + 6 =$ **888**
$9 \times 987 + 5 =$ **8,888**
$9 \times 9,876 + 4 =$ **88,888**
$9 \times 98,765 + 3 =$ **888,888**

8.
$4^2 =$ **16**
$34^2 =$ **1,156**
$334^2 =$ **111,556**
$3,334^2 =$ **11,115,556**
$33,334^2 =$ **1,111,155,556**

9.
$5^2 =$ **25**
$35^2 =$ **1,225**
$335^2 =$ **112,225**
$3,335^2 =$ **11,122,225**
$33,335^2 =$ **1,111,222,225**

10. Experiment with your calculator. Find your own number pattern. Share the pattern with your classmates.

114D

ACHIEVEMENT TEST

Reasoning Logically
CHAPTER 3 FORM A

Name _____
Date _____

MATH MATTERS BOOK 1
Chicha Lynch
Eugene Olmstead

SOUTH-WESTERN PUBLISHING CO.

SCORING RECORD	
Possible	Earned
10	

1. Refer to the figure below. Make a statement about line segments *AB* and *BC*. Then check to see whether your statement is true or false.

   Line segments *AB* and *BC* have the same length.

2. Take any two digit number. Multiply it by 3. Then multiply that product by 7. Now multiply that product by 13. Multiply that product by 37. What number do you get? Repeat this process several times with different two-digit numbers. Make a conjecture about the result of the process.

   For any two-digit number *mn*, the result will be *mnm,nmn*.

3. Experiment with several different sets of numbers and make a conjecture about the product of an odd number, an even number, and an odd number.

   The product will always be even.

4. Write two conditional statements using the following two sentences:
   An animal is a sheep. It does not shed its fur.

   If an animal is a sheep, then it does not shed its fur.

   If it does not shed its fur, then an animal is a sheep.

For each of the numbered statements, write *true* or *false*. If false, give a counterexample.

5. If a figure is a square, then the figure is a rectangle.

   true

6. If a number is divisible by 7, then the number is divisible by 14.

   false; 35

3A-1

Name _____   Date _____

Determine whether the following arguments are *valid* or *invalid*.

7. If the bulb is burnt out, then the light will not go on.
   The light will not go on.
   Therefore, the bulb is burnt out.
   invalid

8. If a number is prime, then it has only two factors.
   23 is a prime number.
   Therefore, 23 has only two factors.
   valid

9. In an informal survey of several English classes, it was found that 56 students enjoy reading plays, 70 enjoy novels, and 45 enjoy poetry. 46 enjoy reading both plays and novels, 15 enjoy both novels and poems, 22 enjoy both poems and plays. In all, 18 enjoy all three. How many students like either plays or novels?

   80 students enjoy reading either plays or novels.

10. Mona, Jim, and Sara have different jobs. One is a teacher, one is an accountant, and one is a court reporter. The teacher and Jim are friends. The court reporter plays tennis with Sara. The teacher and Mona went to the accountant for help at tax time. Which person has which job?

   Mona is the court reporter; Jim is the accountant; Sara is the teacher.

3A-2

ACHIEVEMENT TEST

Reasoning Logically
CHAPTER 3 FORM B

Name _____
Date _____

MATH MATTERS BOOK 1
Chicha Lynch
Eugene Olmstead

SOUTH-WESTERN PUBLISHING CO.

SCORING RECORD	
Possible	Earned
10	

1. Refer to the figure at the right. Make a statement about the length and direction of *AB*, *CD*, and *EF*. Then check to see whether your statement is true or false.

   Line segments *AB*, *CD*, and *EF* are the same length and are parallel.

2. Take three one-digit numbers. Use these three numbers to write six different three-digit numbers (such as 123, 132, 231, 213, and so on). Take the average of those six three-digit numbers. Then divide that average by 37. What number do you get? Now, add the three one-digit numbers you chose at first. What number do you get? Repeat the process with different sets of digits. Make a conjecture about the result of the process.

   For any three one-digit numbers, the result of each process will always be the
   same number.

3. Give a counterexample to show why the following conjecture is false:
   The square of any two-digit whole number below 40 is a three-digit whole number.

   false: 39 × 39 = 1,521; 33 × 33 = 1,089

4. Write two conditional statements, using the following two sentences:
   The food contains protein. It is nourishing.

   If the food contains protein, then it is nourishing.

   If it is nourishing, then the food contains protein.

3B-1

Name _____   Date _____

For each of the numbered statements, write *true* or *false*. If false, give a counterexample.

5. If a number is divisible by 5, then it is divisible by 15.

   false; 20

6. If a number is a multiple of 6, then it is a multiple of 3.

   true

Determine whether the following arguments are *valid* or *invalid*.

7. If the animal is a rodent, then it has sharp teeth.
   The animal has sharp teeth.
   Therefore, the animal is a rodent.
   invalid

8. If the substance is flammable, then it will burn.
   The substance is flammable.
   Therefore, it will burn.
   valid

9. In an informal survey of one class, it was found that 75 students prefer rock music, 60 prefer rhythm and blues, and 55 enjoy classical music. 48 enjoy both rock and rhythm and blues; 44 enjoy both rock and classical music; and 32 enjoy both rhythm and blues and classical music. In all, 28 enjoy all three. How many students like only classical music?

   7 students enjoy classical music only.

10. Frank, Kim, and Louise study different languages. One studies French, one studies German, and one studies Russian. The German student and Kim are cousins. The French student plays tennis with Louise and Frank. The Russian student and Frank went to a football rally together. Which student studies which language?

   Frank studies German; Kim studies French; Louise studies Russian.

3B-2

114E

Teacher's Notes

CHAPTER 4 SKILLS PREVIEW

1. Find the prime factorization of 280. $2^3 \times 5 \times 7$

Find the GCF of each pair of numbers.

2. 8, 12 4
3. 4, 10 2
4. 6, 15 3
5. 12, 18 6

Find the LCM of each pair of numbers.

6. 3, 4 12
7. 9, 15 45
8. 8, 20 40
9. 12, 16 48

10. Multiply to find two fractions that are equivalent to $\frac{11}{15}$. Answers may vary. Sample: $\frac{22}{30}, \frac{33}{45}$

11. Divide to find two fractions that are equivalent to $\frac{24}{28}$. Answers may vary. Sample: $\frac{12}{14}, \frac{6}{7}$

Write each fraction in lowest terms.

12. $\frac{15}{25}$ $\frac{3}{5}$
13. $\frac{20}{24}$ $\frac{5}{6}$
14. $\frac{24}{40}$ $\frac{3}{5}$
15. $\frac{14}{21}$ $\frac{2}{3}$

Write each fraction as a decimal.

16. $\frac{31}{100}$ 0.31
17. $\frac{7}{8}$ 0.875
18. $\frac{4}{15}$ $0.2\overline{6}$
19. $\frac{7}{9}$ $0.\overline{7}$

Write each decimal as a fraction in lowest terms.

20. 0.78 $\frac{39}{50}$
21. 0.3 $\frac{3}{10}$
22. 0.001 $\frac{1}{1000}$
23. 4.92 $4\frac{23}{25}$

Replace ● with <, >, or =.

24. $\frac{5}{8} \bullet \frac{3}{5}$ >
25. $\frac{2}{3} \bullet 0.\overline{6}$ =
26. $0.31 \bullet \frac{5}{16}$ <
27. $\frac{5}{6} \bullet 0.8\overline{3}$ =

28. Write 6^{-8} as a fraction. $\frac{1}{6 \times 6 \times 6 \times 6 \times 6 \times 6 \times 6 \times 6} = \frac{1}{1,679,616}$

29. Write $\frac{1}{8 \times 8 \times 8 \times 8 \times 8 \times 8 \times 8 \times 8}$ in exponential form. 8^{-8}

30. Write 2.813×10^{-6} in standard form. 0.000002813

31. Write 0.00684 in scientific notation. 6.84×10^{-3}

Find each answer. Write answers in lowest terms.

32. $\frac{1}{2} + \frac{3}{8}$ $\frac{7}{8}$
33. $6\frac{3}{8} - 2\frac{1}{6}$ $4\frac{5}{24}$
34. $1\frac{1}{2} \times 2\frac{2}{3}$ 4
35. $\frac{5}{12} \div \frac{2}{3}$ $\frac{15}{24}$

Solve.

36. Yoki has read $\frac{5}{6}$ of a book that is 365 pages long. Estimate to find about how many pages she has read. Estimates may vary. Sample: 300 pages

37. Rosa plans to hike $1\frac{3}{5}$ mi each day to raise money for a charity. Her sponsor has pledged $1.25 to the charity for each mile she hikes. If she hikes each day for $6\frac{2}{7}$ weeks, how much money will the charity receive? $88

OVERVIEW

In this chapter, students work with factors, multiples, and equivalent fractions in order to lay the groundwork for studying operations with fractions. They study negative exponents and expand their understanding of scientific notation to include its application to numbers less than 1. Students estimate to solve problems involving fractions and use the strategy of using simpler numbers to help solve a problem.

SPECIAL CONCERNS

In their everyday lives, students will often encounter situations involving fractions. It is important that they learn to interpret statements involving fractions and perform computations with fractions. Students must be able to estimate with fractions in order to determine if their solutions to problems are reasonable. Students may have varying degrees of difficulty with this chapter. Through the use of manipulatives and working cooperatively in solving problems with fractions, the skill level of most students can be improved.

VOCABULARY

composite number
equivalent fractions
factor
greatest common factor (GCF)
least common denominator
least common multiple (LCM)
like fractions

multiple
prime factorization
prime number
reciprocal
repeating decimal
terminal decimal
unlike fractions

MATERIALS

calculators
rectangular pieces of paper

paper strips of the same
size and shape

BULLETIN BOARD

Have students help you put together a Fractions-in-Sports bulletin board. Using old sports magazines and newspapers, cut out pictures of basketball players. Under each one, write the player's name, team, and height expressed as a fraction (for instance, express a height of 6 ft 6 in. as 6 1/2 ft). Arrange the pictures in order of players' heights. Cut out pictures of baseball players and make a card for each that shows the player's team and batting average expressed as a decimal.

INTEGRATED UNIT 1

The skills and concepts involved in Chapters 1–4 are included within the special Integrated Unit 1 entitled "Voting and Citizenship." This unit is in the Teacher's Edition beginning on page 150H. Worksheets for this integrated unit appear in the Enrichment Activities booklet, pages 47–49.

TECHNOLOGY CONNECTIONS

- Calculator Worksheet, 79
- Computer Worksheet, 80
- MicroExam, Apple Version
- MicroExam, IBM Version

- *MathTalk Fractions,* First Byte Software
- *NumberMaze Decimals & Fractions,* Great Wave Software

TECHNOLOGY NOTES

Computer learning-activities programs, such as those listed in Technology Connections, challenge students' skills in working with fractions and decimals, then offer tutorial screens if incorrect answers are entered. Some programs provide game formats that entice students to solve problems in order to progress through various levels of difficulty toward a goal.

CHAPTER 4

EXPLORING FRACTIONS

PLANNING GUIDE

SECTIONS	TEXT PAGES	ASSIGNMENTS		
		BASIC	AVERAGE	ENRICHED
Chapter Opener/Decision Making	116–117			
4–1 Factors and Multiples	118–121	1–21, 26, 32, 39–40, 41	1–22, 26–28, 32–34, 38–40, 41–42	1–4, 9–12, 17–22, 23–40, 41–44
4–2 Equivalent Fractions	122–125	1–16, 26–32, 38–41	1–25, 26–36, 38–42	9–25, 29–37, 38–43
4–3 Fractions and Decimals	126–129	1–24, 32, 39–40, 43–44	1–33, 39–40, 43–44	21–33, 36–42, 43–45
4–4 Zero and Negative Exponents	130–133	1–16, 19–26, 31, 33–36, 41	1–18, 19–31, 33–40, 41–43	13–18, 27–40, 41–45
4–5 Multiplying and Dividing Fractions	134–137	1–17, 19–20, 23–25, PSA 1–6	1–18, 19–22, 23–29, PSA 1–9	5–8, 13–16, 19–22, 26–29, PSA 1–11
4–6 Adding and Subtracting Fractions	138–141	1–8, 17, 19–24, 28–30, PSA 1–6	1–18, 19–27, 28–32, PSA 1–10	9–18, 21–27, 31–34, PSA 1–14
4–7 Problem Solving Skills: Estimating	142–143	1–15	1–15, 1–19	4–19
4–8 Problem Solving Strategies: Solve a Simpler Problem	144–145	1–7	1–9	3–12
Technology	119, 120, 127, 132	✔	✔	✔

ASSESSMENT				
Skills Preview	115	All	All	All
Chapter Review	146–147	All	All	All
Chapter Test	148	All	All	All
Cumulative Review	149	All	All	All
Cumulative Test	150	All	All	All

ADDITIONAL RESOURCES			
RETEACHING	**ENRICHMENT**	**TECHNOLOGY**	**TRANSPARENCY**
4–1	4–1		TM 15
4–2	4–2		TM 16
4–3	4–3		
4–4	4–4		
4–5	4–5		TM 17, 18, 19
4–6	4–6	4–6	
4–7	4–7		
4–8	4–8		

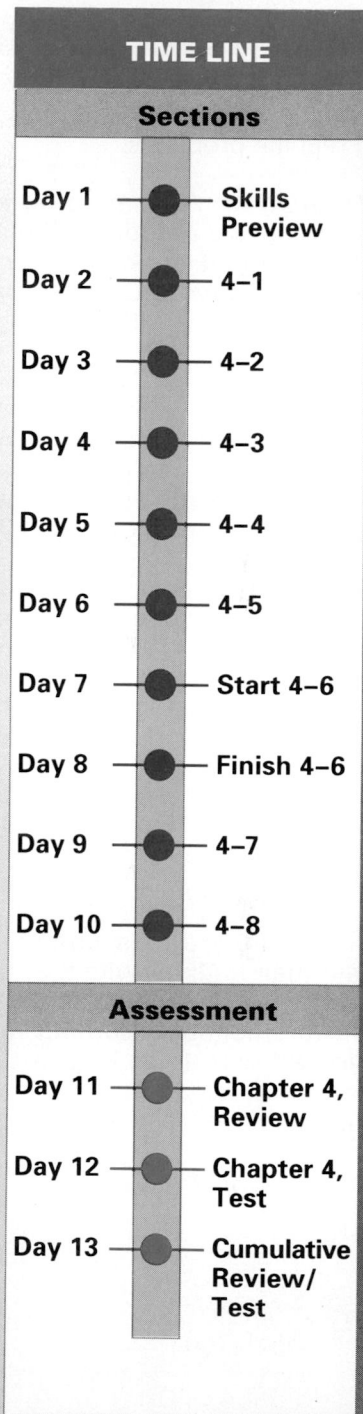

TIME LINE

Sections

Day 1 — Skills Preview

Day 2 — 4–1

Day 3 — 4–2

Day 4 — 4–3

Day 5 — 4–4

Day 6 — 4–5

Day 7 — Start 4–6

Day 8 — Finish 4–6

Day 9 — 4–7

Day 10 — 4–8

Assessment

Day 11 — Chapter 4, Review

Day 12 — Chapter 4, Test

Day 13 — Cumulative Review/ Test

ASSESSMENT OPTIONS

Chapter 4, Test Forms A and B	
Chapter 4, Test	Text, 148
Alternative Assessment	TAE, 148
Chapter 4 MicroExam	

EXPLORING FRACTIONS

THEME Environment

Objective To explore the meaning of fractions and use data in solving real-life problems

Introduction After reading and discussing the introductory material, talk about the data given in the chart. Have students share their experiences with recycling.

Decision Making
Using Data Have students work in small groups to answer these questions and then come together as a large group to discuss their conclusions to Exercises 7–9.

Working Together Brainstorm with the class about methods of getting information about recycling in your community. You may wish to appoint one spokesperson for the class who will contact local agencies with the various groups' questions. Students may also want to find out how recycled materials are used and if recycled products are available in their community. Give each group a chance to share their findings with the class. Sometimes projects such as this lead to students becoming more involved with recycling in their community or organizing school or industrial recycling drives.

The word **fraction** comes from the Latin word *fractio*, which means "act of breaking." A fraction represent a unit broken up into equal parts. Decimals are fractions written as part of the decimal number system. In this chapter, you will compute with fractions, compare decimals and fractions, use negative exponents, and solve problems that involve fractions.

Solid waste is composed of household and industrial wastes. There are three major ways to dispose of solid waste: in landfills, in incinerators, and by recycling. The table shows the amount of solid waste in the United States and how this waste was disposed of over two decades.

	PROFILE OF U.S. WASTE GENERATION, DISPOSAL, AND REUSE			
Year	Pounds of waste generated per person per day	Pounds of waste recycled per person per day	Pounds of waste recovered for energy production	Net pounds of waste disposed of per day
1955	2.77	0.17	0.01	2.59
1970	3.16	0.21	0.01	2.94
1975	3.11	0.23	0.02	2.86
1980	3.35	0.32	0.06	2.96
1981	3.36	0.31	0.05	2.99
1982	3.25	0.30	0.08	2.86
1983	3.37	0.32	0.12	2.92
1984	3.43	0.35	0.15	2.93

DECISION MAKING

Using Data

Use the information in the table to answer each question.

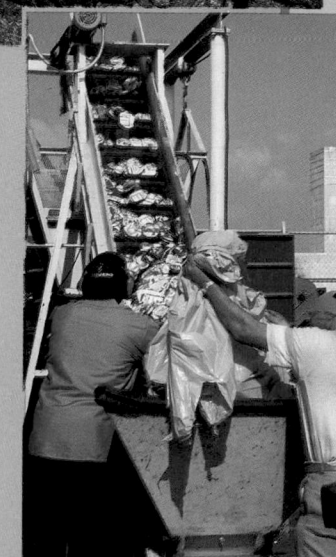

1. There were 203,302,031 people in the United States in 1970. How many pounds of solid waste were generated per day by the entire population in 1970? **642,434,417.96 lb**

2. There were 266,545,805 people in the United States in 1980. How many pounds of solid waste were generated per day by the entire population in 1980? **892,928,446.75 lb**

3. What trend do you see emerging from the data on the total amount of solid waste generated in the United States? **Answers may vary, but in general it is increasing.**

4. What trend do you see emerging from the data on the number of pounds of waste generated per person per day? **Answers may vary, but in general it is increasing.**

5. What trend do you see emerging from the data on the net number of pounds of waste disposed of per person per day? **Answers may vary since it has fluctuated.**

6. What kind of graph would you choose to display these trends? Explain. **Answers will vary. Sample answer: Line graph to show data over time.**

7. What conclusions can you draw about the attitudes of the population that affected these trends? **Answers may vary.**

8. What can you do to encourage recycling at home and in your school? **Answers may vary.**

9. What other conservation methods could you use to save resources? **Answers may vary.**

Working Together

Your group's task is to research the solid-waste management and the recycling programs in your area. Contact the department of waste management in your area, or use newspapers, almanacs, or other reference books. Find the change in the volume of solid waste recycled in your community over the past ten years. If possible, compare the changes in the recycling of different materials, such as newspapers, cans, and glass. Record your results in a chart or graph. Then use the chart or graph to make a poster that will encourage others to recycle. **Answers will vary.**

4-1   Factors and Multiples

Make a chart like this one.

a. Cross out 1.

b. Cross out all multiples of 2 except 2.

c. Cross out all multiples of 3 except 3.

d. Cross out all multiples of 5 except 5.

e. Cross out all multiples of 7 except 7.

1	2	3	4	5	6	7	8	9	10
11	12	13	14	15	16	17	18	19	20
21	22	23	24	25	26	27	28	29	30
31	32	33	34	35	36	37	38	39	40
41	42	43	44	45	46	47	48	49	50
51	52	53	54	55	56	57	58	59	60
61	62	63	64	65	66	67	68	69	70
71	72	73	74	75	76	77	78	79	80
81	82	83	84	85	86	87	88	89	90
91	92	93	94	95	96	97	98	99	100

The numbers that you did not cross out are called *prime numbers.* How many prime numbers are there between 1 and 100? How many of them, other than 2, are even numbers? Explain how you know.
25; none, since even numbers are multiples of 2.

SKILLS DEVELOPMENT

If one whole number is divisible by a second, the second number is a **factor** of the first. For example, since 6 is divisible by 1, 2, 3, and 6, all these numbers are factors of 6.

$$6 \div 1 = 6 \qquad 6 \div 2 = 3 \qquad 6 \div 3 = 2 \qquad 6 \div 6 = 1$$

A **prime number** is a whole number that has exactly two factors, 1 and itself.

$$2 = 1 \times 2 \qquad 3 = 1 \times 3 \qquad 13 = 1 \times 13$$

A whole number that has more than two factors is called a **composite number.** You can write any composite number as a product of prime numbers. This is called the **prime factorization** of the number.

TALK IT OVER

Is the number 1 prime, composite, or neither? Give an argument to support your answer.

The number 1 has only one factor. It is neither prime nor composite.

Example 1

Find the prime factorization of 140.

Solution
Use a *factor tree* to find the prime factors.

► Write the composite number as the product of two factors.
► Repeat with any remaining composite numbers.
► When all numbers are prime, write the prime factorization.

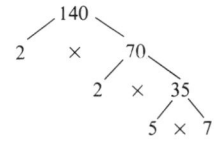

$140 = 2 \times 2 \times 5 \times 7$, or $2^2 \times 5 \times 7$ ◄

TEACHING TIP

If students do not see how to factor a number when attempting to find the prime factorization, suggest they divide the number by 2, 3, 5, 7, and so on until they find a pair of factors to use.

Two whole numbers may have some factors that are the same for both numbers. These are called **common factors.** The greatest of these is called the **greatest common factor (GCF)** of the numbers.

Example 2

Find the GCF of 18 and 60.

Solution 1
To find the GCF using common factors:
► List all the factors of each number.
► Underline the common factors.

factors of 18: <u>1</u>, <u>2</u>, <u>3</u>, <u>6</u>, 9, 18
factors of 60: <u>1</u>, <u>2</u>, <u>3</u>, 4, 5, <u>6</u>, 10, 12, 15, 20, 30, 60

► Identify the greatest common factor.   The GCF of 18 and 60 is 6. ◄

Solution 2
To find the GCF using prime factors:
► Write the prime factorization of each number.

$18 = \underline{2} \times \underline{3}^2$
$60 = \underline{2}^2 \times \underline{3} \times 5$

► Underline the common prime factors.
► Choose the least power of each underlined factor. Then multiply.

$2 \times 3 = 6$
The GCF of 18 and 60 is 6. ◄

When you multiply a number by 1, 2, 3, 4, and so on, you obtain **multiples** of that number. Here are some multiples of 3.

$$1 \times 3 = 3 \qquad 2 \times 3 = 6 \qquad 3 \times 3 = 9$$

Two whole numbers may have some multiples that are the same for both numbers. These are called **common multiples.** The least of these is called the **least common multiple (LCM)** of the numbers.

Example 3

Find the LCM of 18 and 60.

Solution 1
To find the LCM using multiples:
► List multiples of each number.

multiples of 18: 18, 36, 54, 72, 90, 108, 120, 144, 163, <u>180</u>, ...
multiples of 60: 60, 120, <u>180</u>, ...

► Underline common multiples.
► Write the LCM.

The LCM of 18 and 60 is 180. ◄

Solution 2
To find the LCM using prime factors:
► Write the prime factorization of each number.

$18 = 2 \times 3^2$
$60 = 2^2 \times 3 \times 5$

► Choose the greatest power of each prime factor. Then multiply.

$2^2 \times 3^2 \times 5 = 180$
The LCM of 18 and 60 is 180. ◄

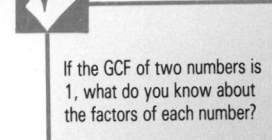

CHECK UNDERSTANDING

If the GCF of two numbers is 1, what do you know about the factors of each number?

Neither is a factor or a multiple of the other.

COMPUTER

This program has the computer find the LCM of two numbers by comparing the common multiples of the numbers.

```
10 LET A = 0:B = 0
20 INPUT "FIRST NUMBER?
   ";A
30 INPUT "SECOND NUMBER?
   ";B
40 FOR X = 1 TO B
50 LET M = A * X
60 FOR Y = 1 TO A
70 LET N = B * Y
80 IF M = N THEN GOTO 110
90 NEXT Y
100 NEXT X
110 PRINT : PRINT "THE LCM
    OF ";A;" AND ";B;" IS
    ";M
```

ASSIGNMENTS

BASIC
1–21, 26, 32, 39–40, 41

AVERAGE
1–22, 26–28, 32–34, 38–40, 41–42

ENRICHED
1–4, 9–12, 17–22, 23–40, 41–44

ADDITIONAL RESOURCES
Reteaching 4–1
Enrichment 4–1
Transparency Master 15

Example 2: Before discussing this example, have students identify factors of 18. **1, 2, 3, 6, 9, 18** Remind students to check to determine that 6 is a factor of both 18 and 60.

Example 3: Before discussing this example, have students identify multiples of pairs of one-digit numbers. Have students name the common multiples in each pair.

Example 4: If students do not understand why the LCM of 10 and 15 is the solution, help them make a chart showing the number of customers, from first to thirtieth. Have them then mark which customers got popcorn, which got lemonade, and which got both.

Additional Questions/Examples
1. Can any two numbers have a common multiple?
   Yes, their product is a common multiple.
2. How can you decide whether to use the GCF or LCM to solve a problem?
   Answers may vary.

Guided Practice/Try These

Have students work in small groups. Point out that the group members should come to an agreement on each answer.

3 SUMMARIZE

Write About Math Have students write paragraphs in their math journal explaining how to find the GCF and LCM of a pair of numbers.

4 PRACTICE

Practice/Solve Problems Encourage students to make two different factor trees for each of Exercises 1–4 to show that the prime factorization is unique.

Extend/Solve Problems Help students solve Exercises 29 and 35, each of which is the first exercise of its type to have variables as factors.

Think Critically/Solve Problems Ask students to explain their answer to Exercise 44 using prime factorization.

5 FOLLOW-UP

Extra Practice

1. Mario buys a lunch ticket every 5 days and a bus pass every 12 days. After how many days will he again have to buy both a lunch pass and a bus pass? **60 days**

2. Carla has three pieces of ribbon: 63 in. of red ribbon, 84 in. of yellow ribbon, and 105 in. of white ribbon. How long are the longest pieces of equal length that can be cut from each ribbon? **21 in.**

COMPUTER

This computer program will list all the factors of any numbers that you input. You can use the program to help you find the GCF for any two numbers.

```
10 INPUT "FIRST NUMBER?
    ";N
20 FOR X = 1 TO N
30 IF N / X = INT (N / X) THEN
    PRINT " ";N / X;
40 NEXT X
50 PRINT:INPUT "SECOND
    NUMBER? ";A
60 FOR Y = 1 TO A
70 IF A / Y = INT (A / Y)
    THEN PRINT " ";A / Y;
80 NEXT Y
```

PRACTICE/ SOLVE PROBLEMS

Example 4

The manager of a popcorn stand gives a free box of popcorn to every 10th customer and a free lemonade to every 15th customer. Which customer will be the first to receive both popcorn and a drink?

Solution

To solve the problem, you need to find the LCM of 10 and 15.

▶ Write the prime factorization of each number.
$10 = 2 \times 5$
$15 = 3 \times 5$

▶ Choose the greatest power of each prime factor. Then multiply.
$2 \times 3 \times 5 = 30$

The 30th customer will receive both popcorn and a drink. ◀

TRY THESE

Find the prime factorization of each number.

1. 56 $2 \times 2 \times 2 \times 7$, or $2^3 \times 7$
2. 88 $2 \times 2 \times 2 \times 11$, or $2^3 \times 11$
3. 117 $3 \times 3 \times 13$, or $3^2 \times 13$
4. 336 $2 \times 2 \times 2 \times 2 \times 3 \times 7$, or $2^4 \times 3$

Find the GCF of each pair of numbers.

5. 8, 12 **4**
6. 6, 15 **3**
7. 6, 8 **2**
8. 30, 40 **10**

Find the LCM of each pair of numbers.

9. 8, 9 **72**
10. 6, 15 **30**
11. 18, 24 **72**
12. 14, 21 **42**

Solve.

13. Every 6th person at a party receives a balloon and every 8th person receives a noisemaker. Which person will be the first to receive both a balloon and a noisemaker? **24th person**

EXERCISES

Find the prime factorization of each number.

1. 104 $2 \times 2 \times 2 \times 13$, or $2^3 \times 13$
2. 108 $2 \times 2 \times 3 \times 3 \times 3$, or $2^2 \times 3^3$
3. 150 $2 \times 3 \times 5 \times 5$, or $2 \times 3 \times 5^2$
4. 300 $2 \times 2 \times 3 \times 5 \times 5$, or $2^2 \times 3 \times 5^2$

Find the GCF of each pair of numbers.

5. 4, 6 **2**
6. 8, 12 **4**
7. 12, 15 **3**
8. 9, 21 **3**
9. 10, 22 **2**
10. 12, 20 **4**
11. 16, 24 **8**
12. 24, 32 **8**

Find the LCM of each pair of numbers.

13. 4, 7 **28**
14. 8, 9 **72**
15. 12, 16 **48**
16. 9, 15 **45**
17. 4, 14 **28**
18. 6, 21 **42**
19. 18, 20 **180**
20. 15, 27 **135**

21. Every 4th person in line at the theater will receive a free matinee pass. Every 6th person will receive a free evening pass. Which person in line will be the first to receive both passes? **12th person**

22. Jane comes to the tennis court every two days. Ellen comes every three days. How many days will pass before the two can play tennis together? **6 days**

Write the prime numbers that fall between each pair of numbers.

23. 2^2 and 3^2 **5, 7**

24. 2^4 and 4^3 **17, 19, 23, 29, 31, 37, 41, 43, 47, 53, 59, 61**

25. 4^3 and 2^5 **37, 41, 43, 47, 53, 59, 61**

Find the GCF.

26. 6, 12, 15 **3**

27. 4, 6, 8 **2**

28. 8, 12, 16 **4**

29. $3a^2b$, $6ab^2$ **3ab**

30. $8a^3b^4$, $4a^2b^3$ **$4a^2b^3$**

31. $12a^7b^3$, $15a^6b^5$ **$3a^6b^3$**

Find the LCM.

32. 2, 3, 5 **30**

33. 4, 6, 10 **60**

34. 3, 7, 12 **84**

35. $3ab^2$, $4a^2b$ **$12a^2b^2$**

36. $2a^4b$, $5ab^4$ **$10a^4b^4$**

37. $6ab$, $3a^2b^2$ **$6a^2b^2$**

Solve.

38. On December 1, Jenny, Dan, and Marita went jogging together. Jenny goes jogging every third day, Dan every fourth day, and Marita every fifth day.
   a. On what day will they next go jogging together? **January 30**
   b. How many more times will Jenny have jogged than Marita by the time they jog together again? **8 more times**

39. At a rummage sale, Christine bought 36 books, 27 cassette tapes, and 15 board games and shared them equally with her friends. With how many friends did she share? **2 friends**

40. Clarence shared $85.50 with each of five friends. What is each person's share? **$14.25**

Find each answer.

41. If the GCF of 2 and another number is 1, describe what you know about the other number. **It is odd.**

42. What is the GCF of two prime numbers? **1**

43. Describe the LCM of two prime numbers. **It is the product of the two numbers.**

44. Three road blinkers are turned on simultaneously. One blinks every 6 seconds; another every 10 seconds; and the third every 12 seconds. How many times per hour do they blink at once? **60 times**

EXTEND/SOLVE PROBLEMS

WRITING ABOUT MATH

In your journal, write a brief paragraph that explains the differences between the GCF and the LCM of two numbers. Then outline the method you prefer to use to find the GCF and the method you prefer to find the LCM. Make a note of any differences in methods.

THINK CRITICALLY/SOLVE PROBLEMS

Extension Have students find the GCF of each pair of factors in Exercises 13–20. Then ask them to compare the product of the GCF and the LCM of a pair of numbers with the product of the pair of numbers. **They will be the same.** Ask students to use prime factorization to explain why. **The factors of the two products are the same.**

Section Quiz Find the prime factorization of each number.
1. 90 **$2 \times 3 \times 3 \times 5$**
2. 100 **$2 \times 2 \times 5 \times 5$**
3. 231 **$3 \times 7 \times 11$**

Find the GCF of each pair of numbers.
4. 7, 21 **7**   5. 25, 35 **5**
6. 11, 17 **1**

Find the LCM of each pair of numbers.
7. 5, 8 **40**   8. 14, 21 **42**
9. 8, 40 **40**

10. Fiona went to a charity fair. She bought 12 paperback books, 16 cassette tapes, and 20 old magazines. She planned to share these items equally with her friends. With how many friends could she share? **3 friends**

CHALLENGE

Have students draw a Venn diagram to show sets A and B, where A = the set of factors of 20 and B = the set of factors of 28. Then use the diagram to find the GCF of 20 and 28.
See margin for Venn diagram.
GCF = 4

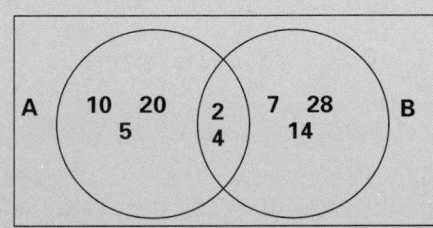

WARM-UP

Find the GCF and LCM of each pair of numbers
1. 24, 36 **12, 72**
2. 10, 30 **10, 30**

1 MOTIVATE

Explore Have students work in small groups to discuss the concept of a fractional part of a whole. Provide them with small plastic circles and triangles or similar manipulatives so they can illustrate concepts such as *2/3 of the objects are green,* and *1/2 of the objects are circles.* Students should understand that, in discussing parts of a whole, we need to consider those parts that are identical in size and shape; in discussing parts of a set, we need to consider only the number of elements with a common feature.

2 TEACH

Use the Pages/Skills Development Have students read this part of the section and then discuss the examples. Explain how fraction bars are used to show equivalent fractions. Ask students how 2/4 is

---

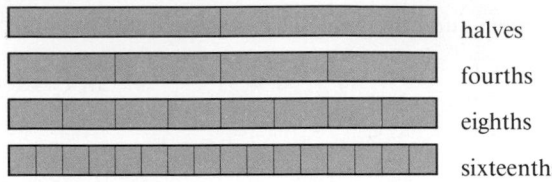

4-2 Equivalent Fractions

EXPLORE

Use fraction bars. Find the bar that is divided into halves. Use one half to cover part of the fourths bar. How many fourths does one half cover?

halves

fourths

eighths

sixteenths

Now use one half to cover the eighths bar, then the sixteenths bar. How many parts of each bar does one half cover?

SKILLS DEVELOPMENT

Fractions that represent the same amount are **equivalent fractions.**

The terms of a fraction, such as $\frac{2}{3}$, tell you what the fraction means.

$$\frac{\text{number of parts being described}}{\text{number of parts that make up the whole}} = \frac{2}{3} \begin{array}{l} \leftarrow \textbf{numerator} \\ \leftarrow \textbf{denominator} \end{array}$$

For example, a fraction, such as $\frac{2}{3}$, can be used to show equal parts of a whole object or a part of a set of objects.

Equal parts of a whole object   Equal parts of a set of objects.

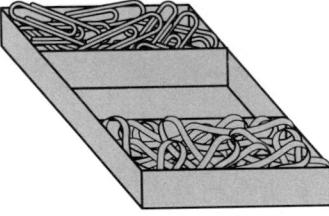   

2 out of 3 sections are full. So $\frac{2}{3}$ of the box is full.

2 out of 3 boxes are full. So $\frac{2}{3}$ of the boxes are full.

An equivalent fraction can be found for any fraction by multiplying or dividing the numerator and the denominator by the same nonzero number.

Example 1

a. Multiply to find four fractions equivalent to $\frac{1}{2}$.

b. Divide to find four fractions equivalent to $\frac{24}{48}$.

122   CHAPTER 4   Exploring Fractions

TEACHING TIP

You can show students how to use cross-products to check if two fractions are equivalent.
For example, 3/4 = 12/16 because 3 × 16 = 48 and 4 × 12 = 48. However, 2/3 ≠ 6/10 because 2 × 10 = 20, but 3 × 6 = 18.

Solution

a. $\frac{1}{2} = \frac{1 \times 2}{2 \times 2} = \frac{2}{4}$ $\qquad \frac{1 \times 3}{2 \times 3} = \frac{3}{6}$ $\qquad \frac{1 \times 4}{2 \times 4} = \frac{4}{8}$ $\qquad \frac{1 \times 5}{2 \times 5} = \frac{5}{10}$

b. $\frac{24}{48} = \frac{24 \div 2}{48 \div 2} = \frac{12}{24}$ $\qquad \frac{24 \div 3}{48 \div 3} = \frac{8}{16}$ $\qquad \frac{24 \div 4}{48 \div 4} = \frac{6}{12}$ $\qquad \frac{24 \div 6}{48 \div 6} = \frac{4}{8}$ ◄

A fraction is said to be in **lowest terms** when the GCF of its numerator and denominator is 1.

Example 2

Write $\frac{24}{48}$ in lowest terms.

Solution

► Find the GCF of the $24 = 2^3 \times 3$
numerator and the denominator. $48 = 2^4 \times 3$ The GCF is $2^3 \times 3 = 24$.

► Divide both the numerator and $\frac{24 \div 24}{48 \div 24} = \frac{1}{2}$ ◄
the denominator by the GCF.

It is easier to compare fractions when they have the same denominator. The **least common denominator (LCD)** of two or more fractions is the least common multiple of their denominators.

Example 3

Compare. Replace ● with $<$, $>$, or $=$: $\frac{5}{6}$ ● $\frac{7}{8}$.

Solution

► Find the LCD. $6 = 2 \times 3$ and $8 = 2^3$
    The LCD is $2^3 \times 3 = 24$.

► Write the fractions using the LCD $\frac{5}{6} = \frac{5 \times 4}{6 \times 4} = \frac{20}{24}$ and $\frac{7}{8} = \frac{7 \times 3}{8 \times 3} = \frac{21}{24}$
as the denominator of each.

► Compare. $\frac{20}{24} < \frac{21}{24}$, so $\frac{5}{6} < \frac{7}{8}$. ◄

TRY THESE

Multiply to find two fractions equivalent to the given fraction.
Answers will vary. Samples are given.

1. $\frac{2}{3}$ $\frac{4}{6}, \frac{6}{9}$
2. $\frac{3}{4}$ $\frac{6}{8}, \frac{9}{12}$
3. $\frac{7}{8}$ $\frac{14}{16}, \frac{21}{24}$
4. $\frac{3}{10}$ $\frac{6}{20}, \frac{9}{30}$

Divide to find two fractions equivalent to the given fraction.
Answers will vary. Samples are given.

5. $\frac{12}{16}$ $\frac{6}{8}, \frac{3}{4}$
6. $\frac{8}{12}$ $\frac{4}{6}, \frac{2}{3}$
7. $\frac{8}{24}$ $\frac{4}{12}, \frac{1}{3}$
8. $\frac{20}{30}$ $\frac{10}{15}, \frac{2}{3}$

Write each fraction in lowest terms.

9. $\frac{2}{4}$ $\frac{1}{2}$
10. $\frac{4}{12}$ $\frac{1}{3}$
11. $\frac{6}{10}$ $\frac{3}{5}$
12. $\frac{15}{20}$ $\frac{3}{4}$

Compare. Replace ● with $<$, $>$, or $=$.

13. $\frac{7}{8}$ ● $\frac{6}{8}$ $>$
14. $\frac{3}{5}$ ● $\frac{2}{3}$ $<$
15. $\frac{5}{6}$ ● $\frac{10}{12}$ $=$

ASSIGNMENTS

BASIC
1–16, 26–32, 38–41

AVERAGE
1–25, 26–36, 38–42

ENRICHED
9–25, 29–37, 38–43

ADDITIONAL RESOURCES
Reteaching 4–2
Enrichment 4–2
Transparency Master 16

related to 1/2. **Both numerator and denominator have been multiplied by 2.** Ask how 1/2 is related to 8/16. **Both the numerator and denominator have been divided by 8.**

Example 1: Stress that there are many fractions, not just one, that are equivalent to a given fraction. Have students find one more fraction for each part.

Example 2: Remind students that a fraction is in lowest terms when the numerator and denominator have no common factors. Have students also find the GCF using factors if it seems easier.

Example 3: Make sure that students know that, in comparing two fractions with the same denominator, the fraction with the greater numerator is the greater.

Additional Questions/Examples
1. If the GCF of the numerator and denominator of a fraction is 1, what is true about the fraction? **It is in lowest terms.**
2. When you multiply or divide the numerator and denominator of a fraction by the same number, by what number are you actually multiplying or dividing the fraction? **1**

AT-RISK STUDENTS

Have students make their own sets of paper fraction bars to help them determine when two fractions are equivalent or to find one fraction that is equivalent to another.

123

3. Write a fraction for the number of primes in the first hundred counting numbers. **1/4**

Guided Practice/Try These Have students compare answers in Exercises 1–8.

3 SUMMARIZE

Key Questions
1. What are equivalent fractions? **fractions that represent the same amount**
2. How do you find a fraction equivalent to a given fraction? **Multiply or divide the numerator and denominator by the same number.**
3. How do you use equivalent fractions to compare two given fractions? **Write both fractions as equivalent fractions with the same denominator. Then compare numerators.**

4 PRACTICE

Practice/Solve Problems Students should be able to do these problems mentally. For each of Exercises 5–8 ask students how they found a factor by which to divide.

Extend/Solve Problems In Exercise 37, ask students what they have to find before answering part a. **the total number of representatives** Have students explain how they could use part a to find the answer to part b. Then have them suggest a second way to solve part b. **Find the number present and compare with the total number.**

Think Critically/Solve Problems Have students explain how they answered Exercises 38–41. For instance, if they used rounding in

---

EXERCISES

PRACTICE/ SOLVE PROBLEMS

Multiply to find two fractions equivalent to the given fraction. **Answers will vary. Samples are given.**

1. $\frac{1}{3}$ $\frac{2}{6}, \frac{3}{9}$ 2. $\frac{3}{8}$ $\frac{6}{16}, \frac{9}{24}$ 3. $\frac{5}{12}$ $\frac{10}{24}, \frac{15}{36}$ 4. $\frac{5}{6}$ $\frac{10}{12}, \frac{15}{18}$

Divide to find two fractions equivalent to the given fraction. **Answers will vary. Samples are given.**

5. $\frac{12}{18}$ $\frac{6}{9}, \frac{2}{3}$ 6. $\frac{36}{54}$ $\frac{18}{27}, \frac{12}{18}$ 7. $\frac{30}{36}$ $\frac{15}{18}, \frac{10}{12}$ 8. $\frac{24}{42}$ $\frac{12}{21}, \frac{4}{7}$

Write each fraction in lowest terms.

9. $\frac{2}{4}$ $\frac{1}{2}$ 10. $\frac{6}{8}$ $\frac{3}{4}$ 11. $\frac{15}{45}$ $\frac{1}{3}$ 12. $\frac{8}{64}$ $\frac{1}{8}$

13. $\frac{18}{20}$ $\frac{9}{10}$ 14. $\frac{9}{24}$ $\frac{3}{8}$ 15. $\frac{8}{12}$ $\frac{2}{3}$ 16. $\frac{12}{15}$ $\frac{4}{5}$

Compare. Replace ● with <, >, or =.

17. $\frac{3}{8}$ ● $\frac{2}{3}$ <   18 $\frac{4}{5}$ ● $\frac{2}{3}$ >   19. $\frac{3}{5}$ ● $\frac{6}{10}$ =

20. $\frac{2}{3}$ ● $\frac{3}{4}$ <   21. $\frac{5}{6}$ ● $\frac{7}{8}$ <   22. $\frac{7}{8}$ ● $\frac{2}{3}$ >

Solve. Write the answer in lowest terms.

23. There are 3 people in a minivan that can seat 9 people. What fraction of the seats are occupied? $\frac{1}{3}$

24. A football team is made up of 30 players. Six players were absent from practice. What fraction of the team was absent? $\frac{1}{5}$

25. A large glass can hold 16 fl oz of liquid. If it is filled with 12 fl oz of water, what fraction of the glass is filled with water? $\frac{3}{4}$

EXTEND/ SOLVE PROBLEMS

Order the following sets of fractions from least to greatest.

26. $\frac{1}{8}, \frac{7}{8}, \frac{3}{8}$ $\frac{1}{8}, \frac{3}{8}, \frac{7}{8}$ 27. $\frac{1}{3}, \frac{5}{6}, \frac{2}{3}$ $\frac{1}{3}, \frac{2}{3}, \frac{5}{6}$ 28. $\frac{1}{2}, \frac{3}{10}, \frac{2}{5}$ $\frac{3}{10}, \frac{2}{5}, \frac{1}{2}$

The manager of a parking garage recorded the types of vehicles housed on the premises over the last two years. Use this data for Exercises 29–36. Write your answers in lowest terms.

Vehicles	Last Year	This Year
full-size cars	52	40
compact cars	42	51
vans	6	9

Use the data in the column marked *last year.*

29. What fraction of the total were compact cars? $\frac{21}{50}$

30. What fraction of the total were full-size cars? $\frac{13}{25}$

31. What fraction of the total number of vehicles were vans? $\frac{3}{50}$

124 CHAPTER 4 Exploring Fractions

32. Compact cars and full-size cars together made up what fraction of the total number of vehicles? $\frac{47}{50}$

Use the data in the column marked *this year*.

33. What fraction of the total were compact cars? $\frac{51}{100}$

34. What fraction of the total were full-size cars? $\frac{2}{5}$

35. What fraction of the total number of vehicles were vans? $\frac{9}{100}$

36. Compact cars and full-size cars together make up what fraction of the total number of vehicles? $\frac{91}{100}$

37. A student council consists of eight representatives from each of four classes. Eight representatives were absent from a meeting.

a. What fraction of the council was absent? $\frac{1}{4}$

b. What fraction of the council was present? $\frac{3}{4}$

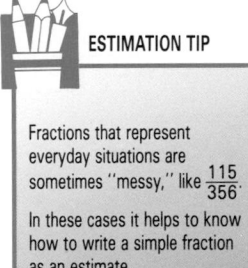

ESTIMATION TIP

Fractions that represent everyday situations are sometimes "messy," like $\frac{115}{356}$.

In these cases it helps to know how to write a simple fraction as an estimate.

$\frac{115}{356}$ is close to $\frac{120}{360}$.

So, $\frac{115}{356}$ is about $\frac{120}{360}$, or $\frac{1}{3}$.

Solve.

Each student is reading a book for a book report. For Exercises 38–41, write a fraction to estimate what part of each student's book has been read so far. **Answers may vary. Samples are given.**

Student	Pages in Book	Pages Read
Suzanne	464	117
Jack	680	512
Jennifer	720	430
Frank	520	345

THINK CRITICALLY/ SOLVE PROBLEMS

38. $\frac{1}{4}$

39. $\frac{3}{4}$

40. $\frac{4}{7}$

41. $\frac{3}{5}$

42. The denominator of the lowest-terms fraction equivalent to $\frac{48}{84}$ is 5 more than a number. What is the number? **2**

43. The sum when 3 is added to 2 times a number is the numerator of the lowest-terms fraction that is equivalent to $\frac{30}{54}$. What is the number? **1**

4-2   Equivalent Fractions   **125**

CHALLENGE

Show students the following:
Given a/b and c/d.   If $ad = bc$, then $a/b = c/d$.
If $ad < bc$, then $a/b < c/d$.
If $ad > bc$, then $a/b > c/d$.
Have students use these facts to solve Exercises 17 – 22 without using equivalent fractions.

Exercise 38, they might answer 100/500 = 1/5; if they used compatible numbers, they might answer 120/480 = 1/4. Once students have written the fractions in Exercises 42 and 43 in lowest terms, they might use the guess-and-check strategy to solve the problem.

5 FOLLOW-UP

Extra Practice
1. Write an equivalent fraction.
   a. 3/4 **for example, 6/8**
   b. 28/36 **for example, 14/18**
2. Find the value of *n* so that the following are equivalent fractions: 2/3 = n/45 **30**
3. Write a fraction to compare the number of boys in your class with the total number of students. Then write a fraction to compare the number of girls with the total number of students. **Answers may vary.**

Extension Have students find values of *n* such that n/36 is greater than, equal to, and less than 3/4. **Answers may vary. Possible answers: 30, 27, 1**

Section Quiz Multiply to find two fractions equivalent to the given fraction.
1. 3/5 **for example, 6/10, 9/15**
2. 7/10 **for example, 14/20, 21/30**
Divide to find two fractions equivalent to the given fraction.
3. 50/100 **for example, 1/2, 25/50**
4. 36/48 **for example, 18/24, 3/4**
Write each fraction in lowest terms.
5. 32/56 **4/7**   6. 17/34 **1/2**
Compare. Replace ● with <, >, or =.
7. 3/8 ● 5/8 **<**   8. 15/30 ● 1/3 **>**

126

4-3   Fractions and Decimals

EXPLORE   Think about the names given to U.S. coins. Why is a 25-cent piece called a *quarter?* (What is it a quarter of?) **$1.00**

None of our other coins in general use today have names that are the same as names of fractions. But suppose they did. Which coins might be called a *tenth*, a *twentieth*, and a *hundredth*? **dime, nickel, penny**

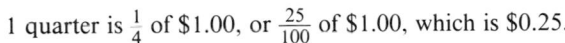

SKILLS DEVELOPMENT   Since fractions and decimals both show parts of a whole, any fraction can be expressed in decimal form. For example:

1 quarter is $\frac{1}{4}$ of $1.00, or $\frac{25}{100}$ of $1.00, which is $0.25.

1 dime is $\frac{1}{10}$ of $1.00, or $\frac{10}{100}$ of $1.00, which is $0.10.

1 nickel is $\frac{1}{20}$ of $1.00, or $\frac{5}{100}$ of $1.00, which is $0.05.

1 penny is $\frac{1}{100}$ of $1.00, or $\frac{1}{100}$ of $1.00, which is $0.01.

To write a decimal as a fraction, recall the place value of each digit.

Example 1

Write each decimal as a fraction or mixed number in lowest terms.
a. 0.3   **b.** 0.84   **c.** 0.375   **d.** 2.06

Solution

Decimal	Read	Write fraction with denominator of 10, 100, or 1,000.	Write in lowest terms.
a. 0.3	3 tenths	$\frac{3}{10}$	$\frac{3}{10}$
b. 0.84	84 hundredths	$\frac{84}{100}$	$\frac{84}{100} = \frac{84 \div 4}{100 \div 4} = \frac{21}{25}$
c. 0.375	375 thousandths	$\frac{375}{1,000}$	$\frac{375}{1,000} = \frac{375 \div 125}{1,000 \div 125} = \frac{3}{8}$
d. 2.06	2 and 6 hundredths	$2\frac{6}{100}$	$\frac{6}{100} = \frac{6 \div 2}{100 \div 2} = \frac{3}{50};$ $2\frac{6}{100} = 2\frac{3}{50}$ ◀

Remember that a bar can indicate division. Therefore, you can write a fraction as a decimal by dividing the numerator by the denominator. When the remainder of this division is 0, the decimal is a **terminating decimal.**

Example 2

Write each fraction or mixed number as a decimal.

a. $\frac{3}{8}$ b. $6\frac{1}{2}$

COMPUTER TIP

To make a computer act like a calculator, just type ? PRINT followed by the computation you want the computer to do.

Solution

Divide. Annex zeros if necessary.

a. $\frac{3}{8} \rightarrow$

$$
\begin{array}{r}
0.375 \\
8\overline{)3.000} \\
\underline{2\ 4} \\
60 \\
\underline{56} \\
40 \\
\underline{40} \\
0
\end{array}
$$

Since the remainder is 0, the decimal terminates.

$\frac{3}{8} = 0.375$

b. Write $\frac{1}{2}$ as a decimal.

$$
\frac{1}{2} \rightarrow
\begin{array}{r}
0.5 \\
2\overline{)1.0} \\
\underline{1\ 0} \\
0
\end{array}
$$

Since the remainder is 0, the decimal terminates.
Now, write $6\frac{1}{2}$ in decimal form.

$$6\frac{1}{2} = 6 + \frac{1}{2}$$
$$= 6 + 0.5$$
$$= 6.5 \blacktriangleleft$$

When the remainder of the division is not zero and a block of digits in the quotient repeats, the decimal is a **repeating decimal.**

Example 3

Write each fraction as a decimal.

a. $\frac{1}{3}$ b. $\frac{1}{11}$

Solution

Divide. Annex zeros if necessary.

a.
$$
\frac{1}{3} =
\begin{array}{r}
0.33\ldots \\
3\overline{)1.00} \\
\underline{9} \\
10 \\
\underline{9} \\
1
\end{array}
$$

The remainder is not zero but the digit 3 repeats in the quotient. Place a bar over the 3.

$\frac{1}{3} = 0.\overline{3}$

b.
$$
\frac{1}{11} =
\begin{array}{r}
0.0909\ldots \\
11\overline{)1.0000} \\
\underline{99} \\
100 \\
\underline{99} \\
1
\end{array}
$$

The remainder is not zero but the digits 0 and 9 repeat as a block in the quotient. Place a bar over 09.

$\frac{1}{11} = 0.\overline{09}$ $\blacktriangleleft$

Example 4

Replace ● with <, >, or =.

a. $0.3 \bullet \frac{2}{5}$ b. $\frac{3}{8} \bullet 0.375$ c. $0.812 \bullet \frac{7}{8}$

Solution

Before comparing, write each fraction as a decimal.

a. $0.3 \bullet \frac{2}{5}$
$0.3 \bullet 0.4$
$0.3 < 0.4$

b. $\frac{3}{8} \bullet 0.375$
$0.375 \bullet 0.375$
$0.375 = 0.375$

c. $\frac{7}{8} \bullet 0.812$
$0.875 \bullet 0.812$
$0.875 > 0.812$ $\blacktriangleleft$

4-3 Fractions and Decimals **127**

Use a place-value chart as needed to supplement this example.

Example 2: Remind students that, to divide one whole number by another, it is necessary to place a decimal point after the whole number in the dividend and annex zeros.

Example 3: Have students do their own divisions to see what they discover about the remainders. Have them compare their results with those in the example.

Example 4: Ask students what the advantage is in changing the fractions to decimals in order to compare rather than changing the decimals to fractions. **It is easier to compare two decimals than to compare two fractions.**

Additional Questions/Examples
Fractions, such as 1/8, 1/20, 1/50, and 3/40, are all equivalent to terminating decimals. When fractions are in lowest terms, the only prime factors of their denominators are 2 or 5. Fractions, such as 1/3, 1/7, 1/9, and 1/22, are equivalent to repeating decimals.

1. What kinds of fractions result in repeating decimals? **those fractions whose denominators lack 2 or 5 as their only prime factors**

2. Do you think the decimal equivalent of 1/32 is a repeating or terminating decimal? Why? **terminating; fractions whose denominators have prime factors of only 2 or 5 are terminating decimals. The only prime factor of 32 is 2.**

3. Do you think the decimal equivalent of 4/7 is terminating or repeating? Why? **The only prime factor of the denominator is 7, not 2 or 5.**

4. Find the decimal equivalent of 4/7. **0.$\overline{571428}$**

5. How would you write 1.3434... using a bar? **1.$\overline{34}$**

6. How would you use a bar to write 1.34333...? **1.34$\overline{3}$**

Guided Practice/Try These Have students work independently and then compare results with a partner. Have them use a place-value chart for Exercises 1–8 if needed.

3 SUMMARIZE

Talk It Over Have students define the terms *terminating decimal* and *repeating decimal* and tell how to write a repeating decimal.

4 PRACTICE

Practice/Solve Problems Ask students to tell what they did to compare the pairs of numbers in Exercises 29–31.

Extend/Solve Problems If students have difficulty with Exercises 34–35, suggest they write each fraction or decimal on a small slip of paper, write the decimal equivalent of each fraction above it, and then arrange the slips of paper in order from least to greatest.

MIXED REVIEW

Find each answer. Use rounding to estimate and compare to see if the answer is reasonable.

1. 4,562 − 796 **3,766**
2. $412.36 − $381.53
3. 1,698 + 245 **1,943**
4. $122.19 + 296.34

    2. **$30.83**
    4. **$418.53**

Solve.

5. Randi bought a bicycle for $263.75, a helmet for $39.95, and a lock for $27.50. There was no tax. How much did she spend in all? **$331.20**

TRY THESE

Write each decimal as a fraction or mixed number in lowest terms.

1. 0.9 **$\frac{9}{10}$**
2. 0.75 **$\frac{3}{4}$**
3. 0.512 **$\frac{64}{125}$**
4. 2.95 **$2\frac{19}{20}$**
5. 0.016 **$\frac{2}{125}$**
6. 3.8 **$3\frac{4}{5}$**
7. 1.02 **$1\frac{1}{50}$**
8. 7.806 **$7\frac{403}{500}$**

Write each fraction or mixed number as a terminating decimal.

9. $\frac{3}{10}$ **0.3**
10. $\frac{61}{100}$ **0.61**
11. $\frac{912}{1,000}$ **0.912**
12. $\frac{4}{5}$ **0.8**
13. $\frac{3}{4}$ **0.75**
14. $\frac{3}{15}$ **0.2**
15. $4\frac{5}{8}$ **4.625**
16. $2\frac{7}{8}$ **2.875**
17. $1\frac{1}{2}$ **1.5**
18. $2\frac{1}{4}$ **2.25**
19. $6\frac{3}{8}$ **6.375**
20. $4\frac{2}{5}$ **4.4**

Write each fraction as a repeating decimal.

21. $\frac{2}{3}$ **0.$\overline{6}$**
22. $\frac{1}{6}$ **0.1$\overline{6}$**
23. $\frac{2}{11}$ **0.$\overline{18}$**
24. $\frac{5}{18}$ **0.2$\overline{7}$**
25. $1\frac{2}{9}$ **1.$\overline{2}$**
26. $2\frac{11}{12}$ **2.91$\overline{6}$**
27. $6\frac{7}{15}$ **6.4$\overline{6}$**
28. $4\frac{7}{30}$ **4.2$\overline{3}$**

Replace ● with <, >, or =.

29. 0.2 ● $\frac{2}{5}$ **<**
30. $\frac{1}{8}$ ● 0.125 **=**
31. $\frac{5}{6}$ ● 0.93 **<**

EXERCISES

PRACTICE/ SOLVE PROBLEMS

Write each decimal as a fraction or mixed number in lowest terms.

1. 0.7 **$\frac{7}{10}$**
2. 0.892 **$\frac{223}{250}$**
3. 0.932 **$\frac{233}{250}$**
4. 4.62 **$4\frac{31}{50}$**
5. 0.058 **$\frac{29}{500}$**
6. 4.3 **$4\frac{3}{10}$**
7. 8.05 **$8\frac{1}{20}$**
8. 3.901 **$3\frac{901}{1000}$**

Write each fraction or mixed number as a terminating decimal.

9. $\frac{9}{10}$ **0.9**
10. $\frac{89}{100}$ **0.89**
11. $\frac{417}{1000}$ **0.417**
12. $\frac{3}{5}$ **0.6**
13. $\frac{1}{4}$ **0.25**
14. $\frac{18}{45}$ **0.4**
15. $\frac{3}{8}$ **0.375**
16. $\frac{9}{25}$ **0.36**
17. $1\frac{2}{5}$ **1.4**
18. $8\frac{1}{25}$ **8.04**
19. $4\frac{7}{8}$ **4.875**
20. $3\frac{7}{28}$ **3.25**

Write each fraction as a repeating decimal.

21. $\frac{5}{6}$ **0.8$\overline{3}$**
22. $\frac{4}{11}$ **0.$\overline{36}$**
23. $\frac{5}{9}$ **0.$\overline{5}$**
24. $\frac{8}{15}$ **0.5$\overline{3}$**
25. $\frac{8}{9}$ **0.$\overline{8}$**
26. $\frac{1}{15}$ **0.0$\overline{6}$**
27. $\frac{1}{3}$ **0.$\overline{3}$**
28. $\frac{21}{22}$ **0.9$\overline{54}$**

Replace ● with >, <, or =.

29. 0.85 ● $\frac{8}{25}$ **>**
30. 0.625 ● $\frac{5}{8}$ **=**
31. $\frac{7}{12}$ ● 0.578 **>**

MAKING CONNECTIONS

Have students find instances where decimals are used in various sports to express scores or ratings of players. Have them discuss the meaning of any score or rating, such as a batting average, expressed in this way.

Solve.

32. Mary Lee jogged $1\frac{4}{5}$ mi. Twana jogged 1.6 mi. Who jogged a greater distance? **Mary Lee**

33. During basketball practice, Christine shot free throws. She made 60 out of 75 shots. She said she made $\frac{4}{5}$ of her shots. Her coach said she made 0.8 of her shots. Who is correct? Explain.
Christine and her coach are both correct since $\frac{60}{75} = \frac{4}{5} = 0.8$

**EXTEND/
SOLVE PROBLEMS**

Order each set of numbers from least to greatest.

34. $\frac{1}{8}$, 0.28, $\frac{1}{5}$, 0.75, $\frac{4}{15}$, 10.031
$\frac{1}{8}, \frac{1}{5}, \frac{4}{15}$, 0.28, 0.75, 10.031

35. 0.561, 0.25, 0.781, $\frac{4}{5}$, $\frac{3}{7}$, $\frac{7}{8}$
0.25, $\frac{3}{7}$, 0.561, 0.781, $\frac{4}{5}$, $\frac{7}{8}$

Let $a = 1$, $b = 2$, and $c = 3$. Replace ● with <, >, or =.

36. $\frac{b}{c}$ ● 0.45   **>**

37. $\frac{a}{c}$ ● 0.125   **>**

38. 0.3 ● $\frac{a}{b}$ **<**

Solve.
José biked 12.5 mi. Betty biked $12\frac{1}{4}$ mi. Mona biked $12\frac{3}{5}$ mi.

39. Who traveled the greatest distance? **Mona**

40. Who traveled the least distance? **Betty**

Shiroh has collected $\frac{8}{9}$ of his goal for a fund-raising drive. Greg has collected 0.83 of his goal. Otis has collected $\frac{16}{17}$ of his goal.

41. Who is closest to meeting his goal? **Otis**

42. Who is farthest from meeting his goal? **Greg**

**THINK CRITICALLY/
SOLVE PROBLEMS**

43. **Answers may vary.
Sample: $\frac{3}{4}, \frac{6}{8}, \frac{9}{12}, \frac{75}{100}$**

Find each answer.

43. Name at least four fractions that have the same value as 0.75.

44. What are the greatest and the least decimals that can be written using each of the digits 3, 4, 5, and 7 just once? **greatest 754.3, least 3.457; Accept also: greatest 7543, least 0.3457**

Solve.

45. Maria has finished reading $\frac{4}{5}$ of a new novel, while Allison has finished 0.832 of it. Melanie has read an amount that is somewhere in between. Express that amount in decimal form.
between 0.8 and 0.832

CHALLENGE

Show students this method for finding the fractional equivalent of $0.\overline{45}$. Have them use this technique to find the fractional equivalents of the following :

$$100n = 45.\overline{45}$$
$$-\quad n = 0.\overline{45}$$
$$99n = 45$$
$$n = 45/99 = 5/11$$

1. $0.\overline{12}$ **4/33**   2. $0.\overline{54}$ **6/11**   3. $0.\overline{1}$ **1/9**

Think Critically/Solve Problems
Ask students to make up a problem similar to Exercise 44.

5 FOLLOW-UP

Extra Practice
1. Caleb bought a package of meat weighing 0.88 lb. Write the decimal as a fraction in lowest terms. **22/25**
2. Pat mowed 5/8 of the lawn before it rained. Write the fraction as a terminating decimal. **0.625**
3. Twelve out of the 18 students in one class read the school newspaper. Express this number as a fraction, then write the fraction as a decimal. **0.6̄**
4. Ramon completed 7/8 of the assignment and Kirk completed 0.88 of it. Who completed more? **Kirk**
5. Sharif swam 3 5/8 laps, Luwanda swam 3.7 laps, and Kari swam 3 2/3 laps. Who swam the greatest distance? **Luwanda** Who swam the least distance? **Sharif**

Extension Have students research the won-lost records of their favorite sports teams. By expressing the number of wins to the number of games played as a decimal, students can rank the teams.

Section Quiz Write each decimal as a fraction or mixed number in lowest terms.
1. 0.65 **13/20**
2. 3.375 **3 3/8**
Write each fraction or mixed number as a decimal.
3. 2 3/4 **2.75**
4. 3/7 **0.$\overline{428571}$**
5. 36/5 **7.2**
6. 4/15 **0.2$\overline{6}$**
7. Francie picked 33 3/4 lb of apples and Shari picked 33.7 lb. Which girl picked more apples? **Francie**

SPOTLIGHT

OBJECTIVE
• Use zero and negative exponents and scientific notation

VOCABULARY
negative exponent, nonzero number, positive exponent

WARM-UP

Write in exponential form.
1. $4 \times 4 \times 4 \times 4$ **4^4**
2. $10 \times 10 \times 10 \times 10 \times 10$ **10^5**

Write in standard form.
3. 4^6 **4,096** 4. 10^6 **1,000,000**

1 MOTIVATE

Explore Have students work with partners to discover the patterns in this activity. Then have small groups compare and discuss their answers. Ask what they can conclude about the value of 3^0 and 3^{-2}.
$3^0 = 1$, $3^{-2} = 1/3^2 = 1/9$

2 TEACH

Use the Pages/Skills Development
Have students read this part of the section and then discuss the examples. Be sure students understand that the definitions of zero and negative exponents follow the laws of exponents. For example, $2/2^2 = 2^{1-2} = 2^{-1}$ according to the laws of exponents. But $2/2^2 = 1/2$ so $1/2$ must be the same as 2^{-1}. Also point out that 0^0 is undefined.

Example 1: Stress that 4^{-2} is another name for a fraction. Have students practice writing such

EXPLORE

Describe the patterns formed by the numbers at the right. Then continue the pattern to show 0.00001, 0.000001, and 0.0000001 in exponential form.

$10,000 = 10^4$
$1,000 = 10^3$
$100 = 10^2$
$10 = 10^1$
$1 = 10^0$

Now try to show a similar pattern using 3 as the base.

$$0.00001 = \frac{1}{100,000} = \frac{1}{10^5} = 10^{-5}$$
$$0.000001 = \frac{1}{1,000,000} = \frac{1}{10^6} = 10^{-6}$$
$$0.0000001 = \frac{1}{10,000,000} = \frac{1}{10^7} = 10^{-7}$$

$$0.1 = \frac{1}{10} = \frac{1}{10^1} = 10^{-1}$$
$$0.01 = \frac{1}{100} = \frac{1}{10^2} = 10^{-2}$$
$$0.001 = \frac{1}{1,000} = \frac{1}{10^3} = 10^{-3}$$
$$0.0001 = \frac{1}{10,000} = \frac{1}{10^4} = 10^{-4}$$

See Additional Answers.

SKILLS DEVELOPMENT

Any nonzero number to the zero power has a value of 1.

$$a^0 = 1$$

Although this fact surprises many people, it is easy to show why it must be true. For instance, this is how you can show that 10^0 must be equal to 1.

You know that $100 \div 100 = 1$, so $10^2 \div 10^2 = 1$.
By the laws of exponents, $10^2 \div 10^2 = 10^{2-2} = 10^0$.
Therefore, $10^0 = 1$.

If a nonzero number has a negative exponent, you can write it as a fraction with a numerator of 1 and a denominator that has a positive exponent.

$$a^{-n} = \frac{1}{a^n}$$

Example 1

Write as a fraction.

a. 4^{-2}　　　　b. 6^{-3}　　　　c. 2^{-4}

Solution

a. 4^{-2}
$= \frac{1}{4^2}$
$= \frac{1}{4 \times 4}$
$= \frac{1}{16}$

b. 6^{-3}
$= \frac{1}{6^3}$
$= \frac{1}{6 \times 6 \times 6}$
$= \frac{1}{216}$

c. 2^{-4}
$= \frac{1}{2^4}$
$= \frac{1}{2 \times 2 \times 2 \times 2}$
$= \frac{1}{16}$ ◀

TEACHING TIP

Point out to students that if the decimal to be expressed in scientific notation is less than 1, the power of 10 will be negative.

Example 2

Write in exponential form.

a. $\dfrac{1}{11 \times 11 \times 11}$

b. $\dfrac{1}{3 \times 3}$

Solution

a. $\dfrac{1}{11 \times 11 \times 11}$

$= \dfrac{1}{11^3} = 11^{-3}$

b. $\dfrac{1}{3 \times 3}$

$= \dfrac{1}{3^2} = 3^{-2}$ ◀

Negative exponents are useful when you need to write very small numbers in scientific notation.

Example 3

Write 3.416×10^{-4} in standard form.

Solution

$3.416 \times 10^{-4} = 3.416 \times \dfrac{1}{10^4} = 3.416 \times \dfrac{1}{10,000}$

$= 3.416 \times 0.0001$

$= 0.0003416$ ◀

Example 4

Write 0.016 in scientific notation.

Solution

$0.0016 = 1.6 \times 0.001 = 1.6 \times \dfrac{1}{1,000}$

$= 1.6 \times \dfrac{1}{10^3}$

$= 1.6 \times 10^{-3}$ ◀

TRY THESE

Write as a fraction.

1. 6^{-2} **$\dfrac{1}{36}$**

2. 3^{-5} **$\dfrac{1}{243}$**

3. 6^{-4} **$\dfrac{1}{1296}$**

4. 9^{-3} **$\dfrac{1}{729}$**

Write in exponential form.

5. $\dfrac{1}{3 \times 3 \times 3 \times 3 \times 3}$ **3^{-5}**

6. $\dfrac{1}{4 \times 4 \times 4 \times 4}$ **4^{-4}**

Write in standard form.

7. 6.876×10^{-7} **0.0000006876**

8. 4.2×10^{-6} **0.0000042**

Write in scientific notation.

9. 0.00000002195 **2.195×10^{-8}**

10. 0.00012 **1.2×10^{-4}**

11. 0.000009 **9×10^{-6}**

12. 0.00000007 **7×10^{-8}**

CONNECTIONS

The definition of scientific notation applies to both very large and very small numbers. A number written in scientific notation has one factor that is greater than or equal to 1 and less than 10 and another factor that is a power of ten.

MENTAL MATH TIP

Here's a shortcut for writing a number in scientific notation. Count the number of places that the decimal point moves to the left or right. Use that number in the exponent. If the decimal point moves left, the exponent should be positive. If it moves right, the exponent should be negative.

$73,000 = 7.3 \times 10^4$

$0.00073 = 7.3 \times 10^{-4}$

MAKING CONNECTIONS

Have students use science books or articles about scientific topics to find uses of scientific notation involving positive and negative exponents. (They may find, for example, the way in which the mass of the sun or the mass of a proton is expressed.) Ask students why they think scientific notation is used so extensively.

ASSIGNMENTS

BASIC
1–16, 19–26, 31, 33–36, 41

AVERAGE
1–18, 19–31, 33–40, 41–43

ENRICHED
13–18, 27–40, 41–45

ADDITIONAL RESOURCES
Reteaching 4–4
Enrichment 4–4
Transparency Masters, 17, 18, 19

expressions as $1/4^2$ before completing the example.

Example 2: Have students try writing each of these fractions as whole numbers with negative exponents. Remind them that it is helpful to first write the denominator in exponential form.

Example 3: Remind students that they can write 10^{-4} as a decimal by writing 1 and then moving the decimal point 4 places to the left. Students may discover the shortcut of simply moving the decimal point 4 places to the left.

Example 4: Review the meaning of scientific notation. Students should find the mental math tip useful in determining the exponent.

Additional Questions/Examples

1. When you write a fraction, such as, $1/(3 \times 3 \times 3)$, in exponential form, will the exponent be positive or negative? **negative**

2. Will the value of 6^{-3} be greater than 1 or less than 1? **less than 1**

3. When a number less than 1 is written in scientific notation, will the power of 10 be positive or negative? **negative**

Guided Practice/Try These
Have students work independently on these problems and then discuss their answers with a partner.

3 SUMMARIZE

Write About Math Ask students to write the definition of a negative exponent in their math journals and give one or two examples.

4 PRACTICE

Practice/Solve Problems Students should note that when a number is in exponential form, it is simply a base raised to a power.

Extend/Solve Problems Suggest that students having difficulty with Exercises 23–25 write all the numbers in fraction form before they order them from least to greatest.

Think Critically/Solve Problems In Exercises 33–36 suggest that students first write each denominator in exponential form. In Exercise 41 students should discover that since both numbers have the same power of 10 when in exponential form, they can simply subtract and then write the result, 0.002×10^{-27}, in scientific notation.

5 FOLLOW-UP

Extra Practice
1. Write in exponential form:
$$\frac{1}{4 \times 4 \times 4 \times 4 \times 4} \quad \textbf{4}^{\textbf{-5}}$$
2. Write as a fraction: 8^{-3}
$$\frac{1}{8^3} \quad \textbf{or} \quad \frac{1}{512}$$
3. Elias needs 5.3×10^{-3} L of solution for an experiment. Write this in standard form. **0.0053**

132

EXERCISES

PRACTICE/
SOLVE PROBLEMS

Write as a fraction.

1. 2^{-6} $\dfrac{1}{64}$ 2. 7^{-3} $\dfrac{1}{343}$ 3. 3^{-2} $\dfrac{1}{9}$ 4. 3^{-4} $\dfrac{1}{81}$

Write in exponential form.

5. $\dfrac{1}{8 \times 8 \times 8 \times 8 \times 8 \times 8}$ 8^{-6} 6. $\dfrac{1}{5 \times 5 \times 5 \times 5 \times 5 \times 5 \times 5}$ 5^{-7}

7. $\dfrac{1}{2 \times 2}$ 2^{-2} 8. $\dfrac{1}{3 \times 3 \times 3}$ 3^{-3}

COMPUTER TIP

A computer will automatically express any number that has more than nine digits and a decimal point in scientific notation.

Write in standard form.

9. 7.1×10^{-3} **0.0071**   10. 8.62154×10^{-8} **0.0000000862154**

11. 9.1302×10^{-6} **0.0000091302**   12. 4.3276×10^{-7} **0.00000043276**

Write in scientific notation.

13. 0.0025 **2.5×10^{-3}**   14. 0.000000009 **9×10^{-9}**

15. 0.00000000000196 **1.96×10^{-12}**   16. 0.00087 **8.7×10^{-4}**

17. Write the number 0.000000000000001 in exponential form. **10^{-15}**

18. The smallest organisms are the viroids. Each has a diameter of 0.00000000007 in. Write this number using scientific notation. **7×10^{-11}**

EXTEND/
SOLVE PROBLEMS

 Write as a fraction. Use a calculator for the computation.

19. 2^{-20} $\dfrac{1}{1,048,576}$ 20. 7^{-10} $\dfrac{1}{282,475,249}$ 21. 15^{-8} $\dfrac{1}{2,562,890,625}$ 22. 3^{-12} $\dfrac{1}{531,441}$

Order the numbers in each row from least to greatest.

23. 6.2×10^{-5} $\quad$ 6.27×10^{-5} $\quad$ 7.3×10^{-4} $\quad$ $\dfrac{1}{1,000}$

24. $\dfrac{1}{10^8}$ $\quad$ 1.25×10^{-8} $\quad$ 1.25×10^{-9} $\quad$ 10^{-9}

25. 10^{-7} $\quad$ 8.886×10^{-6} $\quad$ 8.668×10^{-8} $\quad$ $\dfrac{1}{10^6}$

26. $\dfrac{1}{4^4}$ $\quad$ $\dfrac{1}{5^4}$ $\quad$ $\dfrac{1}{6^4}$ $\quad$ $\dfrac{1}{7^4}$

Extend/Solve Problems answers (left column):
23. 6.2×10^{-5}, 6.27×10^{-5}, 7.3×10^{-4}, $\dfrac{1}{1,000}$
24. 10^{-9}, 1.25×10^{-9}, 10^{-8}, 1.25×10^{-8}
25. 8.668×10^{-8}, 10^{-7}, $\dfrac{1}{10^6}$, 8.886×10^{-6}
26. $\dfrac{1}{7^4}$, $\dfrac{1}{6^4}$, $\dfrac{1}{5^4}$, $\dfrac{1}{4^4}$
27. 0.001 does not belong. It is equal to $\dfrac{1}{1,000}$, not $\dfrac{1}{10,000}$.
28. 0.0027 does not belong. It is equal to $\dfrac{27}{10,000}$.
30. $\dfrac{1}{20 \times 20 \times 20}$ does not belong. It is equal to $\dfrac{1}{160,000}$.

Write *equal* if all the numbers in each row are equal. If not, indicate which number does not belong and explain why.

27. $\dfrac{1}{10 \times 10 \times 10 \times 10}$ $\quad$ 0.001 $\quad$ $\dfrac{1}{10,000}$ $\quad$ 1^{-4}

28. $\dfrac{1}{3 \times 3 \times 3}$ $\quad$ 0.0027 $\quad$ $\dfrac{1}{27}$ $\quad$ 3^{-3}

29. $\dfrac{1}{2 \times 2 \times 2 \times 2}$ $\quad$ 0.0625 $\quad$ $\dfrac{1}{16}$ $\quad$ 2^{-4} **equal**

30. $\dfrac{1}{20 \times 20 \times 20 \times 20}$ $\quad$ 0.000125 $\quad$ $\dfrac{1}{8,000}$ $\quad$ 20^{-3}

AT-RISK STUDENTS

Have students multiply a number such as 0.000004 by 10 repeatedly on a calculator until they have a number between 1 and 10. Have them record the number of factors of 10 they multiplied by. Point out that when the number is written in standard form, the base will be 10 and the exponent will be negative.

Solve.

31. The diameter of the nucleus of an atom is measured in femtometers. One femtometer equals 10^{-15} meter. A picometer equals 0.000000000001 meter. Which unit is larger? Explain.
Picometer. A picometer = 10^{-12}, which is greater than 10^{-15}.

32. One micrometer equals 0.000001 meter. A microsporidian spore can measure from 2 to 20 micrometers. Write this span in meters using scientific notation. 2×10^{-5} to 2×10^{-6}

Write in exponential form.

33. $\frac{1}{125}$ 5^{-3} 34. $\frac{1}{1,000}$ 10^{-3} 35. $\frac{1}{64}$ 2^{-6} or 4^{-3} or 8^{-2} 36. $\frac{1}{343}$ 7^{-3}

Use mental math to tell which number is greater.

37. 0.0025 or $\frac{1}{2^2}$ $\frac{1}{2^2}$ 38. 9 or $\frac{1}{3^3}$ 9

39. 6.2193×10^{-3} or $6.2193 \times 10,000$ $6.2193 \times 10,000$

40. 7.987×10^3 or $7,987$ **They are equal.**

Solve.

THINK CRITICALLY/ SOLVE PROBLEMS

41. The mass of a proton is 1.673×10^{-27} kg. The mass of a neutron is 1.675×10^{-27} kg.

  a. Which one is heavier? **neutron**

  b. How much heavier is it? 2×10^{-30} kg

Replace ● with $<$, $>$, or $=$.

42. 7.192×10^{-2} ● 10^{-2} $>$ 43. 3.96×10^4 ● 3.961×10^{-4} $>$

44. x^{-3} ● 10^0, where x is greater than zero but less than one $>$

45. x^{-3} ● 10^0, where x is greater than one $<$

4. The mass of a spider is 0.000107 kg. Write the number in scientific notation.
1.07×10^{-4}

5. Replace ● with $<$, $>$, or $=$.
3.8×10^{-5} ● 1.1×10^{-4} $<$

6. Write 1/10,000 in exponential form. **10^{-4}**

Extension Have students work in small groups using the laws of exponents to find values such as the following:
1. $5^6 \div 5^8$ **1/25**
2. $6 \div 6^5$ **1/1,296**

Section Quiz
Write in exponential form.
1. $\dfrac{1}{3 \times 3 \times 3 \times 3}$ **3^{-4}**
2. $\dfrac{1}{10 \times 10}$ **10^{-2}**
Write as a fraction.
3. 6^{-7} **$1/6^7$**
4. 10^{-4} **$1/10^4$**
Write in scientific notation.
5. 0.000034 **3.4×10^{-5}**
6. 0.00203 **2.03×10^{-3}**
Write as a decimal.
7. 1.8×10^{-3} **0.0018**
8. 4.002×10^{-6} **0.000004002**
9. Write 0.0000000001 in exponential form **10^{-10}**

Get Ready rectangular pieces of paper

Additional Answers
See page 576.

Additional answers for odd-numbered exercises are found in the Selected Answers portion of the page.

CHALLENGE

Have students find the following products or quotients using scientific notation:
1. $(48 \times 10^{-2})^2$ **2.304×10^{-1}**
2. $\dfrac{3.3 \times 10^{-2}}{4.8 \times 10^{-3}}$ **6.875**
3. $0.00034 \times 0.00000012$ **4.08×10^{-11}**
4. $\dfrac{(0.00023)(4,100)}{0.005}$ **1.886×10^2**

 MOTIVATE

Explore Have students work in small groups to do the paper-folding activity. Then ask students to fold paper to show 1/2 × 1/2, 1/2 × 3/4, and 2/3 × 9/10.

 TEACH

Use the Pages/Skills Development Have students read this part of the section and then discuss the examples. Discuss the steps in multiplying two fractions or mixed numbers. Ask students why all mixed numbers must first be written as fractions. **The rule for multiplying fractions applies only to numbers written in fraction form.**
   Example 1: Have students identify each step in part b of the example. Point out that any fraction whose value is 1 or greater must be rewritten as a mixed number.

134

4-5 Multiplying and Dividing Fractions

EXPLORE   Try this activity.

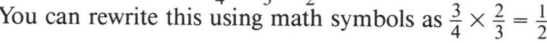

a. Fold a small rectangular piece of paper into thirds using horizontal folds. Shade $\frac{2}{3}$ of the paper yellow.

b. Then fold the paper into fourths using vertical folds and shade $\frac{3}{4}$ of the yellow part blue.

c. There are 6 green sections, so $\frac{6}{12}$, or $\frac{1}{2}$, of the paper is green. You have found that $\frac{3}{4}$ of $\frac{2}{3}$ is $\frac{1}{2}$. You can rewrite this using math symbols as $\frac{3}{4} \times \frac{2}{3} = \frac{1}{2}$.

Now, try folding paper and shading sections to find $\frac{1}{2}$ of $\frac{1}{2}$, $\frac{1}{2}$ of $\frac{3}{4}$, and $\frac{2}{3}$ of $\frac{9}{10}$.

SKILLS DEVELOPMENT   To multiply by a fraction or a mixed number, follow these steps.

Step 1: Write all whole numbers and mixed numbers as fractions.
Step 2: Multiply the numerators.
Step 3: Multiply the denominators.
Step 4: Write the product in lowest terms.

WRITING ABOUT MATH

Describe in your journal how you would write a whole number as a fraction and how you would write a mixed number as a fraction.

Example 1

a. Multiply: $\frac{1}{2} \times \frac{3}{5}$ b. Multiply: $1\frac{1}{3} \times 3\frac{3}{10}$

Solution

a. $\frac{1}{2} \times \frac{3}{5} = \frac{1 \times 3}{2 \times 5} = \frac{3}{10}$

   ┌─ Divide 4 and 10 by their GCF, 2.
b. $1\frac{1}{3} \times 3\frac{3}{10} = \frac{\overset{2}{\cancel{4}}}{\underset{1}{\cancel{3}}} \times \frac{\overset{11}{\cancel{33}}}{\underset{5}{\cancel{10}}} = \frac{2 \times 11}{1 \times 5} = \frac{22}{5} = 4\frac{2}{5}$ ◄
   └─ Divide 3 and 33 by their GCF, 3.

Two numbers whose product is 1 are called **reciprocals.** Each pair of numbers is an example of reciprocals.

$$\frac{2}{3} \times \frac{3}{2} = 1 \qquad 6 \times \frac{1}{6} = 1 \qquad 1\frac{2}{3} \times \frac{3}{5} = \frac{5}{3} \times \frac{3}{5} = 1$$

TEACHING TIP

If students encounter terms they are not sure of (for example, *mixed number*, *product,* or *numerator),* have them look up the terms in the glossary and then write these definitions—along with an example or two—in their math journals.

To divide fractions and mixed numbers, follow these steps.

Step 1: Write all whole numbers and mixed numbers as fractions.
Step 2: Multiply by the reciprocal of the divisor.
Step 3: Write the product in lowest terms.

Example 2

a. Divide: $\frac{3}{8} \div \frac{1}{4}$

b. Divide: $4\frac{1}{2} \div 1\frac{1}{5}$

Solution

The reciprocal of $\frac{1}{4}$ is $\frac{4}{1}$.

a. $\frac{3}{8} \div \frac{1}{4} = \frac{3}{\overset{2}{\cancel{8}}} \times \frac{\overset{1}{\cancel{4}}}{1} = \frac{3}{2} = 1\frac{1}{2}$

Divide 8 and 4 by their GCF, 4.

The reciprocal of $\frac{6}{5}$ is $\frac{5}{6}$.

b. $4\frac{1}{2} \div 1\frac{1}{5} = \frac{9}{2} \div \frac{6}{5} = \frac{\overset{3}{\cancel{9}}}{2} \times \frac{5}{\overset{\cancel{6}}{2}} = \frac{15}{4} = 3\frac{3}{4}$ ◄

Divide 9 and 6 by their GCF, 3.

Example 3

There were 864 glass bottles in the recycling bin. If $\frac{3}{4}$ of them were made from clear glass, how many clear glass bottles were in the bin?

Solution

$\frac{3}{4}$ of $864 = \frac{3}{4} \times 864 = \frac{3}{\overset{\cancel{4}}{1}} \times \frac{\overset{216}{\cancel{864}}}{1}$

$= 3 \times 216 = 648$

There were 648 clear glass bottles. ◄

Multiply. Be sure the product is in lowest terms.

1. $\frac{1}{3} \times \frac{1}{2}$ **$\frac{1}{6}$**
2. $\frac{3}{8} \times \frac{4}{5}$ **$\frac{3}{10}$**
3. $\frac{3}{4} \times 40$ **30**
4. $\frac{1}{2} \times 1\frac{1}{3}$ **$\frac{2}{3}$**
5. $1\frac{2}{3} \times 2\frac{1}{2}$ **$4\frac{1}{6}$**
6. $5 \times 2\frac{3}{10}$ **$11\frac{1}{2}$**
7. $3\frac{1}{3} \times 1\frac{1}{5}$ **4**
8. $3\frac{1}{2} \times 2\frac{2}{5}$ **$8\frac{2}{5}$**

Divide. Be sure the quotient is in lowest terms.

9. $\frac{3}{8} \div \frac{3}{4}$ **$\frac{1}{2}$**
10. $\frac{4}{5} \div \frac{3}{10}$ **$2\frac{2}{3}$**
11. $3 \div \frac{3}{10}$ **10**
12. $3\frac{1}{2} \div 1\frac{1}{6}$ **3**
13. $1\frac{1}{2} \div \frac{2}{3}$ **$2\frac{1}{4}$**
14. $1\frac{1}{2} \div 1\frac{1}{4}$ **$1\frac{1}{5}$**
15. $\frac{8}{15} \div 4$ **$\frac{2}{15}$**
16. $2\frac{3}{4} \div 1\frac{1}{2}$ **$1\frac{5}{6}$**

Solve.

17. A ride on the roller coaster takes $2\frac{1}{2}$ minutes. Tony rode 7 times. For how many minutes did he ride? **$17\frac{1}{2}$ minutes**

Give the reciprocal of each number. $\frac{2}{3}$, $\frac{1}{10}$, $3\frac{1}{4}$, 7

Why doesn't zero have a reciprocal?

$\frac{3}{2}$, $\frac{10}{1}$, $\frac{4}{13}$, $\frac{1}{7}$

When the product of two numbers is 1, the numbers are reciprocals. Since the product of zero and any number is zero, it cannot have a reciprocal.

ASSIGNMENTS

BASIC
1–17, 19–20, 23–25, PSA 1–6

AVERAGE
1–18, 19–22, 23–29, PSA 1–9

ENRICHED
5–8, 13–16, 19–22, 26–29, PSA 1–11

ADDITIONAL RESOURCES
Reteaching 4–5
Enrichment 4–5

Example 2: Review the steps for dividing one fraction by another after having students provide other examples of reciprocals. Then have the students identify the steps as they are used in solving parts a and b of the example.

Example 3: Ask students how to decide whether to multiply or divide to solve this example. Have students solve the problem on their own and then compare their answer with the one given.

Additional Questions/Examples

1. Describe how you would find the reciprocal of a mixed number. **Write the number as a fraction and then reverse the numerator and denominator.**

2. Describe the reciprocal of a whole number. **a fraction with 1 in the numerator and the whole number in the denominator**

3. In which operation, multiplication or division, is the reciprocal of a fraction used? **division**

Guided Practice/Try These Have students work independently on these examples. In Exercise 17, ask students how they decided whether to multiply or divide to solve.

5-MINUTE CLINIC

Exercise	Student's Error	Error Diagnosis
Divide.		• Student uses the reciprocal of the wrong fraction.
1. 2/3 ÷ 5/6	3/2 × 5/6 = 5/4 = 1 1/4	
2. 1 3/8 ÷ 2/7	11/8 × 2/7 = 11/28	• Student forgets to multiply by the reciprocal of the second fraction.

135

SUMMARIZE

Talk It Over Call on students one at a time to give the steps needed to find the product of two fractions or mixed numbers. Then call on students one at a time to give the steps in finding the quotient of two fractions or mixed numbers. Write the steps on the chalkboard as students give them.

4 PRACTICE

Practice/Solve Problems For students having difficulty with these concepts, talk about the steps in rewriting each problem before the final multiplication is done.

Extend/Solve Problems Ask students to identify the data they need to solve Exercises 19 and 20.

Think Critically/Solve Problems For Exercise 24 encourage students to identify the first factor as the reciprocal of the second before they solve.

Problem Solving Applications Have students identify the units used in the recipe for pizza. Ask what they would do if they wanted to make 4 pizzas or 1 pizza. Talk about why it is sometimes difficult to divide a recipe in half. **Fractional parts can become difficult to measure exactly, and an ingredient such as one egg cannot easily be divided.**

5 FOLLOW-UP

Extra Practice Multiply or divide. Write each answer in simplest form.
1. $10/15 \div 2/5$ **1 2/3**
2. $2 \, 2/9 \times 4 \, 1/2$ **10**
3. $7/16 \times 24/28$ **3/8**
4. $3 \, 3/4 \div 1 \, 1/8$ **3 1/3**

EXERGISES

PRACTICE/
SOLVE PROBLEMS

Multiply. Be sure the product is in lowest terms.

1. $\frac{3}{4} \times \frac{8}{9}$ **$\frac{2}{3}$**
2. $\frac{2}{5} \times \frac{5}{8}$ **$\frac{1}{4}$**
3. $\frac{2}{3} \times 27$ **18**
4. $\frac{3}{4} \times 1\frac{1}{3}$ **1**
5. $1\frac{1}{8} \times 4\frac{5}{7}$ **$5\frac{17}{56}$**
6. $1\frac{3}{5} \times 1\frac{3}{8}$ **$2\frac{1}{5}$**
7. $1\frac{4}{5} \times 3\frac{1}{2}$ **$6\frac{3}{10}$**
8. $5\frac{1}{5} \times 1\frac{1}{2}$ **$7\frac{4}{5}$**

Divide. Be sure the quotient is in lowest terms.

9. $\frac{8}{9} \div \frac{8}{12}$ **$1\frac{1}{3}$**
10. $\frac{3}{10} \div \frac{5}{8}$ **$\frac{12}{25}$**
11. $14 \div \frac{7}{10}$ **20**
12. $1\frac{1}{3} \div 1\frac{1}{6}$ **$1\frac{1}{7}$**
13. $1\frac{1}{5} \div \frac{5}{8}$ **$1\frac{23}{25}$**
14. $3\frac{4}{5} \div 1\frac{9}{10}$ **2**
15. $4\frac{1}{2} \div 1\frac{4}{9}$ **$3\frac{3}{26}$**
16. $2\frac{1}{9} \div 12$ **$\frac{19}{108}$**

Solve.

17. Heather works 4 days each week at the hamburger stand. She works $2\frac{3}{4}$ hours each day. How many hours per week is that?
11 hours

18. Chris bought 15 yd of fabric from which to cut sashes for the dance team. At $\frac{3}{4}$ yd per sash, how many sashes can he cut?
20 sashes

EXTEND/
SOLVE PROBLEMS

USING DATA Use the Data Index on page 546 to find the carbon dioxide emissions. Then solve each problem.

19. How many tons of carbon dioxide would be emitted by 2 cars that get 60 mi/gal? **$34\frac{29}{50}$ tons**

20. How many tons of emissions would these be over the lifetime of a car that gets 26 mi/gal? **$19\frac{129}{200}$ tons**

Marilyn ordered 24 pizzas for the homeroom party.

21. If each person eats $\frac{2}{3}$ of a pizza, how many students can be fed?
36 students

22. Each pizza cost $12.75. If the students share the cost equally, how much will each student pay? **$8.50**

THINK CRITICALLY/
SOLVE PROBLEMS

Use mental math to find each missing number.

23. $1\frac{3}{5} \div \frac{\blacksquare}{5} = 1$ **8**
24. $\frac{4}{\blacksquare} \times 1\frac{1}{4} = 1$ **5**
25. $\blacksquare \div \frac{1}{4} = 1$ **$\frac{1}{4}$**

Solve.

A swim club has 24 members. Which of these statements cannot be true? Explain.

26. One-half are female.
27. Two-thirds are divers.
28. Three-eighths swim daily.
29. Three-fifths are lifeguards.

Exercise 29 cannot be true because the product of $\frac{3}{5} \times 24$ is a mixed number, not a whole number.

136 CHAPTER 4 Exploring Fractions

ESL STUDENTS

Write a multiplication or division problem on a set of cards, one step per card. Describe each step on another set of cards. Give out the cards. Have students holding cards for the solved problem stand in the order in which the steps should be performed. Have each student holding the description of a step find his or her partner.

Problem Solving Applications:

CHANGING RECIPE QUANTITIES

TOMATO AND CHEESE PIZZA Yield: 2 pizzas

DOUGH

Combine and set aside.	Combine in food processor.	With the processor running, add steadily:	Process 1 min. Knead dough for 30 s, divide in half, and refrigerate.
$\frac{1}{4}$ c warm water	$2\frac{3}{4}$ c all-purpose flour	yeast mixture	
$\frac{1}{2}$ t sugar	$1\frac{1}{2}$ t salt	$\frac{3}{4}$ c cold water	
1 t dry yeast		3 T olive oil	

SAUCE

1 c tomato sauce
1 6-oz can tomato paste
1 T olive oil
2 T water
1 t oregano
$\frac{1}{2}$ t salt
$\frac{1}{2}$ t sugar
$\frac{1}{8}$ t crushed dried hot red chili peppers

Warm all ingredients in saucepan.

CHEESE TOPPING

$\frac{3}{4}$ lb sliced or grated mozzarella cheese

$\frac{1}{3}$ to $\frac{1}{2}$ c grated Parmesan cheese

DIRECTIONS

Place $\frac{1}{2}$ dough on each baking sheet, top each with $\frac{1}{2}$ sauce and $\frac{1}{2}$ cheese toppings. Bake in 450°F oven for 16–21 min or until cheese is melted.

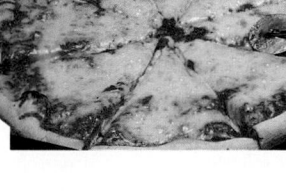

Ken used this recipe for two pizzas.

Solve. **See Additional Answers.**

Suppose Ken cuts the recipe in half to make only one pizza.

1. How much of each ingredient would he use to make the dough?

2. How much of each ingredient would he use to make the sauce?

3. How much of each ingredient would he use to make the cheese topping?

Suppose Ken needs to double the recipe to make four pizzas.

4. How much of each ingredient would he use to make the dough?

5. How much of each ingredient would he use to make the sauce?

6. How much of each ingredient would he use to make the cheese topping?

7. If Ken has $6\frac{3}{4}$ lb mozzarella cheese, how many pizzas can he make with a cheese topping? (Remember that the recipe makes 2 pizzas.)

Explain how to adapt recipes with each number of servings.

8. A recipe serves 12 people. You want to make 8 servings.

9. A recipe serves 4 people. You want to make 6 servings.

Janet wanted to make $\frac{1}{12}$ of a recipe that serves 12 people. The recipe calls for $\frac{1}{2}$ t baking soda.

10. How much baking soda would she have needed for 1 serving?

11. Is it reasonable to think that Janet could make $\frac{1}{12}$ of a serving?

4-5   Multiplying and Dividing Fractions   **137**

5. Jerry filled jars with 2 1/2 lb of granola each. If he started with 82 1/2 lb of granola, how many jars did he fill? **33**

6. Caleb sold 1/2 of a 20-acre lot. He then sold 2/5 of the remaining lot. How much did he sell in all? **14 acres**

Extension Have students apply the order of operations to evaluate the following expressions.

1. 3 1/2 ÷ 7/8 × 3/4   **3**
2. 2 4/5 ÷ (6 2/3 ÷ 10)   **4 1/5**
3. 2 5/8 × (4/7)²   **6/7**
4. 5 1/3 ÷ 1 7/9 ÷ 1 1/2 ÷ 2/7   **7**

Section Quiz

Multiply. Write each answer in the simplest form.

1. 1/3 × 3/5   **1/5**
2. 7/16 × 6/7   **3/8**
3. 6 2/3 × 5 1/4   **35**
4. 4 6/7 × 9 3/8   **45 15/28**

Divide. Write each answer in the simplest form

5. 5/7 ÷ 3/5   **1 4/21**
6. 5/6 ÷ 2   **5/12**
7. 21 ÷ 7/8   **24**
8. 12 3/8 ÷ 6 3/5   **1 7/8**
9. A crew can pave 2/3 mi of highway in one day. How long will it take them to pave 13 1/5 mi?   **19 4/5 days**
10. Ori can run 6 2/3 mi/h. How far can she run in only 15 min?   **1 2/3 mi**

Get Ready paper strips of the same size and shape

Additional Answers
See page 576.

CHALLENGE

A complex fraction is one in which either the numerator or denominator, or both, are fractions or mixed numbers. Have students simplify each of the following complex fractions.

1. $\dfrac{5}{\frac{5}{8}}$ **8**   2. $\dfrac{\frac{17}{20}}{\frac{2}{5}}$ **$\frac{17}{40}$**   3. $\dfrac{\frac{5}{6}}{\frac{2}{3}}$ **$\frac{5}{12}$**

4-6 Adding and Subtracting Fractions

EXPLORE

Model $\frac{4}{5} - \frac{2}{5}$. Cut two identical strips of paper. Divide each strip into 5 equal parts. On one strip, fold back one fifth to show $\frac{4}{5}$. Then, to show how to subtract $\frac{2}{5}$ from $\frac{4}{5}$, fold back two more fifths. Compare the strips. The two parts remaining on the folded strip represent the difference, $\frac{2}{5}$. How would you change the paper strips to model subtraction of other fractions?

 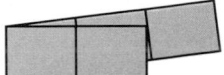

SKILLS DEVELOPMENT

To add or subtract **like fractions,** those fractions with denominators that are alike, follow these steps.

Step 1: Add or subtract the numerators.
Step 2: Keep the same denominator.
Step 3: Write the answer in lowest terms.

Example 1
a. Add: $\frac{3}{8} + \frac{3}{8}$ **b.** Subtract: $\frac{7}{9} - \frac{1}{9}$

Solution

a. $\frac{3}{8} + \frac{3}{8} = \frac{3+3}{8} = \frac{6}{8} = \frac{3}{4}$ **b.** $\frac{7}{9} - \frac{1}{9} = \frac{7-1}{9} = \frac{6}{9} = \frac{2}{3}$ ◄

To add or subtract fractions with unlike denominators, first rename them as like fractions so that the denominators are the same. Then add or subtract and write the answer in lowest terms.

Example 2
a. Add: $\frac{1}{2} + \frac{3}{10}$ **b.** Subtract: $\frac{5}{6} - \frac{2}{3}$

Solution

a.
$$\frac{1}{2} = \frac{5}{10}$$
$$+ \frac{3}{10} = \frac{3}{10}$$
$$\frac{8}{10} = \frac{4}{5}$$

b.
$$\frac{5}{6} = \frac{5}{6}$$
$$-\frac{2}{3} = \frac{4}{6}$$
$$\frac{1}{6}$$ ◄

To add or subtract mixed numbers with unlike denominators, first rename the mixed numbers using like denominators. Then add or subtract and write the answer in lowest terms.

138 CHAPTER 4 Exploring Fractions

138

Example 3

a. Add: $1\frac{11}{12} + 2\frac{5}{6}$

b. Subtract: $4\frac{3}{5} - 3\frac{1}{3}$

Solution

a.
$$1\frac{11}{12} = 1\frac{11}{12}$$
$$+ 2\frac{5}{6} = 2\frac{10}{12}$$
$$3\frac{21}{12} = 4\frac{9}{12} = 4\frac{3}{4}$$

b.
$$4\frac{3}{5} = 4\frac{9}{15}$$
$$-3\frac{1}{3} = 3\frac{5}{15}$$
$$1\frac{4}{15} \blacktriangleleft$$

Example 4

On Saturday, Jamie spent 6 hours at a family picnic and $2\frac{3}{4}$ hours at a ball game. How many more hours did Jamie spend at the picnic than at the ball game?

Solution

Subtract.
$$6 = 5\frac{4}{4}$$
$$-2\frac{3}{4} = 2\frac{3}{4}$$
$$3\frac{1}{4}$$

Jamie spent $3\frac{1}{4}$ more hours at the picnic than at the ball game. ◄

TRY THESE

Add or subtract. Write each answer in lowest terms.

1. $\frac{1}{3} + \frac{1}{3}$ $\frac{2}{3}$
2. $\frac{7}{8} - \frac{3}{8}$ $\frac{1}{2}$
3. $\frac{8}{12} + \frac{1}{12}$ $\frac{3}{4}$
4. $\frac{7}{9} - \frac{4}{9}$ $\frac{1}{3}$
5. $\frac{1}{3} + \frac{2}{5}$ $\frac{11}{15}$
6. $\frac{5}{6} - \frac{2}{3}$ $\frac{1}{6}$
7. $\frac{1}{5} + \frac{1}{2}$ $\frac{7}{10}$
8. $\frac{1}{2} - \frac{3}{8}$ $\frac{1}{8}$
9. $5\frac{1}{2} - 3\frac{2}{5}$ $2\frac{1}{10}$
10. $2\frac{3}{5} + 1\frac{1}{10}$ $3\frac{7}{10}$
11. $6\frac{1}{2} + 5\frac{4}{3}$ $12\frac{5}{6}$
12. $4\frac{1}{6} - 2\frac{1}{3}$ $1\frac{5}{6}$

Solve.

13. Elena found a cone from a Scotch pine that is $1\frac{1}{4}$ in. long and a cone from a white pine that is $4\frac{3}{8}$ in. long. How much longer was the cone from the white pine than the cone from the Scotch pine? $3\frac{1}{8}$ in

EXERCISES

Add or subtract. Write your answers in lowest terms.

PRACTICE/ SOLVE PROBLEMS

1. $\frac{1}{8} + \frac{5}{8}$ $\frac{3}{4}$
2. $\frac{5}{6} - \frac{1}{6}$ $\frac{2}{3}$
3. $\frac{3}{10} - \frac{1}{10}$ $\frac{1}{5}$
4. $\frac{7}{12} + \frac{1}{12}$ $\frac{2}{3}$
5. $\frac{1}{6} - \frac{1}{8}$ $\frac{1}{24}$
6. $\frac{3}{10} + \frac{1}{2}$ $\frac{4}{5}$
7. $\frac{1}{3} + \frac{2}{5}$ $\frac{11}{15}$
8. $\frac{3}{4} - \frac{1}{3}$ $\frac{5}{12}$
9. $3\frac{1}{5} - 1\frac{1}{10}$ $2\frac{1}{10}$
10. $1\frac{1}{2} + 3\frac{1}{5}$ $4\frac{7}{10}$
11. $5\frac{3}{5} + 3\frac{3}{5}$ $9\frac{1}{5}$
12. $9\frac{1}{2} - 6\frac{2}{3}$ $2\frac{5}{6}$
13. $7\frac{3}{8} + 1\frac{1}{6}$ $8\frac{13}{24}$
14. $5\frac{1}{6} - 3\frac{1}{6}$ 2
15. $9 - 4\frac{7}{12}$ $4\frac{5}{12}$
16. $1\frac{5}{6} + 1\frac{1}{6}$ 3

4-6 Adding and Subtracting Fractions **139**

The LCD of the two fractions is the LCM of the two denominators.

Example 3: It may be helpful to use models to illustrate these problems so that you can emphasize that the fractional parts and the whole-number parts must be added separately.

Example 4: A model consisting of circles divided into fourths can be used to illustrate the renaming required in this example. Show that 6 is the same as 5 + 1 = 5 + 4/4 = 5 4/4.

Additional Questions/Examples

1. If the sum of two fractions is 18/12, what steps should you take to write the fraction in simplest form? **First divide the numerator and denominator by 6, then write 3/2 as a mixed number.**

2. What is the first step you would take to find 5 1/3 – 2 3/4? What is the second step? **Write 5 1/3 and 2 3/4 as 5 4/12 and 2 9/12, respectively. Rewrite 5 4/12 as 4 16/12.**

3. What is the solution to Exercise 2? **2 7/12**

Guided Practice/Try These
It may be helpful for students to discuss the steps in solving Exercises 1–12.

3 SUMMARIZE

Write About Math Have students summarize the steps for adding and subtracting mixed numbers in their math journals.

4 PRACTICE

Practice/Solve Problems Ask students to describe how they decided whether to add or subtract in solving Exercises 17 and 18.

Extend/Solve Problems Ask students to describe how they go about finding the LCD for three fractions, such as those in each of Exercises 19–22. **Find the LCM of the three denominators.**

Think Critically/Solve Problems Students can use the guess-and-check strategy in solving Exercises 31–33.

Problem Solving Applications Begin by asking students to explain how many beats each note would get per measure in the sample (the half note gets two beats, the quarter note one beat, and so on). This will help students understand how each note is named. In solving the first six problems, students may guess and check, replacing each note with its fractional equivalent and adding to see if their guess is correct.

Solve.

17. Ming caught a trout that weighed $14\frac{3}{4}$ lb and a walleye that weighed $7\frac{7}{8}$ lb. How much more did the trout weigh? $6\frac{7}{8}$ **lb**

18. Jamie spent $\frac{3}{8}$ of her free time on Tuesday morning and $\frac{1}{6}$ of her free time on Tuesday afternoon at the pool. What fraction of her free time did she spend at the pool on Tuesday? $\frac{13}{24}$ **of her free time**

EXTEND/ SOLVE PROBLEMS

Add. Write each answer in lowest terms.

19. $\frac{1}{2} + \frac{1}{3} + \frac{1}{4}$ $1\frac{1}{12}$

20. $\frac{1}{6} + \frac{1}{3} + \frac{1}{2}$ **1**

21. $\frac{3}{8} + \frac{1}{6} + \frac{3}{4}$ $1\frac{7}{24}$

22. $1\frac{3}{5} + 3\frac{1}{2} + 2\frac{7}{10}$ $7\frac{4}{5}$

23. **Answers may vary. Sample answer is given.**
Yard wastes, glass, metals, textiles, other, and household hazardous waste total
$$\frac{40}{100} = \frac{4}{10}.$$

 USING DATA Use the Data Index on page 546 to find the average mix of refuse. Then solve each problem.

23. What other wastes have the same combined weight as the weight of paper and paperboard and wood/fiber combined?

24. What fraction describes the combined weight of household hazardous waste, rubber and leather, and plastics? $\frac{57}{500}$

The school band practiced the following amounts of time during a two-week period. Express the answers to Exercises 25–27 in hours and minutes.

Week 1: $2\frac{1}{2}$ h, $1\frac{2}{3}$ h   Week 2: $3\frac{1}{2}$ h, $2\frac{3}{4}$ h

25. How long did the band practice during Week 1? **4 h 10 min**

26. How long did the band practice during Week 2? **6 h 15 min**

27. How long did the band practice for the two-week period? **10 h 25 min**

THINK CRITICALLY/ SOLVE PROBLEMS

Complete.

28. $\frac{7}{8} + \frac{\blacksquare}{6} = 1\frac{1}{24}$ **1**

29. $1\frac{3}{4} - \frac{\blacksquare}{10} = 1\frac{1}{20}$ **7**

30. $3\frac{2}{3} - \blacksquare\frac{5}{\blacksquare} = \frac{5}{6}$ **2, 6**

Use the clues to find each fraction in Exercises 31–34.

31. The sum of the digits of the numerator and denominator is 9. Their difference is 7. $\frac{1}{8}$

32. The sum of the digits of the numerator and the denominator is 11. Their difference is 1. $\frac{5}{6}$

33. The sum of the digits of the numerator and the denominator is 5. Their difference is 3. $\frac{1}{4}$

34. Find the sum of the fractions you found for Exercises 31–33. $1\frac{5}{24}$

AT-RISK STUDENTS

Help students put several pieces of paper together on which they can make a large fraction number line to be placed on the floor. Have them walk on the number line to illustrate addition and subtraction of like fractions.

Problem Solving Applications:

MUSIC FOR FRACTION LOVERS

The notes in a piece of music
are given the following names.

○ whole note   ♩ $\frac{1}{2}$ note   ♪ $\frac{1}{8}$ note

♩ $\frac{1}{4}$ note   ♪ $\frac{1}{16}$ note

Two eighth notes ♪♪   Two sixteenth notes ♪♪

are written as ♫.   are written as ♫.

The time signature in a piece of music is written in fraction form.
The numerator tells the number of beats in one bar. The
denominator tells the kind of note that gets one beat.

4 beats to a bar

a quarter note gets one beat

bar   bar   bar   bar

The total value of the notes in each bar must equal 1.

1 whole note $\frac{1}{4} + \frac{1}{4} + \frac{1}{2} = 1$ $\frac{1}{2} + \frac{1}{8} + \frac{1}{16} + \frac{1}{16} + \frac{1}{4} = 1$

Find the missing note that would make each pair equal in value.

1. ♩ = ♪♪ ?
2. ♩ = ♪♪♪ ?
3. ♩ = ♫ ♫ ?

4. ○ = ♩ ?
5. ♪ = ♪ ?
6. ♩ = ♫ ?

Each piece of music is written in $\frac{4}{4}$ time. What type of note is missing
from each bar?

7.

8.

9.

10.

11.

12.

13.

14.

1. ♪ , eighth note
2. ♪ , sixteenth note
3. ♪ , eighth note
4. ♩ , half note
5. ♪ , sixteenth note
6. ♪ , eighth note
7. ♩ , quarter note
8. ♪ , eighth note
9. ♩ , half note
10. ♩ , quarter note
11. ♩ , quarter note;
    ♩ , quarter note;
    ♩ , half note;
    ♩ , half note
12. ♩ , quarter note;
    ♩ , quarter note;
    ♩ , half note
13. ♩ , quarter note;
    ♪ , eighth note;
    ♩ , quarter note
14. ♩ , quarter note;
    ♪ , eighth note;
    ♩ , quarter note

4-6   Adding and Subtracting Fractions   **141**

5   **FOLLOW-UP**

Extra Practice Predict which of
Exercises 1–3 have sums greater
than 1. **2 and 3** Then find the
sum to check your prediction.
1. 1/2 + 1/3   **5/6**
2. 11/12 + 3/8   **1 7/24**
3. 5/6 + 1/2   **1 1/3**
Find each sum or difference.
4. 9/10 – 2/10   **7/10**
5. 3/4 + 11/12   **1 2/3**
6. 7 1/2 – 3 4/5   **3 7/10**
7. 3 1/2 + 9 3/8   **12 7/8**
8. Marianne ran 9/10 mi, walked
1/5 mi, and jogged 3/4 mi. How
far did she travel?   **1 17/20 mi**
9. Find the missing number.
    ■ + 2/3 = 1 4/14   **13/21**

Extension Have students number
10 small squares of paper with the
digits 1–10. Have them arrange the
squares in the following pattern to
find the greatest and least possible
answers: ■/■ + ■/■
greatest: 10/1 + 9/2 = 14 1/2;
least: 1/9 + 2/10 = 14/45

Section Quiz Find each sum.
1. 2/3 + 3/5   **1 4/15**
2. 7 1/5 + 3 1/10   **10 3/10**
3. 7 7/8 + 12 1/4   **20 1/8**
4. 21 5/6 + 10 1/4   **32 1/12**
Find each difference.
5. 1 – 3/5   **2/5**
6. 1 7/20 – 3/5   **3/4**
7. 5/8 – 1/2   **1/8**
8. 11 4/9 – 7 2/3   **3 7/9**
9. The school band practiced
3 1/2 h Wednesday and 2 3/5 h
on Thursday.
a. How long did they practice
in all?   **6 1/10 h**
b. How much longer did they
practice on Wednesday than
on Thursday?   **9/10 h**

WARM-UP

Mark has mowed about 5/8 of the front lawn. If the yard measures 60 ft by 20 ft, about how many square feet has he mowed? **750 ft²**

1 MOTIVATE

Introduction The estimation techniques that students learned for use with whole numbers can also be used with fractions. Review the techniques of rounding, using compatible numbers, and using front-end estimation.

2 TEACH

Use the Pages/Problem Have students read this part of the section and then discuss the problem. Ask why an estimate is acceptable. **The exact number of hours is not needed.**

Use the Pages/Solution Discuss each of the methods shown for estimating the total number worked. Ask why Daryl's answer is probably close to the answer. **because some of the numbers are rounded up and some are rounded down**

142

4-7 Problem Solving Skills:

ESTIMATING WITH FRACTIONS AND DECIMALS

▶ READ
▶ PLAN
▶ SOLVE
▶ ANSWER
▶ CHECK

You often estimate to see if your answer is reasonable, but sometimes you don't have to find the exact answer—an estimate is all you need.

You have already used some estimation techniques, such as rounding, finding compatible numbers, and front-end estimation, when working with whole numbers and decimals. Now you will learn how to use these techniques when estimating with fractions.

PROBLEM

Garth worked $7\frac{3}{4}$h on Monday, $9\frac{1}{2}$h on Tuesday, $9\frac{1}{3}$h on Wednesday, $7\frac{1}{3}$h on Thursday, and $8\frac{1}{2}$h on Friday. He receives overtime pay for every hour he works over 40 hours. Should he receive overtime pay for this week?

 CHECK UNDERSTANDING

How can you be sure from Garth's answer that he worked *more than* 40 hours?

Answers will vary. Discussion should take into account that Garth's estimate did not include the fractional parts. So if the whole numbers alone total 40, then the fractional parts will bring the total over 40 hours.

SOLUTION

Garth and his friends, Jeff and Daryl, estimated to see if he is eligible for overtime pay.

Garth used front-end estimation.	Jeff used numbers close to 0, $\frac{1}{2}$, and 1.	Daryl used rounding.
$7\frac{3}{4} \rightarrow 7$	$7\frac{3}{4} \rightarrow 8$	$7\frac{3}{4} \rightarrow 8$
$9\frac{1}{2} \rightarrow 9$	$9\frac{1}{2} \rightarrow 9\frac{1}{2}$	$9\frac{1}{2} \rightarrow 10$
$9\frac{1}{3} \rightarrow 9$	$9\frac{1}{3} \rightarrow 9\frac{1}{2}$	$9\frac{1}{3} \rightarrow 9$
$7\frac{1}{3} \rightarrow 7$	$7\frac{1}{3} \rightarrow 7\frac{1}{2}$	$7\frac{1}{3} \rightarrow 7$
$8\frac{1}{2} \rightarrow 8$	$8\frac{1}{2} \rightarrow 8\frac{1}{2}$	$8\frac{1}{2} \rightarrow 9$
$\overline{40}$	$\overline{43}$	$\overline{43}$

All the estimation methods indicate that Garth worked more than 40 hours and so is eligible for overtime pay.

TEACHING TIP

Have students use fraction strips to help them round fractions. By placing a strip that represents 2 2/3 on a number line, for instance, students can readily see that the fraction can be rounded to 3.

PROBLEMS

Use estimation to check Pilar's math answers. Write *R* for a reasonable answer and *NR* for an unreasonable answer.

1. $4\frac{1}{8} + 5\frac{1}{2} = 9\frac{5}{8}$ **R**
2. $8\frac{7}{8} - 2\frac{3}{4} = 6\frac{5}{8}$ **NR**
3. $1\frac{4}{5} \times 1\frac{2}{9} = 2\frac{1}{5}$ **R**
4. $9\frac{1}{8} \div 3\frac{1}{4} = 2\frac{21}{26}$ **R**
5. $4 \times 3\frac{9}{10} = 12\frac{3}{5}$ **NR**
6. $7\frac{1}{2} - 3\frac{11}{12} = 3\frac{7}{12}$ **R**
7. $7\frac{1}{2} + 3\frac{5}{8} = 11\frac{1}{8}$ **R**
8. $3\frac{2}{3} + 4\frac{1}{12} = 7\frac{1}{12}$ **NR**
9. $9\frac{1}{4} \div 3\frac{1}{8} = 2\frac{24}{25}$ **R**

$$\frac{2}{9} \times 446$$
$$\downarrow$$
$$\frac{2}{9} \times 450 = \frac{2 \times 450}{9}$$
$$= \frac{900}{9}$$
$$= 100$$

Estimate to solve each problem.

10. Donna has read about $\frac{2}{9}$ of a book that is 446 pages long. About how many pages has she read? **about 100 pages**

11. Raúl wants to cut four boards, each $3\frac{5}{12}$-ft long, from a board whose length is 13 ft. Is this possible? **no**

12. Marti wants to buy a loaf of bread for $0.99, 3 cans of beans for $1.25, ground beef for $6.95, apples for $2.38, and frozen items for $5.84. The sales tax was $1.05. She has $20. Does she have enough money to pay for these items? **yes**

13. Angela had $50. She purchased 2 cassette tapes at $7.98 each and a sweater for $24.89. The sales tax was $2.45. Does she have enough money left to purchase a compact disc for $19.98? **no**

14. Keith needs fourteen pieces of twine that each measure $1\frac{5}{6}$-ft long. He has a $28\frac{1}{2}$-ft roll of twine. Does he need to buy more? **no**

15. A party balloon costs $0.89. A party favor costs $1.98. Tom has $35. Can he buy 8 balloons and 16 party favors, not including tax? If not, give at least two combinations of items he could buy with his money. **No. Answers will vary. Sample answers are given. 6 balloons and 12 favors; 8 balloons and 8 favors**

16. A portion of a park trail is 5,885 meters long. Darlene's footstep is 0.62 meter long. She estimates she would take about 1,000 steps if she walked the trail. Do you agree with her estimate? Why or why not? **No, because 6,000 ÷ 0.6 is 10,000.**

17. Jeanne is buying a coat that is marked $\frac{1}{4}$-off the original price of $149. The sales tax on the coat comes to $7.45. She has $135. Does she have enough money to pay for the coat? **yes**

18. Leon swam $3\frac{1}{10}$ lengths of the pool. Jim swam $1\frac{9}{10}$ times as far. Estimate how many lengths Jim swam. **about 6 lengths**

19. Jackie walked 3.8 km on Monday. She walked half that distance on Tuesday and 5.8 km on Wednesday. Estimate the mean distance she walked per day. **about 4 km**

CHALLENGE

You are flying from Seattle to Tucson. You leave Seattle at 10 a.m. The flight to Los Angeles takes 2 2/3 h. You are on the ground in Los Angeles for 1 1/4 h and the flight to Tucson takes 1 1/3 h. Allowing for a 1-h time difference, about what time is it when you arrive in Tucson?
about 4 p.m.

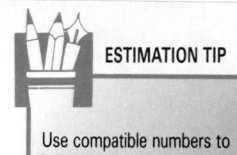

ESTIMATION TIP

Use compatible numbers to help you estimate the answer to Exercise 10.

ASSIGNMENTS

BASIC
1–15

AVERAGE
1–15, 18–19

ENRICHED
4–19

ADDITIONAL RESOURCES
Reteaching 4–7
Enrichment 4–7

---

3 SUMMARIZE

Key Questions

1. How can you determine if an estimate made by rounding might be greater or less than the actual answer? **by noting whether you rounded most of the fractions up or down**

2. How can you decide whether to round a fraction to zero or to one? **If it is less than 1/2, round to 0; if 1/2 or greater, round to 1.**

4 PRACTICE

1. Have students explain how they determined the reasonableness for each of Problems 1–9.

2. Ask students for another way to estimate the answer to Problem 10. **For example, 1/3 of 450 is 150, or 1/9 of 450 is 50.**

5 FOLLOW-UP

Extra Practice
A crew has to paint 88 2/3 mi of center line down a highway. They painted 23 1/5 mi on Monday and 21 3/5 mi on Tuesday. Estimate to determine if they have painted more than half the full distance.
yes

143

OBJECTIVE
* Use the strategy of solving a simpler problem

VOCABULARY
estimating, variable

WARM-UP

Jo spends 1 1/2 h in science lab on each of two days a week. How many hours does she spend in lab during a 16-week semester? **48 h**

1 MOTIVATE

Introduction Ask students whether problems involving whole numbers, fractions, or decimals are the easiest to solve. Elicit the idea that it is often easiest to see which operation to use in a problem involving whole numbers.

2 TEACH

Use the Pages/Problem Have students read this part of the section and then discuss the problem. Ask them to tell what information they are given in the problem.

Use the Pages/Solution Ask the students what they could do to make the numbers easier. Use the students' suggested numbers or those given in the solution of the simpler problem to think through the operations needed to solve the problem.

144

4-8   Problem Solving Strategies:
SOLVE A SIMPLER PROBLEM

► READ
► PLAN
► SOLVE
► ANSWER
► CHECK

Sometimes you can find the answer to a problem by solving another problem that is like it but has simpler numbers. Using simpler numbers can help you focus on choosing the correct operation(s).

PROBLEM

In January, the Willistown Recycling Center paid the Save Our World Club $0.16 per pound for $95\frac{1}{2}$ lb of newspaper and $0.32 per pound for $84\frac{3}{4}$ lb of aluminum cans. How much did the recycling center pay the club?

SOLUTION

Solve a simpler problem. Substitute whole numbers for the decimals and fractions.

Simpler Problem
Suppose the recycling center paid $0.10 per pound for 10 lb of newspaper.
    Multiply: $10 \times 10 = 100$ They paid 100 cents, or $1.00.

Suppose the recycling center paid $0.10 per pound for 20 lb of aluminum cans.
    Multiply: $10 \times 20 = 200$ They paid 200 cents, or $2.00.

Add the amounts to find the total paid: $1.00 + $2.00 = $3.00

Original Problem
Follow the same operations used to solve the simpler problem.
    Multiply: $0.16 \times 95\frac{1}{2} = 15.28 \rightarrow \$15.28$
    Multiply: $0.32 \times 84\frac{3}{4} = 27.12 \rightarrow \$27.12$
    Add: $15.28 + 27.12 = 42.40 \rightarrow \$42.40$

The Willistown Recycling Center paid the Save Our World Club $42.40 during the month of January.

Check the answer by estimating.

At $0.16 per pound, 100 lb of newspaper is $16; 75 lb is $12.
At $0.32 per pound, 100 lb of aluminum cans is $32; 75 lb is $24.

Since the actual number of pounds is between 75 and 100, the answer must be between $36 and $48. So $42.40 is reasonable.

READING MATH

Read each problem carefully and describe what each variable represents. Then, choose the expression that tells how to solve the problem.

1. Andy rode his bike *a* miles in April, *b* miles in May, and *c* miles in June. What was the average number of miles he rode each month?

   a. $a + b - c$

   b. $a \times b \times c$

   c. $\dfrac{a + b + c}{3}$

2. Jenny bought a sweater for *a* dollars, a pair of shoes for *b* dollars, a pocketbook for *c* dollars, a blouse for *d* dollars, and slacks for *e* dollars. She paid the clerk with *f* dollars. How much change did she receive?

   a. $f + a + b + c + d + e$

   b. $f - a + b + c + d + e$

   c. $f - (a + b + c + d + e)$

TEACHING TIPS

It is often helpful for students to work with partners or in small groups when solving problems such as these.

PROBLEMS

Solve.

1. The recycling center paid $0.32 for each pound of aluminum cans. The student council received $18.56 for its returned cans. How many pounds of cans was that? **58 lb**

2. A dripping faucet wasted $\frac{1}{4}$ c of water per day. How much water was wasted over $14\frac{2}{3}$ days? **$3\frac{2}{3}$ c**

3. Mrs. Martinez teaches five classes each day, Monday through Friday. Each class is $1\frac{1}{2}$ h long. How many hours does Mrs. Martinez teach in 36 weeks? **1,350 h**

4. The price of cashews is $7.50 per pound. If you buy 0.46 lb at the grocery and 0.38 lb at the specialty store, how much will you pay altogether? **$6.30**

5. A car averages 31.5 mi/gal. How many miles did the car travel if it used 8.8 gal? **277.2 mi**

6. A bicycle costs $\frac{4}{5}$ as much as a moped. The moped costs $\frac{1}{6}$ as much as a car. If the car costs $7,200, how much does the bicycle cost? **$960**

7. Recyclers paid $0.15 per pound for newspapers. How much did a homeowners' group receive for $75\frac{3}{5}$ lb of newspapers in June and $56\frac{1}{2}$ lb of newspapers in July? **$19.82**

A sponsor for the Hike-A-Thon has promised to pay Anita's favorite charity $0.12 for each mile she hikes. She plans to hike $17\frac{1}{4}$ mi each day for a total of 138 miles.

8. How many days will it take Anita to complete 138 miles? **8 days**

9. How much money will Anita's sponsor pay to her charity? **$16.56**

A lawn mower costs $475.98. Next week it will be sold at $\frac{1}{3}$-off. The Garden Club decides to wait until next week to buy the mower at the sale price.

10. How much will the Garden Club save by buying the mower at the sale price? **$158.66**

11. What is the sale price? **$317.32**

12. What is the total amount of money the Garden Club will pay if the sales tax is 0.06 of the sales price? **$336.36**

PROBLEM SOLVING TIP

Remember to round amounts of money to the nearest cent.

4-8   Problem Solving Strategies: Solve a Simpler Problem   **145**

MAKING CONNECTIONS

Have students find at least one problem from another subject area or from everyday life that they could solve by using simpler numbers.

3 SUMMARIZE

Talk It Over Have students describe how they would go about simplifying a problem that seems difficult to understand.

4 PRACTICE

Call on students to tell how they would simplify each of Problems 1–5. Then have students try to solve the problems on their own.

5 FOLLOW-UP

Extra Practice

1. To accommodate her wheelchair, Roberta had every doorway in her house made 1 2/3 ft wider. If the new doorways are 4 1/6 ft wide, how wide were the doorways originally? **2 1/2 ft**

2. Erica estimates her driving costs at $0.48/mi. How much does it cost her each week to drive to work 7 1/2 mi a day, 5 days a week? **$18**

4 CHAPTER REVIEW

Introduction The Chapter Review emphasizes the major concepts, skills, and vocabulary presented in this chapter and can be used for diagnosing students' strengths and weaknesses. Page references direct students back to appropriate sections for additional review and reteaching.

Using Pages 146–147 Suggest that students scan the Chapter Review and ask questions about any section they find confusing. Exercises 1–5 review vocabulary. Some students may prefer to find the GCF and the LCM by using factors and multiples.

In Exercise 40, students will have to find the waste generated per person per day for each of the four years and multiply by the number of people and the number of days per year for each year; those four figures will then have to be added. Students may prefer to write 150,000,000 in scientific notation and use that form when multiplying, in order to simplify their work.

Informal Evaluation Have students explain how they arrived at any incorrect answers. They will probably find their own mistakes and give you some clues as to the nature of their errors. Make sure students understand this material before administering the Chapter Test.

Follow-Up Have students simplify each expression.
1. 1/2 + 3/8 ÷ 6 **9/16**
2. 2 2/3 × 3/4 − 1 5/6 **1/6**

146

Choose the word from the list that completes each statement.

1. A ___?___ number has only two factors, 1 and itself. **c**
2. When writing a fraction as a decimal, the answer is a ___?___ decimal when the remainder is zero. **a**
3. When writing a fraction as a decimal, the answer is a ___?___ decimal when the remainder is not zero and a block of digits repeats in the quotient. **e**
4. A fraction with a numerator of 1 and a denominator written in exponential form can be written as a whole number with a ___?___ exponent. **b**
5. Two numbers whose product is 1 are ___?___. **d**

a. terminating
b. negative
c. prime
d. reciprocals
e. repeating

SECTION 4–1   FACTORS AND MULTIPLES   (pages 118–121)

► To find the GCF of a pair of numbers, use the prime factorization to find the least power of each common prime factor, then multiply.
► To find the LCM of a pair of numbers, write the prime factorization of each number, then choose the greatest power of each prime factor, and multiply.

Find the GCF of each pair of numbers.

6. 16, 24 **8**   7. 12, 18 **6**   8. 15, 25 **5**   9. 36, 40 **4**

Find the LCM of each pair of numbers.

10. 7, 13 **91**   11. 10, 25 **50**   12. 12, 15 **60**   13. 27, 36 **108**

SECTION 4–2   EQUIVALENT FRACTIONS   (pages 122–125)

► You can multiply or divide the numerator and the denominator of a fraction by the same nonzero number to find an equivalent fraction.

Write each fraction in lowest terms.

14. $\frac{16}{20}$ **$\frac{4}{5}$**   15. $\frac{7}{14}$ **$\frac{1}{2}$**   16. $\frac{30}{50}$ **$\frac{3}{5}$**   17. $\frac{24}{36}$ **$\frac{2}{3}$**

SECTION 4–3   FRACTIONS AND DECIMALS   (pages 126–129)

► To write a fraction as a decimal, divide the numerator by the denominator.
► To compare a decimal and a fraction, write the fraction as a decimal.

Write each fraction as a decimal.

18. $\frac{4}{5}$ **0.8**   19. $\frac{5}{16}$ **0.3125**   20. $\frac{7}{9}$ **$0.\overline{7}$**   21. $\frac{3}{11}$ **$0.\overline{27}$**

SECTION 4-4 ZERO AND NEGATIVE EXPONENTS (pages 130-133)

▶ To write a number in scientific notation, express it as a product in which one factor is greater than or equal to 1 and less than 10 and another factor is a power of ten.

▶ A number with a negative exponent can be expressed as a fraction.

Write in exponential form.

22. $\dfrac{1}{3 \times 3 \times 3 \times 3 \times 3}$ **3^{-5}**

23. $\dfrac{1}{4 \times 4}$ **4^{-2}**

24. $\dfrac{1}{9 \times 9 \times 9 \times 9}$ **9^{-4}**

Write in standard form.

25. 8.63×10^{-5} **0.0000863**

26. 4.11108×10^{-12} **0.00000000000411108**

Write in scientific notation.

27. 0.00258
2.58×10^{-3}

28. 0.0000034
3.4×10^{-6}

29. 0.657
6.57×10^{-1}

30. 0.0000009
9×10^{-7}

SECTIONS 4-5 and 4-6 OPERATIONS WITH FRACTIONS (pages 134-141)

▶ To multiply fractions, multiply the numerators, then the denominators.

▶ To divide fractions, multiply by the reciprocal of the divisor.

▶ To add or subtract fractions, rename them, if necessary, so that the denominators are the same; then add or subtract the numerators.

Compute. Write the answer in lowest terms.

31. $\dfrac{3}{8} \times 4$ **$1\dfrac{1}{2}$**

32. $2\dfrac{1}{2} \times 2\dfrac{3}{4}$ **$6\dfrac{7}{8}$**

33. $\dfrac{3}{8} + \dfrac{1}{6}$ **$\dfrac{13}{24}$**

34. $\dfrac{2}{3} - \dfrac{1}{2}$ **$\dfrac{1}{6}$**

35. $\dfrac{2}{3} \div 8$ **$\dfrac{1}{12}$**

36. $1\dfrac{3}{5} + 2\dfrac{9}{10}$ **$4\dfrac{1}{2}$**

37. $2\dfrac{1}{5} \div 1\dfrac{5}{6}$ **$1\dfrac{1}{5}$**

38. $9\dfrac{1}{4} - 2\dfrac{11}{12}$ **$6\dfrac{1}{3}$**

SECTIONS 4-7 and 4-8 PROBLEM SOLVING (pages 142-145)

▶ To estimate when working with fractions, it is often helpful to use numbers close to 0, $\dfrac{1}{2}$, and 1.

▶ Sometimes you can try simpler numbers to help you focus on choosing the operation(s) needed to solve a problem.

Estimate to solve.

39. Jodie needs 6 pieces of string that each measure $13\dfrac{1}{2}$ in. She has a roll of string that is 78 in. long. Does she need to buy more? **yes**

USING DATA Use the table on page 116 to answer the following question.

40. How many pounds of waste would 150,000,000 people have generated for the years 1980 through 1984? Leap years were 1980 and 1984. **918,627,000,000 lb**

Study Skills Tip Point out the relationships between adding and subtracting with fractions and between multiplying and dividing with fractions to show students that there are basically only two algorithms for the four operations.

CHAPTER TEST

Introduction The Chapter Test uses a variety of questioning techniques to assess students' mastery of the major objectives of Chapter 4. If you prefer, you may use the Skills Preview (page 115) as an alternative form of the Chapter Test. The items on this test and the Skills Preview correspond in content and level of difficulty.

Alternative Assessment
Critical Thinking Find the missing number.
1. $2/3 \times \blacksquare = 6/7$ **1 2/7**
2. $\blacksquare - 2\ 1/2 = 3\ 5/6$ **6 1/3**
3. $13\ 4/5 + \blacksquare = 20\ 1/2$ **6 7/10**
4. $6\ 2/3 \div \blacksquare = 10\ 2/3$ **5/8**

1. Find the prime factorization of 200. **$2^3 \times 5^2$**

Find the GCF of each pair of numbers.

2. 6, 8 **2**
3. 4, 6, **2**
4. 9, 15 **3**
5. 24, 36 **12**

Find the LCM of each pair of numbers.

6. 7, 8 **56**
7. 4, 6, **12**
8. 6, 8 **24**
9. 8, 12 **24**

10. Multiply to find two fractions that are equivalent to $\frac{11}{12}$. **Answers may vary. Sample: $\frac{22}{24}, \frac{33}{36}$**

11. Divide to find two fractions that are equivalent to $\frac{24}{36}$. **Answers may vary. Sample: $\frac{12}{18}, \frac{2}{3}$**

Write each fraction in lowest terms.

12. $\frac{25}{75}$ **$\frac{1}{3}$**
13. $\frac{36}{42}$ **$\frac{6}{7}$**
14. $\frac{12}{15}$ **$\frac{4}{5}$**
15. $\frac{18}{30}$ **$\frac{3}{5}$**

Write each fraction as a decimal.

16. $\frac{71}{100}$ **0.71**
17. $\frac{5}{8}$ **0.625**
18. $\frac{11}{15}$ **$0.7\overline{3}$**
19. $\frac{5}{9}$ **$0.\overline{5}$**

Write each decimal as a fraction in lowest terms.

20. 0.32 **$\frac{8}{25}$**
21. 0.8 **$\frac{4}{5}$**
22. 0.004 **$\frac{1}{250}$**
23. 3.88 **$3\frac{22}{25}$**

Replace ● with <, >, or =.

24. $\frac{7}{8}$ ● $\frac{11}{12}$ **<**
25. $\frac{1}{6}$ ● $0.\overline{3}$ **<**
26. 0.58 ● $\frac{7}{12}$ **<**
27. $\frac{3}{16}$ ● 0.1875 **=**

28. Write 7^{-4} as a fraction **$\frac{1}{7 \times 7 \times 7 \times 7} = \frac{1}{2,401}$**

29. Write $\frac{1}{4 \times 4 \times 4 \times 4 \times 4}$ in exponential form. **4^{-5}**

30. Write 7.392×10^{-8} in standard form. **0.00000007392**

31. Write 0.000138 in scientific notation. **1.38×10^{-4}**

Find each answer. Write answers in lowest terms.

32. $\frac{3}{4} + \frac{1}{8}$ **$\frac{7}{8}$**
33. $12\frac{3}{8} - 9\frac{1}{3}$ **$3\frac{1}{24}$**
34. $3\frac{1}{2} \times 1\frac{1}{3}$ **$4\frac{2}{3}$**
35. $\frac{2}{3} \div \frac{5}{6}$ **$\frac{4}{5}$**

Solve.

36. Don has read $\frac{3}{5}$ of a book that is 348 pages long. Estimate to find about how many pages he has read. **Estimates may vary. Sample: 210 pages**

37. Mai Ling plans to hike $2\frac{3}{4}$ mi each day to raise money for a charity. Her sponsor has pledged \$1.50 to the charity for each mile she hikes. If she hikes each day for $5\frac{3}{7}$ weeks, how much money will the charity receive? **\$156.75**

CHAPTER 4 ● CUMULATIVE REVIEW

The following data show math test scores for one class.

80 70 84 95 68 90 76 85 100 65
89 97 71 69 73 88 91 76 85

1. To find out whether you are in the top half of the class, would you use the mean or the median? **median**

2. Would you be in the top half of the class if you had one of the 85's? **yes**

Use the bar graph for Exercises 3–6.

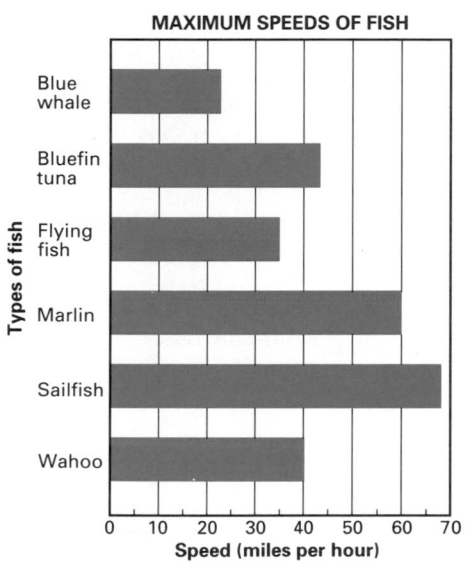

MAXIMUM SPEEDS OF FISH

3. Which is the fastest fish? **sailfish**

4. How many miles per hour faster is the marlin than the flying fish? **25**

5. Which fish can swim over 40 miles per hour? **bluefin tuna, marlin, sailfish, wahoo**

6. The speed of the sailfish is about how many times faster than the speed of the flying fish? **about two times**

Evaluate each expression.
Let $a = 3$ and $b = 6$.

7. $a + 2b$ **15**

8. $4b \div a$ **8**

9. $\frac{5b}{2a}$ **5**

10. $4a \times 2b$ **144**

11. $b - 1a$ **3**

12. $\frac{12a}{b}$ **6**

13. Determine whether the following argument is *valid* or *invalid*.

If a number is divisible by 8, then the number is divisible by 4.
96 is divisible by 8.

Therefore, 96 is divisible by 4. **valid**

14. Pick a number. Double it. Add 3. Then add the original number. Divide by 3. Subtract 1. What number results? Repeat the process several times. Make a conjecture about the result.
The result is always the original number.

15. Write two conditional statements using these two sentences: **See Additional Answers.**
The fruit is a plum.
The fruit has a pit.

Find the prime factorization for each number.

16. 120 **$2^3 \times 3 \times 5$**

17. 768 **$2^8 \times 3$**

Find the GCF of each pair of numbers.

18. 12, 15 **3**

19. 14, 24 **2**

Find the LCM of each pair of numbers.

20. 6, 8 **24**

21. 18, 24 **72**

Write in lowest terms.

22. $\frac{42}{48}$ **$\frac{7}{8}$**

23. $\frac{21}{28}$ **$\frac{3}{4}$**

Cumulative Review **149**

4 CUMULATIVE REVIEW

Introduction The purpose of this Cumulative Review is to maintain previously taught skills and concepts and to apply them to the material presented in this chapter. At least one major objective of each chapter is included in the review.

Item Analysis The table below correlates the Cumulative Review items with the chapter and section that are being reviewed.

Section	Items
1–6	3–6
1–9	1–2
2–2	7–12
3–2	14
3–3	13, 15
4–1	16–21
4–2	22–23

Additional Answers
See page 576.

4 CUMULATIVE TEST

Introduction The Cumulative Test uses a standardized-test format of multiple-choice questions to assess retention of previously learned concepts and test-taking skills. Test results may be used to diagnose students' strengths and weaknesses.

Item Analysis The table below correlates the Cumulative Test items with their corresponding objectives.

Section	Items
1–1	1–2
1–9	3
2–2	4–7
3–3	8–9
4–1	10–12
4–2	13
4–3	14
4–5	16
4–6	15

1. To determine the baseball player with the best batting average, which is the best way to collect data?
   A. personal interviews
   B. records of performance
   C. questionnaire
   D. telephone interviews

2. Which is the best method of choosing a sample to gather data about the quality of Trendy Tires?
   A. personal interviews
   B. records of performance
   C. questionnaire
   D. telephone interviews

3. What is the range of this set of data?
   62   16   49   104   30   21
   A. 104
   B. 78
   C. 62
   D. none of these

4. Evaluate $a + 3b$. Let $a = 2$ and $b = 3$.
   A. 9   B. 8   C. 11   D. 12

5. Evaluate $6b \div a$. Let $a = 2$ and $b = 3$.
   A. 4   B. 9   C. 6   D. 8

6. Evaluate $\frac{6b}{3a}$. Let $a = 2$ and $b = 4$.
   A. 4   B. 1   C. 2   D. 3

7. Evaluate $6a \times 3b$. Let $a = 2$ and $b = 4$.
   A. 24   B. 0   C. 1   D. 144

8. Which is a counterexample for the following statement?

   If a whole number x is divisible by 3, then x is divisible by 9.
   A. 27   B. 81   C. 9   D. 39

9. Which conlusion can be drawn from these statements?

   If yesterday was Wednesday, then I went to school.
   Yesterday was Wednesday.

   A. I did not go to school.
   B. I went to school.
   C. Yesterday was Wednesday.
   D. none of these

10. Which is the prime factorization of 288?
    A. $2^3 \times 3^2 \times 4$ B. $2 \times 3^3 \times 4$
    C. $2^3 \times 4 \times 9$ D. $2^5 \times 3^2$

11. Which is the GCF of 15 and 24?
    A. 7360   B. 209
    C. 3   D. 2

12. Which is the LCM of 6 and 10?
    A. 60   B. 30
    C. 3   D. 2

13. Express $\frac{24}{32}$ in lowest terms.
    A. $\frac{12}{16}$ B. $1\frac{1}{4}$
    C. $\frac{4}{5}$ D. $\frac{3}{4}$

14. Which is the decimal for $\frac{7}{30}$?
    A. $0.2\overline{3}$ B. 4.285
    C. 0.32   D. $1.\overline{37}$

15. Add. $\frac{3}{8} + \frac{1}{6}$
    A. $\frac{2}{3}$ B. $\frac{4}{14}$
    C. $\frac{2}{7}$ D. $\frac{13}{24}$

16. Divide. $\frac{1}{2} \div \frac{7}{8}$
    A. $\frac{2}{4}$ B. $\frac{2}{3}$
    C. $\frac{4}{7}$ D. $\frac{1}{3}$

Name _____ Date _____

Factors and Multiples

A **prime number** is a number that has exactly two factors, 1 and itself.
A **composite number** is a number that has more than two factors.
Any composite number can be written as the product of prime factors.
The prime factorization of 36, for example, is $2 \times 2 \times 3 \times 3$, or $2^2 \times 3^2$.

► Example 1 _____

Find the greatest common factor (GCF) of 27 and 36.

Solution 1

List the factors of each number and underline the common factors.
27: 1, 3, 9, 27
36: 1, 2, 3, 4, 6, 9, 12, 18, 36
The greatest common factor is 9.

Solution 2

Write the prime factorization of each number.
$27 = 3 \times 3 \times 3$
$36 = 2 \times 2 \times 3 \times 3$
Compare. The GCF includes all the common prime factors. Underline the common factors.
$27 = 3 \times \underline{3} \times 3$
$36 = 2 \times 2 \times \underline{3} \times 3$
Choose the least power of each common factor and multiply. The GCF of 27 and 36 is 9.

► Example 2 _____

Find the least common multiple (LCM) of 27 and 36.

Solution 1

List the multiples of each number until you find a common multiple.
27: 27, 54, 81, 108, 135
36: 36, 72, 108
The least common multiple is 108.

Solution 2

Write the prime factorization of each number.
$27 = 3 \times 3 \times 3$
$36 = 2 \times 2 \times 3 \times 3$
Compare. Underline all the factors of 27 and any factors of 36 that are not factors of 27.
$27 = 3 \times \underline{3} \times \underline{3}$
$36 = \underline{2} \times \underline{2} \times 3 \times 3$
Choose the greatest power of each prime factor and multiply. The LCM is 108.

EXERCISES

Find the prime factorization for each number.

1. 45 $\underline{3 \times 3 \times 5}$ or $3^2 \times 5$
2. 60 $\underline{2 \times 2 \times 3 \times 5}$ or $2^2 \times 3 \times 5$
3. 20 $\underline{2 \times 2 \times 5}$ or $2^2 \times 5$
4. 35 $\underline{5 \times 7}$
5. 25 $\underline{5 \times 5}$ or 5^2
6. 28 $\underline{2 \times 2 \times 7}$ or $2^2 \times 7$

Find the GCF of each pair of numbers.

7. 45, 20 **5**
8. 35, 28 **7**
9. 27, 15 **3**
10. 28, 60 **4**
11. 42, 63 **21**

Find the LCM of each pair of numbers.

12. 25, 60 **300**
13. 20, 35 **140**
14. 25, 45 **225**
15. 17, 51 **51**
16. 10, 12 **60**

Name _____ Date _____

Perfect, Abundant, and Deficient Numbers

The **proper factors** include all the factors except the number itself.

If the sum of the factors of a number is **less than** the number itself, the number is called **deficient**. If the sum is **equal to** the number itself, the number is called **perfect**. If the sum is **greater than** the number itself, the number is called **abundant**.

EXERCISES

1. Complete this chart for numbers 1–30.

Number	Proper factors	Sum of factors	Number	Proper factors	Sum of factors
1	none	—	16	1, 2, 4, 8	15
2	1	1	17	1	1
3	1	1	18	1, 2, 3, 6, 9	21
4	1, 2	3	19	1	1
5	1	1	20	1, 2, 4, 5, 10	22
6	1, 2, 3	6	21	1, 3, 7	11
7	1	1	22	1, 2, 11	14
8	1, 2, 4	7	23	1	1
9	1, 3	4	24	1, 2, 3, 4, 6, 8, 12	36
10	1, 2, 5	8	25	1, 5	6
11	1	1	26	1, 2, 13	16
12	1, 2, 3, 4, 6	16	27	1, 3, 9	13
13	1	1	28	1, 2, 4, 7, 14	28
14	1, 2, 7	10	29	1	1
15	1, 3, 5	9	30	1, 2, 3, 5, 6, 10, 15	42

2. Which of the numbers in the chart are deficient?
2, 3, 4, 5, 7, 8, 9, 10, 11, 13, 14, 15, 16, 17, 19, 21, 22, 23, 25, 26, 27, 29

3. Which numbers in the chart are perfect?
6 and 28

4. Which numbers in the chart are abundant?
12, 18, 20, 24, 30

5. Find a deficient number greater than 30.
Answers may vary. Sample: 32

6. Find an abundant number greater than 30.
Answers may vary. Sample: 36

Name _____ Date _____

Equivalent Fractions

Fractions that represent the same amount are called **equivalent fractions**. Equivalent fractions can be formed
► by multiplying the numerator and denominator of a fraction by the same nonzero number;
► by dividing the numerator and denominator of a fraction by the same nonzero number.

► Example 1 _____

Multiply to find two fractions that are equivalent to $\frac{3}{10}$.

Solution

Multiply the numerator and denominator by 2.
$\frac{3 \times 2}{10 \times 2} = \frac{6}{20}$
Multiply the numerator and denominator by 3.
$\frac{3 \times 3}{10 \times 3} = \frac{9}{30}$
Two fractions equivalent to $\frac{3}{10}$ are $\frac{6}{20}$ and $\frac{9}{30}$.

► Example 2 _____

Write $\frac{40}{55}$ in lowest terms.

Solution

Find the GCF of 40 and 55.
$40 = 2 \times 2 \times 2 \times 5$
$55 = 5 \times 11$
The GCF is 5.
Divide both the numerator and denominator by 5.
$\frac{40 \div 5}{55 \div 5} = \frac{8}{11}$
$\frac{40}{55}$ in lowest terms is $\frac{8}{11}$.

EXERCISES

Answers will vary. Sample answers are given.

Multiply to find two fractions that are equivalent to the given fraction.

1. $\frac{2}{5}$ $\frac{4}{10}, \frac{6}{15}$
2. $\frac{3}{7}$ $\frac{6}{14}, \frac{9}{21}$
3. $\frac{5}{8}$ $\frac{10}{16}, \frac{15}{24}$

Answers will vary. Sample answers are given.

Divide to find two fractions that are equivalent to the given fraction.

4. $\frac{20}{36}$ $\frac{10}{18}, \frac{5}{9}$
5. $\frac{18}{24}$ $\frac{9}{12}, \frac{6}{8}$
6. $\frac{24}{48}$ $\frac{12}{24}, \frac{8}{16}$

Write each fraction in lowest terms.

7. $\frac{16}{20}$ $\frac{4}{5}$
8. $\frac{25}{35}$ $\frac{5}{7}$
9. $\frac{24}{33}$ $\frac{8}{11}$
10. $\frac{27}{36}$ $\frac{3}{4}$
11. $\frac{9}{33}$ $\frac{3}{11}$
12. $\frac{14}{35}$ $\frac{2}{5}$
13. $\frac{20}{45}$ $\frac{4}{9}$
14. $\frac{18}{48}$ $\frac{3}{8}$

15. Suppose that when two fractions are each rewritten in lowest terms, they produce the same fraction. What is true about the two fractions?
They are equivalent.

Name _____ Date _____

Exploring Equivalent Fractions

Materials: index cards

Write the digits from 1 to 9, each on a separate index card. Shuffle the cards and lay them face down.

Choose four of the cards without looking at the faces. On a sheet of paper on which you have drawn two fraction bars, use the cards to make different pairs of fractions in which the numerator is less than the denominator. Try also to make a pair of equivalent fractions.

EXERCISES

1. From any four digits, how many different fractions can be made in which the numerator is less than the denominator?
six

2. What pairs of equivalent fractions are possible?
$\frac{1}{2}, \frac{3}{6}$; $\frac{1}{2}, \frac{4}{8}$; $\frac{1}{3}, \frac{2}{6}$; $\frac{2}{3}, \frac{4}{6}$; $\frac{2}{6}, \frac{1}{3}$; $\frac{2}{8}, \frac{3}{4}$; $\frac{6}{8}, \frac{3}{4}$

3. What is the largest number of equivalent fraction pairs that could be produced from a random draw of four cards? Write the pairs.
Two pairs; some possibilities are $\frac{1}{4}, \frac{2}{8}$ and $\frac{4}{8}, \frac{3}{6}$; $\frac{4}{8}$ and $\frac{3}{6}, \frac{4}{8}$

4. Choose any four cards for numbers greater than 9 to be added to the deck. Choose numbers that will increase the number of equivalent fraction pairs that would be possible. Create several new pairs of equivalent fractions.
Answers will vary. Good choices: 12, 20, 24, 30.

150A

Name _____ Date _____

Fractions and Decimals

Both fractions and decimals are used to show parts of a whole. Any number that can be written as a fraction can also be written as a decimal. Many decimals can be written as fractions.

▶ **Example 1** _____

Write each decimal as a fraction or mixed number in lowest terms.

a. 0.72   b. 4.008

Solution

Use the place value of the digits to help you write each fraction.

$0.72 = \frac{72}{100}$ $4.008 = 4\frac{8}{1000}$

Write the fractions in lowest terms.

$\frac{72}{100} = \frac{18}{25}$ $4\frac{8}{1000} = 4\frac{1}{125}$

a. $0.72 = \frac{18}{25}$ b. $4.008 = 4\frac{1}{125}$

▶ **Example 2** _____

Write each fraction or mixed number as a decimal.

a. $\frac{4}{5}$ b. $2\frac{5}{9}$

Solution

Divide each numerator by the denominator.

$5\overline{)4.0}$ gives $\frac{.8}{40}$ ... $\frac{40}{0}$ $9\overline{)5.00}$ gives $\frac{.55...}{45}$ ... $\frac{50}{45}$ ... $\frac{45}{5}$

a. $\frac{4}{5} = 0.8$ b. $\frac{5}{9} = 0.555...$

A repeating decimal can be expressed with a bar over the repeating digit. $0.555... = 0.\overline{5}$

EXERCISES

Write each decimal as a fraction or as a mixed number in lowest terms.

1. 0.9 $\frac{9}{10}$
2. 3.2 $3\frac{1}{5}$
3. 0.41 $\frac{41}{100}$
4. 0.06 $\frac{3}{50}$
5. 0.505 $\frac{101}{200}$
6. 3.007 $3\frac{7}{1000}$
7. 0.009 $\frac{9}{1000}$
8. 30.03 $30\frac{3}{100}$

Write each fraction or mixed number as a terminating decimal.

9. $\frac{3}{4}$ **0.75**
10. $5\frac{11}{20}$ **5.55**
11. $5\frac{6}{25}$ **5.24**
12. $\frac{7}{40}$ **0.175**
13. $\frac{3}{50}$ **0.06**
14. $1\frac{7}{10}$ **1.7**
15. $4\frac{1}{40}$ **4.025**
16. $3\frac{7}{20}$ **3.35**

Write each fraction as a repeating decimal.

17. $\frac{2}{9}$ **0.$\overline{2}$**
18. $\frac{5}{12}$ **0.41$\overline{6}$**
19. $\frac{4}{11}$ **0.$\overline{36}$**
20. $\frac{8}{15}$ **0.5$\overline{3}$**
21. $\frac{5}{6}$ **0.8$\overline{3}$**
22. $\frac{7}{9}$ **0.$\overline{7}$**
23. $\frac{14}{15}$ **0.9$\overline{3}$**
24. $\frac{6}{11}$ **0.$\overline{54}$**

Name _____ Date _____

Babylonian Numeration

The ancient Babylonians used a number system with the base 60. There were two different symbols, ▼ for 1 and ◄ for 10, which could be repeated to represent numbers from 1 to 59. Thus,

represented 4   represented 30   represented 34

Representing numbers 10 and greater was difficult without a placeholder, such as zero. The symbols ◄▼▼ might represent 12 or 12×60 or $12 \times (60)^2$. The value could only be guessed from the context. For example, if a number represented how many sheep were in a flock, one might assume 12×60 was meant. At first a space was left open as a placeholder; later the symbol ◄◄ was used as a placeholder. Thus, the number 3,623 in our decimal system would have been written as

1×60^2 0×60^1 23×60^0
3600   0   23   = 3,623

If the placeholder symbol began a number, the number was a fraction. Thus,

0×60^0 12×60^{-1} 30×60^{-2}

represented the fraction

$0 + \frac{12}{60} + \frac{30}{3600} = 0 + \frac{1}{5} + \frac{1}{60}$

EXERCISES

Write decimal equivalents for each Babylonian numeral.

1.   **42**
2.   **3,620**
3.   **63**

Write fractional equivalents for each numeral.

4. $0 + \frac{5}{60} + \frac{2}{3600}$
5. $0 + \frac{20}{60} + \frac{42}{3600}$

Name _____ Date _____

Zero and Negative Exponents

Study the patterns in this list. They can help you to write numbers in standard form or in scientific notation.

Exponential Form		Standard Form	
10^5	=	100,000	(5 zeros)
10^4	=	10,000	(4 zeros)
10^3	=	1,000	(3 zeros)
10^2	=	100	(2 zeros)
10^1	=	10	(1 zero)
10^0	=	1	(no zero)

A positive exponent tells you how many zeros appear after 1 in the standard form of the number.

10^{-1}	=	0.1	(1 decimal place)
10^{-2}	=	0.01	(2 decimal places)
10^{-3}	=	0.001	(3 decimal places)
10^{-4}	=	0.0001	(4 decimal places)
10^{-5}	=	0.00001	(5 decimal places)

A negative exponent tells you how many decimal places are in the standard form of the number.

▶ **Example 1** _____

Write 6.31×10^{-5} in standard form.

Solution

Write the number that is multiplied by the power of 10.
6.31
Then move the decimal point 5 places to the left (since the exponent is −5). Use zeros as placeholders.
0.0000631
In standard form,
$6.31 \times 10^{-5} = 0.0000631$

▶ **Example 2** _____

Write 0.000457 in scientific notation.

Solution

Write the number in standard form. Then move the decimal to the right to get a number greater than or equal to 1 but less than 10.
0.000457
Count the number of places you moved the decimal point.
0.000457 ← 4 places to the right
Use that number of places to write the correct negative exponent of 10: 10^{-4}. In scientific notation,
$0.000457 = 4.57 \times 10^{-4}$

EXERCISES

Write in scientific notation.

1. 0.000035   **3.5×10^{-5}**
2. 0.001302   **1.302×10^{-3}**
3. 0.0000072   **7.2×10^{-6}**
4. 0.00412   **4.12×10^{-3}**
5. 0.000018   **1.8×10^{-5}**
6. 0.00000089   **8.9×10^{-7}**

Write as a decimal.

7. 2.54×10^{-4} **0.000254**
8. 7.9×10^{-7} **0.00000079**
9. 1.3×10^{-5} **0.000013**
10. 8.01×10^{-8} **0.0000000801**

Name _____ Date _____

Metric Prefixes

In the metric system, prefixes are used to name measures that are multiples of decimal parts of metric units.

Prefix	Multiple	Prefix	Decimal Part
exa-	10^{18}	deci-	10^{-1}
peta-	10^{15}	centi-	10^{-2}
tera-	10^{12}	milli-	10^{-3}
giga-	10^9	micro-	10^{-6}
mega-	10^6	nano-	10^{-9}
kilo-	10^3	pico-	10^{-12}
hecto-	10^2	femto-	10^{-15}
deka-	10^1	atto-	10^{-18}

These prefixes can be combined with any metric units, including seconds.

▶ **Example 1** _____

What part of a second is 6 microseconds?

Solution

The table shows that the decimal part that matches the prefix *micro* is 10^{-6}.
6 microseconds = 6×10^{-6} s
= 0.000006 s

▶ **Example 2** _____

How many meters are in 7 petameters?

Solution

The table shows that *peta* means 10^{15}.
7 petameters = 7×10^{15} m
= 7,000,000,000,000,000 m

EXERCISES

Use the table to find each of the following.

1. What part of a meter is 3 picometers?
   0.000000000003 m
2. What part of a liter is 9 nanoliters?
   0.000000009 L
3. How many tons are in 5 megatons?
   5,000,000 t
4. Write 6 teragrams in grams.
   6,000,000,000,000 g
5. How many liters are in 12 gigaliters?
   12,000,000,000 L
6. What part of a meter is 52 femtometers?
   0.000000000000052 m
7. Write 24 picograms as a number of grams.
   0.000000000024 g
8. Write 85 microliters as a number of liters.
   0.000085 L

150B

Name _____ Date _____

Multiplying and Dividing Fractions

To multiply by a fraction or mixed number,
► Write all whole numbers and mixed numbers as fractions;
► Multiply the numerators; then multiply the denominators;
► Write the product in lowest terms.

► **Example 1**

Multiply: $\frac{3}{8} \times \frac{2}{3}$

Solution

Multiply the numerators and denominators. $\frac{3 \times 2}{8 \times 3}$

Divide out any common factors; then multiply numerators and denominators. $\frac{\overset{1}{3} \times \overset{1}{2}}{\underset{4}{8} \times \underset{1}{3}} = \frac{1}{4}$

To divide fractions and mixed numbers,
► Write all whole numbers and mixed numbers as fractions;
► Multiply the dividend by the reciprocal of the divisor;
► Write the product in lowest terms.

► **Example 2**

Divide: $1\frac{1}{4} \div 1\frac{3}{4}$

Solution

Write each mixed number as a fraction. $1\frac{1}{4} \div 1\frac{7}{8} = \frac{5}{4} \div \frac{15}{8}$

Multiply $\frac{5}{4}$ by the reciprocal of $\frac{15}{8}$. Divide by any common factors of a numerator and a denominator. $= \frac{\overset{1}{5}}{\underset{1}{4}} \times \frac{\overset{2}{8}}{\underset{3}{15}} = \frac{2}{3}$

EXERCISES

Multiply. Write the product in lowest terms.

1. $\frac{2}{3} \times \frac{4}{5}$ $\frac{8}{15}$ 2. $\frac{5}{7} \times \frac{3}{10}$ $\frac{3}{14}$ 3. $1\frac{2}{3} \times 2\frac{1}{5}$ $3\frac{1}{3}$ 4. $1\frac{4}{5} \times 6\frac{2}{3}$ 12

Divide. Write the quotient in lowest terms.

5. $\frac{2}{2} \div \frac{2}{5}$ $2\frac{1}{2}$ 6. $1\frac{1}{4} \div 1\frac{7}{8}$ $\frac{2}{3}$ 7. $\frac{3}{4} \div \frac{1}{9}$ $6\frac{3}{4}$ 8. $\frac{8}{9} \div 1\frac{1}{3}$ $\frac{2}{3}$

Reteaching • SECTION 4-5   71

Name _____ Date _____

Conjectures About Products

When two whole numbers are multiplied, the product is greater than either factor. Compare the products of whole numbers and fractions or pairs of fractions to see what conjecture can be made.

EXERCISES

Find each product.

1. a. $4 \times 2 = $ __8__   2. a. $4 \times \frac{1}{2} = $ __2__   3. a. $4 \times \frac{1}{8} = $ __$\frac{1}{2}$__

   b. $4 \times 1 = $ __4__   b. $4 \times \frac{1}{4} = $ __1__   b. $4 \times \frac{1}{16} = $ __$\frac{1}{4}$__

Write four exercises of your own, in which a whole number is multiplied by a fraction between 0 and 1. **Answers will vary.**

4. _____   5. _____

6. _____   7. _____

8. In Exercises 2–7, compare each product with its set of factors. Make a conjecture.

   Answers may vary. The product is always less than the whole number factor.

   _____

Complete this chart. Use a calculator to find the decimal equivalent of each product, to the nearest thousandth.

		Product	Decimal Equivalent
9.	$\frac{2}{3} \times \frac{1}{2} =$ (0.667 × 0.5)	$\frac{1}{3}$	0.333
10.	$\frac{4}{5} \times \frac{5}{6} =$ (0.8 × 0.8$\overline{3}$)	$\frac{2}{3}$	0.667
11.	$\frac{1}{8} \times \frac{3}{4} =$ (0.125 × 0.75)	$\frac{3}{32}$	0.094
12.	$\frac{3}{10} \times \frac{5}{8} =$ (0.3) × 0.625)	$\frac{3}{16}$	0.1885

13. In Exercises 9–12, compare each product with its set of factors. Make a conjecture.

    Answers may vary. In each instance, the product is always less than either of the fractional factors.

72   Enrichment • SECTION 4-5

Name _____ Date _____

Adding and Subtracting Fractions

To add or subtract fractions with the same denominator, you merely add or subtract the numerators: $\frac{1}{7} + \frac{3}{7} = \frac{1+3}{7} = \frac{4}{7}$

To add or subtract fractions with unlike denominators, you must rename the fractions so that the denominators are the same.

► **Example 1**   ► **Example 2**

Add: $\frac{1}{3} + \frac{4}{9}$ Subtract: $\frac{4}{5} - \frac{1}{2}$

Solution   **Solution**

Rename so that the denominators are the same. Then add.   Rename so that the denominators are the same. Then subtract.

$\begin{array}{r} \frac{1}{3} = \frac{3}{9} \\ +\frac{4}{9} = \frac{4}{9} \\ \hline \frac{7}{9} = \frac{7}{9} \end{array}$ $\begin{array}{r} \frac{4}{5} = \frac{8}{10} \\ -\frac{1}{2} = \frac{5}{10} \\ \hline \frac{3}{10} \end{array}$

For mixed numbers, add or subtract the whole number parts and the fraction parts. Sometimes, you must rename.

► **Example 3**   ► **Example 4**

Add: $2\frac{2}{3} + 3\frac{1}{2}$ Subtract: $5\frac{1}{3} - 2\frac{2}{3}$

Solution   **Solution**

Rename and add the fraction parts. Add the whole number parts.   Rename and subtract the fraction parts. Subtract the whole number parts.

$\begin{array}{r} 2\frac{2}{3} = 2\frac{4}{6} \\ +3\frac{1}{2} = 3\frac{3}{6} \\ \hline 5\frac{7}{6} = 6\frac{1}{6} \end{array}$ $\begin{array}{r} 5\frac{1}{3} = 4\frac{4}{3} \\ -2\frac{2}{3} = 2\frac{2}{3} \\ \hline 2\frac{2}{3} \end{array}$

EXERCISES

Add or subtract. Write your answers in lowest terms.

1. $\frac{2}{5} + \frac{1}{5}$ $\frac{3}{5}$ 2. $\frac{4}{9} - \frac{2}{9}$ $\frac{2}{9}$ 3. $\frac{5}{12} - \frac{1}{12}$ $\frac{1}{3}$ 4. $\frac{2}{11} + \frac{10}{11}$ $1\frac{1}{11}$

5. $\frac{3}{4} + \frac{1}{3}$ $1\frac{1}{12}$ 6. $\frac{8}{9} - \frac{1}{3}$ $\frac{5}{9}$ 7. $\frac{4}{5} + \frac{3}{10}$ $1\frac{1}{10}$ 8. $\frac{4}{9} - \frac{1}{18}$ $\frac{7}{18}$

9. $4\frac{1}{3} - \frac{4}{9}$ $3\frac{8}{9}$ 10. $5\frac{4}{5} + 2\frac{2}{10}$ 8 11. $6\frac{1}{8} - 3\frac{3}{4}$ $2\frac{3}{8}$ 12. $3\frac{1}{3} - 2\frac{1}{4}$ $1\frac{1}{12}$

Reteaching • SECTION 4-6   73

Name _____ Date _____

Fractions in Ancient Egypt

To write fractions, the Egyptians used unit fractions almost exclusively, although there was a special symbol for $\frac{2}{3}$.

A **unit fraction** is a fraction with a numerator of one, such as $\frac{1}{2}$, $\frac{1}{3}$, or $\frac{1}{10}$. The fraction $\frac{3}{4}$, for example, was written as the sum of two different unit fractions: $\frac{1}{2} + \frac{1}{4}$. It was important that the unit fractions be different. For example, $\frac{3}{8} = \frac{1}{8} + \frac{1}{8} + \frac{1}{8}$, but the acceptable form was $\frac{1}{4} + \frac{1}{8}$.

► **Example**

Find two or more unit fractions with a sum of $\frac{2}{7}$.

Solution

Try subtracting a unit fraction that is close in value to $\frac{2}{7}$.

$\frac{2}{7} - \frac{1}{6} = \frac{12-7}{42}$ or $\frac{5}{42}$
                        ↑
                  not a unit
                  fraction

$\frac{2}{7} - \frac{1}{4} = \frac{8-7}{28}$ or $\frac{1}{28}$

Therefore, $\frac{2}{7} = \frac{1}{4} + \frac{1}{28}$.

EXERCISES

Write each fraction as a sum of different unit fractions. **Answers may vary.**

1. $\frac{3}{4}$ $\frac{1}{2} + \frac{1}{4}$ 2. $\frac{3}{8}$ $\frac{1}{4} + \frac{1}{8}$ 3. $\frac{3}{5}$ $\frac{1}{2} + \frac{1}{10}$

4. $\frac{5}{6}$ $\frac{1}{2} + \frac{1}{3}$ 5. $\frac{7}{12}$ $\frac{1}{2} + \frac{1}{12}$ 6. $\frac{2}{5}$ $\frac{1}{3} + \frac{1}{15}$

7. $\frac{4}{9}$ $\frac{1}{3} + \frac{1}{9}$ 8. $\frac{5}{7}$ $\frac{1}{2} + \frac{1}{7} + \frac{1}{14}$ 9. $\frac{3}{10}$ $\frac{1}{5} + \frac{1}{10}$

10. $\frac{2}{3}$ $\frac{1}{3} + \frac{1}{3}$ 11. $\frac{4}{11}$ $\frac{1}{3} + \frac{1}{33}$ 12. $\frac{7}{15}$ $\frac{1}{3} + \frac{1}{8} + \frac{1}{120}$

13. $\frac{7}{8}$ $\frac{1}{2} + \frac{1}{4} + \frac{1}{8}$ 14. $\frac{7}{9}$ $\frac{1}{2} + \frac{1}{4} + \frac{1}{36}$ 15. $\frac{17}{20}$ $\frac{1}{2} + \frac{1}{4} + \frac{1}{10}$

74   Enrichment • SECTION 4-6

150C

Name _____ Date _____

Problem Solving Skills: Estimating with Fractions and Decimals

There are several techniques for estimating with fractions and decimals. Use the method that is most convenient for the numbers and operations involved.

▶ **Example 1**

A delivery service offers a lower rate if the combined weight of the packages is at least 25 pounds. Mr. Rivera's company is sending packages with the following weights. Will the delivery qualify for the lower rate?

$$2\frac{3}{4} \quad 2\frac{5}{8} \quad 3\frac{7}{3} \quad 4\frac{1}{2} \quad 3\frac{1}{3}$$
$$4\frac{1}{4} \qquad 5\frac{1}{3}$$

Solution

Use front-end digits. Begin by adding only the whole number parts.
$$2 + 2 + 3 + 4 + 3 + 4 + 5 = 23$$
Then, look at fraction parts. Four fractions are greater than or equal to one-half.
$$\frac{1}{2} + \frac{1}{2} + \frac{1}{2} + \frac{1}{2} = 2$$

Mr. Rivera has at least 25 lb in all.

▶ **Example 2**

Juliet wants to buy a coat that costs $77.85. The store is selling it for one-fourth off the regular price. About how much will Juliet pay for the coat?

Solution

Round the regular price to a number that it is easy to take one-fourth of.

$77.85 is close to $80.
$$\frac{1}{4} \times 80 = 20$$

If the coat is selling for $20 off, Juliet will be able to buy it for about $80 − 20 or $60.

EXERCISES

Estimate to solve each problem.

1. Jeff bought milk for $2.65, a loaf of bread for $1.25 and three cans of soup for $.79 each. He has $10. Does he have enough money to buy a package of cheese that costs $3.99?

   no

2. Louise earns $4 an hour babysitting. She wants to earn enough money to buy a set of CDs that costs $22.99. About how many hours must she work to earn enough money?

   about 6 hours

3. Pens cost $.79 each and markers cost $1.25 each. Denise has $10 and wants to buy 7 pens and 5 markers. Does she have enough money? If not, give at least two combinations of items she could buy with her money.

   no; possible answers: 7 pens,

   3 markers; 6 pens, 4 markers

4. Franco wants to buy a CD player that is marked down by $\frac{1}{4}$. The original price was $172. The sales tax comes to $7.74. Franco has $140. Is that enough to buy the CD player?

   yes

Name _____ Date _____

Shopping

The Westons have four children who are all going to Sleep-Away Camp this summer. They decided to buy whatever clothes the children need through a mail-order catalog.

Each child goes through his or her closet and makes a list of what is needed for camp. The parents have budgeted $150 per child for camp clothes. The list at the right gives the price for each item.

Item	Price
T-shirt	$8.95
shorts	9.50
slacks	12.75
swimsuit	21.85
socks	12.25/6 pairs

EXERCISES

Estimate the total of each child's list. **Answers may vary.**

1. Eleanor:
   5 T-shirts
   4 shorts
   2 slacks
   2 swimsuits
   12 pairs of socks

   Total: **$180**

2. Marylou:
   5 T-shirts
   3 shorts
   1 slacks
   2 swimsuits
   6 pairs of socks

   Total: **$144**

3. James:
   4 T-shirts
   3 shorts
   3 slacks
   2 swimsuits
   6 pairs of socks

   Total: **$160**

4. Jonathan:
   6 T-shirts
   4 shorts
   2 slacks
   3 swimsuits
   18 pairs of socks

   Total: **$222**

5. Which children went over the spending limit on their lists?

   Eleanor, Jonathan, and James

6. Revise the lists of the two children named in Exercise 5 so that their orders will be less than $150.

   Answers will vary. _____

   _____

   _____

Name _____ Date _____

Problem Solving Strategies: Solve a Simpler Problem

One way to approach a problem that contains fractions and/or decimals is to substitute simpler, whole numbers in the problem.

▶ **Example**

Tanya worked 2 h on Thursday, $2\frac{1}{4}$ h on Friday, and $7\frac{3}{4}$ h on Saturday. She gets paid $5.40 per hour. How much did she earn?

Solution

Begin by restating the problem with simpler numbers.

Tanya worked 2 h, 2 h, and 8 h. She earns $5 an hour. How much did she earn?

To solve, add the number of hours and multiply the sum by the hourly rate.
$$2 + 2 + 8 = 12$$
$$12 \times \$5 = \$60$$

Now, perform the same operations using the numbers in the problem:
$$2 + 2\frac{1}{2} + 7\frac{3}{4} = 2 + 2\frac{1}{4} + 7\frac{3}{4}$$
$$= 11\frac{5}{4} \text{ or } 12\frac{1}{4}$$

Multiply $5.40 by $12\frac{1}{4}$.
$$\$5.40 \times \$12.25 = \$66.15$$

Tanya earned $66.15.

EXERCISES

Solve.

1. Three investors had combined their money to buy some stock. To split the profits fairly, one investor received $\frac{2}{5}$. Another received $\frac{7}{10}$ of what was left and the third received the rest. The total profits were $4,356. What was each investor's share?

   $1742.40; $1829.52; $794.08

2. Helga bought four turtleneck shirts for $12.98 each. The sales tax was 0.05 of the total. How much did she pay in all for the shirts?

   $54.52

3. Joe works $3\frac{3}{4}$ h every Friday and $6\frac{1}{2}$ h on Saturday. How many hours does he work in six weeks?

   $61\frac{1}{2}$ h

4. Mrs. Chang's company will pay her $0.26 a mile for driving her car on business and pay any tolls. Last week, she drove 231 miles and paid a $2.25 bridge toll six times. How much will the company pay her back?

   $73.56

Name _____ Date _____

Solving Nonroutine Problems

Some problems that look very difficult can be solved by looking at a simpler version of the problem first. You may observe a pattern that will help you solve the original problem.

▶ **Example**

Find the sum:

$$\frac{1}{1 \times 2} + \frac{1}{2 \times 3} + \frac{1}{3 \times 4} + \cdots + \frac{1}{99 \times 100}$$

Solution

Find the value of the first fraction, then the sum of the first two, the sum of the first three and see if there is a pattern.

$$\frac{1}{1 \times 2} = \frac{1}{2}$$

$$\frac{1}{1 \times 2} + \frac{1}{2 \times 3} = \frac{2}{3}$$

$$\frac{1}{1 \times 2} + \frac{1}{2 \times 3} + \frac{1}{3 \times 4} = \frac{3}{4}$$

The pattern is that the sum can be found by taking the two factors of the last fraction to be added and using the first as the numerator and the second as the denominator. Therefore:

$$\frac{1}{1 \times 2} + \frac{1}{2 \times 3} + \frac{1}{3 \times 4} + \cdots + \frac{1}{99 \times 100} = \frac{99}{100}$$

EXERCISES

1. There are 12 people at a party. Each one will shake hands with every other person one time. How many handshakes will take place?

   66

2. Triangular numbers are numbers for which dots can be arranged as shown at the right. What is the tenth triangular number?

   55

1   3   6   10

3. What is the sum of the first twelve odd numbers?

   144

150D

Name _____ Date _____

Calculator Activity: Fractions

The $\boxed{a^{b}/_{c}}$ key on a scientific calculator makes it easy to work with fractions. Check to see how your calculator handles fractions.

▶ **Example 1** _____

Use a calculator to write $\frac{16}{24}$ in lowest terms.

Solution

Enter: 16 $\boxed{a^{b}/_{c}}$ 24 $\boxed{=}$ 2 ⌐ 3

The display indicates that $\frac{16}{24} = \frac{2}{3}$.

▶ **Example 2** _____

Use a calculator to solve $8\frac{1}{3} + 6\frac{3}{4} - 1\frac{3}{8}$.

Solution

Enter: 8 $\boxed{a^{b}/_{c}}$ 3 $\boxed{a^{b}/_{c}}$ 5 $\boxed{+}$ 6 $\boxed{a^{b}/_{c}}$ 3 $\boxed{a^{b}/_{c}}$ 4 $\boxed{-}$ 1 $\boxed{a^{b}/_{c}}$ 3 $\boxed{a^{b}/_{c}}$ 8 $\boxed{=}$
13 ⌐ 39 ⌐ 40

The display indicates that the answer is $13\frac{39}{40}$.

▶ **Example 3** _____

Use a calculator to solve $3\frac{4}{5} \times (2\frac{3}{8} + 3\frac{1}{3}) \div 4\frac{3}{9}$.

Solution

Enter: 3 $\boxed{a^{b}/_{c}}$ 4 $\boxed{a^{b}/_{c}}$ 5 $\boxed{\times}$ $\boxed{(}$ 2 $\boxed{a^{b}/_{c}}$ 3 $\boxed{a^{b}/_{c}}$ 8 $\boxed{+}$
3 $\boxed{a^{b}/_{c}}$ 1 $\boxed{a^{b}/_{c}}$ 3 $\boxed{)}$ $\div$ 4 $\boxed{a^{b}/_{c}}$ 3 $\boxed{a^{b}/_{c}}$ 9 $\boxed{=}$ 5 ⌐ 3 ⌐ 520

The display indicates that the answer is $5\frac{3}{520}$.

EXERCISES

Show a calculator key sequence that will help you write each fraction in lowest terms.

1. $\frac{36}{54}$ **36 $\boxed{a^{b}/_{c}}$ 54 $\boxed{=}$**   2. $\frac{75}{215}$ **75 $\boxed{a^{b}/_{c}}$ 215 $\boxed{=}$**

3. $\frac{90}{225}$ **90 $\boxed{a^{b}/_{c}}$ 225 $\boxed{=}$**   4. $\frac{105}{120}$ **105 $\boxed{a^{b}/_{c}}$ 120 $\boxed{=}$**

Solve. Use a calculator with a fraction key.

5. $2\frac{3}{4} + 5\frac{7}{9} =$ **$8\frac{19}{36}$**   6. $15\frac{8}{15} - 7\frac{3}{5} =$ **$7\frac{14}{15}$**   7. $25\frac{2}{3} \times 18\frac{5}{9} =$ **$476\frac{7}{27}$**

8. $99\frac{1}{2} \div 6\frac{4}{5} =$ **$14\frac{43}{68}$**   9. $7\frac{2}{3} + 5\frac{3}{5} - 2\frac{1}{8} =$ **$11\frac{17}{120}$**   10. $9\frac{7}{9} \times 6\frac{3}{7} + 4\frac{1}{2} =$ **$13\frac{61}{63}$**

11. $8\frac{2}{3} \times (6\frac{7}{8} + 5\frac{4}{9}) \div 10\frac{1}{2} =$ **$10\frac{97}{210}$**   12. $(5\frac{1}{3} - 2\frac{2}{9}) \times (6\frac{1}{4} + 4\frac{9}{10}) =$ **$28\frac{89}{180}$**

Name _____ Date _____

Computer Activity: Adding by Multiplying

You can always get a common denominator for two fractions by multiplying the denominators of the fractions. This computer program uses that concept and the horizontal form of adding fractions to find the sum of two fractions.

```
10  INPUT "ENTER THE FIRST NUMERATOR:
        ";N1: PRINT
20  INPUT "ENTER THE FIRST DENOMINATOR:
        ";D1: PRINT
30  INPUT "ENTER THE SECOND NUMERATOR:
        ";N2: PRINT
40  INPUT "ENTER THE SECOND DENOMINATOR:
        ";D2: PRINT
50  NS = N1 • D2 + D1 • N2
60  DS = D1 • D2
70  PRINT "SUM = ";NS;"/";DS
```

EXERCISES

1. RUN the program using the following pairs of fractions. For each RUN, what is the value of NS in line 50?

   a. $\frac{3}{5} + \frac{2}{3}$ **19**   b. $\frac{3}{4} + \frac{9}{10}$ **66**   c. $\frac{3}{8} + \frac{3}{4}$ **36**

2. For each pair of fractions, predict what the value of NS in line 50 will be. RUN the program to check your answer.

   a. $\frac{4}{5} + \frac{3}{4}$ **31**   b. $\frac{5}{7} + \frac{2}{9}$ **59**   c. $\frac{5}{6} + \frac{4}{11}$ **79**

   d. $\frac{9}{10} + \frac{2}{3}$ **47**   e. $\frac{12}{13} + \frac{3}{5}$ **99**   f. $\frac{7}{12} + \frac{6}{11}$ **149**

3. When will the denominator of the sum (DS in line 60) be the same as the LCD?

   When the denominators have no common factors except 1.

4. Will this work for subtraction of fractions? What adjustment can you make to the program to find out? RUN the adjusted program with a–f above to check your answer.

   50 NS = N1 • D2 − D1 • N2

   A. $\frac{1}{20}$ B. $\frac{31}{63}$ C. $\frac{31}{66}$ D. $\frac{7}{30}$ E. $\frac{21}{65}$ F. $\frac{5}{121}$

ACHIEVEMENT TEST

Exploring Fractions

CHAPTER 4 FORM A

Name _____

Date _____

MATH MATTERS BOOK 1

Chicha Lynch
Eugene Olmstead

SOUTH-WESTERN PUBLISHING CO.

SCORING RECORD	
Possible	Earned
37	

1. Find the prime factorization of 340.   **$2^2 \times 5 \times 17$**

Find the GCF of each pair of numbers.

2. 9, 12   **3**   3. 8, 20   **4**   4. 36, 42   **6**   5. 15, 20   **5**

Find the LCM of each pair of numbers.

6. 4, 5   **20**   7. 9, 12   **36**   8. 15, 18   **90**   9. 12, 20   **60**

10. Multiply to find two fractions that are equivalent to $\frac{3}{4}$.   **$\frac{9}{12}, \frac{18}{24}$**

11. Divide to find two fractions that are equivalent to $\frac{32}{48}$.   **$\frac{16}{24}, \frac{8}{12}$**

Write each fraction in lowest terms.

12. $\frac{12}{18}$ **$\frac{2}{3}$**   13. $\frac{14}{42}$ **$\frac{1}{3}$**   14. $\frac{16}{40}$ **$\frac{2}{5}$**   15. $\frac{20}{28}$ **$\frac{5}{7}$**

Write each fraction as a decimal.

16. $\frac{19}{100}$ **0.19**   17. $\frac{5}{8}$ **0.625**   18. $\frac{2}{9}$ **$0.\overline{2}$**   19. $\frac{3}{11}$ **$0.\overline{27}$**

Write each decimal as a fraction in lowest terms.

20. 0.65   **$\frac{13}{20}$**   21. 0.7   **$\frac{7}{10}$**   22. 0.0001   **$\frac{1}{10,000}$**   23. 4.84   **$4\frac{21}{25}$**

Replace ◯ with <, >, or =.

24. $\frac{3}{8}$ Ⓒ $\frac{2}{5}$ 25. $\frac{1}{3}$ Ⓔ 0.03   26. 0.44 Ⓖ $\frac{7}{16}$ 27. $\frac{8}{9}$ Ⓖ 0.87

28. Write 8^{-7} as a fraction.   **$\frac{1}{8 \times 8 \times 8 \times 8 \times 8 \times 8 \times 8} = \frac{1}{2,097,152}$**

29. Write $\frac{1}{3 \times 3 \times 3 \times 3 \times 3}$ in exponential form.   **3^{-5}**

30. Write 2.45×10^{-5} in standard form.   **0.0000245**

31. Write 0.00000531 in scientific notation.   **5.31×10^{-6}**

Name _____ Date _____

Find each answer.

32. $\frac{1}{3} + \frac{4}{7}$ **$\frac{19}{21}$**   33. $5\frac{1}{6} - 2\frac{5}{8}$ **$2\frac{13}{24}$**   34. $4\frac{1}{2} \times 2\frac{2}{3}$ **12**

35. $\frac{7}{12} \div \frac{3}{4}$ **$\frac{7}{9}$**

Solve.

36. Margo has driven $\frac{3}{4}$ the length of a highway that is 578 mi long. Estimate to find about how many miles she has traveled.

   Estimates may vary. Sample: 450 mi

37. Ralph plans to walk $2\frac{1}{5}$ mi each day to raise money for a charity. One sponsor has pledged to donate \$1.40 for each mile he walks. If Ralph walks each day for 25 days, how much money will the sponsor donate to the charity?

   \$77

150E

ACHIEVEMENT TEST

Exploring Fractions
CHAPTER 4   FORM B

MATH MATTERS BOOK 1
Chicha Lynch
Eugene Olmstead

SOUTH-WESTERN PUBLISHING CO.

Name _____

Date _____

SCORING RECORD	
Possible	Earned
37	

1. Find the prime factorization of 480. $2^5 \times 3 \times 5$

Find the GCF of each pair of numbers.

2. 9, 21   **3**   3. 6, 8   **2**   4. 18, 24   **6**   5. 36, 42   **6**

Find the LCM of each pair of numbers.

6. 8, 9   **72**   7. 6, 8   **24**   8. 12, 15   **60**   9. 10, 35   **70**

10. Multiply to find two fractions that are equivalent to $\frac{5}{6}$.   **Answers may vary. Samples:** $\frac{10}{12}, \frac{15}{18}$

11. Divide to find two fractions that are equivalent to $\frac{42}{72}$.   **Answers may vary. Samples:** $\frac{21}{36}, \frac{14}{24}$

Write each fraction in lowest terms.

12. $\frac{15}{25}$ **$\frac{3}{5}$**   13. $\frac{42}{48}$ **$\frac{7}{8}$**   14. $\frac{16}{18}$ **$\frac{8}{9}$**   15. $\frac{28}{80}$ **$\frac{7}{20}$**

Write each fraction as a decimal.

16. $\frac{63}{100}$ **0.63**   17. $\frac{1}{8}$ **0.125**   18. $\frac{8}{15}$ **$0.5\overline{3}$**   19. $\frac{1}{11}$ **$0.\overline{09}$**

Write each decimal as a fraction in lowest terms.

20. 0.4   **$\frac{2}{5}$**   21. 0.35   **$\frac{7}{20}$**   22. 0.009   **$\frac{9}{1,000}$**   23. 4.48   **$4\frac{12}{25}$**

Replace $\bigcirc$ with <, >, or =.

24. $\frac{5}{6} \bigcirc \frac{7}{8}$ <   25. $\frac{5}{16} \bigcirc 0.03$ >   26. $\frac{5}{12} \bigcirc 0.4$ >   27. $0.5625 \bigcirc \frac{9}{16}$ =

28. Write 7^{-6} as a fraction. $\dfrac{1}{7 \times 7 \times 7 \times 7 \times 7 \times 7} = \dfrac{1}{117,649}$

29. Write $\frac{1}{5 \times 5 \times 5 \times 5 \times 5}$ in exponential form. 5^{-5}

30. Write 8.32×10^{-4} in standard form.   **0.000832**

31. Write 0.00000147 in scientific notation. 1.47×10^{-6}

Copyright © 1993 by South-Western Publishing Co.
MIO1AG

4B-1

Name _____   Date _____

Find each answer.

32. $\frac{3}{14} + \frac{2}{7}$ **$\frac{1}{2}$**   33. $10\frac{3}{4} - 7\frac{2}{3}$ **$3\frac{1}{12}$**   34. $5\frac{1}{4} \times 2\frac{2}{3}$ **14**

35. $\frac{8}{9} \div \frac{2}{3}$ **$1\frac{1}{3}$**

Solve.

36. Max has read $\frac{2}{5}$ of a book that contains 246 poems.
   Estimate to find about how many poems he has read.
   Estimates may vary. Sample: 100 poems

37. Flora plans to walk $2\frac{1}{4}$ mi each day to raise money for a charity.
   One sponsor has pledged to donate $1.35 for each mile she walks. If Flora walks
   each day for 40 days, how much money will the sponsor donate to the charity?
   $121.50

4B-2

150F

Teacher's Notes

Voting and Citizenship

Objective: Students will develop skills of number and operation, observation, analysis, inference, and evaluation in the process of obtaining, organizing, displaying, and reasoning about demographic, economic, and political data concerning their community.

UNIT OVERVIEW

Through active study of their local community, students will become more aware of their roles as citizens and their capabilities as mathematically literate voters. The unit consists of four lessons, each of which uses or builds upon data obtained from local sources.

LESSON 1 **A Snapshot of...**   will result in a colorful graphic bulletin board displaying information about the students' own town or city. Students will gather existing data, both current and historical, on population, housing, income, employment, voter registration, and other topics of interest. Then they will organize and display the information gathered by using at least one each of a pictograph, line graph, bar graph, and multiple bar or line graph. Students will also write short descriptive paragraphs that summarize numerical information. To report the results, students will apply problem solving skills and employ whole and fraction number operations to express and compare data in different forms.

LESSON 2 **What's Up Pop(ulation)?** includes the design, execution, and analysis of a poll or survey using appropriate sampling techniques. Students will decide on the topic for investigation based on critical local issues. Students will use inductive reasoning to make inferences about the data collected.

LESSON 3 **Believe It or Not?** using data collected in Lessons 1 and 2, students will write conditional (if-then) statements about their community. Then groups of students will design and produce a poster to demonstrate either accurate or misleading use of data. All students will judge whether valid or invalid reasoning was used.

LESSON 4 **Multicultural Munchies** will highlight fractional operation skills through the use of recipe conversions. After studying the local population, students will develop a sense of the cultural and ethnic diversity of their community. Then students will celebrate that diversity by collecting ethnic recipes from family and friends. They will convert recipes to serve one-half, twice, and three times as many people. They might end with a party in which they prepare and eat various specialties using their converted recipes.

TIME 12-20 days

INTRODUCTION AND MOTIVATION
- For several days prior to the unit introduction, have students collect newspaper articles and features, advertising for employment, housing, or businesses, and possibly brochures from real estate agents or the local Chamber of Commerce about their community or city.
- After students share the gathered materials, ask small groups to discuss:
  1. Which information is meaningful or useful to them as citizens?
  2. Which information is accurate? Which might be misleading?
  3. What would a stranger to the community think or know about the place using the collected information?
- Tell students that in this unit they will use their computation and data-gathering skills, their problem solving skills, and their graphing and reasoning skills to create a true picture of their community.

OBJECTIVES

Students will:
- compile data from local government sources
- use problem solving to select appropriate scales for their graphs
- construct a pictograph, bar graph, line graph, and multiple bar or line graph to portray local data
- find the mean, median, mode, and range to describe data
- use fractions to express and compare data
- round numbers and use scientific notation to describe and graph data

TIME

5 to 8 class periods (not always on consecutive days. Time, in and out of class, should be allowed for data collection and graph preparation.)

MATERIALS

graph paper, markers, 11 X 17 white paper, poster board in different colors; if possible, a local and/or state publication that includes a town profile and census data, obtained by the teacher in advance

INTRODUCING THE LESSON

1. Summarize the discussion from the activity used to motivate the unit.
2. Lead the class in brainstorming a list of topics or questions related to their community about which they would like to find more information.
3. Have students help you categorize the topics or questions under the headings: Population, Income, Housing; Budget and Taxes; Voting; Employment.

FACILITATING THE LESSON

1. Arrange students in groups of 2 to 4 students. Assign (or have groups choose) one of the topics above, making sure that each topic is chosen at least once. Instruct each group that they will gather data to answer their questions concerning the assigned topic. Explain that they will eventually construct a pictograph, a line graph, a bar graph, and a multiple line or bar graph to illustrate the data.
2. As a whole class, generate a list of local and state agencies, offices, publications, or libraries where data can be gathered about the topics.
3. Provide groups with copies of Worksheet 1 , Group Decision Sheet. Have each group complete the worksheet indicating Research Topic, Question(s) to be Answered, Type of Graph, Type of Data Needed, and Possible Sources.
4. Explain that each graph will be accompanied by a short paragraph that summarizes the data using measures such as mean, median, mode, and range. Emphasize that data can be summarized by using comparisons expressed as fractions.
5. Allow students time to gather and organize data, construct their graphs, and write their paragraphs. Circulate to monitor group work and check progress. Remind students to express large numbers in rounded form and/or scientific notation.

SUMMARIZING THE RESULTS

Have each group prepare a final large display copy of their graphs and typewritten copies of their paragraphs for a bulletin board display entitled: A Snapshot Of... Each group should present their data in an oral report no longer than 3 minutes. Encourage students to summarize what was different about the true facts about their community compared to their original perceptions from the newspaper and advertising.

EXTENDING THE LESSON

- Display the completed bulletin board in the school or in a local public building.
- Invite a public official to review the display and discuss the town statistics.

ASSESSING THE LESSON

Check students' graphs and paragraphs for accuracy of mathematical data, appropriate use of scales, and choice of appropriate graphs. Have each student write a short essay summarizing one key finding about their community in each category: Population; Income; Housing; Budget and Taxes; Voting; Employment.

OBJECTIVES

Students will
- design and conduct an opinion poll to obtain information about a current local issue
- use appropriate sampling techniques
- report their opinion poll results using a frequency table and graphs
- use inductive reasoning to make predictions about the findings

TIME

3 to 5 class periods

MATERIALS

copies of local newspapers covering a 2 to 3 week span; access to telephones or access to typewriter/word-processor and copy machine

INTRODUCING THE LESSON

1. Discuss possible subjects for a poll. Use the information from Lesson 1 and/or recent newspaper articles or editorials on issues of local concern.
2. Together, develop a list of possible opinion poll questions that can be answered yes or no, dealing with the suggested subjects. Ask students to try out the questions on 2 or 3 adults before the next class in order to determine which questions reveal "hot topics" and/or elicit emotional responses.

FACILITATING THE LESSON

1. In groups of 4 to 5, have students choose the questions that provoked the most interest from respondents. After each group reports their results, help the class come to consensus about the issue that will be investigated and the questions to be used.
2. Have the groups discuss possible sampling procedures. Ask each group to justify their suggested methods.
3. Have the class develop an opinion-poll form to be filled out by each respondent. It should include a space for the respondent's name or identification number. As a class, fill in the questions to be asked. Encourage the students to think about the form of the questions so that the respondents can easily indicate their answers and so that the results will be easy to tally. The class should decide whether respondents will be identified by any subgroups. Depending on the issue being investigated, students may want to compare answers of males vs. females; those with or without school-age children; long-term vs. short-term residents; or registered vs. non-registered voters.
4. Give students one week to conduct the poll. Then use class time to compile results. Have groups decide how best to tally and organize the data. Suggest frequency tables and one type of graph for reporting the data.
5. Each group should use inductive reasoning to write predictions as to the outcome of the local issue based on the data gathered from the poll.

SUMMARIZING THE RESULTS

Have each group present an oral report of the poll results, showing their tables and graphs. Their report should include their predictions and their justification for them.

EXTENDING THE LESSON

- Students might write an article for a local newspaper explaining their opinion poll, its results, and their predictions.
- Some students may wish to attend a public hearing, town meeting, or some other forum to make their results known.
- If time permits, students can repeat the project with improvements on their methodology and/or questions based on their experience.

ASSESSING THE LESSON

Have students write a report that is a critical analysis of the survey questions, sampling methods, and predictive value of the results. The report should include suggestions for improvement of the activity.

OBJECTIVES

Students will
- apply deductive reasoning skills
- write conditional statements about data
- design posters to demonstrate the use of conditional statements
- determine whether arguments based on the conditional statements are valid or invalid
- give counterexamples to prove conditional statements are false

TIME

2 to 4 class periods

MATERIALS

different types of magazines such as fashion, sports, hobby; posterboard; markers

INTRODUCING THE LESSON

1. Have students look for magazine ads and write an if-then statement for several ads. For example, a blue jeans ad featuring a young man surrounded by beautiful women might be translated into the conditional: If you wear Brand X jeans, you will attract many beautiful women. Ask students to share one or two ads they found and their conditional statements. Explain that they will be writing and illustrating ads using the data from previous lessons about their town.
2. Remind students that a conditional statement is false if a counterexample can be found that satisfies the hypothesis (if statement), but not the conclusion (then statement). Have students give counterexamples for their conditionals.
3. Review valid and invalid arguments using the conditional. For example, a valid argument based on the conditional above would state: John wears Brand X jeans. Therefore, John attracts many beautiful women. An invalid argument would state: John attracts many beautiful women. Therefore, John wears Brand X jeans. Emphasize that the conditional can be false, yet have a valid argument deduced logically from it.

FACILITATING THE LESSON

1. Instruct students to use the data from Lesson 1 or Lesson 2 to write conditional statements about their town or city. From these they will develop and illustrate an advertising poster that promotes their community. After reminding them of the introductory activity, encourage them to write some statements that are accurate and some that are misleading, reflecting valid and invalid arguments.
2. Form 'advertising companies' of 2 to 4 students to chose a slogan to illustrate. Tell the groups that they will share their posters with the class and have an opportunity to judge whether others' posters show true or false statements or valid or invalid arguments. Give students class time to do the posters.
3. Hang the posters around the room. Allow students to judge whether each poster presents an accurate or misleading statement or argument. Have students use Worksheet 2 Believe It or Not? as a judging sheet. For each poster, they should write the conditional that the ad implies in the first column. In the second column they should explain why they think the poster ad is accurate (valid) or misleading (invalid). They may write a counterexample to demonstrate that the conditional is false.

SUMMARIZING THE RESULTS

Sample class opinion as to which posters are an accurate portrayal of the community. Have volunteers justify their reasoning about the posters. Add the accurate posters to the bulletin board displaying information about the community.

EXTENDING THE LESSON

- Invite representatives of local organizations or businesses that try to promote the community to speak with the class. Have the visitors bring any promotional materials they use. Later, have students use their deductive reasoning abilities to critique the materials.

ASSESSING THE LESSON

Check students' judging sheets for understanding and application of deductive reasoning techniques.

LESSON 4   Multicultural Munchies

OBJECTIVES

Students will
- become aware of the ethnic and cultural heritage of their community
- collect recipes that reflect the cultural diversity of the community
- use fractional operations to convert recipes to one half, twice, and three times the quantity

TIME

2 to 3 class periods

MATERIALS

international cookbooks; ingredients to make recipes (optional)

INTRODUCING THE LESSON

1. Ask students what they know or what they discovered in Lesson 1 about the ethnic and cultural diversity of the community. You might ask: Has there been any recent migration into this community? Are there special organizations that reflect diverse groups in the community?
2. Have students share information about special holidays, celebrations, or customs from their own experience.

FACILITATING THE ACTIVITY

1. After the diverse heritage of the community has been identified, show students an ethnic recipe from an international cookbook. Ask students to bring in recipes from their family or friends that reflect their background. If the class does not represent a true picture of the diversity of the community, assign some students to locate recipes for the unrepresented groups in the international cookbooks you have collected or in cookbooks found in the school or local library.
2. After numerous recipes have been collected, arrange students into groups so that each group represents one ethnic or cultural part of the community. Try to place students in groups that differ from their own ethnic or cultural background. Have each group member fill in a copy of Worksheet 3, Recipe Conversions. Each group member should do the computations to convert a different recipe to one half, twice, and three times the quantity.
3. Have each group choose one recipe to prepare for a class collection and an ethnic food-tasting party. Have them convert that recipe into enough servings to feed the entire class.

SUMMARIZING THE RESULTS

For a finished product, each group should print or type the recipe chosen for the class party neatly on one page and illustrate it, if desired. The pages will be collated into a booklet, copies of which will be distributed on the day of the food-tasting celebration. The ethnic recipes may be prepared at home and brought to school, prepared in the school during class time, if possible, or prepared as a special event after school.

EXTENDING THE LESSON

1. Students may wish to invite other classes, parents, or the entire school to their ethnic food celebration.
2. A more complete cookbook of all of the recipes that students have collected could be created and made available to the local library or sold to earn money for a school function.
3. Invite a speaker to the class to answer questions about food service as a career and what role mathematics plays in that career.

ASSESSING THE LESSON

Monitor students' recipe conversions as they work in groups. Worksheets as well as the completed booklet pages can be used to assess students' practical application of fractional operations.

Names _____ Date _____

of Group _____

Members _____

_____

Group Decision Sheet

Research Topic:

Questions	Type of Graph*	Type of Data	Possible Sources

*There must be one each of pictograph, bar graph, line graph, and multiple bar or line graph.

Special Unit: Voting and Citizenship • Worksheet 1

150M

Name _____ Date _____

Believe It or Not?

Poster	Conditional Statement	Comments
1.		
2.		
3.		
4.		
5.		
6.		
7.		
8.		
9.		
10.		

Recipe Conversions

Calculate $\frac{1}{2}$ of the recipe ingredients.

Calculate twice the ingredients.

Calculate three times the ingredients.

Teacher's Notes

CHAPTER 5 SKILLS PREVIEW

Complete.

1. 48 in. = ■ ft **4**

2. 5 gal = ■ qt **20**

3. 8 lb = ■ oz **128**

4. 3 mi = ■ yd **5,280**

5. 490 cm = ■ m **4.9**

6. 2 L = ■ mL **2,000**

7. 958 mg = ■ g **0.958**

8. 1 km = ■ m **1,000**

Write each answer in simplest form.

9. 4 lb 8 oz
   +7 lb 9 oz
   ‾‾‾‾‾‾‾‾‾
   12 lb 1 oz

10. 5 gal
   −3 gal 3 qt
   ‾‾‾‾‾‾‾‾‾‾
   1 gal 1 qt

11. 1 c 8 fl oz
       × 5
   ‾‾‾‾‾‾‾‾
   10 c

12. 2 ft 8 in. ÷ 4
   8 in.

Complete.

13. 150 mm + 450 mm = __?__ cm **60**

14. 2 km − 250 m = __?__ km **1.75**

15. 3.28 cm × 2 = __?__ mm **65.6**

16. 8.4 m ÷ 6 = __?__ cm **140**

Solve.

17. Will one 40-inch roll of tape be enough tape to make 3 strips of tape, each 1 ft 2 in. long? **No**

Find the perimeter of each figure.

18. 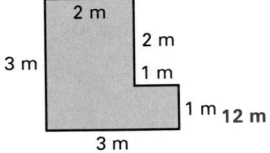 **12 m**

19. 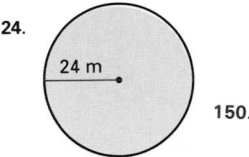 **28 in.**

Use the formula to find the perimeter for a rectangular figure with the given dimensions.

20. length: 78.34 m
    width: 29.87 m **216.42 m**

21. length: 375.4 mi
    width: 291.3 mi **1,333.4 mi**

Find the circumference. Use 3.14 or $\frac{22}{7}$ for π, as appropriate.

22. 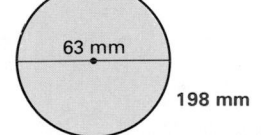 63 mm **198 mm**

23. 15.5 ft **97.3 ft**

24. 24 m **150.7 m**

Draw a picture to help you solve.

25. A triangular garden is enclosed by a fence that has 10 posts on each side. How many posts are there in all? **27 posts**

Introduction The purpose of this Skills Preview is to assess students' abilities on all the major objectives of Chapter 5. Test results may be used

- to determine those topics which may need only to be reviewed and those topics which need to be more carefully developed;
- for class placement;
- in prescribing for individual differences.

If you prefer, you may use the Skills Preview as an alternative form of the Chapter Test (page 180) to evaluate mastery of chapter objectives. The items on the Skills Preview and the Chapter Test correspond in content and level of difficulty.

5

MEASUREMENT AND GEOMETRY

OVERVIEW

In this chapter, students review and extend their understanding of the metric and customary systems of measurement with emphasis on changing units within each system. The decimal system of numeration is used to establish an efficient place-value method for changing from one metric unit to another. Measurement is reinforced in sections on perimeter of polygons and circumference. Estimation techniques, formulas for perimeter, and the draw-a-picture strategy are also developed.

SPECIAL CONCERNS

Although there is no formal strategy for relating metric units to customary units, students should be encouraged to note simple relationships. For example, a centimeter is equivalent to about a half inch, a liter equivalent to about a quart, and a kilogram to about 2 pounds.

As students work with the formula for circumference, it is important to point out that, since π is an irrational number, it is incorrect to say that π "equals" 3.14 or 22/7. These numbers are merely approximations of π.

VOCABULARY

centimeter (cm)	diameter	milligram (mg)
circle	kilogram (kg)	milliliter (mL)
circumference	kilometer (km)	millimeter (mm)
customary units	liter (L)	perimeter
gram (g)	meter (m)	radius

MATERIALS

calculators	tape measures or yardsticks
string	metric and customary rulers
erasers	paper clips

BULLETIN BOARD

Have students draw and then cut out large irregularly-shaped polygons and circles from colored construction paper or oaktag. Display these on the bulletin board. Encourage students to choose a figure, measure its sides, diameter, or radius, and then find its perimeter or circumference.

INTEGRATED UNIT 2

The skills and concepts involved in Chapters 5–8 are included within the special Integrated Unit 2 entitled "United States and World Travel." This unit is in the Teacher's Edition beginning on page 298F. Worksheets for this integrated unit appear in the Enrichment Activities booklet, pages 93–94.

TECHNOLOGY CONNECTIONS

- Computer Worksheet, 95
- Calculator Worksheet, 96
- MicroExam, Apple Version
- MicroExam, IBM Version

- *Elastic Lines: The Electronic Geoboard, Explorer Metros, The Factory,* Sunburst Communications

TECHNOLOGY NOTES

Computer software programs, such as those listed in Technology Connections, challenge students to create geometric shapes for which they can find the perimeters, allow students to hone their measurement skills, and encourage them to draw pictures to solve problems, thus expanding their sense of visual discrimination and spatial perception.

CHAPTER 5

MEASUREMENT AND GEOMETRY

PLANNING GUIDE

SECTIONS	TEXT PAGES	ASSIGNMENTS			
		BASIC	AVERAGE	ENRICHED	
Chapter Opener/Decision Making	152–153				
5–1 Customary Units of Measurement	154–157	1–10, 15, 17–18, 21, 24	1–16, 17–23, 24–25	9–16, 21–23, 24–26	
5–2 Metric Units of Measurement	158–161	1–16, 17–18, 21–22, 25–26, 31	1–16, 17—28, 31	15–16, 17–30, 31–32	
5–3 Working with Measurements	162–165	1–17, 19–20, 23–24, 29	1–18, 19–26, 20–30	9–18, 19–28, 29–32	
5–4 Perimeter	166–169	1–5, 6–9, 12, 15, 18, PSA 1–3	1–5, 6–16, 18–22, PSA 1–5	6–17, 18–23, PSA 1–6	
5–5 Problem Solving Skills: Use a Formula	170–171	1–9	1–12	1–13	
5–6 Circumference	172–175	1–13, 15–16, 19–21	1–14, 15–21, 22–23	15–26	
5–7 Problem Solving Strategy: Draw a Picture	176–177	1–15	1–18	9–18	
Technology	161, 168, 169	✔	✔	✔	

ASSESSMENT					
Skills Preview	151	All	All	All	
Chapter Review	178–179	All	All	All	
Chapter Test	180	All	All	All	
Cumulative Review	181	All	All	All	
Cumulative Test	182	All	All	All	

ADDITIONAL RESOURCES

RETEACHING	ENRICHMENT	TECHNOLOGY	TRANSPARENCY
5–1	5–1		TM 20
5–2	5–2		TM 21
5–3	5–3		
5–4	5–4	5–4	
5–5	5–5		
5–6	5–6	5–6	TM 22
5–7	5–7		

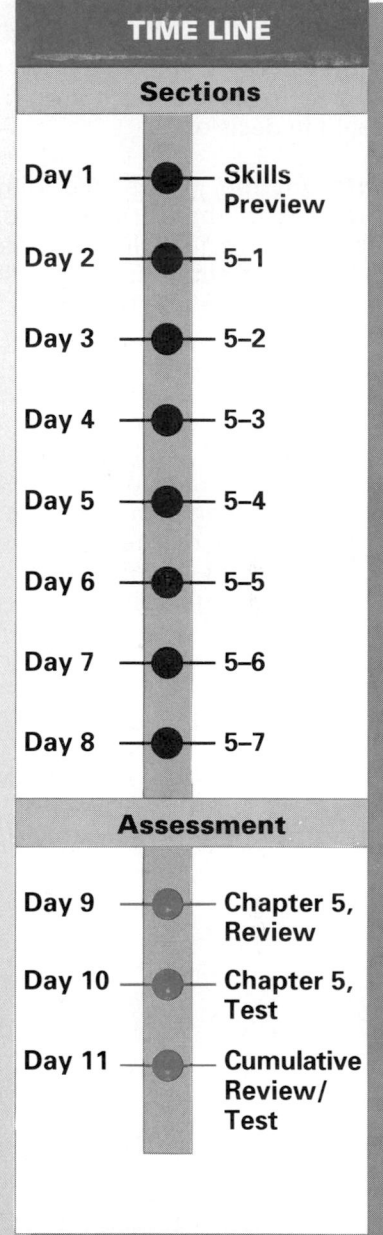

TIME LINE

Sections

Day 1	●	Skills Preview
Day 2	●	5–1
Day 3	●	5–2
Day 4	●	5–3
Day 5	●	5–4
Day 6	●	5–5
Day 7	●	5–6
Day 8	●	5–7

Assessment

Day 9	●	Chapter 5, Review
Day 10	●	Chapter 5, Test
Day 11	●	Cumulative Review/ Test

ASSESSMENT OPTIONS

Chapter 5, Test Forms A and B	
Chapter 5, Test	Text, 180
Alternative Assessment	TAE, 180
Chapter 5, MicroExam	

5 MEASUREMENT AND GEOMETRY

THEME Architecture Around the World

Architecture involves the planning and designing of structures such as buildings, bridges, and towers. Architecture can also involve the designing and positioning of the furniture within a building. Landscape architecture involves arranging the plantings on a piece of property.

The architect seeks to design a structure that is both functional and pleasing to the eye. The style of a structure reflects the cultural preferences of the society that exists at the time during which the structure was produced.

In this chapter, you will solve problems that involve units of measurement, estimating measurements, and finding perimeter and circumference.

152 CHAPTER 5 Measurement and Geometry

Specific styles of architecture became popular during certain periods in history. Some of these styles of Western architecture are listed in the table. The "c." preceding a date is an abbreviation for *circa,* a Latin word meaning "around." This indicates that the date given is an approximation, not an exact date.

STYLES OF WESTERN ARCHITECTURE	
Style	**Date**
Archaic	c. 750–500 B.C.
Classical	c. 500–323 B.C.
Early Byzantine	330–726
Iconoclastic Age	726–843
Byzantine	843–1453
Italian Gothic	c. 1200–1400
Italian Renaissance	1402–1520
Baroque	c. 1600–1750
Classicist	1750–1830
Gothic Revival	c. 1790–1930

DECISION MAKING

Using Data

Use the information in the chart above to answer the following questions.

1. How many years did the Italian Gothic period span?
   approximately 200 years
2. How many years were there between the end of the Classical period and the start of the Baroque period? **approximately 1,923 years**
3. What name do you think might apply in the future to the style of architecture that will be popular from 2000 to 2100? **Answers will vary.**

Working Together

Research the styles of architecture that have been popular from 1930 to the present. Identify what is meant by the *International Style* and by *post-modernism.* Find out about the work of famous architects, such as Mies van der Rohe, Le Corbusier, and Philip Johnson. Collect pictures to illustrate the work of the style periods and the architects. Decide which style and architect you prefer. Discuss your choices with your group.

SPOTLIGHT

OBJECTIVES
- Use customary units of measure to solve problems
- Change to larger/smaller customary units

MATERIALS NEEDED
calculators, tape measures or yardsticks

VOCABULARY
customary units

WARM-UP

Measure.
Use a tape measure or yardstick to find your height in inches. How many feet is that? **Answers will vary.**

1 MOTIVATE

Explore Have students answer the questions independently. Discuss their results.

2 TEACH

Use the Pages/Skills Development
Have students read this part of the section and then discuss the examples. Allow students the option of using calculators.

Examples 1 and 2: For each example, have students tell whether they are changing from a smaller unit to a larger unit or vice versa. Be sure they know whether the number of units will increase or decrease as they multiply or divide.

Example 3: Use this example to demonstrate the need for changing units to solve problems.

5-1   Customary Units of Measurement

EXPLORE   What is the longest bridge you have ever been across? Guess its length, in feet. Write down your guess.

USING DATA   Use the Data Index on page 546 to find the list of longest bridge spans in the world. Is the bridge you named listed in the table?

a. If so, find the actual length of the bridge, in feet. How close was your guess to the actual length?

b. If not, compare your guess to the lengths of the longest bridge spans. Was your guess close to any of these lengths? Was it much greater than or much less than these? How do the data suggest that you adjust your guess?

SKILLS DEVELOPMENT   The table below lists some of the **customary units** that are used to measure length, weight, and capacity. Equivalent measurements are also given.

Measure	Units	Equivalents
length	inch (in.)	
	foot (ft)	1 ft = 12 in.
	yard (yd)	1 yd = 3 ft, or 36 in.
	mile (mi)	1 mi = 5,280 ft, or 1,760 yd
weight	ounce (oz)	
	pound (lb)	1 lb = 16 oz
	ton (T)	1 T = 2,000 lb
capacity	fluid ounce (fl oz)	
	cup (c)	1 c = 8 fl oz
	pint (pt)	1 pt = 2 c, or 16 fl oz
	quart (qt)	1 qt = 2 pt, or 4 c
	gallon (gal)	1 gal = 4 qt, 8 pt, or 16 c

154   CHAPTER 5   Measurement and Geometry

TEACHING TIP

Discuss with students how to remember, when changing units, when to multiply and when to divide. **When a smaller unit is changed to a larger unit, the number of units decrease. When a larger unit is changed to a smaller unit, the number of units increase.**

To change from a larger unit of measure to a smaller unit, multiply the number of larger units by the number of smaller units equivalent to one larger unit.

Example 1

Change 517 yd to feet.

Solution

1 yd = 3 ft

Multiply 517 by 3 to find how many feet are in 517 yd.

$$517 \times 3 = 1{,}551$$

So 517 yd = 1,551 ft. ◄

To change from a smaller unit of measure to a larger one, divide the number of smaller units by the number of smaller units equivalent to one larger unit.

Example 2

Change 45 c to quarts.

Solution

4 c = 1 qt

Divide 45 by 4 to find how many quarts make 45 cups.

$$45 \div 4 = 11\tfrac{1}{4}$$

So 45 c = $11\tfrac{1}{4}$ qt. ◄

Example 3

Ramona has a dress pattern that calls for 7 yd of lace. She already has 4 ft of lace. How many feet of lace will she need to buy?

Solution

1 yd = 3 ft, so   **Change 7 yd to the equivalent number of feet.**
7 yd = 21 ft.

21 ft − 4 ft = 17 ft   **Subtract 4 ft from 21 ft.**

Ramona needs to buy 17 ft of lace. ◄

TRY THESE

Change the larger unit of measure to the smaller unit.

1. 5 ft to inches   60 in.
2. 1 gal to cups   16 c
3. $3\tfrac{1}{4}$ lb to ounces   52 oz
4. $2\tfrac{1}{2}$ gal to quarts   10 qt

JUST FOR FUN

Did you know that a 10-gallon hat can hold only $\tfrac{3}{4}$ gal of liquid? How much liquid would ten 10-gallon hats hold?
$7\tfrac{1}{2}$ gal

Additional Questions/Examples
Change

1. 26 in. to feet and inches.
   2ft 2 in.
2. 90 in. to yards and inches.
   2 yd 18 in.
3. 35 fl oz to cups and ounces.
   4 c 3 oz
4. The world's highest waterfall is Angel Falls in Venezuela, with a height of 3,212 ft. How many yards and feet is that?
   1,070 yd 2 ft

Guided Practice/Try These
Observe students as they complete these exercises independently. Discuss any questions or problems they might have. For Exercises 3, 4, 6, and 8, students may prefer working with decimals rather than fractions.

3 SUMMARIZE

Write About Math Have students write in their math journals how to change from a larger unit of measure to a smaller unit and from a smaller unit to a larger one.

5. 1,760 yd to miles **1 mi**

6. 30 in. to feet **$2\frac{1}{2}$ ft**

7. 48 fl oz to pints **3 pt**

8. 3,000 lb to tons **$1\frac{1}{2}$ T**

Solve.

9. Brooke has jogged 520 yards. How many more yards must she jog before she has jogged 1 mile? **1,240 yd**

EXERCISES

**PRACTICE/
SOLVE PROBLEMS**

Complete.

1. 32 oz = ■ lb **2**

2. 4 yd = ■ ft **12**

3. $3\frac{1}{2}$ T = ■ lb **7,000**

4. 112 in. = ■ ft ■ in. **9, 4**

5. 5 qt = ■ gal ■ qt **1, 1**

6. 40 fl oz = ■ pt ■ c **2, 1**

7. 46 ft = ■ yd ■ ft **15, 1**

8. 46 oz = ■ lb ■ oz **2, 14**

Name the best customary unit for expressing the measure of each.

9. capacity of a pitcher **fl oz**

10. weight of a moving van **T**

11. amount of water needed to fill a bathtub **gal**

12. distance from one city to another **mi**

13. weight of two plums **oz**

14. amount of fabric needed to make a dress **yd**

Solve.

15. A living room is 24 ft long. How many yards long is it? **8 yd**

16. A watermelon weighs 5 lb. How many ounces does it weigh? **80 oz**

**EXTEND/
SOLVE PROBLEMS**

Choose the estimate that best expresses the measure of each.

17. length of a classroom

   a. 30 in.   (b.) 30 ft   c. 30 yd

18. capacity of a water glass

   a. 12 pt   b. 12 qt   (c.) 12 fl oz

19. length of a newborn baby

   (a.) 19 in.   b. 19 ft   c. 19 yd

20. weight of a bag of potatoes

   a. 5 oz   b. 5,000 oz   (c.) 5 lb

---

4 PRACTICE

Practice/Solve Problems For Exercise 6, make sure that, after students divide to find the number of pints, they write the remainder in terms of cups, not ounces.

Extend/Solve Problems Students should understand that each estimate they choose in Exercises 17–20 must be a reasonable one in which to express the object named.

Think Critically/Solve Problems For Exercises 24 and 25, students must realize that Janice already has enough jars to store 12 of the 18 qt of beans.

5 FOLLOW-UP

Extra Practice
Which unit—inches, feet, or miles—would you use to measure each of the following?

1. pencil **inches**
2. pin **inches**
3. fence **feet**
4. highway **miles**
5. knife **inches**
6. flagpole **feet**
7. Lydia's car weighs 3,124 lb. How much more than a ton is that? **1,124 lb**
8. Paolo has a 40-gal fish tank. So far, he's poured 125 qt of water into the tank. How many more quarts will he need to fill the tank? **35 qt**
9. Harvey bikes 6 mi to school each day. How many feet will he bike in a week? **316,800**

MAKING CONNECTIONS

Assign each student one of the customary measures to research. Ask them to report on the origin of the word used to identify the unit, and how and when the unit was standardized. Some students may enjoy researching less familiar measures; for example, *troy weights,* used in jewelry making, and *bushels* and *pecks.*

Solve.

21. A cement company has 1,700 lb of sand on its lot. A supplier delivers 4 T of sand. How many pounds of sand are now on the lot? **9,700 lb**

22. Jeffrey ran 545 yd in the morning and 750 yd in the afternoon. How many more yards must he run later in the day if he wants to have run a total of 3 mi that day? **3,985 yd**

23. Paul knows that his size $10\frac{1}{2}$ shoe is about 12 in. long. By walking slowly along the length of a room with one foot touching the other, heel to toe, Paul can estimate the length of the room. Paul's son, Davey, wears a shoe that is half as long as Paul's. If Davey uses his father's method of measuring the length of a 24-ft-long room, how many steps will Davey take? **48 steps**

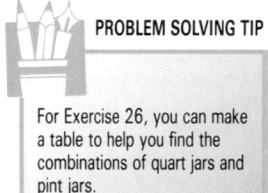

PROBLEM SOLVING TIP

In Exercise 23, how many steps would Paul need to take to measure the room?

THINK CRITICALLY/ SOLVE PROBLEMS

Janice wants to can 18 qt of beans. She already has 9 quart jars and 6 pint jars. She knows she will need to buy more jars.

24. If she buys only quart jars, how many will she need?
6 quart jars

25. If she buys only pint jars, how many will she need?
12 pint jars

26. List five combinations of quart jars and pint jars that Janice might buy.

Quart Jars	Pint Jars
1	10
2	8
3	6
4	4
5	2

PROBLEM SOLVING TIP

For Exercise 26, you can make a table to help you find the combinations of quart jars and pint jars.

Extension Have students solve the following problem.

If you had only a cup measure, a pint measure, a quart measure, and a half-gallon measure, and could fill each measure only once, how could you obtain 1 cup, 2 cups, 3 cups, . . . 15 cups? **1c = c; 2c = pt; 3c = pt, c; 4c = qt; 5c = qt, c; 6c = qt, pt; 7c = qt, pt, c; 8c = 1/2 gal; 9c = 1/2 gal, c; 10c = 1/2 gal, pt; 11c = 1/2 gal, pt, c; 12c = 1/2 gal, qt; 13c = 1/2 gal, qt, c; 14c = 1/2 gal, qt, pt; 15c = 1/2 gal, qt, pt, c**

Section Quiz Complete.
1. 18 in. = ■ ft ■ in. **1, 6**
2. 1/4 mi = ■ ft **1,320**
3. 1/2 pt = ■ c **1**
4. 7,056 yd = ■ mi ■ yd **4, 16**
5. Jan caught 17 fish in the morning and 12 fish in the afternoon. If each fish weighed an average of 1 lb 5 oz, what was the total weight? **38 lb 1 oz**

Get Ready calculators

WARM-UP

Multiply.
1. 10×32.6 **326**
2. 100×1.54 **154**
3. 100×0.78 **78**
4. $1,000 \times 2.6$ **2,600**

1 MOTIVATE

Explore Have students name the basic metric units used to measure length, mass, and capacity. **meter, gram, liter** Discuss their answers to the questions.

2 TEACH

Use the Pages/Skills Development Have students read this part of the section and then discuss the examples. Discuss how the chart can help students understand how metric units are related.

Example 1 and 2: Guide students through the examples and solutions. Demonstrate the following "shortcut" for changing metric units.

To change 36.8 m to centimeters, write 36.8 on the chart.

km	hm	dam	m	dm	cm	mm
			3	6.	8	

5-2 Metric Units of Measurement

EXPLORE Which metric unit of measurement would you use to express the answer to each of these questions?

a. About how much does a person weigh? **kilogram**

b. About how far might you travel in a day? **kilometer**

c. How much juice would there be in a glass? **milliliter**

SKILLS DEVELOPMENT The basic unit of *length* for metric measure is the **meter (m)**. Other commonly used units include the **kilometer (km)**, the **centimeter (cm)**, and the **millimeter (mm)**.

about 1 km

about 1 m wide

about 1 mm at tip

The most commonly used metric units of *mass* are the **kilogram (kg)**, the **gram (g)**, and the **milligram (mg)**.

about 1 kg

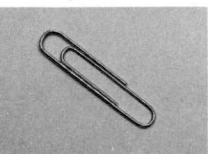
about 1 g

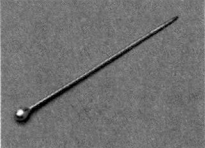
about 100 mg

The most commonly used metric units of *capacity* are the **liter (L)** and the **milliliter (mL)**.

about 1 L

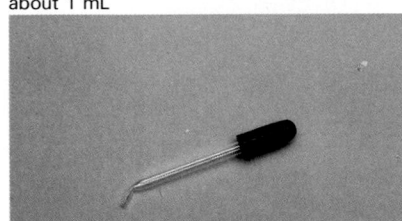
about 1 mL

Today, the meter is defined as being equal to 1,650,763.73 wavelengths in a vacuum of the orange-red radiation of the element krypton 86.

158 CHAPTER 5 Measurement and Geometry

TEACHING TIP

Continue to display the reminders suggested in the Teaching Tip in Section 5-1. Encourage students to refer to them as they complete the activities in this section. The "shortcut" presented on this page will be helpful to all students. Suggest that students prepare a metric place-value chart on a sheet of paper for use throughout this section.

This chart can help you see the relationships between the prefixes used to name metric units. It is important to understand these relationships in order to change values from one metric unit to another.

thousands	hundreds	tens	ones	tenths	hundredths	thousandths
kilo-	hecto-	deka-	no prefix	deci-	centi-	milli-
kilometer (km) 1,000 m	hectometer (hm) 100 m	dekameter (dam) 10 m	meter (m) 1 m	decimeter (dm) 0.1 m	centimeter (cm) 0.01 m	millimeter (mm) 0.001 m
kilogram (kg) 1,000 g	hectogram (hg) 100 g	dekagram (dag) 10 g	gram (g) 1 g	decigram (dg) 0.1 g	centigram (cg) 0.01 g	milligram (mg) 0.001 g
kiloliter (kL) 1,000 L	hectoliter (hL) 100 L	dekaliter (daL) 10 L	liter (L) 1 L	deciliter (dL) 0.1 L	centiliter (cL) 0.01 L	milliliter (mL) 0.001 L

Use the chart to help you change units. Just as with customary units, to change from a larger metric unit to a smaller one, multiply the number of larger units by the number of smaller units equivalent to one larger unit.

Example 1

Change 5 m to centimeters.

Solution
1 m = 100 cm
5 × 100 = 500   Multiply 5 by 100 to find how many centimeters are in 5 m.

So 5 m = 500 cm. ◄

Just as with customary units, to change from a smaller metric unit to a larger one, divide the number of smaller units by the number of smaller units equivalent to one larger unit.

Example 2

Change 5,642 g to kilograms.

Solution
1,000 g = 1 kg
5,642 ÷ 1,000 = 5.642   Divide 5,642 by 1,000 to find how many kilograms make 5,642 g.

So 5,642 g = 5.642 kg. ◄

Example 3

A pharmacist has 4.5 L of cough syrup. How many 150-mL bottles can the pharmacist fill with the cough syrup?

Solution
1 L = 1,000 mL, so   Change 4.5 L to the equivalent number of
4.5 L = 4,500 mL.   milliliters.

4,500 mL ÷ 150 mL = 30   Divide 4,500 mL by 150 mL.

The pharmacist can fill thirty 150-mL bottles. ◄

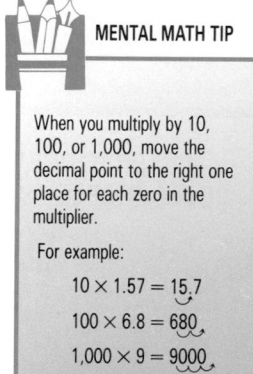

MENTAL MATH TIP

When you multiply by 10, 100, or 1,000, move the decimal point to the right one place for each zero in the multiplier.

For example:

10 × 1.57 = 15.7

100 × 6.8 = 680

1,000 × 9 = 9000

MENTAL MATH TIP

When you divide by 10, 100, or 1,000, move the decimal point to the left one place for each zero in the divisor.

For example:

15.7 ÷ 10 = 1.57

680 ÷ 100 = 6.80

9,000 ÷ 1,000 = 9.000

BASIC
1–16, 17–18, 21–22, 25–26, 31

AVERAGE
1–16, 17–28, 31

ENRICHED
15–16, 17–30, 31–32

ADDITIONAL RESOURCES
Reteaching 5–2
Enrichment 5–2
Transparency Master 21

The decimal point to the right of the meters column indicates that the unit of measurement is the meter. To change to centimeters, just move the decimal point to the right of the centimeter column, adding a zero.

km	hm	dam	m	dm	cm	mm
			3	6	8	0.

Thus, 36.8 m = 3,680 cm.

To change 36.8 m to kilometers, move the decimal point to the right of the kilometer column, adding two zeros.

km	hm	dam	m	dm	cm	mm
0.	0	3	6	8		

Thus, 36.8 m = 0.0368 km.

This method of changing units works because the metric units are related to one another in the same way as are the units of our decimal numeration system.

Example 3: Use this example to demonstrate the need for changing units to solve problems.

Additional Questions/Examples
Which unit would you use to measure each of the following?
1. length of a pencil   **centimeter**
2. mass of an apple   **gram**
3. capacity of a baby-food jar
   milliliter

ESL STUDENTS

Encourage ESL students to share what they know about the metric system with the class. You may wish to pair ESL students with native English-speaking students who are having difficulty understanding the metric system.

4. Drew drives 5.5 km to work each day. Toby drives 7,832 m to work each day. Who drives farther? How much farther?
Toby; 2,332 m farther

Guided Practice/Try These Have students work in small groups. Point out that all group members must come to an agreement about each answer.

Write About Math Have students explain in their math journals how our decimal place-value system is like the metric system of measurement.

4 PRACTICE

Practice/Solve Problems Be sure that students understand that, to change larger to smaller units, they must multiply; to change small units to larger units, they must divide.

Extend/Solve Problems Be sure that students note when different units of measure are given in the same problem and recognize the need to change units.

Think Critically/Solve Problems Discuss the calculator sequence necessary for Exercises 31 and 32.

5 FOLLOW-UP

Extra Practice
1. Marie walked 4 km. How many meters is that? **4,000 m**
2. Florie jogged 3,567 m. How many more meters must she jog to have jogged 5 km? **1,433 m**
3. Bernie divided 8 L of iced tea into jars that hold 500 mL. How many jars did he use? **16**

READING MATH

In this textbook, you are learning about the most commonly used units in the metric system. You may come across other metric units as you read newspapers, magazines, or other textbooks.

Use a dictionary to find the meaning of each of these metric units. Suggest a situation for which each of these units is likely to be used.

1. nanosecond
2. megahertz
3. micrometer
4. gigameter
5. kilowatt

See Additional Answers.

TRY THESE

Change the larger unit of measure to the smaller one.

1. 7 m to centimeters **700 cm**   **2.** 2 L to milliliters **2,000 mL**

3. 12.75 g to milligrams **12,750 mg**   **4.** 5.5 km to meters **5,500 m**

Change the smaller unit of measure to the larger one.

5. 5,000 m to kilometers **5 km**   **6.** 8,000 mL to liters **8 L**

7. 3,500 g to kilograms **3.5 kg**   **8.** 1,254 mg to grams **1.254 g**

Solve.

9. The Dixie Luncheonette had 12 L of milk. If 9.25 L were used, how many liters of milk were left? How much is that in milliliters?
2.75 L; 2,750 mL

EXERCISES

PRACTICE/ SOLVE PROBLEMS

Complete.

1. 8 L = ■ mL **8,000**   **2.** 125 cm = ■ m **1.25**

3. 2 km = ■ m **2,000**   **4.** 388 mm = ■ m **0.388**

5. 4,783 mL = ■ L **4.783**   **6.** 307 g = ■ kg **0.307**

7. 4,000 g = ■ kg **4**   **8.** 255 mL = ■ L **0.255**

Name the best metric unit for expressing the measure of each.

9. length of a car **m**   **10.** mass of a feather **mg**

11. capacity of a gasoline tank **L**   **12.** height of a child **cm**

13. mass of a bag of groceries **kg**   **14.** capacity of a teacup **mL**

Solve.

15. A container holds 7 L of liquid. How many milliliters does it hold? **7,000 mL**

16. It is 0.85 km from Jason's house to the library. How far is that distance in meters? **850 m**

EXTEND/ SOLVE PROBLEMS

Choose the most reasonable unit of measure. Write *km, m, cm,* or *mm.*

17. The length of an airport landing strip is about 1 ■. **km**

18. The length of a pen is about 14 ■. **cm**

19. Janine's height is about 164 ■. **cm**

20. The length of a minivan is about 5 ■. **m**

Write *kg, g,* or *mg.*

21. The mass of a nickel is about 5 ■. **g**

22. A box of raisins has a mass of 500 ■. **mg**

23. A large pot can hold about 0.75 ■. **kg**

24. A grain of rice has a mass of about 1 ■. **g**

Write *L* or *mL.*

25. A fish tank has a capacity of about 60 ■. **L**

26. A baby's bottle can hold about 250 ■. **mL**

Solve.

27. Steve began an 8-km hike on Monday by walking 454 m. He walked 2 km on Tuesday and 533 m on Wednesday. How many meters farther must he walk to complete the hike? **5,013 m**

28. Maria bought 1,224 cm of ribbon at $0.75 per meter. How much did Maria pay for the ribbon, to the nearest cent? **$9.18**

29. Mike had 2 L of milk. After he used 250 mL to bake muffins and 500 mL to bake bread, how much milk did he have left? **1,250 mL**

30. The running time on the videotape of a movie that Dina rented was 108 min. Dina began watching the movie at 7:45 p.m. At what time did the movie end? **9:33 p.m.**

■ Gina's father is a jeweler who cuts gold chain of different sizes for necklaces. On Gina's 16th birthday, she received a 10-g gold necklace. On her 17th birthday, she received a 20-g gold necklace; on her 18th birthday, a 40-g gold necklace; on her 19th birthday, an 80-g gold necklace; and so on.

31. After receiving her necklace for her 21st birthday, Gina decided to determine the value of the gold in all her necklaces. If the price of gold was $20 per gram, find the value of all her necklaces. Make a table to help you. **See Additional Answers.**

32. By Gina's 25th birthday, the price of gold had risen to $30 per gram. Find the value of the gold in all of Gina's necklaces, including the one she received for her 25th birthday. (Hint: Make a table or write a calculator sequence to help you find the answer.) **See Additional Answers.**

COMPUTER

This computer program will compute the total value of the gold in Exercise 31 quite quickly. Which line would you change to reflect the change in the price of gold specified in Exercise 32?

```
10 INPUT "WHAT BIRTHDAY?
   ";B: PRINT
20 FOR X = 0 TO B − 16
30 G = 10 * 2 ↑ X
40 V = 20 * G
50 S = S + V
60 NEXT X
70 PRINT : PRINT "GINA'S
   GOLD IS WORTH $";S;"."
```

In line 40, the "20" would be changed to "30."

**THINK CRITICALLY/
SOLVE PROBLEMS**

5-2   Metric Units of Measurement   **161**

---

4. Troy bought 3,215 g of potting soil. If the soil sells for $0.95 per kilogram, how much did Troy spend, to the nearest cent? **$3.05**

Extension Have students refer to the Data Index on page 546 to locate data from which to create problems involving metric units. Students should then exchange problems and solve.

Section Quiz Complete.
1. 16 L = ■ mL   **16,000**
2. 345 cm = ■ m   **3.45**
3. 38 mg = ■ g   **0.038**
4. Kelly drank 250 mL of milk. She used another 250 mL to make dinner, 500 mL in a cake she baked, and 400 mL for breakfast the next day. Did she use more or less than 1 L of milk? How much more or less? **more; 400 mL more**
5. In a book, each line of print is 18 cm long. Each page has 48 lines of print. What is the total length, in kilometers, of the lines of print on 524 pages? **4.52736 km**

SPOTLIGHT

OBJECTIVES
- Write customary and metric units of measure
- Change units of measure when performing operations with measurements
- Compare units of measure in solving problems

VOCABULARY
capacity, equivalent

WARM-UP

Look at your ruler and notice the length of 10 cm. Without measuring, find at least five things in the classroom that are about 10 cm long.

1 MOTIVATE

Explore/Working Together Have students work in groups of four for this activity.

2 TEACH

Use the Pages/Skills Development Have students read this part of the section and then discuss the examples.

Examples 1 and 2: Point out that writing answers in simplest form is easiest when students have memorized basic equivalences of customary measurements: 16 oz = 1 lb, 12 in. = 1 ft, 1 gal = 4 qt.

Example 3: Point out that, while the steps to the solution may be performed in a different order, a step is saved if students change units of measure in the last step.

Example 4: Point out that units of measure must be alike before they can be compared.

162

5-3 Working with Measurements

EXPLORE/ WORKING TOGETHER

Work in groups of three. Decide if the measurements in each set are equivalent. If not, determine which is larger.

1. 5,000 cm; 5 m **5,000 cm**
2. 48 oz; 4 lb **4 lb**
3. 6 ft; 2 yd **equivalent**
4. 3 L; 300 mL **3 L**
5. 5.3 kg; 5,300 g **equivalent**
6. $7\frac{1}{2}$ qt; 15 pt **equivalent**
7. 3 ft 6 in.; 36 in. **3 ft 6 in.**
8. 100 oz; 6 lb 4 oz **equivalent**

SKILLS DEVELOPMENT

There are many times that you need to perform operations that involve measurements. In carpentry, for example, cutting wood to an appropriate length often involves adding or subtracting lengths. Similarly, in cooking, adjusting a recipe may involve adding or multiplying weights or liquid capacities. When you add or multiply with customary measurements, you may need to change your answer to its simplest form.

Example 1

Write each answer in simplest form.

a. 3 lb 9 oz
   + 1 lb 9 oz

b. 4 ft 9 in.
        × 3

Solution

a. 3 lb 9 oz
   + 1 lb 9 oz
   ─────────
    4 lb 18 oz   18 oz is more than 1 lb, so simplify.

   4 lb + 18 oz = 4 lb + (1 lb + 2 oz) = 5 lb 2 oz

b. 4 ft 9 in.
        × 3
   ─────────
   12 ft 27 in.   Be sure to multiply both the number of inches and the number of feet by 3.

   12 ft 27 in. = 12 ft + (2 ft + 3 in.) = 14 ft 3 in. ◄

When subtracting or dividing with customary measurements, you may need to change a measurement before you can perform the operation.

TEACHING TIP

Have students keep a chart of customary-unit equivalents handy, so that they can perform operations with these units quickly. Point out that repeated experiences with changing units will make memorization of these equivalents easy.

Example 2

Write each answer in simplest form.

a. 8 gal 1 qt
      −6 gal 2 qt

b. 7 yd 1 ft ÷ 2

Solution

a. 8 gal 1 qt   *1 qt is less than 2 qt, so you*
      −6 gal 2 qt   *cannot subtract in this form.*

8 gal 1 qt = (7 gal + 4 qt) + 1 qt = 7 gal 5 qt

   8 gal 1 qt   →   7 gal 5 qt
 −6 gal 2 qt   −6 gal 2 qt
                        1 gal 3 qt ◄

b. 7 yd 1 ft ÷ 2   *1 ft is not divisible by 2, so you need to "borrow" from the number of yards.*

7 yd 1 ft = (6 yd + 3 ft) + 1 ft = 6 yd 4 ft

                          3 yd 2 ft
2)7 yd 1 ft   →   2)6 yd 4 ft ◄

You also may need to change measurements when performing operations with metric measurements.

Example 3

a. Complete: 65 cm + 72 cm = ___?___ m

b. 5 L ÷ 8 = ___?___ mL

Solution

a. 65 cm   100 cm = 1 m
      +72 cm   137 ÷ 100 = 1.37
       137 cm   137 cm = 1.37 m   So, 65 cm + 72 cm = 1.37 m. ◄

b.   1 L = 1,000 mL
       5 × 1,000 = 5,000   625 mL
       5 L = 5,000 mL   8)5,000 mL   So, 5 L ÷ 8 = 625 mL. ◄

TALK IT OVER

In Example 3, is it possible to perform the steps in each solution in a different order? Explain.

Yes; in a, you can first change cm to m and then add. In b, you can first divide and then change L to mL.

Example 4

For a craft project, Linnette needs twenty pieces of ribbon that are each 1 ft 4 in. long. She can buy a 25-ft roll of the ribbon for $4.98. Will this be enough ribbon?

Solution

Multiply 1 ft 4 in. by 20 to find the total amount of ribbon that she needs.

   1 ft 4 in.   20 ft 80 in. = 20 ft + (6 ft 8 in.)
    × 20   = 26 ft 8 in.
  20 ft 80 in.

Compare the amount of ribbon on the roll to the amount Linnette needs.

   25 ft ___?___ 26 ft 8 in.
      25 ft < 26 ft 8 in.   No, the 25-ft roll is not enough. ◄

5-MINUTE CLINIC

Exercise	Student's Error	Error Diagnosis
Write the answer in simplest form.		• Student fails to rename and subtract feet.
9 yd − 3 yd 2 ft	9 yd − 3 yd 2 ft 6 yd 2 ft	

Additional Questions/Examples
Write the answer as a mixed number.
1. 4 ft 5 in. + 1 ft 4 in. **5 3/4 ft**
2. 4 × (3 qt 1 pt) **14 qt**
3. 9 lb 4 oz ÷ 2 **4 5/8 lb**
4. Naomi has a 2-lb box of apricots. She plans to make 6 loaves of apricot bread. Each loaf takes 5 oz of apricots. Will she have enough apricots? **yes**

Guided Practice/ Try These Have students work in small groups. Point out that all group members must come to an agreement on each answer.

Talk It Over
1. Why is it important to change measurements to like units before performing operations? **It is not possible to perform operations with units that are not alike.**
2. When might it be important to change smaller units to larger ones? **When purchasing things that are sold by the larger unit, such as paint sold by the gallon or food sold by the pound.**

4 PRACTICE

Practice/Solve Problems Ask students to express each of the answers to Exercises 9–16 in terms of both of the given units.

Extend/Solve Problems In Exercises 20–22 students should make sure, when renaming units, to use the correct number of smaller units equivalent to each larger unit.

Think Critically/Solve Problems Students should undertand that, to do Exercises 29–32, they need to use inverse operations.

5 FOLLOW-UP

Extra Practice Complete.
1. 3.2 kg + 500 g = ■ kg **3.7**
2. 5 × 70 mm = ■ cm **35**
3. 7.2 L ÷ 6 = ■ mL **1,200**
4. 2,700 mm ÷ 3 = ■ cm **90**
5. 2.8 kg − 0.75 kg = g **2,050**
Write each answer in simplest form.
6.
```
  3 gal  3 qt  1 pt
+ 1 gal  4 qt  2 pt
  6 gal        1 pt
```
7.
```
  5 yd  2 ft  5 in.
− 2 yd  3 ft  7 in.
  2 yd  1 ft 10 in.
```

Extension Have students make up problems of their own that are similar to Exercises 9–16, 19–22, or 29–32. They should then exchange papers with a partner and solve each other's problems.

Section Quiz
Write each answer in simplest form.
1.
```
  5 lb 13 oz
+ 3 lb  4 oz
  9 lb  1 oz
```
2.
```
  9 gal 2 qt
− 3 gal 4 qt
  5 gal 2 qt
```

MIXED REVIEW

Find the mean, the median, and the mode for each set of data. **372, 384, 384**
1. 178, 235, 679, 384, 384
2. 13, 98, 45, 67, 67, 98, 17, 28, 98 **59, 67, 98**
3. 31, 56, 94, 79, 85, 62, 31, 92, 39, 31 **60, 59, 31**

Evaluate each expression.
Let $a = 4$ and $b = 5$.
4. ab **20**
5. $a^2 + b$ **21**
6. $3a - b$ **7**

Solve.
7. Bus tickets cost $1.15 each. How much do 8 bus tickets cost? **$9.20**

TRY THESE

Write each answer in simplest form.
1.
```
  4 c 6 fl oz
+ 1 c 4 fl oz
  6 c 2 fl oz
```
2.
```
  1 yd 2 ft
      × 5   8 yd 1 ft
```
3.
```
  7 lb 2 oz
−     10 oz
  6 lb 8 oz
```
4. $5\overline{)6\text{ ft }3\text{ in.}}$ **1 ft 3 in**

Complete.
5. 200 mL + 750 mL = ___?___ L **0.95**
6. 3 kg − 145 g = ___?___ g **2,855**
7. 4.15 cm × 10 = ___?___ mm **415**
8. 7.2 m ÷ 6 = ___?___ cm **120**

9. In her laboratory, Angela has three beakers with a capacity of 750 mL each and a fourth beaker with a capacity of 1.25 L. Do the four beakers together have the capacity to hold 3.5 L of a saline solution? **Yes**

EXERCISES

PRACTICE/ SOLVE PROBLEMS

Write each answer in simplest form.
1.
```
  6 lb 12 oz
+ 4 lb  9 oz
  11 lb 5 oz
```
2.
```
  9 ft 2 in.
− 8 ft 8 in.
       6 in.
```
3.
```
  8 yd
− 2 yd 1 ft
  5 yd 2 ft
```
4.
```
  5 gal 2 qt
+ 1 gal 3 qt
  7 gal 1 qt
```
5.
```
  2 c 3 fl oz
        × 4
  9 c 4 fl oz
```
6.
```
  1 lb 8 oz
       × 6
  9 lb
```
7. 5 yd 1 ft ÷ 2 **2 yd 2 ft**
8. 1 ft 8 in. ÷ 4 **5 in.**

Complete.
9. 2.4 kg + 600 g = ___?___ kg **3**
10. 4 × 65 mm = ___?___ cm **26**
11. 1.6 L ÷ 5 = ___?___ mL **320**
12. 4 m − 250 cm = ___?___ cm **150**
13. 1.25 cm × 14 = ___?___ mm **175**
14. 120 mL + 860 mL = ___?___ L **0.98**
15. 1.9 kg − 1.25 kg = ___?___ g **650**
16. 2,500 mm ÷ 4 = ___?___ cm **62.5**

164 CHAPTER 5 Measurement and Geometry

CHALLENGE

A micron equals 1 millionth of a meter (1×10^{-6}).
Write the number of microns in each measurement.
1. 1 m **1,000,000 or 1×10^6**
2. 2 cm **20,000 or 2×10^4**
3. 4 mm **4,000 or 4×10^3**
4. 7 km **7,000,000,000 or 7×10^9**

17. Don needs two pounds of raisins to make some loaves of bread for a family get-together. He has three 12-oz boxes of raisins. Will this be enough? Explain. **Yes; 2 lb = 32 oz and 3 × 12 oz = 36 oz**

18. Erica is planning to install three new cabinets in her kitchen. She would like to get two cabinets that are each 2 ft 9 in. wide and a third cabinet that is 3 ft 6 in. wide. Will these three cabinets fit side-by-side along a wall space that is 8 ft 6 in. wide? **No; the cabinets will occupy 9 ft**

Write each answer in simplest form.

19.　2 gal 2 qt 1 pt
　　　+1 gal 3 qt 1 pt
　　　　　　4 gal 2 qt

20.　　4 yd 1 ft 3 in.
　　　−1 yd 2 ft 9 in.
　　　　　　2 yd 1 ft 6 in.

21.　2 yd 1 ft 4 in.
　　　−1 yd 1 ft 10 in.
　　　　　　2 ft 6 in.

22.　　4 gal
　　　−1 gal 3 qt 1 pt
　　　　　　2 gal 1 pt

Customary units are often written using mixed numbers. For instance, since a measurement of 6 in. is half of one foot, a measurement of 2 ft 6 in. might be written $2\frac{1}{2}$ ft. In Exercises 23–28, write the answer in mixed number form.

23. 3 ft 4 in. + 1 ft 4 in. $4\frac{2}{3}$ **ft**

24. 6 lb 2 oz − (1 lb 10 oz) $4\frac{1}{2}$ **lb**

25. 5 × (2 qt 1 pt) $12\frac{1}{2}$ **qt**

26. (9 yd 1 ft) ÷ 2 $4\frac{2}{3}$ **yd**

27. 3 lb 4 oz + $1\frac{1}{2}$ lb $4\frac{3}{4}$ **lb**

28. $4\frac{2}{3}$ ft − 10 in. $3\frac{5}{6}$ **ft**

EXTEND/ SOLVE PROBLEMS

Replace each ■ with the measurement that makes a true statement.

29. 7 lb 6 oz + ■ = 18 lb 2 oz
10 lb 12 oz

30. ■ − 3 ft 5 in. = 1 ft 7 in. **5 ft**

31. 5 × ■ = 2 c 4 fl oz **4 oz**

32. ■ ÷ 6 = 1 ft 2 in. **7 ft**

THINK CRITICALLY/ SOLVE PROBLEMS

Complete.
3. 3.4 kg + 700 g = ■ kg **4.1**
4. 2.30 cm × 12 = ■ mm **276**
5. In her laboratory, Mala has four beakers, each with a capacity of 500 mL, and another with a capacity of 1.25 L. Can the three beakers accommodate 4.2 L of liquid? **no**

Get Ready erasers, paper clips

5-4   Perimeter

EXPLORE   Take any book from your desk. Find the distance around the front cover using an eraser or a paper clip as a unit of measure.

a. How many units long is the book's cover? How many units wide?
Answers will vary.

b. What was the fastest way to find the distance around the cover?
Measure one length and one width. Find the sum, then double it.

c. If another book's cover measures the same distance around, but has a different length and width, what might be its dimensions?
Answers will vary.

SKILLS DEVELOPMENT

The distance around a plane figure is called its **perimeter.** You can find the perimeter of a plane figure by adding the lengths of its sides.

Example 1

Find the perimeter of each figure.

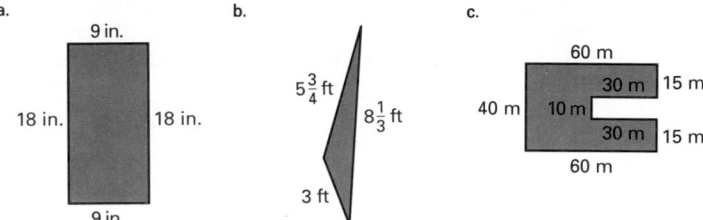

Solution

a. Add the lengths of the four sides.
$$9 + 18 + 9 + 18 = 54$$
The perimeter is 54 in.

b. Add the lengths of the three sides.

$$
\begin{array}{rcl}
3 & \to & 3 \\
5\frac{3}{4} & \to & 5\frac{9}{12} \\
+\ 8\frac{1}{3} & \to & +\ 8\frac{4}{12} \\
\hline
& & 16\frac{13}{12} \to 17\frac{1}{12}
\end{array}
$$

The perimeter is $17\frac{1}{12}$ ft.

c. Add the lengths of the eight sides.
$$40 + 60 + 15 + 30 + 10 + 30 + 15 + 60 = 260$$

The perimeter is 260 m.   ◄

Example 2

Find the perimeter of a
rectangular swimming pool that
is 12.4 m long and 5.6 m wide.

Solution

The swimming pool is in the shape of a rectangle, so it has two sides
that measure 12.4 m and two sides that measure 5.6 m. Add the lengths
of the sides.

$$12.4 + 12.4 + 5.6 + 5.6 = 36$$

The perimeter of the swimming pool is 36 m. ◄

TRY THESE

Find the perimeter of each figure.

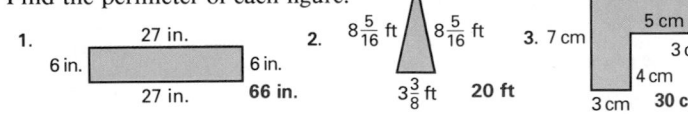

1. 27 in. / 6 in. / 27 in. / 6 in.

2. $8\frac{5}{16}$ ft / $8\frac{5}{16}$ ft / $3\frac{3}{8}$ ft / **20 ft**

3. 8 cm / 5 cm / 7 cm / 3 cm / 4 cm / 3 cm / **30 cm**

4. Find the perimeter of a square courtyard that measures $16\frac{1}{3}$ yd
on each side. **$65\frac{1}{3}$ yd**

EXERCISES

1. Find the perimeter of each figure drawn on the grid if each unit
of the grid represents 1 ft. **See Additional Answers.**

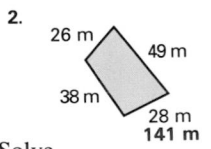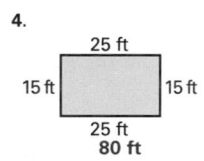

Find the perimeter of each figure.

2. 26 m / 49 m / 38 m / 28 m / **141 m**

3. 35 yd / 23 yd / 32 yd / **90 yd**

4. 25 ft / 15 ft / 15 ft / 25 ft / **80 ft**

Solve.

5. Find the perimeter of a rectangular building foundation that has
two sides that measure 45 ft and two sides that measure $37\frac{1}{2}$ ft.
165 ft

5-4   Perimeter   **167**

✓ CHECK UNDERSTANDING

In Example 2, the pool is
described as being *rectangular*.
What do you know about the
sides of an object that is
described as rectangular? What
if it is described as *square*?

If it's a rectangle, two sides
are one length and two
sides are a different length.
If it's a square, all four
sides are the same length.

PRACTICE/ SOLVE PROBLEMS

2. The perimeter of a square is 24
in. Find the length of each side.
6 in.

3. One side of a parallelogram
measures 4.5 m. The perimeter
is 26.4 m. Find the measures of
the other three sides.   **4.5 m,
8.7 m, 8.7 m**

4. Find the perimeter.

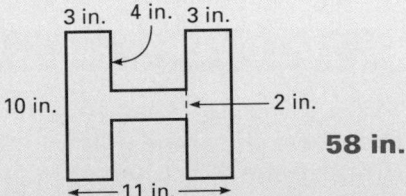

3 in. / 4 in. / 3 in. / 10 in. / 2 in. / 11 in.   **58 in.**

Guided Practice/Try These Have the
students complete these exercises
independently.

③ SUMMARIZE

Key Questions

1. If you know the length of a rectangle and its perimeter, how
can you find its width?   **subtract twice the length and
divide the difference by 2**

2. If you know the length of each
side of a square, how can you
find its perimeter?   **add or
multiply by 4**

5-MINUTE CLINIC

Exercise	Student's Error	Error Diagnosis
Find the perimeter. 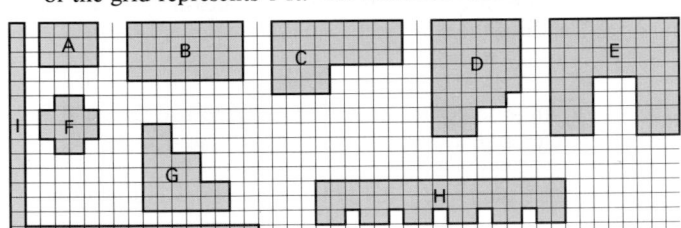 2 in. / 3 in. / 8 in. / 5 in.	$2 + 8 + 5 + 3 = 18$ in.	• Student adds only the dimensions given, instead of first determining, then including, the missing measurements.

Practice/Solve Problems Each of exercises 1–4 provide needed practice in reading measurements from diagrams.

Extend/Solve Problems For each of Exercises 15–17, students must find the length of a missing side when given the perimeter.

Think Critically/Solve Problems Logical thinking is required to find the lengths of the missing sides of the polygon in Exercises 19–22.

Problem Solving Applications These applications contain multi-step problems involving finding the perimeter of a region and then determining the cost of enclosing the region. Students should be alert to the error of multiplying the dimensions of the figure, rather than adding them, to find the perimeter.

Extra Practice Ask students to draw five different polygons on graph paper. Explain that they must draw along the lines of the graph paper, but may make their figures as intricate as possible. Have students exchange papers and find the perimeters of each other's figures.

Extension Have students use one or two shapes to make a tiling pattern. They may use shapes such as an equilateral triangle—a triangle with sides of equal length—and a regular hexagon—a six-sided figure having sides of equal length. Ask students to exchange papers, measure one or more sides of the pattern as necessary, then find the perimeter.

168

EXTEND/ SOLVE PROBLEMS

COMPUTER

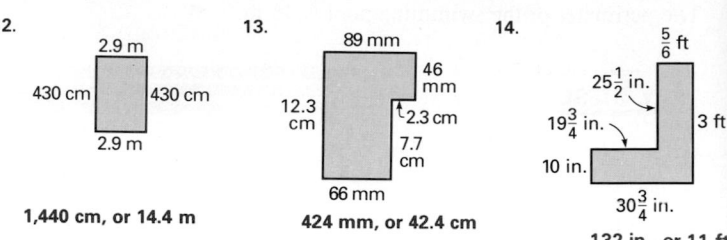

To find the perimeter of a rectangle, you may want to use this program.

```
10 INPUT "ENTER THE
   LENGTH; ";L : PRINT
20 INPUT "ENTER THE
   WIDTH; ";W: PRINT
30 LET P = 2 * L + 2 * W
40 PRINT "THE PERIMETER IS
   ";P; "."
```

To find the perimeter of a triangle, you could make line 10:

```
10 INPUT "ENTER THE
   LENGTHS; ";A,B,C?
```

and delete line 20. How would you then have to change line 30 to get the perimeter?

30 Let P = A + B + C

THINK CRITICALLY/ SOLVE PROBLEMS

WRITING ABOUT MATH

Study the dimensions of the figure used in Exercises 19–22. Write a description of what such a figure could represent. What would be the perimeter of this figure? Why might someone need to know it? **See Additional Answers.**

USING DATA Use the Data Index on page 546 to find the table showing dimensions of noted rectangular structures. Find the perimeter, in meters, of each of these structures.

6. Izumo Shrine **43.6 m**
7. Step Pyramid of Zosar **468 m**
8. Kibitsu Shrine (main) **64.8 m**
9. Palace of the Governors **214 m**
10. Bakong Temple, Roluos **280 m**
11. Kongorinjo hondo **83.4 m**

Find the perimeter of each figure.

12.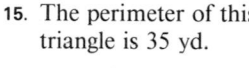
1,440 cm, or 14.4 m

13.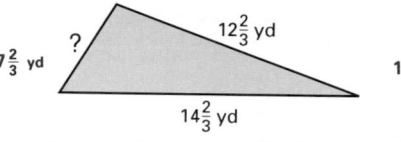
424 mm, or 42.4 cm

14.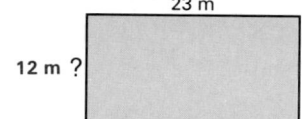
132 in., or 11 ft

Given the perimeters, find the missing dimensions.

15. The perimeter of this triangle is 35 yd.
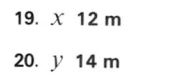

16. The perimeter of this rectangle is 70 m.
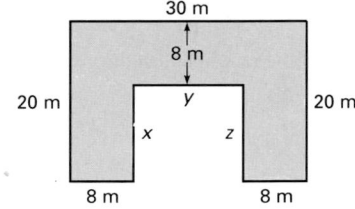

17. A triangular rose garden has two sides that measure 4.8 m each. Find the length of the third side if the perimeter is 14.5 m. **4.9 m**

18. One way to find the perimeter of a square is to add the lengths of its sides. Describe a different method that can be used. **Multiply the length of one side by 4.**

Use the figure for Exercises 19–22. Find each unknown dimension.

19. x **12 m**
20. y **14 m**
21. z **12 m**
22. Find the perimeter. **124 m**

Solve.

23. One side of a rectangle with a perimeter of 45.8 m measures 17 m. Find the lengths of the other three sides. **17 m, 5.9 m, 5.9 m**

CHALLENGE

Have students solve the following problem: Ribbon costs $0.45 per foot. Find the cost of the ribbon on this package. Allow 36 in. for the bow and any overlap. **$5.40**

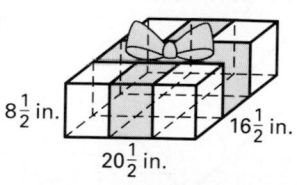

Problem Solving Applications:
SOLVING PROBLEMS THAT INVOLVE PERIMETER

Problems involving perimeter arise in many different practical situations. For example, before you install a fence, you must know how much fencing you need and how much it will cost.

PROBLEM

a. Determine about how much fencing would be needed to fence in the property shown.

b. If the cost of fencing is $8.92/m, find about how much it would cost to fence in the property shown.

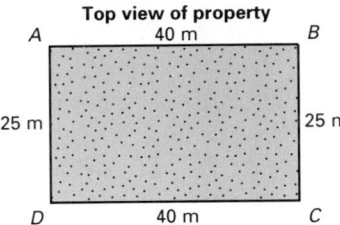
Top view of property

SOLUTION

a. 40 m + 25 m + 40 m + 25 m = 130 m

About 130 m of fencing would be needed to fence in the property shown.

b. Since the perimeter of the property is 130 m, and 1 m of fence costs $8.92, multiply 130 by $8.92.

The cost of fencing would be 130 × $8.92, or about $1,159.60.

Solve.

1. If the cost of fencing material is $2.55 per meter, how much fencing would be needed to enclose a rectangular dog run that measures 7.5 m by 3.7 m? What would be the cost of the fencing? **22.4 m; $57.12**

2. A rectangular swimming pool measures 15 m by 20 m. Find the cost of enclosing the pool if fencing costs $5.25/m. **$367.50**

3. A court is shown in the diagram. What is the cost of placing a railing around the court if railing costs $6.29/m? **$163.54**

4. The path walked in a golf game is shown in the diagram. How long would it take to walk the course at 5 km/h? Each unit represents 100 m. **1.74 hours**

5. Find the cost of enclosing a rectangular courtyard that is 12.4 m long and 5.6 m wide with fencing that costs $7.29/m. **$262.44**

6. A rectangular Japanese garden measures 27.4 m by 43.5 m. Calculate the cost of placing a low decorative fence around it if fencing costs $2.69/m. **$381.44**

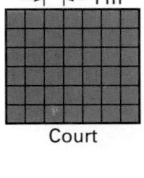

COMPUTER TIP

If you add these lines to the program on p. 168, you can run it to check your results.

```
25 INPUT "ENTER THE COST:
   ";C:?
35 H = P * C
50 ? "THE COST IS $ ";H;"."
```

→| |←1 m
Court

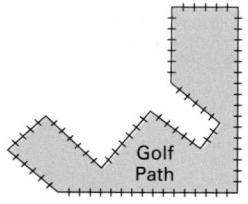
Golf Path

5-4   Perimeter   **169**

Section Quiz

Find the perimeter.

1.
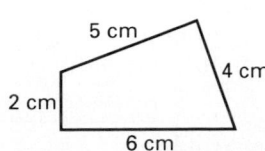
5 cm
4 cm
2 cm
6 cm

17 cm

2.
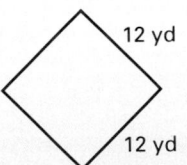
12 yd
12 yd

48 yd

3.
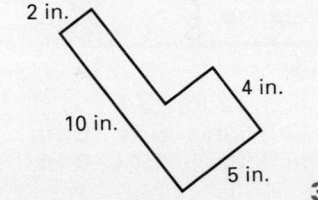
2 in.
4 in.
10 in.
5 in.

30 in.

4.
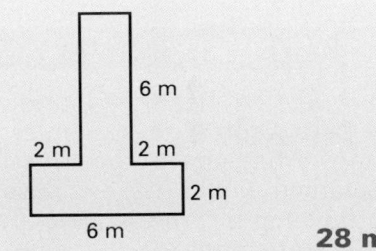
6 m
2 m   2 m
2 m
6 m

28 m

5. If molding costs $1.25 a foot, what would it cost to frame a painting that measures 4 ft by 3 ft? **$17.50**

Additional Answers
See page 577.

Additional answers for odd-numbered exercises are found in the Selected Answers portion of the page.

5-5 Problem Solving Skills:
USE A FORMULA

► READ
► PLAN
► SOLVE
► ANSWER
► CHECK

You have solved problems involving perimeter by adding the lengths of the sides to find the distance around a figure. In this lesson, you will learn how to use a formula to find the perimeter of a rectangle.

PROBLEM

The base of the Parthenon in Greece is a rectangular shape about 69.5 m long and 30.9 m wide. Find the perimeter of its base.

SOLUTION

Since a rectangle has two pairs of equal sides, you can find the perimeter by doubling the given length and doubling the given width, then adding the results.

Choose variables: P = perimeter
l = length
w = width

Write a formula: $P = 2l + 2w$. Then replace l and w with the given length and width.

$$P = 2l + 2w$$
$$P = 2(69.5) + 2(30.9)$$
$$P = \quad 139 \quad + \quad 61.8$$
$$P = \qquad 200.8$$

The perimeter of the base of the Parthenon is about 200.8 m.

PROBLEMS

1. The base of a building is in the shape of a triangle, with sides that measure a feet, b feet, and c feet.
   a. Write a formula to find the perimeter of the base.
   b. Use your formula to find the perimeter if $a = 65$, $b = 97$, and $c = 46$. **$P = a + b + c$; $P = 208$ ft**

2. The base of a building is shaped like a square with sides of length s meters.
   a. Write a formula to find the perimeter of the base.
   b. Use your formula to find the perimeter if each side of the square is 32 m long. **$P = 4 \times s$; $P = 128$ m**

Architects plan and design many types of buildings. They draw plans for houses, such as the plan shown below. Refer to this plan for Problems 3–13.

READING MATH

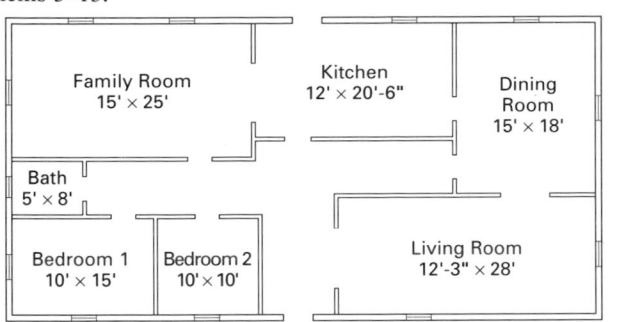

In building plans like the one shown on this page, the symbol ' stands for *feet* and the symbol " stands for *inches*.

Read *12' × 20'6"* as "twelve feet by twenty feet, six inches." This means that the width of the room is 12 ft and the length is $20\frac{1}{2}$ ft.

Suppose the architect wants to install decorative molding around the ceilings in some of the rooms. Use a formula to find how much molding would be needed for each room.

3. family room **80 ft**   **4.** kitchen **65 ft**   **5.** dining room **66 ft**

6. bedroom 1 **50 ft**   **7.** bedroom 2 **40 ft** **8.** living room **$80\frac{1}{2}$ ft**

Suppose the decorative molding costs $1.79 per foot.

9. How much would the molding cost for each of the six rooms listed above?   **See Additional Answers**.

10. What would be the total cost of the molding for the six rooms?   **$682.89**

The bathroom will be tiled as shown at the right. In Problems 11 and 12, allow a 3-ft opening for the bathroom door.

11. How many feet of cap tile are needed for the four walls of the bathroom? How many feet of base tile are needed? **23 ft, 23 ft**

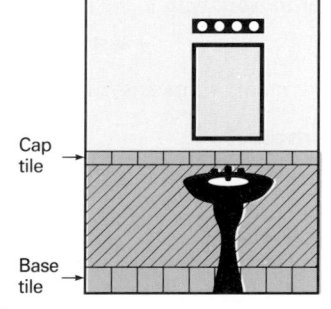

Cap tile →

Base tile →

12. At $1.59 per foot for cap tile and $2.89 per foot for base tile, how much will these bathroom tiles cost? **$103.04**

13. *CRITICAL THINKING* The new owners of this house want to build a deck that they can walk onto from the kitchen. The deck will be 12 ft wide and will run lengthwise across half the back of the house. How many feet of railing are needed for the deck? Allow a $2\frac{1}{2}$-ft opening for the steps down to the yard. **$51\frac{3}{4}$ ft**

sides of different lengths.
P = a + b + c + d
2. What formula could you write for finding the perimeter of an isosceles triangle? **P= 2a + b**

4 **PRACTICE**

Problems Suggest that students draw diagrams for the problems as necessary. Have students identify the formula they used for each problem.

5 **FOLLOW-UP**

Extra Practice
1. A 5-sided polygon has 3 equal sides. Write a formula for the perimeter of the polygon.
P = 3a + b + c
2. The length of each side of a square garden is 46 m. Fencing comes in 25-m rolls. How many rolls of fencing are needed to enclose the garden? **8 rolls**

Get Ready string, metric rulers, calculators

Additional Answers
See page 577.

CHALLENGE

1. Have students write a formula for the Challenge problem given for Section 5–4.
P = 2 [4 (8.5) + 2 (16.5) + 2 (20.5)] + 35 or
P = 8 (8.5) + 4 (16.5) + 4 (20.5) + 35
2. Find the perimeter of a figure with sides that measure 6 yd 6 in., 5 yd 2 ft., 7 ft 3 in., and 4 yd 10 in. **55 ft 7 in. or 18 yd 1 ft 7 in.**

5-6 Circumference

EXPLORE/ WORKING TOGETHER

How can you find the distance around a circle?

Work with a partner. Carefully fit a piece of string around the circle below. Now, stretch out the string and use a centimeter ruler to measure the amount of string that you fit around the circle.

What is the distance around the circle? **about 10 cm**

SKILLS DEVELOPMENT

A **circle** is the set of all points in a plane that are a fixed distance from a given point in the plane. The point is called the **center** of the circle. The fixed distance is called the **radius** of the circle. The distance across a circle is twice its radius. This distance is called the **diameter** of the circle. The distance around a circle is called its **circumference.**

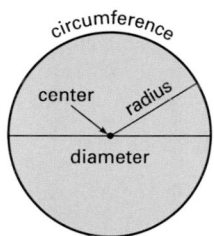

For a circle of any size, the quotient when the circumference (C) is divided by the diameter (d) is always equal to the same number. This number is represented by the Greek letter π (pi).

$$\frac{C}{d} = \pi, \qquad \text{or} \qquad C = \pi d$$

The number π is an *irrational number.* This means that it is a decimal that never ends, and that its digits never repeat. To make it easier to work with π, we use the following approximations.

$$\pi \approx 3.14 \qquad \text{or} \qquad \pi \approx \frac{22}{7}$$

The symbol $\approx$ means *is approximately equal to.*

Example 1

Find the circumference. Use 3.14 for π.

Solution
$C = \pi d$
$C \approx 3.14 \times 5$
$C \approx 15.70$

The circumference is approximately equal to 15.70 in. ◀

5 in.

To find the circumference of a circle when you know the radius (r), use the formula $C = 2\pi r$.

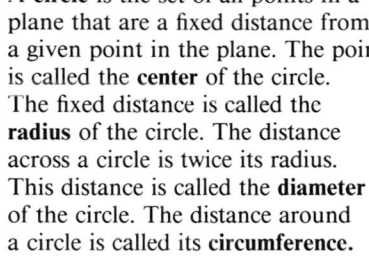

MATH: WHO, WHERE, WHEN

Srinivasa Ramanujan (1887–1920) was a mathematician who was largely self-taught. It was only when he was 27 that he went to England from his native India to study mathematics formally. He is credited with having developed over 6,000 mathematical theorems, one of which was a formula for calculating a decimal approximation for π. Ramanujan's formula was, remarkably, much like that used by modern mathematicians in 1987 when they used high-speed computers to determine the value of π to more than 100 million decimal places!

172 CHAPTER 5 Measurement and Geometry

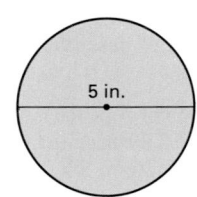

Example 2

Find the circumference. Use $\frac{22}{7}$ for π.

7 mm

Solution

$$C = 2\pi r$$
$$C \approx 2 \times \frac{22}{7} \times \overset{1}{\cancel{7}}$$
$$C \approx 44 \quad \frac{}{1}$$

The circumference is approximately equal to 44 mm. ◄

Example 3

A circular trampoline has a radius of 35 in. Find how much rubber stripping is needed to go around the rim of the trampoline.

Solution

The rim of the trampoline is shaped like a circle. The amount of rubber stripping needed is equal to the circumference of this circle. Since $r = 35$, use the formula $C = 2\pi r$.

$$C = 2\pi r$$
$$C \approx 2 \times \frac{22}{\cancel{7}} \times \overset{5}{\cancel{35}}$$
$$C \approx 220 \text{ in., or } 18\frac{1}{3} \text{ ft}$$

About $18\frac{1}{3}$ ft of rubber stripping is needed. ◄

TRY THESE

Find the circumference. Use 3.14 for π. If necessary, round your answer to the nearest tenth.

1.

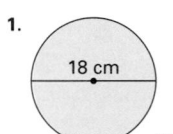

18 cm

56.5 cm

2.

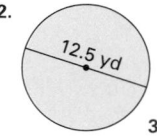

12.5 yd

39.3 yd

3.

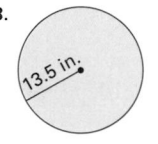

13.5 in.

84.8 in.

Find the circumference. Use $\frac{22}{7}$ for π.

4.

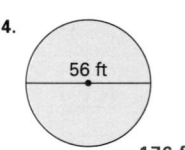

56 ft

176 ft

5.

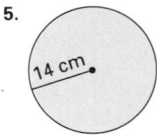

14 cm

88 cm

6.

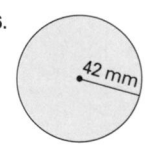

42 mm

264 mm

Solve. Use 3.14 or $\frac{22}{7}$ for π, as appropriate. If necessary, round your answer to the nearest tenth.

7. Find the circumference of a patio umbrella with a radius of 1.4 ft.
8.8 ft

8. The diameter of the face-off circle used in hockey is 6 m. Find the circumference. **18.8 m**

✓ **CHECK UNDERSTANDING**

In the solution of Example 2, why is the equal sign in the circumference formula replaced by the symbol $\approx$?

Because $\frac{22}{7}$ is an approximation of pi.

✓ **CHECK UNDERSTANDING**

In the solution of Example 3, why has $\frac{22}{7}$, and not 3.14, been used for π?

Because the radius, 35, is divisible by 7.

ASSIGNMENTS

BASIC
1–13, 15–16, 19–21

AVERAGE
1–14, 15–21, 22–23

ENRICHED
15–26

ADDITIONAL RESOURCES
Reteaching 5–6
Enrichment 5–6
Transparency Master 22

Additional Questions/Examples
1. A wheel has a diameter of 7 cm. About how many times must the wheel turn to travel 1 m? **about 4 1/2 times**
2. What happens to the circumference of a circle if the diameter is doubled? **The circumference is doubled.**
3. What is the perimeter of the track? **343.5 m**

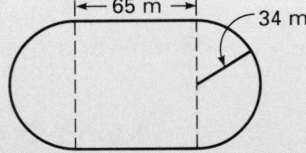

← 65 m →   34 m

Guided Practice/Try These Allow students to use calculators when they do calculations using 3.14 for π. Remind them to label their answers with the correct units. For Exercises 1–3 make sure answers are rounded to the nearest tenth. For Exercises 7 and 8, have students tell which value of π they used and why.

5-MINUTE CLINIC

Exercise	Student's Error	Error Diagnosis
Find the circumference of a circle with a radius of 6 in.	$3.14 \times 6 = 18.84$	• Student confuses the radius with the diameter and so, multiplies using the radius, 6, instead of the diameter, 12.

173

3 SUMMARIZE

Key Questions

1. How do you find the circumference of a circle when you know the diameter? **Use the formula $C = \pi d$.**
2. How do you find the circumference of a circle when you know the radius? **Use the formula $C = 2\pi r$.**
3. When would you use 22/7 for π? **When the radius or the diameter is divisible by 7 or when dimensions are given as fractions or mixed numbers.**
4. When would you use 3.14 for π? **When the radius or diameter is not divisible by 7.**

4 PRACTICE

Allow students to use calculators for all exercises.

Practice/Solve Problems For Exercises 9–12 students may use either $C = 2\pi r$ or $C = \pi d$, depending on the dimensions that are given.

Extend/Solve Problems Discuss the techniques students used to solve Exercises 15–17. For each of Exercises 20–21 students should note that the lengths of the diameter of the circle and the side of the square are equal.

Think Critically/Solve Problems Allow students to work in pairs. For Exercise 26, students must find the distance the front wheels travel in 100 turns, and divide this result by the distance the rear wheels travel in one turn.

PRACTICE/SOLVE PROBLEMS

Find the circumference of a circle with the given diameter. Use 3.14 or $\frac{22}{7}$ for π, as appropriate.

1. $d = 19$ cm
   59.7 cm
2. $d = 5.4$ mm
   17 mm
3. $d = 13.5$ ft
   42.4 ft
4. $d = 56$ yd
   175.8 or 176 yd

Find the circumference of a circle with the given radius. Use 3.14 or $\frac{22}{7}$ for π, as appropriate.

5. $r = 21$ in.
   132 in.
6. $r = 5.8$ yd
   36.4 yd
7. $r = 12.2$ mm
   76.6 mm
8. $r = 17$ cm
   106.8 cm

Copy and complete the chart.

	radius	diameter	Circumference	
9.	15 cm	■	■	30 cm; 94.2 cm
10.	■	100 ft	■	50 ft; 314 ft
11.	500 in.	■	■	1,000 in.; 3,140 in.
12.	■	0.1 mm	■	0.05 mm; 0.3 mm

 Solve.

13. The face of the clock known as Big Ben is about 7.1 m in diameter. What is the circumference of the clock face? **22.3 m**

14. The first Ferris wheel, named after its designer, George Ferris, was built in 1893. The diameter of that first wheel was 76 m. What was its circumference? **238.6 m**

ESTIMATION TIP

Whenever you need to use π in a calculation, an easy way to estimate the answer is to use 3 as an approximate value for π.

For example, to estimate the circumference of a circle with a diameter of 10 cm, think:

$C = \pi d$
$C \approx 3 \times 10$
$C \approx 30$

The circumference is about 30 cm.

EXTEND/SOLVE PROBLEMS

READING MATH

What is a *semicircle*? Find out the meaning of the prefix *semi–*. Then, find the meanings of these words.

1. semifinal
2. semiprecious
3. semiannual
4. semiprivate

See Additional Answers.

Find the distance around each figure. (Hint: Look for *semicircles*.)

15.
    ←—5 cm—→
    31.4 cm

16. 8 in. ←6 in.→
    34.8 in.

17.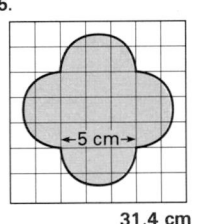
    ←——12 cm——→
    37.68 cm

18. Find the distance a runner would cover in running one lap around this track. **170 m**
    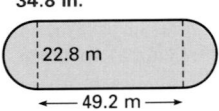
    22.8 m
    ←——49.2 m——→

19. Which is greater, the perimeter of a square that measures 55.5 mm on each side or the circumference of a circle with a radius of 55.5 mm? **circumference of circle; perimeter is 222 mm, circumference is 348.5 mm**

MAKING CONNECTIONS

Have students research information about the approximations of a value for pi that were developed by ancient people such as the Egyptians, Greeks, and Babylonians. Have students prepare short reports about how the estimation for pi was developed and how accurate it was.

Use the diagram for Exercises 20 and 21.

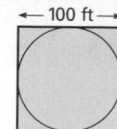
←— 100 ft —→

20. Find the perimeter of the square. **400 ft**

21. Find the circumference of the circle. **314 ft**

A circular jogging course is shown. Jogger A
uses the outer edge of the track. Jogger B
uses the inner edge of the track.

**THINK CRITICALLY/
SOLVE PROBLEMS**

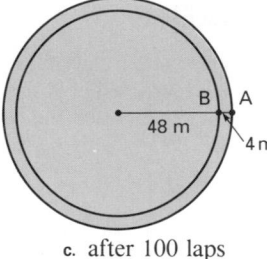

B   A
48 m
4 m

22. How many meters does each jogger
cover if each jogs one complete lap?
Jogger A, 326.6 m; Jogger B, 301.4 m

23. If the two joggers begin at the same
time and at the same points on
their tracks, how much farther will
Jogger A have jogged than Jogger B
after each number of laps?

Note: Answers will vary
depending on value of pi
used in calculation. Answers
here use pi = 3.14

a. after 5 laps   b. after 10 laps   c. after 100 laps
126 m   252 m   2,520 m

24. The diameter of a car tire is 70.4 cm. How many times must the
wheel turn to travel 1 km? **452.4 times**

The radius of the front wheels of a car is 42.3 cm.
The radius of the rear wheels is 47.9 cm. Use these data for
Exercises 27 and 28.

25. Find how far the car travels for one turn of the rear wheels. **301 cm**

26. For 100 turns of the front wheels, how many times do the rear
wheels turn? **88.3 times**

---

5 **FOLLOW-UP**

Extra Practice

1. Find the circumference, to the
nearest tenth, of a circle having
a diameter of 18.3 cm.
57.5 cm

2. Find the distance around the
figure. **16.3 in.**

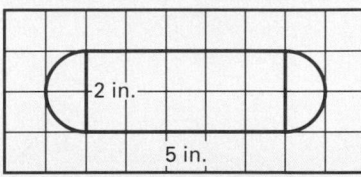
2 in.
5 in.

3. A circle has a circumference of
51 in. What is its radius, to the
nearest inch? **8 in.**

4. The inside of a 4-inch-thick pipe
has a diameter of 12 in. What is
the circumference of the outer
surface of the pipe to the nearest tenth? **63 in.**

Extension The computer program below calculates the
circumference of a circle by
inputting the radius. Have students modify the program to
calculate the circumference by
inputting the diameter.

```
10 PRINT "CIRCUMFERENCE OF
   CIRCLE"
20 INPUT "RADIUS IS"; R
30 C = 2*R*π
40 PRINT "CIRCUMFERENCE IS";
   C
```

Section Quiz

1. Find the circumference, to the
nearest tenth, of a circle having
a radius of 4.5 cm. **28.3 cm**

2. Find the circumference, to the
nearest inch, of a circle having
a diameter of 45 in. **141 in.**

3. Find the diameter of a circle with
circumference 314 ft. **100 ft**

Additional Answers
See page 577.

WARM-UP

Find the next three numbers.
1. 1, 4, 9, 16,...**25, 36, 49**
2. 2, 4, 8, 16,...**32, 64, 128**
3. 3, 9, 27, 81,...**243, 729, 2,187**

1 MOTIVATE

Introduction Ask students to suggest instances where drawing a picture has helped them to solve a problem.

2 TEACH

Use the Pages/Problem Have students read this part of the section and then discuss the problem. Then have them work in small groups to solve the problem.

Use the Pages/Solution Have students compare their solutions with the one presented in the text.

3 SUMMARIZE

Talk About It Have students discuss which types of problems are best solved by drawing a picture.

5-7   Problem Solving Strategies:
DRAW A PICTURE

► READ
► PLAN
► SOLVE
► ANSWER
► CHECK

Sometimes a problem is easier to understand and solve if you draw a picture to help you organize and visualize the data.

PROBLEM

Keisha wants to tack four photographs onto the bulletin board in her bedroom. What is the least number of thumbtacks that she will need if she tacks all the corners of every photograph?

SOLUTION

Keisha begins thinking about the problem. She knows that, if she tacks each photograph individually, she will need 4 thumbtacks for each photograph, or 16 in all. She draws this picture to show the arrangement.

Arrangement 1

Then Keisha considers overlapping the corners of the photographs in different ways. She draws these three arrangements of the photographs to help her figure out how many thumbtacks she will need.

Arrangement 2

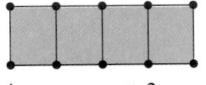

This takes 4 thumbtacks for the first photograph and 2 for each of the other three photographs—10 thumbtacks in all.

Arrangement 3

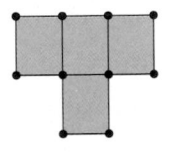

This takes 4 thumbtacks for the first photograph and 2 for each of the other three—10 in all.

TALK IT OVER

Why do you think fewer thumbtacks are needed for Arrangement 4 than for the other arrangements?

Arrangement 4

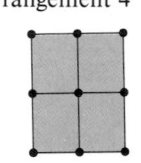

This takes only 9 tacks in all!

TEACHING TIP

Have students work together in pairs or in small groups. Students should take turns reading the problems aloud and drawing the pictures.

PROBLEMS

What is the least number of thumbtacks that would be needed to tack each number of photographs to a bulletin board?

1. 9 photographs **16 thumbtacks**
2. 10 photographs **18 thumbtacks**
3. 11 photographs **19 thumbtacks**
4. 12 photographs **20 thumbtacks**
5. 13 photographs **22 thumbtacks**
6. 14 photographs **23 thumbtacks**
7. 15 photographs **24 thumbtacks**
8. 16 photographs **25 thumbtacks**

What is the greatest number of photographs that could be tacked to a bulletin board with each number of thumbtacks?

9. 4 thumbtacks **1 photograph**
10. 9 thumbtacks **4 photographs**
11. 16 thumbtacks **9 photographs**
12. 25 thumbtacks **16 photographs**
13. 36 thumbtacks **25 photographs**
14. 49 thumbtacks **36 photographs**

15. Record your answers for Exercises 9–14 in a table like the one below. Look for a pattern in the numbers of photographs and the numbers of thumbtacks. Then complete the table by predicting the number of photographs that could be displayed using 64 thumbtacks. **49 photos**

Number of Photographs	1	4	9	16	25	36	49
Number of Thumbtacks	4	9	16	25	36	49	64

Suppose you notice that your neighbor, Kathy, has begun to build a fence around her square garden plot. You see that she has installed eight fence posts along each side of the garden. Each of the posts has a diameter of six inches. The posts have been placed so that the distance from one post to the next is one yard.

16. How many posts are there? **28 posts**

17. What is the perimeter of the garden? **33 yd 1 ft**

18. Suppose Kathy has told you that she would like to plant her square garden so that there are four rows of tomato plants with the same number of plants in each row. Each row will be 100 ft long. If she places the plants four feet apart and starts planting at one corner of the garden, how many plants will she need to buy? **104 plants**

CHALLENGE

Challenge students to solve the following problem by drawing a family tree and using logical thinking.

How many cousins does Inez have? Each pair of her grandparents has three children, and each of these children has three children.   **12 cousins**

4 PRACTICE

Problems Encourage all students to try all the problems. For Problems 1–14, some students might find it beneficial to model the problem with pieces of paper and counters. For Problem 17, remind students to give their answers in yards and feet.

5 FOLLOW-UP

Extra Practice

1. A tour train consists of an engine car, a passenger car, a snack car, and a caboose. The passenger car is three times longer than the snack car. The snack car is twice as long as the engine car. The caboose is half as long as the engine car. How long is each car if the snack car is 8 m?   **engine car: 4 m, passenger car: 24 m, caboose: 2 m**

2. The boards used to build a dock are each 7 1/2 in. wide. They are nailed side by side with spaces of 1 in. between. How wide is the dock after 10 boards have been nailed in place?   **84 in. or 7 ft**

5 CHAPTER REVIEW

Introduction The Chapter Review emphasizes the major concepts, skills, and vocabulary presented in this chapter and can be used for diagnosing students' strengths and weaknesses. Page references direct students back to appropriate sections for additional review and reteaching.

Using Pages 178–179 Allow students to quickly scan the Chapter Review and ask questions about any section they find confusing. Exercises 1–7 review key vocabulary.

Informal Evaluation Have students explain how they arrived at any incorrect answers. They will probably find their own mistakes and give you some clues as to the nature of their errors. Make sure students understand this material before administering the Chapter Test.

Follow-Up Have students research the history of linear measurement. Use their information as the basis for a bulletin board display with illustrations. Include a description of those early measurement units based on the parts of the body, such as *span, cubit, girth, yard, foot, fathom*. Also include the origin and development of the metric system from its development in France to its present-day usage as part of a much larger system called Le Système Internationale d'Unités.

1. The distance across a circle is the circle's ___?___. **f**
2. The distance around a plane figure is its ___?___. **d**
3. The height of a doorway could be expressed in ___?___. **a**
4. The fixed distance from any point on a circle to the circle's center is the ___?___ of the circle. **e**
5. The amount of water in a cup could be expressed in ___?___. **b**
6. The mass of a basket of apples could be expressed in ___?___. **g**
7. The distance around a circle is its ___?___. **c**

a. meters
b. milliliters
c. circumference
d. perimeter
e. radius
f. diameter
g. kilograms

SECTIONS 5–1 AND 5–2 CUSTOMARY AND METRIC UNITS   (pages 154–161)

▶ To change from a larger unit of measure to a smaller unit, multiply.
▶ To change from a smaller unit of measure to a larger unit, divide.

Complete.

8. 4 yd = ■ ft **12**
9. 38 oz = ■ lb ■ oz **2, 6**
10. 5 pt = ■ c **10**

11. 3 mi = ■ yd **5,280**
12. 28 in. = ■ ft ■ in. **2, 4**
13. 10 ft = ■ yd ■ ft **3, 1**

14. 16 kg = ■ g **16,000**
15. 900 cm = ■ m **9**
16. 290 mm = ■ m **0.290**

17. 4 km = ■ m **4,000**
18. 8 m = ■ cm **800**
19. 4.6 L = ■ mL **4,600**

SECTION 5–3 WORKING WITH MEASUREMENTS   (pages 162–165)

▶ When you add or multiply with customary measurements, you may need to simplify your answer.
▶ When you subtract or divide with customary measurements, you may need to change a measurement.
▶ You may also need to change measurements when performing operations with metric measurements.

Write each answer in simplest form.

20. 6 gal 2 qt
   +2 gal 6 qt
   _____
   10 gal

21. 4 ft 3 in.
   −2 ft 10 in.
   _____
   1 ft 5 in.

Complete.

22. 84 cm + 92 cm = ___?___ m **1.76**
23. 8 km ÷ 4 = ___?___ m **2,000**

SECTION 5-4 PERIMETER (pages 166–169)

▶ The **perimeter** of a plane figure is the distance around it.
▶ You can find the perimeter of a figure by adding the lengths of its sides.

Find the perimeter of each figure.

24.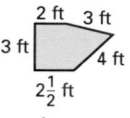
2 ft 3 ft
3 ft 4 ft
$2\frac{1}{2}$ ft
$14\frac{1}{2}$ ft

25.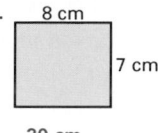
8 cm
7 cm
30 cm

26.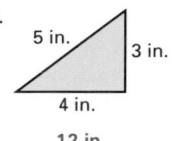
5 in. 3 in.
4 in.
12 in.

27.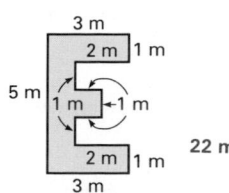
3 m
2 m 1 m
5 m
1 m 1 m
2 m 1 m
3 m
22 m

SECTION 5-5 PROBLEM SOLVING SKILLS: USE A FORMULA (pages 170–171)

▶ You can use the formula $P = 2l + 2w$ to find the perimeter of a rectangle.

Use the formula to find the perimeter of a rectangular figure with the given dimensions.

28. length: 5.6 km
width: 2.9 km **17 km**

29. length: 16.5 mi
width: 12.4 mi **57.8 mi**

SECTION 5-6 CIRCUMFERENCE (pages 172–175)

▶ The formulas $C = \pi d$ and $C = 2\pi r$ can be used to find the circumference of a circle.

Find the circumference.
Use 3.14 or $\frac{22}{7}$ for π,
as appropriate.

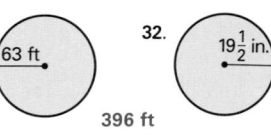

30. 42.7 m
134.1 m

31. 63 ft
396 ft

32. $19\frac{1}{2}$ in.
$122\frac{1}{2}$ in.

SECTION 5-7 PROBLEM SOLVING STRATEGIES (pages 176–177)

▶ Sometimes a problem is easier to understand and solve if you draw a picture.

33. A square garden is enclosed by a fence that has 12 posts on each side. How many posts are there in all? **44 posts**

USING DATA Use the Data Index on page 546 to find the table showing dimensions of noted rectangular structures. Find the perimeter, in meters, of the base of each structure.

34. Cleopatra's Needle **9.4 m**

35. Ziggurat of Ur **210 m**

36. Great Pyramid of Cheops **922.4 m**

37. Ta Keo Templean. **450 m**

USING DATA Use the chart on page 153 to answer the question.

38. About how much longer than the Italian Gothic period was the Byzantine period? **about 400 years**

Review **179**

Complete.

1. 72 in. = ■ ft 6

2. 9 gal = ■ qt 36

3. 10 lb = ■ oz 160

4. 2 mi = ■ yd 3,520

5. 280 mm = ■ m 0.280

6. 5 L = ■ mL 5,000

7. 589 mg = ■ g 0.589

8. 2 km = ■ m 2,000

Write each answer in simplest form.

9.   6 c 4 fl oz
   +2 c 10 fl oz
   ‾‾‾‾9 c 6 fl oz‾‾

10.   6 lb
   −4 lb 5 oz
   ‾‾‾1 lb 11 oz‾‾

11. 2 ft 6 in.
   × 3
   ‾‾‾7 ft 6 in.‾‾

12. 13 gal 2 qt ÷ 3
   4 gal 2 qt

Complete.

13. 230 cm + 370 cm = __?__ m 6

14. 12 m − 750 cm = __?__ m 4.5

15. 3.21 cm × 3 = __?__ mm 96.3

16. 400 m ÷ 8 = __?__ cm 5,000

Solve.

17. Will one 112-inch ball of string be enough to make twelve pieces of string, each $\frac{3}{4}$ ft long? yes

Find the perimeter of each figure.

18. 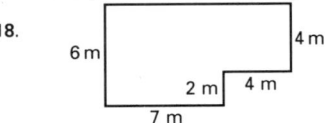 34 yd

19. 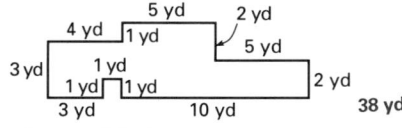 38 yd

Use the formula to find the perimeter for a rectangular figure with the given dimensions.

20. length: 7.35 m
   width: 4.81 m 24.32 m

21. length: 25.4 mi
   width: 16.9 mi 84.6 mi

Find the circumference. Use 3.14 or $\frac{22}{7}$ for π, as appropriate.

22.

49 mm

154 mm

23.

6.6 ft

41.4 ft

24.

17 m

106.8 m

Draw a picture to help you solve.

25. A rectangular garden is enclosed by a fence that has 10 posts on each of its longer sides and 8 posts on each of its shorter sides. How many posts are there in all? 32 posts

1. Organize these data into a stem-and-leaf plot.

90	68	78	62	92	73
87	98	96	70	85	95
84	67	79	86	64	76

6	8 7 2 4
7	8 9 0 3 6
8	7 4 6 5
9	0 8 6 2 5

Which measure of central tendency would you use to represent the middle of each set of data?

2. average salary of part-time help
   mean or median

3. salary received by most part-time help
   mode

4. speed of vehicles on a highway **mean**

Complete.

5. $7.8 = \blacksquare \times 1$ **7.8**

6. $179 + 612 = \blacksquare + 179$ **612**

7. $68 \times \blacksquare = 24 \times 68$ **24**

8. $(8 + 2) + 3 = \blacksquare + (2 + 3)$ **8**

9. $(5 \times \blacksquare) \times 4 = 5 \times (9 \times 4)$ **9**

10. $36 \times \blacksquare = 0$ **0**

11. Determine whether the following argument is *valid* or *invalid*.

    If an animal is a snake, then it is a reptile.
    Squeezy is a reptile.
    _____
    Therefore, Squeezy is a snake. **invalid**

12. Is the following conditional statement *true* or *false*? If false, give a counterexample.

    If a number is divisible by 8, then it is divisible by 16. **false; 24**

Find each answer.

13. $\frac{1}{6} \times 18$ **3**

14. $\frac{9}{16} \div 3$ **$\frac{3}{16}$**

15. $19 - 4\frac{5}{6}$ **$14\frac{1}{6}$**

16. $\frac{5}{8} + \frac{2}{3}$ **$1\frac{7}{24}$**

17. $\frac{4}{9} \times \frac{3}{8}$ **$\frac{1}{6}$**

18. $\frac{3}{4} \div \frac{5}{8}$ **$1\frac{1}{5}$**

Complete.

19. $5 \text{ mi} = \blacksquare \text{ ft}$ **26,400**

20. $4.5 \text{ m} = \blacksquare \text{ mm}$ **4,500**

21. $96 \text{ oz} = \blacksquare \text{ lb}$ **6**

22. $1,572 \text{ m} = \blacksquare \text{ km}$ **1.572**

23. $60 \text{ oz} = \blacksquare \text{ lb} \blacksquare \text{ oz}$ **3; 12**

24. $6,250 \text{ cm} = \blacksquare \text{ m}$ **62.50**

Find the perimeter of each figure.

25.

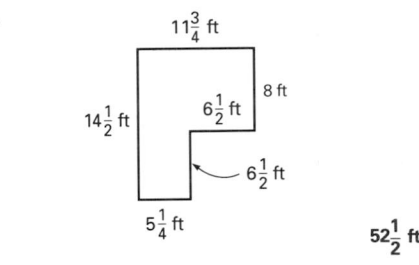

$52\frac{1}{2}$ ft

26.

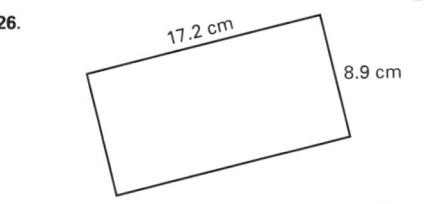

52.2 cm

27.

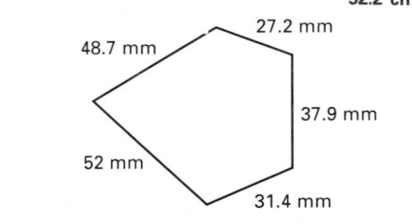

197.2 mm

Introduction The purpose of this Cumulative Review is to maintain previously taught skills and concepts and to apply them to the material presented in this chapter. At least one major objective of each chapter is included in the review.

Item Analysis The table below correlates the Cumulative Review items with the chapter and section that are being reviewed.

Section	Items
1–2	1
1–9	2–4
2–9	5–10
3–2	12
3–3	11
4–5	13–14, 17–18
4–6	15–16
5–1	19, 21, 23
5–2	20, 22, 24
5–4	25–27

5 CUMULATIVE TEST

Introduction The Cumulative Test uses a standardized-test format of multiple-choice questions to assess retention of previously learned concepts and test-taking skills. Test results may be used to diagnose students' strengths and weaknesses.

Item Analysis The table below correlates the Cumulative Test items with the chapter and section that are being tested.

Section	Items
1–9	1–2
2–7	3–4
2–9	5–6
3–3	7–8
4–5	11–12
4–6	9–10
5–1	13
5–2	14
5–7	15

In a set of data, the least number is decreased by 4 and the greatest number is increased by 4.

1. How is the range affected?
   A. It is increased by 8.
   B. It is decreased by 8.
   C. It remains the same.
   D. none of these

2. How is the mean affected?
   A. It is increased.
   B. It is decreased.
   C. It remains the same.
   D. none of these

3. Multiply. $5^9 \times 5^4$.
   A. 25^{13} B. 5^{13}
   C. 5^5 D. 25^5

4. Use the quotient rule to divide. $10^{10} \div 10^7$.
   A. 10^{17} B. 17^{10}
   C. 100^{17} D. 10^3

5. Complete.
   $(9 + 3) + 5 = \blacksquare + (3 + 5)$
   A. 17   B. 3
   C. 5   D. 9

6. $(7 \times \blacksquare) \times 3 = 7 \times (5 \times 3)$
   A. 17   B. 5
   C. 15   D. 7

7. Which conclusion can be drawn from these statements?

   If a person lives in Dallas, then that person lives in Texas.
   Malerie lives in Dallas.

   A. Malerie does not live in Dallas.
   B. Malerie lives in Houston.
   C. Malerie lives in Texas.
   D. none of these

8. Which is a counterexample for the following statement?

   If two numbers are odd, then their sum is odd.
   A. 3 and 5   B. 3 and 6
   C. 4 and 5   D. none of these

9. Add. $\frac{5}{6} + \frac{1}{8}$
   A. $\frac{6}{14}$ B. $\frac{3}{7}$
   C. $\frac{23}{24}$ D. $1\frac{1}{24}$

10. Subtract. $\frac{9}{10} - \frac{2}{5}$
    A. $1\frac{2}{5}$ B. $\frac{7}{10}$
    C. $\frac{2}{3}$ D. $\frac{1}{2}$

11. Multiply. $\frac{3}{4} \times \frac{8}{15}$
    A. $\frac{2}{5}$ B. $\frac{24}{45}$
    C. $\frac{15}{32}$ D. $1\frac{13}{32}$

12. Divide. $\frac{3}{8} \div \frac{1}{4}$
    A. $\frac{3}{32}$ B. $\frac{1}{2}$
    C. $\frac{1}{3}$ D. $1\frac{1}{2}$

13. Complete. 5 gal = $\blacksquare$ pt
    A. 10 pt   B. 20 pt
    C. 40 pt   D. 50 pt

14. Complete. 4,568 cm = $\blacksquare$ m
    A. 0.4568 m   B. 456.8 m
    C. 45.68 m   D. 4.568 m

15. A garden has 6 rows of equal numbers of pepper plants. Each row is 40 ft long. The plants are placed 5 ft apart starting at the corner of the garden. How many pepper plants are in the garden?
    A. 54 plants   B. 60 plants
    C. 48 plants   D. 240 plants

Customary Units of Measurement

Follow these rules for changing from one customary unit of measure to another:

- To change from a *larger* unit of measure to a *smaller* one, **multiply** the number of larger units by the number of smaller units that is equivalent to the larger unit.
- To change from a *smaller* unit of measure to a *larger* one, **divide** the number of smaller units by the number of smaller units that is equivalent to the larger unit.

Customary Units of Measure

Measure	Units	Equivalents
length	inch (in.)	
	feet (ft)	1 ft = 12 in.
	yard (yd)	1 yd = 3 ft, or 36 in.
	mile (mi)	1 mi = 5,280 ft, or 1,760 yd
weight	ounce (oz)	
	pound (lb)	1 lb = 16 oz
	ton (T)	1 T = 2,000 lb
capacity	fluid ounce (fl oz)	
	cup (c)	1 c = 8 fl oz
	pint (pt)	1 pt = 2 c, or 16 fl oz
	quart (qt)	1 qt = 2 pt, or 4 c
	gallon (gal)	1 gal = 4 qt, 8 pt, or 16 c

▶ **Example 1**

Change 108 feet to yards.

Solution

3 ft = 1 yd

Divide the total number of feet by 3.
108 ÷ 3 = 36

There are 36 yd in 108 ft.

▶ **Example 2**

Change 3 pounds to ounces.

Solution

You know that 1 lb = 16 oz.

Multiply 16 by 3.
16 × 3 = 48

There are 48 oz in 3 lb.

EXERCISES

Complete.

1. 9 ft = **108** in.
2. 12 ft = **4** yd
3. 2 mi = **10,560** ft
4. 6 ft = **72** in.
5. 8 yd = **24** ft
6. 18 in. = **1** ft **6** in.
7. 64 oz = **4** lb
8. 24 fl oz = **3** c
9. 4 qt = **8** pt
10. 48 pt = **96** c
11. 5 yd = **15** ft
12. 8 ft = **2** yd **2** ft
13. 1 T = **2,000** lb
14. 0.5 T = **1,000** lb
15. 24 oz = **1** lb **8** oz

Other Customary Units of Measure

The difference between a foot and a yard is not very great. There is a vast difference between a yard and a mile, as you well know. For this reason, it became convenient to use units of measure between the yard and the mile. The *chain*, the *rod*, and the *furlong* are such units. The chain has 100 links.

1 chain = 66 ft   1 rod = $5\frac{1}{2}$ yd   1 furlong = 40 rods

Remember that, to change from a smaller to a larger unit of measure, you divide. To change from a larger to a smaller unit, you multiply.

▶ **Example 1**

Change 440 yd to furlongs.

Solution

First, change 440 yards to rods.
440 yd ÷ 5.5 yd = 80 rods

Then change rods to furlongs.
80 rods ÷ 40 rods = 2 furlongs

So, 440 yd = 2 furlongs.

▶ **Example 2**

Change 1 mi to furlongs.

Solution

First, change 1 mi to rods.
1 rod = 5.5 yd
1 mi = 1,760 yd
1,760 ÷ 5.5 = 320 rods

Then, change rods to furlongs.
320 rods ÷ 40 rods = 8 furlongs

So 1 mi = 8 furlongs.

EXERCISES

1. How many feet are in 1 rod?
   16.5 ft

2. How many yards are in 1 furlong?
   220 yd

3. How many rods are in 1 chain?
   4 rods

4. How many miles are in 1,280 rods?
   4 mi

5. How many chains are in 1 mile?
   80 chains

6. How many links are in 88 yd?
   400 links

7. How many furlongs are in 10 chains?
   1 furlong

Metric Units of Measure

Follow these rules for changing from one metric unit of measure to another:

- To change from a *larger* unit of measure to a *smaller* one, **multiply** the number of larger units by the number of smaller units equivalent to 1 larger unit.
- To change from a *smaller* unit of measure to a *larger* one, **divide** the number of smaller units by the number of smaller units equivalent to 1 larger unit.

Metric Units of Measure

thousands	hundreds	tens	ones	tenths	hundredths	thousandths
kilo-	hecto-	deka-	no prefix	deci-	centi-	milli-
kilometer (km) 1,000 m	hectometer (hm) 100 m	dekameter (dam) 10 m	meter (m) 1 m	decimeter (dm) 0.1 m	centimeter (cm) 0.01 m	millimeter (mm) 0.001 m
kilogram (kg) 1,000 g	hectogram (hg) 100 g	dekagram (dag) 10 g	gram (g) 1 g	decigram (dg) 0.1 g	centigram (cg) 0.01 g	milligram (mg) 0.001 g
kiloliter (kL) 1,000 L	hectoliter (hL) 100 L	dekaliter (daL) 10 L	liter (L) 1 L	deciliter (dL) 0.1 L	centiliter (cL) 0.01 L	milliliter (mL) 0.001 L

▶ **Example 1**

Change 3.5 g to milligrams.

Solution

1 g = 1,000 mg
3.5 × 1,000 = 3,500

So, there are 3,500 milligrams in 3.5 grams.

▶ **Example 2**

Change 400 m to kilometers.

Solution

1 km = 1,000 m or 0.001 km = 1 m
400 × 0.001 = 0.400

So, 400 m = 0.4 km.

EXERCISES

Complete.

1. 4 L = **4,000** mL
2. 740 g = **0.74** kg
3. 5 mm = **0.005** m
4. 525 mm = **52.5** cm
5. 4,990 mL = **4.99** L
6. 25 m = **0.025** km
7. 2 kg = **2,000** g
8. 4 km = **4,000** m
9. 500 mL = **0.5** L
10. 15 cm = **150** mm
11. 45 L = **45,000** mL
12. 0.18 kg = **180** g
13. 1,800 mm = **1.8** m
14. 5,400 m = **5.4** km
15. 6.8 m = **680** cm
16. 17 g = **17,000** mg
17. 750 mm = **0.75** m
18. 9,100 g = **9.1** kg

Metric Magic

You can use a metric place-value table to change metric units.

Metric Prefixes

kilo-	hecto-	deka-	no prefix	deci-	centi-	milli-
1	5					
1	5	0	0			
		4	2	0	0	
	0	4	2	0	0	

▶ **Example 1**

Change 1.5 km to meters.

Solution

Put a decimal point after the "kilo-" prefix in the chart. Enter 1.5 km in the chart. In the next row, put a decimal point after the "no prefix" label.

Copy the numbers from the first row to the second without changing the columns. Fill in any empty spaces in front of the new decimal point with zeros.

So, 1.5 km = 1,500 m.

▶ **Example 2**

Change 4,200 mL to hectoliters.

Solution

Put a decimal point after the "milli-" prefix in the chart, and enter 4,200 mL. In the next row, put a decimal point after the "hecto-" prefix.

Copy the numbers from the first row to the second row without changing the columns. Fill in any spaces after the new decimal point with zeros.

So, 4,420 mL = 0.042 hL.

EXERCISES

Make yourself a table and use it to change from one unit to another.

1. 5 km = **500** dam
2. 6.3 L = **0.0063** kL
3. 7 L = **70** dL
4. 543 cm = **5.43** m
5. 5,343 mg = **5.343** g
6. 0.48 kg = **4.8** hg
7. 5,876 mm = **0.05876** hm
8. 3.2 km = **320,000** cm
9. 18 hL = **1,800,000** mL
10. 720,000 cg = **7.2** kg
11. 18,400 dL = **1,890** L
12. 1,000 km = **1,000,000,000** mm

182A

Name _____ Date _____

Working with Measurements

When adding or multiplying measurements, you may need to change the units of the sum or the product.

▶ **Example 1** _____

a.
```
  4 lb  6 oz
+2 lb 11 oz
```
b.
```
3 ft 8 in.
      ×3
```

Solution

a.
```
  4 lb  6 oz
+2 lb 11 oz
  6 lb 17 oz   ← 17 oz is more than 1 lb. Simplify.
  6 lb 17 oz = 7 lb 1 oz
```
b.
Solution
```
3 ft 8 in.
      ×3
9 ft 24 in.
9 ft 24 in. = 11 ft
```

When subtracting or dividing measurements, you may need to change a measurement before you can perform the operation.

▶ **Example 2** _____

a. Write the answer in simplified form.
```
 9 gal 2 qt
−2 gal 5 qt
```
b. Complete. 6 L ÷ 8 = __?__ mL

Solution

Rename 9 gal 2 qt as 8 gal 6 qt.
```
 8 gal 6 qt
−2 gal 5 qt
 6 gal 1 qt
```

Solution

Change 6 L to mL
1 L = 1,000 mL
6 L = 6,000 mL
6,000 mL ÷ 8 = 750 mL

EXERCISES

Write each answer in simplified form.

1.
```
 7 lb 13 oz
+3 lb  7 oz
11 lb 4 oz
```
2.
```
10 ft  7 in.
+4 ft 11 in.
15 ft 6 in.
```
3.
```
10 yd 7 ft
−4 yd 9 ft
 5 yd 1 ft
```
4.
```
12 gal 3 qt
−8 gal 5 qt
 3 gal 2 qt
```

5.
```
3 c 12 fl oz
        ×4
15 c
```
6.
```
9 lb 7 oz
      ×5
47 lb 3 oz
```
7. 7 yd 1 ft ÷ 2   **3 yd 2 ft**

8. 7 c 11 fl oz ÷ 3   **2 c 6 1/3 oz**

Complete.

9. 3.5 kg + 500 g = **4** kg

10. 1.8 L ÷ 5 = **360** mL

11. 2.75 cm × 12 = **330** mm

12. 6 m − 350 cm = **250** cm

Reteaching • SECTION 5-3   85

Name _____ Date _____

Estimating Measurements

There are several different ways of estimating measurements.

▶ **Example 1** _____

Estimate by rounding.
984 m − 519 m

Solution

984 → 1,000
−519 → −500   and 1,000 − 500 = 500
So, 984 m − 519 m ≈ 500 m.

▶ **Example 2** _____

Estimate using compatible numbers.
2,582 kg ÷ 63

Solution

2,582 ÷ 63
2,400 ÷ 60 = 40   24 is divisible by 6.
So, 2,582 kg ÷ 63 ≈ 40 kg.

▶ **Example 3** _____

Estimate by using front-end estimation.
7.86 lb + 3.4 lb + 9.1 lb

Solution

7 + 3 + 9 = 19
and 0.8 + 0.4 + 0.1 = 1.3
So, 7.86 lb + 3.4 lb + 9.1 lb is slightly more than 20 lb.

▶ **Example 4** _____

Estimate by clustering.
528.5 cm + 526.2 cm + 521.3 cm + 526.4 cm

Solution

The four numbers are close to 525.
525 × 4 = 2,100
So, 528.5 cm + 526.2 cm + 521.3 cm + 526.4 cm ≈ 2,100 cm.

EXERCISES

Estimate by rounding.

1. 3.86 ft × 47   **200 ft**   2. 52,304 m − 28,127 m **20,000 m** 3. 56 in. + 89 in. **150 in.**

Estimate by using compatible numbers.

4. 475 yd ÷ 79   **6 yd**   5. $78\frac{2}{3}$ ft ÷ $7\frac{1}{2}$ **10 ft**   6. 2,369 km ÷ 624   **4 km**

Estimate by using front-end estimation.

7. 8,756 mL − 3,922 mL
slightly less than 5,000 mL

8. 821 km + 956 km + 1,201 km
slightly more than 2,900 km

Estimate by using clustering.

9. 273 bolts + 286 bolts + 329 bolts
900 bolts

10. 5,734 yd + 6,302 yd + 5,887 yd + 5,991 yd
24,000 yd

Use estimation to place the decimal point in the answer.

11. 586 m ÷ 40 = 1465   **14.65**   12. 9.23 m × 56 = 51688   **516.88**

86   **Enrichment** • SECTION 5-3

Name _____ Date _____

Perimeter

The **perimeter** of a plane figure is the distance around it. To find the perimeter of any plane figure, find the sum of the lengths of the sides.

▶ **Example** _____

Find the perimeter of the figure shown below.

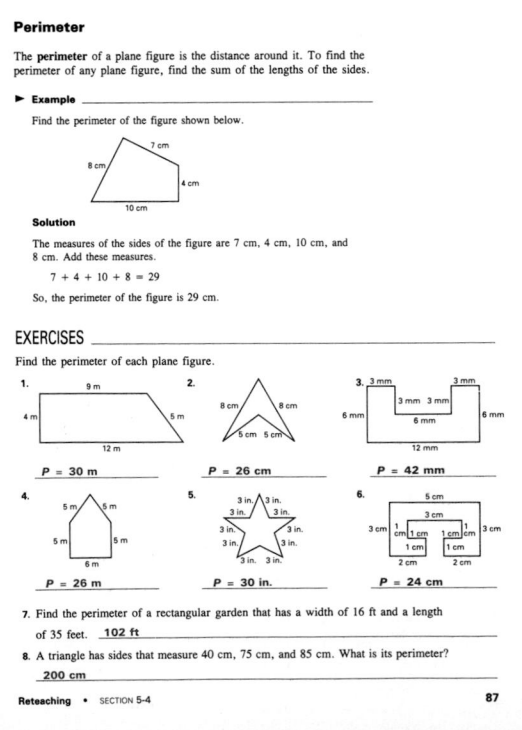

Solution

The measures of the sides of the figure are 7 cm, 4 cm, 10 cm, and 8 cm. Add these measures.

7 + 4 + 10 + 8 = 29

So, the perimeter of the figure is 29 cm.

EXERCISES

Find the perimeter of each plane figure.

1.   **P = 30 m**

2.   **P = 26 cm**

3.   **P = 42 mm**

4.   **P = 26 m**

5.   **P = 30 in.**

6.   **P = 24 cm**

7. Find the perimeter of a rectangular garden that has a width of 16 ft and a length of 35 feet.
102 ft

8. A triangle has sides that measure 40 cm, 75 cm, and 85 cm. What is its perimeter?
200 cm

Reteaching • SECTION 5-4   87

Name _____ Date _____

Thinking About Perimeter

The perimeter of this figure is 24 cm.

Each of the following figures is constructed of squares the same size as above. Find the perimeter of each.

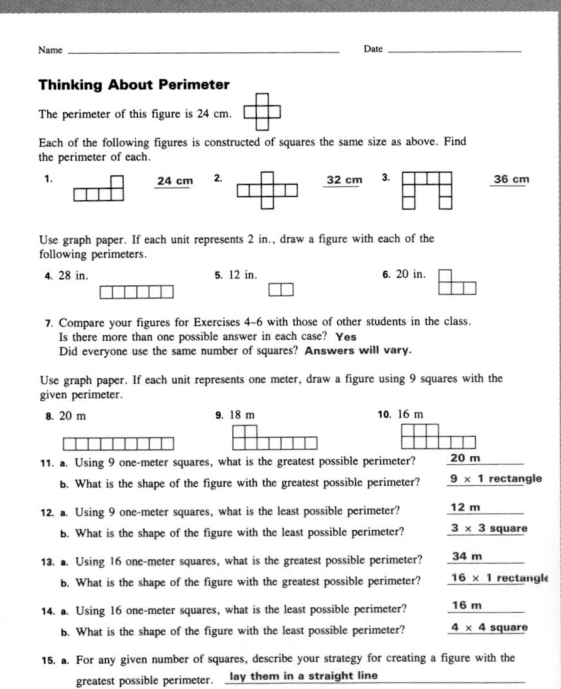

1.   **24 cm**   2.   **32 cm**   3.   **36 cm**

Use graph paper. If each unit represents 2 in., draw a figure with each of the following perimeters.

4. 28 in.   5. 12 in.   6. 20 in.

7. Compare your figures for Exercises 4–6 with those of other students in the class.
Is there more than one possible answer in each case? **Yes**
Did everyone use the same number of squares? **Answers will vary.**

Use graph paper. If each unit represents one meter, draw a figure using 9 squares with the given perimeter.

8. 20 m   9. 18 m   10. 16 m

11. a. Using 9 one-meter squares, what is the greatest possible perimeter?   **20 m**
    b. What is the shape of the figure with the greatest possible perimeter?   **9 × 1 rectangle**

12. a. Using 9 one-meter squares, what is the least possible perimeter?   **12 m**
    b. What is the shape of the figure with the least possible perimeter?   **3 × 3 square**

13. a. Using 16 one-meter squares, what is the greatest possible perimeter?   **34 m**
    b. What is the shape of the figure with the greatest possible perimeter?   **16 × 1 rectangle**

14. a. Using 16 one-meter squares, what is the least possible perimeter?   **16 m**
    b. What is the shape of the figure with the least possible perimeter?   **4 × 4 square**

15. a. For any given number of squares, describe your strategy for creating a figure with the greatest possible perimeter.   **lay them in a straight line**
    b. For any given number of squares, describe your strategy for creating a figure with the least possible perimeter.   **make as close to a square as possible**

88   **Enrichment** • SECTION 5-4

Name _____ Date _____

Problem Solving: Use a Formula

You can find the perimeter of a rectangle by doubling the given length and doubling the given width and adding the results. You can find the perimeter of a square by multiplying the length of the given side by 4.

► **Example 1** _____

Find the perimeter of the rectangle shown.

10 cm
6 cm

Solution

The length is 10 cm and the width is 6 cm. Use a formula.

$P = 2l + 2w = 2(10) + 2(6)$
$P = 20 + 12 = 32$

The perimeter is 32 cm.

EXERCISES _____

Use a formula to find the perimeter of each figure.

1.
4 m
6 m
P = 20 m

2.
10 in.
3 in.
P = 26 in.

3.
20 cm
7 cm
P = 54 cm

4. Square
7 mm
P = 28 mm

5. Rectangle
6 yd
15 yd
P = 42 yd

6. Square
3 m
P = 12 m

Reteaching • SECTION 5-5

89

Name _____ Date _____

Rectangles, Squares, and Factors

Suppose that you want to cut sheets of paper into squares of the same size, with no paper left over. You have sheets of paper that are of the sizes shown below.

I — 56 cm / 32 cm
II — 90 cm / 54 cm
III — 120 cm / 84 cm
IV — 128 cm / 96 cm

You could cut 16-centimeter squares from Sheet IV, since 16 is a common factor of both 96 and 128.

16 cm
16 cm

How many such squares could you cut? __48__

Which sheets could be cut into

1. 3-cm squares? **II, III**
2. 4-cm squares? **I, III, IV**
3. 8-cm squares? **I, IV**
4. 9-cm squares? **II**
5. 12-cm squares? **III**
6. 18-cm squares? **II**

90

Enrichment • SECTION 5-5

Name _____ Date _____

Circumference

The **circumference** of a circle is the distance around it. The circumference of any circle, divided by the diameter, is always equal to the number π (pi), which has the approximate value 3.14 or $\frac{22}{7}$.
You can find the circumference of a circle when you know its diameter, using the formula $C = \pi d$.

► **Example 1** _____

Find the circumference of the circle shown.

10 m

Solution

$C = \pi d$
$C = 3.14 \times 10$
$C = 31.4$

So, the circumference is 3.14 m. ◄

► **Example 2** _____

Find the circumference of the circle shown.

7 ft

Solution

$C = 2\pi r$
$C = 2 \times \frac{22}{7} \times 7$
$C = 44$

So, the circumference is 44 ft. ◄

Because the diameter is twice the radius, if you know the radius of a circle, you can find the circumference of the circle using the formula $C = 2\pi r$.

EXERCISES _____

Find each circumference. Use 3.14 for π.

1. 12 m — **37.68 m**
2. 24 in. — **75.36 in.**
3. 9 mm — **28.26 mm**

Find each circumference. Use $\frac{22}{7}$ for π.

4. 21 cm — **66 cm**
5. 28 in. — **88 in.**
6. 0.7 m — **2.2 m**

Find each circumference. Use 3.14 or $\frac{22}{7}$ for π. Round your answer to the nearest tenth.

7. 40 m — **251.2 m**
8. 4.2 cm — **13.2 cm**
9. 28 in. — **176 in.**

10. 7.5 ft — **47.1 ft**
11. 6.3 yd — **19.8 yd**
12. 23 mm — **72.2 mm**

Reteaching • SECTION 5-6

91

Name _____ Date _____

Inside or Outside Around

A school has a track that is 20 ft wide. The track runs around a playing field that is 300 ft long and 160 ft wide. At each end of the field the track goes around a semicircular area. The distance around each semicircle is equal.

300 ft
160 ft
20 ft

What is the approximate perimeter of the field, measured along the inside of the track?

The distance around the field along the inside of the track is twice the length of the field (300 ft) plus the distance around both semicircles.

$P = 300 + 300 + \pi d$
$P \approx 300 + 300 + 3.14 \times 160$
$P \approx 600 + 502.4$
$P \approx 1,102.4$

So, the perimeter of the field is about 1,102 ft.

Suppose that two runners ran once around the track, one along the inside and one along the outside of the 20-ft wide track.

1. What is the diameter of each semicircle for the outside track?
   200 ft

2. What is the circumference of each semicircle for the outside track?
   about 314 ft

3. What is the length of each straight side of the outside track?
   300 ft

4. What is the approximate total distance run by the runner who runs along the outside of the track?
   about 1,228 ft

5. How much farther does that runner run than does the runner who goes around the inside of the track?
   about 126 ft

6. About how many laps around would you have to run along the inside track to run 1 mile?
   about 5 laps

92

Enrichment • SECTION 5-6

Name _____ Date _____

Problem Solving Strategy: Draw a Picture

Drawing a picture can often help you to solve a problem by providing
a way to visualize the information you are given.

PROBLEM

Flo plans to put a fence around her triangular rose garden. Each side is
12 ft long. She plans to set a rosebush at each corner and along each side.
The bushes will be 24 in. apart. How many bushes must she plant?

SOLUTION

Draw a picture of the triangular plot and indicate where
the bushes are to be planted.

Find the number of bushes. There will be one bush at
each of the three corners. To find the number of bushes
in between these end bushes, divide 12 ft by 24 in.
Remember to convert inches to feet.

There are six 2-ft intervals along each side. So, there will be 5 bushes
in between each corner on each of the three sides. To this number add
3, the number of corner bushes.

$3 \times 5 = 15$ $15 + 3 = 18$

So, Flo needs 18 bushes.

PROBLEMS

1. Jim and Edna are planting a rectangular garden with 6 rows of equal numbers of
pepper plants. Each row will be 60 ft long. They set the plants 3 ft apart and
start planting at one corner of the garden. How many plants will they need?
126

2. Ferris is fencing a rectangular garden plot. He sets 10 fence posts along each
longer side and 8 posts along each shorter side. Each post is 6 in. in diameter.
The posts are 4 yd apart. How many posts will he use? What is the perimeter of
the garden plot?
32 posts; 134 yd

3. Marie is sewing a large diamond-shaped patch on the back of a jacket. Each side of
the diamond is the same length. She wants to sew small sequins along the border
of the patch 1 in. apart. Each sequin is $\frac{1}{4}$ in. in diameter. She wants 8 sequins
along each side. How many sequins will she need? What is the perimeter of the
patch?
28; 36 in.

Reteaching • SECTION 5-7   93

Name _____ Date _____

Tacky Arrangements

Draw a picture to help you solve these problems. **Check students' drawings.**

1. Ron wants to pin up a series of triangular baseball pennants in a row along one
wall of his bedroom. They will hang with the narrow corner down. It takes
3 tacks to pin up one pennant. What is the least number of tacks it will take to
pin up 5 pennants if no pennant is tacked to any other pennant? **15 tacks**

2. If Ron overlaps the top corners of each pennant, how many tacks will he use? **11 tacks**

3. Katina is covering a bulletin board with 6-sided
figures. She will have to overlap each figure in
this way:

If she pins 7 of these figures in a row, how many pins will she use? **36 pins**

Make up your own problem about arranging shapes in a row.

94   Enrichment • SECTION 5-7

Name _____ Date _____

Computer Activity: Dimension Options

Suppose you are given a specific amount of materials to build a
storage container whose dimensions must be whole numbers. Will any
perimeter value enable you to do this?

This program asks you to INPUT the value of the perimeter. When you
do, it then lists all possible whole number length-width combinations
that you can get for that perimeter.

```
10  INPUT "WHAT IS THE PERIMETER ? ";P: PRINT
20  PRINT "LENGTH","WIDTH"
30  FOR L = 1 TO P
40  FOR W = 1 TO P
50  IF 2 * L + 2 * W = P THEN PRINT L,W
60  NEXT W
70  NEXT L
```

EXERCISES

1. List five perimeter values that will not list any combinations. Describe the set of
values needed for the perimeter to list combinations.
P must be an even number.
_____
_____

2. List five perimeter values that will enable you to form a square (length = width).
Describe the set of values needed for the perimeter to enable you to get a square.
P must be a multiple of 4.
_____
_____
_____

3. Predict the dimension combinations for each perimeter. RUN the program to
check your answers.

a. 24   1 2 3 4 5 6 ... 11
         11 10 9 8 7 6 ... 1

b. 38   1 2 3 4 5 6 7 ... 18
         18 17 16 15 14 13 12 ... 1

c. 46   1 2 3 4 5 ... 22
         22 21 20 19 18 ... 1

d. 60   1 2 3 4 5 ... 29
         29 28 27 26 25 ... 1

Challenge: Predict the number of combinations for any given perimeter P. $\frac{P}{2} - 1$, P must be even.

Technology • SECTION 5-4   95

Name _____ Date _____

Calculator Activity: Circumference

Scientific calculators have a $\boxed{\pi}$ key which gives a more precise
calculation of pi than 3.14 or $\frac{22}{7}$.

▶ **Example 1**

Find the circumference of a circle with radius 10.4 cm. Round to the nearest hundredth.

Solution

Use the formula for circumference.
$C = 2\pi r$
$C = 2 \times \pi \times 10.4$
Then use a calculator: $2 \boxed{\times} \boxed{\pi} \boxed{\times} 10.4 \boxed{=} 65.34512719$
Rounded to the nearest hundredth, the circumference is about 65.35 cm.

▶ **Example 2**

Find the diameter of a circle with circumference 49.32 in. Round to the nearest hundredth.

Solution

Use the formula for circumference.
$C \approx \pi d$
$49.32 \approx \pi d$
$\frac{49.32}{\pi} \approx \frac{\pi d}{\pi}$
$\frac{49.32}{\pi} \approx d$
Then use a calculator: $49.32 \boxed{\div} \boxed{\pi} \boxed{=} 15.69904359$
Rounded to the nearest hundredth, the diameter is about 15.70 in.

EXERCISES

Find the circumference to the nearest hundredth. HINT: Use the fraction key for
Exercises 7 and 8.

1. $d = 24$ cm	2. $d = 11.5$ in.	3. $d = 125$ yd	4. $d = 52.5$ ft
75.40 cm	**36.13 in.**	**392.70 yd**	**164.93 ft**

5. $r = 6.3$ cm	6. $r = 18.7$ ft	7. $r = 2\frac{1}{2}$ in.	8. $r = 8\frac{3}{4}$ yd
39.60 cm	**117.50 ft**	**15.71 in.**	**54.98 yd**

Find the diameter or radius to the nearest hundredth. HINT: Use the $\boxed{(}$ and $\boxed{)}$ keys
for Exercise 10.

9. $C \approx 83$ m; find the diameter.	10. $C \approx 75$ in.; find the radius.
26.42 m	**11.94 in.**

96   Technology • SECTION 5-6

182D

ACHIEVEMENT TEST

Measurement and Geometry

CHAPTER 5 FORM A

MATH MATTERS BOOK 1

Chicha Lynch
Eugene Olmstead

SOUTH-WESTERN PUBLISHING CO.

Name _____

Date _____

SCORING RECORD	
Possible	Earned
25	

Complete.

1. 36 in. = **3** ft
2. 3 gal = **12** qt
3. 7 lb = **112** oz
4. 87 ft = **29** yd
5. 58 m = **5,800** cm
6. 3L = **3,000** mL
7. 872 mg = **0.872** g
8. 2 km = **2,000** m

Write each answer in simplest form.

9. 3 lb 9 oz
   +8 lb 8 oz
   12 lb 1 oz

10. 4 gal
    −2 gal 2 qt
    1 gal 2 qt

11. 2 c 7 fl oz
    ×5
    14 c 3 fl oz

12. 6 ft 10 in. ÷ 2
    3 ft 5 in.

Complete.

13. 180 mm + 500 mm = **68** cm
14. 3 km − 150 m = **2.85** km
15. 820 cm × 2 = **16,400** mm
16. 8.4 m ÷ 2 = **420** cm

17. Will one 72-inch roll of ribbon be enough to make 4 strips of ribbon 1 ft 5 in. long?
    yes

Find the perimeter of each figure.

18.

12 m

19.

24 in.

Copyright © 1993 by South-Western Publishing Co.
M101AG

5A-1

Name _____ Date _____

Use the formula to find the perimeter of a rectangular figure with the given dimensions.

20. length: 82.48 m
    width: 32.42 m
    229.8 m

21. length: 473.3 yd
    width: 203.4 yd
    1,353.4 yd

Find the circumference. Use 3.14 or $\frac{22}{7}$ for π, as appropriate.

22. (74 mm)
    232.36 mm

23. (14.5 ft)
    91.06 ft

24. (38 in.)
    238.64 in.

Draw a picture to help you solve.

25. A triangular garden has 12 posts on each side. How many posts are there in all?
    33 posts

5A-2

ACHIEVEMENT TEST

Measurement and Geometry

CHAPTER 5 FORM B

MATH MATTERS BOOK 1

Chicha Lynch
Eugene Olmstead

SOUTH-WESTERN PUBLISHING CO.

Name _____

Date _____

SCORING RECORD	
Possible	Earned
25	

Complete.

1. 84 in. = **7** ft
2. 11 gal = **44** qt
3. 8 lb = **128** oz
4. 57 ft = **19** yd
5. 342 mm = **0.342** m
6. 6L = **6,000** mL
7. 483 mg = **0.483** g
8. 3 km = **3,000** m

Write each answer in simplest form.

9. 7 c 5 fl oz
   +1 c 12 fl oz
   10 c 1 fl oz

10. 6 lb
    −3 lb 6 oz
    2 lbs 10 oz

11. 2 ft 5 in.
    ×3
    7 ft 3 in.

12. 15 gal 2 qt ÷ 2
    7 gal 3 qt

Complete.

13. 460 mm + 340 mm = **80** cm
14. 14 m − 800 cm = **6** m
15. 2.74 cm × 3 = **82.2** mm
16. 560 m ÷ 8 = **7,000** cm

17. Will one 146-inch ball of twine be enough to make 8 shorter pieces each $1\frac{3}{4}$ ft long?
    no

Find the perimeter of each figure.

18.

48 yd

19.

58 in.

Copyright © 1993 by South-Western Publishing Co.
M101AG

5B-1

Name _____ Date _____

Use the formula to find the perimeter of a rectangular figure with the given dimensions.

20. length: 8.24 m
    width: 6.3 m
    29.08 m

21. length: 37.5 yd
    width: 18.4 yd
    111.8 yd

Find the circumference. Use 3.14 or $\frac{22}{7}$ for π, as appropriate.

22. (56 mm)
    175.84 mm

23. (5.8 ft)
    36.42 ft

24. (18 cm)
    113.04 cm

Draw a picture to help you solve.

25. A rectangular garden is enclosed by a fence that has 12 posts along each of its longer sides and 10 posts along each of its shorter sides. How many posts are there in all?
    40 posts

5B-2

Teacher's Notes

CHAPTER 6 SKILLS PREVIEW

Find each absolute value.

1. $|3|$ 3
2. $|-8|$ 8
3. $|-21|$ 21
4. $|6|$ 6

Replace each ● with $<$, $>$, or $=$.

5. -16 ● 16 $<$
6. -14 ● -12 $<$
7. 18 ● -21 $>$
8. 0 ● -1 $>$
9. -10 ● -1 $<$
10. -16 ● 5 $<$

Add.

11. $-8 + 4$ -4
12. $8 + (-12)$ -4
13. $-6 + (-8)$ -14
14. $5 + (-43)$ -38
15. $-7 + (-7)$ -14
16. $12 + (-12)$ 0

Subtract.

17. $7 - (-8)$ 15
18. $-6 - 4$ -10
19. $-5 - (-9)$ 4
20. $-4 - 3$ -7
21. $9 - 12$ -3
22. $4 - (-8)$ 12

Multiply.

23. 9×6 54
24. -6×8 -48
25. $-4 \times (-5)$ 20
26. $7 \times (-8)$ -56
27. $-3 \times (-4)$ 12
28. -5×9 -45

Divide.

29. $16 \div 4$ 4
30. $-24 \div 3$ -8
31. $-56 \div (-8)$ 7
32. $63 \div (-7)$ -9
33. $-28 \div (-4)$ 7
34. $-12 \div 4$ -3

Write each rational number as a ratio of two integers.

35. 3 $\frac{12}{4}$
36. 0.9 $\frac{9}{10}$
37. $-3\frac{5}{6}$ $-\frac{23}{6}$
38. 2.6 $\frac{26}{10}$
39. -5.2 $-\frac{52}{10}$
40. $4\frac{7}{8}$ $\frac{39}{8}$
41. 0.17 $\frac{17}{100}$
42. 1 $\frac{8}{8}$

Solve. **Answers may vary.**

43. A number is three less than negative eight. What is the number? -11

44. A number is nine more than negative twenty-two. What is the number? -13

45. Six teams will play in a tournament. Each team will play every other team once. How many games will be played? **15 games**

6 SKILLS PREVIEW

Introduction The purpose of this Skills Preview is to assess students' abilities on all the major objectives of Chapter 6. Test results may be used
- to determine those topics which may need only to be reviewed and those topics which need to be more carefully developed;
- for class placement;
- in prescribing for individual differences.

If you prefer, you may use the Skills Preview as an alternative form of the Chapter Test (page 216) to evaluate mastery of chapter objectives. The items on the Skills Preview and the Chapter Test correspond in content and level of difficulty.

CHAPTER

6

INTEGERS

OVERVIEW

In this chapter, students are introduced to the concept of negative integers. Everyday uses of integers and operations with integers are explored. The concepts of common fractions and decimal fractions are extended to rational numbers. Students write rational numbers as the ratio of two integers. They use the number line to compare and order rational numbers. Problem solving sections focus on choosing the operation and choosing an appropriate strategy to solve problems.

SPECIAL CONCERNS

It is important for students to understand what negative integers represent and not just to memorize the rules. Therefore, emphasis should be placed on the everyday uses of negative integers.

The concept of absolute value is used when performing operations with integers and rational numbers. Be sure that students learn to distinguish between the absolute value of a number and the number itself.

VOCABULARY

absolute value
addition property of opposites
closed
integer

negative integers
opposites
positive integers
rational numbers

MATERIALS

integer chips or two-color counters
calculators

BULLETIN BOARD

Divide the bulletin board into four sections labeled *Addition, Subtraction, Multiplication,* and *Division.* Below each heading, list the rules for performing the operation with integers. Provide index cards with basic, average, and enriching exercises and word problems involving each operation. Place these cards in envelopes labeled *Easy, Average,* and *Difficult.* Have students take a card from an appropriate level, solve the problem, and then tack the card under the appropriate section on the bulletin board.

INTEGRATED UNIT 2

The skills and concepts involved in Chapters 5–8 are included within the special Integrated Unit 2 entitled "United States and World Travel." This unit is in the Teacher's Edition beginning on page 298F. Worksheets for this integrated unit appear in the Enrichment Activities booklet, pages 93–94.

TECHNOLOGY CONNECTIONS

- Calculator Worksheet, 113
- Computer Worksheet, 114
- MicroExam, Apple Version
- MicroExam, IBM Version

- *A.I., An Experience with Artificial Intelligence,* Scholastic
- *Teasers by Tobbs with Integers,* Sunburst Communications

TECHNOLOGY NOTES

Computer software programs, such as those listed in Technology Connections, build critical-thinking skills and problem solving ability by leading students to understand how computers evaluate multiple options at great speeds and by confronting students with exercises having missing integers, challenging them to choose the options needed to complete the exercises.

6 INTEGERS

PLANNING GUIDE

SECTIONS	TEXT PAGES	ASSIGNMENTS		
		BASIC	AVERAGE	ENRICHED
Chapter Opener/Decision Making	184–185			
6-1 Exploring Integers	186–189	1–13, 17, 19–22, 26, PSA 1–12	1–18, 19–22, 26–29, PSA 1–15	11–18, 23–25, 26–30, PSA 5–17
6-2 Adding Integers	190–193	1–13, 14–16, 22	1–13, 14–21, 22–23	9–13, 14–21, 22–25
6-3 Subtracting Integers	194–197	1–14, 15–17, 21–24, 30–32	5–14, 15–29, 30–41, 45	8–14, 21–29, 30–46
6-4 Multiplying Integers	198–201	1–12, 14–16, 20, 24, 26–27	5–9, 10–25, 26–33	10–36
6-5 Dividing Integers	202–205	1–17, 19–22, 27–39, 42–54	1–18, 19–40, 42–58	13–18, 23–41, 48–59
6-6 Problem Solving Skills: Choose the Operation	206–207	1–18	1–18	1–18
6-7 Rational Numbers	208–211	1–14, 18–31, 36–37	1–17, 18–35, 36–39	9–17, 18–35, 36–41
6-8 Problem Solving/Decision Making: Choose a Strategy	212–213	1–10	1–10	1–10
Technology	187, 188, 191, 201, 212, 213	✔	✔	✔

ASSESSMENT				
Skills Preview	183	All	All	All
Chapter Review	214–215	All	All	All
Chapter Test	216	All	All	All
Cumulative Review	217	All	All	All
Cumulative Test	218	All	All	All

	ADDITIONAL RESOURCES			
	RETEACHING	**ENRICHMENT**	**TECHNOLOGY**	**TRANSPARENCY**
	6–1	6–1		TM 23–25
	6–2	6–2	6–2	TM 26
	6–3	6–3		TM 27
	6–4	6–4		TM 28
	6–5	6–5	6–5	TM 29
	6–6	6–6		TM 23
	6–7	6–7		
	6–8	6–8		

ASSESSMENT OPTIONS

Chapter 6, Test Forms A and B	
Chapter 6, Test	Text, 216
Alternative Assessment	TAE, 216
Chapter 6, MicroExam	

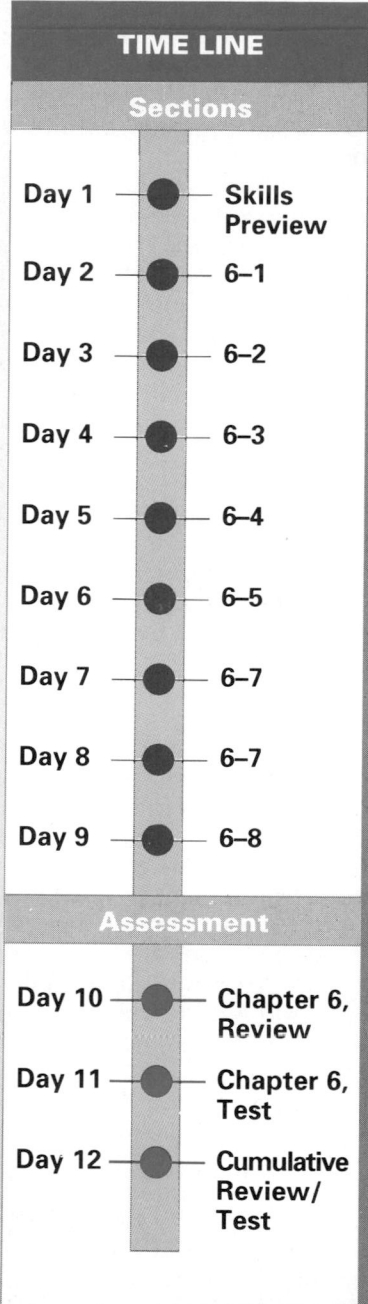

TIME LINE

Sections

Day 1 — Skills Preview
Day 2 — 6–1
Day 3 — 6–2
Day 4 — 6–3
Day 5 — 6–4
Day 6 — 6–5
Day 7 — 6–7
Day 8 — 6–7
Day 9 — 6–8

Assessment

Day 10 — Chapter 6, Review
Day 11 — Chapter 6, Test
Day 12 — Cumulative Review/ Test

CHAPTER 6 EXPLORING INTEGERS

THEME Earth Science

Until now, the only numbers you have worked with in this textbook have been positive numbers, numbers greater than zero. The earth science topics in this chapter introduce you to negative numbers, numbers with values less than zero. You can use negative numbers to express measurements such as depths below sea level and temperatures less than zero degrees.

The earth is shaped more or less like a ball. The inside of the ball is made up of layers that differ in composition and in physical properties, such as density and temperature. The very center of the earth is a solid core of iron and nickel.

184   CHAPTER 6   Exploring Integers

184

LAYERS OF THE EARTH		
Layer	**Approximate Thickness (km)**	
Core: Inner	1,300	
Outer	2,250	
Mantle: Lower	2,650	
Upper	180	
Crust	10–40	

DECISION MAKING

Using Data

Use the information in the table to find the answers to the following questions. Give your answers as negative numbers.

1. We live on the surface of the crust. If the surface is at a depth of 0 km, at what depth does the lower mantle begin? (Give your answer as a range of numbers.) –190 km to –220 km

2. What is the greatest depth at which the bottom of the lower mantle lies? –2,870 km

3. At about what depth is the center of the earth? (Round your answer to the nearest hundred kilometers.) –6,400 km

Working Together

Make a model of the earth's interior based on the data in the table. Your model should be a cross section of the earth. Be sure to label each layer.

Use earth science books or encyclopedias to find out the temperatures at the core and at each of the other layers of the earth. Compare the temperatures. Compare them to temperatures you have experienced.

The earth is enclosed by the atmosphere, which begins at the surface of the earth and rises far out into space. Like the interior of the earth, the atmosphere has a layered structure. If you were a space scientist responsible both for launching space vehicles and for the safety of humans aboard, what would you want to know about the structure and composition of the atmosphere? How could a model like the one of the earth's interior help you?

Answers will vary. Sample answer: A space scientist might want to know the thickness, composition, and temperature of the different layers and how these factors would affect the path of a space vehicle's travel and the functioning of the space vehicle; a model helps to make the structure of the atmosphere clear and is useful when you need to recall important facts, such as the location of boundaries between layers.

OBJECTIVES
- Use integers to represent points on a number line
- Give the absolute value of an integer
- Compare and order integers

VOCABULARY
absolute value, integer, negative integers, opposites, positive integers

WARM-UP

Compare.

1. 54 ● 52 >
2. 1,003 ● 1,000 >
3. 2.4 ● 2.45 <
4. 0.003 ● 0.001 >
5. 93.04 ● 93.40 <

1 MOTIVATE

Explore Allow approximately five minutes for students to complete this activity independently. List their pairs of antonyms on the board. Suggest that, just as words can have opposites, numbers can have opposites.

2 TEACH

Use the Pages/Skills Development Have students read this part of the section and then discuss the examples.

Examples 1 and 2: Emphasize the idea that the positive and negative numbers, such as +4 and –4, are *opposites* that have the same *absolute value,* 4. Students are often confused by absolute-value notation. Point out that |–6| is read "the absolute value of negative six." Thus, |–6| = 6 is read, "the absolute value of negative six equals six."

186

6-1 Exploring Integers

EXPLORE

For most English words that indicate direction, there are antonyms, or words that mean the opposite. *Above* and *below* and *forward* and *backward* are examples of antonyms.

Make a list of antonyms for these words: *in, over, east, north,* and *up.*

Add other related direction words to your list.

SKILLS DEVELOPMENT

An **integer** is any number in the following set.
$$\{\ldots, -3, -2, -1, 0, 1, 2, 3, \ldots\}$$

Integers that are greater than zero are **positive integers.** Integers that are less than zero are **negative integers.** Zero is neither positive nor negative. A positive number can be written with a positive sign (+), but it does not need to have a sign. A negative number must be written with a negative sign (−).

Integers can be shown as points on a number line. On a horizontal number line, positive numbers are to the right of 0; negative numbers are to the left of 0.

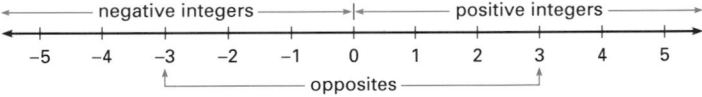

Two numbers that are the same distance from 0, but in opposite directions, are **opposites.** The opposite of a number n is written $-n$.

The distance a number is from zero on a number line is the **absolute value** of the number. The integer 3 is 3 units from 0. The integer -3 is also 3 units from 0. So, opposite integers have the same absolute value. The absolute value of 3 is written $|3|$.

TALK IT OVER

Integers can be used to represent real-life situations. For example, in a football game, 40-yd gain can be represented by +40 and a 40-yd loss by −40.

Discuss other situations that could be represented by positive and negative integers.

Example 1

Write the integer for each lettered point. Then give its opposite.

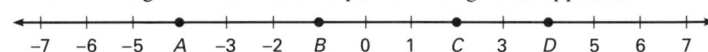

186 CHAPTER 6 Exploring Integers

TEACHING TIP

To avoid confusion with the < and > symbols, tell students to think of the symbol as an arrow which always points to the smaller number.

Solution

a. Point A represents −4. Its opposite is 4.
b. Point B represents −1. Its opposite is 1.
c. Point C represents 2. Its opposite is −2.
d. Point D represents 4. Its opposite is −4. ◄

Example 2

Find each absolute value.   a. |−7| b. |7| c. |0|

Solution

a. Since −7 is 7 units from zero,
   the absolute value of −7 is 7. |−7| = 7
b. Since 7 is 7 units from zero,
   the absolute of 7 is 7. |7| = 7
c. The absolute value of 0 is 0. |0| = 0 ◄

COMPUTER TIP

The ABS function will give you the absolute value of any number, N. Just type: ? ABS(N) and press ENTER.

If integers are graphed on a horizontal number line, the number to the right is the greater number.

Example 3

Replace ● with <, >, or =.
a. 3 ● 2   b. −3 ● −4   c. −4 ● 3   d. 2 ● −3

Solution

Use a number line to help you compare integers.

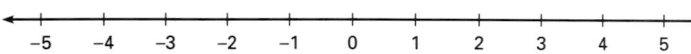

−5 −4 −3 −2 −1 0 1 2 3 4 5

a. Since 3 is to the right of 2, 3 > 2.
b. Since −3 is to the right of −4, −3 > −4.
c. Since −4 is to the left of 3, −4 < 3.
d. Since 2 is to the right of −3, 2 > −3. ◄

TRY THESE

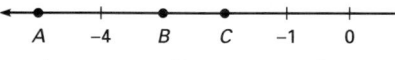

**MATH:
WHO, WHERE, WHEN**

The symbols we use for positive and negative first appeared in print about 500 years ago. The positive sign, +, may have been an abbreviated form of the Latin word *et*, meaning "and." The negative sign, −, probably came from the mark called a macron that was over the Latin word *minus*, meaning "less."

Write the integer for each lettered point. Then give its opposite.

A −4 B C −1 0 D 2 3 E 5

1. *A*   2. *B*   3. *C*   4. *D*   5. *E*
−5 and 5   −3 and 3   −2 and 2   1 and −1   4 and −4

Find each absolute value.

6. |−8| 7. |3| 8. |−45| 9. |−13| 10. |72|
   8   3   45   13   72

Replace each ● with <, >, or =.

11. 0 ● 5 <   12. −7 ● −9 >   13. −4 ● 2 <

Example 3: Comparing two negative numbers can be confusing to students. They may feel that −7 is greater than −6 because 7 > 6. Stress the importance of using a number line when in doubt about comparing integers.

Additional Questions/Examples
Name an integer for each situation.
1. Jim withdrew $50 from his savings account. **−50**
2. April earned $12 babysitting. **12**
Name the opposite of each integer.
3. −21 **21**   4. +15 **−15**
5. −1 **1**
Name each absolute value.
6. |−5| **5**   7. |23| **23**
Order from least to greatest.
8. −3, 8, 0, 3, −2 **−3, −2, 0, 3, 8**
9. 15, −12, −10, 7, 0 **−12, −10, 0, 7, 15**

Guided Practice/Try These Have students work in small groups. Group members should come to an agreement on each answer. For each of Exercises 6–10 have students name another integer with the same absolute value. **8, −3, 45, 13, −72**

AT-RISK STUDENTS

Distribute blank number lines. Have students label the zero point in the middle of one line. Have them locate the two points that are eight units from zero. **8, −8** Ask them to explain ways to distinguish these points. **left and right of zero, greater than or less than zero, negative and positive**

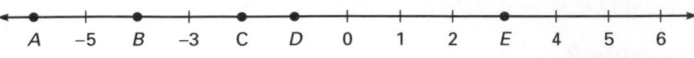
PRACTICE/ SOLVE PROBLEMS

COMPUTER TIP

This program will let you check Exercise 26. You can use your knowledge of conditionals in logic to change the program to check Exercises 27–30.

```
10 INPUT "ENTER 2
   INTEGERS: ";A ,B: PRINT
20 IF A > B AND ABS (A) >
   ABS (B) THEN PRINT
   "TRUE": END
30 PRINT "FALSE"
```

Write the integer for each lettered point. Then give its opposite.

A −5 B −3 C D 0 1 2 E 4 5 6

1. A −6 and 6
2. B −4 and 4
3. C −2 and 2
4. D −1 and 1
5. E 3 and −3

Find each absolute value.

6. $|-2|$ **2**   7. $|-34|$ **34**   8. $|0|$ **0**   9. $|17|$ **17**   10. $|-20|$ **20**

Replace each ● with $<$, $>$, or $=$.

11. $0 ● -2$ **>**   12. $-4 ● -10$ **>**   13. $-2 ● 2$ **<**
14. $-8 ● -3$ **<**   15. $-9 ● -8$ **<**   16. $6 ● -7$ **>**

Write an integer to represent the quantity in each situation.

17. The top of Little Bear, a peak in Colorado, is 4,278 m above sea level. **4,278**

18. The lowest point on the earth's surface is 10,859 below sea level. **−10,859**

EXTEND/ SOLVE PROBLEMS

Draw a number line. At the zero point, write "Today." Label a point to represent each of the following days on the number line. Write an integer to represent each day. **See Additional Answers.**

19. X, yesterday
20. Y, a week from today
21. H, the day after tomorrow
22. Q, the day before yesterday

Write the integers in order from least to greatest.

23. $-9, -13, 14, 0, 7$
    −13, −9, 0, 7, 14
24. $6, -5, 5, 7, -9$
    −9, −5, 5, 6, 7
25. $-8, 14, 8, 9, -15$
    −15, −8, 8, 9, 14

THINK CRITICALLY/ SOLVE PROBLEMS

Decide whether each statement is *true* or *false*. Give an example to support your choice. **See Additional Answers.**

26. For all integers a and b, if $a > b$, then $|a| > |b|$.

27. For all integers a and b, if $|a| = |b|$, then $a = b$.

28. For all negative integers a and b, if $a > b$, then $|a| > |b|$.

29. For all positive integers a and b, if $|a| > |b|$, then $a > b$.

30. If $a > b$ and $b > c$, decide whether each statement is *true* or *false*.
    a. $a > b > c$ b. $b > c > a$ c. $c > a > b$ d. $a > c$

Problem Solving Applications:
COMPARING TEMPERATURES

► READ
► PLAN
► SOLVE
► ANSWER
► CHECK

Some thermometers measure temperature in degrees Celsius (°C); others measure temperature in degrees Fahrenheit (°F).

Write a suitable temperature in degrees Celsius for each activity. **Answers will vary.**

1. shoveling snow **−10°C to −1°C**
2. picnicking in park **25°C to 30°C**
3. complaining about heat wave **35°C to 40°C**
4. wearing jacket outdoors **1°C to 10°C**

Choose the more reasonable temperature for each situation.

5. boiling water: 100°C or 100°F **100°C**
6. room temperature: 76°C or 76°F **76°F**
7. swimming in July: 85°C or 85°F **85°F**
8. removing frost from windshield: 15°C or 15°F **15°F**

Which temperature is warmer?

9. 0°C or −32°F **0°C**
10. 92°C or 180°F **92°C**
11. 32°C or 198°F **198°F**
12. 100°C or 198°F **100°C**
13. 20°C or 80°F **80°F**
14. −10°C or 10°F **−10°C**

15. The highest temperature ever recorded in the United States was 134°F in Death Valley, California. Do you think the integer used to record the equivalent temperature in degrees Celsius is greater or less than 134? Explain. **Less. By comparing the thermometer scales, one can see that the equivalent temperature, in degrees Celsius, is about 56°C.**

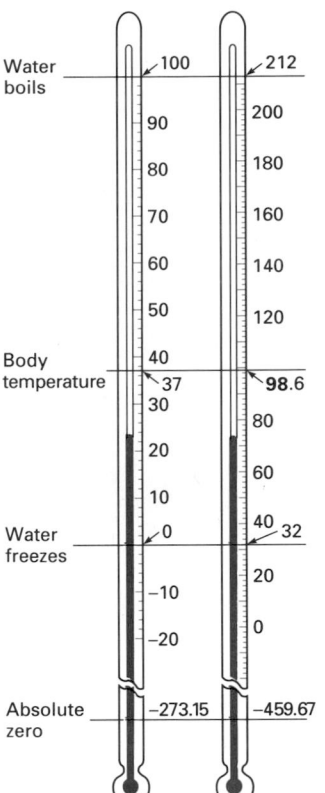

CELSIUS FAHRENHEIT

	Celsius	Fahrenheit
Water boils	100	212
	90	200
	80	180
	70	160
	60	140
	50	120
	40	
Body temperature	37	98.6
	30	80
	20	60
	10	40
Water freezes	0	32
	−10	20
	−20	0
Absolute zero	−273.15	−459.67

Use the table for Exercises 16 and 17.

16. Write the cities in order of their temperatures, from lowest to highest. **See below.**
17. Use an almanac. Find the *highest* recorded temperatures for these cities and order them from lowest to highest. **Honolulu, Key West, (tie between Hartford and Boston), Madison, (tie between Atlanta and Helena)**

16. **Helena, Madison, Hartford, Boston, Atlanta, Key West, Honolulu**

RECORD LOW TEMPERATURES		
		High
Hartford, CT	−26°F	102°F
Key West, FL	41°F	95°F
Atlanta, GA	−8°F	105°F
Honolulu, HI	53°F	94°F
Boston, MA	−12°F	102°F
Helena, MT	−42°F	105°F
Madison, WI	−37°F	104°F

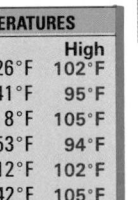

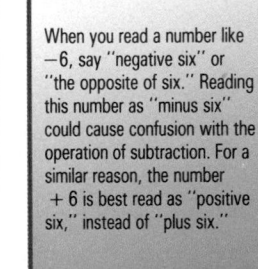

READING MATH

When you read a number like −6, say "negative six" or "the opposite of six." Reading this number as "minus six" could cause confusion with the operation of subtraction. For a similar reason, the number +6 is best read as "positive six," instead of "plus six."

6-1   Exploring Integers   **189**

---

7. an integer that is greater than −15 **Answers will vary.**
8. an integer that is neither negative nor positive **0**
9. an integer with the same absolute value as |19| **−19**

Extension Have students decide whether each statement is *true* or *false*.

1. |36| = 36 **true**
2. |−18| > 7 **true**
3. −6 < |−7| **true**
4. −24 > −15 **false**
5. |−51| = |51| **true**
6. −18 > 3 **false**
7. |78| > |−92| **false**
8. −5 < −12 **false**
9. 1 > −17 **true**
10. −45 > −55 **true**

Section Quiz Write an integer for each situation.

1. $10 more **10**
2. loss of $25 **−25**
3. 10 degrees below zero **−10**

Write the opposite of each integer.

4. −12 **12**
5. 16 **−16**
6. −45 **45**

Write the absolute value.

7. |5| **5**
8. |−7| **7**
9. |−23| **23**

Replace each ● with <, >, or =.

10. −12 ● −4 **<**
11. 26 ● 0 **>**
12. −60 ● −59 **<**
13. 0 ● −4 **>**

Get Ready calculators

Additional Answers
See page 577.

Additional answers for odd-numbered exercises are found in the Selected Answers portion of the page.

CHALLENGE

Have students describe the solutions of each sentence.
1. | a | = 5 **a = 5 or −5**
2. | t | = 0 **t = 0**
3. | r | < 4 **r = all numbers between −4 and 4**
4. | b | > 4 **b = all numbers less than −4 and all numbers greater than 4**

1 MOTIVATE

Explore/Work Together Have students work in pairs or small groups for this activity. Allow time for them to share their results.

2 TEACH

Use the Pages/Skills Development Have students read this part of the section and then discuss the examples.

Example 1: Use these exercises to show students how to add integers using a number line.

Example 2: Stress the rules for adding integers. However, you may wish to "prove" that the rules work by using a number line.

Example 3: Use these exercises to reinforce the *addition property of opposites*.

Example 4: Remind students to look for combinations of addends that will help them find the sum.

190

6-2 Adding Integers

EXPLORE/ WORKING TOGETHER

YARDAGE—TEAM A		
Play	Gain (Yd)	Loss (Yd)
1	5	
2	5	
3	7	
4		– 6
5	10	
6	12	
7		– 3
8	8	

In a football game, Team A gained control of the ball on the 50-yd line. Team A's gains and losses of yardage in the next eight plays are shown in the table. At the end of the eight plays, how many yards did Team A still have to cover to reach Team B's end zone and score a touchdown? **12 yd**

Show on a diagram of a football field how Team A's position changed with each of the eight plays.

Discuss with your group how you might use a number line instead of a diagram of a football field to show how Team A's position changed with the eight plays. Hint: Let the 0 on the number line represent the field's 50-yd line.

SKILLS DEVELOPMENT

You can use a number line to show the addition of two integers. Start at 0. Move to the right to add a positive integer. Move to the left to add a negative integer.

Example 1

Use a number line to help you add.
a. $4 + 3$ **b.** $-2 + (-4)$ **c.** $5 + (-4)$ **d.** $-4 + 2$

Solution

a.
```
      +4      +3
  ‹─┼─┼─┼─┼─┼─┼─┼─┼─┼─›
   -1 0 1 2 3 4 5 6 7 8 9
```

b.
```
          -4      -2
  ‹─┼─┼─┼─┼─┼─┼─┼─┼─┼─›
   -7 -6 -5 -4 -3 -2 -1 0 1 2
```

Begin at 0. Move right 4 places, then 3 places more. The stopping place is +7.
$$4 + 3 = 7$$

Begin at 0. Move left 2 places, then 4 places more. The stopping place is –6.
$$-2 + (-4) = -6$$

TEACHING TIP

Remind students that, when they add two positive or negative integers, they can determine the sign of the sum before doing the addition. They should record the sign first so as not to forget it in their answer. When adding integers with unlike signs, they should first record the sign of the number with the greater absolute value before subtracting.

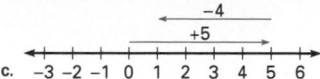

c.
Begin at 0. Move right 5 places. Then move left 4 places. The stopping place is +1.

$$5 + (-4) = 1$$

d.
Begin at 0. Move left 4 places. Then move right 2 places. The stopping place is −2.

$$-4 + 2 = -2 \quad \blacktriangleleft$$

You can also add two integers by using some simple rules. The rule that you use depends on whether the integers have the *same* sign or *different* signs. Study these rules to learn how to add integers correctly.

▶ To add integers with the *same* signs, add the absolute values of the integers. Give the sum of the two integers the sign of the addends.

▶ To add integers with *different* signs, subtract the absolute values. Give the sum the sign of the addend with the greater absolute value.

Example 2

Add.

a. $-7 + (-3)$ **b.** $-8 + 5$ **c.** $9 + (-5)$

Solution

a. Find the absolute values: $|-7| = 7; |-3| = 3$
Since the signs are the same, add the absolute values: $7 + 3 = 10$
The addends are both negative. So, the sum is negative:
$-7 + (-3) = -10$

b. Find the absolute values: $|-8| = 8; |5| = 5$
Since the signs are different, subtract the absolute values: $8 - 5 = 3$
The negative addend has the greater absolute value. So, the sum is negative: $-8 + 5 = -3$

c. Find the absolute values: $|9| = 9; |-5| = 5$
Since the signs are different, subtract the absolute values: $9 - 5 = 4$
The positive addend has the greater absolute value. So, the sum is positive: $9 + (-5) = 4$ ◀

When opposites are added, the sum is always 0. This is called the **addition property of opposites.**

$$a + (-a) = 0 \qquad -a + a = 0$$

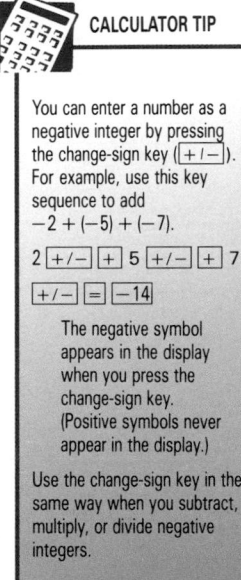

CALCULATOR TIP

You can enter a number as a negative integer by pressing the change-sign key ($\boxed{+/-}$). For example, use this key sequence to add
$-2 + (-5) + (-7)$.

$2 \boxed{+/-} \boxed{+} 5 \boxed{+/-} \boxed{+} 7$
$\boxed{+/-} \boxed{=} \boxed{-14}$

The negative symbol appears in the display when you press the change-sign key. (Positive symbols never appear in the display.)

Use the change-sign key in the same way when you subtract, multiply, or divide negative integers.

6-2   Adding Integers   **191**

Additional Questions/Examples
Which is greater?
1. −4 + (−8) or −8 + 4
 −8 + 4
2. 5 + (−5) or 0 + (−5)
 5 + (−5)
3. −7 + 5 or 6 + (−10)
 −7 + 5
4. Give an example of how you would use the integers −8 and 5 in a problem situation.
 Answers will vary. Possible answer: The temperature fell eight degrees and then rose five degrees.

Guided Practice/Try These Observe students as they complete the exercises independently. Provide guidance as needed.

❸ SUMMARIZE

Write About Math Have students explain in their math journals how to add integers with like signs and those with unlike signs.

AT-RISK STUDENTS

Help students relate adding integers to combining positive and negative electric charges. Each positive charge cancels a negative one. To show −7 + 5, students could draw the following.

Example 3

Add.
a. $7 + (-7)$ b. $-3 + 3$

Solution
Find the absolute values. Then subtract the absolute values.
a. $|7| = 7$ b. $|-3| = 3$
   $|-7| = 7$ $|3| = 3$
   $7 - 7 = 0$ $3 - 3 = 0$

The integer 0 is neither positive nor negative.

$$7 + (-7) = 0 \qquad\qquad -3 + 3 = 0 \quad \blacktriangleleft$$

Sometimes you have to add more than two integers.

Example 4

Add.
a. $4 + 9 + 16$ b. $-3 + (-6) + (-14)$ c. $12 + (-4) + 8 + (-2)$

Solution
a. $4 + 9 + 16 = 20 + 9 = 29$ } The signs are the same. Add the absolute values. Use the sign of the addends.

b. $-3 + (-6) + (-14) = -3 + (-20) = -23$

c. $12 + (-4) + 8 + (-2)$ { The signs are different. Find the sums of integers with the same signs. Then add the sums according to the rule for adding integers with different signs.
   $= 20 + (-6) = 14$ ◄

TRY THESE

Use a number line to help you add.

1. $6 + 3$ 2. $4 + (-3)$ 3. $-7 + (-2)$ 4. $4 + 2$

Add. Use the rules for adding integers.

5. $-6 + 3$ **–3**   6. $9 + (-7)$ **2**   7. $-8 + (-5)$ **–13**   8. $-4 + (-4)$ **–8**

9. $4 + (-4)$ **0**   10. $-6 + 6$ **0**   11. $-5 + (-5)$ **–10**   12. $-9 + 9$ **0**

13. $4 + (-12) + (-8)$ **–16**   14. $-9 + 6 + 12$ **9**

15. $14 + (-9) + 3 + 4$ **12**   16. $-20 + (-4) + (-7) + 2$ **–29**

Solve.

17. A football team lost 5 yd, gained 10 yd, gained 22 yd, lost 16 yd, and then lost 15 yd. What was the net loss or net gain in yards?
    net loss: 4 yd
18. A football team gained 4 yd, lost 10 yd, lost 2 yd, gained 14 yd, and then gained 2 yd. What is the net loss or net gain?
    net gain: 8 yd

EXERCISES

Add. Use a number line or the rules for adding integers.

1. 5 + 5 **10**   2. −5 + (−4) **−9**   3. −6 + (−2) **−8**   4. 6 + (−2) **4**

5. −7 + (−3) **−10** 6. −7 + 3 **−4**   7. 3 + (−3) **0**   8. −9 + (−9) **−18**

9. 14 + 8 + 6 + (−20) **8**   10. −7 + 12 + (−7) + 7 **5**

11. 3 + (−9) + (−6) + 12 **0**   12. −4 + (−8) + 9 + (−3) **−6**

Solve.

13. The price of a share of stock at Friday's close was $45. The price rose $5 on Monday, fell $2 on Tuesday, rose $4 on Wednesday, rose $3 on Thursday, and fell $6 on Friday. What was the price of a share of stock then? **$49**

Replace ● with <, >, or =.

14. −8 + 3 ● −3 + (−2) **=**   15. 4 + (−8) ● −6 + ▷

16. 7 + 2 ● −1 + 10 **=**   17. −6 + 3 ● −5 + (−1) **>**

18. −7 + 8 ● 10 + (−8) **<**   19. 8 + (−6) ● −4 + 10 **<**

On Monday, Jessica had $264 in her checking account. On Tuesday, she wrote checks for $69 and $92. On Thursday, she deposited $21.

20. Use paper and pencil to show how to find the balance in the account. **See Additional Answers.**

21. Use a calculator to find the balance in the account. **Check students' work.**

Find the pair of integers.

22. Two negative integers are 6 units apart. Their sum is −8. **−1 and −7**

23. The sum of a positive integer and a negative integer is 2. They are 14 units apart. **8 and −6**

24. A positive integer and a negative integer are 16 units apart. The absolute value of the positive integer is 3 times the absolute value of the negative integer. **12 and −4**

25. A positive integer and a negative integer are 48 units apart. The absolute value of the negative integer is half the absolute value of the positive integer. **32 and −16**

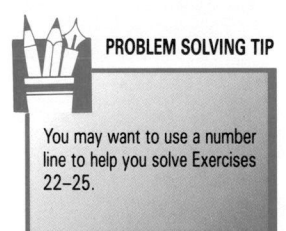

PROBLEM SOLVING TIP

You may want to use a number line to help you solve Exercises 22–25.

6-2   Adding Integers   **193**

Section Quiz Add.

1. −6 + 4 **−2**   2. 6 + 3 **9**
3. 8 + (−5) **3**   4. −5 + (−8) **−13**
5. 7 + (−7) **0**   6. 0 + (−5) **−5**
7. 10 + (−2) **8**   8. −10 + 4 **−6**
9. 6 + (−3) + (−2) + 5 + (−8) **−2**
10. −7 + (−4) + 3 + 5 + (−6) **−9**
11. In a dart game, Kia gained 7 points, lost 5 points, and lost 3 points. Nama lost 10 points, gained 9 points, and lost 4 points. Who won the game? **Kia**

Get Ready integer chips or two-color counters

Additional Answers
See page 578.

CHALLENGE

Students will have to add in order to find the path through the maze that gives the sum −16.

start
−5	4	−9	2	
	−6	3	0	
		6	8	−5	−16	end

−5 + 4 + (−6) + 3 + (−9) + 2 + 0 + (−5) = −16

WARM-UP

Ed entered the elevator on the twelfth floor. He went down 6 floors, then up 4, then down 8, and up 3 floors. On which floor was he then? **fifth**

1 MOTIVATE

Explore/Work Together Provide each pair of students with 20 positive integer chips and 20 negative integer chips. Observe them as they work through parts a–d, providing guidance as needed. Then, take students through the modeling of the two subtraction examples before allowing them to complete part e independently.

2 TEACH

Use the Pages/Skills Development Have students read this part of the section and then discuss the examples.

Example 1: After students model the examples, write on the board the number sentences for each of the subtraction problems completed thus far:

6-3 Subtracting Integers

EXPLORE/ WORKING TOGETHER

You can use integer chips to represent integers. A $\boxed{+}$ chip represents $+1$. A $\boxed{-}$ chip represents -1.

$$\boxed{+}\ \boxed{+}\ \boxed{+} = 3 \text{ and } \boxed{-}\ \boxed{-}\ \boxed{-} = -3$$

a. Use integer chips to model these integers.

$$6 \qquad 4 \qquad -5 \qquad -7$$

You can use either $\boxed{+}$ or $\boxed{-}$ integer chips to model addition of integers with the same signs. For example, use $\boxed{-}$ chips to model $-3 + (-5)$.

$$\begin{array}{ccc} -3 & + & -5 \\ \boxed{-}\ \boxed{-}\ \boxed{-} & & \boxed{-}\ \boxed{-}\ \boxed{-}\ \boxed{-}\ \boxed{-} \end{array}$$

$$\boxed{-}\ \boxed{-}\ \boxed{-}\ \boxed{-}\ \boxed{-}\ \boxed{-}\ \boxed{-}\ \boxed{-}$$

$$-3 + (-5) = -8$$

b. Use integer chips to model these sums.

$$-4 + (-3)\ \textbf{-7} \qquad 6 + 9\ \textbf{15} \qquad -7 + (-5)\ \textbf{-12}$$

You can use both $\boxed{+}$ and $\boxed{-}$ integer chips to model addition of integers with different signs.

c. The combination of $\boxed{+}$ and $\boxed{-}$ always equals zero. Explain why.

$$\boxed{+}\text{--}\boxed{-} = 0$$

Make combinations of zero to model addition of integers with different signs. For example, make pairs of $\boxed{+}$ and $\boxed{-}$ chips to add $3 + (-2)$.

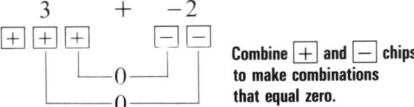

After making all possible combinations, one $\boxed{+}$ chip remains. $3 + (-2) = 1$

d. Use integer chips to model these sums. Remember that a pair of $\boxed{+}$ and $\boxed{-}$ chips equals zero.

$$-7 + 5\ \textbf{-2} \qquad 6 + (-2)\ \textbf{4} \qquad 5 + (-5)\ \textbf{0}$$

You can use integer chips to model subtraction of integers. For example, use $\boxed{-}$ chips to subtract $-5 - (-3)$. Show 5 $\boxed{-}$ chips.

Subtract 3 of the chips. $\boxed{-}\ \boxed{-}\ \boxed{\diagdown}$
$\boxed{\diagdown}\ \boxed{\diagdown}$

Two $\boxed{-}$ chips remain. $\quad -5 - (-3) = -2$

AT-RISK STUDENTS

Students having difficulty subtracting integers should write out the addition instead of trying to compute it mentally. For example:

$$4 - (-8) = 4 + 8 = 12 \qquad 6 - 12 = 6 + (-12) = -6$$

Sometimes you must add combinations of zero to get enough $+$ or $-$ chips to subtract. For example, subtract $3 - 5$.

Show 3. $\boxed{+}$ $\boxed{+}\text{---}\boxed{-}$
          $\boxed{+}$ $\boxed{+}\text{---}\boxed{-}$ Add two combinations of zero.
          $\boxed{+}$

Subtract 5 $\boxed{\not{+}}$ $\boxed{\not{+}}\text{---}\boxed{-}$
$+$ chips. $\boxed{\not{+}}$ $\boxed{\not{+}}\text{---}\boxed{-}$
             $\boxed{\not{+}}$

Two $-$ chips remain. $3 - 5 = -2$

e. Use integer chips to model these differences.

$-6 - (-4)$ **-2** $9 - 5$ **4** $4 - 5$ **-1**

Using integer chips is one way to subtract integers.

SKILLS DEVELOPMENT

Example 1

Subtract.
a. $-4 - (-3)$ b. $6 - (-2)$

Solution
a. $-4 - (-3)$
   Show -4. Then subtract $\boxed{-}\,\boxed{\not-}\,\boxed{\not-}\,\boxed{\not-}$
   3 $-$ chips. One $-$ chip remains. $-4 - (-3) = -1$

b. $6 - (-2)$
   Show 6. Then add two pairs $\boxed{+}\,\boxed{+}\,\boxed{+}$
   of $+$ and $-$ chips. $\boxed{+}\,\boxed{+}\,\boxed{+}$
   Subtract 2 $-$ chips. $\boxed{+}\,\boxed{\not-}\,\boxed{+}\,\boxed{\not-}$
   Eight $+$ chips remain. $6 - (-2) = 8$ ◄

Notice how subtraction and addition of integers are related.

$$-4 - (-3) = -1 \qquad 6 - (-2) = 8$$
$$-4 + 3 = -1 \qquad 6 + 2 = 8$$

These examples illustrate the following rule for subtracting integers.
► To subtract an integer, add its opposite.

Example 2

Subtract.
a. $-7 - (-6)$ b. $-3 - 8$

Solution
a. $-7 - (-6)$ b. $-3 - 8$
   $-7 + 6$ ← Add the opposite of -6, $-3 + (-8)$ ← Add the opposite
   -1 which is 6. -11 of 8, which is -8. ◄

BASIC
1–14, 15–17, 21–24, 30–32

AVERAGE
5–14, 15–29, 30–41, 45

ENRICHED
8–14, 21–29, 30–46

ADDITIONAL RESOURCES
Reteaching 6–3
Enrichment 6–3
Transparency Master 27

$-5 - (-3) = -2$ $3 - 5 = -2$
$-6 - (-4) = -2$ $9 - 5 = 4$
$4 - 5 = -1$ $-4 - (-3) = -1$
$6 - (-2) = 8$

Example 2: Use the example to demonstrate that subtracting integers is the same as adding their opposites.

Example 3: Students may use a number line to model their answer.

Additional Questions/Examples
Will the answer be *positive* or *negative*?
1. $-12 - (-6)$ **negative**
2. $-12 - 6$ **negative**
3. $0 - (-5)$ **positive**
4. $0 - 5$ **negative**
Give the absolute value.
5. $|7|$ **7**
6. $|-7|$ **7**
7. $|-4|$ **4**
Would you *add* or *subtract* to make each sentence true?
8. $3 \bullet -7 = -4$ **add**
9. $-4 \bullet -6 = 2$ **subtract**
10. $-8 \bullet -4 = -4$ **subtract**

Guided Practice/Try These Have students work in pairs. Point out that each pair should come to an agreement on each answer.

5-MINUTE CLINIC

Exercise	Student's Error	Error Diagnosis
Subtract: $7 - (-4)$	$7 + (-4) = 3$	• Student remembers that subtraction of integers involves addition, but forgets to add the opposite integer, 4.

Key Questions *True* or *false*? Give an example to support your answer. (Examples will vary. Samples are given.)
1. The difference between two positive integers is always positive. **false; 3 – 6 = –3**
2. The difference between two negative integers is always negative. **false; –4 – (–8) = 4**

4 **PRACTICE**

Practice/Solve Problems students review the rules for adding integers before they begin these exercises.

Extend/Solve Problems For Exercises 15–20 point out that students are not to find the differences, but are only to tell whether the answer will be positive or negative. For Exercises 25–28, students should note that they are to work with the absolute value of each number.

Think Critically/Solve Problems For Exercises 36–44 students may find it helpful to check their work by using integer chips or number lines.

5 **FOLLOW-UP**

Extra Practice
Subtract.
1. $6 - 8$ **–2**
2. $-7 - (-4)$ **–3**
3. $-9 - (-12)$ **3**
4. $-4 - 0$ **–4**
5. $0 - 6$ **–6**
6. $0 - (-8)$ **8**
7. The highest temperature ever recorded in Huron, North Dakota, was 112°F. The lowest temperature was –39°F. By how many degrees do these temperatures differ? **151°F**

196

Sometimes you need to subtract integers in order to solve problems.

Example 3
The highest recorded temperature in Colorado is 118°F. The lowest recorded temperature is $-61°$F. By how many degrees do these temperatures differ?

Solution
Subtract.
$118 - (-61)$
  $118 + 61$ ← **Add the opposite of –61, which is 61.**
   179

The highest and lowest recorded temperatures differ by 179°F. ◄

TRY THESE

Use integer chips to model these differences. **See Additional Answers.**

1. $4 - 1$	2. $-2 - 2$	3. $-1 - 2$
4. $3 - (-6)$	5. $-2 - 1$	6. $-5 - 4$

Subtract. Use the rule for subtracting integers.

7. $6 - 3$ **3**	8. $2 - 6$ **–4**	9. $-8 - (-4)$ **–4**
10. $-9 - 3$ **–12**	11. $5 - (-4)$ **9**	12. $7 - 8$ **–1**

Solve.

13. The highest recorded temperature in Texas is 120°F. The lowest is $-23°$F. By how many degrees do these temperatures differ? **143°F**

EXERCISES

PRACTICE/ SOLVE PROBLEMS

Use integer chips to model these differences. **See Additional Answers.**

1. $12 - (-6)$ **18** 2. $-7 - (-10)$ **3** 3. $4 - (-2)$ **6** 4. $15 - (-6)$ **21**

Subtract. Use the rule for subtracting integers.

5. $8 - (-3)$ **11**	6. $-17 - (-11)$ **–6**	7. $-22 - 3$ **–25**
8. $-8 - (-10)$ **2**	9. $40 - (-24)$ **64**	10. $-16 - (-6)$ **–10**
11. $-10 - 13$ **–23**	12. $17 - 21$ **–4**	13. $-12 - (-3)$ **–9**

14. *USING DATA* Use the Data Index on page 546 to locate information about the highest elevation and the lowest point on the earth. Find the distance between these two points. **30,340 ft**

Without subtracting, tell whether the answer will be positive or negative.

15. $-16 - (-4)$ **neg.**   **16.** $5 - 18$ **neg.**   **17.** $12 - (-7)$ **pos.**

18. $-38 - (-39)$ **pos.**   **19.** $-72 - 71$ **neg.**   **20.** $38 - 14$ **pos.**

Replace ● with $<$, $>$, or $=$.

21. $-3 - (-2) ● -8 - (-5)$ **>**   **22.** $-14 - (-5) ● 14 - (-5)$ **<**

23. $4 - 18 ● 12 - 5$ **<**   **24.** $5 - 9 ● 9 - 13$ **=**

25. $|-6| - |-9| ● |5| - |-10|$ **>**   **26.** $|12| - |-9| ● |-10| - |-5|$ **<**

27. $|-4| - |5| ● |9| - |-9|$ **<**   **28.** $|-8| - |-9| ● |8| - |-9|$ **=**

Solve.

29. 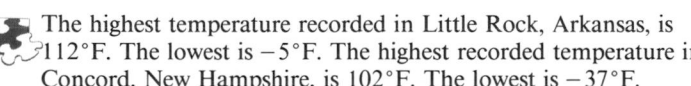 The highest recorded temperature in Louisiana is $114°F$. The high varies from the lowest recorded temperature by $130°F$. What is the lowest recorded temperature in Louisiana?

−16°F

Replace each ● with $+$ or $-$.

30. $8 ● -9 = -1$ **+**   **31.** $-8 ● -9 = 1$ **−**   **32.** $-5 ● -10 = 5$ **−**

33. $6 ● -8 = -2$ **+**   **34.** $-9 ● 4 = -5$ **+**   **35.** $-16 ● -4 = -20$ **+**

Find the integer represented by ■.

36. $7 - ■ = 2$ **5**   **37.** $-4 - ■ = 5$ **−9**   **38.** $■ - (-3) = -8$ **−11**

39. $■ - 7 = -10$ **−3**   **40.** $-9 - ■ = 0$ **−9**   **41.** $■ - 2 = 14$ **16**

42. $30 - ■ = -40$ **70**   **43.** $0 - ■ = 3$ **−3**   **44.** $-6 - ■ = -13$ **7**

The highest temperature recorded in Little Rock, Arkansas, is $112°F$. The lowest is $-5°F$. The highest recorded temperature in Concord, New Hampshire, is $102°F$. The lowest is $-37°F$.

45. Find the differences between the high and low temperatures for Little Rock and for Concord. **Little Rock: 117°F; Concord: 139°F**

46. Which of the two cities had the greater temperature difference? **Concord**

6-3 Subtracting Integers **197**

EXTEND/ SOLVE PROBLEMS

MIXED REVIEW

Find the range, mean, median and mode for each set of data.

1. 63 67 43 40 51
78 89 89
49, 65, 65, 89

2. 225 275 190 211
263 259 205 204
85, 229, 218, none

Express each as a decimal.

3. $\frac{5}{16}$ **4.** $\frac{5}{6}$ **5.** $\frac{3}{5}$
0.3125 0.8$\overline{3}$ 0.6

6. $\frac{1}{8}$ **7.** $\frac{7}{8}$ **8.** $\frac{9}{25}$
0.125 0.875 0.36

9. Find the circumference of a circle with a diameter of 10 centimeters. **31.4 cm**

THINK CRITICALLY/ SOLVE PROBLEMS

WRITING ABOUT MATH

Would your answer to Exercise 46 be the same if the temperatures were recorded in degrees Celsius instead of in degrees Fahrenheit? Write a paragraph explaining your answer. **Yes. Explanations will vary.**

Extension

1. Give a value for ■ that will make ■ + 4 negative. **Any number less than −4 will work.**

2. Give a value for ■ that will make 3 − ■ positive. **Any number less than 3 will work.**

3. Give a value for ■ that will make 5 − (■ + 4) negative. **Any number greater than 1 will work.**

Section Quiz

1. $4 - 8$ **−4**
2. $3 - 9$ **−6**
3. $12 - (-4)$ **16**
4. $-9 - 3$ **−12**
5. $8 - (-5)$ **13**
6. $-7 - 7$ **−14**
7. $5 - (-3)$ **8**
8. $-5 - (-3)$ **−2**
9. $0 - (-7)$ **7**
10. $0 - 3$ **−3**
11. Oscar parachuted from a plane that was flying at an altitude of 13,000 ft. He landed in a valley 48 ft below sea level. How far did Oscar parachute? **13,048 ft**

Get Ready integer chips or two-color counters

Additional Answers
See page 578.

CHALLENGE

Label a number line from −12 to 13. Above each point, write a letter of the alphabet, starting with *A* above −12. Have students use the integers for the letters to make equations like the following:
Example: HID → −5 + (−4) = −9
1. MAY **0 − (−12) = 12**   **2.** HOUR **−5 + 2 + 8 = 5**
3. WHEEL **10 − (−5) + (−8) + (−8) = −1**

1 MOTIVATE

Explore Guide students through the activity. Remind them that multiplication is repeated addition.

2 TEACH

Use the Pages/Skills Development Have students read this part of the section and then discuss the examples.

Example 1: Have students use integer chips to model each part of the example. Compare parts b and c. Have students explain why it would be cumbersome to model -8×4. **In order to model -8×4, you would have to start by modeling 0 by using pairs of 32 positive and 32 negative chips. Then, taking 4 positive chips away 8 times leaves 32 negative chips. So $-8 \times 4 = -32$.**

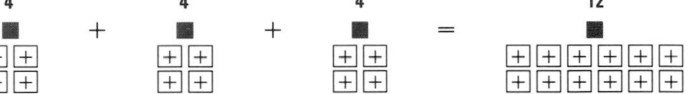

6-4 Multiplying Integers

EXPLORE You have used integer chips to model addition and subtraction of integers. Now you can use integer chips to model multiplication of integers.

Make three groups of four ⊞ chips.

Put the three groups of four ⊞ chips together to make one large group of ⊞ chips.

Write the number sentence that the modeling represents.

You know that multiplication is repeated addition. So, you can write a multiplication sentence to replace the addition sentence. First, complete this sentence.

■ groups of ■ ⊞ chips = 1 group of ■ ⊞ chips

Now use integers to complete this multiplication sentence.

$$\blacksquare \times \blacksquare = \blacksquare$$

Repeat these steps using ⊟ integer chips. Make three groups, each containing four ⊟ chips. Then, put the three groups of ⊟ chips together to make one large group of ⊟ chips. Write the number sentence that the modeling represents.

$$\blacksquare + \blacksquare + \blacksquare = \blacksquare$$

Replace the addition sentence with a multiplication sentence. Use integers to complete the sentence.

$$\blacksquare \times \blacksquare = \blacksquare$$

SKILLS DEVELOPMENT Several techniques can be used to multiply integers. In some cases, you can use integer chips. Sometimes you use the commutative property of multiplication. Sometimes you look for patterns.

Example 1

Multiply.
a. 3×4 b. $4 \times (-8)$ c. -8×4 d. $-4 \times (-3)$

Solution

a. Use integer chips.

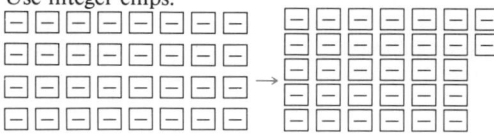

3 groups of 4 ⊞ chips = 1 group of 12 ⊞ chips
  $3 \times 4 = 12$

b. Use integer chips.

⊟⊟⊟⊟⊟⊟⊟⊟
⊟⊟⊟⊟⊟⊟⊟⊟ → ⊟⊟⊟⊟⊟⊟⊟⊟
⊟⊟⊟⊟⊟⊟⊟⊟   ⊟⊟⊟⊟⊟⊟⊟⊟
⊟⊟⊟⊟⊟⊟⊟⊟   ⊟⊟⊟⊟⊟⊟⊟⊟
                  ⊟⊟⊟⊟⊟⊟⊟⊟

4 groups of 8 ⊟ chips = 1 group of 32 ⊟ chips
  $4 \times (-8) = -32$

c. You know that $4 \times (-8) = -32$. ← **Use the commutative property.**
   So, $-8 \times 4 = -32$.

d. Use a pattern. Begin with two factors whose product you know.
   Then extend the pattern.
   $(-4)(3) = -12$ **Look for the pattern. Each product is**
   $(-4)(2) = -8$ **4 greater than the preceding product.**
   $(-4)(1) = -4$ **Extend the pattern.**
   $(-4)(0) = 0$
   $(-4)(-1) = ? \rightarrow (-4)(-1) = 4$
   $(-4)(-2) = ? \rightarrow (-4)(-2) = 8$
   $(-4)(-3) = ? \rightarrow (-4)(-3) = 12$ ◄

Notice the sign of the product in each part of Example 1. Compare
with the signs of the factors. These results illustrate the following
rules for multiplying integers.

► The product of two integers with the *same signs* is positive.
► The product of two integers with *different signs* is negative.

Example 2

Multiply.
a. 8×4 b. $-9 \times (-6)$ c. -6×5 d. $4 \times (-2)$

Solution

a. $8 \times 4 = 32$ **The product is positive because the signs of the factors are the same.**

b. $-9 \times (-6) = 54$ **The product is positive because the signs of the factors are the same.**

c. $-6 \times 5 = -30$ **The product is negative because the signs of the factors are different.**

d. $4 \times (-2) = -8$ **The product is negative because the signs of the factors are different.** ◄

5-MINUTE CLINIC

Exercise	Student's Error	Error Diagnosis
Multiply: $4 \times (-6)$	$4 \times (-6) = 24$	• Student forgets to include the sign in the product.
$-7 \times (-7)$	$-7 \times (-7) = -49$	• Student confuses the rules for multiplying integers.

ASSIGNMENTS

BASIC
1–12, 14–16, 20, 24, 26–27

AVERAGE
5–9, 10–25, 26–33

ENRICHED
10–36

ADDITIONAL RESOURCES
Reteaching 6–4
Enrichment 6–4
Transparency Master 28

List the four examples and their products on the board. Have students note the signs of the factors and products. Elicit the two rules for multiplying integers.
  Example 2: Have students complete the exercises orally, stating the rules used.
  Example 3: Use this example to review the use of the commutative property when multiplying more than two factors.
  Example 4: Have students create a similar problem of their own.

Additional Questions/Examples
1. Complete this multiplication chart.

	+	−
+	+	−
−	−	+

2. 6×7 **42**   3. $4 \times (-7)$ **−28**
4. -6×5 **−30** 5. $-3 \times (-6)$ **18**
6. 0×-2 **0**   7. -4×0 **0**

Guided Practice/Try These
Complete these exercises orally. For Exercises 1–8 students are to analyze each exercise to determine which method they could use. Have students state the multiplication sentence used to solve Exercise 12.

1. Explain the two rules for multiplying integers.
2. For what types of exercises can the commutative property be used to multiply integers? **for exercises in which there are more than two factors**

4 **PRACTICE**

Practice/Solve Problems Ask students to write the multiplication sentence they used to solve Exercise 9.

Extend/Solve Problems Remind students to carefully consider the use of parentheses or brackets in Exercises 20–23.

Think Critically/Solve Problems Remind students that, for Exercises 48–53, they must first determine the missing factor and then determine the sign of that factor. Make sure students understand Exercise 36.

5 **FOLLOW-UP**

Extra Practice
Multiply.
1. -4×1 **-4**
2. $-4 \times (-1)$ **4**
3. 7×3 **21**
4. -2×8 **-16**
5. $5 \times (-7)$ **-35**
6. $-9 \times (-6)$ **54**
7. $0 \times (-10)$ **0**
8. $6 \times (-5)$ **-30**
9. $3 \times (-4) \times (-2) \times 1$ **24**
10. $-5 \times (-1) \times (-3) \times 0$ **0**

Extension
Calculate. Remember $(-4)^2$ means $-4 \times (-4)$. The first one is done for you.
1. $(-4)^2 = -4 \times (-4) = 16$
2. $(-2)^2$ **4**
3. $(-5)^2$ **25**
4. 5^2 **25**
5. $(-3)^2$ **9**
6. $(-2)^3$ **-8**
7. $(-3)^3$ **-27**

TALK IT OVER

What is the sign of the product of any three positive integers? of any three negative integers? Explain.
positive, negative

Example 3
Multiply: $-2 \times 3 \times (-5)$

Solution
$$\begin{aligned} -2 \times 3 \times (-5) &= -2 \times (-5) \times 3 \\ &= 10 \times 3 \\ &= 30 \blacktriangleleft \end{aligned}$$

You may multiply factors in any order to make it easier to multiply mentally.

Sometimes you need to multiply integers in order to solve problems.

Example 4
The temperature has been falling at a constant rate of 2°F per hour. Describe the temperature 4 hours ago.

Solution
The rate at which the temperature has been falling can be represented by a negative number, -2. The four hours that have passed can be represented by a negative number, -4.

$$-2 \times (-4) = 8 \qquad \textbf{Multiply.}$$

Four hours ago, the temperature was 8°F higher. ◄

TRY THESE

Multiply. Decide whether to use integer chips, the commutative property, or a pattern. **See Additional Answers.**

1. $-4 \times (-5)$
2. $6 \times (-2)$
3. -2×6
4. $-7 \times (-3)$

Multiply. Use the rules for multiplying integers.

5. -7×8 **-56**
6. $-6 \times (-6)$ **36**
7. -4×9 **-36**
8. $2 \times (-9)$ **-18**
9. $-4 \times (-3) \times (-2)$ **-24**
10. $-4 \times 3 \times (-2)$ **24**
11. $4 \times 3 \times (-2)$ **-24**
12. A football team lost 7 yd on each of the last 3 plays. What was the total yardage on the plays? **-21 yd**

EXERCISES

PRACTICE/ SOLVE PROBLEMS

Multiply.

1. $4 \times (-8)$ **-32**
2. $3 \times (-7)$ **-21**
3. -8×8 **-64**
4. $-9 \times (-8)$ **72**
5. -5×8 **-40**
6. $-4 \times 5 \times (-2)$ **40**
7. $4 \times 5 \times (-2)$ **-40**
8. $-4 \times (-5) \times (-2)$ **-40**

9. The temperature has been falling at a constant rate of 5°F per hour. Describe the temperature 3 hours ago. **15°F higher**

200 CHAPTER 6 Exploring Integers

Write the multiplication sentence that describes each word phrase. Then find the product.

10. five times negative eight $5 \times (-8) = -40$

11. negative two multiplied by negative six $-6 \times (-2) = 12$

12. negative seven times four $-7 \times 4 = -28$

13. the product when three is multiplied by the sum of negative one and two $(-1 + 2) \times 3 = 3$

Replace each ● with $<$, $>$, or $=$.

14. $-4 \times (-8)$ ● 4×8 $=$

15. $2 \times (-5)$ ● -3×3 $<$

16. $-4 \times (-6)$ ● -4×6 $>$

17. $7 \times (-8)$ ● -9×6 $<$

18. -3×1 ● $3 \times (-1)$ $=$

19. $3 \times (-9)$ ● -4×7 $>$

Multiply.

20. $5 \times (-2) \times (-4) \times (-1)$ -40

21. $5 \times (-7) \times 2 \times (-3)$ 210

22. $10 \times 5 \times (-3) \times 0$ 0

23. $6 \times (-3) \times (-1) \times (-1)$ -18

24. By Friday, the price of a share of stock had fallen $1 for each day of the past work week. How did the price compare to the previous Friday's price? **It was $5 less.**

25. The temperature on Mars drops 15°C each hour at night. Use integers to show the temperature drop for 5 h.
$5 \times (-15) = -75$, **a drop of 75°C**

Find the integer represented by ■.

26. $-6 \times$ ■ $= -30$ 5

27. $-15 \times$ ■ $= 75$ -5

28. ■ $\times (-13) = -52$ 4

29. $9 \times$ ■ $= -81$ -9

30. $-24 \times$ ■ $= 144$ -6

31. $5 \times$ ■ $= -60$ -12

Give the next four integers to continue the pattern. Then state the rule. **Rules may vary. Sample rules are given.**

32. $1, -2, 4, -8, 16, -32, \ldots$ **$64, -128, 256, -512$ Multiply by -2.**

33. $-3, -6, -12, -24, -48, \ldots$ **$-96, -192, -384, -768$ Multiply by 2.**

34. $0, 1, -1, 2, -2, 3, -3, \ldots$ **$4, -4, 5, -5$ Add 1, subtract 2, add 3, subtract 4, add 5, and so on.**

35. Find two integers with a sum of -7 and a product of 12.
-3 and -4

36. Create a rule for predicting the sign of an even power and an odd power of a negative integer. **See Additional Answers.**

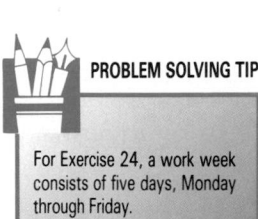

PROBLEM SOLVING TIP

For Exercise 24, a work week consists of five days, Monday through Friday.

THINK CRITICALLY/ SOLVE PROBLEMS

COMPUTER TIP

A RUN of this program will give you data from which to create a rule for Exercise 36.

```
10 FOR X = 1 TO 8
20 PRINT "(-2) ^";X, (-2) ^ X
30 NEXT X
```

8. $(-5)^3$ -125
9. $-2 \times (-5)^2$ -50
10. $3 \times (-4)^2$ 48
11. $-4 \times (-4)^3$ 256
12. $(-2)^2 \times (-4)^2$ 64
13. $(-2)^3 \times (-3)^3$ 216

Section Quiz
Complete.
1. The product of two negative integers is a ___?___ integer. **positive**
2. The product of a negative integer and a positive integer is a ___?___ integer. **negative**
3. The product of two positive integers is a ___?___ integer. **positive**
Multiply.
4. $5 \times (-3)$ **-15**
5. -6×4 **-24**
6. $-4 \times (-5)$ **20**
7. -3×0 **0**
8. $1 \times (-5)$ **-5**
9. $-8 \times (-8)$ **64**
10. $9 \, x \, (-2) \, x \, (-1)$ **18**

Additional Answers
See page 578.

CHALLENGE

Write *true* or *false*. Give an example to support your answer. **Examples will vary.**
If a is a positive integer and b is a negative integer, then,
1. $a + b < a$ and $a + b > b$ **true**
2. $a \times b < a$ and $a \times b > b$ **false**
3. $(a \times b) \times (a \times b) > 0$ **true**

6-5 Dividing Integers

EXPLORE   You can use what you know about multiplying integers to explore dividing integers.

Recall that multiplication undoes division and division undoes multiplication.

$9 \times 5 = 45$, and $45 \div 5 = 9$
$5 \times 9 = 45$, and $45 \div 9 = 5$

Complete.

a. $-9 \times 5 = -45$, and $-45 \div 5 = \blacksquare$ **−9**

b. $9 \times (-5) = -45$, and $-45 \div (-5) = \blacksquare$ **9**

c. $-9 \times (-5) = 45$, and $45 \div (-5) = \blacksquare$ **−9**

SKILLS DEVELOPMENT   Division and multiplication are related operations. The rules for finding the sign of the quotient of two integers are the same as the rules for finding the sign of the product of two integers.

► The quotient of two integers with the *same* signs is positive.

► The quotient of two integers with *different* signs is negative.

Because division and multiplication are inverse operations, you can multiply to check your answer to division.

Example 1

Find each quotient. Then check by multiplying.
a. $56 \div 7$ **b.** $-42 \div (-7)$ **c.** $48 \div (-6)$ **d.** $-72 \div 8$

Solution

a. $56 \div 7 = 8$
   Check: $8 \times 7 = 56$

The quotient is positive because the signs of the factors are the same.

b. $-42 \div (-7) = 6$
   Check: $6 \times (-7) = -42$

The quotient is positive because the signs of the factors are the same.

c. $48 \div (-6) = -8$
   Check: $-8 \times (-6) = 48$

The quotient is negative because the signs of the factors are different.

d. $-72 \div 8 = -9$
   Check: $-9 \times 8 = -72$

The quotient is negative because the signs of the factors are different. ◄

202   CHAPTER 6   Exploring Integers

TEACHING TIP

To ensure that they have the correct sign in the quotient when they divide integers, have students first write whether the quotient will be positive or negative. Have them recheck the sign when they complete the division.

Example 2

The price of a stock decreased $20 over four days. If the rate of decrease was spread equally over the four days, how did the price of the stock change in one day?

Solution

The total amount of decrease in the price can be represented by -20. Divide to find the amount of decrease in one day.

$-20 \div 4 = -5$

The price decreased $5 in one day. ◄

TRY THESE

Divide. Then check by multiplying.

1. $8 \div (-4)$ **-2** 2. $-15 \div 3$ **-5**   3. $24 \div 8$ **3**   4. $-21 \div (-7)$ **3**

5. $-56 \div 7$ **-8** 6. $64 \div (-8)$ **-8** 7. $-48 \div (-6)$ **8.** 8. $49 \div (-7)$ **-7**

9. $42 \div 6$ **7**   10. $-35 \div (-5)$ **7** 11. $28 \div (-4)$ **-7** 12. $-27 \div 9$ **-3**

Solve.

13. The price of a share of stock fell $10 over five days. If the rate of decrease was spread equally over the five days, how much did the price of the stock decrease in one day? **decreased by $2**

14. A borer was being used to drill a well. It drilled to a depth of 200 ft in four minutes. If the rate of drilling was constant, how far had it drilled in one minute? **50 ft**

15. The temperature fell 12°F over a four-hour period. If the decrease was spread equally over the four hours, how much did the temperature change in each hour? **decreased 3°F**

EXERCISES

Divide. Then check by multiplying.

PRACTICE/
SOLVE PROBLEMS

1. $-36 \div 4$ **-9** 2. $-35 \div (-5)$ **7** 3. $12 \div 3$ **4**   4. $40 \div (-8)$ **-5**

5. $35 \div (-7)$ **-5** 6. $-54 \div 9$ **-6**   7. $-56 \div (-8)$ **7** 8. $-36 \div 6$ **-6**

9. $14 \div 2$ **7**   10. $18 \div (-9)$ **-2** 11. $-4 \div (-2)$ **2** 12. $-16 \div 4$ **-4**

13. $-25 \div 5$ **-5** 14. $72 \div (-6)$ **-12** 15. $-60 \div (-5)$ **12** 16. $-36 \div 12$ **-3**

MAKING CONNECTIONS

Have students develop a class list of how integers are used in other subjects (measuring temperature in science), how their parents use integers (deposits and withdrawals from bank accounts), and how people in business use integers (profit and loss).

ASSIGNMENTS

BASIC
1–17, 19–22, 27–39, 42–54

AVERAGE
1–18, 19–40, 42–58

ENRICHED
13–18, 23–41, 48–59

ADDITIONAL RESOURCES
Reteaching 6–5
Enrichment 6–5
Transparency Master 29

Additional Questions/Examples

1. What signs would you use with the numbers 3 and 7 to make their product positive? **Both should be + or both should be –.**

2. Write the two related multiplication sentences for the numbers in Exercise 1. **3 × 7 = 21, –3 × (–7) = 21**

3. Write two related division sentences for each multiplication sentence in Exercise 2. **21 ÷ 3 = 7, 21 ÷ 7 = 3; 21 ÷ (–3) = –7, 21 ÷ (–7) = –3**

Guided Practice/Try These Observe students as they complete these exercises independently. Discuss any questions or problems.

3 SUMMARIZE

Write About Math Have students explain in their math journals why the rules for multiplying integers also apply to dividing integers.

4 PRACTICE

Practice/Solve Problems Be sure that students check each exercise by multiplying. Then have students

17. The temperature decreased 16°F over 8 h. If the decrease was spread equally over the 8 h, how did the temperature change in each hour? **decreased 2°F**

write a division sentence for each word problem.

Extend/Solve Problems For Exercises 19–26 suggest that students first determine the number and then the sign. Before students complete Exercises 35–38 review with them how to find the mean of a set of data. Point out that one must find the sum of the set of integers before dividing to find the mean.

Think Critically/Solve Problems Students will guess and check for Exercises 55–58. Discuss their reasoning for Exercises 55 and 59.

5 FOLLOW-UP

Extra Practice Find each quotient.
1. –30 ÷ (–5) **6**
2. 45 ÷ (–9) **–5**
3. 56 ÷ (–8) **–7**
4. –36 ÷ (–6) **6**
Find the mean of each set of integers.
5. –10, –4, 5, 9 **0**
6. 30, –6, –8, –12, 6 **2**
Find the integer represented by ■.
7. 48 ÷ ■ = –8 **–6**
8. ■ ÷ (–20) = 6 **–120**

Extension Have students complete these exercises.
1. 2.4 ÷ (–0.3) **–8**
2. –8.2 ÷ (–4) **2.05**
3. 9.63 ÷ (–0.003) **–3,210**
4. –1/2 ÷ (–3/4) **2/3**
5. 1 1/2 ÷ (–2 1/4) **–2/3**

17. The temperature decreased 16°F over 8 h. If the decrease was spread equally over the 8 h, how did the temperature change in each hour? **decreased 2°F**

18. A stock portfolio decreased $99 in value. If the decrease was spread equally over 3 days, how did the value of the portfolio change each day? **decreased $33**

EXTEND/ SOLVE PROBLEMS

MATH IN THE WORKPLACE

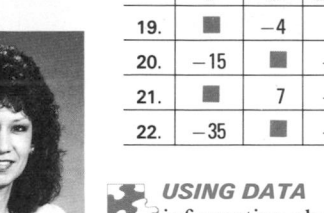

Ruth Gonzalez is a geophysical mathematician. Her work involves developing the mathematics for computer programs used in the exploration and production of oil and gas reservoirs. These programs perform calculations and produce diagrams that help provide information about the earth's subsurface.

Ms. Gonzalez has a bachelor's and master's degree from the University of Texas and a doctorate in mathematical sciences from Rice University.

Copy and complete the table for variables a and b.

	a	b	$a \div b$			a	b	$a \div b$
19.	■	–4	2 **–8**		23.	■	–8	–7 **56**
20.	–15	■	–5 **3**		24.	–72	■	4 **–18**
21.	■	7	–3 **–21**		25.	■	–16	–4 **64**
22.	–35	■	–5 **7**		26.	–100	■	25 **–4**

USING DATA Use the Data Index on page 546 to find information about record holders from the earth sciences.

27. About how many times deeper is the Marianas Trench than the Grand Canyon? **about 6.7 times deeper**

28. About how many times older are the oldest slime molds than the oldest birds? **about 11 times older**

Replace each ● with <, >, or = .

29. $-18 \div (-3)$ ● $18 \div 3$ **=**
30. $10 \div (-5)$ ● $-15 \div (-3)$ **<**
31. $-24 \div (-6)$ ● $-25 \div 5$ **>**
32. $54 \div (-6)$ ● $-54 \div 6$ **=**
33. $-3 \div 1$ ● $-3 \div (-1)$ **<**
34. $49 \div (-7)$ ● $-48 \div 6$ **>**

Find the mean of each set of integers.

35. $-12, -8, 5$ **–5**
36. $-18, 16, -14, -13, 9$ **–4**
37. $-9, -7, 15, 16, 12, -15, 14, -2$ **3**
38. $11, 7, -8, 5, -7, -16, -3, -2, 4, 9$ **0**

Solve.

39. The temperature decreased 20°F over 4 hours. If the decrease was spread equally over the 4 hours, how did the temperature change in 30 minutes? **decreased 2.5°F**

CHALLENGE

Have students complete these exercises. Remind them to follow the standard order of operations that they learned for whole numbers.
1. $3 \times (-5 - 1) + 6 \times (2 - 3 - 4)$ **–48**
2. $-6 + 10 \div (-2) + (-6) - 8$ **–25**
3. $[-64 + (-4)2] \div (-12 - 12)$ **3**
4. $15 - [(-15 \div 3) \times (-1)3] + 16 \div 4$ **4**

40. A stock portfolio had a decrease in price of $120. If the decrease was spread equally over two work weeks, how much did the price of the portfolio change in one day? **$12**

41. The price of a share of stock increased $8.50. If the increase was spread equally over five days, how much did the price increase in one day? **$1.70**

Find the integer represented by ■.

42. $18 \div ■ = -3$ **-6**

43. $-20 \div ■ = -10$ **2**

44. $■ \div (-15) = 3$ **-45**

45. $■ \div (-10) = -5$ **50**

46. $-39 \div ■ = -13$ **3**

47. $■ \div 11 = -2$ **-22**

Replace each ● with $+$, $-$, $\times$, or $\div$. **Sample answers are given.**

48. $-7 ● -5 ● 10 = -20$ **-, ×**

49. $4 ● -3 ● -2 = -14$ **-, ×**

50. $-1 ● -1 ● -1 = -3$ **+, +**

51. $5 ● -5 ● 5 = -125$ **×, ×**

52. $6 ● -6 ● -6 = -7$ **÷, +**

53. $-4 ● 6 ● 3 = -8$ **×, ÷**

Solve.

54. The average daily temperature decreased 31°F in October. If the decrease had been spread equally over the month, how would the temperature have changed in one week? **decreased 7°F**

55. The sum of two integers is always an integer. Is the quotient of two integers always an integer? Explain. **No. For example, the answer to 6 ÷ 5 is not an integer.**

56. The product of two integers is -36. The quotient is -4. What are the integers? **-3 and 12 or 3 and -12**

57. The product of two integers is -16. The quotient is -1. What are the integers? **4 and -4**

58. The sum of two integers is 20. The quotient is -5. What are the integers? **25 and -5**

59. You can multiply to check your answer to a division problem. Using this knowledge, explain why zero can be divided by a nonzero number, but no number can be divided by zero. **See Additional Answers.**

THINK CRITICALLY/
SOLVE PROBLEMS

6-5 Dividing Integers **205**

205

206

WARM-UP

Evaluate.
1. $4 + (-5)$ **–1**   2. $-8 + (-3)$ **–11**
3. $-7 + 9$ **2**   4. $15 - (-9)$ **24**
5. $5 \times (-3)$ **–15**   6. $-8 \times (-4)$ **32**
7. $63 \div (-9)$ **–7**   8. $-32 \div (-8)$ **4**

1 MOTIVATE

Introduction Read and discuss the introductory material with students.

2 TEACH

Use the Pages/Problems Have students read this part of the section and then discuss the problem. Discuss the key words that suggest which operation is required. **more than, less than**

Use the Pages/Solution Review the solutions presented in the text.

3 SUMMARIZE

Talk It Over Have students prepare a class list of words that indicate relationships between numbers; for example, *greater than*, *more than*, *less than*, *product of*.

206

6-6 Problem Solving Skills:
CHOOSE THE OPERATION

► READ
► PLAN
► SOLVE
► ANSWER
► CHECK

In order to solve many problems that involve integers, it is necessary to decide how the numbers in a problem are related. This relationship between the numbers is the key to deciding what operation to use in solving the problem.

PROBLEMS

a. A number is eight more than negative three. What is the number?

b. A number is eleven less than seven. What is the number?

SOLUTIONS

a. The number is greater than negative three. Therefore, add eight to negative three to find the number.

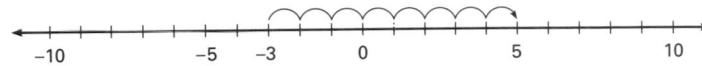

$-3 + 8 = 5$ Check: Five is eight more than negative three.
The number is five.

b. The number is less than seven. Therefore, subtract eleven from seven to find the number.

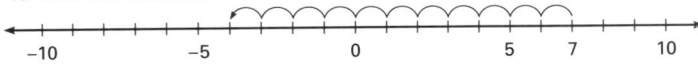

$7 - 11 = -4$ Check: Negative four is eleven less than seven.
The number is negative four.

PROBLEMS

Choose the operation required. Then solve the problem.

1. A number is six less than negative four. What is the number? **–10**

2. A number is nine more than negative four. What is the number? **5**

3. Eight is nine more than some number. What is the number? **–1**

4. A number is the product of six and negative fourteen. What is the number? **–84**

---

5. A number is eighteen more than the sum of six and negative six. What is the number? **18**

6. Negative seventy-seven is seven times some number. What is the number? **−11**

7. A number is the product of negative nine and negative twenty-one. What is the number? **189**

8. A number is four more than negative nine. What is the number? **−5**

9. Negative one hundred sixty-eight is fourteen times some number. What is the number? **−12**

10. A number is the product of five and negative twelve. What is the number? **−60**

11. A number is sixteen more than negative eight. What is the number? **8**

12. A number divided by negative thirteen equals negative twelve. What is the number? **156**

Solve.

13. David has $24 in his checking account. If he writes checks for $6, $13, and $36, what will be the balance in his account? **−$31**

14. The Falcons received a penalty of −15 yd after they made a 33 yd gain. What was the net gain or loss for the team? **18 yd gain**

15. Maria ate a breakfast that supplied her with 300 calories of energy. Then she ran for an hour, which used up 14 calories of energy per minute. Give Maria's net calorie intake after running. **−540 calories**

16. Swimming uses up 10 calories of energy per minute. If Ivan wants to expend the same number of calories in swimming that he takes in by eating a 350-calorie snack, how long would he have to swim? **35 min**

17. Juanita had a balance of $16 in her bank account. In the next few days she made a deposit of $8, wrote a check for $9, and made two more deposits of $10 and $15. How much money is in her account now? **$40**

18. Carlos played a board game in which a player can move forward or backward from the starting line. In the first round Carlos moved forward 8 squares, then back 6 squares, then back 4 squares, then back 9 squares, and then forward 6 squares. At the end of the round, what was Carlos' position relative to the starting line? **5 behind starting line**

4   PRACTICE

Problems Encourage all students to try to solve all the problems. You may wish to have some or all of them work in pairs or small groups.

5   FOLLOW-UP

Extra Practice Solve.
1. A number is seven less than negative three. What is the number? **−10**
2. A number is eleven more than negative eight. What is the number? **3**
3. A number is the product of negative twelve and negative 3. What is the number? **36**
4. A number divided by negative fifteen equals twenty-four. What is the number? **−360**
5. Jason wants to burn 456 calories of energy exercising. If the exercises he does help him burn 12 calories of energy per minute, how long should Jason exercise? **38 minutes**

5-MINUTE CLINIC

Exercise	Student's Error	Error Diagnosis
A number is eleven less than seven. What is the number?	11 − 7 = 4	• Student does not think about what the statement means and so reverses the order of the numbers in the problem.

207

1 MOTIVATE

Explore Have students explore the question, *Is the sum of any two integers always an integer?* Continue with the difference, the product, and the quotient of any two integers. Discuss the students' observations. **Addition, subtraction, and multiplication of integers always results in an integer. Because division sometimes results in remainders, the quotient of two integers is not always an integer.**

2 TEACH

Use the Pages/Skills Development
Have students read this part of the section and then discuss the examples.

Example 1: Use these exercises to demonstrate how to write a rational number as a ratio of two integers.

208

6-7 Rational Numbers

EXPLORE

You already know how to find the quotient of two integers.

$$\frac{10}{2} = 5 \qquad \frac{15}{-3} = -5 \qquad \frac{-18}{3} = -6 \qquad \frac{-24}{-6} = 4$$

Is the quotient of any two integers always an integer? **no**

SKILLS DEVELOPMENT

When you add, subtract, or multiply integers, the result is always another integer. Study these examples.

$$4 + (-3) = 1 \qquad -15 + 14 = -1$$
$$-10 + (-4) = -14 \quad 6 - (-5) = 11$$
$$8 - 9 = -1 \qquad -12 - (-14) = 2$$
$$(6)(5) = 30 \qquad (-4)(2) = -8$$
$$(3)(-2) = -6 \qquad (-8)(-2) = 16$$

However, the quotient of two integers is not always another integer.

$$0 \div 5 = \frac{0}{5} = 0 \qquad 12 \div 3 = \frac{12}{3} = 4 \qquad 3 \div 4 = \frac{3}{4} \qquad 11 \div 8 = \frac{11}{8}$$

Quotient is an integer. Quotient is a fraction.

When an operation on a set of numbers always results in a number that is in the same set, the set of numbers is said to be **closed** under that operation. You have seen that the set of integers is closed under addition, subtraction, and multiplication, but not under division. To obtain a set of numbers that is closed under all four operations, you need to consider the set of *rational numbers*.

A **rational number** is any number that can be expressed in the form $\frac{a}{b}$, where a is any integer and b is any integer except 0. That is, a rational number can be expressed as a ratio of two integers. So all simple fractions are rational numbers, as are all terminating and repeating decimals. Any integer can be expressed as a fraction, such as $4 = \frac{4}{1} = \frac{8}{2}$, so all the integers are rational numbers. The quotient of any two rational numbers is always another rational number.

Example 1

Write each rational number as a ratio of two integers.

a. 4   b. $0.\overline{4}$ c. -0.87 d. $2\frac{2}{5}$ e. -2.3

Solution

a. $4 = \frac{8}{2}$ b. $0.\overline{4} = \frac{4}{9}$ c. $-0.87 = -\frac{87}{100}$

d. $2\frac{2}{5} = \frac{12}{5}$ e. $-2.3 = -\frac{23}{10}$ ◄

TEACHING TIP

To help students to see that more than one rational number may be equivalent to another, have students first write at least two fractions equivalent to a given fraction.

Example 2

Name the point that corresponds to each rational number.

a. 0.8 **b.** $-\frac{3}{5}$ **c.** -1.2 **d.** 2.5

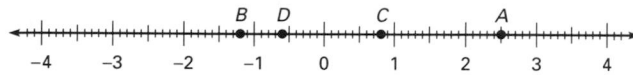

Solution

a. Point C corresponds to 0.8.

b. Point D corresponds to $-\frac{3}{5}$. $\left(-\frac{3}{5} = -\frac{6}{10}\right)$

c. Point B corresponds to -1.2. $\left(-1.2 = -1\frac{2}{10}\right)$

d. Point A corresponds to 2.5. $\left(2.5 = 2\frac{5}{10}\right)$ ◄

Any rational number can be graphed as a point on a number line.

Example 3

Graph each rational number on a number line.

a. $2\frac{1}{6}$ **b.** $1\frac{2}{3}$ **c.** $-\frac{5}{6}$ **d.** $-2\frac{1}{3}$

Solution

The denominator of each rational number is either 3 or 6. Mark the number line in sixths and represent each number by a point on the number line. ◄

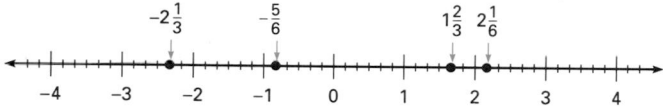

You can use a number line to order rational numbers. If one rational number lies to the right of another on the number line, then it is greater than the other rational number. If one rational number lies to the left of another on the number line, then it is less than the other rational number.

Example 4

Order these rational numbers from least to greatest: $3, -4.1, \frac{1}{2}, 4.5, -2\frac{1}{2}$

Solution

Graph the points on a number line.

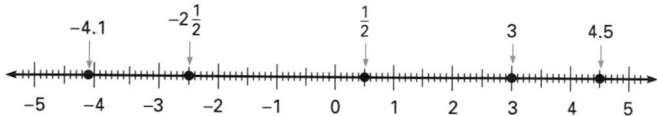

From least to greatest the order is $-4.1, -2\frac{1}{2}, \frac{1}{2}, 3, 4.5$. ◄

ASSIGNMENTS

BASIC
1–14, 18–31, 36–37

AVERAGE
1–17, 18–35, 36–39

ENRICHED
9–17, 18–35, 36–41

ADDITIONAL RESOURCES
Reteaching 6–7
Enrichment 6–7

Example 2: Point out that fifths and tenths can all be graphed on a number line divided into tenths.

Example 3: Point out that, because thirds can be expressed as sixths, these rational numbers can be graphed on the same number line. However, it is not common practice to graph numbers, such as 4/5 or 3/4, on that same number line.

Reinforce the idea that there is, theoretically, only one number line. For convenience sake, one may arbitrarily divide units into any fractional parts when graphing a set of numbers.

Example 4: Use this example to demonstrate how to use the number line to order rational numbers.

Additional Questions/Examples
Express each rational number in lowest terms.

1. −14/−2 **7**   **2.** −6/−8 **3/4**
3. −15/5 **−3**   **4.** −30/45 **−2/3**

Express each rational number as a decimal.

5. 1/4 **0.25**   **6.** −2/3 **$-0.\overline{6}$**
7. 4/−5 **−0.8**   **8.** −1/−2 **0.5**

5-MINUTE CLINIC

Exercise	Student's Error	Error Diagnosis
Compare: −3/4, −1/2	−3/4 > −1/2	• Student may have neglected the sign of the numbers when comparing the fractions. • Student may have confused the < and > signs.

9. What rational number is three and one-quarter units greater than –4? **–3/4 or –0.75**

Guided Practice/Try These Have students work in small groups. Point out that all group members should come to an agreement on each answer.

3 SUMMARIZE

Write About Math Have students explain in their math journals the difference between the set of integers and the set of rational numbers.

4 PRACTICE

Practice/Solve Problems Help the students determine the fractional parts of the number lines they will use for Exercises 9–12. Students should write equivalent fractions for the rational numbers in Exercises 13–16.

Extend/Solve Problems Students should be aware that more than one answer is possible for Exercises 32–35.

Think Critically/Solve Problems You may want to help students in testing the statements in Exercises 36–41.

5 FOLLOW-UP

Extra Practice Graph the rational numbers on a number line.
1. 2/5, –7/10, –4/5, 3/5
2. –3 1/2, –5/4, 3/4, 4/2
Order the numbers from least to greatest.
3. 4/5, 7/10, –9/20   **–9/20, 7/10, 4/5**
4. –2/3, 5/12, –1/4   **–2/3, –1/4, 5/12**

210

TRY THESE

Answers may vary. Sample answers are given.

Write each rational number as a ratio of two integers.

1. 6 $\frac{12}{2}, \frac{6}{1}$
2. 0.6 $\frac{6}{10}, \frac{3}{5}$
3. –0.75 $-\frac{75}{100}, -\frac{3}{4}$
4. –1.4 $-\frac{14}{10}, -\frac{7}{5}$

Name the point that corresponds to each rational number.

5. $-\frac{7}{10}$ **B**
6. 2.4 **D**
7. –2.3 **A**
8. 0.7 **C**

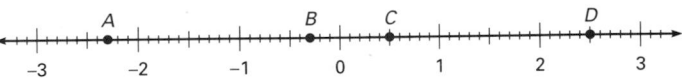

Graph each rational number on a number line.

9. $-1\frac{1}{2}$
10. –0.9
11. $2\frac{1}{5}$ **Check students' graphs.**

12. Order these rational numbers from least to greatest: 2.7, –2.1, –0.3, 0.3 **–2.1, –0.3, 0.3, 2.7**

EXERCISES

PRACTICE/ SOLVE PROBLEMS

Write each rational number as a ratio of two integers.
Answers may vary. Sample answers are given.

1. 7 $\frac{7}{1}, \frac{14}{2}$
2. 0.7 $\frac{7}{10}, \frac{14}{100}$
3. $3\frac{1}{3}$ $\frac{10}{3}, \frac{20}{6}$
4. 1.9 $\frac{19}{10}, \frac{190}{100}$

Name the point that corresponds to each rational number.

5. $-\frac{3}{10}$ **B**
6. 2.5 **D**
7. 0.5 **C**
8. –2.3 **A**

Graph the rational numbers on a number line.

9. $1\frac{1}{3}, -2\frac{2}{3}, -\frac{2}{3}, 3\frac{2}{3}$
10. $-\frac{1}{4}, 2\frac{1}{2}, -1\frac{3}{4}, -2\frac{1}{2}$ **See Additional Answers.**
11. $\frac{2}{5}, -\frac{1}{5}, 1\frac{3}{5}, -1\frac{4}{5}$
12. $\frac{7}{2}, -\frac{5}{2}, 1\frac{1}{2}, -1\frac{1}{2}$

Order the numbers from least to greatest.

13. $\frac{7}{10}, \frac{7}{20}, \frac{3}{5}$ $\frac{7}{20}, \frac{3}{5}, \frac{7}{10}$
14. $\frac{15}{40}, \frac{7}{32}, \frac{9}{20}, \frac{5}{16}$ $\frac{7}{32}, \frac{5}{16}, \frac{15}{40}, \frac{9}{20}$
15. $\frac{3}{4}, \frac{2}{3}, \frac{7}{12}$ $\frac{7}{12}, \frac{2}{3}, \frac{3}{4}$
16. $-0.8, -\frac{13}{16}, -\frac{7}{8}, -\frac{13}{20}$ $-\frac{7}{8}, -\frac{13}{16}, -0.8, -\frac{13}{20}$

MAKING CONNECTIONS

Have students use the Data Index on page 546 to find data about waterfalls. Have them graph the data using rational numbers.

Solve.

17. In the all-county school chorus, 0.28 of the 200 singers are altos. Express the number of altos as a ratio of two integers. **Answers will vary. Possible answers are $\frac{28}{100}$ and $\frac{56}{200}$.**

18. Which of the rational numbers is equivalent to $\frac{7}{4}$?
$\frac{21}{12}, \frac{14}{11}, 1.75, -1.70$ **$\frac{21}{12}$, 1.75**

19. Which of the rational numbers is equivalent to $-\frac{3}{5}$?
$0.8, -0.6, \frac{3}{10}, -\frac{18}{30}$ **$-0.6, -\frac{18}{30}$**

Replace ● with <, >, or =.

20. $-\frac{4}{5}$ ● $\frac{5}{6}$ **<** 21. $-\frac{5}{8}$ ● $-\frac{3}{4}$ **>** 22. $\frac{8}{15}$ ● $\frac{2}{3}$ **<** 23. $\frac{27}{40}$ ● $-\frac{5}{8}$ **>**

24. $\frac{1}{3}$ ● 0.3 **>** 25. $\frac{2}{3}$ ● $0.\overline{6}$ **=** 26. $-\frac{7}{18}$ ● $-\frac{1}{3}$ **<** 27. 0.55 ● $-\frac{11}{20}$ **>**

Write a rational number for each description.

28. decrease in price of $4.50 **−4.5** 29. gain of $3\frac{1}{5}$ inches **$3\frac{1}{5}$ or 3.2**

30. loss of $4\frac{1}{2}$ pounds **$-4\frac{1}{2}$ or $-4\frac{1}{2}$** 31. loss of $32\frac{1}{8}$ ounces **$-32\frac{1}{8}$**

Write a rational number for each description. Use a number line to help you. **Answers will vary. Samples are given.**

32. number halfway between 2 and 3 **2.5**

33. number halfway between 0 and $-1\frac{1}{2}$ **$-\frac{3}{4}$**

34. number four and one-half units less than 1 **$-3\frac{1}{2}$**

35. number two and one-quarter units greater than -3 **$-\frac{3}{4}$**

Write *true* or *false* for each statement. If you think the statement is false, give a counterexample using rational numbers that proves the statement false.

Suppose that p and q are both positive rational numbers.

36. $p + q$ is positive **true** 37. $q - p$ is positive **false: 1.5 − 2.5 = −1**

Suppose that p and q are both negative rational numbers.

38. $p - q$ is negative **false: $-\frac{1}{2} - (-2) = 1\frac{1}{2}$** 39. $p(p + q)$ is positive **true**

Suppose that p is positive and q is negative.

40. pq is less than p **true** 41. $p + q$ is greater than p **false: $\frac{3}{4} + -\frac{3}{4} = 0$**

6-7 Rational Numbers **211**

EXTEND/ SOLVE PROBLEMS

MIXED REVIEW

Write in scientific notation.

1. 9,432 2. 0.000035
9.432 × 10³ **3.5 × 10⁻⁵**
3. 0.00073 4. 25,038
7.3 × 10⁻⁴ **2.5038 × 10⁴**
Complete.

5. 72 c = ■ qt **18**
6. 19 ft = ■ in. **228 in.**
7. 9 mL = ■ L **0.009**
8. 200 m = ■ cm **20,000**

Solve.

9. A garden has 3 rows of corn plants with the same number of plants in each row. Each row is 30 feet long. The plants are placed 3 feet apart starting at the corner of the garden. How many corn plants are in the garden? **33**

THINK CRITICALLY/ SOLVE PROBLEMS

5. What rational number is halfway between −1 and −2? **−1.5 or −1 1/2**

6. What rational number can be used to represent a loss of 15 3/4 lb? **−15.75 or −15 3/4**

7. Five of the twenty-five students in Roy's math class wear glasses. Express the ratio of students who wear glasses to the total number of students in lowest terms. Find the decimal equivalent of the fraction. **5/25 = 1/5 = 0.2**

Extension Demonstrate the following technique for finding a rational number that is exactly halfway between two given rational numbers; for example, 1/2 and 7/8.

1. Find the difference. 7/8 − 1/2 = 3/8

2. Multiply by 1/2. 1/2 × 3/8 = 3/16

3. Add the result to the lesser number, 1/2. 1/2 + 3/16 = 11/16

Section Quiz Graph the rational numbers in Exercises 1 and 2 on a number line. **Check students' work.**

1. 3/8, −1/2, 0.5, −1/4
2. −2/3, 5/6, 11/12, 1/6

Order the numbers from least to greatest.

3. 25/40, −5/32, −11/20, 9/16
−11/20, −5/32, 9/16, 25/40

4. What rational number is four and three-quarter units greater than −2? **2 3/4 or 2.75**

5. Ruth's Nursery sold 150 of the 450 rose bushes in stock on opening day. Express the ratio of rose bushes sold to the total in stock as a fraction in lowest terms. Find the decimal equivalent of the fraction. **1/3; 0.$\overline{3}$**

Get Ready calculators

Additional Answers
See page 578.

WARM-UP

Inez numbered the pages of a report she wrote from 1 to 100. How many times did she write the digit 1? **21 times**

212

6-8

► READ
► PLAN
► SOLVE
► ANSWER
► CHECK

Problem Solving/ Decision Making:
CHOOSE A STRATEGY

PROBLEM SOLVING TIP

Here is a checklist of the problem solving strategies that you have studied so far in this book.

Find a pattern
Logical reasoning
Solve a simpler problem
Draw a picture

In this book you have been studying a variety of problem solving strategies. Experience in applying these strategies will help you decide which will be most appropriate for solving a particular problem. Sometimes only one strategy will work. In other cases, any one of several strategies will offer a solution. There may be times when you will want to use two different approaches to a problem in order to be sure that the solution you found is correct. For certain problems, you will need to use more than one strategy in order to find the solution.

PROBLEMS

Solve. Name the strategy you used to solve the problems.
Strategies may vary. One possible strategy is suggested.

1. The members of the geology club went on a one-week trip. They traveled 16 mi the first day, then traveled $1\frac{1}{2}$ times farther each succeeding day. How far would they travel by the end of the third day? the fifth day? the seventh day? **36 mi, 81 mi, 182.25 mi; draw a picture**

COMPUTER TIP

For Problem 3, you may want to RUN this program several times and make a chart of the results to help discover the pattern.

```
10. INPUT "HOW MUCH
    MONEY? ";M: PRINT
20  FOR X = 3 TO M STEP 2
30  W = (M − 1) / 2
40  NEXT X
50  PRINT "$";M;"
    DEPOSITED IN WEEK
    ";W;"."
```

2. Suppose five athletes enter a swimming competition. They will compete in pairs until each athlete has competed against every other athlete one time. How many competitions will there be?
10 competitions in all; draw a picture

3. Stefanie plans to deposit $3 into a savings account the first week, $5 the second week, $7 the third week, and so on until she reaches her goal of depositing $25 in a week. In which week will she make a $25 deposit? **the twelfth week; find a pattern**

212 CHAPTER 6 Exploring Integers

CHALLENGE

A farmer wants to use fences to divide a rectangular field into 16 sections. What is the least number of straight fences needed to do this? Hint: Find a pattern. **5 fences**
See margin for diagram.

4. Peter, José, Dwayne, Yolanda, Dolores, and Alma are three brother-and-sister pairs who play doubles tennis. Brother and sister cannot play in the same pair. Peter and Yolanda play Dwayne and Dolores on one day. Dwayne and Yolanda play José and Alma on another day. Name each brother-and-sister pair. **Peter and Dolores; José and Yolanda; Dwayne and Alma; logical reasoning**

5. Some baseball teams will play in a tournament. Each team will play every other team once. How many games will be played if there are 8 teams? 10 teams? 25 teams? **28 games; 45 games; 300 games; find a pattern**

6. Some players enter an elimination tournament. Once beaten by another player, they cannot play another round. How many matches will be played if there are 4 players? 8 players? 16 players? **3 matches; 7 matches; 15 matches; draw a picture**

7. The members of the Explorer's Club traveled 26 mi on the first day of their hike. On each of the following days, they hiked only half as far as they had the day before. How many miles would they cover on the second day? On the third day? On the fourth day? **13 mi; 6.5 mi; 3.25 mi; find a pattern**

8. Find the next row of the pattern.
$$\frac{L}{\frac{M}{Q}} \quad \frac{N}{\frac{O}{R}} \quad \frac{L}{\frac{O}{P}} \quad \frac{N}{\frac{M}{R}} \frac{T}{U} \quad \frac{V}{W} \quad \frac{T}{W} \quad \frac{V}{U'}$$
find a pattern

9. The producers of a TV game show award money to contestants who answer questions correctly. If a question is answered correctly by the first contestant, that person will receive $300. If the first contestant misses and the second contestant answers correctly, the amount is doubled, and so on. How much money could someone win who answered correctly and was the third contestant? the fifth contestant? the twelfth contestant? **$1,200; $4,800; $614,400; find a pattern**

10. Geologists examined rock formations at a road-cut site in Ridgeway. Ridgeway is 8 mi due west of Mountain Pass. Dalton is 6 mi due west of Haber, which is 9 mi south and 4 mi west of Mountain Pass. How many miles and in which direction is Dalton from the rock formations in Ridgeway? **9 mi south and 2 mi west; draw a picture**

6-8   Problem Solving/Decision Making: Choose a Strategy   **213**

ASSIGNMENTS

BASIC	1–10
AVERAGE	1–10
ENRICHED	1–10
ADDITIONAL RESOURCES	Reteaching 6–8 Enrichment 6–8

5 FOLLOW-UP

Extra Practice

1. Eric's time in a running race was 2 min 42.4 s. This was 2.3 s slower than Topo's time. Harry's time of 2 min 57.6 s was 0.5 s faster than Will's. Les finished the race in 2 min 50.8 s. List the racers in order of finish from first to last, and give their times.
   Topo: 2 min 40.1 s;
   Eric: 2 min 42.4 s;
   Les: 2 min 50.8 s;
   Harry: 2 min 57.6 s;
   Will: 2 min 58.1 s

2. Yuri started working in 1990 with a salary of $24,000. He received a $1,500 raise each year after that. Betty started working in 1992 with a salary of $25,500. She gets an annual raise of $1,800. In what year will Yuri and Betty be earning the same salary? **1997**

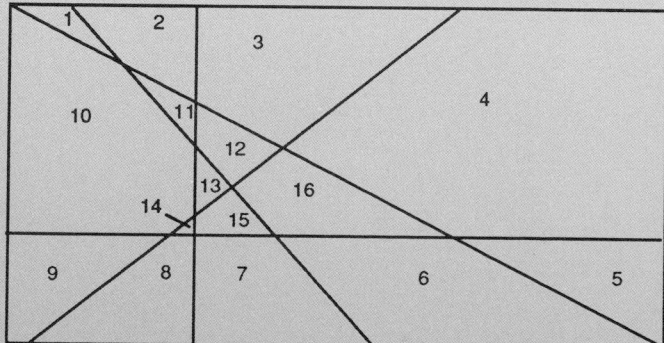

1. A number greater than zero is a(an) __?__ number. **e**
2. A number less than zero is a(an) __?__ number **c**
3. Two numbers are __?__ when they are the same distance from 0, but in different directions. **d**
4. The distance a number is from zero is called its __?__. **a**
5. A number that can be expressed as $\frac{a}{b}$, where a is any integer and b is any integer except 0, is a(an) __?__ number. **b**

a. absolute value
b. rational
c. negative
d. opposites
e. positive

SECTION 6–1   EXPLORING INTEGERS   (pages 186–189)

▶ Opposite integers have the same absolute value.
▶ If two numbers are graphed on a horizontal number line, the number that is to the right is the greater number.

Find each absolute value.

6. $|-23|$ **23**
7. $|-17|$ **17**
8. $|17|$ **17**
9. $|45|$ **45**
10. $|38|$ **38**

Replace each ● with $<$, $>$, or $=$.

11. $23 ● -23$ **>**
12. $-8 ● -7$ **<**
13. $-2 ● 0$ **<**

14. **USING DATA**   Use the table on page 185 to help you solve. Express the distance from the crust to the center of the earth as a negative number. **−6,400 km**

SECTIONS 6–2 and 6–3   ADDING and SUBTRACTING INTEGERS (pages 190–197)

▶ To add integers with the same signs, add the absolute values. Give the sum the sign of the addends.
▶ To add integers with different signs, subtract the absolute values. Give the sum the sign of the addend with the greater absolute value.
▶ To subtract an integer, add its opposite.

Add.

15. $18 + 15$ **33**
16. $18 + (-15)$ **3**
17. $-18 + 15$ **−3**
18. $-18 + (-15)$ **−33**

Subtract.

19. $-34 - 18$ **−52**
20. $36 - 38$ **−2**
21. $-23 - (-4)$ **−19**
22. $-10 - (-4)$ **−6**

---

6 CHAPTER REVIEW

Introduction The Chapter Review emphasizes the major concepts, skills, and vocabulary presented in this chapter and can be used for diagnosing students' strengths and weaknesses. Page references direct students back to appropriate sections for additional review and reteaching.

Using Pages 214–215 Allow students to quickly scan the Chapter Review and ask questions about any section they find confusing. Exercises 1–5 review key vocabulary.

Informal Evaluation Have students explain how they arrived at any incorrect answers. They will probably find their own mistakes and give you some clues as to the nature of their errors. Make sure students understand this material before administering the Chapter Test.

Follow-Up   Have students research the history and development of the concept of integers.

SECTIONS 6–4 and 6–5   MULTIPLYING and DIVIDING INTEGERS (pages 198–205)

▶ The product or quotient of two integers with the *same* sign is positive.
▶ The product or quotient of two integers with *different* signs is negative.

Multiply.

23. 7×9 **63**

24. -8×4 **−32**

25. $-6 \times (-8)$ **48**

26. $9 \times (-3)$ **−27**

27. $0 \times (-14)$ **0**

28. -6×7 **−42**

29. $-7 \times (-8)$ **56**

30. $4 \times (-6)$ **−24**

Divide.

31. $81 \div 9$ **9**

32. $-49 \div 7$ **−7**

33. $18 \div (-3)$ **−6**

34. $-72 \div (-8)$ **9**

35. $-36 \div 4$ **−9**

36. $-16 \div (-8)$ **2**

37. $56 \div (-7)$ **−8**

38. $-48 \div (-6)$ **8**

SECTION 6–7   RATIONAL NUMBERS   (pages 208–211)

▶ A **rational number** is any number that can be expressed in the form $\frac{a}{b}$, where a is any integer and b is any integer except 0.
▶ A rational number can be graphed as a point on a number line.

Write each rational number as a ratio of two integers.

39. 6 $\frac{12}{2}$

40. 0.6 $\frac{6}{10}$

41. $5\frac{2}{3}$ $\frac{17}{3}$

42. -3.3 $-\frac{33}{10}$

Name the point that corresponds to each rational number.

43. -0.6 **C**

44. $\frac{2}{5}$ **A**

45. -1.5 **D**

46. 3.5 **B**

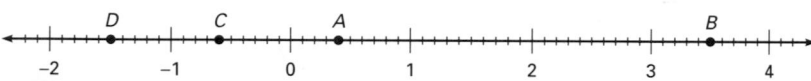

SECTIONS 6–6 and 6–8   PROBLEM SOLVING   (pages 206–207 and 212–213)

▶ Sometimes only one strategy will work in solving a problem. In other cases, you can use any one of several strategies.

▶ The relationship between the numbers in a problem is often the key to the operation needed to solve the problem.

Choose the operation. Then solve the problem.

47. A number is nine more than negative three. What is the number? **6**

Choose the strategy to help you solve.

48. Twelve baseball teams will play in a tournament. Each team will play every other team once. How many games will be played? **66 games**

Review   **215**

Study Skills Tip Have students write the rules for adding and subtracting integers on one index card, and the rules for multiplying and dividing integers on a second index card. Encourage them to memorize the rules by referring to the cards whenever they have some free time.

6 CHAPTER TEST

Introduction The Chapter Test uses a variety of questioning techniques to assess students' mastery of the major objectives of Chapter 6. If you prefer, you may use the Skills Preview (page 183) as an alternative form of the Chapter Test. The items on this test and the Skills Preview correspond in content and level of difficulty.

Alternative Assessment
Write About Math
1. Why can the same rules be used for multiplying and dividing integers?
2. Describe the set of integers.
3. Describe the set of rational numbers.

Find each absolute value.

1. $|-6|$ **6**
2. $|12|$ **12**
3. $|-3|$ **3**
4. $|-18|$ **18**

Replace each ● with $<, >, =$.

5. $12 ● -12$ **>**
6. $-13 ● -18$ **>**
7. $-14 ● 2$ **<**
8. $-5 ● 0$ **<**
9. $-9 ● -5$ **<**
10. $-12 ● 7$ **<**

Add.

11. $-8 + 3$ **-5**
12. $9 + (-11)$ **-2**
13. $-7 + (-13)$ **-20**
14. $2 + (-4)$ **-2**
15. $-12 + (-4)$ **-16**
16. $20 + (-20)$ **0**

Subtract.

17. $5 - (-9)$ **14**
18. $-3 - 7$ **-10**
19. $-8 - (-10)$ **2**
20. $-6 - 8$ **-14**
21. $7 - 13$ **-6**
22. $6 - (-2)$ **8**

Multiply.

23. $5 × 7$ **35**
24. $-4 × 8$ **-32**
25. $-6 × (-9)$ **54**
26. $8 × (-8)$ **-64**
27. $-2 × (-4)$ **8**
28. $-3 × 9$ **-27**

Divide.

29. $25 ÷ 5$ **5**
30. $-36 ÷ 9$ **-4**
31. $-42 ÷ (-7)$ **6**
32. $-54 ÷ 6$ **-9**
33. $-48 ÷ (-8)$ **6**
34. $-27 ÷ 3$ **-9**

Write each rational number as a ratio of two integers.

35. 5 $\frac{10}{2}$
36. 0.4 $\frac{4}{10}$
37. $-4\frac{3}{4}$ $-\frac{19}{4}$
38. -5.1 $-\frac{51}{10}$
39. -7 $-\frac{35}{5}$
40. $2\frac{9}{10}$ $\frac{29}{10}$
41. -0.24 $-\frac{24}{100}$
42. 0.014 $\frac{14}{1,000}$

Solve.

43. A number is fourteen more than negative three. What is the number? **11**

44. A number is six less than negative fourteen. What is the number? **-20**

45. Ten players enter an elimination tournament. Once a player is beaten by another player, he or she cannot play another round. How many matches will be played? **9 matches**

Find the mean and median for each set of data.

1. 4 7 9 9 11 **8; 9**

2. 17 29 17 12 50 17 17 25 **23; 17**

3. 6 7 9 19 19 35 35 67 **24.6; 19**

4. 150 123 128 149 123 19 **115.3; 125.5**

5. The school newspaper wants to know what its readers think of its new movie review feature. They question the first ten students to enter the school. What kind of sampling does this situation represent? **convenience**

Complete.

6. $4(8 + 3) = (4 \times 8) + (\blacksquare \times 3)$ **4**

7. $8(\blacksquare - 2) = (8 \times 3) - (8 \times 2)$ **3**

8. $\blacksquare(4 + 3) = (6 \times 4) + (6 \times 3)$ **6**

9. Determine whether the following argument is *valid* or *invalid*.

   If a figure is a rectangle, then it has four sides.
   Figure *ABCD* has four sides.
   Therefore, figure *ABCD* is a rectangle.
   invalid

10. Pick a number. Add 1. Multiply by 2. Add 8. Divide by 2. Subtract the original number. What number results? Repeat the process with several other numbers. Make a conjecture about the result.
    The result is always 5.

11. Write 0.16 as a fraction in lowest terms. $\frac{4}{25}$

12. Write $\frac{3}{24}$ as a terminating decimal. **0.125**

Add or subtract. Write your answer in lowest terms.

13. $\frac{5}{8} + \frac{2}{3}$ $1\frac{7}{24}$

14. $8 - 2\frac{1}{3}$ $5\frac{2}{3}$

15. $\frac{7}{9} + \frac{1}{3}$ $1\frac{1}{9}$

16. $\frac{7}{12} - \frac{3}{8}$ $\frac{5}{24}$

Solve.

17. The length of one side of a regular octagon measures 12.7 cm. Find the perimeter. **101.6 cm**

18. Use the formula to find the perimeter of a rectangular figure with these dimensions: length: 5.2 mi, width: 2.5 mi **15.4 mi**

Find each circumference. Round your answer to the nearest tenth or whole number.

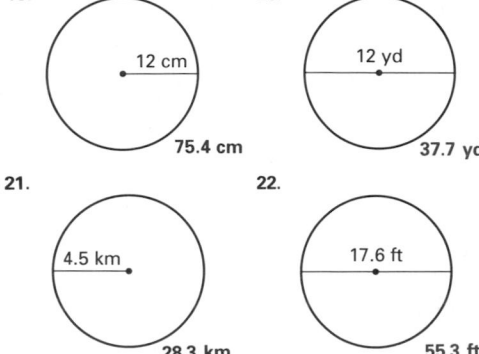

19. 12 cm **75.4 cm**

20. 12 yd **37.7 yd**

21. 4.5 km **28.3 km**

22. 17.6 ft **55.3 ft**

Find each answer.

23. $-64 \div 8$ **–8**

24. $-7 + (-6)$ **–13**

25. $-6 - (-8)$ **2**

26. $-63 \div 9$ **–7**

27. negative twenty-four divided by negative eight **3**

28. the product of fourteen and negative seven **–98**

Introduction The purpose of this Cumulative Review is to maintain previously taught skills and concepts and to apply them to the material presented in this chapter. At least one major objective of each chapter is included in the review.

Item Analysis The table below correlates the Cumulative Review items with the chapter and section that are being reviewed.

Section	Items
1–1	5
1–9	1–4
2–9	6–8
3–2	10
3–3	9
4–2	11
4–3	12
4–6	13–16
5–4	17
5–5	18
5–6	19–22
6–2	24
6–3	25
6–4	28
6–5	23, 26–27

6 CUMULATIVE TEST

Introduction The Cumulative Test uses a standardized-test format of multiple-choice questions to assess retention of previously learned concepts and test-taking skills. Test results may be used to diagnose students' strengths and weaknesses.

Item Analysis The table below correlates the Cumulative Test items with the chapter and section that are being tested.

Section	Items
1–1	1
1–9	2
2–8	4
2–10	3
3–3	5–6
4–3	7–10
5–6	11–12
6–2	13
6–3	14, 17
6–4	16
6–5	15

1. To decide which shirt to buy in a store, Will asked the opinions of the other shoppers in the clothing department. What type of sampling did he use?
   A. systematic   B. cluster
   C. convenience   D. random

2. What is the mean number of days in a month for a year that is not a leap year?
   A. 30   B. 29   C. 31   D. 30.4

3. Complete.
   $6(\blacksquare - 3) = (6 \times 5) - (6 \times 3)$
   A. 12   B. 3   C. 5   D. 6

4. Simplify. $4 \times 8 + 10 \div 2$
   A. 37   B. 21   C. 40   D. 80

5. Which conclusion can be drawn from these statements?

   If people live in Puerto Rico, then they live on an island.
   Marie lives in Puerto Rico.
   A. Marie lives on an island.
   B. Marie does not live in Puerto Rico.
   C. Marie lives in Hawaii.
   D. none of these

6. Which conclusion can be drawn from these statements?

   If the animal is a hare, then it hops.
   The animal hops.
   A. The animal is not a hare.
   B. The animal is a hare.
   C. The animal is a kangaroo.
   D. none of these

7. Express 0.312 as a fraction in lowest terms.
   A. $\frac{8}{25}$ B. $\frac{39}{125}$ C. $\frac{156}{500}$ D. $\frac{312}{1000}$

8. Which is the decimal for $\frac{7}{8}$
   A. 1.142   B. 7.8
   C. 0.625   D. 0.875

9. Which is the decimal for $\frac{3}{20}$?
   A. 6.6   B. 3.2
   C. 0.15   D. 1.05

10. Express 2.125 as a fraction in lowest terms.
    A. $2\frac{1}{8}$ B. $2\frac{1}{125}$
    C. $\frac{125}{1000}$ D. $2\frac{1}{4}$

11. Which is the circumference of a circle with a diameter of 39 yd? Round your answer to the nearest tenth or whole number.
    A. 122.5 yd   B. 244.9 yd
    C. 61.2 yd   D. 192.2 yd

12. Which is the circumference of a circle with a radius of 10.8 in.? Round your answer to the nearest tenth or whole number.
    A. 53.4 in.   B. 21.6 in.
    C. 17 in.   D. 67.8 in.

13. Add. $-8 + (-13)$
    A. -5 B. -21 C. -27 D. 12

14. Subtract. $5 - (-4)$
    A. 9   B. 1   C. -9 D. -1

15. Divide. $63 \div (-9)$
    A. 54   B. -7 C. 9   D. -54

16. Multiply. $9 \times (^-3)$
    A. 6   B. -3 C. -27 D. 12

17. Subtract. $-10 - 6$
    A. -4 B. 4
    C. 16   D. none of these

Name _____ Date _____

Exploring Integers

An integer is any number in the following set:

$$\{..., -3, -2, -1, 0, 1, 2, 3, ...\}$$

Integers can be shown on a number line.

Numbers that are **opposites** are the same distance from zero but in opposite directions. For example, -3 and 3 are opposites. The absolute value of an integer is its distance from 0.

$$|-4| = 4 \qquad |4| = 4 \qquad |-2| = 2$$

► **Example 1** _____

Write the integer represented by each of the points A, B, and C on the number line. Then give its opposite.

Solution

On the number line, $A = -3$, $B = -1$, and $C = 2$.

The opposite of -3 is 3.
The opposite of -1 is 1.
The opposite of 2 is -2.

Example 2 _____

Find each absolute value.

a. $|-11|$ b. $|32|$ c. $|0|$

Solution

The absolute value of an integer is its distance from 0. It is always a positive number.

a. $|-11| = 11$ b. $|32| = 32$ c. $|0| = 0$

EXERCISES _____

Write the integer represented by each labeled point on the number line. Then give its opposite.

1. A **-7; 7**
2. B **-4; 4**
3. C **-2; 2**
4. D **1; -1**
5. E **3; -3**
6. F **5; -5**
7. G **7; -7**
8. H **9; -9**

Find each absolute value.

9. $|-21|$ **21**
10. $|16|$ **16**
11. $|200|$ **200**
12. $|-121|$ **121**

Name _____ Date _____

Above and Below Sea Level

When measuring height and depth of locations on the surface of the earth, sea level is considered to be the starting point. Thus, on a number line representing elevations and depths, sea level corresponds to the 0 point. Positive integers represent heights above sea level. Negative integers represent depths below sea level.

EXERCISES _____

Graph each distance on the number line to the right.

1. Lake Huron, USA & Canada, 579 ft
2. Death Valley, CA, -282 ft
3. Valdes Peninsula, Argentina, -131 ft
4. New Orleans, LA, -8 ft
5. Ozark, AK, 396 ft
6. Ohio River, USA, 455 ft
7. Amsterdam, Netherlands, 5 ft
8. London, England, 149 ft
9. Paris, France, 164 ft
10. Prague, Czechoslovakia, 662 ft
11. Lake Eyre, Australia, -52 ft
12. Caspian Sea, Asia, -92 ft

When you have completed plotting the locations on the graph, answer these questions.

13. Which place listed has the lowest elevation?
    Death Valley

14. Which place on the graph is about the same distance above sea level as the Valdez Peninsula is below sea level?
    London, England

Name _____ Date _____

Adding Integers

To add integers on a number line:
- start with zero;
- move to the right for a positive integer;
- move to the left for a negative integer.

► **Example 1** _____

Add $-4 + 5$.

Solution

Use a number line.

The sum is 1. $-4 + 5 = 1$

Example 2 _____

Add $-2 + -1$.

Solution

Use a number line.

The sum is -3. $-2 + -1 = -3$

EXERCISES _____

Add, using the given number lines.

1. $-3 + -4$ **7**

2. $5 + -9$ **-4**

3. $-6 + 4$ **-2**

Draw a number line and add.

4. $-4 + -4$ **-8**
5. $12 + -8$ **4**
6. $-5 + -8$ **-13**
7. $-2 + -9$ **-11**
8. $6 + -3$ **3**
9. $-3 + -7$ **-10**

Name _____ Date _____

Adding on a Nomograph

At the right are three number lines labeled x, y, and z.

These three lines form a nomograph which can be used to find the sum of two integers.

► **Example** _____

Add $-4 + 6$ on the nomograph.

Solution

Find -4 on the number line labeled x and 6 on the number line labeled y. Use a ruler to draw a line between those two points. Use a sharp pencil and draw carefully.

The line you have drawn will cross the number line labeled z at 2, which is the sum of $-4 + 6$.

EXERCISES _____

1. Describe how the number lines are positioned.
   The three lines are parallel.

2. Describe how the scales of the lines are related.
   Each interval on the z line has a value exactly twice that of the corresponding
   interval on the x or y line.

Use the nomograph to find the following sums.

3. $-9 + 8$ **-1**
4. $-5 + -3$ **-8**
5. $-3 + 7$ **4**
6. $6 + -5$ **1**
7. $-2 + -7$ **-9**
8. $2 + -8$ **-6**

9. List some advantages and disadvantages of using a nomograph to add integers.
   Answers may vary. To add with the nomograph, one does not need to
   know any rules. However, the sums one can represent are limited by the size
   of the scales.

Name _____ **Date** _____

Subtracting Integers

You can use integer chips to model the subtraction of integers.

► **Example** _____

To model −7 − 2, start with 7 ⊟ chips.

⊟ ⊟ ⊟ ⊟
⊟ ⊟ ⊟

You cannot take away two ⊞ chips.

Therefore, add two combinations of ⊞ and ⊟ chips so that you can subtract two ⊞ chips.

⊞ ⊞ ⊟ ⊟ ⊟
⊞ ⊞ ⊟ ⊟

When you take away two ⊞ chips, the result is 9 ⊟ chips.

Therefore, −7 − 2 = −9.

Notice that the result is the same as when you add the opposite of 2.

$$-7 - 2 = -7 + (-2) = -9$$

EXERCISES _____

Choose the correct answer.

1. Which exercise will give the same result as 8 − 3?

   a. 8 + (−3)   **b.** −8 + 3   **c.** 8 + 3

2. To subtract −9 from 20,

   a. add −20 to 9   **b.** add 9 to 20   **c.** add −9 to 20

Subtract.

3. 3 − 7   **−4**   4. −7 − (−6)   **−1**   5. −5 − 17   **−22**

6. 15 − (−8)   **23**   7. −21 − (−8)   **−13**   8. 19 − (−7)   **26**

9. −14 − 6   **−20**   10. −3 − (−32)   **29**   11. −6 − 6   **−12**

12. −9 − (−9)   **0**   13. 34 − 43   **−9**   14. 16 − (−1)   **17**

Name _____ **Date** _____

Subtracting with a Nomograph

At the right are three number lines labeled x, y, and z.

These three lines form a nomograph, which can be used to find the sum or difference of two integers.

To add using the nomograph, carefully draw a line (use a ruler) from one addend on line x to the other addend on line y. The sum will be found where the line intersects line z.

Since addition and subtraction are related operations, you can also use this nomograph to subtract.

EXERCISES _____

1. Describe how the nomograph could be used to subtract integers.

   Descriptions will vary. To find −4 − 5 carefully draw a line from the −4 on

   line z to 5 on line x. Extend the line so that it intersects line y. The difference

   will be found where the line you drew intersects line y.

   _____

   _____

   _____

Use the nomograph to find each difference.

2. 4 − 9   **−5**   3. −3 − (−4)   **1**   4. −4 − 2   **−6**

5. −5 − (−3)   **−2**   6. 7 − (−3)   **10**   7. 3 − 6   **−3**

8. −8 − (−1)   **−7**   9. 6 − (−4)   **10**   10. −7 − 3   **−10**

11. Name a difference that cannot be found using the nomograph.

   Answers will vary. _____

Name _____ **Date** _____

Multiplying Integers

Look at these patterns in multiplying integers.

4 × 2 = 8	1 × (−2) = −2	−8 × 3 = −24
4 × 1 = 4	0 × (−2) = 0	−8 × 2 = −16
4 × 0 = 0	−1 × (−2) = 2	−8 × 1 = −8
4 × (−1) = −4	−2 × (−2) = 4	−8 × 0 = 0
4 × (−2) = −8	−3 × (−2) = 6	−8 × (−1) = 8
4 × (−3) = −12	−4 × (−2) = 8	−8 × (−2) = 16

These patterns suggest the following rules for multiplying integers.
• The product of two integers having the same signs is positive.
• The product of two integers having opposite signs is negative.

► **Example 1** _____   ► **Example 2** _____

Find the product of −9 × 8.   Find the product of −7 × −6.

Solution   **Solution**
Since the signs of the factors are opposite, the sign of the product is negative.   Since the signs of the factors are the same, the sign of the product is positive.

$$-9 \times 8 = -72$$ $$-7 \times -6 = 42$$

EXERCISES _____

Use the rules for multiplying integers to find each product.

1. −8 × 5   **−40**   2. 7 × (−11)   **−77**   3. −9 × (−7)   **63**

4. 3 × 12   **36**   5. −8 × (−10)   **80**   6. −4 × 9   **−36**

7. −6 × −6   **36**   8. 8 × (−6)   **−48**   9. −2 × (−30)   **60**

10. −3 × 5 × −9   **135**   11. −6 × 5 × 8   **−240**

12. −5 × −6 × −5   **−150**   13. −4 × −7 × 2   **56**

Solve.

14. The temperature was falling 3°F every hour. How many degrees had the temperature dropped after 8 hours? Write the change in temperature as an integer.

   24°F; −24°F

15. A diving bell is descending at a rate of 40 m per second. How many meters will it have descended after 15 seconds? Write the change in depth as an integer.

   600 m; −600 m

Name _____ **Date** _____

Magic Squares

A magic square is an arrangement of numbers in rows and columns so that the sum along any row, column, or diagonal is the same. In the magic square at the right, the sum along any row, column, or diagonal is −9.

−6	−1	−2
1	−3	−7
−4	−5	0

EXERCISES _____

1. Complete the square on the right by adding 2 to each entry in the magic square at the left.

−6	−1	−2
1	−3	−7
−4	−5	0

−4	1	0
3	−1	−5
−2	−3	2

2. Is the new square a magic square? If so, what is the sum of any row, column, or diagonal?

   yes; −3 is the sum of each row, column, and diagonal.

3. Complete the square on the left. Multiply each entry in the magic square at the left by the same integer.

−1	0	−5
−6	−2	2
1	−4	−3

3	0	15
18	6	−6
−3	12	9

4. Is the new square a magic square? If so, what is the sum of any row, column, or diagonal?

   yes; the sums are 18

5. What do you predict will happen if you begin with a magic square, multiply each entry by 2, and then add −1 to the result? Test your guess on the magic square at the top of the page.

   The result will also be a magic square.

218B

Name _____ Date _____

Dividing Integers

Division and multiplication are related operations. The rules for dividing integers are the same as the rules for multiplying integers.

Since $3 \times -4 = -12$, $-12 \div 3 = -4$ and $-12 \div -4 = 3$.

Since $-5 \times -6 = 30$, $30 \div -5 = -6$ and $30 \div -6 = -5$.

The rules for dividing integers are as follows:

The quotient of two integers having the same signs is positive.

The quotient of two integers having opposite signs is negative.

▶ **Example 1** _____

Divide $-36 \div 6$. Then check by multiplying.

Solution

The signs of the dividend and divisor are different. So, the quotient will be negative.

$-36 \div 6 = -6$

Check: $6 \times (-6) = -36$

▶ **Example 2** _____

Divide $-60 \div (-12)$. Then check by multiplying.

Solution

The signs of the dividend and the divisor are the same. So, the quotient will be positive.

$-60 \div (-12) = 5$

Check: $-12 \times 5 = -60$

EXERCISES

Find each quotient. Then check by multiplying.

1. $14 \div (-7)$ __−2__
2. $-32 \div (-4)$ __8__
3. $-63 \div 7$ __−9__
4. $-54 \div (-9)$ __6__
5. $-72 \div 8$ __−9__
6. $-18 \div (-3)$ __6__
7. $-66 \div 11$ __−6__
8. $28 \div (-4)$ __−7__
9. $-56 \div (-7)$ __8__
10. $156 \div (-13)$ __−12__
11. $-112 \div 8$ __−14__
12. $-352 \div (-22)$ __16__

Solve.

13. A company lost $400 in March, $500 in April, and $300 in May. Write an integer for the average loss per month.

__−$400__

14. Jeremy rode his bike for 4 hours. He covered 36 miles in that time. Write an integer for his rate.

__9 mph__

Name _____ Date _____

Patterns

Number patterns can be created using integers.

▶ **Example 1** _____

Find the next three terms in this pattern.

$4, -2, 3, -3, 2, -4, \ldots$

Solution

Subtract each term from the term that follows it to see if there is a pattern.

$4 \quad -2 \quad 3 \quad -3 \quad 2 \quad -4,$
$-6 \quad 5 \quad -6 \quad 5 \quad -6 \quad 5$

The pattern appears to be add -6, add 5, then repeat the same two steps.

Therefore, the next three terms are:

$4, -2, 3, -3, 2, -4,\ \mathbf{1},\ \mathbf{-5},\ \mathbf{0}$
$-6 \quad 5 \quad -6 \quad 5 \quad -6 \quad 5 \quad -6 \quad 5$

▶ **Example 2** _____

Find the next three terms in this pattern.

$3, -12, 48, -192, 768, \ldots$

Solution

Notice that the terms increase very quickly. This fact suggests that they may be related by multiplication.
Divide each term by the term preceding it to check.

$-12 \div 3 = -4;\ 48 \div -12 = -4;$
$-192 \div 48 = -4$

The pattern is to multiply each term by -4 to find the term that comes next. The next three terms are:

$3, -12, 48, -192, 768,\ \mathbf{-3072},\ \mathbf{12{,}288},\ \mathbf{49{,}152}$

EXERCISES

State the rule for generating the next term. Then find the next three terms in each pattern.

1. $10, 6, 2, -2, -6, \ldots$
   __Add −4; −10, −14, −18__

2. $-37, -29, -21, -13, \ldots$
   __Add 8; −5, 3, 11__

3. $3, 13, 8, 18, 13, 23, \ldots$
   __Add 10, add −5; 18, 28, 23__

4. $-40{,}000, 20{,}000, -10{,}000, \ldots$
   __Divide by −2; 5000, −2500, 1250__

5. $2, -6, 18, -54, \ldots$
   __Multiply by −3; 162; −468, 1458__

6. $-44, -22, 88, 44, -176, \ldots$
   __Divide by 2; multiply by −4; −88, 352, 176__

Make up two patterns of your own using integers and any operation or combination of operations. Switch patterns with a partner and see if you can continue each other's patterns. **Answers will vary.**

7. _____

8. _____

Name _____ Date _____

Problem Solving Skills: Choosing the Operation

In solving word problems, you must understand how the numbers in the problem are related. Then, the operation needed to solve the problem becomes clear.

▶ **Problem 1** _____

A number is seven less than negative eight. What is the number?

Solution

Restate the problem, using a blank for the number you are looking for and symbols for the other information.

A number is seven less than negative eight.

$\blacksquare = -8 \underset{\text{less than}}{\overline{}} 7$

$\blacksquare = -15$

Check. $-15 - 7 = -8$

So, -15 *is* less than -8.

▶ **Problem 2** _____

Negative thirty-six is four times some number. What is the number?

Solution

Negative thirty-six is four times some number.

$-36 = 4 \times \blacksquare$

Since -36 is the product, you are looking for one of the factors. Divide to find the other factor.

$-36 \div 4 = -9$

Check. $4 \times (-9) = -36$

So, -36 *is* four times -9.

EXERCISES

Choose the operation required. Then solve the problem.

1. A number is five more than negative six. What is the number?
   __−1__

2. A number is eight less than negative twenty-one. What is the number?
   __−29__

3. A number is three less than negative two. What is the number?
   __−5__

4. Twelve is ten more than some number. What is the number?
   __2__

5. A number is the product of four and negative seven. What is the number?
   __−28__

6. A number divided by negative fifteen is negative four. What is the number?
   __60__

7. Negative forty-eight is six times some number. What is the number?
   __−8__

8. A number is three more than the sum of negative nine and negative eleven. What is the number?
   __−17__

Name _____ Date _____

Creating Problems

Given two numbers and an operation, many different problems can be constructed.

▶ **Example** _____

Create a problem using the numbers 3 and -2 and the operation of multiplication.

Solution 1

Write a problem involving temperature.

The temperature dropped 2°F every hour. What was the change in temperature after 3 hours?

Solution 2

Write a problem about a bank account.

Geraldine's bank charges $2 every time her balance drops below $500. What is the total charge to her account if her balance dropped below $500 three different times in one month?

EXERCISES

For each operation and pair of numbers, create two different problems.

Answers may vary. Sample answers are given.

Addition, 56 and 72

1. __Dave bought a bicycle helmet for $56 and rain gear for $72. What was the total cost?__

2. __Louise drove 56 miles in the morning and 72 miles in the afternoon. How many miles did she drive in all?__

Subtraction, 120 and 98

3. __Martha read 120 books this summer. Last summer, she read 98 books. How many more did she read this year than last?__

4. __Steve saved $120 this year. He then spent $98 on a new bike. How much money does he have left?__

Division, 60 and 4

5. __José gets paid $4 an hour. How many hours must he work to earn $60?__

6. __Sixty campers are to be organized into 4 troops. How many campers will be in each troop?__

Name _____ Date _____

Rational Numbers

A rational number is any number that can be expressed in the form, $\frac{a}{b}$ where a is any integer and b is any integer except 0. Rational numbers can be graphed on a number line.

► **Example 1**

Write each rational number as a ratio of two integers.

a. -4　　b. 0.06　　c. -2.7

Solution

a. $-4 = \frac{-4}{1}$, or $\frac{-8}{2}$

b. $0.06 = \frac{6}{100}$

c. $-2.7 = \frac{-27}{10}$

Example 2

Graph the rational numbers on a number line.

$\frac{1}{2}, -\frac{3}{4}, -1\frac{1}{4}, -3\frac{3}{4}, 2\frac{1}{2}$

Solution

Since all the denominators are either 2 or 4, divide the number line into fourths.

$-3\frac{3}{4}$　　$-1\frac{1}{4}$　$-\frac{3}{4}$　$\frac{1}{2}$　　$2\frac{1}{2}$

$-4\ -3\ -2\ -1\ \ 0\ \ 1\ \ 2\ \ 3\ \ 4$

EXERCISES

Write each rational number as a ratio of two integers.

1. 12　$\frac{12}{1}$

2. 0.34　$\frac{34}{100}$, or $\frac{17}{50}$

3. -3.1　$\frac{-31}{10}$

4. -5　$\frac{-5}{1}$

5. 1.003　$\frac{1003}{1000}$

6. -2.13　$\frac{-213}{100}$

Name the point on the number line below that corresponds to each rational number.

7. 3.1　E

8. -2.2　A

9. 0.6　D

10. -0.2　C

11. -1.8　B

$A\ \ B$　　　C　　D　　　　E

-3　-2　-1　　0　　1　　2　　3

Order these numbers from least to greatest.

12. $\frac{3}{5}, \frac{-4}{5}, \frac{-2}{5}$　$\frac{-4}{5}, \frac{-2}{5}, \frac{3}{5}$

13. $\frac{-1}{4}, \frac{-3}{4}, \frac{-1}{2}$　$\frac{-3}{4}, \frac{-1}{2}, \frac{-1}{4}$

14. $\frac{3}{10}, -\frac{7}{20}, \frac{4}{5}$　$-\frac{7}{20}, \frac{3}{10}, \frac{4}{5}$

15. $-0.7, -\frac{11}{16}, -\frac{4}{5}, -\frac{11}{20}$　$-\frac{4}{5}, -0.7, -\frac{11}{16}, -\frac{11}{20}$

Name _____ Date _____

Betweenness

Rational numbers have the property of **betweenness.** That is, between any two rational numbers, there is always another rational number.

► **Example 1**

Find a rational number between $\frac{1}{7}$ and $\frac{1}{8}$.

Solution

There are many, many rational numbers between these two fractions. To determine one of them, you can find the rational number halfway between them by finding the average of the two numbers.

$\frac{1}{7} + \frac{1}{8} = \frac{8+7}{56} = \frac{15}{56}$

$\frac{15}{56} \div 2 = \frac{15}{56} \times \frac{1}{2} = \frac{15}{104}$

So, $\frac{15}{104}$ is the rational number halfway between $\frac{1}{7}$ and $\frac{1}{8}$.

Example 2

Find a rational number between 0.23 and 0.24.

Solution

There are many, many rational numbers between these two decimals. Increase the number of decimal places in each to help you find a decimal between them.

$0.23 = 0.230$
$0.24 = 0.240$

A rational number between 0.23 and 0.24 is 0.231.

EXERCISES

Find a rational number between each pair of rational numbers.

Answers will vary. Sample answers are given.

1. 2 and $2\frac{1}{2}$　$2\frac{1}{4}$

2. 0.3 and 0.32　0.31

3. 1.2 and -1.2　0

4. $\frac{3}{5}$ and $\frac{4}{5}$　$\frac{7}{10}$

5. $-1\frac{1}{3}$ and $-1\frac{1}{4}$　$-1\frac{7}{24}$

6. 0.5 and 0.51　0.505

7. Read the following statement. Decide whether or not you think it is true. Explain your thinking.

Between any two rational numbers, there are an infinite number of rational numbers.

This is a true statement. There is always a rational number between any two

rational numbers. This process can be repeated over and over. Therefore, the

number is infinite.

Name _____ Date _____

Problem Solving/Decision Making: Choose a Strategy

There are many problems for which more than one strategy may be used. There are also times when you will want to use one strategy to solve the problem and another to check your solution. The strategies you have studied up to this point are these:

Find a pattern　　　　Solve a simpler problem
Logical reasoning　　　Draw a picture

PROBLEMS

Name the strategy you would use to solve each problem. Then solve.

1. Suppose that six chess players enter a competition. The players must compete in pairs until each player has played one game against every other player one time. How many games must there be?
   draw a picture; 15 games

2. After opening a savings account, Margo deposited $5 into her savings account the first week. In the second week she deposited $8, and in the third week $12. If she keeps up this schedule, in which week will she deposit $30?
   find a pattern; the sixth week

3. Eva and her family visited friends in Walbridge. Walbridge is 9 miles due east of Farmdale. Martinsville is 13 miles due east of Duniston, which is 15 miles south and 5 miles west of Farmdale. How many miles and in which direction is Martinsville from Walbridge?
   draw a picture; 15 mi south and 8 mi west

4. The members of the nature club went hiking each day for a week. They hiked 8 miles the first day; each day thereafter they hiked $1\frac{1}{5}$ time farther each succeeding day. How far did they travel by the end of the fourth day? the sixth day? the seventh day?
   draw a picture; 11.4 mi; 14.4 mi; 16.2 mi

5. Some players enter a Ping-Pong tournament. Once a player is beaten by another player, he or she is eliminated from play. How many games must be played to determine a winner if there are 5 players?
   draw a picture; 4 games

6. Ann, Bill, Carlo, and Diane are each taking a science course. The courses offered are biology, chemistry, physics, and physiology. Ann has taken both physics and biology, but needs chemistry before she can take physiology. Neither of the boys has taken physics yet. Bill says that he found biology and chemistry really important prerequisites for the course he is taking now. The student taking biology also sings bass in the choir. Which science course is each student taking now?
   logical reasoning; Diane, physics; Anne, chemistry; Carlo, biology; Bill, physiology

---

Name _____ Date _____

Determining Stock Prices

The table below gives data on selected stocks. Included are the opening price of a share of stock (in dollars) for a given week, the closing price at the end of the week, and the net change in price for each stock during that period.

The net change in a share of stock is found by using this formula.

Net change = Closing price − Opening price

You can use this formula to find net change, closing price, or opening price, given any two of the three data. Fractional parts of a dollar are always given in eighths, fourths, or halves.

EXERCISES

Complete the table.

	WEEKLY STOCK QUOTATIONS		
Stock Name	Opening Price	Closing Price	Net Change
1. Fullco	$52\frac{1}{8}$	54	$+1\frac{7}{8}$
2. EcoProd	16	$15\frac{3}{8}$	$-\frac{5}{8}$
3. MetInk	$82\frac{1}{4}$	$83\frac{7}{8}$	$+1\frac{5}{8}$
4. ToyWorld	$26\frac{5}{8}$	$24\frac{3}{8}$	$-2\frac{1}{4}$
5. CompCo	$114\frac{3}{8}$	$111\frac{3}{4}$	$-2\frac{5}{8}$
6. PaperProd	$27\frac{1}{8}$	$28\frac{1}{2}$	$+1\frac{3}{8}$
7. PowerCo	$121\frac{3}{4}$	$117\frac{3}{8}$	$-4\frac{3}{8}$
8. Skinsft	$12\frac{7}{8}$	$15\frac{1}{4}$	$2\frac{3}{8}$
9. CleanCo	$24\frac{1}{4}$	$23\frac{1}{8}$	$-1\frac{1}{8}$
10. QuikCarry	$42\frac{5}{8}$	$45\frac{1}{2}$	$2\frac{7}{8}$

218D

Name _____   Date _____

Computer Activity: Gain and Loss

Problems that involve a repetitive gain and loss can create interesting patterns and results. For example: A football team is on their own 30-yard line. On each play, they gain 10 yards but get a 5-yard penalty. How many plays will it take the team to score a touchdown?

This program will ask you to INPUT the gain, the loss, and the total distance (in this case, 70 yards). It will then print a list showing how far the team advances on each play.

```
10   INPUT "WHAT IS THE GAIN? ";G        INPUT gain
     : PRINT
20   INPUT "WHAT IS THE LOSS? ";L        INPUT loss
     : PRINT
30   INPUT "WHAT IS THE TOTAL DISTANCE? ";D   INPUT distance
     : PRINT
40   PRINT "HOW MANY","HOW FAR"          PRINTS column heads
50   FOR X = 1 TO D
60   AG = AG − L + G                     Determines actual gain
70   PRINT X,AG
80   IF AG > = D THEN GOTO 100           When actual gain equals
90   NEXT X                              or exceeds distance, the
100  END                                 program exits the loop
```

EXERCISES

1. RUN the program for a distance of 20, then 40, then 60 yards. Use the results to predict the number of plays needed for the touchdown. RUN to check your answer.

   3 plays; 7 plays; 11 plays; 13 plays

2. Repeat Exercise 1 when the gain is 5 and the loss is 3.

   33 plays for 70 yards

3. Will the results change if the loss comes before the gain? Change line 60 to AG = AG − L + G and line 100 to END to check.

   yes

Technology • SECTION 6-2   113

Name _____   Date _____

Calculator Activity: Finding Missing Integers

The $\boxed{+/-}$ key and working backwards can help you find unknown integers.

▶ **Example**

Use a calculator to find each unknown integer.

a. $\square + (-6) = 8$ b. $\square - (-4) = -10$ c. $7 \times \square = -42$ d. $\square \div -3 = -11$

Solution

a. Since addition and subtraction are inverse operations, subtract −6 from 8 to find the unknown integer.
   $8 \boxed{-} 6 \boxed{+/-} \boxed{=} 14$

b. Add −10 and −4 to find the unknown integer.
   $10 \boxed{+/-} \boxed{+} 4 \boxed{+/-} \boxed{=} -14$

c. Divide −42 by 7 to find the unknown integer.
   $42 \boxed{+/-} \boxed{\div} 7 \boxed{=} -6$

d. Multiply −11 and −3 to find the unknown integer.
   $11 \boxed{+/-} \boxed{\times} 3 \boxed{+/-} \boxed{=} 33$

EXERCISES

Use a calculator to find each unknown integer.

1. $11 + \square = -17$ __−28__   2. $\square + (-6) = -4$ __2__
3. $\square - 16 = -8$ __8__   4. $\square \times 6 = -36$ __−6__
5. $\square \div 8 = -10$ __−80__   6. $14 \times \square = -98$ __−7__
7. $36 + \square = 24$ __−12__   8. $-15 \times \square = 90$ __−6__
9. $\square - (-62) = 87$ __25__   10. $\square \div 28 = -252$ __−7,056__
11. $\square - 150 = -245$ __−95__   12. $128 + \square = -12$ __−140__
13. $\square \times -55 = -605$ __11__   14. $\square - (-36) = -45$ __−81__
15. $\square \div 28 = -12$ __−366__   16. $\square \div -16 = 32$ __−512__
17. $212 + \square = -14$ __−226__   18. $108 \times \square = -540$ __−5__
19. $-77 + \square = -144$ __−67__   20. $\square - 37 = -12$ __25__
21. $\square - (-18) = 36$ __18__   22. $-12 \times \square = -144$ __12__
23. $\square \div -52 = -6$ __312__   24. $\square \div 45 = -6$ __−270__

114   Technology • SECTION 6-5

ACHIEVEMENT TEST

Exploring Integers

CHAPTER 6   FORM A

MATH MATTERS BOOK 1
Chicha Lynch
Eugene Olmstead

SOUTH-WESTERN PUBLISHING CO.

Name _____
Date _____

SCORING RECORD	
Possible	Earned
45	

Find each absolute value.

1. $|7|$ __7__   2. $|-9|$ __9__   3. $|-18|$ __18__   4. $|5|$ __5__

Compare by writing $<$, $>$, or $=$.

5. -8 __<__ 8 6. -22 __<__ -20 7. 12 __>__ -2
8. 0 __>__ -7 9. -15 __<__ -3 10. -11 __<__ 4

Add.

11. $-7 + 5$ __−2__   12. $8 + (-24)$ __−16__   13. $-5 + (9)$ __4__
14. $9 + (-37)$ __−28__   15. $-12 + (-12)$ __−24__   16. $4 + (-4)$ __0__

Subtract.

17. $6 - (-7)$ __13__   18. $-5 - 8$ __−13__   19. $-12 - (-14)$ __2__
20. $-3 - 5$ __−8__   21. $8 - 15$ __−7__   22. $4 - (-9)$ __13__

Multiply.

23. 7×8 __56__   24. -5×8 __−40__   25. $-6 \times (-4)$ __24__
26. $9 \times (-6)$ __−54__   27. $-8 \times (-9)$ __72__   28. -5×6 __−30__

Divide.

29. $18 \div 9$ __2__   30. $-63 \div 21$ __−3__   31. $-65 \div (-5)$ __13__
32. $128 \div (-8)$ __−16__   33. $-36 \div (-4)$ __9__   34. $-4 \div 4$ __−1__

6A-1

Name _____   Date _____

Write each rational number as a ratio of two integers.

35. $4\frac{16}{4}$ __$\frac{16}{4}$__   36. 0.7 __$\frac{7}{10}$__   37. $-4\frac{2}{3}$ __$-\frac{14}{3}$__   38. 7.1 __$\frac{71}{10}$__
39. -6.7 __$-\frac{67}{100}$__   40. $3\frac{8}{9}$ __$\frac{35}{9}$__   41. -0.23 __$-\frac{23}{100}$__   42. 1 __$\frac{7}{7}$__

Solve.

43. A number is nine less than negative ten. What is the number?
    −19

44. A number is seven more than negative sixteen. What is the number?
    −9

45. Five teams will play in a tournament. Each team will play every other team once. How many games will be played?
    10 games

6A-2

ACHIEVEMENT TEST

Exploring Integers
CHAPTER 6   FORM B

MATH MATTERS BOOK 1
Chicha Lynch
Eugene Olmstead

SOUTH-WESTERN PUBLISHING CO.

Name _____

Date _____

SCORING RECORD	
Possible	Earned
45	

Find each absolute value.

1. $|5|$ ___5___   2. $|-3|$ ___3___   3. $|-23|$ ___23___   4. $|9|$ ___9___

Compare by writing $<$, $>$, or $=$.

5. -6 ___$<$___ 8 6. -11 ___$<$___ -10 7. 22 ___$>$___ -22

8. 0 ___$>$___ -4 9. -14 ___$<$___ -7 10. -10 ___$<$___ 5

Add.

11. $-9 + 5$ ___-4___   12. $7 + (-27)$ ___-20___   13. $-6 + (8)$ ___2___

14. $17 + (-47)$ ___-30___   15. $-15 + (-12)$ ___-27___   16. $6 + (-6)$ ___0___

Subtract.

17. $3 - (-7)$ ___10___   18. $-9 - 5$ ___-14___   19. $-12 - (-16)$ ___4___

20. $-6 - 2$ ___-8___   21. $9 - 11$ ___-2___   22. $4 - (-5)$ ___9___

Multiply.

23. 7×9 ___63___   24. -9×8 ___-72___   25. $-6 \times (-3)$ ___18___

26. $8 \times (-5)$ ___-40___   27. $-9 \times (-9)$ ___81___   28. -4×6 ___-24___

Divide.

29. $16 \div 8$ ___2___   30. $-46 \div 23$ ___-2___   31. $-60 \div (-5)$ ___12___

32. $144 \div (-12)$ ___-12___   33. $-39 \div (-3)$ ___13___   34. $-6 \div 6$ ___-1___

Copyright © 1993 by South-Western Publishing Co.
M IO1AG

6B-1

Name _____   Date _____

Write each rational number as a ratio of two integers.

35. 5 ___$\frac{10}{2}$___   36. 0.9 ___$\frac{9}{10}$___   37. $-3\frac{1}{3}$ ___$\frac{-10}{3}$___   38. 5.3 ___$\frac{53}{10}$___

39. -6.7 ___$\frac{-67}{100}$___   40. $3\frac{5}{9}$ ___$\frac{32}{9}$___   41. 0.33 ___$\frac{33}{100}$___   42. 1 ___$\frac{8}{8}$___

Solve.

43. A number is fifteen less than negative two. What is the number?
___-17___

44. A number is nine more than negative eight. What is the number?
___1___

45. Eleven players enter a tournament. Once a player is eliminated by another player, he or she cannot play another round. How many tournament matches will be played?
10 matches

6B-2

CHAPTER 7 SKILLS PREVIEW

Tell whether the equation is *true, false,* or an *open* sentence.

1. $9 = 6 - w$ open

2. $-15 = 5(4 - 7)$ true

Solve each equation.

3. $-2k = -9$ $4\frac{1}{2}$

4. $t - 3 = -8$ -5

5. $6.3 = y + 8.5$ -2.2

6. $-3 = \frac{n}{-7}$ 21

7. $8c + 11 = -13$ -3

8. $14 = 3z - 4$ 6

9. $\frac{8}{9}n = -\frac{2}{3}$ $-\frac{3}{4}$

10. $\frac{5}{4}x + 6 = 1$ -4

11. $3(x + 3) = 15$ 2

12. $8c - 9 = 6c + 5$ 7

13. Find the height of a triangle with a base measuring 16 cm and an area of 360 cm². The formula for the area of a triangle is $A = \frac{1}{2}bh$ where A = area, b = length of base, and h = height. 45 cm

14. Solve the formula $r = t - n$ for t. $t = r + n$

Simplify.

15. $\frac{1}{3}y + \frac{1}{3}y + 4y + \frac{1}{3}y$ $5y$

16. $4(x - 3) + 2(x + 5) - 3x$ $3x - 2$

Graph each open sentence on a number line.

17. $x \geq -3$

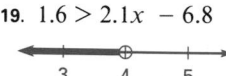

18. $-1 = 3y + 5$

19. $1.6 > 2.1x - 6.8$

20. $\frac{m}{-2} + 3 \leq 2$

Write an equation and solve.

21. Ms. Flynn bought a melon for $0.79 and three grapefruit. The total cost of the fruit was $1.84. How much did each grapefruit cost? $3g + 0.79 = 1.84$; $0.35

22. Lisa bought a package of postcards. She hung 4 on her bulletin board, then gave 3 to her friend José. Finally, she sent half of the remaining cards to pen pals. She had 7 cards left. How many postcards were in the package she bought? 21 cards

23. Midori earned $120.90 before taxes. Her hourly wage is $7.80. How many hours did she work? $15\frac{1}{2}$ hours

24. Lena paid $(a + 3b)$ cents for a pint of cream and $(2a + b)$ cents for 2 quarts of milk. Write and simplify an expression for the total cost of her purchases. $2(2a + b) + (a + 3b)$; $5a + 5b$

7 SKILLS PREVIEW

Introduction The purpose of this Skills Preview is to assess students' knowledge of all the major objectives of Chapter 7. Test results may be used
- to determine those topics which may need only to be reviewed and those topics which need to be more carefully developed;
- for class placement;
- in prescribing for individual differences.

If you prefer, you may use the Skills Preview as an alternative form of the Chapter Test (page 264) to evaluate mastery of chapter objectives. The items on the Skills Preview and the Chapter Test correspond in content and level of difficulty.

OVERVIEW

In this chapter, students learn to solve one–step equations by adding or subtracting the same number to or from an equation or by multiplying or dividing the equation by the same number. They also learn to write an equation to fit a particular situation and to solve two-step equations. Students evaluate formulas and solve them for a specified variable. They also learn to solve and graph one- and two-step inequalities. Finally, the strategy of working backward to solve a problem is introduced.

SPECIAL CONCERNS

This chapter provides an introduction to key concepts involved in first-year algebra. Students discover the idea of modeling an equation with a balance scale; that is, whatever operation is performed on one side of the balance (equation), must be performed on the other side in order to maintain the balance. Students may also find it useful to use algebra tiles to illustrate the solutions of equations.

VOCABULARY

combining like terms	one-step equation	time
equation	open sentence	true equation
false equation	rate	two-step equation
formula	real numbers	unlike terms
inequality	reciprocal	
inverse operations	solution	
like terms	terms	

MATERIALS

algebra tiles	small slips of paper

BULLETIN BOARD

Make a bulletin board showing a large balance scale and two or three equations underneath it. Ask students to find formulas or equations used in other areas of study, such as in science or business courses, and add them to the display with a short description of how the formula or equation is used. Have students provide worked-out solutions for some of their display examples.

INTEGRATED UNIT 2

The skills and concepts involved in Chapters 5–8 are included within the special Integrated Unit 2 entitled "U. S. and World Travel." This unit is in the Teacher's Edition beginning on page 298F. Worksheets for this integrated unit appear in the Enrichment Activities booklet, pages 93–95.

TECHNOLOGY CONNECTIONS

- Calculator Worksheets, 137–138
- Computer Worksheets, 139–140
- MicroExam, Apple Version
- MicroExam, IBM Version

- *Algebra Concepts*, Venture Educational Systems
- *Algebra Shop*, Scholastic

TECHNOLOGY NOTES

Computer algebra software, such as those listed in Technology Connections, provide students with practice in the use of basic algebra terminology and symbols. The students can input their own values for variables in expressions and equations. Some programs use game-like formats, allowing students to manipulate simulations of algebra tiles on the screen.

EQUATIONS AND INEQUALITIES

PLANNING GUIDE

SECTIONS	TEXT PAGES	ASSIGNMENTS			
		BASIC	AVERAGE	ENRICHED	
Chapter Opener/Decision Making	220–221				
7–1 Introduction to Equations	222–225	1–14, 19–27, 29–30, 35, 38	1–28, 31–36, 38–39	7–28, 29–37, 38–41	
7–2 Using Addition or Subtraction to Solve an Equation	226–229	1–12, 17, 18–23, 28, 31	1–17, 18–29, 31–33	18–30, 31–34	
7–3 Problem-Solving Skills: Writing an Equation	230–231	1–8	1–8	1–8	
7–4 Using Multiplication or Division to Solve an Equation	232–235	1–15, 19, 21–28, 35, PSA 1–5	1–20, 23–36, 40–41, PSA 1–5	21–36, 37–41, PSA 1–5	
7–5 Using More Than One Operation to Solve an Equation	236–239	1–15, 17–18, 22–23, 25	1–18, 22–23, 25–26	11–18, 19–23, 24–27	
7–6 Using a Reciprocal to Solve an Equation	240–243	1–13, 14–19, PSA 1–3	1–13, 14–22, PSA 1–5	14–22, 23, PSA 1–5	
7–7 Working with Formulas	244–247	1–4, 9–13, 14–16	1–13, 14–18, 19	5–13, 14–18, 19–20	
7–8 Problem-Solving Strategies	248–249	1–8	1–8	1–8	
7–9 Combining Like Terms	250–253	1–17, 20, 22–25, 28–31, 41, 46	1–27, 28–35, 40–43, 46	28–45, 46–47	
7–10 Graphing Open Sentences	254–257	1–12, 15–16, 18–20, 23, 26–27	1–17, 18–25, 26–31	7–8, 13–35	
7–11 Solving Inequalities	258–261	1–16, 19, 26, 28	1–17, 19–21, 24, 26, 28–30	10–17, 18–25, 26–32	
Technology	237, 253, 259, 261	✔	✔	✔	

ASSESSMENT				
Skills Preview	219	All	All	All
Chapter Review	262–263	All	All	All
Chapter Test	264	All	All	All
Cumulative Review	265	All	All	All
Cumulative Test	266	All	All	All

ADDITIONAL RESOURCES

RETEACHING	ENRICHMENT	TECHNOLOGY	TRANSPARENCY
7–1	7–1		
7–2	7–2		
7–3	7–3		
7–4	7–4		
7–5	7–5		TM 30
7–6	7–6	7–6	
7–7	7–7		
7–8	7–8		
7–9	7–9		
7–10	7–10	7–10	TM 31, 32
7–11	7–11		

ASSESSMENT OPTIONS

Chapter 7, Test Forms A and B	
Chapter 7, Test	Text, 264
Alternative Assessment	TAE, 264
Chapter 7, MicroExam	

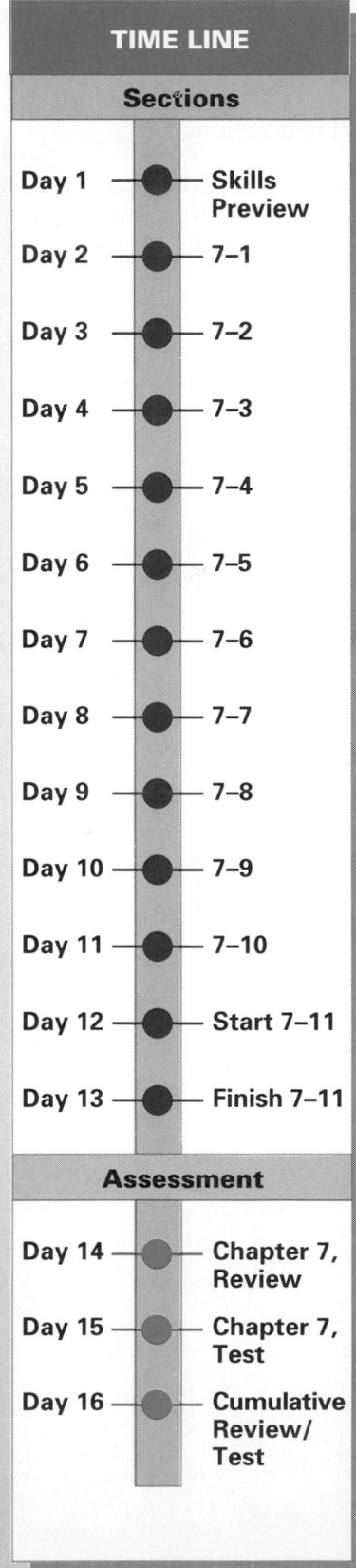

TIME LINE

Sections

Day 1	Skills Preview
Day 2	7–1
Day 3	7–2
Day 4	7–3
Day 5	7–4
Day 6	7–5
Day 7	7–6
Day 8	7–7
Day 9	7–8
Day 10	7–9
Day 11	7–10
Day 12	Start 7–11
Day 13	Finish 7–11

Assessment

Day 14	Chapter 7, Review
Day 15	Chapter 7, Test
Day 16	Cumulative Review/Test

7 CHAPTER OPENER

Objective To explore how data can be used as a basis for making real-life decisions

Introduction After reading and discussing the introductory material, have the students use numbers to give examples of equations and inequalities. Then introduce the formula given here for computing a bowler's handicap. Students may need to research the meaning of *handicap* and how it is used in various sports.

Decision Making Using Data Have students work in small groups to find answers to these questions. Have them share their answers to Exercise 3 with the class.

Working Together Brainstorm possible sources for formulas used in sports and ideas for kinds of formulas students might research. Give each group a chance to present its problems to the class.

CHAPTER 7

EQUATIONS AND INEQUALITIES

THEME Recreation

Algebra is a branch of mathematics in which symbols are used to represent numbers. Unlike symbols in arithmetic, algebraic symbols can vary in value. Only when numbers are substituted for them do algebraic symbols take on specific values. Their variability gives them a wide range of applications in solving problems, particularly problems involving equations and inequalities.

Because formulas use letters to represent numbers, they are studied in algebra. Formulas are often used to compute statistics in sports. You might use a formula like the one that follows to determine a bowler's handicap.

formula:
$h = (0.8)(200 - a)$

meaning of letters:
h = bowler's handicap
a = average score

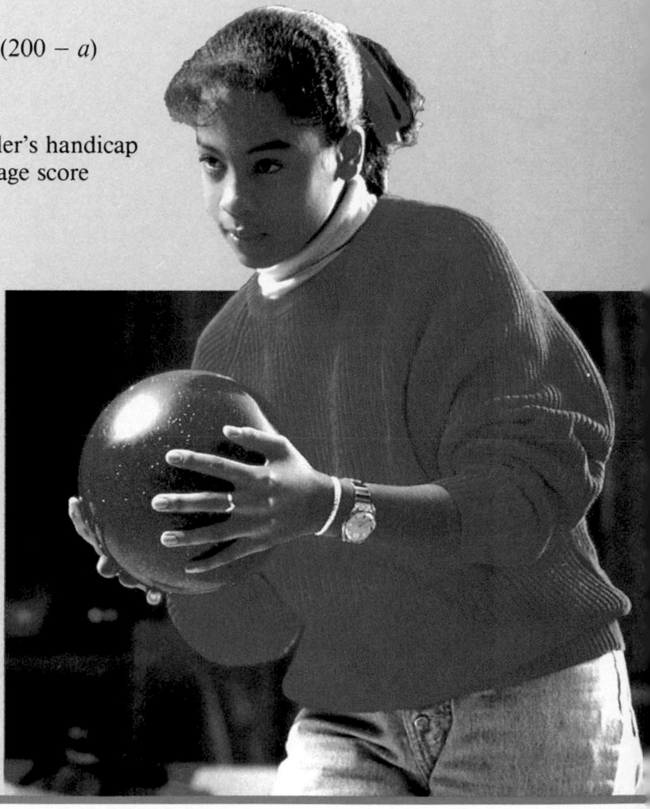

BOWLING RESULTS			
Name	Game 1	Game 2	Game 3
		Score	
Christy	85	121	145
Mark	114	102	126
Evan	96	97	95
Rox	156	177	186

DECISION MAKING

Using Data

Use the table to answer the following questions.

1. Calculate each bowler's handicap. Round to the nearest whole number. Christy, 66; Mark, 69; Evan, 83; Rox, 22

2. A bowler's handicap was 40. Describe a method for finding the bowler's average. What was the average? Answer will vary; guess and check; 150

3. Why is it harder to use the formula to find an average than it is to use it to find a handicap? The unknown quantity (the average) is on the right side of the formula. The formula does not give it directly.

Working Together

Your group's task is to find at least five formulas that are used to calculate sports statistics. Use your own knowledge, library reference materials, or interviews with people involved in sports. Write the formulas, clearly identifying the meaning of the letters in each. Then write two problems involving each formula. Exchange your problems with another group.

OBJECTIVE
* Determine if an equation is true, false, or open

VOCABULARY
equation, true equation, false equation, open sentence, solution of an equation

WARM-UP

Find the value of each expression for n = 3.
1. $-4(-2 + n)$ **-4**   2. $-8n$ **-24**
3. $2n + (-9)$ **-3**   4. $(-12)^2 \div n$
                                    48

1 MOTIVATE

Explore Have students work in small groups to solve this problem. Point out that we do not know what Pat weighs, so we can let her weight be whatever number between 120 and 134 we need to balance the team. It may be helpful to have students write the names and weights on a slip of paper and use the pieces of paper to form balanced teams. Talk about the fact that Pat's weight could vary—that is, we could assign different values to it to make our teams balance. Allow time for students to compare answers.

2 TEACH

Use the Pages/Skills Development Have students read this part of the section and then discuss the examples.

7-1 Introduction to Equations

EXPLORE

Answers will vary. One possible answer might be that Mario, Jill, William, Ed, and Pat make up one team. Pat must then weigh 120 pounds to make both teams have the same total weight, 631 pounds.

Tug-of-war teams are to be chosen so that the total weight on both sides is equal. Students participating are Mario, who weighs 135 lb; Julie, 120 lb; Gwen, 125 lb; Pao, 143 lb; Heather, 104 lb; Mike, 139 lb; Ed, 112 lb; Jill, 122 lb; William, 142 lb; and Pat, who weighs between 120 lb and 134 lb.

How would you make up the teams so that each side has the same total weight? What is Pat's weight in your arrangement? Check with other students. Did they make the same arrangements?

SKILLS DEVELOPMENT

An **equation** is a statement that two numbers or expressions are equal.

This equation is **true.**	This equation is **false.**	This equation is neither true nor false.
$12 - 7 = 5$	$13 - 7 = 5$	$x - 7 = 5$

An **open sentence** is a sentence that contains one or more variables. It can be true or false, depending upon what values are substituted for the variables. A value of the variable that makes an equation true is called a **solution of the equation.**

Open sentence: $x - 7 = 5$
Solution: $x = 12$

TEACHING TIP

Students commonly think that false equations are not equations. It may help to write a variety of expressions and true and false equations on the chalkboard or a worksheet and have students identify the equations. Once they have done so, they can determine which are true and which are false.

Example 1

Tell whether the equation is *true, false,* or an *open sentence.*

a. $3(8 - 2) = 20$ **b.** $2y - 6 = 10$ **c.** $-9 + 5 = 2(7 - 9)$

Solution

a. $3(8 - 2) = 20$
  $3(6) \stackrel{?}{=} 20$
  $18 \stackrel{?}{=} 20$
  The equation is *false.*

b. $2y - 6 = 10$
  The equation contains a variable, so it is an *open sentence.*

c. $-9 + 5 = 2(7 - 9)$
  $-4 \stackrel{?}{=} 2(-2)$
  $-4 = -4$
  The equation is *true.* ◀

Example 2

Which of the values 5 and 6 is a solution of the equation $3x - 1 = 17$?

Solution

Substitute 5 for x in the equation.

$3x - 1 = 3(5) - 1$ ⟵ **Remember the order of operations: Multiply first, then subtract.**
  $= 15 - 1$
  $= 14$

So 5 is not a solution.

Substitute 6 for x in the equation.

$3x - 1 = 3(6) - 1$
  $= 18 - 1$
  $= 17$

So 6 is a solution. ◀

Example 3

Use mental math to solve the equation $x + 9 = 14$.

Solution

Think: What number added to 9 equals 14?
You know that $5 + 9 = 14$, so $x = 5$. ◀

Example 4

Phil, who weighs 191 lb, will have a tug of war against Andy, who weighs 98 lb, and another person. Andy's teammate will be either Anna (97 lb) or Tonio (93 lb). Use the equation $98 + x = 191$. Find the partner Andy should team with so that the weights will be equal on both sides.

Solution

$$98 + x = 191$$

Substitute Anna's weight for x.
  $98 + 97 \stackrel{?}{=} 191$
  $195 \neq 191$ ⟵ **≠ means "is not equal to."**

Substitute Tonio's weight for x.
  $98 + 93 \stackrel{?}{=} 191$
  $191 = 191$

Andy should team up with Tonio. ◀

READING MATH

Sometimes the solution of an equation is said to satisfy the equation. An everyday meaning of the word satisfy is "to meet the needs or requirements of." So a solution of an equation like $3x - 1 = 17$ meets a certain requirement. What is that requirement?

When used as a value of the variable, the number makes a true equation.

Discuss the meaning of *equation* and have students pick out equations from among several equations and expressions you write on the chalkboard. Introduce the terms *true equation, false equation,* and *open sentence.*

Example 1: Have students complete this example on their own and explain their choices before looking at the given solution.

Example 2: Point out that, if you replace the variables in an equation with numbers, you will get either a true equation or a false equation. If the equation is true, that number is a solution of the equation. It may be helpful to substitute other values for x in this equation to help make the point clear.

Example 3: Students can use their knowledge of number facts to solve problems such as this one.

Example 4: Have students try to find a solution of the equation $98 + x = 191$ on their own before discussing the given solution.

TRY THESE

Tell whether each equation is *true, false,* or an *open sentence.*

1. $-3(5 - 7) = 6$ **true**
2. $n - 4 = 4$ **open**
3. $-6 + 3 = \frac{12}{7 - 3}$ **false**

Which of the given values is a solution of the equation?

4. $3 - x = -1$; 2, 4, -4 **4**
5. $2m - 1 = 9$; 3, 4, 5   **5**

Use mental math to solve the following equations.

6. $t + 18 = 25$ **7**
7. $h - 3 = -1$ **2**
8. $p + 5 = -2$ **-7**

Solve.

9. On a seesaw, Art and Beth balance Craig. Art weighs 113 lb and Craig weighs 219 lb. Use the equation $113 + x = 219$ to find whether Beth weighs 96 lb or 106 lb.   **Beth weighs 106 lb.**

EXERCISES

PRACTICE/ SOLVE PROBLEMS

Tell whether each equation is *true, false,* or an *open sentence.*

1. $8 = 9 - y$ **open**
2. $-11 = -15 - 4$ **false**
3. $24 \div (-6) = -4$ **true**
4. $4(2 + 5) = 30 - (4 - 6)$ **false**
5. $\frac{-3 + 11}{2} = 5 - (4 - 3)$ **true**
6. $8 + (7 - 5) = 3 - x + 2$ **open**

Which of the given values is a solution of the equation?

7. $b - 5 = -1$; 4, 6   **4**
8. $9 + c = 9$; 0, 1   **0**
9. $d + 7 = -1$; $-6, -8$ **-8**
10. $1 - e = 4$; $-3, -5$ **-3**
11. $-6 = 6 + h$; 0, -12 **-12**
12. $8 = f - 4$; 4, 12   **12**
13. $2k = 2$; 0, 1, -1 **1**
14. $8m - 2 = -10$; -8, 1, -1 **-1**
15. $17 = 2n$; $8\frac{1}{2}, 9\frac{1}{2}, \frac{1}{2}$ **$8\frac{1}{2}$**
16. $19 = 10 - 2p$; 5.5, 4.5, -4.5 **-4.5**
17. $\frac{p}{4} = 12$; 3, 8, 48   **48**
18. $7 + \frac{1}{2}k = 9$; 0, 2, 4   **4**

Use mental math to solve each equation.

19. $x + 5 = 7$ **2**
20. $n - 4 = 3$ **7**
21. $2 = 10 - k$ **8**
22. $3 + h = 0$ **-3**
23. $-4m = 20$ **-5**
24. $-3 = 18 \div m$ **-6**
25. $-9 + p = -5$ **4**
26. $6 = 12 + y$ **-6**

224 CHAPTER 7   Equations and Inequalities

MAKING CONNECTIONS

Students will encounter true and false sentences in other subjects such as science, English, and history. Have them write two true statements about a topic of their choice and then write two false statements about the same topic.

Solve.

27. Zoa scored 13 more points than Tad did. Zoa scored 41 points. Use the equation $41 - T = 13$ and these values for x: 22, 25, 28. Find T, the number of points that Tad scored. **28 points**

28. Pete's earnings equaled the sum of Doug's and Sam's earnings. Pete earned $82 and Sam earned $39. Use the equation $39 + x = 82$ and these values for x: 37, 43, 45. Find x, the amount that Doug earned. **$43**

Which of the given values is a solution of the equation?

EXTEND/
SOLVE PROBLEMS

29. $4b = -3 + 5$; $1, \frac{1}{2}, -1\frac{1}{2}$

30. $-\frac{1}{4} + c = -\frac{1}{2}$; $\frac{1}{2}, -\frac{1}{4}; \frac{1}{4} - \frac{1}{4}$

31. $2a + (-2) = 0$; $0, 1, -1$ **1**

32. $-9 + 3 = 3z - 15$; $0, 1, -1$
none of these

33. $(5 - 7)d = \frac{-12}{9 - 5}$; $-1.5, 1, 1.5$ **1.5**

34. $4.6f - 2.3 = 1 - 7.9$; $1, 2, -1$ **–1**

Solve.

35. A stack of nickels is worth $1.35. Use the equation $0.05n = 1.35$ and these values for n: 26, 27, 28. Find n, the number of nickels in the stack. **27 nickels**

36. The sum of Mike's and Hedda's ages equals the sum of Nan's and Sol's ages. Hedda is 26, Nan is 33, and Sol is 41. Use the equation $M + 26 = 33 + 41$ and these values for M: 48, 39. Find M, Mike's age. **48 years**

37. A radio was on sale for $48, which was $\frac{2}{3}$ of its regular price. Use the equation $\frac{2}{3}r = 48$ and these values for r: 32, 72, 81. Find r, the regular price. **$72**

Use mental math to find two solutions of the equation.

THINK CRITICALLY/
SOLVE PROBLEMS

38. $x^2 = 25$ **5, –5**

39. $h^2 - 4 = 5$ **3, –3**

40. Verify that 5 is a solution of the equation $-2 + k = 3$. Do you think any other value of k is a solution? Explain.

40. Since $-2 + 5 = 3$, a solution of the equation is 5. No other value of k can satisfy the equation. Possible explanation: The only way to get from -2 to 3 on a number line by a single addition is to move 5 units to the right, that is, to add 5. So no value of k except 5 satisfies $-2 + k = 3$.

USING DATA   In baseball, a pitcher's earned run average (ERA) gives the average number of earned runs the pitcher gives up per game. Use the Data Index on page 546 to find ERA statistics, and solve.

41. Ed Plank's ERA is 0.46 greater than that of another pitcher. Use the values 1.82 and 1.88 and the equation $P - p = 0.46$ to find the other pitcher's name (P = Ed Plank's ERA; p = the other pitcher's ERA). **Addie Joss**

Think Critically/Solve Problems
Ask students why there are two solutions to each of the equations in Exercises 38 and 39. (The variables are to be squared, so both a positive and a negative number is the solutiuon to each.)

5 FOLLOW-UP

Extra Practice

1. Give an example of a true equation. **Answers will vary.**

2. Give an example of a false equation. **Answers will vary.**

3. Give an example of an open sentence. **Answers will vary.**

4. Kathi earned $43 more than Kurt. Kurt earned $83. Find x, the amount Kathi earned. Use the equation $x - 43 = 83$ and these values for x: 120, 126, 128. **126**

Extension   Challenge students to write an equation to fit the following problem and then use the guess-and-check strategy to find the answer. For three weeks, Robin exercised by running. She ran 5 mi more the second week than she did the first and 5 mi more the third week than the second. If she ran 78 mi in all during the third week, how far did she run the first week? **$x + (x + 5) + (x + 10) = 78$; $x = 21$**

Section Quiz   Tell whether each equation is *true, false,* or *open*.

1. $-8 = 3 - 11$ **true**
2. $2 = 7 + x$ **open**
3. $5(2 + 7) = 20 - (6 - 3)$ **false**
4. $-12 \div 6 = 2$ **false**

Which of the given values is a solution of the equation?

5. $b - 6 = 12$ 6, 18 **18**
6. $2m - 3 = -15$ $-9, -6$ **–6**
7. $21 - 3x = 30$ $-3, 3$ **–3**

CHALLENGE

Challenge the students to write an equation relating the number of students, s, to the number of classes, c.

classes	1	2	3	c
students	30	60	90	s

$s = 30c$

7-2 Using Addition or Subtraction to Solve an Equation

EXPLORE

A double-pan balance has a 16-g weight on the left side and weights of 11 g and 5 g on the right side.

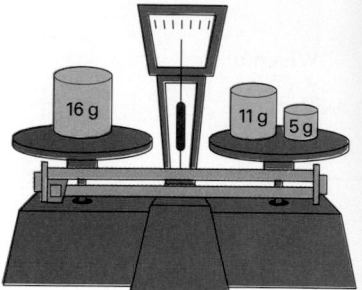

Yes. Both pans hold the same total weight.

1. Is the scale in balance? How can you tell?

Yes; replace with 2-g and 3-g weights. Yes; replace with 4-g, 5-g, and 2-g weights.

2. Suppose you could choose any whole-number gram weight up to 10 grams. Could you replace the 5-g weight with two other weights without losing balance? Give an example. Could you replace the 11-g weight with three other weights without losing balance? Give an example.

3. What will happen if you add a 4-g weight to the left side? What can you do on the right side to bring the scale into balance?
It will tip down; add 4 g.

4. What will happen if you remove the 11-g weight from the right side? What can you do on the left side to bring the scale into balance? It will tip up; replace the 16-g weight with a 5-g weight.

SKILLS DEVELOPMENT

When you **solve an equation,** you find all the values of the variable that make the equation true. In solving an equation, it helps to think of the equation as a balance scale.

"Undo" each operation performed on the variable by using the inverse operation. Keep the equation in balance by performing the same operation on both sides.

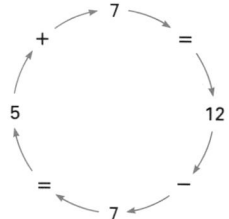

Addition and subtraction are inverse operations.

Continue until the variable is alone on one side of the equals sign.

Check your solution by substituting it in the original equation.

TEACHING TIP

If you do not have balance scales available, you can make simple scales using a hanger and small paper cups. Attach one cup to each end of the hanger and then place objects in the cups to illustrate the concept of balance.

Example 1

Solve $x + 7 = 11$. Check the solution.

Solution

$$x + 7 = 11$$

Undo addition
with subtraction.
$$x + 7 - 7 = 11 - 7$$
$$\uparrow \qquad\quad \uparrow$$

**Keep the equation
in balance.**

$$x + 0 = 4$$
$$x \quad = 4$$

Check.
$$x + 7 = 11 \leftarrow \text{original equation}$$
$$(4) + 7 \overset{?}{=} 11$$
$$11 = 11 \checkmark \quad \text{\textbf{The solution checks.}}$$

The solution is 4. ◄

Example 2

Solve $-3 = n - 5$. Check the solution.

Solution

$$-3 = n - 5$$

Undo subtraction
with addition.
$$-3 + 5 = n - 5 + 5$$
$$\uparrow \qquad\qquad \uparrow$$

**Keep the equation
in balance.**

$$2 = n + 0$$
$$2 = n$$

Check.
$$-3 = n - 5$$
$$-3 \overset{?}{=} (2) - 5$$
$$-3 = -3 \checkmark$$

The solution is 2. ◄

Example 3

Eight less than a certain number equals -2 times 3. To model this
situation, Nancy let $x =$ "a certain number." Then she wrote the
equation $x - 8 = -2(3)$. Solve the equation.

Solution

$$x - 8 = -2(3)$$
$$x - 8 = -6$$
$$x - 8 + 8 = -6 + 8$$
$$x = 2$$

Check.
$$x - 8 = -2(3)$$
$$(2) - 8 \overset{?}{=} -6$$
$$-6 = -6$$

The solution is 2. ◄

CHECK UNDERSTANDING

Would you add or subtract to
solve each equation? Explain.
1. $x - 7 = 3$
2. $t + 6 = -2$
3. $-5 = m + 1$

Add to undo -7.
Subtract to undo $+6$.
Subtract to undo $+1$.

balance by performing the same
operation on both sides of it.

Example 1: Ask students what
must be done to undo 7 added to
x (subtract 7). Remind them that
the same thing must be done to
both sides of the equation. Point
out that the reason for performing
the inverse operation is to get the
variable alone on one side of the
equation. In that way, what is not
known (the variable) can be
expressed in terms of numbers
that are known. Stress the importance of checking the solution in
the original equation.

Example 2: Students may find
Example 2 more difficult than
Example 1 because the variable is
to the right of the equals sign.

Example 3: Read the first sentence in Example 3 to the
students and talk about the equation they would write to model
the situation. Compare their
results with the given solution.

Additional Questions/Examples
1. Write an equation that you
could solve by subtracting 8
from both sides.
Example: $x + 8 = 3$

5-MINUTE CLINIC

Exercise	Student's Error	Error Diagnosis
Solve: $x - 7 = 15$	$x - 7 = 15$ $x = 15$	• Student adds 7 to only one side of the equation.
	$x - 7 = 15$ $x = 8$	• Student subtracts 7 from both sides, instead of adding 7.

2. Write an equation that you could solve by adding –3 to both sides.
Example: $x - (-3) = 5$

Guided Practice/Try These Ask students what they would do to solve each equation. Then have them solve each one and check the solutions. In Exercise 3, they can solve by first adding z to both sides and then adding 3 to both sides.

3 SUMMARIZE

Talk It Over Ask students to describe the steps they would use to solve an equation involving addition and then to solve one involving subtraction.

4 PRACTICE

Practice/Solve Problems In Exercises 1–6 be sure that students see that the result of the inverse operation they choose is to get the variable alone on one side of the equals sign.

Extend/Solve Problems In Exercises 18–27 point out that, before trying to undo any operation, it is important to simplify the equation. Thus the equation in Exercise 18 should be written as $x - 8 = 2$ before adding 8 to both sides.

Think Critically/Solve Problems Ask students to share their techniques for solving Exercises 31 and 32. Remind students that each of Exercises 33 and 34 has more than one solution.

228

TRY THESE

Solve each equation. Check the solution.

1. $y + 8 = -3$ **–11**
2. $4 = h - 6$ **10**
3. $5 - z = -3$ **8**
4. $4 + t = -5$ **–9**
5. $x - 7 = 2$ **9**
6. $2 + y = -8$ **–10**
7. $35 = x - 15$ **50**
8. $t + 18 = 56$ **38**
9. $r + 7 = -16$ **–23**
10. $k - 45 = 22$ **67**
11. $-15 + a = 19$ **34**
12. $-3 = w + 2$ **–5**

13. Two more than a certain number equals negative ten divided by negative two. To model this situation, Lucy let r equal "a certain number." Then she wrote the equation $r + 2 = -10 \div (-2)$. Solve the equation. **3**

14. Forty is 12 less than a certain number. To model this situation, Eileen let n equal "a certain number." Then she wrote the equation $40 = n - 12$. Solve the equation. **52**

15. Five more than a certain number equals half of -10. To model this situation, Chi let m equal "a certain number." Then he wrote the equation $m + 5 = \frac{1}{2}(-10)$. Solve the equation. **–10**

EXERCISES

PRACTICE/ SOLVE PROBLEMS

Write the number that must be added or subtracted to solve the equation. Do not solve.

1. $n - 5 = 6$ **add 5**
2. $y + 9 = -3$ **subtract 9**
3. $-8 + k = 3$ **add 8**
4. $2.7 = m - 0.4$ **add 0.4**
5. $p - \frac{4}{5} = 1\frac{1}{5}$ **add $\frac{4}{5}$**
6. $-4.6 = 35 + c$ **subtract 35**

Solve each equation. Check the solution.

7. $c - 6 = -3$ **3**
8. $d + 9 = -2$ **–11**
9. $4 = e - 4$ **8**
10. $3 + f = -5$ **–8**
11. $g - 1.2 = 2.8$ **4**
12. $0 = k + 7$ **–7**
13. $\frac{3}{4} + m = -1$ **$-1\frac{3}{4}$**
14. $-4.5 = n - 2.7$ **–1.8**
15. $-6 = y - 6$ **0**
16. $t + 1\frac{5}{8} = 2\frac{3}{4}$ **$1\frac{1}{8}$**

Solve.

17. Tai has 87 rock samples and wants to collect a total of 100. To model the situation, he let x equal "the number of samples needed." Then he wrote the equation $87 + x = 100$. Solve the equation. **13**

Solve each equation. Check the solution.

EXTEND/ SOLVE PROBLEMS

18. $x - 8 = -2(4 - 5)$ **10**

19. $2 + 3 \times 4 = k + 5$ **9**

20. $\frac{8(9)}{3(2)} = n + 17$ **−5**

21. $6 + p + 4 = -2 + 8$ **−4**

22. $2(9 - 5) = c - 36 + 14$ **30**

23. $x + 2.5 + 4.7 = 8.1 - 5.3$ **−4.4**

24. $x + 21{,}677 = 13{,}291$ **−8,386**

25. $m - 7.48 = 25.793$ **33.273**

26. $465.91 = 209.8 + y$ **256.11**

27. $53{,}809 = e - 84{,}277$ **138,086**

Solve.

28. After Mark scored -5 he had a score of 12. To model the situation, he let s equal "my score before this turn." Then he wrote the equation $s + (-5) = 12$. Solve the equation to find his score before this turn. **17 points**

29. Today's average temperature of $-9°C$ is $8°C$ higher than yesterday's average. Ann let y equal "yesterday's average temperature." Then she wrote the equation $y + 8 = -9$. Solve the equation to find yesterday's average temperature. **−17°C**

30. Death Valley's elevation is 14,776 ft lower than the top of Mount Whitney, 14,494 ft above sea level. Let d equal Death Valley's elevation. Solve the equation $d + 14{,}776 = 14{,}494$ to find Death Valley's elevation. **−282 ft**

31. Write an equation involving addition that has 4 as its solution. Answers may vary. Sample: $x + 3 = 7$

THINK CRITICALLY/ SOLVE PROBLEMS

32. Write an equation involving subtraction that has -7 as its solution. Answers may vary. Sample: $h + 8 = 1$

Solve each equation.

33. $x^2 - 2.9 = 6.1$ $x = \pm 3$

34. $62.8 = n^2 + 13.8$ $n = \pm 7$

7-2   Using Addition or Subtraction to Solve an Equation   **229**

5 FOLLOW-UP

Extra Practice Solve each of the following equations.

1. $x - 3 = -8$ **−5**

2. $-4 + y = 7$ **11**

3. $5x = 5$ **1**

4. $43.2 = a + 12.9$ **30.3**

5. In the equation $-2 = c - 5$, will c be positive or negative? How do you know? **positive, because −5 < −2**

Extension Have students write simple number puzzles such as the following: *I am thinking of a number. If I add 12 to my number, I get 3. What is my number?* Then have students exchange puzzles and write and solve equations to solve the puzzles.

Section Quiz Write the inverse of each step.

1. add -4 **subtract −4**

2. subtract 1.2 **add 1.2**

Write the number that must be added or subtracted to solve. Then solve.

3. $-7 + m = -2$ **add 7; 5**

4. $2.1 = t - 3.4$ **add 3.4; 5.5**

Solve and check.

5. $-5 + y = -5$ **0**

6. $b + 4.3 = 6.2$ **1.9**

7. $t - 1 \frac{1}{2} = 4 \frac{3}{5}$ **6 1/10**

8. $0 = c + 4$ **−4**

9. Rand had read 38 books of the 50 books he wants to read on a suggested reading list. To model the situation, he let x equal the number of books he still wants to read. Then he wrote the equation $38 + x = 50$. Solve the equation. **12**

230

SPOTLIGHT

OBJECTIVE
• Write an equation needed to solve a problem

VOCABULARY
unknown quantity, variable

WARM-UP

Write an expression for each phrase.
1. a number increased by 4
   $n + 4$
2. 12 less than a number   **$n - 12$**

1 MOTIVATE

Introduction Talk about the steps that must be used in writing an equation to solve a problem. Stress that it is very important to indicate what the variable represents.

2 TEACH

Use the Pages/Problem Have students read this part of the section and then discuss the problem. Ask students to tell how they would find a goal amount of money if they know the amount they have and the amount they need to reach the goal. Use this information to help construct the equation for this example.

Use the Pages/Solution Have the students check the solution and talk about whether it is a sensible answer.

230

7-3 Problem Solving Skills:
WRITING AN EQUATION

► READ
► PLAN
► SOLVE
► ANSWER
► CHECK

Often you can solve a problem by writing an equation and then finding the solution of the equation. Read the problem carefully. Think about how the numbers in the problem are related. Choose a variable to represent the unknown quantity. Then, write an equation representing the situation described in the problem.

PROBLEM

Members of the Chess Club are earning money to buy new chess sets. Their goal is to earn $225. Today's accounting shows that they have earned $157 so far. How much more must they earn in order to reach their goal?

SOLUTION

Given: The target amount is $225.
           Their current total is $157.
Find: how much more is needed

TEACHING TIP

One student may interpret a problem as one involving addition. Another student may use subtraction to solve the same problem. If the equations they write are equivalent, then both are correct.

Let a represent the amount needed to reach $225. ⟵ The variable represents the unknown quantity.

The $157 earned so far plus the amount needed equals $225. ⟵ This sentence describes how the numbers in the problem are related.

$$157 + a = 225$$ ⟵ The equation translates the problem situation into numbers and variables.
$$157 - 157 + a = 225 - 157$$
$$a = 68$$ ⟵ The solution of the equation is 68.

Since a represents the amount needed, the club needs to earn $68 more.

PROBLEMS

Choose the letter of the equation that represents the problem situation correctly. Do not solve the problem.

1. After Harriet spent $21 to rent roller skates, she had $23 left. How much did she have to begin with? Let x represent the amount she had to begin with.
   a. $x + 21 = 23$ b. $x - 23 = 21$ (c.) $x - 21 = 23$

2. Stock in the ABC Corporation rose $2\frac{3}{4}$ points to a value of $17\frac{1}{8}$ points. Find the value before the increase. Let v represent the original value.
   (a.) $v + 2\frac{3}{4} = 17\frac{1}{8}$ b. $v - 2\frac{3}{4} = 17\frac{1}{8}$ c. $v + 17\frac{1}{8} = 2\frac{3}{4}$

Solve each problem by writing an equation.

3. Sandra gave her brother 16 baseball cards, leaving her with 42 cards. How many cards did she have to begin with? **58 cards**

4. If a certain number is decreased by 56, the result is 36. Find the number. **92**

5. A sweater was on sale for $47. The sale price was $16 lower than the regular price. Find the regular price. **$63**

6. Spike rode his bike for $3\frac{1}{2}$ h on Saturday. If he rode $1\frac{1}{4}$ h in the morning, how long did he ride the rest of the day? **$2\frac{1}{4}$ h**

7. On her first turn in a game Verna scored -18. After her second turn her total score was 2. How many points did she score on her second turn? **20 points**

8. Janine has read 230 pages of a novel. That is 39 more pages than Harrison has read. How many pages has Harrison read? **191 pages**

 WRITING ABOUT MATH

Write two different word problems that could be solved using the equation $8 + n = 14$.

Answers may vary. Samples are given.
1. Rex is running in a 14-km race. He has run 8 km so far. How far does he have to go?
2. The sum of Jake's and Geri's ages is 14. Jake is 8. How old is Geri?

3 SUMMARIZE

Key Questions
1. In a problem, what does the variable represent? **the unknown quantity**
2. How does an equation help to solve a problem? **It shows the relationship between numbers and a variable.**

4 PRACTICE

Problems Have students identify the key words they used to help construct their equations. After they write each equation, have them translate it back into words to see if it matches the problem.

5 FOLLOW-UP

Extra Practice
Write an equation and solve.
1. Rita collected $48 more than Juanita for the walkathon. If Rita collected $117, how much did Juanita collect? **$69**
2. If a number is increased by 4 3/4, the result is 16 2/3. Find the number. **11 11/12**

Get Ready small slips of paper

MAKING CONNECTIONS

Have students make up their own problems. For instance, they might write problems involving study time for two classes or the numbers of papers written by two students. Then have students exchange problems and write equations for each other's problems.

SPOTLIGHT

OBJECTIVE
- Use multiplication or division to solve an equation

MATERIALS NEEDED
small slips of paper

VOCABULARY
inverse operations

WARM-UP

Evaluate each expression for n = –12.
1. $3n$ **–36**   2. $-4n$ **48**
3. $-n/-2$ **–6**   4. $n \div 6 + 8$ **6**

1 MOTIVATE

Introduction Ask students to explain how they found the original number in each case. Elicit the fact that, even though students chose different numbers, in each case they could undo multiplication by 7 and division by 3 with division by 7 and multiplication by 3.

2 TEACH

Use the Pages/Skills Development Have students read this part of the section and then discuss the examples. Bring out the idea that, just as addition and subtraction undo each other, so do multiplication and division. To emphasize this point, have students identify how to undo multiplication by 4 and division by –6.
   Example 1: Ask students to identify what is being done to m (multiplying by 4). Discuss the solution and how to undo it

232

7-4 Using Multiplication or Division to Solve an Equation

EXPLORE

a. Choose a number.

b. Multiply the number by 7.

c. Write the result on a slip of paper.

d. Exchange slips with a classmate.

e. Find your classmate's original number.

f. Choose a new number.

g. Divide the number by 3.

h. Write the result on a slip of paper.

i. Exchange slips with a classmate.

j. Find your classmate's original number.

SKILLS DEVELOPMENT

Like addition and subtraction, multiplication and division are inverse operations. You can solve equations involving multiplication or division by using the inverse operation.

Example 1

Solve $4m = -52$. Check the solution.

Solution

$$4m = -52$$

Undo multiplication with division.

$$\frac{4m}{4} = \frac{-52}{4}$$

   ↑   ↑
Keep the equation in balance.

$$(1)m = -13$$
$$m = -13$$

Check. $4m = -52$
$$4(-13) \stackrel{?}{=} -52$$
$$-52 = -52 \checkmark$$

The solution is -13. ◄

Multiplication and division are inverse operations.

TEACHING TIP

If students estimate the answer before solving an equation, it will give them a quick check of whether or not their answer is reasonable. Checking, of course, is important, whether or not estimation is used.

Example 2

Solve $\frac{k}{8} = -2$. Check the solution.

Solution

$$\frac{k}{8} = -2$$

Undo division
with multiplication. $8\left(\frac{k}{8}\right) = 8(-2)$

$$(1)k = -16$$
$$k = -16$$

Check. $\frac{k}{8} = -2$

$$\frac{(-16)}{8} \stackrel{?}{=} -2$$
$$-2 = -2 \checkmark \qquad \text{The solution is } -16. \blacktriangleleft$$

Example 3

Ted needs 60 sandwiches to serve at his party. He can fit 12 sandwiches on a serving platter. How many platters are needed?

Solution

Write and solve an equation that represents the situation.

Let p = number of platters needed

$$12p = 60$$
$$\frac{12p}{12} = \frac{60}{12}$$
$$p = 5 \qquad \text{Ted needs 5 serving platters.} \blacktriangleleft$$

TRY THESE

Solve each equation. Check the solution.

1. $\frac{p}{-3} = 9$ **–27**

2. $51 = 15e$ **3.4**

3. Kirk earned $83.75 working for an hourly wage of $6.70. How many hours did he work? **$12\frac{1}{2}$ h**

EXERCISES

Write the number by which you must multiply or divide to solve the equation. Do not solve.

**PRACTICE/
SOLVE PROBLEMS**

1. $-8y = -10$ **divide by –8**

2. $5.2h = -15.6$ **divide by 5.2**

3. $\frac{c}{2} = -2$ **multiply by 2**

4. $19 = -3.5m$ **divide by –3.5**

5. $-3 = \frac{d}{16}$ **multiply by 16**

6. $\frac{n}{-9} = -8$ **multiply by –9**

(dividing by 4). Again, stress the importance of checking the solution in the original equation.

Example 2: Remind students to multiply both sides of the equation by 8 to keep the equation in balance.

Example 3: If students suggest the equation $p = 60/12$, note that it will also give the correct solution.

Additional Questions/Examples

1. When you look at an equation, how can you tell whether addition, subtraction, multiplication, or division will be needed to solve? **You look at what is done to the variable and undo it.**

2. Can you think of a way to solve $4a = 12$ using multiplication? **Multiply by the reciprocal of 4, 1/4.** Is this way equivalent to dividing both sides by 4? **yes** Why? **Dividing by a number is the same as multiplying by its reciprocal.**

Guided Practice/Try These
Have students work in small groups. Point out that all group members should come to an agreement on each answer.

Write About Math Have students write a paragraph in their math journals summarizing how to go about solving an equation involving multiplication or division.

4 PRACTICE

Practice/Solve Problems It may help students to begin Exercises 7–18 by identifying the number by which they will multiply or divide both sides of the equation. Then, encourage them to write out the step in which they show the indicated multiplication or division on both sides before they complete the solution.

Extend/Solve Problems In Exercises 21–34 encourage students to simplify each equation before solving for the variable. In Problem 36 it may help students to draw a time line showing the number of measurements made. The first can be marked 11°F and the last, –1°F.

Think Critically/Solve Problems Be sure students understand that division by zero is impossible.

Problem Solving Applications Be sure students see that they do not need to know the actual prices of the stock in order to solve the problem.

5 FOLLOW-UP

Extra Practice
1. Given $n/-3 = 21$, will n be positive or negative? Explain.

Solve each equation. Check the solution.

7. $228 = 4x$ **57**
8. $\frac{h}{-7} = -5$ **35**
9. $28 = \frac{k}{14}$ **392**
10. $-12p = 3$ **$-\frac{1}{4}$**
11. $-15 = \frac{y}{-2.2}$ **33**
12. $6k = 6$ **1**
13. $30r = 18$ **0.6**
14. $\frac{w}{8} = -21$ **-168**
15. $\frac{m}{-3} = \frac{4}{9}$ **$-\frac{4}{3}$**
16. $1 = \frac{h}{5}$ **5**
17. $3.7x = 8.51$ **2.3**
18. $-y = 6$ **-6**

Solve.

19. A rectangular garden with an area of 105 ft² has a length of 12 ft. Find the width. **$8\frac{3}{4}$ ft**

20. Board games are on sale for $8.95 each. During the sale, members of the Seniors Club spent $107.40 on board games. How many games did club members buy? **12 games**

EXTEND/ SOLVE PROBLEMS

Solve and check.

21. $-4(-5) = 8n$ **2.5**
22. $\frac{K}{3} = 2(6 - 8 + 12)$ **60**
23. $\frac{p}{4 - 10} = \frac{4 + 5}{8 - 5}$ **-18**
24. $\frac{64.8}{3} = -2.4x$ **-9**
25. $6 + 18 \div 9 = (9 - 29)y$ **-0.4**
26. $4(-2.1) = \frac{n}{12(-3.5)}$ **352.8**
27. $56.8g = 1,209.84$ **21.3**
28. $13.6 = \frac{e}{-9.7}$ **131.92**
29. $\frac{h}{123} = 0.88$ **108.24**
30. $2,142 = -2,520p$ **-0.85**
31. $8.036 = \frac{y}{-0.45}$ **-3.6162**
32. $2.7x = 8.1(-9.3)$ **-27.9**
33. $7.2 = 3(y - 2.3)$ **4.7**
34. $12(n + \frac{2}{9}) = \frac{11}{3}$ **$\frac{1}{12}$**

PROBLEM SOLVING TIP

For Exercise 36, ask yourself how many *changes* in temperature the forecaster measured.

Solve.

35. Mary Beth bought 4 pens priced at $0.79 each and 2 pens priced at $0.97 each. Find the average price she paid per pen. **$0.85**

36. A weather forecaster made 7 temperature measurements on Sunday afternoon. The first reading was 11°F and the last was –1°F. What was the average temperature change between measurements? **-2°F**

THINK CRITICALLY/ SOLVE PROBLEMS

Describe all solutions of each equation. **See Additional Answers.**

37. $2n = 0$
38. $0n = 2$
39. $0n = 0$

40. What is the result if you add 0 to each side of an equation? **The original equation remains unchanged.**
41. What is the result if you multiply each side of an equation by 0? **The equation becomes 0 = 0.**

AT-RISK STUDENTS

Make a worksheet by mixing up equations such as $n/4 = 48$ in one column, directions such as "Multiply both sides by 4" in a second column, and equivalent equations, such as $n = 4 \times 48$, in a third column. Have students match each equation in the first column to a direction in the second and an equivalent equation in the third.

Problem Solving Applications:
STOCK MARKET REPORTS

▶ READ
▶ PLAN
▶ SOLVE
▶ ANSWER
▶ CHECK

Most stock market reports tell the value at which a stock opens (value of the stock in the morning), what change in value happens that day, and the value at which a stock closes (value at the end of the day).

Marilyn likes to keep track of what is happening in the stock market. Part of her hobby is to keep track of the stock prices and the changes that occur during a day. One day she spilled juice on part of a report. However, she could still read some information. From what she could read she was able to find the opening price, change for the day, and closing price of each stock listed.

Show how Marilyn could use an equation to represent each situation. Then solve the equation.

1. ABC stock closed at $14\frac{1}{2}$. It had opened at 12 that day. What was the change for the day? $12 + x = 14\frac{1}{2}$; $+2\frac{1}{2}$

2. Innercomp Communications Systems stock opened at 33. It closed at 28. What was the change for the day? $33 + x = 28$; -5

3. Big Company stock has a change of $+2$ in one day. The value of all Mr. Ruiz's shares in Big Company increased by $300 that day. How many shares of Big Company stock does Mr. Ruiz own? $2x = 300$; 150 shares

4. Bruns Corporation had a change of -5. If its stock closed at 93, at what price did the stock open that day? $x + (-5) = 93$; 98

5. An investor sold 120 shares of Williams Wardrobes for $180 less than was paid for those shares. What was the change in price of a share between the time of purchase and the time of sale? $\frac{180}{120} = x$; $-1\frac{1}{2}$

7-4   Using Multiplication or Division to Solve an Equation   **235**

---

negative because it is the product of a positive and a negative number

2. Given $5n = 105$, will n be greater than or less than 105? **less than**

3. Solve and check.
   a. $-4w = -9$ **2 1/4**
   b. $n/3.2 = -4.8$ **−15.36**

4. Sun Li bought $356.50 worth of shirts to sell at her boutique. If she bought 23 shirts, how much did each one cost? **$15.50**

Extension Have students investigate exchange rates for different currencies and then write problems using the data that involve multiplication and division equations.

Section Quiz Write the inverse of each step.
1. Divide by -4. **Multiply by −4.**
2. Multiply by 0.4. **Divide by 0.4.**
Write the number by which you must multiply or divide to solve.
3. $5x = 35$ **Divide by 5.**
4. $n/-2 = 4.3$ **Multiply by −2.**
Solve and check.
5. $45n = -90$ **−2**
6. $2.4 = j/3.2$ **7.68**
7. $g/2.4 = -100$ **−240**
8. $2(y-3) = -20$ **−7**
9. A triangular yard has an area of 45 ft². If the base measures 9 ft, find the height. **10 ft**

Get Ready algebra tiles

Additional Answers
See page 578.

Additional answers for odd-numbered exercises are found in the Selected Answers portion of the page.

CHALLENGE

Have students solve each equation for b.

1. $a + b = 5$ $b = 5 - a$
2. $a - b = 12$ $b = a - 12$
3. $r = b/a$ $b = ar$
4. $mb = a$ $b = a/m$

WARM-UP

Evaluate.
1. $5n = -32$
  -6 2/5
2. $24 = n + 28$
  -4
3. $a/{-4.1} = 8$
  -32.8
4. $m - 17.3 = 1.9$
  19.2

7-5

Using More Than One Operation to Solve an Equation

EXPLORE You can use algebra tiles to model equations.

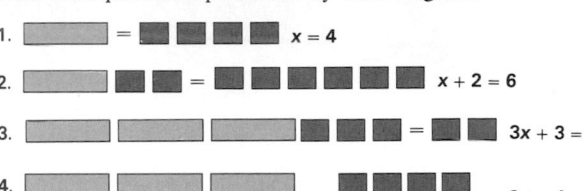
represents x.   represents 1.
represents $2x + 3 = 5$.

Write the equation represented by each diagram.

1. $x = 4$

2. $x + 2 = 6$

3. $3x + 3 = 2$

4. $3x + 4 = 8$

Use tiles to model the equation.   **See Additional Answers.**

5. $x + 1 = 5$
6. $x + 4 = 3$
7. $3x = 6$
8. $2x + 2 = 4$
9. $3x + 1 = 7$
10. $4x + 3 = 9$

11. Model the equation $x + 4 = 7$. Describe how you can use tiles to solve the equation. What is the solution? **See Additional Answers for Exercises 11–15.**

12. Model the equation $3x = 9$. Describe how you can use tiles to solve the equation. What is the solution?

Model the equation $2x + 5 = 9$.

13. Show how to subtract 5 from both sides.

14. Show how to divide both sides into two equal groups.

15. What is the solution of the equation? How do you know?

Use tiles to solve each equation.

16. $2x + 1 = 7$ **3**
17. $3x + 2 = 8$ **2**
18. $4x + 1 = 9$ **2**
19. $2x + 5 = 7$ **1**

MATH: WHO, WHERE, WHEN

Methods of solving equations by performing the same operation on both sides had been developed by Arab and Hindu mathematicians before the ninth century. Around A.D. 830, the Arab mathematician Al-Khowarizmi wrote a book describing methods of solving equations, including equations of the types studied in this chapter. Al-Khowarizmi was a teacher who had earlier worked at the court of the Moslem ruler Harun-al-Rashid in Baghdad. The word *algebra* is derived from the word *al-jabr*, which appeared in the title of his book.

Until now you have been able to solve every equation by applying a single inverse operation. Such an equation is called a **one-step equation.**

Some equations involve two operations. To solve these **two-step equations,** undo addition and subtraction first. Then undo multiplication and division.

Example 1

Solve $4x + 7 = 23$. Check the solution.

Solution

$$4x + 7 = 23$$

Undo the addition. $\quad 4x + 7 - 7 = 23 - 7$

$$4x = 16$$

Undo the multiplication. $\quad \dfrac{4x}{4} = \dfrac{16}{4}$

$$x = 4$$

Check.
$$4x + 7 = 23$$
$$4(4) + 7 \stackrel{?}{=} 23$$
$$16 + 7 \stackrel{?}{=} 23$$
$$23 = 23 \ \checkmark$$

The solution is 4. ◄

PROBLEM SOLVING TIP

Double-check to be sure you are substituting your solution in the original equation.

Example 2

Solve $7(x + 4) = 21$. Check the solution.

Solution

$$7(x + 4) = 21$$

Multiply each term in parentheses by 7. $\quad 7x + 28 = 21$

Undo addition with subtraction. $\quad 7x + 28 - 28 = 21 - 28$

Simplify.
$$7x + 0 = -7$$
$$7x = -7$$

Undo multiplication with division. $\quad \dfrac{7x}{7} = \dfrac{-7}{7}$

$$x = -1$$

Check.
$$7(x + 4) = 21$$
$$7(-1 + 4) \stackrel{?}{=} 21$$
$$7(3) \stackrel{?}{=} 21$$
$$21 = 21 \ \checkmark$$

The solution is −1. ◄

CALCULATOR

You can solve equations using your calculator.

To solve
$3.2x - 4.8 = -17.92$, use this key sequence.

17.92 $\boxed{+/-}$ $\boxed{+}$ 4.8 $\boxed{\div}$ 3.2 $\boxed{=}$

On some calculators, you may have to press the $\boxed{=}$ key after you enter the 4.8.

The solution is −4.1.

make errors and results in a more tedious computation, it is better to begin by undoing addition or subtraction.

Example 2: Point out that it is possible to begin solving this equation in one of two ways—by multiplying to eliminate the parentheses or by first dividing both sides of the equation by 7 and then subtracting.

Example 3: Ask students what each person's share will be ($n/3$). From this, students should be able to see how to write the equation.

Additional Questions/Examples
1. Solve the equation $2x - 5 = 11$ by adding and then dividing. **8**
2. Solve the equation $2x - 5 = 11$ by dividing and then adding. **8**
3. Which method is easier? Why do you think we generally start the solution of a two-step equation by adding or subtracting? **Answers may vary.**
4. In the equation $2(x - 3) = 8$, why is it possible to first divide and then add? **Because the left side of the equation is a product of which 2 is a factor; thus, both sides of the equation are divisible by 2.**

AT-RISK STUDENTS

Continue to allow students to feel free to use manipulatives to solve two-step equations as long as necessary. After students solve an equation using the manipulatives, have them write out the steps to show how the solution is accomplished.

238

Guided Practice/Try These
Have students work in small groups. Point out that all group members should come to an agreement on each answer.

3 SUMMARIZE

Talk It Over Call on students to describe the steps they used in solving Exercises 1–3 in Try These.

4 PRACTICE

Practice/Solve Problems To help students who are having difficulty, ask them to identify the steps they would use in solving each of Exercises 5–16.

Extend/Solve Problems Remind students to do any numerical simplification before adding or subtracting to isolate the variable. They will also have to use the distributive property in Exercise 19 in order to simplify $5p - 3p$.

Think Critically/Solve Problems Ask students to share their ideas for creating two-step equations meeting the requirements in Exercises 26 and 27.

5 FOLLOW-UP

Extra Practice
Solve and check.
1. $9x - 15 = 39$ **6**
2. $10 = -4 + c/2$ **28**
3. $d/31.4 + 312.9 = 412.7$
   3,133.72
4. $2c + 5 = 3c - 8$ **13**
Write an equation and solve.
5. Bonita is saving $20 per week and already has $200. In how many weeks will she have saved $300? **5**

Example 3
Arnie, Megan, and Piero picked apples, which they shared equally. Piero ate 2 apples, leaving him with 7. How many apples did Arnie, Megan, and Piero pick?

Solution
Write and solve an equation that represents the situation.

Let n = total number of apples picked.

Piero's share minus 2 left Piero with 7 apples.

$$\frac{n}{3} - 2 = 7$$

Undo the subtraction. $\quad \frac{n}{3} - 2 + 2 = 7 + 2$

$$\frac{n}{3} = 9$$

Undo the division. $\quad 3\left(\frac{n}{3}\right) = 3(9)$

$$n = 27$$

The three picked 27 apples. ◄

TRY THESE

Solve each equation. Check the solution.

1. $3x - 5 = 16$ **7** 2. $\frac{n}{-6} + 8 = 5$ **18**

3. $1 = -9y + 7$ **$\frac{2}{3}$** 4. $2(y - 6) = 18$ **15**

5. A bicycle can be rented for $4.50 per hour plus a $3 service fee. Pete paid $25.50 to rent a bike. For how many hours did he rent it? **5 h**

EXERCISES

PRACTICE/ SOLVE PROBLEMS

Write the first step in solving the equation. Do not solve.

1. $\frac{y}{-7} - 3 = 4$ **Add 3.** 2. $\frac{1}{2} = 8x + 5$ **Subtract 5.**

3. $8 + 3n = -1$ **Subtract 8.** 4. $9 = -6 + \frac{k}{2}$ **Add 6.**

Solve each equation. Check the solution.

5. $6p - 7 = 5$ **2** 6. $11 = -2x + 5$ **−3** 7. $\frac{k}{-2} + 4 = -8$ **24**

8. $3 = 3h + 6$ **−1** 9. $7 + \frac{e}{6} = 0$ **−42** 10. $34 = \frac{f}{-2} + 47$ **26**

11. $9n - 1.5 = 10.2$ **1.3** 12. $\frac{3}{5} + 3c = 3$ **$\frac{4}{5}$** 13. $\frac{x}{7} - 15 = -4$ **77**

14. $5(k + 3) = 30$ **3** 15. $-7(p - 9) = 14$ **7** 16. $26 = 8(m + 7)$ **−3.75**

Greg's age of 15 is 6 years more than Marv's age divided by 4.

17. Let M = Marv's age. Write an equation relating Marv's age to Greg's. $\frac{M}{4} + 6 = 15$

18. Solve the equation you wrote in Exercise 17. Check your solution. **36**

Solve each equation. Check the solution.

**EXTEND/
SOLVE PROBLEMS**

19. $\frac{-45}{-9} + (5p - 3p) = 4 + 7 \times 3$ **10**

20. $\frac{y}{15.6} + 319.7 = 334$ **223.08**

21. $47.3k - 8.77 = -131.75$ **−2.6**

It costs $24 per day plus $0.08 per mile to rent a car. Jeanne's charge for a one-day rental was $42.32.

22. Let m = number of miles driven. Write an equation representing Jeanne's charge. **0.08 m + 24 = 42.32**

23. Solve the equation you wrote in Exercise 22. Check the solution. **229**

Solve.

**THINK CRITICALLY/
SOLVE PROBLEMS**

24. $9\left(x - \frac{1}{3}\right) = -21$ **−2**

25. $-28 = -4(2y - 3)$ **5**

26. Write a two-step equation involving addition and division that has -12 as its solution. **Answers may vary. Sample: $\frac{t}{3} + 5 = 1$**

27. Write a two-step equation involving subtraction and multiplication that has 3 as its solution. **Answers may vary. Sample: $2f - 5 = 1$**

7-5   Using More Than One Operation to Solve an Equation   **239**

Extension Have students investigate local cab rates (most are a fixed charge plus a charge per mile.) Have students write an equation that relates the number of miles traveled, m, the rate per mile, and the fixed charges, with the cost of the ride, c. Then have them substitute values for c and solve.

Section Quiz Write the first step in solving the equation.
1. $4a - 3 = 12$ **add 3**
2. $c/-5 + 3 = 7$ **subtract 3**
Solve and check.
3. $v/6 - 2 = 7$ **54**
4. $4 - 3x = 16$ **−4**
Write an equation and solve.
5. Robine is $3y$ older than 4 times Karim's age. If Robine is 19, how old is Karim? **4y**

Additional Answers
See page 578.

CHALLENGE

Using square blue tiles for negative integers, square white tiles for positive integers, and rectangular tiles for variables, students can model solutions of equations such as $2x - 5 = -1$: *Add 5 white tiles to each side. Remove pairs of positive and negative tiles, because each pair equals zero. Divide groups into two equal parts.* **Solution: 2**

SPOTLIGHT

OBJECTIVE
- Use the reciprocal of a number to solve an equation

VOCABULARY
reciprocal

WARM-UP

Find the reciprocal of each.
1. 2/3 **1 1/2**   2. -9/7 **–7/9**
3. 3 2/3 **3/11**   4. 1 **1**

1 MOTIVATE

Explore/Working Together Use a different example so students understand why 1/12 of a 12–h job is completed in 1h. For example, if Jeff had 12 apples to eat and ate 1 apple per hour, he would eat 1/12 of the total per hour.

2 TEACH

Use the Pages/Skills Development Have students read this part of the section and then discuss the examples.

Example 1: Make the point that dividing each side of the equation 3/8k = –27 by 3/8, would be the same as multiplying by the reciprocal of 3/8. Call on a student to explain why 8/3 (–27) = –72.

Example 2: Students may suggest that it is easier to add before writing the mixed numbers as improper fractions. The important point is to undo the subtraction

240

7-6 Using a Reciprocal to Solve an Equation

EXPLORE/ WORKING TOGETHER

Jeff has many different hobbies. His favorite hobby is building model airplanes. The table belows gives the amount of time he spent working on some of his models. Study the table carefully.

Plane	Number of hours to build	Fraction of job completed per hour
B-29	12	$\frac{1}{12}$
Tigercat	5	■ $\frac{1}{5}$
Stinger	$\frac{15}{4}$	■ $\frac{4}{15}$
Shark	$\frac{11}{2}$	■ $\frac{2}{11}$

a. Work with a partner to complete the table. Find what fraction of the job Jeff completed per hour. Assume he worked at a steady rate.

b. What relationship do you notice between the numbers in the second and third columns? **They are reciprocals of each other.**

c. For each model, find the product of the numbers in the second and third columns. **Each product equals 1.**

SKILLS DEVELOPMENT

When an equation involves a variable that is multiplied by a fraction, you can use the reciprocal of the fraction to solve the equation.

Example 1
Solve $\frac{3}{8}k = -27$. Check the solution.

240   CHAPTER 7   Equations and Inequalities

TEACHING TIP

Sometimes the steps in solving a problem must be done in a prescribed order (for example, using the order of operations), but other times, more than one order can be used. When students suggest alternative solutions, help them explore whether or not these solutions will work and which one is easier.

Solution

$$\frac{3}{8}k = -27$$

Multiply both sides by the reciprocal of $\frac{3}{8}$.

$$\frac{8}{3}\left(\frac{3}{8}\right)k = \frac{8}{3}(-27)$$

$$(1)k = -72$$

$$k = -72$$

Check.

$$\frac{3}{8}k = -27$$

$$\frac{3}{8}(-72) \stackrel{?}{=} -27$$

$$-27 = -27 \checkmark \quad \text{The solution is } -72. \quad \blacktriangleleft$$

CONNECTIONS

Recall that the product of a fraction and its reciprocal is 1. You can find the reciprocal of a fraction by interchanging the numerator and denominator.

Example 2

Solve $-\frac{9}{2}n - 6 = 12$. Check the solution.

Solution

$$-\frac{9}{2}n - 6 = 12$$

Undo the subtraction.

$$-\frac{9}{2}n - 6 + 6 = 12 + 6$$

$$-\frac{9}{2}n = 18$$

Multiply both sides by the reciprocal of $-\frac{9}{2}$.

$$-\frac{2}{9}\left(-\frac{9}{2}n\right) = -\frac{2}{9}(18)$$

$$n = -4$$

Check.

$$-\frac{9}{2}n - 6 = 12$$

$$-\frac{9}{2}(-4) - 6 = 12$$

$$\frac{36}{2} - 6 = 12$$

$$18 - 6 = 12$$

$$12 = 12 \checkmark$$

The solution is -4. $\blacktriangleleft$

Example 3

Copies of a math book are $1\frac{1}{8}$ in. thick. How many can be stored on a shelf $20\frac{1}{4}$ in. wide?

Solution

Write and solve an equation that represents the situation.

Let n = the number of copies.

$$1\frac{1}{8}n = 20\frac{1}{4}$$

$$\frac{9}{8}n = \frac{81}{4} \qquad \text{Write the mixed numbers as fractions.}$$

$$\frac{8}{9}\left(\frac{9}{8}n\right) = \frac{8}{9}\left(\frac{81}{4}\right)$$

$$n = 18$$

Eighteen books can be stored on the shelf. $\blacktriangleleft$

CHECK UNDERSTANDING

How does multiplying by the reciprocal simplify the process of solving an equation?

It converts the number in front of the variable to 1 in a single step.

7-6   Using a Reciprocal to Solve an Equation   **241**

ASSIGNMENTS

BASIC
1–13, 14–19, PSA 1–3

AVERAGE
1–13, 14–22, PSA 1–5

ENRICHED
14–22, 23, PSA 1–5

ADDITIONAL RESOURCES
Reteaching 7–6
Enrichment 7–6

before undoing the multiplication.

Example 3: To help students develop an equation, ask them to write a sentence to check if the answer is 3 (1 1/8 × 3 = 20 1/4). Then ask what equation we can write if n is the number of books. Have students check the answer, 18 books. Students should be able to reason that the answer is reasonable, since it is slightly less than 20 1/4.

Additional Questions/Examples
1. How do you find the reciprocal of a mixed number? **Write it as a fraction, interchanging the numerator and denominator.**
2. Is the reciprocal of a negative number positive or negative? **negative** Explain. **The product must be positive so both numbers must be negative.**
3. Will the solution of 3/4 x = 9 be *greater than* or *less than* 9? Explain. **greater than, since you multiply by a number less than 1 to get 9**

AT-RISK STUDENTS

Make a worksheet or a set of index cards showing the mixed numbers and their reciprocals used in this section. Have students match numbers to their reciprocals on the worksheets or match pairs of cards showing numbers and reciprocals. They can then use the worksheet or cards to help them in solving the problems.

Guided Practice/Try These
Have students describe the steps they will use in solving each of the given equations.

3 SUMMARIZE

Write About Math Have students summarize the process of using reciprocals to solve an equation such as $2/3c = -8$. Ask them to write their summaries in a paragraph.

4 PRACTICE

Practice/Solve Problems It may help students to work with partners in solving some of Exercises 1–12. For Exercise 13 asking students how far the snail crawled in 1h, 2h, and 3h may help them write an equation to solve the problem.

Extend/Solve Problems It may be simpler if students do the addition or subtraction in Exercises 14–20 before they write the mixed numbers as fractions. In Exercise 21 they must use the distributive property first or else divide both sides by 9. Check student' equations in Exercise 22 before they complete the solution. It may help them to draw a picture.

Think Critically/Solve Problems Have students experiment with some numbers to help them arrive at a conclusion for Exercise 23.

Problem Solving Applications Have students work with partners or in small groups to solve these problems.

TRY THESE

Solve each equation. Check the solution.

1. $\frac{5}{6}y = 35$ **42**   2. $\frac{10}{3}g + 5 = 8$ **$\frac{9}{10}$**   3. $\frac{4}{5}m = \frac{3}{4}$ **$\frac{15}{16}$**

4. Marci planted a seedling plum tree 28 in. tall. The tree grew at an average rate of $8\frac{1}{4}$ in. per year. After how many years was the tree 127 in. tall? **12 years**

EXERCISES

PRACTICE/ SOLVE PROBLEMS

Solve each equation. Check the solution.

1. $\frac{3}{4}x = -6$ **-8**   2. $9 = -\frac{2}{3}x$ **$-\frac{27}{2}$**   3. $\frac{4}{5}x = 8$ **10**

4. $-\frac{3}{7}h = 9$ **-21**   5. $1 = \frac{4}{5}c$ **$\frac{5}{4}$**   6. $\frac{4}{3}e = 16$ **12**

7. $\frac{2}{5}x - 3 = 9$ **30**   8. $8 = -\frac{3}{7}x + 4$ **$-\frac{28}{3}$**   9. $\frac{5}{4}x - 6 = 14$ **16**

10. $\frac{1}{9}m = 1\frac{1}{3}$ **12**   11. $\frac{1}{3} = -\frac{1}{2}t$ **$-\frac{2}{3}$**   12. $0.4 = \frac{2}{9}k$ **1.8**

13. A snail crawled at a rate of $8\frac{3}{4}$ inches per hour. How long did it take the snail to crawl $41\frac{7}{8}$ inches? **$4\frac{11}{14}$ h**

EXTEND/ SOLVE PROBLEMS

Solve each equation. Check the solution.

14. $3 = \frac{7}{8}p + \frac{1}{4}$ **$3\frac{1}{7}$**   15. $\frac{8}{3}f - \frac{3}{4} = 7\frac{1}{4}$ **3**

16. $\frac{8}{5}y - 1\frac{1}{3} = -\frac{2}{3}$ **$\frac{5}{12}$**   17. $\frac{7}{9}k - 6\frac{1}{3} = 1\frac{1}{3}$ **$10\frac{2}{7}$**

18. $0 = \frac{15}{4}n + 1\frac{1}{2}$ **$-\frac{2}{5}$**   19. $\frac{9}{4}c + 2\frac{1}{4} = 3\frac{3}{4}$ **$\frac{2}{3}$**

20. $\frac{16}{15}x + \frac{1}{3} = -\frac{1}{3}$ **$-\frac{5}{8}$**   21. $9\left(\frac{3}{4}m - 1\right) = -5\frac{5}{8}$ **$\frac{1}{2}$**

22. Bricks $2\frac{1}{4}$ in. thick are stacked atop each other to make a wall. A $9\frac{1}{2}$-in. wood border decorates the top of the wall. If the wall plus the border is $65\frac{3}{4}$ in. high, how many rows of bricks are there? **25 rows**

THINK CRITICALLY/ SOLVE PROBLEMS

23. If the product of two numbers is 1, the numbers are reciprocals of one another. What can you say about two numbers whose quotient is 1? Two numbers whose product is 0? **They are equal; one or the other or both are 0.**

MAKING CONNECTIONS

Have students write a report about the stock market, including a sample of listings of stock prices. Have them note the fractions used and explain the meaning of those fractions.

Problem Solving Applications:

BREAKING THE CODE

The young of various animals have been given special names.

In the table, each name is given a code number.

To find the names of the young of various animals in the questions that follow, you need to solve the equations, then use the codes given in the table. Here is an example.

What is a young goose called?

Solve the equation $-6 = 2g - 4$ and then use the code.

$$-6 = 2g - 4$$
$$-6 + 4 = 2g - 4 + 4$$
$$-2 = 2g$$
$$-1 = g$$

From the table, -1 is the code number of "gosling." So a young goose is called a gosling.

CODE	
Code number	Name of young
−4	leveret
−3	foal
−2	calf
−1	gosling
0	parr
1	squab
2	piglet
3	cygnet
4	chick
5	lamb
6	duckling

Solve each equation. Then use the code to determine the name of the young of each animal given.

1. What is a young donkey called?
   Solve the equation $7d + 6 = -15$ and use the code. **$d = -3$; foal**

2. What is a young elephant called?
   Solve the equation $-10 = 2e - 6$ and use the code. **$e = -2$; calf**

3. What is a young pigeon called?
   Solve the equation $3p + (-12) = -9$ and use the code. **$p = 1$; squab**

4. What is a young swan called?
   Solve the equation $5s - 15 = 0$ and use the code. **$s = 3$; cygnet**

5. What is a young turkey called?
   Solve the equation $1.5t - (-3) = 9$ and use the code. **$t = 4$; chick**

7-6   Using a Reciprocal to Solve an Equation   **243**

5 FOLLOW-UP

Extra Practice
Have students solve the following equations. Have them first tell whether each answer will be *greater than* or *less than* the number on the right side of the equals sign.
1. $3/4 b = 3$ **greater than; 4**
2. $11/3 x = 22$ **less than; 6**
3. $7/4 m + 4 = 25$ **less than; 12**
Solve.
4. $-5/6 m = 2\ 1/2$ **−3**
5. $-5/2 z = 17\ 1/2$ **−7**
6. $3/4 n + 1/4 = 1\ 3/4$ **2**

Extension The equation used to convert a temperature in degrees Celsius to degrees Fahrenheit is $5/9(F - 32) = C$. Have students find Fahrenheit temperatures for specific Celsius temperatures such as 100°C, 0°C, 25°C, and so on. Then have them use reciprocals to solve for F in terms of C. **212; 32; 77; $F = 9/5 C + 32$**

Section Quiz
Solve and check.
1. $2/5 n - 23 = 11$ **85**
2. $7/8 d - 5/8 = 2$ **3**
3. $8/3 p + 1/2 = -2\ 1/4$ **−1 1/32**
4. A supermarket display consists of a stand that is 8 3/4 in. high plus rows of cans each 3 1/2 in. high. If the total height of the display is 40 1/4 in., how many rows of cans are there? **9**

WARM-UP

Evaluate each expression for n = 3.
1. $3n - 5$ **4**   2. $-n - 4$ **-7**
3. $-2(n - 6) - 12$ **-6**
4. $3.2n + 4.16$ **13.76**

1 MOTIVATE

Explore Before students read the questions in the Explore activity, ask them how they think they could find out the cost of a kite tail. Then let students work with partners or in small groups to answer the questions posed.

2 TEACH

Use the Pages/Skills Development Have students read this part of the section and then discuss the examples. Point out that a formula involves two or more quantities represented by variables and that formulas are often used to describe physical situations.

Example 1: The distance formula in Example 1 is one that students will encounter often in mathematics and science. Introduce the formula by asking how far a person would have gone after 1 h, 2 h, and 3 h, traveling at a rate

244

7-7   Working with Formulas

EXPLORE

Several members of the Kite Klub ordered fancy kite tails from a catalog. All the kite tails have the same price, and the mail-order house charges a fixed amount for shipping any order.

Member	Number of Kite Tails	Total Cost of Order
Allie	4	$18.50
Bob	2	$10.50
Mi-Su	7	$30.50
Greg	5	$22.50

a. Use the information in the table. Notice that Allie ordered 2 more kite tails than Bob. How much do 2 more tails add to the cost of an order? How much does one kite tail cost? **$8; $4**

b. How much does shipping cost per order? **$2.50**

c. Suppose you want to find the total cost of ordering a certain number of kite tails. You can begin by multiplying the number of tails by the cost per tail. What would you do next? **Add the shipping cost**

d. Write a formula for the total cost, T, of ordering n kite tails.
   $T = 4n + 2.50$

SKILLS DEVELOPMENT

A formula is an equation stating a relationship between two or more quantities. For example, the number of square units in the area (A) of a rectangle is equal to the number of units of length (l) multiplied by the number of units of width (w).

Here is the formula for the area of a rectangle: $A = lw$

Sometimes you can evaluate a variable in a formula directly, using the given information.

$A = lw$
$A = 9 \times 7$
$A = 63$

The area is 63 cm².

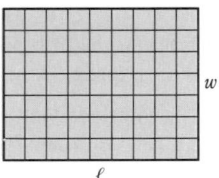

7 cm

9 cm

At other times, however, you must use your knowledge of equations to solve for a variable in a formula.

244   CHAPTER 7   Equations and Inequalities

TEACHING TIP

If students have difficulty solving formulas for one of the variables when the formula contains more than two variables, have them substitute numbers for the variables for which they are not solving and keep track of all the steps in the solution. Then they can duplicate the steps in working with the variables.

Example 1

A car traveled 120 mi in $2\frac{1}{2}$ h. Find its average rate.

Solution

Use the formula $d = rt$, where
d = distance, r = rate, and
t = time.

Substitute the known values of
the variables into the formula.

Solve the equation.

$$d = rt$$
$$120 = r\left(2\frac{1}{2}\right)$$
$$120 = r\left(\frac{5}{2}\right)$$
$$120\left(\frac{2}{5}\right) = r\left(\frac{5}{2}\right)\left(\frac{2}{5}\right)$$
$$48 = r$$

The car's average rate was 48 mi/h. ◄

Example 2

The formula $F = \frac{9}{5}C + 32$ relates Fahrenheit and Celsius temperatures,
where F = Fahrenheit temperature and C = Celsius temperature. Use
the formula to convert $113°F$ to a Celsius temperature.

Solution

$$F = \frac{9}{5}C + 32$$
$$113 = \frac{9}{5}C + 32$$
$$113 - 32 = \frac{9}{5}C + 32 - 32$$
$$81 = \frac{9}{5}C$$
$$\left(\frac{5}{9}\right)81 = \left(\frac{5}{9}\right)\frac{9}{5}C$$
$$45 = C$$

The temperature is $45°C$. ◄

Sometimes you must solve a formula for a variable that is *not* alone
on one side of the equals sign.

Example 3

Solve the formula $d = rt$ for r.

Solution

To solve a formula for a given variable, rewrite the formula so that the
variable stands alone on one side of the equals sign.

$$d = rt$$
$$\frac{d}{t} = \frac{rt}{t} \qquad \text{Divide both sides by } t$$
$$\qquad \text{to get } r \text{ alone on one side.}$$
$$\frac{d}{t} = r \quad ◄$$

The new formula indicates that rate can be found by dividing distance
by time.

ASSIGNMENTS

BASIC
1–4, 9–13, 14–16

AVERAGE
1–13, 14–18, 19

ENRICHED
5–13, 14–18, 19–20

ADDITIONAL RESOURCES
Reteaching 7–7
Enrichment 7–7

of 50 mi/h. Then develop the formula.

Example 2: Discuss temperatures
with which students are familiar (for
instance, the boiling point of water)
and then introduce the formula.
Have students complete the example and compare their solutions
with those in the text.

Example 3: Point out that in
Example 1 students used the
same formula, $d = rt$, substituting
given values for the variables into
the formula. In Example 3 they
are solving the formula for r in
terms of d and t, which results in
an expression for r in terms of
distance (d) and time (t).

Additional Questions/Examples
To help students understand the
concept of a formula, have them
work in small groups. Write a sentence such as "the selling price of
an item is equal to the cost plus
the markup." Have students identify variables and write a formula
for the relationship.

Guided Practice/Try These Students should note that they will
need to use the formula $d = rt$ in
Exercise 1.

3 SUMMARIZE

Key Questions
1. What is a formula? **an equation stating a relationship between two or more quantities**
2. How do you use a formula such as *d* = *rt* to find *r* when you know *d* and *t*? **Solve for *r* and substitute, or substitute and then solve for *r*.**

4 PRACTICE

Practice/Solve Problems In working Exercises 9–13 have students describe how they solve the equation for the indicated variable. For example, in Exercise 9 they might say, "*t* is multiplied by *p* and *r*." Then ask how the students would find the value of *t*.

Extend/Solve Problems In the formulas for Exercises 14–17 more than one step is required to isolate the variable. Have students identify how they would proceed in each case.

Think Critically/Solve Problems Suggest that students think through how they solved Exercise 19 in trying to answer Exercise 20.

5 FOLLOW-UP

Extra Practice The formula used to determine a player's batting average is *a* = *h/n*, where *a* is the average, *h* is the number of hits, and *n* is the number of times at bat. Use this formula for Exercise 1 that follows.

TRY THESE

Solve.

1. Jolene jogged $8\frac{1}{8}$ mi at an average rate of $6\frac{1}{2}$ mi/h. How long did she jog? **$1\frac{1}{4}$ h**

2. A formula for the perimeter of a rectangle is $P = 2l + 2w$, where P = perimeter, l = length, and w = width. A rectangle has a width of 7 cm and a perimeter of 46 cm. Find the length of this rectangle. **16 cm**

3. The formula for the area of a rectangle is $A = lw$, where A = area, l = length, and w = width. Solve this formula for w. **$w = \frac{A}{l}$**

EXERCISES

PRACTICE/ SOLVE PROBLEMS

Solve.

1. The formula for the area of a rectangle is $A = lw$, where l = length and w = width. A rectangle has an area of 336 cm² and a width of 14 cm. Find the length of this rectangle. **24 cm**

2. The formula for the area of a triangle is $A = \frac{1}{2}bh$, where b = length of base and h = height. A triangle has a base measuring 13 cm and an area of 91 cm². Find the height. **14 cm**

3. The formula for mass is $M = DV$, where D = density and V = volume. The largest gold nugget ever found had a mass of 216,160 g. The density of gold is 19.3 g/cm³. Calculate the volume of the nugget. **11,200 cm³**

4. The formula for the volume of a rectangular prism is $V = lwh$, where l = length, w = width, and h = height. The volume of a rectangular prism is 121.5 cm³. The length of the prism is 12 cm and the height is 4.5 cm. Find the width. **2.25 cm**

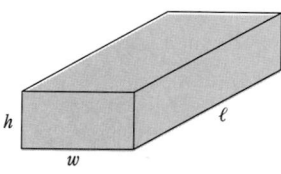

5. Crickets increase their rate of chirping as the temperature increases. One formula for estimating temperature from cricket chirps is $F = \frac{n}{4} + 32$, where F = Fahrenheit temperature and n = number of chirps per minute. Find the chirping frequency at 70°F. **152 chirps per minute**

MAKING CONNECTIONS

Have some students find examples of formulas used in business and science and share their findings with the class. Make a bulletin-board display of the formulas and how they are used. Then have other students use the formulas to make up other problems to solve.

6. Juanita sells magazine subscriptions. She earns a salary of $840 per month plus $3 for each subscription she sells. Her earnings can be computed using the formula $E = 840 + 3s$, where E = earnings and s = number of subscriptions sold. One month she earned $1,932. Find the number of subscriptions she sold.
364 subscriptions

USING DATA The earned run average (ERA) of a baseball pitcher can be computed using the formula $E = \frac{9R}{I}$, where E = ERA, R = number of runs given up, and I = number of innings pitched. Use the Data Index on page 546 to find the ERA statistics. Use them to solve Exercises 7 and 8.

7. Find the number of runs Rube Waddell gave up in 150 innings.
36 runs

8. Jeff Tesreau pitched 1,800 innings. How many runs did he give up? **486 runs**

Solve each formula for the indicated variable.

9. $I = prt$, for t $t = \frac{I}{pr}$

10. $E = mc^2$, for m $m = \frac{E}{c^2}$

11. $A = p + i$, for i $i = A - p$

12. $P = 3s$, for s $s = \frac{P}{3}$

13. $K = 273 + C$, for C $C = K - 273$

Solve each formula for the indicated variable.

14. $P = 2l + 2w$, for w

15. $A = \frac{1}{2}bh$, for h

16. $F = \frac{9}{5}C + 32$, for C

17. $y = mx + b$, for x

18. $T = p + prt$, for r

19. A traffic court judge fined speeding offenders $75 plus $2 for every mile per hour they exceeded the speed limit. Calculate the fine for driving at the given rate.

a. 45 mi/h; speed limit: 35 mi/h **$95**

b. 70 mi/h; speed limit: 55 mi/h **$105**

c. 92 mi/h; speed limit: 65 mi/h **$129**

20. A motorist drove r mi/h on a highway posting a speed limit of s mi/h.

a. If $r > s$, by how much did the motorist exceed the speed limit?

b. Refer to Exercise 19 for the method of calculating a fine. Write a formula for computing the fine, f. $f = 2(r - s) + 75$

---

14. $P = 2l + 2w$
$P - 2l = 2w$
$\frac{1}{2}(P - 2l) = w$

15. $A = \frac{1}{2}bh$
$2A = bh$
$\frac{2A}{b} = h$

16. $F = \frac{9}{5}C + 32$
$F - 32 = \frac{9}{5}C$
$\frac{5}{9}(F - 32) = C$

17. $y = mx + b$
$y - b = mx$
$\frac{y - b}{m} = x$

18. $T = p + prt$
$T - p = prt$
$\frac{T - p}{pt} = r$

EXTEND/ SOLVE PROBLEMS

THINK CRITICALLY/ SOLVE PROBLEMS

1. Jake would like to maintain a batting average of 0.25. If he knows he will be at bat 40 times, how many hits does he need? **10**

2. The voltage, V, across any part of a circuit is the product of the current, I, and resistance, R. Write a formula to express this relationship and solve it for I. **$V = IR$; $I = V/R$**

3. Solve the formula $T = 2\pi r^2 + 2\pi rh$ for h.
$h = \dfrac{T - 2\pi r^2}{2\pi r}$

Extension The formula for the number of kilowatt-hours used by an appliance is $k = hw/1{,}000$. In the formula, h is the number of hours the appliance is used and w is the number of watts, or units of power, it uses. Have one group of students find out how many watts their hair dryers use and determine the number of kilowatt-hours a hair dryer would use if it was turned on for 20 min a day for a year. Then have them find the average cost per kilowatt-hour in your community to determine the cost of running the hair dryer. Have other groups of students choose different appliances to investigate in the same way.

Section Quiz The formula for finding the area of a rectangle is $A = lw$.

1. Find the length of a rectangle if its width is 8.3 m and its area is 62.25 m². **7.5 m**

2. Find the width if the length is 24 ft and the area is 432 ft². **18 ft**

3. If $T = cx^2$, find c for $T = 22.05$ and $x = 2.1$. **5**

7-8 Problem Solving Strategies:
WORK BACKWARD

▶ READ
▶ PLAN
▶ SOLVE
▶ ANSWER
▶ CHECK

Some problems tell you the outcome of a series of steps and ask you to find the fact that sets those steps in motion. To solve such a problem, start at the end and work backward to the beginning.

PROBLEM

Tanya purchased a $930 clarinet using her $5.20/h earnings as a cashier. While she was saving, her uncle donated $150 toward her goal. How many hours did Tanya work to earn the total?

SOLUTION

?	×	$5.20	+	$150	=	$930
number of hours		hourly wages		uncle's contribution		total

Multiply.   Add.

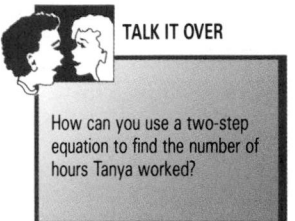

TALK IT OVER

How can you use a two-step equation to find the number of hours Tanya worked?

Let w = the number of hours she worked.

$$5.20w + 150 = 930$$
$$5.20w + 150 - 150 = 930 - 150$$
$$5.20w = 780$$
$$\frac{5.20w}{5.20} = \frac{780}{5.20}$$
$$w = 150$$

The diagram shows the steps and operations leading to the $930 total. To solve, work backward from $930.

Start with the total.	$930
Subtract the contribution.	$\dfrac{-\ 150}{\$780}$
Divide by the hourly wage.	$780 ÷ $5.20 = 150

Tanya worked 150 h.

Check: $(150 × \$5.20) + \$150 = \$930$

PROBLEMS

1. During April the price of an apple tripled. In May the price dropped $0.18. In June it halved. In July the price rose $0.06 to $0.30 per apple. Find the price at the beginning of April. **$0.22**

2. A mountain climber ascended at a rate of 1,200 ft/h, fell and slid 600 ft down, got up and ascended for $2\frac{1}{3}$ h at a rate of 900 ft/h. The total gain in altitude was 4,500 ft. How long did the mountaineer climb at a rate of 1,200 ft/h? **$2\frac{1}{2}$ h**

3. Marcus spends 55 min each morning getting ready for work. He walks 7 min to a cafeteria where he spends 27 min eating breakfast. Then he walks 4 min to a subway station where he catches a subway for a 12-min ride to work. At what time must he get up in order to get to his job at 8:30 a.m.? **6:45 a.m.**

4. Lee wants to save $900 for his vacation. When he has saved four times as much as he already has saved, he'll need just $44 more to reach his goal. How much has he saved? **$214**

5. Julie multiplied her age by 2, subtracted 5, divided by 3, and added 9. The result was 20. How old is she? **19**

6. Abbie, Bill, Colin, and Diane divided up a flat of petunia plants. Abbie took half. Bill took half of what was left. Colin took a third of what was left. Diane took the remaining 6 plants. How many petunia plants were there to begin with? **36 plants**

7. A new car carried a sticker price of $11,586. The price consisted of the base price plus $1,360 in options, multiplied by 1.05 for sales tax. To this amount a $180 dealer preparation fee and $66 license fee were added. Find the base price. **$9,440**

8. On Monday, the price of a share of Heavy Metal Industries fell $2\frac{1}{4}$ points. On Tuesday the price rose $5\frac{1}{2}$ points, but it fell $\frac{7}{8}$ on both Wednesday and Thursday. On Friday the stock gained $1\frac{3}{4}$ points to close the week at $28\frac{3}{8}$. Find the price of a share of Heavy Metal at the start of trading on Monday. **$25\frac{1}{8}$**

CHALLENGE

Challenge students to write equations to solve each of the problems in this section.

ASSIGNMENTS

BASIC
1–8

AVERAGE
1–8

ENRICHED
1–8

ADDITIONAL RESOURCES
Reteaching 7–8
Enrichment 7–8

to find out how much Tanya has to earn and, thus, how many hours she has to work.

3 SUMMARIZE

Write About Math Have students describe in their math journals how they would decide when to leave home for baseball practice if they knew what time they had to be at practice and had errands taking 15 minutes, 20 minutes, and 10 minutes to do before they left.

4 PRACTICE

Problems Have students work with partners in solving these problems.

5 FOLLOW-UP

Extra Practice
1. A ball bounces on the floor. After each bounce, it is 2/3 as high as it was on the previous bounce. On the fifth bounce, it is 2 ft above the floor. How high was the ball on the first bounce? **10 1/8 ft**
2. Begin with a number. Add 3, divide by 2, subtract 4, and multiply by 5. The answer is –5. What is the number? **3**

Get Ready algebra tiles

WARM-UP

Use the distributive property to rewrite each expression.
1. $3 \times 4 + 5 \times 4$ **(3 + 5)4**
2. $2.3 \times 7 + 2.3 \times 3$ **2.3 (7 + 3)**
3. $3a + 3b$ **3(a + b)**

1 MOTIVATE

Explore Provide algebra tiles and let students work in groups on this activity. Before they begin, be sure all students can represent terms, such as $2x$ and $5x$, using the tiles. Ask students what they conclude from the activity.

2 TEACH

Use the Pages/Skills Development
Have students read this part of the section and then discuss the examples. Write $3xy$ and $x + y$ on the chalkboard and ask students to describe how they are different. Use this to introduce the concept of the terms of an expression. Have students identify the terms in the expression given in the Skills Development. Write $7a$ on the chalkboard and have students give three like terms and three unlike terms.

7-9 Combining Like Terms

EXPLORE Model each expression using algebra tiles. Use your model to find a simpler way to write each expression. Recall that ▭ represents x and ▪ represents 1. **See Additional Answers.**

1. $3x + 2x$
2. $6x - x$

SKILLS DEVELOPMENT The parts of a variable expression that are separated by addition or subtraction signs are called **terms.** The expression $7a + 2b + 5a - 3b^2$ contains four terms, $7a$, $2b$, $5a$, and $3b^2$. The terms $7a$ and $5a$ are called **like terms** because they have identical variable parts. The terms $2b$, $5a$, and $3b^2$ are called **unlike terms** because they have different variable parts.

Example 1
Identify the terms as like or unlike.

a. $3x$, $3y$ b. $7k$, $-3k$ c. $2p$, $2pt$

Solution
a. The terms are unlike because the variable parts, x and y, are different.
b. The terms are like because the variable parts, k and k, are identical.
c. The terms are unlike because the variable parts, p and pt, are different. ◄

You simplify an expression when you perform as many of the indicated operations as possible. You can use the distributive property to simplify a variable expression that contains like terms. This process is called **combining like terms.** Unlike terms cannot be combined.

Example 2
Simplify. a. $3m + 5m$ b. $9x - x + 5y + 2y^2$ c. $4k - 3h + 2k$

Solution
a. Use the distributive property. $3m + 5m = (3 + 5)m$
$= 8m$

b. Use the distributive property. *Recall that $x = 1x$.*
$9x - x + 5y + 2y^2 = (9 - 1)x + 5y + 2y^2$
$= 8x + 5y + 2y^2$

c. Rewrite using the commutative property. $4k - 3h + 2k = 4k + 2k - 3h$
Use the distributive property. $= (4 + 2)k - 3h$
$= 6k - 3h$ ◄

TEACHING TIP

Point out that students can sometimes check the way in which they simplified an expression by substituting the same value for the variable in both the given expression and in their answer.

Example 3

Jesse worked $2p$ hours in the morning, $5r$ hours in the afternoon, and $4p$ hours in the evening. Write and simplify an expression for the total number of hours he worked.

Solution

Total times worked = sum of morning, afternoon, and evening hours

$$= 2p + 5r + 4p$$
$$= 6p + 5r \quad \blacktriangleleft$$

Some equations have a variable term on both sides of the equals sign. Combining like terms is an important part of the process of solving these equations.

Example 4

Solve $5x - 7 = 3x + 13$. Check the solution.

Solution

	$5x - 7 = 3x + 13$
Subtract $3x$ from both sides.	$5x - 3x - 7 = 3x - 3x + 13$
Simplify.	$2x - 7 = 13$
Add 7 to both sides.	$2x - 7 + 7 = 13 + 7$
	$2x = 20$
Divide both sides by 2.	$\frac{2x}{2} = \frac{20}{2}$
	$x = 10$
Check.	$5x - 7 = 3x + 13$
	$5(10) - 7 = 3(10) + 13$
	$50 - 7 = 30 + 13$
	$43 = 43 \checkmark$

The solution is 10. $\blacktriangleleft$

TRY THESE

Identify the terms as like or unlike.

1. $-4a, -5a$
   like

2. $3mn, 3mn^2$
   unlike

3. $6x, -4x$
   like

4. $7a, 7ab$
   unlike

Simplify.

5. $2y + 13y$ **15 y**

6. $8x + 5z + 4 - 3z$ **8x + 2z + 4**

7. $7n + 3m - m + 2n$ **9n + 2m**

8. $e + 5e + e^2 - 3e$ **3e + e²**

7-9   Combining Like Terms   **251**

Example 1: Have students discuss this example and then compare their solutions to those in the text. See if students can explain why parts *a* and *c* contain unlike terms before they read the solution in the text.

Example 2: Point out that *simplifying* means combining all the like terms. Review the properties used to simplify the expression.

Example 3: Make sure that students understand how the expression for the total time worked was derived.

Example 4: Use algebra tiles to illustrate the solution of this equation. Then discuss the steps given in the text. Ask students how they used the distributive property in the second step.

Additional Questions/Examples

1. Are x and x^2 like terms? Explain. **No, x squared has the value of x multiplied by itself.**

2. How many terms are in the expression $-3x - 2y + 5y + 8x$? Are any like terms? **4; −3x and 8x, 2y and 5y are like terms.**

3. Simplify the following expression: $-3x - 2y + 5y + 8x$ **5x + 3y**

9. On four successive days, Meg's driving times were $18k$, $11h$, $13k$, and $9h$ hours. Write and simplify an expression for the total amount of time she drove. **$18k + 11h + 13k + 9h$; $31k + 20h$**

Solve and check.

10. $4n + 7 = n - 2$ **-3**

11. $9x = 6x - 15$ **-5**

12. $7c - 3 = 3c + 5$ **2**

13. $8m = 2m + 18$ **3**

EXERCISES

PRACTICE/SOLVE PROBLEMS

Identify the terms as like or unlike.

1. $9n$, $3m$ **unlike**
2. $-2p$, $-2p^2$ **unlike**
3. x, $-2x$ **like**
4. $5y$, 5 **unlike**
5. $-3h$, $-3h$ **like**
6. xy, xz **unlike**
7. $-8a$, $8a$ **like**
8. $3s$, $6s$ **like**
9. $-k$, k **like**

Simplify.

10. $4m + 8m$ **12m**
11. $8s + 5t - 3s$ **$5s + 5t$**
12. $x + 11x$ **12x**
13. $7y + 3y - 4$ **$10y - 4$**
14. $2 + 9p - 6p$ **$2 + 3p$**
15. $a + 2a + 3a$ **6a**
16. $3w + 2w + 5 - 7$ **$5w - 2$**
17. $14n - 6x + (-5n) + 8x$ **$9n + 2x$**
18. $2c + 8c^2 + 11c + 9a$ **$13c + 8c^2 + 9a$**
19. $-3x - 3x + 3xy$ **$-6x + 3xy$**

20. Emilio spent $5a$ min on his math homework, $3a$ min on history, $8b$ min on social studies, and $7b$ min on science. Write and simplify an expression for the total amount of time he spent on homework. **$5a + 3a + 8b + 7b$; $8a + 15b$**

21. In four successive basketball games, Ruth scored $11s$, $9p$, $7s$, and $6k$ points. Write and simplify an expression for the total number of points she scored. **$11s + 9p + 7s + 6k$; $18s + 9p + 6k$**

Solve each equation. Check the solution.

22. $5x - 7 = 2x + 2$ **3**
23. $3n - 4 = 2n + 3$ **7**
24. $4n + 9 = n - 3$ **-4**
25. $8r - 11 = 6r - 6$ **$\frac{5}{2}$**
26. $3n - 10 = -2n + 5$ **3**
27. $n - 3 = 5n + 1$ **-1**

EXTEND/SOLVE PROBLEMS

Simplify.

28. $4.6x + 3.5y - 6.9x + y$ **$-2.3x + 4.5y$**
29. $\frac{1}{2}x + \frac{1}{2}x + 3x$ **4x**
30. $\frac{2}{3}m + \frac{3}{4}n + \frac{5}{8}n$ **$\frac{2}{3}m + \frac{11}{8}n$**
31. $7 + 14.2a - 8 + 3.8a$ **$18a - 1$**

32. $z + z + z + z^2 + z$ **4z + z²**

33. $297k + 403k$ **700k**

34. $3(x - 5) + 2(x + 2)$ **5x – 11**

35. $-4n + 5(3n) - 2(-2n)$ **15n**

36. $6(-3y) + x + 2(2x + y)$
5x – 16y

37. $8(a + b + c) + 2(a + 2b + 3c)$
10a + 12b + 14c

38. $9.177x + 8.591y + 13.962x - 12.413y$ **23.139x – 3.822y**

39. $173x + 23(14x - 27y) + 45(22x + 12y)$ **1,485x – 81y**

40. Benito bought 3 quarts of milk costing $(a + 2b)$ cents per quart and 2 loaves of bread costing $(3a + b)$ cents per loaf. Write and simplify an expression for the total cost of his purchases.
3(a + 2b) + 2(3a + b); 9a + 8b

41. Write and simplify an expression for the combined areas of the two rectangles.
4(x + y) + 3(2x – y); 10x + y

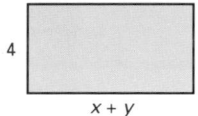
4

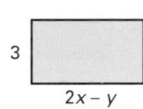
3

$x + y$ $2x - y$

Solve each equation. Check the solution.

42. $7(x - 2) = 5(x + 4)$ **17**

43. $3n + 5(n - 2) = 3(n - 5)$ **–1**

44. $6a + 4a + 5 = 3 + 4(a + 5)$ **3**

45. $\frac{3}{4}(x + 12) = \frac{5}{8}(x + 8)$ **–32**

46. Choose three numbers. Copy and complete the table. Describe the unusual results you obtain. **Each result will be 3, regardless of the number chosen.**

	First number	Second number	Third number
Add 3.	■	■	■
Multiply by 4.	■	■	■
Subtract 6.	■	■	■
Divide by 2.	■	■	■
Subtract twice the original number.	■	■	■

47. Follow these steps to discover why you obtained the above result. Let n = the chosen number. Write and simplify the expression that results when you apply the given operation.

a. Add 3 to n. **n + 3**

b. Multiply the result obtained in a by 4. **4n + 12**

c. Subtract 6 from the result obtained in b. **4n + 6**

d. Divide the result obtained in c by 2. **2n + 3**

e. Subtract twice the original number from the result obtained in d. **3**

f. Explain why you obtained the unusual results in the table above.
The variable n is eliminated from the calculations in the final step, leaving 3 no matter what the value of n.

THINK CRITICALLY/ SOLVE PROBLEMS

COMPUTER

This program will give unusual results, too. Run it several times and compare the result with the number you entered. Can you "predict" the result for any number entered? Why is this true? Can you change lines 20 and 30 to create a different result?

```
10 INPUT "ENTER A
   NUMBER: ";N: PRINT
20 LET X =
   (3 * (N – 2) – 12) / 6
30 LET R = X + 3
40 PRINT "THE RESULT IS
   ";R: PRINT
50 INPUT "TRY IT AGAIN? Y
   OR N: ";X$: PRINT
60 IF X$ = "Y" THEN GO TO
   10
```

See Additional Answers.

Simplify.

3. $5a + 8a$ **13a**

4. Condra spent $4a$ hours driving one week, $5b$ hours the next week, and $9a$ hours the third week. Write and simplify an expression for the total number of hours she spent driving.
4a + 5b + 9a = 13a + 5b

Solve and check.

5. $4x + 2 = x - 10$ **–4**

6. $8r + 1 = 4r + 1$ **0**

7. $3(-2y) + 5 = 9y - 40$ **3**

Extension Have students work in small groups to make up puzzles similar to the one in Exercise 46, except give students the goal that each puzzle should end with the number with which it started. Then have students use the puzzle to determine other students' ages or classroom numbers, for example.

Section Quiz
Identify the terms as *like* or *unlike*.

1. $5h, -2h$ **like**

2. $3x, 3xy$ **unlike**

Simplify.

3. $6m + 7m$ **13m**

4. $x - 9x$ **–8x**

5. $a + b + 4a$ **5a + b**

6. $3m + 7m - 2m$ **8m**

7. $4x + 2y - 4x$ **2y**

8. $2 - 3n + (-4m) - 8$
–3n – 4m – 6

Solve and check.

9. $4a - 5 = 2a + 3$ **4**

10. $2 - 5x = x - 10$ **2**

Additional Answers
See page 579.

7-10   Graphing Open Sentences

EXPLORE

INEQUALITY SYMBOLS	
Symbol	**Meaning**
>	"is greater than"
<	"is less than"
≥	"is greater than or equal to"
≤	"is less than or equal to"

Match each statement with a symbolic representation.

1. The temperature is less than 17°F. **b**
2. No more than 17 tickets remain. **d**
3. No one under 17 admitted. **c**
4. The blouse cost at least $17. **c**
5. More than 17 students attended. **a**
6. She finished in under 17 minutes. **b**

   a. $x > 17$
   b. $x < 17$
   c. $x \geq 17$
   d. $x \leq 17$

SKILLS DEVELOPMENT

A mathematical sentence that contains one of the symbols $<$, $>$, $\leq$, or $\geq$ is called an **inequality**.

$$5 < 7 \qquad\qquad 7 \geq y - 2$$
5 is less than 7.   7 is greater than or equal to $y - 2$.

In Section 7-1 you learned that an open sentence is a sentence that contains one or more variables. An open sentence may be either an equation or an inequality. By itself, an open sentence like $x + 7 = 11$ or $k > 3$ is neither true nor false. When you substitute a number for the variable, however, you can determine whether the result is true or false. Any value of the variable that makes the sentence true is called a **solution of the open sentence**.

equation: $x + 7 = 11$ inequality: $k > 3$
solution: 4   some solutions: 4
                                  $7\frac{1}{3}$
                                  π (3.141592...)

254   CHAPTER 7   Equations and Inequalities

How many solutions are there for the inequality $k > 3$? Every integer greater than 3 is a solution. In fact, any rational number, such as $7\frac{1}{3}$, is a solution. The number π, which is approximately equal to 3.14, is also a solution.

Numbers such as π that are non-terminating, non-repeating decimals are called **irrational numbers.** There are many irrational numbers beside π that are greater than 3. Together the set of rational numbers and the set of irrational numbers make up the set of **real numbers.** So we say that the solutions of $k > 3$ are *all the real numbers that are greater than 3.*

Since solutions of open sentences are real numbers, you can graph them on a number line.

Example 1

Graph the equation $m + 9 = 14$.

Solution

$$m + 9 = 14$$
$$m + 9 - 9 = 14 - 9$$
$$m = 5$$

The equation has one solution. Graph the solution on a number line by drawing a solid dot at the point 5.

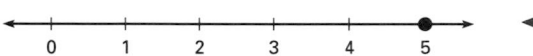

Example 2

Graph the inequality $p > -2$.

Solution
The solution consists of all real numbers greater than -2. Graph the solution by drawing a solid arrow beginning at -2 and pointing to the right. To indicate that -2 is not part of the solution, draw an open circle at the point -2.

Example 3

Graph the inequality $k \leq 4$.

Solution
The solution consists of all real numbers less than or equal to 4. Graph the solution by drawing a solid arrow beginning at 4 and pointing to the left. To indicate that 4 is part of the solution, draw a solid dot at the point 4.

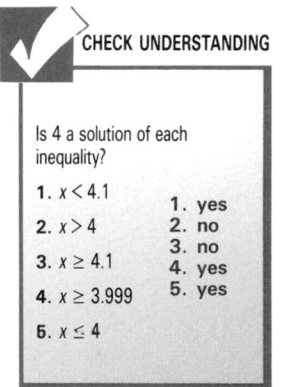

CHECK UNDERSTANDING

Is 4 a solution of each inequality?

1. $x < 4.1$
2. $x > 4$
3. $x \geq 4.1$
4. $x \geq 3.999$
5. $x \leq 4$

1. yes
2. no
3. no
4. yes
5. yes

ASSIGNMENTS

BASIC
1–12, 15–16, 18–20, 23, 26–27

AVERAGE
1–17, 18–25, 26–31

ENRICHED
7–8, 13–35

ADDITIONAL RESOURCES
Reteaching 7–10
Enrichment 7–10
Transparency Masters 31, 32

Example 1: Students should understand that the solid dot represents the one integer that is the solution of the equation.

Example 2: Have students give some numbers that are solutions of this inequality and then draw the graph. Ask students again why an open circle is used in the graph.

Example 3: Ask students if 4 is a solution of this equation. Have them refer to the graph.

Additional Questions/Examples

1. Write an inequality involving the symbol < that is equivalent to "$x > 4$."
   $4 < x$
2. Write an inequality equivalent to "x is not less than 3." $x \geq 3$
3. Write an inequality equivalent to "x is not greater than or equal to -5." $x < -5$

Guided Practice/Try These Ask students which open sentence will have a graph with an open circle and why. Then ask which will have a solid dot and why.

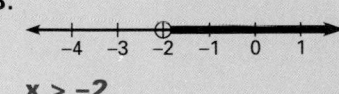

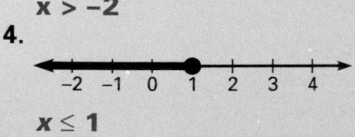

3 SUMMARIZE

Talk It Over Ask students to describe in their own words how they would go about graphing an inequality.

4 PRACTICE

Practice/Solve Problems Before they begin graphing Exercises 1–8 have students decide if the starting point of each graph will be included or not. In each of Exercises 15–17 ask students to indicate the key words they used in determining the appropriate inequality symbol.

Extend/Solve Problems Again, in Exercises 23–25 ask students to indicate the key words they used in interpreting the inequality.

Think Critically/Solve Problems Suggest that students try actual numbers as they determine how to draw the graph. Thus, in Exercise 29 you might ask if –1 is included. **No, because it is not less than or equal to –2.**

5 FOLLOW-UP

Extra Practice Give three numbers that are solutions of each inequality and one number that is not.
1. $x < -2$ **for example, –3, –4, –5; 0**
2. $a \geq 0$ **for example, 0, 1, 2; –1**

Write an open sentence whose solution is shown by the graph.
3.

x > –2

4.

x ≤ 1

256

1.

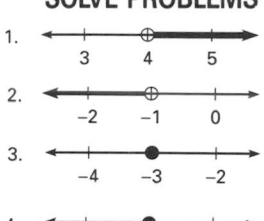

2.

3.

Graph each open sentence on a number line.

1. $x - 7 = 3$ **2.** $h \geq -5$ **3.** $e < 1$

EXERCISES

PRACTICE/ SOLVE PROBLEMS

1.

2.

3.

4.

5.

6.

7.

8.

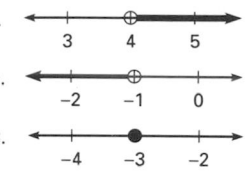
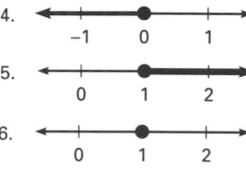
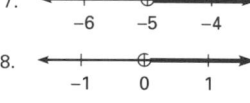

Graph each open sentence on a number line.

1. $x > 4$ **2.** $m < -1$
3. $h + 5 = 2$ **4.** $y \leq 0$
5. $p \geq 1$ **6.** $-5 = e - 6$
7. $x > -5$ **8.** $0 < w$

Tell whether the given number is a solution of the inequality.

9. $n < 4$: 0; 4; 5; –4
yes; no; no; yes

10. $y \geq 6$: 0; –5; –6; –7
no; no; no; no

11. $x \leq -1$: 1; 0; –1; –2
no; no; yes; yes

12. $m > -2$: –5; –1; –2; –4
no; yes; no; no

13. $k \leq -\frac{1}{2}$: –1; $-\frac{3}{4}$; –0.5; 0
yes; yes; yes; no

14. $e > -3.4$: –2; –4; –3.45; –3.40 yes; no; no; no

Write an inequality to describe the situation.

15. Car repairs cost more than $300. Let r equal the repair cost. **r > 300**

16. Joaquin's height is greater than or equal to 2 m. Let h equal Joaquin's height. **h ≥ 2**

17. Every frame in the store was priced at under $25. Let c equal the cost of a frame. **c < 25**

EXTEND/ SOLVE PROBLEMS

Write an open sentence for each graph. **Answers may vary.**

18.

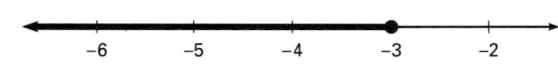

x ≤ –3

19.

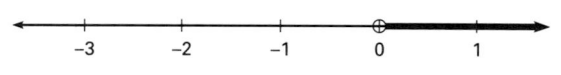

x > 0

20.

x < –1

AT-RISK STUDENTS

Use the technique of writing inequalities, such as $x < 4$, on index cards, one symbol to a card. To test if a number is a solution to a given inequality, students can place a card with that number over the variable, read the resulting inequality, and then decide whether or not it is true.

21.

$x = 2$

22.

$x \geq 3$

Write an inequality to describe the situation.

23. Gregor hit at least 8 foul shots in every game. Let f equal the number of foul shots that he hit. **$f \geq 8$**

24. Sue planned to spend no more than \$2,000 on advertising. Let a equal the amount she planned to spend. **$a \leq 2,000$**

25. Police estimated the size of the crowd as not less than 50,000 people. Let p equal the number of people present. **$p \geq 50,000$**

Graph on a number line.

26. all values of n that are greater than -3 and less than 3

27. all values of n that are greater than or equal to -2 and less than 2

28. all values of n that are greater than -3 and greater than 1

29. all values of n that are less than 0 and less than or equal to -2

USING DATA Use the table on page 221.
Let a = average of the four scores in Game 1
  b = average of the four scores in Game 2
  c = average of the four scores in Game 3

Write *true* or *false*.

30. $a \geq b$ false

31. $c > a$ true

32. $b < c$ true

33. $b < 125$ true

34. $c \leq 138$ true

35. $a > 113$ false

7-10 Graphing Open Sentences **257**

**THINK CRITICALLY/
SOLVE PROBLEMS**

26.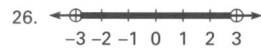

27.

28.

29.

Write an inequality to describe the situation.

5. Aaron's age is less than or equal to 17 years. Let a equal Aaron's age. **$a \leq 17$**

6. Yolanda plans to spend no less than \$135 on a bicycle. Let s equal the amount she plans to spend. **$s \geq 135$**

7. Graph on a number line: all values of n that are less than 0 and greater than -4.

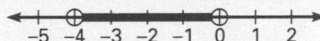

Extension Have students work in groups to write sentences about their lives that could be described with inequalities. For example, "I watch no more that 3 h of television per week."

Section Quiz Graph the solution of each open sentence on a number line.
1. $n \leq -3$

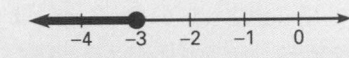

2. $w < -1$

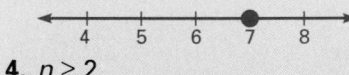

3. $-2 + a = 5$

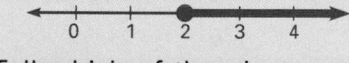

4. $n \geq 2$

Tell which of the given numbers, if any, is a solution of the inequality.
5. $a > 2$: $-4, -2, 0, 2$ **none**
6. $m \geq 4$: $-2, 0, 4, 6$ **4, 6**
7. $x \leq -1$: $-4, -2, -1, 0$ **$-1, -2, -4$**
8. $w < 5$: $-4, 0, 2, 5$ **$-4, 0, 2$**
9. Write an inequality: Lunches cost at least \$4. Let c equal the cost of lunch. **$c \geq 4$**

WARM-UP

Replace ● with <, >, ≤ or ≥ .
1. $-3 < 0$, so $-3 + 2 ● 0 + 2$ **<**
2. $4 > 1$, so $4 - 8 ● 1 - 8$ **>**
3. $2 ≥ -1$, so $3 × 2 ● 3 × (-1)$ **>**
4. $-1 ≤ 0$, so $-2 × (-1) ● -2 × 0$ **≥**

1 MOTIVATE

Explore Have students work with partners on the Explore activity. When they have finished, ask each group to share its discoveries with the class.

2 TEACH

Use the Pages/Skills Development
Have students read this part of the section and then discuss the examples. Talk about the rules for solving equations. (You can add, subtract, multiply, or divide both sides of an equation by the same number without changing the equality.) Ask students if they feel that will be true for inequalities. Use the students' experience with the Explore activity to elicit the idea that when multiplying or dividing both sides by a negative number, the inequality will change. Discuss the rule and

258

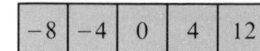

7-11 Solving Inequalities

EXPLORE

-8	-4	0	4	12

The integers above are ordered from least to greatest. Write the 5 integers that result when you perform the indicated operation on each of the above integers.

1. add 6 **-2, 2, 6, 10, 18**
2. add -4 **-12, -8, -4, 0, 8**
3. subtract 8 **-16, -12, -8, -4, 4**
4. subtract -3 **-5, -1, 3, 7, 15**
5. multiply by 5 **-40, -20, 0, 20, 60**
6. multiply by -1 **8, 4, 0, -4, -12**
7. divide by 2 **-4, -2, 0, 2, 6**
8. divide by -4 **2, 1, 0, -1, -3**

9. List the exercises in which the operations you performed resulted in a set of integers ordered least to greatest.
Exercises 1, 2, 3, 4, 5, and 7
10. List the exercises in which the operations you performed resulted in a set of integers ordered greatest to least. **Exercises 6 and 8**

SKILLS DEVELOPMENT

Most of the equations you have solved have had only one solution. As you discovered in the last section, an inequality may have an infinite number of solutions. To find the solutions of an inequality containing several operations, you follow the rules for solving equations. The single exception to the rules is the following.

▶ When multiplying or dividing both sides of an inequality by a negative number, reverse the direction of the inequality.

Example 1

Solve and graph $k + 3 ≥ -4$.

Solution

$$k + 3 ≥ -4$$

Undo the addition. $k + 3 - 3 ≥ -4 - 3$

Simplify. $k ≥ -7$

Draw the graph.

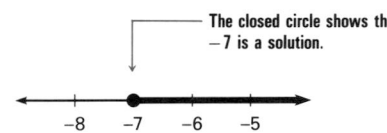

The solution is all numbers greater than or equal to -7. ◀

258 CHAPTER 7 Equations and Inequalities

Example 2

Solve and graph $-3n - 5 < 1$.

Solution

$$-3n - 5 < 1$$

Undo the subtraction. $\quad -3n - 5 + 5 < 1 + 5$

Simplify. $\quad\quad\quad\quad\quad\quad -3n < 6$

Undo the multiplication. $\quad \dfrac{-3n}{-3} > \dfrac{6}{-3} \leftarrow$ { **Dividing by a negative number reverses the direction of the inequality.**

Simplify. $\quad\quad\quad\quad\quad\quad n > -2$

Draw the graph.

The open circle shows that -2 is not a solution.

◀

Example 3

The pilot of a small plane wanted to stay at least 1,400 ft beneath the storm clouds, which are at 11,000 ft. After takeoff, for how long could the pilot ascend at a rate of 640 ft/min?

Solution

Write and solve an inequality that represents the situation.

Let m = number of minutes ascending at 640 ft/min.

$$640m + 1,400 \le 11,000$$

$$640m + 1,400 - 1,400 \le 11,000 - 1,400$$

$$640m \le 9,600$$

$$m \le 15$$

The pilot can climb for up to 15 min. ◀

 CALCULATOR

You can solve inequalities using your calculator.

Solve as though you were solving an equation, but remember to use the correct inequality symbol instead of an equals sign in your answer.

To solve $5.1x - 20.8 \le -3.46$, use this key sequence.
3.46 $\boxed{+/-}$ $\boxed{+}$ 20.8 $\boxed{\div}$ 5.1 $\boxed{=}$ 3.4

On some calculators, you have to press the $\boxed{=}$ key after you enter the 20.8.

The solution is $x \le 3.4$.

illustrate with examples such as $-1 < 3$, but $(-2)(-1) > (-2)\,3$.

Example 1: Ask students to identify the first step in solving this inequality before you discuss the solution in the text.

Example 2: Call on students to identify the first step in solving $-3n - 5 < 1$. Once 5 has been added to both sides, point out that the multiplication still needs to be undone. Remind students that multiplying or dividing by a negative number reverses the inequality. Have students check their answers by substituting -2 (to show they have the number that makes both sides equal) and another number, such as 0.

Example 3: It may help students to draw a picture to illustrate this example.

Additional Questions/Examples
1. How can you check your answer when solving an inequality? **In the original inequality, check the value that would make an equation and one value that should be a solution.**
2. Which of the following is equivalent to $3x < -12$?
   a. $x < -4$ **b.** $x > -4$ **a**

3. Which of the following is equivalent to $-2x < 6$?

a. $x < -3$ **b.** $x > -3$ **b**

Guided Practice/Try These Have students work independently and then compare their graphs with their partners' graphs. Encourage students to check their answers.

3 SUMMARIZE

Write About Math Ask students to write a paragraph in their math journals summarizing the rules for solving inequalities.

4 PRACTICE

Practice/Solve Problems Ask students what inequality symbols describe "at least" in Exercise 16 ($\geq$) and "no more than" in Exercise 17 ($\leq$).

Extend/Solve Problems For added practice, students can graph the solutions to Exercises 18–23. In Exercise 24, students might solve using the guess-and-check strategy, but encourage them to write an inequality to solve. In Exercise 25, a diagram may prove helpful.

Think Critically/Solve Problems In Exercises 26 and 27, have students give their problems to other students to solve. Remind students that, in Exercises 28–32 the word *and* means that both conditions must be met by the integers. (Thus, 0 is not a solution in Exercise 28 because, although 0 > −8, 0 is not less than −4.)

5 FOLLOW-UP

Extra Practice
Solve and graph each inequality.
1. $x + 7 \geq 7$ **$x \geq 0$**

260

1.

2.

3.

4.

TRY THESE

Solve and graph each inequality.

1. $x - 9 \geq -5$ **$x \geq 4$**   **2.** $5y < -45$ **$y < -9$**

3. $-2p \leq 10$ **$p \geq 5$**   **4.** $-7 < 2k + 3$ **$k > -5$**

Solve.

5. Members of the Music Club hope to raise at least $1,800 from the sale of tickets to the spring concert. They estimate that 450 people will attend the concert. How much should they charge for tickets? **$4 or more**

EXERCISES

PRACTICE/ SOLVE PROBLEMS

Solve and graph each inequality. **See Additional Answers.**

1. $w - 6 < -2$ **$w < 4$**   **2.** $3p \geq -12$ **$p \geq -4$**   **3.** $x + 9 \leq 3$ **$x \leq -6$**

4. $5n > 0$ **$n > 0$**   **5.** $\frac{y}{-3} \geq 1$ **$y \leq -3$**   **6.** $4 < \frac{2}{3}h$ **$h > 6$**

7. $15 + t > 17$ **$t > 2$**   **8.** $12 \leq -4e$ **$e \leq -3$**   **9.** $\frac{1}{7}m > -1$ **$m > -7$**

10. $13 \leq c + 8$ **$c \geq 5$**   **11.** $-x < 2$ **$x > -2$**   **12.** $k - (-3) > -2$ **$k > -5$**

13. $6w + 2 > -4$ **$w > -1$**   **14.** $4 + 5y < 29$ **$y < 5$**   **15.** $7n + 4 \geq -10$ **$n \geq -2$**

Solve.

16. Marcia needs to score at least 270 points on 3 tests to earn an A. On her first 2 tests she scored 88 and 87. What must she score on the final to receive an A? **at least 95**

17. Multiply a number by -2. Then subtract 5 from the product. The result is no more than 7. Find the number. **$n \geq -6$**

EXTEND/ SOLVE PROBLEMS

Solve.

18. $4\left(x - \frac{1}{2}\right) < 6$ **$x < 2$**   **19.** $2y - 5y \geq 12$ **$y \leq -4$**

20. $-2 + 5(2) > -2(x - 1)$ **$x > -3$**   **21.** $\frac{2}{3}x - \frac{3}{4}x \leq -1$ **$x \geq 12$**

22. $-1 < 4n - 5n$ **$n < 1$**   **23.** $15(e + 4) \leq 5(12 - 3e)$ **$e \leq 0$**

24. Ted and Tom worked more than 14 h painting a garage. Ted worked 2 h more than Tom. Find the least whole number of hours each of them might have worked. **Tom, 6 h; Ted, 8 h**

25. On an ocean dive, Tracy wants to stay at least 7 m above her deepest previous depth, 40 m. How long can she descend at a rate of 2.2 m/min? **15 min**

 **COMPUTER**

This program will check your answer for Exercise 28. What happens if the number is both < -4 and > -8? What would you change in line 20 to have it check your answers for Exercises 29–32?

```
10 FOR X = -10 TO 10
20 IF X < -4 AND X > -8
     THEN PRINT X
30 NEXT X
```

See Additional Answers.

26. Write a problem that could be solved using the inequality $t - 5 > 9$. **Answers will vary. Sample: If Tanya spends $5, she'll still have more than $9 left. How much does she have?**

27. Write a problem that could be solved using the inequality $2x + 4 < 30$.

Find all integers, if any, that are solutions of both inequalities.

28. $x < -4$ and $x > -8$ **−7, −6, −5**

29. $x \geq 3$ and $x < 9$ **3, 4, 5, 6, 7, 8**

30. $x < -2$ and $x > -5$ **−4, −3**

31. $x > 0$ and $x < -6$ **none**

32. $x \geq 4$ and $x \leq 4$ **4**

7-11 Solving Inequalities **261**

THINK CRITICALLY/ SOLVE PROBLEMS

27. Answers will vary. Sample: Anthony plans to continue going to school until he's twice his present age, then to travel for 4 years. He will still be under 30 years of age. How old is he?

CHALLENGE

Have students solve and graph each inequality.

1. $x^2 \geq 9$
   $x \geq 3$ or $x \leq -3$
2. $a^2 < 4$
   $-2 < a < 2$
3. $z^2 > 0$
   $z > 0$ or $z < 0$
4. $m^2 \leq 0$
   $0 = m$
   See margin for graphs.

---

2. $x/6 < -1$ $x < -6$

3. $-2/3\, a \leq -4$ $a \geq 6$

4. $2x - 7 > -3$ $x > 2$

5. $-2(3 + x) > 8$ $x < -7$

6. $-2 \leq 6n - 8n$ $n \leq 1$

7. When this number is divided by −3, the result is at least 30. Find the number. $n \leq -90$

Extension Have students work in small groups to find the two smallest consecutive integers whose sum is greater than 125. Have them write an inequality for solving the problem rather than using the guess-and-check strategy.

Section Quiz Solve and graph each inequality.

1. $3x - 1 \leq 8$ $x \leq 3$

2. $-2x > 4$ $x < -2$

3. $4 + 3y \geq -8$ $y \geq -4$

4. $m/-4 < 1\,3/4$ $m > -7$

5. A number is multiplied by −3. Then 7 is added. The result is more than −5. Find the number.
   $-3\,n + 7 > -5, n < 4$

Additional Answers
See page 579.

1.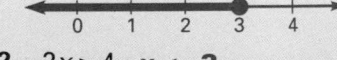

2.

3.

4.

261

7 CHAPTER REVIEW

Introduction The Chapter Review emphasizes the major concepts, skills, and vocabulary presented in this chapter and can be used for diagnosing students' strengths and weaknesses. Page references direct students back to appropriate sections for additional review and reteaching.

Using Pages 262–263 Allow students to quickly scan the Chapter Review and ask questions about any section they find confusing. Exercises 1–3 review key vocabulary. For Exercises 23–28 students should have a clear understanding of the difference between graphing sentences with < or > and with ≤ or ≥.

Informal Evaluation Have students explain how they arrived at answers that are not correct. They may find their own mistakes or may give you some clues as to the nature of their errors. Make sure students understand this material before administering the Chapter Test.

Follow-Up Have students solve the following problem by writing and solving an equation: The cost of a coat is $40 more than the cost of a jacket. The cost of two coats and three jackets is $580. Find the cost of a jacket. **$100**

Choose a phrase from the list to complete each statement.

1. An equation that is neither true nor false is ___?___ . **b**
2. Multiplication is ___?___ of division. **c**
3. The product of a fraction and ___?___ equals 1. **a**

    a. its reciprocal
    b. an open sentence
    c. the inverse

SECTION 7–1   INTRODUCTION TO EQUATIONS   (pages 222–225)

▶ An equation is a statement that two numbers or expressions are equal.
▶ An open sentence is a sentence that contains one or more variables.

State whether the equation is *true, false,* or an *open sentence.*

4. $k - 7 = -1$ **open**   5. $-4 + 6 = -6 + 4$ **false**   6. $-2(3 + 5) = -16$ **true**

SECTIONS 7–2 AND 7–4   SOLVE EQUATIONS   (pages 226–229 and 232–235)

▶ Undo an operation by using the inverse operation.

Solve.

7. $f + 22 = -16$ **−38**   8. $c + 1\frac{1}{2} = 2\frac{1}{2}$ 9. $1.4 = e - 1.4$ **2.8**

10. $36 = -18k$ **−2**   11. $\frac{m}{7} = 4$ **28**   12. $3.42 = 2p$ **1.71**

SECTIONS 7–3 AND 7–8   PROBLEM SOLVING   (pages 230–231 and 248–249)

▶ To solve a problem by writing an equation, decide how the numbers in the problem are related. Choose a variable to represent the unknown quantity. Write an equation representing the problem situation.

USING DATA   Solve using the table on page 221.

13. The sum of Val's, Mark's, and Evan's scores in Game 1 equaled 307. Write an equation and solve to find Val's score.
    V + 114 + 96 = 307; V = 97

SECTION 7–5   EQUATIONS: MORE THAN ONE OPERATION   (pages 236–239)

▶ Undo addition and subtraction first, then multiplication and division.

Solve.

14. $7r + 12 = 5$ **−1**   15. $6(s - 3) = 12$ **5**   16. $3.2x - 2.1 = 4.3$ **2**

SECTION 7–6   USING A RECIPROCAL TO SOLVE AN EQUATION (pages 240–243)

► To solve an equation in which the variable is multiplied by a fraction, multiply both sides of the equation by the reciprocal of the fraction.

Solve.

17. $\frac{3}{8}m = 21$ **56**

18. $\frac{5}{6}k = 10$ **12**

SECTION 7–7   WORKING WITH FORMULAS   (pages 244–247)

► When working with a formula, substitute for known values of the variables. Then solve the resulting equation.

19. Find the time it takes to walk $19\frac{1}{8}$ mi at $4\frac{1}{4}$ mi/h. $4\frac{1}{2}$ **h**

SECTION 7–9   COMBINING LIKE TERMS   (pages 250–253)

► You combine like terms as part of the process of solving an equation with a variable term on both sides.

Simplify.

20. $2n + 5 = n - 4$ **–9**

21. $2n = 4n + 10$ **–5**

22. $3y - 6 = -4y + 7$ $\frac{13}{7}$

SECTION 7–10   GRAPHING SOLUTIONS   (pages 254–257)

► To indicate that a number is a solution of an equality or inequality, draw a solid dot at the corresponding point on a number line. To indicate that the number is not a solution, draw an open circle at the point.

Graph each open sentence on a number line.

23. $m < -1$

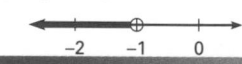

24. $-7 = p - 2$

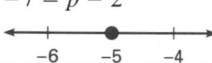

25. $m \geq 2$

SECTION 7–11   SOLVING INEQUALITIES   (pages 260–261)

► Follow all steps for solving equations except for the following situation: When multiplying or dividing both sides of an inequality by a negative number, reverse the direction of the inequality.

Solve and graph each inequality.

26. $x - 4 < -3$ $x < 1$

27. $-3n \geq 12$ $n \leq -4$

28. $\frac{h}{-6} + 5 > -2$ $h < 42$

Review   **263**

Study Skills Tip Remind students that the rules for solving equations and inequalities are identical except that the sign of the inequality must be reversed when multiplying or dividing both sides of an inequality by a negative number.

Introduction The Chapter Test uses a variety of questioning techniques to assess students' mastery of the major objectives of Chapter 7. If you prefer, you may use the Skills Preview (page 219) as an alternative form of the Chapter Test. The items on this test and the Skills Preview correspond in content and level of difficulty.

Alternative Assessment The formula for estimating temperature from the number of cricket chirps is $F = n/4 + 32$, where F stands for the temperature in degrees Fahrenheit and n stands for the number of chirps per minute.
1. Compute F for $n = 40$, 60, and 80. **42°, 47°, 52°**
2. Compute n for $F = 50°$ and 60°. **72, 112**
3. Can n be any value? **n can be any whole number.**
4. For what values of F will n be less than 0? Is the formula valid for those values of F? **$n < 32°$; no, because then n would be a negative number, which would be an impossibility**

Tell whether the equation is *true, false,* or an *open sentence.*

1. $v + 8 = -6$ **open sentence**
2. $3(4 - 5) = 3$ **false**

Solve each equation.

3. $e - 5 = 2$ **7**
4. $-6d = 420$ **-70**
5. $m + 4.4 = 12.2$ **7.8**
6. $\frac{x}{3} = -5$ **-15**
7. $5m - 3 = 17$ **4**
8. $19 = 2c + 3$ **8**
9. $-\frac{5}{6}y = \frac{1}{5}$ **$-\frac{6}{25}$**
10. $\frac{3}{2}x + 4 = 2$ **$-\frac{4}{3}$**
11. $2(y - 3) = 14$ **10**
12. $5d + 5 = 2d - 4$ **-3**

Solve.

13. Convert a temperature of 167°F to the Celsius scale. Use the formula $F = \frac{9}{5}C + 32$ (F = Fahrenheit, C = Celsius). **75°C**
14. Solve the formula $P = 24 + d$ for d. **$d = P - 24$**

Simplify.

15. $x + 2x + 3x^2 + 5x$ **$8x + 3x^2$**
16. $7(a + b - c) - 3(a + 2b - 4c)$ **$4a + b + 5c$**

Graph each open sentence on a number line.

17. $f \le -2$

18. $2x - 5 = -7$

19. $2 < 0.5x + 1.5$

20. $-4p + 9 \le -3$

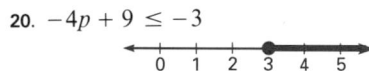

Write an equation and solve.

21. For a bike-a-thon, Wendy had pledges that totaled $34.35 for each kilometer she rode. She also collected $64.55 in other donations. If she collected $751.55 in all, how far did she ride?
$34.35d + 64.55 = 751.55$; 20 km

22. Mr. Cintron spent $5.60 in the delicatessen. Then he spent $\frac{1}{2}$ of the money he had left in the bakery. Finally, he spent $10.80 in the hardware store. He had $7.50 left. How much did he have before making his purchases? **$42.20**

23. Marcia earns $6.90 per hour. Her weekly salary before taxes is $241.50. How many hours does she work per week? **35 hours**

24. On four trips to the supermarket, Lou drove $3s$ miles, $4p$ miles, $5s$ miles, and $2p$ miles. Write and simplify an expression for the number of miles he drove. **$3s + 4p + 5s + 2p$; $8s + 6p$**

1. If the average of three chemistry test scores was 80, could all three test scores be above 80? Would they all have to be equal to 80? **no; no**

Simplify.

2. $24 ÷ (3 + 5) − 1$ **2**

3. $64 ÷ 4 − 5 × 3$ **1**

4. $5 + (7 − 2)^2$ **30**

5. $3 × 5 − (5 + 1)$ **9**

6. Determine whether the following argument is *valid* or *invalid*.

   If the fiber is wool, then it is a natural fiber.
   The fiber is a natural fiber.
   Therefore, the fiber is wool. **invalid**

7. Write two conditional statements from the following two sentences:
   See Additional Answers.
   The sun has set.
   The sky will grow dark.

8. Teams A, B, and C played three other teams, X, Y, and Z, but not necessarily in that order. Team A did not play team X. Team B did not play teams X or Y. Which teams played each other?
   See Additional Answers.

Find the prime factorization of each number.

9. 40 **$2^3 × 5$**

10. 84 **$2^2 × 3 × 7$**

11. 64 **2^6**

12. 120 **$2^3 × 3 × 5$**

Write each in exponential form.

13. $\frac{1}{12 × 12 × 12}$ **12^{-3}**

14. $\frac{1}{7 × 7 × 7 × 7 × 7}$

7-5

Find each circumference. Round your answer to the nearest tenth or whole number.

15.
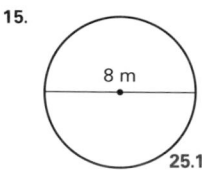
8 m
25.1 m

16.
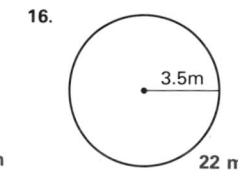
3.5m
22 m

17.
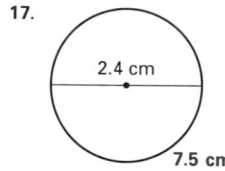
2.4 cm
7.5 cm

18.
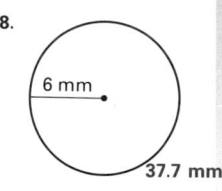
6 mm
37.7 mm

19. Find the perimeter of the figure.

31 cm

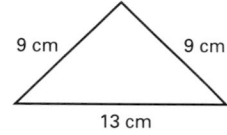
9 cm   9 cm
13 cm

20. Use the formula to find the perimeter of a rectangular figure with these dimensions.
    length: 4.5 m
    width: 10.1 m **29.2 m**

Find each answer.

21. $−3 × (−4)$ **12**

22. $−72 ÷ 6$ **−12**

23. $−36 ÷ 12$ **−3**

24. $6 × (−25)$ **−150**

25. $−6 + −2$ **−8**

26. $7 − (−3)$ **10**

Solve each equation or inequality.

27. $−18 > −3t$ **$t > 6$**

28. $24 = 2m − 26$ **$m = 25$**

29. $15 ≤ 2d + 10$ **$d ≥ 2\frac{1}{2}$**

30. $7x = −14$ **$x = −2$**

Cumulative Review **265**

7 CUMULATIVE REVIEW

Section Value The purpose of this Cumulative Review is to maintain previously taught skills and concepts and to apply them to the material presented in this chapter. At least one major objective of each chapter is included in the review.

Item Analysis The table below correlates the Cumulative Review items with the chapter and section that is being reviewed.

Section	Items
1–9	1
2–8	2–5
3–3	6–7
3–5	8
4–1	9–12
4–4	13–14
5–4	19
5–5	20
5–6	15–18
6–2	25
6–3	26
6–4	21, 24
6–5	22–23
7–4	30
7–6	28
7–10	27, 29

7 CUMULATIVE TEST

Introduction The Cumulative Test uses a standardized-test format of multiple-choice questions to assess retention of previously learned concepts and test-taking skills. Test results may be used to diagnose students' strengths and weaknesses.

Item Analysis The table below correlates the Cumulative Test items with the chapter and sections that is being tested.

Sections	Items
1–9	1
2–10	2–3
3–3	4
4–4	5–9
5–1	10–11
5–5	12
6–7	13–14
7–2	15
7–4	16–17
7–6	18

1. The mean of three test scores was 85. Two of the scores were 78 and 86. What was the third score?
   (A) 91   B. 85   C. 88   D. 100

2. Complete.
   $$\blacksquare(2 + 9) = (5 \times 2) + (5 \times 9)$$
   A. 2   (B) 5   C. 9   D. 24

3. Complete.
   $$3(6 - 1) = (3 \times \blacksquare) - (3 \times 1)$$
   A. 15   B. 1   C. 3   (D) 6

4. Which is a counterexample for the following statement?

   If a number is divisible by 5, then it is divisible by 50.
   A. 100   (B) 75   C. 150   D. 2000

5. Express 3.2×10^{-2} in standard form.
   A. 3,200   B. 0.0032
   C. 3.200   (D) 0.032

6. Express 5^{-4} as a fraction.
   A. $\frac{1}{20}$ (B) $\frac{1}{625}$ C. $\frac{1}{256}$ D. $-\frac{1}{20}$

7. Compare. $0.6 \; \blacksquare \; \frac{3}{7}$
   A. <   (B) >
   C. =   D. none of these

8. Express $\frac{1}{2 \times 2 \times 2 \times 2 \times 2 \times 2 \times 2}$ in exponential form.
   (A) 2^{-7} B. 2^7 C. 7^2 D. 7^{-2}

9. Express 0.000045 in scientific notation.
   A. 45×10^5 (B) 4.5×10^{-5}
   C. $4^5 \times 10$ D. 4.5×10^{-6}

10. Complete. $36 \text{ ft} = \blacksquare \text{ yd}$
    (A) 12 yd   B. 3 yd
    C. 48 yd   D. 432 yd

11. Complete. $6 \text{ yd} = \blacksquare \text{ in.}$
    (A) 216 in.   B. 72 in.
    C. 3 in.   D. 0.5 in.

12. Use the formula to find the perimeter of a rectangular figure whose length is 340 mi and whose width is 216 mi.
    (A) 1,112 mi   B. 556 mi
    C. 73,440 mi   D. 1,090 mi

13. Which group of rational numbers is in order from least to greatest.
    A. $0.23, -0.21, -0.7$
    (B) $-0.7, -0.21, 0.23$
    C. $-0.7, 0.23, -0.21$
    D. $-0.21, 0.23, -0.7$

14. Which group of rational numbers is in order from least to greatest?
    A. $\frac{11}{16}, \frac{5}{8}, -\frac{13}{24}$ B. $\frac{5}{8}, \frac{11}{16}, -\frac{13}{24}$
    C. $-\frac{13}{24}, \frac{11}{16}, \frac{5}{8}$ (D) $-\frac{13}{24}, \frac{5}{8}, \frac{11}{16}$

15. Solve. $x + 18 = 15$
    A. $x = 33$ B. $x = 3$
    (C) $x = -3$ D. none of these

16. Solve. $-6z = 24$
    A. $z = 4$ (B) $z = -4$
    C. $z = 144$ D. none of these

17. Solve. $8 = \frac{x}{2}$
    A. $x = 4$ B. $x = 2$
    (C) $x = 16$ D. none of these

18. Solve. $5x - 3 = 12$
    A. $x = 1\frac{4}{5}$ B. $x = 5$
    C. $x = -3$ (D) none of these

Name _____ Date _____

Introduction to Equations

An **equation** is a statement that two numbers or expressions are equal.

An **open sentence** is a sentence containing one or more variables. These are open sentences.
$$x + 1 = 4 \qquad y - 2 = 3$$

An open sentence can be either true or false, depending on what value is substituted for the variables. A value of the variable that makes the equation true is called a **solution** of the equation.

▶ **Example 1**

Tell whether the equation is *true*, *false*, or is an *open sentence*.

a. $4(3 + 2) = 20$ **b.** $2x - 4 = 6$ **c.** $-4 + 1 = 2(6 - 8)$

Solution

a. $4(3 + 2) = 20$
$4(5) = 20$
$20 = 20$

The equation is true.

b. $2x - 4 = 6$

The equation contains a variable.

The equation is an open sentence.

c. $-4 + 1 = 2(6 - 8)$
$-3 = 2(-2)$
$-3 \neq -4$
$\underset{\text{means "is not equal to"}}{}$

This equation is false.

▶ **Example 2**

Use mental math to solve the equation $-2 + x = 4$.

Solution

Think: What number added to -2 equals 4? You know that $-2 + 6 = 4$, so $x = 6$.

EXERCISES

Tell whether the equation is *true*, *false*, or an *open sentence*.

1. $4(9 - 3) = 22$ **2.** $2x - 3 = 5$ **3.** $12 - 4 = 2(10 - 2)$
   false   open sentence   false

4. $-9 + 3 = 2(3 - 6)$ **5.** $3y + 4 = 16$ **6.** $-2(4 - 6) = -16 + 20$
   true   open sentence   true

Use mental math to solve the following equations.

7. $x + 4 = 9$ **8.** $y - 2 = 7$ **9.** $c + 9 = 17$ **10.** $d + 8 = -2$
   5   9   8   −10

11. $2a = 18$ **12.** $-11 = p + 8$ **13.** $-6 + r = -1$ **14.** $7 = 13 + z$
   9   −19   5   −6

Name _____ Date _____

Money Matters

On the planet Zorb, money has different shapes. The value of each coin is recognized by its shape.

◯ is the coin with the least value.

Study the following chart of relative values.

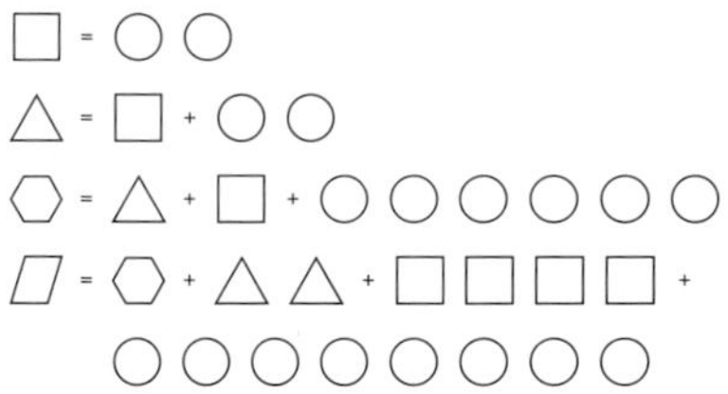

▢ = ◯◯

△ = ▢ + ◯◯

⬡ = △ + ▢ + ◯◯◯◯◯◯

▱ = ⬡ + △△ + ▢▢▢▢ + ◯◯◯◯◯◯◯◯◯◯

Write the value of each coin shown below in terms of only one other type of coin. Use the least possible number of coins in each case.

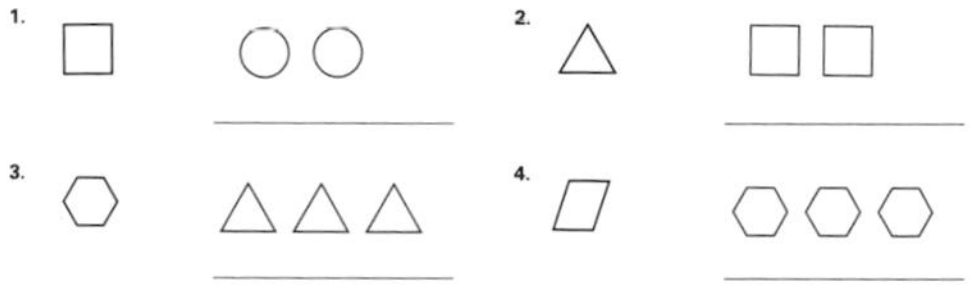

1. ▢   ◯◯   **2.** △   ▢▢

_____   _____

3. ⬡   △△△   **4.** ▱   ⬡⬡⬡

_____   _____

Write the value of each coin shown below in terms of other coins using any combination you wish.

5. ⬡   **6.** ▱

_____   _____

Answers will vary.   Answers will vary.

Name _____ Date _____

Using Addition or Subtraction to Solve an Equation

To solve an equation, find all values of the variable that make the equation true. To do so, you must perform operations on the equation so that you can *get the variable alone on one side of the equals sign*.

Many equations contain addition or subtraction.
$$x + 7 = 10 \qquad y - 3 = 8$$

To solve such equations, you must reverse, or "undo," the addition or subtraction in the equation. Undo addition by subtracting. Undo subtraction by adding. In doing so, you must keep the equation in balance by adding or subtracting the same number from both sides of the equation.

▶ **Example 1**

Solve $x + 7 = 10$.

Solution

You want to get x alone on the left side of the equals sign. Undo the addition in the equation by subtracting 7 from both sides of the equation.

$x + 7 = 10$
$x + 7 - 7 = 10 - 7$
$x + 0 = 3$
$x = 3$

CHECK: Substitute the value 3 into the original equation.

$3 + 7 = 10$
$10 = 10$

▶ **Example 2**

Solve $y - 3 = 8$.

Solution

Undo the subtraction by adding 3 to both sides of the equation.

$y - 3 = 8$
$y - 3 + 3 = 8 + 3$
$y + 0 = 11$
$y = 11$

CHECK: Substitute the value 11 into the original equation.

$11 - 3 = 8$
$8 = 8$

EXERCISES

Solve each equation. Check the solution.

1. $x + 5 = 7$ $x = 2$ **2.** $y - 3 = 10$ $y = 13$ **3.** $x + 3 = 10$ $x = 7$

4. $n - 4 = 9$ $n = 13$ **5.** $t - 6 = -6$ $t = 0$ **6.** $x + 4 = 3$ $x = -1$

7. $p + 7 = -7$ $p = -14$ **8.** $r - 21 = -9$ $r = 12$ **9.** $-y + 4 = 1$ $y = 3$

10. $-6 = n - 5$ $n = -1$ **11.** $8 + f = -8$ $f = -16$ **12.** $7 - y = -13$ $y = 20$

Name _____ Date _____

More Addition and Subtraction Equations

Some addition and subtraction equations involve decimals and fractions.

▶ **Example 1**

Solve the equation $m + 1.47 = 12.5$.

Solution

Undo the addition by subtracting 1.47 from both sides of the equation.

$m + 1.47 - 1.47 = 12.5 - 1.47$
$m = 11.03$

CHECK: Substitute 11.8 for m in the original equation.
$11.03 + 1.47 = 12.5$
$12.50 = 12.5$

▶ **Example 2**

Solve the equation $w - 3\frac{1}{4} = 18\frac{1}{2}$.

Solution

Undo the subtraction by adding $3\frac{1}{4}$ to each side of the equation.

$w - 3\frac{1}{4} = 18\frac{1}{2} + 3\frac{1}{4}$
$w = 21\frac{3}{4}$

CHECK: Substitute $21\frac{3}{4}$ for w in the original equation.
$21\frac{3}{4} - 3\frac{1}{4} = 18\frac{2}{4}$
$18\frac{2}{4} = 18\frac{1}{2}$

EXERCISES

Solve each equation. Check your solution.

1. $x - 12.6 = 9.5$ **2.** $y - 7.3 = 8.9$ **3.** $t + 7.3 = 17.2$
   22.1   16.2   9.9

4. $b + 1.46 = 32.4$ **5.** $14.8 + z = 22.5$ **6.** $r - 143.08 = 471.13$
   30.94   7.7   614.21

7. $a + 3\frac{1}{8} = 12\frac{1}{2}$ **8.** $x - 4\frac{5}{6} = 5\frac{1}{3}$ **9.** $12\frac{5}{9} + y = 14\frac{2}{18}$
   $9\frac{3}{8}$ $10\frac{1}{6}$ $1\frac{5}{9}$

10. $k + 13.678 = 44.7$ **11.** $m - 54.08 = 33.3$ **12.** $5.555 + x = 9.444$
   30.022   87.38   3.889

266A

Name _____   Date _____

Problem Solving Skills: Writing an Equation

Many problems can be solved by writing an equation to relate the numbers in the problem and then solving the equation.

► **Example 1**

If 5 more than a number is 17, what is the number?

Solution

Let x represent the number. You know that 5 more than the number equals 17. So,

$$x + 5 = 17$$
$$x + 5 - 5 = 17 - 5$$
$$x = 12$$

CHECK: $12 + 5 = 17$

► **Example 2**

If 3 less than a number is 12, what is the number?

Solution

Let n represent the number. You know that 3 less than the number equals 12. So,

$$n - 3 = 12$$
$$n - 3 + 3 = 12 + 3$$
$$n = 15$$

CHECK: $15 - 3 = 12$

EXERCISES

Choose a variable, write an equation for each problem. Then solve the problem.

1. If a number is 11 more than 6, what is the number?
$n = 6 + 11; 17$

2. If 3 less than a number is 10, what is the number?
$n - 3 = 10; 13$

3. If 14 more than a number is 12, what is the number?
$n + 14 = 12; -2$

4. If a number is 5 less than 23, what is the number?
$n = 23 - 5; 18$

5. If a number plus 4 equals two times 6, what is the number?
$n + 4 = 2(6); 8$

6. If a number minus 10 equals four times 15, what is the number?
$n - 10 = 4(15); 70$

7. Mark bought two books. He spent $14. He had $12 left. How much did Mark have before he bought the books?
$n - 14 = 12; \$26$

8. Together, José and Tommy ate 7 slices of pizza. If Tommy ate 2 slices, how many did José eat?
$n + 2 = 7; 5$ slices

9. On Tuesday Sarah rode her bike 3 miles before lunch. By evening she had ridden 11 miles in all. How far did she ride after lunch?
$3 + n = 11; 8$ miles

Reteaching • SECTION 7-3   119

Name _____   Date _____

Equations in Buying and Selling

Equations are useful in solving problems about costs and prices of consumer goods.

► **Example**

Margo wants to buy a car. The current model of the car she prefers costs $11,176, which is $945 more than the cost of last year's model, and includes $431 in costs for extras. What was the price of last year's model?

Solution

Let x represent the cost of last year's model. Write an equation that reflects the facts. You know that the cost of the older model, plus $945 and an additional $431 in extra costs equals the cost of the current model.

$$x + \$945 + \$431 = \$11,176$$
$$x + \$1,376 = \$11,176$$
$$x + \$1,376 - \$1,376 = \$11,176 - \$1,376 = \$9,800$$

The cost of last year's model was $10,661.

EXERCISES

Write the equation needed to solve the problem. Then solve it.

1. A model railroad costs $584. If all the deluxe items were purchased, a buyer would pay an additional $205, but would get a bonus of $50 deducted from the total price. If Angie and her father buy the train and the extras, what would they pay, excluding any taxes?
$\$584 + \$205 - \$50 = x; \$739$

2. Bill is shopping for a used car. Six months ago he would have paid $920 less than he will now have to pay for the same car. The price he will pay this year is $8,400. What price would he have paid last year?
$x + \$920 = \$8,400; \$7,480$

3. Fran bought a used car. She had a $350 stereo and CD player installed. The dealer gave her a $200 rebate on the total price. If she paid a total of $9,700, what was the original price of the car?
$x + \$350 - \$200 = \$9,700; \$9,550$

4. Write an original problem of your own about costs and pricing that can be solved using an equation.

_____
_____
_____

120   Enrichment • SECTION 7-3

Name _____   Date _____

Using Multiplication or Division to Solve an Equation

Multiplication and division are inverse operations. Therefore, you can solve an equation involving multiplication by dividing. Likewise, you can solve an equation involving division by multiplying. Remember, both sides of the equation must be multiplied or divided by the same number.

► **Example 1**

Solve $3x = 6$.

Solution

Undo the multiplication.

$\dfrac{3x}{3} = \dfrac{6}{3}$ Divide each side by 3 to balance the equation.

$1x = 2$

$x = 2$

CHECK: $3(2) = 6$

► **Example 2**

Solve $\frac{n}{2} = 4$.

Solution

Undo the division.

$2\left(\frac{n}{2}\right) = 2(4)$ Multiply each side by 2 to balance the equation.

$1n = 8$

$n = 8$

CHECK: $\frac{8}{2} = 4$

EXERCISES

Solve each equation and check.

1. $3x = 12$
$x = 4$

2. $\frac{n}{3} = 7$
$n = 21$

3. $10m = 80$
$m = 8$

4. $\frac{t}{7} = 7$
$t = 49$

5. $16 = 4y$
$y = 4$

6. $6b = 72$
$b = 12$

7. $5 = \frac{x}{9}$
$x = 45$

8. $\frac{x}{8} = 0$
$y = 0$

9. $\frac{c}{2} = -5$
$c = -10$

10. $8a = 64$
$a = 8$

11. $18y = -36$
$y = -2$

12. $\frac{x}{-5} = 8$
$x = -40$

13. $\frac{y}{3} = 8$
$y = 24$

14. $-12x = 48$
$x = -4$

15. $-4n = -16$
$n = 4$

16. $-2y = 0$
$y = 0$

17. $\frac{x}{8} = 7$
$z = 56$

18. $16c = 96$
$c = 6$

19. $-18d = 234$
$d = -13$

20. $\frac{w}{5} = -21$
$w = -105$

Reteaching • SECTION 7-4   121

Name _____   Date _____

Balanced Expressions

Find a value for the variable so that each scale is balanced.

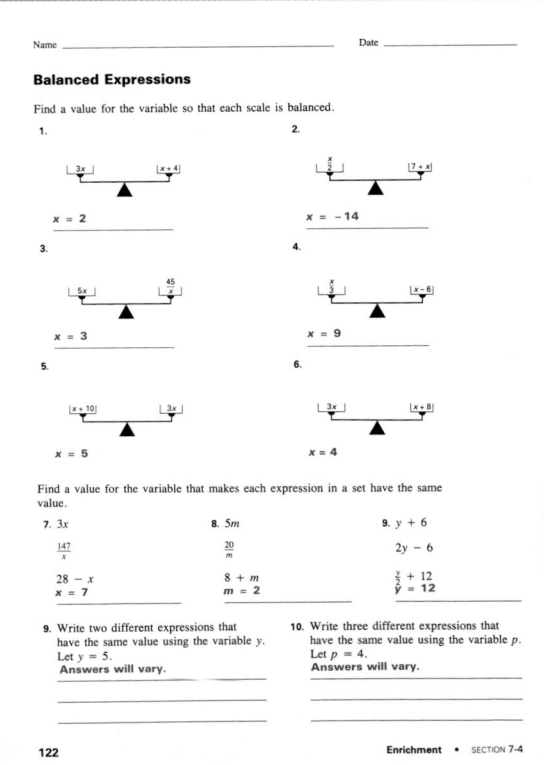

1.
$x = 2$

2.
$x = -14$

3.
$x = 3$

4.
$x = 9$

5.
$x = 5$

6.
$x = 4$

Find a value for the variable that makes each expression in a set have the same value.

7. $3x$
$\frac{147}{x}$
$28 - x$
$x = 7$

8. $5m$
$\frac{20}{m}$
$8 + m$
$m = 2$

9. $y + 6$
$2y - 6$
$\frac{x}{2} + 12$
$y = 12$

9. Write two different expressions that have the same value using the variable y. Let $y = 5$.
Answers will vary.

_____
_____

10. Write three different expressions that have the same value using the variable p. Let $p = 4$.
Answers will vary.

_____
_____

122   Enrichment • SECTION 7-4

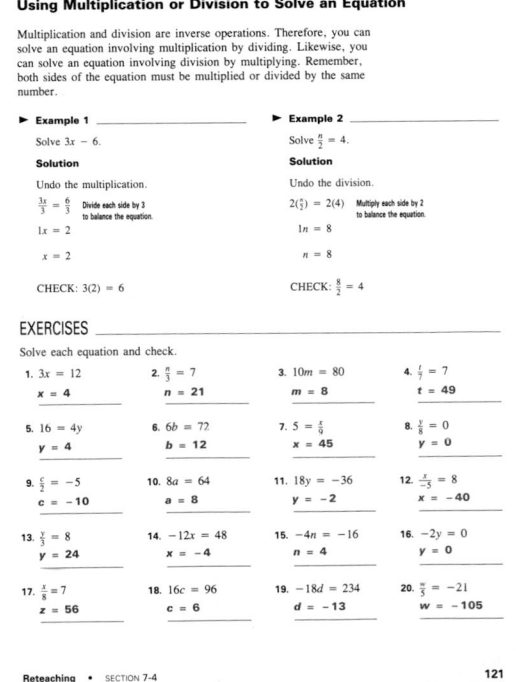

Name _____ Date _____

Using More Than One Operation to Solve an Equation

Some equations, called **two-step equations**, require more than one inverse operation to solve. Follow these steps:

- First, undo addition or subtraction.
- Then undo multiplication or division.

▶ **Example 1** _____

Solve $3x - 4 = 5$.

Solution

Undo the subtraction.
$$3x - 4 + 4 = 5 + 4$$
$$3x = 9$$
$$\frac{3x}{3} = \frac{9}{3}$$
$$x = 3$$

CHECK: $3(3) - 4 = 5$
$$9 - 4 = 5$$
$$5 = 5$$

▶ **Example 2** _____

Solve $\frac{y}{-5} + 3 = 13$.

Solution

Undo the addition.
$$\frac{y}{-5} + 3 - 3 = 13 - 3$$
$$\frac{y}{-5} = 10$$

Undo the division.
$$-5\left(\frac{y}{-5}\right) = -5(10)$$
$$y = -50$$

CHECK: $\frac{-50}{-5} + 3 = 13$
$$10 + 3 = 13$$
$$13 = 13$$

EXERCISES _____

Solve each equation. Check the solution.

1. $5n + 4 = 29$
   $n = 5$

2. $\frac{m}{7} - 3 = 8$
   $m = 77$

3. $\frac{x}{2} + 6 = 4$
   $x = -4$

4. $8y - 7 = 17$
   $y = 3$

5. $9t - 5 = -14$
   $t = -1$

6. $12 + \frac{n}{2} = 18$
   $n = 12$

7. $15 + 5y = 20$
   $y = 1$

8. $-3 - 7x = -17$
   $x = 2$

9. $\frac{y}{-4} - 6 = 5$
   $y = -44$

Write an equation for the following problem and solve.

10. Jeff bought two tapes on sale at the music store. They both cost the same amount. Then Jeff bought a tape-head cleaner for $6. Altogether Jeff spent $22. How much did Jeff pay for one tape?
    $2x + 6 = 22$; Jeff paid $8 for one tape.

Reteaching • SECTION 7-5   123

Name _____ Date _____

Writing a Problem to Fit an Equation

Here is a word sentence and the equation that models it:

Four more than a number is nine. $n + 4 = 9$

You can create a problem for which a given equation is a model.

▶ **Example 1** _____

Write a word sentence for which the equation $x - 4 = 12$ is a model.

Solution

Four less than a number is 12.

You can also write a word problem about a situation for which the equation is a model.

▶ **Example 2** _____

Write a word problem for which the equation $2x + 4 = 20$ is a model.

Solution

Four more than twice Flora's age is equal to 20.

EXERCISES _____

Write a word sentence for which each of the following equations is a model. **Answers will vary.**

1. $b + 3 = 9$

2. $2a + 7 = 15$
   _____

3. $x - 14 = 7$
   _____

4. $3(x - 4) = 15$
   _____

5. Write a word problem for one of the equations in Exercises 1–4.
   Answers will vary.

124   **Enrichment** • SECTION 7-5

Name _____ Date _____

Using a Reciprocal to Solve an Equation

Some equations have variables that are multiplied by a fraction. To solve such an equation you can multiply the fraction by its reciprocal. Recall that reciprocals are two fractions whose product is 1.

▶ **Example 1** _____

Solve $\frac{3x}{4} = 6$.

Solution

Multiply each side by the reciprocal of $\frac{3}{4}$.
$$\frac{4}{3}\left(\frac{3x}{4}\right) = \frac{4}{3}(6)$$
$$x = 8$$

CHECK: $\frac{3(8)}{4} = 6$
$$6 = 6$$

▶ **Example 2** _____

Solve $\frac{2x}{3} + 4 = 12$.

Solution

First, undo the addition. Subtract 4.
$$\frac{2x}{3} + 4 - 4 = 12 - 4$$
$$\frac{2x}{3} = 8$$

Then, multiply each side by the reciprocal of $\frac{2}{3}$.
$$\frac{3}{2}\left(\frac{2x}{3}\right) = \frac{3}{2}(8)$$
$$x = 12$$

CHECK: $\frac{2(12)}{3} + 4 = 12$
$$8 + 4 = 12$$
$$12 = 12$$

EXERCISES _____

Solve. Check your solutions.

1. $\frac{5y}{6} = 25$
   $y = 30$

2. $\frac{3t}{5} = 30$
   $t = 50$

3. $\frac{7x}{10} = -14$
   $x = -20$

4. $\frac{8n}{-9} = 24$
   $n = -27$

5. $\frac{2t}{-3} = -18$
   $t = 27$

6. $\frac{-5m}{7} = 100$
   $n = -140$

7. $\frac{8n}{-9} = \frac{16}{18}$
   $n = -1$

8. $\frac{7x}{10} = -\frac{21}{20}$
   $x = -\frac{3}{2}$

Solve.

9. $\frac{5x}{7} + 3 = 13$
   $x = 14$

10. $\frac{3m}{4} - 8 = 4$
    $m = 16$

11. $\frac{2x}{-3} + 5 = 7$
    $x = -3$

12. $\frac{4y}{7} + 8 = 4$
    $y = -7$

13. $\frac{9p}{-10} + 2 = 2$
    $p = 0$

14. $\frac{5x}{8} + 10 = -5$
    $x = 24$

Reteaching • SECTION 7-6   125

Name _____ Date _____

Writing Equations

To solve many problems, you need to write equations to represent the situation in the problem. Sometimes, the equation can have variables on both sides.

▶ **Example 1** _____

Write an equation to represent this problem: Four times a number is 3 more than a number.

Solution

Let n be the number. The verb is means "equals." Four times the number can be written $4n$, and 3 more than the number can be written $n + 3$. The relationship stated in the problem is now as follows:
$$4n = n + 3$$

▶ **Example 2** _____

Two thirds of a number, increased by 5, is equal to the sum of the number and 1.

Solution

Let n be the number. Two thirds of the number, increased by 5, can be written $\frac{2}{3}n + 5$. The sum of the number and 1 can be written $n + 1$.

The problem can be represented by this equation:
$$\frac{2}{3}n + 5 = n + 1$$

EXERCISES _____

Write an equation to represent each problem. Use n for the variable.

1. Twice the difference between a number and three equals five more than four times the number.
   $2(n - 3) = 4n + 5$

2. Three more than four times a number is equal to three less than six times the number.
   $4n + 3 = 6n - 3$

3. The difference when two is subtracted from twice a number equals 3 times the sum of the number and 4.
   $2n - 2 = 3(n + 4)$

4. Three fourths of a number, minus 2 is equal to twice the sum of a number and four.
   $\frac{3n}{4} - 2 = 2(n + 4)$

5. Rob's age is four less than three times the age of his little sister Sue. In five years his age will be twice Sue's age.
   $(3n - 4) + 5 = 2(n + 5)$

6. Sheri's mother is five times as old as Sheri. In ten years, her mother will be eight more than two times as old as Sheri.
   $5n + 10 = 2(n + 10) + 8$

126   **Enrichment** • SECTION 7-6

Working with Formulas

An equation that states a relationship between two quantities is a **formula.** Formulas are commonly used in many fields, including mathematics, the sciences, statistics, and banking. Here are some of these formulas.

Volume of a prism	Rate/Distance/Time	Perimeter of a rectangle	Area of a triangle
$V = lwh$	$d = rt$	$p = 2l + 2w$	$A = \frac{1}{2}bh$
$V =$ volume, $l =$ length	$d =$ distance, $r =$ rate	$p =$ perimeter	$A =$ area, $b =$ base
$w =$ width, $h =$ height	$t =$ time	$l =$ length, $w =$ width	$h =$ height

▶ **Example 1**

Find the volume of a rectangular prism 12 ft long, 2 ft wide, and 5 ft high.

Solution

Substitute the values into the formula for volume. Solve for V.
$V = l \times w \times h$
$V = 12(2)(5)$
$V = 120$

Volume is expressed in cubic units. So, the volume is 120 ft³.

▶ **Example 2**

How long would it take a car to go 135 mi at an average rate of 45 mi/h?

Solution

Substitute the values into the formula for distance. Solve for t.
$d = rt$
$135 = 45t$
$\frac{135}{45} = \frac{45t}{45}$
$3 = t$

It would take the car 3 h to go a distance of 135 mi.

EXERCISES

Use the formulas given above to solve these problems.

1. How far would a car go in 4 h at an average speed of 55 mi/h?

   220 mi

2. Find the area of a triangle with a base of 3 cm and a height of 8 cm.

   12 cm²

3. Find the perimeter of a rectangle 4 m long and 3.5 m wide.

   15 m

4. Find the volume of a rectangular prism 4 in. long, 5 in. wide, and 7 in. high.

   140 in.³

5. How long would it take you to walk 10 mi if you walk at an average rate of 4 mi/h?

   $2\frac{1}{2}$ h

6. What is the height of a rectangular prism if the volume is 210 cm², the width, 6 cm, and the length, 7 cm?

   5 cm

Cube It

A formula states the relationship between two or more quantities.

1. How many faces are there in a cube? **6**
2. How many edges are there in a cube? **12**

3. Write a formula for the area of one face of a cube with edge of length x.

   $A = x^2$

4. Substitute 4 cm for x.

   $A = 4^2$; $A = 16$ cm²

5. Write a formula for the entire surface area of a cube with edge of length x.

   $S = 6x^2$

6. Substitute 5 in. for x.

   $S = 6(5^2)$; $S = 150$ in.²

7. Write a formula for the perimeter of one face of a cube with edge of length x.

   $P = 4x$

8. Substitute 0.5 km for x.

   $P = 4(0.5)$; $P = 2$ km

9. Write a formula for the length of all the edges of a cube with edge of length x.

   $E = 12x$

10. Substitute 6 ft for x.

   $E = 12(6)$; $E = 72$ ft

Students choice of letter for area, surface area, perimeter, and length may vary.

Problem Solving Strategies: Work Backward

Some problems give you the final result of a series of steps and ask you to find the starting condition. To solve such a problem you can work backward to the beginning.

▶ **Example**

Kim spent a total of $18 at the movies. She bought 3 tickets, a bucket of popcorn for $3.50, and a drink for $1.00. How much did she spend for each ticket?

Solution

Everything Kim spent must add up to $18. First, subtract the cost of the drink and the popcorn from that total.

Subtract the cost of the drink.	$18 − $1.00 = $17
Subtract the cost of the popcorn.	$17 − $3.50 = $13.50
The cost of three tickets is $13.50. Let t be the cost of one ticket.	$3t = $13.50
Divide by 3 to find the cost of a ticket.	$t = \frac{\$13.50}{3} = \$4.50$

Kim spent $4.50 for each ticket.

EXERCISES

Solve by working backward.

1. Mario had $5 left after shopping at the mall. He spent $12 at the record shop, $23 at the sports store, and $5 at the food mart. How much did Mario have when he went to the mall?

   $45

2. Patty works at a pet store. Last week she earned $45. She makes $5 an hour. Mrs. Sanchez paid her an extra $10 for clipping her garden hedge. How many hours did Patty work at the pet store?

   7 h

3. David, Chen, and Leslie raised money for the school in a walk-a-thon. Altogether they made $46. Leslie made $10. Chen made twice as much as David. How much did David make?

   $12

4. Willie had 55 baseball cards after trading with his friends. He traded 5 cards for one special card. Lee gave him 11 cards for a Mickey Mantel card. How many cards did Willie start with?

   49 cards

5. Classes start at 8:30 a.m. at Lee's school. It takes him 15 minutes to ride his bike to school, but he likes to arrive 20 minutes early. What time should he leave home?

   7:55 a.m.

6. After Julie went to sleep, the temperature rose 6° F before falling twice as far as it had risen. The next morning, the temperature was 14° F. What was the temperature when Julie went to sleep?

   20° F

Writing Two-Step Equations with a Given Solution

You can write an equation from a given solution. In doing so, you are working backward from the answer.

▶ **Example**

Write a two-step equation having the solution $x = 3$. You can use addition or subtraction combined with multiplication or division.

Solution

$x = 3$

Multiply both sides by 3.
$3x = 3(3)$ or $3x = 9$

Now subtract 4 from both sides.
$3x − 4 = 9 − 4$ or $3x − 4 = 5$

The desired equation is $3x − 4 = 5$.

CHECK: Reverse the steps to solve the equation.
$3x − 4 + 4 = 5 + 4$
$3x = 9$
$\frac{3x}{3} = \frac{9}{3}$
$x = 3$ ← The solution from which you began

EXERCISES

For each solution given, write two equations, each solvable by a two-step process involving either addition or subtraction, and multiplication or division. Check your equations by solving them. **Answers will vary. Samples are given.**

1. $x = −1$ $2x − 8 = −10$; $\frac{x}{3} + 4 = 4\frac{1}{3}$

2. $y = 4$ $5y + 12 = 32$; $4y − 32 = −16$

3. $z = −7$ $\frac{z}{3} + 8 = −29$; $2z + 19 = 5$

4. $a = \frac{1}{3}$ $6a + 14 = 16$; $\frac{a}{3} + 4 = \frac{37}{9}$

5. $b = 1.6$ $3b + 1.2 = 6$; $4b − 12.3 = −7.1$

6. $x = −0.5$ $2x + 12.1 = 11.1$; $\frac{x}{5} + 3 = 2.9$

Name _____ Date _____

Combining Like Terms

Those parts of a variable expression that are separated by addition or subtraction signs are called **terms.**
The expression $3x + 5y - 2x + x^2$ contains 4 terms.

$3x$ and $-2x$ are **like terms.** The variable parts are identical.
$5y$, x^2, and $3x$ are **unlike terms.** The variable parts are different.

You can simplify expressions by combining like terms.

▶ **Example 1**

Simplify $3a - 4b + 5a$.

Solution

Rewrite using the commutative property. $3a - 4b + 5a = 3a + 5a - 4b$
Use the distributive property. $3a + 5a - 4b = (3 + 5)a - 4b$
                                                    $= 8a - 4b$

You can solve equations by combining like terms.

▶ **Example 2**

Solve $4x - 16 = 2 - 2x$.

Solution

Add $2x$ to each side. $4x + 2x - 16 = 2 - 2x + 2x$ CHECK. $4(3) - 16 = 2 - 2(3)$
Simplify. $6x - 16 = 2$ $12 - 16 = 2 - 6$
Add 16 to each side. $6x - 16 + 16 = 2 + 16$ $-4 = -4$
                                   $6x = 18$
Divide both sides by 6. $\frac{6x}{6} = \frac{18}{6}$
                                    $x = 3$

EXERCISES

Simplify by combining like terms.

1. $x + 3y + 2x + 5$ 2. $5a + 8 - 2a + 1$ 3. $3x + 7y^2 - 3x + 2y^2$
   $3x + 3y + 5$ **$3a + 9$**   **$9y^2$**

4. $m + m^3 + 2m + m^3$ 5. $7x + 3xy - 2xy + x$ 6. $5x^2y + 2xy^2 + 4x^2y$
   $3m + 2m^3$ **$8x + xy$**   **$9x^2y + 2xy^2$**

Solve and check.

7. $5y + 3 - 4y = 5$ 8. $m + 3m + 2 + 2m = 14$ 9. $3t + 17 + t = 1$
   $y = 2$ **$m = 2$**   **$t = -4$**

10. $5x + 4 - 2x = 2 + 2$ 11. $4y - 2y + 7 - 2y = y$ 12. $4n = 2n + 6$
    $x = 0$ **$y = 7$**   **$n = 3$**

Name _____ Date _____

A—MAZE—ing!

Can you combine like terms? Find your way from the top of the maze to the bottom by moving to a square containing an expression that is equivalent to the expression in the first square and the last.

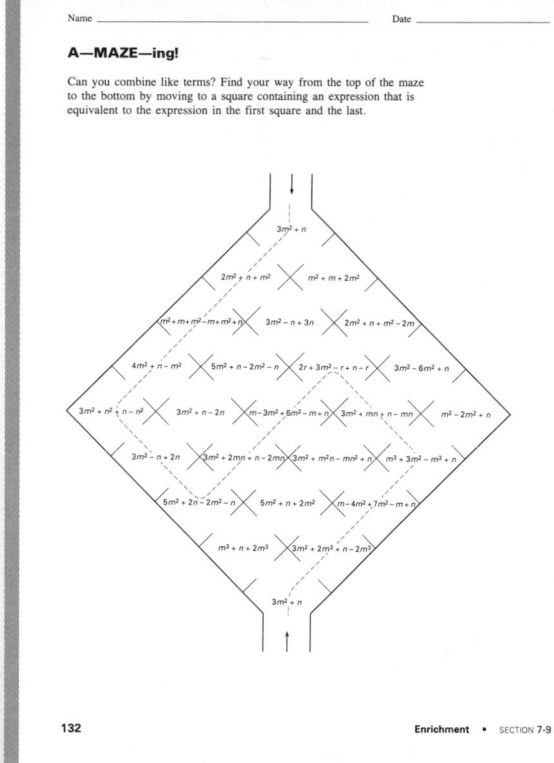

Name _____ Date _____

Graphing Open Sentences

An open sentence can be an equation or an inequality and can be graphed on a number line. A solid dot means the number is included. An open circle means the number is not included.

▶ **Example 1**

a. $x = 5$ b. $x < 5$

   Solution   **Solution**
   The graph of $x = 5$ is a solid dot on 5 on   The graph of $x < 5$ is an open circle on
   a number line.   5 with an arrow pointing to numbers less
                                                   than 5.

▶ **Example 2**

a. $x > 5$ b. $x \le 5$

   Solution   **Solution**
   The graph of $x > 5$ is an open circle on   The graph of $x \le 5$ is a solid dot on 5
   5 with an arrow pointing to the numbers   with an arrow pointing to the numbers
   greater than 5.   less than 5.

▶ **Example 3**

$x \ge 5$

   Solution
   The graph of $x \ge 5$ is a solid dot on 5 with an arrow pointing to the numbers greater than 5.

EXERCISES

Graph each open sentence.

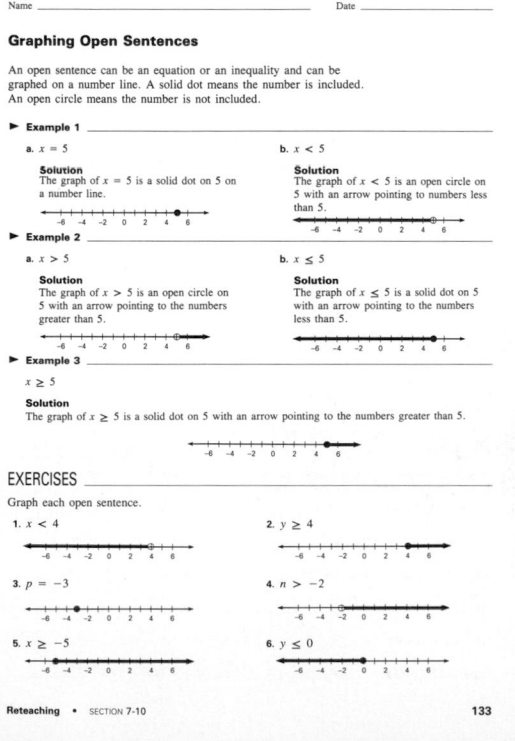

1. $x < 4$ 2. $y \ge 4$

3. $p = -3$ 4. $n > -2$

5. $x \ge -5$ 6. $y \le 0$

Name _____ Date _____

Absolute Value Equations

You know that the absolute value of an integer and its opposite is the same number. For example,
$|2| = 2$ and $|-2| = 2$

So, the equation $|x| = 2$ has two solutions: $x = 2$ and $x = -2$. The graph of the equation is at the right.

As with other equations and inequalities, to solve equations involving absolute value, you must undo addition and subtraction first, and then undo multiplication and division.

▶ **Example 1**   ▶ **Example 2**

Solve $|x| + 4 = 7$.   Solve $4|y| = 24$.
Graph the solution.   Graph the solution.

Solution   **Solution**
$|x| + 4 - 4 = 7 - 4$ $\frac{4|y|}{4} = \frac{24}{4}$
$|x| = 3$ $|y| = 6$
So, $x = 3$ or $x = -3$.   So, $y = 6$ or $y = -6$.

CHECK: $|3| + 4 = 3 + 4 = 7$ CHECK: $4|6| = 4(6) = 24$
$|-3| + 4 = 3 + 4 = 7$ $4|-6| = 4(6) = 24$

EXERCISES

Solve each equation and graph the solution.

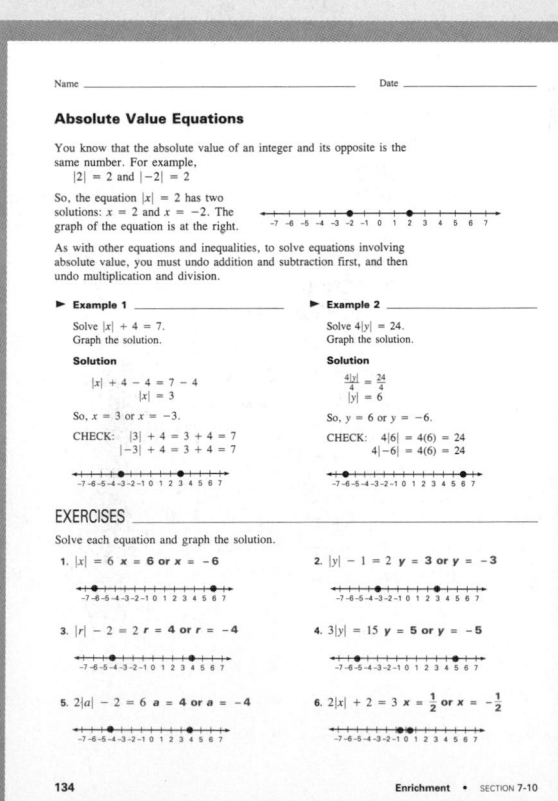

1. $|x| = 6$ **$x = 6$ or $x = -6$**

2. $|y| - 1 = 2$ **$y = 3$ or $y = -3$**

3. $|r| - 2 = 2$ **$r = 4$ or $r = -4$**

4. $3|y| = 15$ **$y = 5$ or $y = -5$**

5. $2|a| - 2 = 6$ **$a = 4$ or $a = -4$**

6. $2|x| + 2 = 3$ **$x = \frac{1}{2}$ or $x = -\frac{1}{2}$**

Name _____ Date _____

Solving Inequalities

Inequalities can be solved in almost the same way as equations.
However, when you multiply or divide by a negative number, you
must reverse the sign.

► Example 1 _____

Solve $\frac{x}{3} + 3 < 5$.

Solution

Subtract 3 from each side.

$\frac{x}{3} + 3 - 3 < 5 - 3$

$\frac{x}{3} < 2$

Multiply each side by 4.

$\frac{x}{3}(3) < 2(3)$

$x < 6$

Graph the solution.

► Example 2 _____

Solve $-3y - 4 \leq 2$.

Solution

Add 4 to each side.

$-3y - 4 + 4 \leq 2 + 4$

$-3y \leq 6$

Divide each side by -3.

$\frac{-3y}{-3} \geq \frac{6}{-3}$

$y \geq -2$

Graph the solution.

EXERCISES

Solve each inequality and graph the solution.

1. $4s + 2 < 10$

   $s < 2$

2. $3t + 5 \geq 2$

   $t \geq -1$

3. $\frac{3y}{4} - 2 > 1$

   $y > 4$

4. $\frac{x}{-3} < -2$

   $x > 6$

5. $5n + 3 \geq -12$

   $n \geq -3$

6. $-6m + 2 \geq 14$

   $m \leq -2$

Name _____ Date _____

Compound Inequalities

Pairs of inequalities connected by *and* or *or* form **compound inequalities**.
Solutions of compound inequalities can be graphed on a number line.

► Example 1 _____

Graph the solutions of $x > -1$ *and* $x \leq 3$ on one number line.

Solution

First, graph each inequality on a separate number line.

The graph of $x > -1$ is

The graph of $x \leq 3$ is

To satisfy a compound inequality connected by *and*, the solution must satisfy both inequalities.
Graph these inequalities on the same number line by showing that range in which the two
graphs overlap.

► Example 2 _____

Graph the solution of $x < -1$ *and* $x \geq 1$ on the same number line.

Solution

The graph of $x < -1$ is

The graph of $x \geq 1$ is

To satisfy a compound inequality with *or*, the solution must satisfy at least one of the
inequalities. That is, the solution can be on either graph. Graph both inequalities on the same
number line. Any solution of either inequality will satisfy the compound inequality.

EXERCISES

Graph these compound inequalities.

1. $x > 0$ and $x \leq 4$

2. $x \leq -3$ or $x \geq 3$

3. $x < -2$ or $x > 0$

4. $x < -2$ or $x \geq 2$

5. $x > 1$ and $x < 3$

6. $x \leq -5$ or $x > -1$

Name _____ Date _____

Calculator Activity: Solving Equations and Checking Solutions

You can use a calculator to help you solve equations and then to check
your solutions.

► Example 1 _____

Use a calculator to solve the equation $0.75 + m = -1$. Check your
solution.

Solution

To solve this equation, you must subtract $\frac{3}{4}$ from both sides.

$0.75 + m = -1$ — Use your calculator to evaluate $-1 - \frac{3}{4}$.

$0.75 - 0.75 + m = -1 - 0.75$ — $1 \boxed{+/-} \boxed{+} 0.75 \boxed{+/-} \boxed{=} -1.75$

$m = -1 - 0.75$

$m = -1.75$

To check this solution, replace m with -1.75 in the original equation.

$\frac{3}{4} + m = -1$ — Use your calculator to evaluate $0.75 + (-1.75)$.
If the solution is correct, the result will be -1.

$0.75 + (-1.75) = -1$ — $0.75 \boxed{+} 1.75 \boxed{+/-} \boxed{=} -1$

$-1 = -1$ — The result is -1. The solution is $-1\frac{3}{4}$.

EXERCISES

Use a calculator to solve each equation and then to check each solution.

1. $n - 5.8 = 6.4$ **12.2**

2. $8.7 = f - 8.7$ **17.4**

3. $441 = 7x$ **63**

4. $\frac{j}{-12} = -14$ **168**

5. $w - 3.5 = -16.75$ **−13.25**

6. $-7.36 = m - 5.4$ **−1.96**

7. $-3.2y = -0.8$ **0.25**

8. $\frac{a}{3.76} = -2.3$ **−8.648**

9. $z - 37.2 = -37.2$ **0**

10. $0.8b = -0.875$ **−1.09375**

11. $p - 8.6 = -11.34$ **−2.74**

12. $t - 0.04 = -0.04$ **0**

Name _____ Date _____

Solving Equations and Checking Solutions *(continued)*

You can also use your calculator to solve and check two-step equations.

► Example 2 _____

Use a calculator to solve the equation $34 = \frac{x}{-2} + 47$.
Then check the solution.

Solution

To solve this equation, first subtract 47 from both sides. Then divide
both sides by -2. This is how it would look on the calculator.

$34 \boxed{-} 47 \boxed{=} -13$

$\boxed{\times} 2 \boxed{+/-} \boxed{=} 26$

To check this solution, replace x with 26 in the original equation.

$34 = \frac{26}{-2} + 47$

Use a calculator to evaluate $\frac{26}{-2} + 47$. If the solution is correct, the result
will be 34.

$26 \boxed{\div} 2 \boxed{+/-} \boxed{+} 47 \boxed{=} 34$

The result is 34. The solution is 26.

EXERCISES

Determine the two steps you need to solve each equation. Then use a calculator to
perform the two steps. Check your solutions on the calculator.

13. $15k - 45 = 30$ **5**

14. $0.8m + 18 = +4$ **−17.5**

15. $6.5t + 3.6 = -9.4$ **−2**

16. $\frac{b}{-6} + 3 = -15$ **108**

17. $0.75 + (-4m) = -16$ **4.1875**

18. $\frac{14n}{-7} = 28$ **−14**

19. $12(n + 0.1875) = 26.25$ **2**

20. $47.3x - 8.77 = -131.75$ **−2.6**

21. $-10 = 2e - 6$ **−2**

22. $9w - 1.5 = 10.2$ **1.3**

23. $0.6 + 3c = 3$ **0.8**

24. $\frac{x}{-12.4} + 18.6 = 27.6$ **−111.6**

Name _____ Date _____

Computer Activity: Spreadsheet Solutions

When an equation involves equal expressions, a spreadsheet can be used to find the solution of the equation. By entering each expression and possible values for the variable you can determine the solution by finding what value of the variable gives both expressions the same value. It will be helpful to observe patterns that emerge from the different values that you enter using this "guess and check" method.

The following spreadsheet shows how the solution was found for the equation $3x - 1 = 2x + 4$.

	A	B	C	D	E	F
1:						
2:		Equation $3x - 1 = 2x + 4$				
3:						
4:		Value of x:		$3x - 1$		$2x + 4$
5:		-2		-7		0
6:		-1		-4		2
7:		0		-1		4
8:		1		2		6
9:		2		5		8
10:		3		8		10
11:		4		11		12
12:		5		14		14
13:		6		17		16

C5: (Value) 3*A5 − 1

EXERCISES

1. What is the formula for E12?

   2 * A12 + 4

2. Why did the second set of values start above 2 instead of below −2?

   The differences get smaller as x increases.

3. What in the pattern of values in column C and column E indicates that the solution will be close to 2?

   The columns differ by only 3 when x = 2.

4. What is the solution of the equation $3x - 1 = 2x + 4$?

   5

Name _____ Date _____

Spreadsheet Solutions (continued)

Use the following spreadsheet for Exercises 5–7.

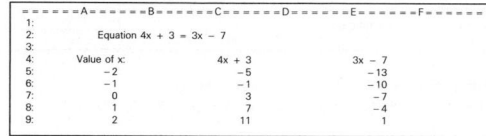

	A	B	C	D	E	F
1:						
2:		Equation $4x + 3 = 3x - 7$				
3:						
4:		Value of x:		$4x + 3$		$3x - 7$
5:		-2		-5		-13
6:		-1		-1		-10
7:		0		3		-7
8:		1		7		-4
9:		2		11		1

5. Would the additional values you check be above 2 or below −2? Why?

   Below −2 because the differences get smaller as x decreases.

6. Based on the differences between the values in column C and column E, how far from 2 or −2 will the solution be?

   The differences at −2 is 8 and it decreases 1 each time x decreases so the solution is 8 away from −2.

7. What is the solution of the equation $4x + 3 = 3x - 7$? **−10**

Use the following spreadsheet for Exercises 8–10.

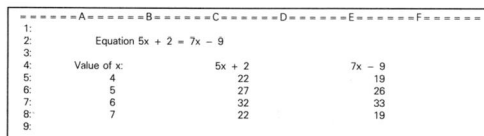

	A	B	C	D	E	F
1:						
2:		Equation $5x + 2 = 7x - 9$				
3:						
4:		Value of x:		$5x + 2$		$7x - 9$
5:		4		22		19
6:		5		27		26
7:		6		32		33
8:		7		22		19
9:						

8. What do you notice about the differences between the values in column C and column E?

   They go from 3 to 1 to 1 to 3. They never equal 0.

9. What does this pattern indicate about the solution? Why?

   The solution is between 5 and 6. The differences equal 0 before they get larger again.

10. What is the solution of the equation $5x + 2 = 7x - 9$? **$5\frac{1}{2}$**

ACHIEVEMENT TEST Name _____

Equations and Inequalities Date _____

CHAPTER 7 FORM A

MATH MATTERS BOOK 1

Chicha Lynch
Eugene Olmstead

SOUTH-WESTERN PUBLISHING CO.

SCORING RECORD	
Possible	Earned
24	

Tell whether the equation is *true*, *false*, or an *open* sentence.

1. $8 = 5 + x$ **open**   2. $-12 = 6(-3 + 1)$ **true**

Solve each equation.

3. $-3y = -9$ **3**   4. $v - 4 = -9$ **−5**   5. $8.4 = z - 2.2$ **10.6**

6. $-4 = \frac{n}{-8}$ **32**   7. $4c + 5 = -7$ **−3**   8. $16 = 3x - 2$ **6**

9. $\frac{2}{3}n = -\frac{8}{9}$ **$-\frac{4}{3}$**   10. $\frac{4}{5}x + 5 = 1$ **−5**

11. $4(x + 5) = 28$ **2**   12. $9c - 4 = 6c + 5$ **3**

13. Find the height of a triangle with a base measuring 18 cm and an area of 90 cm². The formula for the area of a triangle is $A = \frac{1}{2}bh$, where A = area, b = length of base, and h = height.

   10 cm

14. Solve the formula $m = n + p$ for n. **$n = m - p$**

Simplify.

15. $\frac{1}{4}y + \frac{1}{4}y + 3y + \frac{1}{4}y + \frac{1}{4}y$ **4y**   16. $3(x - 2) + 3(x + 4) - 4x$ **2x + 6**

Graph each open sentence on a number line.

17. $x \geq -4$

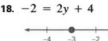

18. $-2 = 2y + 4$

19. $0.6 > 2.4x - 4.2$

20. $\frac{m}{-3} + 4 \geq 5$

Name _____ Date _____

Write an equation and solve.

21. Ms. Oliveda bought a grapefruit for $0.89 and three oranges. The total cost of the fruit was $2.15. How much did each orange cost?

   $3x + 0.89 = 2.15$; $0.42

22. Mel bought a box of holiday greeting cards. He sent 8 to relatives and loaned a friend 2. Then he mailed away half of the remaining cards to friends. He had 6 cards left. How many greeting cards were in the box he bought?

   22 cards

23. Jo's salary before taxes is $132.00. She worked 16 hours. What is her hourly wage?

   $8.25

24. Sal paid $(x + 2y)$ cents for a quart of milk and $(2x + y)$ cents for each of three cartons of orange juice. Write and simplify an expression for the total cost of Sal's purchases.

   $(x + 2y) + 3(2x + y)$; $7x + 5y$

266G

TEST B

Equations and Inequalities

CHAPTER 7   FORM B

MATH MATTERS BOOK 1

Chicha Lynch
Eugene Olmstead

SOUTH-WESTERN PUBLISHING CO.

Name _____

Date _____

SCORING RECORD	
Possible	Earned
24	

Tell whether the equation is *true, false,* or an *open* sentence.

1. $4 = 6 + x$ __open__

2. $-16 = 4(2 - 6)$ __true__

Solve each equation.

3. $-4y = -12$ __3__

4. $v + 12 = 7$ __−5__

5. $2.4 = z - 3.4$ __5.8__

6. $-4 = \frac{n}{-9}$ __36__

7. $3c + 5 = -7$ __−4__

8. $15 = 3x - 3$ __6__

9. $\frac{4}{5}n = -\frac{8}{15}$ __−$\frac{2}{3}$__

10. $\frac{5}{8}x - 4 = 1$ __8__

11. $5(x + 6) = 5$ __−5__

12. $8c - 11 = 3c + 4$ __3__

13. Find the height of a triangle with a base measuring 21 cm and an area of 315 cm². The formula for the area of a triangle is $A = \frac{1}{2}bh$, where A = area, b = length of base, and h = height.

    __30 cm__

14. Solve the formula $m = p - n$ for n.   __$n = p - m$__

Simplify.

15. $x + 3y + 5x - 4y + y$

    __$6x$__

16. $3(2a - b + c) - 3(a + 2b - 3c)$

    __$3a - 9b + 12c$__

Graph each open sentence on a number line.

17. $f \le -3$

    −4   −3   −2

18. $2x - 6 = -8$

    −2   −1   0

19. $3 < 0.6x + 1.8$

    1   2   3

20. $-3p + 6 \le -3$

    2   3   4

7B-1

Name _____   Date _____

Write an equation and solve.

21. Joan bought a can of soup for $0.79 and three bananas. The total cost of the groceries was $1.42. How much did each banana cost?

    __$3b + 0.79 = 1.42;\ \$0.21$__

22. Juan bought a packet of postcards. He sent 4 to relatives and loaned his brother 4. Then he mailed away a third of the remaining cards to friends. He had 8 cards left. How many postcards were in the packet he bought?

    __20 cards__

23. Bo's weekly salary before taxes is $104. His hourly wage is $6.50. How many hours per week does he work?

    __16 h__

24. Sal paid $(3x + 2y)$ cents for a half-gallon of orange juice and $(2x - y)$ cents for each of two cans of soup. Write and simplify an expression for the total cost of Sal's purchases.

    __$(3x + 2y) + 2(2x - y);\ 7x$__

7B-2

TEST B

266H

CHAPTER 8 SKILLS PREVIEW

Write each ratio in two other ways.

1. 2 to 7 **2:7; $\frac{2}{7}$**

2. $\frac{3}{5}$ **3 to 5; 3:5**

3. 6:3 **$\frac{6}{3}$; 6 to 3**

4. 5 to 6 **$\frac{5}{6}$; 5:6**

5. $\frac{1}{4}$ **1 to 4; 1:4**

6. 3:4 **3 to 4; $\frac{3}{4}$**

7. 4:2 **$\frac{4}{2}$; 4 to 2**

8. $\frac{6}{1}$ **6 to 1; 6:1**

9. $\frac{1}{7}$ **1 to 7; 1:7**

10. 8:4 **8 to 4; $\frac{8}{4}$**

11. 9 to 10 **$\frac{9}{10}$; 9:10**

12. 5:15 **5 to 15; $\frac{5}{15}$**

Find the unit rate.

13. $42 to 7 h **$6/h**

14. 630 mi:15 h **42 m/h**

15. 304 m to 9.5 s **32 m/s**

16. 150 words:5 min **30 words/1 min**

17. $\frac{392 \text{ revolutions}}{7 \text{ min}}$ **56 revolutions/ min**

18. 200 students to 4 buses **50 students/ bus**

19. 144 cans:12 cartons **12 cans/carton**

20. $\frac{1{,}512 \text{ m}}{18 \text{ s}}$ **84 m/s**

21. 273 mi to 4.2 h **65 mi/h**

Write three ratios equivalent to the given ratios.

22. $\frac{1}{5}$ **$\frac{2}{10}$, $\frac{3}{15}$, $\frac{4}{20}$**

23. 4:12 **2:6; 1:3; 8:24**

24. $\frac{50}{60}$ **$\frac{100}{120}$, $\frac{150}{180}$, $\frac{200}{240}$**

25. $\frac{1}{6}$ **$\frac{2}{12}$, $\frac{3}{18}$, $\frac{4}{24}$**

26. 15:45 **$\frac{1}{3}$, $\frac{3}{9}$, $\frac{5}{15}$**

27. $\frac{3}{5}$ **$\frac{6}{10}$, $\frac{9}{15}$, $\frac{12}{20}$**

28. 12 to 18 **2 to 3; 4 to 6; 6 to 9**

29. 28:42 **2:3; 4:6; 6:9**

30. $\frac{30}{80}$ **$\frac{60}{160}$, $\frac{120}{320}$, $\frac{180}{480}$**

Solve each proportion.

31. $\frac{7}{8} = \frac{x}{24}$ **x = 21**

32. $\frac{9}{x} = \frac{3}{2}$ **x = 6**

33. $\frac{x}{10} = \frac{12}{15}$ **x = 8**

34. $\frac{5}{3} = \frac{15}{x}$ **x = 9**

35. $\frac{26}{65} = \frac{2}{x}$ **x = 5**

36. $\frac{21}{x} = \frac{7}{2}$ **x = 6**

37. $\frac{8}{9} = \frac{32}{x}$ **x = 36**

38. $\frac{3}{x} = \frac{42}{56}$ **x = 4**

39. $\frac{x}{5} = \frac{90}{150}$ **x = 3**

40. The length of a living-room floor in a scale drawing is 8 in. The actual length of the floor is 20 ft. What is the scale of the drawing? **1 in.:2$\frac{1}{2}$ ft**

41. The actual width of a door is 3 feet. If the scale of a drawing of the room is 1 in.:12 ft, what is the drawing length of the width of the door? **$\frac{1}{4}$ in.**

42. The scale on a map is 1 in.:30 mi. If the map distance from Acton to Parker is 3 in., what is the actual distance between these two towns? **90 mi**

43. On a scale drawing, the width of a parking lot is 6 in. The actual width of the lot is 108 ft. What is the scale of the drawing? **1 in.:18 ft**

Introduction The purpose of this Skills Preview is to assess students' abilities on all the major objectives of Chapter 8. Test results may be used
- to determine those topics which may need only to be reviewed and those topics which need to be more carefully developed;
- for class placement;
- in prescribing for individual differences.

If you prefer, you may use the Skills Preview as an alternative form of the Chapter Test (page 296) to evaluate mastery of chapter objectives. The items on the Skills Preview and the Chapter Test correspond in content and level of difficulty.

8

EXPLORING RATIO AND PROPORTION

OVERVIEW

In this chapter, students explore the concepts involved in comparing like quantities using ratios and equivalent ratios. They learn to read, write, and determine rates and unit rates. The application of ratios to proportions and the use of scale drawings are also included. Students also learn to solve problems using proportions and organized lists.

SPECIAL CONCERNS

Students may possibly have trouble with the new vocabulary in Sections 1, 2, 3, and 6. You may wish to suggest that students list each new term, its definition, and an example which illustrates the term on an index card. Encourage students to carry these cards with them and to review them whenever they have a few minutes to spare.

VOCABULARY

cross-products	ratio
equivalent ratios	term
proportion	scale drawing
rate	unit rate

MATERIALS

calculators	graph paper

BULLETIN BOARD

Display a scale drawing of a house plan on the bulletin board with the scale given. Have students determine the actual dimensions of each room.

INTEGRATED UNIT 2

The skills and concepts involved in Chapters 5–8 are included within the special Integrated Unit 2 entitled "U. S. and World Travel." This unit is in the Teacher's Edition beginning on page 298F. Worksheets for this integrated unit appear in the Enrichment Activities booklet, pages 93–95.

TECHNOLOGY CONNECTIONS

- Calculator Worksheet, 155
- Computer Worksheet, 156
- MicroExam, Apple Version
- MicroExam, IBM Version

- *Semcalc*, Sunburst Communications
- Spreadsheet Software (various publishers)

TECHNOLOGY NOTES

Computer software programs, such as those listed in Technology Connections, allow students to work with units (such as gallons per mile or miles per hour) encouraging them to analyze calculations involving units rates. Spreadsheet software can be used to set up tables of rates and ratios that enable students to quickly change quantities from one unit to another.

 EXPLORING RATIO AND PROPORTION

PLANNING GUIDE

SECTIONS	TEXT PAGES	ASSIGNMENTS			
		BASIC	AVERAGE	ENRICHED	
Chapter Opener/Decision Making	268–269				
8-1 Ratios	270–273	1–12, 17, 18, PSA 1–3	1–12, 13–17, 18–19, PSA 1–3	11–12, 13–17, 18–21, PSA 1–3	
8-2 Rates	274–277	1–8, 11, 13–15, 18, PSA 1–4	1–12, 13–18, 19, PSA 1–6	13–18, 19–20, PSA 1–6	
8-3 Equivalent Ratios	278–281	1–15, 19, 22, 27	1–20, 22–23, 27–28	19–21, 22–26, 27–29	
8-4 Solving Proportions	282–285	1–12, 16, 18–20, 33	4–17, 18–32, 33–34	13–17, 18–32, 34–36	
8-5 Problem Solving Skills: Using Proportions to Solve Problems	286–287	1–7	1–7	1–7	
8-6 Scale Drawings	288–291	1–7, 10–11, 17, PSA 1–6, 12–13	1–8, 9–14, 16–17, PSA 1–10, 12–13	5–8, 9–15, 16–17, PSA 4–13	
8-7 Problem Solving Strategies: Make an Organized List	292–293	1–8	1–8	1–8	
Technology	277, 282, 284, 293	✔	✔	✔	

ASSESSMENT				
Skills Preview	267	All	All	All
Chapter Review	294–295	All	All	All
Chapter Test	296	All	All	All
Cumulative Review	297	All	All	All
Cumulative Test	298	All	All	All

| | ADDITIONAL RESOURCES | | | |
	RETEACHING	ENRICHMENT	TECHNOLOGY	TRANSPARENCY
	8–1	8–1		TM 33
	8–2	8–2		
	8–3	8–3		
	8–4	8–4		
	8–5	8–5	8–5	
	8–6	8–6		
	8–7	8–7		

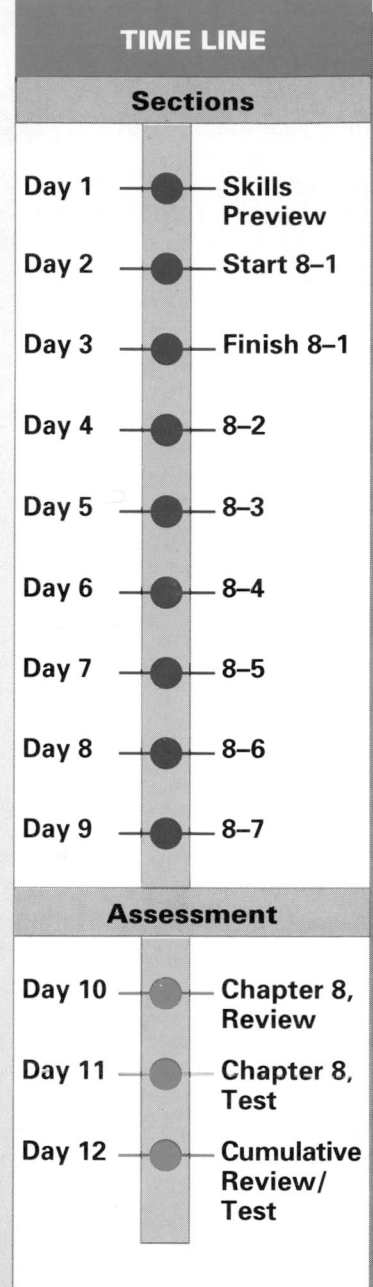

TIME LINE

Sections

Day 1 — Skills Preview
Day 2 — Start 8–1
Day 3 — Finish 8–1
Day 4 — 8–2
Day 5 — 8–3
Day 6 — 8–4
Day 7 — 8–5
Day 8 — 8–6
Day 9 — 8–7

Assessment

Day 10 — Chapter 8, Review
Day 11 — Chapter 8, Test
Day 12 — Cumulative Review/ Test

ASSESSMENT OPTIONS

Chapter 8, Test Forms A and B	
Chapter 8, Test	Text, 296
Alternative Assessment	TAE, 296
Chapter 8 MicroExam	

Objective To explore how rates and ratios can be used to determine real-life statistics

Introduction Ask students to examine the information or data in the table to determine what factors are important in determining a baseball player's batting average. **number of hits divided by the number of times at bat** Discuss the chart, making sure that students understand each column; then ask the following questions.

- If you were researching this topic for your own personal interest, what information would this data provide?
- What questions might this data answer for you?
- What decision might you make after analyzing the data?

Decision Making Using Data After discussing the questions ask, *How did the data in the chart help you answer the questions?* **They provided the numbers that were needed to find and compare batting averages.**

Working Together Have students share their charts. Discuss how this data might affect a baseball manager's decision in organizing a lineup. Ask students to describe how they would determine a lineup and other factors they would consider.

CHAPTER 8
EXPLORING RATIO AND PROPORTION

THEME Hobbies

Many of our hobbies and sports pastimes involve comparing two numbers or quantities. Batting averages, football turnover ratios, and model building all incorporate the concept of ratios. In this chapter you will solve problems that involve ratios, rates, proportions, and scale drawings.

Statisticians calculate a baseball player's batting average by using this ratio.

$$\text{batting average} = \frac{\text{number of hits}}{\text{number of times at bat}}$$

The ratio is expressed as a decimal rounded to the nearest thousandth.

This table gives statistics for some leading American League batters at the end of a recent season.

PLAYER	TEAM	HITS	AT BATS
Boggs	Boston	187	619
Brett	Kansas City	179	544
Thomas	Chicago	63	191
Polonia	California	135	403
Mattingly	New York	101	394
Sierra	Texas	170	608
Gallego	Oakland	80	389

DECISION MAKING

Using Data

Use the information in the table to answer the following questions.

1. Who had the greatest number of hits? **Boggs**

2. Who had the highest batting average? **Polonia**

3. What was Boggs's batting average for the baseball season? **.302**

4. Who had the lower batting average, Mattingly or Sierra? How much lower was it? **Mattingly; .024 lower**

5. At the end of the season, who had the higher batting average, Brett or Thomas? **Thomas**

Working Together

Your group's task is to find the players with the top batting averages for a recent baseball season. For your research, use almanacs, sports magazines, or newspapers. Organize your results in a chart. Compare your group's chart with those made by other groups. Did you include both National League and American League players? Were some of the players selected by more than one group?

In 1901, batter Napoleon Lajoie was the American League batting champion with a .422 batting average. Since then, most batting champions have not batted over .366. Why do you think this is so? What possible changes could influence the drop in averages over time? **Answers will vary.**

Exploring Ratio and Proportion **269**

8-1 Ratios

Work with a partner. Supply the missing word in this analogy:

Hot is to cold as warm is to ___?___ .

Each partner should make up four analogies. Exchange papers with your partner and find the missing words in the analogies. Discuss the following questions in your group. **Answers will vary.**

1. What is meant by an analogy?

2. How does the relationship of the words in each of your analogies show comparison?

3. The word *analogy* comes from the Greek *analogos*, which means "in due ratio." How is a ratio of one number to another like the relationship between one word and another in an analogy?

A **ratio** is the quotient of two numbers and is used to compare one number to the other. The order of the numbers in a ratio is important.

Example 1

Write each ratio in two other ways.

a. 1 to 3　　　　　　b. $\frac{3}{5}$　　　　　　c. 9:10

Solution

a. 1 to 3 can be written as $\frac{1}{3}$ or as 1:3.　← Read: "one to three"

b. $\frac{3}{5}$ can be written as 3 to 5 or as 3:5.　← Read: "three to five"

c. 9:10 can be written as 9 to 10 or as $\frac{9}{10}$.　← Read: "nine to ten"　◄

Example 2

Write each ratio as a fraction in lowest terms.

a. 2 cm to 6 cm　　　　　　b. 50 cm to 2 m

Solution

a. When a ratio is written as a fraction, write it in lowest terms.

$$\frac{2}{6} = \frac{2 \div 2}{6 \div 2} = \frac{1}{3}$$

b. To compare two measurements, use the same unit for both measures. If necessary, rename the larger unit. Then write the ratio.

1 m = 100 cm, so 2 m = 200 cm.

$$\frac{50}{200} = \frac{50 \div 50}{200 \div 50} = \frac{1}{4} \quad ◄$$

270　CHAPTER 8　Exploring Ratio and Proportion

Example 3

Carlene grows geraniums. She feeds the plants once a week with a mixture of liquid plant food and water. This mixture calls for three parts plant food to eight parts water. How much food should Carlene mix with 40 fl oz of water? What is the ratio of plant food to water?

Solution

For 8 fl oz of water, Carlene would need 3 fl oz of plant food. Then, for each additional 8 fl oz of water, she would need an additional 3 fl oz of plant food. Use this pattern to make a table. Extend the table until you arrive at 40 fl oz of water.

CHECK UNDERSTANDING

In Example 3, how much plant food should Carlene mix with 56 fl oz of water? What would be the ratio of plant food to water?
21 fl oz; $\frac{21}{56}$**, or** $\frac{3}{8}$

	+3	+3	+3	+3	
fluid ounces of plant food	3	6	9	12	15
fluid ounces of water	8	16	24	32	40
	+8	+8	+8	+8	

Carlene should mix 15 fl oz of plant food with 40 fl oz of water. The ratio of plant food to water is $\frac{15}{40}$, or $\frac{3}{8}$. ◄

TRY THESE
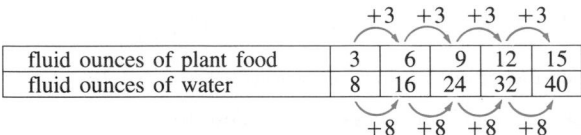

Write each ratio in two other ways.

1. 1 to 2   **1:2;** $\frac{1}{2}$
2. $\frac{4}{8}$ **4 to 8; 4:8**
3. 10 to 9 $\frac{10}{9}$**; 10:9**
4. 9:13 $\frac{9}{13}$**; 9 to 13**
5. $\frac{6}{7}$ **6 to 7; 6:7**
6. 7:10 $\frac{7}{10}$**; 7 to 10**
7. $\frac{8}{9}$ **8 to 9; 8:9**
8. 4 to 1 $\frac{4}{1}$**; 4:1**

Write the ratio of the first quantity to the second as a fraction in lowest terms.

9. 8 cm to 3 cm $\frac{8}{3}$
10. 5 kg to 2 kg $\frac{5}{2}$
11. 25 qt to 5 qt $\frac{5}{1}$
12. 2 h to 6 h $\frac{1}{3}$
13. \$1.10 : 55¢ $\frac{2}{1}$
14. 7 qt to 4 gal $\frac{7}{16}$
15. 65 min to 3 h $\frac{13}{36}$
16. 5 m : 2.5 cm $\frac{200}{1}$

Solve.

17. Andres mixes 2 parts peat moss with 5 parts potting soil for planting his hybrid petunias. How much peat moss should be mixed with 30 parts of potting soil? What is the ratio of peat moss to potting soil?   **12 parts peat moss,** $\frac{12}{30}$**, or** $\frac{2}{5}$

terms of a ratio are reversed, that ratio will take on a different meaning. For example, the ratio 1 to 3 is not the same as 3 to 1.

Example 2: Stress the importance of comparing like units of measure when writing ratios.

Example 3: Write the fraction 1/8 on the chalkboard. Ask students to name a fraction equivalent to 1/8. Point out that the table represents equivalent fractions. Discuss the concept of whole and part. Tell students that a ratio may represent any of three kinds of comparisons: *part/part*, *part/whole*, and *whole/part*. Elicit that the table in the solution represents the *part/part* comparison.

Additional Questions/Examples
Write the ratio in two other ways.
1. 2 ft to 5 yd   **2:15; 2/15**
2. 22 cm to 7 cm   **22:7; 22/7**
3. A school has 384 students and 16 teachers. Find the ratio of students to teachers.
384:16 or 24:1
4. Robert used 7 gallons of gasoline in traveling 154 miles. Find the ratio of miles to gallons of gasoline.   **154:7 or 22:1**

5-MINUTE CLINIC

Exercise	Student's Error	Error Diagnosis
Write a ratio of 6 grams to 22 grams as a fraction.	22/6	• Student reverses the order of the terms of the ratio. The ratio should be 6/22.

Guided Practice/Try These Have students work in small groups. Point out that all group members should come to an agreement on each answer. Be sure that students remember to compare like units of measure.

3 SUMMARIZE

Write About Math Have students describe the meaning of ratio in their math journals. Have them include a statement of the ratio in words and then write the ratio in three ways, using symbols.

4 PRACTICE

Practice/Solve Problems For Exercises 1–10 remind students to keep in mind the proper order of each ratio.

Extend/Solve Problems For Exercises 13–14 have students find a pattern for the numbers in the charts. **For Exercise 13 part A increases by 5 and part B increases by 6. For Exercise 14 part A increases by 3 and part B increases by 4.**

Think Critically/Solve Problems For Exercises 18–21 students should rewrite the ratios as fractions with like denominators and then compare.

Problem-Solving Applications Ratios provide a useful way to compare numbers and related quantities. Have students think of other situations in which ratios might be used; for example, at the supermarket, in recipes, and on a map.

PRACTICE/ SOLVE PROBLEMS

Write each ratio in two other ways.

1. 6 to 9   6:9; $\frac{6}{9}$ **2.** 10:3   10 to 3; $\frac{10}{3}$ **3.** $\frac{8}{1}$ 8 to 1; 8:1

4. 5:4   5 to 4; $\frac{5}{4}$ **5.** $\frac{1}{3}$ 1 to 3; 1:3   **6.** 2 to 3   2:3; $\frac{2}{3}$

Write each ratio as a fraction in lowest terms.

7. 5 cm to 2 cm $\frac{5}{2}$ **8.** 4 min to 20 min $\frac{1}{5}$

9. 1.5 m to 30 cm $\frac{5}{1}$ **10.** 16 gal to 12 qt $\frac{16}{3}$

 Solve. Write the ratio as a fraction in lowest terms.

11. Molly planted marigold seedlings in a window box. There were 21 yellow and 12 orange marigold plants. What is the ratio of yellow to orange plants? $\frac{7}{4}$

12. A recipe calls for 4 parts cornstarch to 7 parts water. How much cornstarch should be mixed with 28 parts water? What is the ratio of cornstarch to water?   16 parts cornstarch; $\frac{16}{28}$, or $\frac{4}{7}$

EXTEND/ SOLVE PROBLEMS

Copy and complete each of the following *ratio tables*.

13. The ratio of part A to part B is 5 to 6.

part A	5	10	?	?	?
part B	?	?	18	24	30

15; 20; 25
6; 12

14. The ratio of part A to part B is 3 to 4. The ratio of part B to part C is 1 to 3.

part A	?	6	?	12	?
part B	4	?	12	?	20
part C	?	?	?	48	?

3; 9; 15
8; 16
12; 24; 36; 60

What is the ratio of the width to length in each rectangle? Express the ratio as a fraction in lowest terms.

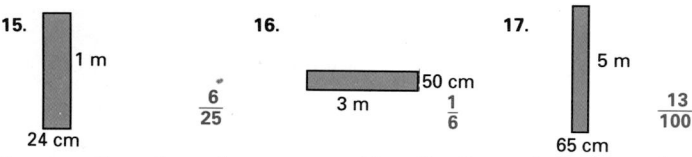

15. 1 m / 24 cm $\frac{6}{25}$

16. 3 m / 50 cm $\frac{1}{6}$

17. 5 m / 65 cm $\frac{13}{100}$

THINK CRITICALLY/ SOLVE PROBLEMS

Compare the ratios using < or >. Describe the method you used to compare. **See Additional Answers.**

18. 1 to 3 and 2 to 9 $\frac{1}{3} > \frac{2}{9}$ **19.** 6 to 11 and 5 to 13 $\frac{6}{11} > \frac{5}{13}$

20. 3 to 4 and 10 to 12 $\frac{3}{4} < \frac{10}{12}$ **21.** 4 to 9 and 7 to 12 $\frac{4}{9} < \frac{7}{12}$

Problem Solving Applications:

RATIOS IN SPORTS

Football and baseball are two sports in which ratios are used to make comparisons.

For football, statisticians keep track of the percent of successful field goals made by an individual kicker. The kicking ratio is the ratio of successful field goals to the total number of field goals attempted.

$$\text{kicking ratio} = \frac{\text{successful field goal kicks}}{\text{attempted field goals}}$$

For example, if a player had 17 successful kicks out of 21 attempted field goals, that player's *kicking ratio* is $\frac{17}{21}$, or .810. Note that the ratio is expressed as a three-place decimal rounded to the nearest thousandth.

In baseball, a player's *batting average* is calculated using the following ratio.

$$\text{batting average} = \frac{\text{number of hits}}{\text{number of times at bat}}$$

A batting average also is expressed as a decimal rounded to the nearest thousandth.

 USING DATA Use the information in the chart on page 268 to answer Exercises 1–2.

1. Brett and Thomas had almost the same batting average for the season. For each, compare the number of hits to the number of times at bat. Tell how their averages can be very close even though Thomas's number of at-bats is much lower than Brett's.

2. What was Polonia's batting average for the season? **.335**

3. If a player successfully kicks 15 field goals in 20 attempts, then his kicking ratio is ____?____ . **.750**

8–1 Ratios **273**

 CHECK UNDERSTANDING

The ratio of 4 to 2 is a whole-number ratio.

$\frac{4}{2} = \frac{2}{1}$, and $\frac{2}{1} = 2$

The first number is divisible by the second.

Rewrite each ratio as a fraction with a denominator of 1.

1. 20 to 4 $\frac{20}{4} = \frac{5}{1}$

2. 50 to 2 $\frac{50}{2} = \frac{25}{1}$

3. 48 to 8 $\frac{48}{8} = \frac{6}{1}$

4. 21 to 3 $\frac{21}{3} = \frac{7}{1}$

1. Possible answer: The average is a comparison of two numbers. Even though Thomas had fewer at-bats than Brett, the ratio of his at-bats to his hits is about the same as Brett's.

---

5 **FOLLOW-UP**

Extra Practice Write a ratio to compare each of the following.
1. 4 ounces to 7 ounces **4:7**
2. 3 nickels to a quarter **15:25**
3. 9 inches to 3 feet **9:36**
4. 5 pints to 1 1/2 gallons **5:12**
5. 4 hours to 15 minutes **240:15**
6. Last season, the Bulls won 11 games and lost 15 games. There were no tie games. What is the ratio of wins to games played? **11:26**

Extension Have students complete and explain what is meant by this statement: 1 is to 5 as 12 is to ■. **60; The two ratios represent equivalent fractions.**

Section Quiz Write a ratio to compare each of the following.
1. 5 out of 8 people live in a suburban area. **5:8**
2. 36 people out of 54 own CD players. **36:54**
3. 3 pounds to 45 ounces **48:45**
4. 9 quarts out of 24 gallons **9:96**
5. 3 hours to 25 seconds **10,800:25**
6. Complete the ratio table.

Part A	2	4	6	8	10
Part B	5	10	15	20	25

Get Ready calculators

Additional Answers
See page 579.
Additional answers for odd-numbered exercises are found in the Selected Answers portion of the page.

8-2 Rates

EXPLORE/ WORKING TOGETHER

Work in small groups. Look through newspapers or magazines and cut out articles or advertisements that give examples of rates. Look for advertisements that specify two different units, such as 10 ears of corn for $1.00 or 35 miles per gallon of gasoline.

1. Display the group's advertisements and explain how they may relate to each other.

2. Make a conjecture about why retailers may offer customers a discount, such as 5 cans of motor oil for $9, when the price of a single can is $1.99.

3. Look at the advertisements and discuss how rates are related to ratios.

SKILLS DEVELOPMENT

A **rate** is a ratio that compares two different kinds of quantities. For example, suppose a typist can type 275 words in 5 minutes. The typist's rate is a ratio comparing the total number of words typed (275) to the amount of time it takes to type that number of words (5 minutes).

Example 1

Simplify the rate *275 words in 5 minutes.*

Solution

$$\text{words} \rightarrow \frac{275}{5} = \frac{275 \div 5}{5 \div 5} = \frac{55}{1} \leftarrow \text{minutes}$$

The rate is 55 words in 1 minute. ◄

A rate that has a denominator of 1 unit is called a **unit rate.** The rate *55 words in 1 minute* is an example of a unit rate. You can write this unit rate as 55 words per minute, or 55 words/min.

274 CHAPTER 8 Exploring Ratio and Proportion

READING MATH

How is a rate related to ratio? What makes a rate different from a ratio?

A rate and ratio both compare quantities, but ratios compare two numbers of the same unit while rates involve two different units.

ASSIGNMENTS

BASIC
1–8, 11, 13–15, 18, PSA 1–4

AVERAGE
1–12, 13–18, 19, PSA 1–6

ENRICHED
13–18, 19–20, PSA 1–6

ADDITIONAL RESOURCES
Reteaching 8–2
Enrichment 8–2

Example 2

The Hortons drove to a sports car show in another town. Their car consumed 12 gallons of gasoline on the 390-mile trip. How many miles per gallon did they average?

Solution

$$\text{miles} \rightarrow \frac{390}{12} = \frac{390 \div 12}{12 \div 12} = \frac{32.5}{1} \leftarrow \text{gallons}$$

The Hortons' car averaged 32.5 miles per gallon of gasoline. You can write this as 32.5 mi/gal. ◄

TRY THESE

Find the unit rate.

1. $\frac{\$63}{7 \text{ h}}$ **$9/h**

2. $\frac{702 \text{ mi}}{13 \text{ h}}$ **54 mi/h**

3. $\frac{299 \text{ meters}}{11.5 \text{ seconds}}$ **26 m/s**

4. $\frac{140 \text{ students}}{5 \text{ teachers}}$ **28 students/teacher**

5. $\frac{450 \text{ words}}{9 \text{ min}}$ **50 words/min**

6. $\frac{190 \text{ revolutions}}{5 \text{ min}}$ **38 revolutions/min**

7. 80 km in 2 h **40 km/h**

8. $120 for 15 days **$8/d**

9. 65 mi in 2 h **32.5 mi/h**

10. 125 words in 5 min **25 words/min**

11. 608 mi in 8 h **76 mi/h**

12. 360 km on 6 gal **60 km/gal**

Solve.

13. Darlene works in a bakery. She makes 520 muffins in an 8-hour shift. How many muffins does she make in one hour? **65 muffins**

14. Chuck read 45 pages in 60 minutes. What is his reading rate in pages per minute? **0.75 page/min**

8–2   Rates   **275**

Example 2: Students should understand that the unit rate is what is asked for (miles per *one* gallon). Have students point out key words that will help them set up and solve the problem. Students' responses should include the word *average.* Tell students that the word *average* indicates division.

Additional Questions/Examples
Find each unit rate.
1. $92/4 h **$23/h**
2. 138 mi/3 gal **46 mi/gal**
3. 72 mi in 6 h **12 mi/h**
4. 360 words in 4 min
   90 words/min

Guided Practice/Try These Have students work in small groups. Point out that all group members should come to an agreement on each answer.

3 SUMMARIZE

Key Questions Ask students to name a ratio that is not a rate and a rate that is not a ratio. Students should understand that a rate is always a ratio, but a ratio is not always a rate. For example, 5 in. to 12 in. is a ratio, but it is not a rate.

276

4 PRACTICE

Practice/Solve Problems For Exercises 1–10 emphasize that a rate is a comparison of two different quantities.

Extend/Solve Problems For Exercises 13–17 encourage students to use calculators. Tell students that, when comparing rates to determine the lower unit rate, they must be sure to write the terms of both ratios in the same order.

Think Critically/Solve Problems For Exercise 20 have students explain what each rate actually means. **48 mi/gal means that for every 48 miles traveled, one gallon of gas was used. 0.02 gal/mi means that for every mile, 0.02 gal was used.**

Problem-Solving Applications Discuss the definition of *unit price* with students and help them understand how a unit price is found. For each of Exercises 5 and 6, have students describe the method they used to find the answer.

5 FOLLOW-UP

Extra Practice Find the unit rate.
1. 10 ounces/$2.50   **4 oz/dollar**
2. $16/2 lb   **$8/lb**
Find each unit rate.
3. 450 miles on 18 gallons
    25 mi/gal
4. 385 miles in 7 hours
    55 mi/h
5. 96 hours in 6 days
    16 h/day

PRACTICE/ SOLVE PROBLEMS

Find the unit rate.

1. $\frac{300 \text{ mi}}{6 \text{ h}}$ **50 mi/h**
2. $\frac{\$42}{5 \text{ h}}$ **$8.40/h**
3. $\frac{440 \text{ words}}{8 \text{ min}}$ **55 words/min**
4. $\frac{\$12.15}{9 \text{ gal}}$ **$1.35/gal**
5. $\frac{\$20}{4 \text{ lb}}$ **$5/lb**
6. $\frac{264 \text{ mi}}{22 \text{ gal}}$ **12 mi/gal**
7. $260 for 4 rooms   **$65/room**
8. $272 for 4 tires   **$68/tire**
9. 6,000 km in 12 days
    500 km/day
10. 420 ft in 14 min   **30 ft/min**

Solve.

11. Fiona drove 550 mi in 10 h. What was her average speed in miles per hour?   **55 mi/h**

12. Elaine bought 7 yd of fabric for $69.65. What did the fabric cost per yard?   **$9.95/yd**

EXTEND/ SOLVE PROBLEMS

Which is the lower rate?

13. 2,133 km in 6 h   or   1,498 km in 7 h   **1,498 km in 7 h**

14. 495 km on 45 L   or   175 km on 25 L   **175 km on 25 L**

15. 144 students in 3 buses   or   208 students in 4 buses   **144 students in 3 buses**

16. 88 h in 4 weeks   or   108 h in 5 weeks   **108 h in 5 weeks**

17. $3.96 for 36 oz   or $6.30 for 42 oz   **$3.96 for 36 oz**

Solve.

18. Will Raymond earn more money working 10 h per week at $4.75/h or 12 h per week at $4.25/h?   **12 h at $4.25/h**

THINK CRITICALLY/ SOLVE PROBLEMS

19. Jerome's car uses 18.75 gal of gasoline to travel 900 mi. What is the rate of fuel consumption in miles per gallon? What is the rate in gallons per mile?

    48 mi/gal; about 0.02 gal/mi

20. The answers to Exercise 19 are each unit rates. Explain how they differ from one another.   **They compare the same two quantities, but the order in which the quantities are compared is reversed.**

Problem Solving Applications:

UNIT PRICE

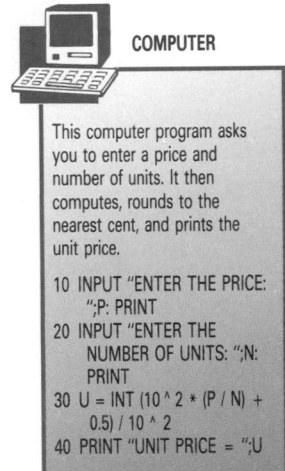

A rate that identifies the cost of an item per unit is called the **unit price.** Many retailers and grocers post the unit price of an item as well as the total price. Sometimes, the retailer will offer a lower unit price to consumers when they buy a larger quantity.

Price-conscious consumers often calculate the unit price to determine the better buy. To find the unit price:

► Write a ratio of the price to the number of units.
► Rewrite the ratio so that the denominator represents 1 unit.
► Round the numerator to the nearest cent, if necessary.

Find the unit price. Round to the nearest cent, if necessary.

1. As the office manager, Megan buys supplies for a doctor's office. One brand of copy paper sells for $11.98 for 500 sheets. What is the unit price? **$0.02/sheet**

2. A large pizza costs $14.98 and serves 6 people. What is the price per serving? **$2.50/serving**

3. The manager at the Fruit Stand is selling a 3-lb bag of apples for $2.19. What is the unit price? **$0.73/lb**

4. The Deli King is selling 1 dozen bagels for $4.20. What does one bagel cost? **$0.35**

5. A one-gallon container of skim milk sells for $1.99. A half gallon sells for $.98. Which is the better buy? **the half gallon**

6. Pasta Palace is selling a 3-lb package of spaghetti for $2.45. The cost of a 32-oz package is $1.70. There is a 25-cents-off coupon on the 32-oz package which is redeemable at the checkout stand. Which package is the better buy? **the 32-oz package**

COMPUTER

This computer program asks you to enter a price and number of units. It then computes, rounds to the nearest cent, and prints the unit price.

```
10 INPUT "ENTER THE PRICE:
   ";P: PRINT
20 INPUT "ENTER THE
   NUMBER OF UNITS: ";N:
   PRINT
30 U = INT (10 ^ 2 * (P / N) +
   0.5) / 10 ^ 2
40 PRINT "UNIT PRICE = ";U
```

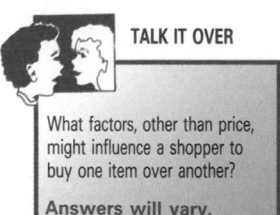

TALK IT OVER

What factors, other than price, might influence a shopper to buy one item over another?

Answers will vary.

8–2 Rates **277**

Extension Have students work in small groups to solve and check the solution to this problem: On vacation, Robert drove as follows: Monday, 676 mi in 13 h; Tuesday, 195 mi in 5 h; Wednesday, 72 mi in 2 h; Thursday, 132 mi in 3 h; Friday, 660 mi in 12 h; Saturday, 864 mi in 16 h; and Sunday, 200 mi in 4 h. Find his average rate per day and per week to the nearest mile per hour.
52 mi/h, 39 mi/h, 36 mi/h, 44 mi/h, 55 mi/h, 54 mi/h, 50 mi/h; 51 mi/h

Section Quiz Find each unit rate.
1. 480 words/12 min
   40 words/min
2. 385 mi/7 h **55 mi/h**
3. 15 laps/5 min **3 laps/min**
4. $7.85/5 lb **$1.57/lb**
5. 1,054 km in 17 days
   62 km/day
6. 792 ft in 66 min **12 ft/min**
7. Joseph's pulse rate is 50 beats every 30 seconds. What is his pulse rate per minute?
   100 beats per minute
8. Marge bought 8 yd of yarn for $2.56. How much did the yarn cost per yard? **$0.32**

SPOTLIGHT

OBJECTIVES
- To find a ratio or a group of ratios equivalent to a given ratio
- To write a ratio in lowest terms

VOCABULARY
equivalent ratios

WARM-UP

Find the greatest common factor (GCF) for each.
1. 45, 60 **15**   2. 16, 36 **4**
3. 18, 30 **6**   4. 81, 18 **9**

1 MOTIVATE

Explore/Working Together Elicit from students that the ratio for the recipe is 2:7. Students should realize that 10 fl oz of cranberry juice are needed for 35 fl oz of pineapple juice and that 42 fl oz of pineapple juice are needed for 12 fl oz of cranberry juice.

2 TEACH

Use the Pages/Skills Development Have students read this part of the section and then discuss the examples. Write 1/3 and 3/9 on the chalkboard. Have students recall ways to determine if these fractions are equivalent. Point out that equivalent ratios can be expressed as equivalent fractions.

Example 1: Emphasize the importance of multiplying the numerator and the denominator of a ratio by the same number. Students should see that, if each

---

EXPLORE/ WORKING TOGETHER

Work with a partner.

Carmine is making fruit punch for a party. The recipe calls for 2 parts cranberry juice to 7 parts pineapple juice.

1. Copy and complete the table to show the ratios of cranberry juice to pineapple juice, in fluid ounces.

parts of cranberry juice	2	4	?	?	?	?	14	?
parts of pineapple juice	?	?	21	28	35	42	?	56

cranberry juice: 6, 8, 10, 12, 16; pineapple juice: 7, 14, 49

2. How much cranberry juice is needed for 35 fl oz of pineapple juice? **10 fl. oz**

3. How many fluid ounces of pineapple juice are needed for 12 fl oz of cranberry juice? **42**

4. Look at your table. What patterns do you notice? **Answers will vary.**

5. How could you have found the answers to questions 2 and 3 without using a table? **Answers will vary.**

SKILLS DEVELOPMENT

Two ratios that represent the same comparison are called **equivalent ratios.** One way to find a ratio equivalent to a given ratio is to multiply both terms of the given ratio by the same nonzero number.

Example 1

Find three ratios equivalent to 2:3.

Solution
Write the ratio as a fraction. Multiply the numerator and the denominator by the same number.

$$\frac{2}{3} = \frac{2 \times 2}{3 \times 2} = \frac{4}{6}$$

$$\frac{2}{3} = \frac{2 \times 3}{3 \times 3} = \frac{6}{9}$$

$$\frac{2}{3} = \frac{2 \times 4}{3 \times 4} = \frac{8}{12}$$

The ratios 4:6, 6:9, and 8:12 are equivalent to 2:3. ◄

Another way to find a ratio equivalent to another ratio is to *divide* both terms of the given ratio by the same nonzero number.

ESL STUDENTS

Pair ESL students with other students. Since reading comprehension is crucial to mathematical understanding, have students read and model each ratio for Exercises 1–6.

Example 2

Find three ratios equivalent to 20 to 30.

Solution

Write the ratio as a fraction.

$$\frac{20}{30} = \frac{20 \div 2}{30 \div 2} = \frac{10}{15} \qquad \frac{20}{30} = \frac{20 \div 5}{30 \div 5} = \frac{4}{6} \qquad \frac{20}{30} = \frac{20 \div 10}{30 \div 10} = \frac{2}{3}$$

The ratios 10 to 15, 4 to 6, and 2 to 3 are equivalent to 20 to 30. ◄

To determine if two given ratios are equivalent, you write each ratio as a fraction in lowest terms.

Example 3

Is each pair of ratios equivalent?

a. 10:12, 15:18 **b.** 8 to 6, 12 to 8

Solution

a. $\frac{10}{12} = \frac{10 \div 2}{12 \div 2} = \frac{5}{6}$

$\frac{15}{18} = \frac{15 \div 3}{18 \div 3} = \frac{5}{6}$

Yes, the ratios 10:12 and 15:18 are equivalent.

b. $\frac{8}{6} = \frac{8 \div 2}{6 \div 2} = \frac{4}{3}$

$\frac{12}{8} = \frac{12 \div 4}{8 \div 4} = \frac{3}{2}$

No, the ratios 8 to 6 and 12 to 8 are not equivalent. ◄

Example 4

Jane answered 5 out of 6 questions correctly on the lesson quiz and 25 out of 30 correctly on the chapter test. Was the ratio of the number of correct answers to the number of questions the same for both the quiz and the test?

Solution

Write the ratios as fractions in simplest form.

5 out of 6 questions → $\frac{5}{6}$

25 out of 30 questions → $\frac{25}{30} = \frac{25 \div 5}{30 \div 5} = \frac{5}{6}$

Yes, the ratios of Jane's scores— 5 out of 6 correct on the lesson quiz and 25 out of 30 correct on the chapter test— were the same. ◄

8–3 Equivalent Ratios **279**

5-MINUTE CLINIC

Exercise	Student's Error	Error Diagnosis
Find a ratio equivalent to 14/56.	$\frac{14 \div 2}{56 \div 7} = \frac{7}{8}$	• Student forgets to divide both terms of the ratio by the same number.

term of a ratio were multiplied by a different number, the resulting ratio would not be equivalent to the original ratio.

Example 2: Reinforce the idea that, to find an equivalent ratio having smaller numbers, always divide each term of the ratio by the same number.

Example 3: Make sure students understand that, to write any fraction in lowest terms, they must divide the numerator and the denominator by the GCF.

Example 4: Have students explain why dividing each term in the ratio 25/30 by their GCF will result in a fraction in lowest terms. **Dividing each term by the GCF results in a fraction whose numerator and denominator have no common factor.**

Additional Questions/Examples
Find three ratios equivalent to each.
1. 5/12 **10/24, 15/36, 20/48**
2. 15/60 **1/4, 3/12, 5/20**
3. 1/3 **2/6, 3/9, 4/12**
4. Are the ratios 8:13 and 32:52 equivalent? **yes**

280

Guided Practice/Try These

Have students work in small groups. Point out that there are an infinite number of equivalent ratios for each of the ratios in Exercises 1–3.

Write About Math Have students describe in their math journals how to find equivalent ratios.

4 PRACTICE

Practice/Solve Problems For those ratios in Exercises 10–18 that are not equivalent, have students explain why they are not equivalent and then write two ratios that are equivalent.

Extend/Solve Problems For Exercises 22–24 you may wish to have students substitute other values for the variables and then write equivalent ratios for these ratios.

Think Critically/Solve Problems For Exercises 27–29 have students explain how to write three-term ratios in lowest terms. **find the GCF of all three numbers, then divide each by that common factor**

5 FOLLOW-UP

Extra Practice Write three equivalent ratios for each.
1. 5/6 **10/12, 15/18, 20/24**
2. 12/32 **6/16, 3/8, 24/64**
3. 6/21 **2/7, 12/42, 18/63**
4. 16/44 **4/11, 8/22, 32/88**
Are the ratios equivalent? Write *yes* or *no*.
5. 1 to 4, 16 to 64 **yes**
6. 3:2, 12:9 **no**
7. 30:6, 90:18 **yes**
8. 4 to 6, 20 to 18 **no**

TRY THESE

Use multiplication to find three equivalent ratios. **Possible answers are given.**

1. $\frac{1}{4}$ **$\frac{2}{8}$, $\frac{4}{16}$, $\frac{8}{32}$**

2. 4:6 **8:12; 16:24; 32:48**

3. 2 out of 7 **$\frac{4}{14}$, $\frac{6}{21}$, $\frac{8}{28}$**

Use division to find three equivalent ratios.

4. $\frac{30}{50}$ **$\frac{15}{25}$, $\frac{6}{10}$, $\frac{3}{5}$**

5. 18 to 24 **3 to 4; 9 to 12; 6 to 8**

6. 15:75 **3:15; 5:25; 1:5**

Are the ratios equivalent? Write *yes* or *no*.

7. 1 to 3, 25 to 75 **yes**
8. 3:2, 15:5 **no**
9. $\frac{7}{3}$, $\frac{42}{18}$ **yes**
10. 30:98, 10:15 **no**
11. $\frac{4}{5}$, $\frac{12}{18}$ **no**
12. 3 to 4, 6 to 8 **yes**

13. Julio answered 51 out of 60 questions correctly on a history test. He answered 17 out of 20 questions correctly on a math quiz. Was the ratio of correct answers to questions the same for both the test and the quiz? **Yes; $\frac{51}{60} = \frac{17}{20}$.**

EXERCISES

PRACTICE/ SOLVE PROBLEMS

Write three equivalent ratios for each given ratio.

1. $\frac{6}{7}$ **$\frac{12}{14}$, $\frac{18}{21}$, $\frac{24}{28}$**

2. 15 to 18 **75 to 90; 5 to 6; 60 to 72**

3. 3:4 **15:20; 6:8; 9:12**

4. 16:20 **8 to 10; 4 to 5; 64 to 80**

5. 5 to 3 **10 to 6; 15 to 9; 20 to 12**

6. $\frac{7}{8}$ **$\frac{14}{16}$, $\frac{21}{24}$, $\frac{28}{32}$**

7. $\frac{8}{9}$ **$\frac{16}{18}$, $\frac{24}{27}$, $\frac{32}{36}$**

8. 12:14 **6:7; 24:28; 36:42**

9. 3:10 **6 to 20; 9 to 30; 12:40**

Are the ratios equivalent? Write *yes* or *no*.

10. 4 to 5, 12 to 15 **yes**
11. 8:2, 16:9 **no**
12. $\frac{14}{20}$, $\frac{170}{100}$ **no**
13. 24 to 40, 144 to 80 **no**
14. $\frac{2}{3}$, $\frac{12}{18}$ **yes**
15. 5:8, 15:32 **no**
16. $\frac{8}{14}$, $\frac{24}{44}$ **no**
17. $\frac{6}{7}$, $\frac{36}{42}$ **yes**
18. 2:5, $\frac{100}{250}$ **yes**

19. Janna appeared on the show *Trivia Bowl*. She answered 7 out of 10 questions correctly during the first round and 28 out of 40 questions correctly in the second round. Was the ratio of the number of correct answers to the number of trivia questions the same for both rounds? **yes; 7 to 10 = 28 to 40**

20. The school reading club took a survey. They discovered that in the winter 17 out of 21 readers regularly read mystery novels, but in the spring 68 out of 84 readers chose adventure novels. Was the ratio of the number of readers of mystery novels the same as the ratio for readers of adventure novels? **yes; $\frac{17}{21} = \frac{68}{84}$**

MIXED REVIEW

1. Solve: $\frac{2}{5}x = 6$ **$x = 15$**

2. Solve: $3x + 5 = 14$ **$x = 3$**

3. Estimate the quotient: 4,507 ÷ 23   **about 200**

4. Find the product: $\frac{3}{4} \times \frac{5}{8}$ **$\frac{15}{32}$**

5. Find the sum: $\frac{2}{9} + \frac{7}{12}$ **$\frac{29}{36}$**

21. The local car dealership was running a clearance sale in November. They sold 6 out of 9 midsize cars and 24 out of 45 economy cars. Was the ratio for midsize cars sold the same as the ratio for economy cars sold? Write a number statement to justify your answer. **no; $\frac{6}{9} = \frac{2}{3}$, $\frac{24}{45} = \frac{8}{15}$**

For Exercises 22–24, answers will vary. Possible answers are given.

22. Write a ratio equivalent to $a:b$, if $a = 5$ and $b = 4$. **30:24**

23. Write a ratio equivalent to x to y, if $x = 45$ and $y = 50$. **9 to 10**

24. Write a ratio equivalent to $\frac{d}{e}$, if $d = 17$ and $e = 25$. **$\frac{51}{75}$**

Jewelry that contains gold is measured in karats (k), with pure gold being 24 k. If you have a 10-k gold ring, the ratio of the mass of pure gold in the ring to the mass of the ring is 10:24.

25. Express the gold content of a gold ring marked 18 k as a ratio in lowest terms. **3:4**

26. Express the gold content of a gold earring marked 14 k as a ratio in lowest terms. **7:12**

It is possible to write ratios with three numbers. For example, 30:85:60 means 30 compared to 85 compared to 60.

27. Write the ratio 30:85:60 in lowest terms. **6:17:12**

28. Write two 2-number ratios using the ratio 30:85:60.
Possible answer: 30:85; 85:60

29. Write two equivalent 3-number ratios for 30:85:60.
Possible answer: 6:17:12; 60:170:120

EXTEND/ SOLVE PROBLEMS

THINK CRITICALLY/ SOLVE PROBLEMS

8–3 Equivalent Ratios **281**

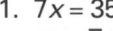

WARM-UP

Solve each equation.

1. $7x = 35$
   $x = 5$
2. $15x = 60$
   $x = 12$
3. $1/3x = 8$
   $x = 24$
4. $4/7x = 16$
   $x = 28$

1 MOTIVATE

Explore/Working Together Ask students what the ratio 2/7 represents. **The ratio of the number of boys who made the team to the number of those who tried out.** Students should explain that, to solve this proportion using an equation, they must set up two ratios, use the cross-products of the ratios to write an equation, and then solve for the missing number.

2 TEACH

Use the Pages/Skills Development Have students read this part of the section and then discuss the examples. Write the word *proportion* on the chalkboard. To check for student understanding, have students explain what the word means.

Example 1: Have students suggest two other ways to determine

8-4 Solving Proportions

EXPLORE/ WORKING TOGETHER

Work with a partner.

Twenty-eight boys were accepted on the football team. Two out of every seven boys who tried out were accepted on the team. How many boys tried out for the football team?

1. Write a ratio to represent how many boys were accepted on the football team.

2. Explain how you might solve this problem using an equation.

3. Tell how to find an equivalent ratio for $\frac{2}{7}$ using the number 28 as the numerator of the ratio.

SKILLS DEVELOPMENT

A **proportion** is an equation that states that two ratios are equivalent.

You can write a proportion two ways.

$$\frac{2}{3} = \frac{6}{9} \qquad 2:3 = 6:9$$

Read: "Two is to three as six is to nine."

 **COMPUTER**

Identify what A, B, C, and D represent in this program. What happens if the cross-products are not equal?

```
10 INPUT "ENTER A,B,C,D:
   ";A,B,C,D
20 IF A * D = B * C THEN
   GOTO 40
30 PRINT "NOT A
   PROPORTION"
35 END
40 PRINT "PROPORTION"
```

See Additional Answers.

The numbers 2, 3, 6, and 9 are called the **terms** of the proportion. If a statement is a proportion, the **cross-products** of the terms are equal.

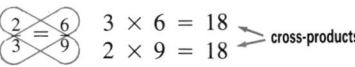
$$3 \times 6 = 18$$
$$2 \times 9 = 18 \quad \longrightarrow \text{cross-products}$$

Example 1

Tell whether the statement is a proportion.

$$\frac{3}{4} \stackrel{?}{=} \frac{21}{28}$$

Solution
Find the cross-products.

$$3 \times 28 = 84$$
$$4 \times 21 = 84$$

$$84 = 84, \text{ so } \frac{3}{4} = \frac{21}{28}$$

Yes, the statement is a proportion. ◄

Example 2

Tell whether the statement is a proportion.

$$2:7 \overset{?}{=} 6:10$$

Solution

Write each ratio as a fraction. Find the cross products.

$$\frac{2}{7} \overset{?}{=} \frac{6}{10} \quad \begin{array}{l} 2 \times 10 = 20 \\ 7 \times 6 = 42 \end{array}$$

$20 \neq 42$, so $2:7 \neq 6:10$

No, the statement is not a proportion. ◀

Sometimes one term of a proportion is unknown.
You can use cross-products to *solve* this type of proportion.

Example 3

Solve the proportion.

$$\frac{n}{16} = \frac{15}{24}$$

Solution

Write the cross-products.

$$n \times 24 = 15 \times 16$$
$$24n = 240$$
$$\frac{24n}{24} = \frac{240}{24} \leftarrow \text{Divide both sides of the equation by 24.}$$
$$n = 10$$

Check by substituting the value for n. Then determine if the two ratios are equivalent.

$$\frac{10}{16} = \frac{10 \div 2}{16 \div 2} = \frac{5}{8}$$
$$\frac{15}{24} = \frac{15 \div 3}{24 \div 3} = \frac{5}{8} \qquad \text{So } n = 10. \ ◀$$

Example 4

A baseball pitcher pitched 12 losing games. The ratio of wins to losses was 3:2. How many games did the pitcher win?

Solution

Let n represent the number of games the pitcher won.
Write and solve a proportion.

$$\begin{array}{l} \text{wins} \rightarrow \\ \text{losses} \rightarrow \end{array} \frac{3}{2} = \frac{n}{12} \begin{array}{l} \leftarrow \text{wins} \\ \leftarrow \text{losses} \end{array}$$
$$3 \times 12 = 2 \times n$$
$$36 = 2n$$
$$\frac{36}{2} = \frac{2n}{2}$$
$$18 = n$$

Check: $\frac{3}{2} = \frac{3}{2}$ $\frac{18}{12} = \frac{18 \div 6}{12 \div 6} = \frac{3}{2}$

The pitcher won 18 games. ◀

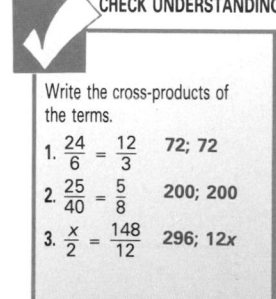

CHECK UNDERSTANDING

Write the cross-products of the terms.

1. $\frac{24}{6} = \frac{12}{3}$ **72; 72**

2. $\frac{25}{40} = \frac{5}{8}$ **200; 200**

3. $\frac{x}{2} = \frac{148}{12}$ **296; 12x**

whether the statement is a proportion. Student responses should include a ratio table, equal ratios, and multiplying or dividing the terms of the ratio by a common factor.

Example 2: Have students first solve the example using mental math and then check their results using cross-products.

Example 3: Students should recall that multiplication and division are inverse operations. Have them solve the proportion.

Example 4: Emphasize that, when solving a proportion, both ratios must compare the same subjects in the same order.

Additional Questions/Examples
Tell whether each statement is a proportion. Replace ● with = or ≠.
1. 11:2 ● 22:4 **=**
2. 3:5 ● 15:9 **≠**
3. 1:4 ● 4:8 **≠**
4. 6:9 ● 18:27 **=**
Solve each proportion.
5. $x/6 = 35/42$ **5**
6. $63/21 = 18/x$ **6**
7. $6/7 = 54/x$ **63**
8. A team scores 48 points in 45 minutes. At that rate, how many points will they score in 1 1/2 hours? **96 points**

5-MINUTE CLINIC

Exercise	Student's Error	Error Diagnosis
Solve. 2:x = 4:10	x/2 = 4/10 x = 4/5	• Students set up the terms of the proportion in the incorrect order. The correct proportion is 2/x = 4/10, so x = 5.

284

Guided Practice/Try These
Have students work in small groups to complete these exercises.

3 SUMMARIZE

Key Questions Ask students: How do you know whether two ratios form a proportion? **They form a proportion if the cross-products of their terms are equal.**

4 PRACTICE

Practice/Solve Problems For Exercises 1–6 have students use mental math to determine whether each statement is a proportion.

Extend/Solve Problems For Exercises 18–29 remind students to check their solution by substituting the solution into the original proportion to see whether the ratios are equal.

Think Critically/Solve Problems For Exercise 34 emphasize that it is important to express the rates in the same units before solving the proportions. Have students explain the steps they used in solving Exercise 35.

5 FOLLOW-UP

Extra Practice Solve each proportion.
1. $x:4 = 5:20$ **1**
2. $x:24 = 18:108$ **4**
3. $10:15 = 4:x$ **6**
4. $4:6 = 38:x$ **57**
5. Find the value of a if the ratios 3:10 and a:60 are equivalent. **18**
6. Find the value of b if the ratios 6:b and 16:40 are equivalent. **15**

TRY THESE

Tell whether each statement is a proportion. Write *yes* or *no*.

1. $\frac{3}{9} \overset{?}{=} \frac{9}{27}$ **yes**
2. $\frac{2}{5} \overset{?}{=} \frac{12}{11}$ **no**
3. $4:12 \overset{?}{=} 7:36$ **no**
4. $7:9 \overset{?}{=} 49:63$ **yes**

Solve each proportion.

5. $\frac{x}{6} = \frac{30}{36}$ **$x = 5$**
6. $\frac{4}{3} = \frac{a}{45}$ **$a = 60$**
7. $n:5 = 16:40$ **$n = 2$**
8. $12:3 = y:18$ **$y = 72$**
9. $\frac{x}{8} = \frac{21}{56}$ **$x = 3$**
10. $3:15 = b:45$ **$b = 9$**

11. If a cyclist travels 10 mi in 2 h, how far will the cyclist travel in 5 h? **25 mi**

EXERCISES

PRACTICE/ SOLVE PROBLEMS

Tell whether each statement is a proportion. Write *yes* or *no*.

1. $\frac{2}{3} \overset{?}{=} \frac{4}{6}$ **yes**
2. $\frac{4}{8} \overset{?}{=} \frac{1}{4}$ **no**
3. $6:8 \overset{?}{=} 3:4$ **yes**
4. $5:6 \overset{?}{=} 40:48$ **yes**
5. $\frac{8}{5} \overset{?}{=} \frac{15}{24}$ **no**
6. $2:7 \overset{?}{=} 6:21$ **yes**

CALCULATOR

You can use a calculator to solve a proportion.

Solve: $9:m = 15:10$

Write the ratios as fractions.
$$\frac{9}{m} = \frac{15}{10}$$

Then use cross-products to find the answer.

Enter: 9 $\boxed{\times}$ 10 $\boxed{\div}$ 15 $\boxed{=}$

The solution is 6.
$9:6 = 15:10$

Write the key sequence for each proportion.
1. $\frac{2.4}{3.6} = \frac{x}{1.8}$
2. $m:22.5 = 7:9$
3. $\frac{4.4}{z} = \frac{11}{2.5}$

See Additional Answers.

Solve each proportion. Check your answers.

7. $\frac{1}{2} = \frac{n}{16}$ **$n = 8$**
8. $\frac{h}{15} = \frac{20}{75}$ **$h = 4$**
9. $\frac{3}{10} = \frac{x}{60}$ **$x = 18$**
10. $r:28 = 7:4$ **$r = 49$**
11. $b:12 = 2:24$ **$b = 1$**
12. $\frac{6}{15} = \frac{z}{40}$ **$z = 16$**
13. $\frac{1}{8} = \frac{n}{32}$ **$n = 4$**
14. $3:2 = s:50$ **$s = 75$**
15. $8:y = 2:4$ **$y = 16$**

Solve.

16. An artist makes green paint by mixing blue paint and yellow paint in the ratio 1:2. How many parts of yellow paint would be mixed with 6 parts of blue paint to make green paint? **12 parts yellow**

17. Robert swims 30 laps in 20 minutes. How many laps can he swim in 40 minutes? **60 laps**

Solve each proportion.

18. $\frac{13}{10} = \frac{52}{x}$ **x = 40** 19. $\frac{78}{m} = \frac{18}{12}$ **m = 52** 20. $56:32 = x:4$
x = 7

21. $9:21 = 15:m$
m = 35
22. $z:39 = 4:12$ **z = 13** 23. $\frac{x}{81} = \frac{5}{9}$ **x = 45**

24. $\frac{11}{44} = \frac{12}{n}$ **n = 48** 25. $6:x = 42:77$ **x = 11** 26. $\frac{0.004}{0.12} = \frac{x}{1.8}$ **x = 0.06**

27. $\frac{4.5}{6} = \frac{a}{20}$ **a = 15** 28. $\frac{0.9}{3.6} = \frac{1.2}{b}$ **b = 4.8** 29. $\frac{2.4}{32} = \frac{y}{16}$ **y = 1.2**

30. The basketball team scored 38 points in 30 minutes. At that rate, how many points are they likely to score in 1 hour? **76 points**

31. Jan can run 525 yards in 7.5 minutes. Express her average speed in yards per minute. **70 yd/min**

32. **USING DATA** Refer to the Data Index on page 546 to answer this question: The ratio of the length of a 2-penny nail to the length of a 20-penny nail is 1:4. How long is a 20-penny nail? **10 cm**

Answers will vary. Possible answer: $\frac{9}{27} = \frac{35}{105}$

33. Write a proportion using the numbers 9, 27, 35, and 105.

34. The track-team star runs 0.5 miles in 4 minutes. What is the star's rate in feet per minute? **660 ft/min**

35. The width and length of a rectangular plot are in the ratio $3:8$. The perimeter of the lot is 440 ft. What are the dimensions of the width and length? **w = 60 ft; l = 160 ft**

36. The cost of a tool shed and the cost of a lawn mower are in the ratio $5:2$. The total cost of the shed and the mower is $2,100. What is the cost of each? **Shed, $1,500; mower, $600**

8–4 Solving Proportions   **285**

**EXTEND/
SOLVE PROBLEMS**

**THINK CRITICALLY/
SOLVE PROBLEMS**

CHALLENGE

Find the value of x that will make each proportion true.

1. $x:2 = 18:x$ **x = 6** 2. $x:25 = 1:x$ **x = 5**

3. $x:54 = 6:x$ **x = 18** 4. $7:x = x:28$ **x = 14**

WARM-UP

Is each a proportion, yes *or* no*?*
1. 5 : 4 = 24 : 20 **no**
2. 3 : 8 = 15 : 40 **yes**
3. 12 : 7 = 36 : 21 **yes**
4. 10 : 3 = 20 : 3 **no**

1 MOTIVATE

Introduction Emphasize the need to interpret carefully the information given in the problem. Use the concept of cross-products to write an equation to find the missing number.

2 TEACH

Use the Pages/Problem Have students read this part of the section and then discuss the problem. Students should see that this problem involves writing a proportion using the *part : whole* ratio. Ask students to identify the ratio for *part : whole*. **The part is 3 and the whole is 20**.

8-5 Problem Solving Skills:

USING PROPORTIONS TO SOLVE PROBLEMS

► READ
► PLAN
► SOLVE
► ANSWER
► CHECK

Before you begin to solve a problem, you need to decide which mathematical ideas or skills will help you find the solution. How would you solve this problem?

PROBLEM

Sue's hobby is making ceramic masks. Out of every 20 masks she makes, 3 are clown masks. If Sue makes 240 masks, how many of them will be clown masks?

SOLUTION

You know that 3 out of every 20 masks are clown masks. This tells you that the ratio of clown masks to all the masks is 3:20.

If Sue makes 240 masks, the ratio of clown masks to all masks will still be 3:20. So, you can solve the problem by finding a ratio equivalent to 3:20 in which the second term is 240.

To find an equivalent ratio, you can write and solve a proportion. Remember that the two sides of the proportion must compare the quantities in the same order. Let a variable such as x represent the number of clown masks.

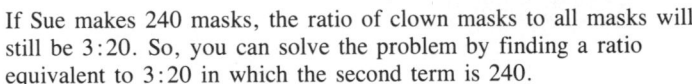

$$\text{number of clown masks} \rightarrow \frac{x}{240} = \frac{3}{20} \leftarrow \text{number of clown masks} \atop \text{total number of masks}$$

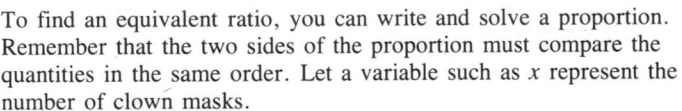

$$\frac{3}{20} = \frac{3 \times 12}{20 \times 12} = \frac{36}{240} \leftarrow \text{Write a ratio equivalent to } \frac{3}{20}.$$

$$x = 36 \leftarrow \text{number of clown masks in 240}$$

Notice that an alternative way to solve the proportion would be to use cross-products. Therefore, whenever Sue makes 240 masks, 36 of them will be clown masks.

The words *ratio* and *proportion* did not appear in the problem, yet you found that you could use a proportion to solve it.

286   CHAPTER 8   Exploring Ratio and Proportion

CHALLENGE

A profit of $2,400 is divided among three people in the ratio 8 : 5 : 2. How much profit does each person receive? **$1,280; $800; $320**

PROBLEMS

1. Evan's hobby is carving wooden statuettes. Out of every 25 statuettes he makes, 8 of them are birds. If Evan carves 200 statuettes, how many of them will be birds? **64 statuettes**

2. Cindy makes model airplanes. She paints 8 out of every 20 models she makes. If Cindy makes 65 model planes, how many of them will she paint? **26 model airplanes**

3. Marta uses color film for 48 out of every 72 photographs she takes. If she takes 156 photographs, how many will be on color film? **104 photographs**

MENTAL MATH TIP

You may be able to solve certain problems mentally, rather than by writing a proportion. For Problem 1, ask yourself how many groups of 25 statuettes there are in 200. Then how many groups of 8 birds would there be?

4. The students in the Art Club used 3 L of paint for 36 pictures. How many liters of paint would they need for 60 pictures? **5 L**

5. Out of every 30 students, 23 belong to an after-school club. How many students out of every 120 students belong to an after-school club? **92 students**

6. Ben makes ceramic pots. He decorates 11 out of every 15 pots he makes and leaves the rest undecorated. If Ben makes 60 pots, how many of them will be undecorated? **16 pots**

7. The Crafts Club used 4 lb of clay to make 20 sculptures. If 5 lb more clay had been used, how many sculptures could have been made? **45 sculptures**

Use the Pages/Solution Have students check the solution by using cross-products and then comparing their results.

3 SUMMARIZE

Talk About It Have students discuss the advantages of solving problems using equivalent ratios and using cross-products to write and solve an equation.

4 PRACTICE

Students should note words and phrases that signal the operations needed to solve problems. For Exercise 7 point out that the word *more* indicates addition; therefore they should add 5 lb to the 4 lb to obtain the numerator of the ratio needed to solve the problem.

5 FOLLOW-UP

Extra Practice The aerobics club exercises 45 minutes a day. How many hours will the club spend exercising in the next 12 days? **9 hours**

Get Ready inch graph paper, centimeter graph paper

WARM-UP

Solve each proportion.
1. $x:5 = 9:45$ $x = 1$
2. $12:21 = x:7$ $x = 4$
3. $7:x = 42:54$ $x = 9$
4. $4:36 = 2:x$ $x = 18$

1 MOTIVATE

Explore/Working Together Elicit from students that this procedure demonstrates that each partner's first design was not lost just because the size of the shape was enlarged. Point out that large images can be reduced in size by reproducing the part of the design that 4 small squares cover on 1 small square of graph paper.

2 TEACH

Use the Pages/Skills Development Have students read this part of the section and then discuss the examples. Have them compare the methods of finding the actual and drawing lengths in scale drawings. Students should conclude that the methods of finding the missing length (actual or scale) are the same.

8-6   Scale Drawings

EXPLORE/ WORKING TOGETHER

Work with a partner. Each partner should create a simple design on a section of graph paper. (Be sure that your design is drawn only on the grid lines of the graph paper.) Exchange designs with your partner. Copy your partner's design, but make the following change. For every small square that your partner's design covers, your design should cover 4 small squares that are arranged in the shape of a large square. Discuss the results with your group.

SKILLS DEVELOPMENT

A **scale drawing** is a drawing that represents a real object. All lengths in the drawing are proportional to the actual lengths in the object. The ratio of the size of the drawing to the size of the actual object is called the **scale** of the drawing.

To find the actual length or the drawing length, set up and solve a proportion.

Example 1

The scale of this drawing is 1 in. : 2 ft. Find the actual length of the mural.

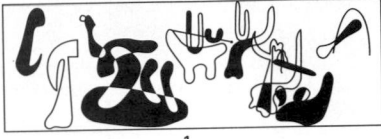

$2\frac{1}{2}$ in.

Solution
Write and solve a proportion. Let n represent the actual length.

$$\text{drawing length (inches)} \rightarrow \frac{1}{2} = \frac{2\frac{1}{2}}{n} \leftarrow \text{actual length (feet)}$$
$$1 \times n = 2 \times 2\frac{1}{2} \quad \leftarrow \text{Write the cross-products.}$$
$$n = 5$$

The actual length of the mural is 5 ft. ◀

Example 2

The scale of a drawing is 1 in. : 3 ft. Find the drawing length that would be used to represent an actual length of 12 ft.

Solution
Write and solve a proportion.

Let n represent the length in the drawing.

$$\text{drawing length (inches)} \rightarrow \frac{1}{3} = \frac{n}{12} \leftarrow \text{actual length (feet)}$$
$$1 \times 12 = 3n \quad \leftarrow \text{Write the cross-products.}$$
$$\frac{12}{3} = \frac{3n}{3}$$
$$4 = n$$

The drawing length would be 4 in. ◀

MAKING CONNECTIONS

Have students find information about what architectural blueprints look like, what scales are used to make them, and why scale drawings are an important part of an architect's work.

Example 3

The length of the bedroom in this scale drawing is 3 in. The actual length is 15 ft.
a. What is the scale of the drawing?
b. What is the actual width of the bedroom?

Solution

a. Use the information given to write a ratio. Then simplify the ratio.

drawing length (inches) → $\dfrac{3}{15} = \dfrac{3 \div 3}{15 \div 3} = \dfrac{1}{5}$
actual length (feet) →

The scale of the drawing is 1 in. : 5 ft.

b. Use the scale to write and solve a proportion. Let w represent the actual width of the bedroom.

drawing length (inches) → $\dfrac{1}{5} = \dfrac{1\frac{2}{3}}{w}$
actual length (feet) →

$1 \times w = 5 \times 1\frac{2}{3}$ ← Write the cross-products.

$w = 8\frac{1}{3}$ The actual width of the bedroom is $8\frac{1}{3}$ ft. ◄

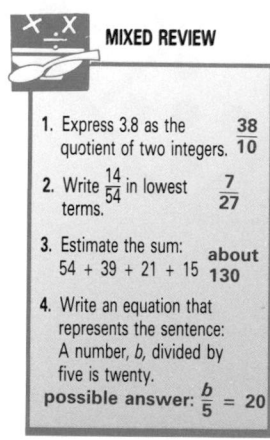

TRY THESE

Find the actual length for each drawing length given.

1. Drawing length: 8 in.
   Scale: 1 in. : 2 ft **16 ft**

2. Drawing length: 4 cm
   Scale: 1 cm : 0.5 m **2 m**

3. Drawing length: 12 in.
   Scale: 1 in. : 6 yd **72 yd**

4. Drawing length: 5 in.
   Scale: 1 in. : 7 mi **35 mi**

Find the drawing length for each actual length given.

5. 35 ft
   Scale: 1 in. : 7 ft **5 in.**

6. 32 mi
   Scale: 1 in. : 8 mi **4 in.**

7. 4.5 km
   Scale: 1 cm : 0.5 km **9 cm**

8. 30 yd
   Scale: 1 in. : $2\frac{1}{2}$ yd **12 in.**

9. Jane made a scale drawing of her living room. The length of the living room in her drawing is 4 in. The actual length is 20 ft. What scale did Jane use for her drawing? **1 in. : 5 ft**

EXERCISES

Find the actual or drawing length.

1. Scale: 1 cm : 6 m
   Drawing length: 3 cm
   Actual length: ? **18 m**

2. Scale: 1 cm : 12 km
   Drawing length: ?
   Actual length: 120 km **10 cm**

3. Scale: 1 in. : 22 ft
   Drawing length: ?
   Actual length: 330 ft **15 in.**

4. Scale: 1 in. : 7 mi
   Drawing length: 9 in.
   Actual length: ? **63 mi**

**PRACTICE/
SOLVE PROBLEMS**

MIXED REVIEW

1. Express 3.8 as the quotient of two integers. $\dfrac{38}{10}$

2. Write $\dfrac{14}{54}$ in lowest terms. $\dfrac{7}{27}$

3. Estimate the sum:
   54 + 39 + 21 + 15 **about 130**

4. Write an equation that represents the sentence: A number, b, divided by five is twenty.
   possible answer: $\dfrac{b}{5} = 20$

ASSIGNMENTS

BASIC
1–7, 10–11, 17, PSA 1–6, 12–13

AVERAGE
1–8, 9–14, 16–17, PSA 1–10, 12–13

ENRICHED
5–8, 9–15, 16–17, PSA 4–13

ADDITIONAL RESOURCES
Reteaching 8–6
Enrichment 8–6

Example 1: Point out that the scale: 1 in. : 2 ft is used as a ratio to find the actual length of n. Point out that the order of inches to feet in one ratio must be maintained in the second ratio.

Example 2: Challenge students to find the drawing length by using inches as unit/measure for the actual length, instead of feet. Tell students to use proportions to change units. 12 in. : 1 ft = x : 3 ft; 12 in. : 1 ft = x : 12 ft. Elicit from students that 3 ft is equal to 36 in. and 12 ft is equal to 144 in. Have students compare their results with the example.

Example 3: Point out that, to determine the scale of a drawing, one actual length and the corresponding scale length must be given.

Additional Questions/Examples
Find the actual or drawing length for the scale given.
1. scale length: 7 in.
   scale: 1 in. = 12 ft
   actual length : ■ **84 ft**
2. scale length: 11 in.
   scale: 1 in. = 48 mi
   actual length: ■ **528 mi**

3. actual length: 112 cm
scale: 2 mm = 28 cm
scale length: ■ **8 mm**

4. actual length: 45 yd
scale: 4 in. = 15 yd
scale length: ■ **12 in.**

Guided Practice/Try These
Have students work in small groups. Point out that all group members should come to an agreement on each answer.

③ SUMMARIZE

Write About Math Have students explain in their math journals what a scale is and how it is used to find actual dimensions.

Making Connections A garden in the shape of a regular pentagon (5 equal sides) has sides that each measure 3 in. on a scale drawing. Find the perimeter of the garden if the scale is 1 in. : 2 ft. **30 ft**

④ PRACTICE

Practice/Solve Problems Remind students that the ratio of the units of the scale is the basis for finding either the drawing or actual dimensions.

Extend/Solve Problems For Exercise 14 point out that students should first use the given scale to determine the actual dimensions of the bedroom and then compare the dimensions of the walls with those of the bed.

Think Critically/Solve Problems For Exercises 16–17 have students present their scale drawings to the class and discuss how they determined the scale for each.

Find the actual or drawing length.

5. Scale: 1 cm:5 m
Drawing length: ?
Actual length: 55 m   **11 cm**

6. Scale: 1 cm:8.5 m
Drawing length: 6 cm
Actual length: ?   **51 cm**

7. A scale model of a square-based statue is constructed using the scale of 1 cm:0.1 m. If the length of the base of the model is 14 cm, what is the length of the base of the statue?   **1.4 m**

8. If the actual height of the statue in Exercise 7 is 3.5 m, find the height of the scale model.   **35 cm**

EXTEND/ SOLVE PROBLEMS

9. A movie company made a large model of a domino. The actual width of the domino is 15 mm. If the width on the scale model is 6 m, what is the scale of the model?   **1 m:2.5 mm**

Actual and drawing lengths are given in the table below. Express the scale in the form 1 cm:__?__ km or 1 in.:__?__ mi.

	ACTUAL LENGTH	DRAWING LENGTH	
10.	8 km	2 cm	1 cm:4 km
11.	50 mi	5 in.	1 in.:10 mi
12.	100 km	25 cm	1 cm:4 km
13.	$416\frac{1}{2}$ mi	$8\frac{1}{2}$ in.	1 in.:49 mi

14. The scale 1 in.:3 ft was used to make a scale drawing of Anne's bedroom. In the drawing, the bedroom is 5 in. long and 2 in. wide. Could a bed 72 in. long and 42 in. wide fit along one of the walls?   **Y**

15. The actual dimensions of a deck are 10 ft by 15 ft. If the dimensions of a scale drawing of the deck are 2 in. by 3 in., what is the scale of the drawing?   **1 in.:5 ft**

THINK CRITICALLY/ SOLVE PROBLEMS

16. Make a scale drawing of a room in your home. Measure the length and width of the room to the nearest inch. Choose a scale for your drawing. Make and cut out a scale model for each piece of furniture. Position the cutouts on the scale drawing. Be sure to write the scale on your drawing.   **Check students' drawings.**

17. Make a scale drawing of your classroom. Measure the room to the nearest inch. Decide on a scale to use. Show where windows, desks, and tables are located. Be sure to include the scale.
Check students' drawings.

► READ
► PLAN
► SOLVE
► ANSWER
► CHECK

Problem Solving Applications:

MAPS

Cartography is the art of representing a geographical area graphically, usually on a flat surface such as a map. It is an ancient discipline that dates from the prehistoric depiction of hunting and fishing territories.

The discovery of the New World led to the need for new techniques in map making, and by the seventeenth and eighteenth centuries printed maps had increased in accuracy. Today, cartography involves the use of aerial photographs as a base for some maps.

 Use a customary ruler to measure each map distance. Then use the scale 1 in. : 15 mi to find the actual distance.

1. school to library   **3.75 mi**

2. town pool to post office   **15 mi**

3. fire station to general store   **22.5 mi**

4. hospital to school $11\frac{1}{4}$ **mi**

5. Which two locations are farthest apart?   **school and post office**

6. Which two locations are closest together?   **school and the library**

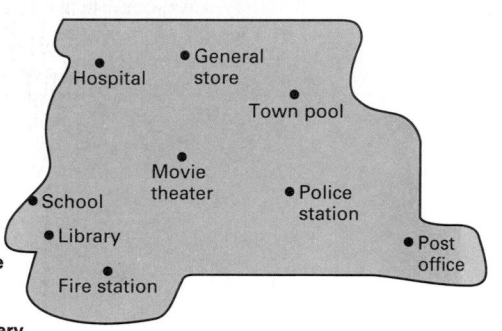

Copy and complete the table.

	Scale	Distance on a Map	Actual Distance
7.	1 cm represents 10 km	22 cm	? **220 km**
8.	1 : 1500	? **10 m**	15,000 m
9.	1 in. : 200 mi	? $1\frac{1}{4}$ **in.**	250 mi
10.	1 cm : 0.4 km	20 cm	? **8 km**

11. On a map, 1 cm represents 50 km. If the actual distance between two cities is 230 km, what is the distance on the map?   **4.6 cm**

12. Create a map of your neighborhood. Choose an appropriate scale and include your favorite places to visit.   **Check students' drawings.**

13. How is a map like a scale drawing?   **They both are drawings made so actual dimensions can be determined by using a proportion.**

Problem Solving Applications
You may wish to have students work together in small groups.

5 FOLLOW-UP

Extra Practice   Find the actual or scale length.
Use the scale 1 in. : 12 ft.
1. actual: 180 ft, scale: ■ **15 in.**
2. scale: 7 in., actual: ■ **84 ft**
3. actual: 144 ft, scale: ■ **12 in.**
4. scale: 2 ft, actual: ■ **288 ft**
5. actual: 1,296 in., scale: ■ **9 in.**
6. The actual length of a porch deck is 6 ft. The length of the porch in a scale drawing is 2 in. What is the scale used for the drawing?   **1 in. : 3 ft**

Extension   Ask students to explain how they can determine if an object is actually smaller or larger than the scale drawing. **If the terms of the ratio have the same unit of measure and the number of units in the first term is larger than in the second, the object is smaller. If the first term is smaller than the second term, the object is larger.**

Section Quiz   Find the drawing length, using the scale 1 cm : 8 m for each.
1. 24 m   **3 cm**
2. 120 m   **15 cm**
3. 176 m   **22 cm**
4. 528 m   **66 cm**
Find the actual length, using the scale 1 in. : 32 mi for each.
5. 5 in.   **160 mi**
6. 3 ft   **1,152 mi**
7. 11 in.   **352 mi**
8. 2 2/3 ft   **1,024 mi**
9. In a scale drawing, a living room measures 4 in. × 7 in. What are the actual dimensions of the room if the scale is 2 in.:5 ft?   **10 ft × 17 1/2 ft**

SPOTLIGHT

OBJECTIVE
- To solve problems by making an organized list

MATERIALS NEEDED
calculators

VOCABULARY
organized list

WARM-UP

Give the LCM for each.
1. 5, 8 **40**
2. 4, 16 **16**
3. 10, 25 **50**
4. 4, 8, 12 **24**

1 MOTIVATE

Introduction This lesson is designed to use the strategy of making an organized list. However, students may wish to use other strategies, such as finding a pattern or using logical reasoning.

2 TEACH

Use the Pages/Problem Have students read this part of the section and then discuss the problem. Be sure that students understand that each condition restricts the way in which the possible frames can be made.

Use the Pages/Solution Stress the importance of deciding which condition is the best to use to begin investigating. Point out that, although students begin the table by investigating condition d, as they complete the table, they

292

► READ
► PLAN
► SOLVE
► ANSWER
► CHECK

Often a problem calls for finding the number of ways in which something can be done. You may decide to list all the possible ways and then count them. Depending on how you choose to organize your list, though, the process could be either long and confusing or clear and systematic.

PROBLEM

Ramón makes triangular frames for collages. He has decided that the frames he makes this month will all satisfy these conditions:

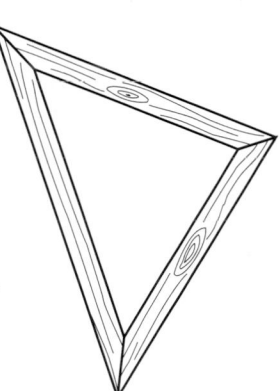

a. The length of each side is a whole number of centimeters.
b. The total length of the three sides will be 100 cm.
c. No side will be more than 50 cm long.
d. Two of the sides will have lengths in the ratio of 4:3.

List all the different frame sizes Ramón can use. How many such sizes are there?

SOLUTION

You could try every possible grouping of three numbers that add up to 100 and then check whether each grouping satisfies all of Ramón's conditions. However, this would take a very long time and you would probably overlook some groupings. Instead, you can organize your list of frame sizes in a way that saves you from considering most of the impossible groupings.

From conditions **a** and **d** you can reason that one side of any frame must have a length that is a multiple of 4. Since no side can be longer than 50 cm, begin by listing a frame with a side 48 cm long.

Lengths of sides in ratio of 4:3		Length of third side
48 cm	36 cm	16 cm
44 cm	33 cm	23 cm
40 cm	30 cm	30 cm
36 cm	27 cm	37 cm
32 cm	24 cm	44 cm
28 cm	21 cm	51 cm

$\frac{48}{36} = \frac{4}{3}$;

$100 - (48 + 36) = 16$

This frame would not satisfy condition c.

CHALLENGE

James invited Robert, Andy, Steve, Vinny, Paul, and Anthony to his house for a chess tournament. How many different combinations of two players could they have? **21**

If you continue listing frames whose first two sides are shorter than 28 cm, the third side will always violate condition **c.** Thus, by making your list in a logical and organized way, you have quickly found the only frame sizes Ramon could make.

PROBLEMS

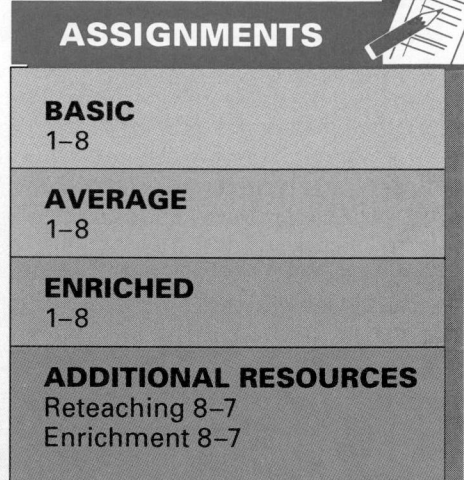

1. Suppose Ramón changed only condition **d** to have two sides of the frame in the ratio of 3:2. Complete the following chart to find how many different frames he could then make.   **7 frames**

Lengths of sides in ratio of 3:2		Length of third side	
■ cm 48	32 cm	■ cm	20
■ cm 45	■ cm 30	25 cm	
42 cm	■ cm 28	■ cm	30
■ cm 39	■ cm 26	■ cm	35
■ cm 36	■ cm 24	■ cm	40
■ cm 33	■ cm 22	■ cm	45
■ cm 30	■ cm 20	■ cm	50

Make an organized list to find all the possible answers for each problem.

2. How many different 4-digit numbers can be made with the digits 2, 3, 4, and 5? No digit may be used more than once in any number.   **24 numbers**

3. In how many different ways can you make 50 cents using nickels, dimes, or nickels and dimes?   **6 ways**

4. In how many different ways can you make a one-dollar-and-fifteen-cent bus fare with nickels, dimes, and at least two quarters?   **14 ways**

5. How many different 4-digit odd numbers can be made with the digits 5, 6, 7, and 0? No digit may be used more than once in any number.   **8 odd numbers**

6. How many 4-digit even numbers can be made with the digits 4, 5, 6, and 8? No digit may be used more than once in any number.   **12 even numbers**

7. In a student-council election, there are 3 candidates for president: Mack, Billie, and Sue; 3 candidates for vice-president: Marge, Lu, and Dana; and 2 candidates for secretary-treasurer: Steve and Mel. In how many different combinations could these candidates be elected?   **18 ways**

8. How many different 3-digit numbers can be made from the digits 6, 7, 8, and 9? No digit may be used more than once in any number.   **24 ways**

COMPUTER

This program will let you determine how many numbers can be made from a given number of digits when no digits repeat. Can you use the program to figure out how you could do it on a calculator?

```
10 LET P = 1
20 INPUT "HOW MANY
   DIGITS? ";N: PRINT
30 FOR X = N TO 1 STEP
   – 1
40 LET P = P * X
50 NEXT X
60 PRINT "THERE ARE ";P;"
   NUMBERS."
```

See Additional Answers.

are taking account of conditions b and c at the same time.

3 SUMMARIZE

Write About Math Have students explain in their math journals how an organized list may be helpful in solving problems.

4 PRACTICE

Have students explain how they organized the given information to help solve the problems.

5 FOLLOW-UP

Extra Practice Make an organized list to answer each problem.

1. How many different signals can be made using four different colored flags hanging one below the other on a flagpole? Each color may be used only once.   **24**

2. How many different 5-digit numbers can you write using the digits 1, 2, 3, 4, and 5? No digit should be used more than once in any number.   **120**

Additional Answers
See page 580.

8 CHAPTER REVIEW

Introduction The Chapter Review emphasizes the major concepts, skills, and vocabulary presented in this chapter and can be used for diagnosing students' strengths and weaknesses. Page references direct students back to appropriate sections for additional review and reteaching.

Using Pages 294–295 Allow students to quickly scan the Chapter Review and ask questions about any section they may still find confusing. Exercises 1–5 review key vocabulary. For Exercises 7–10 encourage the use of division as well as multiplication.

Informal Evaluation Have students explain how they arrived at any incorrect answers. They will probably find their own mistakes or give you some clues as to the nature of their errors. Make sure students understand this material before administering the Chapter Test.

Follow-Up A copying machine can make a copy of a page in a reduced size. If a photograph that measures 8 in. x 10 in. is copied at 13/20 of the original size, what space will the copy of the photograph occupy? **5.2 in. × 6.5 in.**

1. A statement that two ratios are equal is called a ___?___ . **c**

2. A ___?___ is a ratio of two quantities with different units of measure. **b**

3. A drawing of an object in which the lengths in the drawing are proportional to the actual lengths is called a ___?___ drawing. **e**

4. A quotient of two numbers that is used to compare the numbers is called a ___?___ . **a**

5. The cost per unit of a given item is the ___?___ . **d**

a. ratio
b. rate
c. proportion
d. unit price
e. scale

SECTION 8–1   RATIO   (pages 270–273)

▶ A ratio compares two numbers by division.
▶ A ratio can be written in three ways:

$$1:2 \qquad 1 \text{ to } 2 \qquad \frac{1}{2}$$

Write each ratio in two other ways.

6. $9:10$ $\frac{9}{10}$; 9 to 10

7. $\frac{1}{3}$ 1 to 3; 1:3

8. 4 to 5 $\frac{4}{5}$; 4:5

SECTION 8–2   RATES   (pages 274–277)

▶ A rate is a ratio that is used to compare two different kinds of quantities.
▶ A unit rate is the ratio of a number to 1.

Find the unit rate.

9. $\frac{\$150}{5 \text{ h}}$ $30/h

10. $\frac{48 \text{ revolutions}}{6 \text{ s}}$ 8 revolutions/s

11. $\frac{112.5 \text{ m}}{7.5 \text{ s}}$ 15 m/s

SECTION 8–3   EQUIVALENT RATIOS   (pages 278–281)

You can find equivalent ratios in either of these ways.

▶ Multiply both terms of the ratio by the same nonzero number.
▶ Divide both terms of the ratio by the same nonzero number.

Write two equivalent ratios for each given ratio. **Possible answers are given.**

12. $3:9$ 1:3; 6:18

13. $20:200$ 1:10; 2:20

14. $\frac{5}{50}$ $\frac{1}{10}$; $\frac{10}{100}$

15. 16 to 20 4 to 5; 32 to 40

SECTION 8–4 SOLVING PROPORTIONS (pages 282–285)

▶ A proportion is an equation that states that two ratios are equivalent.
▶ To solve a proportion, use the cross-products to write an equation.

Solve each proportion.

16. $\frac{6}{4} = \frac{y}{1.4}$ *y = 2.1*

17. $\frac{16}{8} = \frac{x}{3}$ *x = 6*

18. $\frac{x}{15} = \frac{2.2}{3.3}$ *x = 10*

SECTION 8–5 PROBLEM SOLVING SKILLS (pages 286–287)

▶ You can solve problems by writing and solving a proportion.

19. Stan collects postage stamps. For every 3 United States stamps in his collection, he has 7 foreign stamps. If Stan has 420 foreign stamps, how many United States stamps does he have? **180 stamps**

SECTION 8–6 SCALE DRAWINGS (pages 288–291)

▶ A scale drawing represents a real object. All lengths in the drawing are proportional to the actual lengths in the object.

Find the actual or scale length.

20. Scale: ? **1 in.:20 mi**
Drawing length: 4 in.
Actual length: 80 mi

21. Scale: 1 cm:16 m
Drawing length: 21 cm
Actual length: ? **336 m**

22. Scale: 1 cm:4 km
Drawing length: 6 mm
Actual length: ? **24 km**

SECTION 8–7 PROBLEM SOLVING STRATEGIES (pages 292–293)

▶ An organized list helps you solve problems by showing information in a way that accounts for all possibilities and avoids repetition.

23. In how many different ways can you make $1.00 using quarters and nickels or nickels and at least five dimes? **10 ways**

24. *USING DATA* Refer to the data index on page 546 to find the chart for baseball at-bats. Write a ratio of Gallego's hits to his at-bats.
80:389; .206

Review **295**

8 CHAPTER TEST

Introduction The Chapter Test uses a variety of questioning techniques to assess students' mastery of the major objectives of Chapter 8. If you prefer, you may use the Skills Preview (page 267) as an alternative form of the Chapter Test. The items on this test and the Skills Preview correspond in content and level of difficulty.

Alternative Assessment
Critical Thinking Use local newspaper or magazine advertisements to find information and set up ratios to compare different sizes of the same product. Is the claim, "The larger the quantity, the lower the price" always true? Explain and give examples to justify your explanations.

Write each ratio in two other ways.

1. 5 to 4 $\frac{5}{4}$; **5:4**
2. $\frac{6}{7}$ **6 to 7; 6:7**
3. 7:11 $\frac{7}{11}$; **7 to 11**
4. 8 to 10 **8:10;** $\frac{8}{10}$
5. $\frac{1}{3}$ **1 to 3; 1:3**
6. 2:3 $\frac{2}{3}$; **2 to 3**
7. 9:7 $\frac{9}{7}$; **9 to 7**
8. $\frac{5}{2}$ **5 to 2; 5:2**
9. $\frac{4}{7}$ **4 to 7; 4:7**
10. 8:4 **8 to 4;** $\frac{8}{4}$
11. 7 to 21 **7:21;** $\frac{7}{21}$
12. 4:28 **4 to 28;** $\frac{4}{28}$

Find the unit rate.

13. $72 to 8 h **$9/h**
14. 636 mi:12 h **53 mi/h**
15. 342 m to 9.5 s **36 m/s**
16. 220 words:5 min **44 words/min**
17. $\frac{252 \text{ revolutions}}{6 \text{ min}}$ **42 revolutions/min**
18. 120 students to 4 teachers **30 students/teacher**
19. 72 boxes:6 cases **12 boxes/case**
20. 228 m/12 h **19 m/h**
21. 852 mi to 12 h **71 mi/h**

Write three ratios equivalent to each given ratio.

22. $\frac{1}{6}$ **$\frac{2}{12}$; $\frac{4}{24}$; $\frac{3}{18}$**
23. 2:10 **4:20; 6:30; 8:40**
24. $\frac{100}{200}$ **$\frac{200}{400}$; $\frac{300}{600}$; $\frac{400}{800}$**
25. $\frac{1}{8}$ **$\frac{2}{16}$; $\frac{3}{24}$; $\frac{4}{32}$**
26. 16:24 **2:3; 4:6; 8:12**
27. $\frac{2}{5}$ **$\frac{4}{10}$; $\frac{6}{15}$; $\frac{8}{20}$**
28. 11 to 22 **1 to 2; 22 to 44; 33 to 66**
29. 18:90 **1:5; 2:10; 3:15**
30. $\frac{30}{70}$ **$\frac{60}{140}$; $\frac{90}{210}$; $\frac{120}{280}$**

Solve each proportion.

31. $\frac{22}{10} = \frac{x}{5}$ **x = 11**
32. $\frac{x}{25} = \frac{4}{10}$ **x = 10**
33. $\frac{1.5}{x} = \frac{10}{20}$ **x = 3**
34. $\frac{28}{32} = \frac{56}{x}$ **x = 64**
35. $x:1.5 = 8:3$ **x = 4**
36. $\frac{15}{21} = \frac{40}{x}$ **x = 56**
37. $\frac{52}{39} = \frac{x}{3}$ **x = 4**
38. $\frac{5}{8} = \frac{x}{56}$ **x = 35**
39. $\frac{x}{2.8} = \frac{5}{20}$ **x = 0.7**

40. The length of a bedroom floor in a scale drawing is 6 in. The actual length of the floor is 15 ft. What is the scale of the floor plan? **1 in.:2$\frac{1}{2}$ ft**

41. The actual height of a door on a train is 6 ft. If the scale of a model train is 1 in.:4 ft, what is the height of the door in the model? **1$\frac{1}{2}$ in.**

42. The scale on a map is 1 in.:50 mi. If the map distance from Morton to Curryville is 3.5 in., what is the actual distance between these two towns? **175 mi**

43. On a scale drawing, the width of a corner lot is 8 in. The actual width of the lot is 200 ft. What is the scale of the drawing? **1 in.:25 ft**

The following data show the temperature readings, in degrees Celsius, at ten locations inside three kilns.

Kiln A: 240 250 230 260 250 250 220
240 250 210

Kiln B: 260 250 240 270 250 250 230
260 250 240

Kiln C: 220 240 280 275 260 210 255
240 230 285

1. Find the mean, median, mode, and range for each. **Kiln A: 240, 245, 250, 50**
Kiln B: 250, 250, 250, 40
Kiln C: 249.5, 247.5, 240, 75

Write in standard form.

2. 356×10^4 **35,600**   3. 6.324×10^5
6,324,000

4. 82.64×10^5 5. 2.0001×10^4
8,264,000   **20,001**

Write each answer in exponential form.

6. $4^5 \times 4^2$ **4^7**   7. $b^{14} \div b^7$ **b^7**

8. $25^{10} \div 25^2$ **25^8**   9. $p^6 \times p^3$ **p^9**

10. Sonia, Theo, Jay, and Kim are not in the same first period class. English, math, French, and computer lab are taught in first period. Theo has English first period. Sonia does not have French. Jay and Sonia have computer lab during second period. Which class is each student in during first period?
See Additional Answers.

11. Determine whether the following argument is *valid* or *invalid*.

If the water is too cold, then you will not be able to swim.
The water is too cold.

Therefore, you will not be able to swim.
valid

Replace ● with <, >, or =.

12. 0.47 ● $\frac{7}{13}$ **<**   13. $\frac{5}{9}$ ● 0.54 **>**

14. 0.6 ● $\frac{3}{7}$ **>**   15. $\frac{1}{8}$ ● 0.125 **=**

Complete.

16. 0.6 m = ■ mm **600** 17. 62 g = ■ kg
0.062

18. 64 in. = ■ ft ■ in. **5, 4**

19. 100 oz = ■ lb ■ oz **6, 4**

20. Find the perimeter.

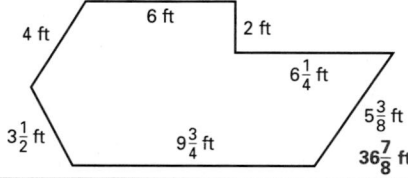

Find each answer.

21. $-4 + (-21)$ **−25** 22. $9 - (-13)$ **22**

23. $5 \times (-8)$ **−40** 24. $27 \div (-3)$ **−9**

Solve each inequality. Graph the solution set on a number line. **See Additional Answers.**

25. $-6 + x \geq 2$ 26. $-e < 5$

27. $t + 2 \leq -1$ 28. $x - 2 < -6$

Write each ratio in two other ways.

29. $16:20$ **16 to 20; $\frac{16}{20}$** 30. 2 to 5 **$\frac{2}{5}$; 2:5**

Solve each proportion.

31. $\frac{3}{6} = \frac{9}{x}$ **$x = 18$** 32. $5:8 = x:24$
$x = 15$

33. $\frac{5}{x} = \frac{30}{42}$ **$x = 7$** 34. $\frac{5}{6} = \frac{x}{48}$ **$x = 40$**

Are the ratios equivalent? Write *yes* or *no*.

35. $10:15, 12:18$ **yes** 36. $\frac{15}{10}, \frac{10}{15}$ **no**

Introduction The purpose of this Cumulative Review is to maintain previously taught skills and concepts and to apply them to the material presented in this chapter. At least one major objective of each chapter is included in the review.

Item Analysis The table below correlates the Cumulative Review items with the chapter and section that are being reviewed.

Section	Items
1–9	1
2–6	2–9
3–3	11
3–5	10
4–3	12–15
5–1	18–19
5–2	16–17
5–5	20
6–2	21
6–3	22
6–4	23
6–5	24
7–9	25–28
8–1	29–30
8–3	35–36
8–4	31–34

8 CUMULATIVE TEST

Introduction The Cumulative Test uses a standardized-test format of multiple-choice questions to assess retention of previously learned concepts and test-taking skills. Test results may be used to diagnose students' strengths and weaknesses.

Item Analysis The table below correlates the Cumulative Test items with the chapter and section that are being tested.

Section	Items
1–9	1
3–3	2
4–4	3
5–5	5
5–6	4
6–2	6
6–3	7
6–4	8
6–5	9
7–2	11
7–6	12
7–9	10
8–2	16
8–3	14–15
8–4	13

1. If you had 250 total points for three tests, what score would you need on the next test to receive an 85 average?

   A. 50   B. 95   C. 130   (D) 90

2. Which conclusion can be drawn from these statements?

   If the plant is a violet, its flowers are small.
   The flowers are small.

   A. the plant is a violet.
   B. The plant is not a violet.
   C. The plant is sick.
   (D) none of these

3. Express 3^{-5} as a fraction.

   (A) $\frac{1}{243}$ B. $\frac{1}{125}$ C. $\frac{3}{3,125}$ D. $\frac{3}{243}$

4. Find the circumference. Round your answer to the nearest tenth.

   A. 75.6 m
   B. 37.7 m
   (C) 80.4 m
   D. 40.1 m

   12.8 m

5. Find the perimeter of a rectangular figure whose length is 7.2 km and whose width is 5.4 km.

   (A) 25.2 km   B. 12.6 km
   C. 18 km   D. 38.88 km

6. Add. $-16 + (-8)$

   (A) -24 B. -8
   C. -8 D. 128

7. Subtract. $24 - (-6)$

   A. -4 B. -18
   (C) 30   D. none of these

8. Multiply. $-6 \times (-2)$

   A. 3   B. -8 C. -4 (D) 12

9. Divide. $-15 \div 3$

   A. -12 (B) -5 C. -18 D. 45

10. Which graph shows the solution set?
    $-180 \le 60x$

    A.
    (B)
    C.

    D. none of these

11. Solve. $4 + r = 4$

    (A) $r = 0$ B. $r = -1$
    C. $r = 1$ D. none of these

12. Solve. $40 = 6x - 8$

    A. $x = \frac{1}{3}$ B. $x = 6\frac{2}{3}$
    (C) $x = 8$ D. none of these

13. What is the value of y in the proportion $\frac{32}{8} = \frac{16}{y}$?

    A. 3   B. 5   (C) 4   D. 6

14. Which of the following ratios is equivalent to $12:9$?

    A. $9:12$ B. $4:2$
    C. $6:9$ (D) $4:3$

15. Which of the following ratios is equivalent to $26:130$?

    A. $5:1$ (B) $1:5$
    C. $1:6$ D. none of these

16. What is the unit rate for $\frac{\$54}{9 \text{ h}}$?

    A. $\frac{\$54}{9}$ (B) $\frac{\$6}{1 \text{ h}}$
    C. $\frac{1}{6}$ D. none of these

Name _____ Date _____

Ratio and Proportion

A **ratio** is the quotient of two numbers. A ratio is used to compare one number to the other. Each number is called a **term** of the ratio. Ratios can be expressed in three ways.

► **Example 1** _____

Write a ratio to compare the number of vowels with the number of consonants in the word, *tumbler* in three ways.

Solution

Write a word expression for the ratio.

Vowels to Consonants

The number of vowels is 2; the number of consonants, 5. The ratio of 2 to 5 can be written in three ways.

2 to 5 or 2:5 or $\frac{2}{5}$

► **Example 2** _____

Write the ratio of 2 cups to 1 quart as a fraction in lowest terms.

Solution

To write a ratio that compares these quantities, each must be expressed in the same unit. Use cups.

2 cups = 2 cups
1 quart = 4 cups

The ratio is $\frac{2}{4}$, which in lowest terms is $\frac{1}{2}$.

EXERCISES _____

Write each ratio in two other ways.

1. $\frac{1}{5}$ **1 to 5** / **1:5**
2. $\frac{3}{7}$ **3 to 7** / **3:7**
3. 11 to 7 $\frac{11}{7}$; **11:7**
4. 8:11 $\frac{8}{11}$; **8 to 11**
5. 4:9 $\frac{4}{9}$; **4 to 9**
6. 3 to 1 $\frac{3}{1}$; **3:1**

Write each ratio as a fraction in lowest terms.

7. 5 m to 20 m $\frac{1}{4}$
8. 7 g to 3 g $\frac{7}{3}$
9. 10L:15L $\frac{2}{3}$
10. 4 km:6 km $\frac{2}{3}$
11. 8 kg to 36 kg $\frac{2}{9}$
12. 5 pt:40 pt $\frac{1}{8}$
13. 50 cm:1 m $\frac{1}{2}$
14. 250 mL:3L $\frac{1}{12}$
15. 750 g:2 kg $\frac{3}{8}$
16. 3 qt to 1 gal $\frac{3}{4}$
17. 75¢ to $3 $\frac{1}{4}$
18. 6 m:5 cm $\frac{120}{1}$

Reteaching • SECTION 8-1   141

Name _____ Date _____

Gears

A gear is a tooth-edged wheel that is used to transmit motion. In the diagram at the right, gear X is being turned in a clockwise direction. Since gear X has more teeth than gear Y, gear Y will turn around faster.

The ratio of the number of teeth in gear X to the number of teeth in gear Y is 24:12, or 2:1. Gear Y will turn around twice as fast as gear X, so the ratio of the speed of gear X to the speed of gear Y is 1:2.

EXERCISES _____

Use the diagram of gears A, B, C, D, and E for the exercises below.

Write a ratio in lowest terms comparing the number of teeth in each pair of gears.

1. A to C **4:1**
2. D to B **1:2**
3. E to C **6:1**
4. D to E **1:5**
5. A to E **2:3**
6. A to D **10:3**

Suppose each given pair of gears is connected. Write a ratio in lowest terms comparing the speed of each pair.

7. A to C **1:4**
8. D to B **2:1**
9. E to C **1:6**

10. In the diagram at the top of the page, in which direction will gear Y move, clockwise or counterclockwise? **counterclockwise**

11. Suppose that seven gears are intermeshed. If the first gear is set in motion in a clockwise direction, in which direction will the seventh gear move? **clockwise**

142   **Enrichment** • SECTION 8-1

Name _____ Date _____

Rates

A **rate** is a ratio that is used to compare two different kinds of quantities.

A **unit rate** is a rate that has a denominator of 1 unit.

25 miles per gallon and 40 words per minute are everyday examples of unit rates.

► **Example 1** _____

Find the unit rate: 35 gal in 7 s.

Solution

Write the rate as a fraction in lowest terms.
$\frac{35 \text{ gal}}{7 \text{ s}} = \frac{35 \div 7}{7 \div 7} = \frac{5}{1}$

The rate is 5 gal/s.

► **Example 2** _____

Find the unit rate: $430 for a 4-month membership.

Solution

Write as a fraction. Divide by $\frac{4}{4}$.
$\frac{\$430}{4 \text{ mo}} = \frac{430 \div 4}{4 \div 4} = \frac{107.5}{1}$

The unit rate is $107.50/month.

EXERCISES _____

Find the unit rate.

1. $\frac{540 \text{ words}}{5 \text{ min}}$ **108 words/min**
2. $\frac{285 \text{ miles}}{15 \text{ gal}}$ **19 mi/gal**
3. $\frac{60 \text{ pages}}{5 \text{ hours}}$ **12 p/h**
4. $\frac{\$85}{17 \text{ h}}$ **$5/h**
5. $\frac{90 \text{ ft}}{3 \text{ s}}$ **30 ft/s**
6. $\frac{12 \text{ mi}}{4 \text{ min}}$ **3 mi/min**
7. 35 km in 5 h **7 km/h**
8. $160 for 20 weeks **$8/wk**
9. 85 mi in 2 h **42.5 mi/h**
10. 450 words in 9 min **50 words/min**
11. 60 mi on 4 gal **15 mi/gal**
12. 40 mi in 4 h **10 mi/h**

Solve.

13. Polly assembled 56 dolls in 8 hours. How many dolls can she assemble in one hour? **7 dolls/h**

Reteaching • SECTION 8-2   143

Name _____ Date _____

Averaging Rates

Suppose that you drove a distance of 100 miles at an average rate of 45 miles per hour, but averaged 55 miles per hour driving the same distance back. What was your average rate for the whole round-trip distance?

Here is a formula you can use to find the overall average rate for two trips:
$A = \frac{2 \times R \times r}{R + r}$

In this formula A is the average rate, R represents the average rate over the first part of the journey, and r represents the average rate over the second part of the journey. To find the average rate, substitute values for R and r and then compute.
$A = \frac{2 \times 45 \times 55}{45 + 55}$
$= \frac{4,950}{100}$
$= 49.5$

So, the average rate over the entire distance of 200 miles was 49.5 mi/h.

Use the formula to solve.

1. Mark rode his bike from home to school at an average rate of 12 mi/h. His average on the way home was 6 mi/h. What was his average for the total trip?

8 mi/h

2. Mario walked the length of the park at a rate of 6 mi/h. He then walked back over the same route at a rate of 4 mi/h. What was his average rate for the total distance he walked?

4.8 mi/h

3. An airplane averaged 550 mi/h on a trip from Atlanta to San Francisco. On the return trip it averaged 625 mi/h. What was the average speed for the entire round trip?

585.1 mi/h

4. Jessica hiked from her campsite up a five-mile trail at a rate of 2 mi/h. She hiked back to the campsite at a rate of 3 mi/h. What was her average rate for the distance she hiked?

2.4 mi/h

5. Philia Moran drove to work this morning at a rate of 45 mi/h. On the trip back home this evening, she averaged only 30 mi/h.

36 mi/h

6. Lt. Sanders flew his usual helicopter route from Melville to the Corvale airport at an average rate of 100 mi/h. On the way back, he averaged only 88 mi/h. What was his overall average speed on the entire trip?

93.6 mi/h

144   **Enrichment** • SECTION 8-2

298A

Equivalent Ratios

Name _____ Date _____

When ratios represent the same comparison, they are called **equivalent ratios**. Finding equivalent ratios is similar to finding equivalent fractions.

▶ **Example 1**

Write three ratios equivalent to 4:15.

Solution

Write the ratio as a fraction. Then, multiply or divide the numerator and denominator by the same nonzero number.

$\frac{4}{15} = \frac{4 \times 2}{15 \times 2} = \frac{8}{30}$

$\frac{4}{15} = \frac{4 \times 3}{15 \times 3} = \frac{12}{45}$

$\frac{4}{15} = \frac{4 \times 4}{15 \times 4} = \frac{16}{60}$

So, the ratios 8:30, 12:45, and 16:60 are equivalent to 4:15.

▶ **Example 2**

Is this pair of ratios equivalent?
$\frac{12}{18} \quad \frac{6}{9}$

Solution

$\frac{12}{18} = \frac{12 \div 6}{18 \div 6} = \frac{2}{3}$

$\frac{6}{9} = \frac{6 \div 3}{9 \div 3} = \frac{2}{3}$

$\frac{2}{3} = \frac{2}{3}$

Yes, the ratios $\frac{12}{18}$ and $\frac{6}{9}$ are equivalent.

EXERCISES

Write three equivalent ratios for each of the following. **Answers may vary. Possible answers are given.**

1. 2:7   **4:14; 6:21**

2. 11:12   **22:24; 33:36**

3. 48:36   **8:6; 4:3; 24:18**

4. $\frac{5}{8}$ **$\frac{10}{16}$, $\frac{15}{24}$**

5. $\frac{16}{3}$ **$\frac{32}{6}$, $\frac{48}{9}$**

6. $\frac{9}{10}$ **$\frac{18}{20}$, $\frac{27}{30}$**

Are the ratios equivalent? Write *yes* or *no*.

7. $\frac{2}{5}, \frac{10}{25}$ **yes**

8. $\frac{3}{7}, \frac{4}{21}$ **no**

9. 10:12, 15:18   **yes**

10. $\frac{4}{12}, \frac{10}{30}$ **yes**

11. 5:14, 20:70   **no**

12. $\frac{16}{24}, \frac{4}{6}$ **yes**

Reteaching • SECTION 8-3   **145**

Batting Averages

Name _____ Date _____

When a ratio is expressed in numbers that have few common factors, it can be more convenient to express the ratio as a decimal. This is exactly what happens with batting averages in baseball.

Player	Hits	At Bats
George	27	111
Aaron	27	105
Enrique	31	109
Peter	20	62
Otis	34	115
José	32	123
Phil	24	98
Leon	37	135
Bob	30	93

▶ **Example**

Leon made 37 hits in 135 at bats. Find his batting average.

Solution

Write the ratio of hits to at bats.

$\frac{hits}{at \ bats} = \frac{37}{135}$

Use a calculator to change the fraction form to a decimal. Round the answer to three decimal places. $37 \div 135 = 0.274$

Leon's batting average is 0.274.

EXERCISES

Use the chart above to find the batting average for each of these players. Use a calculator.

1. George   **0.243**
2. Otis   **0.296**
3. Phil   **0.245**
4. Bob   **0.323**
5. Aaron   **0.257**
6. Enrique   **0.284**
7. Peter   **0.323**
8. José   **0.260**

Answer these questions.

9. Did the player with the most hits have the highest batting average? Explain.
   No, Leon had the most hits.

10. Which two players have the same batting average? Why?
    Bob and Peter; the ratios of hits to at bats for each is the same.

11. Otis' hits include 13 singles, 7 doubles and 11 home runs. What is the ratio of doubles and home runs to his total hits?
    18:31 or 0.581

12. Explain why batting average alone may not show who is the team's most valuable hitter.
    Answers may vary. Sample: A player may have a lower batting average but drive in more runs than other players.

13. Steve is a new player who has gotten 3 hits in his first 5 at bats. Find his batting average.
    0.600

14. Is Steve the best hitter on the team? Explain your answer.
    Answers may vary. Steve has not played long.

146   Enrichment • SECTION 8-3

Solving Proportions

Name _____ Date _____

A **proportion** is an equation that states that two ratios are equivalent. A proportion can be written in two ways:

$5:15 = 3:9$ or $\frac{5}{15} = \frac{3}{9}$

The **cross-products** of the terms of a proportion are equal. You can use this fact to determine whether a statement is a proportion or to find the unknown term in a proportion.

▶ **Example 1**

Tell whether the statement is a proportion.
$\frac{8}{22} = \frac{20}{55}$

Solution

Find the cross-products.

$8 \times 55 = 440$

$20 \times 22 = 440$

So the statement

$\frac{8}{22} = \frac{20}{55}$

is a proportion.

▶ **Example 2**

Solve the proportion
$\frac{4}{12} = \frac{n}{36}$

Solution

Find the cross-products and solve for n.

$4 \times 36 = 12 \times n$

$144 = 12n$

$12 = n$

Check by substituting the value 12 for n.

$\frac{4}{12} = \frac{4 \div 4}{12 \div 4} = \frac{1}{3}$

$\frac{12}{36} = \frac{12 \div 12}{36 \div 12} = \frac{1}{3}$

So n = 12.

EXERCISES

Tell whether each statement is a proportion. Write = or ≠.

1. $\frac{15}{55} \ (\neq) \ \frac{10}{50}$

2. $\frac{4}{18} \ (=) \ \frac{16}{72}$

3. $\frac{60}{9} \ (=) \ \frac{40}{6}$

4. $\frac{2}{3} \ (=) \ \frac{44}{66}$

5. $\frac{26}{52} \ (\neq) \ \frac{50}{75}$

6. $\frac{5}{6} \ (=) \ \frac{75}{90}$

Solve.

7. $3:4 = 9:x$ **x = 12**

8. $6:x = 4:20$ **x = 30**

9. $35:15 = x:3$ **x = 7**

10. $\frac{14}{21} = \frac{24}{x}$ **x = 36**

11. $\frac{x}{20} = \frac{100}{125}$ **x = 16**

12. $\frac{18}{45} = \frac{x}{25}$ **x = 10**

Reteaching • SECTION 8-4   **147**

Proportional Representation

Name _____ Date _____

The Rooftop Stargazers' Club has 400,000 members nationwide. The club decided to hold a national convention of 200 delegates from six regions. The number of delegates from each region will be proportional to the number of members in the region.

REGION	MEMBERS
Southwest	40,000
Northwest	64,000
North Central	104,000
South Central	28,000
Southeast	68,000
Northeast	96,000

▶ **Example**

How many delegates will the Southwest send?

Solution

The proportion to use is:

$\frac{region's \ members}{total \ members} = \frac{number \ of \ delegates}{200}$

For the Southwest,

$\frac{40,000}{400,000} = \frac{x}{200}$

$x = 20$

Therefore, the Southwest will send 20 delegates.

EXERCISES

Find the number of delegates each region will send. Use a calculator.

1. Northwest   **32**
2. North Central   **52**
3. South Central   **14**
4. Southeast   **34**
5. Northeast   **48**

6. About how many members does one delegate represent?
   2,000 members

7. Another way of achieving fair representation is to decide that there will be one delegate for every 1,000 members. If that system was used, how many delegates would be needed?   **400**

8. Compare the system of proportional representation described in the example with that described in Exercise 7. Describe the advantages and disadvantages of each.
   Answers will vary.

148   Enrichment • SECTION 8-4

298B

Name _____ Date _____

Problem Solving Skills: Solving Problems with Proportions

Some problems can be solved by writing and solving a proportion.
Follow these steps to solve such problems.

► **Steps**

1. Read the problem. Ask yourself:
   What information is given?
   What must I find?

2. Write a word statement of the proportion described in the problem.

3. Choose a letter to represent what you are looking for. Substitute in the word statement.

4. Solve the proportion.

5. Substitute the answer in the original problem to see if the answer is reasonable.

► **Problem**

A certain color of paint is formed by mixing 2 parts yellow to 3 parts blue. How much blue paint is needed to mix with 24 mL of yellow paint?

Solution

After reading the problem, write a word statement of the given proportion:

$$\frac{2 \text{ parts yellow}}{3 \text{ parts blue}} = \frac{24 \text{ mL yellow}}{\text{How many mL blue?}}$$

Let b = the amount of blue paint needed.

Write the proportion. $\frac{2}{3} = \frac{24}{b}$

Solve. $2b = 72$

$b = 36$

Therefore, 36 mL of blue paint are needed.

Does this answer make sense? Yes, there is more blue paint than yellow paint. The ratio of yellow to blue is 24:36, or 2:3 in lowest terms.

EXERCISES

Solve each problem.

1. Diana is making a mosaic pattern that uses 2 blue tiles for every 5 white tiles. How many blue tiles does she need if she has 60 white tiles?

   24 tiles

2. A necklace is made of 12 red beads and 18 orange beads. How many red beads would be needed to make the same necklace with 90 orange beads?

   60 beads

3. Tom found out that 3 L of fruit punch serves 8 people. He wants to make enough for 48 people. How many liters of punch does he need?

   18 L

4. A recipe uses 4 c of rice to every 5 c of beans. How many cups of beans are needed for 10 c of rice?

   12.5 c of beans

Name _____ Date _____

Estimating Wildlife Populations

You can use a proportion to estimate an entire population if you take a sample of that population and know that a part of that sample has certain characteristics. Sue James, a naturalist, wants to estimate total populations for several types of animals living in a wildlife preserve.

► **Example**

Sue sets traps and captures 20 raccoons in different parts of the preserve. She tags each raccoon and then releases them. A week later, she captures 15 raccoons and notes that 6 out of the 15 have tags. How can she use this ratio to estimate the total number of raccoons living on the preserve?

Solution

Sue assumes that the raccoons she tagged are now mixed in with the other raccoons on the preserve. So, she writes and solves this proportion:

$$\frac{\text{total number of raccoons tagged}}{\text{total population of raccoons}} = \frac{\text{number of tagged raccoons captured}}{\text{total number of raccoons captured}}$$

$$\frac{20}{r} = \frac{6}{15} \qquad \text{where } r = \text{the total population of raccoons}$$

$$6r = 300$$

$$r = 50$$

Sue estimates that there are a total of 50 raccoons in the preserve.

EXERCISES

1. What has Sue assumed about the sample population of 15 raccoons?

   The ratio of tagged raccoons to the

   total raccoon population is the

   same for the total population as it

   is in the sample.

2. Describe how Sue could increase the accuracy of her estimate.

   Repeat the sampling several more

   times and average the results. She

   might also take a larger sample

   number of raccoons in her first

   sampling.

Solve, using a proportion.

3. Sue captures, tags, and releases 50 field mice. Two weeks later, she captures 35 mice. She finds that 10 have tags. Estimate the total number of field mice.

   175 field mice

4. Ninety geese are captured, tagged, and released. Later, a sample of 25 are captured. Of these, 9 geese have tags. Estimate the total number of geese.

   250 geese

5. Sue estimates the total deer population on the preserve to be 200. A random sample of 50 deer shows that 8 of them have a certain disease. Estimate the number of deer out of the total number on the preserve that have the disease.

   32 deer

Name _____ Date _____

Scale Drawings

A **scale drawing** is a representation of a place or object. All the measurements shown in the drawing are proportional to the actual measurements of the place or object. The ratio of the actual dimensions of the object to the dimensions of the drawing is called the **scale** of the drawing. You can use scale ratios to solve problems.

► **Example**

At the right is a scale drawing of a room. The scale is 1 in.:4 ft. Find the actual length of the wall that includes the door.

Solution

Write and solve a proportion, using the scale of the drawing as one ratio. Let x = the actual length of the wall.

$$\frac{1 \text{ in.}}{4 \text{ ft}} = \frac{2 \text{ in.}}{x}$$

$$1 \times x = 2 \times 4$$

$$x = 8$$

The actual length of the wall is 8 ft.

EXERCISES

Solve for x to find either the actual or drawing measurement.

1. scale: 1 in.:3 yd
   drawing: x = __**15 in.**__
   actual: 45 yd

2. scale: 2 cm:5 m
   drawing: 8 cm
   actual: x = __**20 m**__

3. scale: 1 in.:20 mi
   drawing: 4 in.
   actual: x = __**80 mi**__

4. scale: 3 cm:2 km
   drawing: x = __**15 cm**__
   actual: 10 km

5. scale: 1 cm:6 m
   drawing: x = __**3 cm**__
   actual: 18 m

6. scale: 1 cm:20 cm
   drawing: 7 cm
   actual: x = __**140 cm**__

7. scale: 1 cm:2 mm
   drawing: 5 cm
   actual: x = __**10 mm**__

8. scale: 2 cm:1 km
   drawing: x = __**38 cm**__
   actual: 19 km

9. scale: 2 in.:3.5 mi
   drawing: x = __**6 in.**__
   actual: 10.5 mi

Name _____ Date _____

Scale Drawings in Art

A graphic artist is a person who designs and produces pictures that, for example, appear in book illustrations, magazine advertising, signs, business cards, menus and packaging. Scale drawings can be useful when artists make sketches for large designs.

► **Example**

An artist made this sketch of a design which will appear on a large billboard. What will be the actual diameter of the circle?

Solution

The scale on the drawing states that 1 in. represents 30 ft. A ruler shows that the diameter of the circle in the drawing is $1\frac{1}{2}$ in. Write and solve a proportion.

Let d = the diameter, in feet.

$$\frac{1 \text{ in.}}{30 \text{ ft}} = \frac{1.5 \text{ in.}}{d}$$

$$d = 45 \text{ ft}$$

The diameter of the circle will be 45 ft.

"Save The Planet"

Scale: 1 in.: 30 ft

EXERCISES

Solve.

1. A business card measures $3\frac{1}{2}$ in. by 2 in. Brandon wants to make a scale model that is 7 in. long. How wide should the model be?

   4 in.

2. A restaurant is enlarging its menus, which now measure 6 in. by 8 in., so that the longer side will measure 12 in. What will be the length of the shorter side?

   9 in.

3. Odetta must design a new label for canned peaches. The diameter of the can is 7 cm. The height of the can is 10 cm. She plans a scale drawing with dimensions $2\frac{1}{2}$ times those of the actual label. What dimensions should she use for the drawing? (Use 3.14 for π.)

   17.5π cm × 25 cm or 55 cm × 25 cm

4. In a scale drawing for a poster, the width of the space to be used for a photo is 15 mm. The scale is 1 mm:40 cm. What will be the width of the photo in the actual poster?

   60 cm

298C

Name _____ Date _____

Problem Solving Strategy: Make an Organized List

Some problems ask you to find out in how many different ways items can be arranged. Many of these problems can be solved by making an organized list. The list must be clearly set up, then carefully filled in.

▶ **Example**

In how many ways can you make 15¢ using any combination of pennies, nickels and dimes?

Solution

Make a list that lets you record the number of each kind of coin. Each row must give a total of 15¢. Work in a certain order. For example, begin by using the least amount of pennies and work up to using all pennies. The completed list shows that 6 combinations are possible.

Pennies	Nickels	Dimes
0	1	1
0	3	0
5	0	1
5	2	0
10	1	0
15	0	0

EXERCISES

Copy and complete the list to find all possible answers for each problem.

1. How many different 3-digit numbers can you make using the digits 9, 2 and 1? No digit should be used more than once in any number.

Hundreds	Tens	Ones
9	2	1
9	1	2

Number of ways: **6**

2. There are 4 runners, A, B, C, and D, in a race. In how many different ways can the runners finish first, second, third, or fourth.

First	Second	Third	Fourth
A	B	C	D
A	B	D	C

Number of ways: **24**

Make an organized list to find all possible answers for each problem.

3. How many different 3-digit numbers can you make using the digits 8, 4, and 0? No digit should be used more than once in any number.

4

4. How many different ways can you make 12¢ using any combination of pennies, nickels and dimes?

2

Name _____ Date _____

Permutations

A **permutation** is any arrangement of a group of items in a particular order. The number of permutations of the items can be found by making an organized list. It can also be found by using multiplication.

▶ **Example**

Find the number of permutations of the letters A, B, and C, if these are to be written side by side.

Solution 1

Make an organized list.

First position	Second	Third
A	B	C
A	C	B
B	A	C
B	C	A
C	A	B
C	B	A

There are 6 different possible permutations.

Solution 2

Use multiplication.

The number of letters possible for the first position is 3.

Once the first letter is chosen, there are 2 possible letters for the second position.

After the second letter is chosen, there is only one possibility for the third position.

Therefore, the total number of possible permutations is

$3 \times 2 \times 1$, or 6.

The method shown in Solution 2 uses the fundamental counting principle:

To find the number of permutations of a group of items, multiply together the number of possibilities for each position.

EXERCISES

Find the number of permutations in each situation. Use a calculator if you wish.

1. In how many ways can the digits 3, 4, 7, and 9 be arranged from left to right to form a four-digit whole number?

24 ways

2. Juanita has 6 spelling bee trophies. In how many different ways can she arrange them from left to right on her shelf?

720 ways

3. Janet is planting five rows of vegetables in her garden side by side. She will fill each row with a different vegetable. In how many different ways can she arrange the rows of vegetables?

120 ways

4. In how many ways can the class president and vice president be chosen from a slate of 10 candidates?

90 ways

Name _____ Date _____

Calculator Activity: Solving Proportions

A calculator can help you solve proportions.

▶ **Example**

Gina mixes the juice of 8 lemons with 3.5 quarts of water when she makes lemonade. How many lemons will she need if she uses 10 quarts of water?

Solution

Write and solve a proportion. Let x represent the number of lemons.

$$\text{lemons} \rightarrow \frac{8}{3.5} = \frac{x}{10} \leftarrow \text{water}$$

Use a calculator to solve the proportion.

$8 \boxed{\times} 10 \boxed{\div} 3.5 \boxed{=} 22.9$

Gina will need about 23 whole lemons.

EXERCISES

Use a calculator to solve each proportion. Round your answer to the nearest tenth.

1. $\frac{7}{12} = \frac{n}{18}$ **n = 10.5**

2. $\frac{32.6}{15.3} = \frac{y}{7.1}$ **y = 15.1**

3. $\frac{2\frac{1}{2}}{5} = \frac{w}{7\frac{3}{4}}$ **w = 3.9**

4. $\frac{90.6}{120} = \frac{x}{60}$ **x = 45.3**

5. $\frac{m}{24.6} = \frac{7.3}{43.5}$ **m = 4.1**

6. $\frac{p}{7\frac{1}{2}} = \frac{3\frac{1}{2}}{5\frac{3}{4}}$ **p = 4.6**

7. $\frac{275}{6.5} = \frac{350}{z}$ **z = 8.3**

8. $\frac{500}{60} = \frac{b}{2.4}$ **b = 20**

9. $\frac{1.2}{7\frac{1}{4}} = \frac{3\frac{4}{5}}{a}$ **a = 23.8**

10. $\frac{15.7}{5\frac{1}{2}} = \frac{d}{15.7}$ **d = 44.8**

11. $\frac{4\frac{1}{2}}{5\frac{3}{4}} = \frac{5\frac{1}{4}}{n}$ **n = 6.7**

12. $\frac{0.45}{8.13} = \frac{x}{32.7}$ **x = 1.8**

Solve.

13. Elsa drove 245 m in 4.5 h. How many hours will it take her to drive 650 m?

about 11.9 h

14. William bought 6 apples for $1.98. How much will 20 apples cost?

$6.60

Name _____ Date _____

Computer Activity: Properties of Proportions

Proportions have interesting properties that enable you to put the terms of a proportion in different positions.

This program will print out cross products for a proportion.

```
10  READ A,B,C,D              DATA is read in order given
20  PRINT A • D;"   ";B • C    Cross products are printed
30  DATA 2,3,4,6              Four terms of the proportion
```

EXERCISES

1. RUN the program. Record the cross products and determine if the pairs of ratios form a proportion.

$\frac{2}{3} = \frac{4}{6}$ **12**   **12**   Proportion ? **Yes**

2. Change line 30 so the DATA is entered 3, 2, 6, 4 (This interchanges A with B and C with D). Repeat Exercise 1.

$\frac{3}{2} = \frac{6}{4}$ **12**   **12**   Proportion ? **Yes**

3. Change line 30 so the DATA is entered 2, 4, 3, 6 (This interchanges the original B and C). Repeat Exercise 1.

$\frac{2}{4} = \frac{3}{6}$ **12**   **12**   Proportion ? **Yes**

4. Repeat Exercises 1-3 for the following sets of DATA.

a. 3, 5, 9, 15

$\frac{3}{5} = \frac{9}{15}$ **45**   **45**   **Yes**

$\frac{5}{3} = \frac{15}{9}$ **45**   **45**   **Yes**

$\frac{3}{9} = \frac{5}{15}$ **45**   **45**   **Yes**

b. 2, 5, 8, 20

$\frac{2}{5} = \frac{8}{20}$ **40**   **40**   **Yes**

$\frac{5}{2} = \frac{20}{8}$ **40**   **40**   **Yes**

$\frac{2}{8} = \frac{5}{20}$ **40**   **40**   **Yes**

5. If $\frac{a}{b} = \frac{c}{d}$, describe your conclusion about $\frac{b}{a}$ and $\frac{d}{c}$.

When the ratios of the original proportion are inverted, the result is also a proportion.

6. If $\frac{a}{b} = \frac{c}{d}$, describe your conclusion about $\frac{a}{c}$ and $\frac{b}{d}$.

When the means are switched in a proportion, the result is also a proportion.

ACHIEVEMENT TEST

Exploring Ratio and Proportion
CHAPTER 8 FORM A

MATH MATTERS BOOK 1
Chicha Lynch
Eugene Olmstead

SOUTH-WESTERN PUBLISHING CO.

Name _____

Date _____

SCORING RECORD	
Possible	Earned
43	

Write each ratio in two other ways.

1. 3 to 8 $3:8; \frac{3}{8}$ 2. $\frac{5}{6}$ **5 to 6; 5:6**   3. 5:2   **5 to 2; $\frac{5}{2}$**

4. 4 to 9 $\frac{4}{9}$; 4:9   5. $\frac{1}{5}$ **1:5; 1 to 5**   6. 4:5   **4 to 5; $\frac{4}{5}$**

7. 3:1   **3 to 1; $\frac{3}{1}$**   8. $\frac{11}{12}$ **11 to 12; 11:12**   9. $\frac{1}{9}$ **1 to 9; 1:9**

10. 7:4   **7 to 4; $\frac{7}{4}$**   11. 6 to 10   **6:10; $\frac{6}{10}$**   12. $\frac{3}{4}$ **3 to 4; 3:4**

Find the unit rate.

13. $\frac{\$36}{4 \text{ h}}$ **\$9/h**   14. $\frac{770 \text{ mi}}{14 \text{ h}}$ **55 mi/h**   15. $\frac{374 \text{ m}}{8.5 \text{ s}}$ **44 m/s**

16. $\frac{300 \text{ words}}{5 \text{ min}}$ **60 words/min**   17. $\frac{364 \text{ cycles}}{7 \text{ min}}$ **52 cycles/min**   18. $\frac{513 \text{ mi}}{19 \text{ gal}}$ **27 mi/gal**

19. $\frac{288 \text{ bottles}}{12 \text{ cases}}$ **24 bottles/case**   20. $\frac{14,300 \text{ mi}}{22 \text{ h}}$ **650 mi/s**   21. $\frac{252 \text{ m}}{6 \text{ s}}$ **42 m/s**

Write three ratios equivalent to the given ratio.

22. $\frac{1}{4}$ $\frac{2}{8}, \frac{3}{12}, \frac{4}{16}$ 23. $\frac{1}{6}$ $\frac{2}{12}, \frac{3}{18}, \frac{4}{24}$ 24. $\frac{30}{40}$ $\frac{60}{80}, \frac{90}{120}, \frac{3}{4}$

25. $\frac{1}{8}$ $\frac{2}{16}, \frac{3}{24}, \frac{4}{32}$ 26. $\frac{6}{30}$ $\frac{3}{15}, \frac{2}{10}, \frac{1}{5}$ 27. $\frac{2}{7}$ $\frac{4}{14}, \frac{6}{21}, \frac{8}{28}$

28. 12:20   **6:10, 3:5, 24:40**   29. 14:35   **2:5, 28:70, 42:105**   30. 10:18   **5:9, 20:36, 30:54**

Solve each proportion.

31. $\frac{6}{7} = \frac{x}{35}$ **x = 30**   32. $\frac{8}{x} = \frac{4}{6}$ **x = 12**   33. $\frac{9}{14} = \frac{27}{x}$ **x = 42**

34. $\frac{6}{4} = \frac{x}{32}$ **x = 48**   35. $\frac{13}{65} = \frac{x}{5}$ **x = 1**   36. $\frac{28}{16} = \frac{7}{x}$ **x = 4**

37. $\frac{8}{32} = \frac{9}{x}$ **x = 36**   38. $\frac{3}{42} = \frac{x}{56}$ **x = 4**   39. $\frac{16}{6} = \frac{80}{x}$ **x = 30**

8A-1

Name _____

Date _____

Solve.

40. The width of a dining room floor in a scale drawing is 6 in. The actual width is 30 ft. What is the scale of the floor plan?

   1 in.:5 ft _____

41. Carla made a scale drawing of her classroom. The length of the classroom in the drawing is 5 in. The actual length is 40 ft. What scale did Carla use for her drawing?

   1 in.:8 ft _____

42. The scale of a drawing is 1 in.:4 ft. Find the drawing length that would be used to represent an actual length of 24 ft.

   6 in. _____

43. The scale on a map is 0.5 in.:10 mi. If the map shows that the distance from Barkley to Winston is 4.5 in., what is the actual distance between those two towns?

   90 mi _____

8A-2

ACHIEVEMENT TEST

Exploring Ratio and Proportion
CHAPTER 8 FORM B

MATH MATTERS BOOK 1
Chicha Lynch
Eugene Olmstead

SOUTH-WESTERN PUBLISHING CO.

Name _____

Date _____

SCORING RECORD	
Possible	Earned
43	

Write each ratio in two other ways.

1. 4 to 9 $4:9; \frac{4}{9}$ 2. $\frac{5}{8}$ **5 to 8; 5:8**   3. 4:3   **4 to 3; $\frac{4}{3}$**

4. 2 to 7 $\frac{2}{7}$; 2:7   5. $\frac{1}{8}$ **1:8; 1 to 8**   6. 8:6   **8 to 6; $\frac{8}{6}$**

7. 7:1   **7 to 1; $\frac{7}{1}$**   8. $\frac{12}{14}$ **12 to 14; 12:14**   9. $\frac{1}{6}$ **1 to 6; 1:6**

10. 9:5   **9 to 5; $\frac{9}{5}$**   11. 3 to 10   **3:10; $\frac{3}{10}$**   12. $\frac{2}{3}$ **2 to 3; 2:3**

Find the unit rate.

13. $\frac{\$63}{7 \text{ h}}$ **\$9/h**   14. $\frac{930 \text{ mi}}{15 \text{ h}}$ **62 mi/h**   15. $\frac{361 \text{ m}}{9.5 \text{ s}}$ **38 m/s**

16. $\frac{225 \text{ words}}{5 \text{ min}}$ **45 words/min**   17. $\frac{384 \text{ cycles}}{8 \text{ min}}$ **48 cycles/min**   18. $\frac{528 \text{ mi}}{22 \text{ gal}}$ **24 mi/gal**

19. $\frac{312 \text{ cans}}{26 \text{ cases}}$ **12 cans/case**   20. $\frac{4,500 \text{ mi}}{6 \text{ h}}$ **750 mi/h**   21. $\frac{448 \text{ m}}{8 \text{ s}}$ **56 m/s**

Write three ratios equivalent to the given ratio.

22. $\frac{1}{2}$ $\frac{2}{4}, \frac{3}{6}, \frac{5}{10}$ 23. $\frac{1}{5}$ $\frac{2}{10}, \frac{3}{15}, \frac{4}{20}$ 24. $\frac{30}{20}$ $\frac{60}{40}, \frac{6}{4}, \frac{3}{2}$

25. $\frac{1}{7}$ $\frac{2}{14}, \frac{3}{21}, \frac{4}{28}$ 26. $\frac{6}{42}$ $\frac{1}{7}, \frac{12}{84}, \frac{18}{126}$ 27. $\frac{2}{9}$ $\frac{4}{18}, \frac{6}{27}, \frac{8}{36}$

28. 10:32   **5:16, 20:64, 30:96**   29. 14:42   **2:6, 7:21, 1:3**   30. 36:24   **9:6, 12:8, 6:4**

Solve each proportion.

31. $\frac{6}{8} = \frac{x}{32}$ **x = 24**   32. $\frac{7}{12} = \frac{42}{x}$ **x = 72**   33. $\frac{3}{14} = \frac{x}{56}$ **x = 12**

34. $\frac{x}{48} = \frac{5}{8}$ **x = 30**   35. $\frac{5}{x} = \frac{35}{49}$ **x = 7**   36. $\frac{27}{15} = \frac{9}{x}$ **x = 5**

37. $\frac{84}{144} = \frac{x}{12}$ **x = 7**   38. $\frac{0.6}{1.8} = \frac{1}{x}$ **x = 3**   39. $\frac{x}{2} = \frac{2}{8}$ **x = 0.5**

8B-1

Name _____

Date _____

Solve.

40. The width of a dining room floor in a scale drawing is 7 in. The actual width is 35 ft. What is the scale of the floor plan?

   1 in.:5 ft _____

41. Carla made a scale drawing of her classroom. The length of the classroom in the drawing is 8 in. The actual length is 48 ft. What scale did Carla use for her drawing?

   1 in.:6 ft _____

42. The scale of a drawing is 1 in.:5 ft. Find the drawing length that would be used to represent an actual length of 40 ft.

   8 in. _____

43. The scale on a map is 0.5 in.:15 mi. If the map shows that the distance from Conley to Dunwall is 4.5 in., what is the actual distance between those two towns?

   135 mi _____

8B-2

United States and World Travel

Objective: In the context of solving travel problems, students will enhance their consumer and mathematics skills by estimating and computing with whole numbers, integers, and rational numbers; expressing mathematical information using variables, tables, and graphs; using formulas; and operating with measures.

UNIT OVERVIEW

Using up-to-the-minute real-world information from newspapers, travel bureaus, and other sources, students will research and compare options, analyze advantages and disadvantages, and make and defend realistic decisions while planning travel experiences. The three lessons focus on both consumer and mathematical literacy.

LESSON 1 **And the Winner Is...** asks students to imagine that they have won a vacation to any city outside of the U.S. They must select two very different cities and compile facts about the cities. Students will write travel brochures or design travel posters including such information as mean and range of temperatures, air fares, hotel rates, and dinner prices. They will express at least one set of data in a stem-and-leaf plot. To represent temperature variation or changes in elevation, students will add and subtract integers. They will round large population figures and may express them in scientific notation. They will add measures of time to determine total travel time and use a formula to convert Fahrenheit to Celsius temperatures.

LESSON 2 **Bargain Getaway** narrows the travel focus to a city within the U.S., and how best to get there. After students choose a destination at least 1000 miles away from their town or city, they must determine and compare the transportation costs of using a car, a bus, a train, or an airplane. Students will employ number operations, comparisons, and estimates. They will write a paper summarizing their findings with algebraic expressions, inequalities, graph or table presentations, and generalizations using inductive reasoning.

LESSON 3 **Making Tracks** asks students to plan an itinerary for a car trip to one of the U.S. National Parks. First, students select a National Park and describe it in terms of popularity (number of visitors annually), area, perimeter, and other numerical data. They will prepare a map of the park using scale drawings and metric measurements. Students will "buy" a used car based on comparisons of price, reliability, gas mileage, and other factors. They will use the distance formula to estimate time on the road and then compute the total cost, including food and lodging, for their trip.

TIME 8–16 days (not consecutively since students will need time to gather materials)

INTRODUCTION AND MOTIVATION
- Before beginning the unit, save several weeks' copies of a Sunday newspaper travel section. Also, gather old and new travel brochures.
- Have students work in groups of 3 to 5 to examine the brochures and newspapers and discuss among themselves vacations they have taken. Then ask:
  1. Where have you gone on vacation?
  2. How would you go about choosing a vacation spot?
  3. Which places are the best buys?
  4. How would you get there?
  5. What would the total costs be?
- Tell students that they will be gathering and using information like that in the newspapers and brochures to become better consumers of services.

LESSON 1   And the Winner Is . . .

OBJECTIVES

Students will:
- collect data and compute the mean and range of data
- construct a stem-and-leaf plot
- add and subtract integers to find the range of temperatures and/or elevations
- use a formula to convert Fahrenheit to Celsius temperature
- add and subtract time measures

TIME

3 to 6 class periods (allow time for outside information gathering)

MATERIALS

travel brochures, magazines, newspaper travel sections, almanacs, *The Times Books World Weather Guide* (optional), posterboard, markers, construction paper

INTRODUCING THE LESSON

1. Tell students to pretend they are all winners of a special vacation to any two cities in the world outside of the U.S. and, as such, will collect and compare real information on two world cities.
2. Remind students of the introductory activity in which they discussed vacation spots. Arrange students in pairs. Ask each pair to select two cities to investigate further. Provide research materials mentioned above. Emphasize that the two cities must be different in character. For example, if one city is in the Northern hemisphere, choose the second from the Southern hemisphere.

FACILITATING THE LESSON

1. Lead the class in brainstorming the type of information they will seek about their cities. Hand out **Worksheet 1, Vital Statistics**. Have students be certain to include at least the items listed in their report.
2. Help student pairs to prepare a list of where to write or phone to obtain some of the information needed on their chosen cities. One suggestion might be that they contact the country's embassy in the U.S.
3. Allow students time to gather information. Then give students class time to work on computations and conclusions. Check work, and give assistance where needed.
4. Provide time and materials for students to create travel posters or brochures. Their finished products should include short paragraphs in which the information is embedded, as well as illustrations, charts, and a stem-and-leaf plot.

SUMMARIZING THE RESULTS

Hang student posters and brochures. Allow class members to make a "world tour" by circulating around the room to read each other's work. Then ask students to find the best travel buy. Before they circulate again to make their choices, discuss possible criteria and computations they might use to determine the best buy. Have students write a final paragraph justifying their choice.

EXTENDING THE LESSON

- Select two of the students' choices for best buy. Arrange a formal debate among students for or against each choice.
- Invite speakers who have lived in or traveled to some of the cities studied by the students to visit the classroom and share their experiences and photographs.
- Ask students to carry out computations to respond to such questions as:
  1. What are the advantages and disadvantages of going on a packaged tour of a city compared to visiting it on your own and making your own arrangements?
  2. If the whole class was going to one of the cities, what might change?

ASSESSING THE LESSON

Check students' computations for accuracy and their choices of best buys for reasonable justification. Look for originality, as well as accuracy in their posters and brochures. You may wish to give a final assignment in which you propose a particular package tour to one of their chosen cities, and ask them to compare its value to a best-buy trip already selected.

298G

OBJECTIVES

Students will
- use outside resources to find transportation costs
- invesigate and compare transportation costs by car, van, plane, train, and bus
- use fractions and whole numbers in comparisons and estimates of relative costs
- use tables or graphs to display comparative data
- make generalizations, using inductive reasoning, about relative costs for travel

TIME

2 to 4 class periods

MATERIALS

almanacs, U.S. atlas or state maps, train (Amtrak) schedules, newspapers with travel ads, long distance bus company schedules, publications from automobile travel clubs, EPA gas mileage estimates for automobiles

INTRODUCING THE LESSON

1. Ask students to guess which method of travel within the U.S. would be the best bargain: travel by car, train, plane, or bus. Give them the opportunity to look through some sample travel ads before responding. In your discussion, emphasize that there may be no one right answer. The costs depend on where they go, how long they stay, and how wise they are as consumers.
2. Explain to students that they will choose a destination at least 1000 miles away from their home and investigate the relative costs of different methods of travel. Emphasize that this lesson concerns transportation costs *only*. Allow individuals to make tentative choices of their destinations; then form groups of 2 to 3 students with the same choice.

FACILITATING THE LESSON

1. Help groups of students decide where to write or call for travel information. Hand out **Worksheet 2, Which Way To Go?,** to help students organize their comparisons.
2. After students have collected car, train, bus, and plane information, give them class time to make calculations for traveling by each method. Remind them to include transportation to and from the airport or bus and train stations and parking fees, if applicable. For car travel, they may choose to estimate costs for their family vehicle or for a rented vehicle. Remind them to include tolls.

SUMMARIZING THE RESULTS

Each group must submit a paper summarizing their results and including a generalization about which mode of travel might be the best buy throughout the U.S. Emphasize that credit will be given for their use of algebraic expressions, including inequalities, to describe the relationship of different cost factors. Groups may also decide to display specific comparison data using a graph or table. You may wish to have students present their papers orally.

EXTENDING THE LESSON

- Post a map of the United States and attach markers to show where each group has traveled. Ask students to make up mathematical questions or problems about the location such as: How much farther is the longest than the shortest distance traveled?
- Have students use their generalizations about relative travel costs to write an editorial on the subject for a local publication.

ASSESSING THE LESSON

Evaluate students' thoroughness of research, accuracy of computations, clarity of comparison data, reasonableness and justification for their generalizations and use of appropriate algebraic expressions and/or inequalities in their papers.

OBJECTIVES

Students will
- evaluate different advertised used cars and pretend to buy one of them
- study one U.S. National Park in depth
- plan a car trip to a national park
- use the distance formula to help determine an itinerary for the trip
- using metric and customary measures, draw a map to scale and prepare another scale drawing

TIME

3 to 6 class periods (allow time for outside information gathering)

MATERIALS

almanacs, U.S. maps, travel articles on U.S. National Parks, consumer magazine issues focusing on used cars; local newspaper classified sections showing car ads, publications from automobile travel clubs, poster boards, markers, graph paper, rulers (metric and customary measures), *The National Parks: Lesser-Known Areas,* available from U.S. Government Printing Office, Washington, D.C. 20402 ($1.50, GPO Stock #024-005-009-11-6; optional)

INTRODUCING THE LESSON

1. Divide the class into two large groups. Provide one group with the previously-collected resources concerning U.S. National Parks. Have them consider such questions as:
   - Which park has attractions you would like to see?
   - Which is the least crowded?
   - What types of accomodations are available?
   - What kinds of wildlife might you see there?
   - Which would be a likely destination for a vacation by car, two weeks in length?

   Provide the second group with ads for used cars and consumer magazines focusing on buying used cars. Ask this group to consider:
   - Which car is affordable?
   - Which would be reliable on a two-week trip?
   - How many miles per gallon will you average on a long trip?
   - What other factors would you use to decide which car to buy?

2. Have two or three people from each group give their responses to the questions. Tell the class that this lesson will include both of the activities that they just sampled. They will choose a U.S. National Park to visit, and they will pretend to buy a used car to get them there. They will have to describe the park and justify their choice of car in a short essay. Other final products will include some statistics, a map, and some information on wildlife.

FACILITATING THE LESSON

1. Give all students the opportunity to look over the information on U.S. National Parks and make a tentative choice of one to visit. Then list all the tentative choices on the chalkboard and arrange students who wish to visit the same place in groups of 2 to 5.

2. Students should bring in additional ads for used cars from magazines, newspapers, or even local bulletin boards. Give groups class time to analyze the advertised cars with respect to the information in consumer magazines. They should consider repair records, gas consumption and other operating costs, and safety records, as well as asking price when they evaluate the cars. Groups must decide which car to buy and write a one-page essay carefully explaining their reasoning and citing statistics that justify their choice.

3. Instruct the groups that they will produce:
   - using customary measures, a map to scale of their park and information on the number of acres or square miles, and an estimate of its perimeter.
   - using metric measures, a scale drawing of an animal or animal track that might be seen in the park.
   - statistics about the park, such as the number of yearly visitors and the cost of various types of lodging.
   - an estimate of the number of days' driving needed to reach their destination, based on the distance, driving speed, and number of hours of driving each day, and an estimate of the total cost of the trip, considering the number of nights they need to pay for lodging, the cost of meals and snacks along the way, gasoline expense, possible cost of tolls on the highways, entrance fees to the park and park attractions, and, possibly, photographs and souvenirs.

SUMMARIZING THE RESULTS

Help groups decide where to find the needed information. For example, students may need to look in wildlife books or an encyclopedia to find size information in order to draw an animal or animal track.

4. Students may use **Worksheet 3, Making Tracks Checklist/Plan,** to help them organize their information.

EXTENDING THE LESSON

Encourage groups to exhibit their creativity in the way they bring together the information required in this lesson. Some might choose to use a large poster board to display the map, the animal drawing, the essay, and the statistics about the park and travel expenses. Others may design a "package tour" in a brochure format. Set aside class time for sharing the results and discussing questions such as:
- What have you learned from this activity?
- How has mathematics made it easier for you to plan wisely?
- Is there anything you would do differently if you repeated this project?

ASSESSING THE LESSON

1. Ask every group to return to their choice of car and research and report on additional information such as insurance cost or the advantages and disadvantages of financing the purchase.
2. Do an in-depth examination of one or mre of the U.S. National Parks studied by the students. Find out more about the wildlife. Are there any endangered plants or animals in those parks? Why are they endangered? What is being done to protect them? How do wildlife experts go about counting or estimating the number of a certain species of animal? What rules should people visiting a park follow to protect wildlife and the environment? If possible, invite a wildlife expert to discuss questions such as those above. Invite a park ranger to speak about his or her career.

Evaluate students' work in progress and their final products for accuracy of maps and drawing, reasonable use of data, and mathematical justification for their car choice. You might ask students to wite a short essay which includes mathematics facts to support or reject (a) a camping trip or (b) flying to the destination as cost-effective travel alternatives.

Name _____ Date _____

Vital Statistics

	City 1:	City 2:
Population (rounded)		
Elevation		
Monthly rainfall		
Extreme temperatures/ range of variation		
Average temperature in July		
Air fare and flight information (departures, arrivals, change planes, waiting time between flights)		
Cost of taxi, bus, or shuttle to and from airport		
Hotel rates (shared room for two)		
Dinner cost (for one)		
Sales tax and/or any other special taxes, charges or visitor fees		
Cost of special attractions, sightseeing		

Which Way To Go?

By plane
By bus
By train
By car

Name _____ Date _____

Making Tracks   Checklist/Plan

Destination: _____

Leave: _____

	Daily Travel			Daily Costs		
Day	Number of miles	Location for stopover	Lodging cost	Food cost	Sight-seeing cost	Gas and tolls

Totals

Costs at Park

       Entrance fee _____

       Lodging _____

       Attractions _____

       Food _____

       Other _____

Return: _____

	Daily Travel			Daily Costs		
Day	Number of miles	Location for stopover	Lodging cost	Food cost	Sight-seeing cost	Gas and tolls

Totals

Teacher's Notes

Teacher's Notes

Teacher's Notes

CHAPTER 9 SKILLS PREVIEW

Write each as a percent.

1. 0.3 30%

2. $\frac{1}{4}$ 25%

3. 0.007 0.7%

Find the percent of each number.

4. 18% of 30 5.4

5. 40% of 500 200

6. $12\frac{1}{2}$% of 70 8.75

7. 3.2% of 45 1.44

8. 15% of 120 18

9. 33% of 512 168.96

Find each percent.

10. What percent of 50 is 30? 60%

11. What percent of 56 is 40? 71%

12. 65 is what percent of 450? 14%

13. 4.38 is what percent of 6? 73%

14. What percent of 20 is 7? 35%

15. 12 is what percent of 15? 80%

Find the percent of increase or decrease. Round to the nearest tenth.

16. Original amount: $535
New amount: $642 20% increase

17. Original amount: $145
New amount: $174 20% increase

18. Original amount: 125
New amount: 120 4% decrease

19. Original amount: 80
New amount: 70 $12\frac{1}{2}$% decrease

20. Original amount: $64
New amount: $40 37.5% decrease

21. Original amount: $30
New amount: $39 30% increase

Find the unknown number.

22. 17 is 50% of what number? 34

23. $37\frac{1}{2}$% of what number is 72? 192

24. 12 is 75% of what number? 16

25. 150% of what number is 36? 24

26. 12 is 40% of what number? 30

27. 5% of what number is 12.8? 256

Find the discount and the sale price. Round to the nearest cent.

28. Regular price: $179.89
Percent of discount: 20% D = $35.98
S = $143.91

29. Regular price: $45.23
Percent of discount: 15% D = $6.78
S = $38.45

Find the interest and the amount.

30. Principal: $320
Rate: 5%/year
Time: 1 year
I = $16
A = $336

31. Principal: $7,200
Rate: 8.5%/year
Time: 24 months
I = $1,224
A = $8,424

32. Principal: $819
Rate: 13%/year
Time: 6 months
I = $53.24
A = $872.24

Introduction The purpose of this Skills Preview is to assess students' abilities on all the major objectives of Chapter 9. Test results may be used

- to determine those topics which may need only to be reviewed and those topics which need to be more carefully developed;
- for class placement;
- in prescribing for individual differences.

If you prefer, you may use the Skills Preview as an alternative form of the Chapter Test (page 336) to evaluate mastery of chapter objectives. The items on the Skills Preview and Chapter Test correspond in content and level of difficulty.

OVERVIEW

In this chapter, students are introduced to the concept of percent. Students write percents as decimals and fractions, and write decimals and fractions as percents. Percent of increase and decrease, as well as discount, sale price, and simple interest are also introduced. Problem solving sections focus on choosing a computation method and choosing a strategy.

SPECIAL CONCERNS

Students often have difficulty differentiating among the three types of percent problems—finding the percent of a number, finding what percent one number is of another, and finding a number when a percent of it is known. Help them to conceptualize what is being done in each case so that they can determine whether an equation or a proportion should be used to solve a given percent problem.

VOCABULARY

commission   percent of increase
commission rate   principal
discount   rate
interest   sale price
percent   time
percent of decrease

MATERIALS

graph paper   crayons or markers
8 cards and a paper bag   calculators
  for each pair of students

BULLETIN BOARD

Have students bring in clippings from newspapers, magazines, or advertisements that specify interest paid, interest earned, sales prices, discount prices, percent of increase, percent of decrease, or commission in terms of percents. Display the clippings under the appropriate headings on the bulletin board. Ask students to create word problems using data from the clippings and to pin their problems up on the bulletin board. Encourage students to solve these problems in their spare time.

INTEGRATED UNIT 3

The skills and concepts involved in Chapters 9–11 are included within the special Integrated Unit 3 entitled "On Your Own." This unit is in the Teacher's Edition beginning on page 434J. Worksheets for this integrated unit appear in the Enrichment Activities booklet, pages 137–139.

TECHNOLOGY CONNECTIONS

- Calculator Worksheets, 175-176
- Computer Worksheets, 177–178
- MicroExam, Apple Version
- MicroExam, IBM Version

- *Decimal Dungeon*, Unicorn Software
- *Safari Search*, Sunburst Communications

TECHNOLOGY NOTES

Computer software programs, such as those listed in Technology Connections, build students' critical-thinking skills and problem solving abilities. Programs that are written in game format provide students with an opportunity to practice their skills while having fun, thereby further enhancing their computational skills.

EXPLORING PERCENT

PLANNING GUIDE

SECTIONS	TEXT PAGES	ASSIGNMENTS			
		BASIC	AVERAGE	ENRICHED	
Chapter Opener/Decision Making	300–301				
9–1 Exploring Percent	302–305	1–45, 47–50, 54–57	1–46, 47–53, 54–57	25–46, 47–53, 54–60	
9–2 Finding the Percent of a Number	306–309	1–22, 23–31, 38–39, 41–42	10–22, 23–40, 41–42	16–22, 29–40, 41–45	
9–3 Finding What Percent One Number Is of Another	310–313	1–10, 12, 13–14, PSA 1–2	1–12, 13–15, 16, PSA 1–2, 5	5–12, 13–15, 16–17, PSA 1–5	
9–4 Percent of Increase and Decrease	314–317	1–15, 21, 22	3–10, 13–15, 16–21, 22	5–8, 13–15, 16–21, 22–23	
9–5 Finding a Number When a Percent of It Is Known	318–321	1–13, 15–17	3–14, 15–21	5–14, 18–21, 22–24	
9–6 Problem Solving Skills: Choose a Computation Method	322–323	1–7	1–7	1–7	
9–7 Discount	324–327	1–7, 9–12, 13, PSA 1–7	5–8, 9–12, 13–14, PSA 1–7, 9	7–8, 9–12, 13–15, PSA 1–9	
9–8 Simple Interest	328–331	1–14, 16–17, 20–23	1–15, 16–18, 20–28	8–15, 16–19, 20–29	
9–9 Problem Solving/Decision Making: Choose a Strategy	332–333	1–14	1–14	1–14	
Technology	306, 307, 311, 319, 328	✔	✔	✔	

ASSESSMENT					
Skills Preview	299	All	All	All	
Chapter Review	334–335	All	All	All	
Chapter Test	336	All	All	All	
Cumulative Review	337	All	All	All	
Cumulative Test	338	All	All	All	

ADDITIONAL RESOURCES			
RETEACHING	**ENRICHMENT**	**TECHNOLOGY**	**TRANSPARENCY**
9–1	9–1		TM 34
9–2	9–2		
9–3	9–3		
9–4	9–4		
9–5	9–5		
9–6	9–6		
9–7	9–7		
9–8	9–8	9–8	
9–9	9–9		

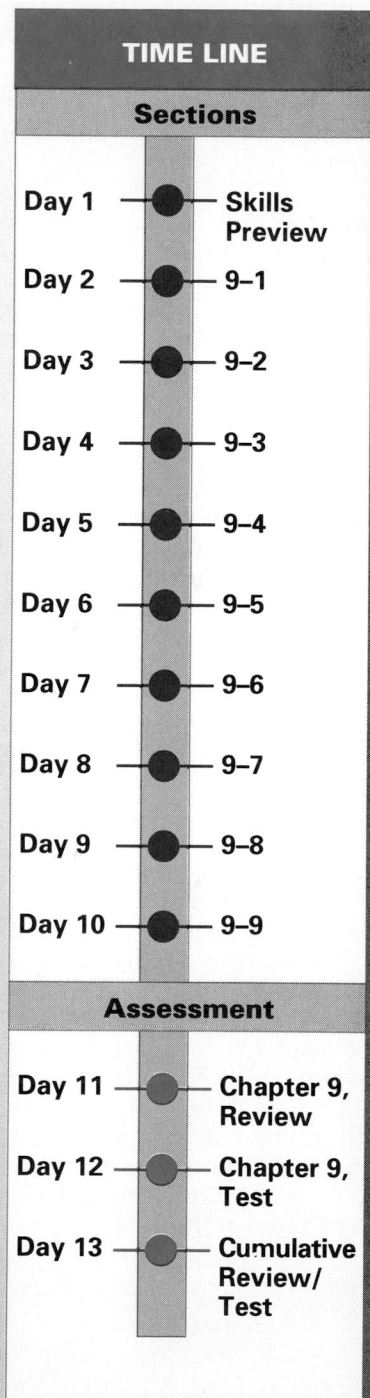

TIME LINE

Sections

Day 1 — Skills Preview
Day 2 — 9–1
Day 3 — 9–2
Day 4 — 9–3
Day 5 — 9–4
Day 6 — 9–5
Day 7 — 9–6
Day 8 — 9–7
Day 9 — 9–8
Day 10 — 9–9

Assessment

Day 11 — Chapter 9, Review
Day 12 — Chapter 9, Test
Day 13 — Cumulative Review/ Test

ASSESSMENT OPTIONS

Chapter 9, Test Forms A and B	
Chapter 9, Test	Text, 336
Alternative Assessment	TAE, 336
Chapter 9, MicroExam	

9 EXPLORING PERCENT

THEME Shopping

Objective To explore how data involving percents can be used as a basis for making real-life decisions

Introduction Discuss and list on the board the decisions that must be made when purchasing a new or used car—make, model, year, options, and financing. Have students decide whether they would buy a new car or a used car. Have them explain their reasoning.

Decision Making Using Data Have students examine the depreciation schedule. After discussing the two questions, ask whether any students who wanted to buy a new car have changed their minds.

Working Together Have students work in groups of three. Clarify each student's role, if necessary. Point out that most buyers finance a new car. Suggest that each group first decide how much money they would be willing to use for a down payment and how much they would be willing to pay each month in financing. Then they must decide on the car, its options, and price. The "car dealers," of course, must try to get the best deal for the company, while the "buyer" must give the "bankers" some proof of the ability to pay. When the groups are ready, have them each role-play their car-buying situation for the class.

Buying a car involves some of the most important—and most difficult—purchasing decisions most people ever make. Often the first decision is whether to buy a new or a used car. There are other factors to consider as well. What make and model should be purchased? What options should be chosen? How will the car be paid for? If a car loan is needed, what is the best financing arrangement?

In this chapter, you will solve problems that involve percents, discount, and simple interest.

This chart shows what happens to the value of a new car as it gets older. The car's value drops dramatically during the first three years of ownership.

NEW AUTOMOBILE DEPRECIATION SCHEDULE

Year	1	2	3	4	5–7	8+
% Decrease*	30–32	24–26	18–20	7–8	3–5	1–2

*The percent given in each column represents a decrease in the value of the car from the year before.

DECISION MAKING

Using Data

Use the information in the chart to answer the following questions.

Answers will vary.

1. Why do you think the percent of decrease in a car's value drops so dramatically after the first three years?

2. Suppose you wanted to buy a new car. You already own a car that is about three years old. Would you sell the car independently or would you try to trade it in at the dealership? What factors would you consider before making the decision? Does the depreciation schedule assist you?

Working Together

Your group's task will be to research and "buy" a new car. One person will be the car buyer, another will be the car dealer, and the third will be the banker. Look through newspapers, car magazines, and consumer magazines and decide which make and model car to buy. The car dealer should look for ads with the different prices and options for the model the car buyer has chosen. The banker should meet with both the dealer and the buyer and decide whether or not the bank will loan the buyer some of the money to buy the car. The banker should explain the terms and conditions of the bank loan. The buyer should analyze all the information and decide if this is the car to buy.

Make a list of all of the factors to consider before making your decision. Compare your lists with those of other groups. What is the most reasonable thing to do?

9-1 Exploring Percent

EXPLORE/ WORKING TOGETHER

Work with a partner. Outline a ten-unit-by-ten-unit square on grid paper. Shade 25 one-unit squares.

a. Write a ratio that compares the number of shaded squares to the total number of squares.   **25:100**

b. Write the ratio as a fraction and a decimal. $\frac{25}{100}$; **0.25**

c. Make a similar model to show the ratio 32:100.

d. What does the ratio 32:100 represent?
   32 out of 100 squares are shaded

SKILLS DEVELOPMENT

A ratio that compares a number to 100 is called a **percent.** Percent means *per one hundred*. The symbol % is used to show percent.

This grid has 35 out of the 100 squares shaded blue.

The ratio $\frac{35}{100}$ means 35 per 100, so 35% of the grid is shaded blue.

Example 1

Write each percent as a decimal.

a. 45%   **b.** 5.9%   **c.** $7\frac{1}{2}\%$

Solution

a. $45\% = \frac{45}{100} = 0.45$

b. $5.9\% = \frac{5.9}{100} = \frac{5.9 \times 10}{100 \times 10} = \frac{59}{1,000} = 0.059$

c. $7\frac{1}{2}\% = \frac{7\frac{1}{2}}{100} = \frac{7.5}{100} = \frac{7.5 \times 10}{100 \times 10} = \frac{75}{1,000} = 0.075$ ◄

Example 2

Write each percent as a fraction.

CHECK UNDERSTANDING

How is 0.2 different from 0.2%?

0.2 is equivalent to 20%; 0.2% is the decimal equivalent to 0.002.

a. 35%   **b.** 0.2%   **c.** $\frac{3}{4}\%$

Solution

Write as a fraction with a denominator of 100 and simplify.

a. $35\% = \frac{35}{100} = \frac{7}{20}$

b. $0.2\% = \frac{0.2}{100} = \frac{0.2 \times 10}{100 \times 10} = \frac{2}{1000} = \frac{1}{500}$

c. $\frac{3}{4}\% = \frac{\frac{3}{4}}{100} = \frac{3}{4} \div 100 = \frac{3}{4} \times \frac{1}{100} = \frac{3}{400}$ ◄

Example 3

Write each decimal as a percent.

a. 0.6   **b.** 0.025   **c.** $0.66\frac{2}{3}$

Solution

a. $0.6 = 0.60 = \frac{60}{100} = 60\%$

b. $0.025 = \frac{25}{1,000} = \frac{25 \div 10}{1,000 \div 10} = \frac{2.5}{100} = 2.5\%$

c. $0.66\frac{2}{3} = \frac{66\frac{2}{3}}{100} = 66\frac{2}{3}\%$ ◄

Example 4

Write each fraction as a percent.

a. $\frac{3}{5}$ **b.** $\frac{5}{8}$ **c.** $\frac{1}{12}$

Solution

a. $\frac{3}{5} = \frac{3 \times 20}{5 \times 20} = \frac{60}{100} = 60\%$

b. $\frac{5}{8} = 0.625 = \frac{625}{1,000} = \frac{625 \div 10}{1,000 \div 10} = \frac{62.5}{100} = 62.5\%$

c. $\frac{1}{12} = 0.08\frac{1}{3} = \frac{8\frac{1}{3}}{100} = 8\frac{1}{3}\%$ ◄

Example 5

A basketball team won 15 out of 25 games. What percent of the total games played did the team win?

Solution

Write the ratio as a fraction.

$\frac{15}{25}$ ← games won / ← total games played

Write a fraction that has a denominator of 100, and then write the percent.

$\frac{15}{25} = \frac{15 \times 4}{25 \times 4} = \frac{60}{100} = 60\%$

The basketball team won 60% of the total games played. ◄

TRY THESE

Write each percent as a decimal.

1. 38%   **0.38**   **2.** 6.3%   **0.063**   **3.** $8\frac{1}{4}\%$ **0.0825**

Write each percent as a fraction.

4. 32% $\frac{8}{25}$ **5.** 15% $\frac{3}{20}$ **6.** $\frac{1}{2}\%$ $\frac{1}{200}$

9–1   Exploring Percent   **303**

MENTAL MATH TIP

Here is a shortcut for changing a percent to a decimal or a decimal to a percent.

Since percent means per hundred, 45% means 45 per hundred, or 45/100, or 45 ÷ 100.

So a quick way to change 45% to a decimal is to divide 45 by 100 by moving the decimal point 2 places to the left.

45% = 0.45

A quick way to change a decimal such as 0.68 to a percent is to multiply by 100 by moving the decimal point 2 places to the right.

0.68 = 68%

ASSIGNMENTS

BASIC
1–45, 47–50, 54–57

AVERAGE
1–46, 47–53, 54–57

ENRICHED
25–46, 47–53, 54–60

ADDITIONAL RESOURCES
Reteaching 9–1
Enrichment 9–1
Transparency Master 34

Example 3: Have students explain the solutions.

Example 4: Elicit from students the fact that in part b, since the denominator has only factors of 2 or 5, the fraction is a terminating decimal; whereas, in part c, because some of the factors of the denominator are not 2 or 5, the decimal is non-terminating.

Example 5: Use this example to demonstrate how to solve problems involving percent.

Additional Questions/Examples
Use these fractions for the following exercises.

| 2/5 | 1/2 | 4/7 | 7/8 | 5/12 |
| 3/4 | 8/15 | 7/9 | 9/10 | 2/3 |

1. Which fractions can be written as terminating decimals?
   2/5, 1/2, 3/4, 9/10, 7/8
2. Which fractions can be written as fractional percents?   **4/7, 5/12, 8/15, 7/9, 2/3**

5-MINUTE CLINIC

Exercise	Student's Error	Error Diagnosis
Write 45 1/3 % as a decimal.	45 1/3% = 4.5 1/3	• Student counts the fraction as one decimal place when moving the decimal point.
	45 1/3% = 0.45 1/3%	• Student failed to drop the percent sign.

303

Write each decimal as a percent.

7. 0.4 **40%**   **8.** 0.032 **3.2%**   **9.** $0.56\frac{1}{4}$ $56\frac{1}{4}\%$

Write each fraction as a percent.

10. $\frac{7}{10}$ **70%**   **11.** $\frac{3}{8}$ **37.5%**   **12.** $\frac{5}{12}$ $41\frac{2}{3}\%$

Solve.

13. John's batting average is 0.325. What percent of the times he has been at bat has he had a hit?   **32.5%**

Guided Practice/Try These

Guided Practice/Try These You may wish to have students work in pairs. Encourage calculator use. Observe students and discuss any questions or problems they may have.

3 SUMMARIZE

Write About Math Have students explain in their math journals how to do the following.
1. Write 36%, 7.4%, and 8 3/4 % as decimals.
2. Write 52%, 0.6%, and 5/8% as fractions.
3. Write 0.7, 0.036, and 5.38 as percents.
4. Write 4/5, 7/8, and 1/15 as percents.

4 PRACTICE

Practice/Solve Problems For each of Exercises 37–44 encourage students to calculate the decimal equivalent by dividing and then to use a calculator to check.

Extend/Solve Problems For Exercises 47 and 49, students should first find the percent, then the decimal. For Exercises 48 and 50, they should first find the decimal, then the fraction.

Think Critically/ Solve Problems Ask students to suggest other types of questions similar to Exercises 58–60.

EXERCISES

PRACTICE/ SOLVE PROBLEMS

Write each percent as a decimal.

1. 50% **0.5**   **2.** 42% **0.42**   **3.** 17% **0.17**   **4.** 95% **0.95**

5. 8.5% **0.085**   **6.** 16.4% **0.164**   **7.** 88.1% **0.881**   **8.** 7.2% **0.072**

9. $13\frac{1}{2}\%$ **0.135**   **10.** $3\frac{1}{4}\%$ **0.0325**   **11.** $5\frac{3}{4}\%$ **0.0575**   **12.** $2\frac{1}{3}\%$ $0.02\frac{1}{3}$

Write each percent as a fraction.

13. 55% $\frac{11}{20}$ **14.** 19% $\frac{19}{100}$ **15.** 20% $\frac{1}{5}$ **16.** 64% $\frac{16}{25}$

17. 0.4% $\frac{1}{250}$ **18.** 2.4% $\frac{3}{125}$ **19.** 0.7% $\frac{7}{1000}$ **20.** 0.5% $\frac{1}{200}$

21. $\frac{3}{8}\%$ $\frac{3}{800}$ **22.** $5\frac{1}{2}\%$ $\frac{55}{1000}$ **23.** $16\frac{1}{4}\%$ $\frac{13}{80}$ **24.** $1\frac{3}{4}\%$ $\frac{7}{400}$

Write each decimal as a percent.

25. 0.2 **20%**   **26.** 0.55 **55%**   **27.** 0.79 **79%**   **28.** 0.09 **9%**

29. 0.015 **1.5%**   **30.** 0.113 **11.3%**   **31.** 0.005 **0.5%**   **32.** 0.061 **6.1%**

33. $0.12\frac{1}{2}$ $12\frac{1}{2}\%$ **34.** $0.09\frac{1}{4}$ $9\frac{1}{4}\%$ **35.** $0.16\frac{2}{3}$ $16\frac{2}{3}\%$ **36.** $0.03\frac{1}{2}$ $3\frac{1}{2}\%$

Write each fraction as a percent.

37. $\frac{12}{100}$ **12%**   **38.** $\frac{4}{5}$ **80%**   **39.** $\frac{3}{4}$ **75%**   **40.** $\frac{1}{2}$ **50%**

41. $\frac{7}{8}$ **87.5%**   **42.** $\frac{5}{6}$ $83\frac{1}{3}\%$ **43.** $\frac{5}{12}$ $41\frac{2}{3}\%$ **44.** $\frac{1}{3}$ $33\frac{1}{3}\%$

Solve.

45. Michelle attended three out of every 20 away football games. What percent of the away football games did she attend?   **15%**

46. Jonathan received a score of 80% on a history quiz. What fraction of the questions did he answer correctly? $\frac{4}{5}$

CONNECTIONS

Only about a quarter of the earth's surface is covered by land. The rest is covered by water. The table below shows what fraction of the total land is occupied by each continent. Write each fraction as a percent.

Continent	Fraction of Earth's Land	
Africa	$\frac{1}{5}$	20%
Antarctica	$\frac{19}{200}$	9.5%
Asia	$\frac{3}{10}$	30%
Australia	$\frac{1}{20}$	5%
Europe	$\frac{13}{200}$	6.5%
North America	$\frac{4}{25}$	16%
South America	$\frac{3}{25}$	12%

304   CHAPTER 9   Exploring Percent

MAKING CONNECTIONS

Have students survey their classmates on a subject of their choosing, such as their favorite book, rock group, song, or sport. Have them record the results of their surveys as fractions, decimals, and percents of the total number surveyed. Have them prepare graphs to illustrate the results of their surveys.

Copy and complete the chart.

	Fraction	Decimal	Percent
47.	$\frac{1}{20}$	■ 0.05	■ 5%
48.	■ $\frac{3}{10}$	■ 0.30	30%
49.	$\frac{6}{15}$	■ 0.4	■ 40%
50.	■ $\frac{3}{20}$	■ 0.15	15%

Solve.

51. Sally received $60 for her birthday. She bought a game for $45. What percent of her money did she spend on the game? **75%**

52. In a referendum, 18 people out of 200 voted against adding a slide to the town pool. What percent voted against the slide? **9%**

53. Over the summer, Peter read 9 out of 20 library books. What percent of the books did he read? **45%**

In Exercises 54–57, decide whether it makes sense to rewrite the fraction or the decimal as a percent. Write *yes* or *no*. If your answer is *yes*, then write the percent.

54. Renee hiked a distance of 3.8 kilometers. **no**

55. Two thirds of the people surveyed did their grocery shopping after work or on the weekend. **yes; $66\frac{2}{3}$%**

56. By the end of the day, Sharon had finished only 0.7 of the job. **yes, 70%**

57. Harold added one quarter pound of ground meat to the recipe. **no**

Solve.

58. What percent of 1 meter is 1 centimeter? **1%**

59. What percent of 1 yard is 1 foot? **$33\frac{1}{3}$%**

60. What percent of 1 year is 1 month? **$8\frac{1}{3}$%**

9–1 Exploring Percent **305**

WARM-UP

Write as both a decimal and a fraction.
1. 85% **0.85, 17/20**
2. 33 1/3% **0.$\overline{3}$, 1/3**
3. 60% **0.6, 3/5**

1 MOTIVATE

Explore Have students work in pairs. Discuss the techniques they used to answer part c.

2 TEACH

Use the Pages/Skills Development Have students read this part of the section and then discuss the examples.

Example 1: Demonstrate how to translate the question into an equation.

Discuss Solution 1, obtained by writing the percent as a decimal, and Solution 2, obtained by writing the percent as a fraction. Have half the class use a decimal to solve the equation and have the other half use a fraction. Compare the two methods. Discuss which method students prefer and why.

Example 2: Use this example to introduce using a proportion to find the percent of a number.

306

9-2 Finding the Percent of a Number

EXPLORE

Salespeople at Milo's Appliance Center earn a commission, or a percent of the price of each appliance they sell. New employees earn 3% and a manager earns 12%.

This 10-by-10 grid represents the price of a large-screen television. The shaded part represents the manager's commission.

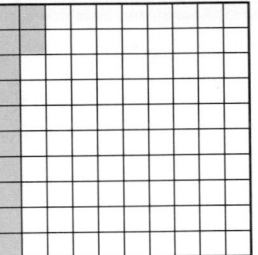

a. What is the ratio of the number of shaded parts to the total number of squares? **12:100**

b. What fraction of the squares is shaded? $\frac{12}{100} = \frac{3}{25}$

c. Suppose the price of the television was $2,000. How much money would the manager receive as a commission? **$240**

SKILLS DEVELOPMENT

To find a percent of a number, you can write and solve an equation.

CALCULATOR TIP

% KEY
You can use the percent key (%) on your calculator to find the percent of a number.

To find 40% of 360, you can use this key sequence:

360 [×] 40 [%]

On some calculators, you have to press the [=] key to get the answer.

Example 1

What number is 60% of 45?

Solution 1
Find the percent of a number by writing an equation using a decimal.

What number is 60% of 45?

$$x = 0.60 \times 45$$
$$x = 27$$

Let x represent the unknown number

Solution 2
Find the percent of a number by writing an equation using a fraction.

What number is 60% of 45?

$$x = \frac{3}{5} \times 45$$
$$x = 27$$

So 60% of 45 is 27. ◄

Another way to find the percent of a number is to write and solve a proportion. You write the proportion using this relationship.

$$\frac{\text{part}}{\text{whole}} = \frac{\text{part}}{\text{whole}}$$

One of the ratios in this proportion represents the percent. The "whole" referred to in this ratio is 100.

Example 2

Find $5\frac{1}{2}\%$ of 30.

Solution

Write $5\frac{1}{2}\%$ as the ratio $\frac{5\frac{1}{2}}{100}$. Use this ratio to write a proportion.

$$\begin{array}{c}\text{part} \rightarrow \\ \text{whole} \rightarrow\end{array} \frac{5\frac{1}{2}}{100} = \frac{x}{30} \begin{array}{c}\leftarrow \text{part} \\ \leftarrow \text{whole}\end{array}$$

Then solve the proportion.

$$\frac{5\frac{1}{2}}{100} = \frac{x}{30}$$

$$5\frac{1}{2} \times 30 = 100x$$

$$165 = 100x$$

$$\frac{165}{100} = \frac{100x}{100}$$

$$1.65 = x \qquad \text{So } 5\frac{1}{2}\% \text{ of } 30 \text{ is } 1.65. \blacktriangleleft$$

Example 3

Keiko bought a pocket camera for 20% off the original price of $45. How much money did she save?

Solution

Write and solve an equation.

What amount is 20% of $45?

$$x = \frac{1}{5} \times 45$$

$$x = 9 \qquad \text{Keiko saved \$9. } \blacktriangleleft$$

TRY THESE

Find the percent of each number using an equation.

1. 40% of 75 **30**
2. 22% of 50 **11**
3. 25% of 48 **12**
4. 90% of 30 **27**

Find the percent of each number using a proportion.

5. 16% of 125 **20**
6. 30% of 70 **21**
7. $7\frac{1}{2}\%$ of 60 **4.5**
8. 15% of 85 **12.75**

Solve.

9. What number is 35% of 25? **8.75**

10. Larry bought stereo speakers on sale for 32% off the original price of $735. How much money did he save? **$235.20**

ASSIGNMENTS

BASIC
1–22, 23–31, 38–39, 41–42

AVERAGE
10–22, 23–40, 41–42

ENRICHED
16–22, 29–40, 41–45

ADDITIONAL RESOURCES
Reteaching 9–2
Enrichment 9–2

Example 3: Once students find the amount Keiko saved, have them find the sale price of the camera. **$36**

Additional Questions/Examples

1. The manager of a camera shop marked up the price of a $45 camera by 15%. What is the marked-up price? **$51.75**
2. What proportion would you use to find 20% of 50?
   20/100 = x/50
3. Use the proportion you wrote for Exercise 2 to find 20% of 50.
   10

Use mental math.

4. If you know that 20% of 50 is 10, what is 10% of 50? What is 40% of 50? **5; 20**
5. If you know that 10% of 600 is 60, what is 1% of 600? 2% of 600? 20% of 600? 40% of 600?
   6; 12; 120; 240

Guided Practice/Try These   Observe students as they complete these exercises independently, making sure they set up their equations and proportions correctly. Watch for students who forget to change the percent to a decimal or fraction.

5-MINUTE CLINIC

Exercise	Student's Error	Error Diagnosis
What number is 35% of 400?	35 × 400 = 14,000	• Student forgets to replace the percent with a decimal or fraction before multiplying.

308

SUMMARIZE

Talk It Over

1. Which method would you use to find 8 1/2% of 125? Explain. **Answers will vary.**

2. Which method would you use to find 0.1% of 250? Explain. **Answers will vary.**

4 PRACTICE

Practice/Solve Problems Students should use fractions to represent the percents in Exercises 7–9.

Extend/Solve Problems Discuss the estimation techniques students will have to use for Exercises 23–28. Exercises 35–37 involve percents less than 1%. Exercises 38 and 39 deal with discounts and mark-ups. Some students may have difficulty with Exercise 40. You may have to point out that they may use either the equation 315 = 0.07 × x or the proportion 7/100 = 315/x to solve the problem.

Think Critically/Solve Problems For Exercise 45, point out that students must find the lowest and highest trade-in values after the second year.

5 FOLLOW-UP

Extra Practice

Find the percent of each number using an equation.
1. 45% of 315 **141.75**
2. 12 1/2% of 75 **9.375**
3. 93 3/4% of 200 **187.5**

Find the percent of each number, using a proportion.
4. 35% of 350 **122.5**
5. 75% of 98 **73.5**
6. 2/3 of 1,200 **8**
7. A video recorder is on sale for

EXERCISES

PRACTICE/ SOLVE PROBLEMS

Find the percent of each number using an equation.

1. 87% of 80 **69.6**
2. 25% of 600 **150**
3. 20% of 93 **18.6**
4. 72% of 90 **64.8**
5. 60% of 55 **33**
6. 25% of 10 **2.5**
7. $33\frac{1}{3}$% of 189 **63**
8. $66\frac{2}{3}$% of 210 **140**
9. $87\frac{1}{2}$% of 560 **490**

Find the percent of each number using a proportion.

10. 30% of 210 **63**
11. 80% of 25 **20**
12. 40% of 125 **50**
13. 60% of 80 **48**
14. 45% of 60 **27**
15. 75% of 36 **27**
16. $12\frac{1}{2}$% of 92 **11.5**
17. $8\frac{1}{2}$% of 100 **8.5**
18. $66\frac{2}{3}$% of 60 **40**

Solve.

19. What number is 83% of 20? **16.6**

20. Mario plans to make a 15% down payment on a new mountain bike that sells for $355. What will his down payment be? **$53.25**

21. The regular price of a compact disc player is $210. It is now on sale for 22% off its original price. What is the amount of the savings? **$46.20**

22. A gallon of gasoline costs $1.50. Of that amount, 32% goes to state, local, and highway taxes. How much money per gallon goes to taxes? **$0.48**

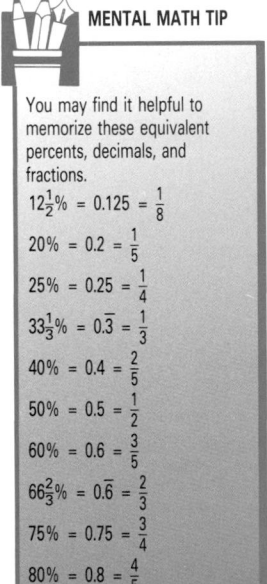

MENTAL MATH TIP

You may find it helpful to memorize these equivalent percents, decimals, and fractions.

$12\frac{1}{2}$% = 0.125 = $\frac{1}{8}$

20% = 0.2 = $\frac{1}{5}$

25% = 0.25 = $\frac{1}{4}$

$33\frac{1}{3}$% = $0.\overline{3}$ = $\frac{1}{3}$

40% = 0.4 = $\frac{2}{5}$

50% = 0.5 = $\frac{1}{2}$

60% = 0.6 = $\frac{3}{5}$

$66\frac{2}{3}$% = $0.\overline{6}$ = $\frac{2}{3}$

75% = 0.75 = $\frac{3}{4}$

80% = 0.8 = $\frac{4}{5}$

MAKING CONNECTIONS

Students can practice using percents by solving statistical problems, such as those that apply to social studies. For example, have students use data on population to write and solve percent problems.

Estimate.

23. 42% of 495 **200**   **24.** 70% of 189 **140**   **25.** 21% of 204 **40**

26. 32% of 63 **20**   **27.** 11% of 375   **28.** 49% of 93 **45**

  37 or 38

Find the percent of each number.

29. 20% of 4,225 **845**   **30.** 2% of 1,550 **31**   **31.** 6% of 16,450 **987**

32. 9.5% of 60 **5.7**   **33.** 54.3% of 300 **162.9**   **34.** 12.5% of 200 **25**

35. 0.1% of 150 **0.15**   **36.** $\frac{3}{4}$% of 80 **0.6**   **37.** 0.1$\frac{3}{4}$% of 12 **0.021**

38. Mavis bought a $90 blazer on sale for 54% off. How much did she spend on the blazer? **$41.40**

39. The manager at Milo's Appliance Center marked up the price of a $129 vacuum cleaner by 43%. How much was the final cost of the vacuum cleaner? **$184.47**

40. Seven percent of Paul's monthly salary is deducted from his payroll check each month for the company's savings plan. He saves $315 per month. What is Paul's monthly salary? **$4,500**

USING DATA Use the Automobile Depreciation Schedule on page 301 for Exercises 41–45. Some of your answers should be given as a range of numbers.

Consider a new car that is worth $10,000.

41. How much does the car depreciate, in dollars, in the first year?
  $3,000 to $3,200
42. What is the trade-in value of the $10,000 car after the first year?
  $6,800 to $7,000
43. Find the maximum dollar amount of depreciation in the second year. **26% of $7,000 = $1,820**

44. Find the minimum dollar amount of depreciation in the second year. **24% of $6,800 = $1,632**

45. Find the trade-in value of the car at the end of the second year. Write an explanation of how you found the trade-in value of the car.

45. **$5,032 to $5,320. The lowest value is found by subtracting 26% of $6,800 from $6,800; the highest value is found by subtracting 24% of $7,000 from $7,000.**

EXTEND/ SOLVE PROBLEMS

ESTIMATION TIP

One way to estimate a percent of a number is to use compatible numbers. For example, this is how you might estimate 34% of 1,098.

Think:
34% is close to 33$\frac{1}{3}$%, or $\frac{1}{3}$.

1,098 is close to 1,200.

$\frac{1}{3}$ × 1,200 = 400

So 34% of 1,098 is about 400.

THINK CRITICALLY/ SOLVE PROBLEMS

25% off the regular price of $199. What is the sale price? **$149.25**

Extension Have students use the distributive property to calculate percents greater than 100%. For example, 145% of 600 = (100% of 600) + (45% of 600) = 600 + 270 = 870.

Section Quiz Calculate. Round answers to the nearest tenth.
1. 25% of 40 **10**
2. 94% of 625 **587.5**
3. 8.4% of 230 **19.3**
4. 15 1/4% of 96 **14.6**
5. 9 1/3% of 240 **22.4**
6. In 20 tries at the basket, Sheena sank 65% of her shots. How many baskets did she make? **13**
7. A coat that originally cost $250 was on sale for 30% off. What was the sale price of the coat? **$175**

Get Ready 8 cards and a paper bag for each pair of students, calculators

CHALLENGE

Wanda has six coins. One third of the coins are dimes. The dimes represent 25% of the total value of the coins. What coins does Wanda have?

WARM-UP

Write each ratio as a percent.
1. 7/20 **35%**   2. 4/20 **20%**
3. 19/20 **95%**   4. 13/65 **20%**
5. 15/25 **60%**   6. 36/10 **360%**

1 MOTIVATE

Explore/Working Together Provide each pair of students with 8 cards and a paper bag. Make sure they understand what to do. Discuss their explanations for finding the percent of a total number.

2 TEACH

Use the Pages/Skills Development Have students read this part of the section and then discuss the examples.

Example 1: Use this example to demonstrate how to translate the question into an equation: What percent (*b*) one number (13) is of another number (65).

Example 2: Use this example to demonstrate how to use a proportion to find what percent one number is of another number.

310

9-3 Finding What Percent One Number Is of Another

EXPLORE/ WORKING TOGETHER

Work with a partner. Write the digits 1 to 8 on 8 different cards and mix up the cards in a paper bag. One partner should select a card from the bag 20 times while the other partner creates a tally chart for the number of times each card is taken from the bag. Remember to put each card back in the bag before taking another.

Copy and complete the tally chart by writing a ratio for the number of times out of 20 that each card was selected. Then calculate the percents from the ratios.

	1	2	3	4	5	6	7	8
Tally								
Ratio								
Percent								

Explain how you might find the tally count if you knew only the percent and the total number of times cards were drawn from the bag.

SKILLS DEVELOPMENT

To find what percent one number is of another, you can write and solve an equation or a proportion.

Example 1

What percent of 65 is 13?

Solution
Write an equation. Let b = the percent.
What percent $\times$ 65 = 13?

$$b \times 65 = 13$$
$$65b = 13$$
$$\frac{65b}{65} = \frac{13}{65}$$
$$b = \frac{13}{65}$$
$$b = 0.2 \qquad \text{So 13 is 20\% of 65.} \blacktriangleleft$$

Example 2

20 is what percent of 10?

Solution
Let the unknown percent be $x\%$, and write this as $\frac{x}{100}$. Use this as the basis for writing a proportion.

$$\begin{array}{c} \text{part} \to \\ \text{whole} \to \end{array} \frac{x}{100} = \frac{20}{10} \begin{array}{c} \leftarrow \text{part} \\ \leftarrow \text{whole} \end{array}$$

$$10x = 2{,}000$$

$$\frac{10x}{10} = \frac{2{,}000}{10}$$

$$x = 200 \qquad \text{So 20 is 200\% of 10.} \quad \blacktriangleleft$$

Example 3

The sale price of the VCR Alicia bought was $360. The sales tax she paid was $23.40. What was the sales tax rate?

Solution

Let the sales tax rate be $s\%$. Write and solve a proportion.

$$\begin{array}{c} \text{part} \to \\ \text{whole} \to \end{array} \frac{s}{100} = \frac{23.40}{360} \begin{array}{c} \leftarrow \text{part} \\ \leftarrow \text{whole} \end{array}$$

$$360s = 2{,}340$$

$$\frac{360s}{360} = \frac{2{,}340}{360}$$

$$s = 6.5 \qquad \text{So the sales tax rate was 6.5\%.} \quad \blacktriangleleft$$

COMPUTER TIP

You can change the program on page 307 to find what percent one number is of another. Below is the adapted program.

```
10 INPUT "ENTER NUMBER
   PART: ";P: PRINT
20 INPUT "ENTER NUMBER
   WHOLE: ";W: PRINT
30 PRINT W;"X = 100*";P
35 PRINT W;"X = ";100 * P
40 X = (100 * P) / W
50 PRINT "X = ";X: PRINT
60 INPUT "RUN AGAIN? Y
   OR N: "; X$: PRINT
70 IF X$ = "Y" THEN
   GOTO 10
```

TRY THESE

Find each percent.

1. What percent of 25 is 15? **60%**
2. What percent of 64 is 32? **50%**
3. What percent of 56 is 14? **25%**
4. What percent of 25 is 45? **180%**
5. What percent of 3 is 12? **400%**
6. What percent of 125 is 87.5? **70%**
7. 48 is what percent of 32? **150%**
8. 3 is what percent of $4\frac{1}{2}$? **$66\frac{2}{3}$%**
9. 12 is what percent of 72? **$16\frac{2}{3}$%**
10. 34 is what percent of 80? **42.5%**
11. 250 is what percent of 625? **40%**
12. 36 is what percent of 10? **360%**

Solve.

13. A 6-volume CD anthology was marked down from $120 to $96. What percent of the original price was the sale price? **80%**

14. Four concert tickets cost $110. Ellison paid $4.73 in sales tax. What was the sales tax rate? **4.3%**

9-3   Finding What Percent One Number Is of Another   **311**

5-MINUTE CLINIC

Exercise	Student's Error	Error Diagnosis
What percent of 600 is 45?	$x \times 600 = 45$ $x/100 = 45/600$ $x = 0.075$ $x = 7.5$	• Student forgets to include the percent sign in the answer.

ASSIGNMENTS

BASIC
1–10, 12, 13–14, PSA 1–2

AVERAGE
1–12, 13–15, 16, PSA 1–2, 5

ENRICHED
5–12, 13–15, 16–17, PSA 1–5

ADDITIONAL RESOURCES
Reteaching 9–3
Enrichment 9–3

Example 3: Have half the students use an equation and the other half use a proportion to solve this problem. Have the two groups compare their results.

Additional Questions/Examples
1. Which ratio represents a percent less than 1%, 1/10, 1/100, or 1/1,000? **1/1,000**
2. Which ratio represents a percent greater than 100%, 46/45, 45/45, or 45/50? **46/45**
3. 60 is what percent of 150? **40%**
4. What percent of 80 is 24? **30%**
5. What percent of $96 is $7.68? **8%**

Guided Practice/Try These   Observe students as they complete the exercises independently. Discuss any problems or questions they may have.

3 SUMMARIZE

Key Questions
1. What kind of proportion is used to solve percent problems? **part/whole = part/whole**
2. How would you use a proportion to find 12% of 50? **12/100 = x/50; x = 6**

3. What equation would you use to find 12% of 50?
$0.12 \times 50 = x$

4. How would you use a proportion to find what percent 18 is of 120?
$x/100 = 18/120; x = 15$

5. What equation would you use to find what percent 18 is of 120?
$x \times 120 = 18$

6. How would you use a proportion to find a number if you know that 90% of the number is 450?
$90/100 = 450/x; x = 500$.

7. What equation would you use to solve Exercise 6?
$0.90 \times x = 450$

4 PRACTICE

Allow calculator use for all exercises.

Practice/Solve Problems Point out that in Exercise 11, the ratio given is for the number of days it rained. The question, however, asks for the percent of the days it did *not* rain.

Extend/Solve Problems Tell students that in Exercise 15, they are asked to round the percent.

Think Critically/Solve Problems These questions allow students to explore what happens to the percent if one of the other numbers doubles.

Problem Solving Applications Observe students to make sure that, for Exercise 3, they base their percents on the total number of games played.

5 FOLLOW-UP

Extra Practice Find each percent to the nearest whole number.
1. What percent of 600 is 450? **75%**
2. 60 is what percent of 75? **80%**

EXERCISES

PRACTICE/
SOLVE PROBLEMS

Find each percent.

1. What percent of 16 is 4? **25%**
2. What percent of 25 is 24? **96%**
3. What percent of 200 is 68? **34%**
4. What percent of 500 is 450? **90%**
5. What percent of 770 is 77? **10%**
6. What percent of 120 is 18? **15%**
7. 56 is what percent of 160? **35%**
8. 1 is what percent of 40? **2.5%**
9. 15 is what percent of 15? **100%**
10. 2.4 is what percent of 200? **1.2%**

Solve.

11. Last April it rained 9 out of 30 days. What percent of the days did it not rain last April? **70%**

12. Ellen purchased a snowboard for $625. The sales tax she paid was $50. What was the sales tax rate? **8%**

EXTEND/
SOLVE PROBLEMS

13. One weekend at a mountain resort, 7 out of 10 guests skied on Saturday and 69 out of 100 guests skied on Sunday. What percent of guests skied on Saturday? on Sunday? **70%; 69%**

14. Refer to Exercise 13. On which day did the greater percent of guests ski? **Saturday**

TALK IT OVER

Why is the sum of the answers to Exercise 15 greater than 100%?

See Additional Answers.

15. Out of a total annual income of $24,000, the Davis family saves $1,500. To the nearest tenth of a percent, what percent of their total income do they save? What percent do they spend? Write and solve an equation to answer each question.
$24,000x = 1,500$; $x = 0.0625 \approx 6.3\%$ $24,000y = 22,500$; $y = 0.9375 \approx 93.8\%$
They save about 6.3% and spend about 93.8%.

THINK CRITICALLY/
SOLVE PROBLEMS

Refer to Exercise 15 to complete Exercises 16–17.

16. Assume that the amount of money saved by the Davis family doubles but that their income remains the same. Does the percent saved also double? **Yes. It is 12.5%.**

17. Assume that the Davises' income doubles but the amount they save remains the same. What happens to the percent saved?

It is halved to 3.1%.

CHALLENGE

A test has been divided into four parts. Part I is worth half as much as Part II. Part II is worth four times as much as Part III. Part IV is worth 37%. What percent are each of Parts I, II and III worth? HINT: Work backwards. **Part I = 18%, Part II = 36%, Part III = 9%**

► READ
► PLAN
► SOLVE
► ANSWER
► CHECK

Problem Solving Applications:

SPORTS AND STANDINGS

The performances of players or teams can be compared by using percents. This chart gives information about field goals for five top-scoring basketball players at Ashton High School.

Player	Field Goals Attempted	Field Goals Made
Gilson	993	560
Johnson	1,307	790
McNance	1,062	585
Parker	937	557
Williams	1,057	588

1. For each player, find the percent of field goals made out of field goals attempted. Round each answer to the nearest tenth of a percent. **56.4%; 60.4%; 55.1%; 59.4%; 55.6%**

2. List the players in order, from the one with the highest percent to the one with the lowest.
Johnson, Parker, Gilson, Williams, McNance

This chart shows the won–lost records of the baseball teams in the Western Division of the American League partway through a recent season.

Team	Games Won	Games Lost
California	55	58
Chicago	66	48
Kansas City	60	52
Minnesota	68	47
Oakland	64	50
Seattle	60	53
Texas	57	54

3. Write as a percent each of the seven ratios of games won to games played. **48.7%; 57.9%; 53.6%; 59.1%; 56.1%; 53.1%; 51.4%**

4. Which team is in first place? Which team is in last place?
Minnesota; California

The chart below gives passing records for five of the top quarterbacks in the National Football League during a recent season.

Quarterback	Passes Attempted	Passes Completed
Eason	448	276
Esiason	469	273
Krieg	375	225
Marino	623	378
O'Brien	482	300

5. Rank the quarterbacks in order of percent of passes completed, from least to greatest. **Esiason, Krieg, Marino, Eason, O'Brien**

9–3   Finding What Percent One Number Is of Another   **313**

AT-RISK STUDENTS

Discuss with students the method they prefer to use when working with percents: writing and solving an equation or writing and solving a proportion.

---

3. What percent of 266.7 is 80? **30%**

4. 35 is what percent of 388.9? **9%**

5. Dinner cost Tony $19. He left the waiter a tip of $2.85. What percent of the bill did Tony leave as a tip? **15%**

6. Inez was absent from school on 9 of the 21 school days in April. What percent of the days in April was she in school? **57%**

Extension Have each student work with a number cube marked 1 to 6. Have each toss the cube 50 times, tallying the results. Have students write their tallies as ratios, express each ratio as a percent, and use the percents to make predictions for 100, 150, 200, and 250 tosses. Then have groups of five students combine their results and compare their individual predictions with the totals.

Section Quiz
1. What percent of 280 is 70? **25%**
2. What percent of 650 is 260? **40%**
3. 55 is what percent of 275? **20%**
4. 60 is what percent of 200? **30%**
5. A baseball stadium has a capacity of 36,000. What percent of the stadium was filled if only 9,000 people attended a game? **25%**
6. On a school bus, 22 of the 40 students are girls. What percent are boys? **45%**

Get Ready calculators

Additional Answers
See page 580.
Additional answers for odd-numbered exercises are found in the Selected Answers portion of the page.

9-4 Percent of Increase and Decrease

EXPLORE/ WORKING TOGETHER

Work in your group to complete these sentences. For each of the following sentences, first simplify the fraction and then write the equivalent percent.

a. When 5 is added to 25, the whole, 25, has been increased by $\frac{5}{25}$, or ___?___. So 5 added to 25 represents an increase of ___?___%. $\frac{1}{5}$; 20

b. When 5 is subtracted from 30, the whole, 30, has been decreased by $\frac{5}{30}$, or ___?___. So 5 subtracted from 30 represents a decrease of ___?___%. $\frac{1}{6}$; $16\frac{2}{3}$

Discuss the following with your group.

c. It is true that $25 + 5 = 30$ and $30 - 5 = 25$.
It is false that 5 added to 25 and 5 subtracted from 30 represent the same percent of increase and decrease.

SKILLS DEVELOPMENT

The **percent of increase** tells what percent the amount of increase is of the original amount.

The **percent of decrease** tells what percent the amount of decrease is of the original amount.

To find the percent of increase or decrease, express the ratio of the amount of increase or decrease to the original amount as a percent.

Example 1

Find the percent of increase.
Original number: 70
New number: 84

Solution
Find the amount of increase.
$$84 - 70 = 14$$

Write a ratio.
$$\frac{\text{amount of increase}}{\text{original number}} = \frac{14}{70}$$

Find the percent.
$$\frac{14}{70} = 0.2 = 20\%$$

So the percent of increase is 20%. ◄

Example 2

Find the percent of decrease. $200 to $175

Original amount: $200

New amount: $175

Solution

Find the amount of decrease.

Write a ratio.

$$\frac{\text{amount of decrease}}{\text{original number}} = \frac{\$25}{\$200} = \frac{1}{8}$$

Find the percent.

$$\frac{1}{8} = 12\frac{1}{2}\%$$

So the percent of decrease is $12\frac{1}{2}\%$. ◄

Example 3

At the beginning of Barbara's exercise program she walked 40 minutes a day. After six weeks she walked 55 minutes a day. Find the percent of increase.

Solution

Find the amount of increase.

$$55 - 40 = 15$$

Write a ratio.

$$\frac{15}{40} = \frac{3}{8} = 0.375$$

Find the percent.

$$0.375 = 37.5\%$$

So the percent of increase from 40 to 55 is 37.5%. ◄

Example 4

In May, the round-trip airfare from Denver to Newark was $440. In January, the airlines reduced the fare to $330 round trip. Find the percent of decrease.

Solution

Find the amount of decrease.

$$\$440 - \$330 = \$110$$

Write a ratio.

$$\frac{\$110}{\$440} = \frac{1}{4}$$

Find the percent.

$$\frac{1}{4} = 25\%$$

So the percent of decrease is 25%. ◄

9–4   Percent of Increase and Decrease   **315**

5-MINUTE CLINIC

Exercise	Student's Error	Error Diagnosis
Find the percent of increase from 202 to 305.	305 − 202 = 103 103/305 = 33.8%	• Student expresses the percent of increase as the ratio of the increase to the new number, rather than to the original number. Correct: 103/202 = 51%

315

Allow students to use calculators. Point out that students are to round all answers to the nearest tenth.

Practice/Solve Problems Before solving Exercise 14 have students estimate their answer. Be sure they understand why "about 50%" is a good estimate.

Extend/Solve Problems Discuss students' techniques for solving Exercise 20. (Some may find the percent of decrease at Computer Discount and compare that with 40%; others may find 40% of $750 and compare that with $479. The latter technique will give them the answer to both questions in one step.)

Think Critically/Solve Problems For Exercise 23 explain that students are to find the percent of increase in population in each region and compare the results.

5 **FOLLOW-UP**

Extra Practice
Find the percent of increase. Round to the nearest tenth.
1. Original price: $340
   New price: $400   **17.6%**
2. Original weight: 20 lb
   New weight: 35 lb   **75%**
Find the percent of decrease. Round to the nearest tenth.
3. Original fare: $315
   New fare: $255   **19%**
4. Original price: $41,599
   New price: $39,700   **4.6%**
5. Walter bought a house for $130,000. Twelve years later, he sold the house for $540,000. What was the percent of increase?   **315%**

316

TRY THESE

Find the percent of increase.

1. Original price of refrigerator: $600
   New price of refrigerator: $630 **5%**

2. Original number of vacation days: 14
   New number of vacation days: 21 **50%**

3. Original weight: 140 lb
   New weight: 161 lb **15%**

4. Original salary: $450
   New salary: $513 **14%**

5. Original fare: $75
   New fare: $99 **32%**

6. Original number of employees: 375
   New number of employees: 600 **60%**

Find the percent of decrease.

7. Original population: 840
   New population: 504 **40%**

8. Original price: $250
   New price: $175 **30%**

9. Original beats per minute: 80
   New beats per minute: 72 **10%**

10. Original airfare: $680
    New airfare: $646 **5%**

11. Original score: 185
    New score: 148 **20%**

12. Original weight: 200 lb
    New weight: 144 lb **28%**

Solve.

13. The Jacksons bought a house for $145,000 four years ago. They sold the house this year for $174,000. What is the percent of increase? **20%**

14. The original price of a gold bracelet is $325. The sale price is $260. What is the percent of decrease? **20%**

EXERCISES

PRACTICE/ SOLVE PROBLEMS

Find the percent of increase.

1. Original rent: $500
   New rent: $550 **10%**

2. Original price: $0.25
   New price: $1.50 **500%**

3. Original price: $120
   New price: $132 **10%**

4. Original weight: 125 lb
   New weight: 135 lb **8%**

5. Original number: 75
   New number: 105 **40%**

6. Original number: 32
   New number: 128 **300%**

MAKING CONNECTIONS

Have students clip sale advertisements from their local newspaper. Ask them to use the information in their ad to create a problem that deals with percent of increase or percent of decrease. Have them exchange papers and solve each other's problems.

Find the percent of decrease.

7. Original number of players: 80
   New number of players: 60 **25%**

8. Original population: 320
   New population: 208 **35%**

9. Original price: $500
   New price: $410 **18%**

10. Original fare: $800
    New fare: $750 **6 $\frac{1}{4}$%**

11. Original salary: $900
    New salary: $765 **15%**

12. Original price: $680
    New price: $578 **15%**

Solve.

13. The regular price of a handheld video game is $40. The sale price is $34. What is the percent of decrease? **15%**

14. Molly bought a used car for $5,500. She sold it two years later for $2,475. What is the percent of decrease? **55%**

15. After a workout, Bill's pulse rate rose from 70 beats per minute to 91 beats per minute. What is the percent of increase? **30%**

Find the percent of increase or decrease.

EXTEND/ SOLVE PROBLEMS

16. $1500 to $300 **80% decrease**

17. 16,000 to 19,840 **24% increase**

18. 0.45 to 0.27 **40% decrease**

19. $360 to $378 **5% increase**

Solve.

20. Daniel is shopping for a new computer monitor that sells for $750 retail. The monitor is on sale at Computer Discount for $479. Byteland is having a 40% off sale on the same monitor. Which is the better buy? How much could Daniel save off the original price? **Byteland; $300**

21. A leather jacket was priced at $550 in the store. The jacket cost the owner of the store $250. What is the owner's percent of increase over the owner's cost? **120%**

22. An $8 item is marked up 50%. Then the new price is marked down 50%. Is the final price $8? Explain.
    No, $6. The markdown is 50% of 12.

THINK CRITICALLY/ SOLVE PROBLEMS

23. *USING DATA* Use the Data Index on page 546 to find the table of United States Resident Population Change by Region. Use a calculator. Determine which regions of the United States had the greatest and least percent of increase in population from 1980 to 1990. Northeast, about 3.5% increase (least); West, about 22.2% increase (greatest)

9–4   Percent of Increase and Decrease   **317**

6. Tabatha is shopping for a new TV that usually sells for $1,200. The set is on sale at TV Land for $989. Video Plus is selling the same TV for 30% off. Which store has the better buy? by how much? **Video Plus; $149**

Extension Have students study the following calculator keystrokes to find the new number when there is a percent increase or decrease.

% Increase
Original: 25
Increase: 45%

25 [x] 145 [%] 36.25

% Decrease
Original: 25
Decrease: 45%

25 [x] 55 [%] 13.75

Have students use their calculators to find the new number to the nearest tenth.

1. Original: 150
   Increase: 38% **207**
2. Original: 575
   Decrease: 74% **149.5**
3. Original: 977
   Increase: 55% **1,514.4**

Section Quiz
Find the percent of increase. Round to the nearest tenth.
1. Original salary: $23,500
   New salary: $25,400 **8.1%**
2. Original number: 456
   New number: 584 **28.1%**
Find the percent of decrease. Round to the nearest tenth.
3. Original price: $57.99
   New price: $46.40 **20%**
4. Original weight: 206 lb
   New weight: 175 lb **15%**

Get Ready calculators

CHALLENGE

Talbot High School has an enrollment of 500 students. During the school year, 45 students moved out of the district and 60 new students moved in. What was the overall percent of increase or decrease? **3% increase**

9-5 Finding a Number When a Percent of It Is Known

EXPLORE/ WORKING TOGETHER

The manager of a discount clothing outlet estimated that 75% of the number of items sold were sold to people under 20 years old. Last year the outlet sold about 52,500 items. About how many items were sold to people under 20 years old?

You can draw a triangle diagram to model the parts of any percent problem.

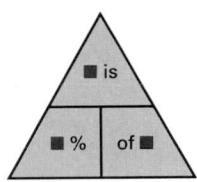

a. Draw and complete a triangle diagram to represent the situation above. **x is 75% of 52,500**

b. Use the diagram to write an equation. **x = 75% × 52,500**

c. Solve the equation. **39,375**

d. Answer the question.
About 40,000 items were sold to people under 20 years old.

SKILLS DEVELOPMENT

To find a number when a percent of it is known, you can write and solve an equation or a proportion.

Example 1

4.2% of what number is 2.31?

Solution
Write an equation. Let n represent the whole amount.
4.2% of what number is 2.31?

$$4.2\% \times \quad n \quad = 2.31$$
$$0.042n = 2.31$$
$$\frac{0.042n}{0.042} = \frac{2.31}{0.042}$$
$$n = 55 \qquad \text{So 4.2\% of 55 is 2.31.} \quad \blacktriangleleft$$

Example 2

$\frac{3}{8}\%$ of what number is 30?

Solution

Write and solve an equation. Let n represent the whole amount.

$\frac{3}{8}\%$ of what number is 30?

$\frac{3}{8}\% \times \quad n \quad = 30$

$\frac{3}{800}n = 30$

$\frac{800}{3}\left(\frac{3}{800}n\right) = \frac{800}{3}(30)$

$n = 8,000$

So $\frac{3}{8}\%$ of 8,000 is 30. ◄

Example 3

20 is 25% of what number?

Solution

Write 25% as $\frac{25}{100}$. Use this as the basis for writing a proportion. Let x represent the whole amount.

$\text{part} \rightarrow \frac{25}{100} = \frac{20}{x} \leftarrow \text{part}$
$\text{whole} \rightarrow \qquad\qquad \leftarrow \text{whole}$

$25x = 2,000$

$\frac{25x}{25} = \frac{2,000}{25}$

$x = 80$

So 20 is 25% of 80. ◄

Example 4

Joanne took Annemarie to a restaurant for lunch. Joanne left a tip of $5.25, 15% of the bill. How much was the bill for lunch?

Solution

Write and solve an equation. Let n represent the whole amount.

$5.25 is 15% of what number?

$\$5.25 = 15\% \times \quad n$

$5.25 = 0.15n$

$\frac{5.25}{0.15} = \frac{0.15n}{0.15}$

$35 = n$

So the bill for lunch was $35. ◄

COMPUTER TIP

If you want to solve these problems using a proportion, use the adaptation of the program on p. 307 to find a number when a percent of it is known.

```
10 INPUT "ENTER % PART:
   ";P: PRINT
20 INPUT "ENTER NUMBER
   PART: ";N: PRINT
30 PRINT P;"X = 100*";N
35 PRINT P;"X";" = ";100
   * N
40 X = (100 * N) / P
50 PRINT "X = ";X: PRINT
60 INPUT "RUN AGAIN? Y
   OR N: ";X$: PRINT
70 IF X$ = "Y" THEN
   GOTO 10
```

ASSIGNMENTS

BASIC
1–13, 15–17

AVERAGE
3–14, 15–21

ENRICHED
5–14, 18–21, 22–24

ADDITIONAL RESOURCES
Reteaching 9–5
Enrichment 9–5

Examples 1–3: Use these examples to introduce the third type of problem, finding the number when a percent of it is known. Ask students to use the triangle diagram to explain how to write equations for questions.
Example 4: Stress the use of the triangle diagram for solving word problems.

Additional Questions/Examples

1. Alex is driving from North Carolina to Florida in two days. He increased the distance he drove on the second day over the distance he drove the first day by 20%. If he drove 360 miles the second day, how many miles had he driven the first day? **20% of what number is 360 less that number? 20% × n = 360 − n; n = 300**

2. The price of a cassette deck was increased by 12%. If the original price was $145, what is the new price? **145 + 12% × 145 = n; n = $162.40**

Guided Practice/Try These serve students as they complete these exercises independently. Discuss any questions or problems they may have. Stress the use of the triangle diagram for Exercise 7.

5-MINUTE CLINIC

Exercise	Student's Error	Error Diagnosis
Solve. 152 is 38% of what number?	$152 \times 0.38 = 57.76$	• Student automatically multiplies when seeing a percent problem and so uses an incorrect equation. Correct: $0.38 \times n = 152$ $\qquad\qquad\qquad n = 400$

Talk It Over Draw the triangle diagram on the board. Ask volunteers to make up percent problems. Use a triangle diagram to solve each one. Continue until all three types of percent problems have been covered.

4 PRACTICE

Allow students to use calculators.

Practice/Solve Problems Exercise 14 is similar to Exercise 2 in **Additional Questions/Examples.** If necessary, help students set up the equation.
34 + 20% × n = n

Extend/Solve Problems Exercises 15–20 deal with finding the amount of increase in volume when the percent of increase is given.

Think Critically/Solve Problems Exercise 21 is a multiple-step problem. If necessary, review the steps with students. Exercises 22–24 involve a computer program written in BASIC.

5 FOLLOW-UP

Extra Practice
1. 56% of what number is 140?
   250
2. 5.4 is 12% of what number? **45**
3. 33 1/3% of what number is 33?
   99
4. 3 is 0.6% of what number?
   500
5. 3/4% of what number is 3.6?
   480
6. A painting that sells for $545 was marked down by 15%. What is the sale price of the painting? **$463.25**

TRY THESE

Find each answer.

1. 25% of what number is 6? **24**
2. 42 is 60% of what number? **70**
3. 31.5 is 75% of what number? **42**
4. 29% of what number is 145? **500**
5. $26\frac{2}{3}$% of what number is 21.6? **81**
6. 80 is 20% of what number? **400**

7. A real estate agent received a $9,900 commission, which was 6% of the selling price. At what price did the agent sell the house? **$165,000**

EXERCISES

PRACTICE/ SOLVE PROBLEMS

Find each answer.

1. 35% of what number is 54.6? **156**
2. 4.68 is 18% of what number? **26**
3. $33\frac{1}{3}$% of what number is $29\frac{1}{3}$? **88**
4. 2 is 0.8% of what number? **250**
5. 8.3 is 10% of what number? **83**
6. 88 is 22% of what number? **400**
7. 68% of what number is 17? **25**
8. 15 is 18% of what number? **$83\frac{1}{3}$**
9. 20% of what number is 65? **325**
10. 3 is 0.1% of what number? **3,000**
11. $\frac{1}{2}$% of what number is 3? **600**
12. 18 is $2\frac{1}{2}$% of what number? **720**

13. Suppose that 50,400 people attended a football game at the Cotton Bowl Stadium in Dallas, Texas. If the people who attended the game filled 70% of the seats in the stadium, what is the seating capacity of the stadium? **72,000 seats**

14. The manager of a construction company totals the costs of labor and materials and then adds 16%. The company has just been awarded a bid for a recreational center. If the added cost is $32,000, what is the cost of labor and materials? **$200,000**

MIXED REVIEW

1. Write 19.5% as a fraction and a decimal. **$\frac{39}{200}$, 0.195**
2. Find 5% of 15.6. **0.78**
3. What percent of 84 is 7? **$8\frac{1}{3}$%**
4. 11 is what percent of 88? **12.5%**

Multiply or divide. Be sure the answer is in lowest terms.

5. $6\frac{2}{3} \times 2\frac{1}{3}$ **$15\frac{5}{9}$**
6. $2\frac{2}{3} \div 8$ **$\frac{1}{3}$**
7. $6\frac{3}{10} \times 1\frac{2}{3}$ **$10\frac{1}{2}$**
8. Find the perimeter of a square patio that measures $22\frac{1}{3}$ yd on a side. **$89\frac{1}{3}$**

Do you know why ice floats? Any object will float if it is less dense, or less concentrated, than the fluid in which it is immersed. The **density** of a material is found by dividing its mass by its volume.

When water freezes, its volume increases. Its mass remains the same, so its density decreases. Therefore, ice is less dense than water and so ice will float.

Assume that, when water freezes, the volume of ice that is formed is 112% of the volume of the water. Find the volume of water that formed each amount of ice.

15. 112 cm³ **100 cm³**   16. 168 cm³ **150 cm³**   17. 672 cm³ **600 cm³**

18. 95.2 cm³ **85 cm³**   19. 336 cm³ **300 cm³**   20. 476 cm³ **425 cm³**

21. **USING DATA** Use the Data Index on page 546 to find state sales tax rates.
During a recent vacation, Sharon bought these souvenirs.

Flag $3.69   Postcard set $6.25
Slides $16.25   Wood carvings $36.95

If Sharon paid $66.77 for the souvenirs, including sales tax, in which state did she make the purchase? **Nevada**

The computer program is written in BASIC.

22. Describe what the computer program does.

23. How would you use the program for a sales tax rate of 8.5%?
Enter 8.5 for R in Line 20.

24. Explain why .5 is added to the product of P and R in Line 50. (Hint: The INT function on a computer, INT(X), always gives the greatest integer that is less than or equal to *x*.)

to round dollar amounts to the nearest cent

```
10 PRINT "SALES TAX AND
   TOTAL AMOUNTS"
20 INPUT "ENTER THE
   PERCENT TAX RATE";R
30 PRINT "PRICE",
   "PERCENT TAX RATE",
   "TAX", "TOTAL PRICE"
40 READ P
50 TAX = INT(P*R + .5)/100:
   TOTAL = P + TAX
60 PRINT "$"P, R"%", "$"TAX,
   "$"TOTAL
70 IF P = 45.89 THEN END
80 GOTO 40
90 DATA 28.79, 56.87, 112.13,
   67.34, 81.08, 45.89
```

EXTEND/ SOLVE PROBLEMS

THINK CRITICALLY/ SOLVE PROBLEMS

22. **computes sales tax and totals for six given prices**

9–5   Finding a Number When a Percent of It Is Known   **321**

9-6 Problem Solving Skills:
CHOOSE A COMPUTATION METHOD

► READ
► PLAN
► SOLVE
► ANSWER
► CHECK

Many problems can be solved using different computation methods such as mental math, estimation, paper and pencil, or a calculator.

When you encounter a problem, first read it carefully. Then, develop a plan. Ask:
► Do you need an exact answer or an estimate?
► Which computation method is appropriate?

Try to solve the problem using mental math. If this method is not appropriate, try pencil and paper or a calculator.

PROBLEM

Stacey works as a clerk. She earns $6.73 per hour. One week she worked 10.5 hours. How much money did she earn?

SOLUTION

Which computation method is appropriate?

Pencil and paper

$$\begin{array}{r} \$6.73 \\ \times \ \ 10.5 \\ \hline 3365 \\ 000 \\ 673 \\ \hline \$70.665 \approx \$70.67 \end{array}$$

Can the computation be done quickly with pencil and paper?

Estimation

$$\begin{array}{r} \$6.70 \\ \times \ \ \ 10 \\ \hline \$67.00 \end{array}$$

Is an exact answer needed?

Calculator

6.73 $\boxed{\times}$ 10.5 $\boxed{=}$ $\boxed{70.665}$ Can the computation
                           ↓   be done quickly
                        $70.67   with a calculator?

Using a calculator is probably the best method to solve this problem.

322   CHAPTER 9   Exploring Percent

PROBLEMS

Indicate which computation method you would use to solve each problem.

1. Dawn bought mascara for $7.00, nail polish for $2.25, and shampoo for $6.00. How much did she spend? **mental math; $15.25**

2. Justin is hanging a wallpaper border in his rectangular-shaped bathroom. The bathroom is 6 feet 3 inches long and 4 feet 6 inches wide. What is the perimeter of his bathroom? **mental math; $21\frac{1}{2}$ ft**

3. Penny stopped at the fruit stand on her way home from school. She spent $5.27 for watermelon, $2.39 for pears, $6.19 for green peppers, and $3.05 for apples. How much money did she spend? **calculator; $16.90**

4. Justin wants to buy 4 new custom rims for his car tires. Each rim sells for $129.99. He has $500. Does he have enough money? **estimate; no**

5. Bonnie is dividing her collection of 335 baseball cards evenly among herself and her 4 brothers. How many baseball cards will each person receive? **mental math; 67**

6. Mel wants to buy 3 compact discs. Each disc sells for $14.98, including tax. If Mel has $45, does he have enough money? **estimation; yes**

7. The bowling team has decided to buy new shoes for the upcoming tournament. There are 6 people on the team. Bowling shoes cost $58.95 per pair plus 8% sales tax. How much money does the team need to buy the shoes? **calculator; $382.00 or $382.02, depending on method used**

ASSIGNMENTS

BASIC
1-7

AVERAGE
1-7

ENRICHED
1-7

ADDITIONAL RESOURCES
Reteaching 9-6
Enrichment 9-6

---

4 PRACTICE

Problems Some students may opt to use paper and pencil, rather than mental math, for Problem 2. For Problem 5, make sure students understand that the divisor is 5, not 4.

---

5 FOLLOW-UP

Extra Practice Indicate which computation method you would use to solve each problem. Then solve.

1. Henrietta is making aprons as Christmas gifts. She needs 1 1/2 yd of fabric for each apron. If she is making 15 aprons, how much fabric should she buy? **mental math; 22.5 yd**

2. Juanita used 1 2/3 c of milk and 3/4 c of water to make a dish for dinner. How much liquid did she use? **paper and pencil; 2 5/12 c**

Get Ready calculators

9-7 Discount

VIDEO VIBES
Special Sale
27" Remote
COLOR TV
NOW **$50** OFF
REGULAR PRICE $459.99

EXPLORE

Drucilla is shopping for a new TV set. She finds this ad in Sunday's paper. As a Video Vibes club member, Drucilla gets 12% off the regular price of store merchandise on Tuesdays. If she wants the TV set advertised, should she buy it at the sale price, or should she wait until Tuesday to buy it?

a. What is 12% of $459.99? **$55.20**

b. How much would Drucilla save by waiting until Tuesday to make the purchase? **$5.20**

SKILLS DEVELOPMENT

If an item is on sale, you can save money by buying it at less than the regular price.

The **discount** is the amount that the regular price is reduced.

The **sale price** is the regular price less the discount.

WRITING ABOUT MATH

Write a paragraph in your journal explaining the difference between the terms *regular price*, *discount*, and *sale price*.

Example 1

Find the discount and sale price of an item that sells for $298.95 with an 18% discount.

Solution
Find the discount.

$$\text{discount} = \text{percent of discount} \times \text{regular price}$$

$$\begin{aligned} \text{discount} &= 18\% \times \$298.95 \\ &= 0.18 \times \$298.95 \\ &= \$53.811, \text{ or } \$53.81 \end{aligned}$$

The discount is $53.81.

Find the sale price.

$$\begin{aligned} \text{sale price} &= \text{regular price} - \text{discount} \\ \text{sale price} &= \$298.95 - \$53.81 \\ &= \$245.14 \end{aligned}$$

The sale price is $245.14. ◄

324   CHAPTER 9   Exploring Percent

Example 2

 Use a calculator to find the sale price if it is $15\frac{1}{2}$% off the regular price of $359.62.

Solution

Since discount = percent of discount × regular price, you can use this key sequence.

359.62 ⨉ 15.5 % `55.7411`

+/− + 359.62 = `303.8789`

The sale price is $303.88. ◄

Example 3

Kayla went to a 25%-off sale at Lacy's Department Store. The regular price of a leather belt is $48. Find the sale price.

Solution

Think: $25\% = \frac{1}{4}$

$\frac{1}{4}$ of 48 = 12

$48 − $12 = $36

The sale price of the belt is $36. ◄

Try These

Find the discount and the sale price. Round your answers to the nearest cent.

1. regular price: $415.38
   percent of discount: 15%
   D = $62.31 S = $353.07
2. regular price: $63.45
   percent of discount: 25%
   D = $15.86 S = $47.59
3. regular price: $89.95
   percent of discount: 50%
   D = $44.98 S = $44.98
4. regular price: $369.99
   percent of discount: 20%
   D = $74 S = $295.99
5. regular price: $115.18
   percent of discount: 3%
   D = $3.46   S = $111.72
6. regular price: $3,880
   percent of discount: 12.5%
   D = $485 S = $3,395
7. regular price: $1,385
   percent of discount: 5%
   D = $69.25 S = $1,315.75
8. regular price: $8,175
   percent of discount: 16%
   D = $1,308 S = $6,867

Solve.

9. Francie bought a $49.95 sweater on sale for 20% off. What was the sale price? **$39.96**

10. Bruce went to a 50% end-of-ski-season sale. How much did he pay for $1,580 worth of regular-priced merchandise? **$790**

ASSIGNMENTS

BASIC
1–7, 9–12, 13, PSA 1–7

AVERAGE
5–8, 9–12, 13–14, PSA 1–7, 9

ENRICHED
7–8, 9–12, 13–15, PSA 1–9

ADDITIONAL RESOURCES
Reteaching 9–7
Enrichment 9–7

Example 3: Use this example to develop awareness of the fact that sometimes fractions can be used, instead of decimals, to find the sale price.

Additional Questions/Examples

1. The price of a vest was reduced 15%. After two weeks, it was reduced an additional 8%. If the regular price was $34.98, what was the sale price two weeks later? **$27.35**
2. Are the computations done in Exercise 1 the same as taking 23% off the regular price? **no**
3. Which reduction, the one in Exercise 1 or the one in Exercise 2, represents a greater savings? **23% off the regular price**
4. A pair of skis was discounted by 35%. Judd paid cash and received an additional 5% off the sale price. If he paid $249.50 for the skis, what was the regular price? HINT: Work backward. **x = sale price; 249.50 = x − 0.05x; x = 262.63; y = regular price; 262.63 = y − 0.35y; y = $404.05.**

Guided Practice/Try These Allow calculator use.

EXERCISES

Find the discount and the sale price. Round your answers to the nearest cent.

1. regular price: $150.50
   percent of discount: 10%
   D = $15.05 S = $135.45
2. regular price: $389.55
   percent of discount: 20%
   D = $77.91 S = $311.64
3. regular price: $689.99
   percent of discount: 18.5%
   D = $127.65 S = $562.34
4. regular price: $1,450.39
   percent of discount: 25%
   D = $362.60 S = $1,087.79
5. regular price: $511.09
   percent of discount: 13%
   D = $64.44 S = $446.65
6. regular price: $48.88
   percent of discount: 3%
   D = $1.47 S = $47.41

Solve.

7. A $410 ski jacket is on sale for 75% off. What is the sale price? **$102.50**

8. The regular price of a Caribbean cruise is $895. For a limited time, the price is being reduced by 28%. How much would you save if you traveled at the reduced rate? **$250.60**

The Pants Palace discounts all merchandise that has been on the racks for six weeks. Any merchandise remaining another four weeks is discounted again by taking a percent of the first discount price. Complete the table to show the final sale price.

	Regular Price	First Discount	Second Discount	Sale Price
9.	$89.00	10%	15%	See Additional Answers. ■
10.	$384.50	5%	20%	■
11.	$25.75	25%	10%	■
12.	$2,853.27	30%	5%	■

13. During a special sale, a guitar was marked 20% off and sold for $85. A banjo was marked down 25% and sold for $72. Which instrument had the greater amount of discount? **the banjo**

14. James received a 25% discount on a suit during a sale. He also received a 5% discount on the sale price for paying cash. If he paid $120.50 for the suit, what was the regular price of the suit? **$169.12**

No. The 35% reduction represents a greater saving. The additional 15% is based on a sale price lower than the original price.

15. A blouse that is reduced by 20% one week is marked down an additional 15% after two weeks. Does this blouse cost the same as if it had been reduced by 35% at the start of the sale? Explain.

Problem Solving Applications:

COMMISSIONS ON SALES

Many salespeople work on commission. This means that they earn an amount of money that is a percent of their total sales. The percent is called the **commission rate,** and the amount of money is called the **commission.** Often, their income is made up of a combination of salary and commission.

Solve. Express dollar answers to the nearest cent.

1. Corretta is a real estate agent. Her commission rate is 3%. Find her commission on a house that sold for $85,000. **$2,550**

2. Laura's commission rate for selling cosmetics is 15%. If her sales totaled $225 one week, how much commission did she receive? **$33.75**

3. One week, Alberto sold $42,000 worth of computers. His commission was $1,260. Find his commission rate. **3%**

4. Shirley receives a base salary of $135 a week and a commission rate of 2% of sales. What did she earn in a week in which she sold $1,859 worth of cosmetics? **$172.18**

5. Rosemarie receives commissions at the rate of 12% for the first $15,000 of sales and 30% for sales in excess of $15,000. What is her commission on $27,000 of sales? **$5,400**

6. Jackson sells two cars. Each car is priced at $9,245.80, and his commission rate is $2\frac{1}{2}$%. How much commission does he receive? **$462.29**

7. Robin sells cars on commission. She receives 5% for new cars and 4% for used cars. Today she sold two used cars totaling $7,120 and one new car selling for $6,895. How much commission did she earn? **$629.55**

8. Lemuel sells furniture. He receives a base salary of $150 a week and a commission on sales. One week he sold $6,800 worth of furniture. His combined salary and commission was $320. What is his commission rate? **2.5%**

9. Juanita receives a commission of 25% on sales of merchandise in excess of $2,000. One week her sales were $7,500. What was the amount of her commission for that week? **$1,375**

MIXED REVIEW

Find each answer. Round decimal quotients to the nearest thousandth.

1. 7 ÷ 12 **0.583**

2. 0.03 × 650 **19.5**

3. 1.45 × 62 **89.9**

4. 745 ÷ 640 **1.164**

5. −18 + 7 **−11**

6. 42 − (−8) **50**

7. 9 × (−12) **−108**

8. −250 ÷ (−10) **25**

9. −25 + (−16) **−41**

10. If the cost of fencing is $4.60 per meter, how much would it cost to enclose a rectangular garden that measures 9.2 m by 4.6 m? **$126.96**

9–7   Discount   **327**

3. Mrs. Yamoto received an 18% discount on a dishwasher. If she paid $360.39, what was the regular price? **$439.50**

4. A store plans to decrease its prices by 25% in January, then increase them by 25% in February. Will the final prices be higher or lower than before the changes? What percent higher or lower? **lower; 6.25% lower**

Extension Provide students with mail-order catalogs. Tell them to assume that they have $250 to spend and that all prices that appear in the catalogs are now reduced by 15%. Have students list the items they will buy, the regular prices, the discounts, and the sale prices. Have them add the amount of sales tax appropriate to your area. Remind them that their totals cannot go over $250.

Section Quiz Find the discount and the sale price. Round answers to the nearest cent.

1. regular price: $224.75
   percent of discount: 15%
   d = $33.71; sp = $191.04

2. regular price: $4,837.50
   percent of discount: 16.5%
   d = $798.19; sp = $4,039.31

3. Bill bought a $57 pair of shoes for 20% off. What was the sale price? **$45.60**

4. The regular price of a $89.98 sweat suit is being reduced by 35%. How much would you save if you bought the sweat suit at the reduced price? **$31.49**

Get Ready calculators

Additional Answers

See page 580.

CHALLENGE

Georgia has a choice of how she will be paid. She may take a salary of $10,000 per year plus a commission of 15% on everything she sells, or she may take only a commission of 20%. What must her total sales be to make the second offer worthwhile? **over $200,000**

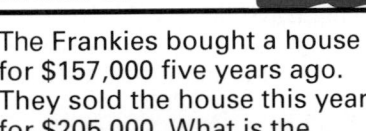

WARM-UP

The Frankies bought a house for $157,000 five years ago. They sold the house this year for $205,000. What is the percent of increase? **30.6%**

1 MOTIVATE

Explore Have students share their results. Point out the importance of applying the order of operations in computing the entries in the Amount column.

2 TEACH

Use the Pages/Skills Development Have students read this part of the section and then discuss the examples. After discussing the new vocabulary, have students tell what part of a year 3, 4, 6, 8, 9, and 10 months represent. **1/4, 1/3, 1/2, 2/3, 3/4, and 5/6**

EXPLORE

a. Follow the pattern. Copy and complete the table.

Number of Years	Amount	
1	$100 + 1 × 5% of $100 = ■	**$105**
2	$100 + 2 × 5% of $100 = ■	**$110**
3	$100 + 3 × 5% of $100 = ■	**$115**
4	$100 + 4 × 5% of $100 = ■	**$120**
5	$100 + 5 × 5% of $100 = ■	**$125**

b. What would be the entry in the Amount column for 8 years?
$100 + 8 × 5% of $100 = $140

c. What do you think the amount would be for 6 months? **$102.50**

SKILLS DEVELOPMENT

Interest is money that is paid for the use of money over a period of time. For example, if you put money into a savings account, a bank will pay you interest for the use of that money over a given period of time. If you borrow money from a bank, the bank will charge you interest for the use of their money for a given period of time.

The amount of money that is earning interest or that you are borrowing at interest is called the **principal. Simple interest** is paid only on the principal.

The **rate** is the percent of the principal charged for the use of the money over a given period of time, usually a year.

The **time** is the number of time periods during which the principal remains in the bank account or has not been paid back. Time is usually expressed in years or parts of a year.

The interest (I) earned or paid on a given principal (p) at a given rate (r) over a time period (t) is given by a formula.

$$I = prt$$

Example 1

Find the interest on $3,000 for 2 years at a rate of 8% per year.

Solution
Use the formula $I = prt$ to solve.
Substitute the values into the formula.
$I = \$3,000 × 8\% × 2$
$I = \$3,000 × 0.08 × 2$
$I = \$480$ The interest is $480. ◄

COMPUTER TIP

The spreadsheet is a type of computer application software that is widely used today. A spreadsheet is made up of rows and columns into which you can enter information. The computer will perform calculations on the numerical information in a spreadsheet according to formulas that you enter. If you have spreadsheet software available, you can use its power to investigate various "what if" situations.

328   CHAPTER 9   Exploring Percent

TEACHING TIP

Demonstrate that the simple-interest formula may be used to calculate any one of the four variable quantities, given any three. Display the formulas on a chart for student reference.

$I = prt$ $p = I/rt$ $r = I/pt$ $t = I/pr$

Example 2

Find the interest and the total amount in the account for a
$6,300 deposit for 4 years at a rate of 5.4% per year.

Solution
Find the interest.

$I = prt$
$I = \$6,300 \times 5.4\% \times 4$
$I = \$6,300 \times 0.054 \times 4$
$I = \$1,360.80 \leftarrow$ interest

Find the total amount by adding the interest to the principal.
$\$6,300 + \$1,360.80 = \$7,660.80$

So the total amount in the account after 4 years
is $7,660.80. ◄

When you repay money borrowed, the **amount due** is equal to the
principal plus the interest accrued.

Example 3

Kendra borrowed $4,125 from a bank at an annual interest rate of
18.5%. The term of the loan was 36 months. Find the simple interest
and the total amount due on the loan.

Solution
Find the interest.

$I = prt$
$I = \$4,125 \times 0.185 \times 3 \leftarrow$ **Change 36 months to 3 years.**
$I = \$2,289.375 \approx \$2,289.38$

Find the total amount due.
$\$4,125 + \$2,289.38 = \$6,414.38$

Kendra paid $2,289.38 in simple interest and $6,414.38
over the life of the loan. ◄

TRY THESE

Find the interest.

1. Principal: $1,000
   Rate: 8%/year
   Time: 2 years **$160**

2. Principal: $895
   Rate: 17.5%/year
   Time: 3 years **$469.88**

3. Principal: $1,300
   Rate: 5.4%/year
   Time: 1 year **$70.20**

4. Principal: $475
   Rate: 6%/year
   Time: 6 months
   $14.25

5. Principal: $900
   Rate: 7.5%/year
   Time: 24 months
   $135

6. Principal: $3,450
   Rate: 13%/year
   Time: 12 months
   $448.50

9-8   Simple Interest   **329**

Examples 1 and 2: Point out
that the rate, given as a percent,
must be changed to a decimal in
order to do the computation.
Example 3: Point out that the
time periods must be consistent
with the rate at which the interest
is computed. Since the interest is
computed annually, 36 months
must be changed to 3 years.

Additional Questions/Examples
Find the missing information.
1. interest: $120
   principal: ■ **$1,200**
   rate: 2% per month
   time: 5 months
2. interest: $2,301.88
   principal: $6,350
   rate: 14 1/2 % per year
   time: ■ **2.5 years**
3. interest: $4,200
   principal: $10,000
   rate: 14% per year
   time: ■ **3 years**

Guided Practice/Try These Have
students complete these exercises independently. Discuss any
questions or problems they may
have.

5-MINUTE CLINIC

Exercise	Student's Error	Error Diagnosis
Find the interest on a $4,000 loan for 18 months at a rate of 7.5%.	$I = 4,000 \times 0.075 \times 18$ $I = \$5,400$	• Student forgets to write the time period as a fractional part of a year.

329

7. Principal: $650
Rate: 4%/year
Time: 5 years
$130; $780

8. Principal: $980
Rate: 7.5%/year
Time: 3 years
$220.50; $1,200.50

9. Principal: $1,500
Rate: 15%/year
Time: 36 months
$675; $2,175

Solve.

10. Clint borrowed $6,400 from a bank at 14% interest per year. How much simple interest will he owe after 4 years? **$3,584**

11. Gina opened a savings account with a deposit of $1,100. What was the total amount of money in her account after 2 years if she earned simple interest at 6% per year and she made no other deposits or withdrawals? **$1,232**

EXERCISES

PRACTICE/
SOLVE PROBLEMS

Find the interest and the amount due.   See Additional Answers.

	Principal	Annual Rate	Time	Interest	Amount Due
1.	$410	5%	6 months	■	■
2.	$35	7%	2 years	■	■
3.	$1,575	12.5%	3 years	■	■
4.	$4,853	8%	48 months	■	■
5.	$3,000	5.4%	5 years	■	■
6.	$8,957	4%	12 months	■	■
7.	$505.50	10.5%	1 year	■	■
8.	$2,200.75	11%	8 years	■	■
9.	$775	15%	36 months	■	■
10.	$8,000	16.5%	5 years	■	■
11.	$10,100	6.5%	24 months	■	■
12.	$69.75	8.5%	4 years	■	■

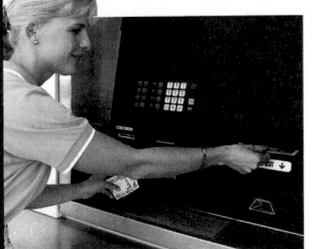

Solve.

13. Bruce purchased a certificate of deposit for $2,000. The certificate paid 6% simple interest per year. How much money will Bruce have in his account after 2 years? **$2,240**

14. Erica borrows $10,000 for 4 years at 11.5% simple interest per year. How much money does she repay the bank after 4 years? **$14,600**

330   CHAPTER 9   Exploring Percent

---

3 SUMMARIZE

Write About Math In their math journals, have students define simple interest and explain how it is calculated.

4 PRACTICE

Allow calculator use.

Practice/Solve Problems In Exercises 1–12 be sure students realize that their answers for the interest column are amounts of interest, not rates of interest.

Extend/Solve Problems For Exercises 16–17 students must write the time periods as fractional parts of the year. Exercise 19 requires multiple steps.

Think Critically/Solve Problems Watch for students who have difficulty with the time period in Problem 28. Remind them that they can express the time in fractional parts of a year

5 FOLLOW-UP

Extra Practice
Complete the table.

Principal	Rate	Time	Interest
1. $2,400	6%	24 mo	**$288**
2. **$500**	10%	3 y	$150
3. $5,725	**11.4%**	6 mo	$326.33
4. $3,748	15%	26 wk	**$281.10**

5. Randy lent his sister $1,200 for 2 years. She promised to pay him $120 interest. What interest rate did that amount represent? **5%**

Extension Have students solve the following problem.
Kara paid for a car in 48 monthly installments of $356.40 each. The price of the car was $10,800. How much interest did Kara pay? If simple interest was charged, what was the annual interest rate?
$6,307.20; 14.6%

330

CHALLENGE

Rubin deposited $2,000 in a savings account which pays 8.5% simple interest per year. If he makes no additional deposits or withdrawals, how much interest will he earn in 5 years? **$850**

15. Rowland opened a money market account with a deposit of $3,000. The account pays 8.2% annual interest. How much interest will be credited to his account after 3 years if he makes no other deposits or withdrawals? **$738**

Solve.

16. Teresa invested $2,280 at 9% per year for 3 months. How much interest did she earn? **$51.30**

17. Rob loaned $625 to his niece. She agreed to repay the loan in 9 months, at a rate of 13% per year. How much interest will Rob receive? **$60.94**

18. Patsy borrowed $6,800 at $14\frac{1}{8}$% simple interest per year for 2 years. How much money will she have to repay the bank? **$8,721**

19. Chuck borrowed $18,000 to expand his courier business. He took out two different loans. One loan was for $10,000 at 12% per year and the other loan was for $8,000 at $15\frac{1}{4}$% per year. Both loans were for 5 years. How much money must he repay at the end of 5 years? **$30,100**

Find the principal, rate, time, or interest:

	Principal	Annual Rate	Time	Interest
20.	$1,300	■ 14%	$2\frac{1}{2}$ years	$455
21.	■ $900	12.5%	4 years	$450
22.	$1,500	12%	■ $\frac{1}{2}$ y	$90
23.	$4,500	10%	■ $1\frac{1}{2}$ y	$675
24.	■ $6,750	11%	26 weeks	$371.25
25.	$3,000	■ 12%	6 months	$180
26.	$3,500	11%	18 months	■ $577.50

Solve.

27. Janice opened a savings account with a deposit of $900. She made no other deposits or withdrawals. She received $90 in interest after one year. What was the annual interest rate? **10%**

28. Norm purchased a certificate of deposit for $4,800 at $6\frac{1}{4}$% interest per year. The term is for 2 years and 5 months. How much interest did he receive on his deposit at the end of the term? **$725**

29. Suppose you earn an annual salary of $32,000. You are offered a 10.5% raise for a 4-year period, or you can have a 3% raise each year for 4 years. Which is the better offer?
10.5% = $141,440 over 4 years
3% = $137,892.34 over 4 years

EXTEND/ SOLVE PROBLEMS

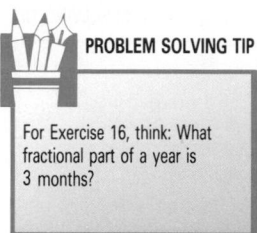

PROBLEM SOLVING TIP

For Exercise 16, think: What fractional part of a year is 3 months?

THINK CRITICALLY/ SOLVE PROBLEMS

Section Quiz Find the interest and the total amount.

	Principal	Rate	Time
1.	$625	3%	9 mo
	$14.06; $639.06		
2.	$2,870	5.5%	3 y
	$473.55; $3,343.55		
3.	$8,000	7.6%	48 mo
	$2,432; $10,432		
4.	$12,500	16.8%	24 mo
	$4,200; $16,700		

5. Paulo borrowed $5,000 for 4 years at 9 1/4% simple interest per year. How much money must he repay the bank after 4 years? **$6,850**

Get Ready calculators, graph paper

Additional Answers
See page 580. Additional answers for odd-numbered exercises are found in the Selected Answers portion of the page.

MAKING CONNECTIONS

Ask students to bring in newspaper advertisements for bank loans and for car loans. Have them compare the interest rates and terms offered.

9-9

► READ
► PLAN
► SOLVE
► ANSWER
► CHECK

Problem Solving/ Decision Making:
CHOOSE A STRATEGY

 **PROBLEM SOLVING TIP**

Here is a checklist of the problem solving strategies that you have studied so far in this book.

Solve a simpler problem
Find a pattern
Guess and check
Work backward
Make an organized list
Make a table
Use logical reasoning
Act it out

In this book you have been studying a variety of problem solving strategies. Experience in applying these strategies will help you decide which will be most appropriate for solving a particular problem. Sometimes, only one strategy will work. In other cases, any one of several strategies will offer a solution. There may be times when you will want to use two different approaches to a problem in order to be sure that the solution you found is correct. For certain problems, you will need to use more than one strategy in order to find the solution.

PROBLEMS

Solve. Name the strategy you used to solve the problem.
Strategies may vary. One possible strategy is suggested.

1. Six months ago, Joe received an 8% pay raise. His salary was then $256. Two months ago, Joe took an 8% pay cut. What was Joe's salary after the pay cut? How does that amount compare with his salary six months ago? (Always round Joe's salary to the nearest dollar.) **Work backward; original salary: $237; current salary: $236**

2. The Walters are having an outdoor barbecue. They plan to set up square card tables joined in a row to seat 36 people. How many card tables will be needed to seat 36 people if just one person sits on each side of a table? **Find a pattern; 17 tables**

3. The annual membership fee of the Movie-Goers Club is $15. Admission to each movie is $3.75 for members and $5.50 for nonmembers. How many movies would someone have to see before it would be worth becoming a member? **Use logical reasoning; 9 movies**

4. A game is played by drawing three of these number cards from a bag. Each player's score is the sum of the digits on the cards picked. How many possible scores are there for each pick?

Make an organized list. There are 10 scores but only 9 different ones, since 13 occurs twice.

6 2 4 0 7

Use this information for Problems 5 and 6.

These are pentominoes.   These are *not* pentominoes.

5. Find how many unique pentominoes are possible.
   Guess and check. There are 12 pentominoes.
6. How many pentominoes can be folded to form open-top boxes?
   Act it out. 8 can be folded into open-top boxes.
7. A construction foreman determined that 3 workers can do half a job in 20 days. Working at the same rate, how long would it take 12 workers to do the whole job? **Make a table; 10 days**

8. Complete the pattern.

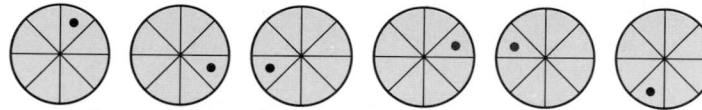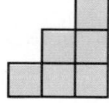

9. A grandmother divided her figurine collection among her four grandchildren. She gave the oldest grandchild half of her collection, the second oldest one fourth of her collection, the third oldest one fifth of her collection, and the youngest 48 figurines. How many figurines were there in her collection altogether?
   Work backward; 960 figurines
10. Six cubes make a 3-step staircase.

    How many cubes are needed
    to make a 9-step staircase?

    Find a pattern; 45 cubes

In each of these puzzles a letter represents a digit. Different letters represent different digits. Find the numbers that solve each puzzle.

See Additional Answers.

11.	12.	13.	14.
A	A B	T W O	H O C U S
× A	+ B A	+ T W O	+ P O C U S
BA	C A C	F O U R	P R E S T O

ASSIGNMENTS

BASIC
1–14

AVERAGE
1–14

ENRICHED
1–14

ADDITIONAL RESOURCES
Reteaching 9–9
Enrichment 9–9

4 PRACTICE

Problems Encourage all students to try all problems. You may wish to have students work in pairs or small groups. Students will need graph paper for Problems 5 and 6. Ask students which strategy worked for Problems 11–14.

5 FOLLOW-UP

Extra Practice
1. David gave one present to each of his brothers and sisters. He gave as many presents to his brothers as he did to his sisters. One of his sisters, Edith, also gave one present to each. She gave half as many presents to sisters as she did to brothers. How many brothers and sisters are there?
   3 sisters and 4 brothers
2. Nita started riding in a bike rally at 9:00 a.m. and pedaled at a rate of 12 mi/h. Her brother Petri started two hours later and pedaled at a rate of 18 mi/h. At what time will Petri catch up to his sister? **3 p.m.**

Additional Answers
See page 580.

CHALLENGE

1. How many 3-digit numbers can be made with the digits 0, 1, 2, and 3? **24**

2. A health-food store sells brown rice in 16-kg, 17-kg, 23-kg, 30-kg, and 40-kg packages. How can a customer get exactly 100 kg of brown rice? **by buying two 16-kg and four 17-kg packages**

9 CHAPTER REVIEW

Introduction The Chapter Review emphasizes the major concepts, skills, and vocabulary presented in this chapter and can be used for diagnosing students' strengths and weaknesses. Page references direct students back to appropriate sections for additional review and reteaching.

Using Pages 334–335 Allow students to quickly scan the Chapter Review and ask questions about any section they find confusing. Exercises 1–5 review key vocabulary.

Informal Evaluation Have students explain how they arrived at any incorrect answers. They will probably find their own mistakes and give you some clues as to the nature of their errors. Make sure students understand this material before administering the Chapter Test.

1. A __?__ is a ratio that compares a number to 100. **d**
2. The percent of the regular price you save when you buy an item on sale is the __?__. **c**
3. The __?__ is the amount you save when you buy an item on sale. **a**
4. The percent of sales received as income is __?__. **b**
5. The __?__ is the amount saved or the amount borrowed. **e**

a. amount of discount
b. commission rate
c. discount rate
d. percent
e. principal

SECTION 9–1   EXPLORING PERCENTS   (pages 302–305)

► A ratio that compares a number to 100 is called a percent.
► Percent means *per one hundred*.

Write each as a percent.

6. $\frac{7}{8}$ **87.5%**   7. $\frac{3}{5}$ **60%**   8. $\frac{23}{100}$ **23%**

SECTION 9–2   FINDING THE PERCENT OF A NUMBER   (pages 306–309)

► To find a percent of a number, you can write and solve an equation or a proportion.

What number is 75% of 80?   4% of 15 is what number?

$$x \qquad = 75\% \times 80$$

$$\text{part} \rightarrow \frac{4}{100} = \frac{n}{15} \leftarrow \text{part} \\ \text{whole} \rightarrow \qquad\qquad \leftarrow \text{whole}$$

Find the percent of each number.

9. 75% of 64 **48**   10. 7% of 2,000 **140**   11. 64.3% of 200 **128.6**

SECTION 9–3   FINDING WHAT PERCENT ONE NUMBER IS OF ANOTHER   (pages 310–313)

► To find what percent one number is of another, you can write and solve an equation or a proportion.

12. What percent of 80 is 12? **15%**   13. What percent of 86 is 215? **250%**

14. 5 is what percent of 30? **$16\frac{2}{3}\%$**   15. 1 is what percent of 40? **2.5%**

SECTION 9–4   PERCENT OF INCREASE AND DECREASE   (pages 314–317)

► The percent of increase tells what percent the amount of increase is of the original amount.
► The percent of decrease tells what percent the amount of decrease is of the original amount.

Find the percent of increase or decrease.

16. 96 to 108 **12.5% increase**   17. 950 to 836 **12% decrease**   18. 800 to 716 **10.5% decrease**

SECTION 9–5 FINDING A NUMBER WHEN A PERCENT OF IT IS KNOWN
(pages 318–321)

▶ To find a number when a percent of it is known, use an equation or a proportion.

30 is 50% of what number?

$$30 = 50\% \times n$$

25% of what number is 12?

$$\begin{array}{c} \text{part} \to \\ \text{whole} \to \end{array} \frac{25}{100} = \frac{12}{n} \begin{array}{c} \leftarrow \text{part} \\ \leftarrow \text{whole} \end{array}$$

Find the unknown number.

19. 60 is 40% of what number? **150**

20. 70% of what number is 147? **210**

SECTION 9–7 DISCOUNT
(pages 324–327)

▶ The discount is the amount that the regular price is reduced.
▶ The sale price is the regular price minus the discount.

Find the discount and the sale price. Round to the nearest cent.

21. Regular price: $899.98
Discount: 25% **D = $225 S = $674.98**

22. Regular price: $510.20
Discount: 10% **D = $51.02 S = $459.18**

SECTION 9–8 SIMPLE INTEREST
(pages 328–331)

▶ The principal is the amount of money in a bank account or the amount of money to be borrowed from the bank.
▶ The rate is the percent of principal earned or paid per year.
▶ The time is the number of months or years the money remains in the bank account or the length of the loan.

Find the interest.

23. Principal: $600
Rate: 7%/year
Time: 2 years **I = $84**

24. Principal: $598
Rate: 3.4%/year
Time: 36 months **I = $61**

25. Principal: $9,000
Rate: 14%/year
Time: 4 years **I = $5,040**

SECTIONS 9–6 and 9–9 PROBLEM SOLVING
(pages 322–323, 332–333)

▶ Sometimes more than one strategy will be needed in order to find the solution to a problem.

Choose a strategy. Solve the problem.

26. The sum of two numbers is 212. If the smaller number is subtracted from the greater number, the result is 14. What are the numbers?
Guess and check; 99 and 113

USING DATA Use the Automobile Depreciation Schedule on page 301.

27. Find the minimum dollar amount of depreciation on a 2-year-old car that cost $12,000 when it was new. **$2,880**

Review **335**

Follow-Up Have students research the history and development of the percent symbol.

Study Skills Tip If students continue to find it difficult to differentiate among the three types of percent problems, work with them to conceptualize what is needed to solve each problem, displaying the work for each problem on the chalkboard, one next to another. Review the work done for each, pointing out the similarities, but highlighting the differences.

CHAPTER TEST

9

Introduction The Chapter Test uses a variety of questioning techniques to assess students' mastery of the major objectives of Chapter 9. If you prefer, you may use the Skills Preview (page 299) as an alternative form of the Chapter Test. The items on this test and the Skills Preview correspond in content and level of difficulty.

Alternative Assessment
Write About Math Imagine that you are interested in buying a CD player regularly priced at $289 that is on sale for 25% off. Describe how you would use mental math to estimate the sale price.

Write each as a percent.

1. $\frac{16}{100}$ **16%**   2. $\frac{13}{20}$ **65%**   3. 0.05 **5%**

Find the percent of each number.

4. 17% of 50 **8.5**   5. 35% of 800 **280**   6. $13\frac{1}{2}$% of 80 **10.8**

7. 5.8% of 70 **4.06**   8. 20% of 140 **28**   9. 62% of 410 **254.2**

Find each percent.

10. What percent of 65 is 29.9? **46%**   11. What percent of 95 is 49.4? **52%**

12. 84 is what percent of 350? **24%**   13. 2.61 is what percent of 3? **87%**

14. What percent of 50 is 10? **20%**   15. 27 is what percent of 9? **300%**

Find the percent of increase or decrease.

16. Original amount: $90
New amount: $144 **60% increase**

17. Original amount: $300
New amount: $405 **35% increase**

18. Original amount: 500
New amount: 320 **36% decrease**

19. Original amount: 75
New amount: 60 **20% decrease**

20. Original amount: $280
New amount: $126 **55% decrease**

21. Original amount: $72
New amount: $99 **37.5% increase**

Find the unknown number.

22. 441 is 9% of what number? **4,900**   23. 60 is 40% of what number? **150**

24. 2% of what number is 3? **150**   25. 25% of what number is 36? **144**

26. 8 is 4% of what number? **200**   27. 25% of what number is 48? **192**

Find the discount and the sale price. Round to the nearest cent.

28. Regular price: $419
Discount percent: 12% **D = $50.28**
S = $368.72

29. Regular price: $53.69
Discount percent: 25% **D = 13.42**
S = $40.27

Find the interest and the amount.

30. Principal: $170
Rate: 6%/year
Time: 2 years **I = $20.40**
A = $190.40

31. Principal: $8,100
Rate: 7.5%/year
Time: 36 months **I = $1,822.50**
A = $9,922.50

32. Principal: $517
Rate: 12%/year
Time: 6 months
I = $31.02
A = $548.02

Use the pictograph for Exercises 1–5.

STAR VIDEO RENTALS
(for June 16)

Drama ⊡ ⊡ ⊡ ⊡ ⊡

Comedy ⊡ ⊡ ⊡ ⊡ ⊡ ⊡ ⊡

Foreign ⊡ ⊡

Sci Fi ⊡ ⊡ ⊡

Sports ⊡ ⊡

Key: ⊡ represents 10 rentals

1. How many sports videos were rented? **20**

2. Which group had the most rentals? **comedy**

3. Suppose 35 Do-It-Yourself videos were rented. How many video symbols would be needed to represent this number $3\frac{1}{2}$

4. What was the total number of videos rented for June 16? **175**

5. The number of drama rentals was how many times greater than the number of foreign rentals? **three**

Write each answer in exponential form.

6. $a^3 \times a^5$ a^8 7. $n^{12} \div n^4$ n^8

8. Write two conditional statements using the following two sentences:
   See Additional Answers.
   Clara can read French.
   She can translate this letter.

9. Write 0.000025 in scientific notation.
   2.5×10^{-5}

10. A fence that encloses a square garden has 10 fence posts on each side of the garden. There is a post in each corner. How many fence posts are there in all?
    36 posts

Replace ● with $<$, $>$, or $=$.

11. -16 ● -13 $<$ 12. -15 ● -11 $<$

13. 3 ● -8 $>$ 14. 0 ● -3 $>$

Solve.

15. $\frac{x}{3} = 12$ $x = 36$ 16. $4d - 18 = -18$
                                      $d = 0$

17. $-y + 3 = 19$ 18. $\frac{5}{3}k = 25$ $k = 15$
    $y = -16$

Write each ratio in two other ways.

19. 4 to 1 $\frac{4}{1}$, 4:1   20. 8:2 8 to 2, $\frac{8}{2}$

21. $\frac{19}{20}$ 19 to 20, 19:20

22. Write the unit rate for $\frac{125 \text{ words}}{5 \text{ min}}$. $\frac{25 \text{ words}}{1 \text{ min}}$

23. All merchandise in Lucy's Gift Shop is marked 20% off. Write this reduction as a fraction. $\frac{1}{5}$

24. Find 75% of 48. **36**

25. 16 is what percent of 25? **64%**

26. A $450 television set is on sale at 10% off. Find the discount and the sale price.
    D = $45; S = $405

27. Find the interest for one year on a $2,500 deposit if the interest rate is $8\frac{1}{2}$% per year. **I = $212.50**

Cumulative Review **337**

CUMULATIVE TEST

Introduction The Cumulative Test uses a standardized-test format of multiple-choice questions to assess retention of previously learned concepts and test-taking skills. Test results may be used to diagnose students' strengths and weaknesses.

Item Analysis The table below correlates the Cumulative Test items with the chapter and section that are being tested.

Sections	Items
1–9	1
2–7	2
3–3	3
4–3	4
5–2	5
5–6	6
6–2	7
7–4	8
7–10	9
8–2	12
8–3	11
8–4	10
8–6	13
9–1	14
9–2	15
9–3	16
9–7	17

1. If the mean score on a math test was 80, what would a student who was absent have to score on a makeup test to keep the same mean?
 A. 85   B. 90
 C. 80   D. none of these

2. Multiply. $x^8 \times x^2$
 A. x^6 B. x^{16} C. x^4 D. x^{10}

3. Which conclusion can be drawn from these statements?

 If Willa is Scandinavian, then she is not Italian.
 Willa is Scandinavian.
 A. Willa is not Italian.
 B. Willa is from Rome.
 C. Willa is Scandinavian.
 D. none of these

4. Which is the decimal for $2\frac{3}{8}$?
 A. 2.6   B. 2.375
 C. 0.375   D. 2.38

5. Complete. 716 mg =   kg.
 A. 7.16 kg   B. 0.716 kg
 C. 0.0716 kg   D. 0.000716 kg

6. Find the circumference. Round your answer to the nearest tenth or whole number.
 A. 34.5 cm
 B. 8.6 cm
 C. 27 cm
 D. 84.8 cm

 (circle with 5.5 cm radius)

7. Add. $25 + (-24)$
 A. 49   B. 1   C. −49   D. −1

8. Solve. $3 = \frac{t}{4}$
 A. $t = 16$ B. $t = 15$
 C. $t = 11$ D. none of these

9. Solve. $5x + 11 > -4$
 A. $x > -3$ B. $x < -3$
 C. $x > 1\frac{2}{5}$ D. none of these

10. Solve. $3:4 = b:16$
 A. 7   B. 3   C. 12   D. 6

11. Which of the following ratios is equivalent to $12:8$?
 A. $12:2$ B. $36:24$
 C. $6:2$ D. none of these

12. Which is the unit rate for 87 mi on 6 gal?
 A. $\frac{13 \text{ mi}}{\text{gal}}$ B. $\frac{15.4 \text{ mi}}{\text{gal}}$
 C. $\frac{14.5 \text{ mi}}{\text{gal}}$ D. none of these

13. The scale of a map is 1 in. : 10 mi. What is the actual distance between two cities if the distance on the map is $7\frac{1}{2}$ in.?
 A. 70 mi   B. $7\frac{1}{2}$
 C. 75 mi   D. none of these

14. Express $\frac{2}{5}$ as a percent.
 A. 20%   B. 40%
 C. 25%   D. 50%

15. What is $33\frac{1}{3}$% of 58.5?
 A. 175.5   B. 195
 C. 1.95   D. none of these

16. 8 is what percent of 12?
 A. $66\frac{2}{3}$%   B. 80%
 C. 75%   D. $33\frac{1}{3}$%

17. An $800 refrigerator is on sale at 15% off. What is the discount?
 A. $785   B. $120
 C. $680   D. none of these

Name _____ Date _____

Exploring Percent

A **percent** is a ratio comparing a number to 100. The ratio $\frac{47}{100}$, for example, means 47 out of 100 or 47 per 100. It is written as 47%.

▶ **Example 1** _____

Write $12\frac{1}{2}\%$ as a decimal.

Solution

$12\frac{1}{2}\% = \frac{12\frac{1}{2}}{100} = 12\frac{1}{2} \times \frac{1}{100} = \frac{25}{2} \times \frac{1}{100} = \frac{25}{200} = \frac{1}{8}$

▶ **Example 2** _____

Write $\frac{7}{12}$ as a percent.

Solution

First, express the fraction as a decimal: $\frac{7}{12} = 0.58333\ldots$

Next, move the decimal point two places to the right.

$0.58333\ldots$ or 58.33% or $58\frac{1}{3}\%$

EXERCISES

Write each percent as a decimal.

1. 40% **0.4**   2. 61% **0.61**   3. 9% **0.09**   4. 90% **0.9**

5. 65.2% **0.652**   6. 4.5% **0.045**   7. 1.25% **0.0125**   8. $3\frac{2}{3}\%$ **0.03666...**

Write each percent as a fraction.

9. 45% $\frac{9}{20}$ 10. 30% $\frac{3}{10}$ 11. 0.2% $\frac{1}{500}$ 12. 23% $\frac{23}{100}$

13. 6.5% $\frac{13}{200}$ 14. $\frac{3}{4}\%$ $\frac{3}{400}$ 15. $2\frac{1}{2}\%$ $\frac{1}{40}$ 16. $6\frac{1}{4}\%$ $\frac{1}{16}$

Write each decimal as a percent.

17. 0.7 **70%**   18. 0.19 **19%**   19. 0.04 **4%**   20. 0.012 **1.2%**

21. 0.765 **76.5%**   22. 0.001 **0.1%**   23. 0.043 **4.3%**   24. $0.05\frac{1}{2}$ **$5\frac{1}{2}$%**

Write each fraction as a percent.

25. $\frac{73}{100}$ **73%**   26. $\frac{2}{5}$ **40%**   27. $\frac{2}{3}$ **$66\frac{2}{3}\%$**   28. $\frac{9}{10}$ **90%**

29. $\frac{1}{4}$ **25%**   30. $\frac{1}{6}$ **$16\frac{2}{3}\%$**   31. $\frac{7}{12}$ **$58\frac{1}{3}\%$**   32. $\frac{3}{1000}$ **0.3%**

Name _____ Date _____

Grading with Percents

Isabelle took four spelling tests this month. These were her results.

Quiz A: 20 out of 25 correct
Quiz B: 11 out of 15 correct
Quiz C: 15 out of 18 correct
Quiz D: 11 out of 12 correct

EXERCISES

Answer these questions about Isabelle's spelling scores.

1. Make a guess: On which quiz did she receive her best score?
   Answers will vary.

Write each of Isabelle's scores as a percent. Use a calculator. Round to the nearest whole percent.

2. $\frac{20}{25}$ **80%**   3. $\frac{11}{15}$ **73%**   4. $\frac{15}{18}$ **83%**   5. $\frac{11}{12}$ **92%**

6. Can you predict the best score based on the number of correct answers? Explain your thinking.
   No; the ratio of correct answers to total questions gives the score.

7. On which of Isabelle's quizzes did a correct answer count the most? Explain.
   Quiz D; there are fewer questions, so each answer counts for more.

8. Suppose Isabelle's next quiz had 100 spelling words. Describe how you would find her percent score.
   The number of words she got correct will be the same as her percent score.

9. Suppose a spelling test had 50 words on it. Describe a way of using mental math to determine the percent score.
   Multiply the number of correct answers by two.

Name _____ Date _____

Finding the Percent of a Number

To find a percent of a number, you can write and solve an equation or a proportion.

▶ **Example** _____

What number is 20% of 750?

Solution 1

Write an equation. Use a decimal or a fraction for the given percent.

$x = 0.20 \times 750$

$x = 150$

or $x = \frac{1}{5} \times 750$

$x = 150$

Solution 2

Write a proportion.

$\frac{part}{whole} = \frac{part}{whole}$

$\frac{20}{100} = \frac{x}{750}$

Solve the proportion.

$20 \times 750 = 100x$

$150 = x$

EXERCISES

Find the percent of each number, using an equation.

1. 27% of 700   2. 30% of 120   3. 75% of 60
   189   **36**   **45**

4. 80% of 98   5. $2\frac{1}{2}\%$ of 40   6. 6.7% of 500
   78.4   **1**   **33.5**

7. 0.5% of 34   8. 14.8% of 2,000   9. $33\frac{1}{3}\%$ of 360
   0.17   **296**   **120**

Find the percent of each number, using a proportion.

10. 40% of 700   11. 25% of 4,800   12. 55% of 165
   280   **1,200**   **90.75**

13. 50% of 492   14. 11% of 300   15. $12\frac{1}{2}\%$ of 72
   246   **33**   **9**

16. 7.5% of 2,000   17. $12\frac{1}{2}\%$ of 64   18. 0.4% of 20
   150   **8**   **0.08**

Name _____ Date _____

Royalties

The author of a book usually receives payment in the form of a royalty. This is a percent of the sale price of each book. The amount the author earns depends on the number of copies of the book sold.

Sometimes, the percent of the royalty can change, based on the number of books sold.

▶ **Example** _____

Ms. Williams signed a contract with a publishing company that will produce her book. The royalty agreement is summarized below. How much would Ms. Williams earn if the book sells 115,000 copies?

Sales price: $24.50
First 50,000 copies   —3% of sales price
Next 50,000 copies   —4% of sales price
Number over 100,000—5% of sales price

Solution

The chart shows that Ms. Williams will earn 3% of the sales price of the first 50,000 copies, 4% of the sales price of the second 50,000 copies and 5% of the sales price of all sales over 100,000.

$50,000 \times \$24.50 = \$1,225,000$
$\$1,225,000 \times 0.03 = \$36,750$
$\$1,225,000 \times 0.04 = \$49,000$
$15,000 \times \$24.50 = \$367,500$
$\$367,500 \times 0.05 = \$18,375$
$\$36,750 + 49,000 + 18,375 = \$104,125$

So, Ms. Williams would earn royalties of $104,125 in all.

EXERCISES

Assume that all these authors have the same publishing contract as does Ms. Williams. Find their royalties. Use a calculator.

1. Mr. Peterson's book sold 40,000 copies at $16.95 each.
   $20,340

2. Mrs. Smith's book sold 45,000 copies at $18.50 each.
   $24,975

3. Mrs. Lopez's book sold 60,000 copies at $12.50 each.
   $23,750

4. Mr. Brown's book sold 80,000 copies at $14 each.
   $37,800

5. Mrs. Ferrara's book sold 90,000 copies at $10.75 each.
   $33,325

6. Ms. O'Brien's book sold 75,000 copies at $25.50 each.
   $63,750

7. Ms. Hawkins' book sold 105,000 copies at $35.40 each.
   $132,750

8. Mr. Wong's book sold 180,000 copies at $9.95 each.
   $74,625

Finding What Percent One Number Is of Another

▶ **Example**

What percent of 300 is 270?

Solution 1

Write an equation.

Let p = the percent.

$p \times 300 = 270$

$300p = 270$

$\frac{300p}{300} = \frac{270}{300}$

$p = 0.9$

So, 270 is 90% of 300.

Solution 2

Write a proportion.

Let the unknown percent be $x\%$ and write it as $\frac{x}{100}$.

$\frac{x}{100} = \frac{270}{300}$

$300x = 27,000$

$\frac{300x}{300} = \frac{27,000}{300}$

$x = 90$

So, 270 is 90% of 300.

EXERCISES

Find each percent.

1. What percent of 80 is 4?   **5%**

2. What percent of 56 is 7?   **12.5%**

3. What percent of 60 is 15?   **25%**

4. What percent of 100 is 5?   **5%**

5. What percent of 72 is 27?   **37.5%**

6. What percent of 600 is 12?   **2%**

7. 36 is what percent of 144?   **25%**

8. 9 is what percent of 3?   **300%**

9. 24 is what percent of 3?   **800%**

10. 8 is what percent of 1,000?   **0.8%**

11. 100 is what percent of 50?   **200%**

12. 140 is what percent of 224?   **62.5%**

13. 20 is what percent of 10,000?   **0.2%**

14. 16 is what percent of 80?   **20%**

15. Marta bought a raincoat for $110. The sales tax was $7.70. What was the sales tax rate?
7%

Budget Analysis

The Franklin family is working together to see how they are spending their money. The amount of money they have available after taxes is $25,000.

They have compiled these totals for the past twelve months:

Housing: $11,400   Clothing: $1,400
Utilities: $492   Vacation: $750
Food: $5,720   Transportation: $1,560
Gifts: $400   Household purchases: $550
Entertainment: $820   Miscellaneous expenses: $108
Savings: $500   Medical bills: $1,300

Answer these questions about the Franklin family's budget.

1. What percent of their total budget goes to housing?
45.6%

2. What percent goes to food?
22.9%

3. What percent of their total budget is saved?
2%

4. What parts of their budget do you think the Franklins have the most control over?
Answers will vary. Clothing, gifts, entertainment, vacation, household purchases and miscellaneous expenses are most likely answers.

5. Next year, the Franklins are determined to save 5% of their income after taxes. How much money should they save? How much more is this than they have saved during the past twelve months?
$1,250; $750

6. Which categories would you adjust and by how much so that the Franklins will be able to increase their savings by the amount you determined in Exercise 5?
Answers will vary; the total amount of adjustment must be $750.

Percent of Increase and Decrease

To find the percent of increase or decrease, express the ratio of the amount of increase or decrease to the original amount as a percent.

▶ **Example 1**

Find the percent of increase.
Original number: 56
New number: 63

Solution

Find the amount of increase.

$63 - 56 = 7$

Write a ratio.

$\frac{\text{amount of increase}}{\text{original number}} = \frac{7}{56}$

Find the percent for this ratio.

$\frac{7}{56} = 0.125 = 12.5\%$

So, the percent of increase is 12.5%.

▶ **Example 2**

Find the percent of decrease.
Original number: 240
New number: 18

Solution

First, find the amount of the decrease.

$240 - 18 = 222$

Write a ratio.

$\frac{\text{amount of decrease}}{\text{original number}} = \frac{222}{240}$

Write a percent for this ratio.

$\frac{222}{240} = 0.925 = 92.5\%$

So, the percent of decrease is 92.5%.

EXERCISES

Find the percent of increase. Round to the nearest tenth.

1. Original price: $300
New price: $360
20%

2. Original weight: 120 lb
New weight: 140 lb
$16\frac{2}{3}$%

3. Original rent: $500
New rent: $575
15%

4. Original number: 4,500
New number: 4,590
2%

Find the percent of decrease. Round to the nearest tenth.

5. Original price: $50
New price: $35
30%

6. Original number: 400
New number: 352
12%

7. Original salary: $60
New salary: $45
25%

8. Original size: 80 cm
New size: 48 cm
40%

Enlarging and Reducing

Many duplicating machines can produce a copy whose size is reduced or enlarged to a given percent of the original size. Documents, drawings, or photos can only be reduced or enlarged by whole-number percents.

▶ **Example 1**

Mrs. Wales was preparing a layout for a page in the PTA newsletter. She wanted to use a drawing that was 6 in. high in a space that was 2.5 in. high.

Solution

Determine the reduction by finding what percent 2.5 is of 6.

$\frac{2.5}{6} = 41\frac{2}{3}\%$

Since the machine only handles whole-number percents, it must be set for 42%.

▶ **Example 2**

Mr. Jones has a logo on his newsletter that is 4 in. long. He wants to enlarge it to a length of 6 in. to use on an advertising poster.

Solution

Determine the percent of enlargement by finding what percent 6 is of 4.

$\frac{6}{4} = 150\%$

The machine must be set for a 150% enlargement.

EXERCISES

Solve. Round to the nearest whole-number percent.

1. A photo that is 5 in. wide must fit in a space that is 2 in. wide. At what percent should the copy machine be set?
40%

2. A trademark that is 8 cm in length must be enlarged to a length of 14 cm. What percent enlargement should be used?
175%

3. A drawing that is 15 in. long must be reduced to fill a space 12 in. long. What percent reduction should be used?
80%

4. A picture that is 2.5 cm wide must be enlarged to fill a space 4 cm wide. What percent enlargement should be used?
160%

5. A line of calligraphy that is 5 cm long must be enlarged to fill a space 7 cm long. What percent enlargement is needed?
140%

6. A blueprint that is 20 in. wide must be reduced to fit on a sheet of paper 8.5 in. wide. What percent reduction should be used?
42%

338B

Name _____   Date _____

Finding a Number When a Percent of It Is Known

To find a number when a percent of it is known, write and solve an equation or a proportion.

► **Example** _____

12 is 25% of what number?

Solution 1

Write an equation.
Let n = the number
$$0.25 \times n = 12$$

Solve the equation.

$$0.25n = 12$$

$$\frac{0.25n}{0.25} = \frac{12}{0.25}$$

$$n = 48$$

So, 12 is 25% of 48.

Check: $48 \times 0.25 = 12$.

Solution 2

Write a proportion.
Let n = the number
$$\frac{part}{whole} \rightarrow \frac{12}{n} = \frac{25}{100}$$

Solve the proportion.

$$12 \times 100 = 25 \times n$$

$$1200 = 25n$$

$$48 = n$$

So, 12 is 25% of 48.

EXERCISES

Find each answer.

1. 45% of what number is 135?   **300**

2. 33 is 75% of what number?   **44**

3. 15% of what number is 84?   **560**

4. 54 is 40% of what number?   **135**

5. 562 is 50% of what number?   **1,124**

6. 100 is 2% of what number?   **5,000**

7. 105 is 30% of what number?   **350**

8. 34% of what number is 238?   **700**

9. 22 is $\frac{1}{2}$% of what number?   **4,400**

10. 50 is 0.2% of what number?   **25,000**

11. 40 is 200% of what number?   **20**

12. 49 is 700% of what number?   **7**

Name _____   Date _____

Percent Game

Copy the percent triangles below and cut out a copy of each triangle. Then mix up the triangles in a pile in the center of a table. Each player should have a blank sheet of paper.

In turn, each player should pick one of the triangles from the pile. The player should then copy the information given on the triangle and write in the missing number. The other player should use a calculator to check the first player's answer. Players should alternate picking triangles and checking answers until the pile has been used up. Each correct answer gets one point. A tie game is possible.

For the next game, each player may create six new triangles and the game can begin anew.

Name _____   Date _____

Problem Solving Skills: Choose a Computation Method

After carefully reading a problem, ask yourself:
    Do you need an exact answer or an estimate?
    Which computation method is appropriate?
        First, consider mental math. If that method is
        not appropriate, try pencil and paper or a calculator.

► **Example 1** _____

Eleanor wants to buy some new music cassettes. She has $20 to spend. The cassettes she wants cost $6.99, $7.95, and $5.99. Does she have enough money?

Solution

For this problem, only an estimate will be needed.

Round each price. Then find the estimated sum
    $7 + $8 + $6 = $21
Eleanor does not have enough money for all three tapes.

► **Example 2** _____

Otis is selling tickets for the school play. Each ticket costs $2. He has sold 35 tickets. How much money has he collected?

Solution

For this problem, an exact answer is necessary. The computation can be done using mental math.
    $35 \times $2 = 70$

So, Otis collected $70.

EXERCISES

Indicate which computation method you would use to solve each problem. Then solve.

1. At the used-book fair, Loretta bought 6 paperbacks for $.50 each and 11 hardcover books for $1 each. What was her total cost?
    mental math; $14

2. Steven earns $7.25 as a grocery clerk. Last week he worked 14 hours. How much did he earn?
    pencil and paper or calculator; $101.50

3. There are 221 students signed up for the end-of-year picnic. A bus can carry 42 students. If 6 buses are available, will all 221 students go by bus to the picnic?
    estimation; yes

4. Toby's restaurant ordered $435.78 worth of supplies from a catalogue. Toby must pay 5.5% sales tax on this amount. How much sales tax must he pay?
    calculator; $23.97

Name _____   Date _____

Computational Shortcuts

Here are some shortcuts that you can use to make some computations easier.

Shortcut 1 A quick way to multiply by 5 uses the fact that $5 = \frac{10}{2}$.
To multiply by 5, annex a zero (multiply by 10) and divide by 2.
    Example: $67 \times 5 \rightarrow 2)\overline{670}$ → 335
    $67 \times 5 = 335$

Shortcut 2 A quick way to multiply by 25 uses the fact that $25 = \frac{100}{4}$.
To multiply by 25, annex two zeros (multiply by 100) and divide by 4.
    Example: $38 \times 25 \rightarrow 4)\overline{3800}$ → 950
    $38 \times 25 = 950$

Shortcut 3 To find a total cost when a percent has been added to the original price, use the distributive property.

Total cost = original price + percent added
    Total = $A + (p\% \times A) = A(1 + p)\%$

The total cost of a $14 book to which a tax of 5% is added would be
    Total = $14(1 + 0.05) = $14(1.05) = 14.70

EXERCISES

Use one of the shortcuts described above to find each answer.

1. 793×5 **3965**

2. 4256×5 **21,280**

3. $34,671 \times 5$ **173,355**

4. 79×25 **1975**

5. 351×25 **8775**

6. 1298×25 **32,450**

7. James received a salary increase of 4%. His weekly salary was $240. What will his new salary be?
    $249.60

8. Linda gets 15% off on things she buys at the book store. What will she pay for a $25 book?
    $21.25

9. The price of a coat is $42. What is the cost plus 3% sales tax?
    $43.26

10. What is the cost of a car listed at $20,000 with a 6% add-on tax?
    $21,200

11. How could you use the distributive property to find the total price when a percent of the original price has been subtracted from the original price?
    Total price = original price − percent of original price
    Total = $A(1 − p)\%$

Name _____ Date _____

Discount

Many stores offer a discount at one time or another to encourage people to buy their merchandise. The **discount** is the percent of the regular price by which that price is reduced. The sale price is the regular price less the discount.

► Example 1

The regular price of a wool coat at Sterling Clothes is $84.50. Sterling is offering a 15% discount this week. What will the sale price be after the discount?

Solution

First, find the amount of the discount.
$84.50
× 0.15
12.675 or $12.68 to the nearest cent
Then, subtract the discount from the regular price.
$84.50 − 12.68 = $71.82

So, the sale price will be $71.82.

► Example 2

The employees at Book Market get a 10% discount on any purchase. Dave works at Book Market and wants a book that costs $14.95. How much will he pay for it?

Solution

First, find the amount of the discount. Use mental math:

$14.95 × 0.1 = $1.495 or $1.50
Now, subtract the discount from the regular price.
$14.95 − 1.50 = $13.45

So, Dave will pay $13.45.

EXERCISES

Find the discount and the sale price. Round your answers to the nearest cent.

1. Regular price: $225.75
   Percent of discount: 30%
   $67.73; $153.02

2. Regular price: $42.78
   Percent of discount: 4%
   $1.71; $41.07

3. Regular price: $25.60
   Percent of discount: 25%
   $6.40; $19.20

4. Regular price: $6,000
   Percent of discount: 8.5%
   $510; $5,490

5. Regular price: $340
   Percent of discount: 20%
   $68; $272

6. Regular price: $459.99
   Percent of discount: 40%
   $184.00; $275.99

7. Regular price: $65.40
   Percent of discount: 5%
   $3.27; $62.13

8. Regular price: $36.37
   Percent of discount: 1%
   $.36; $36.01

Name _____ Date _____

Commission Decision

Matt has just been offered a job selling motorcycles. He has a choice with regard to how he will get paid. He can receive a salary of $150 per week plus 6% commission on sales, or he can just receive 12% commission on sales. To help him decide which plan is best, he decides to make a table to compare the amount he would make under each plan.

EXERCISES

Complete the table.
Let Plan A = $150 + 6% of sales. Let Plan B = 12% of sales.

	Amount of sales	Plan A	Plan B
1.	$600	$186	**$72**
2.	$1,200	**$222**	**$144**
3.	$1,500	**$240**	**$180**
4.	$2,000	**$270**	**$240**
5.	$2,500	**$300**	**$300**
6.	$3,000	**$330**	**$368**
7.	$3,500	**$360**	**$420**
8.	$4,000	**$390**	**$480**

9. Under what conditions would Plan A be best for Matt?
   if the sales he makes are below $2,500

10. Under what conditions would Plan B be best for Matt?
    if the sales he makes are more than $2,500

11. Matt found out that the previous sales representative sold about two motorcycles a week. The average cost of a motorcycle is $750. At that rate, which plan would be best. Why?
    Plan A is best; on sales of $1,500, Plan A pays $240 while Plan B only pays $180.

12. How many motorcycles a week must be sold to make Plan B the best?
    at least 4 at $750 each or at a more expensive price

Name _____ Date _____

Simple Interest

Interest (*I*) is money that is paid for the use of money over a period of time. The amount of money earning interest, or that has been borrowed at interest, is called the **principal** (*p*). The **rate** (*r*) is the percent of the principal charged for the use of the money over a given period of time. The **time** (*t*) is the number of time periods during which the principal earns interest or has not been paid back. The formula for interest earned or to be paid is:
$$I = prt$$

► Example 1

Find the interest due on a loan of $500, at a rate of 12% per year for 6 months.

Solution

Identify the principal, rate and time.
$p = \$500,\ r = 12\%,\ t = 0.5$
(6 months is a half year)
Substitute these amounts in the formula.
$I = prt$
$I = \$500 × 0.12 × 0.5$
$I = \$30$
The interest due on the loan is $30.

► Example 2

Find the total amount in a savings account that pays $6\frac{1}{2}\%$ interest per year after 6 years. The original amount invested was $350.

Solution

Identify the principal, rate, and time. Then use the interest formula.
$p = \$350,\ r = 6\frac{1}{2}\%,\ t = 6$
$I = \$350 × 0.065 × 6$
$I = \$136.50$
To find the total amount in the savings account, add the amount of interest to the principal.
$350 + 136.50 = \$486.50$
So, the total amount after 6 years is $486.50.

EXERCISES

Find the interest and the total amount.

	Principal	Rate	Time	Interest	Total Amount
1.	$640	5%	6 mo	**$16**	**$656**
2.	$3,000	4%	2 y	**$240**	**$3,240**
3.	$475	11%	3 y	**$156.75**	**$631.75**
4.	$700	7.5%	2 y	**$105**	**$805**
5.	$58.50	$5\frac{1}{4}\%$	$2\frac{1}{2}$ y	**$7.68**	**$66.18**
6.	$200	9.5%	3 mo	**$4.75**	**$204.75**
7.	$860	12%	4 mo	**$34.40**	**$894.40**
8.	$138	18%	2 y	**$4.14**	**$142.14**
9.	$620	6.65%	2 y	**$82.46**	**$702.46**

Name _____ Date _____

Finding the Average Percent

To find an average of two or more different percents, you must find each percent separately. Then you must average them in a special way.

► Example

In the first regional game, Greg was successful in 75% of his free throws. In all, he had 16 attempts. In the second regional game, he had 20 attempts and made 90% of them successfully. What was Greg's average percent of free throws for both games? Round to the nearest whole number.

Solution

First find how many free throws the player actually made during each game.
75% of 16: 0.75 × 16 = 12 free throws
90% of 20: 0.90 × 20 = 18 free throws

The total number of free throw attempts in both games:

$16 + 20 = 36$ free throws

Total number of successful throws in both games:

$12 + 18 = 30$

Now, find what percent 30 is of 36.
$\frac{30}{36} = \frac{5}{6} = 0.8\overline{3}$ or about 83%

Greg's average of free throws for both games was about 83%.

EXERCISES

Solve. Round to the nearest whole-number percent.

1. During the first week of practice for a gymnastics tournament, Ester had strong dismounts on 60% of the 15 times she practiced her free-exercise routine. During the second week, she managed strong dismounts 70% of the 20 times she did her routine. What was her average for successful dismounts for the two weeks?
   about 66%

2. During the opening game of the football season, Joey, the first team quarterback, completed 30% of 30 attempted passes. During the second week's game, he made 42% of 50 attempts. What was his completed-pass average for the two games?
   about 38%

3. During the first marking period, Laura's average was 85% on 8 tests. During the second marking period, her average was 80% on 6 tests. In the final marking period, her average was 70% on 5 tests. What was Laura's final average for all three marking periods?
   about 79%

338D

Name _____ Date _____

Problem Solving/Decision Making: Choose a Strategy

Sometimes only one problem solving strategy will solve a problem. At other times, more than one approach is possible. And sometimes, you may need to use multiple strategies to solve a problem. The strategies you have learned already are these:

Solve a simpler problem   Find a pattern   Guess and check
Work backward   Make an organized list   Make a table
        Use logical reasoning   Act it out

► **Example**

Lou is in training for a race. On Monday, she ran 2 km, on Tuesday she ran 2.6 km, and on Wednesday she ran 3.2 km. If she continues to increase her distance in this way, how many kilometers must she run on Saturday?

Solution

Find a pattern. The difference between the distance run on any day and that run on the previous day is 0.6 km. Her running distance from Wednesday to Saturday will be as follows:

Wednesday: 3.2 km
Thursday: 3.2 km + 0.6 km = 3.8 km
Friday: 3.8 km + 0.6 km = 4.4 km
Saturday: 4.4 km + 0.6 km = 5.0 km

EXERCISES

Name the strategy you used to solve the problem. Then solve.
Strategies may vary. One possible strategy is suggested.

1. Jo, May, and Eva are friends. One plays the flute, one plays the piano and one plays the violin. Eva is the only one who cannot carry her instrument. Neither Eva nor May play flute. What instrument does each friend play?

   Make a table; Eva, piano; Jo, flute; May, violin

2. Mario bought a bicycle helmet today for $29. That left him with $6. Before he bought the helmet, his dad had given him $10 for his birthday. How much had Mario saved before his birthday?

   Work backwards; $25

3. The Stars beat the Tars by 5 points. If the Tars had scored twice as many points as they did, they would have beaten the Stars by 5 points. How many points did each team score?

   Guess and check; Stars 15, Tars 10

4. George has six coins, worth a total of 56¢. What coins does he have?

   Make an organized list; 1 quarter, 2 dimes, 2 nickels, 1 penny

---

Name _____ Date _____

Polyominoes

A **polyomino** is a set of squares connected along their edges.

Dominoes are examples of polyominoes composed of two squares. There is only one possible domino.

A triomino is a set of three squares connected along their edges.

1. How many different triominoes are possible? Draw your examples on the grid below. **2**

A tetromino is a set of four squares connected along their edges.

2. How many different tetrominoes are possible? Draw your examples on the grid below. **5**

A hexomino is a set of six squares connected along their edges. There are 35 different hexominoes. **Answers will vary. Possible answers are shown.**

3. Draw four different hexominoes on the grid below. Compare with your classmates. How many different hexominoes has your class found?

Name _____ Date _____

Calculator Activity: Compound Interest

A savings account earns interest which is **compounded** or added to the total, or principal, at the end of each interest period. In this way, if the money is left in the account and not withdrawn, the interest starts earning interest.

► **Example 1**

A bank pays 10% interest per year. You deposit $100. How much money will you have after 3 years if you do not make any withdrawals or additional deposits?

Solution

First year: $100 + 10% of $100 = $100 + $10 = $110
Second year: $110 + 10% of $110 = $110 + $11 = $121
Third year: $121 + 10% of $121 = $121 + $12.10 = $133.10

You will have $133.10 after three years.

The formula $P_n = P_o(1 + r)^n$ can be used to calculate compound interest.

Let P_o = original principal on deposit
Let r = interest rate per period
Let n = number of periods

► **Example 2**

Use a calculator to find the total amount on deposit if $1000 was deposited in a CD account for 5 years at an interest rate of 8.5%.

Solution

Use the formula $P_n = P_o(1 + r)^n$. Write the numbers in place of the variables.
$P_n = $1000(1 + .085)^5$
Then, use a calculator.
1000 ⊠ ⊂ 1 ⊞ .085 ⊃ x^y 5 ⊜ 1503.65669
After 5 years, $1503.66 would be in the account.

EXERCISES

Use a calculator and the compound-interest formula to find the total amount on deposit.

1. principal: $1500
   interest rate: 7.9%
   periods: 4
   $2033.19

2. principal: $5000
   interest rate: 5.6%
   periods: 10
   $8622.02

3. principal: $10,000
   interest rate: 8.3%
   periods: 8
   $18,924.64

Name _____ Date _____

Compound Interest (continued)

► **Example 3**

Joseph deposits $150 in his savings account on the last day of each month. The bank pays interest at a rate of 0.7% per month compounded monthly. How much will Joseph have in his account after 1 year if he does not withdraw any money?

Solution

Use the formula $T = D \times \left[\dfrac{(1 + r)^n - 1}{r}\right]$

Let T = total value
Let D = amount of deposit
Let r = rate of interest
Let n = number of periods

$T = 150 \times \left[\dfrac{(1 + .007)^{12} - 1}{.007}\right]$

Then use your calculator.

1.007 x^y 12 ⊟ 1 ⊜ ⊞ .007 ⊠ 150 ⊜ 1870.942755

Joseph will have $1870.94 in his account after 1 year.

EXERCISES

4. How much money did Joseph actually deposit into his account in the example above?

   $1800

5. How much interest did Joseph earn on his money after 1 year?

   $70.94

Find the total value.

6. Six monthly deposits of $250 at an interest rate of .65% per month, compounded monthly.

   $1524.59

7. Twelve monthly deposits of $750 at an interest rate of .72% per month, compounded monthly.

   $9365.09

8. Five yearly deposits of $10,000 at an interest rate of 10.5% per year, compounded annually.

   $61,661.60

9. Twenty-five yearly deposits of $5,000 at an interest rate of 15% per year, compounded annually.

   $1,063,965.09

Computer Activity: Speculating with a Spreadsheet

Name _____ Date _____

An electronic spreadsheet has many features that allow you to speculate "what if?" very quickly and accurately. One of the most important features is that whenever you change the data in one cell, the data in any other cell that uses that data is automatically changed at the same time.

This spreadsheet shows interest options from different banks.

	A	B	C	D	E	F	G
	Bank	Principal	Rate	Time	Interest		
1:							
2:							
3:	A	3000	.05	2	300		
4:	A	3000	.055	5	825		
5:	A	3000	.08	10	2400		
6:	A	3000	.09	15	4050		
7:							
8:	B	3000	.04	2	240		
9:	B	3000	.046	5	690		
10:	B	3000	.07	10	2100		
11:	B	3000	.085	15	3825		
12:							
13:	C	3000	.045	2	270		
14:	C	3000	.06	5	900		
15:	C	3000	.075	10	2250		
16:	C	3000	.09	15	4050		
17:							
18:							

E15: (Value) +B15•C15•D15

1. Explain the formula used to get the contents of E15.
   interest formula: principal × rate × time

2. Tell the bank(s) you would choose to:
   a. get the best return after 15 years   **A or C**
   b. invest money for only 5 years   **C**
   c. expect a 0.105 rate for 20 years   **C**
   d. make a 2-year and a 5-year investment   **A (2 yr); C (5 yr)**
   e. earn 70% of the principal in 10 years   **B**
   f. use if these were the loan rates   **B**

3. In 10 years, what interest would you expect to earn in the given bank on the given principal?
   a. A, 7000 **5600**   b. B, 9000 **6300**   c. C, 6000 **4500**

Speculating with a Spreadsheet (continued)

Name _____ Date _____

Use the changes shown on the portion of the spreadsheet below for Exercises 4–8.

	A	B	C	D	E	F	G
	Bank	Principal	Rate	Time	Interest		
1:							
2:							
3:	A	6000	.05	2	600		
4:	A	3000	.11	5	1650		
5:	A	3000	.08	20	4800		
6:	A	6000	.18	30	32400		
7:							

	Row	What was doubled	What happened to interest
4.	3	**principal**	**doubled**
5.	4	**rate**	**doubled**
6.	5	**time**	**doubled**
7.	6	**principal, rate, time**	**increases 8 fold**

8. What conclusion is suggested?   **Interest doubles for each factor that is doubled.**

Use the changes shown on the portion of the spreadsheet below for Exercises 9–13.

	A	B	C	D	E	F	G
	Bank	Principal	Rate	Time	Interest		
1:							
2:							
13:	C	5000	.04	3	600		
14:	C	5000	.06	3	900		
15:	C	5000	.08	3	1200		
16:	C	5000	.1	3	1500		
17:							

If you could

9. invest twice as much money, what rate would give you almost $2,500 in interest?   **8%**

10. find a bank to loan you $5,000 at 2% interest, how much interest would you owe after 3 years?   **$300**

11. get a 7% interest rate on $5,000, how much interest would you earn after 3 years?   **$1050**

12. only invest your $5,000 for 18 months, what rate would you need to earn $900 in interest?   **12%**

13. invest half as much money for the 3 years, what rate would you need to earn $1,200 in interest?   **16%**

ACHIEVEMENT TEST

Exploring Percent
CHAPTER 9 FORM A

MATH MATTERS BOOK 1
Chicha Lynch
Eugene Olmstead

SOUTH-WESTERN PUBLISHING CO.

Name _____
Date _____

SCORING RECORD	
Possible	Earned
32	

Write each as a percent.
1. 0.7 **70%**   2. $\frac{7}{20}$ **35%**   3. 0.004 **0.4%**

Find the percent of each number.
4. 22% of 45 **9.9**   5. 30% of 210 **63**   6. 42% of 360 **151.2**
7. 4.6% of 80 **3.68**   8. 12% of 35 **4.2**   9. 55% of 128 **70.4**

Find each percent.
10. What percent of 140 is 28? **20%**   11. What percent of 65 is 29.9? **46%**
12. 45 is what percent of 360? **12.5%**   13. 14 is what percent of 80? **17.5%**
14. What percent of 48 is 6? **12.5%**   15. 20 is what percent of 50? **40%**

Find the percent of increase or decrease.
16. Original amount: $450
New amount: $639
42% increase

17. Original amount: $720
New amount: $864
20% increase

18. Original amount: $200
New amount: $175
12.5% decrease

19. Original amount: $124
New amount: $93
25% decrease

20. Original amount: $60
New amount: $42
30% decrease

21. Original amount: $3,200
New amount: $3,488
9% increase

Find the unknown number.
22. 364 is 7% of what number? **5,200**   23. 300 is 80% of what number? **375**
24. 4% of what number is 33? **825**   25. 15% of what number is 75? **500**
26. 96 is 24% of what number? **400**   27. 32% of what number is 896? **2,800**

Name _____   Date _____

Find the discount and the sale price. Round to the nearest cent.
28. Regular price: $149.99
Discount percent: 25%
D = $37.50; S = $112.49

29. Regular price: $88.50
Discount percent: 24%
D = $21.24; S = $67.26

Find the interest and the amount due.
30. Principal: $500
Rate: 7%
Time: 1 year
I = $35; A = $535

31. Principal: $3,200
Rate: 6.5%
Time: 6 months
I = $104; A = $3,304

32. Principal: $679
Rate: 14.2%
Time: 24 months
I = $192.84; A = $871.84

ACHIEVEMENT TEST

Exploring Percent
CHAPTER 9 FORM B

MATH MATTERS BOOK 1
Chicha Lynch
Eugene Olmstead

SOUTH-WESTERN PUBLISHING CO.

Name _____

Date _____

SCORING RECORD	
Possible	Earned
32	

Write each as a percent.

1. 0.9 __**90%**__ 2. $\frac{17}{20}$ __**85%**__ 3. 0.055 __**5.5%**__

Find the percent of each number.

4. 18% of 60 __**10.8**__ 5. 25% of 460 __**115**__ 6. 36% of 980 __**352.8**__

7. 5.2% of 90 __**4.68**__ 8. 15% of 70 __**10.5**__ 9. 40% of 90? __**36**__

Find each percent.

10. What percent of 68 is 27.2? __**40%**__ 11. What percent of 96 is 43.2? __**45%**__

12. 52 is what percent of 1,040? __**5%**__ 13. 2.8 is what percent of 8? __**35%**__

14. What percent of 360 is 9? __**2.5%**__ 15. 150 is what percent of 60? __**250%**__

Find the percent of increase or decrease.

16. Original amount: $560
    New amount: $630
    __**12.5% increase**__

17. Original amount: $840
    New amount: $966
    __**15% increase**__

18. Original amount: $400
    New amount: $380
    __**5% decrease**__

19. Original amount: $375
    New amount: $345
    __**8% decrease**__

20. Original amount: $960
    New amount: $624
    __**35% decrease**__

21. Original amount: $4,750
    New amount: $5,130
    __**8% increase**__

Find the unknown number.

22. 9 is 6% of what number? __**150**__ 23. 574 is 70% of what number? __**820**__

24. 3% of what number is 33? __**1,100**__ 25. 18% of what number is 1,125? __**6,250**__

26. 112 is 14% of what number? __**800**__ 27. 12% of what number is 108? __**900**__

9B-1

Name _____

Date _____

Find the discount and the sale price. Round to the nearest cent.

28. Regular price: $158.50
    Discount percent: 20%
    __**D = $31.60; S = $126.90**__

29. Regular price: $89.60
    Discount percent: 15%
    __**D = $13.44; S = $76.16**__

Find the interest and the amount due.

30. Principal: $900
    Rate: 8%
    Time: 1 year
    __**I = $72; A = $972**__

31. Principal: $4,150
    Rate: 4.5%
    Time: 24 months
    __**I = $373.50;**__
    __**A = $4,523.50**__

32. Principal: $1,800
    Rate: 13.5%
    Time: 3 months
    __**I = $60.75;**__
    __**A = $1,860.75**__

9B-2

338G

Teacher's Notes

CHAPTER 10 SKILLS PREVIEW

Give all the names that apply to each polygon or polyhedron.

1.
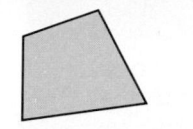
quadrilateral

2. 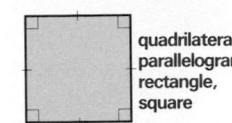 right triangle

3.
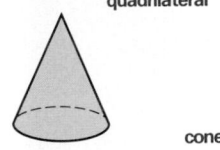 quadrilateral, parallelogram, rectangle, square

4.
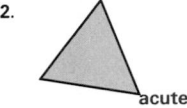
cone

5.
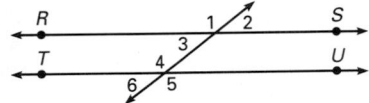
hexagonal prism

6.
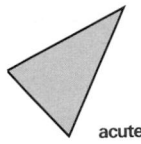
triangular pyramid

Classify each angle as *acute, right, straight*, or *obtuse*.

7. $15°$ acute **8.** $180°$ straight **9.** $90°$ right **10.** $98°$ obtuse

11. In the figure below, lines *RS* and *TU* are parallel. Identify the numbered angles that have equal measures. $\angle 1 = \angle 4 = \angle 5; \angle 2 = \angle 3 = \angle 6$

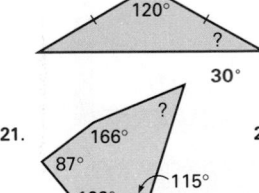

Classify each triangle as *right, acute,* or *obtuse*.

12.
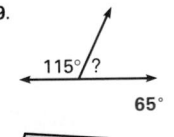
acute

13.
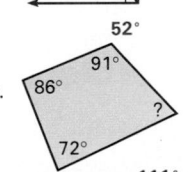
right

14.
obtuse

15.
acute

16. The measure of $\angle ABC$ is $98°$. Ray *BD* bisects $\angle ABC$. What is the measure of $\angle DBC$? $49°$

17. Use a protractor to draw an angle that measures $118°$. Construct its bisector. Check students' drawings.

Find the unknown angle measure.

18.
$120°$?
$30°$

19.
$115°$?
$65°$

20.
$38°$
?
$52°$

21.
$166°$
$87°$
$122°$? $115°$
$50°$

22.
$102°$?
$75°$ $68°$
$115°$

23.
$91°$
$86°$
$72°$?
$111°$

Introduction The purpose of this Skills Preview is to assess students' abilities on all the major objectives of Chapter 10. Test results may be used
* to determine those topics which may need only to be reviewed and those topics which need to be more carefully developed;
* for class placement;
* in prescribing for individual differences.

If you prefer, you may use the Skills Preview as an alternative form of the Chapter Test (page 380) to evaluate mastery of chapter objectives. The items on the Skills Preview and the Chapter Test correspond in content and level of difficulty.

CHAPTER 10

GEOMETRY OF SIZE AND SHAPE

OVERVIEW

In this chapter, students recognize and identify polygons and three-dimensional figures based on their shapes, number of sides, and angles. In addition, students learn to describe the basic geometric figures: angle, point, line, line segment, ray, and plane. Practical applications are introduced through learning to measure, construct, and bisect angles and line segments with a protractor and compass.

SPECIAL CONCERNS

The chapter contains technical vocabulary which students must understand in order to complete the activities. As students complete each section, you may wish to suggest they make a study sheet that includes vocabulary, definitions, and examples. Encourage students to refer to their study sheet whenever needed and to review the information whenever possible.

VOCABULARY

acute angle	bisector	degree
adjacent angle	collinear points	diagonal
alternate exterior angle	compass	edge
alternate interior angle	complementary angle	endpoints
altitude	cone	equiangular
angle	construction	equilateral
bases	coplanar points	face
bisect	cylinder	*(Continues on p. 341)*

MATERIALS

calculators	protractors
coins	scissors
compasses	square sheet of paper
envelopes	straightedge
geoboards, bands (optional)	strips of paper
graph paper	straws (optional)
grid paper	tape
posterboard	

BULLETIN BOARD

Display several pictures of buildings or artifacts that contain geometric figures such as planes, lines, points, angles, and polygons. Invite students to bring in additional pictures to be displayed. Beneath the pictures list such questions as, *How are points used? Which polygons can you identify? Which lines are parallel and/or perpendicular?* Whenever time permits, ask the questions with respect to one of the buildings or works of art displayed.

INTEGRATED UNIT 3

The skills and concepts involved in Chapters 9–11 are included within the special Integrated Unit 3 entitled "On Your Own." This unit is in the Teacher's Edition beginning on page 434J. Worksheets for this integrated unit appear in the Enrichment Activities booklet, pages 137–139.

TECHNOLOGY CONNECTIONS

- Calculator Worksheet, 199
- Computer Worksheets, 200
- MicroExam, Apple Version
- MicroExam, IBM Version

- *Geometry Concepts*, Ventura Educational Software
- *The Geometric Supposer: Triangles*, Sunburst Communications

TECHNOLOGY NOTES

Computer software programs, such as those listed in Technology Connections, lead students to discover basic geometric properties. Some programs provide databases of geometric terms and allow students to draw and rotate 2- and 3-dimensional figures on the screen. Other programs enable students to construct angle bisectors and line bisectors employing simulated compasses and straightedges.

CHAPTER 10

GEOMETRY OF SIZE AND SHAPE

PLANNING GUIDE

SECTIONS	TEXT PAGES	ASSIGNMENTS			
		BASIC	**AVERAGE**	**ENRICHED**	
Chapter Opener/Decision Making	340–341				
10-1 Points, Lines, and Planes	342–345	1–12, 16–22, 24–26	4–13, 14–23, 24–31	9–13, 14–23, 24–31	
10-2 Angles and Angle Measure	346–349	1–15, 18–20, 24–25	1–17, 18–22, 24–26	11–17, 21–23, 24–27	
10-3 Parallel and Perpendicular Lines	350–353	1–11, 13–14	4–12, 13–15, 16	12, 13–15, 16–17	
10-4 Triangles	354–357	1–8, 10	1–3, 8–10, 14–16	8, 10, 11–16	
10-5 Polygons	358–361	1–6, 7, 9–11, 15–16	1–6, 7–14, 15–18	9–14, 15–20	
10-6 Three-Dimensional Figures	362–365	1–10, 11–13, 21, 24	1–10, 14–21, 22, 24	7–10, 14–20, 22–25	
10-7 Problem Solving Skills: Make a Conjecture	366–367	1–6	1–10	1–10	
10-8 Problem Solving Strategies: Act It Out	368–369	1–10	1–13	1–13	
10-9 Constructing and Bisecting Angles	370–373	1–7	4–7, 8, 10	5–7, 8–9, 10	
10-10 Constructing Line Segments and Perpendiculars	374–377	1–10	4–9, 10–12, 13–14	10–16	
Technology	343, 352, 360, 361, 367	✔	✔	✔	

ASSESSMENT				
Skills Preview	339	All	All	All
Chapter Review	378–379	All	All	All
Chapter Test	380	All	All	All
Cumulative Review	381	All	All	All
Cumulative Test	382	All	All	All

ADDITIONAL RESOURCES			
RETEACHING	**ENRICHMENT**	**TECHNOLOGY**	**TRANSPARENCY**
10–1	10–1		TM 35
10–2	10–2		TM 36
10–3	10–3		TM 37, 38
10–4	10–4		TM 39
10–5	10–5	10–5	TM 40–45
10–6	10–6		TM 46–53
10–7	10–7		
10–8	10–8		
10–9	10–9		TM 54, 55
10–10	10–10		

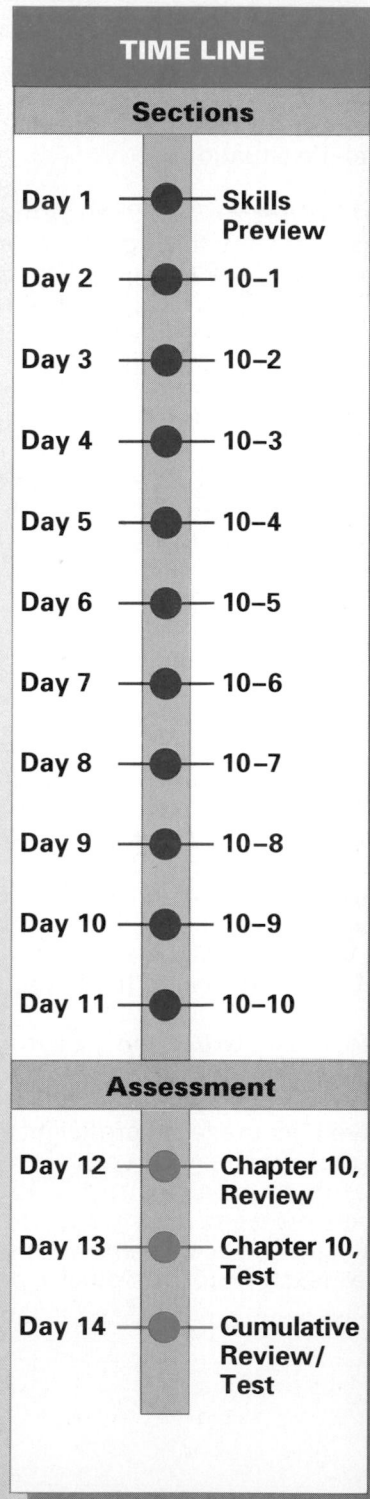

TIME LINE

Sections

Day 1	Skills Preview
Day 2	10–1
Day 3	10–2
Day 4	10–3
Day 5	10–4
Day 6	10–5
Day 7	10–6
Day 8	10–7
Day 9	10–8
Day 10	10–9
Day 11	10–10

Assessment

Day 12	Chapter 10, Review
Day 13	Chapter 10, Test
Day 14	Cumulative Review/ Test

ASSESSMENT OPTIONS

Chapter 10, Test Forms A and B	
Chapter 10, Test	Text, 380
Alternative Assessment	TAE, 380
Chapter 10 MicroExam	

Objective To explore how geometry can be used as a basis for acquiring information needed in real-life situations

Introduction Ask students to examine how the signals for various letters are related. Encourage them to specifically recognize similarities and differences. Then discuss the following:
• How many different angles are used in Semaphore code to signal the letters *a-z*? What are the angles? **4: 90°, 45°, 180°, 135°**
• Do you think it would be a good decision to use shapes and angles as international code symbols? Why or why not?
• Who might use Semaphore code signals? When and why?

**Decision Making
Using Data** Before completing Exercises 3, 4, and 5, have volunteers draw an example of angles that measure 90°, 45°, and 180°. (Students' drawings can be approximate.) Point out that the angles referred to are the angles at which the signaler's arms are held with respect to the body.

Working Together Students might enjoy competing in groups to answer signals. Once students have become fairly proficient with signaling, have several groups watch as one group signals a math problem. The group to signal back the correct answer first is the next group to signal a problem.

CHAPTER 10 GEOMETRY OF SIZE AND SHAPE

THEME Road Signs and Other Signs

Since ancient times, every culture has communicated with the aid of signs and symbols. In the past, piles of stones or notches in tree trunks marked the trails of wilderness travelers. Today, signs of different geometric shapes tell drivers to stop, yield, or slow down for a curve. You will learn the meanings of many symbols as you study the geometry of size and shape.

Some signals form a code. Semaphore, for example, is a system of signals used in the U.S. Navy. Semaphore code makes use of two flags, which are held in different positions to signal the letters of the alphabet. When preceded by the signal for *numeral*, the positions in which the flags are held to represent each of the first ten letters represent the numbers from 1 through 9 and 0.

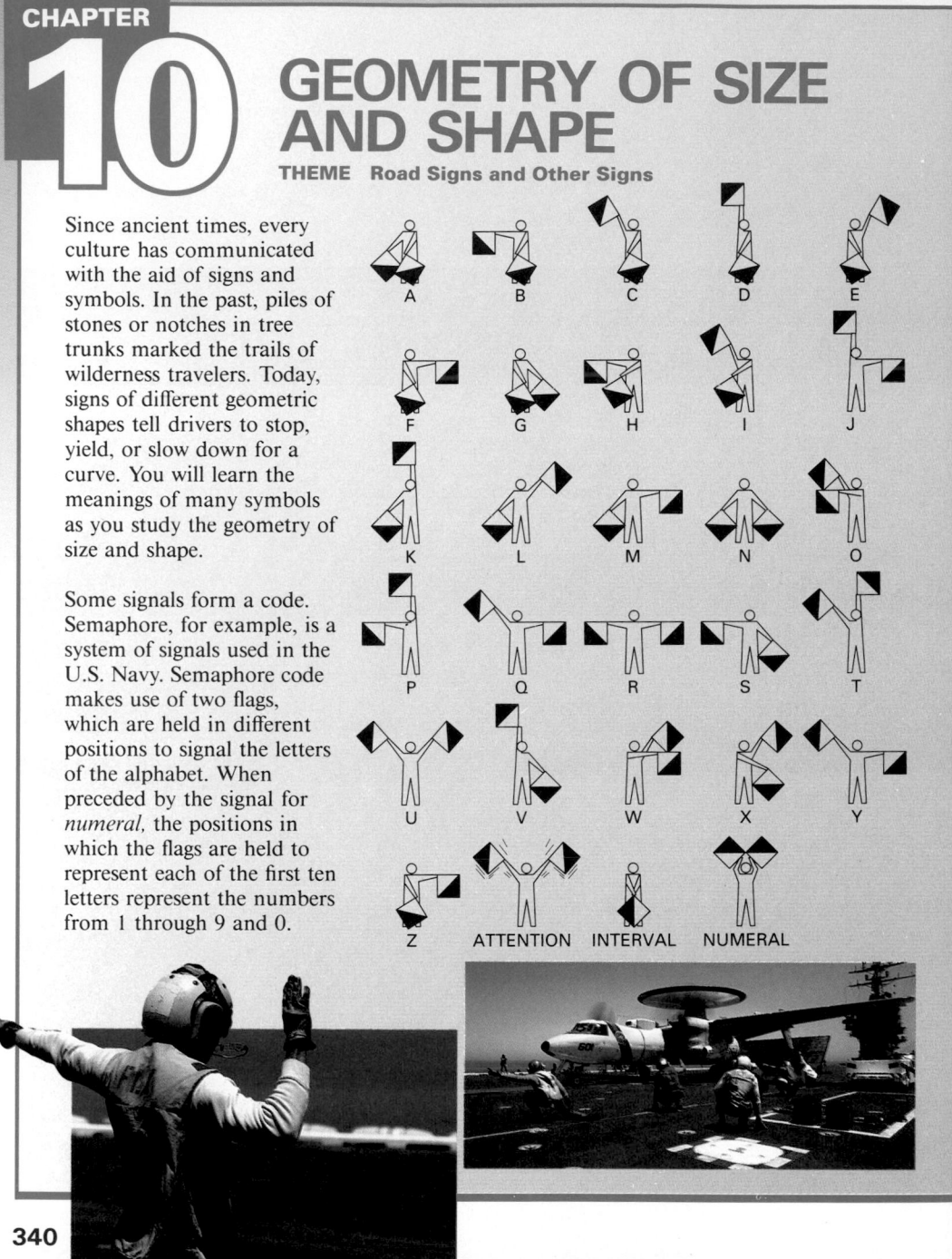

340

DECISION MAKING

Using Data

Use the semaphore chart to answer the questions.

1. What is the shape of each flag? square

2. What is the shape of the sections of each flag? triangle

3. For which letters are the signaler's arms at a 90° angle?
   I, J, N, P, U

4. For which letters are the signaler's arms at a 45° angle?
   H, O, T, W

5. For which letters are the signaler's arms at a 180° angle? D, L, R

Working Together

In small groups, look at the semaphore code to discover ways of learning it easily. Are there patterns that would allow you to learn the signals for several letters at a time instead of individually? What pattern do you detect in the sequence for the letters from A through D? What other patterns can you find?

Make a set of semaphore flags for your group, using light-colored and dark-colored triangles. Practice signaling and reading the words *add, subtract, multiply,* and *divide* and the numbers from 1 through 99. Then each group can signal math problems to other groups, which can signal the answers in return. Begin with problems like 3 + 5 and go on to problems like 53 × 49.

(Continued from page 339A)
intersect
isosceles
line
line segment
midpoint
obtuse
obtuse angle
parallel
perpendicular
perpendicular bisector
plane
point
polygon
polyhedron
prism
protractor
pyramid
quadrilateral
ray
regular polygon
right angle
scalene
sides
sphere
straight angle
straightedge
supplementary angle
transversal
vertex
vertical angle

342

10-1   Points, Lines, and Planes

EXPLORE   Geometric shapes appear in many places. You can see many of them in the superstructures of buildings and bridges. What geometric shapes do you see in the photograph?

SKILLS DEVELOPMENT   Here is a chart of some basic geometric figures. Note how each is drawn, named, and symbolized.

Figure	Name	Description
• *A*	point *A*	A **point** is a location in space. Although a point has no dimension, it is usually represented by a dot.
←•—•→ *B* *C*	line *BC* ($\overleftrightarrow{BC}$) or line *CB* ($\overleftrightarrow{CB}$)	A **line** is a set of points that extends without end in two opposite directions. Two points determine a line. Points that are on the same line are said to be **collinear points**.
•—• *D* *E*	line segment *DE* ($\overline{DE}$) or line segment *ED* ($\overline{ED}$)	A **line segment** is a part of a line that consists of two **endpoints** and all points between them.
•—•→ *F* *G*	ray *FG* ($\overrightarrow{FG}$)	A **ray** is a part of a line that starts at one endpoint and extends without end in one direction.
A *B* *C* (angle)	angle *ABC* (∠*ABC*) or angle *CBA* (∠*CBA*) or angle *B* (∠*B*).	An **angle** is the figure formed by two rays that share a common endpoint. The endpoint is called the **vertex** of the angle. The rays are called the **sides** of the angle.
4 (angle)	angle 4 or ∠4	If there is a number between the rays that form an angle, the angle can be named by that number.
m *x* *y* *z* (plane)	plane *XYZ* or plane *m*	A **plane** is a flat surface that extends without end in all directions. It is determined by three noncollinear points. Points that are on the same plane are said to be **coplanar points**.

CHECK UNDERSTANDING

When you use three letters to name an angle, which letter must be in the middle?

Why would it be incorrect to also refer to the angle at the right as ∠A?

the letter that represents the vertex of the angle; point *A* is not at the vertex of the angle.

342   CHAPTER 10   Geometry of Size and Shape

Example 1

Write the symbol for each figure.

a. 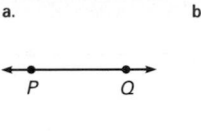　b. 　c. 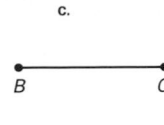　d.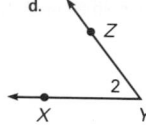

Solution

a. $\overrightarrow{PQ}$ or $\overrightarrow{QP}$　　　　b. $\overrightarrow{DE}$

c. $\overline{BC}$ or $\overline{CB}$　　　　d. $\angle XYZ$ or $\angle ZYX$ or $\angle Y$ or $\angle 2$ ◀

A **polygon** is a closed plane figure that is formed by joining three or more line segments at their endpoints. Each line segment joins exactly two others and is called a **side of the polygon.** The point at which two sides meet is a **vertex of the polygon.** Some polygons have special names.

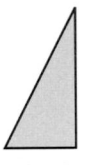　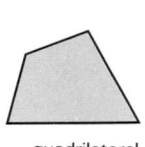　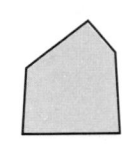　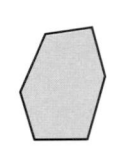　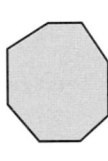

triangle　quadrilateral　pentagon　hexagon　octagon
(3 sides)　(4 sides)　(5 sides)　(6 sides)　(8 sides)

A polygon is named by the letters at its vertices. You list the vertices in order.

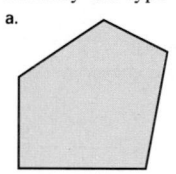　　　　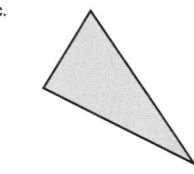

triangle *DEF* or △*DEF*　　quadrilateral *ABCD*

Example 2

Identify the type of polygon.

a. 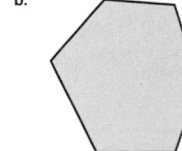　b.　c.

Solution

a. There are 5 sides. It is a pentagon.

b. There are 6 sides. It is a hexagon.

c. There are 3 sides. It is a triangle. ◀

READING MATH

The prefixes in the names of polygons tell you how many sides and angles the polygon has. Use a dictionary to find out what number these prefixes signify: *hepta-, nona-, deca-, dodeca-, icosa-.* What do you think is the definition of a *nonagon*? What do you think is the name of a polygon with twelve sides? **9-sided polygon; dodecagon**

COMPUTER

Complete this program by writing lines 40–60. With these lines, the program can be RUN for all polygons from 3-sided to 6-sided.

```
10  INPUT "ENTER THE
    NUMBER OF SIDES IN
    THE POLYGON: ";S:
    PRINT
20  INPUT "ENTER THE
    NAME OF THE POLYGON:
    ";N$: PRINT
30  IF S = 3 AND N$ =
    "TRIANGLE" THEN
    GOTO 80
70  PRINT "SORRY, NOT
    QUITE RIGHT.":PRINT :
    GOTO 90
80  PRINT "GREAT JOB":
    PRINT
90  INPUT "RUN AGAIN? Y
    OR N:";X$: PRINT
100 IF X$ = "Y" THEN
    GOTO 10
```

See Additional Answers.

ASSIGNMENTS

BASIC
1–12, 16–22, 24–26

AVERAGE
4–13, 14–23, 24–30

ENRICHED
9–13, 14–23, 24–31

ADDITIONAL RESOURCES
Reteaching 10–1
Enrichment 10–1
Transparency Master 35

Example 1: Have students identify, and write the symbol for, each geometric figure.　**a. line, b. ray, c. line segment, d. angle**
Example 2: Point out that when identifying polygons, the order of the letters is important.

Additional Questions/Examples
1. What is the vertex of an angle? **the common endpoint shared by the two rays that form the angle**
2. What are coplanar points? **points that are on the same plane**
3. Is an angle a plane figure? Explain. **Yes, an angle is named by three distinct points. Thus, an angle will determine a plane.**

Guided Practice/Try These　Point out that when naming angles as done in Exercise 4, the middle letter must always be the letter at the vertex of the angle. For Exercises 7–9 encourage students to count the sides and use their knowledge of prefixes to name the polygons.

343

Write the symbol for each figure.

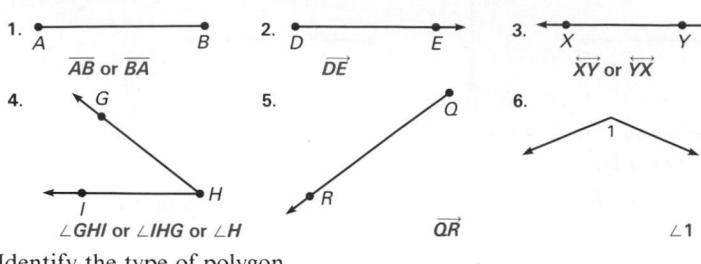

1. $\overrightarrow{AB}$ or $\overrightarrow{BA}$
2. $\overrightarrow{DE}$
3. $\overleftrightarrow{XY}$ or $\overleftrightarrow{YX}$

4. ∠GHI or ∠IHG or ∠H
5. $\overrightarrow{QR}$
6. ∠1

Identify the type of polygon.

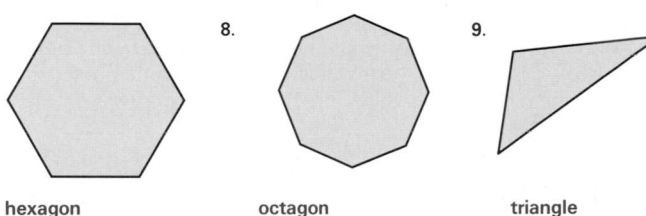

7. hexagon
8. octagon
9. triangle

EXERCISES

Write the symbol for each figure.

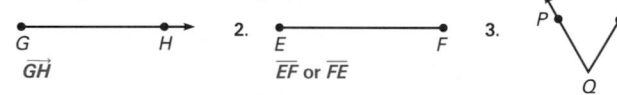

1. $\overrightarrow{GH}$
2. $\overline{EF}$ or $\overline{FE}$
3. ∠PQR or ∠RQP or ∠

Draw a representation of each figure. **Check students' drawings.**

4. $\overleftrightarrow{MN}$ 5. $\overrightarrow{PQ}$ 6. plane m 7. $\overline{PR}$ 8. ∠XYZ

9. pentagon   10. hexagon   11. quadrilateral   12. octagon

USING DATA Refer to the Data Index on page 546 to answer this question.

13. What two-dimensional figure is suggested by the shape of the Mosque of Omar? **The base suggests an octagon. The dome suggests a semicircle.**

Use this figure for Exercises 14–15.

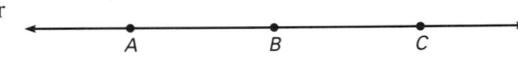

14. Use symbols to name the line in at least four ways. **Answers may vary. At least 4 of these: AC, AB, BC, CA, CB, BA**

15. Use symbols to name two rays with B as an endpoint. **$\overrightarrow{BA}$, $\overrightarrow{BC}$**

Talk It Over Have a class discussion about each of the following: How are lines, line segments, and rays similar? How are they different? How are the sides of an angle and the sides of a polygon the same? How are they different?

Practice/Solve Problems For each of Exercises 2 and 3, you may wish to have students give all possible symbols. For Exercises 9–12 remind students to use their knowledge of prefixes.

Extend/Solve Problems For Exercises 16–23 point out that some descriptions can be matched to more than one lettered term.

Think Critically/Solve Problems For Exercises 24–29 you may wish to have students correct the false statements to make them true. For Exercise 30 some students might want to use straws to act out the problem.

Extra Practice Write a symbol for each figure.

1.

∠XZY, ∠YZX, or ∠Z

2.

plane *TUV* or plane *n*

3.
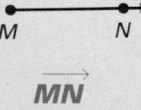

$\overrightarrow{MN}$

344

ESL STUDENTS

Have students work in pairs to draw a picture that includes coplanar points, collinear points, rays, lines, and line segments. Have each pair write directions on how to draw their picture. They must use symbols in their directions. Then have student pairs exchange directions and follow each others' directions.

Use the terms in the box at the right. Write the letters of all the terms that match each description.

| a. ray |
| b. line |
| c. polygon |
| d. line segment |
| e. point |
| f. plane |
| g. angle |

16. It is a location in space.   e

17. It extends without end in two   b
opposite directions.

18. It has one vertex.   g

19. It extends without end in only   a
one direction.

20. It can be named with two endpoints. d

21. It is determined by three noncollinear points.   f

22. It can be named by two points.   a, b, d

23. It has three or more vertices.   c

Refer to the figure at the right, then write *true* or *false* for each.

24. Point M is on $\overleftrightarrow{NO}$. true

25. Point M is on $\overleftrightarrow{PQ}$ but not on   false

26. $\overrightarrow{MO}$ can also be named $\overrightarrow{OM}$. false

27. $\overleftrightarrow{PQ}$ and $\overrightarrow{PM}$ name the same figure. true

28. $\overleftrightarrow{MN}$ has two endpoints. false

29. $\overleftrightarrow{MQ}$ has no endpoints. true

Suppose that you have eight straws that you can use to represent eight line segments.

30. What polygon and what combinations of polygons could you model using all eight straws each time? Name all the possibilities.

31. Refer to your answers to Exercise 30. What is the total number of vertices in each polygon or combination of polygons?
8 vertices in each combination

30. octagon
   2 quadrilaterals
   2 triangles and
   1 quadrilateral
   1 pentagon and
   1 triangle

THINK CRITICALLY/ SOLVE PROBLEMS

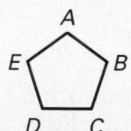 **MATH: WHO, WHERE, WHEN**

Archie Alexander (1867–1958) is among the most innovative of United States bridge builders. One of the first African Americans to graduate from the University of Iowa with a degree in engineering, Alexander went on to continue his studies in bridge design at the University of London.

While he implemented the construction of many other civil engineering (construction) projects, Alexander's greatest achievement was the River Bridge at Mt. Pleasant, Iowa. When it was built, this bridge was known to be the longest and most efficient in the state of Iowa.

Draw a figure to represent each.
4. Two rays, one with endpoint C and one with endpoint D, where C and D lie on the same line, line CD.

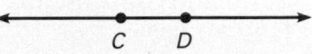

5. A polygon with five vertices labeled $ABCDE$.

Extension Provide students with the following definition and example: Locus—a set of points that satisfies given conditions. For example, this circle represents the locus of points that are 1 cm from S.

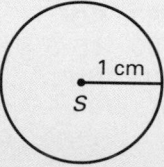

Have students draw a circle showing the locus of points that are 2 cm from center point, S. Then have them shade in the region that is the locus of all points less than 2 cm from S.

Get Ready protractors, paper strips

Additional Answers
See page 581.

Additional answers for odd-numbered exercises are found in the Selected Answers portion of the page.

MAKING CONNECTIONS

Have students use their knowledge of geometric figures to map the locations of their school, home, library, a local park, train or bus station, and a shopping center. Have students use points to represent each place, and lines to connect them. When their maps are complete, have students tell which points, if any, are collinear.

10-2 Angles and Angle Measures

EXPLORE Cut out two paper strips. Attach them at one end with a paper fastener. Notice that the two strips in the diagram are positioned to make a "square angle." Make a matching angle with your strips.

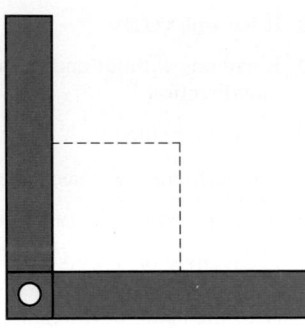

Now use the strips to make an angle larger than the square angle. Then make an angle that is smaller than the square angle. Describe all the differences you observe between the larger and the smaller angles.

SKILLS DEVELOPMENT The size of an angle is commonly measured in the unit called the **degree.** The number of degrees in the angle's measure indicates the amount of openness between the sides of the angle. Angles are measured with the instrument called a **protractor.** On the protractor, one scale shows degree measures from 0° to 180° in a clockwise direction; the other scale shows these degree measures in a counterclockwise direction.

Example 1

Use a protractor to measure $\angle PQR$.

Solution
Place the center of the protractor on vertex Q. Place the 0° line of the protractor on $\overrightarrow{QR}$. Locate the position of $\overrightarrow{QP}$ on the counterclockwise scale.

Scale shows 48°

R

Q is at center of protractor

Q

0° line

The measure of $\angle PQR$ is 48°. This is written $m\angle PQR = 48°$. ◄

A protractor can also be used to draw angles.

Example 2

Use a protractor to draw $\angle JKL$ so that $m\angle JKL = 120°$.

Solution
Step 1
Draw a ray from point K through point L.

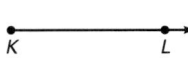

K L

346   CHAPTER 10   Geometry of Size and Shape

Step 2
Place the center of the protractor on the vertex, K. Place the 0° line of the protractor on $\overrightarrow{KL}$. Locate the 120° point on the protractor. Mark point J at 120°.

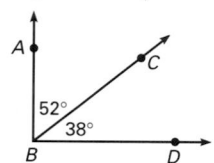

Step 3
Remove the protractor. Draw $\overrightarrow{KJ}$.

$m\angle JKL = 120°$ ◄

Angles are classified according to their measures, as listed in the chart.

Name of Angle	Measure of Angle
acute	between 0° and 90°
right	90°
obtuse	between 90° and 180°
straight	180°

Certain pairs of angles are related in important ways.

Two angles are called **complementary angles** if the sum of their measures is 90°. In the figure at the right, $\angle ABC$ is the *complement* of $\angle CBD$.

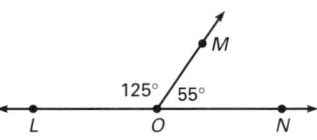

Two angles are called **supplementary angles** if the sum of their measures is 180°. In the figure at the right, $\angle LOM$ is the *supplement* of $\angle MON$.

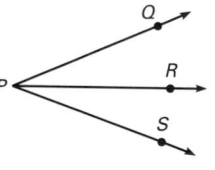

Adjacent angles are angles that have a common vertex and a common side, but have no interior points in common. In the figure at the right, $\angle QPR$ is adjacent to $\angle RPS$.

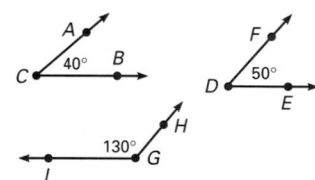

Example 3

Use the figures at the right to name the following.
a. all acute angles
b. all obtuse angles
c. a pair of complementary angles
d. a pair of supplementary angles

5-MINUTE CLINIC

Exercise	Student's Error	Error Diagnosis
Draw $\angle ABC$ so that m $\angle ABC$ is 45°.	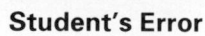	• Student read the wrong scale on the protractor. Correct solution:

Example 2: Have students practice drawing $\angle JKL$ on their own. Check their drawings before going on.

Example 3: Ask students why none of the angles shown are classified as adjacent. **None have a common vertex.**

Additional Questions/Examples
1. Which of the following will always be complementary angles? **a**
   a. two adjacent angles whose exterior sides form a right angle
   b. two angles that are both adjacent and acute
   c. two adjacent angles whose exterior sides form a straight angle
2. At what times would the hands of a clock form a right angle? **12:16, 12:49, 1:22, 1:54, 2:27, 3:00, 3:33, 4:05, 4:38, 5:11, 5:44, 6:16, 6:49, 7:22, 7:54, 8:27, 9:00, 9:33, 10:06, 10:38, 11:11, 11:44**

Guided Practice/Try These For Exercises 1–3 remind students to position their protractors with the center on each vertex.

For Exercises 4, 6, and 8 suggest that students give the complement of each angle. For Exercises 5 and 7 have them give the supplement of the angle.

Write About Math Have students copy and complete the following chart.

Measure	Angle	Type	Name
50°	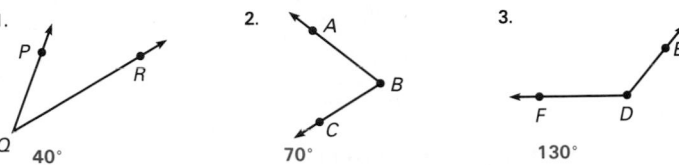	acute	∠ABC
105°			∠DEF
180°			∠XYZ
90°			∠MNO

4 PRACTICE

Practice/Solve Problems After students complete Exercises 1–8, have them identify as many supplementary angles as they can, using the angles marked off on the protractor. **Answers will vary. Samples: ∠ JOM and ∠ MOP; ∠ PON and ∠ NOJ**

Extend/Solve Problems For Exercises 21–23 point out that all angles will have the letter A as the middle letter.

Think Critically/Solve Problems For each of Exercises 26 and 27, suggest students draw a picture to help them solve each problem.

5 FOLLOW-UP

Extra Practice

1. Use a protractor to draw and label an example of each of the following.
   a. an acute angle
   b. an obtuse angle
   c. complementary angles
   d. supplementary angles
   Check students' drawings.

348

Solution

a. Angles *C* and *D* both have measures less than 90°. So, ∠*C* and ∠*D* are acute angles.

b. The measure of angle *G* is greater than 90°. So ∠*G* is an obtuse angle.

c. $m\angle C + m\angle D = 40° + 50° = 90°$, so ∠*C* and ∠*D* are a pair of complementary angles.

d. $m\angle D + m\angle G = 50° + 130° = 180°$, so ∠*D* and ∠*G* are a pair of supplementary angles. ◄

TRY THESE

Use a protractor to measure each angle.

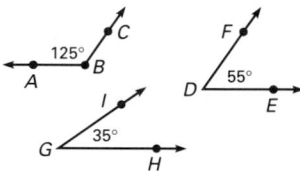

1. 40°
2. 70°
3. 130°

Use a protractor to draw an angle of the given measure.
Check students drawings.

4. 40°   5. 110°   6. 44°   7. 135°   8. 19°

Use the figures at the right to name the following.

9. an acute angle ∠*IGH*

10. a pair of complementary angles ∠*FDE* and ∠*IGH*

11. an obtuse angle ∠*ABC*

12. a pair of supplementary angles ∠*FDE* and ∠*ABC*

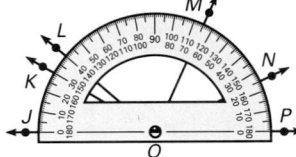

EXERCISES

PRACTICE/ SOLVE PROBLEMS

Find the measure of each angle. Classify the angle.

1. ∠*NOP* — 25°, acute
2. ∠*KOP* — 150°, obtuse
3. ∠*LOP* — 140°, obtuse
4. ∠*MOP* — 65°, acute
5. ∠*JOM* — 115°, obtuse
6. ∠*JOL* — 40°, acute
7. ∠*JOK* — 30°, acute
8. ∠*JON* — 155°, obtuse

Use a protractor to draw an angle of the given measure.
Check students' drawings.

9. 10°   10. 140°   11. 65°   12. 175°   13. 28°

Use the figures at the right to name the following.

14. a right angle ∠*LMN*

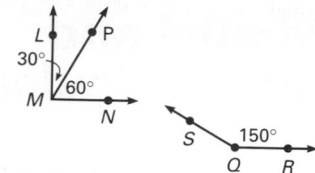

15. a pair of adjacent angles
∠*LMP* and ∠*PMN*

16. a pair of supplementary angles
∠*LMP* and ∠*SQR*

17. a pair of complementary angles
∠*LMP* and ∠*PMN*

For each of Exercises 18–20, classify the angle, estimate its size, and then measure the angle. Estimates will vary. acute, 83°

18.

19.

20.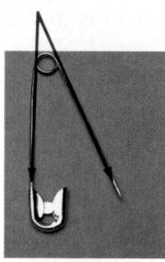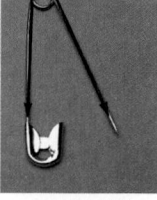

obtuse, 108°   acute, 56°   acute, 25°

Use the figure at the right for Exercises 21–23.
The ⌐ symbol is used to show a right angle.

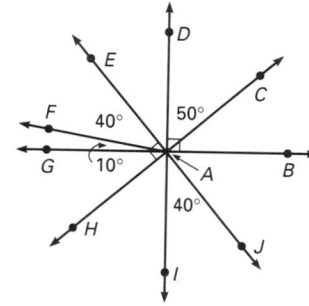

21. Name all the angles with
a measure of 40°.
∠*CAB*, ∠*EAD*, ∠*EAF*, ∠*GAH*, ∠*IAJ*

22. Name all the angles with
a measure of 50°.
∠*CAD*, ∠*HAI*, ∠*JAB*, ∠*EAG*

23. Name all the right angles.
∠*GAD*, ∠*DAB*, ∠*BAI*, ∠*IAG*,
∠*EAC*, ∠*CAJ*, ∠*JAH*, ∠*HAE*

Determine whether each statement is *true* or *false.*

24. The supplement of an obtuse angle is an acute angle. true

25. The complement of an acute angle is a right angle. false

Solve.

26. The measure of an angle is 50° more than its supplement. Find the measure of both angles. 115°, 65°

27. Suppose that ∠*A* and ∠*B* are complementary angles and that ∠*A* and ∠*C* are supplementary angles. What is the relationship between the measure of ∠*B* and the measure of ∠*C*? *m*∠*C* = *m*∠*B* + 90°

EXTEND/ SOLVE PROBLEMS

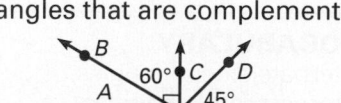

ESTIMATION TIP

To estimate the measure of an angle, think visually using 90°, 45°, and 180° as benchmarks. Knowing that a right angle measures 90°, you can estimate the size of an angle that is half as big, or 45°. In the same way, you can estimate the measure of a larger angle by deciding if it is just a little larger than a right angle, or if it is closer to a *straight angle,* which measures 180°.

THINK CRITICALLY/ SOLVE PROBLEMS

2. Measure and classify each angle.

a.
70° acute

b.
125° obtuse

3. Name all the pairs of adjacent angles that are complementary.

∠ *AOH* and ∠ *HOG*; ∠ *GOF* and ∠ *FOE*; ∠ *EOD* and ∠ *DOC*; ∠ *COB* and ∠ *BOA*

Extension Have students find and measure the angles that are part of of familiar objects located in the classroom.

Section Quiz Use the following drawing to identify each angle described. **Answers will vary.**

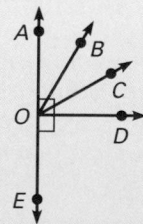

1. adjacent complementary angles ∠ *AOC* and ∠ *COD*

2. adjacent supplementary angles ∠ *AOC* and ∠ *COE*

3. adjacent angles that are neither supplementary nor complementary ∠ *BOC* and ∠ *COD*

4. two adjacent angles, one of which is acute and one of which is obtuse ∠ *BOC* and ∠ *COE*

10-3 Parallel and Perpendicular Lines

EXPLORE/ WORKING TOGETHER

Look at Figures A and B. In each diagram, one line crosses two other lines. Measure the angles formed by the lines in each diagram. What is the same in both diagrams? What is different?

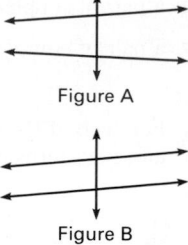

Figure A

Figure B

SKILLS DEVELOPMENT

Any two lines in a plane are related in exactly one of these two ways.

The two lines might **intersect** at a single point.

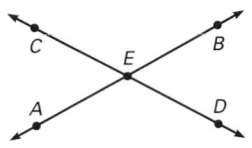

$\overleftrightarrow{AB}$ and $\overleftrightarrow{CD}$ intersect at point E.

The two lines might be **parallel** and have no points in common.

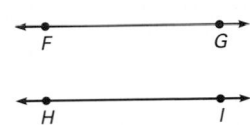

$\overleftrightarrow{FG}$ is parallel to $\overleftrightarrow{HI}$. Write this as $\overleftrightarrow{FG} \parallel \overleftrightarrow{HI}$.

If two lines intersect to form right angles, then the lines are said to be **perpendicular** to each other.

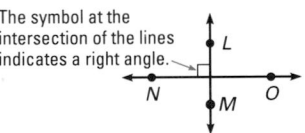

The symbol at the intersection of the lines indicates a right angle.

$\overleftrightarrow{LM}$ is perpendicular to $\overleftrightarrow{NO}$. Write this as $\overleftrightarrow{LM} \perp \overleftrightarrow{NO}$.

When two lines intersect, the angles that are not adjacent to each other are called **vertical angles.** In the figure at the right, $\angle 1$ and $\angle 3$ form a pair of vertical angles, and $\angle 2$ and $\angle 4$ form another pair.

Vertical angles have the same measure. So, in this figure, $m\angle 1 = m\angle 3$ and $m\angle 2 = m\angle 4$.

The double marks indicate that angles 2 and 4 are equal in measure.

The single marks indicate that angles 1 and 3 are equal in measure.

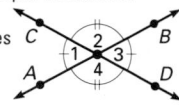

✏ WRITING ABOUT MATH

Use what you know about straight angles and supplementary angles to write a convincing argument explaining why vertical angles must be equal in measure. Use a figure like the one below to illustrate your reasoning.

Answers will vary. Sample answer: Each angle has the same supplementary angle.

350 CHAPTER 10 Geometry of Size and Shape

Example 1

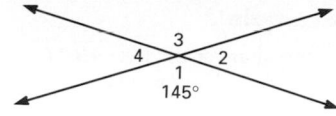

Use the figure at the right.

a. Find $m\angle 3$.   **b.** Find $m\angle 2$.

Solution

a. $\angle 1$ and $\angle 3$ are not adjacent to each other, so they are vertical angles. Vertical angles have the same measure.

$m\angle 1 = m\angle 3$ $m\angle 3 = 145°$

b. $\angle 2$ and $\angle 3$ are supplementary angles, so the sum of their measures is 180°.

$m\angle 2 + 145° = 180°$
$m\angle 2 = 180° - 145°$
$m\angle 2 = 35°$ ◄

A **transversal** is a line that intersects two or more lines in a plane at different points. In the figure at the right, $\overleftrightarrow{AB}$ is a transversal intersecting $\overleftrightarrow{CD}$ and $\overleftrightarrow{EF}$.

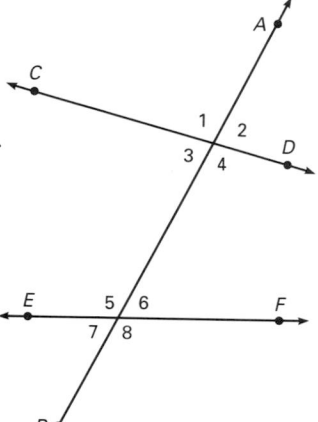

A transversal such as $\overleftrightarrow{AB}$ forms several special pairs of angles with the lines it intersects.

Angles that are in the same position relative to the transversal and the lines are called **corresponding angles**. In this figure, there are four pairs of corresponding angles.

$\angle 1$ and $\angle 5$ $\angle 2$ and $\angle 6$
$\angle 3$ and $\angle 7$ $\angle 4$ and $\angle 8$

Alternate interior angles are pairs of angles that are *interior* to the lines but on *alternate* sides of the transversal. In the figure above, there are two pairs of alternate interior angles.

$\angle 3$ and $\angle 6$ $\angle 4$ and $\angle 5$

Alternate exterior angles are pairs of angles that are *exterior* to the lines and on *alternate* sides of the transversal. In the figure above, there are two pairs of alternate exterior angles.

$\angle 1$ and $\angle 8$ $\angle 2$ and $\angle 7$

When a transversal intersects a pair of parallel lines, there are special relationships among these pairs of angles.

► Corresponding angles have the same measure.
► Alternate interior angles have the same measure.
► Alternate exterior angles have the same measure.

TALK IT OVER

In the definition of a transversal, why do you think it is important to include the phrase *at different points?*

Answers will vary. If the three or more lines all intersected at one point, the special pairs of corresponding angles would not be formed.

Example 2: Emphasize that the special relationships between corresponding and alternate angles exist only when a transversal intersects parallel lines. Point out that, in the first diagram shown next to the definition of transversal, the alternate and corresponding angles are not equal to each other.

Example 3: You may wish to discuss the vertical angles in this figure as well.

Additional Questions/Examples

1. Can parallel lines intersect? Why or why not? **No, because they have no points in common.**

2. In this diagram, what information would you need to determine whether $\overleftrightarrow{AB} \perp \overleftrightarrow{CD}$ or whether $\overleftrightarrow{CD} \parallel \overleftrightarrow{EF}$?

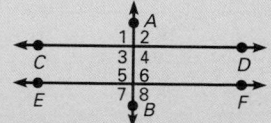

Answers will vary. Possible answers:
m $\angle$ 1; m $\angle$ 3 and m $\angle$ 6

Guided Practice/Try These Have students work in small groups to complete Exercises 1–6. Encourage them to share and discuss their reasons for each answer.

Key Questions
1. What are vertical, corresponding, alternate interior, and alternate exterior angles?
2. What are transversals and parallel and perpendicular lines? Draw and label an example of each.

4 **PRACTICE**

Practice/Solve Problems For Exercise 12 you may wish to discuss two methods students could use to solve this problem—identifying complementary angles or identifying alternate angles.

Extend/Solve Problems For Exercises 14 and 15, point out to students that they should write and solve an equation for each exercise.

Think Critically/Solve Problems For Exercises 16 and 17, emphasize that students should never assume that lines are parallel or perpendicular solely on how the lines appear. They should, instead, reason from known facts.

5 **FOLLOW-UP**

Extra Practice Use this figure for Exercises 1–4.

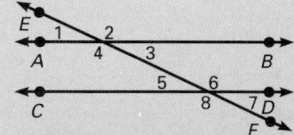

1. Which angles have the same measure as ∠5?
   ∠1, ∠3, ∠7
2. Which angles have the same measure as ∠6? **∠2, ∠ 8, ∠4**
3. Identify two alternate exterior angles.
   ∠2 and ∠8 or ∠1 and ∠7
4. Identify the pairs of corresponding angles. **∠2 and ∠6; ∠1 and ∠ 5; ∠3 and ∠7; ∠4 and ∠8**

352

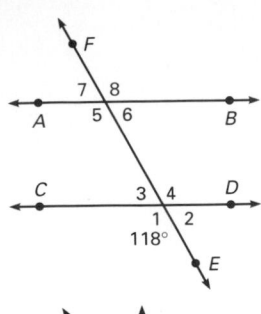

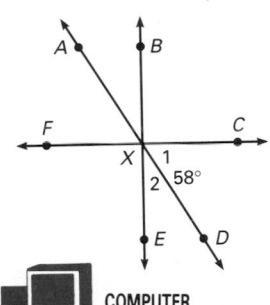

COMPUTER

This program will print the measure of the complement or the supplement of an angle based on your input.

```
10 INPUT "WHAT IS THE
   MEASURE OF THE ANGLE?
   ";M: PRINT
20 INPUT "ARE THE ANGLES
   COMPLEMENTARY OR
   SUPPLEMENTARY?
   C OR S: ";A$: PRINT
30 IF A$ = "C" THEN GOTO
   50
40 PRINT "THE MEASURE OF
   THE SUPPLEMENT IS
   ";180 − M: END
50 PRINT "THE MEASURE OF
   THE COMPLEMENT IS";
   90 − M
```

PRACTICE/
SOLVE PROBLEMS

Example 2

In the figure at the left, $\overleftrightarrow{AB} \parallel \overleftrightarrow{CD}$. Find the measure of each angle.
a. $m\angle 8$ b. $m\angle 7$

Solution
a. ∠8 and ∠1 are a pair of alternate exterior angles, so $m\angle 8 = m\angle 1$.
   $m\angle 8 = 118°$
b. ∠7 and ∠8 are a pair of supplementary angles, so the sum of their measures is 180°.
   $m\angle 7 + m\angle 8 = 180°$
   $m\angle 7 + 118° = 180°$
   $m\angle 7 = 180° − 118° = 62°$ ◄

Example 3

In the figure at the left, $\overleftrightarrow{BE} \perp \overleftrightarrow{CF}$. Find $m\angle 2$.

Solution
Because $\overleftrightarrow{BE} \perp \overleftrightarrow{CF}$, ∠CXE is a right angle, with a measure of 90°. So, ∠1 and ∠2 are a pair of complementary angles.
   $m\angle 1 + m\angle 2 = 90°$
   $58° + m\angle 2 = 90°$
   $m\angle 2 = 90° − 58° = 32°$ ◄

TRY THESE

Use the figure at the right to find the measure of each angle.

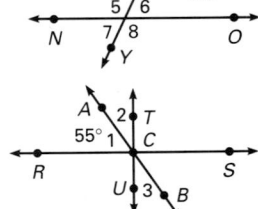

1. ∠1 **120°**   2. ∠4 **60°**

In the figure at the right, $\overleftrightarrow{LM} \parallel \overleftrightarrow{NO}$. Find the measure of each angle.

3. ∠5 **115°**   4. ∠6 **65°**

In the figure at the right, $\overleftrightarrow{RS} \perp \overleftrightarrow{TU}$. Find the measure of each angle.

5. ∠2 **35°**   6. ∠3 **35°**

EXERCISES

Use the figure at the right to find the measure of each angle.

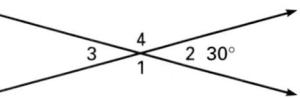

1. ∠3 **30°** 2. ∠1 **150°** 3. ∠4 **150°**

In the figure at the right, $\overleftrightarrow{AB} \parallel \overleftrightarrow{CD}$. Find the measure of each angle.

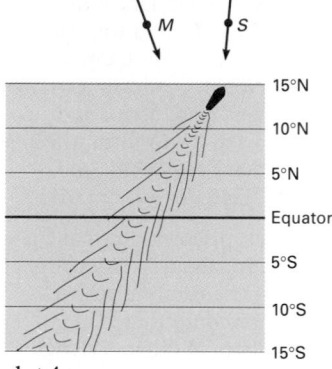

4. ∠1 70°   **5.** ∠2 110°

6. ∠3 110°   **7.** ∠4 70°

8. ∠5 95°   **9.** ∠6 95°

10. ∠7 85°   **11.** ∠8 95°

12. Assume that, over a small area, the surface of the earth can be considered a plane. If a freighter going north crosses the equator at a 59° angle and continues on in the same direction, what will be the angle of the freighter's path as it crosses the 5°N latitude line? **59°**

PROBLEM SOLVING TIP

Remember that lines of latitude are parallel to the equator.

Find the measure of ∠1, ∠2, ∠3, and ∠4.

13.

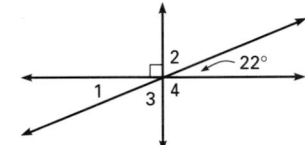

$m\angle 1 = 22°$
$m\angle 2 = 68°$
$m\angle 3 = 68°$
$m\angle 4 = 90°$

EXTEND/ SOLVE PROBLEMS

PROBLEM SOLVING TIP

In Exercises 14 and 15, write an equation that represents the relationship between the angles whose measures are labeled.

Equation for Exercise 14:
$3x - 34 = 2x$

Equation for Exercise 15:
$3x + 30 + 2x = 180$

14. Find $m\angle ATS$ and $m\angle STB$ in the figure below.

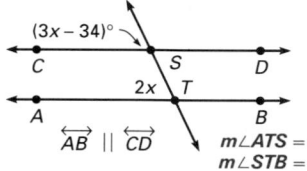

$\overleftrightarrow{AB} \parallel \overleftrightarrow{CD}$
$m\angle ATS = 68°$
$m\angle STB = 112°$

15. Find $m\angle QRN$ and $m\angle PRN$ in the figure below.

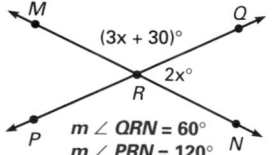

$m\angle QRN = 60°$
$m\angle PRN = 120°$

16. Do you think that $\overleftrightarrow{AB}$ and $\overleftrightarrow{CD}$ in the figure below are parallel? Explain.

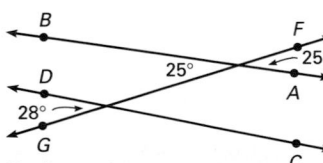

No, because the corresponding angles shown are not equal.

17. In the figure below, $m\angle BAC = m\angle DCE$. Which line segments or lines do you think are parallel? Explain.

$\overleftrightarrow{AB} \parallel \overleftrightarrow{CD}$ because ∠BAC and ∠DCE are corresponding angles.

THINK CRITICALLY/ SOLVE PROBLEMS

10-3   Parallel and Perpendicular Lines   **353**

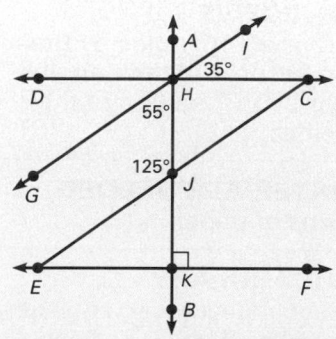
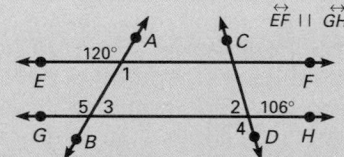

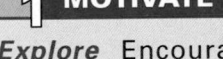
10-4 Triangles

EXPLORE

TALK IT OVER

Suppose that the strips of paper had lengths of 3 in., 5 in., 8 in., and 9 in. Can you predict which combination of strips could be used to form a triangle?

3 in., 8 in., and 9 in. or 5 in., 8 in., and 9 in.

Cut four thin strips of paper that have lengths 2 cm, 5 cm, 6 cm, and 9 cm. Using the strips, try to form a triangle with sides of 5 cm, 6 cm, and 9 cm. Now try to form a triangle with sides 2 cm, 5 cm, and 9 cm. Then try to form a triangle with sides 2 cm, 5 cm, and 6 cm. What do you discover?

Use your discovery to make a conjecture about how lengths of sides of triangles are related. Test your conjecture using paper strips of different lengths.

You cannot make a triangle using 2-cm, 5-cm, and 9-cm strips.

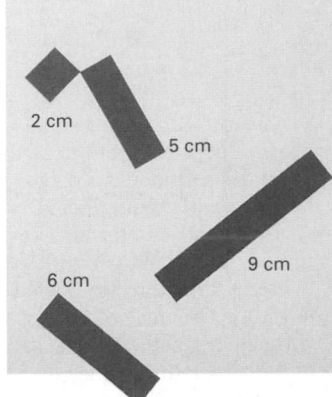

SKILLS DEVELOPMENT

Triangles can be classified by the measures of their angles.

right triangle: A triangle that has one right angle.

acute triangle: A triangle that has three acute angles.

obtuse triangle: A triangle that has one obtuse angle.

Example 1

Classify each triangle as *acute, obtuse,* or *right.*

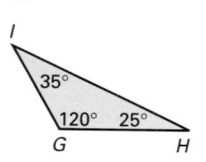 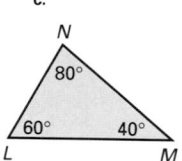

Solution
a. Angle *E* is a right angle. So, △*DEF* is a right triangle.
b. The measure of angle *G* is greater than 90°. So, △*GHI* is an obtuse triangle.
c. The measure of each of the three angles is less than 90°. So, △*LMN* is an acute triangle. ◄

354 CHAPTER 10 Geometry of Size and Shape

Triangles can also be classified by the lengths of their sides.

isosceles triangle: A triangle that has two sides of equal length.

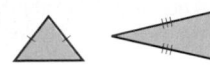

equilateral triangle: A triangle that has all three sides of equal length.

scalene triangle: A triangle that has no sides of equal length.

The marks on the figures indicate which sides are of equal length.

Example 2

Classify each triangle as *isosceles, equilateral,* or *scalene.*

a.    b. 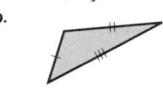   c.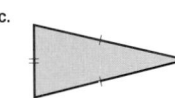

Solution

a. All three sides are of equal length. The figure is an equilateral triangle.
b. No sides are of equal length. The figure is a scalene triangle.
c. Two sides are of equal length. The figure is an isosceles triangle. ◀

The sum of the measures of the angles of a triangle is 180°.

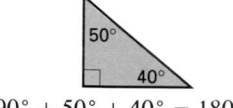

$90° + 50° + 40° = 180°$

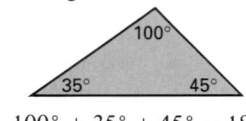

$100° + 35° + 45° = 180°$

The three angles of an equilateral triangle have equal measures. That is, an equilateral triangle is also **equiangular.** Each angle of an equilateral triangle measures 60°.

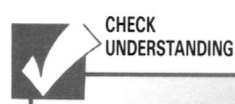

Example 3

a. The angles of a scalene triangle have the measures shown in the figure at the right. What is the value of x?
b. The measure of one acute angle of a right triangle is 30°. What is the measure of the other acute angle?

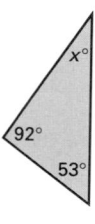

Solution

a. The sum of the angles of a triangle → 180
   Known angle measures $92° + 53°$ → $\underline{-145}$
   35

   The measure of the third angle is 35°, so $x = 35$.

Example 2: Point out that when students classify triangles according to their sides, the actual lengths can be unknown, but the relationships between the lengths must be known.

Example 3: Point out that an equilateral triangle will always have angles that measure 60°, but the length of the equal sides can change.

Example 4: Point out that the measure of a base angle of an isosceles triangle can always be found if the measure of the vertex angle is known.

Additional Questions/Examples

1. Can an obtuse triangle have more than one obtuse angle? Why or why not? **no, because two obtuse angles would equal more than 180°**

2. Can a triangle have sides that measure 4 cm, 6 cm, and 12 cm? Why or why not? **no, because the sum of any two sides of a triangle must always be greater than the third side: 4 + 6 is not greater than 12**

Guided Practice/Try These
Point out that for Exercises 9–11 students must perform at least two operations—addition and subtraction—to solve each.

3 SUMMARIZE

Talk It Over Discuss the similarities and differences between the various triangles in this lesson. **Answers will vary. Sample answers: An equilateral triangle is always an acute triangle. An acute triangle always has 3 acute angles, but a right triangle has only 1 right angle.**

4 PRACTICE

Practice/Solve Problems For Exercises 1–3 you may wish to have students draw diagrams to help them visualize and classify each triangle.

Extend/Solve Problems For Exercise 10 suggest that students first list the different triangles in the figures, then classify them. They may need to copy the figure and outline the triangles one at a time as they list them.

Think Critically/Solve Problems For each of Exercises 11–16 students may need to draw diagrams to determine if the statement is true or false.

5 FOLLOW-UP

Extra Practice Classify each triangle according to its angles.

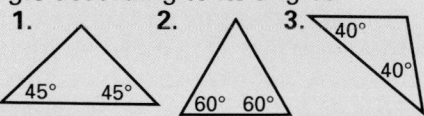

1. 45° 45° **right** **2.** 60° 60° **acute** **3.** 40° 40° **obtuse**

Classify each triangle according to its size.

4. **isosceles** **5.** 5.4 in. 5 in. 3 in. **scalene** **6.** 3 cm 3 cm 3 cm **equilateral**

356

---

b. The sum of the angles of a triangle → 180
The measure of the right angle is 90°.
Known angle measures 90° + 30° → $\underline{-120}$
60
The measure of the other acute angle is 60°. ◄

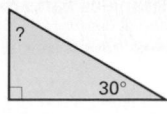

In an isosceles triangle the sides and angles have special names.

Notice that the legs of an isosceles triangle are of equal length. The base angles, which are opposite the legs, also have equal measures.

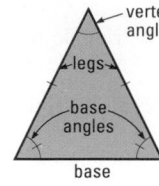

Example 4

The measure of the vertex angle of an isosceles triangle is 120°. What is the measure of each of the base angles?

Solution
The sum of the angles of a triangle → 180
Known angle measure $→ \underline{-120}$
60

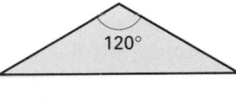

Since the base angles of an isosceles triangle are equal in measure, divide 60 by 2: 60 ÷ 2 = 30. The measure of each base angle is 30°. ◄

TRY THESE

Classify each triangle as *acute*, *obtuse*, or *right*.

1. right **2.** acute **3.** right **4.** obtuse

Classify each triangle as *isosceles*, *equilateral*, or *scalene*.

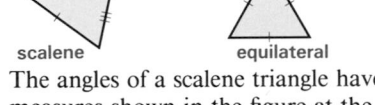

5. scalene **6.** equilateral **7.** isosceles **8.** scalene

9. The angles of a scalene triangle have the measures shown in the figure at the right. What is the value of *x*? **27**

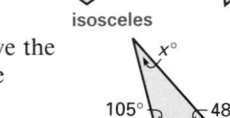
105°   48°   x°

10. ∠*ABC* and ∠*BAC* are the base angles of isosceles triangle *ABC*, and the measure of ∠*ABC* is 50°. What is the measure of ∠*BCA*? **80°**

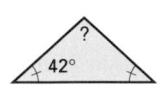
50° B   A   C

11. The measure of one base angle of an isosceles triangle is 42°. What is the measure of the vertex angle? **96°**

42°

EXERCISES

Classify each triangle as *acute, obtuse,* or *right.*

1. △ABC with $m\angle A = 35°$, $m\angle B = 70°$, and $m\angle C = 75°$. acute

2. △LMN with $m\angle M = 40°$, $m\angle L = 90°$, and $m\angle N = 50°$ right

3. △HIJ with $m\angle I = 104°$, $m\angle H = 46°$, and $m\angle J = 30°$ obtuse

Classify each triangle as *isosceles, equilateral,* or *scalene.*

4.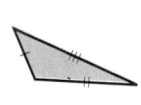

5.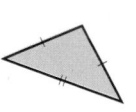

6.

7.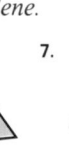

scalene   isosceles   equilateral   scalene

Use the figures at the right for Exercises 8 and 9.

8. The angles of a scalene triangle have the measures shown. What is the value of x? 47°

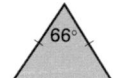

9. The measure of the vertex angle of an isosceles triangle is 66°. What is the measure of each of the base angles? 57°

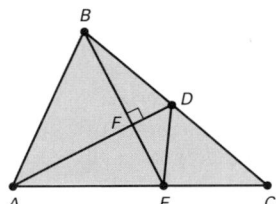

10. Name all the triangles in the figure at the right. Classify each triangle by its angles. △ABC, acute; △ADC, obtuse; △AFE, right; △ABE, acute; △ADE, obtuse; △ABF, right; △BDF, right; △BDE, obtuse; △BCE, obtuse; △CDE, acute; △DEF, right

Determine whether each statement is *true* or *false.*

11. A scalene triangle can also be an obtuse triangle. true

12. An equilateral triangle can also be a right triangle. false

13. An isosceles triangle can also be an obtuse triangle. true

14. An equilateral triangle must also be an acute triangle. true

15. Some right triangles are also isosceles. true

16. All equilateral triangles are also equiangular. true

10-4 Triangles **357**

Is it possible to form:

7. a triangle with lengths of 3 in., 12 in., and 7 in.? **No, the sum of the lengths of the two shorter sides is not greater than the length of the longest side — 3 + 7 < 12.**

8. a triangle with angles of 60°, 60°, and 45°? **No, the sum of these angles is less than 180°.**

Extension List the following types of triangles on the chalkboard. Have students state whether each triangle is possible. If it is *possible,* have them state the measures of its angles. If it is *impossible,* have them tell why.

1. an equilateral obtuse triangle **impossible, because, if it is equilateral, the meaure of each angle is 60°**

2. an acute right triangle **impossible, because a right triangle has one 90°-angle, which is not acute**

3. a right isosceles triangle **possible, angles measure 90°, 45°, 45°**

Section Quiz Classify each triangle according to its sides and angles.

1. a triangle with one right angle and no sides of equal length **right scalene triangle**

2. a triangle with three acute angles and just two sides of equal length **acute isosceles triangle**

3. a triangle with three acute angles and three sides of equal length **acute equilateral triangle**

Get Ready set of tangram pieces

SPOTLIGHT

OBJECTIVES
- To find the sum of the angles of a polygon
- To identify types of quadrilaterals

MATERIALS NEEDED
set of tangram pieces

VOCABULARY
bases, diagonal, quadrilateral, regular polygon

WARM-UP

Find the missing number.
1. $120 + 25 + 70 + ■ = 360$ **145**
2. $45 + ■ + 100 + 110 = 360$ **105**
3. $■ × 180 = 1,080$ **6**
4. $180 × ■ = 540$ **3**

1 MOTIVATE

Explore Students can work in small groups to complete this activity. You may wish to review the names of the polygons discussed on page 359 before students begin this activity.

2 TEACH

Use the Pages/Skills Development Have students read this part of the section and then discuss the examples.

Example 1: Point out that diagonals can be drawn from any of the vertices of the polygon. Have students trace the polygon and draw diagonals from a vertex other than the one shown in the solution. Have them compare their drawing with the one in the text.

10-5 Polygons

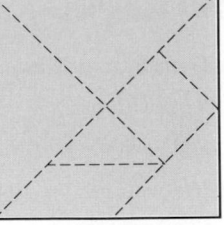

EXPLORE

A tangram is an ancient Chinese puzzle made up of seven pieces. The pieces can be combined to form different figures. Use a set of tangram pieces or trace the figure to make your own.

(If you trace the figure, cut apart your tracing along the lines to form the seven tangram pieces.

a. Use any number of pieces to form each of these figures.

 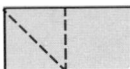  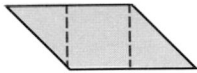

b. Record which pieces were used to make each figure.

c. Compare the pieces you used to make each figure with a classmate. Is there more than one way to make each figure? **yes** Discuss.

SKILLS DEVELOPMENT

A **diagonal** of a polygon is a line segment that joins two vertices and is not a side. Polygons can be separated into nonoverlapping triangular regions by drawing all the diagonals from one vertex.
The sum of the measures of the angles of a polygon is the product of the number of triangles formed and 180°.

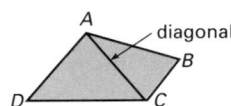

The sum of the angles of △ ABC equals 180°.
The sum of the angles of △ ACD equals 180°.

CHECK UNDERSTANDING

What pattern do you see that relates the number of sides of a polygon and the number of triangles formed by diagonals from one vertex?

The number of triangles is always 2 less than the number of sides.

Example 1

Find the sum of the angle measures for this polygon.

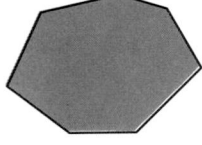

Solution
The polygon has 7 sides. When all the diagonals are drawn from one vertex, there are 5 triangular regions. Since $5 × 180° = 900°$, the sum of the angle measures is 900°. ◄

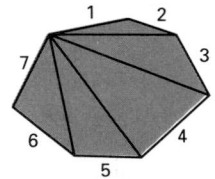

AT-RISK STUDENTS

You may wish to provide additional manipulative activities for these students. Have them work together in small groups using geoboards and rubber bands of two colors. Have students form polygons with rubber bands of one color, then use rubber bands of a second color to show all diagonals from one vertex.

Example 2

Find $m\angle D$ in the polygon at the right.

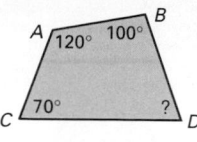

Solution

Since the polygon can be separated into two triangles, the sum of the angle measures is 360°.

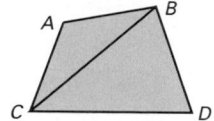

$$m\angle A + m\angle B + m\angle C + m\angle D = 360°$$
$$120° + 100° + 70° + m\angle D = 360°$$
$$290° + m\angle D = 360°$$
$$m\angle D = 70° \quad \blacktriangleleft$$

Recall that a four-sided polygon is called a *quadrilateral*. Some quadrilaterals have special names.

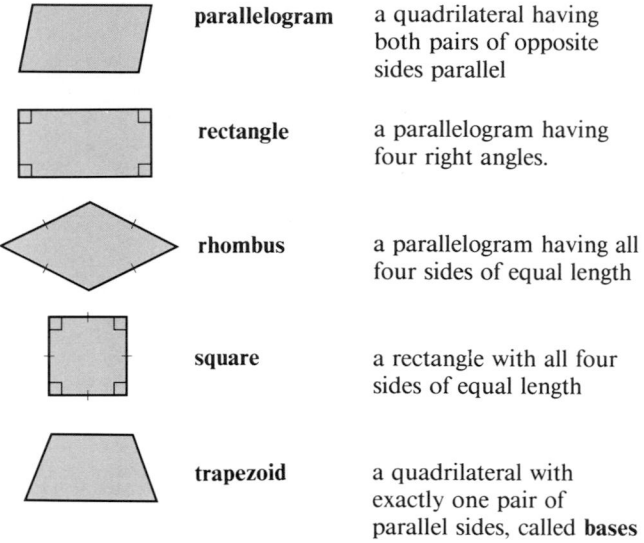

parallelogram — a quadrilateral having both pairs of opposite sides parallel

rectangle — a parallelogram having four right angles.

rhombus — a parallelogram having all four sides of equal length

square — a rectangle with all four sides of equal length

trapezoid — a quadrilateral with exactly one pair of parallel sides, called **bases**

Example 3

Give all the names that apply to the figure at the right.

Solution

It is a four-sided figure that has both pairs of opposite sides parallel. So, the figure is a polygon, a quadrilateral, and a parallelogram. ◄

A polygon is **regular** if all its sides are of equal length and all its angles are of equal measure. That is, a regular polygon is equilateral and equiangular.

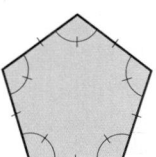

ASSIGNMENTS

BASIC
1–6, 7, 9–11, 15–16

AVERAGE
1–6, 7–14, 15–18

ENRICHED
9–14, 15–20

ADDITIONAL RESOURCES
Reteaching 10–5
Enrichment 10–5
Transparency Masters 40–45

Example 2: Point out that, when diagonal BC is drawn on the polygon, the sum of the angles in each of the two resulting triangles is 180°.

Example 3: Remind students that a closed plane figure with three or more sides is a polygon. You may wish to discuss why the figure shown is neither a rectangle, a rhombus, nor a trapezoid.

Additional Questions/Examples

1. Explain how to find the measure of an unknown angle in an 8-sided polygon, where the measures of seven angles are known. **Solve the equation: sum of measures of seven angles + a = 1,080°**

2. How can you find the sum of the measures of the angles of a regular polygon? **Draw all possible diagonals from one vertex and multiply the number of diagonals by 180°.**

Guided Practice/Try These For Exercises 3–5 point out that students must first find the sum of the measures of all angles in each polygon before finding the unknown angle measure.

360

3 SUMMARIZE

Write About Math In their journals, have students write a procedure for each of the following.
1. how to find the sum of the measures of all angles of a polygon
2. how to find the measure of an unknown angle in a polygon where the measures of all but one angle are given
3. how to determine if a polygon is regular

4 PRACTICE

Practice/Solve Problems For Exercises 1–6 you may wish to have students draw diagrams of several polygons listed, including all the diagonals from one vertex of each. Review the meanings of the prefixes used in the names of the polygons.

Extend/Solve Problems For Exercise 7 be sure that students know what is meant by *regular octagon*. For Exercise 8 students must first calculate the sum of the measures of all angles in the regular hexagon. For Exercises 9–12 point out that a diagonal extends from one vertex to another and that a diagonal is not a side.

Think Critically/Solve Problems For Exercises 15–18 students may need to review the definition of quadrilateral.

5 FOLLOW-UP

Extra Practice
1. Find the sum of the angle measures for polygon *ABCDEFG*.
900°

Find the sum of the angle measures for each polygon.

1.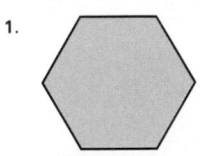
720°

2.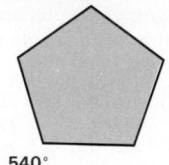
540°

Find the unknown angle measure for each polygon.

3.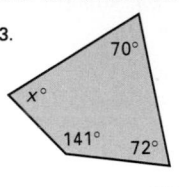
70°
x°
141° 72°
77°

4.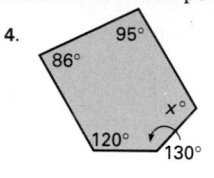
95°
86°
120° 130°
x°
109°

5.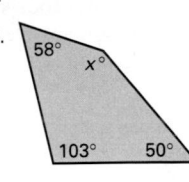
58° x°
103° 50°
149°

Give all the names that apply to the figure.

6.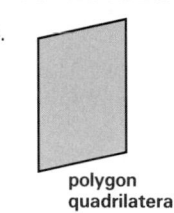
polygon
quadrilateral
parallelogram

7.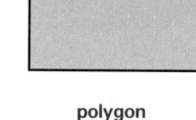
polygon
quadrilateral
parallelogram
rectangle

8.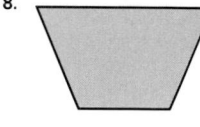
polygon
quadrilateral
trapezoid

COMPUTER

This program tells you the sum of the angle measures of a polygon when you enter the number of sides. Can you explain how the formula in line 20 was derived?

```
10 INPUT "HOW MANY
   SIDES? ";S: PRINT
20 LET D = (S – 2) * 180
30 PRINT "THE SUM OF THE
   MEASURES OF THE
   ANGLES IS ";D;"."
```

See Additional Answers.

EXERCISES

PRACTICE/ SOLVE PROBLEMS

Copy and complete the chart.

Polygon	Number of sides	Number of triangles formed by drawing all possible diagonals from one vertex	Sum of the angle measures
triangle	3	1	180
quadrilateral	4	2	2 × 180 = 360°
1. pentagon	▪ 5	▪ 3	▪
2. hexagon	▪ 6	▪ 4	▪
3. heptagon	7	▪ 5	▪
4. octagon	▪ 8	▪ 6	▪
5. nonagon	9	▪ 7	▪
6. decagon	▪ 10	▪ 8	▪

1. 3 × 180 = 540°
2. 4 × 180 = 720°
3. 5 × 180 = 900°
4. 6 × 180 = 1,080°
5. 7 × 180 = 1,260°
6. 8 × 180 = 1,440°

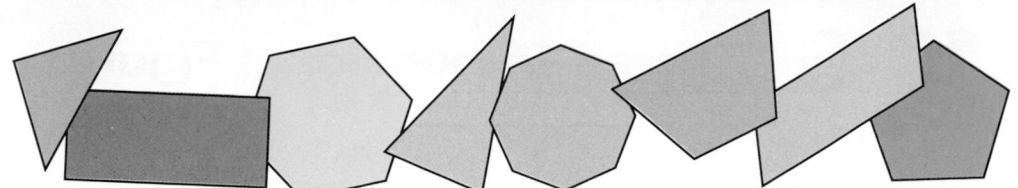

7. The perimeter of a regular octagon is 29.92 cm. What is the length of each side? **3.74 cm**

**EXTEND/
SOLVE PROBLEMS**

8. What is the measure of each angle of a regular hexagon? **60°**

Trace each polygon. Then draw all its diagonals.

9.

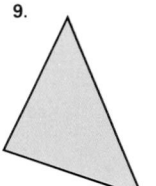

no diagonals

10.

11.

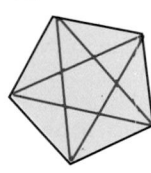

12.

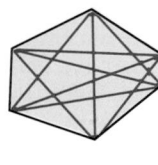

13. Copy and complete using your results from Exercises 9–12.

number of sides	3	4	5	6	7	8	9
number of diagonals	▪0	▪2	▪5	▪9	14	20	27

14. Refer to the table you completed in Exercise 13. What is the pattern that shows the relationship between the number of sides of a figure and the number of diagonals? Use the pattern to predict the number of diagonals of a decagon. **See Additional Answers.**

Determine whether each statement is *true* or *false*.

15. Some quadrilaterals are trapezoids. **true**

16. A rhombus is an equilateral parallelogram. **true**

17. Not every square is a parallelogram. **false**

18. Every parallelogram is also a rectangle. **false**

19. Is it possible for a hexagon to have six sides of equal length, but not to be a regular hexagon? Explain.

20. Is it possible for a hexagon to have six angles of equal measure, but not to be a regular hexagon?

COMPUTER

Can you complete and use the chart to compute the number of diagonals in a 10-sided or 11-sided polygon quicker than the computer? Have a friend use this program while you challenge the computer!

```
10 D = 0
20 INPUT "HOW MANY
   SIDES? ";N: PRINT
30 FOR X = 0 TO N − 1
40 D = D + X − 1
50 NEXT X
60 PRINT "THERE ARE ";D;"
   DIAGONALS."
```

**THINK CRITICALLY/
SOLVE PROBLEMS**

19. Yes. Consider this figure.

20. Yes. Consider this figure.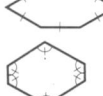

---

2. What is the measure of each angle in regular polygon *ABCDE*? **108°**

3. Find the unknown angle measure in the polygon below. **140°**

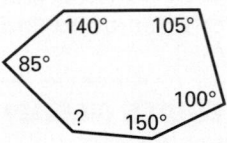

Complete by writing *sometimes, always,* or *never.*

4. A rhombus is ? a square. **sometimes**

5. A square is ? a rhombus. **always**

6. The bases of a trapezoid are ? parallel. **always**

7. A trapezoid is ? a parallelogram. **never**

Extension Have students examine newspapers, magazines, mail-order catalogs, and college catalogs for examples of logos based on geometric shapes. Encourage students to also look for logos in stores and on labels when they are shopping. For example, they can find logos at gasoline stations, supermarkets, and other chain stores. Have students copy or cut out logos and display them on a bulletin board.

Section Quiz
Identify each quadrilateral.

1. It is a rectangle with all sides of equal length. **square**

2. It has just one pair of parallel sides. **trapezoid**

3. It has four right angles. **rectangle**

4. It is a parallelogram with four right angles. **rectangle**

Get Ready graph paper

Additional Answers
See page 581.

CHALLENGE

Direct students to Exercises 13 and 14 in this section. Challenge them to write a formula that expresses the relationship between the number of sides of any polygon with *n* sides and the number of all its diagonals.

$$\frac{1}{2} n(n-3) \text{ or } \frac{n(n-3)}{2}$$

10-6 Three-Dimensional Figures

EXPLORE

Which of these patterns could be folded along the dotted lines to form a cube?

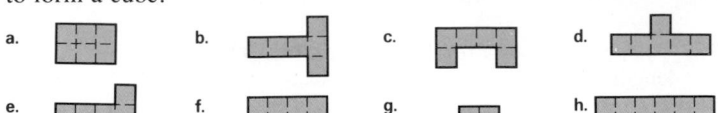

b and e

There are eleven different patterns of six squares that can be folded to form a cube. Sketch the eleven patterns on a piece of graph paper.
See Additional Answers.

SKILLS DEVELOPMENT

A **polyhedron** (plural: *polyhedra*) is a three-dimensional figure in which each surface is a polygon. The surfaces of a polyhedron are called its **faces.** Two faces meet, or intersect, at an **edge.** Three or more edges intersect at a **vertex** (plural: *vertices*).

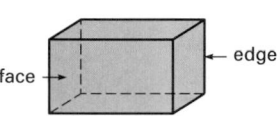

Each of these polyhedra is a prism. A **prism** has two identical, parallel faces, which are called **bases.** The other faces are parallelograms. A prism is named by the shape of its base.

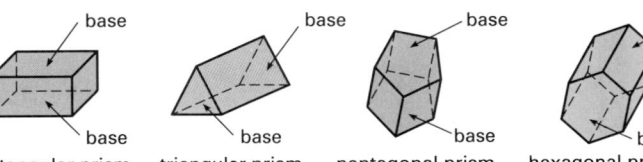

rectangular prism   triangular prism   pentagonal prism   hexagonal prism

A rectangular prism with all edges of the same length is called a **cube.**

Example 1

Identify the polyhedron.

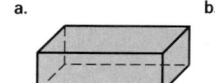 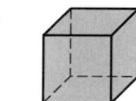 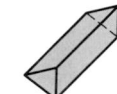

Solution
a. The bases are identical rectangles. It is a rectangular prism.
b. The bases are identical hexagons. It is a hexagonal prism.
c. It is a rectangular prism with all edges of the same length. It is a cube.
d. The bases are identical triangles. It is a triangular prism. ◀

A **pyramid** is a polyhedron with only one base. The other faces are triangles. A pyramid is named by the shape of its base.

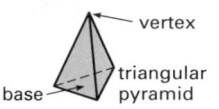
vertex

base triangular pyramid

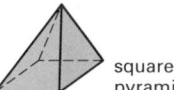
square pyramid

hexagonal pyramid

Example 2

Identify the pyramid.

a.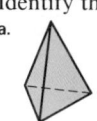
b.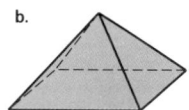
c.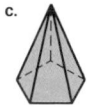

a. The base is a triangle. It is a triangular pyramid.
b. The base is a rectangle. It is a rectangular pyramid.
c. The base is a pentagon. It is a pentagonal pyramid. ◄

Some three-dimensional figures have a curved surface.

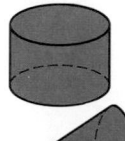 A **cylinder** has two identical, parallel, circular bases.

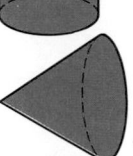 A **cone** has one circular base and one vertex.

 A **sphere** is the set of all points in space that are the same distance from a given point, called the center of the sphere.

Example 3

Identify the figure.

a.
b.
c.

a. It has a curved surface and one circular base. It is a cone.
b. It has a curved surface with no bases, and all points are the same distance from the center. It is a sphere.
c. It has a curved surface with two circular bases. It is a cylinder. ◄

Example 4

Identify the three-dimensional figure that would be formed if the pattern at the right were folded along the dotted lines.

Solution
The figure would have two triangular bases and three other faces that are rectangles. It would be a triangular prism. ◄

CHECK UNDERSTANDING

Why are cylinders, cones, and spheres *not* called polyhedra?

These figures are not made up of polygons.

AT-RISK STUDENTS

Provide students with posterboard, scissors, and tape. Have them work in small groups to create patterns that can be folded to make various prisms and pyramids. Then have them construct the solids and identify each.

ASSIGNMENTS

BASIC
1–10, 11–13, 21, 24

AVERAGE
1–10, 14–21, 22, 24

ENRICHED
7–10, 14–20, 22–25

ADDITIONAL RESOURCES
Reteaching 10–6
Enrichment 10–6
Transparency Masters 46–53

Example 3: Point out that three-dimensional figures with curved surfaces can also be identified by their number of bases: sphere—0; cone—1; cylinder—2.

Example 4: You may wish to have students trace the pattern and construct the prism to prove the solution is correct.

Additional Questions/Examples
1. What plane figure would you see if you looked down at a triangular prism from above? What would you see if you looked down at a cyclinder from above?   **triangle, circle**
2. What are the three dimensions of a three-dimensional figure?   **length, width, and height**

Guided Practice/Try These
Exercises 1–5 orally. Have volunteers explain which properties they used to identify each figure.

3 SUMMARIZE

Write About Math Have students list in their journals the properties that identify each of the three-dimensional figures discussed in this section.

Practice/Solve Problems For Exercises 1–4 point out that students must show bases in their drawings. For Exercises 7–10 suggest that students identify the bases to help them name the figures.

Extend/Solve Problems For Exercise 20 you may wish to have students write the formula to express the generalization they explained. $F + V - E = 2$

Think Critically/Solve Problems Students must use spatial visualization skills to solve Exercises 24 and 25.

Extra Practice Complete this table.

Figure	Shape and No. of Bases	Name of Figure
1.	rectangle 2	rectangular prism
2.	triangle 2	triangular prism
3.	circle 2	cylinder
4.	circle 1	cone
5.	none	sphere

TRY THESE

Identify each figure.

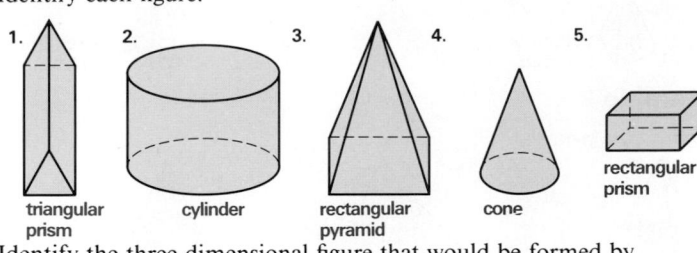

1. triangular prism 2. cylinder 3. rectangular pyramid 4. cone 5. rectangular prism

Identify the three-dimensional figure that would be formed by folding the pattern.

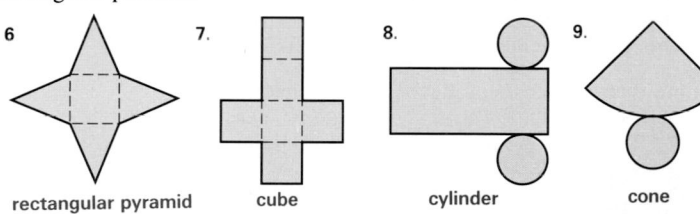

6. rectangular pyramid 7. cube 8. cylinder 9. cone

EXERCISES

PRACTICE/ SOLVE PROBLEMS

Make a drawing of each three-dimensional figure.
Check students' drawings.

1. triangular prism
2. rectangular pyramid
3. cube
4. cone

Identify each three-dimensional figure.

5. It has one base that is an octagon. The other faces are triangles. octagonal pyramid
6. Its two bases are identical, parallel pentagons. The other faces are parallelograms. pentagonal prism

Identify the three-dimensional figure that would be formed by folding each pattern along the dotted lines.

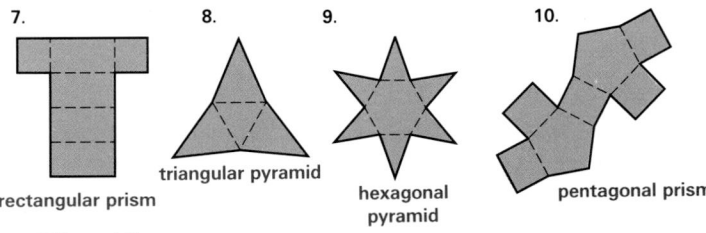

7. rectangular prism 8. triangular pyramid 9. hexagonal pyramid 10. pentagonal prism

Name three everyday objects that have the given shape. **Answers will vary.**

11. rectangular prism   12. cylinder   13. sphere

Copy and complete this table.

Polyhedron	Number of Faces (F)	Number of Vertices (V)	Number of Edges (E)	F + V – E
14. triangular prism	■ 5	■ 6	■ 9	■ 2
15. rectangular prism	■ 6	■ 8	■ 12	■ 2
16. pentagonal prism	■ 7	■ 10	■ 15	■ 2
17. triangular pyramid	■ 4	■ 4	■ 6	■ 2
18. rectangular pyramid	■ 5	■ 5	■ 8	■ 2
19. pentagonal pyramid	■ 6	■ 6	■ 10	■ 2

20. Use your results from Exercises 14–19. Make a generalization about the relationship between the faces, edges, and vertices of a polyhedron. Write a brief paragraph to explain your generalization. **Answers will vary but should include that, for any polyhedron, the sum of the faces and vertices minus the edges equals 2.**

21. The offices of the U.S. Department of Defense are located in Arlington, Virginia, in a building called the Pentagon. Why do you think the building was given this name?
It is built in the shape of a pentagonal prism.

Name the solid that would be formed form each pattern.

6.

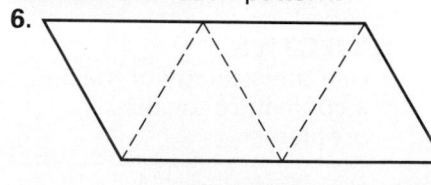

triangular pyramid

7.

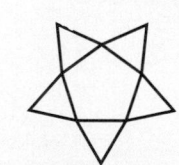

pentagonal pyramid

8.

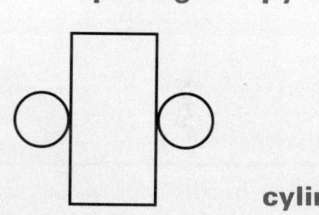

cylinder

Extension Have students work in small groups to discuss and then describe the paths made if they rolled a cylinder, a cone, or a sphere along a flat, plane surface. Have them compare the paths and directions of each. **A cylinder rolls in a straight path in one direction either forward or backward. A sphere also rolls in a straight path but in many directions. A cone rolls in a circular path.**

Section Quiz
Complete the following table of prisms and pyramids.

THINK CRITICALLY/ SOLVE PROBLEMS

22. Each base of a prism is a polygon with *n* sides. Write a variable expression that represents the total number of faces of the prism. **$n + 2$**

23. The base of a pyramid is a polygon with *t* sides. Write a variable expression that represents the total number of edges of the pyramid. **$2t$**

24. Suppose that a cylinder is cut in half along a plane that is perpendicular to the bases. What is the shape of the flat face of each half of the cylinder? **rectangle**

25. Suppose that a cube is cut in half along a plane that passes through the diagonal of one of its faces. What is the shape of each half of the cube? **triangular prism**

Poly-hedron	No. of Faces	No. of Vertices	No. of Edges
hexagonal prism	8	12	18
hexagonal pyramid	7	7	12
pentagonal prism	7	10	15
pentagonal pyramid	6	6	10

Additional Answers
See page 581.

CHALLENGE

Have students work in groups to draw the top, bottom, and all the side views of the figure shown after it has been rolled to the right one time.

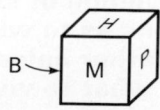

Top: ⊡
Bottom: ⊡

10-7 Problem Solving Skills:

MAKE A CONJECTURE

► READ
► PLAN
► SOLVE
► ANSWER
► CHECK

Sometimes you can solve a problem by making a conjecture. A **conjecture** is a guess that seems reasonable based on knowledge that you have. After making a conjecture, test it to find out whether it is the correct solution to the problem.

PROBLEM

In triangle *ABC*, point *D* is the midpoint of $\overline{AC}$ and point *E* is the midpoint of $\overline{BC}$. How would the line segment joining points *D* and *E* relate to side *AB* of the triangle?

SOLUTION

Make a conjecture. From the figure, it looks as though $\overline{DE}$ would be parallel to $\overline{AB}$. Also, it looks as though the length of $\overline{DE}$ would be about two thirds the length of $\overline{AB}$.

Test both parts of your conjecture. Trace the triangle onto a sheet of paper and draw line segment *DE*. Then use a protractor to measure $\angle CDE$ and $\angle BAD$, which are corresponding angles. If $\overline{DE}$ is parallel to $\overline{AB}$, $\angle CDE$ and $\angle BAD$ should be equal in measure.

Use a ruler to measure the length of $\overline{DE}$ and the length of $\overline{AB}$. Compare the measurements.

When the conjecture is tested, you find that $\overline{DE} \parallel \overline{AB}$. The length of $\overline{DE}$ is one half the length of $\overline{AB}$.

PROBLEMS

For each problem, use the question in part **a** to make a conjecture. Then use a ruler or protractor to answer the question in part **b**.

1. For parallelogram *ABCD*, find $m\angle C$.
   a. Conjecture: What is the relationship between the *opposite angles* of a parallelogram?
   b. What is $m\angle C$? **78°**

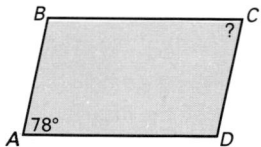

366 CHAPTER 10 Geometry of Size and Shape

2. For parallelogram *PQRS*, find the length of $\overline{RS}$.
   a. Conjecture: What is the relationship between the *opposite sides* of a parallelogram?
   b. What is the length of $\overline{RS}$? **2 cm**

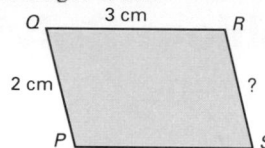

3. For rectangle *WXYZ*, where $\overline{WY}$ measures $1\frac{3}{4}$ in., find the length of $\overline{XZ}$.
   a. Conjecture: What is the relationship between the diagonals of a rectangle?
   b. What is the length of $\overline{XZ}$? **$1\frac{3}{4}$ in.**

4. For rhombus *JKLM*, find $m\angle 1$.
   a. Conjecture: What is the relationship between the diagonals of a rhombus?
   b. What is $m\angle 1$? **90°**

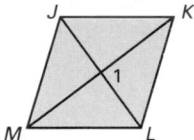

5. For isosceles trapezoid *EFGH*, find $m\angle H$. (In an isosceles trapezoid the nonparallel sides are equal in length.)
   a. Conjecture: What is the relationship between the base angles of an isosceles trapezoid?
   b. What is $m\angle H$? **59°**

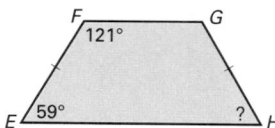

6. For isosceles trapezoid *MNOP*, where $\overline{MO}$ measures 3.5 cm, find the length of $\overline{NP}$.
   a. Conjecture: What is the relationship between the diagonals of an isosceles trapezoid?
   b. What is the length of $\overline{NP}$? **3.5 cm**

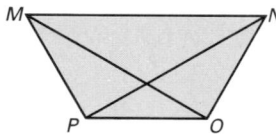

Test each of these conjectures to determine whether it is *true* or *false*.

7. The diagonals of a parallelogram are equal in length. **false**

8. Opposite angles of a trapezoid are equal in measure. **false**

9. Each diagonal of a rhombus separates the other diagonal into two segments of equal length. **true**

10. The diagonals of a parallelogram are perpendicular to each other. **false**

COMPUTER TIP

Geometric drawing software can help you to draw geometric figures quickly and easily. With such software a simple command will usually give you a quick reading of the length of a line segment or the measure of an angle. If you have this kind of software available, you might find it helpful in testing conjectures about geometric figures.

3 SUMMARIZE

Talk It Over Have students explain what a conjecture is and how it can be used to solve problems.

4 PRACTICE

For Problems 7–10 suggest that students draw diagrams to test each conjecture.

5 FOLLOW-UP

Extra Practice For isosceles triangle *ABC*, find $m\angle B$.

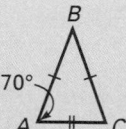

a. What is the relationship among the angles of an isosceles triangle? **Two angles are equal. Their sum, added to the measure of the third angle, equals 180°.**
b. What is $m\angle B$? **40°**

Get Ready two pennies

 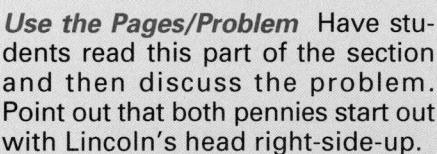

10-8 Problem Solving Strategies:
ACT IT OUT

► READ
► PLAN
► SOLVE
► ANSWER
► CHECK

The most efficient way to solve some problems is by acting out the situation described in the problem. This strategy can be especially helpful in solving problems that involve geometric shapes.

PROBLEM

Eileen wondered whether, if one penny was placed next to another and then was rolled halfway around the other, Lincoln's head would be right-side-up or upside down.

SOLUTION

Act it out. Take two pennies, place them as shown, and then roll the right penny halfway around the left penny. Lincoln's head is right-side-up.

Starting Position

Ending Position

PROBLEMS

Solve.

1.

1. Susan and Kathy were making geometric shapes with toothpicks. Susan challenged Kathy to make three squares by moving exactly three toothpicks in the pattern shown here. Act it out. Draw a picture of the three squares.

2. Starting with the original pattern in Exercise 1, Kathy challenged Susan to make three squares by moving exactly four toothpicks. Act it out. Draw a picture of the three squares.

Trace and cut out square *ABCD*. Label the corners as shown. What figure is formed as each of the following moves is made? Check by acting it out.

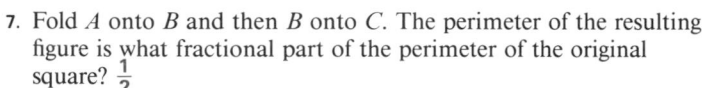

3. *A* folded onto *B*
   rectangle
4. *A* folded onto *D*
   isosceles right triangle
5. *A* folded onto *B*, then *B* folded onto *C*
   square
6. *A* folded onto *D*, then *B* folded onto *C*
   isosceles right triangle

Use your cutout of square *ABCD* to solve.

7. Fold *A* onto *B* and then *B* onto *C*. The perimeter of the resulting figure is what fractional part of the perimeter of the original square? $\frac{1}{2}$

8. Fold *A* onto the midpoint of side *AB*. If the area of the resulting figure is 12 cm², what was the area of the original square? 16 cm²

Use coins or cardboard cutouts of coins to act out each problem. In how many ways can you make change for each amount of money?

9. 25¢
   12 ways

10. 28¢
    12 ways

11. $1.00, using 7 coins

12. 65¢, using 8 coins only 1 way: 5 dimes 3 nickels

11. a. 2 quarters + 5 dimes
    b. 3 quarters + 1 dime + 3 nickels

13. A heavy log is being moved by rolling it on cylinders. If the circumference of each of the cylinders is 6 ft, how far will the log move for each revolution of the cylinders? Use cardboard or paper to make a model. 6 ft

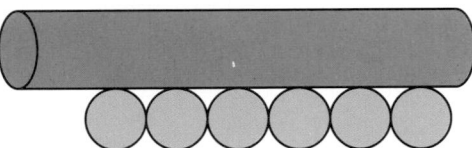

ASSIGNMENTS

BASIC
1–10

AVERAGE
1–13

ENRICHED
1–13

ADDITIONAL RESOURCES
Reteaching 10–8
Enrichment 10–8

3 SUMMARIZE

Talk It Over Have students discuss types of problems for which acting it out is a sensible, and perhaps necessary, strategy to use.

4 PRACTICE

For Exercise 13 students can use two cylinders, each with a 6-in. circumference, to model the problem.

5 FOLLOW-UP

Extra Practice
Nancy is taller than John, but not as tall as Joe. John is shorter than Bill, who is not as tall as Nancy. Arrange the four people in order of size from shortest to tallest.
John, Bill, Nancy, Joe

Get Ready protractor, compass, straightedge

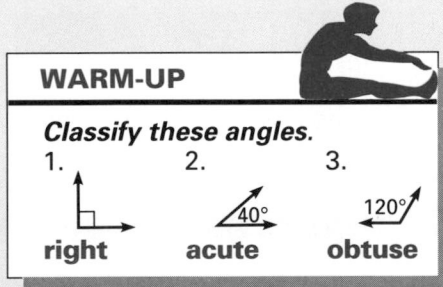
10-9 Constructing and Bisecting Angles

EXPLORE

a. Trace ∠ABC onto a piece of paper.

b. Fold the paper so that $\overrightarrow{BA}$ lies directly on top of $\overrightarrow{BC}$.

c. Unfold the paper. Along the fold line, draw $\overrightarrow{BD}$.

d. If $m\angle ABC = 120°$, what is $m\angle ABD$? Explain.

e. Devise a method for locating $\overrightarrow{BE}$ so that $m\angle ABE = 30°$.

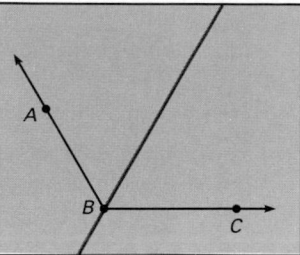

SKILLS DEVELOPMENT

A **construction** is a drawing of a geometric figure that is made using only two tools: a **compass** and an unmarked **straightedge**. (You can use the edge of a ruler as a straightedge, but you must ignore the markings on the ruler.)

Two of the most fundamental constructions are copying an angle and bisecting an angle.

Example 1

Copy ∠ABC using a compass and straightedge.

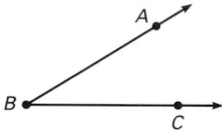

Solution

Step 1
Place the point of the compass at B and draw an arc that intersects both $\overrightarrow{BA}$ and $\overrightarrow{BC}$. Label the intersection points D and E.

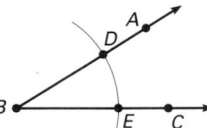

Step 2
Use a straightedge to draw any ray $\overrightarrow{FG}$. Place the point of the compass at F. With the same setting used in Step 1, draw an arc that intersects $\overrightarrow{FG}$. Label the intersection point H.

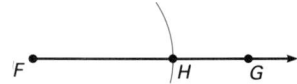

Step 3
Place the point of the compass at *D* and the point of the pencil at *E*. Keep that setting and place the point of the compass at *H*. Draw an arc that crosses the first arc. Label the intersection point *I*.

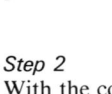

Step 4
Use a straightedge to draw $\overrightarrow{FI}$.

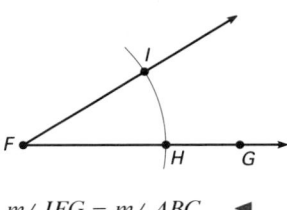

$m\angle IFG = m\angle ABC$ ◄

To *bisect* means "to divide into two equal parts." The **bisector of an angle** is the ray that separates the angle into two adjacent angles of equal measure.

Example 2

Bisect $\angle ABC$ using a compass and a straightedge.

Solution

Step 1
With the compass point at *B*, draw an arc that intersects $\overrightarrow{BA}$ and $\overrightarrow{BC}$. Label the intersection points *P* and *Q*.

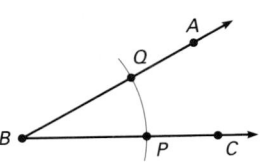

Step 2
With the compass point at *P*, adjust the compass so that the opening is a little greater than half of arc *PQ*. Draw an arc inside the angle, as shown.

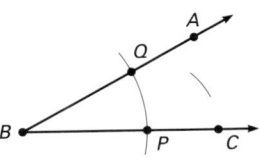

Step 3
Place the point of the compass at *Q*. With the same compass opening you used in Step 2, draw an arc that intersects the first arc at *T*.

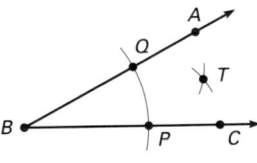

Step 4
Draw $\overrightarrow{BT}$. Ray *BT* is the bisector of $\angle ABC$.

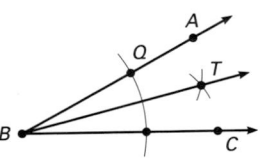

$m\angle ABT = m\angle TBC$ ◄

 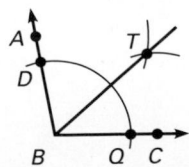

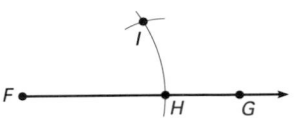 

ASSIGNMENTS

BASIC
1–7

AVERAGE
4–7, 8, 10

ENRICHED
5–7, 8–9, 10

ADDITIONAL RESOURCES
Reteaching 10–9
Enrichment 10–9
Transparency Masters 54, 55

Additional Questions/Examples
1. If you draw the bisector of $\angle BAC$ in the triangle shown, what will the measure of $\angle DAC$ be? Explain.

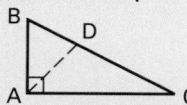

   45°, because bisecting means dividing an angle in half. Since $m \angle BAC = 90°$, $1/2 \times 90° = 45°$.
2. In these triangles, can the measure of $\angle 1$ be equal to the measure of $\angle 2$? Why or why not?

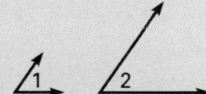

   Yes, the angles can be equal in measure because the length of the rays that make up the sides of an angle do not affect the measure of the angle.

Guided Practice/Try These Have students explain how they constructed each angle for Exercises 1–3. Have them explain their procedure for constructing bisectors in Exercises 4–6.

3 SUMMARIZE

Key Questions
1. What is the procedure for copying an angle?
2. What is the procedure for constructing the bisector of an angle?

4 PRACTICE

Practice/Solve Problems For Exercise 7 have students use their protractors first to draw the angle and then to check measures in the bisected angle.

Extend/Solve Problems Students can use their protractors to first measure all the angles they constructed and then to check that the angles were accurately bisected.

Think Critically/Solve Problems Students can work in pairs to discuss methods for completing Exercise 10.

Problem Solving Applications Point out that students must round each number of degrees to an approximate whole number of degrees, and that the total number of degrees should equal 360.

5 FOLLOW-UP

Extra Practice Trace each angle. Then copy it using a compass and straightedge.

1.
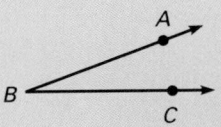

2.
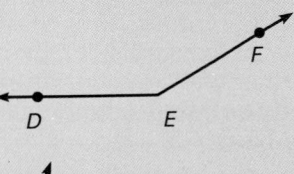

3.
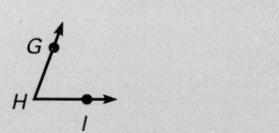

Trace each angle. Then copy it using a compass and straightedge.
Check students' drawings.

1. 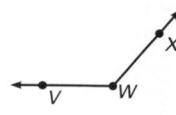   2. 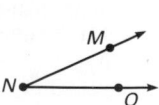   3.

Trace each angle. Then bisect it using a compass and straightedge.
Check students' drawings.

4. 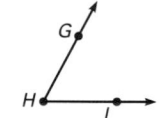   5.    6.

EXERCISES

PRACTICE/ SOLVE PROBLEMS

Trace each angle. Then copy it using a compass and straightedge.
Check students' drawings.

1. 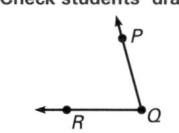   2. 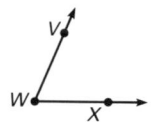   3.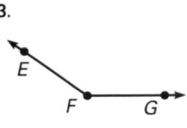

Trace each angle. Then bisect it using a compass and straightedge.
Check students' drawings.

4.   5. 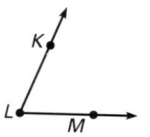   6.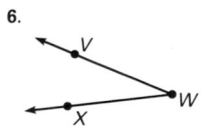

7. Use a protractor to draw an acute angle that measures 54°. Construct the bisector of this angle. **Check students' drawings.**

EXTEND/ SOLVE PROBLEMS

8. Draw any obtuse angle. Then use a compass and straightedge to divide the angle into four equal parts. **Check students' drawings.**

9. Bisect any of the angles that resulted from your construction in Exercise 8. **Check students' drawings.**

THINK CRITICALLY/ SOLVE PROBLEMS

10. Use a compass and protractor to construct a triangle with the same angle measures and with sides of the same length as the triangle at the right. **Check students' drawings.**

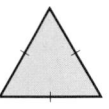

Problem Solving Applications:

CONSTRUCTING CIRCLE GRAPHS

When data are expressed as parts of a whole, you can display the data in a **circle graph.** The entire circle represents 100% of the data. The different parts are represented as **sectors** of the circle. Since there are 360° in a circle, the sum of the **central angles** formed by the sectors must be 360°.

Example

Draw a circle graph to represent the following data. In a survey, 350 people named their favorite movie: *Star Wars,* 119; *Indiana Jones and the Temple of Doom,* 98; *Gone With the Wind,* 77; *It's a Wonderful Life,* 56.

Solution

Step 1 Find the decimal for each part of the whole.

Star Wars:	*Gone With the Wind:*
119 ÷ 350 = 0.34	77 ÷ 350 = 0.22
Indiana Jones:	*Wonderful Life:*
98 ÷ 350 = 0.28	56 ÷ 350 = 0.16

Step 2 Star Wars: 0.34 × 360° = 122.4° ≈ 122°
Indiana Jones: 0.28 × 360° = 100.8° ≈ 101°
Step 3 Gone With the Wind: 0.22 × 360° = 79.2° ≈ 79°
It's a Wonderful Life: 0.16 × 360° = 57.6° ≈ 58°
Step 4 Draw a circle. Use a protractor to construct the central angle for each sector. Label each sector and give the graph a title. ◄

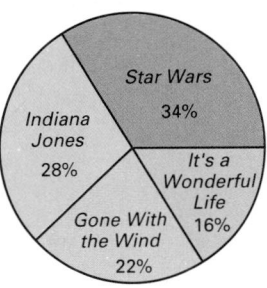

Construct a circle graph to represent each set of data. **Check students' constructions.**

1. The content of cheddar cheese is as follows: protein, 26%; water, 40%; fat, 34%. (*Hint:* Write each percent as a decimal.)

2. A concrete mix consists of 1.5 L of water, 3.8 L of sand, 1 L of cement, and 4.2 L of gravel.

3. A recipe for chili con carne contains the following ingredients. hamburger, 520 g; kidney beans, 280 g; tomato sauce, 150 g; tomato paste, 150 g; seasoning, 60 g; water 440 g.

4. During a sale, an appliance store sold 70 refrigerators, 35 stoves, 60 dishwashers, and 85 color TVs.

See Additional Answers.

---

4. Bisect each angle you constructed in Exercises 1–3.
   Check students' drawings.
Use a protractor to draw an angle of the given measure. Then construct the bisector of each angle.
5. 120°
6. 90°
7. 60°
Check students' drawings.

Extension Use a compass and a protractor to construct a line *EF* parallel to line *AB.*

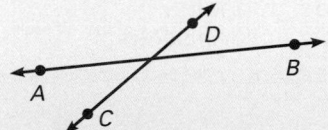

Check students' drawings.

Section Quiz Trace each angle. Then copy it using a compass and straightedge.

1.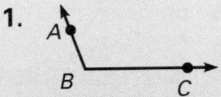

2.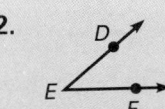

Check students' drawings.
Use a protractor to draw an angle of the given measure Then use a compass to construct the bisector of each angle.
4. 75°
5. 110°
6. 45°
Check students' drawings.

Get Ready compass, straightedge

Additional Answers
See page 581.

374

10-10 Constructing Line Segments and Perpendiculars

EXPLORE

Trace $\overline{AB}$ on a sheet of paper. Fold the paper so that point A of the line segment meets point B. Open the folded paper. Label the crease $\overleftrightarrow{CD}$. Label the point where $\overleftrightarrow{CD}$ intersects $\overline{AB}$ point M. What can you say about $\overline{AM}$ and $\overline{MB}$?

$\overline{AM}$ and $\overline{MB}$ are equal in length.

SKILLS DEVELOPMENT

Copying a line segment and bisecting a line segment are two more important constructions.

Example 1

Construct a line segment that is equal in length to $\overline{AB}$. Use a compass and straightedge.

Solution
Step 1
Use a straightedge to draw $\overrightarrow{CD}$.

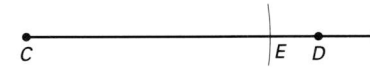

Step 2
Put the compass point at A and the point of the pencil at B.

Step 3
Place the point of the compass at C. With the same setting used in Step 2, draw an arc. Label the intersection point E.

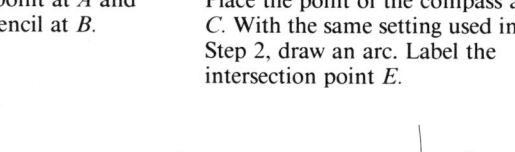

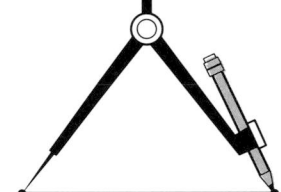

$CE = AB$ ◄

The **midpoint** of a line segment is the point that separates it into two line segments of equal length. The **perpendicular bisector** of a line segment is a line, ray, or line segment that is perpendicular to a line segment at its midpoint.

Example 2

Construct the perpendicular bisector of $\overline{JK}$.

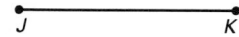

Solution

Step 1
Place the point of the compass at *J*. Open the compass a little more than half the length of $\overline{JK}$. Draw one arc above $\overline{JK}$ and another below $\overline{JK}$.

Step 2
Place the point of the compass at *K*. With the same setting used in Step 1, draw arcs above and below $\overline{JK}$. Label the points of intersection *L* and *N*. Use a straightedge to draw $\overleftrightarrow{LN}$. Label the midpoint of $\overline{JK}$ point *M*.

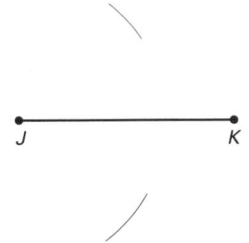

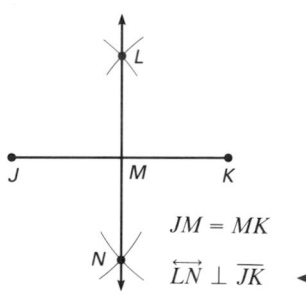

$JM = MK$

$\overleftrightarrow{LN} \perp \overline{JK}$ ◀

Example 3

Construct a line perpendicular to $\overleftrightarrow{RS}$ from point *A*.

Solution

Step 1
With the point of the compass at *A*, draw an arc that intersects $\overleftrightarrow{RS}$ at *B* and *C*.

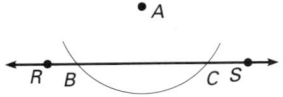

Step 2
With the point of the compass at *B*, open the compass a little more than half the length of $\overline{BC}$. Draw an arc below $\overleftrightarrow{RS}$.

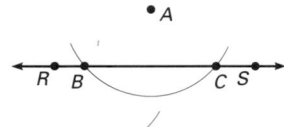

Step 3
Place the point of the compass at *C*. With the same setting used in Step 2, draw an arc below $\overleftrightarrow{RS}$ that intersects the arc drawn in Step 2. Label the point of intersection *D*.

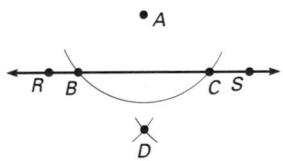

Step 4
Draw $\overleftrightarrow{AD}$.

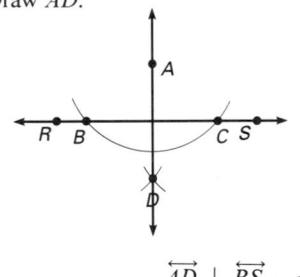

$\overleftrightarrow{AD} \perp \overleftrightarrow{RS}$ ◀

BASIC
1–10

AVERAGE
4–9, 10–12, 13–14

ENRICHED
10–16

ADDITIONAL RESOURCES
Reteaching 10–10
Enrichment 10–10

results: finding the midpoint of a line segment and drawing a line perpendicular to the line segment.

Example 3: Have students try an alternate form of this construction with Point *B* to the left of Point *R* and Point *C* to the right of Point *S*. Discuss whether the resulting construction is the same. **It is.**

Additional Questions/Examples
1. How can you tell if two lines are perpendicular? **They form right angles.**
2. What is the difference between *drawing* a line segment equal in length to another line segment and *constructing* a line segment equal in length to another line segment? **You can draw line segments using a ruler. To *construct* a line segment you may use only a compass and a straightedge.**

Guided Practice/Try These After students complete Exercises 1–3 allow them to use rulers to measure each constructed line segment to check that it is equal in length to the corresponding traced line segment.

5-MINUTE CLINIC

Exercise	**Student's Error**	**Error Diagnosis**
From point *P*, construct a line, *PQ*, perpendicular to line segment *AB*.	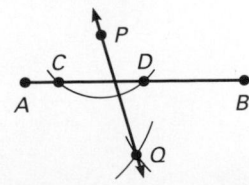	• Student placed compass on point *B*, rather than on point *D*, to form the second arc.

Trace each line segment. Then construct a line segment equal in length. **Check students' drawings.**

1.   2.   3.

Trace each line segment. Then construct its perpendicular bisector. **Check students' drawings.**

4.   5.   6.

Trace each figure. Then construct a line perpendicular to line *AB* from point *P*. **Check students' drawings.**

7.   8.   9.

EXERCISES

PRACTICE/ SOLVE PROBLEMS

Trace each line segment. Then construct a line segment equal in length. **Check students' drawings.**

1.   2.   3.

Trace each line segment. Then construct its perpendicular bisector. **Check students' drawings.**

4.   5.   6.

Trace each figure. Then construct a line perpendicular to line *CD* from point *K*. **Check students' drawings.**

7.   8.   9.

---

3 SUMMARIZE

Key Questions
1. What is the procedure for constructing a line segment equal in length to a given line segment?
2. What is the procedure for constructing the perpendicular bisector of a line segment? for constructing the bisector of an angle?
3. What is the procedure for constructing a line perpendicular to a given line from a point not on the line?

4 PRACTICE

Practice/Solve Problems For Exercises 1–3 have students measure their constructed segments with a ruler to check for accuracy. For Exercises 4–9 have students measure the angles formed by the perpendicular bisectors with a protractor.

Extend/Solve Problems For Exercises 10–12 emphasize that, although the altitude students construct is perpendicular to line *PS*, the altitude does not necessarily bisect line *PS*.

Think Critically/Solve Problems For Exercises 13–15 be sure students use rulers only to measure the indicated distances from the points to the vertices and not to perform any of the constructions.

5 FOLLOW-UP

Extra Practice Use a ruler to draw a line segment of the given length. Then use a compass to construct line segments equal in length to each one you drew.
1. 3 cm
2. 40 mm
Check students' drawings and constructions.

376

An **altitude of a triangle** is a line segment from a vertex of the triangle perpendicular to the opposite side or to a line containing that side. You can use the construction described in Example 3 to construct any altitude of a triangle.

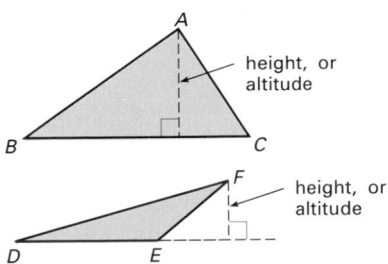

Trace each triangle. Construct an altitude of the triangle from point R to $\overline{PS}$.

10. 11. 12.

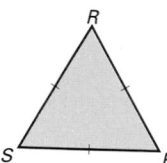

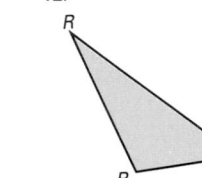

13. Trace △*ABC*. Bisect each side of △*ABC*. The bisectors will meet at a point. Label the point *P*. Measure the distance from point *P* to each vertex. What do you observe? **The distance from point *P* to each vertex is the same.**

14. Trace △*DEF*. Bisect each side of the triangle and label the point where the bisectors intersect point *Q*. Measure the distance from point *Q* to each vertex. What do you observe? **The distances are the same.**

15. Trace △*GHI*. Bisect each side of the triangle and label the point where the bisectors intersect point *R*. Measure the distance from point *R* to each vertex. What do you observe? **The distances are the same.**

16. Compare your answers to Exercises 13–15. How is the position of the intersection point of the perpendicular bisectors of the sides related to the type of triangle? **Acute triangle—inside the triangle; right triangle—at the midpoint of the side opposite the right angle; obtuse triangle—outside the triangle.**

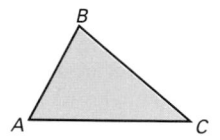

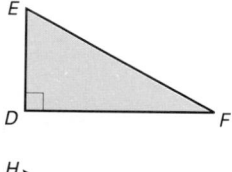

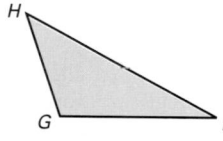

10-10 Constructing Line Segments and Perpendiculars **377**

CHALLENGE

Have partners work together to discover how to construct a line perpendicular to a given line segment through a point *on* the given line. Then challenge students to construct a rectangle with sides equal in length to segments *AB* and *XY*.

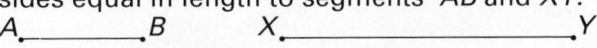

Draw a line segment of the given length with a ruler. Then construct lines of equal length and construct their perpendicular bisectors.

3. 2.5 cm

4. 45 mm

Check students' drawings and constructions.

5. Draw line *AB*. Then draw point *C* above line *AB*. Construct a line *CD* perpendicular to line *AB* from point *C*. **Check students' constructions.**

6. Use a ruler to draw three different triangles all named *ABC* with *AB* = 4 cm and *AC* = 5 cm. Construct any altitude for each triangle. (You may wish to suggest that students draw isosceles, acute, and obtuse triangles.) **Check students' constructions.**

Extension Have students work in pairs to discuss and construct the following:

Construct four squares within the given square so that the area of each is 1/4 the area of the large square.

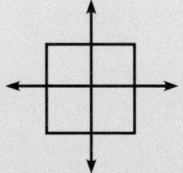

Students need to construct the perpendicular bisectors of each side.

Section Quiz Copy each line segment by using a ruler to measure. Then construct a line segment equal in length to each. Finally, construct the perpendicular bisector to each constructed line segment.

1. (Display a 3-in. segment, *AB*.)

2. (Display a 4-in. segment, *DE*.)

Check students' drawings and constructions.

3. Have students construct a line perpendicular to a given line segment, *GH*, from point *K*.

10 CHAPTER REVIEW

Introduction The Chapter Review emphasizes the major concepts, skills, and vocabulary presented in this chapter and can be used for diagnosing students' strengths and weaknesses. Page references direct students back to appropriate sections for additional review and reteaching.

Using Pages 378–379 For Exercises 1–7 you may also wish to have students draw the figure listed in the second column. For Exercises 9 and 10, students need a protractor. For Exercise 11 students need to know how angles 1 and 2 are related. For Exercise 17 suggest that students trace the octagon and draw all the diagonals from one vertex.

Informal Evaluation Make a frequency chart for all student errors in each section of the Chapter Review. Discuss which sections have the most errors and why the students think this is so. You may wish to review key concepts in these sections. Be sure students understand the procedures and the concepts of the chapter before they take the Chapter Test.

CHAPTER 10 ● REVIEW

Match the letters of the words at the right with the descriptions at the left.

1. has two identical, parallel bases that are triangles and three other faces that are parallelograms g
2. the sum of their measure is 180° b
3. has two sides of equal length e
4. flat surface that extends without end in all directions a
5. parallelogram with four sides of equal length f
6. the sum of their measures is 90° d
7. has one base and three other faces that are triangles c

a. plane
b. supplementary angles
c. triangular pyramid
d. complementary angles
e. isosceles triangle
f. rhombus
g. triangular prism

SECTION 10-1 POINTS, LINES, AND PLANES (pages 342–345)

- ► Basic geometric figures include the **point,** the **line,** the **line segment,** the **ray,** the **angle,** and the **plane.**
- ► A **polygon** is a closed plane figure formed by joining three or more line segments at their endpoints.

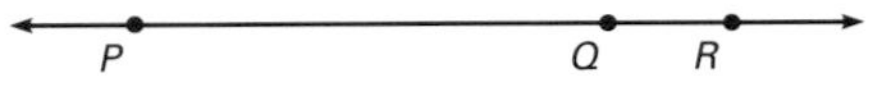

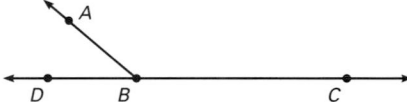

8. Name the line in four ways.
Answers will vary. Any four of these: $\overleftrightarrow{PQ}$, $\overleftrightarrow{PR}$, $\overleftrightarrow{RP}$, $\overleftrightarrow{RQ}$, $\overleftrightarrow{QP}$, $\overleftrightarrow{QR}$

SECTION 10-2 ANGLES AND ANGLE MEASURES (pages 346–349)

- ► Angles are classified by their measure as **acute, right, obtuse,** or **straight.**
- ► **Complementary angles** are two angles whose measures add to 90°.
- ► **Supplementary angles** are two angles whose measures add to 180°.

9. Measure and classify $\angle ABC$.
140°; obtuse
10. Find $m\angle DBA$.
40°

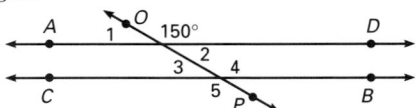

SECTION 10-3 PARALLEL AND PERPENDICULAR LINES (pages 350–353)

- ► Two lines in a plane either intersect at a single point or are parallel and have no points in common.
- ► When a transversal intersects two parallel lines, **vertical angles** along with several other pairs of angles are formed: **corresponding angles, alternate interior angles,** and **alternate exterior angles.**

11. Name a pair of alternate interior angles. $\angle 2$ and $\angle 3$
12. Determine $m\angle 2$ in the figure. 30°

SECTIONS 10–4 AND 10–5　TRIANGLES AND POLYGONS　　(pages 354–361)

▶ Triangles are classified by the measures of their angles and sides.
▶ Polygons are named by the number of their sides.

Classify each triangle as *acute, obtuse,* or *right.*

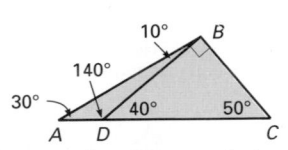

13. △*CBD*
right

14. △*CBA*
obtuse

15. △*ADB*
obtuse

Name each polygon and give
the sum of the angle measures.

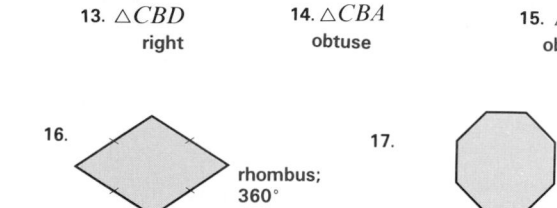

16.
rhombus;
360°

17.
octagon;
1,080°

SECTION 10–6　THREE-DIMENSIONAL FIGURES　　(pages 362–365)

▶ Some three-dimensional figures are **polyhedra** —figures in which each
surface is a polygon. Others have curved surfaces.

Draw the figure. **Check students' drawings.**

18. rectangular prism　　　**19.** rectangular pyramid　　　**20.** cylinder　　　**21.** cone

SECTIONS 10–7 AND 10–8　PROBLEM SOLVING　　(pages 366–369)

▶ Acting out a problem is a useful problem solving strategy.

22. In how many ways can you make change for 70¢ with 4 coins?
Only 1 way: 2 quarters and 2 dimes

SECTIONS 10–9 AND 10–10　CONSTRUCTIONS　　(pages 370–377)

▶ You can copy and bisect any angle using a compass and straightedge.
▶ With a straightedge and compass, you can construct a line segment
equal in length to a given line segment, a perpendicular bisector of a
given line segment, and a perpendicular to a line from a given point
not on a line.

23. Trace ∠*LMN*. Then bisect it.
Check students drawings.

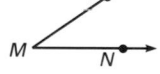

24. Construct a line segment congruent to $\overline{JK}$ and bisect it. 　J•————————•K
Check students' drawings.

USING DATA　　Refer to the chart of semaphore code signals on page 340.

25. For which letters are the flags held at an obtuse angle? C, E, K, M, Q, S, V, Y

Review　**379**

10 CHAPTER TEST

Introduction The Chapter Test uses a variety of questioning techniques to assess students' mastery of the major objectives of Chapter 10. If you prefer, you may use the Skills Preview (page 339) as an alternative form of the Chapter Test. The items on this test and the Skills Preview correspond in content and level of difficulty.

Alternative Assessment
Making Connections Work with a partner. Look in newspapers and magazines for a photograph of a building with a peaked roof seen straight on. Use a compass to bisect the angles formed where the sloped sides of the roof meet the building. Use a straightedge to draw the angle bisectors. Identify the point at which the angle bisectors cross to form the "new" peak of the roof. Make guesses about the height of the original roof. Then guess the distance by which you "lowered" the roof with your constructions.

Give all the names that apply to each polygon or polyhedron.

1.

pentagon

2.
trapezoid, quadrilateral

3.
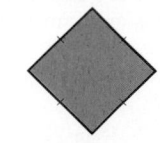
quadrilateral, parallelogram, rhombus

4.
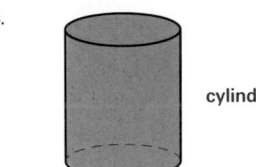
cylinder

5.

pentagonal prism

6.
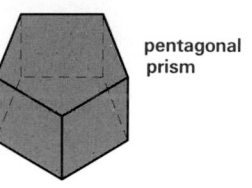
rectangular pyramid

Classify each angle as *acute, right,* or *obtuse.*

7. 150° obtuse   8. 95° obtuse   9. 32° acute   10. 90° right

Use the figure at the right for Exercises 11 and 12.

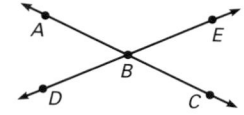

11. Name a pair of vertical angles.
   ∠ABE and ∠DBC or ∠ABD and ∠CBE
12. Name a pair of adjacent angles.
   ∠ABD and ∠ABE or ∠ABE and ∠CBE

Classify each triangle as *isosceles, equilateral,* or *scalene.*

13.
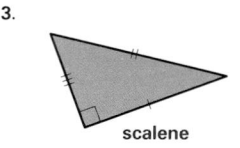
scalene

14.
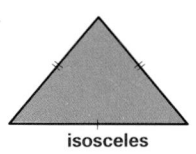
isosceles

15
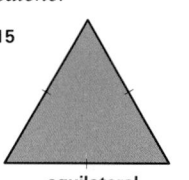
equilateral

16.
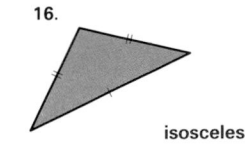
isosceles

17. Trace $\overleftrightarrow{JK}$ with point *L.* Construct a line perpendicular to $\overleftrightarrow{JK}$ from point *L.*
   Check student's work.
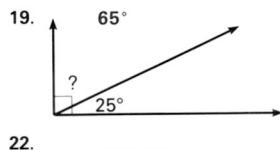

Find the unknown angle measure.

18.
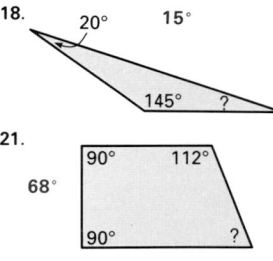
20° 15°
145° ?

19.
65°
?
25°

20.
75°
? 105°

21.
90° 112°
68°
90° ?

22.
124° ? 85°
90°
143° 168° 110°

23.
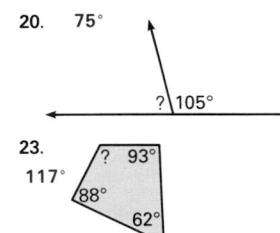
? 93°
117°
88°
62°

1. A quality control manager inspects every seventh knapsack sewn. Which term describes this sample? **systematic**

Write in scientific notation.

2. 8,794   **8.794 × 10³**   3. 15,096   **1.5096 × 10⁴**

4. Write a statement about the width and the height of the figure below. Then check to see if your statement is true or false. **The height and width are the same.** false.

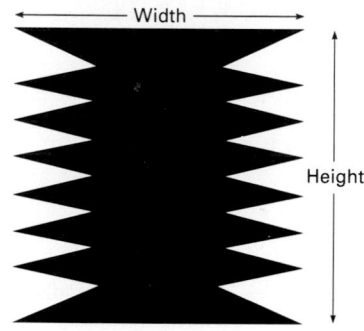
← Width →

Height

5. Write two conditional statements using the following two sentences:

   Rhea helped us study.
   We passed the test. **See Additional Answers.**

Write each fraction as a decimal.

6. $\frac{7}{8}$ **0.875**   7. $\frac{7}{9}$ **0.7̄**   8. $\frac{5}{16}$ **0.3125**

9. Find the perimeter of a rectangular figure whose length is 10.5 cm and whose width is 40.2 cm. **101.4 cm**

Replace ● with <, >, or =.

10. $5 - 16$ ● $10 - (-1)$ **<**

11. $-4 + (16)$ ● $8 + 4$ **=**

12. $13 - (-5)$ ● $45 - 64$ **<**

Solve each equation.

13. $2\frac{1}{2}x = 5$ **x = 2**

14. $r - 18 = 23$ **r = 41**

15. $150 = 10x + 30$ **x = 12**

16. $\frac{b}{7} = -15$ **b = -105**

Solve each proportion.

17. $\frac{4}{5} = \frac{12}{x}$ **x = 15**

18. $8 : x = 6 : 15$ **x = 20**

19. Write $\frac{16}{20}$ and $\frac{45}{60}$ as percents. Which percent is greater? **80%; 75%; 80%**

20. Write a percent equivalent to $\frac{12}{8}$.   **150%**

21. What percent of 50 is 35?   **70%**

22. A jacket that regularly sells for $80 is on sale for $60. What is the discount rate? **25%**

23. Jeff receives a 5% commission on sales of sports equipment. He sold $2,500 worth of equipment last week. What was the amount of his commission? **$125**

Classify each triangle as *acute*, *obtuse*, or *right*.

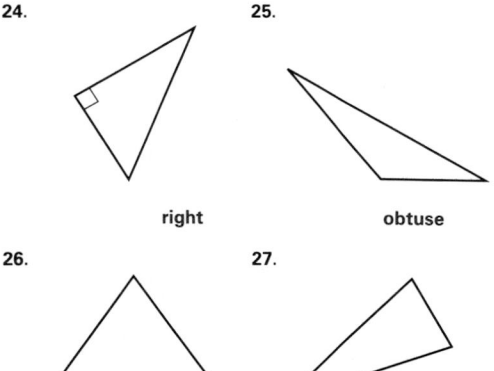

24. **right**

25. **obtuse**

26. **acute**

27. **acute**

Introduction The purpose of this Cumulative Review is to maintain previously taught skills and concepts and to apply them to the material presented in this chapter. At least one major objective of each chapter is included in the review.

Item Analysis The table below correlates the Cumulative Review items with the chapter and section that is being reviewed.

Section	Items
1–1	1
2–6	2–3
3–3	5
3–6	4
4–3	6–8
5–6	9
6–2	11
6–3	12
6–8	10
7–2	14
7–4	15
7–6	13, 16
8–4	17, 18
9–1	19, 20, 22
9–3	21
9–7	23
10–4	24–27

Additional Answers
See page 582.

10 CUMULATIVE TEST

Introduction The Cumulative Test uses a standardized-test format of multiple-choice questions to assess retention of previously learned concepts and test-taking skills. Test results may be used to diagnose students' strengths and weaknesses.

Item Analysis The table below correlates the Cumulative Test items with the chapter and sections that are being tested.

Section	Items
1–9	1
2–5	2
3–3	3
4–3	4
5–1	6
5–2	5
6–3	7
7–7	8
8–3	9
8–6	10
9–2	11
9–5	12
9–7	14
9–9	13
10–2	16
10–9	15

1. If you had a total of 288 test points and an average of 72, how many tests did you take?
   A. 7   B. 8   C. 4   D. 5

2. Divide. 1.836 ÷ 1.2
   A. 153   B. 15.3
   C. 1.53   D. 0.153

3. Which is a counterexample for the following statement?

   If a number is a multiple of 4, then it is a multiple of 8.
   A. 8   B. 16   C. 64   D. 20

4. Which is the decimal for $\frac{5}{6}$?
   A. $0.8\overline{3}$ B. 1.2
   C. 0.8   D. $0.\overline{83}$

5. Complete. 1.73 kg = ■ g
   A. 173 g   B. 17.3 g
   C. 1,730 g   D. 17,300 g

6. Complete. 2 lb 8 oz × 4
   A. 8 lb 10 oz   B. 10 lb 8 oz
   C. 10 lb   D. 9 lb 4 oz

7. Subtract. 25 − (−15)
   A. −40   B. −10
   C. 10   D. 40

8. Stan biked 44 km in $3\frac{2}{3}$ h. What was his average rate of speed? Use the formula $d = rt$.
   A. 12 km/h   B. $161\frac{1}{3}$ km/h
   C. 15 km/h   D. $\frac{2}{3}$ km/h

9. Which of the following ratios is equivalent to 4:5?
   A. 8:10   B. 2:3
   C. 10:20   D. none of these

10. The scale of a drawing is 2 cm:5 m. What is the actual width of a room if the width in the scale drawing is 4 cm?
    A. 2 m   B. 1 cm
    C. 10 m   D. none of these

11. What is 30% of 650?
    A. 195   B. 19.5
    C. 620   D. 1,950

12. 80% of what number is 20?
    A. 60   B. 18
    C. 360   D. 25

13. What is closest to the amount of interest earned in one year on $4,500 if the rate of interest is 8.5% per year?
    A. $380   B. $34
    C. $340   D. $400

14. Risa receives a 4% commission on her furniture sales. Last week she sold $2,800 worth of furniture. What was the amount of her commission?
    A. $400   B. $2,912
    C. $112   D. none of these

15. $\overrightarrow{DB}$ bisects ∠ADC. ∠BDC = 25°. Find m∠ADC.
    A. 25°
    B. 50°
    C. 1.5°
    D. none of these

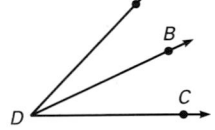

16. Classify ∠GHI.

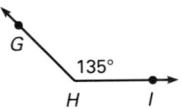

    A. right   B. straight
    C. acute   D. none of these

Name _____ Date _____

Points, Lines, and Planes

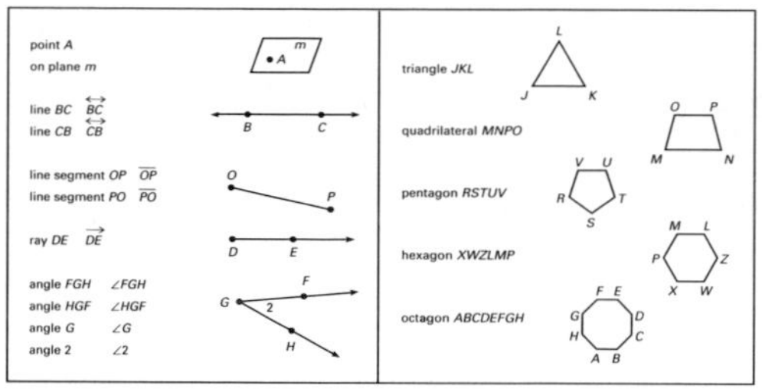

point A on plane m	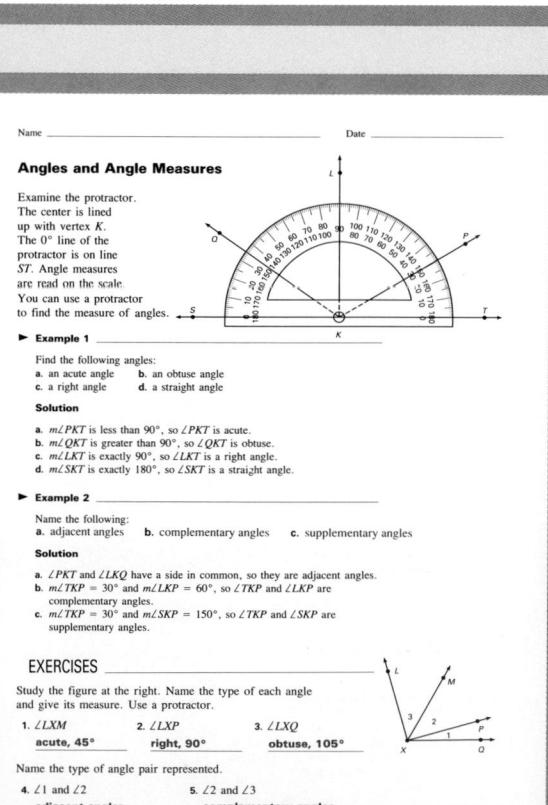	triangle JKL	
line BC $\overrightarrow{BC}$ line CB $\overrightarrow{CB}$		quadrilateral MNPO	
line segment OP $\overline{OP}$ line segment PO $\overline{PO}$		pentagon RSTUV	
ray DE $\overrightarrow{DE}$		hexagon XWZLMP	
angle FGH ∠FGH angle HGF ∠HGF angle G ∠G angle 2 ∠2		octagon ABCDEFGH	

► **Example 1**

Name the endpoints of the line segment in the chart.

Solution

O and P

► **Example 2**

Identify the polygon.

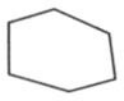

Solution

There are 6 sides. It is a hexagon.

EXERCISES

Write the symbol for each figure.

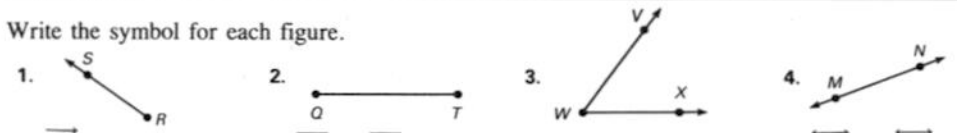

1. $\overrightarrow{RS}$
2. $\overline{QT}$ or $\overline{TQ}$
3. ∠VWX or ∠XWV or ∠W
4. $\overrightarrow{MN}$ or $\overrightarrow{NM}$

Draw a representation of each figure.

5. an octagon
6. a quadrilateral with no two sides equal in length
7. ∠CAT

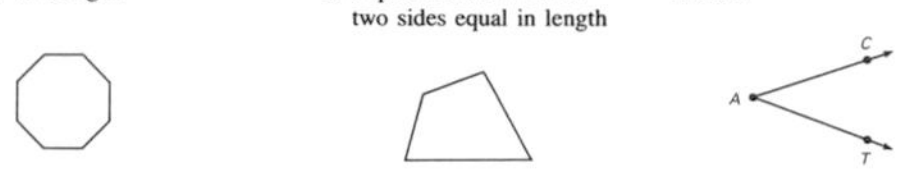

Reteaching • SECTION 10-1

179

Name _____ Date _____

Hidden Polygons

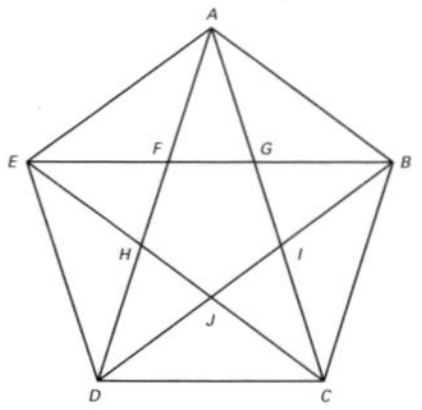

EXERCISES

Use the figure above to answer the questions.

1. What kind of polygon is ABCDE? **pentagon**

2. What polygon is formed by $\overline{GI}$, $\overline{IJ}$, $\overline{JH}$, $\overline{HF}$, and $\overline{FG}$?
   pentagon

3. Name four triangles that have B as one vertex.
   Answers will vary. Samples: ABG, ABI, BIC, BJC

4. Choose a point other than B. Name three different angles having that point as a vertex.
   Answers will vary.

5. Name three triangles that have DJ as one side.
   DJC, DJH, DJE

6. Name four quadrilaterals.
   Answers will vary. Samples: ABCD, EGID, FBJH, EBCD

7. Name one hexagon.
   Answers will vary. Samples: AFEJBG, ICJDEB

8. Name one heptagon (7-sided figure).
   Answers will vary. Sample: AEDCJBG

9. How many triangles are there in the figure?
   35

180

Enrichment • SECTION 10-1

Name _____ Date _____

Angles and Angle Measures

Examine the protractor. The center is lined up with vertex K. The 0° line of the protractor is on line ST. Angle measures are read on the scale. You can use a protractor to find the measure of angles.

► **Example 1**

Find the following angles:
a. an acute angle
b. an obtuse angle
c. a right angle
d. a straight angle

Solution

a. m∠PKT is less than 90°, so ∠PKT is acute.
b. m∠QKT is greater than 90°, so ∠QKT is obtuse.
c. m∠LKT is exactly 90°, so ∠LKT is a right angle.
d. m∠SKT is exactly 180°, so ∠SKT is a straight angle.

► **Example 2**

Name the following:
a. adjacent angles
b. complementary angles
c. supplementary angles

Solution

a. ∠PKT and ∠LKQ have a side in common, so they are adjacent angles.
b. m∠TKP = 30° and m∠LKP = 60°, so ∠TKP and ∠LKP are complementary angles.
c. m∠TKP = 30° and m∠SKP = 150°, so ∠TKP and ∠SKP are supplementary angles.

EXERCISES

Study the figure at the right. Name the type of each angle and give its measure. Use a protractor.

1. ∠LXM
   acute, 45°
2. ∠LXP
   right, 90°
3. ∠LXQ
   obtuse, 105°

Name the type of angle pair represented.

4. ∠1 and ∠2
   adjacent angles
5. ∠2 and ∠3
   complementary angles

Reteaching • SECTION 10-2

181

Name _____ Date _____

Angles of Polygons

Measure and make conjectures about the angles of polygons. You may have to extend the lines in some figures in order to measure the angles.

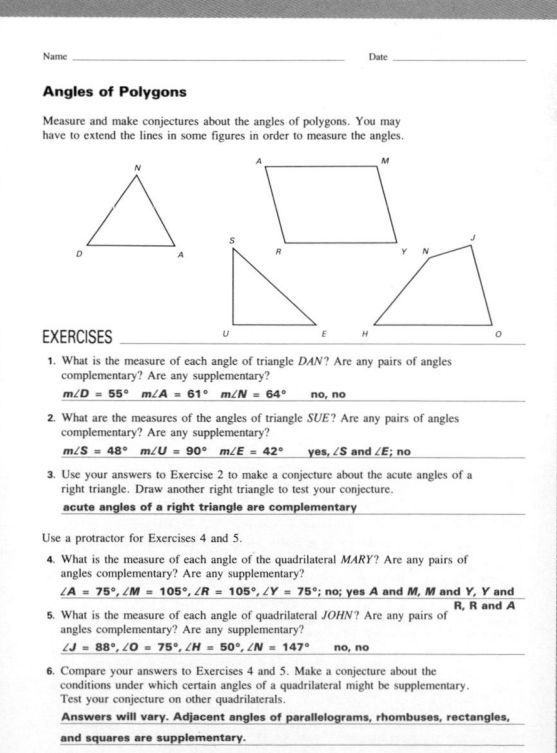

EXERCISES

1. What is the measure of each angle of triangle DAN? Are any pairs of angles complementary? Are any supplementary?
   m∠D = 55°   m∠A = 61°   m∠N = 64°   no, no

2. What are the measures of the angles of triangle SUE? Are any pairs of angles complementary? Are any supplementary?
   m∠S = 48°   m∠U = 90°   m∠E = 42°   yes, ∠S and ∠E; no

3. Use your answers to Exercise 2 to make a conjecture about the acute angles of a right triangle. Draw another right triangle to test your conjecture.
   acute angles of a right triangle are complementary

Use a protractor for Exercises 4 and 5.

4. What is the measure of each angle of the quadrilateral MARY? Are any pairs of angles complementary? Are any supplementary?
   ∠A = 75°, ∠M = 105°, ∠R = 105°, ∠Y = 75°; no; yes A and M, M and Y, Y and R, R and A

5. What is the measure of each angle of quadrilateral JOHN? Are any pairs of angles complementary? Are any supplementary?
   ∠J = 88°, ∠O = 75°, ∠H = 50°, ∠N = 147°   no, no

6. Compare your answers to Exercises 4 and 5. Make a conjecture about the conditions under which certain angles of a quadrilateral might be supplementary. Test your conjecture on other quadrilaterals.
   Answers will vary. Adjacent angles of parallelograms, rhombuses, rectangles, and squares are supplementary.

182

Enrichment • SECTION 10-2

382A

Name _____ Date _____

Parallel and Perpendicular Lines

In the figure at the right, lines *AB* and *CD* are parallel, and line *KL* is a transversal. You can use these facts to identify types of angles and find their measures, given the measure of just one angle.

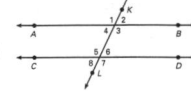

In the figure, the four pairs of vertical angles are ∠1 and ∠3, ∠2 and ∠4, ∠5 and ∠7, and ∠6 and ∠8. The four pairs of corresponding angles are ∠1 and ∠5, ∠2 and ∠6, ∠4 and ∠8, and ∠3 and ∠7.

► **Example**

If m∠3 = 80°, find the measures of the other numbered angles.

Solution

Angles ∠2 and ∠3 are supplementary, so
   m∠2 = 180° − 80° = 100°
Since vertical angles have the same measure,
   m∠3 = m∠1 = 80° and m∠2 = m∠4 = 100°
Since corresponding angles have the same measure,
   m∠1 = m∠5 = 80° and m∠2 = m∠6 = 100°
   m∠4 = m∠8 = 100° and m∠3 = m∠7 = 80°

EXERCISES

Use the figure to find the measure of each angle named.

1. 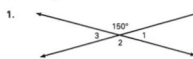   m∠1 = __30°__   m∠2 = __150°__   m∠3 = __30°__

2. 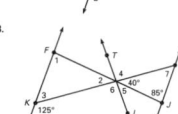   m∠2 = __110°__   m∠3 = __70°__   m∠4 = __110°__

3. FK̅ ∥ H̅J and L̅7 ⊥ H̅J
   m∠2 = __40°__   m∠5 = __50°__   m∠6 = __90°__
   m∠4 = __90°__   m∠1 = __85°__   m∠3 = __55°__
   m∠7 = __55°__

Name _____ Date _____

Angular Reasoning

You can use facts about angle measure to identify pairs of lines as parallel lines.

Suppose you know that two lines are intersected by a transversal and also that at least one of the following is true:
   • the corresponding angles have the same measure;
   • the alternate interior angles have the same measure;
   • the alternate exterior angles have the same measure.

From these facts, you can conclude that the two lines are parallel.

EXERCISES

Use angle measures to determine which lines in the figures are parallel. Line *QT* in the first figure and *JG* in the second figure are straight lines. Explain your reasoning.

1. Are Z̅Y̅ and W̅T̅ parallel?   **∠QYZ ≠ ∠YTW   No, corresponding angles are not equal.**

2. Are R̅S̅ and Q̅T̅ parallel?   **Yes, corresponding angles RST and YTV are equal.**

3. Are S̅V̅ and Y̅X̅ parallel?   **Yes, alternate interior angles STY and TYX are equal.**

4. Are C̅F̅ and I̅H̅ parallel?
   Yes, alternate interior angles FIH and CFI are equal.

5. Are I̅B̅ and F̅E̅ parallel?   **No, corresponding angles BIF and EFG are not equal.**

6. Are A̅I̅ and D̅F̅ parallel?   **No, corresponding angles AIJ and DFI are not equal.**

Name _____ Date _____

Triangles

Triangles can be classified by the measures of their angles and the lengths of their sides.

right triangle (one right angle)   acute triangle (three acute angles)   obtuse triangle (one obtuse angle)

scalene triangle (no sides equal in length)   isosceles triangle (two sides equal in length)   equilateral triangle (three sides equal in length)

► **Example**

Classify the triangle by its sides or by its angles. Then find the measure of the unknown angle.

Solution

No sides are of equal length. There is one obtuse angle of 100°.
The triangle is a scalene obtuse triangle.
Subtract the sum of the measures of the known angles from 180°.
   x = 180° − (50° + 30°)
     = 180° − 80° = 100°

The measure of the unknown angle is 100°.

EXERCISES

Classify each triangle by its sides or by its angles. Then find the measure of ∠x.

1.   2.   3.   4.   5.
acute or isosceles; 77°   right or isosceles; 45°   acute or equilateral; 60°   acute or scalene; 79°   obtuse or isosceles; 135°

6. 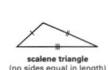   7.   8. 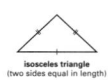   9.   10.
obtuse or scalene; 53°   acute or isosceles; 70°   right or scalene; 19°   obtuse or isosceles; 150°   obtuse or scalene; 17°

Name _____ Date _____

Variable Triangles

To solve some problems involving triangles, you may have to draw triangles and experiment a bit. For example, when the lengths of the sides of a triangle are expressed in terms of a variable, different values of the variable will result in different types of triangles.

► **Example**

The lengths of the sides of a triangle are n, $2n + 2$, and $3n - 2$. Is there a value for n for which an isosceles triangle results?

Solution

Make a chart. Try some values for n.

n	$2n + 2$	$3n - 2$	
1	2(1) + 2 = 4	3(1) − 2 = 1	1, 4, and 1 cannot form a triangle. Try drawing it to see.
2	2(2) + 2 = 6	3(2) − 2 = 4	2, 6, and 4 cannot form a triangle. Draw it to see.
3	2(3) + 2 = 8	3(3) − 2 = 7	3, 8, and 7 form a scalene triangle.
4	2(4) + 2 = 10	3(4) − 2 = 10	4, 10, and 10 are lengths of sides that form an isosceles triangle.

EXERCISES

1. The lengths of the sides of a triangle are given by $4n$, $2n + 10$, and $7n - 15$. Find a value of n such that an equilateral triangle results.   __$n = 5$__

2. What kind of triangle is formed if its angle measures are represented by n, $2n$, and $3n$?   __30°, 60°, 90°   right triangle__

3. Classify a triangle that has angle measures represented by n, $2n$, and $6n$.   __20°, 40°, 120°   obtuse, scalene__

4. Classify a triangle that has angle measures represented by n, $n + 10$, and $n − 10$.   __50°, 60°, 70°; acute, scalene__

5. The lengths of the sides of a triangle are given by $2n − 1$, $n + 5$, and $2n − 8$. Find two values for n that will make the triangle isosceles.   __$n = 6$, $n = 13$__

6. Is there a value for n in Exercise 5 that will make an equilateral triangle?   __no__

382B

Name _____ Date _____

Polygons

The diagram helps you classify quadrilaterals.

| Quadrilateral (4 sides) |
| Parallelogram (both pairs of opposite sides equal) |
| Rectangle (4 right angles) | Square | Rhombus (4 sides equal) |
| Trapezoid (exactly one pair of opposite sides parallel) |

► **Example 1** _____

Give all the names that apply to a rhombus.

Solution
Start with the smallest box that contains the rhombus. A rhombus is a polygon that is also a parallelogram and a quadrilateral.

To find the sum of the measures of the angles of any polygon, count the number of triangles formed by diagonals and multiply by 180°.

► **Example 2** _____

Find the unknown angle measure.

150°   90°

150°   1 2   3   90°

Solution
Copy the pentagon and draw all the diagonals from one vertex. The diagonals form three triangles, so the sum of the angles is $3 \times 180°$, or 540°. Subtract the sum of the known angle measures from 540°. The unknown angle measure is 165°.

EXERCISES

Give all the names that apply to each figure.

1. **polygon, trapezoid, quadrilateral**

2. **polygon, parallelogram, quadrilateral**

3. **polygon, rectangle, parallelogram, quadrilateral**

4. **square, rectangle, parallelogram, quadrilateral, polygon**

Tell the number of triangles that can be formed and the sum of the angles of the polygon. Then find the unknown angle measure.

5. 95°   95°   85°   **2; 360°; 85°**

6. 105° 115°   105° 75°   **3; 540°; 140°**

7. 92° 140° 140° 112°   **4; 720°; 146°**

8. 110° 110° 170° 170° 130° 130° 130°   **6; 1,080°; 130°**

Name _____ Date _____

Tessellations

A **tessellation** is a repeated pattern of polygons or other shapes that covers a plane surface completely without gaps or overlaps.

You can see from the figures at the right that triangles and squares will tessellate a surface.

What other kinds of figures will make tessellations?

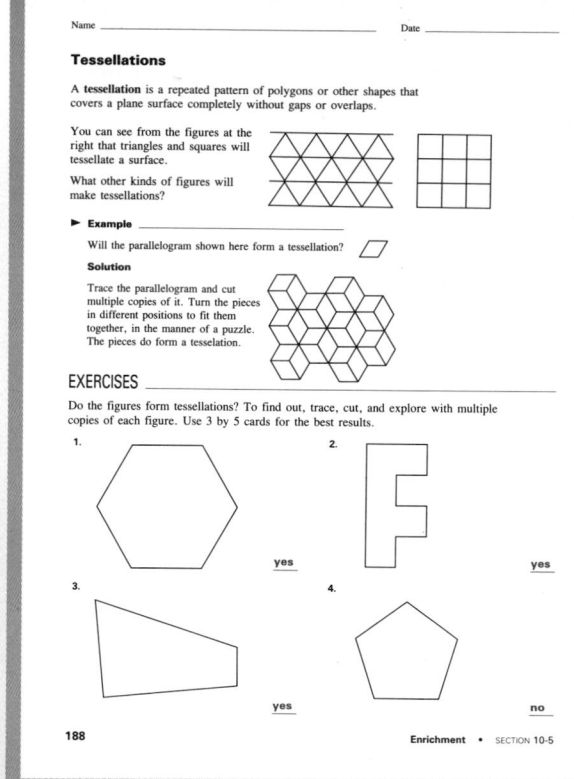

► **Example**

Will the parallelogram shown here form a tessellation?

Solution

Trace the parallelogram and cut multiple copies of it. Turn the pieces in different positions to fit them together, in the manner of a puzzle. The pieces do form a tessellation.

EXERCISES

Do the figures form tessellations? To find out, trace, cut, and explore with multiple copies of each figure. Use 3 by 5 cards for the best results.

1.   **yes**

2.   **yes**

3.   **yes**

4.   **no**

Name _____ Date _____

Three-Dimensional Figures

		triangular	square	rectangular	pentagonal ...
POLYHEDRA (each surface is a polygon)	**PRISMS** (two identical, parallel bases)		(cube)		
	PYRAMIDS (one base and a vertex)		vertex face base		
OTHER 3-DIMENSIONAL FIGURES (some or all surfaces are curved)	**cylinder** (two identical, parallel, circular bases)		**sphere** (all surface points equally distant from the center of the sphere)		
	cone (one circular base, one vertex)				

EXERCISES

Match each figure with its description or with the pattern that can be folded to form it. Identify the figure.

1. **c, sphere**

2. **b, triangular prism**

3. **a, square pyramid**

4. **d, octagonal prism**

5. **f, pentagonal pyramid**

6. **e, cube**

a. a square base and four triangular faces

b.

c. a curved surface having a center equally distant from all points on the surface

d. two octagons as bases

e. four pairs of opposite faces that are squares

f. a single base and five triangular faces

Name _____ Date _____

The Five Regular Polyhedra

A **regular polyhedron** is a polyhedron with faces that are identical regular polygons and edges that are identical.

A regular polyhedron is named by the number of faces it has.

tetrahedron   hexahedron   octahedron   dodecahedron   icosahedron

EXERCISES

Use the figures above to count faces, vertices, and edges to complete the following table.

	Type of polyhedron	Number of faces	Number of vertices	Number of edges
1.	tetrahedron	4	4	6
2.	hexahedron	6	8	12
3.	octahedron	8	6	12
4.	dodecahedron	12	20	30

5. Describe the faces of each regular polyhedron.
   tetrahedron—equilateral triangles; hexahedron—squares;
   octahedron—equilateral triangles; dodecahedron—regular pentagons;
   icosahedron—equilateral triangles

6. Which three regular polyhedra have faces that are equilateral triangles?
   tetrahedron, octahedron, icosahedron

7. Which regular polyhedron has the greatest number of faces? How many does it have?
   icosahedron, 20 faces

8. Which regular polyhedron is a cube?   **hexahedron**

382C

Name _____ Date _____

Problem Solving Skills: Make a Conjecture

To solve a problem, you may need to make a reasonable guess, or conjecture, and then test it.

► **Example**

Triangle ABC is an isosceles triangle. Segment $\overline{BD}$ is perpendicular to $\overline{AC}$. What is the relationship between the length of $\overline{AD}$ and the length of $\overline{DC}$?

Solution

Make a conjecture: It looks as if $\overline{BD}$ bisects $\overline{AC}$. You know from the figure that the length of $\overline{AD}$ is 2 cm. Test your conjecture: Use meterstick to measure $\overline{DC}$. The measure of $\overline{DC}$ is 2 cm. The length of $\overline{DC}$ is equal to the length of $\overline{DC}$. $\overline{BD}$ is not only perpendicular to $\overline{AC}$, it bisects $\overline{AC}$.

PROBLEMS

Use the question in part **a** to make a conjecture. Then use a ruler or protractor to answer the question in part **b**.

1. Triangle XYZ is drawn inside circle C. $\overline{XZ}$ is a diameter of the circle.

   a. Conjecture: what kind of triangle is formed inside a circle if 1 side of the triangle is a diameter of the circle?
   The triangle is a right triangle.

   b. Find $m\angle XYZ$.
   90°

3. For parallelogram $EFGH$, where $m\angle 1 = 76°$, find $m\angle 3$.

   a. Conjecture: What is the relationship between the alternate interior angles created by the diagonals of a parallelogram?
   The alternate interior angles created by a diagonal of a parallelogram have equal measures.

   b. What is the measure of $\angle 3$?
   76°

2. For parallelogram $ABCD$, find the length of $\overline{AC}$, as compared to $\overline{BD}$.

   $\overline{AC}$ = 3.7 cm
   $\overline{BD}$ = 4.7 cm

   a. Conjecture: What is the relationship between the lengths of the diagonals of a parallelogram?
   The two diagonals do not have the same length.

   b. Compare the lengths of $\overline{AC}$ and $\overline{BD}$.
   $\overline{BD} > \overline{AC}$

Name _____ Date _____

Using Conjectures in Comparisons

You can use what you know about geometric figures to make conjectures that can help you to determine which of several comparisons is correct.

► **Example**

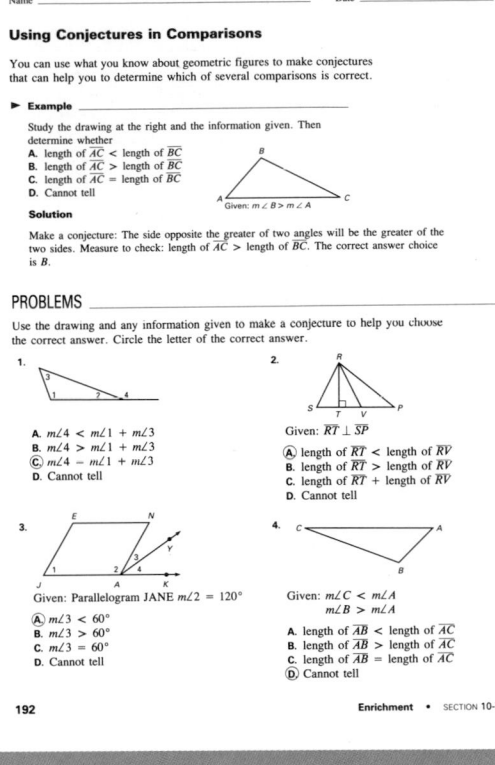

Study the drawing at the right and the information given. Then determine whether
A. length of $\overline{AC}$ < length of $\overline{BC}$
B. length of $\overline{AC}$ > length of $\overline{BC}$
C. length of $\overline{AC}$ = length of $\overline{BC}$
D. Cannot tell

Given: $m\angle B > m\angle A$

Solution

Make a conjecture: The side opposite the greater of two angles will be the greater of the two sides. Measure to check: length of $\overline{AC}$ > length of $\overline{BC}$. The correct answer choice is B.

PROBLEMS

Use the drawing and any information given to make a conjecture to help you choose the correct answer. Circle the letter of the correct answer.

1.

   A. $m\angle 4 < m\angle 1 + m\angle 3$
   B. $m\angle 4 > m\angle 1 + m\angle 3$
   Ⓒ $m\angle 4 = m\angle 1 + m\angle 3$
   D. Cannot tell

2.

   Given: $\overline{RT} \perp \overline{SP}$
   Ⓐ length of $\overline{RT}$ < length of $\overline{RV}$
   B. length of $\overline{RT}$ > length of $\overline{RV}$
   C. length of $\overline{RT}$ + length of $\overline{RV}$
   D. Cannot tell

3.

   Given: Parallelogram JANE $m\angle 2 = 120°$
   Ⓐ $m\angle 3 < 60°$
   B. $m\angle 3 > 60°$
   C. $m\angle 3 = 60°$
   D. Cannot tell

4.

   Given: $m\angle C < m\angle A$
   $m\angle B > m\angle A$
   A. length of $\overline{AB}$ < length of $\overline{AC}$
   B. length of $\overline{AB}$ > length of $\overline{AC}$
   C. length of $\overline{AB}$ = length of $\overline{AC}$
   Ⓓ Cannot tell

Name _____ Date _____

Problem Solving Strategy: Act It Out

You can solve some problems using objects, people, or drawings to "act it out."

► **Example**

In how many ways can three pennies be placed on a 3 by 3 grid so there is only one penny in any row or column?

Solution

Draw a grid like the one shown. Move pennies around to find the solution. Record the solutions on paper. You might find:

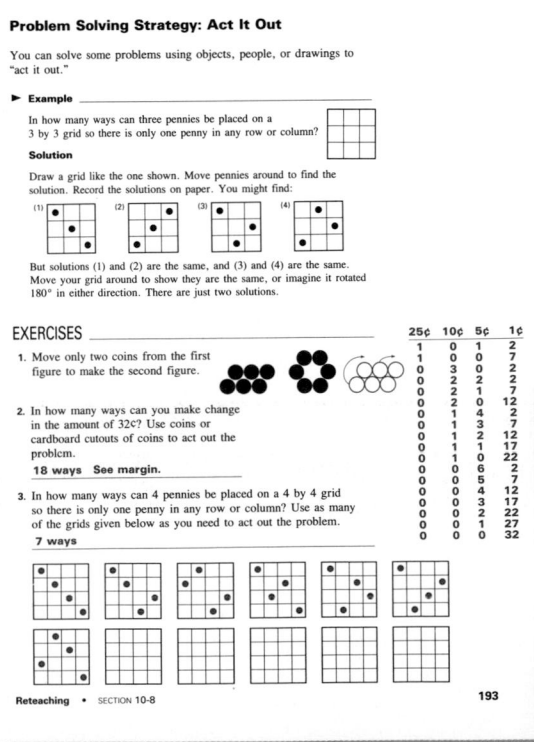

But solutions (1) and (2) are the same, and (3) and (4) are the same. Move your grid around to show they are the same, or imagine it rotated 180° in either direction. There are just two solutions.

EXERCISES

1. Move only two coins from the first figure to make the second figure.

2. In how many ways can you make change in the amount of 32¢? Use coins or cardboard cutouts of coins to act out the problem.
   18 ways   See margin.

25¢	10¢	5¢	1¢
1	0	1	2
1	0	0	7
0	3	0	2
0	2	2	2
0	2	1	7
0	2	0	12
0	1	4	2
0	1	3	7
0	1	2	12
0	1	1	17
0	1	0	22
0	0	6	2
0	0	5	7
0	0	4	12
0	0	3	17
0	0	2	22
0	0	1	27
0	0	0	32

3. In how many ways can 4 pennies be placed on a 4 by 4 grid so there is only one penny in any row or column? Use as many of the grids given below as you need to act out the problem.
   7 ways

Name _____ Date _____

Twisted Surfaces: Möbius Strips

Can a surface have just one edge and one side? If you cut a strip of paper in half lengthwise, will you always get two halves?

You may think you know the answers to these questions, but experimenting with some actual materials may lead you to some surprising discoveries. To solve such problems, you must "act it out."

EXERCISES

You will need paper, tape, and scissors.

1. Take a long strip at least 2 inches in width. Give the strip a half twist. Then tape the ends together. You have made a Möbius strip. It is named for Augustus F. Möbius (1790–1868), who studied it in detail.

2. Color one side of the strip down its length.
   a. What happens? How much is colored? **all of it**
   b. How many sides does a Möbius strip have? **one**
   c. How many edges does it have? **one**

3. Begin with a point midway from the edge. Cut the strip down the middle along its entire length as shown by the dashed line.
   a. How many pieces will you get? **Answers will vary.**
   b. Describe what you found. **One long strip results.**

4. Cut the strip again down the middle.
   a. Predict the outcome. **Answers will vary.**
   b. Describe what you found. **Two pieces are intertwined.**

5. Make a new Möbius strip. Cut a strip lengthwise as in Exercise 3, but $\frac{1}{3}$ of the way from one edge.
   a. Predict the outcome. **Answers will vary.**
   b. Describe what you found. **Two linked strips. One piece is still a Möbius strip.**

382D

Name _____ Date _____

Constructing and Bisecting Angles

You need only compass and straightedge to copy a given angle or to construct the bisector of a given angle.

▶ **Example 1**

Copy ∠ABC, shown at the right.

Solution

Given ∠ABC. Use straightedge to draw $\overrightarrow{DG}$. With compass point on B, draw an arc intersecting the sides of the angle. Label the intersection points X and Y. Keep the same setting, and with compass point on D, draw an arc intersecting $\overrightarrow{DG}$. Label the intersection point E. With compass, measure the distance from X to Y. Keep that setting and place point on E. Draw an arc intersecting the first arc. Label the intersection point Q. Draw $\overrightarrow{DQ}$ to complete the copy of the angle.

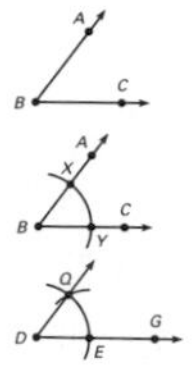

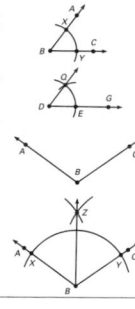

▶ **Example 2**

Bisect ∠ABC, shown at the right.

Solution

Given ∠ABC. With compass point on B, draw an arc that intersects both $\overrightarrow{AB}$ and $\overrightarrow{BC}$. Label the intersection points X and Y, as shown. With compass point on X, draw an arc a little greater than one half of arc XY. With the same setting, place compass point on Y and draw an arc so that it intersects the first one. Label their intersection point Z. Draw $\overrightarrow{BZ}$, which bisects ∠ABC.

EXERCISES

Use a straightedge and compass for the following constructions. Copy each angle in the space to its right. **Check students' constructions.**

1.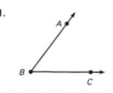

2.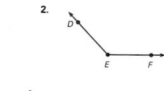

Bisect each angle. **Check students' constructions.**

3.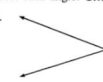

4.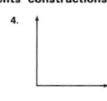

5.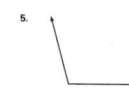

Name _____ Date _____

Using Angles to Construct Triangles

Given the measures of three angles whose sum is 180°, you can construct a triangle.

EXERCISES

Use a straightedge and compass for the following exercises.

1. Construct a triangle by copying these three angles. Let one side of the triangle be 2 in. long. **Check students' drawings.**

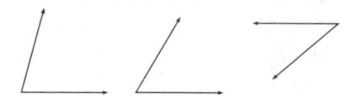

2. Compare your triangle with those of your classmates. Are all the triangles alike? What do you notice about the constructions?

   Answers will vary. Although the triangles have the same shape, they may not

   be the same size.

3. Construct an obtuse triangle by copying these three angles. Compare your triangle with those of your classmates. Are all the triangles alike?

   Students' triangles will have the same shape but will not necessarily be the

   same size.

4. For each triangle you have constructed, bisect each of the angles. What do you notice about the angle bisectors?

   The bisectors of each triangle intersect in a point.

The point of intersection of the angle bisectors of a triangle is called the **incenter**.

5. Use this question to make a conjecture: What do you notice about the point of intersection of the angle bisectors of a triangle? Use your compass to check.

   The point of intersection is the same distance from each side of the triangle,

   but not from each angle.

Name _____ Date _____

Constructing Line Segments and Perpendiculars

Constructing the perpendicular bisector of a line segment can be done with a compass and straightedge.

▶ **Example 1**

Construct the perpendicular bisector of $\overline{AB}$.

Solution

Given $\overline{AB}$. Use a compass opening greater than half the length of $\overline{AB}$. With compass point on A, draw an arc above and below the line segment. Repeat with compass point on B and the same setting, drawing arcs that intersect the first two arcs. Label the intersections X and Y. Draw $\overleftrightarrow{XY}$. $\overleftrightarrow{XY} \perp \overline{AB}$

Label the point M where the perpendicular crossed line $\overline{AB}$. $\overline{AM} = \overline{MB}$. You have also found the midpoint of $\overline{AB}$.

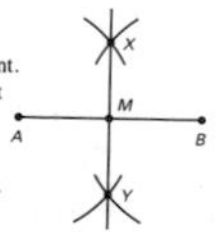

▶ **Example 2**

Construct a line perpendicular to $\overleftrightarrow{CD}$ from T.

Solution

Given point T and $\overleftrightarrow{CD}$. With compass point on T, draw an arc that intersects $\overleftrightarrow{CD}$ at X and Y. Place the compass point on X. With compass open a little more than half the length of $\overline{XY}$, draw an arc below the line. Repeat with compass point on Y and the same setting, intersecting your first arc. Label the point of intersection Z. Draw $\overleftrightarrow{TZ}$. $\overleftrightarrow{TZ} \perp \overleftrightarrow{CD}$.

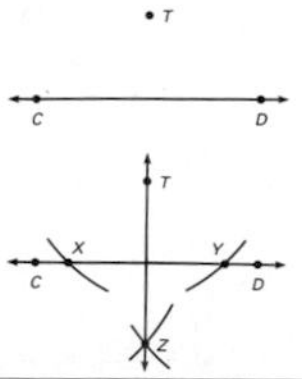

EXERCISES

Use a straightedge and compass for the following constructions. Construct the perpendicular bisector of each line segment. **Check students' drawings.**

1.   2.   3.

Copy each segment onto a separate sheet. Place a point above or below each segment. Construct the perpendicular from the point to the segment. **Check students' drawings.**

4.   5.   6.

Name _____ Date _____

A Surprise Construction

Use compass, straightedge, and the constructions you know to make a surprise figure.

EXERCISES

Use the space at the bottom of the page for your drawings.

1. Draw any line segment AB. Find its midpoint. Label it P.

2. Draw a circle with center at P that passes through A and B.

3. Construct (or extend) the perpendicular bisector of $\overline{AB}$ to intersect the circle. Label the intersection E. Draw $\overline{AE}$ and $\overline{BE}$.

4. Adjust the compass opening to the length of $\overline{AB}$. Use that opening. With compass point at B, draw an arc from A that intersects $\overline{BE}$ at G_1.

5. Use the same opening. With compass point at A, draw an arc from B that intersects $\overline{AE}$ at G_2.

6. Adjust compass opening to the length of $\overline{EG_1}$. With compass point at E, draw a curve from G_1 to G_2.

7. What have you constructed? **egg shape**

8. Try the construction again beginning with a different length for line segment $\overline{AB}$.

   Check students' drawings.

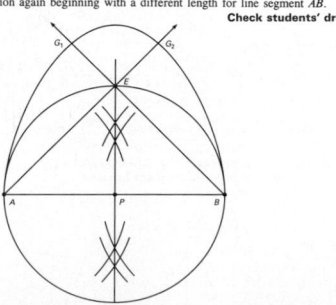

382E

Calculator Activity: Finding Angle Measures

Name _____ Date _____

You can use a calculator to find unknown angle measures for polygons.

► **Example**

Find the unknown angle measure.

96° 96°
124°
x°
103°

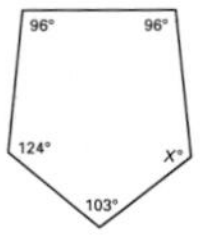

Solution

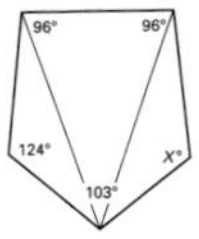

First divide the polygon into triangles. There are three triangular regions.

Use a calculator to find the sum of the angles of the polygon. Then subtract the sum of the given angles.

96° 96°
124°
x°
103°

3×180 M+ $96 +96 +124 +103$ M⁻ MRC
$= 121$

$m\angle x = 121°$

EXERCISES

Use a calculator to find each unknown angle measure.

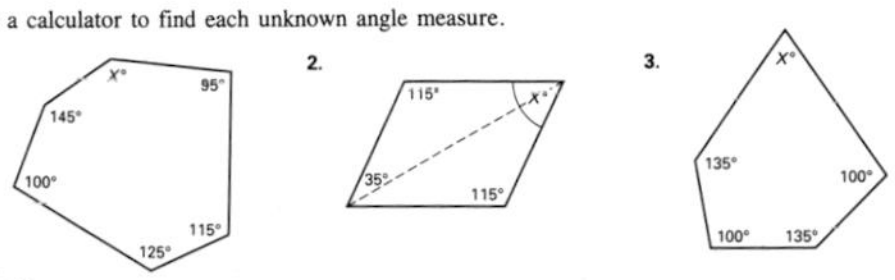

1.
x° 95°
145°
100°
115°
125°

140°

2.
115°
x°
35°
115°

65°

3.
x°
135° 100°
100° 135°

70°

4.
75° 165°
135° x°
90°

85°

5.
x°
69°
107°
127° 141°

96°

Computer Activity: Total Turtle Trip

Name _____ Date _____

Using the computer to draw and investigate properties of geometric shapes can be done quite easily using a computer language called **Logo.** By writing commands and grouping them into **procedures,** you can create and explore with the turtle.

Commands can be given to change the turtle's location or the turtle's direction. The basic ones are listed below.

LOCATION	COMMAND	EXAMPLE	DIRECTION	COMMAND	EXAMPLE
FORWARD	FD	FD 8	RIGHT	RT	RT 75
BACK	BK	BK 29	LEFT	LT	LT 90

When the same command is repeated, use the REPEAT command. For example, to draw a square: REPEAT 4 [FD 50 RT 90].

A procedure can have variables. They are preceded by a colon and allow you to enter different values. For example, in TO SQUARE :LENGTH, the procedure asks you to input the length of the side. To run this procedure, you could type SQUARE 50.

This procedure asks you to input the number of sides in the polygon and the measure of the angle you want to turn.

```
TO POLY :SIDES :ANGLE
  REPEAT :SIDES [FD 40 RT :ANGLE WAIT 25]
END
```
The "WAIT 25" allows the program to run slowly enough for you to see it working.

EXERCISES

1. RUN the procedure for the following values. For example, TO POLY 3 72. Watch the turtle and sketch the result.

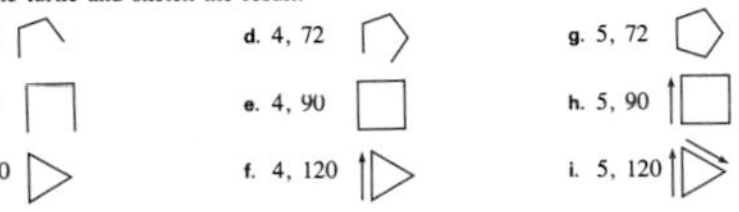

   a. 3, 72
   b. 3, 90
   c. 3, 120
   d. 4, 72
   e. 4, 90
   f. 4, 120
   g. 5, 72
   h. 5, 90
   i. 5, 120

2. How many degrees does a turtle always use in a total trip (start and end at same place)? **360**

3. Predict the correct values for SIDES and ANGLE to have the turtle draw the indicated regular polygon. RUN to check.

   a. decagon   **10**   **36**
   b. octagon   **8**   **45**
   c. hexagon   **6**   **60**
   d. *n*-gon   **n**   **$\frac{360}{n}$**

ACHIEVEMENT TEST

Geometry of Size and Shape
CHAPTER 10 FORM A

MATH MATTERS BOOK 1
Chicha Lynch
Eugene Olmstead

SOUTH-WESTERN PUBLISHING CO.

Name _____
Date _____

SCORING RECORD	
Possible	Earned
23	

Identify each polygon.

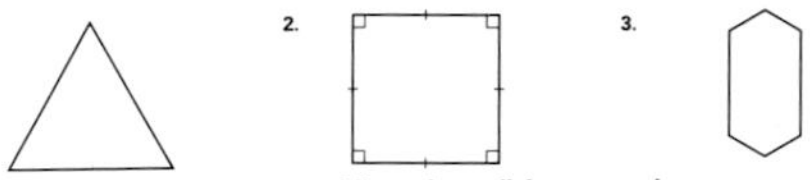

1. **triangle**

2. **quadrilateral, parallelogram, rectangle, square**

3. **hexagon**

Classify each figure.

4. **triangular pyramid**

5. **cone**

6. **rectangular prism**

Classify each angle as *acute, right,* or *obtuse.*

7. 19° **acute**
8. 156° **obtuse**
9. 90° **right**
10. 110° **obtuse**

11. In the figure at the right, lines *RS* and *TU* are parallel. Identify the numbered angles that have equal measures.
$\angle 1 = \angle 3 = \angle 4 = \angle 5; \angle 2 = \angle 6$

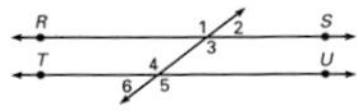

Classify each triangle as *right, acute,* or *obtuse.*

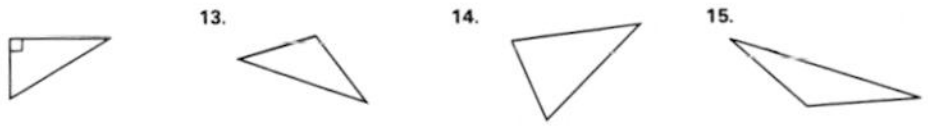

12. **right**
13. **obtuse**
14. **acute**
15. **obtuse**

Name _____ Date _____

16. The measure of $\angle DEF$ is 96°. Ray EG bisects $\angle DEF$. What is the measure of $\angle GEF$? **48°**

17. Use a protractor to draw an angle that measures 110°. Construct its bisector. **Check students' drawings.**

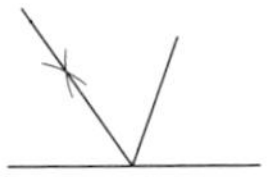

Find the unknown angle measure.

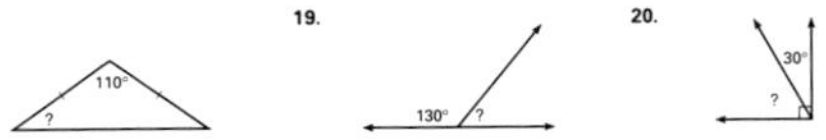

18.
110°
?

35°

19.
130°
?

50°

20.
30°
?

60°

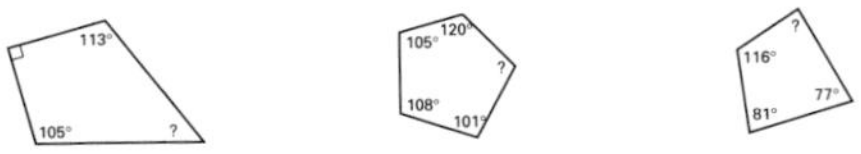

21.
113°
105°
?

52°

22.
105° 120°
108°
101°

106°

23.
116° ?
81° 77°

86°

382F

ACHIEVEMENT TEST

Geometry of Size and Shape

CHAPTER 10 FORM B

Name _____

Date _____

MATH MATTERS BOOK 1

Chicha Lynch
Eugene Olmstead

SOUTH-WESTERN PUBLISHING CO.

SCORING RECORD	
Possible	Earned
23	

Identify each polygon.

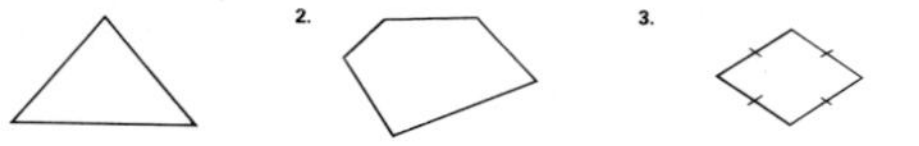

1.

triangle

_____

2.

pentagon

3.

quadrilateral,

parallelogram,

rhombus

Classify each figure.

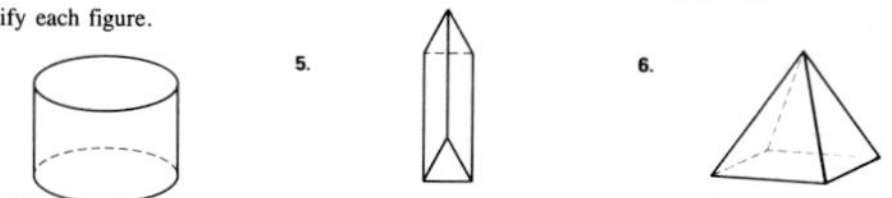

4.

cylinder

5.

triangular prism

6.

rectangular pyramid

Classify each angle as *acute, right,* or *obtuse.*

7. 28°   **acute**

8. 106°   **obtuse**

9. 90°   **right**

10. 140°   **obtuse**

11. In the figure at the right, lines *AD* and *CB* are parallel. Identify the numbered angles that have equal measures.

∠1 = ∠4 = ∠6; ∠2 = ∠3 = ∠5

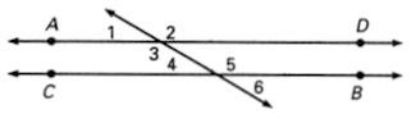

Classify each triangle as *scalene, isosceles,* or *equilateral.*

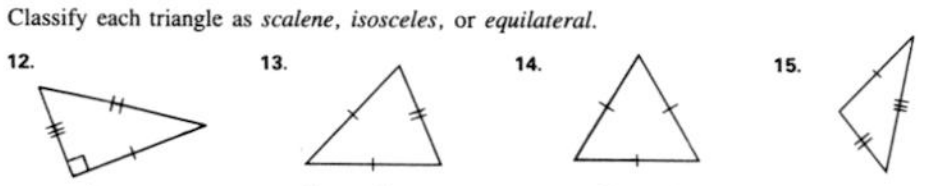

12.

scalene

13.

isosceles

14.

equilateral

15.

scalene

10B-1

Name _____   Date _____

16. Line segment *EF* is 16 cm long. Line segment *CG* bisects $\overline{EF}$ at *G*. Line segment *HJ* bisects $\overline{EG}$ at *K*. How long is line segment *EK*?   **4 cm**

17. Use a protractor to draw an angle that measures 130°. Construct its bisector.   **Check students' work.**

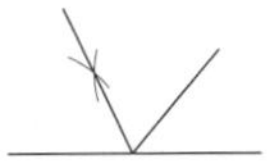

Find the unknown angle measure.

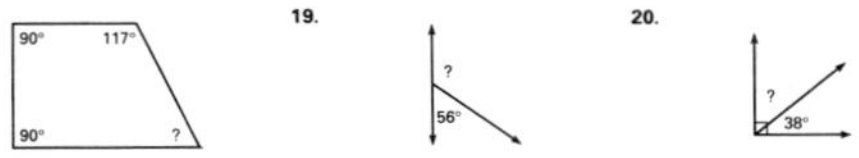

18.

63°

19.

124°

20.

52°

21.

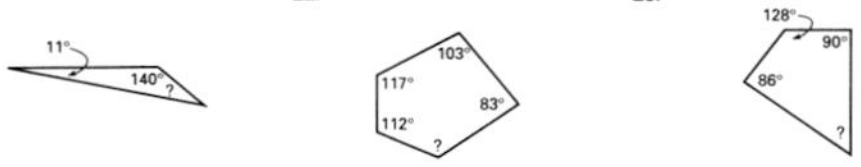

29°

22.

125°

23.

56°

10B-2

Teacher's Notes

CHAPTER 11 SKILLS PREVIEW

Find the area of each figure.

1.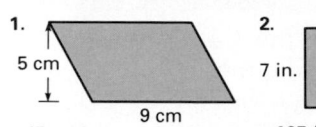
5 cm
9 cm
45 cm²

2.
7 in.
15 in.
105 in.²

3.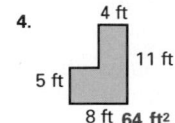
12 m
18 m
108 m²

4.
4 ft
11 ft
5 ft
8 ft **64 ft²**

Find the square root to the nearest tenth.

5. $\sqrt{49}$ **7**   **6.** $\sqrt{4000}$ **20**   **7.** $\sqrt{38}$ **6.2**   **8.** $\sqrt{110}$ **10.5**

Find the unknown length.

9.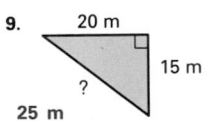
20 m
15 m
?
25 m

10.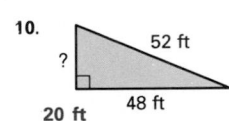
52 ft
?
20 ft
48 ft

11.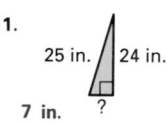
25 in. 24 in.
7 in. ?

Find the volume of each figure. Round to the nearest whole number.

12.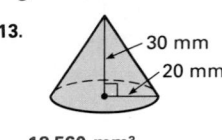
8 in.
8 in.
8 in.
512 in.³

13.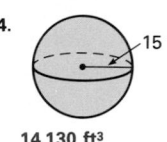
30 mm
20 mm
12,560 mm³

14.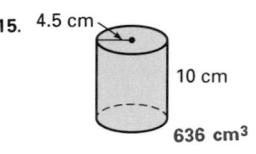
15 ft
14,130 ft³

15.
4.5 cm
10 cm
636 cm³

16. The diameter of a can of tomatoes is 9 cm and the height is 16 cm. Find the volume. Use $\pi \approx 3.14$. Round to the nearest whole number.
1017 cm³

Copy and complete the table.

	radius (r)	diameter (d)	area (A)	
17.	■	30 cm	■	15 cm; 706.5 cm²
18.	8 cm	■	■	16 cm; 200.96 cm²

Find the surface area of each figure.

19.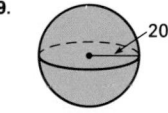
20 ft
5,024 ft²

20.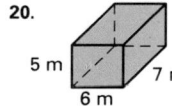
5 m
6 m
7 m
214 m²

21.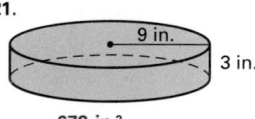
9 in.
3 in.
678 in.²

22.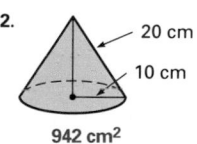
20 cm
10 cm
942 cm²

Solve.

23. The area of a rectangle is 32 in.² The length is twice the width. Find the dimensions of the rectangle. **4 in. wide, 8 in. long**

Skills Preview **383**

CHAPTER 11

AREA AND VOLUME

OVERVIEW

In this chapter, students learn to use formulas for finding the area of rectangles, parallelograms, triangles, and circles; for finding the volume of prisms, cylinders, pyramids, cones, and spheres; and for finding the surface area of three-dimensional figures. Students also learn how to determine squares and square roots, identify rational and irrational numbers, and explore applications of the Pythagorean Theorem. Students apply two problem solving strategies: using a formula and guess and check.

SPECIAL CONCERNS

Although the formulas introduced in this chapter are easily applied, remembering them all may be difficult for some students. You may wish to have students write out each formula and draw a diagram of the figure with which it correlates on individual index cards. Encourage students to review the formulas whenever they have spare time.

VOCABULARY

area	perimeter	square
base	prism	square root
cone	pyramid	surface area
cylinder	Pythagorean	volume
parallelogram	Theorem	
perfect square	sphere	

MATERIALS

calculators	tape
grid paper	scissors
centimeter cubes	24 one-inch tiles
centimeter rulers	

BULLETIN BOARD

Entitle a bulletin board, "Area and Volume Around Us." Divide the board into several sections, one for each of several of the three-dimensional figures discussed in this chapter. Label each section with the name of the figure and include a diagram of the figure along with the formula for finding the volume and/or surface area of that figure. Have students bring pictures of everyday objects shaped like those figures to add to the display.

INTEGRATED UNIT 3

The skills and concepts involved in Chapters 9–11 are included within the special Integrated Unit 3 entitled "On Your Own." This unit is in the Teacher's Edition beginning on page 434J. Worksheets for this integrated unit appear in the Enrichment Activities booklet, pages 137–139.

TECHNOLOGY CONNECTIONS

- Computer Worksheets, 225–226
- Calculator Worksheets, 227–228
- MicroExam, Apple Version
- MicroExam, IBM Version

- *3D Images,* William K. Bradford
- *Geometer's Sketchpad,* Key Curriculum Press

TECHNOLOGY NOTES

Computer geometry programs, such as those listed in Technology Connections, give students the ability to create and manipulate figures on the computer screen while maintaining the figures' geometric definitions. Some programs allow students to view cross-sections of solid figures. Others offer "tools" that enable students to draw lines and locate and label points on those lines.

PLANNING GUIDE

SECTIONS	TEXT PAGES	ASSIGNMENTS BASIC	ASSIGNMENTS AVERAGE	ASSIGNMENTS ENRICHED	
Chapter Opener/Decision Making	384–385				
11–1 Area and Volume	386–389	1–13, 15, 16–18, 22–23	4–15, 16–24, 27, 31	14–15, 16–27, 28–31	
11–2 Area of Rectangles	390–393	1–10, 13, 14, PSA 1–9	1–11, 12–13, 14, PSA 1–9	8–11, 12–13, 14–15, PSA 5–12	
11–3 Area of Parallelograms and Triangles	394–397	1–16, 20–21, 23, 29–30	1–19, 21–26, 29–30	17–19, 21–28, 29–32	
11–4 Problem Solving Skills : Using a Formula for Perim	398–399	1–10	1–15	1–15	
11–5 Squares and Square Roots	400–403	1–32, 37, 38–45, 50	1–37, 38–51	21–37, 38–51, 52–53	
11–6 The Pythagorean Theorem	404–407	1–13, 15–18, 23–26	1–14, 17, 28–30	12–14, 15–27, 28–31	
11–7 Volume of Prisms	408–411	1–6, 9–10, 11–15, 21	1–10, 11–20, 21–23	9–10, 11–17, 20, 21–25	
11–8 Area of Circles	412–415	1–8, 12–14, 15–16, 18–23, 29	5–14, 17–28, 29–34	12–14, 18–28, 29–35	
11–9 Volume of Cylinders	416–419	1–8, 11–12, 14, PSA 1–4	5–10, 11–13, 14, PSA 1–5	9–10, 11–13, 14–16, PSA 1–7	
11–10 Volume of Pyramids Cones, and Spheres	420–423	1–16, 21–24, 28	1–16, 18, 19–20, 21–27, 28	7–20, 21, 25–27, 28–30	
11–11 Surface Area	424–427	1–16, 21	1–17, 18–21	9–15, 18–21, 22	
11–12 Problem Solving	428–429	1, 3–5, 7	1–7	1–8	
Technology	389, 392, 400, 403, 404, 405, 409, 411, 413, 416, 421				

ASSESSMENT					
Skills Preview	383	All	All	All	
Chapter Review	430–431	All	All	All	
Chapter Test	432	All	All	All	
Cumulative Review	433	All	All	All	
Cumulative Test	434	All	All	All	

	ADDITIONAL RESOURCES			
	RETEACHING	**ENRICHMENT**	**TECHNOLOGY**	**TRANSPARENCY**
	11–1	11–1		
	11–2	11–2		
	11–3	11–3		
	11–4	11–4		
	11–5	11–5		
	11–6	11–6	11–6	TM 56
	11–7	11–7		TM 57
	11–8	11–8		
	11–9	11–9		
	11–10	11–10		
	11–11, 11–12	11–11, 11–12	11–11	

ASSESSMENT OPTIONS

Chapter 11, Test Forms A and B	
Chapter 11, Test	Text, 432
Alternative Assessment	TAE, 432
Chapter 11, MicroExam	

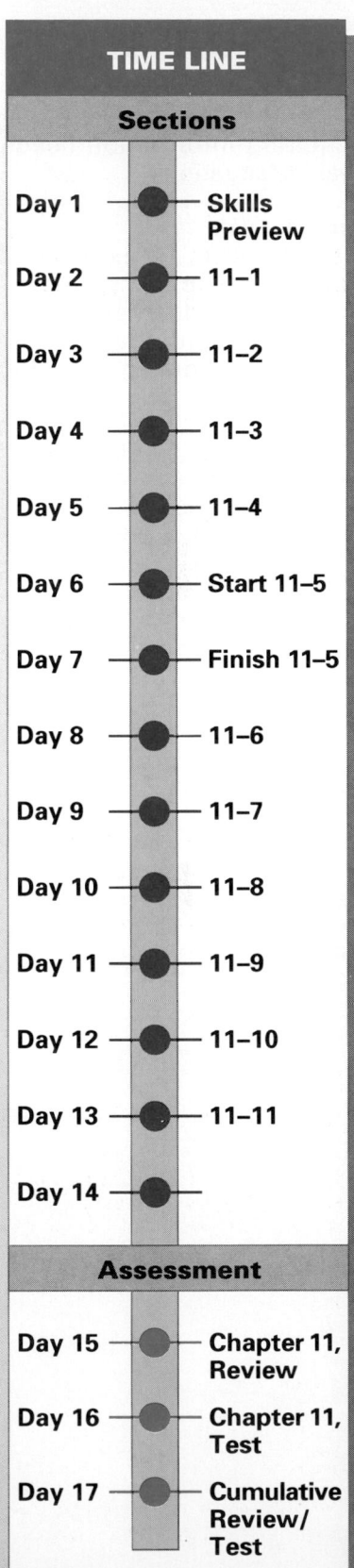

TIME LINE

Sections

Day 1 — Skills Preview
Day 2 — 11–1
Day 3 — 11–2
Day 4 — 11–3
Day 5 — 11–4
Day 6 — Start 11–5
Day 7 — Finish 11–5
Day 8 — 11–6
Day 9 — 11–7
Day 10 — 11–8
Day 11 — 11–9
Day 12 — 11–10
Day 13 — 11–11
Day 14 —

Assessment

Day 15 — Chapter 11, Review
Day 16 — Chapter 11, Test
Day 17 — Cumulative Review/Test

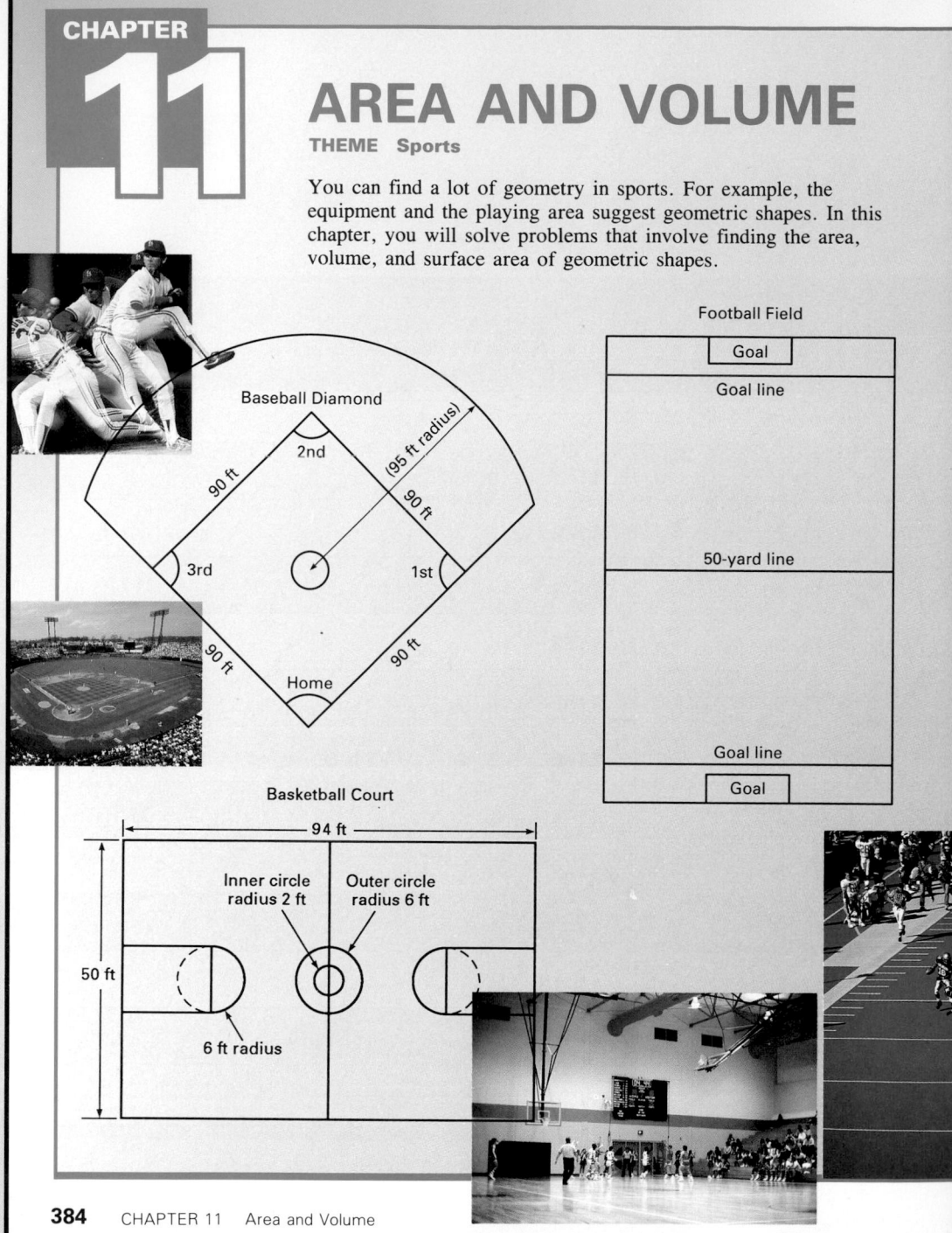

CHAPTER 11 AREA AND VOLUME

THEME Sports

You can find a lot of geometry in sports. For example, the equipment and the playing area suggest geometric shapes. In this chapter, you will solve problems that involve finding the area, volume, and surface area of geometric shapes.

Football Field

Goal

Goal line

50-yard line

Goal line

Goal

Baseball Diamond

2nd

(95 ft radius)

90 ft

90 ft

3rd

1st

90 ft

90 ft

Home

Basketball Court

94 ft

Inner circle radius 2 ft

Outer circle radius 6 ft

50 ft

6 ft radius

DECISION MAKING

Using Data

Use the information in the diagrams to answer the following questions.

1. What is the perimeter of the baseball "diamond"? **360 ft**

2. How great an area would you have to cover to wax the floor of the basketball court? **4,700 ft²**

3. How much clay would you need to cover the tennis court with a 1.5-in. layer? **351 ft³**

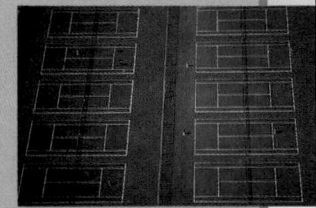

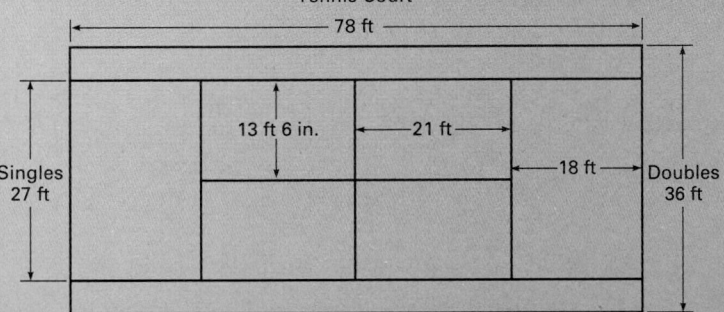

Tennis Court

78 ft

13 ft 6 in. — 21 ft

Singles 27 ft

18 ft

Doubles 36 ft

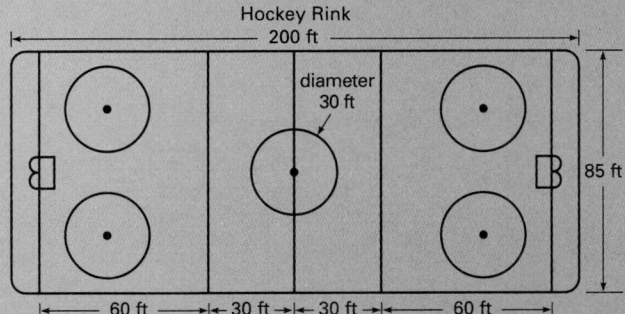

Hockey Rink

200 ft

diameter 30 ft

85 ft

60 ft — 30 ft — 30 ft — 60 ft

Working Together

Your group's task is to choose a sport other than basketball, baseball, tennis, or hockey. Gather information about the field or court area used for that sport. You might choose football or soccer or even swimming. (An Olympic-sized swimming pool can also be thought of as a "field.") Prepare a table of facts about the field and draw a diagram illustrating it and the critical distances and lines on it. Prepare an oral report on your sport to present to the class.

11-1 Area and Volume

EXPLORE

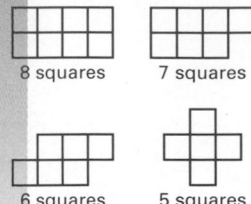

8 squares   7 squares

6 squares   5 squares

This large square, made up of 9 small squares, has a perimeter of 12 units.

Use grid paper to draw other closed figures, each with a perimeter of 12 units. Try figures that are made up of 8, 7, 6, and 5 squares.

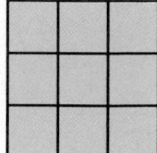

SKILLS DEVELOPMENT

The amount of surface covered by a plane figure is called its **area**.

Example 1

Find the area of this rectangle.

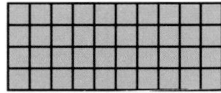

Solution

Count the number of square units. There are 40 squares in all.

The area of the rectangle is 40 square units. ◄

Plane figures with the same area may have different perimeters.

Example 2

Find the perimeter and area of each figure.

a.
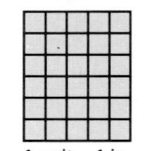

1 unit = 1 in.

b.
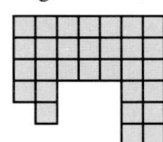

1 unit = 1 in.

Solution

a. To find the perimeter, count the number of units along the sides of the figure. There are 22 units, so the is 22 in.

To find the area, count the number of *square units* covered by the figure. There are 30 square units, so the area is 30 in.²

b. Count the number of units along the sides of the figure. There are 30, so the perimeter is 30 in.

Count the number of *square units*. There are 30, so the area is 30 in.² ◄

TEACHING TIP

Students often fail to express their answers in square or cubic units, as necessary. Post signs, such as these, to serve as reminders.

> Perimeter: Linear Units

> Area: Square Units

> Volume: Cubic Units

The amount of space a three-dimensional figure occupies is called its **volume.** Volume can be measured by counting the number of cubic units that fill a space.

Example 3

Find the volume of this rectangular prism.

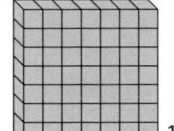

1 unit = 1 cm

Solution

The number of cubic centimeters that fill the space is 49.

The volume of the rectangular prism is 49 cm³. ◄

Sometimes, not all of the cubic units can be seen in the drawing of a three-dimensional figure.

Example 4

Find the volume of the rectangular prism.

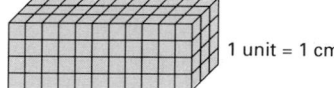

1 unit = 1 cm

Solution

Count the number of cubic units in the top layer.

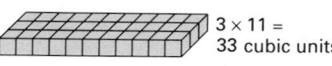

$3 \times 11 =$ 33 cubic units

Count the number of layers.

Multiply to find the volume.

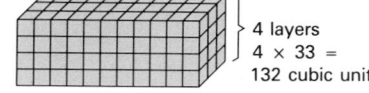

4 layers
$4 \times 33 =$
132 cubic units

The volume of the figure is 132 cubic centimeters, or 132 cm³. ◄

TRY THESE

In each of Exercises 1–4, 1 unit = 1 cm.

1. Find the area of the figure.

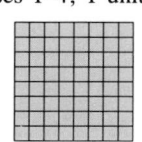

64 cm²

2. Find the perimeter and the area of the figure.

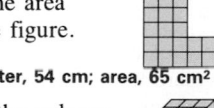

perimeter, 54 cm; area, 65 cm²

3. Find the volume of the rectangular prism.

45 cm³

4. Find the volume of the rectangular prism.

48 cm³

Example 2: Use this example to show how to find areas of irregular figures. Make a connection between this example and the Explore activity by reinforcing the ideas that, just as figures with the same perimeter may have different areas, figures with the same area may have different perimeters.

Example 3: You may wish to have pairs of students build the figure using cubes. Then have them count the cubes to find the volume. Point out that there are different units of measure, such as cubic inches (in.³), cubic feet (ft³), and cubic centimeters (cm³), in which volume can be expressed.

Example 4: Have students count the number of cubes in the top layer of the figure. **33** Ask, "How many layers are there?" **4** "How would you find the number of cubes in all?" **multiply 33 by 4** Have students multiply to find the volume, 132 cm³.

Additional Questions/Examples
1. What would happen to the perimeter of a figure if you doubled the lengths of all of its sides? **The perimeter would be doubled.**
happen if you tripled the lengths of all of its sides? **The perimeter would be tripled.**

2. What would happen to the area of a figure if you doubled the lengths of each of its sides? **The area would become 4 times greater.** What would happen if you tripled the lengths of its sides? **The area would become 9 times greater.**

Guided Practice/Try These Complete these exercises orally. Have volunteers explain how they arrive at each answer.

3 SUMMARIZE

Guided Practice/Try These Have students explain in their math journals the difference between area and volume.

4 PRACTICE

Practice/Solve Problems You might wish to have students use cubes to build each of the figures in Exercises 10–13.

Extend/Solve Problems Students may need cubes to build the figures in Exercises 22–26. They may want to use grid paper to draw a facsimile of the rectangular area for Exercise 27.

Think Critically/Solve Problems Have students share and discuss their conclusions for Exercises 28 and 29.

5 FOLLOW-UP

Extra Practice
Find the area. 1 unit = 1 ft

1.
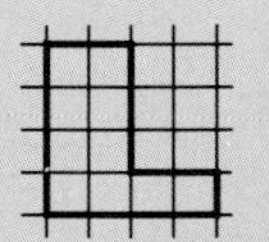
10 ft²

2.
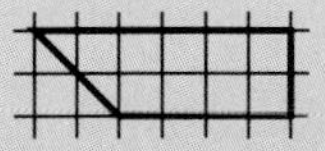
10 ft²

EXERCISES

PRACTICE/ SOLVE PROBLEMS

For each of Exercises 1–9, 1 unit = 1 cm.

Find the area of each rectangle.

1.
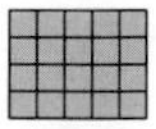
20 cm²

2.
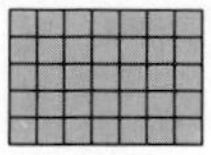
35 cm²

3.
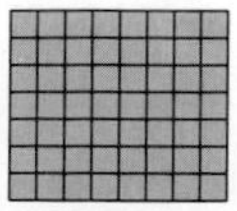
56 cm²

Find the perimeter and the area of each figure.

4.
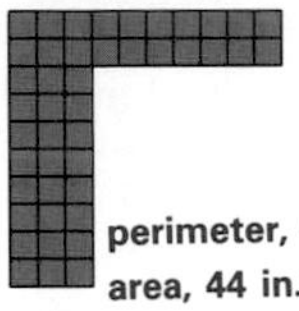
perimeter, 40 in.; area, 44 in.²

5.
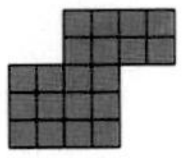
perimeter, 22 in.; area, 20 in.²

6.
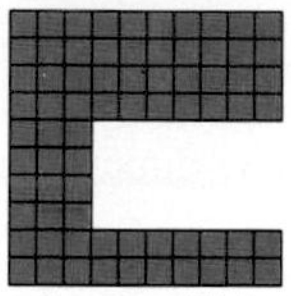
perimeter, 54 in.; area, 72 in.²

Find the volume of each prism.

7.
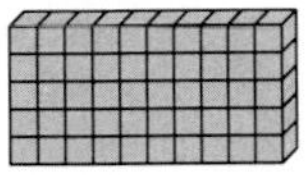
50 cm³

8.
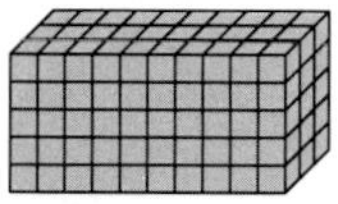
150 cm³

9.
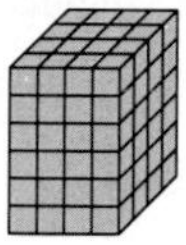
96 cm³

Find the volume of each rectangular prism. Visualize the space or draw a diagram to solve.

10. Length: 5 in.
Width: 1 in.
Height: 2 in. **10 in.³**

11. Length: 6 in.
Width: 3 in.
Height: 1 in. **18 in.³**

12. Length: 4 cm
Width: 3 cm
Height: 2 cm **24 cm³**

13. Length: 7 cm
Width: 2 cm
Height: 1 cm **14 cm³**

Solve.

14. A builder wanted to cover the floor of the foyer of a house with square tiles. The width of the foyer was 6 ft. The length was 8 ft. How many tiles measuring 1 ft on a side would the builder need to cover the floor? **The builder would need 48 tiles.**

15. A manufacturer can pack 3 layers of square boxes in one carton. Each layer holds 4 boxes. How many boxes can be packed in each carton? Draw a rectangular prism to help you solve.
The manufacturer can pack 12 square boxes in each carton.

MIXED REVIEW

Using Data. Use the Data Index on page 546 to find the statistics you need for each answer.

1. Draw a stem-and-leaf plot showing the number of stories in the tallest buildings in the world as of 1990.
2. Draw a bar graph that represents the length, in meters, of the spans of the longest bridges as of 1988.
3. Decide whether a line graph is an appropriate way to express the data in the famous-waterfalls table. Explain your decision.

1. Number of Stories of Tallest Buildings in World

7	2 7
8	0
9	
10	0 2
11	0 0 0

2. Check students' graphs.

3. Answers will vary. Sample: A line graph would be inappropriate. Line graphs are more appropriately used when data for one item or characteristic are compared over a period of time.

MAKING CONNECTIONS

Have students report on how volume might be used in other subject areas, such as science, home economics, or automobile technology. Encourage students to use specific examples of each.

Find the area of each shaded figure. Use 1 unit = 1 ft.

16.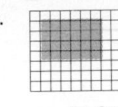
24 ft²

17.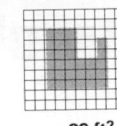
29 ft²

18.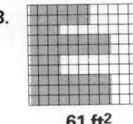
61 ft²

Find the area of each shaded figure. Use 1 unit = 1 cm. Hint: two halves of a square unit can be added to make one square unit.

19.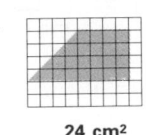
24 cm²

20.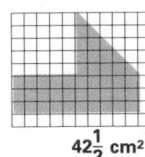
$42\frac{1}{2}$ cm²

21.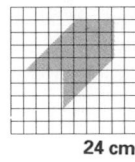
24 cm²

Find the volume of each figure. Use 1 unit = 1 in.

22.
32 in.³

23.
30 in.³

24.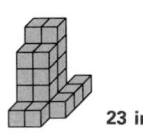
23 in.³

Find the volume of each figure. Use 1 unit = 1 m.

25.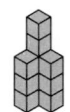
11 m³

26.
14 m³

27. A landscape architect plans to use stones to cover a rectangular area that measures 2 ft by $2\frac{1}{2}$ ft. Each bag of stones will cover 2 ft². How many bags of stones will be needed? Explain.

See Additional Answers.

28. Draw as many rectangles as you can that have whole-number dimensions and an area of 16 square units. Then find the perimeter of each. Write a short paragraph stating your conclusions about the relationship between area and perimeter.

29. How many differently shaped three-dimensional figures can be made using four cubic units?   **5**

30. The volume of a carton is 132 cubic units. If there are 4 layers, how many units are in each layer? Visualize the problem or draw a rectangular prism to solve.   **33 units**

31. Find the volume of this figure. 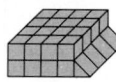   **34 unit³**

THINK CRITICALLY/ SOLVE PROBLEMS

27. 3 bags; Explanations may vary. Sample: Since the area of the region is 5 ft², 3 bags will be needed.

28. Answers will vary. For dimensions 1 × 16, perimeter will be 34; for 2 × 8, perimeter will be 20; for 4 × 4, perimeter will be 16.

Find the volume. 1 unit = 1 cm

3.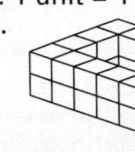
12 cm²

4.
24 cm²

5. What is the area of a rectangle that is 12 in. long and 5 in. wide?   **60 in.²**

6. How many square inches are in a square foot?   **144 in.²**

7. How many square feet are in a square yard?   **9 ft²**

Extension Have students design a floor plan of a house on grid paper. Ask them to find the area of each room, then find the area of the entire house.

Section Quiz
Find the area. 1 unit = 1 in.

1.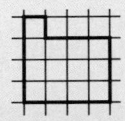
13 in.²

2.
12 in.²

Find the volume. 1 unit = 1 cm

3.
27 cm³

4.
22 cm³

5. The cost of sodding an area with grass is $1.67 per square yard. Find the cost of sodding a rectangular area that measures 28 yd by 20 yd.   **$935.20**

Get Ready   24 one-inch tiles, grid paper

Additional Answers
See page 582. Additional answers for odd-numbered exercises are found in the Selected Answers portion of the page.

CHALLENGE

Have students estimate, then find the answers.
1. If each person were allowed 9 ft² of floor space, how many people could your classroom accommodate?
2. If each person were allowed 45 ft³ of air space, how many people could your classroom accommodate?

WARM-UP

Solve:
1. $24 \times 32 = \blacksquare$ **768**
2. $3.7 \times 2.6 = \blacksquare$ **9.62**
3. $2.04 \times 7.14 = \blacksquare$ **14.5656**
4. $132 \times 3.2 = \blacksquare$ **422.4**

1 MOTIVATE

Explore Have students work in pairs to do this activity. Ask the following question as a hint for students who are having difficulty: "How must the length and width of each rectangle be related to the number 24?" **They must all be factors of 24.**

2 TEACH

Use the Pages/Skills Development Have students read this part of the section and then discuss the examples. Have students work in pairs. You may wish to give each pair a sheet of grid paper. Students should take turns drawing a rectangle and then finding the length, width, and area of the rectangle.

390

11-2 Area of Rectangles

EXPLORE

How many different rectangles can you construct using 24 tiles measuring 1 in. by 1 in.? In each rectangle, you must use all 24 tiles. What are the length and width of each rectangle you construct? Make a list.

SKILLS DEVELOPMENT

You have found the area of a rectangle by counting the number of square units. The rectangle below has 24 one-inch square units. Its area is 24 in.²

The area of the rectangle can also be found by using this formula.

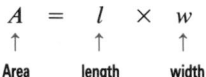

$$A \;=\; l \;\times\; w$$
$$\uparrow \qquad \uparrow \qquad \uparrow$$
Area   length   width

The length of the rectangle measures 6 inches. The width measures 4 inches.

$A = 6 \times 4 = 24$ The area of the rectangle is 24 in.²

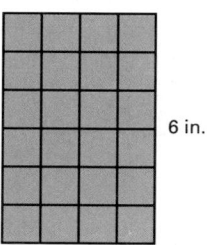

6 in.

4 in.

Example 1

Use the formula $A = l \times w$ to find the area of a rectangle having a length of 16 in. and a width of 7 in.

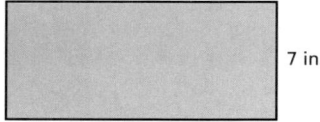

7 in.

16 in.

Solution

$A = l \times w$ ← Substitute 16 for *l* and 7 for *w*.

$A = 16 \times 7$

$A = 112$ The area of the rectangle is 112 in.² ◀

To find the area of an irregular figure, separate it into smaller, rectangular regions.

Example 2

Find the area of the irregular figure at the right. All angles are right angles.

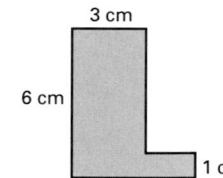

3 cm

6 cm

1 cm

5 cm

390   CHAPTER 11   Area and Volume

TALK IT OVER

A special formula for the area of a square is $A = s^2$, where *s* is the length of a side. Explain how this formula is related to the formula for the area of a rectangle.

A square is a type of rectangle, so the formula for the area of a rectangle can be used, substituting *s* for both *l* and *w*.
$A = l \times w$
$A = s \times s$
$A = s^2$

Solution

Total area	=	Area of region 1	+	Area of region 2
$A =$		5×3	+	5×1
$A =$		15	+	5
$A =$		20		

The area of the irregular figure is 20 cm². ◄

Example 3

The dimensions of a rectangular field on a sod farm are shown.

a. Find the area of the field.
b. At 7.5¢/ft², find the total cost of plowing the field.

96 ft

72 ft

Solution

a. The field is shaped like a rectangle. Use the formula for the area of rectangles.

$A = l \times w$

$A = 96 \times 72$

$A = 6,912$ The area is 6,912 ft².

b. It costs 7.5¢ to plow 1 ft². Write 7.5¢ as $0.075, then multiply.

$6,912 \times 0.075 = 518.4$

The cost of plowing 6,912 ft² is $518.40. ◄

PROBLEM SOLVING TIP

Remember to state your answer to a problem in sentence form.

TRY THESE

Use the formula $A = l \times w$ to find the area of each rectangle.

1.

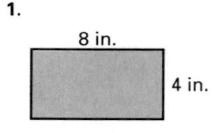

8 in.

4 in.

32 in.²

2.

10 in.

10 in.

100 in.²

3.

19 in.

16 in.

304 in.²

Find the area of each irregular figure. All angles are right angles.

4.

14 cm
4 cm
9 cm
6 cm

86 cm²

5.

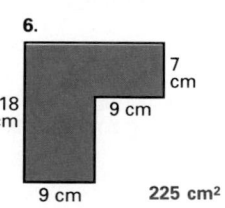

29 cm
10 cm
6 cm   ←12 cm   6 cm

494 cm²

6.

7 cm
18 cm   9 cm
9 cm

225 cm²

A rectangular floor measures 8 yd by 10 yd.

7. Find the area of the floor. **80 yd²**

8. If carpeting costs $12.75/yd², what would be the cost of carpeting the floor? **$1,020**

ASSIGNMENTS

BASIC
1–10, 13, 14, PSA 1–9

AVERAGE
1–11, 12–13, 14, PSA 1–9

ENRICHED
8–11, 12–13, 14–15, PSA 5–12

ADDITIONAL RESOURCES
Reteaching 11–2
Enrichment 11–2

Example 1: Remind students that area is always measured in square units.

Example 2: Discuss this alternative solution to the problem: $3 \times 6 + 2 \times 1 = 20$ cm².

Example 3: You may wish to review how to place the decimal point in the product when multiplying with decimals.

3 SUMMARIZE

Additional Questions/Examples

1. The area of a rectangle is 8 ft². Its length and width, in feet, are whole numbers. What are the possible dimensions of the rectangle? **8 ft by 1 ft or 4 ft by 2 ft**

2. When would you need to know only one dimension of a rectangle in order to find its area? **when the rectangle is a square**

3. What would be the cost of carpeting a room that is 12 ft by 18 ft if carpet costs $7 per square yard? **$168**

Guided Practice/Try These You may wish to have students work in pairs. Have students discuss how they solved Exercises 4–6.

4 PRACTICE

Practice/Solve Problems You may wish to review rounding before students begin Exercises 5–7. For Exercise 11 point out that, when a rectangle is a square, only one dimension is needed to find the area.

Extend/Solve Problems In Exercise 12 students should see that there is more than one way to divide the area into two regions.

Think Critically/Solve Problems For Exercises 14 and 15, students may wish to draw diagrams.

Problem Solving Applications You may wish to have students work in pairs to complete these exercises and then use their calculators to check the accuracy of each solution.

5 FOLLOW-UP

Extra Practice

1 Find the area of a rectangle with *l* = 3.4 in. and *w* = 5.6 in.
19.04 in.²

2. Find the cost of carpet for each room if carpet costs $12.99 per square yard.
   a. Living room: length = 21 ft
   width = 15 ft
   $454.65
   b. Bedroom: length = 12.6 ft
   width = 17.7 ft
   $321.89

3. Identify the figure with the greater area and tell how much greater it is—a square with an area of 312 in.² or a rectangle with a length of 24 in. and a width of 12 in. **The area of the square is 24 in.² greater.**

PRACTICE/SOLVE PROBLEMS

COMPUTER TIP

Here is the way to change the program on page 389 so that it can be used to find the area of rectangles of all sizes:
In line 20, delete the following.
  "PERIMETER"
Delete line 60.
In line 70, delete the following.
  ,P

EXTEND/SOLVE PROBLEMS

PROBLEM SOLVING TIP

In Exercise 13, don't forget to change square feet to square yards.

THINK CRITICALLY/SOLVE PROBLEMS

Find the area of each rectangle.

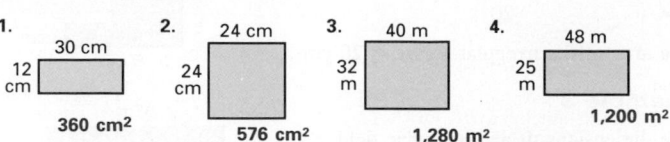

1. 30 cm, 12 cm — **360 cm²**
2. 24 cm, 24 cm — **576 cm²**
3. 40 m, 32 m — **1,280 m²**
4. 48 m, 25 m — **1,200 m²**

Find the area of each plot of land. Round to the nearest square foot.

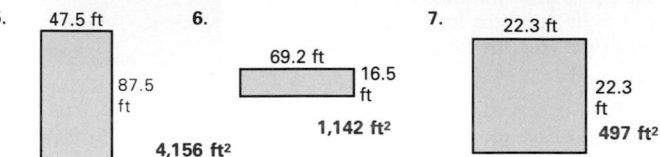

5. 47.5 ft, 87.5 ft — **4,156 ft²**
6. 69.2 ft, 16.5 ft — **1,142 ft²**
7. 22.3 ft, 22.3 ft — **497 ft²**

Find the area of each figure. All angles are right angles.

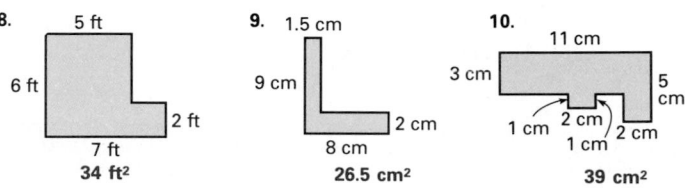

8. 5 ft, 6 ft, 2 ft, 7 ft — **34 ft²**
9. 1.5 cm, 9 cm, 2 cm, 8 cm — **26.5 cm²**
10. 11 cm, 3 cm, 5 cm, 1 cm, 2 cm, 2 cm, 1 cm — **39 cm²**

11. Rose's vegetable garden is square. It measures 24.5 ft on a side. Find its area. **600.25 ft²**

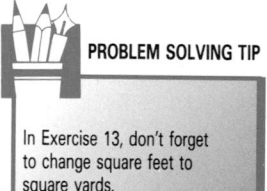

Use this sketch of Ann's living room for Exercises 12 and 13.

12. What is the area of the living room? **648 ft²**

13. If carpeting costs $8.25/yd², what will be the cost of carpeting the entire living room? **$594**

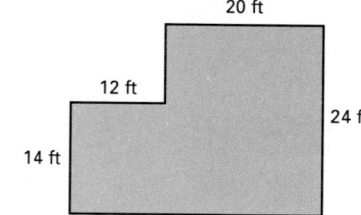

20 ft, 12 ft, 24 ft, 14 ft

14. A rectangular lot 50 ft by 100 ft is surrounded by a sidewalk 5 ft wide. What is the area of the sidewalk? **1,600 ft²**

15. The width of a rectangle is four-fifths its length. If its perimeter is 72 in., what is its area? **320 in.²**

392 CHAPTER 11 Area and Volume

► READ
► PLAN
► SOLVE
► ANSWER
► CHECK

Problem Solving Applications:

BASEBALL STRIKE ZONE

In major league baseball, a pitcher must throw the ball over home plate and between the batter's armpits and knees. The shaded region *ABCD* is called the **strike zone.** You can find the area of the strike zone by using the area formula for a rectangle.

$A = l \times w$
$A = 31 \times 12$
$A = 372$ The area of the strike zone is 372 in.²

Assume that the width of home plate is 12 in. Find the area of each strike zone.

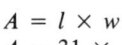
Home Plate

1.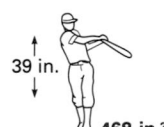
39 in.
468 in.²

2.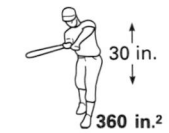
30 in.
360 in.²

3.
27 in.
324 in.²

4.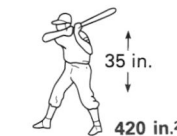
35 in.
420 in.²

The dimensions of the strike zone for each player are listed in the table. Find the area of each strike zone.

	Player	Dimensions	
		l	*w*
5.	Jose Canseco	27 in.	12 in.
6.	Carlton Fisk	32 in.	12 in.
7.	Pete Incaviglia	37 in.	12 in.
8.	Rickey Henderson	34 in.	12 in.
9.	Lance Parrish	28 in.	12 in.

5. 324 in.²
6. 384 in.²
7. 444 in.²
8. 408 in.²
9. 336 in.²

10. Two batting positions are shown.

By how much does the area of the strike zone decrease if this batter crouches? **72 in.²**

Upright 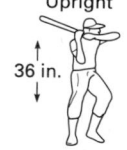 36 in.   Crouch 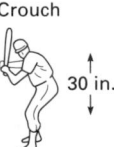 30 in.

Find the decrease in the area of the strike zone if each batter crouches as shown.

11. 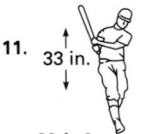 33 in.  28 in.
60 in.²

12. 38 in. 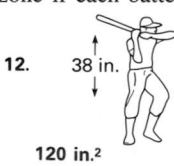 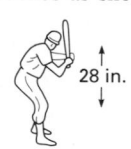 28 in.
120 in.²

11–2   Area of Rectangles   **393**

Extension Have students work in pairs using a yardstick as a "bat." Have partners determine each other's strike zones for both standing and crouching positions. Provide tape measures or rulers for students to measure distances. (Have them assume that "home plate" is 12 in. wide.)

Section Quiz
Find the area of each rectangle.

1.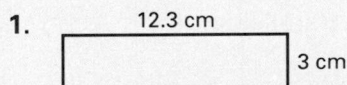
12.3 cm
3 cm
36.9 cm²

2.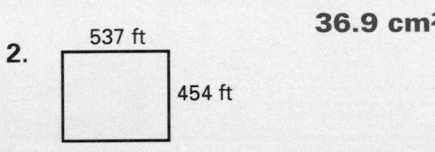
537 ft
454 ft
243,798 ft²

Find the area of each region.

3.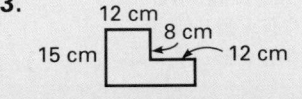
12 cm
8 cm
15 cm
12 cm
264 cm²

4.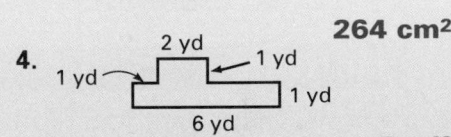
2 yd
1 yd
1 yd
1 yd
6 yd
8 yd²

Find the area of the shaded region of the square shown.

5.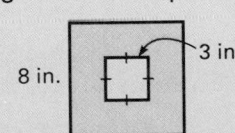
8 in.
3 in.
55 in.²

6. How many 6 in. by 9 in. tiles would be needed to cover a wall that measures 72 in. by 63 in.? **84**

7. How much more would it cost to cover a 12 ft by 10 ft lawn than a 14 ft by 8 ft lawn if a square foot of grass sod costs $3.95? **$31.60**

Get Ready grid paper, scissors, centimeter ruler

393

394

11-3 Area of Parallelograms and Triangles

EXPLORE   Count the square units to find the area of the rectangle and the parallelogram. Compare.
Now copy each figure onto grid paper. Cut out the parallelogram. How can you cut the parallelogram into two pieces that can be rearranged to form the rectangle? Try it!

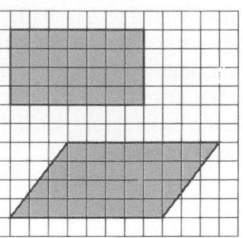

SKILLS DEVELOPMENT   In the Explore activity, you discovered how a parallelogram can be "rearranged" to form a rectangle. Because of this relationship, the area formula for a parallelogram is very much like the area formula for a rectangle.

$$A \;=\; b \;\times\; h$$

↑ Area   ↑ length of base   ↑ height

 Parallelogram

Example 1

Find the area of each parallelogram.

a.
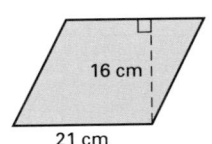
16 cm
21 cm

b.
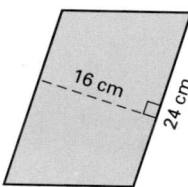
16 cm
24 cm

Solution

a. $A = b \times h$ ← Substitute 21 for *b*
$A = 21 \times 16$ and 16 for *h*.
$A = 336$

The area is 336 cm².

b. $A = b \times h$ ← Substitute 24 for *b*
$A = 24 \times 16$ and 16 for *h*.
$A = 384$

The area is 384 cm². ◄

The area of a triangle is one half the area of a parallelogram with the same base and height.

Area of parallelogram
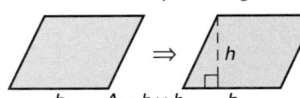
b $A = b \times h$ b

Area of triangle = $\frac{1}{2}$ area of parallelogram
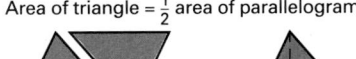
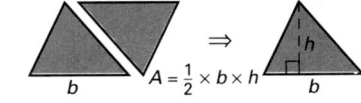
b $A = \frac{1}{2} \times b \times h$ b

394   CHAPTER 11   Area and Volume

Example 2

Find the area of each triangle.

a.

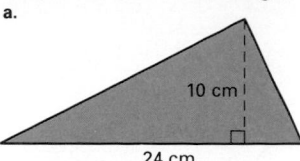

10 cm

24 cm

b.

15 cm

37 cm

Solution

a. $A = \frac{1}{2} \times b \times h$ ←Substitute 24 for b and 10 for h.

$A = \frac{1}{2} \times 24 \times 10$

$A = 120$

The area is 120 cm².

b. $A = \frac{1}{2} \times b \times h$ ←Substitute 37 for b and 15 for h.

$A = \frac{1}{2} \times 37 \times 15$

$A = 18.5 \times 15$

$A = 277.5$

The area is 277.5 cm². ◄

Example 3

The area of a triangle is 450 cm². The length of the base is 36 cm. Find its height.

Solution

$A = \frac{1}{2} \times b \times h$

$450 = \frac{1}{2} \times 36 \times h$ ←Substitute 450 for A and 36 for b.

$450 = 18 \times h$

$\frac{450}{18} = \frac{18h}{18}$

$25 = h$ The height of the triangle is 25 cm. ◄

TRY THESE

Find the area of each parallelogram.

1.

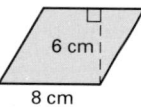

48 cm²

6 cm

8 cm

2.

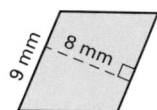

9 mm

8 mm

72 mm²

Find the area of each triangle.

3.

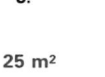

25 m²

5 m

10 m

4.

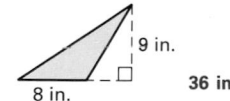

9 in.

8 in.

36 in.²

5. The area of a parallelogram is 72 in.² Its height is 9 in. Find its base.

8 in.

6. The base of a triangle is 40 cm. Its area is 1,200 cm². Find its height.

60 cm.

the triangle in part a and then draw a parallelogram with the same dimensions. Discuss how the area of this parallelogram is related to the area of the triangle. **The area of the parallelogram is twice that of the triangle.**

   Example 3: Have students draw the triangle described and label the base and height.

Additional Questions/Examples

1. How are the methods for finding the areas of triangles, parallelograms, and rectangles the same? How are the methods different? **Finding the area of all three figures is based on finding the product of the measures of the base and height. Areas of triangles are found by multiplying the product of the base and height by 1/2.**

2. Find the area of a parallelogram with a height of 5 in. and a base of 6 in. **30 in.²** Then draw a triangle with the same height and base and find its area.
15 in.²

Guided Practice/Try These

   For Exercises 5 and 6 have students draw and label the figures described.

5-MINUTE CLINIC

Exercise	Student's Error	Error Diagnosis
Find the area. 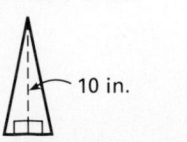 10 in. 4 in.	10 in. × 4 in. = 40 in.²	• Student fails to multiply the product of the base and height by 1/2. Correct Solution: 1/2 × 10 × 4 = 20 in.²

396

3 SUMMARIZE

Write About Math Have students draw diagrams of a triangle and a parallelogram in their math journals. Have them label each figure and supply dimensions for each. Students should then use their dimensions in written descriptions of the method for finding the area of each figure.

4 PRACTICE

Practice/Solve Problems For Exercises 5–16 you may wish to have students use mental math.

Extend/Solve Problems For Exercise 22 remind students that an equal number of tic marks on the sides of the figures mean that those sides are of equal measure.

Think Critically/Solve Problems For Exercise 30 students will need a centimeter ruler. For Exercises 31 and 32, students must do several calculations. You may wish to have them work in small groups to complete these exercises.

5 FOLLOW-UP

Extra Practice
1. Find the area of a parallelogram with base = 12.5 cm and height = 4.2 cm. **52.5 cm²**
2. How much greater in area is a triangle with a base of 12 in. and height of 8 in. than a triangle with a base of 9 in. and a height of 10 in.?
3 in.² greater
3. A can of paint costs $11.20 and covers 160 ft². Find the cost of giving the roof of a garden gazebo two coats of paint. The roof is composed of four triangular areas, each with a base of 20 ft and a height of 8 ft.
$44.80

EXERCISES

PRACTICE/SOLVE PROBLEMS

Find the area of each figure.

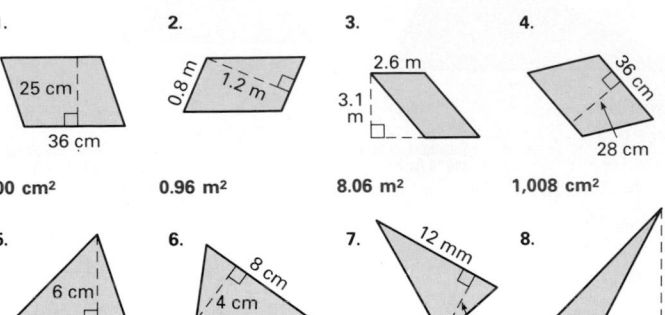

1.
25 cm
36 cm
900 cm²

2.
0.8 m
1.2 m
0.96 m²

3.
2.6 m
3.1 m
8.06 m²

4.
36 cm
28 cm
1,008 cm²

5.
6 cm
8 cm
24 cm²

6.
8 cm
4 cm
16 cm²

7.
12 mm
6 mm
36 mm²

8.
20 cm
9 cm
90 cm²

Find the area of a parallelogram with the given dimensions.

9. $b = 10$ in., $h = 6$ in. **60 in.²**   10. $b = 20$ ft, $h = 20$ ft **400 ft²**

11. $b = 25$ cm, $h = 20$ cm **500 cm²**   12. $b = 60$ mm, $h = 40$ mm **2,400 mm²**

Find the area of a triangle with the given dimensions.

13. $b = 14$ yd, $h = 8$ yd **56 yd²**   14. $b = 20$ in., $h = 10$ in. **100 in.²**

15. $b = 25$ ft, $h = 20$ ft **250 ft²**   16. $b = 28$ cm, $h = 12$ cm **168 cm²**

Solve.

17. A parallelogram has an area of 180 cm² and a height of 10 cm. Find its base. **18 cm**

18. Find the length of a rectangle that has an area of 180 cm² and a width of 10 cm. **18 cm**

19. Draw a diagram to illustrate the answers in Exercises 17 and 18.

19.
18 cm
10 cm

18 cm
10 cm

EXTEND/SOLVE PROBLEMS

Find the area of each shaded region.

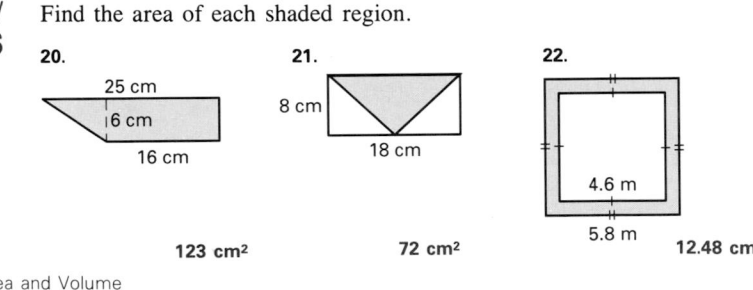

20.
25 cm
6 cm
16 cm
123 cm²

21.
8 cm
18 cm
72 cm²

22.
4.6 m
5.8 m
12.48 cm²

23. Find the amount of aluminum siding needed to cover the side of the house. **38.605 m²**

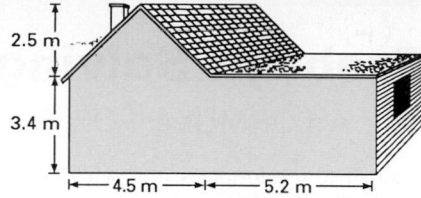
2.5 m
3.4 m
4.5 m
5.2 m

Most sailboats have two basic sails, the jib and the mainsail. Assume that the sails are triangles. Using the data in the diagram, find the area of each sail, to the nearest tenth.

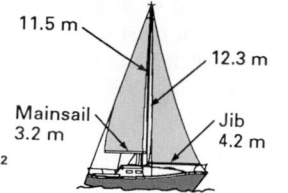
11.5 m
12.3 m
Mainsail
3.2 m
Jib
4.2 m

24. mainsail **18.4 m²**   **25.** jib **25.8 m²**

Find the total area of the sails on each boat, to the nearest tenth.

26.
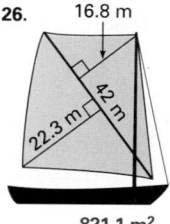
16.8 m
42 m
22.3 m
821.1 m²

27.
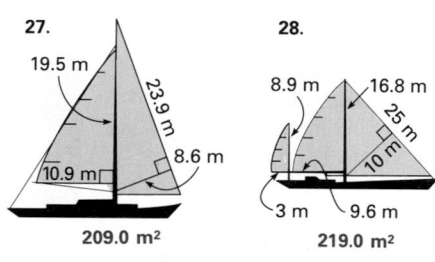
19.5 m
23.9 m
8.6 m
10.9 m
209.0 m²

28.
8.9 m
16.8 m
25 m
10 m
3 m
9.6 m
219.0 m²

Use the triangle at the right for Exercises 29 and 30.

29. What measurements are needed to find the area of this triangle? **height, length of the base**

30. Use a metric ruler. Make the measurements needed and find the area. **5 cm²**

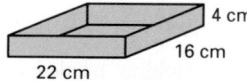
2 cm
5 cm

A box lid is constructed from the pattern shown.

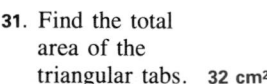
4 cm
16 cm
22 cm

31. Find the total area of the triangular tabs. **32 cm²**

32. Find the total amount of material used to make a box lid. **688 cm²**

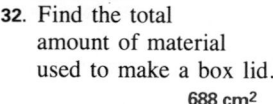
These tabs are used to glue the corners.

THINK CRITICALLY/ SOLVE PROBLEMS

Extension Ask these questions.
1. How does the area of a parallelogram or a triangle change if its height is doubled? **The area of each figure doubles.**
2. How does the area of a parallelogram or a triangle change if both the base and the height are doubled? **The area of each figure quadruples.**

Section Quiz
Find the area of each figure.
1.
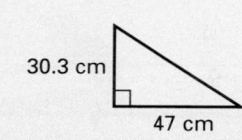
4.3 ft
8 ft
34.4 ft²

2.
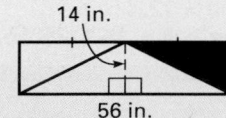
30.3 cm
47 cm
712.05 cm²

3. Find the area of the shaded region of the rectangle.
14 in.
56 in.
196 in.²

4. Which has the greatest area? How much greater is it than the next greatest area?
    rectangle: $l = 5$ m; $w = 7$ m
    parallelogram: $b = 4.5$ m; $h = 8$ m
    triangle: $b = 9$ m; $h = 4$ m
The parallelogram has the greatest area. It is 1 m² greater than the area of the rectangle.

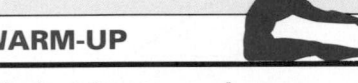
398

11-4 Problem Solving Skills:
USE A FORMULA FOR PERIMETER OR AREA

► READ
► PLAN
► SOLVE
► ANSWER
► CHECK

In order to use formulas in solving problems, you must decide which formula is the one you need. For example, if you want to find the amount of wallpaper that will be needed to cover the walls of a room, you must use a formula for area. If you want to know the amount of fencing that is needed to enclose a field, you will need to use a formula for perimeter.

PROBLEM

Molding is to be installed along the base of each wall of a rectangular room that measures 36 ft by 24 ft. On one of the longer walls, there is a 4-ft-wide door. No molding is needed along the floor at the doorway. Find the length of molding needed for the room.

SOLUTION

The words "along the base of each wall" and "length of molding" suggest that the goal is to find the distance around the room.

Use the formula for perimeter.

$$P = 2l + 2w = 2(36) + 2(24)$$
$$= 72 + 48$$
$$= 120$$

The perimeter of the room is 120 ft.

No molding will be used along the floor at the doorway. Therefore, you subtract the width of the door from the perimeter.

$$120 - 4 = 116$$

The amount of molding needed for the room is 116 ft.

PROBLEMS

Write *P* or *A* to tell if you would use a *perimeter* or an *area* formula.

1. amount of fencing *P*

2. amount of floor covering *A*

---

3. amount of material in a bedspread **A**

4. amount of fringe needed to decorate the edge of a scarf **P**

Choose a formula that can be used to solve the problem. Then solve.

$$P = 2l + 2w \qquad A = l \times w \qquad A = s^2$$
$$A = \frac{1}{2} \times b \times h \qquad P = 4 \times s$$

5. A garden is 7.9 m wide and 14 m long. How much fencing will be needed to enclose it? **$P = 2l + 2w$; 43.8 m**

6. Carpeting costs \$25/yd². How much will it cost to carpet a room that is 9 ft on each side? **$A = s^2$; \$225**

7. A square quilt measures 425 cm on each side. How much binding is needed to bind the edges of the quilt? **$P = 4 \times s$; 1,700 cm**

8. Find the area of the canvas covering the rectangular sides of the tent shown. **$A = (l \times w)$; 16.56 m²**

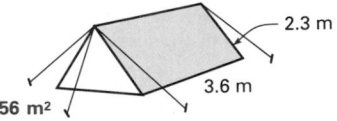
2.3 m
3.6 m

9. A new rectangular park measures 88 yd by 72 yd. Sod costs \$1.10/yd². Find the total cost of covering the entire park with sod. **$A = l \times w$; \$6,969.60**

10. A new garden plot has the dimensions shown. If plastic covering costs \$0.89/m², find the cost of covering the entire plot.
$A = \frac{1}{2} \times b \times h$; \$74.76

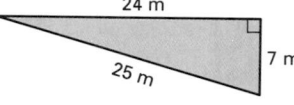
24 m
25 m
7 m

Use the diagram showing the layout of a farm for Exercises 11–15.

11. What is the area of land used to grow hay? **3,120 ft²**

12. It costs 63¢/ft² to fertilize the vegetable garden. What will be the total cost? **\$2,557.80**

13. A fence is to enclose the property around the house. How many feet of fencing are needed? **140 ft**

14. Each apple tree requires 25 ft² of space. How many apple trees can there be in the orchard? **114 trees**

15. How many square feet of the farm are used neither for the house nor for planting? **8,735 ft²**

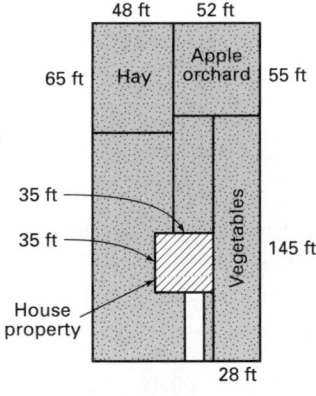
48 ft 52 ft
65 ft Hay Apple orchard 55 ft
35 ft
35 ft Vegetables 145 ft
House property
28 ft

3 SUMMARIZE

Write About Math Have students work in pairs to make up a problem that is best solved by using either the area or perimeter formula. Then have pairs exchange problems and solve each other's problems. Finally, have students write a statement telling how they decided which formula to use to solve the problems.

4 PRACTICE

While students are not told to write formulas to solve Problems 11–15, you may suggest that they use this strategy to solve each one.

5 FOLLOW-UP

Extra Practice
Pose this problem based on the farm layout supplied for Exercises 11–15. *A worker can put up 30 ft of fencing in 1 hour. To the nearest hour, how long would it take the worker to fence in the apple orchard?* **7 h**

Get Ready centimeter cubes, grid paper, calculators

11-5 Squares and Square Roots

EXPLORE

A square field has an area of 100 m². What do you think is the measure of its sides? **10 m**

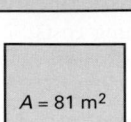

$A = 100$ m²

Suppose another square field has an area of 81 m². What do you think is the measure of its sides? **9 m**

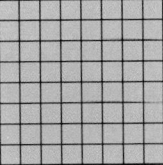

$A = 81$ m²

Another square field has an area of 90 m². What do you think is the measure of its sides?
Answers may vary. Sample answer: greater than 9 m and less than 10 m

SKILLS DEVELOPMENT

The area of the square at the right is 64 square units. The length of each side of the square is 8 units. From the figure you can see why the expression 8^2 is read "8 squared" or "the square of 8."

   The **square** of 8 is 64.   The **square root** of 64 is 8.
      $8^2 = 64$ $\sqrt{64} = 8$

A number a is a square root of another number b if $a^2 = b$. The number 64 has two square roots.

   $64 = (8)^2$ 8 is the positive square root of 64.
            You write $\sqrt{64} = 8$.

   $64 = (-8)^2$ -8 is the negative square root of 64.
            You write $-\sqrt{64} = -8$.

The number 64 is called a **perfect square** because its square roots are integers.

COMPUTER TIP

The square root function, SQR, will give you the positive square root of any non-negative integer, N. Type:
? SQR(N) then press ENTER.

Numbers such as $\frac{4}{9}$ and 0.49 also have two square roots, one positive and one negative.

$$\sqrt{\frac{4}{9}} = \frac{2}{3} \qquad\qquad -\sqrt{\frac{4}{9}} = -\frac{2}{3}$$

$$\sqrt{0.49} = 0.7 \qquad\qquad -\sqrt{0.49} = -0.7$$

The number 0 has only one square root: $\sqrt{0} = 0$.

TEACHING TIP

Emphasize that, when students are computing squares or square roots of decimal numbers, they must have the correct number of decimal places in their answers. Use the following as examples:

$(0.6)^2 = 0.36$ and $(0.06)^2 = 0.0036$

$\sqrt{0.64} = 0.8$ and $\sqrt{0.0064} = 0.08$

Example 1

Find each square.

a. 5^2 **b.** $(-3)^2$ **c.** $\left(-\frac{1}{4}\right)^2$ **d.** $(.09)^2$

Solution

a. $5^2 = 5 \times 5 = 25$

b. $(-3)^2 = (-3) \times (-3) = 9$

c. $\left(-\frac{1}{4}\right)^2 = \left(-\frac{1}{4}\right) \times \left(-\frac{1}{4}\right) = \frac{1}{16}$

d. $(0.09)^2 = 0.09 \times 0.09 = 0.0081$ ◀

Example 2

Find each square root.

a. $\sqrt{\dfrac{9}{16}}$ **b.** $-\sqrt{0.36}$

Solution

a. $\frac{3}{4} \times \frac{3}{4} = \frac{9}{16}$, so $\sqrt{\dfrac{9}{16}} = \frac{3}{4}$

b. $-0.6 \times (-0.6) = 0.36$,

so $-\sqrt{0.36} = -0.6$. ◀

The square root of any rational number that is not a perfect square is a nonterminating, nonrepeating decimal. In Chapter 5 you learned that this type of decimal is called an *irrational number.*

So, numbers like $\sqrt{2}$ and $\sqrt{3}$ cannot be expressed as common fractions or as terminating or repeating decimals. For numbers such as these you can use a table of square roots or a calculator to find an *approximate* value. For those numbers that are not perfect squares, the values in the table are approximations of square roots, rounded to the nearest thousandth.

TABLE OF SQUARE ROOTS

1	1.000	26	5.099	51	7.141	76	8.718
2	1.414	27	5.196	52	7.211	77	8.775
3	1.732	28	5.292	53	7.280	78	8.832
4	2.000	29	5.385	54	7.348	79	8.888
5	2.236	30	5.477	55	7.416	80	8.944
6	2.449	31	5.568	56	7.483	81	9.000
7	2.646	32	5.657	57	7.550	82	9.055
8	2.828	33	5.745	58	7.616	83	9.110
9	3.000	34	5.831	59	7.681	84	9.165
10	3.162	35	5.916	60	6.746	85	9.220
11	3.317	36	6.000	61	7.810	86	9.274
12	3.464	37	6.083	62	7.874	87	9.327
13	3.606	38	6.164	63	7.937	88	9.381
14	3.742	39	6.245	64	8.000	89	9.434
15	3.873	40	6.325	65	8.062	90	9.487
16	4.000	41	6.403	66	8.124	91	9.539
17	4.123	42	6.481	67	8.185	92	9.592
18	4.243	43	6.557	68	8.246	93	9.644
19	4.359	44	6.633	69	8.307	94	9.695
20	4.472	45	6.708	70	8.367	95	9.747
21	4.583	46	6.782	71	8.426	96	9.798
22	4.690	47	6.856	72	8.485	97	9.849
23	4.796	48	6.928	73	8.544	98	9.899
24	4.899	49	7.000	74	8.602	99	9.950
25	5.000	50	7.071	75	8.660	100	10.000

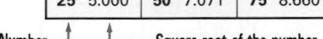
Number ⌐   └── Square root of the number

CHECK UNDERSTANDING

Tell whether each square root is rational or irrational.

1. $\sqrt{169}$

2. $\sqrt{27}$

1. rational, since 169 is a perfect square.
2. irrational, since 27 is not a perfect square.

Example 2: Be sure students understand that in part a, the numerator (3) and denominator (4) of each factor must be the same.

Point out that in part b, it is the negative square root that must be found.

Example 3: Provide additional practice that requires the use of the table of square roots.

Additional Questions/Examples

1. Which integers have square roots that are integers? **Those integers that are perfect squares.**

2. Why is $\sqrt{5}$, given in the table as 2.236, an approximation? **$\sqrt{5}$ is an irrational number. The value given is rounded to three decimal places.**

Guided Practice/Try These
Have students work in groups. Point out that in Exercises 3, 9, and 12, they will be working with negative square roots.

3 SUMMARIZE

Key Questions

1. What are the two square roots of 9? **3 and –3**

2. Why is a number such as 9 a perfect square? **Its square roots are integers.**

3. What is the square root of 0? **0**

4 PRACTICE

Practice/Solve Problems In each of Exercises 5–8 students should take care to place the decimal point correctly in the answer.

Extend/Solve Problems Point out that for Exercise 37 students must find the length of one side of the square stamp.

Think Critically/Solve Problems For Exercises 52 and 53, students must first substitute the information given in each problem for the distance, *d*, before they can solve for *t*.

5 FOLLOW-UP

Extra Practice Find each square.

1. $(0.12)^2$ **0.0144**
2. $(0.004)^2$ **0.000016**
3. $(1.6)^2$ **2.56**
4. $(-0.09)^2$ **0.0081**
5. $(400)^2$ **160,000**
6. $\sqrt{(2.45)^2}$ **2.45**
7. $-\sqrt{34^2}$ **−34**

Extension Find each square root using a calculator. Tell whether each root is a *rational* or an *irrational* number.

8. $\sqrt{99}$ **9.9498743 . . . ; irrational**
9. $\sqrt{234}$ **15.297058 . . . ; irrational**
10. $\sqrt{529}$ **23; rational**

MATH: WHO, WHERE, WHEN

One way of computing square roots is a method developed by the English mathematician Sir Isaac Newton (1642–1727). Here are the steps in this process, known as Newton's Method.

1. Choose a number: 13
2. Estimate $\sqrt{13}$.
   $\sqrt{9} < \sqrt{13} < \sqrt{16}$
   ↓ ↓ ↓
   3 < ? < 4
3. Try $\sqrt{13} \rightarrow 3.8$.
4. Divide: $13 \div 3.8 \approx 3.42$
5. Find the average of your estimate in Step 3 and the quotient from Step 4.
   $\frac{3.8 + 3.42}{2} = 3.61 \approx 3.6$
6. Divide the number 13 by the result in Step 5.
   $13 \div 3.6 = 3.61 \approx 3.6$
7. Check by multiplying.
   $(3.6)^2 = 12.96$
   $(3.61)^2 \approx 13.03$
   Therefore, $\sqrt{13} \approx 3.6$, rounded to the nearest tenth.

Use Newton's method to approximate $\sqrt{27}$ to the nearest tenth.

Example 3

Find $\sqrt{23}$.

Solution 1

Use the table on page 401. Find 23 in the number column. Read to the right to locate the square root, rounded to the nearest thousandth.

$\sqrt{23} \approx 4.796$ ◄ Because the value is an approximation, this symbol, meaning "approximately equal to," is used.

Solution 2

Use a calculator. Enter this key sequence.

23 $\sqrt{}$

Display: 4.7958315
Round the number in the display to the nearest thousandth. $\sqrt{23} \approx 4.796$ ◄

TRY THESE

Find each square.

1. 14^2 **196**   2. 16^2 **256**   3. $(-4)^2$ **16**   4. $\left(\frac{5}{8}\right)^2$ **$\frac{25}{64}$**   5. $(0.6)^2$ **0.36**

Find each square root.

6. $\sqrt{\frac{81}{169}}$ **$\frac{9}{13}$**   7. $\sqrt{\frac{144}{289}}$ **$\frac{12}{17}$**   8. $\sqrt{\frac{16}{121}}$ **$\frac{4}{11}$**   9. $\sqrt{\frac{289}{361}}$ **$\frac{17}{19}$**

10. $-\sqrt{0.25}$ **−0.5**   11. $\sqrt{0.09}$ **0.3**   12. $\sqrt{0.16}$ **0.4**   13. $-\sqrt{1.44}$ **−1.2**

Use the table of square roots on page 401 to find each square root.

14. $\sqrt{87}$ **9.327**   15. $\sqrt{19}$ **4.359**   16. $\sqrt{52}$ **7.211**   17. $\sqrt{7}$ **2.646**

Use a calculator to find each square root. Round to the nearest thousandth.

18. $\sqrt{22}$ **4.690**   19. $\sqrt{53}$ **7.280**   20. $\sqrt{58}$ **7.616**   21. $\sqrt{85}$ **9.220**

EXERCISES

PRACTICE/ SOLVE PROBLEMS

Find each square.

1. 18^2 **324**   2. 17^2 **289**   3. 21^2 **441**   4. 15^2 **225**

5. $(1.8)^2$ **3.24**   6. $(1.7)^2$ **2.89**   7. $(2.1)^2$ **4.41**   8. $(1.5)^2$ **2.25**

9. $\left(\frac{1}{5}\right)^2$ **$\frac{1}{25}$**   10. $\left(\frac{3}{7}\right)^2$ **$\frac{9}{49}$**   11. $\left(-\frac{8}{9}\right)^2$ **$\frac{64}{81}$**   12. $\left(-\frac{18}{23}\right)^2$ **$\frac{324}{529}$**

Find each square root.

13. $\sqrt{256}$ **16**   **14.** $\sqrt{144}$ **12**   **15.** $-\sqrt{100}$ **−10**   **16.** $-\sqrt{0.36}$ **−0.6**

17. $-\sqrt{0.49}$ **−0.7**   **18.** $\sqrt{\frac{1}{49}}$ **$\frac{1}{7}$**   **19.** $\sqrt{\frac{4}{81}}$ **$\frac{2}{9}$**   **20.** $-\sqrt{\frac{16}{121}}$ **$-\frac{4}{11}$**

Use the table of square roots on page 401 to find the square root. Round to the nearest tenth.

21. $\sqrt{12}$ **3.5**   **22.** $\sqrt{90}$ **9.5**   **23.** $\sqrt{84}$ **9.2**   **24.** $\sqrt{72}$ **8.5**

25. $\sqrt{42}$ **6.5**   **26.** $\sqrt{94}$ **9.7**   **27.** $\sqrt{6}$ **2.4**   **28.** $\sqrt{38}$ **6.2**

Use a calculator to find each square root. Round to the nearest thousandth.

29. $\sqrt{153}$ **12.369**   **30.** $\sqrt{527}$ **22.956**   **31.** $\sqrt{976}$ **31.241**   **32.** $\sqrt{369}$ **19.209**

33. $\sqrt{101}$ **10.050**   **34.** $\sqrt{482}$ **21.954**   **35.** $\sqrt{815}$ **28.548**   **36.** $\sqrt{239}$ **15.460**

37. The area of a square stamp is 23 cm². Find the measure of a side. Round your answer to the nearest tenth. **4.8 cm**

EXTEND/ SOLVE PROBLEMS

Find each square.

38. 29^2 **841**   **39.** 34^2 **1,156**   **40.** $(-37)^2$ **1,369**   **41.** $(0.2)^2$ **.04**

42. $(0.16)^2$ **0.0256**   **43.** $(0.8)^2$ **0.64**   **44.** $(0.08)^2$ **0.0064**   **45.** $(0.008)^2$ **0.000064**

46. $(0.0008)^2$ **0.00000064**   **47.** $(\sqrt{8})^2$ **8**   **48.** $(\sqrt{19})^2$ **19**   **49.** $(-\sqrt{24})^2$ **24**

50. The area of a square plot is 72 ft². Find the perimeter of the plot to the nearest tenth. **33.9 ft**

51. A square lot has an area of 500 ft². If fencing costs $10.25 per foot, how much would it cost to enclose the lot? **Answers may vary. $916.76 or $916.79**

The amount of time (t) in seconds it takes an object to fall a distance (d) in meters is expressed in this formula. $t = \sqrt{\frac{d}{4.9}}$

THINK CRITICALLY/ SOLVE PROBLEMS

Use this formula for Exercises 52 and 53. Round each answer to the nearest tenth.

52. An object fell 40 m. How long did it take the object to hit the ground? **2.9 s**

53. A rock falls over a cliff 75 m in height. How long will it take the rock to hit the water at the bottom of the cliff? **3.9 s**

CHALLENGE

Arrange the numbers in order from least to greatest.

$1.73, 1.\overline{7}, 1.\overline{73}, \sqrt{3}, 1.7\overline{3}$

$1.73, \sqrt{3}, 1.\overline{73}, 1.7\overline{3}, 1.\overline{7}$

11-6   The Pythagorean Theorem

EXPLORE

Trace this right triangle. Draw a square on each side of the triangle, so that each side forms one side of a square. Measure a side of each square using a centimeter ruler. Find the area of each square. Look for a relationship between the numbers representing the areas of the squares.

$3^2 + 4^2 = 5^2$ or $9 + 16 = 25$

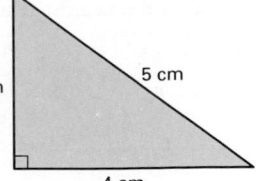

SKILLS DEVELOPMENT

In a right triangle, the side opposite the right angle is called the **hypotenuse.** The other sides are called the **legs.** There is a special relationship between the legs of a right triangle and the hypotenuse.

In any right triangle, the square of the length of the hypotenuse is equal to the sum of the squares of the lengths of the legs.

In right triangle *ABC*, where *c* is the length of the hypotenuse, and *a* and *b* are the lengths of the legs, you can state the relationship as follows.

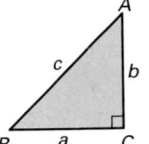

$$c^2 = a^2 + b^2$$

This property is called the **Pythagorean Theorem,** after the Greek mathematician Pythagoras. The Pythagorean Theorem can be used to find the measure of one side of a right triangle if the measures of the other two sides are known.

COMPUTER

You can use this program to check your work when finding the hypotenuse of a right triangle.
```
10 INPUT "ENTER LENGTHS
   OF LEGS: ";A,B:
   PRINT
20 LET C = SQR (A ^ 2 +
   B ^ 2)
30 PRINT "HYPOTENUSE
   IS ";C
```

Example 1

Find the length of the hypotenuse of triangle *ABC*.

Solution
Use the Pythagorean Theorem.

$c^2 = a^2 + b^2$
$c^2 = 5^2 + 12^2$
$c^2 = 25 + 144$
$c^2 = 169$
$c = \sqrt{169}$
$c = 13$

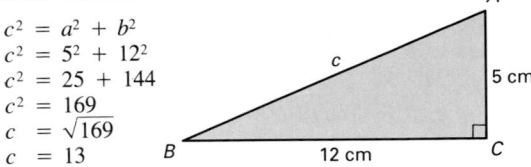

The length of the hypotenuse is 13 cm. ◄

404   CHAPTER 11   Area and Volume

Example 2

In triangle ABC, find the unknown length b to the nearest tenth.

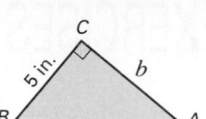

Solution

In triangle ABC, you know that $c^2 = a^2 + b^2$. If you subtract a^2 from each side of the equation, the relationship becomes $b^2 = c^2 - a^2$.

$$b^2 = c^2 - a^2$$
$$b^2 = 8^2 - 5^2$$
$$b^2 = 64 - 25 \quad \text{Use the table of square roots (page 401) or}$$
$$b^2 = 39 \qquad \text{a calculator to find an approximate value}$$
$$b = \sqrt{39} \qquad \text{of } \sqrt{39}.$$
$$b \approx 6.245$$
$$b \approx 6.2 \text{ (nearest tenth)}$$

The length b is about 6.2 in. ◄

Example 3

A ladder 6.1 m in length is placed against a wall. If the foot of the ladder is a distance of 2.5 m from the wall, how far up the wall does the ladder reach? Express your answer as an approximation to the nearest tenth.

Solution

Draw a diagram of the problem. Since the wall meets the ground at a right angle, the ladder serves as the hypotenuse of a right triangle. Use the Pythagorean Theorem.

$$c^2 = a^2 + b^2$$
$$b^2 = c^2 - a^2$$
$$b^2 = (6.1)^2 - (2.5)^2$$
$$b^2 = 37.31 - 6.25$$
$$b^2 = 30.96$$
$$b = \sqrt{30.96} \approx 5.6$$

The ladder reaches about 5.6 m up the wall. ◄

TRY THESE

Use the Pythagorean Theorem to find the unknown length, to the nearest tenth.

1.
5 cm ... c ... C 3 cm B
5.8 cm

2.
A ... c ... B, 9 cm, 5 cm, C
10.3 cm

3.
A, 4 in., 8 in., C a B
6.9 in.

4.
B, a, 40 cm, C 25 cm A
31.2 cm

5. A 15-ft-long wooden brace is built against a wall at a height of 12 ft. How far from the wall is the foot of the brace? **9 ft**

COMPUTER TIP

You can use this program to find a missing leg of a right triangle. Compare this program with the one used to find the hypotenuse on page 404.

```
10 INPUT "ENTER LEG, THEN
   HYPOTENUSE ";A,C
20 LET B = SQR (C ^ 2 -
   A ^ 2)
30 PRINT "OTHER LEG
   IS ";B
```

MATH IN THE WORKPLACE

Here is how carpenters use math to make right angles.

Mark off 3 ft (or 6 ft)
BOARD
Right angle
BOARD
Mark off 4 ft (or 8 ft)
Tape measure
Adjust boards to get a reading of 5 ft (or 10 ft) on the tape measure.

Example 2: Point out that students must subtract a^2 from both sides of the equation in order to isolate and find the value of b^2:
$$c^2 = a^2 + b^2$$
$$c^2 - a^2 = a^2 + b^2 - a^2$$
$$c^2 - a^2 = b^2$$

Example 3: Point out that, since the other measures are given to the nearest tenth, the approximation of the length of side b is also rounded to the nearest tenth.

Additional Questions/Examples
1. Would the Pythagorean Theorem apply to a right triangle if its dimensions were given in fraction form? Why or why not? **Yes. Whether the measurements are expressed as fractions or as their decimal equivalents, the relationship among the sides of the right triangle is the same.**
2. Use the Pythagorean Theorem to find the length of the hypotenuse of a right triangle with legs measuring 6 cm and 9 cm. **about 10.8 cm**

Guided Practice/Try These For Exercise 5 have students draw and label a diagram showing the brace leaning against the wall.

5-MINUTE CLINIC

Exercise	Student's Error	Error Diagnosis
Find the length of the hypotenuse. 12 ... 16	$c^2 = a^2 + b^2$ $= 12 + 16$ $= 28$ $c = 28$ $c \approx 5.3$	• Student forgets to square the measures of the sides. Correct solution: $c^2 = 12^2 + 16^2$ $c^2 = 144 + 256 = 400$ $c = \sqrt{400} = 20$

406

3 SUMMARIZE

Key Questions

1. What is the Pythagorean Theorem used for?
2. How do you find the length of the hypotenuse of a right triangle when given the length of the two legs?
3. How do you find the length of one leg of a right triangle when given the lengths of the other leg and the hypotenuse?

4 PRACTICE

Practice/Solve Problems Allow students to use their calculators or the Table of Square Roots (page 401) to find the square roots when necessary. For Exercise 14 have students draw a diagram.

Extend/Solve Problems Have students draw and label a right triangle for each of Exercises 15–18. Allow students to use calculators or the Table of Square Roots (page 401) to find square roots when necessary

Think Critically/Solve Problems For Exercises 28–30 have students draw and label a diagram of the room described.

5 FOLLOW-UP

Extra Practice Find the unknown length, to the nearest tenth, in each right triangle.

1.

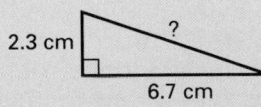

7.1 cm

2.

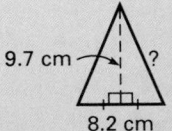

10.5 cm

PRACTICE/ SOLVE PROBLEMS

Use the Pythagorean Theorem to find the unknown length to the nearest tenth.

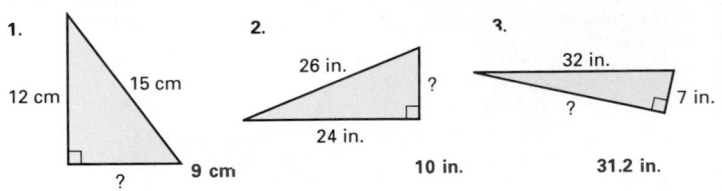

1. 12 cm, 15 cm, ?

2. 26 in., 24 in., 9 cm, ?

3. 32 in., ?, 7 in., 31.2 in., 10 in.

Find the length of the hypotenuse in each right triangle. Round to the nearest tenth.

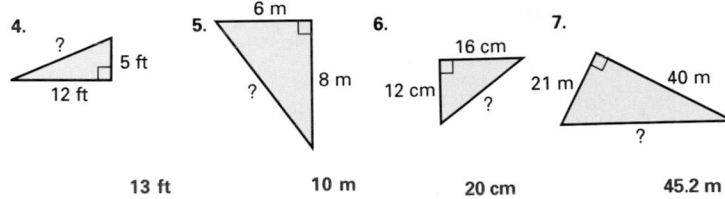

4. ?, 12 ft, 5 ft, 13 ft

5. 6 m, ?, 8 m, 10 m

6. 16 cm, 12 cm, ?, 20 cm

7. 21 m, 40 m, ?, 45.2 m

Find the unknown length in each triangle. Round to the nearest tenth.

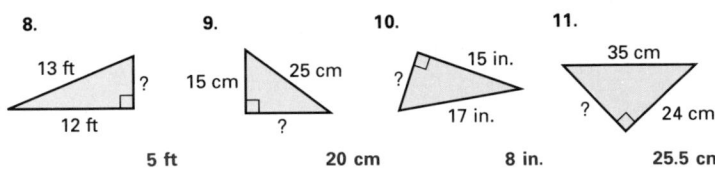

8. 13 ft, ?, 12 ft, 5 ft

9. 15 cm, 25 cm, ?, 20 cm

10. 15 in., ?, 17 in., 8 in.

11. 35 cm, ?, 24 cm, 25.5 cm

Use the Pythagorean Theorem to find each unknown measurement. Round to the nearest tenth.

12. length of the ladder

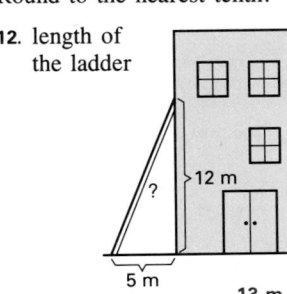

?, 12 m, 5 m, 13 m

13. height of the kite above the ground

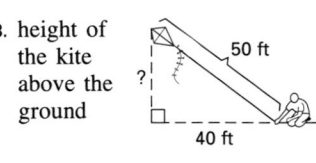

50 ft, ?, 40 ft, 30 ft

Solve.

14. A ladder 7 m long is placed against a wall so that the foot of the ladder is 2.5 m from the wall. Find how high up the wall the ladder reaches. Round to the nearest tenth. **6.5 m**

Calculate the value of the variable. Estimate the square root to the nearest tenth.

15. $c^2 = (1.8)^2 + (3.2)^2$ **3.7**   **16.** $b^2 = (8.6)^2 - (5.3)^2$ **6.8**

17. $a^2 = (7.3)^2 - (4.5)^2$ **5.7**   **18.** $(5.8)^2 + (3.0)^2 = c^2$ **6.5**

Find the unknown length for each. Round to the nearest hundredth.

19. 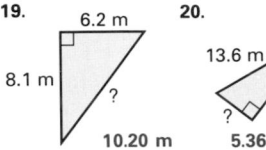 **20.** 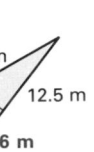 **21.** 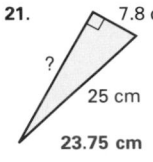 **22.**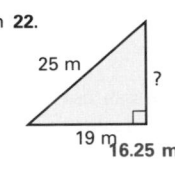

6.2 m   13.6 m   7.8 cm   25 m
8.1 m ?   12.5 m   ?   ?
10.20 m   5.36 m   25 cm   19 m 16.25 m
                   23.75 cm

Find the length of each diagonal to the nearest tenth.

23. 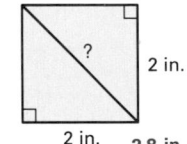 **24.** 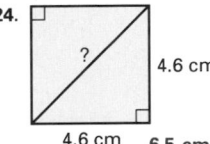 **25.**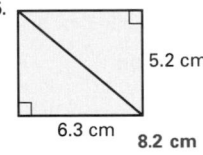

? 2 in.   ? 4.6 cm   5.2 cm
2 in. 2.8 in.   4.6 cm 6.5 cm   6.3 cm 8.2 cm

26. The figure is a diagram of a roof. Find the length of the span (*SP*) of the roof.
7.7 m

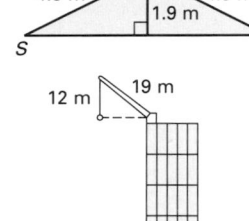

4.3 m   4.3 m
1.9 m
S   P

27. A crane with a 19-m rotating lever drops a load of cement at the end of a 12-m cable so that the load is level with the base of the rotating arm. How far is the load from the edge of the building? **14.7 m**

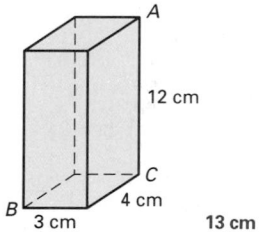

12 m   19 m

In a square room, the distance between opposite corners measures 12.8 m. For Exercises 28–30, find these measurements to the nearest tenth.

28. length of one side **9.1 m**   **29.** perimeter **36.2 m**   **30.** area **81.9 m²**

31. Find the length of $\overline{AB}$, the diagonal of this box.

A
12 cm
C
B   4 cm
3 cm   13 cm

Find the value of the variable in each right triangle with the given dimensions. Round each square root to the nearest tenth.

3. $a = 7.4$ cm $b = 3.6$ cm $c = \blacksquare$
8.2 cm

4. $c = 4.5$ cm $a = 2.4$ cm $b = \blacksquare$
3.8 cm

5. The top of a 15-ft beam leans against a wall at a point that is 6 ft above the floor. How far, to the nearest tenth of a foot, is the base of the beam from the wall?   **13.7 ft**

Extension Provide students with centimeter grid paper. Have them draw a right triangle with legs of 5 cm and 7 cm. Suggest that students make use of the Pythagorean Theorem to find the hypotenuse. **8.6 cm** Then have them draw a right triangle with legs of 10 cm and 14 cm. Have them find the hypotenuse of this triangle. **17.2 cm** Ask students to compare the lengths of the sides of the two triangles. **All dimensions of the larger triangle are twice the length of the dimensions of the smaller triangle.**

Section Quiz
Find the unknown length in each triangle. Round to the nearest hundredth.

1.
3.1 in.
?   8.6 in.

8.02 in.

2.
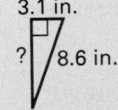
15 in.   ?
18 in.

23.43 in.

3. What is the length, to the nearest tenth, of the hypotenuse of a right triangle with legs that measure 5.6 cm and 7.9 cm?
9.7 cm

Get Ready one-inch cubes

THINK CRITICALLY/ SOLVE PROBLEMS

11-7 Volume of Prisms

EXPLORE

How many one-inch cubes would be needed to build an L-shaped structure that is 3 inches tall and has a floor plan as shown? **48 cubes**

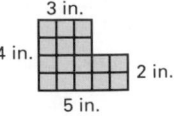

SKILLS DEVELOPMENT

You can use the formula $V = l \times w \times h$ to find the volume of any rectangular prism.

Example 1

Find the volume of a rectangular prism with length 3 ft, width 2 ft, and height 2 ft.

Solution

$V = l \times w \times h$
$V = 3 \times 2 \times 2$ ← Substitute 3 for *l*,
$V = 12$ 2 for *w*, and 2 for *h*.

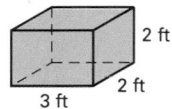

The volume of the rectangular prism is 12 ft³. ◄

Example 2

Find the volume of a cubic carton with edges measuring 6 in.

Solution

The cube is a rectangular prism in which each dimension is the same. $V = l \times w \times h$
$V = 6 \times 6 \times 6$ ← Substitute 6 for
$V = 216$ *l, w,* and *h.*

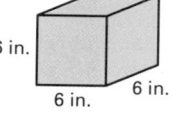

The volume of the cube is 216 in.³ ◄

The area of any rectangular prism is found by multiplying the area of the **base,** $l \times w$, by the height of the prism, *h*.

$$V = B \times h$$
$\uparrow$ $\uparrow$ $\uparrow$
Volume   area   height
of base   of prism

You can generalize this formula to find the volume of any prism.

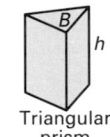 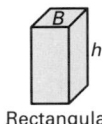 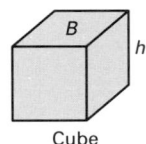 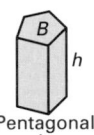 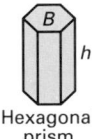

Triangular prism   Rectangular prism   Cube   Pentagonal prism   Hexagonal prism

Example 3

Find the volume of a triangular prism with the dimensions shown.

Solution

First find the area of the base. The base is a triangle with a base of 12 cm and a height of 6 cm.

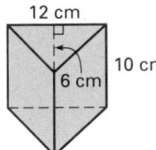

$$B = \frac{1}{2} \times b \times h$$

$$B = \frac{1}{2} \times 12 \times 6 \quad \leftarrow \text{Use the dimensions of the triangle. Substitute 12 for } b \text{ and 6 for } h.$$

$$B = 36$$

The area of the base is 36 cm².

To find the volume of the prism, use the formula $V = B \times h$, where h is the height of the prism.

$$V = B \times h$$

$$V = 36 \times 10 \quad \leftarrow \text{Use the dimensions of the prism. Substitute 36 for } B \text{ and 10 for } h.$$

$$V = 360$$

The volume of the triangular prism is 360 cm³. ◄

Example 4

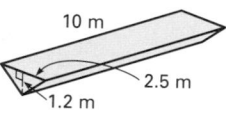

A trench for a pipe is dug with dimensions as shown in the diagram. Find the volume of earth removed in digging the trench.

Solution

The trench is in the shape of a triangular prism. Use the volume formula for prisms. First, calculate the area of one triangular base (B).

$$B = \frac{1}{2} \times b \times h$$

$$B = \frac{1}{2} \times 2.5 \times 1.2 \quad \leftarrow \text{Use the dimensions of the triangular end of the prism. Substitute 2.5 for } b \text{ and 1.2 for } h.$$

$$B = 1.5$$

The area of the end (base) of the trench is 1.5 m².
Now find the volume of the prism.

$$V = B \times h$$

$$V = 1.5 \times 10 \quad \leftarrow \text{Use the dimensions of the entire prism. Substitute 1.5 for } B \text{ and 10 for } h.$$

$$V = 15$$

The volume of earth removed is 15 m³. ◄

TRY THESE

1. Find the volume of a rectangular prism that measures 12 ft by 6 ft by 4 ft. **288 ft³**

2. Find the volume of a cube with edges measuring 7 in. **343 in.³**

3. Find the volume of a triangular prism with dimensions shown. **80 cm³**

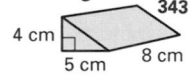

4. A triangular trough has the dimensions shown. Find the volume of the trough. **8.4 m³**

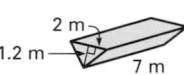

COMPUTER

You may want to use this program to check your work when finding the volume of a rectangular or triangular prism.

```
10 INPUT "ENTER BASE
   AND HEIGHT OF BASE
   OF PRISM: ";B,H: PRINT
20 INPUT "ENTER HEIGHT
   OF PRISM: ";HP: PRINT
30 INPUT "ENTER THE TYPE
   PRISM (TRIANGULAR,
   RECTANGULAR OR
   CUBE) T,R OR C: ";P$:
   PRINT
40 IF P$ = "T" THEN AB =
   1 / 2 * B * H
50 IF P$ = "R" OR P$ =
   "C" THEN AB = B * H
60 V = AB * HP
70 PRINT "AREA OF BASE
   = ";AB:
80 PRINT "VOLUME = ";V
```

3. A rectangular box that is 2 ft high, 3 ft wide, and 6 ft long is half filled. How can you determine the number of cubic feet that are filled? **Multiply length times width times half the height.**

Guided Practice/Try These You may wish to complete these exercises together with the class. For Exercise 2 discuss why only one dimension is given. **The figure is a cube.**

③ SUMMARIZE

Talk It Over Have students discuss similarities and differences between the methods used for finding the volume of triangular prisms and rectangular prisms. **Similarities: For both types of prism, volume is found by multiplying the areas of the base or bases by the height. Differences: The area of the base of a triangular prism is one half the area of a rectangular prism with the same dimensions.**

④ PRACTICE

Practice/Solve Problems You may wish to have students use calculators to check their answers for Exercises 1–10.

Extend/Solve Problems In Exercise 16 students should understand that they must divide the total volume by the volume of one dump-truck load.

Think Critically/Solve Problems For Exercises 21–25 stress the distinction between volume and capacity. Discuss the possibility of taking two different approaches to solving Exercise 25.

EXERCISES

PRACTICE/ SOLVE PROBLEMS Find the volume of each prism. If necessary, round to the nearest tenth.

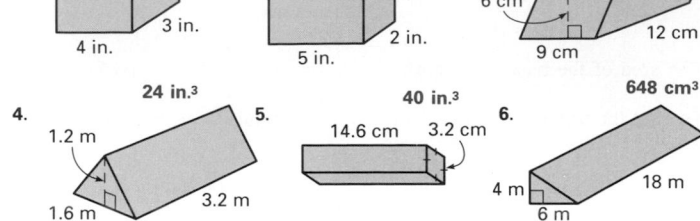

1. 2 in. 3 in. 4 in. — **24 in.³**

2. 4 in. 2 in. 5 in. — **40 in.³**

3. 6 cm 9 cm 12 cm — **648 cm³**

4. 1.2 m 3.2 m 1.6 m — **3.1 m³**

5. 14.6 cm 3.2 cm — **149.5 cm³**

6. 4 m 6 m 18 m — **216 m³**

Which prism in each pair has the greater volume?

7. 4,704 in.³, 5,040 in.³
D has the greater volume.

8. 2,957.4 cm³, 3,003 cm³
G has the greater volume.

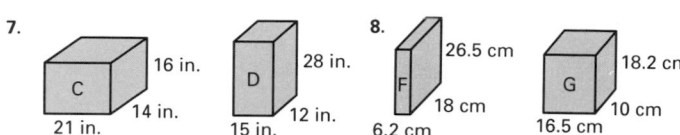

7. C 16 in. 14 in. 21 in. — D 28 in. 12 in. 15 in.

8. F 26.5 cm 18 cm 6.2 cm — G 18.2 cm 10 cm 16.5 cm

9. Find the volume of the cheese. 3.2 cm 2.3 cm 1.5 cm — **5.52 cm³**

10. A bale of hay is 1.5 m long, 0.8 m wide, and 1.4 m high. What is the volume of the bale? **1.68 m³**

EXTEND/ SOLVE PROBLEMS Find the volume. If necessary, round to the nearest tenth.

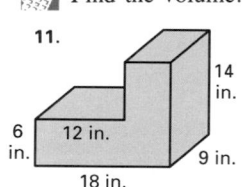
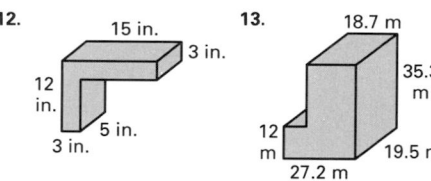

11. 14 in. 6 in. 12 in. 9 in. 18 in. — **1,404 in.³**

12. 15 in. 3 in. 12 in. 3 in. 5 in. — **360 in.³**

13. 18.7 m 35.3 m 12 m 27.2 m 19.5 m — **14,861.1 m³**

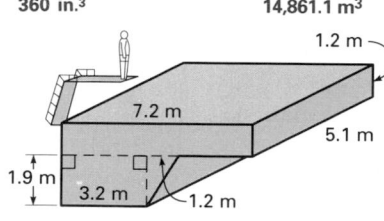

14. Find the volume of water in the swimming pool if the surface of the water is 0.3 m below the top edge of the pool. 1.2 m 7.2 m 5.1 m 1.9 m 3.2 m 1.2 m — **69.87 m³**

15. The cost of heating the air in a room measuring 3.2 m by 4.8 m by 2.9 m is 1.5¢/m³. What is the cost of heating the room? **66.82¢ or 67¢**

16. To construct the basement of a house, a hole with dimensions 10 m by 5 m by 2 m must be dug. A dump truck can haul 25 m³ of earth on each trip. How many trips must the truck make to remove all the earth? **4 trips**

17. There are 10 garbage containers at an apartment site. Each container holds 4.5 yd³ of garbage. If the truck that picks up the garbage has a closed trailer with dimensions 4.5 yd by 2.5 yd by 2 yd, how many trips will it take to haul away the garbage? **2 trips**

18. Before the surface of an ice rink is safe for skating, an area 75.1 m long and 74.8 m wide must be frozen to a depth of 0.5 m. What is the volume of ice that will result? **2,808.74 m³**

19. A cubical tank for storing syrup has sides measuring 1.2 m. What volume of syrup is in the tank when the tank is half full? **0.864 m³**

20. The concrete pillar at the right has a hexagonal base. All the triangular areas are identical. Find the volume of the pillar. **4.46 m³**

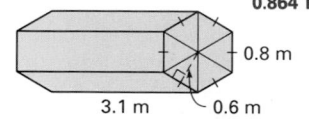

For Exercises 21–25, use these relationships between volume and liquid capacity. 1 L = 1,000 cm³ and 1 mL = 1 cm³

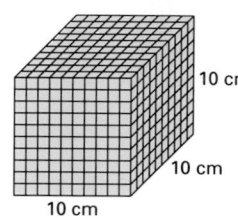

1 L is the capacity of a box that measures 10 cm by 10 cm by 10 cm.

A fish tank is 180 cm long, 150 cm wide, and 70 cm high.

21. What is its volume? **1,890,000 cm³**

22. In liters of water, what is the capacity of the tank? **1,890 L**

23. A goldfish requires about 1,000 cm³ of tank space to survive. How many goldfish can this tank support? **1,890 goldfish**

A fish tank is 42.8 cm long, 19.5 cm wide, and 15.2 cm deep.

24. Find, to the nearest thousandth, the number of liters of water the tank will hold when full. **12.686 L**

25. The tank is filled so that the water is 4.7 cm from the top. How much water is in the tank? **8,763.3 cm³ or 8.7633 L**

COMPUTER TIP

For Exercise 20, you can use the program on page 409. Change line 50 to V = 6 * AB * HP to find the volume of the pillar.

THINK CRITICALLY/ SOLVE PROBLEMS

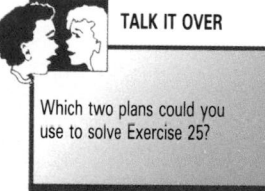
TALK IT OVER

Which two plans could you use to solve Exercise 25?

Find the volume of the water directly, or find the total volume of the tank and subtract the volume of the unfilled space.

11-7   Volume of Prisms   **411**

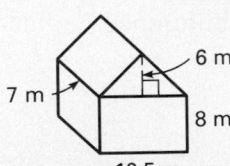
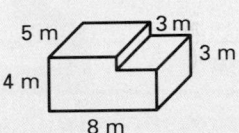
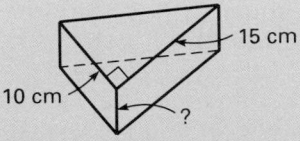

412

11-8 Area of Circles

EXPLORE/WORKING TOGETHER

With a partner, draw several circles of various sizes on a sheet of grid paper. Count the number of square units within each circular region. Estimate the area of each circle in square units. Compare your estimates with those of other classmates.

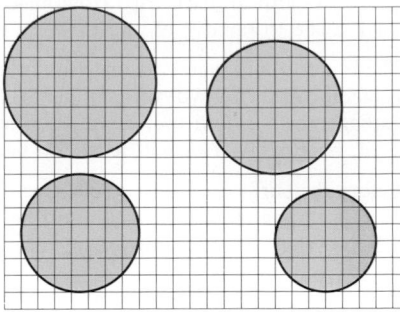

SKILLS DEVELOPMENT

A circle can be cut into equal sections. These sections can then be used to form a figure that is shaped very much like a parallelogram. The length of the base of this "parallelogram" will approximate one half the circumference of the circle. The height of the "parallelogram" will equal the radius of the circle.

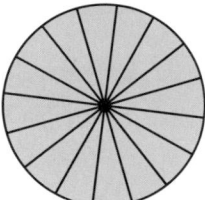 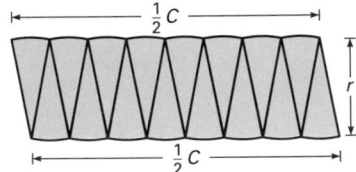

A formula for the area of a circle can then be derived from the formula for the area of a parallelogram.

$$A = b \times h$$
$$A = \tfrac{1}{2}C \times r$$
$$A = \tfrac{1}{2} \times (2\pi r) \times r$$
$$A = \pi r \times r$$
$$A = \pi r^2$$

You can use the formula $A = \pi r^2$ to solve problems involving the area of a circle.

Example 1

Find the area of circle O to the nearest square centimeter.

Use $\pi \approx 3.14$.

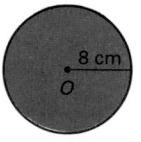

A circle is named by its center point.

412

Solution

$$A = \pi r^2$$
$$A \approx 3.14 \times 8^2 \quad \leftarrow \text{ Substitute 8 for } r.$$
$$A \approx 3.14 \times 64$$
$$A \approx 200.96$$
$$A \approx 201 \quad \leftarrow \text{ Round to the nearest whole number.}$$

The area of the circle is approximately 201 cm². ◄

Example 2

Find the area of a circular garden that has radius 2.6 m. Use $\pi \approx 3.14$. Round your answer to the nearest tenth.

Solution

The garden is shaped like a circle. Use the area formula for circles.

$$A = \pi r^2$$
$$A = \pi \times (2.6)^2 \quad \leftarrow \text{ Substitute 2.6 for } r.$$
$$A \approx 3.14 \times 6.76$$
$$A \approx 21.2264$$
$$A \approx 21.2$$

The area of the garden is approximately 21.2 m². ◄

TRY THESE

Use $\pi \approx 3.14$. Round answers to the nearest tenth.

Find the area of each circle.

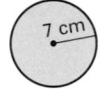 1.   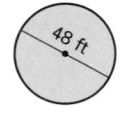 2.   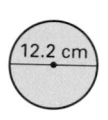 3.   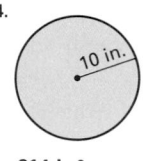 4.

7 cm   48 ft   12.2 cm   10 in.

153.9 cm²   **1,808.6 ft²**   **116.8 cm²**   **314 in.²**

5. The diameter of a circle is 12.6 cm. Find the area of the circle.
124.6 cm²

CHECK UNDERSTANDING

If you know the diameter of a circle, how would you find its radius?

Halve the diameter.

EXERCISES

Use $\pi \approx 3.14$. Round answers to the nearest tenth.

Find the area of each circle.

PRACTICE/ SOLVE PROBLEMS

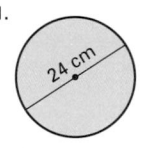 1.   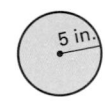 2.   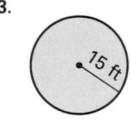 3.   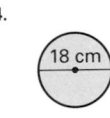 4.

24 cm   5 in.   15 ft   18 cm

452.2 cm²   **78.5 in.²**   **706.5 ft²**   **254.3 cm²**

ASSIGNMENTS

BASIC
1–8, 12–14, 15–16, 18–23, 29

AVERAGE
5–14, 17–28, 29–34

ENRICHED
12–14, 18–28, 29–35

ADDITIONAL RESOURCES
Reteaching 11–8
Enrichment 11–8

out also that the answer is always given in square units.

Example 2: You may wish to review multiplying and rounding decimal numbers. Students can estimate the answer before discussing the solution.

Additional Questions/Examples

1. If you know only the diameter of a circle, what do you need to do to find the area? **You must find the radius so that you can apply the area formula.**

2. What is the relationship between the radius of a circle and the radius of one quarter of that circle? **They are the same.**

3. Which two of the circles with the following dimensions are equal in area? How can you tell?

Circle A: $d = 2.14$ cm
Circle B: $r = 1.07$ cm
Circle C: $d = 2.014$ cm

Circles A and B are equal in area because the radius of circle B is exactly half the diameter of circle A.

Guided Practice/Try These Before students do Exercise 2, 3, and 5, they should calculate the radius of each circle.

AT-RISK STUDENTS

Be sure that students understand that the symbol π stands for a decimal number that, when rounded, is expressed as 3.14.

3 SUMMARIZE

Write About Math In their journals, have students write directions for finding the area of a circle when given the radius and then when given the diameter.

4 PRACTICE

Practice/Solve Problems For Exercises 9–11 suggest that students find the radius of each circle before attempting to match the circles to the areas.

Extend/Solve Problems For Exercises 15 and 16, remind students that the area of every figure is expressed in square units. For Exercise 17 point out to students the need to first find the area of the whole circle.

Think Critically/Solve Problems For Exercise 30 students must realize that the radius of the circle is also the height of each triangle. For Exercises 32–35 you may wish to have students work in pairs.

5 FOLLOW-UP

Extra Practice
1. Find the area, to the nearest hundredth, for circles with the following dimensions:
   a. $r = 2.4$ in. **b.** $d = 34$ m
   c. $r = 12$ m **d.** $d = 16$ ft
   a. 18.09 in.² **b. 907.46 m²**
   c. 452.16 m² **d. 200.96 ft²**

Find the area of a circle with the given radius.

5. 8 ft	**6.** 36 m	**7.** 9.8 in.	**8.** 10.6 m
201.0 ft²	4,069.4 m²	301.6 in.²	352.8 m²

Write the letter of the corresponding circle. The measures are in centimeters.

9. $A = 3.14 \times 100^2$ C

10. $A = 3.14 \times 50^2$ A

11. $A = 3.14 \times 100$ B

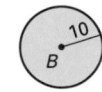 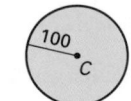

Solve.

12. By how much does the area of circle O exceed the area of circle P?
388.6 ft²

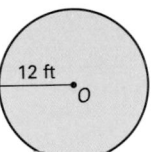 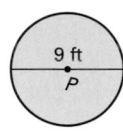

13. Throwing the discus is one of the Olympic field events. The circular discus has a radius of 11 cm. Find its area.   379.9 cm²

14. The discus is thrown by an athlete standing within a circle with a diameter of about 2.8 m. Find the area of the circle.   6.2 m²

EXTEND/ SOLVE PROBLEMS

15. Which has the greater area: a square with 8-in. sides or a circle with a 9-in. diameter?   **square**

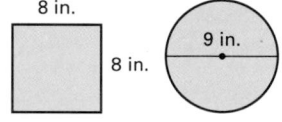

16. What is the difference in area, to the nearest thousandth, of the figures used for Exercise 15?   0.415 in.²

17. The radius of a circle is 1.3 yd. Find the area of one quadrant (one-fourth) of the circle.   about 1.3 yd²

Use this table for Exercises 18–28.

PIZZA MENU				
	Small 10 in.	Medium 14 in.	Large 16 in.	Extra Large 18 in.
BASIC TOMATO SAUCE AND CHEESE	$ 5.00	$ 7.00	$ 8.75	$ 9.25
One choice of topping	5.50	7.75	9.25	10.00
Two choices of toppings	6.00	8.50	9.85	10.45
Three choices of toppings	6.50	8.75	10.00	10.75
Four choices of toppings	7.25	9.25	10.50	12.00

18. What is the cost of 3 small pizzas with 3 choices of toppings?   **$19.50**

Find the area of the pizza of each size.

19. small   **78.5 in.²**   **20.** medium   **153.9 in.²**   **21.** large   **201 in.²**

Find the cost per square inch of each of the following pizzas. Express your answers in cents per square inch.

22. small with 2 choices   **8¢ per square inch**

23. medium with 2 choices   **6¢ per in.² square inch**

24. large with 2 choices   **5¢ per square inch**

25. Which of the pizzas described in Exercises 22–24 costs least per square inch?   **large with 2 choices, Exercise 24**

26. How much greater in size is a large pizza than a small pizza?   **122.5 in.²**

27. Find the area of 4 small pizzas.   **314 in.²**

28. Which is the better buy, 4 small pizzas with 2 choices or 2 large pizzas with 2 choices?   **The better buy is 2 large pizzas with 2 choices.**

Find the area of each shaded region to the nearest tenth.

THINK CRITICALLY/ SOLVE PROBLEMS

29.
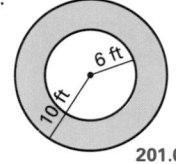
201.0 ft²

30.
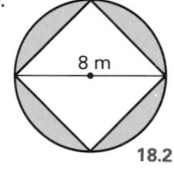
18.2 m²

31.
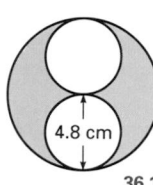
36.1 in.² Answers may vary due to rounding.

Find the area of each figure to the nearest tenth.

32.
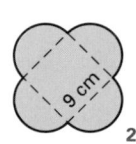
208.2 cm²

33.
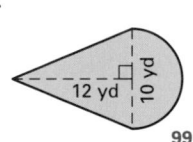
99.3 yd²

34.
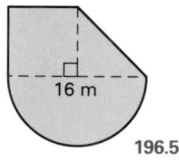
196.5 m²

35. Find the area of the window at the right.
14,142.0 cm²

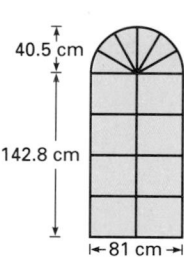

---

2. Find the area of the shaded region in each figure.

a.
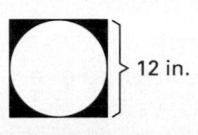
12 in.
30.96 in.²

b.
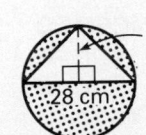
h = 14 cm
28 cm
419.44 cm²

Extension Have students work in small groups to find the area, to the nearest hundredth, of each shaded region.

a. d = 1 in.   **b.** d = 6.4 ft

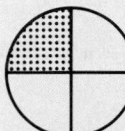

0.20 in.²   **12.06 ft²**

Section Quiz
1. To the nearest hundredth, find the area for circles with the following dimensions:
a. d = 1.04 in.   **b.** r = 23 ft
c. r = 6.12 cm   **d.** d = 4.5 m
a. 0.85 in.²
b. 1,661.06 ft²
c. 117.61 cm²
d. 15.90 m²

2. Find the area of this region to the nearest hundredth.

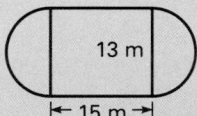
13 m
15 m
327.67 m²

Get Ready calculators

11-9 Volume of Cylinders

EXPLORE

The three-dimensional figure in the diagram below is a cylinder. How is a cylinder different from a prism? How is it like a prism?

Its base is a circle, not a polygon, and has one curved surface instead of several lateral faces. It has two bases.

SKILLS DEVELOPMENT

Recall the general formula for finding the volume of a prism is $V = B \times h$, where B is the area of the base. Since the base of a cylinder is a circle, replace B in this formula with πr^2.

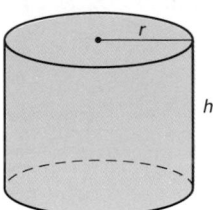

$$V = B \times h$$
$$V = \pi r^2 \times h, \text{ or } V = \pi r^2 h$$

COMPUTER

This program will find the volume of a cylinder when you INPUT the radius and the height. Remember to label your answer in cubic units.

```
10 INPUT "ENTER RADIUS
   AND HEIGHT: ";R,H:
   PRINT
20 LET V = 3.14 * R ^ 2 * H
30 LET A = INT (V * 10 ^ 2
   + 0.5) / 10 ^ 2
40 PRINT "THE VOLUME IS
   ABOUT ";A
```

Example 1

Find the volume of a cylinder with a radius of 4 in. and a height of 40 in. Use $\pi \approx 3.14$.

Solution

Use the formula $V = \pi r^2 \times h$.

$$V = \pi r^2 \times h$$
$$V = \pi \times 4^2 \times 40 \qquad \leftarrow \text{Substitute 4 for } r \text{ and 40 for } h.$$
$$V \approx 3.14 \times 16 \times 40$$
$$V \approx 2,009.6$$

The volume of the cylinder is approximately 2,009.6 in.³ ◄

Example 2

Find the volume of the cylinder below. Use $\pi \approx 3.14$. Round your answer to the nearest tenth.

Solution

Use the formula $V = \pi r^2 \times h$.

The diameter of the cylinder is 16 cm. So, its radius is 8 cm. The height of the cylinder is 36 cm.

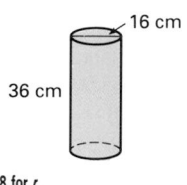

16 cm

36 cm

$$V = \pi r^2 \times h$$
$$V = \pi \times 8^2 \times 36 \qquad \leftarrow \text{Substitute 8 for } r$$
$$V \approx 3.14 \times 64 \times 36 \qquad \text{and 36 for } h.$$
$$V \approx 7,234.56$$
$$V \approx 7,234.6$$

The volume of the cylinder is about 7,234.6 cm³. ◄

Example 3

Find the volume of the can. Use $\pi \approx 3.14$.

Solution

The can has a radius of 5 cm. Use the volume formula for a cylinder.

$$V = \pi r^2 \times h$$
$$V = \pi \times 5^2 \times 14.5 \quad \leftarrow \text{Substitute}$$
$$V \approx 3.14 \times 25 \times 14.5 \quad \begin{array}{l}\text{5 for } r \text{ and}\\ \text{14.5 for } h.\end{array}$$
$$V \approx 1{,}138.25$$
$$V \approx 1{,}138.3$$

The volume of the can is about 1,138.3 cm³. ◄

TRY THESE

Use $\pi \approx 3.14$. Round answers to the nearest tenth.

1. Find the volume of a cylinder with a radius of 3 in. and a height of 28 in.   **791.3 in.³**

Use the dimensions given to find the volume of each cylinder.

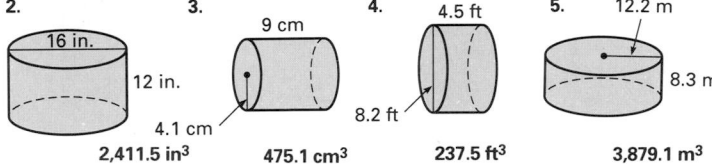

2. **2,411.5 in³**
3. **475.1 cm³**
4. **237.5 ft³**
5. **3,879.1 m³**

Solve.

6. Find the volume of a cylindrical water drum with a height of 4 ft and a radius of 2.5 ft.   **78.5 ft³**

EXERCISES

Use $\pi \approx 3.14$. Round answers to the nearest whole number.

PRACTICE/ SOLVE PROBLEMS

Find the volume of each cylinder.

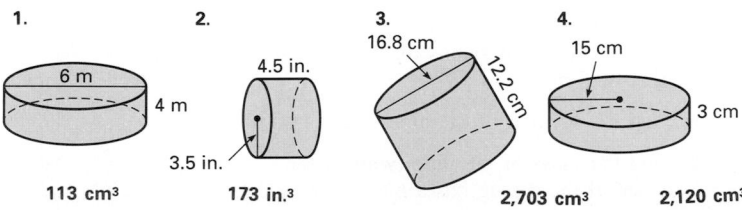

1. **113 cm³**
2. **173 in.³**
3. **2,703 cm³**
4. **2,120 cm³**

MAKING CONNECTIONS

Have students find the volume of a can of frozen orange juice or lemonade concentrate. Students should use centimeter rulers to measure the diameter and height of the can. Have them determine the price per milliliter of the liquid based on the cost and volume of the can.

ASSIGNMENTS

BASIC
1–8, 11–12, 14, PSA 1–4

AVERAGE
5–10, 11–13, 14, PSA 1–5

ENRICHED
9–10, 11–13, 14–16, PSA 1–7

ADDITIONAL RESOURCES
Reteaching 11–9
Enrichment 11–9

Additional Questions/Examples

1. How are the formulas for finding the volume of a prism and for finding the volume of a cylinder the same? How are they different?   **Same: Both require multiplying the area of the base times the height; Different: Base of a cylinder is a circle; base of a prism is a polygon.**

2. The volume of a storage bin is 200 m³. The substance that will be stored inside the bin has a mass of 20 kg/m³. How many grams of the substance can fit into the bin? (Recall 1 kg = 1,000 g.)   **4,000,000 grams**

Guided Practice/Try These For Exercises 2 and 4, remind students to find the radius before finding the volume.

3 SUMMARIZE

Key Questions

1. How can you find the volume of a cylinder when you know the diameter of the base and the height?   **Find the area of the base then multiply the area by the height.**

417

418

2. How can you find the height of a cylinder when you know the area of the base and the volume? **Divide the volume, V, by B, the area of the base.**

4 PRACTICE

Practice/Solve Problems In Exercise 10 you might ask students to find the liquid capacity of the oil drum. **361.571 L**

Extend/Solve Problems For Exercise 13 ask students to explain the difference between engine capacity and liquid capacity. **Engine capacity is given in cubic centimeters, whereas liquid capacity is given in milliliters or liters.**

Think Critically/Solve Problems For Exercises 15 and 16, point out that the units of measure for the height and radius are not the same.

Problem Solving Applications Be sure students do not confuse the shape of a thunderstorm disturbance (cylindrical) with the shape of a tornado (cone). You may wish to point out that not all the problems on this page deal with cylinders. (Problems 4, 6, and 7 deal with rectangular areas.)

5 FOLLOW-UP

Extra Practice
1. Find the volume, to the nearest hundredth, of cylinders with the following dimensions:
   a. $r = 12.4$ in., $h = 10$ in.
   b. $d = 14$ cm, $h = 15$ cm
   a. 4,828.06 in.³
   b. 2,307.9 cm³

Copy and complete the table.

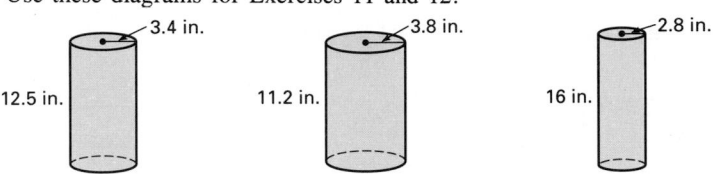

	diameter of base (d)	radius of base (r)	height (h)	Volume (V)	
5.	6 ft	■	8 ft	■	$r = 3$ ft $V = 226$ ft³
6.	18 cm	■	23 cm	■	$r = 9$ cm $V = 5,850$ cm³
7.	■	6.9 in.	4.2 in.	■	$d = 13.8$ in. $V = 628$ in.³
8.	■	2.5 m	4.8 m	■	$d = 5.0$ m $V = 94$ m³

9. The diameter of a cylinder is 24 in. The height is 16 in. Find its volume. **7,235 in.³**

10. An oil drum is 94 cm high. The radius of the base is 35 cm. What is the volume? **361,571 cm³**

EXTEND/SOLVE PROBLEMS

Use $\pi \approx 3.14$. Round answers to the nearest whole number.

Use these diagrams for Exercises 11 and 12.

3.4 in. 12.5 in.   3.8 in. 11.2 in.   2.8 in. 16 in.

11. Estimate which can holds the most. **Answers will vary.**

12. Find the volume of each can. Was your estimate correct?
454 in.³; 508 in.³; 394 in.³

13. A car engine has 8 cylinders, each with a diameter of 7.2 cm and a height of 8.4 cm. The total volume of the cylinders is called the *capacity* of the engine. Find the engine capacity. **2,735 cm³**

THINK CRITICALLY/SOLVE PROBLEMS

Use $\pi \approx 3.14$. Round answers to the nearest whole number.

14. Suppose that the mass of wheat is 120 kg/m³. A silo with a radius of 3.4 m and a height of 7.3 m is filled with wheat. Find the total mass of the wheat in the silo. **31,800 kg Answers may vary due to rounding.**

A section of water pipe has an inner radius of 14.6 cm, an outer radius of 18.2 cm, and a height of 6 m.

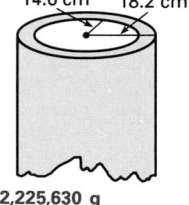

14.6 cm   18.2 cm

15. Find the volume of the pipe. **401,593 cm³**

16. Find the mass of the pipe (without the water) if the piping has a mass of 10 g/cm³.

2,225,630 g

Problem Solving Applications:

THUNDERSTORMS AND TORNADOS

Thunderstorms and tornados are two of nature's most destructive phenomena. After a tornado, the average total property loss is about $24,300,000.

On average, a tornado lasts for less than 30 s. A thunderstorm lasts an average of 2 h.

The atmospheric disturbance created by a thunderstorm is almost cylindrical, as shown at the right.

3.5 km
6.2 km

Use $\pi \approx 3.14$. Round your answer to the nearest tenth.

1. Find the volume of the atmospheric disturbance shown above. **238.5 km³**

2. The smallest thunderstorm on record measured 2.7 km in diameter and its cloud formation had a height of 8.4 km. Find the total volume of the disturbance in the atmosphere. **48.1 km³**

3. The largest thunderstorm on record measured 43.9 km in diameter. It had a height of 20.3 km. Find the total volume this storm occupied in the atmosphere. **30,711.1 km³**

4. A tornado is a narrow funnel-shaped cloud that extends downwards from cumulonimbus clouds. Find the zone of destruction of a tornado that destroys a tract of land 8 mi wide and 6.9 mi long. **55.2 mi²**

Specially equipped aircraft and ground crews record data about thunderstorms. These measurements were taken of two thunderstorms.

Storm A	Storm B
diameter: 9.6 km	diameter: 8.5 km
height: 5.9 km	height: 6.3 km

5. Which storm, A or B, occupied the greater space? **A = 426.8 km³** **B = 375.3 km³ A > B**

6. A tornado in Massachusetts covered a path 0.256 km wide and 27.5 km long. Find the zone of destruction. **7.0 km³**

7. A small tornado swept across a tract of land 0.04 km wide and 7.8 km long. Find the land area that was destroyed. **0.3 km²**

11–9   Volume of Cylinders   **419**

2. Which cylinder has the greater volume?
Cylinder a: $r = 23.3$ cm, $h = 12$ cm
Cylinder b: $B = 1,600$ cm², $h = 10$ cm
Cylinder a: 4,456 cm³ greater

3. Find the volume of the outer part of the cylinder.

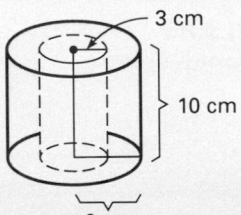

3 cm
10 cm
6 cm

847.8 cm³

Extension Have students work in small groups to solve the following problem. *How many one-liter bottles would be needed to carry the water in a cylinder with a diameter of 20 cm and a height of 65 cm when the water is filled to 6 cm from the top of the cylinder?* **19 one-liter bottles**

Section Quiz
1. Find the volume of cylinders having the following dimensions:
   a. $d = 32$ in., $h = 8$ in.
   b. $r = 7.2$ cm, $h = 15$ cm
   a. 6,430.72 in.³
   b. 2,441.664 cm³
2. If you decrease the diameter of cylinder b, above, by 3 cm, by how much will the volume decrease? **by 911.39 cm³**
3. Jay drinks 8 glasses of water per day. Each glass is a cylinder that is 6 cm in diameter, and it is filled to a height of 12 cm. If 1,000 cm³ = 1 liter, about how many liters of water does Jay drink per day? **2.713 L per day**

Get Ready calculators

WARM-UP

Solve.
1. $1/3 \times 6^2 = \blacksquare$ **12**
2. $4/3 \times 3.14 \times 6 = \blacksquare$ **25.12**
3. $2/3 \times 18.6 = \blacksquare$ **12.4**
4. $1/3 \times 4^2 = \blacksquare$ **5 1/3**

1 MOTIVATE

Explore You may want to perform a demonstration using a rectangular prism and pyramid with equal dimensions. After students agree that the contents of the pyramid would fill 1/3 of the prism, have them discuss how the formulas for finding the volumes of the figures might be related.

2 TEACH

Use the Pages/Skills Development Have students read this part of the section and then discuss the examples. Point out that the three dimensions must be the same for both figures for their volumes to be related.

Example 1: At the chalkboard, have a volunteer draw the figure described, then explain each step of the solution.

11-10 Volume of Pyramids, Cones, and Spheres

EXPLORE

The containers shown are a pyramid and a prism that have identical bases. The containers are equal in height.

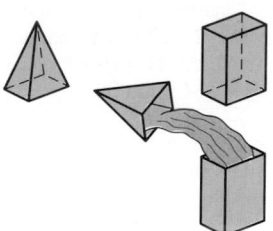

Suppose that the pyramid were filled with water, and the water was then poured into the prism. What part of the prism would be filled? $\frac{1}{3}$

SKILLS DEVELOPMENT

The formula for the volume of a pyramid is related to the formula for the volume of a prism.

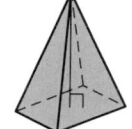

prism: $V = B \times h$
pyramid: $V = \frac{1}{3} \times B \times h$

That is, the volume of a pyramid with a given base is one third the volume of a prism with a base of the same size.

Example 1

Find the volume of a rectangular pyramid with a base of length 6 in. and width 4 in. and with height 8 in.

Solution

Use the volume formula for a pyramid.

$$V = \frac{1}{3} \times B \times h$$
$$V = \frac{1}{3} \times 6 \times 4 \times 8 \quad \leftarrow \text{Substitute 6} \times \text{4 for } B$$
$$\text{and 8 for } h.$$
$$V = 64$$

The volume of the pyramid is 64 in.³ ◄

In a similar way, the formula for the volume of a cone is related to the formula for the volume of a cylinder.

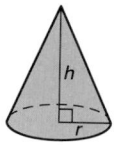 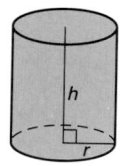

cylinder: $V = \pi r^2 \times h$

cone: $V = \frac{1}{3} \times \pi r^2 \times h$

That is, the volume of a cone with a given radius and height is one third the volume of a cylinder with the same radius and height.

AT-RISK STUDENTS

When finding the volume of different kinds of figures, these students may become confused about which formula to use. Until students have become proficient at applying the correct formulas, you may wish to display a reference poster displaying the various figures and their corresponding formulas.

Example 2

Find the volume of the cone at the right. Use $\pi \approx 3.14$.

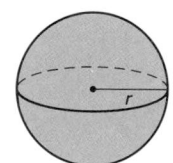

6 cm

10 cm

Solution

$$V = \tfrac{1}{3} \times \pi r^2 \times h$$
$$V = \tfrac{1}{3} \times \pi(10)^2 \times 6 \quad \leftarrow \text{Substitute 10 for } r \text{ and 6 for } h.$$
$$V \approx \tfrac{1}{3} \times 3.14 \times 100 \times 6$$
$$V \approx 628$$

The volume of the cone is approximately 628 cm³. ◄

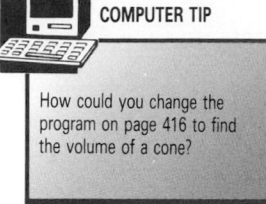

COMPUTER TIP

How could you change the program on page 416 to find the volume of a cone?

The formula for the volume of a sphere is as follows.

sphere: $V = \tfrac{4}{3} \times \pi r^3$

Example 3

Find the volume of a sphere with radius 6 in. Use $\pi \approx 3.14$. Round your answer to the nearest whole number.

Solution

$$V = \tfrac{4}{3} \times \pi r^3$$
$$V = \tfrac{4}{3} \times \pi \times 6^3 \quad \leftarrow \text{Substitute 6 for } r.$$
$$V \approx \tfrac{4}{3} \times 3.14 \times 216$$
$$V \approx 904.32$$
$$V \approx 904$$

The volume of the sphere is approximately 904 in.³ ◄

Example 4

To the nearest whole number, find the volume of a cone-shaped storage bin with a height of 27 m and a radius of 15 m.

Solution

The storage bin is a cone with radius 15 m and height 27 m. Use the volume formula for a cone.

$$V = \tfrac{1}{3} \times \pi r^2 \times h$$
$$V = \tfrac{1}{3} \times \pi(15)^2 \times 27 \quad \leftarrow \text{Substitute 15 for } r \text{ and 27 for } h.$$
$$V \approx \tfrac{1}{3} \times 3.14 \times 225 \times 27$$
$$V \approx 6{,}358.5$$
$$V \approx 6.359 \quad ◄$$

The volume of the bin is approximately 6.359 m³.

11–10   Volume of Pyramids, Cones, and Spheres   **421**

ASSIGNMENTS

BASIC
1–16, 21–24, 28

AVERAGE
1–16, 18, 19–20, 21–27, 28

ENRICHED
7–20, 21, 25–27, 28–30

ADDITIONAL RESOURCES
Reteaching 11–10
Enrichment 11–10

Example 2: Remind students that πr^2 in the volume formula stands for the area of the circular base of the figure.

Example 3: You may wish to suggest that students visualize cutting an orange in half to help them envision the locations of the radius and diameter of a sphere.

Example 4: Students should make sure they do not forget to include the value of π as a factor.

Additional Questions/Examples
1. How can you find the height of a pyramid if you know the volume and the area of the base? **Multiply the volume by 3, then divide the result by the area of the base.**
2. Complete this analogy:
V(pyramid): ■ as
■:V(cylinder)
V(prism); V(cone)

Guided Practice/Try These In Exercise 1 point out that the base of the pyramid is a square. For all exercises have students identify the formula needed before finding the volume.

TEACHING TIP

You may wish to explain to students that a cross-sectional area is the plane figure that results when a solid figure is cut across the center. Sketch and discuss cross-sectional shapes for various figures. **sphere — circle, triangular prism — triangle, rectangular prism — rectangle, cone — circle**

3 SUMMARIZE

Write About Math In their journals, have students write descriptions of the formulas used for finding the volume of the pyramid, the cone, and the sphere. Tell them not to use abbreviations. They should write, for example, "The volume of a pyramid equals one third the length of its base times the width of its base times the height."

4 PRACTICE

Practice/Solve Problems Discuss Exercises 1–4 before students begin them. Have students identify each figure and choose the formula needed to find the volume.

Extend/Solve Problems Students should note that, in Exercises 22–24, the volume of each part of the figure must be determined before the total volume of each figure can be found.

Think Critically/Solve Problems For Exercises 28–30 suggest that students draw diagrams, supply sample values for the dimensions of the figures, and find the volumes before they draw any conclusions.

5 FOLLOW-UP

Extra Practice
1. Find the volume, to the nearest hundredth, for each of the following figures.

a.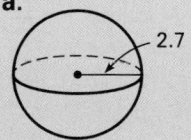

2.7 cm

b.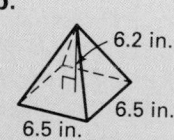

6.2 in.

6.5 in.

6.5 in.

82.41 cm³   87.32 in.³

2. Which has the greater volume? How much greater is it?
Cone: *d* =12.4 cm, *h* = 10 cm
Sphere: *r* = 4 cm
The volume of the cone is 134.39 cm³ greater.

**PRACTICE/
SOLVE PROBLEMS**

1. 65.3 in.³
2. 1,071.8 m³
3. 7,234.6 ft³
4. 232.2 cm³

TRY THESE

Use π ≈ 3.14. Round your answer to the nearest tenth.

Find the volume of each figure.

1.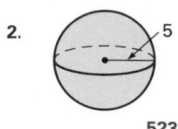
5 cm

6 cm

60.0 cm³

2.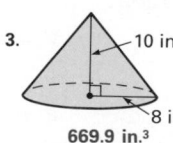
5 in.

523.3 in.³

3.
10 in.

8 in.

669.9 in.³

Solve.

4. A pyramid has a rectangular base with length 8 ft and width 3 ft. Its height is 9 ft. Find the volume of the pyramid. **72 ft³**

EXERCISES

Use π ≈ 3.14. Round your answer to the nearest tenth.

Find the volume of each figure.

1.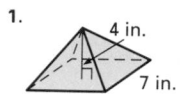
4 in.

7 in.

2.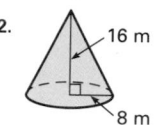
16 m

8 m

3.
12 ft

4.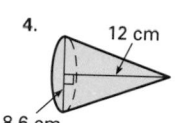
12 cm

8.6 cm

5. Find the volume of a cone 3.8 ft in diameter and 5.1 ft high. **19.3 ft³**

6. Find the volume of a pyramid that is 10.2 m high and has a rectangular base measuring 7 m by 3 m. **71.4 m³**

Find the volume of a sphere with the given radius.

7. 8 in.
2,143.6 in.³

8. 25 cm
65,416.7 cm³

9. 6.8 ft
1,316.4 ft³

10. 10.4 cm
4,709.4 cm³

Find the volume of a sphere with the given diameter.

11. 18 m
3,052.1 m³

12. 9.2 ft
407.5 ft³

13. 13.6 mm
1,316.4 mm³

14. 24.8 cm
7,982.4 cm³

15. A baseball has a radius of 3.6 cm. What is its volume? **195.3 cm³**

16. Find the volume of a cone-shaped funnel with radius 15 cm and height 22 cm. **5,181 cm³**

17. Phosphate is stored in a conical pile. Find the volume of the pile if the height is 7.8 m and the radius is 16.2 m. **2,142.6 m³**

18. A tent is in the form of a pyramid with a base 2 yd wide and 3 yd long, and it has a height of 2.4 yd. Find the volume of the tent. **4.8 yd³**

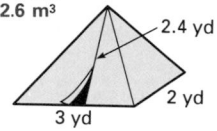
2.4 yd

2 yd

3 yd

USING DATA Use the Data Index on page 546 to locate information on ball sizes for international competition sports. Find the volume of the ball specified in each of Exercises 19 and 20. Round your answer to the nearest tenth.

19. soccer ball **5,572.5 cm³**

20. tennis ball **143.7 cm³**

Use π ≈ 3.14. Unless otherwise stated, round to the nearest tenth.

21. A half sphere is called a *hemisphere*. A hemispherical tank and a conical tank have the dimensions shown. How much greater is the volume of the hemisphere? **25.1 m³**

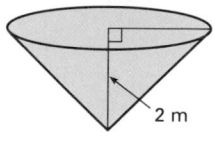 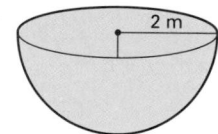

Find the volume of each figure.

22.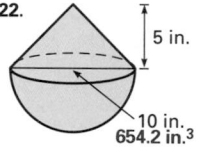
5 in.
10 in.
654.2 in.³

23.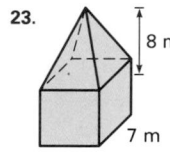
8 m
7 m **473.7 m³**

24.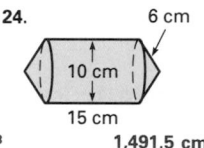
6 cm
10 cm
15 cm
1,491.5 cm³

25. Find the volume of the storage bin. **510.3 yd³**

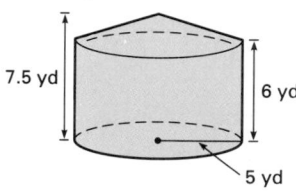
7.5 yd
6 yd
5 yd

26. Find the volume of the barn to the nearest whole number. **3,900 ft³**

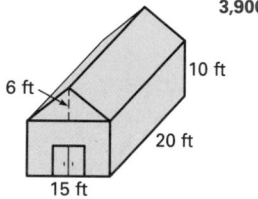
10 ft
6 ft
20 ft
15 ft

27. The thickness of a grapefruit peel is 0.5 cm. If the diameter of an unpeeled grapefruit is 12.8 cm, what part of the grapefruit is peel? Express your answer as a percent. **21.6%**

28. How would doubling the height of a cone affect the volume, if the radius of the base is unchanged? **The volume is doubled.**

29. How would doubling the radius of the base of a cone affect the volume if the height is unchanged? **The volume is multiplied by 4.**

30. How would the volume of a cone be affected if both the height and the radius were doubled? **The volume is multiplied by 8.**

11–10   Volume of Pyramids, Cones, and Spheres   **423**

CHALLENGE

You are designing a box to hold a solid glass sphere. The sphere will just touch each side of the box. If the radius of the sphere is *r*, what percent (to the nearest percent) of the volume of the box will be air? **48%**

11-11 Surface Area

EXPLORE Each of these two-dimensional patterns can be folded to form a polyhedron. Match each pattern to the polyhedron it would form.

1. 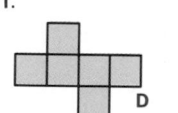 D
2. 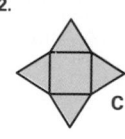 C
3. 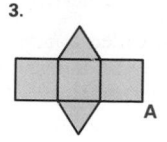 A
4. 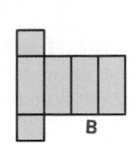 B

A.   B. 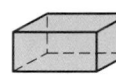   C.    D.

SKILLS DEVELOPMENT

The **surface area** of a prism or pyramid is the sum of the areas of all its faces.

Example 1

Find the surface area of the prism shown at the right.

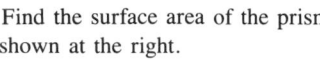

Solution

CHECK UNDERSTANDING

In example 1, why do you multiply Area A, Area B, and Area C by 2 to find the total surface area?

The surfaces opposite Areas A, B, and C are identical to them and have equal areas.

Area of A
$A = l \times w$
$A = 30 \times 12$
$A = 360$

Area of B
$A = l \times w$
$A = 30 \times 15$
$A = 450$

Area of C
$A = l \times w$
$A = 15 \times 12$
$A = 180$

$SA = 2 \times 360 + 2 \times 450 + 2 \times 180$
$SA = 720 + 900 + 360$
$SA = 1{,}980$

The surface area is 1,980 in.² ◄

To find the surface area of a cylinder, add the area of the curved surface to the sum of the areas of the two bases.
$$SA = 2\pi rh + 2\pi r^2$$

Example 2

Find the total surface area of the cylinder.
Round your answer to the nearest
whole number.

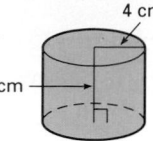

4 cm

9 cm

Solution

First, find the area of the curved surface of the cylinder. When
"unrolled," this surface is a rectangle with length equal to the
circumference of the circle. The width of the rectangle is equal to the
height of the cylinder.

$A = 2\pi r \times h$
$A = 2\pi \times 4 \times 9$
$A \approx 2 \times 3.14 \times 4 \times 9$
$A \approx 226.08$

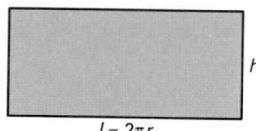

h

$l = 2\pi r$

Next, find the area of a base and multiply it by 2.

$A = \pi r^2$ $2 \times A \approx 2 \times 50.24$
$A = \pi(4)^2$ $2 \times A \approx 100.48$
$A \approx 3.14 \times 16$
$A \approx 50.24$

Add these areas to find the total surface area of the cylinder.

$SA \approx 226.08 + 100.48$
$SA \approx 326.56$
$SA \approx 327$

The total surface area of the cylinder is approximately 327 cm². ◄

The formula for the surface area of a cone
is as follows.

$SA = \pi rs + \pi r^2$

In the formula s represents
the *slant height* of the cone.

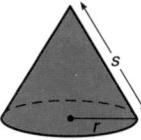

s

r

Example 3

Find the total surface area of the cone at the right. Round your answer
to the nearest whole number.

Solution

$SA = \pi rs + \pi r^2$
$SA = \pi \times 4 \times 10 + \pi \times (4)^2$
$SA \approx 3.14 \times 4 \times 10 + 3.14 \times 16$
$SA \approx 125.6 + 50.24$
$SA \approx 175.84$
$SA \approx 176$

10 cm

4 cm

The surface area of the cone is approximately 176 cm². ◄

that, when unrolled, the curved
surface of a cylinder is actually a
rectangle.
    Example 3: Have students dis-
cuss other ways to write the
formula for the total surface area
of a cone.   **Possible answer:**
$\pi r(s + r)$
    Example 4: Point out that
although the surface area of a
sphere is the amount of space
that covers the outside of the
sphere, the surface-area formula
is based on the radius.

Additional Questions/Examples
1. How would you find the surface area of a rectangular
pyramid?   **Add the areas of
the 4 trianglular and 1 rectangular surfaces.**
2. What is the relationship
between the formula for the
surface area of a sphere and
the diameter and circumference of a cross-section of the
sphere?
$SA = 4\pi r^2 = (2\pi r)2r = C \times d$

5-MINUTE CLINIC

Exercise	Student's Error	Error Diagnosis
Find the total surface area of a rectangular prism with l = 5 in., w = 6 in., and h = 4 in.	$l \times w \times h = 120$ in.³	• Students may confuse surface area formulas with volume formulas. Correct solution: $2 \times 6 \times 4 + 2 \times 4 \times 5 + 2 \times 6 \times 5 = 48 + 40 + 60 = 148$ in.²

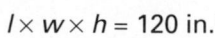

Guided Practice/Try These

Before having students begin these exercises, identify each figure, and discuss the number and shapes of its faces.

3 SUMMARIZE

Write About Math Have students write in their math journals the formulas for the surface area of a prism, a pyramid, a cone, and a sphere. Have them describe each formula in words.

4 PRACTICE

Practice/Solve Problems For Exercises 1–12 have students write out the formula they will use before solving each problem. For Exercises 16 and 17, have students identify the figures described. **rectangular prism and sphere**

Extend/Solve Problems For Exercises 18–20 point out that figure C is a cube. Students should use a calculator to check their answers.

Think Critically/Solve Problems Have students make up an example to fit the conditions of Exercise 22. Then have them apply their formula to find the surface area.

5 FOLLOW-UP

Extra Practice

1. Find the surface area of each figure:

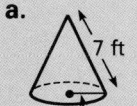
a. 7 ft, 3 ft

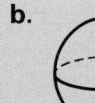
b. 2.2 cm

94.2 ft²   **60.79 cm²**

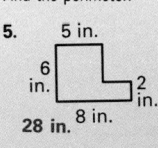
MIXED REVIEW

Tell what kind of graph would best represent the data.

1. average monthly temperature change in one year   **line graph**

2. world's deepest bodies of water   **bar graph**

3. team's budget for the year   **circle graph**

4. Find the median.
   39, 52, 25, 16   **32**

Find the perimeter.

5.   5 in.
     6 in.   2 in.
   28 in. 8 in.

Simplify.

6. $2 + 9 \cdot 6 - 24 \div 3 + 5$ **53**

7. 2^3 **8**   8. $9^2 - 4^3$ **17**

The formula for the surface area of a sphere is as follows.

$$SA = 4\pi r^2$$

In the formula, r represents the radius of the sphere.

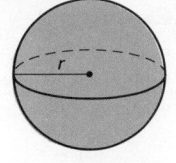

Example 4

Find the surface area of the sphere at the right. Round your answer to the nearest whole number.

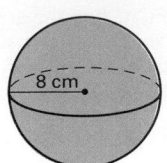
8 cm

Solution

$$SA = 4\pi r^2$$
$$SA = 4 \times \pi \times (8)^2$$
$$SA \approx 4 \times 3.14 \times 64$$
$$SA \approx 803.84$$
$$SA \approx 804$$

The surface area of the sphere is approximately 804 cm². ◄

TRY THESE

Find the surface area of each figure. Use $\pi \approx 3.14$. Round your answer to the nearest whole number.

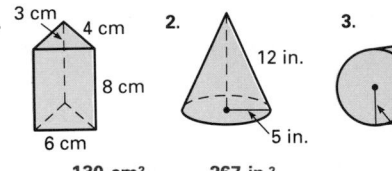
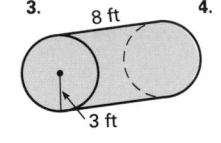
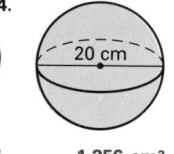

1. 3 cm, 4 cm, 8 cm, 6 cm
2. 12 in., 5 in.
3. 8 ft, 3 ft
4. 20 cm

130 cm²   **267 in.²**   **207 ft²**   **1,256 cm²**

EXERCISES

PRACTICE/ SOLVE PROBLEMS

Find the surface area of each figure. Use $\pi \approx 3.14$. Round your answer to the nearest whole number.

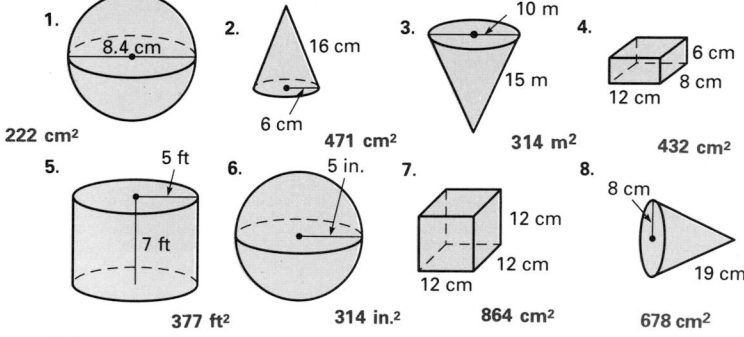

1. 8.4 cm
2. 16 cm, 6 cm
3. 10 m, 15 m
4. 6 cm, 8 cm, 12 cm
5. 5 ft, 7 ft
6. 5 in.
7. 12 cm, 12 cm, 12 cm
8. 8 cm, 19 cm

222 cm²   **471 cm²**   **314 m²**   **432 cm²**

377 ft²   **314 in.²**   **864 cm²**   **678 cm²**

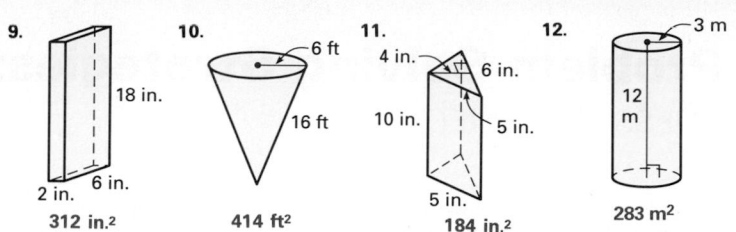

9. 18 in. 6 in. 2 in.
312 in.²

10. 6 ft 16 ft
414 ft²

11. 4 in. 6 in. 10 in. 5 in. 5 in.
184 in.²

12. 3 m 12 m
283 m²

Find the surface area of a sphere with the given diameter. Round your answer to the nearest tenth. Use π ≈ 3.14.

13. 16 cm **803.8 cm²**   **14.** 58 cm **10,563 cm²**   **15.** 9.6 cm **289.4 cm²**

Solve.

16. The dimensions of a gift box are 8 in. by 12 in. by 3 in. How much gift wrap is needed to cover the box? **312 in.²**

17. A major league baseball has a radius of 3.8 cm. What is the surface area of the ball? Round your answer to the nearest hundredth. **181.37 cm²**

Round your answer to the nearest hundredth. Use π ≈ 3.14.

Use the diagrams below for Exercises 18–20.

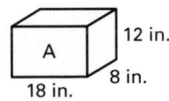 A 12 in. 8 in. 18 in.

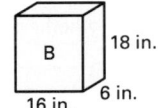 B 18 in. 16 in. 6 in.

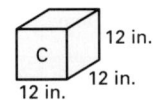 C 12 in. 12 in. 12 in.

18. Find the volume of each box. What do you notice? **The volumes are all the same.**

19. Find the surface area of each box. **912 in.²; 984 in.²; 864 in.²**

20. Which box uses the least amount of material? **Box C**

21. The radius of a sphere is 15 cm. By how much does the surface area of the sphere increase if the radius increases by 2 cm? **803.84 cm²**

22. The base of a rectangular prism is *l* units long and *w* units wide. Its height is *h* units. Write a formula that can be used to find the surface area of the prism. *SA* = 2(*l* × *w* + *l* × *h* + *w* × *h*)

EXTEND/ SOLVE PROBLEMS

THINK CRITICALLY/ SOLVE PROBLEMS

2. A warehouse sets the price of steel cylinders according to the surface area of the cylinders. At $0.05 per square centimeter, how much will 100 cylinders cost if each has a height of 10 cm and a radius of 4 cm? **$1,758.40**

3. The radius of a sphere is 12 in. By how much will the surface area of the sphere increase if the radius increases by 3 in.? **1,017.36 in.³**

Extension Have each student bring in an object from home that is shaped like either a sphere, cone, prism, or cylinder. Have them find the surface area of their object, exchange objects with another student, and then find the surface area of the object they receive. They should then compare answers.

Section Quiz

1. Find the surface area, to the nearest whole number, of each of the following figures:

a.   **b.**

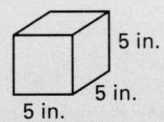 5 in. 5 in. 5 in.

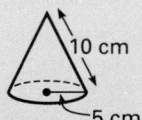 10 cm 5 cm

150 in.²   **236 cm²**

2. How much canvas would you need to make a conical tent with a slant height of 10 ft and a diameter of 12 ft? **301.44 ft²**

3. What is the surface area of a cylinder with a height of 12 m and a diameter of 4 m? **175.84 m²**

Get Ready calculators

SPOTLIGHT

OBJECTIVE
- Use the strategy of guess and check to solve problems

MATERIALS NEEDED
calculators

VOCABULARY
dimensions, guess and check

WARM-UP

Find the value of the variable.
1. $12 + x = 30$ **18**
2. $20 \times a = 180$ **9**
3. $549 \div b = 61$ **9**

 MOTIVATE

Introduction Explain that one way of checking if a guess is correct is by substituting the guess into the formula and then trying to solve the problem.

 **TEACH**

Use the Pages/Problem Have students read this part of the section and then discuss the problem. Before reading the solution, have students discuss ways they would approach this problem. Have students identify what information is given and what information must be found in order to solve the problem.

Use the Pages/Solution Point out that making a list is another problem solving strategy that works well with the guess-and-check strategy.

▶ READ
▶ PLAN
▶ SOLVE
▶ ANSWER
▶ CHECK

To solve some area problems, you may not be able to apply a geometric formula directly. For example, you may know the area of a figure and may need to find the dimensions. An effective approach to such a problem is the use of the strategy, *guess and check*. Make guesses for the length and width of the figure, each time checking to see if the product of the pair of numbers you guessed equals the known area.

PROBLEM

The area of a rectangular rug is 40.5 ft². The length is twice the width. What are the dimensions of the rug?

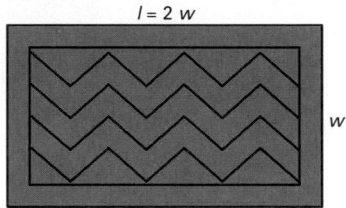

$l = 2w$

w

SOLUTION

Use a guess-and-check approach. After you make each guess for the width, find the corresponding length. Check each pair of your guesses for length and width to see whether the area is 40.5 ft².

	width	length	area	
	w	$l = 2w$	$A = l \times w$	
guess →	10 ft	20 ft	200 ft²	(too large)
guess →	5 ft	10 ft	50 ft²	(too large)
	4 ft	8 ft	32 ft²	(too small)

Notice that 40.5 is between 50 and 32.

The width must be between 4 ft and 5 ft. Since the last digit of 40.5 is 5, try 4.5 ft for the width.

$w = 4.5$
$l = 2 \times 4.5 = 9$
$A = 9 \times 4.5 = 40.5$

The rug is 4.5 ft wide and 9 ft long.

AT-RISK STUDENTS

These students may need additional practice in using guess and check with a familiar problem. Substitute several different values for the dimensions given in the sample problem. Have these students make lists and use guess and check to solve each.

PROBLEMS

1. The area of a triangular tile is 60 in.^{2} The height is 2 in. more than the base. Copy and complete the table to find the base and height of the tile. **base 10 in.; height 12 in.**

base	height	area
b	$h = b + 2$	$\frac{1}{2} \times b \times h$
6	■	■ 8; 24
12	■	■ 14; 84

The number 60 ends in zero, so try ___?___.   **10**

2. The volume of a rectangular box is 500 cm^3. The length is twice the width. The height is 5 cm more than the width. Find the dimensions of the box. **length 10 cm; width 5 cm; height 10 cm**

3. The volume of a pyramid is 98 cm^3. The area of the base is 49 cm^2. Find the height. **6 cm**

4. The area of a rectangular garden is 105 m^2. The width is 8 m less than the length. Find the dimensions of the garden. **length 15 m; width 7 m**

5. The area of a triangular park is 2,400 yd^2. The height of the triangle is three times the base. Find the dimensions of the park. **base 40 yd; height 120 yd**

6. A park in the shape of a parallelogram has a perimeter of 160 m. The longer sides of the park are 10 m longer than the shorter sides. How long are the sides of the park? **longer sides 45 m; shorter sides 35 m**

7. The volume of a cube is 1,728 cm^3. Find the length of an edge of the cube. **12 cm**

8. The distance around the track at the right is 714 yd. The length of the rectangle is twice the diameter of each semicircle. Find the dimensions of the rectangle. **length 200 yd; width 100 yd**

ASSIGNMENTS

BASIC
1, 3–5, 7

AVERAGE
1–7

ENRICHED
1–8

ADDITIONAL RESOURCES
Reteaching 11–12
Enrichment 11–12

3 SUMMARIZE

Key Questions
1. What is the guess-and-check strategy?
2. When is guess and check an appropriate strategy to use?

4 PRACTICE

Before students begin to solve these problems you may wish to review the formulas for finding the area and volume of various 3-dimensional figures. Students should use calculators to complete these problems. For Exercise 8 students may work in pairs to discuss how to approach the problem, which involves higher level thinking skills.

5 FOLLOW-UP

Extra Practice
1. The volume of a cylinder is 942 cm^3. The diameter is 10 cm. What is the height? **12 in.**
2. What is the circumference of a circle that has an area of 113.04 cm^2? **37.68 cm**

11 CHAPTER REVIEW

Introduction The Chapter Review emphasizes the major concepts, skills, and vocabulary presented in this chapter and can be used for diagnosing students' strengths and weaknesses. Page references direct students back to appropriate sections for additional review and reteaching.

Using Pages 430–431 For Exercises 1–6 you may wish to have students draw each figure listed in the box or display examples of the figures and be sure students can identify them. For Exercise 24 have students share their guesses and tell how they arrived at their answers.

Informal Evaluation Have students work in pairs to draw a diagram of a one-floor apartment. Have them include the dimensions and area of each room: kitchen, bathroom, living room, and two bedrooms. Students should also include a closet in each bedroom and find the volume of each. Suggest that students show a circular rug in the bathroom and find its area. Finally, have them show a refrigerator and stove in the kitchen and find the surface area of each.

Match the expression at the left with the name of the figure in the box. Tell whether the expression is used in the formula for *area*, *volume*, or *surface area*. Some words are used more than once.

1. $B \times h$ b; volume
2. $\frac{4}{3} \times \pi r^3$ d; volume
3. $\frac{1}{2} \times b \times h$ a; area
4. $\frac{1}{3} \times B \times h$ c or e; volume
5. $\frac{1}{3} \times \pi r^2 \times h$ c; volume
6. $4 \pi r^2$ d; surface area

a. triangle	d. sphere
b. rectangular prism	e. pyramid
	f. cylinder
c. cone	g. circle

SECTION 11–1   AREA AND VOLUME   (pages 384–387)

► **Area** is the amount of surface enclosed by a plane figure. Area is measured in *square units*.
► **Volume** is the amount of space enclosed by a three-dimensional figure. Volume is measured in *cubic units*.

Find the area.

7.
6 square units

8. 8 square units

Find the volume.

9.
18 cubic units

10. 6 cubic units

SECTIONS 11–2, 11–3   AREA OF POLYGONS   (pages 388–395)

► You can use these formulas to find area:
rectangle: $A = l \times w$ square: $A = s^2$
parallelogram: $A = b \times h$ triangle: $A = \frac{1}{2} \times b \times h$

Find the area of each figure.

11. 9 in. 45 in.² 5 in.

12. 6 cm 24 cm² 8 cm

13. 16 ft² 4 ft

14. 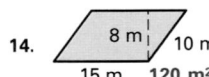 8 m 10 m 15 m 120 m²

SECTIONS 11–5, 11–6   THE PYTHAGOREAN THEOREM   (pages 398–405)

► Since $9^2 = 81$, $\sqrt{81} = 9$. Since $-9^2 = 81$, $-\sqrt{81} = -9$.
► For all right triangles, $c^2 = a^2 + b^2$, where c is the length of the hypotenuse and a and b are the lengths of the legs.

15. Write the positive square roots of 121 and 78 rounded to the nearest tenth.
11; 8.8

16. Find the measure of the hypotenuse of right triangle *ABC* with legs that measure 9 in. and 12 in.   15 in.

SECTION 11-8   AREA OF CIRCLES   (pages 410–413)

► Find the area of a circle with radius r by using the formula $A = \pi r^2$

17. Find the area of a circle with a radius of 7 in. Round to the nearest whole number. **154 in.²**

SECTIONS 11-7, 11-9, 11-10   VOLUME   (pages 406–409, 414–421)

► The **volume** of any prism is given by the formula $V = B \times h$.
► You can use these special formulas to find volume:

pyramid: $V = \frac{1}{3} \times B \times h$ cone: $V = \frac{1}{3} \times \pi r^2 \times h$ sphere: $V = \frac{4}{3} \times \pi r^3$

18. Find the volume of a prism with a height of 7 in. and a base that measures 6 in. by 8 in. Round to the nearest whole number. **336 in.³**

19. Find the volume of a sphere with a radius of 7 in. Round to the nearest whole number. **1,436 in.**

► To find the volume of a cylinder, use the formula $V = \pi r^2 \times h$.

20. Find the volume of a cylinder whose base has a radius of 7 ft and whose height is 10 ft. Round to the nearest tenth. Use $\pi \approx 3.14$. **1,538.6 ft³**

SECTION 11-11   SURFACE AREA   (pages 422–425)

► The **surface area** of a prism or pyramid is the sum of all its faces. These figures have special formulas for surface area:

sphere: $SA = 4\pi r^2$ cone: $SA = \pi rs + \pi r^2$

Find the surface area of each figure.

21.
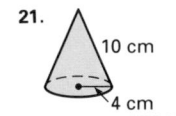
10 cm
4 cm
175.84 cm²

22.
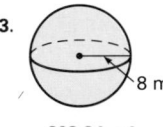
5 in.
4 in.
8 in.
184 in.²

23.
8 m
803.84 m²

SECTIONS 11-4, 11-12   PROBLEM SOLVING   (pages 396–397, 426–427)

► You need to recognize whether the formula for perimeter or for area is the formula you need to solve word problems.

► **Guess and check** is a strategy for solving certain kinds of problems.

24. The area of a triangular arm patch is 20 cm. The base is 3 cm longer than the height. Find the base and height of the patch. **b = 8 cm; h = 5 cm**

USING DATA   Use the diagram on page 385 to answer the following.

25. What is the total playing surface on a hockey rink?   **17,000 ft²**

Review   **431**

Find the area of each figure.

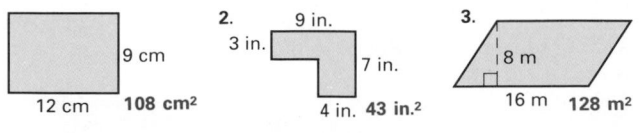
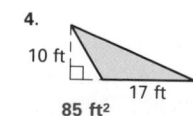

1. 9 cm, 12 cm **108 cm²**

2. 9 in., 3 in., 7 in., 4 in. **43 in.²**

3. 8 m, 16 m **128 m²**

4. 10 ft, 17 ft **85 ft²**

Find the square root to the nearest tenth.

5. $\sqrt{169}$ **13**   6. $\sqrt{1600}$ **40**   7. $\sqrt{91}$ **9.5**   8. $\sqrt{159}$ **12.6**

Find the unknown length.

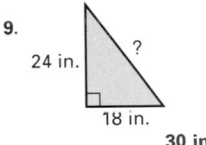
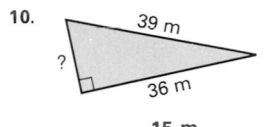
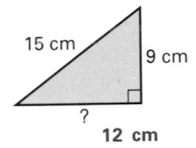

9. 24 in., ?, 18 in. **30 in.**

10. 39 m, ?, 36 m **15 m**

11. 15 cm, 9 cm, ?, 12 cm

Find the volume of each figure. Round to the nearest whole number.

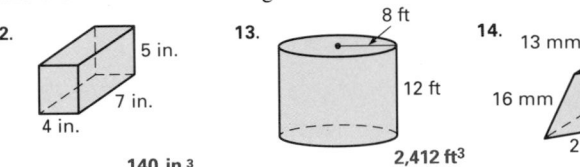
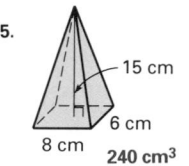

12. 5 in., 7 in., 4 in. **140 in.³**

13. 8 ft, 12 ft **2,412 ft³**

14. 13 mm, 5 mm, 16 mm, 24 mm **960 mm³**

15. 15 cm, 6 cm, 8 cm **240 cm³**

16. A large mound of sand is in the shape of a cone. The diameter is 42 ft and the height is 9 ft. Find the volume. Use $\pi \approx 3.14$. Round to the nearest whole number.
4154 ft³

Copy and complete the table.

	radius (r)	diameter (d)	area (A)	
17.	20 cm	■	■	40 cm; 1,256 cm²
18.	■	10 cm	■	5 cm; 78.5 cm²

Find the surface area of each figure.

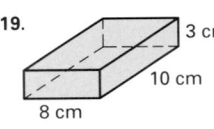
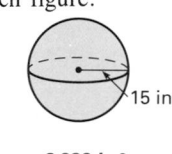
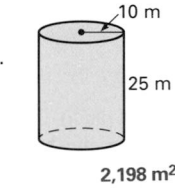
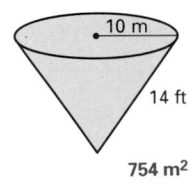

19. 3 cm, 10 cm, 8 cm **268 cm²**

20. 15 in. **2,826 in.²**

21. 10 m, 25 m **2,198 m²**

22. 10 m, 14 ft **754 m²**

23. The area of a rectangle is 84 cm². The length is 5 cm more than the width. Find the dimensions of the rectangle. **7 cm wide, 12 cm long**

432

1. Draw a line graph to show the temperature data. **See Additional Answers.**

AVERAGE MONTHLY TEMPERATURE (BOISE, IDAHO)												
Month	J	F	M	A	M	J	J	A	S	O	N	D
Boise	30	36	41	49	57	66	75	72	63	52	40	32

Complete.

2. $1,468 + \blacksquare = 2,193 + 1,468$ **2,193**

3. $8,762 \times 25 = 25 \times \blacksquare$ **8,762**

4. $(6 + 8) + 4 = \blacksquare + (8 + 4)$ **6**

5. Write two conditional statements using the following two sentences:

Melba is thirteen.
She cannot be twelve.

If Melba is thirteen, then she cannot be twelve.
If Melba is twelve, then she cannot be thirteen.

Replace ● with $<$, $>$, or $=$.

6. $\frac{1}{8}$ ● $\frac{1}{15}$ **>**   7. $\frac{3}{7}$ ● $\frac{3}{8}$ **>**   8. $\frac{5}{8}$ ● $\frac{5}{6}$ **<**

Complete.

9. $17 \text{ ft} = \blacksquare \text{ yd} \blacksquare \text{ ft}$ **5, 2**

10. $7 \text{ mm} = \blacksquare \text{ dm}$ **0.07**

11. $8,025 \text{ g} = \blacksquare \text{ kg}$ **8.025**

12. $19 \text{ pt} = \blacksquare \text{ qt} \blacksquare \text{ pt}$ **9,1**

Divide.

13. $56 \div (-7)$ **-8**   14. $-70 \div (-10)$ **7**

15. $-72 \div 8$ **-9**   16. $24 \div 3$ **8**

Solve each equation.

17. $77 = 10a + 7$ **a = 7**

18. $10.2 = 6t - 1.2$ **t = 1.9**

19. $3m - 2 = 25$ **m = 9**

20. $\frac{x}{-3} + 1 = -1$ **x = 6**

Solve each proportion.

21. $\frac{2.4}{4.8} = \frac{x}{4}$ **x = 2**

22. $2.5:10 = x:40$ **x = 10**

23. What is 40% of 256? **102.4**

24. What is the amount of interest earned in one year on $5,000 if the rate of interest is 7.5% per year? **$375**

Find the unknown angle in each triangle.

25.
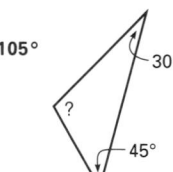

26.
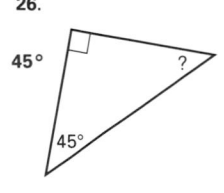

27.
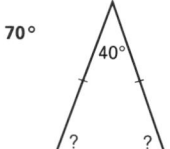

28.
15°
20°
145°
?

Find the area of each figure.

29.
54 m²
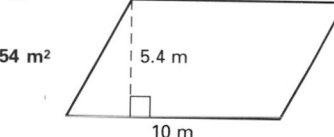
5.4 m
10 m

30.
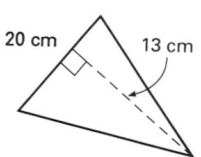
20 cm
13 cm
130 cm²

Introduction The purpose of this Cumulative Review is to maintain previously taught skills and concepts and to apply them to the material presented in this chapter. At least one major objective of each chapter is included in the review.

Item Analysis The table below correlates the Cumulative Review items with the chapter and section that are being reviewed.

Section	Items
1–8	1
2–9	2–4
3–3	5
4–2	6–8
5–1	9, 12
5–2	10–11
6–5	13–16
7–5	17–20
8–4	21–22
9–2	23
9–8	24
10–4	25–28
11–3	29–30

Additional Answers
See page 582.

11 CUMULATIVE TEST

Introduction The Cumulative Test uses a standardized-test format of multiple-choice questions to assess retention of previously learned concepts and test-taking skills. Test results may be used to diagnose students' strengths and weaknesses.

Item Analysis The table below correlates the Cumulative Test items with the chapter and section that are being tested.

Section	Items
1–9	1
2–9	2
3–3	3
4–3	4
5–1	5
6–4	6–7
7–5	9
7–6	8
8–6	10
9–9	11–12
10–2	13–14
11–8	15
11–9	16

1. What is the mean weekly salary of someone who earns $20,000 per year?
   A. $1,667.67   B. $384.62
   C. $9.61   D. none of these

2. Complete.
   $2,934 + \blacksquare = 1,846 + 2,934$
   A. 4,780   B. 1,088
   C. 2,934   D. 1,846

3. Which is a counterexample for the following statement?
   If a number is divisible by 6, then it is divisible by 12.
   A. 18   B. 36
   C. 48   D. 72

4. Which is the fraction for 7.16?
   A. $7\frac{1}{6}$ B. $7\frac{4}{25}$
   C. $\frac{7}{16}$ D. $7\frac{1}{4}$

5. Complete.
   76 oz = $\blacksquare$ lb $\blacksquare$ oz
   A. 4 lb 12 oz   B. 6 lb 4 oz
   C. 9 lb 4 oz   D. 19 lb 0 oz

6. Multiply. $10 \times (-3)$
   A. -7 B. 13
   C. -30 D. -13

7. Multiply. $-20 \times (-4)$
   A. 80   B. -16
   C. 16   D. -80

8. Solve. $\frac{2}{3}t = -18$
   A. $t = 12$ B. $t = 18$
   C. $t = -27$ D. $t = 6$

9. Solve. $4k - 42 = 18$
   A. $k = 24$ B. $k = 15$
   C. $k = -20$ D. $k = -18$

10. The scale of a drawing is 1 in.:3 ft. A room is drawn $4\frac{1}{4}$ in. wide. What is the width of the actual room?
    A. 12 ft   B. $12\frac{3}{4}$ ft
    C. $7\frac{1}{4}$ ft   D. none of these

11. A power tool sells for $125. James paid $5.75 in sales tax. What was the sales tax rate?
    A. 0.4%   B. 6.6%
    C. 21.7%   D. 4.6%

12. What is the interest for one year on a $5,000 deposit if the rate of interest is 9% per year?
    A. $4,500   B. $450
    C. $5,450   D. $4,550

13. $\angle JKL$ is 75°. Find the measure of its complement.
    A. 15°   B. 105°
    C. 75°   D. none of these

14. $\angle GHI$ is 35°. Find the measure of its supplement.
    A. 45°   B. 55°
    C. 145°   D. none of these

15. Find the area of a circle with a radius of 2.6 cm.
    A. 21.2 cm²   B. 16.3 cm²
    C. 5.3 cm²   D. 5.2 cm²

16. Find the volume of a cylinder with a radius of 1.5 in. and a height of 4 in.
    A. 11.3 in.³   B. 18.8 in.³
    C. 28.3 in.³   D. 7.1 in.³

Area and Volume

Name _____ Date _____

To find perimeter, count the number of units of length around the outside of a figure.

1 centimeter, or 1 cm

To find area, count the number of square units needed to cover a figure.

1 square centimeter, or 1 cm²

To find volume, count the number of cubic units that fill the space.

1 cubic centimeter, or 1 cm³

► **Example 1**

Find the area and perimeter of the figure.
1 unit = 1 cm

Solution

There are 5 squares.
The area is 5 cm².

The perimeter is 12 cm.

► **Example 2**

Find the volume of the solid figure.
1 unit = 1 cm

Solution

Count the top layer of cubes.
Count the number of layers.
    4 cubes in top layer
    2 layers
The volume is 2 times 4 cubes, or 8 cubic units. So, the volume is 8 cm³.

EXERCISES

Find the perimeter and area of each figure. 1 unit = 1 cm

1.
P = 16 cm
A = 7 cm²

2.
P = 14 cm
A = 10 cm²

3.
P = 26 cm
A = 12 cm²

4.
P = 22 cm
A = 10 cm²

Find the volume of each solid figure. 1 unit = 1 in.

5.
V = 8 in.³

6.
V = 9 in.³

7.
V = 12 in.³

8.
V = 12 in.³

Shaded Regions

Name _____ Date _____

You can find the area and volume of irregularly shaped regions.

► **Example**

a. Find the area of the shaded regions of the figure. 1 unit = 1 cm

b. Find the volume of the shaded region of the figure. Assume that if a cube is shaded, all the cubes in that row or column are shaded. 1 unit = 1 cm

2 cubes shaded top to bottom

3 cubes shaded front to back

Solution

Each half of the figure is one unit square, or 1 cm². Each shaded region is ½ the area of a square, or ½ cm². So, the area of the two shaded regions is ½ + ½, or 1 cm².

Solution

Count the cubes in the bottom row that is shaded. There are 3. Count the cubes in the column that is shaded. There are 2. Five cubes are shaded. So, the volume of the shaded region is 5 cm³.

EXERCISES

Find the area of the shaded portion of each figure. 1 unit = 1 in.

1.
A = 12 in.²

2.
A = 8 in.²

3.
A = 21 in.²

Find the volume of the shaded region. If a cube is shaded on the outside, then all cubes in that row or column are shaded. 1 unit = 1 cm

4.
V = 7 cm³

5.
V = 19 cm³

6.
V = 32 cm³

Area of Rectangles

Name _____ Date _____

A unit square measures one unit on each side. The area of a figure is the number of unit squares that cover it. To find the area of a rectangle, multiply length times width.

w = width

l = length

Use the formula:
$A = l \times w$

1 square centimeter 1 cm²

1 square inch 1 in.²

► **Example 1**

Find the area of Larry's room.

9 ft
12 ft

Solution

Use the formula $A = l \times w$
$A = 12 \times 9$
$A = 108$

The area of Larry's room is 108 ft².

► **Example 2**

Find the area of Martita's room.

6 ft
6 ft 5 ft
6 ft A₁   A₂ 11 ft
12 ft

Solution

Divide the room into two rectangles. Find the area of each part and add.
$A_1 = 6 \times 6 = 36$
$A_2 = 6 \times 11 = 66$
$A_1 + A_2 = 36 + 66 = 102$

The area of Martita's room is 102 ft².

EXERCISES

Use the floor plan to find the area of the room to the nearest square foot. 1 unit = 2 ft

1. Bedroom 2   160 ft²
2. Bath 2   50 ft²
3. Dining room   148 ft²
4. Master suite/Nursery   240 ft²
5. L-shaped Living room   339 ft²
6. Kitchen   215 ft²

Enclosing the Greatest Area

Name _____ Date _____

Suppose you are making a temporary rectangular pen for your dog. You have 51 feet of fencing to use for three sides of the enclosure. You may use up to 20 feet, the width of your house, for the fourth side.

51 ft of fencing
20 ft
House

EXERCISES

Find the largest rectangular area that you can enclose under each of the following conditions, with minimum waste of fencing. For each possibility, draw a diagram, and explain your reasoning.

1. The fencing can be bent or cut off at 1-foot intervals.
   20 ft / 18 ft / 19 ft / 16 ft
   Not enough fencing to use 20 ft length
   A = 19 × 16 = 304 ft² All the fencing is used.

2. The fencing can be bent or cut off at 2-foot intervals.
   18 ft / 16 ft
   A = 18 × 16 = 288 ft² 1 ft is wasted.

3. The fencing can be bent or cut off at 3-foot intervals.
   18 ft / 15 ft
   Greatest length possible is 18 ft.
   Out of 33 ft left, two 15-ft lengths are possible.
   A = 18 × 15 = 270 ft² 3 ft of fencing is wasted.

4. The fencing can be bent or cut off at 4-foot intervals.
   20 ft / 12 ft
   Greatest length is 20 ft, but of the 31 ft left, you can get two 12-ft lengths at most. A = 20 × 12 = 240 ft² 7 ft is wasted. You can use a 16-ft square, 256 ft², with only 3 ft of fencing wasted.

5. The fencing can be bent or cut off at 6-inch intervals.
   20 ft / 15.5 ft
   Can cut off 20 ft. The 31 ft left can be split into two 15.5-ft widths.
   A = 20 × 15.5 = 310 ft² All the fencing is used.

6. What conclusion can you draw regarding area and maximum use of fencing?
   The less fencing you waste, the larger the area you can enclose.

7. Write a problem like this one for someone else to solve.
   Answers will vary.

434A

Name _____ Date _____

Area of Parallelograms and Triangles

To find the area of a parallelogram or a triangle, you use the measure of the base and the height. The height is always perpendicular to the base. Here are some possible positions of the height and base.

Parallelograms   **Triangles**

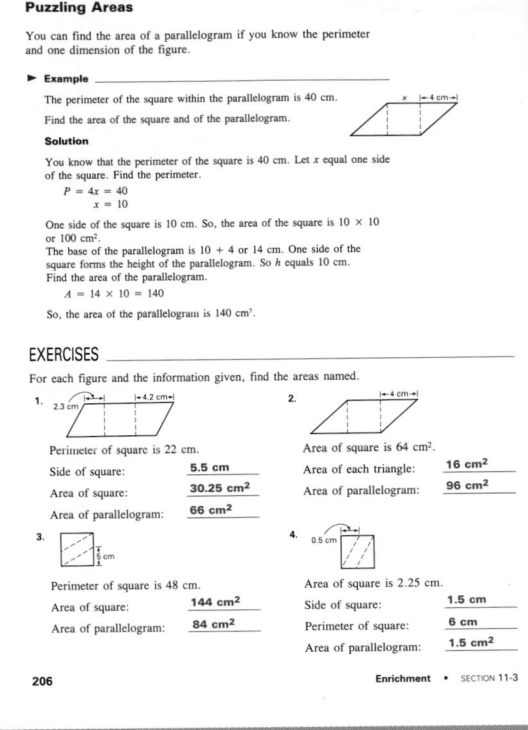

Area of a parallelogram:
$A = b \times h$

Area of a triangle:
$A = \frac{1}{2}(b \times h)$

► **Example 1** _____

Find the area of the parallelogram.

17 m
35 m

Solution

$A = b \times h$
$A = 35 \times 17 = 595$
The area is 595 mm².

► **Example 2** _____

Find the area of the triangle.

5 cm
12 m

Solution

$A = \frac{1}{2}(b \times h)$
$A = \frac{1}{2}(12)(5) = 30$
The area is 30 cm².

EXERCISES

Find the area of each figure to the nearest whole number.

1. 6 in.
   9 in.
   27 in.²

2. 1.2 yd
   19.7 yd
   24 yd²

3. 55 cm
   64 cm
   1760 cm²

4. 11 mm
   13 mm
   143 mm²

5. 15 m
   8.3 m
   125 m²

6. 14.6 ft
   40.5 ft
   296 ft²

7. Find the area of a hair ribbon 18 inches long with its ends cut on the diagonal. The width of the ribbon is 1.5 inches.   **27 in.²**

8. Find the area of the top surface of a napkin folded into a triangle. The folded edge is 13 cm long, and the height is 7 cm.   **45.5 cm²**

Reteaching • SECTION 11-3   205

Name _____ Date _____

Puzzling Areas

You can find the area of a parallelogram if you know the perimeter and one dimension of the figure.

► **Example** _____

The perimeter of the square within the parallelogram is 40 cm. Find the area of the square and of the parallelogram.

x   4 cm

Solution

You know that the perimeter of the square is 40 cm. Let x equal one side of the square. Find the perimeter.
$P = 4x = 40$
$x = 10$

One side of the square is 10 cm. So, the area of the square is 10×10 or 100 cm². The base of the parallelogram is $10 + 4$ or 14 cm. One side of the square forms the height of the parallelogram. So h equals 10 cm. Find the area of the parallelogram.
$A = 14 \times 10 = 140$

So, the area of the parallelogram is 140 cm³.

EXERCISES

For each figure and the information given, find the areas named.

1. 2.3 cm   4.2 cm
   Perimeter of square is 22 cm.
   Side of square: **5.5 cm**
   Area of square: **30.25 cm²**
   Area of parallelogram: **66 cm²**

2. 4 cm
   Area of square is 64 cm².
   Area of each triangle: **16 cm²**
   Area of parallelogram: **96 cm²**

3. 5 cm
   Perimeter of square is 48 cm.
   Area of square: **144 cm²**
   Area of parallelogram: **84 cm²**

4. 0.5 cm
   Area of square is 2.25 cm.
   Side of square: **1.5 cm**
   Perimeter of square: **6 cm**
   Area of parallelogram: **1.5 cm²**

206   Enrichment • SECTION 11-3

Name _____ Date _____

Problem Solving Skills: Use a Formula for Perimeter or Area

Before you can solve a geometric problem of measure, you must recognize which formula you need to use. Sometimes you may need a formula for perimeter. Other times you may need a formula for area.

► **Example** _____

A rectangular flower garden is 9 ft wide and 12 ft long. How much wire fencing will be needed to enclose it?

Solution

The fencing needed is to enclose, or go around, the garden. You need to know "distance around." Find the perimeter of the garden plot.
$P = 2l + 2w$
$= 2 \times 12 + 2 \times 9 = 24 + 18$
$= 42$ ft

So, 42 ft of fencing will be needed to enclose the garden.

EXERCISES

Which would you need, a perimeter or an area formula?

1. amount of lace strip to decorate the edges of a handkerchief   **P**

2. amount of material needed for a square tarpaulin boat cover   **A**

Use one of these formulas to solve each problem.
$P = 2l + 2w$ $A = l \times w$ $A = b \times h$ $A = \frac{1}{2} \times b \times h$

1. The width of one rectangular side of a tent measures 3 m by 4 m. How much canvas would be required for both sides?
   $A = l \times w$; 24 m²

2. Marna's backyard measures 80 yd by 90 yd. If sod costs $1.35/yd², how much will Marna pay for sod to cover the entire yard?
   $A = l \times w$; $108

3. Rhonda's club is making pennants to sell at basketball games. A model for a pennant is shown. How many square feet of felt is needed for 300 pennants?
   8 in.   18 in.
   $A = \frac{1}{2} \times b \times h$; 150 ft²

4. How much wrapping paper will an art dealer need to cover both sides of a large rectangular piece of stained glass? The glass is 7 feet high and 4 feet wide.
   $A = l \times w$; 56 ft²

Reteaching • SECTION 11-4   207

Name _____ Date _____

Using Perimeter Formulas for Polygons

A regular polygon is one whose sides have the same length and whose angles each have the same degree measure. You can use the fact that regular polygons are equilateral to find an unknown side of a polygon.

► **Example** _____

The perimeter of a regular pentagon is 35 cm. Find the length of each side.

Solution

The pentagon has 5 equal sides. Let $n =$ the length of each side. The perimeter can be found by this formula: $P = 5n$. The perimeter equals 35 cm.
$P = 35$
$5n = 35$
$n = 7$

The length of each side of the pentagon is 5 cm.

EXERCISES

Write an equation. Then use it to solve. You may draw a picture to help.

1. The perimeter of a square is 86 mm. Find the length of a side.
   $4n = 86$; $n = 21.5$ mm

2. The perimeter of a regular hexagon is 84 ft. Find the length of a side.
   $6n = 84$; $n = 14$ ft

3. The perimeter of an equilateral triangle is 225 m. Find the length of a side.
   $3n = 225$; $n = 75$ m

4. The perimeter of a regular octagon is 7.68 km. Find the length of each side.
   $8n = 7.68$; $n = 0.96$ km

5. Complete the chart for a perimeter of 72 mm.

Regular polygon	triangle	pentagon	hexagon	octagon	nonagon
Length of a side	24	14.4	12	9	8

Find the missing dimensions.

6. The perimeter of an isosceles triangle is 7.5 m. The length of the base is 3.5 m. Find the length of each of the other sides.   **2 m**

7. The perimeter of a hexagon with four sides of equal length is 13.8 km. The other two sides have lengths of 5 km and 5 km. Find the length of each of the sides that are equal in length.   **0.95 km**

208   Enrichment • SECTION 11-4

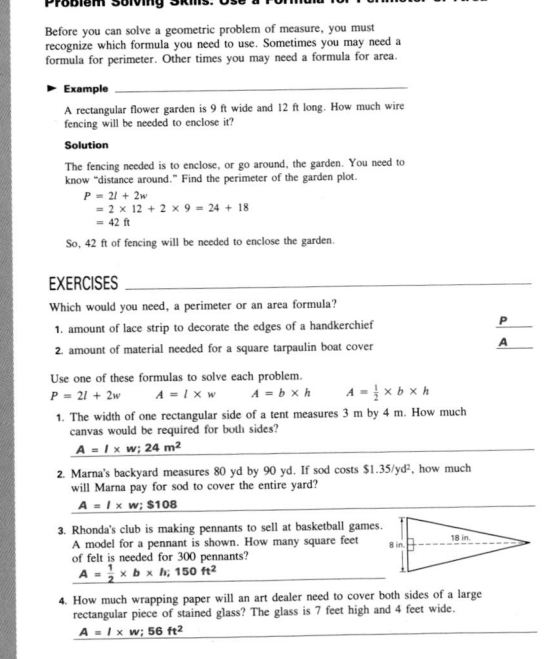

Name _____ Date _____

Squares and Square Roots

The square of 6 is 36, and the square of −6 is 36.

$$6 \times 6 = 6^2 = 36 \quad \text{and} \quad (-6) \times (-6) = (-6)^2 = 36$$

6 is the *positive square root*   −6 is the *negative square root*
of 36. $\sqrt{36} = 6$ of 36. $-\sqrt{36} = -6$

Any number whose square roots are integers is a **perfect square.** In the table of square roots on page 401 of your text, the numbers 1, 4, 9, 16, 25, 36, 49, 64, 81, and 100 are perfect squares. All of the numbers in the table other than these have square roots that are **irrational numbers**—nonterminating, nonrepeating decimals. In the table, the square roots of these numbers are approximations, rounded to the nearest thousandth. The square root of 2, for example, is 1.414.

▶ **Example 1**

Find each square.

a. 5^2 b. $\left(\frac{3}{4}\right)^2$ c. $(0.5)^2$

Solution

a. $5^2 = 5 \times 5 = 25$
b. $\left(\frac{3}{4}\right)^2 = \left(\frac{3}{4}\right) \times \left(\frac{3}{4}\right) = \frac{9}{16}$
c. $(0.5)^2 = 0.5 \times 0.5 = 0.25$

▶ **Example 2**

Find each square root.

a. $\sqrt{0.81}$ b. $-\sqrt{\frac{9}{25}}$ c. $\sqrt{38}$

Solution

a. $0.9 \times 0.9 = 0.81$, so, $\sqrt{0.81} = 0.9$
b. $\left(-\frac{3}{5}\right) \times \left(-\frac{3}{5}\right) = \frac{9}{25}$, so, $-\sqrt{\frac{9}{25}} = -\frac{3}{5}$
c. Find 38 in the *number* column of the table of square roots. Read to the right. The square root of 38, rounded to the nearest thousandth, is 6.164.

EXERCISES

Find each square.

1. 19^2	2. 22^2	3. 25^2	4. $(-16)^2$	5. 39^2	6. $(-28)^2$
361	**484**	**625**	**256**	**1,521**	**784**
7. $(-0.19)^2$	8. $(1.6)^2$	9. $(3.2)^2$	10. $\left(\frac{5}{7}\right)^2$	11. $\left(\frac{8}{9}\right)^2$	12. $\left(\frac{3}{4}\right)^2$
0.0361	**2.56**	**10.24**	**$\frac{25}{49}$**	**$\frac{64}{81}$**	**$\frac{9}{16}$**

Use the table of square roots to find each square root.

13. $\sqrt{1,024}$	14. $\sqrt{676}$	15. $-\sqrt{289}$	16. $\sqrt{0.81}$	17. $\sqrt{0.0049}$	18. $-\sqrt{2.25}$
32	**26**	**−17**	**0.9**	**0.07**	**−1.5**
19. $\sqrt{27}$	20. $-\sqrt{19}$	21. $\sqrt{89}$	22. $-\sqrt{51}$	23. $\sqrt{75}$	24. $-\sqrt{13}$
5.196	**−4.359**	**9.434**	**−7.141**	**8.660**	**−3.606**

Reteaching • SECTION 11-5   **209**

Name _____ Date _____

Hands-On Estimation of Square Roots

You can use base-ten blocks (hundreds, tens, and units) to estimate square roots. (You may trace and cut out multiple copies of these figures.)

▶ **Example 1**

Estimate the square root. $\sqrt{123}$

Solution

123 is formed by 1 hundreds block, 2 tens blocks and 3 units blocks.

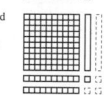

• Arrange the blocks to form the *largest possible square* from the blocks for 123 or for fewer blocks.
• The largest square is formed by 121 blocks in the arrangement at the right. There are 2 units left over.
• Then consider how many more blocks you would need to form the *next largest square.*
  The next largest square is 144 as in the second diagram. You need 23, or 2 tens and 3 units more.
• You already have 2 out of 23 or $\frac{2}{23}$ of the blocks needed to make the next square. So,
  $\sqrt{123} \approx \sqrt{121} + \frac{2}{23}$ or 11.087

▶ **Example 2**

Estimate the square root. $\sqrt{12}$

Solution

The hundreds and tens blocks are too big to use to make squares. Using 12 units, the largest square you can form is a 3 by 3 square of 9. You have 3 left over. The next largest square is 4 by 4 or 16. You need 7 more units. You have 3 out of 7.
$\sqrt{12} \approx \sqrt{9} + \frac{3}{7}$ or 3.429

EXERCISES

Use blocks to estimate each square root, to the nearest thousandth. Use a calculator to check. Find the difference between your estimate and the calculator value. How close is your estimate, in each case, to the calculator value?

1. $\sqrt{13}$ **3.571, 3.606; difference, 0.035**
2. $\sqrt{24}$ **4.889, 4.899; difference, 0.010**
3. $\sqrt{96}$ **9.894, 9.798; difference, 0.007**
4. $\sqrt{148}$ **12.160, 12.166; difference, 0.006**
5. $\sqrt{200}$ **14.138, 14.142; difference, 0.004**
6. $\sqrt{501}$ **22.378, 22.383; difference, 0.005**

210   Enrichment • SECTION 11-5

Name _____ Date _____

The Pythagorean Theorem

The Pythagorean Theorem states a relationship between the legs and the hypotenuse of a right triangle. It states:

In any right triangle, the square of the length of the hypotenuse is equal to the sum of the squares of the lengths of the legs.

That is, in a triangle with sides of lengths *a* and *b* and hypotenuse of length *c*.

$$c^2 = a^2 + b^2$$

You can use this theorem to find the length of the hypotenuse of a right triangle if you know the lengths of the two sides. You can also use the Pythagorean Theorem to find the unknown length of a leg, given the length of the other leg and the length of the hypotenuse.

▶ **Example 1**

Find the length of hypotenuse *c* of triangle *ABC*.

Solution

$c^2 = a^2 + b^2$
$c^2 = 6^2 + 8^2$
$c^2 = 36 + 64$
$c^2 = 100$
$c = 10$

The length of the hypotenuse is 10 cm.

▶ **Example 2**

In triangle *DEF*, find the unknown length *a*, to the nearest tenth.

Solution

$c^2 = a^2 + b^2$

Subtract b^2 from each side.
$c^2 - b^2 = a^2 + b^2 - b^2$
$a^2 = c^2 - b^2$
$a^2 = 16^2 - 9^2 = 256 - 81 = 175$
$a = \sqrt{175} = 13.229$, or 13.2

The length, *a*, is about 13.2 in.

EXERCISES

Use the Pythagorean Theorem to find the unknown length, to the nearest tenth.

1. 9 in., 12 in. → **15 in.**
2. 16 cm, 30 cm → **34 cm**
3. 4.0 cm, 7.5 cm → **8.5 cm**
4. 2 cm, 9 cm → **about 9.2 cm**
5. 12 in., 10 in. → **about 6.6 in.**
6. 3 cm, 14 cm → **about 13.7 cm**

Reteaching • SECTION 11-6   **211**

Name _____ Date _____

Pythagorean Puzzles

Use the Pythagorean Theorem to solve challenging problems.

▶ **Example**

Mayor Charles of Baileyville, on his way to Acton, followed Lake Road 14 kilometers north, Boundary Lane 7 kilometers east, and Anchor Avenue 10 kilometers north. He and the mayor of Acton agreed that a new road should be built connecting the two towns with a straight line. How much shorter would the new road from Baileyville to Acton be than the route Mayor Charles took?

Solution

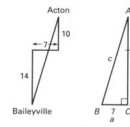

• Draw a diagram. Then redraw the diagram to form a right triangle, with $a = 7$ km; $b = 24$ km.
• Use the Pythagorean Property to find *c*, which is the straight line route between the towns.

$c^2 = a^2 + b^2$
$c^2 = 7^2 + 24^2$
$c^2 = 49 + 576 = 625$
$c = \sqrt{625} = 25$

The new road is 25 km. The original route was $7 + 24$ or 31 km. The new road would be 6 km shorter.

EXERCISES

Use the Pythagorean Theorem to solve.

1. The city bus company plans to change part of the Blue Avenue bus route. At present, after the bus leaves point A on the map, it travels 15 blocks north, then 8 blocks west, and then 15 blocks north again. At that point, the bus turns west again and travels 4 blocks before going 5 blocks north to reach point B. The city is planning an avenue that will go from point A to point B on a diagonal. How many blocks will be cut from the length of the bus route if the new avenue is built?

10 blocks saved

2. A hungry spider is sitting in a room 30 ft long, 12 ft wide and 12 ft high. It is sitting in the middle of an end wall, 1 ft from the ceiling. It spies a fly sitting in the middle of the opposite wall, 1 ft from the floor. What is the shortest distance the hungry spider can travel to catch the sleeping fly? (Hint: Imagine the room opened up like a flattened cardboard box.)

40 ft

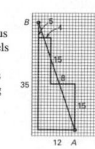

212   Enrichment • SECTION 11-6

Name _____ Date _____

Volume of Prisms

You can find the volume of a prism by counting the number of unit cubes that would fill the space. You can also use a formula. The volume of a prism is equal to the area of its base times its height, so you can use these formulas.

For a rectangular prism: For a triangular prism:
$V = l \times w \times h$ $V = B \times h$ where B = area of base
                             $B = \frac{1}{2} \times b \times h$

▶ **Example 1** _____

Find the volume of the rectangular prism.

Solution

$V = l \times w \times h$
$V = 22 \times 24 \times 12$
$V = 6,336$

The volume of the prism is 6,336 in.³

▶ **Example 2** _____

Find the volume of the triangular prism.

Solution

$V = B \times h$
$B = \frac{1}{2} \times b \times h$
$B = \frac{1}{2} \times 10 \times 7 = 35$
$V = 35 \times 30 = 1,050$

The volume of the prism is 1,050 cm³.

EXERCISES

Complete the chart to find the volume of the figure.

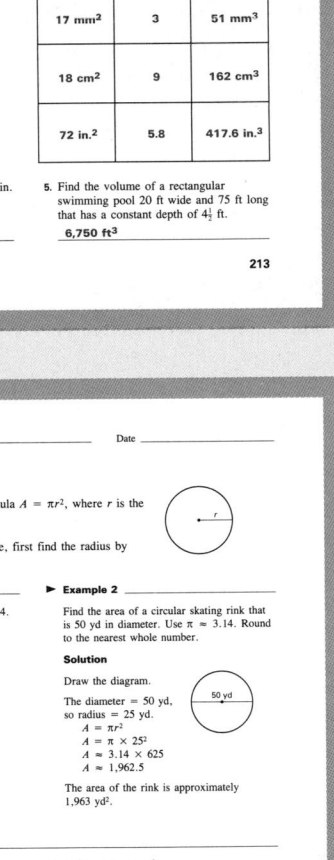

	B	h	Volume
1.	17 mm²	3	51 mm³
2.	18 cm²	9	162 cm³
3.	72 in.²	5.8	417.6 in.³

4. Find the volume of a concrete floor 3 in. thick in a rectangular basement that is 30 ft by 22 ft.

1,980 ft³

5. Find the volume of a rectangular swimming pool 20 ft wide and 75 ft long that has a constant depth of $4\frac{1}{2}$ ft.

6,750 ft³

Name _____ Date _____

Treasure Hunt

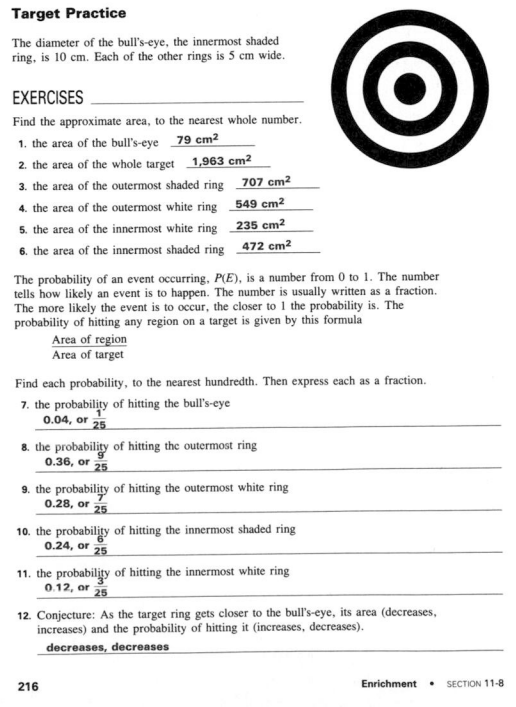

Five treasure chests are lying near a shipwreck. All are rusted shut. One is made of oak, one of maple, one of pine, one of cherry, and one of fir. A legend says that one contains silver, one gold, one platinum, and one diamonds. The fifth contains feathers. Use the following clues to discover which chest holds which treasure.

1. The chest having the greatest volume contains the lightest contents.
2. The least expensive material is in the pine chest.
3. The gold is in the oak chest.
4. The maple chest is greater in volume than the fir chest.
5. The chest with the least volume is the oak chest.
6. The chest filled with platinum has the smallest base area.
7. The chest having the median volume is made of cherry.
8. The diamonds are in the maple chest.

EXERCISES

Find the volume of each chest to the nearest tenth of a cubic inch. Then use the clues to discover what is in each chest.

		Volume	Type of chest	Contents of chest
least volume	C	1,166.4	oak	gold
	A	1,386	fir	silver
	E	1,392	cherry	platinum
	D	1,404.9	maple	diamonds
greatest volume	B	1,605.5	pine	feathers

A **1,386 in.³**   B **1,605.5 in.³**   C **1,166.4 in.³**   D **1,404.9 in.³**   E **1,392 in.³**

Name _____ Date _____

Area of Circles

To find the area of a circle, use the formula $A = \pi r^2$, where r is the radius of the circle. Use $\pi \approx 3.14$.

If you are given the diameter of the circle, first find the radius by taking one-half of the diameter.

▶ **Example 1** _____

Find the area of the circle. Use $\pi \approx 3.14$. Round to the nearest tenth.

Solution

$A = \pi r^2$
$A = \pi \times (2.1)^2$
$A = 3.14 \times 4.41$
$A \approx 13.847$

The area of the circle is approximately 13.8 cm².

▶ **Example 2** _____

Find the area of a circular skating rink that is 50 yd in diameter. Use $\pi \approx 3.14$. Round to the nearest whole number.

Solution

Draw the diagram.

The diameter = 50 yd, so radius = 25 yd.
$A = \pi r^2$
$A = \pi \times 25^2$
$A \approx 3.14 \times 625$
$A \approx 1,962.5$

The area of the rink is approximately 1,963 yd².

EXERCISES

Find the area of each circle. Use $\pi \approx 3.14$. Round to the nearest tenth.

1. **153.9 m²**
2. **0.5 km²**
3. approximately **55.4 mm³**

4. A rotating blade cuts grass in a circular area. If the diameter of the circular area is 2.2 ft, what is the area of the region cut with each rotation?

approximately 3.8 ft²

5. A circular irrigation system on a farm consists of a rotating arm that sprinkles water over a circular region. If the arm is 43 m long, find the area of the region.

approximately 5,805.9 m²

6. A circular sandbox has a diameter of 5.5 ft. What is its area?

approximately 23.7 ft²

7. Which is the better buy, a pizza 10 in. in diameter for $8.00 or a 12-in. square pizza for $9.25?

the square pizza

Name _____ Date _____

Target Practice

The diameter of the bull's-eye, the innermost shaded ring, is 10 cm. Each of the other rings is 5 cm wide.

EXERCISES

Find the approximate area, to the nearest whole number.

1. the area of the bull's-eye   **79 cm²**
2. the area of the whole target   **1,963 cm²**
3. the area of the outermost shaded ring   **707 cm²**
4. the area of the outermost white ring   **549 cm²**
5. the area of the innermost white ring   **235 cm²**
6. the area of the innermost shaded ring   **472 cm²**

The probability of an event occurring, $P(E)$, is a number from 0 to 1. The number tells how likely an event is to happen. The number is usually written as a fraction. The more likely the event is to occur, the closer to 1 the probability is. The probability of hitting any region on a target is given by this formula

$$\frac{\text{Area of region}}{\text{Area of target}}$$

Find each probability, to the nearest hundredth. Then express each as a fraction.

7. the probability of hitting the bull's-eye
   0.04, or $\frac{1}{25}$

8. the probability of hitting the outermost ring
   0.36, or $\frac{9}{25}$

9. the probability of hitting the outermost white ring
   0.28, or $\frac{7}{25}$

10. the probability of hitting the innermost shaded ring
    0.24, or $\frac{6}{25}$

11. the probability of hitting the innermost white ring
    0.12, or $\frac{3}{25}$

12. Conjecture: As the target ring gets closer to the bull's-eye, its area (decreases, increases) and the probability of hitting it (increases, decreases).

    decreases, decreases

434D

Volume of Cylinders

Name _____ Date _____

The base of a cylinder is a circle. You can find the volume of a cylinder by multiplying area of the base by the height.

$V = B \times h$
$V = \pi r^2 \times h$

► **Example**

Find the volume of the cylinder. Round to the nearest cubic centimeter. Use $\pi \approx 3.14$.

Solution

The diameter of the base is 14 cm, so the radius is 7 cm.
$V = \pi r^2 \times h$
$V \approx 3.14 \times 7^2 \times 24 \approx 3,692.64$

The volume is approximately 3,693 cm³.

14 cm
24 cm

EXERCISES

Find the volume of each cylinder. Use $\pi \approx 3.14$. Round to the nearest whole number.

1. 14 in., 7 in. — **4,308 in.³**
2. 8 mm, 3 mm — **226 mm³**
3. 7 ft, 19 ft — **731 ft³**
4. 11.8 m, 18 m — **7,870 m³**
5. 10.5 cm, 3.5 cm — **404 cm³**
6. 20 in., 4 in. — **973 in.³**
7. 12.6 cm, 2.3 cm — **209 cm³**
8. 4.6 cm, 14.7 cm — **977 cm³**

9. The diameter of a cylinder is 36 cm. The height is 48 cm. Find its volume.
48,833 cm³

10. A cylindrical pitcher has a diameter of 10 in. It stands 16 in. high. What is its volume?
1,256 in.³

Changing Dimensions

Name _____ Date _____

What happens to the volume of a cylinder if one or more dimensions change?

EXERCISES

Find each volume to the nearest cubic unit. Use $\pi \approx 3.14$. For Exercises 1–5, consider a cylindrical glass with a diameter of 5 cm and a height of 7.5 cm.

1. Find the volume of the glass.
$V \approx$ **147 cm³**

2. What will the volume of the glass be if its height is doubled?
$V \approx$ **294 cm³**

3. What happens to the volume when height is doubled?
When the height is doubled, the volume doubles.

4. What will the volume of the glass be if the height remains the same, but the diameter is doubled?
$V \approx$ **589 cm³**

5. What happens to the volume when the diameter is doubled?
The volume increases about 4 times.

For Exercises 6–10, use your answers to Exercises 1–5 and consider a cylindrical mug with a diameter of 8 cm and a height of 8 cm. Test each conjecture.

6. Find the volume of the mug.
$V \approx$ **402 cm³**

7. How will the volume change if its diameter remains the same, but its height doubles?
The volume will double. $V \approx$ **804 cm³**

8. How will the volume of the mug change if its diameter doubles, but its height remains the same?
The volume will be 4 times as great. $V \approx$ **1,608 cm³**

9. How will the volume change if the diameter remains the same, but the height triples?
The volume will double. $V \approx$ **1,206 cm³**

10. Predict the volume of a soup mug that is half as tall and twice as wide as the mug you have considered for Exercises 6–9.
The volume will double. $V \approx$ **804 cm³**

Volume of Pyramids, Cones, and Spheres

Name _____ Date _____

Formulas for the volume of a pyramid, a cone, and a sphere are given below. Compare the formula for the volume of a prism with that of a pyramid. Then compare the formula for the volume of a cone with that of a cylinder and that of a sphere.

Rectangular Prism	Rectangular Pyramid	Cylinder	Cone	Sphere
You know	New	You know	New	New
$V = B \times h$	$V = \frac{1}{3} \times B \times h$	$V = B \times h$	$V = \frac{1}{3} \times B \times h$	$V = \frac{4}{3} \times \pi r^3$
$V = l \times w \times h$	$V = \frac{1}{3} \times l \times w \times h$	$V = \pi r^2 \times h$	$V = \frac{1}{3} \times \pi r^2 \times h$	

► **Example**

Find the volume to the nearest whole number. Use $\pi \approx 3.14$.

Solution

a. $V = \frac{1}{3} \times l \times w \times h$
$V = \frac{1}{3} \times 3 \times 4 \times 5$
The volume of the pyramid is approximately 20 in.³

b. $V = \frac{1}{3} \times \pi r^2 \times h$
$V = \frac{1}{3} \times 3.14 \times 4^2 \times 6$
$V = \frac{1}{3} \times 3.14 \times 16 \times 6$
The volume of the cone is approximately 100 in.³

c. $V = \frac{4}{3} \times \pi r^3$
$V = \frac{4}{3} \times 3.14 \times 3^3$
$V \approx \frac{4}{3} \times 3.14 \times 27$
The volume of the sphere is approximately 113 in.³

EXERCISES

Find the volume of each, to the nearest tenth. Use $\pi \approx 3.14$.

1. 10 cm, 8 cm — $V \approx$ **267.9 cm³**
2. 5.3 cm, 7.2 cm — $V \approx$ **91.6 cm³**
3. 14 ft — $V \approx$ **11,488.2 ft³**
4. 12 m, 9 m — $V \approx$ **3,052.1 m³**

5. Find the volume of a cone 4.2 cm in diameter and 6.1 cm high.
$V \approx$ **9.0 cm³**

6. Find the volume of a sphere with a radius of 3 ft.
$V \approx$ **113.0 ft³**

7. Find the volume of a pyramid 12 cm high with a rectangular base measuring 6 cm by 8 cm.
$V \approx$ **192 cm³**

It Matters How You Slice It

Name _____ Date _____

Can you visualize what kind of plane figure, or **cross-section**, will result when you "slice" a three-dimensional figure?

When you slice a sphere horizontally, you get a circle.

When you slice a sphere at any angle, you still get a circle.

EXERCISES

Visualize, sketch the figure, or make a model to answer the questions.

1. Will all the cross-sections of a sphere have the same area?
No

2. Describe how you must slice a sphere in order to obtain circles having equal areas.
You must slice the sphere so that the diameter of each cross section is equal.

3. Name two other three-dimensional figures that could have cross-sections that are circles.
cylinder, cone

4. How would you have to slice these figures so that the cross-sections will be circles?
The slices must be made parallel to the base.

5. What plane figure will you get if you slice a cylinder perpendicular to its base?
a rectangle

6. What plane figure do you get if you slice any prism so that the cross-section is parallel to the base of the figure?
You get a cross-section having the same size and shape as the base.

7. How must you slice a cone so as to obtain a triangular cross-section?
You must slice the cone perpendicular to the base through its vertex.

8. Draw the figures you would get if you sliced a cone perpendicular to its base, but not through its vertex. **Check students' drawings.**

Surface Area

To find the surface area, *SA*, of a prism, find the sum of the areas of all its faces. To find the surface area of a cylinder, add the area of the curved surface to the sum of the areas of the two circular bases. Use these formulas.

rectangular prism: $SA = 2 \times (l \times w + l \times h + w \times h)$ **cylinder:** $SA = 2\pi rh + 2\pi r^2$

To find the surface area of a cone or a sphere, use these formulas.

cone: $SA = \pi rs + \pi r^2$ **sphere:** $SA = 4\pi r^2$
$s = $ slant height

► **Example**

Sketch and label the unfolded faces of each figure. Then use a formula to find the surface area. Use $\pi \approx 3.14$. Round to the nearest whole number.

a. b. c.

Solution

a. $2 \times (2 \times 3) = 12$
$2 \times (2 \times 4) = 16$
$2 \times (3 \times 4) = 24$
$SA = 12 + 16 + 24 = 52$
$SA = 52 \text{ cm}^2$

b. $SA = 2\pi rh + 2\pi r^2$
$2\pi rh = 2 \times 3.14 \times 3 \times 4 \approx 75.36$
$2\pi r^2 = 2 \times 3.14 \times 3^2$
$= 2 \times 3.14 \times 9$
≈ 169.56
$SA \approx 75.36 + 169.56$
≈ 244.9
$SA \approx 245 \text{ cm}^2$

c. $SA = \pi rs + \pi r^2$
$\pi rs \approx 3.14 \times 3 \times 5$
≈ 47.1
$\pi r^2 \approx 3.14 \times 3^2 \approx 28.26$
$SA \approx 47.1 + 28.26$
≈ 75.36
$SA \approx 75 \text{ cm}^2$

EXERCISES

Sketch and label the unfolded faces of each figure on another sheet of paper. Then use a formula to find the surface area. Use $\pi \approx 3.14$. Round to the nearest whole number.

1. **232 in.²**
2. **125.6 cm²**
3. **314 cm²**
4. **170 in.²**

Reteaching • SECTION 11-11   **221**

Painted Surfaces

The cube shown at the right was painted blue on the three faces that you see and painted red on the three faces hidden from your view. It was then sliced into 27 one-centimeter cubes, as shown in the diagram.

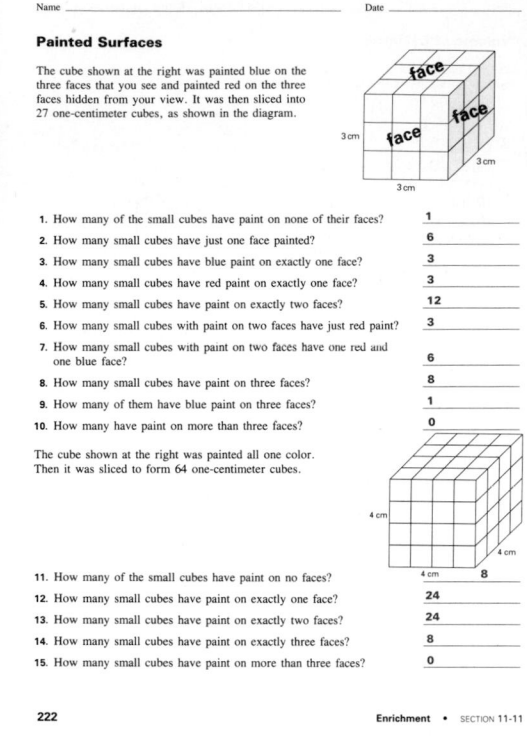

1. How many of the small cubes have paint on none of their faces? **1**
2. How many small cubes have just one face painted? **6**
3. How many small cubes have blue paint on exactly one face? **3**
4. How many small cubes have red paint on exactly one face? **3**
5. How many small cubes have paint on exactly two faces? **12**
6. How many small cubes with paint on two faces have just red paint? **3**
7. How many small cubes with paint on two faces have one red and one blue face? **6**
8. How many small cubes have paint on three faces? **8**
9. How many of them have blue paint on three faces? **1**
10. How many have paint on more than three faces? **0**

The cube shown at the right was painted all one color. Then it was sliced to form 64 one-centimeter cubes.

11. How many of the small cubes have paint on no faces? **8**
12. How many small cubes have paint on exactly one face? **24**
13. How many small cubes have paint on exactly two faces? **24**
14. How many small cubes have paint on exactly three faces? **8**
15. How many small cubes have paint on more than three faces? **0**

222   Enrichment • SECTION 11-11

Problem Solving Strategies: Guess and Check

You can use the guess-and-check strategy to solve problems involving area when you cannot apply a formula directly.

► **Example**

The area of a triangular garden is 48 ft². The height is 4 ft less than the base. Find the dimensions of the garden.

Solution

Construct a table for values of the base, the height, and the area.

base	height	area	
b	$h = b - 4$	$\frac{1}{2} \times b \times h$	
14	10	70 ft²	← too large
10	6	30 ft²	← too small
12	8	48 ft²	

The base of the garden is 12 ft and the height is 8 ft.

PROBLEMS

1. The area of a rectangular room is 420.5 ft². The length is twice the width. Find the dimensions of the room.
   length, 29 ft; width, 14.5 ft

2. The volume of a pyramid is 192 cm³. The area of the base is 32 cm². Find the height.
   height, 18 cm

3. The area of a rectangular plot is 154 yd². The length is 3 yd more than the width. Find the dimensions of the plot.
   length, 14 yd; width, 11 yd

4. The volume of a rectangular box is 1,320 in.³ The length is 3 more than the width. The height is 4 less than the width. Find the dimensions of the box.
   length, 15 in.; width, 12 in.; height, 8 in.

5. A design having the shape of a parallelogram has a perimeter of 96 mm. The longer sides of the parallelogram are 12 mm greater than the shorter sides. How long are the sides of the design?
   longer sides, 30 mm; shorter sides, 18 mm

Reteaching • SECTION 11-12   **223**

Draw, Guess, and Check

Use the guess-and-check strategy to solve geometry problems and other types of problems. Drawing a picture often helps. Sometimes, the only way to solve a problem is with a picture.

► **Example**

How can you cut the 3 by 5 rectangle at the right into three pieces, each of which folds up into a cube with one face open?

Solution

Use graph paper to sketch different arrangements. One solution is shown.

EXERCISES

Solve each problem. You may want to draw a picture to help you.

1. The areas of the sides of a rectangular box are 24, 32, and 48 cm². What is the volume of the box?
   $V = 4 \times 6 \times 8$
   $V = \textbf{192 cm}^3$

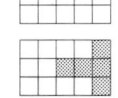

2. A square piece of paper was folded in half vertically to make a figure having a perimeter of 12 in. What was the area of the original square?
   16 in.²

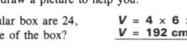

3. A box has length a, width b, and height c, all whole numbers. The length of its diagonal is d. Using only the Pythagorean Theorem, you can show that $d^2 = a^2 + b^2 + c^2$.

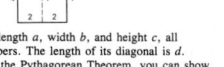
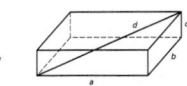

   a. Show that a box having dimensions 4, 5, and 6 units has a diagonal of the same length as a box with dimensions 2, 3, and 8 units.
   b. Find other pairs of boxes whose dimensions are whole numbers and whose diagonals are equal in length.

   a. $4^2 + 5^2 + 6^2 = 2^2 + 3^2 + 8^2$
   $16 + 25 + 36 = 4 + 9 + 64 = 77$
   b. $1^2 + 5^2 + 6^2 = 2^2 + 3^2 + 7^2$
   $2^2 + 4^2 + 9^2 = 1^2 + 6^2 + 8^2$
   other solutions possible

224   Enrichment • SECTION 11-12

434F

Name _____ Date _____

Computer Activity: Pythagorean Pursuit

Suppose you were asked to determine if a triangle is a right triangle. If you know the angle measurements, that is easily done. Simply check to see if one of the angles of the triangle has a measure of 90°.

If you know the side measurements, a, b, c, that is also easily done. Simply use the converse of the Pythagorean Theorem.

Pythagorean Theorem: If a triangle is a right triangle, then $a^2 + b^2 = c^2$.

Converse: If $a^2 + b^2 = c^2$, then a triangle is a right triangle.

The following program will take your INPUT of the measures of the sides and determine if the triangle is a right triangle.

```
10 INPUT "ENTER THE SIDES IN ASCENDING
   ORDER: ";A,B,C: PRINT
20 IF A > B OR A > C OR B > C THEN
   GOTO 10
30 IF C = INT ( SQR (A ^ 2 + B
   ^ 2)) THEN GOTO 50
40 PRINT A;" ";B;" ";C;" DOES NOT FORM
   A RIGHT TRIANGLE."
   : END
50 PRINT A;" ";B;" ";C;" DOES FORM A
   RIGHT TRIANGLE."
```

EXERCISES

1. Why must the measures of the sides be entered in ascending order?

__C represents the longest side, so it must be entered last.__

_____

2. If the measures INPUT were C, A, B or C, B, A, in what order would they have been entered?

__descending order__

3. What does line 20 do?

__Line 20 checks that numbers were entered in ascending order.__

_____

Name _____ Date _____

Pythagorean Pursuit (continued)

4. RUN the program to determine if a triangle with sides of the given measures is a right triangle.

a. 6 m, 8 m, 10 m — __yes__   d. 4 in., 9 in., 7 in. — __no__

b. 9 ft, 15 ft, 12 ft — __yes__   e. 13 cm, 5 cm, 12 cm — __yes__

c. 25 km, 7 km, 21 km — __no__   f. 24 m, 26 m, 10 m — __yes__

Any set of 3 positive integers a, b, c that satisfy the equation $a^2 + b^2 = c^2$ is called a **Pythagorean Triple.** By multiplying each part of a Pythagorean Triple by the same number, you can generate another Pythagorean Triple.

5. For each Pythagorean Triple, generate three others. **Answers may vary.**

a. 3, 4, 5	6, 8, 10	9, 12, 15	15, 20, 25
b. 5, 12, 13	10, 24, 26	15, 36, 39	20, 48, 52
c. 8, 15, 17	16, 30, 34	24, 45, 51	32, 60, 68

The following program generates sets of Pythagorean Triples.

6. RUN the program for M = 2 to 15 and record the results.

```
10   PRINT "A","B","C"
20   FOR M = 3 TO 15 STEP 2
30   LET A = 2 • M:B = M ^ 2 − 1:
     C = M ^ 2 + 1
40   PRINT INT (A), INT (B), INT (C)
50   NEXT M
```

7. For each Pythagorean Triple generated in which a, b, and c are all EVEN numbers, divide a, b, and c by 2 to get other Pythagorean Triples. Record the results.

__6, 8, 10 → 3, 4, 5__

__10, 24, 26 → 5, 12, 13__

__14, 48, 50 → 7, 24, 25__

__18, 80, 82 → 9, 40, 41__

__22, 120, 122 → 11, 60, 61__

__26, 168, 170 → 13, 84, 85__

__30, 224, 226 → 15, 112, 113__

_____

Name _____ Date _____

Calculator Activity: Finding Area and Surface Area

Review these formulas for area and surface area.

Area
rectangle: $A = l \times w$
parallelogram: $A = b \times h$
square: $A = s^2$
triangle: $A = \frac{1}{2} \times b \times h$
circle: $A = \pi r^2$

Surface Area
cylinder: $SA = 2\pi rh + 2\pi r^2$
cone: $SA = \pi rs + \pi r^2$
sphere: $SA = 4\pi r^2$

You can use the memory function of your calculator to find the area or surface area of two- and three-dimensional figures.

► **Example 1**

Find the surface area of the cylinder shown at the right.

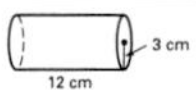

Solution

Use the formula for the surface area of a cylinder.
$SA = 2\pi rh + 2\pi r^2$

Use the memory function to store the results of your calculations as you go along. Use this sequence.
$SA \approx 2 \boxed{\times} 3.14 \boxed{\times} 3 \boxed{\times} 12 \boxed{=} \boxed{M+} 2 \boxed{\times} 3.14 \boxed{\times} 3 \boxed{\times} 3 \boxed{=} \boxed{M+} \boxed{MRC}$ 282.6

The surface area is about 282.6 cm².

► **Example 2**

Find the surface area of the figure. Round to the nearest tenth.

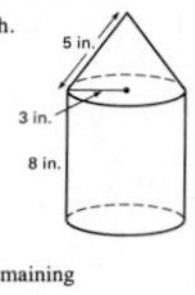

Solution

$SA = \pi rs + \pi r^2$

Since the cone is directly on top of the cylinder, you need not calculate the area of the base of the cone. Thus you can eliminate finding πr^2.
$SA = 2\pi rh + 2\pi r^2$

You need calculate the area of *only one* circular base. The remaining surface area will thus be $2\pi rh + \pi r^2$.

To find the total surface area, add the two formulas and use a calculator.
$SA = \pi rs + 2\pi rh + \pi r^2$
$\approx 3.14 \boxed{\times} 3 \boxed{\times} 5 \boxed{=} \boxed{M+} 2 \boxed{\times} 3.14 \boxed{\times} 3 \boxed{\times} 8 \boxed{=} \boxed{M+} 3.14 \boxed{\times} 3 \boxed{\times} 3 \boxed{=} \boxed{M+} \boxed{MRC}$
226.08

The surface area is about 226.1 in.²

Name _____ Date _____

Finding Area and Surface Area (continued)

EXERCISES

Choose the appropriate formula and use a calculator to find the area of the figure in each of Exercises 1 and 2. Find the area of the shaded region in Exercise 3. Round answers to the nearest tenth.

1. (3.7 cm, 5.2 cm, 6 cm) — __9.6 cm²__

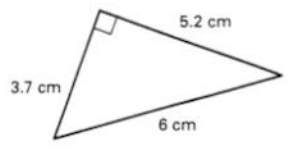

2. (3 ft, 5 ft, 12 ft) — __77.3 ft²__

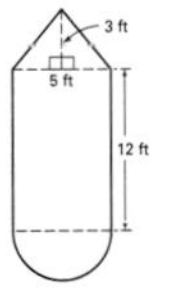

3. (4 m, 4 m) — __24 m²__

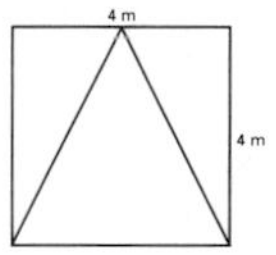

Choose the appropriate formula and use a calculator to find the surface area of each figure. Round answers to the nearest tenth.

4. (14 cm) — __615.8 cm²__

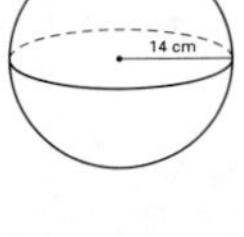

5. (6.8 in., 10.2 in.) — __290.5 in.²__

6. (11 ft, 25 ft) — __527 ft²__

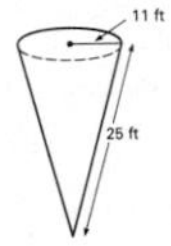

7. What happens to the surface area of the figure in Exercise 4 when the radius is doubled?

__The surface area is 4 times as great.__

434G

ACHIEVEMENT TEST

Area and Volume
CHAPTER 11 FORM A

MATH MATTERS BOOK 1
Chicha Lynch
Eugene Olmstead

SOUTH-WESTERN PUBLISHING CO.

Name _____
Date _____

SCORING RECORD	
Possible	Earned
20	

Find the area of each figure.

1.
4 in.
8 in.

32 in.²

2.
5 m
17 m

85 m²

3.
18 cm
25 cm

225 cm²

Find the square root to the nearest tenth.

4. √64 **8**
5. √900 **30**
6. √58 **7.6**
7. √150 **12.2**

Find the unknown length.

8.
6
8
?

10

9.
16
34
?

30

10.
30
18
?

24

Find the volume of each figure to the nearest whole number.

11.
40 in.
12 in.
12 in.

5,760 in.³

12.
18 cm

24,417 cm³

13.
20 cm
8 cm

1,340 cm³

14. The diameter of a can of green beans is 3 in. and the height is 4.5 in. What formula would you use to find the volume? Find the volume to the nearest whole number. Use 3.14 for π.
V = πr²h; 127 in.³

Copyright © 1993 by South-Western Publishing Co.
MI01AG

11A-1

Name _____   Date _____

Copy and complete the table.

	radius (r)	diameter (d)	area (A)
15.	12.5 cm	25 cm	490.625 cm²
16.	7 cm	14 cm	153.86 cm²

Find the surface area of each figure.

17.
5 ft
4 ft
6 ft
24 ft

145.6 ft²

18.
8 cm
5 cm
30 cm

860 cm²

19.
8 cm
22 cm

1,507.2 cm²

Solve.

20. The area of a rectangle is 147 in.² The length is three times the width. Find the dimensions of the rectangle.
7 in. wide; 21 in. long

11A-2

ACHIEVEMENT TEST

Area and Volume
CHAPTER 11 FORM B

MATH MATTERS BOOK 1
Chicha Lynch
Eugene Olmstead

SOUTH-WESTERN PUBLISHING CO.

Name _____
Date _____

SCORING RECORD	
Possible	Earned
20	

Find the area of each figure.

1.
4 cm
11 cm

44 cm²

2.
18 cm
32 cm

576 cm²

3.
12 ft
14 ft

84 ft²

Find the square root to the nearest tenth.

4. √144 **12**
5. √630 **25.1**
6. √(169/324) **13/18**
7. √23 **4.8**

Find the unknown length.

8.
20
48
?

52

9.
16
34
?

30

10.
89
80
?

39

Find the volume of each figure to the nearest whole number.

11.
15 cm
8 cm
14 cm

1,680 cm³

12.
9 m
2 m

113 m³

13.
28 mm
5 mm
12 mm

840 mm³

14. A cylindrical storage bin has a diameter of 52 ft, and a height of 20 ft. What formula would you use to find the volume? Find the volume.
V = πr²h; 42,452.8 ft³

Copyright © 1993 by South-Western Publishing Co.
MI01AG

11B-1

Name _____   Date _____

Copy and complete the table.

	radius (r)	diameter (d)	area (A)
15.	24 cm	48 cm	1808.64 cm²
16.	13 cm	26 cm	530.66 cm²

Find the surface area of each figure.

17.
30 cm
12 cm
7 cm

1,308 cm²

18.
8 cm

4,571.84 cm²

19.
25 in.
10 in.

942 in.²

Solve.

20. The area of a rectangle is 144 cm². The length is four times the width. Find the dimensions of the rectangle.
6 cm wide; 24 cm long

11B-2

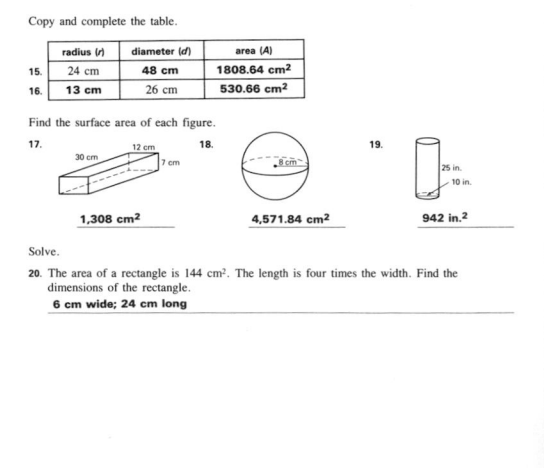

Teacher's Notes

On Your Own

Objective: Students will solve real problems related to living on their own by using operations with whole numbers, fractions, and decimals; computing area, perimeter and volume; comparing costs using ratios, rates, percents, interest, and discounts; and constructing a circle graph.

UNIT OVERVIEW

The four lessons in this unit prepare students for making real-world decisions using current data about rental costs, food prices, and other expenses they will encounter when they are on their own. They will be challenged to compare and contrast, and evaluate alternative purchases while considering both short and long-range planning.

LESSON 1 **Apartment Complexities** involves students in making decisions about living expenses. Students will arrange themselves into 'rooming groups' to research and compare apartment prices. Computations involving rates, ratios, and percents will culminate in the construction of a circle graph displaying their estimated monthly living expenses.

LESSON 2 **Discount Daze** asks students to furnish one room of their apartment by investigating and minimizing the total cost of various necessary items. For one large purchase, they will look for sales, compute discounts and sale price, and determine the relative costs of cash or credit buying. Students will practice computing interest payments using calculators.

LESSON 3 **Wall-to-Wall** taps students' creativity, number sense, and knowledge of geometry while they decorate a room in their apartment. They will calculate wall and floor areas, estimate material needed, and determine comparative rates such as cost per gallon of paint, per yard of wallpaper or carpeting. Finally, students will find out what a contractor would charge for doing the same work and write a paper, using percents to compare costs.

LESSON 4 **Food Fights** has students investigate costs of the same or same quality product in different stores. As they learn supermarket shopping strategies, students will use volumes, and customary and metric measures of capacity and weight; they will also use ratios, rates, and proportions to find comparable values.

TIME 12–25 days (not consecutively since students will need time to gather materials)

INTRODUCTION AND MOTIVATION

- For several days prior to the unit introduction, have students collect and bring to class all receipts or register tapes from family purchases. Then, as a class, read the items (not the cost) and classify them into categories of expenses such as: food, clothing, furniture, or transportation.
- Ask the students to suggest other types of living expenses that a family or a group of young roommates might have. Students should name such things as rent, electric, heating oil or gas, telephone, repairs, insurance, and possibly others.
- Tell students that in this unit they will simulate living on their own. In 'rooming groups,' they will find an apartment, furnish it, decorate it, and buy groceries. Arrange students in groups of four and assign these 'expert' roles: (1) real estate, (2) furnishings (including appliances), (3) decorating, and (4) food.

LESSON 1   Apartment Complexities

OBJECTIVES

Students will:
- compare real estate ads for rental apartments
- select the best apartment for their group after considering several factors
- justify their choice of apartment in a letter to a parent or friend
- construct a circle graph to show the distribution of monthly expenses

TIME

2 to 5 class periods

MATERIALS

local newspapers, real estate listings of available rentals, examples of circle graphs appearing in newspapers and magazines, compasses, protractors

INTRODUCING THE LESSON

1. Give each group a newspaper or other listing of rentals in the surrounding area. Member of each group should pretend to be roommates looking for an unfurnished apartment; they must decide which kind of rental would be appropriate. Have students narrow their choices to two or three reasonable possibilities.
2. After they have narrowed their choices, rooming groups should make a list of questions to ask about the property, such as: What is the size of each room? Are utilities included in the rent? If not, what is the average monthly cost for gas or electric or heating oil? Is there a laundry room and/or parking available? Are pets allowed in the building? Are appliances and/or carpeting included?
   Students should make outside inquiries to find answers to their questions.

FACILITATING THE LESSON

1. Each rooming group should determine the total expenses for each apartment. Remind them to include gas, oil, water, electric bills, and so on, if applicable and not included in the rent. Using the dimensions of each apartment, they should calculate the price per square foot. Then the group should select their apartment. Students may find factors other than least cost that are important to them.
   Students will need two or three copies of **Worksheet 1, How Much Room?** to guide them through the calculations and consideration of other decision factors for each apartment possibility. Each student should write a letter to a parent or friend explaining their choice of apartment
2. Show students some examples of circle graphs from newspaper and magazines. Instruct the groups that they are to determine their total monthly housing costs and then compute the percent of the total that they spend on each item. They will use the percents to construct a circle graph to display their monthly expenses. Encourage students to limit themselves to no more than 6 categories.

SUMMARIZING THE RESULTS

Each group gives an oral presentation of their reasons for choosing their apartment and shows their circle graph. As a class, make a comparison of the data collected. Ask: What was the lowest rent found? What was the highest? What was the difference in dollars between the lowest and highest? What is the percent difference?
   Have students compare the data for monthly expenses in a similar manner. They might ask: On what did most groups spend the most each month? Why?

EXTENDING THE LESSON

- Have students determine how much they would spend over six months or a year for housing expenses. Which of the expenses might vary during different seasons? Why?
- Invite a real estate agent to visit the class and talk about the rental market in the area. He or she might have helpful insights and suggestions for students.

ASSESSING THE LESSON

Use the students' individual letters explaining their apartment choice and their circle graphs of monthly expenses to evaluate both individual and group work.

434K

OBJECTIVES

Students will
- use coupons, discounts, and sales to find best buys for furnishings
- compare the cash and credit cost of one large item
- calculate total amount and total interest to be paid using a credit purchase
- use tables or graphs to display comparative data

TIME

3 to 7 class periods

MATERIALS

newspaper ads, advertising flyers for discount, furniture, and appliance stores, classified section of telephone books, comprehensive mail-order catalogs, calculators

INTRODUCING THE LESSON

1. Ask students to consider, in their rooming groups, the first, most important purchase they would make if they had to furnish their apartment, for example, a refrigerator, or a sofa. Then ask students to suggest one major luxury item they would also like to buy—for example, a large-screen television set, or a computer.
2. On the chalkboard, make a list of the groups' suggestions for the first major purchase—*what they need*— and their most desired purchase—*what they want.* Discuss the difference between needs and wants.

FACILITATING THE LESSON

1. As groups decide which major purchase to make, distribute newspapers, catalogs, and flyers that might have ads for the items. Assist groups as they decide where to find additional information on the cost and financing options for their purchase. If financing is not an option, they should investigate taking out a small bank loan.
2. Hand out calculators, if necessary. Circulate among the groups as they use interest formulas and/or real data from banks, finance companies, or store credit plans to compute the cost of their major purchase. Remind them to include the sales tax. As a final product, each group should present an "advertisement" for their purchase that has a picture, a description, and the figures comparing the cash and credit costs.
3. Hand out **Worksheet 2, Kitchen Costs.** Challenge the groups to find the best buys for the item listed. Tell them that they have $700 to spend. They must buy all of the starred items. If they shop wisely, they may be able to buy some of the optional items. If discounts are not available on some items, encourage students to investigate second-hand sources.

SUMMARIZING THE RESULTS

Post student 'advertisements' on a bulletin board. Have students give one-minute oral reports about purchasing different items—the cost, and the best way to save money. For the kitchen activity, ask whether any groups were able to save enough money to buy optional items. Discuss their shopping strategies.

EXTENDING THE LESSON

- Have each group choose another room to furnish and complete a similar activity.
- If no group has chosen a refrigerator or stove as a major purchase, have students investigate the costs, financing options, and possible discounts for buying either of those major appliances.
- Students may wish to start a campaign within their school or community to encourage donations of basic furniture and appliance items to charity shops or homeless shelters.

ASSESSING THE LESSON

Group worksheets and advertisements can be used to evaluate student performance. Alternatively, you might assign each student a short essay in which they compare a cash and credit purchase for one item. They should apply information gathered in their projects, but use an item different from the one they previously used.

LESSON 3   Wall-to-Wall

OBJECTIVES

Students will
- find the floor and wall area of one room to be decorated in their apartment
- calculate the amount of materials needed to complete the decorating project
- determine the comparative costs for different materials
- compare do-it-yourself costs and contractor costs

TIME

3 to 6 class periods

MATERIALS

wallpaper books, carpet and tile (ceramic or other floor covering) samples (borrowed from merchants), newspaper, catalog, and flyer ads

INTRODUCING THE LESSON

1. Give students some time to "shop around" examining the wallpaper, carpet, and floor covering samples you have previously collected. Ask them to think about decorating one room in their apartment. They will either paint or wallpaper the walls, and carpet or tile the floor.
2. Have the rooming groups discuss which room they will decorate and whether they will paint or wallpaper, and carpet or tile. Have the decorating expert chosen earlier facilitate the group's choice. They will use the actual dimensions of the room to determine the amount of materials they will need. Tell them they will choose real colors and patterns for the materials. Help them to decide what additional information or samples they might need.

FACILITATING THE LESSON

1. After allowing enough time for students to find out the cost of materials and to bring in samples, give the groups class time to make a choice and begin to calculate costs. Distribute **Worksheet 3, Wall-to-Wall** as a guide. Have them indicate their selection and write the reasons for their choice on the back of the sheet.
2. When students have calculated the cost of buying materials and completing the project themselves, ask: How much time do you think the project will take to complete? What alternatives are there to doing the project yourself? Have students investigate the cost of hiring others to do the work.

SUMMARIZING THE RESULTS

Have students write a paper describing the process of (1) choosing their projects, (2) choosing the materials, and (3) comparing their do-it-yourself project with the same work performed by a contractor. Their papers should summarize the relevant mathematical facts and include extra comparative information using percent calculations. Some examples might be the percent increase in cost of a contractor-performed project compared to a do-it-yourself project, or the percent saved by buying one wallpaper pattern over another.

EXTENDING THE LESSON

- Students might use ratio computations to develop a scale diagram or scale model of a room in their apartment showing the true colors and patterns that they have chosen.
- Suggest students undertake a project in their own homes for which they can investigate and report similar figures and comparisons.
- A service project in which students actually do some decorating for a senior citizen housing project, a group home, or a children's shelter is another way for students to put their mathematical knowledge into action.

ASSESSING THE LESSON

Use **Worksheet 3** and the students' final papers to evaluate students' work. If you first provide a checklist of objectives and criteria, you might have each group of students check the work of another group on **Worksheet 3**.

434M

OBJECTIVES

Students will
- compute and compare unit prices of identical items in different supermarkets
- compute and compare unit prices of different brands of the same product
- use ratios and proportions to find comparable values or better values
- make judgments about the relative value of purchasing different sizes or brands

TIME

4 to 7 class periods

MATERIALS

local supermarket ads, manufacturers' and supermarket discount coupons, empty food containers including several from different brands of the same product, register tapes showing the prices of some of the products under consideration

INTRODUCING THE LESSON

1. Display empty food containers and register tapes. Have students notice and list the kinds of numerical information they can obtain from the containers. Tell students that this lesson will focus on price and weight or capacity.
2. Arrange students into new groups of three or more. Explain that they will compare products with one another and compare the same product from different supermarkets. Students may collect empty grocery containers or visit supermarkets to obtain the following data: the item's brand name, size information, and price. To achieve some consistency for comparison, specify the products and sizes; for example, all-purpose flour (5 lb), orange juice (64 oz. container), peanut butter (18 oz.), tuna fish (6 1/8 oz. can), and so on.

FACILITATING THE LESSON

1. Have groups select 5 or 6 items from your list to research and compare.
2. For each product, ask groups to:
   - find the unit price, using both customary and metric measures, if available.
   - compare the prices of the identical product at different supermarkets.
   - compare the prices of different brands of comparable items, including the 'house' or generic brand of the items.
   - use proportions to determine the expected price of a larger quantity, then compare with actual supermarket price
   - list the highest and lowest prices found for each, then compute the totals and the difference in price.
3. Each group should summarize the results in a written paper. The group paper might include a table and/or graph, but must include the unit pricing and proportion problems worked out, and some conclusions about shopping strategies.

SUMMARIZING THE RESULTS

To close the activity, lead a class discussion about students' discoveries and strategies concerning supermarket shopping. Some questions for consideration are:
Is it always better to buy in larger quantities? Explain. Is it preferable to shop at the same market or follow the sales? What factors may affect your decision? How does the use of coupons affect your shopping? Is it always worth using the coupon for a particular product? How do store brand products compare to national brands?

EXTENDING THE LESSON

- Make up a shopping list of ten new items for each group. Then have groups compete to spend the least amount to (hypothetically) buy every item within one week.
- Have students write an essay on how the shape of containers affects the shopper.
- Discuss product serving size and nutritional informtion in conjunction with a nutrition/health unit.

ASSESSING THE LESSON

Monitor students' computations as they work in groups to compute unit prices. Check the group paper. The first suggestion for extending the activity, a shopping competition, can be used as an assessment tool.

Name _____ Date _____

How Much Room?

Apartment Location _____

Room	Dimensions	Square Feet
Living room		
Kitchen or kitchen area		
Dining room or dining area		
Bedroom 1		
Bedroom 2		
Bathroom 1		
Bathroom 2		
Hallways		
Other		

Monthly rent: _____   Total square feet: _____

Cost per square foot: _____

Other Factors
(Parking, closet space, laundry, pets,
utilities, near public transportation, other)

_____

_____

_____

_____

Name _____ Date _____

Kitchen Costs

You have $700 to spend. You must buy the items marked with an * . You may buy others if you save enough money.

Item	Original Price	Sale Price	Discount or Savings
Table*			
4 chairs*			
Table cloth *			
Dishes* (service for 4)			
Eating utensils*			
Pots and pans*			
Toaster*			
Blender*			
Coffee maker*			
Cooking and storage dishes *			
Cooking utensils*			
Garbage pail*			
Mop*			
Broom*			
6 Glasses*			
4 cups or mugs*			
Dust pan*			
Cookbook*			
Toaster oven			
Electric fry pan			
Popcorn popper			
Microwave oven			
Other			

Name _____   Date _____

Wall-to-Wall

Paint   Area of walls _____ square feet

1 gallon covers _____ square feet

gallons needed @ _____ **Total cost** _____

OR

Wall paper Area of walls _____ square feet

1 roll covers _____ square feet

rolls needed @ _____ **Total cost** _____

AND

Carpet   Area of floor_____ square feet

Area of floor_____ square yards

square yards needed @ _____ **Total cost** _____

OR

Tile or other floor covering

Area of floor_____ square feet

Area of floor_____ square yards

boxes of 12 × 12 inch tiles needed @ _____

Total cost _____

OR

boxes of 9 × 9 inch tiles needed @ _____

Total cost _____

Teacher's Notes

CHAPTER 12 SKILLS PREVIEW

Simplify.

1. $x + 3x + (-2x)$ **2x**

2. $4n^2 + (-3n^2) + 2n^2$ **3n²**

3. Three stars are located at the points of a triangular region, *DEF*. Write a simplified expression for the perimeter of the region. **17a**

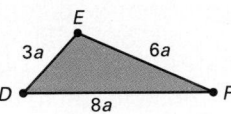

4. $(4n^2 + 2n - 8) + (2n^2 + n - 6)$
  6n² + 3n − 14

5. $(7r^2 + 3r - 9) - (4r^2 + 2r - 10)$
  3r² + r + 1

6. Write a simplified expression for the perimeter of the rectangle.
12x − 8

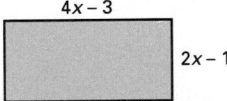

4x − 3

2x − 1

7. $(-3h)(2k)$ **−6hk**

8. $(-4m)(-5n)$ **20mn**

9. $(4t)(2t^3)$ **8t⁴**

10. $(-3c^4)(-2c^2)$ **6c⁶**

11. $\dfrac{14xy}{7x}$ **2y**

12. $\dfrac{-5h^3}{h^3}$ **−5**

13. $\dfrac{12y^3z^4}{-2yz}$ **−6y²z³**

14. $5(p + 1)$ **5p + 5**

15. $-2x(x + 2)$ **−2x² − 4x**

16. $n^2(1 - n - n^2)$ **n² − n³ − n⁴**

Factor each polynomial.

17. $12n + 8$
  4(3n + 2)

18. $6a^2b - 18a$
  6a(ab − 3)

19. $5mn + 10m^2n^2 - 20mnp$
  5mn(1 + 2mn − 4p)

Simplify.

20. $\dfrac{-16xy}{4y}$
  −4x

21. $\dfrac{-24a^2b^3c}{-6ab}$
  4ab²c

22. $\dfrac{35b^4 + 14b^2 - 21b^3}{7b}$
  5b³ + 2b − 3b²

Solve.

23. The area of the rectangle shown is $14x^3y$. Write an expression for the unknown dimension. **2xy**

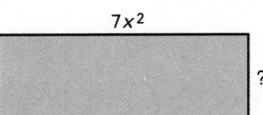

$7x^2$

?

12 SKILLS PREVIEW

Introduction The purpose of this Skills Preview is to assess students' abilities on all the major objectives of Chapter 12. Test results may be used
- to determine those topics which may need only to be reviewed and those topics which need to be more carefully developed;
- for class placement;
- in prescribing for individual differences.

If you prefer, you may use the Skills Preview as an alternative form of the Chapter Test (page 468) to evaluate mastery of chapter objectives. The items on this and the Chapter Test correspond in content and level of difficulty.

12

POLYNOMIALS

OVERVIEW

In this chapter, students are introduced to the concept of polynomials, to adding and subtracting polynomials, multiplying monomials, multiplying a polynomial by a monomial, factoring, and dividing monomials and polynomials by monomials. The vocabulary and techniques introduced are those found in standard introductory algebra texts. Problem solving lessons focus on choosing a particular strategy to solve a problem and on using *chunking* as a way to factor polynomials.

SPECIAL CONCERNS

Students need no previous experience with algebra to complete the activities in this chapter. Errors often occur when students try to simplify the expressions mentally. Encourage them to write out all the steps and to check their solutions.

VOCABULARY

binomial	greatest common	product rule for
chunking	factor	exponents
coefficient	like terms	quotient rule for
common factor	monomial	exponents
constant	polynomial	standard form
degree of a	power of a product	term
polynomial	rule	trinomial
factoring	power rule for exponents	

MATERIALS

calculators   algebra tiles

BULLETIN BOARD

Divide the bulletin board into five sections, one each for addition, subtraction, multiplication, factoring, and division. At the top of the board write the name of the chapter, *Polynomials*. As each topic is studied, write one or more helpful hints and display the hints below each heading. Write exercises on index cards and place in them in an envelope. Encourage students to solve these exercises in their free time and display their solutions on the bulletin board.

INTEGRATED UNIT 4

The skills and concepts involved in Chapters 12–14 are included within the special Integrated Unit 4 entitled "Fast Forward." This unit is in the Teacher's Edition beginning on page 545K. Worksheets for this integrated unit appear in the Enrichment Activities booklet, pages 172–174.

TECHNOLOGY CONNECTIONS

- Computer Worksheet, 245
- Calculator Worksheet, 246
- MicroExam, Apple Version
- MicroExam, IBM Version

- *Tobbs Learns Algebra*, Sunburst Communications
- *Alge-Blaster Plus*, Davidson & Associates

TECHNOLOGY NOTES

Computer algebra programs, such as those listed in Technology Connections, develop students ability to think algebraically, while allowing them to build their problem solving skills. Many such programs enable students to control their own learning by having them choose the level of difficulty at which they will work and then by providing them with tutorials and additional practice.

POLYNOMIALS

PLANNING GUIDE

SECTIONS	TEXT PAGES	ASSIGNMENTS		
		BASIC	AVERAGE	ENRICHED
Chapter Opener/Decision Making	436–437			
12–1 Polynomials	438–441	1–14, 18–21, 29–30	1–17, 18–26, 29–33	4–6, 11–17, 18–28, 29–35
12–2 Adding and Subtracting Polynomials	442–445	1–12, 19–34, 41	1–18, 19–37, 40–43, 44–45	19–28, 36–43, 44–47
12–3 Multiplying Monomials	446–449	1–18, 27–32, 40–43	5–28, 29–43, 44	11–16, 29–43, 44–46
12–4 Multiplying a Polynomial by a Monomial	450–453	1–14, 15–21, 27–28, 30–33	1–14, 15–29, 30–35	7–14, 19–29, 20–37
12–5 Problem Solving/Decision Making: Choose a Strategy	454–455	1–9	1–13	1–13
12–6 Factoring	456–459	1–21, 22–24, 27, PSA 1–7	3–21, 22–30, 31, PSA 1–9	12–21, 22–30, 31, PSA 1–9
12–7 Problem-Solving Skills: Chunking	460–461	1–8	2–12	2–12
12–8 Dividing by a Monomial	462–465	1–16, 20–29, 33–35, 39	1–19, 20–38, 39–40	9–19, 20–32, 36–38, 39–42
Technology	455, 456, 461	✓	✓	✓

ASSESSMENT				
Skills Preview	435	All	All	All
Chapter Review	466–467	All	All	All
Chapter Test	468	All	All	All
Cumulative Review	469	All	All	All
Cumulative Test	470	All	All	All

| | ADDITIONAL RESOURCES | | |
RETEACHING	ENRICHMENT	TECHNOLOGY	TRANSPARENCY
12–1	12–1		TM 30
12–2	12–2		
12–3	12–3		
12–4	12–4		
12–5	12–5		
12–6	12–6	12–6	
12–7	12–7		
12–8	12–8	12–8	

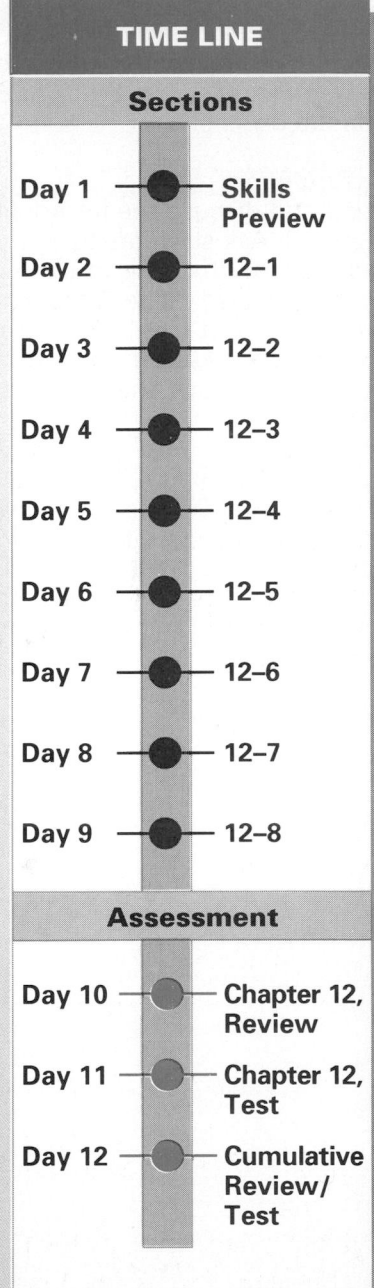

TIME LINE

Sections

Day 1 — Skills Preview
Day 2 — 12–1
Day 3 — 12–2
Day 4 — 12–3
Day 5 — 12–4
Day 6 — 12–5
Day 7 — 12–6
Day 8 — 12–7
Day 9 — 12–8

Assessment

Day 10 — Chapter 12, Review
Day 11 — Chapter 12, Test
Day 12 — Cumulative Review/ Test

ASSESSMENT OPTIONS

Chapter 12, Test Forms A and B	
Chapter 12, Test	Text, 468
Alternative Assessment	TAE, 468
Chapter 12, MicroExam	

POLYNOMIALS

THEME Astronomy

Objective To explore how data can be used as a basis for making real-life decisions

Introduction
Read and discuss the introductory material. Ask students to compare the two arithmetic statements with the two algebraic statements. Elicit the fact that a *variable expression* contains one or more letters that represent numbers.

Decision Making
Using Data Discuss the graph. Help students identify the point at which the first cycle ends and the second cycle begins on the graph. Have students answer the questions orally.

Working Together
Have students work in small groups. Allow opportunities for each group to share its findings with the class.

In arithmetic, you operate on quantities that are represented by specific numbers.

▶ Find the sum of 8 and 11.

▶ A room measures 10 ft by 10 ft. Find its area.

In algebra, you operate on quantities that vary in value and so must be represented by variable expressions.

▶ Find the sum of $3x$ and $6x$.

▶ A room measures y ft by $(y + 5)$ ft. Find its area.

In this chapter you will learn to perform operations on algebraic expressions called *polynomials*.

The measure of how bright a star appears from Earth is called its *apparent magnitude*. The brighter the star, the less its magnitude. Like the quantities in algebraic expressions, the magnitudes of some stars vary. The magnitude of the star Mira, for example, varies as the star expands and contracts over a long period of time. Mira is called a *long-period* variable star.

436   CHAPTER 12   Polynomials

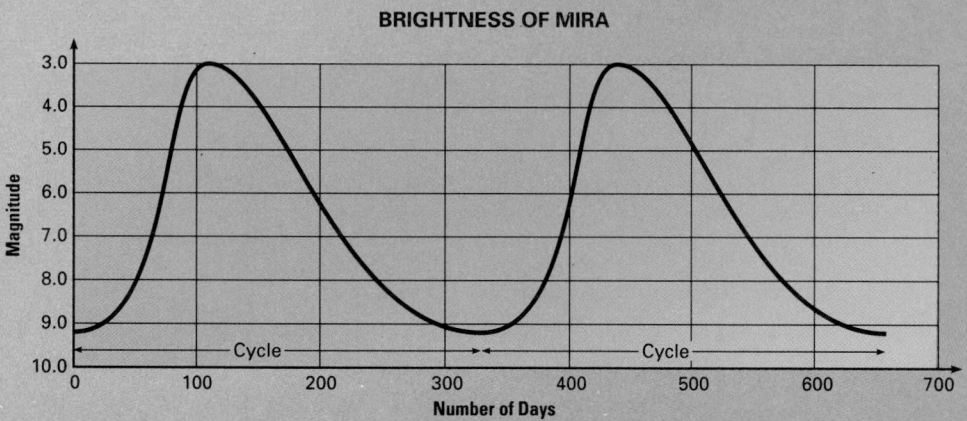

BRIGHTNESS OF MIRA

DECISION MAKING

Using Data

The graph above shows two of Mira's cycles. Use the graph to answer the following questions.

1. What is the maximum magnitude of Mira during a cycle? about 9.2

2. About how many days long is each of Mira's cycles? about 330 days

3. On day d Mira had a magnitude of m. Write an expression to represent the next day on which a magnitude of m occurred.
   $d + 330$

4. The star Sirius has a magnitude of -1.4. Is Sirius brighter or dimmer than Mira? brighter (Sirius is the brightest star in the sky.)

Working Together

Use earth-science books, astronomy books, or encyclopedias to find out about different kinds of variable stars. First, find out how *absolute magnitude* is related to apparent magnitude. Then, look for a Hertzsprung–Russell (H–R) diagram of the stars or some other indicator of the comparative temperature and brightness of the stars. Use this information to make lists of two kinds of stars—those with magnitudes greater than Mira's and those with magnitudes less than Mira's.

For what purpose do you think an astronomer might need the information about a star's variable period and its apparent magnitude?
    Answers may vary; to determine the distance from Earth to the star

12-1   Polynomials

EXPLORE/ WORKING TOGETHER

You can use algebra tiles to model variable expressions. The length of each side of a tile is assigned the value x or 1, as shown in the figure. By calculating the area of each tile, you see that each large square tile represents x^2, each long tile represents x, and each small square tile represents 1.

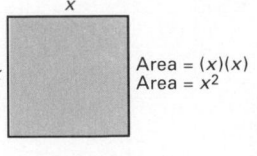

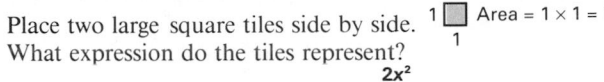

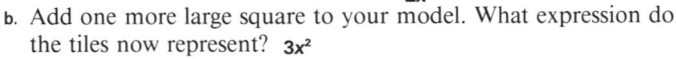

a. Place two large square tiles side by side. What expression do the tiles represent? **2x²**

b. Add one more large square to your model. What expression do the tiles now represent? **3x²**

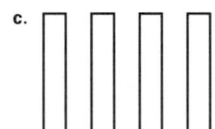

c. Use long tiles to model the expression $4x$.

d. Add three more long tiles to your second model. What expression do the tiles now represent? **7x**

e. Combine the square tiles from the first model with the long tiles from the second model. What expression do the tiles now represent? **3x² + 7x**

SKILLS DEVELOPMENT

Each of these expressions is called a monomial.

$$4 \qquad x \qquad 3y \qquad 2y^2 \qquad 5xy \qquad 6x^2y$$

A **monomial** is an expression that is either a single number, a variable, or the product of a number and one or more variables.

READING MATH

The prefixes *mono–* and *poly–* are from Greek words meaning "one" and "many." The prefixes *bi–* and *tri–* are Latin prefixes meaning "two" and "three." Write these four prefixes across the top of a page in your journal. Under each word, list as many everyday words you can think of that begin with that prefix.

In a monomial such as $3y$, the number 3 is called the **numerical coefficient** or, more simply, the **coefficient**. In monomials such as x, xy, and x^2y, the coefficient is 1. A monomial such as -4, which contains no variable, is called a **constant monomial,** or a **constant**.

The *sum* of two or more monomials is an expression called a **polynomial**. Each monomial is called a **term** of the polynomial. A polynomial with two terms is called a **binomial**. A polynomial with three terms is called a **trinomial**.

binomials: $2x + 3$ $x^2 + x^2y$ $4xy + (-5y)$

trinomials: $x^2 + y + 7$ $2xy^2 + (-5xy) + (-2)$

You can think of a monomial as a polynomial of one term.

Sometimes a polynomial is written with subtractions rather than additions. Think of this kind of polynomial as a sum of monomials, some of which have negative coefficients.

$7z^3 - 2z^2 - z + 6 = 7z^3 + (-2z^2) + (-1z) + 6$ The coefficients are 7, −2, and −1.

A polynomial is written in **standard form** when its terms are arranged in order from greatest to least powers of one of the variables.

Example 1

Write each polynomial in standard form.

a. $8n^2 + 5n + 6n^3 + 9n^4$ b. $7y + 11 + y^2$ c. $5z - 2z^3 - 6 - 4z^2$

Solution

a. Write the terms of the polynomial in order from the greatest power of n to the least power of n.

$8n^2 + 5n + 6n^3 + 9n^4 = 9n^4 + 6n^3 + 8n^2 + 5n$

b. Order the terms from greatest to least power of y. Write the constant last.

$7y + 11 + y^2 = y^2 + 7y + 11$

c. First, write the polynomial as a sum of monomials with negative coefficients.

$5z - 2z^3 - 6 - 4z^2 = 5z + (-2z^3) + (-6) + (-4z^2)$

Then write the standard form.

$-2z^3 + (-4z^2) + 5z + (-6)$, or $-2z^3 - 4z^2 + 5z - 6$ ◄

In a polynomial, terms that are exactly alike, or that are alike *except* for their numerical coefficients, are called *like terms*.

These are examples of like terms: $x^2y, 5x^2y, 4x^2y$

These are not like terms: $xy, x^2y, 2xy^2$

TALK IT OVER

Explain why x^2y and $2xy^2$ are not like terms.

Because the variables x and y in each term do not have the same powers.

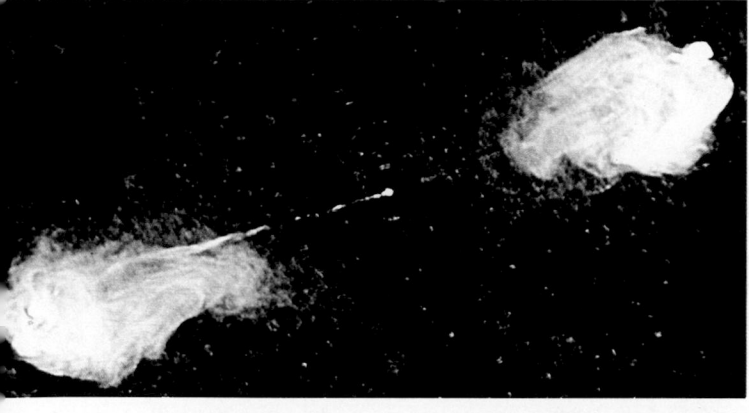

Additional Questions/Examples

1. Which of these are like terms?
5, 2*b*, 2*c*, 3*b*², −6*c*, 4*b*, 5*bc*, 2
5 and 2; 2b and 4b; 2c and −6c

Tell if the expression is a monomial (*m*), a binomial (*b*), or a trinomial (*t*).

2. 4*b*   **m**

3. 3*a* + 6   **b**

4. *a*   **m**

5. 2*xy* − 3*x*   **b**

6. *a* + *b* + *c*   **t**

Write the polynomial in standard form.

7. $b^2 + 3b - 4b^3 + 7$
$-4b^3 + b^2 + 3b + 7$

8. $5p^2 - 6p^3 - 5 + 8p^4$
$8p^4 - 6p^3 + 5p^2 - 5$

Simplify.

9. $5c + 6n + 4c - 2c$
$7c + 6n$

10. $5r^2 - 4r + r - 7r^2$
$-2r^2 - 3r$

11. $6xy + 3xy + 4x^2 - 3x^2 + x + y$
$x^2 + 9xy + x + y$

5-MINUTE CLINIC

Exercise	Student's Error	Error Diagnosis
Simplify: $12s^3 - 2s^2 - 9s^3$	$12s^3 - 2s^2 - 9s^3 = s^3$	• Student combines all variables raised to a power into one term without considering the power to which they are raised.

Guided Practice/Try These Observe students as they complete these exercises independently. Discuss any questions or problems they may have.

SUMMARIZE

Write About Math Have students define each new term in their math journals. Ask them to include explanations of how to simplify a polynomial and to write a polynomial in standard form.

4 PRACTICE

Practice/Solve Problems
Remind students to write polynomials that involve subtractions as a *sum* of monomials with negative coefficients.

Extend/Solve Problems In these exercises, students simplify polynomials *in two variables*. Remind students that terms are *like terms* only if they contain the same variables raised to the same powers.

Think Critically/Solve Problems
In Exercises 29–32 students identify the degree of each polynomial. Discuss students' responses to Exercises 33–35.

5 FOLLOW-UP

Extra Practice
Write in standard form.
1. $x^3 + 2x^2 + 7x + 4x^4$
   $4x^4 + x^3 + 2x^2 + 7x$
2. $t^2 + 4t - 3t^3 - 5$
   $-3t^3 + t^2 + 4t - 5$
3. $6x + 4x^2 - 8x^3 - x^4 + 7$
   $-x^4 - 8x^3 + 4x^2 + 6x + 7$

Tell whether the terms in each pair are *like* or *unlike* terms.

1. $2s, -5s$ **like**

2. $4ab, -3ab$ **like**

3. $7yz, 7xz^2$ **unlike**

4. $d^2t, -d^2t$ **like**

When a polynomial contains two or more like terms, you *simplify* the polynomial by adding these terms. Remember that this process is called *combining like terms*. A polynomial is simplified when all like terms have been combined and only unlike terms remain.

Example 2
Simplify.
a. $7x^2 + 2 + 9x^2$ b. $4d^2 - 6d - 3d^2 - 5d$ c. $-15m + 7m^2 + 9 - 3m$

Solution
a. $7x^2 + 2 + 9x^2 = 7x^2 + 9x^2 + 2$ ← Rearrange, or collect, like terms.
$= (7 + 9)x^2 + 2$ ← Combine like terms by applying the distributive property.
$= 16x^2 + 2$

b. $4d^2 - 6d - 3d^2 - 5d = 4d^2 + (-3d^2) + (-6d) + (-5d)$
$= [4 + (-3)]d^2 + [-6 + (-5)]d$ ← Combine like terms and apply the distributive property.
$= 1d^2 + (-11d)$
$= d^2 - 11d$

c. $-15m + 7m^2 + 9 - 3m = 7m^2 + (-15m) + (-3m) + 9$
$= 7m^2 + [-15 + (-3)]m + 9$
$= 7m^2 + (-18m) + 9$
$= 7m^2 - 18m + 9$ ◄

TRY THESE

Write each polynomial in standard form.

Answers may be written with or without parentheses.
Example: 5. $3c^2 + (-7c)$
or $3c^2 - 7c$

1. $5m + m^3 + 4m^2 + 2m^4$
   $2m^4 + m^3 + 4m^2 + 5m$
2. $5d + d^2 + 14$
   $d^2 + 5d + 14$
3. $-3x - 4x^2 + 2 + 5x^3$ $5x^3 + (-4x^2) + (-3x) + 2$

Simplify.

4. $4x^2 + 7x^3 - 3x^2 - 8$
   $7x^3 + x^2 + (-8)$
5. $5c^2 - 4c - 2c^2 - 3c$
   $3c^2 + (-7c)$
6. $-4y + 2y^2 + 5 + 3y$ $2y^2 + (-y) + 5$

EXERCISES

PRACTICE/ SOLVE PROBLEMS

Write each polynomial in standard form.

1. $n^3 + 2n^2 + 7n + 4n^4$
   $4n^4 + n^3 + 2n^2 + 7n$
2. $4b - 2b^2 + 7$
   $-2b^2 + 4b + 7$
3. $-4z - 5z^2 + 4 + 5z^3$
   $5z^3 + (-5z^2) + (-4z) + 4$
4. $m^2 + 2m - 4m^3 + m^4$
   $m^4 + (-4m^3) + m^2 + 2m$
5. $5n^5 - 8n + 2n^3$
   $5n^5 + 2n^3 + (-8n)$
6. $p^2 + 2p - 3p^3 - 9$
   $-3p^3 + p^2 + 2p + (-9)$

440 CHAPTER 12 Polynomials

Simplify. Be sure your answer is in standard form.

7. $-6a - 3 + 9a$
$3a + (-3)$

8. $-4b + 7 + 8b$
$4b + 7$

9. $-2 + 4c - 9c$
$-5c + (-2)$

10. $n^2 + 2n - 4n^2$
$-3n^2 + 2n$

11. $7r^2 + 4r^2 + r^2 + r$
$12r^2 + r$

12. $14x^2 + 3x^2 + 5x^3 - 6x$
$5x^3 + 17x^2 + (-6x)$

13. $7s^2 - s + 6s^2 - s^2 - s$
$12s^2 + (-2s)$

14. $3a^2 + 4a^3 + 2a^2 - 2a^3$
$2a^3 + 5a^2$

15. $-2 + 18h^2 - 9h^3 + 4h^2$
$-9h^3 + 22h^2 + (-2)$

16. $-7r + 4r^2 + 8r - 12r^2 - 12$
$-8r^2 + r + (-12)$

17. $12s^3 + 4s - 2s^2 - 9s^3 + 7 - 9s$ $3s^3 + (-2s^2) + (-5s) + 7$

EXTEND/
SOLVE PROBLEMS

So far in this lesson, you have been working with polynomials *in one variable.* Each term of the polynomial was either a constant term or contained a power of a single variable. You can extend what you have learned to the simplification of polynomials in two or more variables. Remember that two terms are like only if they contain the same variables raised to the same powers.

Simplify each expression. Be sure to write your answer in standard form.

18. $yz + 8yz$ $9yz$

19. $9x^2y - 4x^2y$
$5x^2y$

20. $yz^3 + 4yz^3 - 3yz^3$
$2yz^3$

21. $2mn - 3mn + 4mn$ $3mn$

22. $5g^2h + 4g^2h - 11g^2h$ $-2g^2h$

23. $-12x^2y^2 + x^2y^2 - z^2 + 4z^2$
$3z^2 + (-11x^2y^2)$

24. $-k - 2km - 9km + 15$
$-k + (-11km) + 15$

If possible, simplify each expression. Otherwise, write *not possible.*

25. $2xy + 3xy + 3x^2 - x^2 - x$
$2x^2 + (-x) + 5xy$

26. $r^3s - 4s + rs^2$ **not possible**

27. $a^3b + a^2b - ab + 4ab^2 + 8ab^3 - 3a^2b + 4ab$
$a^3b + (-2a^2b) + 3ab + 4ab^2 + 8ab^3$

28. $17x^3z + 18 x^3y + 2xy^3 + xy - 4xz^3$ **not possible**

THINK CRITICALLY/
SOLVE PROBLEMS

The *degree of a polynomial* in one variable is the highest power of the variable that appears in the polynomial after it has been simplified. What is the degree of each polynomial?

29. $x^2 + 3x - 5$ **2**

30. $5a - 4 + a^3 - 6a^2$ **3**

31. $m + 4$ **1**

32. $-3x^2 + 2x^3 - x^4 - 7$ **4**

33. Tell whether the statement is *true* or *false:*
If the degree of a polynomial is 4, then the polynomial will have four terms. **false**

34. Explain why the expression $\frac{3x}{2y}$ is not a polynomial.

35. Explain why $3x^4 - 3x^3 + 2x^2 - 5x$ cannot be simplified.

34. The expression is not the sum of monomials, because $\frac{3x}{2y}$ is not a monomial.

35. None of the terms are like terms.

Simplify.

4. $-4 + 14w - 7w - 8$ **$7w - 12$**

5. $7x^2y - 4x^2y$ **$3x^2y$**

6. $-13w + 9w^2 - 5 - 7w + 2w^2$
$11w^2 - 20w - 5$

7. $g^3h + g^2h - gh + 4gh^2 + 3gh^3 - 8 g^2h + 6gh$
$g^3h - 7g^2h + 5gh + 4gh^2 + 3gh^3$

Extension Have students substitute values for the variables in Try These Exercises 4–6 and evaluate both the original and simplified expressions.

Section Quiz
Simplify.

1. $5x + 5 - 8x + 8$ **$-3x + 13$**

2. $7m^2 - 4m^3 - 8m^2 + 6m + 5m^3$
$m^3 - m^2 + 6m$

3. $-4r + 7r^2 + 9r - 1$ **$7r^2 + 5r + (-1)$**

4. $14t^3 + 5t - 3t^2 - 6t^3 + 2 - 8t$
$8t^3 - 3t^2 - 3t + 2$

5. $ab - 5ab + 4a^2b - 3ab^2$
$4a^2b + (-4ab) + (-3ab^2)$

CHALLENGE

Have students replace the variables x and y with any whole number from 0 to 9 to make each sentence true. A number may be used only once in each expression. **Answers may vary. Samples are given.**

1. $xy + 3 = 27$ **4, 6**

2. $(x + 2) \div 3 = y$ **7, 3**

3. $2(x - y) = 4$ **2, 0**

4. $2x + y = 11$ **5, 1**

12-2 Adding and Subtracting Polynomials

EXPLORE Find the term that makes the statement true.

a. $-9d +$ ___?___ $= 5d$ **14d** **b.** $3m + (-7m) +$ ___?___ $= m$ **5m**

c. $-x + (-7x) +$ ___?___ $= -x$ **7x** **d.** $7h^2 +$ ___?___ $+ (-h^2) + 4h^2 = 4h^2$ **$-6h^2$**

SKILLS DEVELOPMENT Just as you can add, subtract, multiply, or divide real numbers, you can perform each of these basic operations with polynomials.

To add two polynomials, write the sum and simplify by combining like terms.

Example 1

Simplify.
a. $3m + (2m - 6)$ **b.** $(2a + 3) + (4a - 1)$
c. $(x^2 + 4x - 2) + (3x^2 + 7)$

Solution *3m and 2m are like terms.*

a. $3m + (2m - 6) = (3m + 2m) - 6 \leftarrow$ *Use the associative property.*

$\qquad = (3 + 2)m - 6 \leftarrow$ *Use the distributive property.*

$\qquad = 5m - 6$

b. $(2a + 3) + (4a - 1) = (2a + 4a) + (3 - 1) \leftarrow$ *2a and 4a are like terms.*

$\qquad = (2 + 4)a + 2$ *3 and -1 are like terms.*

$\qquad = 6a + 2$

c. $(x^2 + 4x - 2) + (3x^2 + 7) = (x^2 + 3x^2) + 4x + (-2 + 7)$

$\qquad = (1 + 3)x^2 + 4x + 5$

$\qquad = 4x^2 + 4x + 5$ ◄

You may also add polynomials by lining up like terms vertically. For example, at the right you see an alternative way to add the polynomials in part **c** of Example 1.

$$\begin{array}{r} x^2 + 4x - 2 \\ +3x^2 \quad\quad + 7 \\ \hline 4x^2 + 4x + 5 \end{array}$$

Subtraction of polynomials is like subtraction of real numbers. To subtract one polynomial from another, add its opposite and simplify. To find the opposite of a polynomial, find the opposite of each term of the polynomial.

Polynomial	*Opposite*
$x + 4$	$-x + (-4)$ or $-x - 4$
$2 + n$	$-2 + (-n)$ or $-2 - n$
$4f^2 - 2f - 7$	$-4f^2 + 2f + 7$
$-3x^2 + 5x + 2$	$3x^2 + (-5x) + (-2)$ or $3x^2 - 5x - 2$

Example 2

Simplify.

a. $9y - (2y - 4)$

b. $(-x + 2) - (-8x - 3)$

c. $(5x^2 - 11) - (3x^2 - 3x + 2)$

Solution

a. $9y - (2y - 4) = 9y + (-2y + 4)$ ← **Add the opposite of (2y − 4).**

$\quad\quad\quad = [9y + (-2y)] + 4$ ← **9y and −2y are like terms.**

$\quad\quad\quad = (9y - 2y) + 4$

$\quad\quad\quad = (9 - 2)y + 4$ ← **Use the distributive property.**

$\quad\quad\quad = 7y + 4$

b. $(-x + 2) - (-8x - 3) = (-x + 2) + (8x + 3)$ ← **Add the opposite of**

$\quad\quad\quad = (-x + 8x) + (2 + 3)$ **(−8x − 3).**

$\quad\quad\quad = (-1 + 8)x + 5$ ← **Write the coefficient of −x as −1.**

$\quad\quad\quad = 7x + 5$

c. $(5x^2 - 11) - (3x^2 - 3x + 2) = (5x^2 - 11) + [-3x^2 + 3x + (-2)]$

$\quad\quad\quad = [5x^2 + (-3x^2)] + 3x + [-11 + (-2)]$

$\quad\quad\quad = (5 - 3)x^2 + 3x + (-11 - 2)$

$\quad\quad\quad = 2x^2 + 3x - 13$ ◄

As with addition, you may subtract polynomials by lining up like terms vertically. This is how you could subtract the polynomials in part **c** of Example 2.

$$\begin{array}{r} 5x^2 \quad\quad - 11 \\ \underline{-3x^2 + 3x - 2} \\ 2x^2 + 3x - 13 \end{array}$$ ← **Add the opposite of each term.**

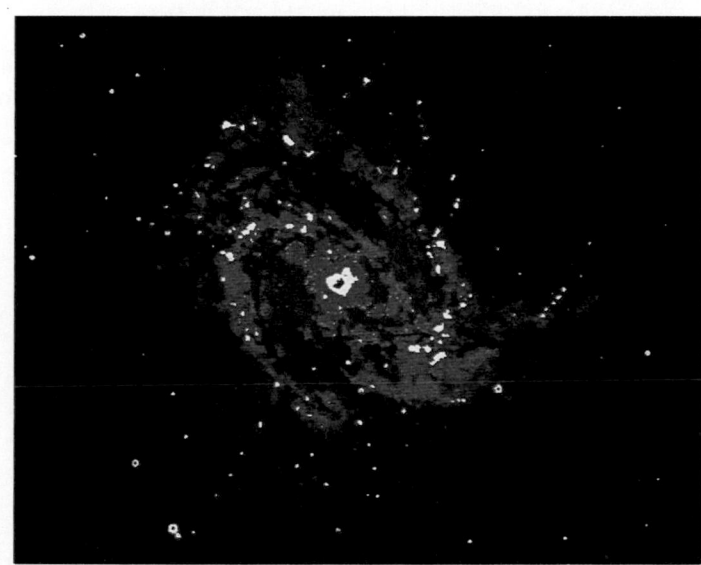

ASSIGNMENTS

BASIC
1–12, 19–34, 41

AVERAGE
1–18, 19–37, 40–43, 44–45

ENRICHED
19–28, 36–43, 44–47

ADDITIONAL RESOURCES
Reteaching 12–2
Enrichment 12–2

Additional Questions/Examples
Write the opposite of each polynomial.

1. $2x - 3$ **−2x + 3**
2. $-t^2 + 5t - 1$ **t² − 5t + 1**
3. $2 + n^2 - n$ **−2 − n² + n**
4. $x^2 + 2x - 5$ **−x² − 2x + 5**
5. $-(x^2 - 3x + 5)$ **x² − 3x + 5**

Simplify.

6. $(4r + 7) + (6r - 5)$ **10r + 2**
7. $(2b^2 - 5) + (b - 10) + (3b^2 + 2b)$
   5b² + 3b − 15
8. $(3a + 4) - (2a + 5)$ **a − 1**
9. $(5m^2 - 9m) - (3m^2 - 4m)$
   2m² − 5m
10. $(-6x - 4) - (-x - 9)$ **−5x + 5**

Guided Practice/Try These Observe students as they complete these exercises independently. Discuss any questions or problems they may have.

5-Minute Clinic

Exercise	Student's Error	Error Diagnosis
Simplify: $(2x^2 - 4x + 3) -$ $(4x^2 - 5x + 3)$	$(2x^2 - 4x + 3) -$ $(4x^2 - 5x + 3) =$ $6x^2 - 9x + 6$	• Student forgets to add the opposite of each term when subtracting polynomials.

3 SUMMARIZE

Key Questions

1. Explain how you would find the sum of $(3x^2 + 4x + 2)$ and $(-7x^2 - 3x + 2)$. **Answers may vary. Possible Answer: line up like terms vertically, then simplify.**

2. What is the sum of the two polynomials in Exercise 1? **$-4x^2 + x + 4$**

3. Explain how you would find the difference $(3x^2 + 4x + 2) - (-7x^2 - 3x + 2)$. Then find the difference. **Line up like terms vertically. Then add the opposite of each term; $10x^2 + 7x$**

4 PRACTICE

Practice/Solve Problems You may wish to have students do the addition or subtraction after lining up the like terms for each expression vertically.

Extend/Solve Problems These exercises involve two variables. To avoid confusion, encourage students to line up like terms vertically, especially for Exercises 27–28, 37, and 40.

Think Critically/Solve Problems In Exercises 44–46 students must analyze the results when the order of a subtraction is reversed.

5 FOLLOW-UP

Extra Practice Simplify.
1. $8c + (3c - 4)$ **$11c - 4$**
2. $(4x^2 - 6) + (-7x^2 + 3)$ **$-3x^2 - 3$**
3. $(7b^2 - 4b + 2) + (-3b^2 - 2b + 5)$ **$4b^2 - 6b + 7$**
4. $(3x - 4y) + (-4x + 3y)$ **$-x - y$**

TRY THESE

Add.

1. $\begin{array}{r} 4x^2 + 3x - 2 \\ + \quad x^2 \quad\;\; + 6 \\ \hline \mathbf{5x^2 + 3x + 4} \end{array}$
2. $\begin{array}{r} 2m^2 + 2m - 5 \\ + \qquad 3m + 4 \\ \hline \mathbf{2m^2 + 5m - 1} \end{array}$
3. $\begin{array}{r} 5y^2 + 2y + 3 \\ 2y^2 \quad\;\; + 4 \\ \hline \mathbf{7y^2 + 2y + 7} \end{array}$

Simplify.

4. $4x + (3x - 5)$ **$7x - 5$**
5. $(4m + 2) + (5m - 3)$ **$9m - 1$**
6. $(2r^2 + 3r - 4) + (2r^2 + 3)$ **$4r^2 + 3r - 1$**
7. $6y - (4y - 3)$ **$2y + 3$**
8. $(-z + 3) - (-4z - 2)$ **$3z + 5$**
9. $(3c^2 - 8) - (2c^2 - 2c + 6)$ **$c^2 + 2c - 14$**

EXERCISES

PRACTICE/ SOLVE PROBLEMS

Simplify.

1. $2y + (4y - 3)$ **$6y - 3$**
2. $3k + (2k + 4)$ **$5k + 4$**
3. $9m + (-2m - 7)$ **$7m - 7$**
4. $(2x^2 - 1) + (3x^2 + 7)$ **$5x^2 + 6$**
5. $(5a^2 + 3) + (-7a^2 - 4)$ **$-2a^2 - 1$**
6. $(2x^2 - 3x + 5) + (4x^2 + 6x - 8)$ **$6x^2 + 3x - 3$**
7. $(a^2 - 2a + 1) + (a^2 - 4)$ **$2a^2 - 2a - 3$**
8. $(2y - 3) + (4y + 5)$ **$6y + 2$**
9. $(3x) - (-2x)$ **$5x$**
10. $2k - (2k + 4)$ **-4**
11. $6z - (2z + 7)$ **$4z - 7$**
12. $(7x - 5) - (5x + 3)$ **$2x - 8$**
13. $(10 - 2k) - (7 - 3k)$ **$3 + k$**
14. $(z^2 - 3z + 10) - (z^2 + 5z - 6)$ **$-8z + 16$**
15. $(8p^2 + 5) - (3p^2 + 2p - 9)$ **$5p^2 - 2p + 14$**
16. $(2a^2 - 3a + 5) - (a^2 - 2a - 4)$ **$a^2 - a + 9$**
17. $(11m^3 + 2m^2 - m) - (-6m^3 + 3m^2 + 4m)$ **$17m^3 - m^2 - 5m$**
18. $(3d^3 - 10d^2 + d + 1) + (5d^3 + 2d^2 - 3d - 6)$ **$8d^3 - 8d^2 - 2d - 5$**

EXTEND/ SOLVE PROBLEMS

Simplify.

19. $(2x - 4y) + (x + 3y)$ **$3x - y$**
20. $(8m - 7n) + (-5m - 4n)$ **$3m - 11n$**
21. $(a + b) + (a + b) + (a + b)$ **$3a + 3b$**
22. $(7r - 3s) + (4r + 2s)$ **$11r - s$**
23. $(8a - 3b) + (5a + 4b)$ **$13a + b$**
24. $(2a + 6b - 1) + (5a + b - 3)$ **$7a + 7b - 4$**
25. $(3x - 4xy + y) + (-4x - 2xy - y)$ **$-x - 6xy$**
26. $(-4m^2n - 3mn^2) + (6m^2n - 3mn^2)$ **$2m^2n - 6mn^2$**
27. $(3r^2t - 2rt^2) + (5r^2t - 4rt)$ **$8r^2t - 2rt^2 - 4rt$**
28. $(-6x^2y - 4x^2) + (xy^2 - 2x^2y)$ **$-8x^2y - 4x^2 + xy^2$**

29. $(4a - 3b) - (a - 4b)$
$3a + b$

30. $(7m - 3n) - (-9m + 4n)$
$16m - 7n$

31. $(2z - 3y) - (5z - 5y)$
$-3z + 2y$

32. $(7x - 2y) - (3x + 4y)$
$4x - 6y$

33. $(8f - 7g - 1) - (-f + 4g + 3)$
$9f - 11g - 4$

34. $(-3k^2g - 4kg^2) - (6k^2g - 2kg^2)$
$-9k^2g - 2kg^2$

35. $(2r^2t^2 + 3rt^2) - (-4r^2t^2 - 7rt^2)$
$10rt^2$

36. $(4xy^2 - 4y^2) - (5xy^2 - x^2y)$
$x^2y - xy^2 - 4y^2$

37. $(4a^2bc - 3abc^2 + 2abc) + (8abc^2 - 5a^2bc - ab^2c)$
$-a^2bc + 5abc^2 + 2abc - ab^2c$

38. $(3a + 9b - 12) - (4a + 10b - 13)$ $-a - b + 1$

39. $(4x - 4xy + 3y) - (4x - 5xy + 3y)$ xy

40. $(12abc - 3ab^2c + 4a^2bc) - (-3ab^2c - 4abc + 4a^2bc)$ $16abc$

Use the diagrams for Exercises 41–43.

41. Write an expression in simplest form for the perimeter of triangle RST. $5x + y$

42. Write an expression in simplest form for the perimeter of quadrilateral $ABCD$. $9x + 2y$

43. Find the difference when the perimeter of triangle RST is subtracted from that of quadrilateral $ABCD$. $4x + y$

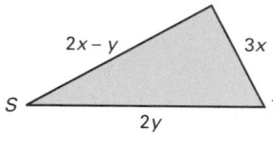

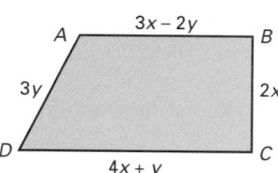

44. Find each difference.

a. $(4n^2 - 2n + 1) - (n^2 + 6n - 5)$ $3n^2 - 8n + 6$
b. $(n^2 + 6n - 5) - (4n^2 - 2n + 1)$ $-3n^2 + 8n - 6$

45. Compare your answers to parts **a** and **b** of Exercise 44. How are they related? **They are opposites of each other.**

46. Suppose that you are working with two polynomials. When you subtract the second polynomial from the first, the result is $x^2 + 3x - 9$. What do you think the result would be if you subtracted the first polynomial from the second? $-x^2 - 3x + 9$

47. Analyze your answers to Exercises 44–46. Make a generalization about what happens when you reverse a subtraction of polynomials.

TALK IT OVER

Is it possible for the sum of two trinomials to be a binomial?

Is it possible for the difference of two trinomials to be a monomial?

Answers will vary. Both situations are possible.
Examples:
$x^2 + 4x + 7 + (x^2 - 4x - 1) = 2x^2 + 6;$
$x^2 + 5x - 6 - (x^2 + 2x - 6) = 3x$

THINK CRITICALLY/ SOLVE PROBLEMS

47. The sign of each term in the difference is opposite when the order of polynomials is reversed in a subtraction.

---

5. $(3w^2 + 2w + 8) - (-3w^2 + 2w - 8)$
$6w^2 + 16$

6. $(2d + 4e) + (3d - 6e) + (-d - 5e)$
$4d - 7e$

7. $4q - (-8q)$ $12q$

8. $(7s - 4) - (6s - 2)$ $s - 2$

9. $(n^2 - 5n + 6) - (3n^2 + 2n - 5)$
$-2n^2 - 7n + 11$

10. $(5p - 2r) - (-7p + 3r)$
$12p - 5r$

11. $(4a + 6b - 3) - (2 + 6a - 2b)$
$-2a + 8b - 5$

12. $(5d^4 - 4d^3 - 3d^2 + 1) - (9d^4 - 4d^3 + 6d^2 + 8d)$
$-4d^4 - 9d^2 - 8d + 1$

Extension Have students check their answers to some of the exercises by substituting values for the variables in the original addition or subtraction and in the simplified answer.

Section Quiz Simplify.

1. $(8m^2 + 3m) - (5m^2 - 6m)$
$3m^2 + 9m$

2. $(4t^2 - 6t) + (t^2 - 3t)$ $5t^2 - 9t$

3. $(8x^3 - 2x) + (3x - 5) - (-4x^3 + 6x)$
$12x^3 - 5x - 5$

4. $(3a^2 + 5a - 2) - (4a^2 - 3a + 1)$
$-a^2 + 8a - 3$

5. $(2c^3 - 20c^2 + c + 10) + (5c^3 + c^2 - 10c + 10)$ $7c^3 - 19c^2 - 9c + 20$

Get Ready 15 long algebra tiles per student

445

12-3 Multiplying Monomials

EXPLORE

Use algebra tiles. Assign the value a to the length of a long tile, and assign the value b to its width.

a. What is the area of each long tile? **$a \times b$ or ab**

b. Use long tiles to model a rectangle with length $3a$ and width $2b$. What is the total area of all the long tiles you used? **$6ab$**

c. Using the area formula $A = l \times w$, give the area of the modeled rectangle as a product of its dimensions. **$3a \times 2b$**

d. Use your model to give the product of $3a$ and $2b$ as a monomial. **$6ab$**

e. Model the following products and give each result as a monomial. What pattern do you see?

e. The product is always found to be the product of the coefficients and $a \times b$, written as a monomial.

$4a \times 3b$ **$12ab$** $a \times 3b$ **$3ab$** $2a \times 2b$ **$4ab$** $3a \times 5b$ **$15ab$**

SKILLS DEVELOPMENT

Recall that, by the commutative property of multiplication, the order in which factors are multiplied does not affect the product: $a \times b = b \times a$.

By the associative property of multiplication, the grouping of factors does not affect the product: $(a \times b) \times c = a \times (b \times c)$.

You can use these properties to find a product of two monomials.

Example 1

Simplify.

a. $(4a)(5b)$ b. $(-3k)(2p)$ c. $\left(-\frac{1}{2}x\right)(-4y)$

Solution

a. $(4a)(5b) = (4)(5)(a)(b)$ ← **Use the commutative and associative properties to regroup the coefficients and variables.**
   $= 20ab$

b. $(-3k)(2p) = (-3)(2)(k)(p) = -6kp$

 **READING MATH**

c. $\left(-\frac{1}{2}x\right)(-4y) = \left(-\frac{1}{2}\right)(-4)(x)(y) = 2xy$ ◄

When multiplying monomials, you often will need to use the laws of exponents, which you learned in Chapter 2.

Recall the *product rule for exponents:* To multiply two powers having the same base, you add the exponents.

Remember that parentheses are often used to show multiplication.

$$a^m \times a^n = a^{m+n}$$

You can use the product rule for exponents, together with the commutative and associative properties of multiplication, to find a product of two monomials having powers of the same base.

Example 2

Simplify.

a. $(-3k)(-2k^3)$ b. $(ab^3)(a^3b^4)$

Solution

a. $(-3k)(-2k^3) = (-3)(-2)(\underline{k \times k^3})$ k means k^1

              $= 6k^{1+3}$

              $= 6k^4$

b. $(ab^3)(a^3b^4) = (a \times a^3)(b^3 \times b^4)$

             $= a^{1+3} \times b^{3+4} \leftarrow$ Use the product rule for each base, a and b.

             $= a^4 b^7 \leftarrow$ Because a and b are unlike bases, you cannot add exponents. ◄

Recall the *power rule for exponents:* To find the power of a monomial that is itself a power, you multiply exponents.

$$(a^m)^n = a^{mn}$$

You can use the power rule, together with the commutative and associative properties for multiplication, to simplify a power of a product such as $(2c^3)^2$.

$$(2c^3)^2 = 2c^3 \times 2c^3$$

$$= (2 \times 2) \times (c^3 \times c^3)$$

$$= 2^2 \times (c^3)^2$$

$$= 4c^6$$

Notice that, when the product $2c^3$ is squared, both 2 and c^3 are squared. This suggests the following rule, called the **power of a product rule:**

► To find the power of a product, find the power of each factor, and multiply.

$$(ab)^m = a^m b^m$$

Example 3

Simplify.

a. $(3z)^2$ b. $(4y^2)^2$ c. $(-2c^2)^3$

Solution

a. $(3z)^2 = (3^2)(z^2)$ b. $(4y^2)^2 = (4)^2 (y^2)^2$ c. $(-2c^2)^3 = (-2)^3 (c^2)^3$

     $= 9z^2$ $= 16y^{2 \times 2}$ $= -8c^{2 \times 3}$

                          $= 16y^4$ $= -8c^6$ ◄

Example 2: Parts a and b illustrate how to multiply two powers having the same base.

Example 3: Parts a–c illustrate how to find the power of a monomial that contains a power. Be sure that students understand why the coefficient of the product is negative in part c.

Additional Questions/Examples

First determine the sign of the product. Then simplify.

1. $(6a)(4b)$ **positive, 24ab**
2. $(-5n)(3r)$ **negative, −15nr**
3. $(-4y)(y^3)$ **negative, −4y⁴**
4. $(3s^2t^4)(4s^3t^2)$
    positive, 12s⁵t⁶
5. $(q^2)^3$ **positive, q⁶**
6. $(2d^3)^2$ **positive, 4d⁶**
7. $(-3w^2)^5$ **negative, −243w¹⁰**
8. $(-2yz^6)^2$ **positive, 4y²z¹²**

Guided Practice/Try These Observe students as they complete the exercises independently. Discuss any problems or questions they may have. Remind students that, in Exercise 2, the coefficient of $-b$ is −1.

5-MINUTE CLINIC

Exercise	Student's Error	Error Diagnosis
Simplify:	$(4c^2)^3 (-2c^3)^4 = -8c^{18}$	• Student forgets that a negative number raised to the fourth power is positive.
$(4c^2)^3 (-2c^3)^4$	$(4c^2)^3 (-2c^3)^4 = 8c^{72}$	• Student multiplies all exponents.

3 SUMMARIZE

3 SUMMARIZE

Write About Math Have students write answers to the following in their math journals.
1. Explain how to multiply the monomials $4c$ and $3d$.
2. Explain how to multiply the monomials $2x^2y$ and $3x^3y^4$.
3. Explain how to simplify $(-2x^3)^3$.
4. If the cube of a negative number is multiplied by the square of a negative number, will the product be positive or negative? **negative**

4 PRACTICE

Practice/Solve Problems Watch for students who have trouble determining the sign of the product in Exercises 19–26. Exercises 27 and 28 involve the areas of a rectangle and a triangle.

Extend/Solve Problems Exercises 29–37 involve the product of monomials with more than one variable. Use Exercises 40–43 to review the formula for the volume of rectangular prisms.

Think Critically/Solve Problems Discuss students' responses to Exercises 44–46.

5 FOLLOW-UP

Extra Practice Simplify.
1. $(5q)(-4h)$ **−20qh**
2. $(-8p)(-1/4\ t)$ **2pt**
3. $(6a^2)(-3a^4)$ **−18a⁶**
4. $(c^2)^3(-4c)^3$ **−64c⁹**
5. $(3a^4b^2)(-3a^5b^3)$ **−9a⁹b⁵**
6. $(-1/3\ x^2y^3)(9x^4y^4)$ **−3x⁶y⁷**
7. $(3d^2e^3)^2(-d^4e^5)^3$ **−9d¹⁶e²¹**
8. $(2x^3y^4)(-3xy^2)^3$ **−54x⁶y¹⁰**
9. $(-5c^2d^3)^2\ (-3cd^4)^3\ (1/5c^2d)^2$ **−27c¹¹d²⁰**

TRY THESE

Simplify.

1. $(3x)(3y)$ **9xy**
2. $(-b)(5d)$ **−5bd**
3. $(4g)(-2h)$ **−8gh**
4. $\left(-\frac{1}{3}a\right)(-3b)$ **ab**
5. $(-2y)(y^2)$ **−2y³**
6. $(2p^2r^3)(5pr^3)$ **10p³r⁶**
7. $(2y)^3$ **8y³**
8. $(3z^3)^2$ **9z⁶**
9. $(-3c^2)^2$ **9c⁴**

EXERCISES

PRACTICE/ SOLVE PROBLEMS

Simplify.

1. $(2a)(3b)$ **6ab**
2. $(-4m)(-3n)$ **12mn**
3. $(6k)(-2m)$ **−12km**
4. $(3x)(-3z)$ **−9xz**
5. $(-2k)(-3p)$ **6kp**
6. $(-6a)\left(\frac{1}{2}b\right)$ **−3ab**
7. $(2c)(3.5d)$ **7cd**
8. $(-3p)(-4.1r)$ **12.3pr**
9. $(2a)^2$ **4a²**
10. $(3m^2)^3$ **27m⁶**
11. $(7y)(y^4)$ **7y⁵**
12. $(5b^3)(-2b)$ **−10b⁴**
13. $(4p^2)(-6p)$ **−24p³**
14. $(-3a^2)(-2a^3)$ **6a⁵**
15. $(5k^4)(-3k^2)$ **−15k⁶**
16. $(2p^3)(-4p^5)$ **−8p⁸**
17. $(3x)^2$ **9x²**
18. $(3y^2)^3$ **27y⁶**
19. $(-2d^3)^3$ **−8d⁹**
20. $(-3x^2)^3$ **−27x⁶**
21. $(9y)(3y)^2$ **81y³**
22. $(3x^2)(2x)^3$ **24x⁵**
23. $(2c)^2(3c)^2$ **36c⁴**
24. $(2a^2)^2$ **4a⁴**
25. $(-4b^4)^3$ **−64b¹²**
26. $(-y^2)^2(-5y^3)^2$ **25y¹⁰**

Write an expression for the area of each figure.

27.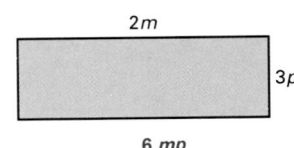

2m

3p

6 mp

28.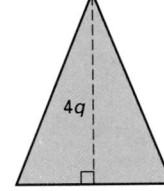

4q

8p   16 pq

MIXED REVIEW

Identify each polygon.

1.
2.

right triangle   quadrilateal

3.
4.

hexagon   octagon

5. Find the area of a triangle whose base is 8 cm and whose height is 4.5 cm.
   18 cm²

MIXED REVIEW

Identify each polygon.

1.
2.

right triangle   quadrilateal

3.
4.

hexagon   octagon

5. Find the area of a triangle whose base is 8 cm and whose height is 4.5 cm.
   18 cm²

Simplify.

29. $(2a^3b^2)(6a^3b)$ **$12a^6b^3$** 30. $(-3x^3y^3)(-4xy^2)$ **$12x^4y^5$** 31. $\left(\frac{-1}{3}p^3r\right)(9p^2r^4)$ **$-3p^5r^5$**

32. $(-2a^2b)^4\left(\frac{1}{2}ab^3\right)^2$ **$4a^{10}b^{10}$** 33. $\left(\frac{1}{3}x^2y\right)^3(3x^5y)^3$ **$x^{21}y^6$** 34. $(2d)^3(4d^2)^3(-d)^2$ **$512d^{11}$**

35. $(2x^3y^6)^4\left(\frac{1}{2}xy^3\right)^2$ **$4x^{14}y^{30}$** 36. $(4z)^2(2z^2)^3\left(\frac{1}{4}z\right)^3$ **$2z^{11}$** 37. $(2p^2r^3)^2(3p^3r)^2$ **$36p^{10}r^8$**

Write an expression for the area of each figure.

38.

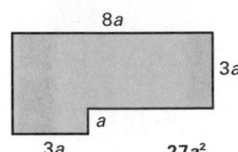
8a
3a
a
3a   27a²

39.

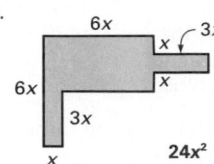 
6x   3x
x   x
6x   x
3x
x   24x²

Write an expression for the volume of each cube.

40.

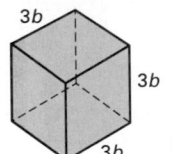
3b
3b
3b   27b³

41.

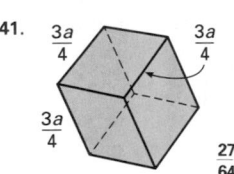
$\frac{3a}{4}$ $\frac{3a}{4}$
$\frac{3a}{4}$ $\frac{27}{64}a^3$

Write an expression for the volume of each prism.

42.

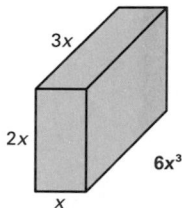
3x
2x
6x³
x

43.

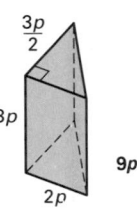
$\frac{3p}{2}$
3p
9p³
2p

44. Which triangle has the greater area? **A**

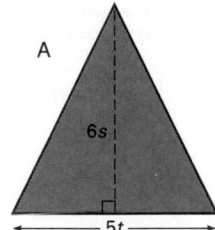
A   6s   5t

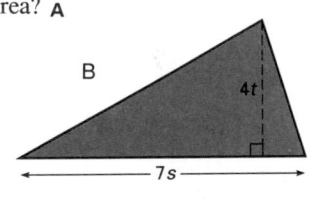
B   4t   7s

45. Give values of s and t for which $30st$ is less than $28st$.
    Any values such that st is negative
46. As a result of answering Exercise 45, did you want to change your answer to Exercise 44 in any way? Explain why or why not.

12-3 Multiplying Monomials **449**

THINK CRITICALLY/ SOLVE PROBLEMS

46. Since there are values of s and t such that $30st < 28st$, it may seem that triangle B can have a greater area than triangle A. However, such values of s and t do not make sense in the diagrams because they would result in a negative number for one of the lengths in each diagram. Thus, triangle A does have the greater area for any meaningful values of s and t.

10. Write an expression for the area of the triangle. **$10x^2$**

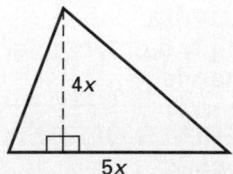
4x
5x

Extension Have students check their answers to the Practice exercises by substituting values for the variables in the original expression and in the simplified expression.

Section Quiz Simplify.
1. $(3b)(4d)$ **$12bd$**
2. $(-5m)(-3n)$ **$15mn$**
3. $(8t)(-2t)$ **$-16t^2$**
4. $(-a^3)(4a^5)$ **$-4a^8$**
5. $(2b^3)(4b^4)$ **$8b^7$**
6. $(5x)^2$ **$25x^2$**
7. $(-4f)^3$ **$-64f^3$**
8. $-(3v^2)^4$ **$-81v^8$**
9. $(4x^2)^3(2x^2)^3$ **$512x^{12}$**
10. $(-p^2)^5(-3p^3)^2$ **$-9p^{16}$**

Get Ready calculators

449

12-4 Multiplying a Polynomial by a Monomial

EXPLORE

a. Figure A shows a rectangle with a length of $2x + 5$ and a width of $2x$. What is the measure of the area of the rectangle expressed as a product? $2x(2x + 5)$

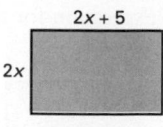
Figure A

b. Figure B shows the same rectangle divided into two smaller rectangles with dimensions as marked. Express the measure of the area of each smaller rectangle as a product. $(2x)(2x), (2x)(5)$

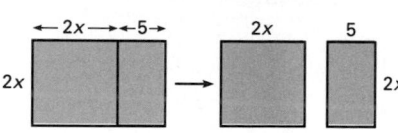
Figure B

c. Now express the measure of the area of the larger rectangle as the sum of the areas of the two smaller rectangles. $(2x)(2x) + (2x)(5)$

d. Compare your answer to a with your answer to c. What property is illustrated? **the distributive property**

SKILLS DEVELOPMENT

You can use the distributive property and the rules for exponents to find the product of a monomial and a polynomial.

Example 1

Simplify.
a. $3a(a + 4)$ **b.** $-2x(x^2 - 3x + 11)$

Solution

a. $3a(a + 4) = 3a(a) + 3a(4)$ ← Use the distributive property.
$= 3a^2 + 12a$ ← Apply the product rule for exponents and the commutative property for multiplication.

b. $-2x(x^2 - 3x + 11) = -2x[x^2 + (-3x) + 11]$
$= -2x(x^2) + [-2x(-3x)] + [-2x(11)]$
$= -2x^3 + 6x^2 + (-22x)$
$= -2x^3 + 6x^2 - 22x$ ◀

Example 2

For a science exhibit, Gail uses a photograph of Saturn's rings that is 12 in. wide and 16 in. long. When she adds a caption below the photograph, the total width will be $(12 + c)$ in. Find an expression for the total area occupied by the photograph and the caption, then simplify the expression.

450   CHAPTER 12   Polynomials

TEACHING TIP

As an aid to multiplying a polynomial by a monomial, students should practice writing the monomial next to each term of the polynomial. For example:
$-5w(2w^2 - 3w - 4) =$
$-5w(2w^2) + (-5w)(-3w) + (-5w)(-4) =$
$-10w^3 + 15w^2 + 20w$

ASSIGNMENTS

BASIC
1–14, 15–21, 27–28, 30–33

AVERAGE
1–14, 15–29, 30–35

ENRICHED
7–14, 19–29, 30–37

ADDITIONAL RESOURCES
Reteaching 12–4
Enrichment 12–4

Example 2: Draw a diagram and work through this example at the board.

Additional Questions/Examples
Simplify.
1. $6x(3x + 2)$ **$18x^2 + 12x$**
2. $-3z(z^2 + 4z - 5)$
   $-3z^3 - 12z^2 + 15z$
3. $2m^2(m^3 - m)$ **$2m^5 - 2m^3$**
4. $3d(-2d + 4)$
   $-6d^2 + 12d$
5. $x(3x^3 - x - 4)$
   $3x^4 - x^2 - 4x$
6. $1/5\,y(5y^2 + 15y - 10)$
   $y^3 + 3y^2 - 2y$

Guided Practice/Try These
Observe students as they complete the exercises independently. Discuss any questions or problems they may have.

Solution
Use the formula for the area of a rectangle.

$A = l \times w$
$A = 16(12 + c)$ ← **Replace *l* with 16 and *w* with (12 + *c*).**
$A = 16(12) + 16(c)$
$A = 192 + 16c$

The total area occupied by the photograph and the caption is $(192 + 16c)$ in.^{2} ◄

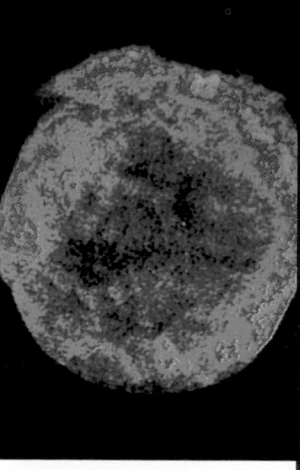

TRY THESE

Simplify.

1. $4x^2(x + 7)$ $4x^3 + 28x^2$
2. $2b(b^2 + 12)$
   $2b^3 + 24b$
3. $3g^2(g^2 - 7)$
   $3g^4 - 21g^2$
4. $2z(3z^2 + z - 8)$
   $6z^3 + 2z^2 - 16z$
5. $6k(2k^2 + k)$
   $12k^3 + 6k^2$
6. $-x(3x^2 - x - 6)$
   $-3x^3 + x^2 + 6x$

Solve.

7. A photograph of a supernova was 12 in. long and 8 in. wide. After Pablo trimmed one side of the photograph, its width was $(8 - t)$ in. Find an expression for the area of the trimmed photograph and then simplify it. $A = 12(8 - t) = 96 - 12t$

12-4   Multiplying a Polynomial by a Monomial   **451**

3 SUMMARIZE

Key Questions
1. Identify the property that is used below:
   $2f(f + 6) = 2f(f) + 2f(6)$
   the distributive property

5-MINUTE CLINIC

Exercise	Student's Error	Error Diagnosis
Simplify: $2x(2x^2 + 6x - 14)$	$2x(2x^2 + 6x - 14) =$ $4x^3 + 6x^2 - 14x$	• Student multiplies only the first term by the coefficient of the monomial.

2. How many terms are in the product of $4x^2 - 2x + 6$ and $3x^2$? **3**

3. True or false: the area of a rectangle whose length is 2 in. greater than its width is given by $2w + 2$. Explain. **False; $l = w + 2$ so area is given by $w(w + 2) = w^2 + 2w$**

4 PRACTICE

Practice/Solve Problems If necessary, review the rules for determining the sign of a product.

Extend/Solve Problems Watch for students who have difficulty with the fractions in Exercises 22 and 23. Allow calculator use with Exercises 28 and 29.

Think Critically/Solve Problems Have students discuss their answers to Exercises 36 and 37.

5 FOLLOW-UP

Extra Practice
Simplify.
1. $3(q^2 - 3q + 4)$ **$3q^2 - 9q + 12$**
2. $-4m(m^2 + 6m - 2)$ **$- 4m^3 - 24m^2 + 8m$**
3. $7s - 3(4s + 6)$ **$-5s - 18$**
4. $3n(-n^3 - 2n + 6)$ **$-3n^4 - 6n^2 + 18n$**
5. $-4p^2(-3p^3 + 2p^2 - 4p + 2)$ **$12p^5 - 8p^4 + 16p^3 - 8p^2$**
6. The width, w, of a rectangle is 3 cm less than its length, l. Write an expression to represent the area and simplify. **$l(l - 3) = l^2 - 3l$**

Extension
Have students create their own problems involving the product of a monomial and a polynomial. Then have them exchange and solve each other's problems.

EXERCISES

PRACTICE/ SOLVE PROBLEMS

Simplify.

1. $2x(x + 3x^2)$ **$2x^2 + 6x^3$**
2. $-4(-y + 2y^2)$ **$4y - 8y^2$**
3. $2a(3a + 1)$ **$6a^2 + 2a$**
4. $10w(-3w + 2)$ **$-30w^2 + 20w$**
5. $2(a^2 + 5a - 1)$ **$2a^2 + 10a - 2$**
6. $-2m(m^2 - 3m - 4)$ **$-2m^3 + 6m^2 + 8m$**
7. $-3(f^2 - 2f - 8)$ **$-3f^2 + 6f + 24$**
8. $4p(p^2 - 3p - 2)$ **$4p^3 - 12p^2 - 8p$**
9. $-r(r^3 - 9r + 6)$ **$-r^4 + 9r^2 - 6r$**
10. $a(4 - a - 5a^2)$ **$4a - a^2 - 5a^3$**
11. $2x(2x^2 - 4x + 3)$ **$4x^3 - 8x^2 + 6x$**
12. $4a^2(-a^2 - 3a + 9)$ **$-4a^4 - 12a^3 + 36a^2$**

Solve.

13. Write an expression for the area of the rectangle. Then simplify the expression. **$10x^2 - 5x$**

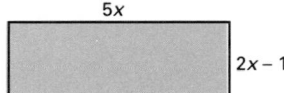
5x
2x − 1

14. Let r represent an even integer. Then $r + 2$ is the next even integer. Write an expression for the product of the two integers. Then simplify the expression. **$r(r + 2) = r^2 + 2r$**

EXTEND/ SOLVE PROBLEMS

Simplify.

15. $4x + 2(x + 1)$ **$6x + 2$**
16. $4(a - 3) - 3a$ **$a - 12$**
17. $-2p - 2(p - 3) + 6$ **$-4p + 12$**
18. $4(m - 1) + 3(m + 5)$ **$7m + 11$**
19. $4(2a - b) - 3(a + b)$ **$5a - 7b$**
20. $4a(2b - a) - 3a(2b + a)$ **$2ab - 7a^2$**
21. $x(2x^2 - x + 2) - x(x^3 + 4x^2 - 3x)$ **$-x^4 - 2x^3 + 2x^2 + 2x$**
22. $\frac{1}{2}x(2x^2 + 6xy - 14y^2)$ **$x^3 + 3x^2y - 7xy^2$**
23. $\frac{1}{3}c^2(3c^2 - 9cd - 18d^2)$ **$c^4 - 3c^3d - 6c^2d^2$**
24. $(4x^2y - 3xy^2 + 7y^3)(4xy)$ **$16x^3y^2 - 12x^2y^3 + 28xy^4$**
25. $(r^3 - 3r^2s + 4s^2)(-2rs)$ **$-2r^4s + 6r^3s^2 - 8rs^3$**

Exercises 26–28 refer to this diagram. The diagram indicates the number of seconds it took a beam of sunlight to travel first to the earth, then on to the moon.

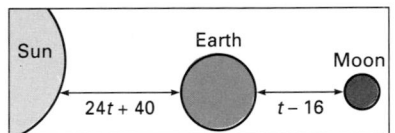
Sun
Earth
Moon
24t + 40
t − 16

26. Write an expression for the total time it took the light to reach the moon. **$(25t + 24)$ s**

CHALLENGE

Have students factor each polynomial into a common monomial factor and a polynomial factor. Have them check their answers by multiplying.
1. $6a - 6b$ **$6(a - b)$**
2. $-3x - 3y - 3z$ **$-3(x + y + z)$**
3. $4gh - 3ht + fh$ **$h(4g - 3t + f)$**

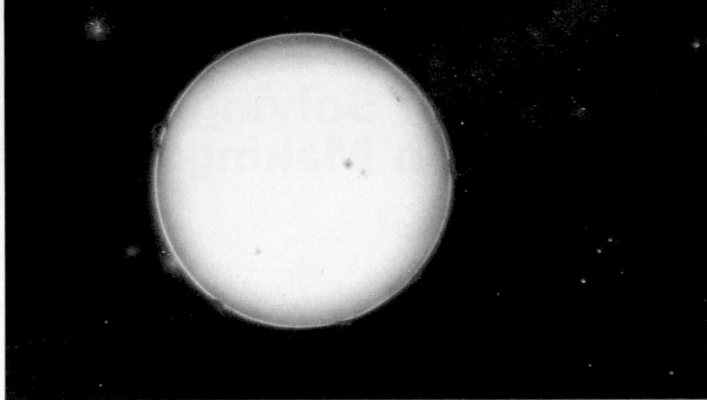

MATH:
WHO, WHERE, WHEN

Annie Jump Cannon (1863–1941) was one of the most well-known and honored astronomers of her day. She contributed much to the study of *variable* stars, stars whose magnitude is not constant, but changes. Ms. Cannon published a 10-volume catalog grouping over 1,000 stars according to their color spectra. Her careful and accurate work in compiling this information has benefited astronomers everywhere.

Section Quiz
Simplify.
1. $4c(c + 5c^2)$ **$4c^2 + 20c^3$**
2. $-3(-d + 6d^2)$ **$3d - 18d^2$**
3. $5s^3(2s^2 - 3s)$ **$10s^5 - 15s^4$**
4. $-6p(p^3 - 4p^2 + p)$
   $-6p^4 + 24p^3 - 6p^2$
5. $3a^2(-a^3 - 4a + 6)$
   $-3a^5 - 12a^3 + 18a^2$
6. Let n represent an integer. Write an expression for the product of that integer and an integer that is 6 greater than 3 times the integer n. Then simplify the expression.
   $n(3n + 6) = 3n^2 + 6n$

27. Suppose that light travels at a rate of c mi/s. Write a product expressing the distance from the sun to the moon. Then simplify the product. **$c(25t + 24) = 25ct + 24c$**

28. Use $c = 186,000$ mi/s and $t = 19$ s to approximate the distance from the sun to the moon. **about 92,814,000 mi**

USING DATA Refer to the Data Index on page 546 to find the table listing the largest moons in the solar system.

29. One moon in our solar system has a diameter of $(w^2v)(w^5v)$ miles, where $w = 2$ and $v = 5$. Which moon is it? **Ganymede**

Write an expression for the area of the shaded region. Then simplify the expression.

**THINK CRITICALLY/
SOLVE PROBLEMS**

30.

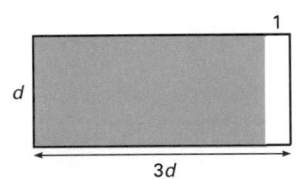

$3d^2 - d$

31.

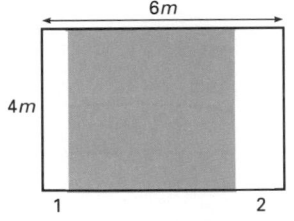

$24m^2 - 12m$

Simplify.

32. $3(x - y) - 2(x - y)$ **$x - y$**

33. $7(x - y) - 6(x - y)$ **$x - y$**

34. $13(x - y) - 12(x - y)$ **$x - y$**

35. $20(x - y) - 19(x - y)$ **$x - y$**

Examine your results for Exercises 32–35. Then, without multiplying, write a simplified expression for each.

36. $112(x - y) - 111(x - y)$ **$x - y$**

37. $50(w + z) - 49(w + z)$ **$w + z$**

12-5

► READ
► PLAN
► SOLVE
► ANSWER
► CHECK

Problem Solving/ Decision Making:

CHOOSE A STRATEGY

 PROBLEM SOLVING TIP

Here is a checklist of the problem solving strategies that you have studied so far in this book.

Solve a simpler problem
Find a pattern
Guess and check
Work backward
Make an organized list
Make a table
Use logical reasoning
Act it out
Draw a picture

In this book you have studied a variety of problem solving strategies. Experience in applying these strategies will help you decide which will be most appropriate for solving a particular problem. Sometimes only one strategy will work. In other cases, any one of several strategies will offer a solution. There may be times when you will want to use two different approaches to a problem in order to be sure that the solution you found is correct. For certain problems, you will need to use more than one strategy in order to find the solution.

PROBLEMS

Solve. Name the strategy you used to solve each problem.
Strategies may vary. One possible strategy is suggested.

1. The sum of two monomials is $11n$. The product of the monomials is $30n^2$. Find the two monomials. **See Additional Answers.**

2. In 1609, Galileo discovered Jupiter's four largest moons—Callisto, Ganymede, Io, and Europa. In how many different orders might Galileo have sighted these moons?
24 different orders; make an organized list

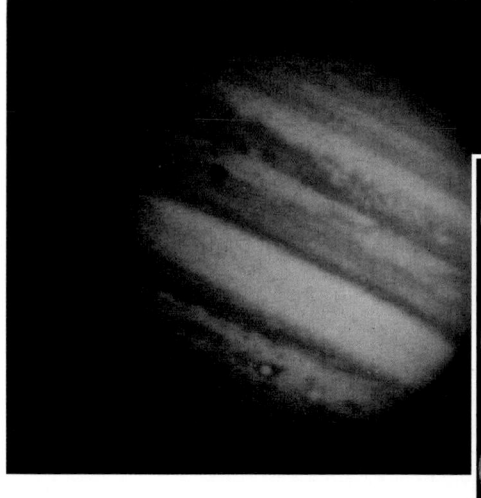

454 CHAPTER 12 Polynomials

3. The product of two monomials is x^6. How many different sums might the monomials have? What are the sums? **four; $1 + x^6$; $x + x^5$; $x^2 + x^4$; $x^3 + x^3 = 2x^3$; make an organized list**

4. The height of a lighthouse is 25 yd plus half its height. How high is the lighthouse? **See Additional Answers.**

5. What is the sum of the first ten odd numbers? What is the sum of the first 75 odd numbers? **See Additional Answers.**

6. Take a certain monomial. Add $7x^3$ to it. Subtract $10x$. Add $4x^3 + 5x$. The result is $20x^3 - 5x$. Find the original monomial. **See Additional Answers.**

7. Akira knows he set his watch to the correct time at 6:00 a.m. When he hears a radio announcer say that it is exactly 10:30 a.m., Akira notices that his watch is 18 min fast. At that rate, how many minutes fast will his watch be by 5:00 p.m.? **See Additional Answers.**

8. Jana is thinking of a number. If she multiplies this number by itself and then adds the number to the result, she gets 420. What is Jana's number? **20; guess and check**

COMPUTER TIP

9. A rectangle has a perimeter of $22x$. One side of the rectangle measures $3x$. Write an expression for the area of the rectangle. **$24x^2$; draw a picture**

10. Study this set of statements.

$$(x^1)^1 = x^1$$
$$(x^{11})^{11} = x^{121}$$
$$(x^{111})^{111} = x^{12,321}$$

Find $(x^{1,111})^{1,111}$ and $(x^{111,111})^{111,111}$. **$x^{1,234,321}$; $x^{12,345,654,321}$; find a pattern**

RUN this program to obtain some data that can help you in solving Problem 5.

```
10 PRINT "ODD NUMBERS",
   "SUM"
20 LET X = -1 : LET S = 0
30 FOR N = 1 TO 7
40 LET X = X + 2; LET
   S = S + X
50 PRINT N,S
60 NEXT N
```

11. On planet S there are four kinds of creatures: Lums, Mads, Nogs, and Ools. Each kind of creature has a standard number of eyes, 6, 8, 14, or 16. Nogs have more than twice as many eyes as Lums. The number of eyes of a Nog is not a perfect square. Mads have more eyes than Ools. How many eyes does each type of creature have? **Lums: 6; Mads: 16; Nogs: 14; Ools: 8; See Additional Answers.**

12. A town is placing two rows of benches in a rectangular park, one row along the north end and one along the south end. Those ends of the park are 200 ft long. Each bench is 8 ft long, and they are to be 4 ft apart. If the rows of benches are to begin and end at a corner of the park, how many benches will be needed? **34 benches; draw a picture. See Additional Answers.**

13. A checkerboard is made up of 8 rows of 8 squares. How many squares of all sizes are there in a checkerboard? **204 squares; find a pattern. See Additional Answers.**

CHALLENGE

Pete decides to visit his friend in the country. It takes him 4 h, walking at a steady rate of 2 1/2 mi/h. After visiting for an hour, his friend drives him home. The road is bumpy and they travel at a rate of only travel 20 mi/h. Pete arrives home at 2:30 p.m. At what time had he left his house in the morning? **9 a.m.**

ASSIGNMENTS

BASIC
1–13

AVERAGE
1–13

ENRICHED
1–13

ADDITIONAL RESOURCES
Reteaching 12–5
Enrichment 12–5

---

the sum of the first two odd numbers, and the first three odd numbers. Look for a pattern. Students should see, in Problem 6, that the best strategy is to work backward. In Problem 13, students may forget the board itself, which is an 8 by 8 square. Encourage all students to try all problems. You may wish to have students work in pairs or small groups.

5 FOLLOW-UP

Extra Practice

1. If Nina's daughter is my son's mother, what is my relationship to Nina if I am male? **Nina's son-in-law**

2. A cup and saucer together weigh twelve ounces. The cup weighs twice as much as the saucer. How much does the saucer weigh? **4 oz**

Additional Answers
See page 583.

Additional answers for odd-numbered exercises are found in the Selected Answers portion of the page.

12-6 Factoring

Suppose that a rectangle has an area of 36 ft². What are all the possible whole-number dimensions that the rectangle might have? Two possibilities are shown on the grid below.
What are all the factors of 36? **1, 2, 3, 4, 6, 9, 12, 18, 36**

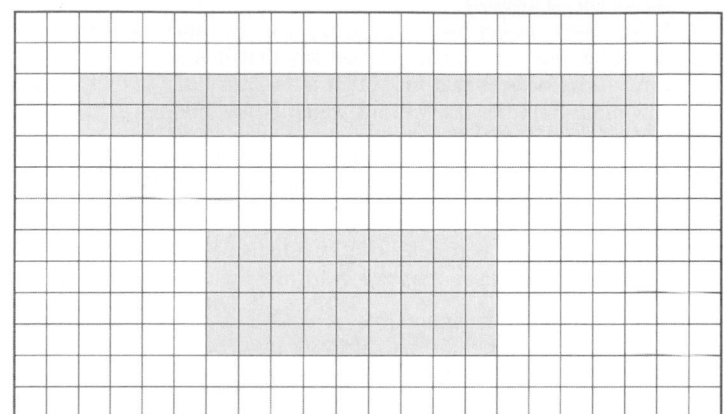

SKILLS DEVELOPMENT

Recall that, to simplify expressions, you can apply the distributive property.

$$2(3x + 4) = 2(3x) + 2(4) = 6x + 8$$

You began with the factors and multiplied them to find the product. You can reverse this process to find what factors were multiplied to obtain the product. The reverse process is called **factoring.**

The **greatest common factor** of two or more monomials is the *common factor* having the greatest numerical factor and the variable or variables of greatest degree.

Example 1

Find the greatest common factor of the monomials.
a. $4x$ and $16xy$ **b.** $-6a^2b$ and $12ab$

Solution
a. Write the factors of $4x$: $(2^2)(x)$
    Write the factors of $16xy$: $(2^4)(x)(y)$
    The GCF of $4x$ and $16xy$ is $(2^2)(x)$, or $4x$.

b. Write the factors of $-6a^2b$: $(2 \times 3)(a^2)(b)$
    Write the factors of $12ab$: $(2^2 \times 3)(a)(b)$
    The GCF of $-6a^2b$ and $12ab$ is $(2 \times 3)(a)(b)$, or $6ab$. ◄

To factor a polynomial, look for the greatest monomial factor common to all the terms. Then write the polynomial as the product of the greatest monomial factor and another polynomial.

Example 2

Factor each polynomial.

a. $3a + 6$

b. $4x^2y - 18x$

Solution

a. Find the GCF of each term of $3a + 6$.

factors of $3a$: $(3)(a)$
factors of 6: (2×3)

The GCF is 3.

Use the GCF to rewrite the polynomial.

$3a + 6 = 3 \times a + 3 \times 2$
$\qquad = 3(a + 2)$ ← **Use the distributive property.**

So, $3a + 6 = 3(a + 2)$.

b. Find the GCF of each term.

factors of $4x^2y$: $(2^2)(x^2)(y)$
factors of $18x$: $(2 \times 3^2)(x)$

The GCF is $2x$.

Use the GCF to rewrite the polynomial.

$4x^2y - 18x = (2x)(2xy) - (2x)(9)$
$\qquad = 2x(2xy - 9)$

So, $4x^2y - 18x = 2x(2xy - 9)$. ◄

Example 3

The formula for the surface area of a rectangular prism is $SA = 2lw + 2wh + 2lh$, where l represents the length, w represents the width, and h represents the height. Use factoring to rewrite the formula.

Solution

Rewrite the formula with the right side as the polynomial in factored form. The GCF of the terms is 2.

$SA = 2lw + 2wh + 2lh$
$SA = 2(lw + wh + lh)$ ◄

Try These

Find the greatest common factor of the monomials.

1. $3y$ and $15yz$ **3y**

2. $4cd^2$ and $20cd$ **4cd**

Factor each polynomial.

3. $4x + 12$ **4(x + 3)**

4. $6y^2z - 9y$ **3y(2yz - 3)**

Example 2: Point out that the polynomial factors must have the same number of terms as the original polynomial.

Explain that the same results can be obtained if a common factor used is not the GCF, but that using the GCF avoids extra steps.

Elicit that students can multiply the two factors to check whether the polynomial was factored correctly. Emphasize the usefulness of this check and encourage students to check routinely.

Additional Questions/Examples
Find the greatest common factor of each pair of terms.

1. $8x^2y$ and $12x^3y^2$ **$4x^2y$**

2. $6c^3d^4$ and $9cd^2$ **$3cd^2$**

Factor each polynomial.

3. $4s^2 + 8s^2t - 16st$
   $4s(s + 2st - 4t)$

4. $5mn^2 - 25m^2n + 10mn$
   $5mn(n - 5m + 2)$

5-MINUTE CLINIC

Exercise	Student's Error	Error Diagnosis
Factor the polynomial: $-3x^2y - 18xy - 12xy^2$	$-3x^2y - 18xy - 12xy^2 =$ $-3xy(x + 18 + 12y)$	• Student forgets to factor out from each term the coefficient of the monomial factor. Correct solution: $-3xy(x + 6 + 4y)$

Guided Practice/Try These You may wish to have students complete these exercises orally.

3 SUMMARIZE

Write About Math Have students explain in their math journals how to factor the polynomial $6q^2h + 4q^2h^2 - 8qh^2$.
$2qh(3q + 2qh - 4h)$

4 PRACTICE

Practice/Solve Problems Use Exercise 10 to point out that if *all* terms in a polynomial are negative, then the common factor should be negative. Watch for students who forget the 1 in the polynomial factor in Exercises 13, 20, and 27.

Extend/Solve Problems Some students may have difficulty with Exercise 29, in which the terms must be re-ordered to find two sets of common factors.

Think Critically/Solve Problems Elicit that, just as one must begin with parentheses first, here one must begin with the innermost set of parentheses and work outward.

Problem Solving Applications Discuss how the polynomial $5n + 12$ was obtained to represent the perimeter of the figure at the top of the page.
    Point out that the radius and the sides of the square in Exercise 9 are not the same length. Have students explain how to find the area of the square. **Find the area of right triangle *AOB* and multiply the result by 4.**

---

5. The mean of four numbers a, b, c, and d can be found using the formula $M = \frac{1}{4}a + \frac{1}{4}b + \frac{1}{4}c + \frac{1}{4}d$. Rewrite the formula by factoring.
$$M = \frac{1}{4}(a + b + c + d)$$

EXERCISES

PRACTICE/SOLVE PROBLEMS

Find the greatest common factor of the monomials.

1. $8y$ and $4xy^2$ **4y**   2. $3c^2$ and $18cd$ **3c**   3. $5ab^2$ and $15a^2b$ **5ab**

4. $7r^2s$ and $9rs$ **rs**   5. $10s^2t$ and $25s^3t^2$ **5s²t**   6. $14x^2y^2$ and $42x^2y$ **14x²y**

Match each set of monomials with their greatest common factor.

7. $2x^2, 2x^2y, 2x^2z$ **d**   a. $2ab$

8. $3y^3z, 3yz, 6y^2z$ **e**   b. $-xy$

9. $6ab^2, -4a^2b, 12a^2b^2$ **a**   c. x^2

10. $-2x^3y, -x^2y, -2xy^2$ **b**   d. **2x²**

11. $ax^2, 2bx^2, 3cx^2, 5dx^2$ **c** e. $3yz$

Factor each polynomial.

12. $35a - 7ab$ **7a(5 − b)**   13. $9 - 18x$ **9(1 − 2x)**   14. $5x^2 - 4x$ **x(5x − 4)**

15. $12ab - 3b^2$ **3b(4a − b)**   16. $35mn - 15m^2n^2$ **5mn(7 − 3mn)**   17. $24x^2 - 6xy$ **6x(4x − y)**

18. $3x^2 + 6x^2y - 18xy$ **3x(x + 2xy − 6y)**   19. $xy + xy^2 - y^2$ **y(x + xy − y)**

20. $8ab - 16a^2b + 32ab^2$ **8ab(1 − 2a + 4b)**   21. $7xy - 14x^2y^2 + 28xyz$ **7xy(1 − 2xy + 4z)**

EXTEND/SOLVE PROBLEMS

Find the missing factor.

22. $8x^5 = (2x^2)(\underline{\quad?\quad})$ **4x³**   23. $-6a^3b^2 = (2ab^2)(\underline{\quad?\quad})$ **−3a²**

24. $21c^2d = (3c^2)(\underline{\quad?\quad})$ **7d**   25. $12xy^2 = (3x)(\underline{\quad?\quad})$ **4y²**

26. $-15x^3y^5 = (-5x^2y^2)(\underline{\quad?\quad})$ **3xy³**

Factor each polynomial. 27. **$\frac{1}{3}xyz(1 − y − xz)$**

27. $\frac{1}{3}xyz - \frac{1}{3}xy^2z - \frac{1}{3}x^2yz^2$ 28. $14f^2g - 42fg^2 - 56g^2h$ **14g(f² − 3fg − 4gh)**

29. $4rs + 4st - 4tu + 4ru$ **4r(s + u) + 4t(s − u)**   30. $12c^3d - 6c^2d^2 - 14c^2d + 4c^2$ **2c²(6cd − 3d² − 7d + 2)**
    (Answers may vary.)

31. To solve a problem with more than one set of grouping symbols, begin working with the expression within the innermost grouping symbol.
$6x^2 - x[3x - 4x(2x - 3)] =$
$6x^2 - x[3x - 8x^2 + 12x] =$
$6x^2 - 3x^2 + 8x^3 - 12x^2 =$
$8x^3 + (6x^2 - 3x^2 - 12x^2) =$
$8x^3 - 9x^2 = x^2(8x - 9)$

THINK CRITICALLY/SOLVE PROBLEMS

31. Explain the order of operations to follow in simplifying this expression. Simplify the expression, then factor.
$6x^2 - x[3x - 4x(2x - 3)]$ **x²(8x − 9)**

► READ
► PLAN
► SOLVE
► ANSWER
► CHECK

Problem Solving Applications:
USING POLYNOMIALS IN GEOMETRY

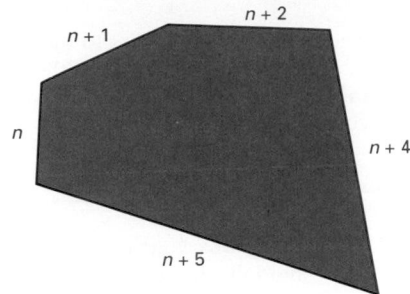

As you continue your study of geometry, you will find that the measures of the sides of a figure are sometimes given as variable expressions. Since the perimeter of a polygon is the sum of the measures of its sides, you can create a polynomial to represent the perimeter.

For instance, the perimeter of the figure above is represented by the polynomial $5n + 12$.

Write an expression for the perimeter of each figure. Then simplify the expression.

1.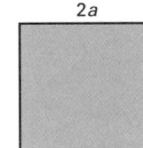
   8a

2.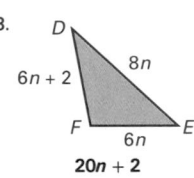
   6n + 12

3.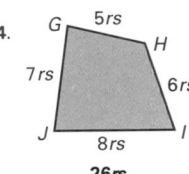
   20n + 2

4.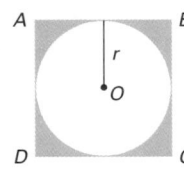
   26rs

5. Write an expression for the area of the rectangle in Exercise 2. **$2n^2 + 6n$**

6. The formula for the area of a trapezoid is $A = \frac{1}{2}hb_1 + \frac{1}{2}hb_2$, where h is the height, b_1 is the length of one base, and b_2 is the length of the other base. Rewrite the formula with the right side as a polynomial in factored form. **$A = \frac{1}{2}h(b_1 + b_2)$**

7. The formula for the surface area of a cylinder with radius r and height h is $SA = 2\pi rh + 2\pi r^2$. Rewrite the formula with the right side as a polynomial in factored form. **$SA = 2\pi r(h + r)$**

Write an expression in factored form for the area of the shaded region.

8.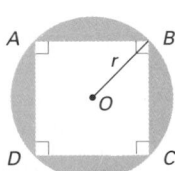
   $r^2(4 - \pi)$

9. (exercise 9 figure)
   $r^2(\pi - 2)$

5 **FOLLOW-UP**

Extra Practice Factor each polynomial.
1. $4ab^3 + 3ab^2$ **$ab^2(4b + 3)$**
2. $7x^2 - 28x$ **$7x(x - 4)$**
3. $16a^2 + 32ab + 48a$
   $16a(a + 2b + 3)$
4. $5m^2n^2 - 5m^2$ **$5m^2(n^2 - 1)$**
5. $d^3 - d^2 + d$ **$d(d^2 - d + 1)$**
6. $5s^2t^3 - 15st^2 - 10s^2t^2$
   $5st^2(st - 3 - 2s)$

Find the missing factor.
7. $10w^6 = 5w^2(\blacksquare)$ **$2w^4$**
8. $-12a^3b^5 = -3a^2b^2(\blacksquare)$ **$4ab^3$**

Extension Have students check their answers to Exercises 12–21 by multiplying the factors.

Section Quiz Factor each polynomial.
1. $27a - 9ab$ **$9a(3 - b)$**
2. $6x^2 - 5x$ **$x(6x - 5)$**
3. $72ab - 56a^2b^2$ **$8ab(9 - 7ab)$**
4. $2w^2 + 4w^2x - 12wx$
   $2w(w + 2wx - 6x)$
5. $6q^2h^2 - 18q^2h^3 - 12q^4h^3$
   $6q^2h^2(1 - 3h - 2q^2h)$
6. $14cd - 21c^3d + 35cd^2$
   $7cd(2 - 3c^2 + 5d)$

WARM-UP

Factor each polynomial.
1. $8x^5 + 7x^3 - 3x^2$
$$x^2(8x^3 + 7x - 3)$$
2. $4a^2b - 24ab + 16a$
$$4a(ab - 6b + 4)$$

1 MOTIVATE

Introduction Read and discuss the introductory material.

2 TEACH

Use the Pages/Problem Have students read this part of the section and then discuss the problem.

Use the Pages/Solution Help students to see how the diagram verifies that the answer is correct.

12-7 Problem Solving Skills:
CHUNKING

► READ
► PLAN
► SOLVE
► ANSWER
► CHECK

The term **chunking** refers to the process of collecting several pieces of information and grouping them together as a single piece of information. Although you might not think much about it, you use chunking every day.

> You chunk the individual letters *d*, *o*, and *g* and think of the word *dog*.

> You chunk your parents, sisters, brothers, and other relatives and talk about your *family*.

In algebra, you are chunking when you think of a polynomial, not as a group of individual terms, but rather as a single variable expression. Sometimes, chunking can help you solve a problem that is different from any problem you have encountered before.

PROBLEM

Factor $n(2 + n) + 5(2 + n)$.

SOLUTION

$n(2 + n) + 5(2 + n) = n\boxed{(2 + n)} + 5\boxed{(2 + n)}$ ← **Think of $(2 + n)$ as a chunk.**

$\qquad\qquad\qquad\qquad = (n + 5)\boxed{(2 + n)}$ ← **Use the distributive property to factor out the chunk.**

So, $n(2 + n) + 5(2 + n) = (n + 5)(2 + n)$.

You can use a diagram like the one below to prove that this answer is correct.

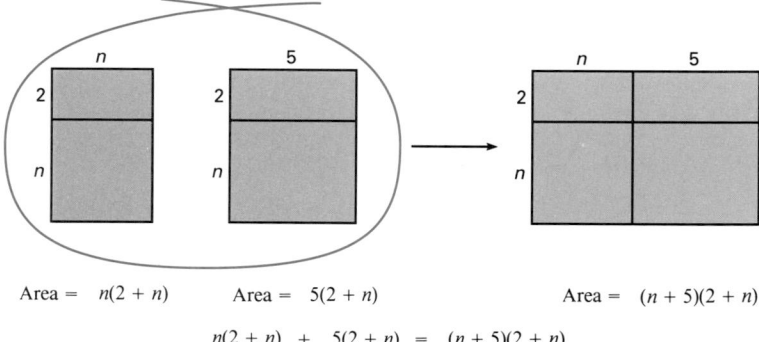

Area = $n(2 + n)$ Area = $5(2 + n)$ Area = $(n + 5)(2 + n)$

$$n(2 + n) \ + \ 5(2 + n) \ = \ (n + 5)(2 + n)$$

460   CHAPTER 12   Polynomials

PROBLEMS

Write the mathematical sentence that each diagram represents.

1.

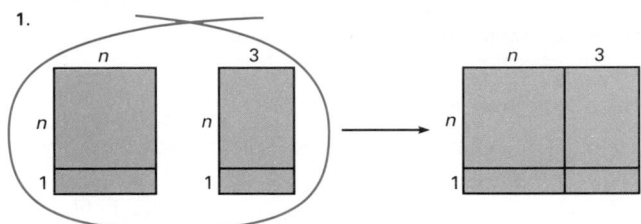

$$n(n + 1) + 3(n + 1) = (n + 3)(n + 1)$$

2.

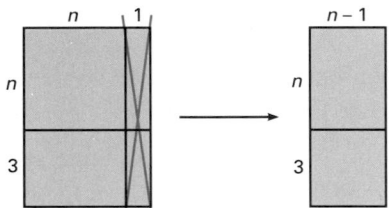

$$n(n + 3) - 1(n + 3) = (n - 1)(n + 3)$$

Use chunking to factor each expression.

3. $a(a - 3) + 2(a - 3)$
$(a + 2)(a - 3)$

4. $(t + 5)t - (t + 5)7$
$(t - 7)(t + 5)$

5. $r(r + 6) + 2(6 + r)$
$(r + 2)(r + 6)$

6. $(k - 1)3 + k(k - 1)$
$(3 + k)(k - 1)$

7. $c^2(c + 1) + c(c + 1) + 5(c + 1)$
$(c + 1)(c^2 + c + 5)$

8. $(z - 5)z^2 - (z - 5)z + (z - 5)2$
$(z - 5)(z^2 - z + 2)$

 **COMPUTER**

After you complete Exercise 11, change 25 to 1, then to 4, then to 9, and then to 16. After each change, use chunking to solve. You can check your results by RUNning this program.

```
10 FOR X = 1 TO 4
20 LET A = X − 2
30 LET B = −X − 2
40 PRINT ''IF (Y + 2) ^ 2 =
   '';X ^ 2;'' THEN
   Y = '';A;'' OR '';B
50 NEXT X
```

Describe how chunking can be used for each of Exercises 9–12.

9. How can you use chunking to solve the equation $2^{m+1} = 32$?
(*Hint:* $32 = 2^?$)

10. How can you use chunking to solve the equation $|d + 1| = 3$?
(*Hint:* Which two integers have an absolute value of 3?)

11. How can you use chunking to solve the equation $(y + 2)^2 = 25$?
(*Hint:* Which two integers squared are equal to 25?)

12. How can you use chunking to solve the equation $\sqrt{b + 4} = 3$?
(*Hint:* The square root of which number is 3?)

9. Use the chunk $(m + 1)$.
Let $m + 1 = 5$. Then
$m = 5 - 1 = 4$.

10. Use the chunk $(d + 1)$.
If $|d + 1| = 3$, then
$d + 1 = 3$ or $d + 1 = -3$.
So $d = 2$ or $d = -4$.

11. Since $(y + 2)^2 = 5^2$,
write these two
equations: $y + 2 = 5$
and $y + 2 = -5$. Then
$y = 3$ and $y = -7$.

12. Use the chunk $(b + 4)$.
If $\sqrt{b + 4} = 3$, then
$b + 4 = 3^2$. So $b = 5$.

ASSIGNMENTS

BASIC
1–8

AVERAGE
2–12

ENRICHED
2–12

ADDITIONAL RESOURCES
Reteaching 12–7
Enrichment 12–7

3 SUMMARIZE

Talk It Over Discuss the following question. *What form must a polynomial take for you to be able to use chunking to factor it?*

4 PRACTICE

Problems You may wish to have students work in pairs or in small groups for Problems 9–12.

5 FOLLOW-UP

Extra Practice Use chunking to factor each expression.
1. $a(a - 5) + 2(a - 5)$
   $(a + 2)(a - 5)$
2. $(t - 3)5 - (t - 3)t$ $(t - 3)(5 - t)$
3. $w^2(w + 4) - 9(w + 4)$
   $(w^2 - 9)(w + 4)$
4. Describe how you can use chunking to solve the equation $(x + 3)^2 = 49$. (Hint: Which two integers squared are equal to 49?) Since $(x + 3)^2 = 49$, write these two equations: $x + 3 = 7$ and $x + 3 = -7$. Then, $x = 4$ and $x = -10$.

5-MINUTE CLINIC

Exercise	Student's Error	Error Diagnosis
Use chunking to factor the expression: $a(a - 3) + 2(a - 3)$	$a(a - 3) + 2(a - 3) =$ $a^2 - 3a + 2a - 6 =$ $a^2 - a - 6$	• Instead of factoring, the student simplifies each term in the polynomial by multiplying. Correct solution: $(a - 3)(a + 2)$

WARM-UP

Multiply.
1. $(4a)(6a)$ **$24a^2$**
2. $(5d)(8d^2)$ **$40d^3$**
3. $(7f^2)(4f^3)$ **$28f^5$**
4. $(6mn)(9m^2)$ **$54m^3n$**

1 MOTIVATE

Explore Have students complete the activity independently. Discuss their results.

2 TEACH

Use the Pages/Skills Development
Have students read this part of the section and then discuss the examples. Write the multiplication $3m \times 2n = 6mn$ on the board and ask a volunteer to write a division sentence that "undoes" the multiplication. ($6mn \div 2n = 3m$ or $6mn \div 3m = 2n$.) Write the division as a ratio: $6mn/2n$ or $6mn/3m$. Elicit that to find the quotient, students divide both coefficients by their GCF and divide the variable parts by their GCF.

12-8 Dividing by a Monomial

EXPLORE You know that division "undoes" multiplication.
$$-4 \times 9 = -36, \quad \text{so} \quad -36 \div 9 = -4.$$

1. Simplify the product $3g \times 5h$. **$15gh$**

2. Write a division equation that "undoes" the multiplication you just performed in a. **$15gh \div 3g = 5h$** or **$15gh \div 5h = 3g$**

3. Write a division equation that "undoes" each of these.
   a. $2a \times 5a = 10a^2$ **$10a^2 \div 5a = 2a$** or **$10a^2 \div 2a = 5a$**
   b. $6y \times 4y^2 = 24y^3$ **$24y^3 \div 6y = 4y^2$** or **$24y^3 \div 4y^2 = 6y$**
   c. $2m \times 7mn = 14m^2n$
   $14m^2n \div 2m = 7mn$ or **$14m^2n \div 7mn = 2m$**

SKILLS DEVELOPMENT You have found the product of two monomials by multiplying the coefficients and multiplying the variables.
$$3m \times 2n = (3 \times 2)(mn) = 6mn$$
Because division is the inverse of multiplication, you can reverse the process to obtain the related division sentence.
$$3m \times 2n = 6mn, \text{ so } 6mn \div 3m = 2n.$$

Now let's look at a different way to find the quotient $6mn \div 3m$.

$$\frac{6mn}{3m} = \frac{6}{3} \times \frac{mn}{m}$$
$$= \frac{\overset{2}{\cancel{6}}}{\underset{1}{\cancel{3}}} \times \frac{mn}{m} \leftarrow \text{Divide both coefficients by their GCF, which is 3.}$$
$$= \frac{\overset{2}{\cancel{6}}}{\underset{1}{\cancel{3}}} \times \frac{\overset{1}{\cancel{m}}n}{\underset{1}{\cancel{m}}} \leftarrow \text{Divide the variable parts by their GCF, which is } m.$$
$$= 2n$$

This division illustrates the following rule for dividing monomials:

► To simplify the quotient of two monomials, find the quotient of any numerical coefficients, then find the quotient of the variables.

Example 1

Simplify.
a. $\dfrac{-4mn}{2m}$ b. $\dfrac{wxz}{wz}$

Solution
a. $\dfrac{-4mn}{2m} = \dfrac{-4}{2} \times \dfrac{mn}{m} = \dfrac{\overset{-2}{\cancel{-4}}}{\underset{1}{\cancel{2}}} \times \dfrac{\overset{1}{\cancel{m}}n}{\underset{1}{\cancel{m}}} = -2 \times n = -2n$

b. $\dfrac{wxz}{wz} = \dfrac{\overset{1}{\cancel{w}}x\overset{1}{\cancel{z}}}{\underset{1}{\cancel{w}}\underset{1}{\cancel{z}}} = \dfrac{x}{1} = x \leftarrow$ **Both coefficients are 1.** ◄

TEACHING TIP

Students often forget the sign when the quotient is negative. To avoid this problem, have students determine and write the sign of the quotient before dividing.

Recall the *quotient rule for exponents:* To divide two powers having the same base, you subtract the exponent of the denominator from that of the numerator.

$$\frac{a^m}{a^n} = a^{m-n}$$

TALK IT OVER

Refer to Example 2b. Explain why the exponent 2 in the denominator is subtracted from the exponent 3 in the numerator but not from the exponent 5.

You can use the quotient rule for exponents to find the quotients of monomials having the same base.

Example 2

Simplify.

a. $\frac{-16x^8}{4x^4}$ **b.** $\frac{-24x^3y^5}{-3x^2y}$

Solution

a. $\frac{-16x^8}{4x^4} = \left(\frac{-16}{4}\right)x^{8-4} = -4x^4$

b. $\frac{-24x^3y^5}{-3x^2y} = \left(\frac{-24}{-3}\right)x^{3-2}y^{5-1}$
    $= 8xy^4$ ◄

The 5 is the exponent of *y*. The 2 is the exponent of *x*. The *x* and the *y* are unlike bases.

Each of the statements is true.

$\frac{18 + 24 + 36}{6} = \frac{78}{6} = 13$ and $\frac{18}{6} + \frac{24}{6} + \frac{36}{6} = 3 + 4 + 6 = 13$

So you can arrive at this conclusion.

$\frac{18 + 24 + 36}{6} = \frac{18}{6} + \frac{24}{6} + \frac{36}{6}$

This example suggests the following general rule.

► When *a*, *b*, and *c* are real numbers, and *c* is not equal to 0,
$$\frac{a+b}{c} = \frac{a}{c} + \frac{b}{c}$$

You can use this rule to divide a polynomial by a monomial.

Example 3

Simplify.

a. $\frac{6a + 9}{3}$

b. $\frac{2x^4 + 8x^3 + 12x^2}{2x^2}$

Solution

a. $\frac{6a + 9}{3} = \frac{6a}{3} + \frac{9}{3}$ ← Divide each term of the polynomial by the divisor.
    $= 2a + 3$

b. $\frac{2x^4 + 8x^3 + 12x^2}{2x^2} = \frac{2x^4}{2x^2} + \frac{8x^3}{2x^2} + \frac{12x^2}{2x^2} = x^2 + 4x + 6$ ◄

✓ **CHECK UNDERSTANDING**

Why is the quotient $\frac{12x^2}{2x^2}$ equal to 6?

$\frac{12}{2} = \frac{\overset{6}{\cancel{12}}}{\cancel{2}} = \frac{6}{1} = 6$

$\frac{x^2}{x^2}\frac{x^2}{x^2} = \frac{1}{1} = 1$

12-8 Dividing by a Monomial **463**

ASSIGNMENTS

BASIC
1–16, 20–29, 33–35, 39

AVERAGE
1–19, 20–38, 39–42

ENRICHED
9–19, 20–32, 36–38, 39–42

ADDITIONAL RESOURCES
Reteaching 12–8
Enrichment 12–8

Example 1: Have students find each quotient. Compare their solutions to those presented in the text.

Example 2: Use these examples to demonstrate how to apply the quotient rule for exponents in dividing a monomial by a monomial.

Example 3: Stress that each term of the polynomial must be divided by the monomial divisor. Remind students that when they divide a polynomial by a monomial, the quotient should have the same number of terms as does the polynomial.

Additional Questions/Examples
Simplify.
1. $18a^2b / -3a$ **–6*ab***
2. $-45x^3y^2 / -9x^2y$ **5*xy***
3. $3st^3 / 9t^2$ ***st* / 3**
4. $-4p^3q^4 / 20\,p^2q^2$ **–*pq*² / 5**
5. $(9a - 3b) / 3$ **3*a* – *b***
6. $(5c^3 - 15c^2 + 10c) / 5c$
    ***c*² – 3*c* + 2**
7. $(3x^6 - 9x^4 - 15x^2) / 3x^2$
    ***x*⁴ – 3*x*² – 5**
8. $(-24m^8n^5 + 16m^3n^2) / 8mn^2$
    –3*m*⁷*n*³ + 2*m*²

Guided Practice/Try These You may wish to have students complete these exercises orally.

5-MINUTE CLINIC

Exercise	Student's Error	Error Diagnosis
Simplify: $\frac{56a^2 - 49a}{-7a}$	$\frac{56a^2 - 49a}{-7a} = 8a - 7$	• Student ignores the negative sign in the divisor. Correct solution: $-8a + 7$

Simplify.

1. $\dfrac{-6xy}{2y}$ $-3x$
2. $\dfrac{abc}{-ac}$ $-b$
3. $\dfrac{8cd}{-4cd}$ -2
4. $\dfrac{12wz}{3z}$ $4w$

5. $\dfrac{24z^6}{6z^5}$ $4z$
6. $\dfrac{27x^7y}{3x^2}$ $9x^5y$
7. $\dfrac{18x^6}{-9x^3}$ $-2x^3$
8. $\dfrac{-21a^7b^8}{-3ab^5}$ $7a^6b^3$

9. $\dfrac{4x+8}{2}$ $2x+4$
10. $\dfrac{6y-12}{3}$ $2y-4$
11. $\dfrac{-14b+21}{7}$ $-2b+3$

12. $\dfrac{3y^4+6y^3+12y^2}{3y}$ y^3+2y^2+4y
13. $\dfrac{4x^5-8x^3-12x^2}{4x^2}$ x^3-2x-3

EXERCISES

**PRACTICE/
SOLVE PROBLEMS**

Simplify.

1. $\dfrac{20cd}{4d}$ $5c$
2. $\dfrac{18gh}{2g}$ $9h$
3. $\dfrac{40ry}{10r}$ $4y$

4. $\dfrac{-12ab}{2a}$ $-6b$
5. $\dfrac{-12pq}{3q}$ $-4p$
6. $\dfrac{-12st}{-4t}$ $3s$

7. $\dfrac{y^3z}{y^2z}$ y
8. $\dfrac{x^3y^2z}{x^2y}$ xyz
9. $\dfrac{cd^6}{d^5}$ cd

10. $\dfrac{4w^2x^2}{-2wx}$ $-2wx$
11. $\dfrac{ab^5}{b^3}$ ab^2
12. $\dfrac{-x^5y^5}{x^3y^4}$ $-x^2y$

13. $\dfrac{2a+12}{2}$ $a+6$
14. $\dfrac{18y+6}{2}$ $9y+3$
15. $\dfrac{9a-18b}{3}$ $3a-6b$

16. $\dfrac{32c-8b-4a}{4}$ $8c-2b-a$
17. $\dfrac{6x^3-9x^2+3x}{3x}$ $2x^2-3x+1$

18. $\dfrac{28x^2y^2-21xy^2+14xy}{7xy}$ $4xy-3y+2$
19. $\dfrac{9a^2b^2c^2-15abc^2}{3abc}$ $3abc-5c$

**EXTEND/
SOLVE PROBLEMS**

Simplify.

20. $\dfrac{28a^8}{7a^2}$ $4a^6$
21. $\dfrac{18b^9}{-3b}$ $-6b^8$
22. $\dfrac{bm^3}{-m^2}$ $-bm$

23. $\dfrac{x^3}{x^2y}$ $\dfrac{x}{y}$
24. $\dfrac{x^2y^2}{y^2z}$ $\dfrac{x^2}{z}$
25. $\dfrac{5z^3}{15z}$ $\dfrac{z^2}{3}$

26. $\dfrac{-18r^6s^2t^3}{-3r^2st^2}$ $6r^4st$
27. $\dfrac{-42q^5r^2s^3}{7qrs^2}$ $-6q^4rs$
28. $\dfrac{-27a^3b^2c^3}{9abc}$ $-3a^2bc^2$

29. $\dfrac{24a^4b-12a^2b^2+18a^3b^3}{3a^2b}$ $8a^2-4b+6ab^2$
30. $\dfrac{35c^3d^2e-50c^4d^3e^2}{5cde}$ $7c^2d-10c^3d^2e$

31. $\dfrac{8z^2xy-12x^2zy+16zx^2y^2-24z^2x^3y^3}{4zxy}$ $2z-3x+4xy-6zx^2y^2$

32. $\dfrac{56a^4b^4c^3-49a^3b^3c^2+35ab^2c^2-28a^2bc^4}{-7abc^2}$ $-8a^3b^3c+7a^2b^2-5b+4ac^2$

---

3 SUMMARIZE

Write About Math Have students describe in their math journals how the arithmetic and algebraic processes of division are alike and how they differ.

4 PRACTICE

Practice/Solve Problems Remind students that in Exercises 13–19 it is important to divide each term of the polynomial by the monomial divisor and write the quotient first as the sum of these divisions.

Extend/Solve Problems Watch for students who have difficulty with Exercises 23–25 in which the quotients are in fractional form.

Think Critically/Solve Problems Ask students how they could apply the chunking technique in solving Exercises 41 and 42.

5 FOLLOW-UP

Extra Practice
Simplify.
1. $25ab / 5a$ $5b$
2. $-14pq / 2q$ $-7p$
3. $-x^5y^3 / -x^3y^2$ x^2y
4. $18s^6t^7 / -6s^6t^2$ $-3t^5$
5. A prism has a volume of $60\ x^2y^4z^3$. Its length is $5xy$ and its width is $6yz$. Find its height. $2xy^2z^2$
6. $(9y^2-3y) / 3y$ $3y-1$
7. $(21x^2-14xz^2+7xz) / 7x$
   $3x-2z^2+z$
8. $(12c^3+3c^2d-9cd^2+21cd^3) / 3c$
   $4c^2+cd-3d^2+7d^3$

Extension Have students check their answers to Exercises 10–19 by multiplying.

CHALLENGE

A triangle has an area of $14\ x^2y$ square units. Its height is $7xy$ units. Find the length of its base. $4x$

Write an expression for the unknown dimension of each rectangle.

33. Area: $25pq$

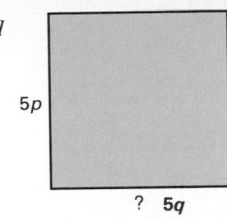

$5p$

? $5q$

34. Area: $36abc$

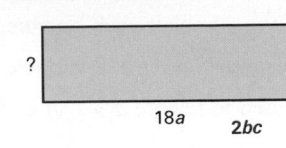

?

$18a$ $2bc$

35. A rectangle has area $36ab$ square units. The width is $4b$ units. Write an expression for the length of the rectangle. **9a units**

36. The area of the rectangle at the right is $28x^3y^2$. Write an expression for the unknown dimension. **7x²y**

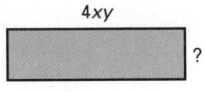

$4xy$

?

37. The area of the square at the right is $25m^4n^6$. Write an expression for the unknown dimension. **5m²n³**

?

38. The product of $3x^2y^5$ and a certain monomial is $21x^7y^9z^2$. Find the missing monomial factor. **7x⁵y⁴z²**

39. Write an expression for the volume of a rectangular prism with a length of $6xz$, a width of $3y^2$, and a height of $5z$. **90xy²z²**

40. A prism with a volume equal to the one in Exercise 39 has a length of $15y$ and a width of $2z$. Write an expression for its height. **3xyz**

THINK CRITICALLY/ SOLVE PROBLEMS

How could you use the distributive property to simplify these expressions?

41. $\dfrac{x^2(x+1) - 4(x+1)}{x+1}$ **x² – 4**

42. $\dfrac{y^2(y+5) - 9(y+5)}{y^2 - 9}$ **y + 5**

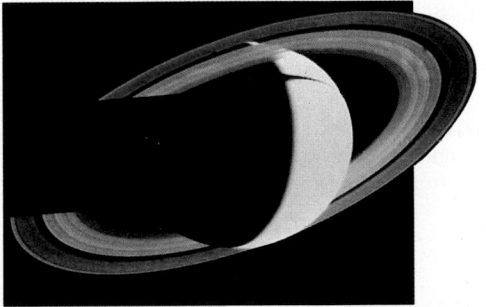

12 CHAPTER REVIEW

Introduction The Chapter Review emphasizes the major concepts, skills, and vocabulary presented in this chapter and can be used for diagnosing students' strengths and weaknesses. Page references direct students back to appropriate sections for additional review and reteaching.

Using Pages 466–467 Allow students to quickly scan the Chapter Review and ask questions about any section they find confusing. Exercises 1–5 review key vocabulary. Use Exercises 11–14 to review the rules for exponents when multiplying polynomials and Exercises 20–23 to review the rules for exponents when dividing polynomials.

Informal Evaluation Have students explain how they arrived at any incorrect answers. They will probably find their own mistakes and give you some clues as to the nature of their errors. Make sure students understand this material before administering the Chapter Test.

Follow-Up Have students illustrate each of the operations (addition, subtraction, multiplication, factoring, or division) introduced in the chapter. Ask them to provide two examples of each operation.

CHAPTER 12 ● REVIEW

Choose a word from the list to complete each statement.

1. Use a(n) ___?___ to indicate the power of a variable. **c**

2. The ___?___ property is used to rewrite $a(b + c)$ as $ab + ac$. **b**

3. $x^3 + x - 3$ is an example of a(n) ___?___. **e**

4. In $5m$, 5 is the numerical ___?___. **a**

5. $4x^2y^2z$ is an example of a(n) ___?___. **d**

a. coefficient
b. distributive
c. exponent
d. monomial
e. trinomial

SECTION 12–1   POLYNOMIALS   (pages 436–439)

► A **monomial** is an expression that is either a single number, a variable such as x or y, or the product of a numerical **coefficient** and one or more variables. A monomial that contains no variable is called a **constant**.
► The sum of two or more monomials is a polynomial. **Like terms** are terms in a polynomial that are exactly alike, or are alike except for their numerical coefficients. One way to simplify a polynomial is to combine like terms.

Simplify.

6. $9n - n$ **8n**

7. $4m + 5m$ **9m**

8. $4x^2 - 3x + 2x^2$ **6x² – 3x**

SECTION 12–2   ADDING AND SUBTRACTING POLYNOMIALS   (pages 440–443)

► To add polynomials, write the sum and then combine like terms.
► To subtract a polynomial, add its opposite and simplify.

Simplify.

9. $(8h + 4) + (5 - 5h)$ **3h + 9**

10. $(13a^2 + 7a) - (4a^2 - a)$ **9a² + 8a**

SECTIONS 12–3 AND 12–4   MULTIPLYING POLYNOMIALS   (pages 444–451)

► To find the product of two monomials, use the commutative and associative properties of multiplication and simplify.
► To multiply two powers having the same base, add the exponents.
► To find the power of a monomial that is itself a power, multiply the exponents.
► To find the power of a product, find the power of each factor and multiply.
► To multiply a polynomial by a monomial, use the distributive property and the rules for exponents.

Simplify.

11. $(2a)(-4b)$ **–8ab**

12. $(x^2y^2)^3$ **x⁶y⁶**

13. $6w(3 - 4w + 2w^2)$
18w – 24w² + 12w³

14. $-k(k^3 - 7k - 5)$
–k⁴ + 7k² + 5k

▶ Experience in applying problem solving strategies will help you decide which one to use to solve a particular problem.

▶ You can group, or "chunk," by grouping pieces of data together.

Solve. Tell what strategy you used.

15. Rhoda is thinking of a number. If she multiplies the number by itself and then subtracts half the number from the result, she gets 248. What number is Rhoda thinking of?
   16; possible strategy: guess and check

Use chunking to factor each expression.

16. $x(x + 4) + 5(x + 4)$ **(x + 5)(x + 4)**   **17.** $(m + 1)4 - m(m + 1)$ **(m + 1)(4 – m)**

SECTION 12–6   FACTORING   **(pages 454–457)**

▶ To **factor** an expression, write the expression as a product of factors.

▶ The greatest common factor of two or more monomials is the common factor having the greatest numerical factor (coefficient) and the variable or variables of greatest degree.

Factor each polynomial.

18. $6x^2y - 24x$ **6x(xy – 4)**   **19.** $ab - a^2b + b^2$ **b(a – a² + b)**

SECTION 12–8   DIVIDING BY A MONOMIAL   **(pages 460–464)**

▶ To simplify the quotient of two monomials, find the quotient of any numerical coefficients and then find the quotient of the variables.

▶ To divide a polynomial by a monomial, divide each monomial term of the polynomial by the monomial.

Simplify.

20. $\dfrac{24mn}{-3m}$ **–8n**   **21.** $\dfrac{-15a^3b^2}{5ab}$ **–3a²b**   **22.** $\dfrac{12c + 15}{3}$ **4c + 5**   **23.** $\dfrac{4x^4 + 6x^3 + 14x^2}{2x^2}$ **2x² + 3x + 7**

USING DATA   For Exercises 24 and 25, use the graph on page 437.

Solve.

24. The magnitude (s) of star X is related to the magnitude (m) of the star Mira by the formula $s = 3m - 6$. Rewrite this formula by factoring. **s = 3(m – 2)**

25. What was the magnitude of star X on the 440th day from the beginning of Mira's first cycle? **3**

Introduction The Chapter Test uses a variety of questioning techniques to assess students' mastery of the major objectives of Chapter 12. If you prefer, you may use the Skills Preview (page 435) as an alternative form of the Chapter Test. The items on this test and the Skills Preview correspond in content and level of difficulty.

Alternative Assessment
Critical Thinking
1. Explain how addition and subtraction of polynomials are similar and how they are different.
2. Explain how multiplication and division of monomials are similar and how they are different.
3. Explain how using a common monomial factor to factor a polynomial is like dividing a polynomial by a monomial.

Simplify.

1. $4m + (-3m) + 2m$ **$3m$**

2. $3x^2 + (-2x^2) + 4x^2$ **$5x^2$**

3. Three meteorite fragments are located at the points of a triangular region, ABC. Write a simplified expression for the perimeter of the region.   **$75k$**

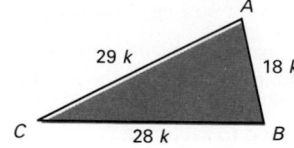

4. $(6z^2 + 4z - 5) + (3z^2 - z + 2)$
   $9z^2 + 3z - 3$

5. $(5p^2 + 4p - 8) - (3p^2 + 2p - 12)$
   $2p^2 + 2p + 4$

6. Write a simplified expression for the perimeter of the triangle.   **$6x - 4$**

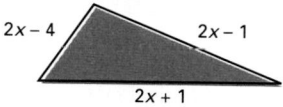

7. $(-2x)(3xy)$ **$-6x^2y$**

8. $(-2m^2p^2)(pq)$ **$-2m^2p^3q$**

9. $(3x)(2x)$ **$6x^2$**

10. $(4k^3)(-2k^4)$ **$-8k^7$**

11. $\dfrac{12ab}{4a}$ **$3b$**

12. $\dfrac{-6p^2}{2p^2}$ **-3**

13. $\dfrac{15t^2m^7}{-3tm^3}$ **$-5tm^4$**

14. $3(y - 2)$ **$3y - 6$**

15. $(-3m)(2m - 1)$
    $-6m^2 + 3m$

16. $a^2(a^3 - 3a + c)$ **$a^5 - 3a^3 + a^2c$**

Factor each polynomial.

17. $14p - 8$
    $2(7p - 4)$

18. $8m^2n + 24m$
    $8m(mn + 3)$

19. $2rs - 12r^2s^2 + 10rst$
    $2rs(1 - 6rs + 5t)$

Simplify.

20. $\dfrac{-15ab}{3a}$
    $-5b$

21. $\dfrac{-36x^3y^2z}{-9xz}$
    $4x^2y^2$

22. $\dfrac{18q^5 + 21q - 9q^3}{3q}$
    $6q^4 + 7 - 3q^2$

Solve.

23. A rectangular garden has an area of $8p^3t^5$ square units. The width is $4pt$ units. Write an expression for the length.
    $2p^2t^4$ units

1. Organize these numbers into a stem-and-leaf plot. **See Additional Answers.**

   72 68 64 49 51 58 54 62 82
   64 53 57 65 76 79 53 71 73

Complete.

2. $(21 + 45) + 16 = 21 + (\blacksquare + 16)$ **45**

3. $5(3 + 8) = (5 \times 3) + (\blacksquare \times 8)$ **5**

4. Is the argument *valid* or *invalid*?

   If the mattress is too soft, then you will sleep uncomfortably.
   You slept uncomfortably.
   Therefore, the mattress is too soft. **invalid**

5. Write a statement about the size of each circle below. Then check to see if your statement is true or false.

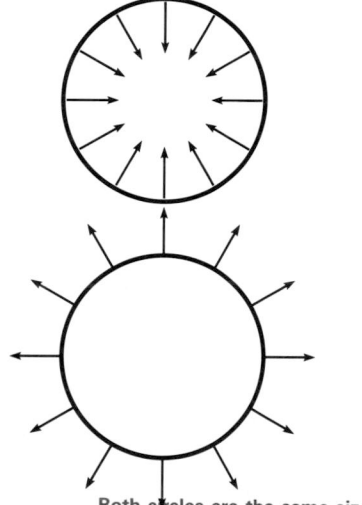

Both circles are the same size.

Replace ● with $<$, $>$, or $=$.

6. $0.18 \bullet \frac{3}{16}$ $<$

7. $\frac{5}{8} \bullet 0.621$ $>$

8. $\frac{3}{4} \bullet \frac{3}{10}$ $>$

9. $0.02 \bullet \frac{1}{5}$ $<$

10. Find the perimeter of a rectangle whose length is 7 in. and whose width is 6.3 in. **26.6 in.**

Subtract.

11. $-27 - (-15)$ **−12**

12. $-21 - 4$ **−25**

13. $4 - (-19)$ **23**

14. $-11 - (-11)$ **0**

Solve each equation.

15. $27 = -4y + 3$ $y = -6$

16. $\frac{k}{-5} = 7$ $k = -35$

17. $38 = 5b$ $b = 7.6$

18. $3n - 4.2 = 1.2$ $n = 1.8$

Write three equivalent ratios for each given ratio.

19. $7{:}3$ **14:6, 21:9, 28:12**

20. 16 to 24 **2 to 3, 4 to 6, 8 to 12**

21. The scale on a map is 1 cm:8 m. If a distance on the map is 9.25 cm, what is the actual distance? **74 m**

22. 78 is 20% of what number? **390**

23. Find the discount and the sale price. Round to the nearest cent.
    regular price: $265
    percent of discount: 25%
    $D = \$66.25, S = \$198.75$

24. Draw an angle with a measure of 68°. Then construct its bisector.
    Check students' drawings.

Use a calculator to find each square root to the nearest thousandth.

25. $\sqrt{58}$ **7.616**

26. $\sqrt{546}$ **23.367**

27. $\sqrt{292}$ **17.088**

Simplify.

28. $5x(-3x^2 + 2x)$ $-15x^3 + 10x^2$

29. $2x^2(y + 3)$ $2x^2y + 6x^2$

Cumulative Review **469**

12 CUMULATIVE REVIEW

Introduction The purpose of this Cumulative Review is to maintain previously taught skills and concepts and to apply them to the material presented in this chapter. At least one major objective of each chapter is included in the review.

Item Analysis The table below correlates the Cumulative Review items with the chapter and section that are being reviewed.

Section	Items
1–2	1
2–9	2
2–10	3
3–1	5
3–3	4
4–2	8
4–3	6–7, 9
5–7	10
6–3	11–14
7–5	15–18
8–3	19–20
8–6	21
9–5	22
9–7	23
10–9	24
11–5	25–27
12–4	28–29

Additional Answers
See page 583.

12 CUMULATIVE TEST

Introduction The Cumulative Test uses a standardized-test format of multiple-choice questions to assess retention of previously learned concepts and test-taking skills. Test results may be used to diagnose students' strengths and weaknesses.

Item Analysis The table below correlates the Cumulative Test items with the chapter and section that are being tested.

Section	Items
1–9	1
2–9	2
3–3	3
4–3	4
5–4	5
6–4	6
7–6	7
8–4	8
9–5	9
9–7	10
10–2	11
10–4	12
11–11	13
11–9	14
12–2	15–16

1. What is the mode for this set of data?
   11.4 9.1 3.1 6.8 7.2 7.0 6.8 7.5 6.8
   A. 7.3   B. 7.0
   C. 6.9   (D.) none of these

2. Complete.
   $9(7 + 4) = (9 \times 7) + (\blacksquare \times 4)$
   (A.) 9   B. 7   C. 4   D. 11

3. Which is a counterexample for the following statement?
   Any number divisible by 5 is divisible by 10.
   A. 10   B. 20   (C.) 15   D. 40

4. Compare. $\frac{11}{15}$ ● 0.7
   A. <   (B.) >
   C. =   D. ≈

5. Find the perimeter.
   A. 925 m
   B. 121.3 m
   C. 78.1 cm
   (D.) 99.7 m

   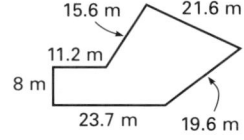

6. Compare. $4 \times (-6)$ ● $6 \times (-4)$
   (A.) =   B. >   C. <   D. ≈

7. Solve. $1\frac{3}{4}n = 17\frac{1}{2}$
   A. $n = 30$ (B.) $n = 10$
   C. $n = 1$ D. $n = 20$

8. Solve. $\frac{3}{4} = \frac{31.5}{x}$
   A. $x = 4.2$ B. $x = 41.4$
   C. $x = 21$ (D.) $x = 42$

9. 18 is $33\frac{1}{3}\%$ of what number?
   A. 6   (B.) 54
   C. 56   D. none of these

10. What is the sale price of an $800 copier if the discount rate is 20%?
    A. $160   B. $600
    C. $960   (D.) none of these

11. Find $m \angle 1$.

    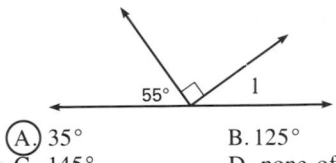

    (A.) 35°   B. 125°
    C. 145°   D. none of these

12. Find the missing angle measure.

    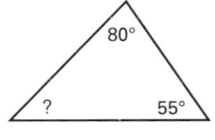

    A. 55°   B. 10°
    (C.) 45°   D. none of these

13. Find the surface area of a sphere with a radius of 2.5 ft.
    A. 25 ft²   B. 19.6 ft²
    (C.) 78.5 ft²   D. 31.4 ft²

14. Which formula would you use to calculate the volume of a cone?
    A. $v = l \times w \times h$
    (B.) $v = \frac{1}{3}Bh$
    C. $v = \pi r^2 h$
    D. $v = Bh$

15. Simplify. $2m + (m - 3)$
    A. $2m^2 - 3$ (B.) $3m - 3$
    C. $2m - 3$ D. none of these

16. Simplify. $(3x - 2) - (4x + 3)$
    A. $x - 1$ B. $7x + 1$
    (C.) $-x - 5$ D. $x + 5$

Polynomials

A **monomial** is an expression that is a single number, a variable, or the product of a number and one or more variables. A **polynomial** is an expression that is the sum of two or more monomials. The monomials are called the **terms** of the polynomials. The following expressions are polynomials:

$$x + (-y) \qquad x^2 + 4x + 4 \qquad 3y^2 + z + (-y) - 2$$

Those terms of a polynomial that are exactly alike or are alike except for their coefficients are called *like terms*.

like terms: $xy^2, 3xy^2, -2xy^2$ \qquad unlike terms: $xy^2, x^2y, 3z, 4x$

You can simplify a polynomial by combining like terms.

▶ **Example** _____

Simplify.
a. $2y^2 + 8y + 3 - 4y + 3y^2 - 4$ \qquad b. $4x^3 - 3x + 2 - 5x + x^3$

Solution

a. $\quad 2y^2 + 8y + 3 - 4y + 3y^2 - 4$
$= 2y^2 + 3y^2 + 8y - 4y + 3 + (-4)$ \qquad Rearrange, or collect, like terms.
$= (2 + 3)y^2 + (8 - 4)y + [3 + (-4)]$ \qquad Use the distributive property
$= 5y^2 + 4y - 1$ \qquad to combine like terms.

b. $4x^3 - 3x + 2 - 5x + x^3 = 4x^3 + x^3 + (-3x) + (-5x) + 2$
$= (4 + 1)x^3 + [-3 + (-5)]x + 2$
$= 5x^3 + (-8)x + 2$
$= 5x^3 - 8x + 2$

EXERCISES

1. $m^2 + m - 6 + 3m - 2$
$\underline{m^2 + 4m - 8}$

2. $5s + 2 - 5s - 7 + 4s$
$\underline{4s - 5}$

3. $t^4 + 3t^3 - 2t^4 + 3 + 2t^3 - t^2 - 1$
$\underline{-t^4 + 5t^3 - t^2 + 2}$

4. $6x^2 + 8x - 8 - 2x^2 - 9x + 4$
$\underline{4x^2 - x + -4}$

5. $-3 + 16z^2 - 8z^3 + 5z^2 + 11 + 4z$
$\underline{-8z^3 + 21z^2 + 4z + 8}$

6. $-7a - 3a^2 + 8a + 12a^2 - 9$
$\underline{9a^2 + a - 9}$

7. $5x^3 + 3x - 4 - 3x^2 + 8x^3 + 6 - 6x$
$\underline{13x^3 - 3x^2 - 3x + 2}$

8. $3y^4 + 3y + 3y^3 - y^2 + y + y^2$
$\underline{3y^4 + 3y^3 + 4y}$

9. $n^2 + 3n + 4n^2 + n^3 - 3n^3 + 1$
$\underline{-2n^3 + 5n^2 + 3n + 1}$

10. $8r^2 + 4r^3 - 3r^2 - r - 4r^2$
$\underline{4r^3 + r^2 - r}$

Degree of a Monomial or a Polynomial

The **degree of a monomial** is the total number of times the variables in the monomial occur as factors. Thus, to find the degree of a monomial, find the sum of the exponents of the variables.

▶ **Example 1** _____

Find the degree of the monomial.
a. $5a^2b$ \qquad b. $x^4y^3z^2$

Solution

a. The power of a is 2. The power of b is 1. The degree of the monomial is $2 + 1$, or 3.

b. The power of x is 4. The power of y is 3. The power of z is 2. The degree of the monomial is $4 + 3 + 2$, or 9.

The **degree of a polynomial** is the greatest of the degrees of its terms after the polynomial has been simplified.

▶ **Example 2** _____

Simplify the polynomial. Then find the degree of the polynomial.
$$xy^4 + 3x^2y^2 + 4x^2y - xy^4$$

Solution

$xy^4 + 3x^2y^2 + 4x^2y - xy^4 = xy^4 - xy^4 + 4x^2y + 3x^2y^2$
$= 4x^2y + 3x^2y^2$

The monomial term of greatest degree is $3x^2y^2$. Its degree is $2 + 2$, or 4. So, the degree of the polynomial is 4.

EXERCISES

Find the degree of each monomial.

1. $4a^3b$ \quad **degree 5**
2. $7n$ \quad **degree 1**
3. $3ab$ \quad **degree 2**
4. xy \quad **degree 2**
5. $6m^3np^2$ \quad **degree 6**
6. $8x^4z$ \quad **degree 5**
7. $12k^2m^3n^2$ \quad **degree 7**
8. $7x^2y^3z$ \quad **degree 6**

Simplify the polynomial. Then find its degree.

9. $3m^2n + 2mn^2 - 4m^2n + mn + m^2n$
$\underline{mn + 2mn^2; \text{ of degree } 3}$

10. $2r^2s^2t + 4r^2s^3 - rs + rs^2 - 2r^2s^2t$
$\underline{4r^2s^3 - rs + rs^2; \text{ of degree } 5}$

11. $6a^2b^2 - 2a^2b^2c + 3b^2c^2 - 3a^2b^2 + 4a^2b^2c$
$\underline{3a^2b^2 + 2a^2b^2c^2 + 3b^2c^2; \text{ of degree } 6}$

Adding and Subtracting Polynomials

You can add and subtract polynomials in vertical form.

▶ **Example 1** _____

Simplify $(3y^3 - 4y^2 + 6) + (2y^3 + 2y^2 - 3y + 6)$.

Solution

Put like terms in the same columns. Write 0 where there is no like term. \qquad Combine like terms.

$$\begin{array}{r} 3y^3 - 4y^2 + 0y + 6 \\ +2y^3 + 2y^2 - 3y + 6 \\ \hline \end{array} \qquad \begin{array}{r} 3y^3 - 4y^2 + 0y + 6 \\ +2y^3 + 2y^2 - 3y + 6 \\ \hline 5y^3 - 2y^2 - 3y + 12 \end{array}$$

To subtract a polynomial from another polynomial, you add the opposite.

▶ **Example 2** _____

Simplify $(4t^4 + t^2 - 5t + 4) - (3t^4 - 3t + 8)$.

Solution

Arrange the problem in vertical columns. \quad Change to an addition problem by changing all signs in the polynomial being subtracted. \quad Combine like terms.

$$\begin{array}{r} 4t^4 + t^2 - 5t + 4 \\ -(3t^4 + 0t^2 - 3t + 8) \\ \hline \end{array} \quad \begin{array}{r} 4t^4 + t^2 - 5t + 4 \\ + -3t^4 - 0t^2 + 3t - 8 \\ \hline \end{array} \quad \begin{array}{r} 4t^4 + t^2 - 5t + 4 \\ + -3t^4 - 0t^2 + 3t - 8 \\ \hline t^4 + t^2 - 2t - 4 \end{array}$$

EXERCISES

Add.

1. $(4y^3 - 2y^2 + y) + (y^3 - 2y^2 + 4y)$
$\underline{5y^3 - 4y^2 + 5y}$

2. $(4m^2 + m - 3) + (2m^2 + 7)$
$\underline{6m^2 + m + 4}$

Subtract.

3. $(6n^2 + 4n + 3) - (3n^2 - 5n + 1)$
$\underline{3n^2 + 9n + 2}$

4. $(5p^3 + 3p^2 - p + 4) - (2p^3 - p^2 + p - 1)$
$\underline{3p^3 + 4p^2 - 2p + 5}$

Simplify.

5. $(4y^2 + y - 5) + (3y^2 - 5y + 6)$
$\underline{7y^2 - 4y + 1}$

6. $(8x^2 - 4x + 2) - (5x^2 - 3x + 4)$
$\underline{3x^2 - x - 2}$

7. $(4m^2 - 3m + 3) - (2m^2 - 3m - 3)$
$\underline{2m^2 + 6}$

8. $(5r^3 - r^2 - 5) + (-6r^3 - r^2 + 3)$
$\underline{-r^3 - 2r^2 - 2}$

More or Less

To add polynomials, combine like terms. To subtract polynomials, change the sign of each term in the polynomial being subtracted and combine like terms.

Solve each problem to find the solution to the riddle:
How do you find the circumference of an apple?

Substitute the letter for the answer in the grid below.

Simplify.

1. $(3x^2 + 2x - 4) - (2x^2 + x + 4)$ \quad $\underline{x^2 + x - 8}$
2. $(4x^4 + 6x^3y - 2x^2y^2) + (2x^4 - 4x^3y + 2x^2y^2)$ \quad $\underline{6x^4 + 2x^3y}$
3. $(3x^3 + 2x^2 - 4x + 5) - (x^2 - 4x + 5)$ \quad $\underline{3x^3 + 6x^2}$
4. $(3x^4 - 5x^3y + 2x^2y^2) - (3x^4 - 4x^3y - x^2y^2)$ \quad $\underline{-x^3y + 3x^2y^2}$
5. $(5x^2 - 6xy + 2y^2) + (x^2 + 3xy)$ \quad $\underline{6x^2 - 3xy + 2y^2}$
6. $(3x^2 + 2xy) - (3x^2 + 5xy - 2y^2)$ \quad $\underline{-3xy + 2y^2}$
7. $(8x^2y - 4xy^2 + 2xy) + (-4x^2y - xy^2 + 2xy)$ \quad $\underline{4x^2y - 5xy^2 + 4xy}$
8. $(9x^3 - x^2) + (-6x^3 + 7x^2)$ \quad $\underline{3x^3 + 6x^2}$
9. $(5xy - 3y^2) - (-6x^2 + 8xy - 5y^2)$ \quad $\underline{6x^2 - 3xy + 2y^2}$
10. $(4x^3 - 7x^2 + 3x - 3) - (x^3 - 5x^2 + 3)$ \quad $\underline{3x^3 - 2x^2 + 3x - 6}$

$3x^3 + 6x^2$	E
$-x^3y + 3x^2y^2$	A
$6x^4 + 2x^3y$	S
$3x^3 - 2x^2 + 3x - 6$	I
$6x^2 - 3xy + 2y^2$	P
$4x^2y - 5xy^2 + 4xy$	L
$x^2 + x - 8$	U

1	2	3		4	5	6	7	8		9	10
U	S	E		A	P	P	L	E		P	I

470A

Name _____ Date _____

Multiplying Monomials

The product rule for exponents is as follows:

$$a^m \times a^n = a^{m+n}$$

So, to multiply two powers having the same base, you add the exponents.

► **Example 1**

Simplify $(4a^2b)(-5ab^3)$.

Solution

$(4a^2b)(-5ab^3) = (4)(-5)(a)(a^2)(b)(b^3)$ ← Use the commutative and associative properties to regroup the coefficients and variables.

$\quad = (-20)(a^{2+1})(b^{1+3})$ ← Use the product rule to add exponents for each like base.

$\quad = -20a^3b^4$

The power rule for exponents is as follows:

$$(a^m)^n = a^{mn}$$

To find the power of a monomial that is itself a power, multiply exponents.

The power of a product rule is as follows:

$$(ab)^m = a^mb^m$$

To find the power of a product, find the power of each factor and multiply.

► **Example 2**

Simplify $(y^2)^3$.

Solution

$(y^2)^3 = y^{2\times3} = y^6$

Check: $(y^2)^3 = y^2 \times y^2 \times y^2$
$\quad = y^{2+2+2}$
$\quad = y^6$

► **Example 3**

Simplify $(4a^2b)^3$.

Solution

$(4a^2b)^3 = (4)^3(a^2)^3(b)^3$
$\quad = 64a^6b^3$

EXERCISES

Simplify.

1. $(2a^3)(5a)$
$10a^4$

2. $(4m^2n)(3mn^3)$
$12m^3n^4$

3. $(-2xy)(4x^2)$
$-8x^3y$

4. $(3xy)(-2x^3y^4)$
$-6x^4y^5$

5. $(-7st^4)(-2s^2t^2)$
$14s^3t^6$

6. $(3w^2y)(6)$
$18w^2y$

7. $(m^5n^2)(mn^4)$
m^6n^6

8. $(-3t^2)(-3st)$
$9st^3$

9. $(4xyz)(2x^2yz^2)$
$8x^3y^2z^3$

10. $(s^3)^3$
s^9

11. $(2x^2)^2$
$4x^4$

12. $(4rt)^2$
$16r^2t^2$

13. $(-3m)^2$
$9m^2$

14. $(-2p^2)^3$
$-8p^6$

15. $(5x^2yz^3)^2$
$25x^4y^2z^6$

Name _____ Date _____

More About Polynomials

A polynomial is the sum of monomials. A monomial may be either a number, a variable, or the *product* of a number and one or more variables. An expression such as $\frac{3x}{y}$ or $\frac{a}{b}$ is not a monomial. So, a polynomial may not have a term that has a variable in the denominator.

Decide whether each expression is a polynomial or not.
Write the expression in the correct column.

	Expressions	Polynomial	Not a polynomial
1.	$3x^2 + \frac{(3+x)}{x}$		$3x^2 + \frac{(3+x)}{x}$
2.	$2a^2b - 2ab^2$	$2a^2b - 2ab^2$	
3.	$5m - \frac{3}{4}n$	$5m - \frac{3}{4}n$	
4.	$4s^2 - \frac{3t}{s}$		$4s^2 - \frac{3t}{s}$
5.	$\frac{5p}{6} + p^2 - 7$	$\frac{5p}{6} + p^2 - 7$	
6.	$\frac{3}{5x} + \frac{5}{2x^3}$		$\frac{3}{5x} + \frac{5}{2x^3}$

A monomial cannot include the square root of a variable. So, $\sqrt{x}$ is not a monomial, and the expression $y + \sqrt{x}$ is not a polynomial.

Decide whether each expression is a polynomial or not.
Write the expression in the correct column.

	Expressions	Polynomial	Not a polynomial
7.	$5mn - \sqrt{m-n}$		$5mn - \sqrt{m-n}$
8.	$4xy + \sqrt{3x}$	$4xy + \sqrt{3x}$	
9.	$4s - \frac{\sqrt{3}}{4}$	$4s - \frac{\sqrt{3}}{4}$	
10.	$8p^2q + \frac{5}{p} - 8$		$8p^2q + \frac{5}{p} - 8$
11.	$6\sqrt{ab}$		$6\sqrt{ab}$
12.	$\sqrt{6ab}$	$\sqrt{6ab}$	

Name _____ Date _____

Multiplying a Polynomial by a Monomial

You can set up the multiplication of a polynomial by a monomial in a manner similar to a multiplication with whole numbers.

► **Example**

Simplify. $6xy(4x^2 + 3y)$

Solution

Write the factors as shown.

$$\begin{array}{r} 4x^2 + 3y \\ \times \quad 6xy \\ \hline \end{array}$$

Multiply the first term in the polynomial by the monomial.

$$\begin{array}{r} 4x^2 + 3y \\ \times \quad 6xy \\ \hline 24x^3y \end{array}$$

Multiply the second term in the polynomial by the monomial.

$$\begin{array}{r} 4x^2 + 3y \\ \times \quad 6xy \\ \hline 24x^3y + 18xy^2 \end{array}$$

EXERCISES

Simplify.

1. $7(2x + 3y)$
$14x + 21y^2$

2. $s(4s^2 - 3s)$
$4s^3 - 3s^2$

3. $m^2(5m + 2)$
$5m^3 + 2m^2$

4. $6m(3mn - 2n)$
$18m^2n - 12mn$

5. $4y(4x^2y + 3xy)$
$16x^2y^2 + 12xy^2$

6. $3p(8pq - 2)$
$24p^2q - 6p$

7. $5(2x^2 + x - 2)$
$10x^2 + 5x - 10$

8. $4m(3m^2 + 2m - 3)$
$12m^3 + 8m^2 - 12m$

9. $3x(x^2 + 3x + 5)$
$3x^4 + 9x^3 + 15x^2$

10. $xy(4x - 3y + 2z)$
$4x^2y - 3xy^2 + 2xyz$

11. $-3t(3s^2t + 5st)$
$-9s^2t^2 - 15st^2$

12. $-mn^2(-m^3 + mn - n^2)$
$m^4n^2 - m^2n^3 + mn^4$

Write an expression for the area of each rectangle. Then simplify the expression.

13. (rectangle with width x, length $x + 5$)
$x(x + 5);\ x^2 + 5x$

14. (square with sides xy and $x + y$)
$x^2y + xy^2$

Name _____ Date _____

Rectangles

Expressions may vary for Exercises 6–8.

Write an expression for the area of each rectangle. Then simplify the expression.

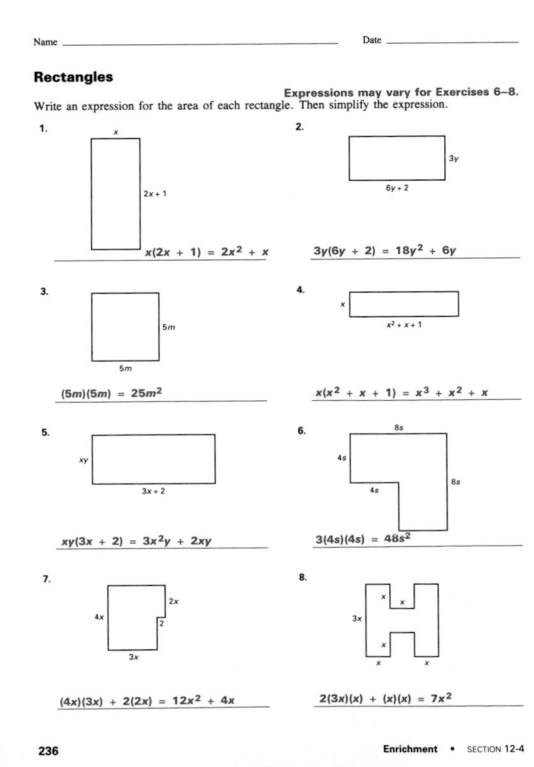

1. $x(2x + 1) = 2x^2 + x$

2. $3y(6y + 2) = 18y^2 + 6y$

3. $(5m)(5m) = 25m^2$

4. $x(x^2 + x + 1) = x^3 + x^2 + x$

5. $xy(3x + 2) = 3x^2y + 2xy$

6. $3(4s)(4s) = 48s^2$

7. $(4x)(3x) + 2(2x) = 12x^2 + 4x$

8. $2(3x)(x) + (x)(x) = 7x^2$

70B

Name _____ Date _____

Problem Solving/Decision Making: Choose a Strategy

Here is a checklist of the strategies you have studied so far.

Solve a simpler problem	Work backward	Use logical reasoning
Find a pattern	Make an organized list	Act it out
Guess and check	Make a table	Draw a picture

▶ **Example** _____

Solve. Name the strategy you used to solve each problem.

A rectangle has a perimeter of 28. Its area is 45 ft². What are the dimensions of the rectangle?

Solution

Since $P = 2l + 2w$ and $A = l \times w$, you need to find two numbers such that their product is 45 and twice their sum is 28. Guess and check.

length	width	perimeter	area
8	6	28	48
10	4	28	40

← The dimensions do not satisfy both conditions. The area is too large.
← The area is too small.

The length of the rectangle is 9 ft. The width is 5 ft.

PROBLEMS _____

Solve. Name the strategy you used to solve each problem. Strategies may vary. One possible strategy is suggested.

1. Leroy began his savings account by depositing $15 in June. Each month thereafter he saved $5 more than he had the previous month. How much will he save in 12 months?

   $510; find a pattern

2. The sum of two monomials is $18n$. Their product is $77\,n^2$. Find the two monomials.

   $11n$ and $7n$; guess and check

3. Mrs. Ramirez wants to hang four new pictures horizontally along one wall of her living room, all in a row. In how many ways can she order the pictures?

   24 ways; make an organized list

4. Nan helped her mother tile the floor. Her mother spent twice as much time working as did Nan. All together, they spent 24 h working. How much time did each person work?

   Nan, 8 h; her mother, 16 h; guess and check

Name _____ Date _____

Tile Table Tops

Mike and Sara are working on a design project. They are to design a table top made of blue and white ceramic tiles. Each tile is 6 in. square. Both will submit a proposed design.

1. Both Mike and Sara are to use n tiles in each column and $n + 1$ tiles in each row. Write an expression for the total number of tiles they will need.

   $n(n + 1)$

2. For the top, each student uses a total of 30 tiles. What are the dimensions of the table top?

   $2\frac{1}{2}$ ft by 3 ft

3. Mike and Sara plan to use twice as many blue tiles as white tiles. Each person spent $29 for 30 tiles. Blue tiles cost $0.10 more than white tiles. What was the cost of each color tile?

   white, $0.90; blue, $1.00

4. How much would they have saved on the cost of the tiles if they had used twice as many white tiles as blue tiles?

   $1.00

Use graph paper to make two possible designs Mike and Sara might create for the table top with their 30 tiles. Remember to use twice as many blue tiles as white tiles.

5.
6.

7. In her design, Sara decided to use a 2-tile-by-2-tile square block. She used two blue tiles and two white tiles. How many possible arrangements could she make?

   6

Mike decided to use a square block 3 tiles by 3 tiles. He uses five blue tiles and four white tiles in his square. Show two possible arrangements of tiles he might have designed.

8.
9.

Designs will vary.

Name _____ Date _____

Factoring

To find the product of a monomial and a polynomial, multiply each term of the polynomial by the monomial. You can reverse this process to find what factors were multiplied to obtain the product. This reverse process is called **factoring**.

To factor a polynomial, find the greatest monomial factor common to all the terms of the polynomial. Then write the polynomial as the product of that greatest common factor and another polynomial.

▶ **Example** _____

Factor the polynomial: $5s^2t + 10st$.

Solution

Write the factors of each monomial term and find the greatest common monomial factor of each term.

factors of $5s^2t$: $(5)(s)(s)(t)$ factors of $10st$: $(2 \times 5)(s)(t)$

GCF: $5st$

Rewrite the polynomial as the product of the greatest common factor and another polynomial. Use the distributive property.

$$5s^2t + 10st = 5st \times s + 5st \times 2$$
$$= 5st(s + 2)$$

EXERCISES _____

Factor each polynomial.

1. $4m + 2n$
   $2(2m + n)$

2. $4x^2 + 3x$
   $x(4x + 3)$

3. $9x^2 + 6x$
   $3x(3x + 2)$

4. $6p^2 + 12p$
   $6p(p + 2)$

5. $5r^2 - 10r$
   $5r(r - 2)$

6. $12m^2 + 4m$
   $4m(3m + 1)$

7. $7s^2t - 14s$
   $7s(st - 2)$

8. $4x^2y + 6xy^2$
   $2xy(2x + 3y)$

9. $3p^2q - 8pq^2$
   $pq(3p - 8q)$

10. $4m^3 + 2m^2 + 6m$
    $2m(2m^2 + m + 3)$

11. $3a^2 + 6a + 3$
    $3(a^2 + 2a + 1)$

12. $2x^3y + 4x^2y - 6xy$
    $2xy(x^2 + 2x - 3)$

13. $3r^3s - r^2s + r$
    $r(3r^2s - rs + 1)$

14. $2x^2 + 4xy - 8y^2$
    $2(x^2 + 2xy - 4y^2)$

15. $8m^3n + 4m^2n - 24mn$
    $4mn(2m^2 + m - 6)$

Name _____ Date _____

Factorable Dimensions

The area of each rectangle shown is written as a polynomial. Factor to find the dimensions of the rectangle. Then label the edges of the rectangle.

1. $A = m^2 + 5m$ m
   $m + 5$

2. $A = 2n^2 + 3n$ n
   $2n + 3$

3. $A = 2p^2 + 6p$ $p + 3$
   $2p$

4. $A = 15s^2 - 9s$ $3s$
   $5s - 3$

5. $A = 14x^2 + 7x$ $2x + 1$
   $7x$

6. $A = 2y^2 + 3y$ y
   $2y + 3$

The volume of each rectangular prism is written as a polynomial. Factor the polynomial to find the unknown dimension of the prism. Express the unknown dimension as a polynomial.

7. $V = 12x^2 + 36x$
   $x^2 + 3x$

8. $V = 24m^2 + 6m$
   $4m^2 + m$

470C

Name _____ Date _____

Chunking

Sometimes a factor can have more than one term. You can think of a factor with more than one term as a single "chunk." You can use the distributive property to factor an expression that contains identical chunks.

► **Example** _____

Factor $3n(n + 3) = 4(n + 3)$.

Solution

Factor each term. Think of $(n + 3)$ as one factor.	Find the greatest common factor of both terms.	Rewrite the expression as the product of the GCF and the sum of the remaining factors.
$3n(n + 3) = (3)(n)(n + 3)$ $4(n + 3) = -(2)(2)(n + 3)$	$3n(n + 3) = (3)(n)(n + 3)$ $4(n + 3) = -(2)(2)(n + 3)$ GCF $= (n + 3)$	$(n + 3)3n + (n + 3)(-4) =$ $(n + 3)(3n - 4)$

EXERCISES _____

Use chunking to factor each expression.

1. $5a(a - 2) + 3(a - 2)$
   $(5a + 3)(a - 2)$

2. $4n(n + 3) - 3(n + 3)$
   $(4n - 3)(n + 3)$

3. $t(t^2 + 4) + 2(t^2 + 4)$
   $(t + 2)(t^2 + 4)$

4. $7x(x + y) - y(x + y)$
   $(7x - y)(x + y)$

5. $2y(y - 3) - 4(y - 3)$
   $(2y - 4)(y - 3)$

6. $4m(5m - 2) - (5m - 2)$
   $(4m - 1)(5m - 2)$

7. $3x(x^2 + 2x - 1) + 4(x^2 + 2x - 1)$
   $(3x + 4)(x^2 + 2x - 1)$

8. $2p(p^2 - 2p + 3) - 3(p^2 - 2p + 3)$
   $(2p - 3)(p^2 - 2p + 3)$

Factor these expressions. (Hint: Factor each expression. Then factor once more.)

9. $4p(p - 2) + 6(p - 2)$
   $2(2p + 3)(p - 2)$

10. $4r^2(2r + 3) - 6r(2r + 3)$
    $2r(2r - 3)(2r + 3)$

Name _____ Date _____

Put It All Together

After using chunking to find the factors of an expression, you can find the product of the factors, using a drawing.

► **Example** _____

Write the expression $2x(x + 2) + 5(x + 2)$ as a trinomial in standard form.

Solution

With $(x + 2)$ as a chunk, use the distributive property to factor the expression.

$2x(x + 2) + 5(x + 2) = (x + 2)(2x + 5)$

Draw a rectangle so that its dimensions represent the factors $2x + 5$ and $x + 2$. Write an expression to represent the area of each region of the rectangle.

	$2x$	5
x	$2x^2$	$5x$
2	$4x$	10

Write the sum of these areas and combine like terms. The result is a trinomial in standard form that represents the area of the large rectangle.

$2x^2 + 5x + 4x + 10 = 2x^2 + 9x + 10$

EXERCISES _____

Use a drawing to help you write each expression as a trinomial in standard form.

1. $x(2x + 4) + 3(2x + 4)$

	$2x$	4
x	$2x^2$	$4x$
3	$6x$	12

$2x^2 + 10x + 12$

2. $3y(2y + 4) + 2(2y + 4)$

	$3y$	2
$2y$	$6y^2$	$4y$
4	$12y$	8

$6y^2 + 16y + 8$

3. $4m(3m + 5) + 3(3m + 5)$

	$4m$	3
$3m$	$12m^2$	$9m$
5	$20m$	15

$12m^2 + 29m + 15$

4. $s(2s + 1) + (2s + 1)$

	$2s$	1
s	$2s^2$	s
1	$2s$	1

$2s^2 + 3s + 1$

Name _____ Date _____

Dividing by a Monomial

The quotient rule for exponents is as follows:
   To divide two powers having the same base, subtract the exponent of the denominator from that of the numerator.

$$\frac{a^m}{a^n} = a^{m-n}$$

Use this rule to find the quotient of two monomials having the same base.

► **Example 1** _____

Simplify $\frac{18x^6}{6x^2}$

Solution

$\frac{18x^6}{6x^2} = \left(\frac{18}{6}\right)x^{6-2} = 3x^4$

When a, b, and c are real numbers, and $c = 0$,

$\frac{a + b}{c} = \frac{a}{c} + \frac{b}{c}$

You can use this general rule to divide a polynomial by a monomial. That is, to divide a polynomial by a monomial, divide each term of the polynomial by the monomial.

► **Example 2** _____

Simplify $\frac{4x^3y^2 + 6x^2y - 2xy}{2xy}$

Solution

$\frac{4x^3y^2 + 6x^2y - 2xy}{2xy} = \frac{4x^3y^2}{2xy} + \frac{6x^2y}{2xy} - \frac{2xy}{2xy}$

$= 2x^{3-1}y^{2-1} + 3x^{2-1}y^{1-1} - 1x^{1-1}y^{1-1}$

$= 2x^2y + 3xy - 1$

EXERCISES _____

Simplify.

1. $\frac{w^3}{w}$
   w^2

2. $\frac{-r^3s}{r^4}$
   $-rs$

3. $\frac{6mn}{3n}$
   $2m$

4. $\frac{4s^2t}{2s}$
   $2st$

5. $\frac{6s^2t^4}{6st^2}$
   st^2

6. $\frac{3xy^2}{xy}$
   $3y$

7. $\frac{15a^2b^2}{3ab}$
   $5ab$

8. $\frac{1.2m^3n}{6m}$
   $0.5m^2n$

9. $\frac{5x^2 + 10x}{5}$
   $x^2 + 2x$

10. $\frac{4m - 8n}{2}$
    $2m - 4n$

11. $\frac{6x^2 - 9x}{x}$
    $6x - 9$

12. $\frac{6m^2 + 4m}{2m}$
    $3m + 2$

13. $\frac{12x^3 - 8x^2}{4x^2}$
    $3x - 2$

14. $\frac{24s^2t + 8st^2}{4st}$
    $6s + 2t$

15. $\frac{4a^2b - 2a}{-2}$
    $-2a^2b + a$

16. $\frac{7p^2 + 14p - 21}{7}$
    $p^2 + 2p - 3$

17. $\frac{12x^3 - 8x^2 + 16x}{4x}$
    $3x^2 - 2x + 4$

18. $\frac{m^3n + 2m^2n^2 - mn^3}{mn}$
    $m^2 + 2mn - n^2$

19. $\frac{3x^3y^2 + 6x^2y^3}{3x^2y}$
    $xy + 2y^2$

20. $\frac{4a^4 + 2a^3 - 2a^2 + 8a}{2a}$
    $2a^3 + a^2 - a + 2$

Name _____ Date _____

Long Division

Dividing a trinomial by a binomial is similar to dividing with whole numbers.

► **Example** _____

Divide $\frac{3x^2 - 2x - 8}{x + 2}$

Solution

Set up the division problem in a similar fashion to whole number division. Remember the four steps in division: Divide, multiply, subtract, bring down.	Divide $3x^2$ by x. Put the quotient $3x$ over the $2x$. Multiply the divisor $x - 2$ by the quotient $3x$ and subtract. Remember to change signs and combine like terms.	Bring down the -8. Divide $-4x$ by x. Put the quotient -4 over the -8. Multiply the divisor $x + 2$ by the quotient -4 and subtract.
$x + 2 \overline{)3x^2 + 2x - 8}$	$\begin{array}{r} 3x \\ x + 2 \overline{)3x^2 + 2x - 8} \\ -(3x^2 + 6x) \\ \hline -4x \end{array}$	$\begin{array}{r} 3x - 4 \\ x + 2 \overline{)3x^2 - 2x - 8} \\ -(3x^2 + 6x) \\ \hline -4x - 8 \\ -(-4x - 8) \end{array}$

EXERCISES _____

Divide.

1. $\frac{x^2 + 7x + 12}{x + 3}$

   $x + 4$

2. $\frac{2x^2 + 6x + 4}{x + 2}$

   $2x + 2$

3. $\frac{3m^2 + 8m - 3}{m + 3}$

   $3m - 1$

4. $\frac{s^2 - 7s + 12}{s - 4}$

   $s - 3$

5. $\frac{2t^2 - 8t + 8}{t - 2}$

   $2t - 4$

6. $\frac{2n^2 + 8n - 20}{n + 4}$

   $2n + 5$

470D

Name _____   Date _____

Computer Activity: GCF According to Euclid

Euclid, a Greek mathematician who lived about 300 BC, is best known for his Geometry. However, he also contributed to other branches of mathematics. In his famous work, *The Elements*, he described a way to find the GCF of two numbers. It is called the **Euclidean Algorithm.**

This program will use the Euclidean Algorithm to find the GCF.

```
10  INPUT "ENTER THE NUMBERS IN
    ASCENDING ORDER: ";S,L
20  IF S > L GOTO 10
30  Q = L / S
40  R = L − INT (Q) ∗ S
50  IF R < > 0 THEN GOTO 70
60  PRINT : PRINT "THE GCF IS ";
    S: END
70  L = S:S = R: GOTO 30
```

Euclidean Algorithm
1. Divide smaller number into larger number.
2. If remainder ≠ 0 then divide remainder into smaller number.
3. Continue Step 2 until remainder = 0. The last divisor = GCF.

EXERCISES

1. RUN the program to find the GCF for each set of numbers.

   a. 15, 40   **5**   c. 124, 620   **124**   e. 195, 520   **65**

   b. 77, 143   **11**   d. 96, 156   **12**   f. 606, 2727   **303**

2. Use the results in Exercise 1 to factor each polynomial.

   a. $15a^2b^4 - 40a^5b^3 + 5ab$
   $5ab(3ab^3 - 8a^4b^2 + 1)$

   b. $77x^3y^4 + 143x^7y^2 - 33x^5y^8$
   $11x^3y^2(7y^2 + 13x^4 - 3x^2y^6)$

   c. $124rs^3t^5 - 620r^5s^4t^7$
   $124rs^3t^5(1 - 5r^4st^2)$

   d. $96c^5d^3 + 156cd - 24c^2d^7$
   $12cd(8c^4d^2 + 13 - 2cd^6)$

   e. $195f^4g^6 - 520fg^4h + 25g^3h^2$
   $5g^3(39f^4g^3 - 104fgh + 5h^2)$

   f. $606j^7k^4 + 2727j^7k^2 - 303j^6k^2$
   $303j^6k^2(2jk^2 + 9j - 1)$

Name _____   Date _____

Calculator Activity: Evaluating Polynomials

You can use a calculator to evaluate polynomials. Check your calculator manual to see exactly how your calculator handles exponents.

► **Example**

Use a calculator to evaluate each expression if $x = -3$ and $y = 2$.

a. $(3x^2 - 2x + 4) - (-2x^2 + x - 4)$ b. $(-3y^2)^4$ c. $\frac{16x^4y^2 - 32x^2y^3 - 24x^3y^3}{8x^2y^2}$

Solution

First, simplify. Next, write the expression using the numbers in place of the variables. Then, use a calculator.

a. $(3x^2 - 2x + 4) - (-2x^2 + x - 4) = (3x^2 - 2x + 4) + (2x^2 - x + 4)$
$$= 5x^2 - 3x + 8$$

Substitute −3 for x.
$5(-3)^2 - 3(-3) + 8$
$5 \boxed{\times} 3 \boxed{+/-} \boxed{y^x} 2 \boxed{=} \boxed{-} 3 \boxed{\times} 3 \boxed{+/-} \boxed{=} \boxed{+} 8 \boxed{+} 62$

b. $(-3y^2)^4 = (-3)^4(y^2)^4 = (-3)^4(y)^8$

Substitute 2 for y.
$(-3)^4(2)^8$
$3 \boxed{+/-} \boxed{y^x} 4 \boxed{=} \boxed{\times} 2 \boxed{y^x} 8 \boxed{=} 20,736$

c. $\frac{16x^4y^2 - 32x^2y^3 - 24x^3y^3}{8x^2y^2} = \frac{16x^4y^2}{8x^2y^2} - \frac{32x^2y^3}{8x^2y^2} - \frac{24x^3y^3}{8x^2y^2}$
$$= 2x^2 - 4y^3 - 3xy$$

Substitute −3 for x and 2 for y.
$2(-3)^2 - 4(2)^3 - 3(-3)(2)$
$2 \boxed{\times} 3 \boxed{+/-} \boxed{y^x} 2 \boxed{=} \boxed{-} 4 \boxed{\times} 2 \boxed{y^x} 3 \boxed{=} \boxed{-} 3 \boxed{\times} 3 \boxed{+/-} \boxed{\times} 2 \boxed{=} 4$

EXERCISES

Use your calculator to evaluate each expression if $x = -4$ and $y = -3$. Remember to simplify first.

1. $(-2x^2 + 4x - 3) + (3x^2 + 7)$ 2. $(-\frac{1}{2}x)(-6y)$
   4   **36**

3. $-3x^2(-x^2 + 4x + 3)$ 4. $(3y^2 - 2y + 7) - (4y^2 + 3y + 5)$
   1,392   **8**

5. $\frac{7x^5y^2 - 49x^2y^4 + 35x^4y^3}{7x^2y}$ 6. $(x^4y^3)^2$
   1,101   **47,775,744**

7. $(4x^2 - 2) + (-3x^2 + 6x + 1)$ 8. $\frac{-8x^3y^3 + 12x^4y^3 - 10x^5y^2}{2x^3y^2}$
   −9   **44**

ACHIEVEMENT TEST   Name _____

Polynomials   Date _____
CHAPTER 12 FORM A

MATH MATTERS BOOK 1
Chicha Lynch
Eugene Olmstead

SOUTH-WESTERN PUBLISHING CO.

SCORING RECORD	
Possible	Earned
18	

Add.

1. $2x + (-4x) + x$ **−x**   2. $3m^2 + (-2m^2) + 4m^2$ **$5m^2$**

Use the figure at the right. Solve.

3. Three stars are located at the points of a triangular region *ABC*. Write an expression for the perimeter of the region.
   19x

Add or subtract.

4. $(6a^2 + 3a - 9) + (a^2 - a + 3)$
   $7a^2 + 2a - 6$

5. $(8t^2 + 4t - 10) - (3t^2 + t - 12)$
   $5t^2 + 3t + 2$

Solve.

6. Write a simplified expression for the perimeter of the rectangle shown at the right.
   20x − 4

Find the product.

7. $(-4m)(3n)$ **−12mn**   8. $(-3b)(-6c)$ **18bc**

Simplify.

9. $(3s)(3s^2)$ **$9s^3$**   10. $(-3x^4)(-5x^3)$ **$15x^7$**

11. $\frac{16ab}{4b}$ **4a**   12. $\frac{-7x^3}{x^3}$ **−7**   13. $\frac{10y^4z^3}{-5y^3z^2}$ **$-2yz^3$**

Find the product.

14. $3(r - 4)$ 15. $-3y(y + 3)$ 16. $n^3(3 - n - n^3)$
   3r − 12   **$-3y^2 - 9y$**   **$3n^3 - n^4 - n^6$**

Name _____   Date _____

Solve.

17. The area of the rectangle shown is $18a^2b$. Write an expression for the unknown dimension.   **3a**

    [rectangle: 6ab, ?]

18. The width of a rectangle is 3 cm less than the length. The perimeter is 42 cm. Find the area of the rectangle.
    l = 12 cm; w = 9 cm; A = 108 cm^2

ACHIEVEMENT TEST

Polynomials

CHAPTER 12   FORM B

MATH MATTERS BOOK 1
Chicha Lynch
Eugene Olmstead

SOUTH-WESTERN PUBLISHING CO.

Name _____

Date _____

SCORING RECORD	
Possible	Earned
18	

Add.

1. $3m + (-2m) + 4m$ **5m**

2. $4y^2 + (-8y^2) + 3y^2$ **$-y^2$**

Use the figure at the right. Solve.

3. Three meteorite fragments are located at the points of a triangular region DEF. Write an expression for the perimeter of the region.

   25m

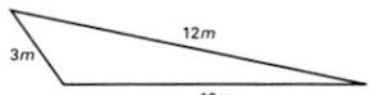

(triangle: $3m$, $12m$, $10m$)

Add or subtract.

4. $(7x^2 + 5x - 6) + (2x^2 - 2x + 4)$

   $9x^2 + 3x - 2$

5. $(6r^2 + 3r - 6) - (2r^2 - r - 18)$

   $4r^2 + 4r + 12$

Solve.

6. Write a simplified expression for the perimeter of the triangle shown at the right.

   $4r + 1$

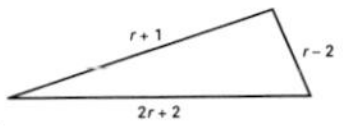

(triangle: $r + 1$, $r - 2$, $2r + 2$)

Find the product.

7. $(-3y)(4xy)$ **$-12xy^2$**

8. $(-2a^2b)(ab^2)$ **$-2a^3b^3$**

Simplify.

9. $(4a)(3a)$ **$12a^2$**

10. $(5m^3)(-3m^5)$ **$-15m^8$**

11. $\dfrac{18st}{6t}$ **$3s$**

12. $\dfrac{-14p^3}{2p^3}$ **-7**

13. $\dfrac{24x^4y^3}{-6xy}$ **$-4x^3y^2$**

Find the product.

14. $5(z - 3)$

    $5z - 15$

15. $-2y(3y + 3)$

    $-6y^2 + 6y$

16. $-x^3(x^2 - 4x - x^3)$

    $-x^5 + 4x^4 + x^6$

12B-1

Name _____   Date _____

Solve.

17. A rectangular garden has an area of $12a^3b^2$ square units. The width is $6ab$ units. Write an expression for the length.   **$2a^2b$**

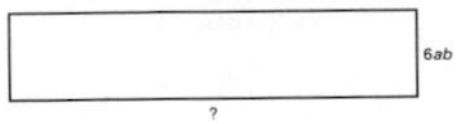

(rectangle: $6ab$, ?)

18. The perimeter of a rectangle is 26 cm. The length is 5 cm greater than the width. Find the area of the rectangle.

   $w = 4$ cm; $l = 9$ cm; $A = 36$ cm^2

12B-2

CHAPTER 13 SKILLS PREVIEW

Find each probability. Use this spinner.

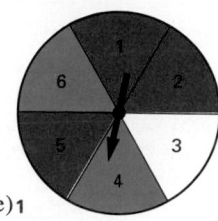

1. $P(\text{red})\frac{1}{2}$

2. $P(\text{not white})$ $\frac{5}{6}$

3. $P(2 \text{ or } 3)\frac{1}{3}$

4. $P(10)$ 0

5. $P(\text{red, white, or blue})$1

Use a tree diagram to find the number of possible outcomes in each sample space.

6. tossing a penny and then spinning a spinner like the one at the right
   8: H1, H2, H3, H4, T1, T2, T3, T4

7. tossing a dime and tossing a number cube
   12: H1, H2, H3, H4, H5, H6, T1, T2, T3, T4, T5, T6

Use the counting principle to find the number of possible outcomes.

8. choosing a sandwich from 3 kinds of bread and 5 sandwich fillings 15

9. tossing a coin five times 10

10. choosing a breakfast from orange juice, tomato juice, or grapefruit juice; with oatmeal or cornflakes; and with whole wheat, rye, or pumpernickel toast 18

11. A bag contains 3 white marbles, 2 blue marbles, and 5 orange marbles. Brian takes one marble from the bag, replaces it, and takes another. Find $P(\text{blue, then white})$. $\frac{3}{50}$

A number cube is tossed 30 times. The outcomes are recorded in this table.

Number on Cube	1	2	3	4	5	6
Number of Outcomes	5	6	4	5	7	3

12. Find $P(4)$. $\frac{1}{6}$

13. Find $P(\text{odd number})$. $\frac{8}{15}$

14. Find $P(1 \text{ or } 3)$. $\frac{3}{10}$

Introduction The purpose of this Skills Preview is to assess students' abilities on all the major objectives of Chapter 13. Test results may be used
- to determine those topics which may need only to be reviewed and those topics which need to be more carefully developed;
- for class placement;
- in prescribing for individual differences.

If you prefer, you may use the Skills Preview as an alternative form of the Chapter Test (page 500) to evaluate mastery of chapter objectives. The items on the Skills Preview and the Chapter Test correspond in content and level of difficulty.

OVERVIEW

In this chapter, students learn the basic concepts of probability. They begin with the definition of probability and then determine sample spaces using tree diagrams. The counting principle is introduced as an alternative means of determining the number of outcomes. Students learn how to compute probabilities of dependent and independent events and to determine probabilities experimentally. Finally, the chapter closes with the problem solving skill of using samples to make predictions and the problem solving strategy of simulating a problem.

SPECIAL CONCERNS

Throughout their lives, students will encounter situations that are described using probability concepts, from the chance of rain to the odds in favor of a candidate winning. Hence, an understanding of what probability means, how it is calculated, and how it is used is very important. People use probability to decide whether buying an ad in a paper is worthwhile or whether an operation will be successful. Probability gives us a framework for dealing with actions whose outcomes we cannot predict with certainty.

VOCABULARY

counting principle
dependent events
equally likely
event
experimental probability
independent events
outcome

probability
random
random sample
sample space
simulate
tree diagram

MATERIALS

coins
marbles
spinners
number cubes
paper cups

BULLETIN BOARD

Make a bulletin board containing clippings or statements involving probability. As the unit progresses, have students contribute to the display and, as their understanding grows, discuss what the statements mean and evaluate their usefulness.

INTEGRATED UNIT 4

The skills and concepts involved in Chapters 12–14 are included within the special Integrated Unit 4 entitled "Fast Forward." This unit is in the Teacher's Edition beginning on page 545K. Worksheets for this integrated unit appear in the Enrichment Activities booklet, pages 172–172.

TECHNOLOGY CONNECTIONS

- Computer Worksheet, 261
- Calculator Worksheet, 262
- MicroExam, Apple Version
- MicroExam, IBM Version

- *Probability Theory,* True BASIC
- *Gears,* Sunburst Communications

TECHNOLOGY NOTES

Computer probability programs, such as those listed in Technology Connections, provide students with simulations that graphically portray probability concepts. Some programs allow students to manipulate the elements of a problem. Many game programs allow for the gathering of data and the building of such problem solving skills as predicting and analyzing.

PLANNING GUIDE

SECTIONS	TEXT PAGES	ASSIGNMENTS		
		BASIC	AVERAGE	ENRICHED
Chapter Opener/Decision Making	472–473			
13–1 Probability	474–477	1–10, 17–22, 31–33	1–16, 23–28, 31–35	11–16, 23–30, 31–38
13–2 Sample Spaces and Tree Diagrams	478–481	1–10, 11, 14	3–10, 11–12, 14–15	7–10, 11–13, 14–16
13–3 The Counting Principle	482–485	1–6, 9–10, 13, PSA 1–3	3–8, 9–11, 13–14, PSA 1–4	3–8, 9–12, 13–14, PSA 1–4
13–4 Independent and Dependent Events	486–489	1–12, 17–19, 28–29	13–16, 17–27, 28–29	13–16, 17–27, 28–31
13–5 Experimental Probability	490–493	1–14, 17, 20–21	8–16, 17–19, 20–21	8–16, 17–19, 20–22
13–6 Problem Solving Skills: Using Samples to Make Predictions	494–495	1–8	1–12	1–12
13–7 Problem Solving Strategies: Simulating the Problem	496–497	1–6	1–9	5–11
Technology	495, 497	✔	✔	✔

ASSESSMENT				
Skills Preview	471	All	All	All
Chapter Review	498–499	All	All	All
Chapter Test	500	All	All	All
Cumulative Review	501	All	All	All
Cumulative Test	502	All	All	All

ADDITIONAL RESOURCES			
RETEACHING	**ENRICHMENT**	**TECHNOLOGY**	**TRANSPARENCY**
13–1	13–1		TM 58
13–2	13–2	13–2	
13–3	13–3	13–3	
13–4	13–4		
13–5	13–5		
13–6	13–6		
13–7	13–7		

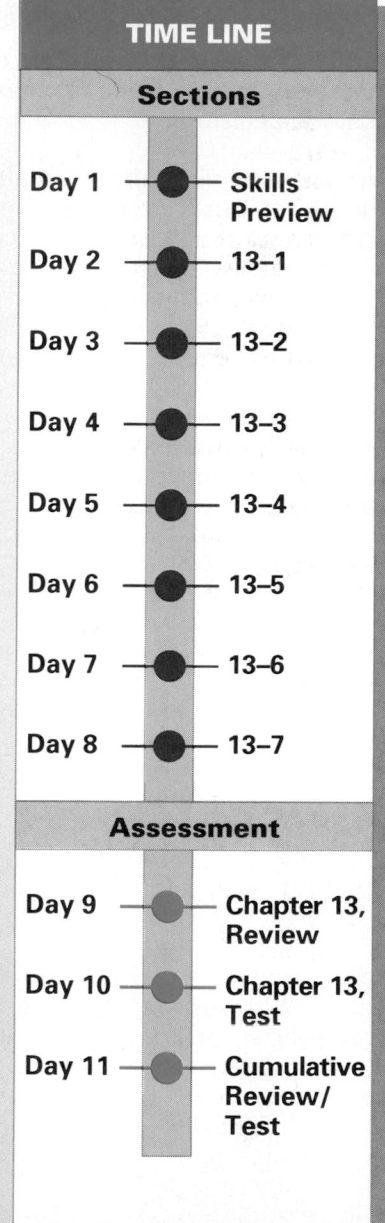

TIME LINE

Sections

Day 1 — Skills Preview
Day 2 — 13–1
Day 3 — 13–2
Day 4 — 13–3
Day 5 — 13–4
Day 6 — 13–5
Day 7 — 13–6
Day 8 — 13–7

Assessment

Day 9 — Chapter 13, Review
Day 10 — Chapter 13, Test
Day 11 — Cumulative Review/Test

ASSESSMENT OPTIONS

Chapter 13, Test Forms A and B	
Chapter 13, Test	Text, 500
Alternative Assessment	TAE, 500
Chapter 13, MicroExam	

PROBABILITY

THEME   Forecasting and Choices

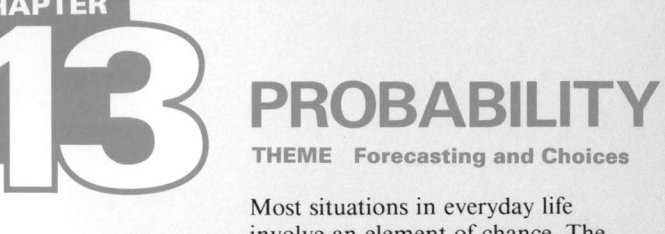

Objective   To explore how data can be used as a basis for making real-life decisions.

Introduction   Talk about situations in which students have heard statements that involve choice or probability. Then discuss the weather data given on page 472. Have students estimate the amount of rainfall, in millimeters, that your city gets in one year. Give students a figure to use to compare with their estimates and also to help make the data in the table more meaningful.

Decision Making
Using Data   Have students work in small groups to find answers to these questions. Have them justify their answers to Exercise 2. In summary, you may wish to point out that students are using the data about past events as a basis for predicting future events.

Working Together   Use as many sources as possible in this activity. Have students compare the accuracy of weather forecasting in different media. For example, is television more reliable than the newspaper, or are they about the same? Allow time for sharing results.

Most situations in everyday life involve an element of chance. The ability to determine the probability of various outcomes is important in helping people decide what to do.

In this chapter you will use a formula to determine the probability of an event.

Raindrops form in clouds when tiny water droplets join together or when larger ice crystals melt. A raindrop must contain at least 1,000 droplets for it to be heavy enough to fall.

GREATEST AVERAGE ANNUAL RAINFALLS		
Continent	Rainfall (in millimeters)	Place
Oceania (Pacific islands)	11,684	Mt. Wai-'ale-'ale, Hawaii
Asia	11,430	Cherrapunji, India
Africa	10,277	Debundseha, Cameroon
South America	8,991	Quibdo, Colombia
North America	6,655	Henderson Lake, British Columbia
Europe	4,648	Crkvice, Yugoslavia
Australia	4,496	Tully, Queensland

DECISION MAKING

Using Data

1. Use the information in the chart on page 472. On the average, how much more rainfall do the residents of Henderson Lake receive than do the residents of Tully? **2,159 mm**

2. In the tropical rain forests in South America, it rains nearly every day. Each year at least 2,030 mm of rain can fall. Decide which of the following ranges best describes the probability of this area suffering from a drought in the next year.
   a. 0%–20%   b. 20%–50%   c. 50%–100%   Answers will vary.

Working Together

Your group's task will be to track the predictions of a particular weather forecaster in the newspaper or on TV for one week. Make a chart showing the weather that was forecast and then the actual weather that occurred for each day during the week.

Based on your data, your group should decide on the probability that your forecaster's next prediction will be accurate.

Probability **473**

474

13-1 Probability

Work with a partner. Toss a number cube 20 times. Keep a tally of the number of times each number lands face up.

1. Predict how many times the number 5 would land face up if you tossed the number cube 40 times.

2. Toss the number cube 40 times and record the times the number 5 lands face up.

3. Compare this number to your prediction. Were you close?

When you toss a number cube, there are six possible **outcomes.** Each outcome is **equally likely** to happen.

An **event** is an outcome or a combination of outcomes.

The **probability** of an event, written $P(E)$, is a number between 0 and 1. It is usually written as a fraction in lowest terms. You can find $P(E)$ by using this formula.

$$P(E) = \frac{\text{number of favorable outcomes}}{\text{number of possible outcomes}}$$

Example 1

Find each probability. Use this spinner.
a. $P(2)$
b. $P(\text{blue})$
c. $P(10)$
d. $P(\text{blue, red, or green})$

Solution

a. $P(2) = \frac{1}{6}$ ← one favorable outcome
   ← six possible outcomes

b. Three of the six sections are blue.

   $P(\text{blue}) = \frac{3}{6} = \frac{1}{2}$

c. No section is labeled 10, so there are no favorable outcomes.

   $P(10) = \frac{0}{6} = 0$ ← The probability of any *impossible event* is 0.

d. All the outcomes are favorable.

   $P(\text{blue, red, or green}) = \frac{6}{6} = 1$ ← The probability of any *certain event* is 1. ◄

474 CHAPTER 13 Probability

Example 2

Find the probability of picking a B or a Y from this set of cards without looking.

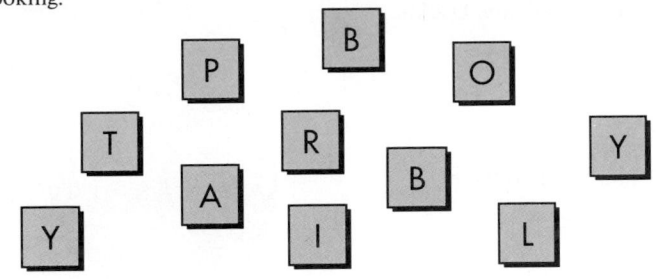

Solution

$P(\text{B or Y}) = \dfrac{4}{11} \leftarrow$ two Bs, two Ys
$\phantom{P(\text{B or Y}) = \dfrac{4}{11}} \leftarrow$ total number of possible outcomes ◀

Example 3

A bag contains 3 purple, 4 green, 4 blue, and 4 white marbles. What is the probability of taking a marble that is not purple out of the bag?

Solution

Find the total number of marbles.

$$3 + 4 + 4 + 4 = 15$$

There are 3 purple marbles, so subtract to find the number of *not purple* marbles.

$$15 - 3 = 12$$

Find the probability.

$P(\text{not purple}) = \dfrac{12}{15} = \dfrac{4}{5}$ ◀

RY THESE

Find each probability. Use this spinner.

1. $P(\text{blue}) \dfrac{1}{2}$

2. $P(\text{not blue}) \dfrac{1}{2}$

3. $P(\text{C or D}) \dfrac{1}{3}$

4. $P(Z) \; 0$

5. $P(\text{red, green, or blue}) \dfrac{5}{6}$

6. $P(\text{A, B, C, D, E, or F}) \; 1$

One of the cards is picked without looking. Find each probability.

7. $P(O) \dfrac{2}{7}$ 8. $P(\text{C, O, or E}) \dfrac{4}{7}$ 9. $P(B) \; 0$ 10. $P(T) \dfrac{1}{7}$

ASSIGNMENTS

BASIC
1–10, 17–22, 31–33

AVERAGE
1–16, 23–28, 31–35

ENRICHED
11–16, 23–30, 31–38

ADDITIONAL RESOURCES
Reteaching 13–1
Enrichment 13–1
Transparency Master 58

events which are not equally likely—for instance, $P(\text{number} \leq 5) \neq P(\text{number} > 5)$.

Example 1: Show students a spinner like the one on page 474 and ask them to identify the outcomes when you spin the spinner. With this spinner, there are two sets of outcomes: the numbers 1, 2, 3, 4, 5, 6, and the colors R, B, G. Ask students to answer the questions and then compare their answers to those in the text.

Example 2: Ask students to identify which of the cards they could draw in Example 2 to have a favorable outcome and then discuss the solution.

Example 3: Have students identify another way to find the number of marbles in the bag that are not purple. Then discuss the solution.

Additional Questions/Examples

1. Can a probability be 0? When would that happen? Give an example. **yes; when the probability of an event is impossible; tossing a 12 on a number cube labeled 1–6**

2. Can a probability be 1? When would that happen? Give an example. **yes; when the probability of an event is certain; tossing a number < 10 with a number cube labeled 1–6**

3. Why is the probability of an event never greater than 1? **because the number of favorable events cannot be greater than the number of events**

Guided Practice/Try These Have students work with partners. For each of Exercises 1–6 have them identify which wedges constitute favorable outcomes before finding the probability.

3 SUMMARIZE

Key Questions
1. What is the difference between an outcome and an event?
2. How would you define the probability of an event?
3. In what forms other than a lowest-terms fraction can probability be expressed? **decimal or percent**

4 PRACTICE

Practice/Solve Problems It may help students to list the outcomes possible when spinning the spinner (1–6 and G, G, G, B, B, R). Remind them to treat each set of outcomes separately—that is, there are 6 outcomes involving numbers and 6 involving colors.

Extend/Solve Problems Challenge students to give the odds against *E* (number of unfavorable outcomes/number of favorable outcomes). Point out that when the odds of an event are written in ratio form *a* to *b*, then *a* + *b* = number of possible outcomes.

Think Critically/Solve Problems Encourage students to justify their choices.

5 FOLLOW-UP

Extra Practice There are 4 red, 3 green, and 5 blue marbles in a bag. Find each probability.
1. *P*(red marble) **1/3**
2. *P*(green marble) **1/4**

476

EXERCISES

PRACTICE/ SOLVE PROBLEMS

Find each probability. Use the spinner.

1. P(blue) $\frac{1}{3}$
2. P(4 or 5) $\frac{1}{3}$
3. P(even number) $\frac{1}{2}$
4. P(number divisible by 2) $\frac{1}{2}$
5. P(not green) $\frac{1}{2}$
6. P(10) **0**
7. P(not red) $\frac{5}{6}$
8. P(6) $\frac{1}{6}$
9. P(red) $\frac{1}{6}$
10. P(number > 0) **1**

One of the cards is drawn without looking. Find each probability.

11. P(vowel) $\frac{1}{3}$
12. P(C) $\frac{1}{3}$
13. P(N or H) $\frac{1}{3}$
14. P(consonant) $\frac{2}{3}$
15. P(S) **0**
16. P(consonant or vowel) **1**

EXTEND/ SOLVE PROBLEMS

A number cube, with faces labeled from 1 through 6, is tossed. Find each probability.

17. P(6) $\frac{1}{6}$
18. P(even number) $\frac{1}{2}$
19. P(prime number) $\frac{1}{2}$
20. P(number > 10) **0**
21. P(2, 3, 5, or 6) $\frac{2}{3}$
22. P(composite number) $\frac{1}{3}$

The *odds in favor of an event* can be found using this formula.

$$\text{odds in favor of } E = \frac{\text{number of favorable outcomes}}{\text{number of unfavorable outcomes}}$$

Use the formula and the spinner to find the odds in favor of each event. Write your answer as a ratio in the form *a* to *b*.

23. B **1 to 4**

24. green **2 to 3**

25. blue or not green **3 to 2**

26. C or D **2 to 3**

27. not A **4 to 1**

28. white **1 to 4**   29. green or A **3 to 2**   30. B, C, or D **3 to 2**

31. **USING DATA** Use the Data Index on page 546 to locate information that will help you answer the following question. Suppose the names of all the Mountain states were put into a hat and one name were to be drawn from the hat without looking. What is the probability that the name drawn would be that of a state larger than Utah? $\frac{3}{4}$

Many times you will see a probability expressed as a percent. For instance, if the probability that it will rain today is $\frac{3}{5}$, a meteorologist usually reports this as a 60% chance of rain. If an event will definitely occur, people often describe it as "100% sure."

THINK CRITICALLY/ SOLVE PROBLEMS

For Exercises 32–37, choose an appropriate probability for each event from probabilities a–e.

32. A name chosen at random from a telephone directory is a woman's name. **c**

a. 0%
b. less than 10%
c. about 50%
d. more than 90%
e. 100%

33. The next car that passes through an intersection was manufactured before 1960. **b**

34. The sun will rise in the west tomorrow. **a**

35. The next person you meet will be someone who sometimes watches television. **d**

36. Water will freeze when its temperature is lowered to −10°F. **e**

37. The next person you meet will be an identical twin. **b**

Write the probability as a percent.

38. A meteorologist says that there is a 30% chance of snow tomorrow. What is the probability that it will *not* snow tomorrow? **70%**

3. *P* (blue or red marble) **3/4**
4. *P* (white marble) **0**
5. What is another way of writing *P* (not green marble)? **P (red or blue)**
6. Find the odds in favor of picking a green marble in this experiment. **3 to 9**

Extension Write a number of outcomes, such as 0, 3, 5, 6, 8, R, T, A, O, on cards. Have each pair of students pick a card. Ask one student to describe an event (such as picking an odd number) and the other student to give the probability of that event.

Section Quiz You are to choose a letter at random from the word MISSISSIPPI. Find each probability.
1. *P*(M) **1/11**
2. *P*(I) **4/11**
3. *P*(S) **4/11**
4. *P*(I or P) **6/11**
5. *P*(vowel) **4/11**
6. *P*(consonant) **7/11**
7. *P*(Q or R) **0**
8. *P*(letter following E in the alphabet) **1**

CHALLENGE

A bag of fruit chews contains 2 cherry chews, 3 lemon chews, and 1 orange chew.
1. How many cherry chews must be added to the bag so that the probability of choosing a cherry chew is 1/2? **2**
2. How many lemon chews must be added to the bag so that the probability of choosing a cherry chew is only 1/5? **4**

477

13-2 Sample Spaces and Tree Diagrams

EXPLORE

TALK IT OVER

If you are only able to guess at the answers, what happens to your chances of getting a passing grade as the number of true–false items increases? Do you think that guessing is ever a good strategy?

Suppose that your history teacher surprises you with a true–false quiz. You haven't done the reading yet, so you have to guess at the answers.

1. List all possible ways to guess at the answer if the quiz has only one item. **true, false**
2. Now suppose that the quiz has just two items. List all the possible ways you might guess at the answers. **true, false; true, true; false, false; false, true**
3. What if the quiz has just three items? List all the possible ways you could guess. How can you make sure you are accounting for all the possibilities? **There are 8 ways to guess: TTT, TTF, TFT, TFF, FTT, FTF, FFT, FFF**

SKILLS DEVELOPMENT

To find a probability, it is important that you account for all outcomes. The term **sample space** is the special name we use for a set of all possible outcomes. Many times it is easy to identify a sample space. For instance, when you toss a number cube, there are six outcomes in the sample space, which you can list as follows.

$$\{1, 2, 3, 4, 5, 6\}$$

At other times, though, a sample space is not easy to identify. In such cases, you might use a **tree diagram** to picture the sample space and then count the outcomes.

Example 1

A penny and a nickel are tossed. Make a tree diagram to picture the sample space. How many possible outcomes are there?

Solution

When each coin is tossed there are two possibilities.

penny → heads or tails
nickel → heads or tails

This tree diagram pictures the outcomes.

There are four possible outcomes. ◀

PENNY	NICKEL	OUTCOMES
H	H →	HH
	T →	HT
T	H →	TH
	T →	TT

478   CHAPTER 13   Probability

You can also use a tree diagram to find a probability.

Example 2

Three coins are tossed. Find P(exactly 2 tails).

Solution

Make a tree diagram to picture the sample space.

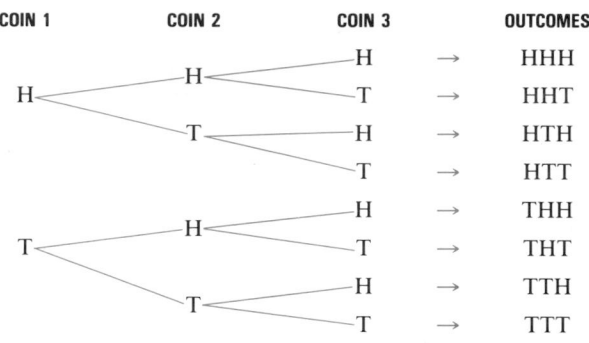

| COIN 1 | COIN 2 | COIN 3 | OUTCOMES |

Look at the list of outcomes in the tree diagram. There are 8 possible outcomes. Only 3 of these outcomes contain *exactly* 2 tails.

HTT, THT, TTH

So P(exactly 2 tails) $= \frac{3}{8}$. ◄

TRY THESE

List each sample space.

1. tossing a coin H, T

2. spinning this spinner
   1, 2, 3, 4, 5, 6, 7, 8

Use a tree diagram to find the number of possible outcomes in the sample space.

3. tossing four coins
   See Additional Answers.

4. tossing a coin and tossing a number cube
   See Additional Answers.

Suppose that you toss a coin and spin the spinner in Exercise 2 above. Find each probability.

5. P(tail and 5) $\frac{1}{16}$

6. P(head and even number) $\frac{1}{4}$

Example 1: Use paper coins to help students visualize the possibilities when tossing two coins. Stress that a tree diagram helps organize the possibilities.

Example 2: Have students work on their own to make a tree diagram for tossing two coins. Then have them compare their results with those in the text. If students are not sure why the word *exactly* is important, point out that, without it, the probability of two tails would include the outcome T T T.

Additional Questions/Examples
1. Why is the outcome HTT different than the outcome THT?
   Because the head occurs on the first toss in HTT but on the second in THT.
2. What are the possible outcomes when you roll a number cube and then toss a coin?
   1H, 1T, 2H, 2T, 3H, 3T, 4H, 4T, 5H, 5T, 6H, 6T

Guided Practice/Try These Ask students why a tree diagram is not necessary in Exercises 1 and 2. **The sample space is easy to find.** Have them work with partners in solving Exercises 3–6.

479

480

3 SUMMARIZE

Write About Math Have students write a paragraph in their math journals describing how to make a tree diagram to show the number of outcomes in a sample space.

4 PRACTICE

Practice/Solve Problems Have students identify the possibilities for each part of the event before making tree diagrams to show the sample spaces.

Extend/Solve Problems Some students may find it helpful to first make a table showing all two-number possibilities in one column and all two-letter possibilities in a second column. They can then use their table to systematically construct the tree diagram.

Think Critically/Solve Problems In Exercises 14–15 students should recognize that spinners 2 and 3 are not divided into equal parts. However, for a spinner, probability depends not on the area for each section but, rather, on the measure of the angle at the point where the spinner is anchored. Thus, for spinner 2 the outcomes are not equally likely; for spinner 3, the intersecting lines form all right angles, so the probability for each letter is the same.

5 FOLLOW-UP

Extra Practice
Two number cubes, marked 1–6, are tossed. List all the possible outcomes.
(1,1)(1,2)(1,3)(1,4)(1,5)(1,6), (2,1)
(2,2)(2,3)(2,4)(2,5)(2,6), (3,1)(3,2)
(3,3)(3,4)(3,5)(3,6), (4,1)(4,2)(4,3)
(4,4)(4,5)(4,6), (5,1)(5,2)(5,3)(5,4)
(5,5)(5,6), (6,1)(6,2)(6,3)(6,4)(6,5)
(6,6)

480

EXERCISES

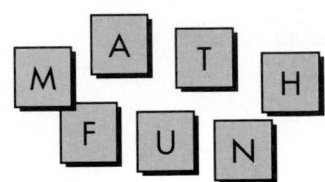

PRACTICE/ SOLVE PROBLEMS

List each sample space.

1. picking a card from those shown here.
   {M, A, T, H, F, U, N}
2. spinning spinner 1 {1, 2, 3}

MIXED REVIEW

Find the sum or difference.

1. $9 - (-12)$ **21**
2. $22 + (-7)$ **15**

Write each decimal as a fraction in lowest terms.

3. 2.75 $2\frac{3}{4}$
4. 0.8 $\frac{4}{5}$
5. 3.15 $3\frac{3}{20}$

Use $I = prt$ to find the total amount due.

6. $p = \$300$
   $r = 5\%$
   $t = 2$ years **$330**

7. $p = \$2,000$
   $r = 8\%$
   $t = 36$ months
   $2,480

8. $p = \$1,750$
   $r = 11\%$
   $t = 6$ months
   $1,846.50

Solve.

9. A rectangular garden measures 7.4 m by 5.5 m. What is the perimeter?

Spinner 1   Spinner 2

Use a tree diagram to find the number of possible outcomes in each sample space. **See Additional Answers.**

3. tossing a penny, a nickel, and a dime

4. picking a card from those shown above and tossing a coin

5. tossing a coin and spinning spinner 1

6. tossing a nickel, tossing a number cube, and spinning spinner 1

Suppose you spin both spinner 1 and spinner 2. Find each probability.

7. $P(3 \text{ and } E)$ $\frac{1}{18}$

8. $P(2 \text{ and consonant})$ $\frac{1}{9}$

9. $P(\text{odd number and vowel})$ $\frac{2}{9}$

10. $P(\text{prime number and C})$ $\frac{1}{9}$

The Highlands Recreation Center has the following rules for coding their membership cards.

Rule 1: Letters and numbers may be used only once in each card.

Rule 2: A membership code must be two numbers followed by two letters.

Use the information given and list the possible codes for membership cards that can be made from the letters and numbers given for each of Exercises 11–13. **See Additional Answers.**

11. 4, 9, 3, A, J, R

12. 1, 6, 7, 5, P, N

13. Choose three letters and four numbers and make up your own membership codes. Make a tree diagram to show all possible outcomes in this sample space. **Check students' diagrams; 72 ways**

For Exercises 14–15, use these three spinners.

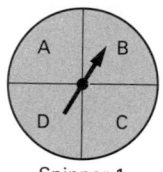
Spinner 1

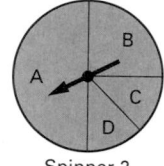
Spinner 2

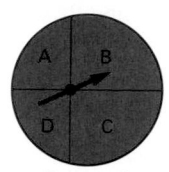
Spinner 3

14. List the sample space for each spinner.
   a. spinner 1 **(A, B, C, D)**
   b. spinner 2 **(A, B, C, D)**
   c. spinner 3 **(A, B, C, D)**

15. Find $P(A)$ for each spinner.
   a. spinner 1 $\frac{1}{4}$
   b. spinner 2 $\frac{1}{2}$
   c. spinner 3 $\frac{1}{4}$

16. Compare your answers to Exercises 14 and 15. How are the spinners alike? How are they different?

16. The sample space is the same on each of the three spinners. However, the probability of the events A, B, C, and D differs.

2. What is the probability that the numbers that land face up will have a sum of 7? **1/6**

3. What is the probability that the numbers that land face up will have a sum greater than 4? **5/6**

4. Describe two different ways to answer Exercise 3. **Find P(sum > 4), or find P(sum ≤ 4) and subtract it from 1.**

Extension Have students work with partners to show the possible outcomes when tossing 1, 2, 3, and 4 coins. Have them predict the number of possible outcomes for 5, 6, and 10 coins.

Section Quiz Use a tree diagram to find the possible outcomes for the sample space.

1. Picking a letter from the word WORK and spinning a spinner with two equal divisions labeled 1 and 2. **(W,1)(W,2), (O,1)(O,2), (R,1)(R,2), (K,1)(K,2)**

2. Find the probability of picking a vowel and spinning an even number in Exercise 1. **1/8**

3. One spinner is divided into three equal divisions labeled 1, 2, 3. Another spinner is divided into eight equal divisions labeled A, B, C, D, E, F, G, H. What is P(even number and vowel) if you spin both spinners? **1/12**

Additional Answers
See page 583.
Additional answers for odd-numbered exercises are found in the Selected Answers portion of the page.

OBJECTIVE
• Use the counting principle to find the number of outcomes in a sample space

VOCABULARY
counting principle

WARM-UP

Find each product.
1. $2 \times 3 \times 5$ **30**
2. $7 \times 8 \times 4$ **224**
3. $8 \times 8 \times 8$ **512**
4. $10 \times 10 \times 10 \times 10$ **10,000**

1 MOTIVATE

Explore Have the students work with partners to make tree diagrams to solve the problem posed in the Explore activity. Involve students in a group discussion to explore ideas for other ways of determining the number of different kinds of sandwiches.

2 TEACH

Use the Pages/Skills Development Have students read this part of the section and then discuss the examples. Use Example 1 to introduce the counting principle.

Example 1: Using the table, students should be able to visualize that for each brand of computer there are two choices of monitor, giving 8 outcomes for monitor and brand of computer. With each of those 8 possibilities, there are then 3 choices of printer, for a total of 24 choices.

482

13-3 The Counting Principle

EXPLORE

Suppose that you and your friend are making sandwiches for lunch. You can use either chicken salad or turkey for the filling. For bread, your choices are pita, whole wheat, and pumpernickel. How many different kinds of sandwiches can you make? (Use only one type of filling and one type of bread for each sandwich.) **6**

SKILLS DEVELOPMENT

TALK IT OVER

When is it not practical to use a tree diagram to show the sample space for an event?

You have probably noticed that, as the number of choices increases, it becomes increasingly less convenient to count outcomes by using a tree diagram. Another method you can use to find the number of outcomes is the **counting principle.** Using the counting principle, you can find the number of outcomes for an activity by multiplying the number of outcomes at each stage of the activity.

Example 1

Benita is configuring a computer system for her home office. This means that she must choose which combination of computer, monitor, and printer to buy. The table shows the possibilities from which she can choose.

Brand of Computer	Type of Monitor	Type of Printer
Plum	color	laser
Ideal	black and white	dot matrix
Sunflower		daisy wheel
AA Tech		

How many different configurations can she choose from?

Solution
There are three stages involved in choosing a configuration: which computer to buy, which monitor to buy, and which printer to buy.

Multiply the number of choices at each stage.

$$\begin{array}{ccccccc} \text{computer} & & \text{monitor} & & \text{printer} & & \\ 4 & \times & 2 & \times & 3 & = & 24 \end{array}$$

Benita can choose from 24 possible configurations. ◄

482 CHAPTER 13 Probability

TEACHING TIP

If students have difficulties understanding the counting principle, apply it to several problems where you can also draw tree diagrams. Draw the tree diagram and then use it to illustrate how the counting principle works.

Example 2

In a game, a player tosses a number cube and chooses one of 26 alphabet cards. Using the counting principle, find P(prime number, T or Q).

Solution

Remember that $P(E) = \frac{\text{number of favorable outcomes}}{\text{number of possible outcomes}}$.

First find the number of possible outcomes.

$$\begin{array}{ccc} \text{number cube} & & \text{alphabet cards} \\ 6 & \times & 26 & = & 156 \end{array}$$

Then find the number of favorable outcomes. There are 3 prime numbers: 2, 3, and 5. There are 2 letters: T and Q.

$$\begin{array}{ccc} \text{prime number} & & \text{letters} \\ 3 & \times & 2 & = & 6 \end{array}$$

P(prime number, T or Q) $= \frac{6}{156} = \frac{1}{26}$ ◄

TRY THESE

Use the counting principle to find the number of possible outcomes.

1. choosing a stereo system from 2 kinds of receivers, 3 kinds of CD players, and 4 kinds of cassette decks **24**

2. making a sandwich using wheat, rye, or pumpernickel bread and peanut butter or jelly as a filling **6**

3. tossing a number cube three times **216**

4. tossing a coin four times **16**

Use the counting principle to find each probability.

5. A number cube is tossed three times. Find $P(6, 6, 6)$. **$\frac{1}{216}$**

6. A coin is tossed five times. Find P(all heads or all tails). **$\frac{1}{16}$**

EXERCISES

Use the counting principle to find the number of possible outcomes.

PRACTICE/ SOLVE PROBLEMS

1. choosing a snack from 3 desserts and 4 kinds of herb tea **12**

2. choosing breakfast from 4 cereals, 5 juices, and 2 kinds of toast **40**

3. choosing a pasta from spaghetti, rigatoni, or linguini and a sauce from plain, meat, mushroom, garden vegetable, or herb **15**

4. choosing a skirt from denim, khaki, or plaid and a blouse from white, tan, red, blue, black, or yellow **18**

13-3   The Counting Principle   **483**

5. tossing a coin three times **8**

6. tossing a number cube two times **36**

Use the counting principle to find the probability.

7. A coin is tossed six times. Find P(all heads). $\frac{1}{64}$

8. A coin is tossed three times and a number cube is tossed. Find P(all tails and even number). $\frac{1}{16}$

4 PRACTICE

Practice/Solve Problems In Exercise 8 have students list the favorable outcomes before attempting to find the probability.

Extend/Solve Problems In Exercises 9–10 students may overlook the "with or without" as offering two additional choices in these problems. Suggest they list a possible choice (for example, *vegetable soup, small, with salt*) to help identify the choices.

Think Critically/Solve Problems Have students work with number cards or paper slips labeled 1, 2, and 3 to help them understand Exercise 13. Emphasize that once a number card is used, it cannot be used again. Exercise 14 can also be modeled using numbered cards or paper slips.

Problem Solving Applications Ask students why a tree diagram is more useful here than is the counting principle. Elicit from students that Kimi wants to consider the actual programs in each of the possible sequences. Have students work in small groups on these problems and then allow time for the groups to compare their solutions.

5 FOLLOW-UP

Extra Practice Use the counting principle to find the number of outcomes.

1. tossing a coin 5 times **32**
2. rolling a number cube, marked 1–6, two times and tossing a coin **72**
3. choosing an outfit from 4 sweaters, 2 pairs of jeans, and 3 pairs of shoes **24**

EXTEND/ SOLVE PROBLEMS

9. A restaurant serves 5 kinds of soup in 3 different sizes, with and without salt. How many different ways can you order soup? **30**

10. A type of telephone is available with or without an answering machine. The phone comes in the colors white, beige, black, slate, and gray. It is available with a programmable memory for either 3, 6, 9, 12, or 20 phone numbers. How many different ways is the telephone available for sale? **50**

11. How many different four-digit numbers can be made using only the digits 1, 2, and 3? (Assume that a digit may be repeated.) $3 \times 3 \times 3 \times 3 = 81$

12. How many different three-letter "words" can be made using any of the letters A through Z? (Assume that a letter may be repeated.) $26 \times 26 \times 26 = 17{,}576$

THINK CRITICALLY/ SOLVE PROBLEMS

13. Three students are going to run a relay race. The track coach has to decide which student will run in each position, first, second, and third.

   a. In how many ways can the coach assign a student to the first position? **3 ways**

   b. Once a student has been assigned to the first position, in how many ways can the coach assign a student to the second position? **2 ways**

   c. Once students have been assigned to the first and second positions, in how many ways can the coach assign a student to the third position? **1 way**

   d. Multiply your answers to **a, b,** and **c.** In how many ways can the coach assign the three students to the three positions? $3 \times 2 \times 1 = 6$; **6 ways**

14. Suppose that a car has room to seat four passengers.

   a. In how many ways can four people be arranged in the four seats if they all are able to drive? **24 ways**

   b. In how many ways can four people be arranged in the four seats if only one of them is able to drive? **6 ways**

484 CHAPTER 13 Probability

► READ
► PLAN
► SOLVE
► ANSWER
► CHECK

Problem Solving Applications:

DECISION MAKING IN TELEVISION PROGRAMMING

Kimi Lee, a programming manager for a television station, must decide what programs will be shown between 7:00 p.m. and 9:00 p.m. on Thursdays. Four half-hour programs are available for this time period: a comedy, a game show, news, and an interview show.

Kimi decides to use a tree diagram to show all the possible ways to fill each half-hour time slot. Then she can choose the program sequence she thinks will be most popular. Here is the beginning of her tree diagram.

7:00 p.m.	7:30 p.m.	8:00 p.m.	8:30 p.m.	POSSIBLE SEQUENCE
	news	comedy	game	INCG
		game	comedy	INGC
	comedy	news	game	ICNG
interview		game	news	ICGN
	game	news	comedy	IGNC
		comedy	news	IGCN

From past popularity ratings of the station's programs, Kimi knows several things:

a. News programs are more popular before than after 8:00 p.m.
b. Many parents of young children do not wish their children to watch this particular comedy.
c. Interview shows draw the largest audience if they come just before or just after a news show.

1. Complete Kimi's tree diagram. How many different possible choices are there for programming between 7:00 p.m. and 9:00 p.m.? **See Additional Answers; 24 possible choices**

2. Decide which branches of the tree diagram you would eliminate because of fact **a.** How many possible choices would then be left? **Answers will vary.**

3. After making your decision in Problem 2, decide which branches of the tree diagram you would eliminate because of facts **b** and **c.** How many possible choices would then be left? **Answers will vary.**

4. Make a recommendation to Kimi about one or two program schedules you think will draw the largest audience. Give reasons for your recommendation. **Answers will vary.**

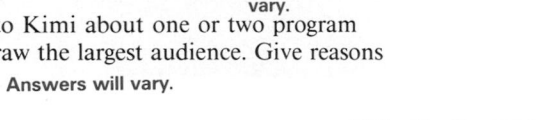

13-3   The Counting Principle   **485**

4. In Exercise 1, what is P(all heads or all tails)? **1/16**
5. In Exercise 3, if you can choose an outfit with or without socks, how many outfits are there? **48**
6. How many different five-digit numbers can be made using the digits from 1 through 5? (Assume that a digit may be repeated.) **3,125**

Extension   Have students work in groups to discover a formula or rule for finding the number of different ways 3 objects, 4 objects, and 5 objects can be arranged in a row. Then have them extend their rule to any number of objects.

Section Quiz   Use the counting principle to find the number of outcomes.
1. rolling a number cube 3 times **216**
2. choosing a sweatshirt from 4 colors and 5 styles **20**
3. rolling a number cube and then tossing a coin twice **24**
4. Find P (odd number and all heads) for Exercise 3. **1/8**

Get Ready   marbles

Additional Answers
See page 583.

SPOTLIGHT

OBJECTIVE
- Find the probabilities of independent and dependent events

MATERIALS NEEDED
marbles

VOCABULARY
dependent events, independent events

WARM-UP

Two number cubes are rolled. Find each probability.
1. *P*(sum of 8) **5/36**
2. *P*(sum of 12) **1/36**
3. *P*(first number is 3) **1/6**

1 MOTIVATE

Explore It may be helpful for students to work with marbles, especially in the second activity, so they can visualize how the situation changes if the marble selected is not replaced. Have students work with partners to complete these activities and then allow time for sharing results.

2 TEACH

Use the Pages/Skills Development
Have students read this part of the section and then discuss the examples. Define *independent events* and have students give some illustrations of such events.

Example 1: Ask students what is in the bag on the first draw and what is in the bag on the second draw. Since the second draw does not depend on the results of the first, the events are independent.

13-4 Independent and Dependent Events

EXPLORE

1a. Check students' diagrams.

1. Suppose that a bag contains one red, one white, and one green marble. A marble is selected from the bag and replaced, and then a marble is selected again.

   a. Make a tree diagram to show the sample space.

   b. Use the tree diagram to find *P*(red, then white). $\frac{1}{9}$

2a. Check students' diagrams.

2. Suppose that a bag contains one red, one white, and one green marble. A marble is selected from the bag, but it is *not* replaced. Then a marble is selected again.

   a. Make a tree diagram to show the sample space.

   b. Use the tree diagram to find *P*(red, then white). $\frac{1}{6}$

SKILLS DEVELOPMENT

Two events are said to be **independent** when the result of the second event does not depend on the result of the first event.

If two events X and Y are independent, the probability that both events will occur is the product of their individual probabilities.
$$P(X \text{ and } Y) = P(X) \times P(Y)$$

Example 1

A bag contains 3 red apples and 2 yellow apples. An apple is taken from the bag and replaced. Then an apple is taken from the bag again. Find *P*(red, then red).

Solution

There are 5 apples in the bag, and 3 of these are red. Because the first apple is replaced, the second selection is independent of the first.

$P(\text{red, then red}) = P(\text{red}) \times P(\text{red})$

$\qquad\qquad = \frac{3}{5} \times \frac{3}{5} = \frac{9}{25}$ ◄

486 CHAPTER 13 Probability

TEACHING TIP

To help students understand the rules illustrated in Examples 1 and 2, have them label the apples R_1, R_2, R_3, Y_1, and Y_2. Then have them draw tree diagrams to show the possible outcomes in each example and use the tree diagrams to find the probabilities.

Two events are **dependent** if the result of the first event *does* affect the result of the second event.

Example 2

A bag contains 3 red apples and 2 yellow apples. An apple is taken from the bag and is *not* replaced. Then an apple is taken from the bag again. Find P(red, then red).

Solution

Because the first apple is not replaced, the second event is dependent on the first.

On the first selection, there are 5 apples in the bag, and 3 of the apples are red.

$$P(\text{red}) = \frac{3}{5}$$

On the next selection, there are only 4 apples in the bag. Assuming that a red apple was removed, only 2 of the apples in the bag are red.

$$P(\text{red after red}) = \frac{2}{4} = \frac{1}{2}$$

Multiply the two probabilities.

$$P(\text{red, then red}) = \frac{3}{5} \times \frac{1}{2} = \frac{3}{10} \blacktriangleleft$$

 **WRITING ABOUT MATH**

Write an outline in your journal explaining the difference between independent and dependent events. Be sure to include an example of each type of event.

TRY THESE

A silverware drawer contains 10 forks, 16 teaspoons, and 4 steak knives. A utensil is taken from the drawer and replaced. Then a utensil is taken from the drawer again. Find each probability.

1. P(fork, then teaspoon) $\frac{8}{45}$

2. P(steak knife, then fork) $\frac{2}{45}$

3. P(fork, then fork) $\frac{1}{9}$

4. P(teaspoon, then teaspoon) $\frac{64}{225}$

Solve.

5. A bag contains 2 red marbles, 3 white marbles, and 5 green marbles. Find P(white, then green) if Jenna selects two marbles from the bag without replacement.

$\frac{1}{6}$

6. A change purse contains 2 quarters, 3 dimes, and 5 pennies. Find P(quarter, then dime) if the coins are selected without replacement. $\frac{1}{15}$

5-MINUTE CLINIC

Exercise	Student's Error	Error Diagnosis
Two marbles are drawn without replacement from a bag containing 3 red and 2 blue marbles. Find P(red, then blue).	$2/3 \times 2/5 = 6/25$ $3/5 + 1/2 = 11/10$	• Student does not recognize that events are dependent. • Student adds rather than multiplies to find the probability.

Example 2: Use this example to help clarify the meaning of dependent events. Ask students to describe other situations that are dependent events. Also have students find the probability of P(red, then yellow). **3/10**

Additional Questions/Examples
1. What information in the statement of the problem can help you determine if events are independent or dependent?
   The statement will tell whether or not replacement is made.
   Are the events *dependent* or *independent*? Explain.
2. rolling two number cubes **independent; The roll of one cube does not affect the roll of the other.**
3. drawing two cards from a deck without replacement. **dependent; The result of the first draw affects the second draw.**

Guided Practice/Try These For each of Exercises 1–2, ask students how many utensils there are for the first selection and then for the second. (The same number available both times indicates that the events are independent.)

3 SUMMARIZE

Talk It Over Have students describe the difference between dependent and independent events and then tell how they find the probability in each case.

4 PRACTICE

Practice/Solve Problems Ask students to determine if the events described for each group of problems are *dependent* or *independent*.

Extend/Solve Problems Elicit from students that another way to state Exercises 17 and 18, respectively, is "find *P*(orange, then orange)" and "find *P*(blue, then blue)." Point out that the word "then" is implied in the statement "one blue tee and (then) one orange tee." (See the Extension activity below.)

Think Critically/Solve Problems In Exercises 28–31 it may help students to list the possible pairs of events and find the probabilities for each. Or they can simply use guess and check as a method of solution.

5 FOLLOW-UP

Extra Practice The letters A–J are written one each on a card, and the cards are placed in a box. Find the probability of each event.
1. P(A, then E) if the first card is replaced **1/100**
2. P(A, then E) if the first card is not replaced **1/90**
3. P(B, then A or E) if the first card is not replaced **1/45**
4. Six red cubes and 5 blue cubes are placed in a box. What is the probability that you select two blue cubes if you reach into the box and select two cubes at random. **2/11**

EXERCISES

PRACTICE/ SOLVE PROBLEMS

MIXED REVIEW

Find each product.
1. $\frac{1}{4} \times \frac{5}{16}$ **$\frac{5}{64}$**
2. $\frac{5}{2} \times \frac{12}{5}$ **6**

Evaluate each expression.
3. $3t - 1$ if $t = (-2)$ **-7**
4. $5 - d$ if $d = (-3)$ **8**
5. $-10z$ if $z = 5$ **-50**

Solve each proportion.
6. $\frac{x}{9} = \frac{35}{45}$ **$x = 7$**
7. $\frac{6}{15} = \frac{m}{80}$ **$m = 32$**
8. A 9-oz tube of toothpaste costs $3.55. Find the unit price to the nearest cent. **$0.39 per ounce**

A game is played by spinning the spinner and then choosing a card without looking. After each turn the card is replaced. Find the probability of each event.

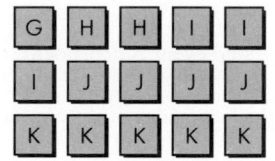

1. P(5, then K) $\frac{1}{30}$
2. P(6, then G) $\frac{1}{75}$
3. P(5, then G) $\frac{1}{150}$
4. P(8, then J) $\frac{8}{75}$
5. P(7, then I) $\frac{3}{50}$
6. P(7, then H) $\frac{1}{25}$
7. P(6, then J) $\frac{4}{75}$
8. P(6, then K) $\frac{1}{15}$
9. P(5, then H) $\frac{1}{75}$
10. P(7, then J) $\frac{2}{25}$
11. P(8, then G) $\frac{2}{75}$
12. P(5, then I) $\frac{1}{50}$

Ten cards are numbered from 1 to 10 and placed in a box. A card is taken from the box and is not replaced. Then a card is taken from the box again. Find the probability of each event.

13. P(4, then 4) **0**
14. P(5, then odd) $\frac{2}{45}$
15. P(6, then prime) $\frac{2}{45}$
16. P(prime, then prime) $\frac{2}{15}$

EXTEND/ SOLVE PROBLEMS

A bucket contains 8 orange golf tees and 6 blue golf tees. Kato reaches into the bucket and selects 2 tees without looking.

17. What is the probability that he selects 2 orange tees? $\frac{4}{13}$
18. What is the probability that he selects no orange tees? $\frac{15}{91}$
19. What is the probability that he selects one blue tee and one orange tee? $\frac{24}{91}$

Two of these cards are chosen at random, one after another, without replacement. What is the probability of each event?

20. P(both vowels) $\frac{5}{33}$
21. P(vowel, then consonant) $\frac{35}{132}$
22. P(both consonants) $\frac{7}{22}$
23. P(A, then T) $\frac{1}{66}$

AT-RISK STUDENTS

These students may have a hard time distinguishing between independent and dependent events. Use manipulatives to illustrate the problem situations in this section and to help students see that in dependent events, the number of choices changes for the second selection.

In a writing contest, 6 nonfiction stories and 4 fiction stories were chosen as finalists. Prizes were awarded on a random basis. The judges put the titles of the winning stories into a box and selected 3 winners at a time. Find each probability.

24. P(3 fiction) $\frac{1}{30}$

25. P(3 nonfiction) $\frac{1}{6}$

26. P(nonfiction, then fiction, then nonfiction) $\frac{1}{6}$

27. P(fiction, then nonfiction, then fiction) $\frac{1}{10}$

A wallet contains three $20 bills, one $10 bill, and two $1 bills. Use this information to give an example for each exercise. **Answers may vary. Sample answers are given.**

28. two independent events whose probability is $\frac{1}{12}$

29. two dependent events whose probability is $\frac{1}{10}$

30. two dependent events whose probability is 0

31. two independent events whose probability is 0

13-4   Independent and Dependent Events   **489**

THINK CRITICALLY/ SOLVE PROBLEMS

28. pulling out a $20 bill, replacing it, and then pulling out a $10 bill

29. pulling out a $20 bill, not replacing it, and then pulling out a $10 bill

30. pulling out a $10 bill, not replacing it, and then pulling out a $10 bill

31. pulling out a $10 bill, replacing it, and then pulling out a $5 bill

OBJECTIVE
• Find experimental probability

MATERIALS NEEDED
number cubes, paper cups

VOCABULARY
experimental probability

WARM-UP

A number cube, marked 1–6, is rolled. Find:
1. $P(1)$ **1/6**   2. $P(8)$ **0**
3. $P(3$ or $4)$ **1/3**
4. $P($number greater than $1)$ **5/6**

1 MOTIVATE

Explore/Working Together Supply groups of students with number cubes so they can complete the experiment described in the Explore activity. Ask how many 5s they would expect if they rolled a cube 60 times (10). Point out that, in the experiment, they could roll all 5s or no 5s. When students compare results, pool their experiments to see if the number of 5s in 1,000 rolls was closer to 1/6 the total number of rolls than the number of 5s in 60 rolls.

2 TEACH

Use the Pages/Skills Development Have students read this part of the section and then discuss the examples. Introduce the concept of experimental probability. Point out that the experimental probability of getting a given number

---

13-5 Experimental Probability

EXPLORE/ WORKING TOGETHER

Work with a partner. Use a number cube with faces labeled from 1 through 6.

1. Use the probability formula to find the probability of rolling a 5 on the first roll. $\frac{1}{6}$

2. Now roll the number cube 60 times, keeping a tally of the results of your *experiment* in a table like this.

Number on Cube	Tally
1	
2	
3	
4	
5	
6	

3. Examine the results of your experiment. How many times did the number 5 actually appear? Compare your results with those of other students in your class.

SKILLS DEVELOPMENT

The probability of an event based on the results of an experiment is called **experimental probability.**

Example 1

This table is a summary of the outcomes from an experiment in which a number cube was tossed 50 times.

Number on Cube	1	2	3	4	5	6
Number of Outcomes	7	9	10	2	11	11

Find each experimental probability.
a. $P(6)$ b. $P(3)$ c. $P($even number$)$ d. $P(4$ or $5)$

CHECK UNDERSTANDING

In the solution of Example 1a, what does the 11 represent in the fraction?

Solution
a. $P(6) = \frac{11}{50}$
b. $P(3) = \frac{10}{50} = \frac{1}{5}$
c. $P($even number$) = \frac{22}{50} = \frac{11}{25}$
d. $P(4$ or $5) = \frac{13}{50}$ ◄

MAKING CONNECTIONS

Ask students to investigate how a manufacturer might use experimental probability for quality-control purposes; for example, in determining the number of defects in a product. See if students can suggest other instances in which experimental probability might be used.

Experimental probabilities are often used when all outcomes are not equally likely. For instance, when you toss a paper cup, there generally are three ways that it could land: up, down, or on its side. So there are three outcomes, but they are not equally likely. You would expect that the cup would most often land on its side.

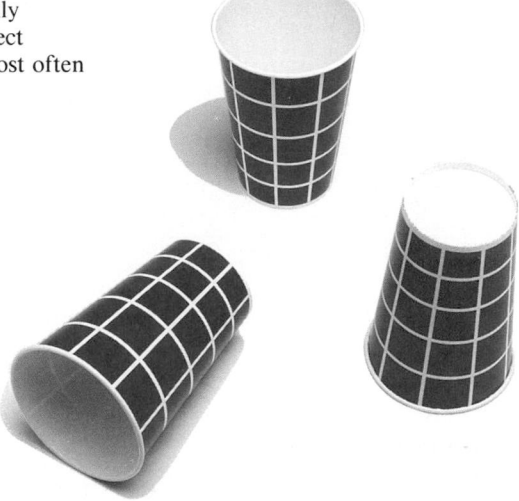

Example 2

A paper cup is tossed 50 times. The results are shown in the table. Find the experimental probability of the cup landing on its side.

Outcome	Tally
up	IIII
down	JHT JHT
side	JHT JHT JHT JHT JHT JHT JHT I

Solution
Counting the tally marks, you see that the paper cup landed on its side 36 times.

$$P(\text{side}) = \frac{36}{50} = \frac{18}{25} \quad \blacktriangleleft$$

You often find an experimental probability using a random sample. Then you use the probability to draw conclusions about the population from which the sample was taken.

Example 3

In a bottle factory, 1,000 bottles were selected at random. Of these, 3 were found to be defective. What is the probability of a bottle being defective?

Solution

$$P(\text{defective}) = \frac{\text{number defective in sample}}{\text{total number in sample}} = \frac{3}{1,000} . \quad \blacktriangleleft$$

CONNECTIONS

In Chapter 1, you learned about *random sampling* as a way of collecting information from people. Example 3 shows a different way that random sampling is used. In this case, the sample is selected from a group of objects. However, the entire group is still called the *population*.

When is a sample considered a random sample?

When the members of the population are given an equal chance of being selected.

13-5   Experimental Probability   **491**

ASSIGNMENTS

BASIC
1–14, 17, 20–21

AVERAGE
8–16, 17–19, 20–21

ENRICHED
8–16, 17–19, 20–22

ADDITIONAL RESOURCES
Reteaching 13–5
Enrichment 13–5

on a roll of a number cube may be different than the theoretical probability of getting that number. Point out that experimental probability can also be used in situations where theoretical probability cannot be readily computed, such as in Example 3.

Example 1: For part a, students should be able to suggest that to find the experimental probability of an event such as getting a 6, they should use the ratio of the number of 6s to the total number of tosses.

Example 2: As a follow-up to this example, ask students how many times they would expect the cup to land on its side if it were tossed 100 times. **72** Also elicit from students that the cup landing upward is least likely.

Example 3: Discuss the situation presented in this example. This type of experimental probability is used often in manufacturing as a means of quality control.

Additional Questions/Examples
1. Do you have any means of predicting how many heads you will get on 100 tosses of a coin? Explain. **You could assume heads and tails are equally likely events each with a probability of 1/2, which may lead you to predict 50**

out of 100 tosses will be heads. Or, you could toss the coin 20 or 50 times to obtain an experimental probability.

2. As the number of coin tosses becomes very large, what do you think is true about the experimental probability of getting heads? **The experimental probability gets very close to the probability of 1/2 that would be found using the definition from Section 13–1.**

Guided Practice/Try These Have the students work with partners to complete these exercises.

③ SUMMARIZE

Write About Math Have students write a paragraph in their math journals explaining how to find the experimental probability of an event, such as getting a 6, when rolling a number cube.

④ PRACTICE

Practice/Solve Problems Ask students to explain the meaning of the statements in Exercises 4–7, describing which outcomes will be considered favorable.

Extend/Solve Problems For Exercises 17 and 18, have students work in small groups and then combine their results. Compare the experimental probabilities found by the groups with those found from the combined data.

Think Critically/Solve Problems Be sure students understand that their answers for Exercises 20–21 will depend on which coin they have used for the tail of their unfair coin.

492

TRY THESE

A number cube is tossed 20 times. The outcomes are recorded in the table. Find each experimental probability.

Number on Cube	1	2	3	4	5	6
Number of Outcomes	3	5	2	5	4	1

1. $P(1)$ $\frac{3}{20}$

2. $P(2)$ $\frac{1}{4}$ 3. $P(3)$ $\frac{1}{10}$

4. $P(4)$ $\frac{1}{4}$ 5. $P(5)$ $\frac{1}{5}$

6. $P(6)$ $\frac{1}{20}$ 7. $P(\text{even number})$ $\frac{11}{20}$

8. $P(\text{odd number})$ $\frac{9}{20}$ 9. $P(2 \text{ or } 4)$ $\frac{1}{2}$

10. On an assembly line, 5,000 randomly selected cars were tested. Of these, 24 cars were found to be defective. What is the probability of finding a car to be defective? $\frac{3}{625}$

EXERCISES

PRACTICE/ SOLVE PROBLEMS

Jessie selected a marble from a bag 80 times. With each selection, she recorded the color and returned the marble to the bag. Use the results in the table to find each experimental probability.

Outcome	Tally	Total
striped	JHT JHT JHT JHT JHT	25
blue	JHT JHT JHT JHT JHT JHT JHT	35
red	JHT JHT IIII JHT	20

1. $P(\text{striped})$ $\frac{5}{16}$

2. $P(\text{blue})$ $\frac{7}{16}$

3. $P(\text{red})$ $\frac{1}{4}$

4. $P(\text{striped or blue})$ $\frac{3}{4}$

5. $P(\text{blue or red})$ $\frac{11}{16}$

6. $P(\text{not blue})$ $\frac{9}{16}$

7. $P(\text{striped or red})$ $\frac{9}{16}$

The table shows the number of muffins sold in the cafeteria during a recent morning. Use these results for Exercises 8–13 to find the experimental probability of each event.

blueberry	cranberry	bran	cinnamon	apple	other
22	6	10	8	4	10

8. P(blueberry) $\frac{11}{30}$

9. P(bran) $\frac{1}{6}$

10. P(cranberry) $\frac{1}{10}$

11. P(not bran) $\frac{5}{6}$

12. P(apple or cinnamon) $\frac{1}{5}$

13. P(not blueberry) $\frac{19}{30}$

14. Fran tossed a bottle cap 25 times. It landed up 14 times and down 11 times. Find the experimental probability that the cup lands up. $\frac{14}{25}$

15. Bruce tossed a thumbtack 20 times. It landed point up 18 times. Find the experimental probability that a tossed tack will not land point up. $\frac{1}{10}$

16. On a production line, 18 defective bolts were found among 2,000 randomly selected bolts. What is the probability of a bolt being defective? $\frac{9}{1,000}$

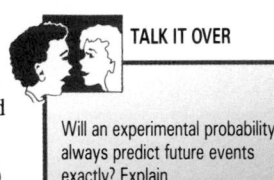
TALK IT OVER

Will an experimental probability always predict future events exactly? Explain.

EXTEND/ SOLVE PROBLEMS

17. Toss a penny and a quarter 60 times and find the experimental probability of the outcomes. Make a table to record your data. **Answers will vary.**

18. Toss a number cube 100 times and find the experimental probability of each of the outcomes. Make a table to record your data. **Answers will vary.**

19. A basketball team won 62 out of 120 games. What is the experimental probability that the team will win their next game? $\frac{31}{60}$

THINK CRITICALLY/ SOLVE PROBLEMS

Make an unfair "coin" by taping a nickel and a penny together. Be sure your coin has one head and one tail.

20. Flip the coin 20 times. Use your results to predict how many tails you will get if you flip the coin 100 times. **Answers will vary.**

21. Flip the coin 80 more times for a total of 100 flips. Was the prediction you made after 20 flips accurate for 100 flips? **Answers will vary.**

22. If you predict the number of tails in 500 flips using your results for 100 flips, do you think your prediction will be more reliable than predicting from 20 flips? Explain. **Yes.**

13-5   Experimental Probability   **493**

WARM-UP

Find each product.
1. $0.35 \times 20,000$ **7,000**
2. 0.16×500 **80**
3. $0.125 \times 350,000$ **43,750**
4. $0.08 \times 6,600$ **528**

1 MOTIVATE

Introduction Review what is meant by a random sample and how poll-takers might go about taking a random sample of a population.

2 TEACH

Use the Pages/Problem Have students read this part of the section and discuss the problem. Elicit from students why it is better to sample 225 voters instead of 10 or 250,000. **10 is too small a sample to be reliable; not practical to sample all 250,000 due to time and expense.**

Use the Pages/Solution Emphasize that finding *P*(Jackson) is a case of determining experimental probability. Talk about writing a probability as a decimal and point out that the given value is rounded.

13-6 Problem Solving Skills:
USING SAMPLES TO MAKE PREDICTIONS

► READ
► PLAN
► SOLVE
► ANSWER
► CHECK

Remember that a random sample is a group chosen from a large population in such a way that each member of the population has an equal chance of being chosen. As you saw in Chapter 1, random sampling is often used to poll people about their opinions or preferences. In this lesson, you will see how the responses of the random sample are used to make predictions about the entire population.

PROBLEM

Members of the Channel 3 election bureau conducted a poll before the mayoral election in Southglenn. There are 250,000 registered voters in this city. The poll-takers asked a random sample of 225 of these voters which of the three candidates they were planning to vote for. The results are shown in the table.

Candidate	Voters Planning to Vote
Miselli	35
Jackson	80
Brimmage	110

Based on this random sample, how many votes might Jackson expect to receive in the election?

TEACHING TIP

In order to help students recall how to find probabilities and predicted numbers, help them write word equations describing what to do. For example, to find the predicted number, students might write the following:

P(event) × total number = predicted number

SOLUTION

Use the sample results to find P(Jackson).

$$P(\text{Jackson}) = \frac{80}{225} = \frac{16}{45} \approx 0.356$$

You can predict how many votes Jackson will receive by finding the product of the probability and the total number of registered voters.

P(Jackson)	×	number of voters	=	predicted number of votes
0.356	×	250,000	=	89,000

So, Jackson might expect to receive about 89,000 votes in the election.

PROBLEMS

CALCULATOR

You can use your calculator to write a probability as a decimal. Just remember to divide the numerator of the fraction by the denominator.

Use a calculator to write each probability as a decimal. Round to the nearest hundredth.

1. $\frac{13}{20}$ **0.65**

2. $\frac{14}{25}$ **0.56**

3. $\frac{18}{65}$ **0.28**

4. $\frac{55}{75}$ **0.73**

A local cable company asked a random sample of 500 of their 50,000 subscribers which kind of programming they preferred. Of those polled in the survey, 150 preferred comedy shows, 80 boxing matches, 225 movies, and 45 sporting events.

1. Find P(boxing). $\frac{4}{25}$

2. Find P(movies). $\frac{9}{20}$

3. Find P(comedy). $\frac{3}{10}$

4. Find P(sporting events). $\frac{9}{100}$

COMPUTER TIP

You may want to use this program to check your answers to Exercises 9–11. Notice how important line 20 is in giving the correct answer.

```
10 INPUT "HOW MANY
   VOTERS? " ;V: PRINT
20 LET N = V * 750
30 PRINT N;" VOTES"
```

Refer to the information given for Problems 1–4. If all of the subscribers were polled, how many people would you expect to select each kind of programming?

5. comedy **15,000**

6. boxing matches **8,000**

7. movies **22,500**

8. sporting events **4,500**

Of the 750,000 registered voters in Centerville, 1,000 were polled about their preferences in the upcoming election. The results of the poll are shown in this table.

Bendel	370
Morales	325
Greenberg	155
Undecided	150

9. Based on the results of the poll, about how many of the registered voters in Centerville would you expect to be undecided? **112,500**

10. Suppose that all the undecided voters vote for Morales. If all the registered voters vote, about how many votes might Morales expect to receive? **356,250**

11. Suppose that Greenberg drops out of the race, and the Greenberg supporters switch their loyalty to Morales. If the undecided voters then split evenly between Bendel and Morales, about how many votes might Bendel expect to receive? **333,750**

12. Would you predict the outcome of the election based on the results of this poll? Explain. **Answers will vary.**

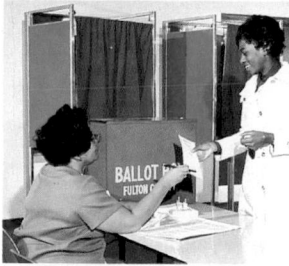

CHALLENGE

Rangers in a national park captured 200 elk, tagged them, and then released them. Three weeks later they captured 50 elk and noted that 18 of them had tags. About how many elk do you think are in the park?

18/50 = 200/n; n = 556; about 556 elk

ASSIGNMENTS

BASIC
1–8

AVERAGE
1–12

ENRICHED
1–12

ADDITIONAL RESOURCES
Reteaching 13–6
Enrichment 13–6

3 SUMMARIZE

Write About Math Write a paragraph explaining how to predict the number of people in a city of 10,000 who might buy a particular product if you know that 40 people out of a sample of 200 bought the product.

4 PRACTICE

Problems Have students explain their reasoning for Problems 10 and 11. When students have answered Problem 12, ask them how they might be able to make a more accurate prediction. **by polling a larger sample**

5 FOLLOW-UP

Extra Practice A poll of 400 commuters showed that 104 bought monthly tickets and 296 bought weekly tickets. Out of 1,500 commuters, about how many could be expected to buy a monthly ticket? **390**

Get Ready coins, number cubes, spinners

WARM-UP

Let a = 2, b = 3, c = 10. Find:
1. b/c **3/10**
2. $(a + b)/(a + c)$ **5/12**
3. ab/bc **1/5**
4. $(a + c)/(b - a)$ **12**

1 MOTIVATE

Introduction Begin by discussing what is meant by a *simulation*. Have students look up the word in the dictionary. Explain that a simulation models the mathematical relationships in the problem to be solved.

2 TEACH

Use the Pages/Problem Have students read this part of the section and then discuss the problem. Ask students to give the important facts about the game show situation. **There are two outcomes for each question and 10 questions in all.**

13-7 Problem Solving Strategies:
SIMULATE THE PROBLEM

► READ
► PLAN
► SOLVE
► ANSWER
► CHECK

Often a problem that seems complicated can be solved by relating the problem to a similar but simpler situation. This strategy is sometimes used in problems about probability. To find probabilities for a complex process you relate it to a simpler process that *simulates*, or acts in the same way as, the complex one.

PROBLEM

Dena Power is taking part in a TV game show. She has to answer 10 true-false questions. To win a prize, she must answer at least 6 of the questions correctly. Design a simulation to predict the probability of Dena winning a prize.

SOLUTION

Each of the 10 questions has two possible outcomes: either Dena answers correctly or she answers incorrectly. So, the game can be simulated by tossing 10 coins.

Let heads represent a question answered correctly and tails represent a question answered incorrectly. Toss 10 coins many times and record the outcomes in a table.

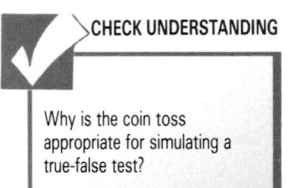

✓ CHECK UNDERSTANDING

Why is the coin toss appropriate for simulating a true-false test?

What simulation would be appropriate for a multiple-choice test with 6 choices? with 4 choices?

Each item on the true-false test has two outcomes and a coin toss has two outcomes.

Tossing a number cube; spinning a 4-part spinner

	Number of heads (correct answers)	Number of tails (incorrect answers)
1st toss	4	6
2nd toss	7	3
3rd toss	5	5
4th toss	6	4
5th toss	7	3

On the first 5 tosses, heads showed on 6 or more coins 3 times. Based on these tosses, the probability of Dena winning a prize is given by the following.

$$P(\text{winning}) = \frac{\text{number of favorable outcomes}}{\text{number of possible outcomes}} = \frac{3}{5}$$

To obtain a more accurate estimate of the probability of Dena winning, the simulation should be continued to find the results for many more tosses.

496   CHAPTER 13   Probability

AT-RISK STUDENTS

Have students solve these problems working with partners. Allow them to use the manipulatives that seem easiest to them.

PROBLEMS

Answers will vary. Check students' simulations.
Use this information to solve Problems 1–4.

A quality control inspector finds that 1 out of every 6 soft drink cans is not sealed properly.

1. Which of the following simulations would you use to predict the probability of getting no defective cans in a group of 7?
   a. toss 7 coins
   b. draw 7 cards from a deck
   c. toss 7 number cubes

2. Describe in detail how you would carry out the simulated experiment in Problem 1.

3. Carry out the simulation 50 times and record the results.

4. Give your estimate of the probability of getting no defective cans.

For Problems 5–11, design a simulation for the situation.

5. In the past, Frank has won 4 out of every 6 video games he has played. What is the probability that Frank will win his next three games?

6. What is the probability that, in a group of 8 people in an elevator, 5 are female?

7. Sandra guesses the answers to the 15 questions on a true-false test. What is the probability that Sandra will get 12 or more answers correct?

8. On a TV station, 7 out of every 10 commercials are for food products. What is the probability that 3 out of the next 4 commercials shown will be for food products?

9. In a school survey, 3 out of every 4 students indicated that they would prefer a rock concert to a dance. If a sample of 10 students is chosen, what is the probability that 8 or more of these students prefer a rock concert?

10. Suppose that 1 out of every 6 motorcycles has racing handlebars, and 1 out of every 2 motorcycles has a sidemarker lamp. What is the probability that the next three motorcycles you see will have neither racing handlebars nor a sidemarker lamp?

11. The entrance tests for a military college consist of two parts, an academic test and an athletic ability test. On the academic test, 4 out of every 6 applicants passes, but on the athletic test, only 1 in every 2 passes. What is the probability that both of the next two applicants will pass both tests?

COMPUTER

The random number generator built into BASIC can be used to simulate the experiments described in the text. For example, to simulate tossing a coin, the following program may be used.

```
10 INPUT "# OF TOSSES:
   ";NTOSS
20 LET H = 0: LET T = 0 :
   REM # HEADS, TAILS
30 FOR I = 1 TO NTOSS
40 LET R = RND(1) : REM
   RANDOM NUMBER
   BETWEEN 0 AND 1
50 IF R < .5 THEN LET
   H = H + 1: PRINT
   "HEADS": GOTO 70
60 LET T = T + 1: PRINT
   "TAILS"
70 NEXT I
80 PRINT "# HEADS = "; H;
   "# TAILS = "; T
```

PROBLEM SOLVING TIP

You may use any of these methods for your simulations.

a. tossing a coin
b. tossing a number cube
c. spinning a spinner

ASSIGNMENTS

BASIC
1–6

AVERAGE
1–9

ENRICHED
5–11

ADDITIONAL RESOURCES
Reteaching 13–7
Enrichment 13–7

Use the Pages/Solution Help students see that tossing a coin simulates the two outcomes of each question; 10 coins are used because there are 10 questions. Point out that if there were 20 questions, 20 coins would be needed; if each question had 3 outcomes, something such as a spinner with three equal parts would be needed.

3 SUMMARIZE

Talk It Over Ask students to summarize the process of simulating an experiment.

4 PRACTICE

Problems Discuss the manipulatives that can be used to simulate Problems 5–9. For instance, in Problem 5 either a number cube or a spinner with six equal divisions could be used.

5 FOLLOW-UP

Extra Practice
Estimate the probability that, of 6 people answering an ad for a job, 4 are female. **Answers will vary.**

CHALLENGE

Have students work in groups to make up problems that can be solved using simulations. Then have groups exchange, then solve, each other's problems.

13 CHAPTER REVIEW

Introduction The Chapter Review emphasizes the major concepts, skills, and vocabulary presented in this chapter and can be used for diagnosing students' strengths and weaknesses. Page references direct students back to appropriate sections for additional review and reteaching.

Using Pages 498–499 Allow students to quickly scan the Chapter Review and ask questions about any section they find confusing. Exercises 1–5 review key vocabulary. In Exercises 8 and 9, it may help students to decide before they begin work whether the problems involve dependent or independent events. Have students work in small groups to discuss their answer to Exercise 18.

Informal Evaluation Have students explain how they arrived at any incorrect answers. They will probably find their own mistakes and give you some clues as to the nature of their errors. Make sure students understand this material before administering the Chapter Test.

Choose a word from the list to complete each statement.

1. The ___?___ of an event is between 0 and 1 and is usually written as a fraction in lowest terms. **c**

2. The set of all outcomes is called the ___?___ . **e**

3. Two events are said to be ___?___ when the result of the second event does not depend on the result of the first event. **b**

4. The probability of an event based on the results of an experiment is called ___?___ . **a**

5. An ___?___ is an outcome or a combination of outcomes. **d**

a. experimental probability
b. independent
c. probability
d. event
e. sample space

SECTION 13–1   PROBABILITY   (pages 474–477)

▶ The **probability** of an event is a number between 0 and 1. You find a probability using this formula:

$$P(E) = \frac{\text{number of favorable outcomes}}{\text{number of possible outcomes}}$$

A number cube, with faces labeled from 1 through 6, is tossed.

6. Find $P(\text{number} < 5)$. $\frac{2}{3}$

7. Find $P(\text{prime number})$. $\frac{1}{2}$

SECTION 13–2   SAMPLE SPACES AND TREE DIAGRAMS   (pages 478–481)

▶ The set of all outcomes is called the **sample space.**

▶ You can make a **tree diagram** to show a sample space.

8. Two coins are tossed. Draw a tree diagram to show the sample space. How many possible outcomes are there?
There are 4 outcomes: HH, HT, TH, TT.

SECTION 13–3   THE COUNTING PRINCIPLE   (pages 482–485)

▶ To find the number of possible outcomes for an activity, multiply the number of choices for each stage of the activity.

Use the counting principle to find the number of possible outcomes.

9. tossing a coin four times 16

10. choosing a snack from 2 desserts and 3 kinds of herb tea 6

11. tossing a number cube three times 216

SECTION 13–4 INDEPENDENT AND DEPENDENT EVENTS (pages 486–489)

▶ Two events are said to be **independent** when the result of the second event does not depend on the result of the first event.

▶ When the result of the first event does affect the result of the second event, the events are **dependent**.

12. A bag contains 3 red, 2 white, and 5 green marbles. One is taken from the bag, replaced, and another is taken. Find P(red, then green). $\frac{3}{20}$

13. A change purse contains 3 quarters, 2 dimes, and 5 nickels. Find P(dime, then nickel) if two coins are selected from the purse without replacement. $\frac{1}{9}$

SECTION 13–5 EXPERIMENTAL PROBABILITY (pages 490–493)

▶ The probability of an event based on the results of an experiment is called **experimental probability**.

A number cube is tossed 20 times. The outcomes are recorded in this table.

Number on Cube	1	2	3	4	5	6
Number of Outcomes	3	5	2	5	2	3

14. Find $P(1)$. $\frac{3}{20}$

15. Find P(even number). $\frac{13}{20}$

16. Find P(3 or 4). $\frac{7}{20}$

SECTIONS 13–6 AND 13–7 PROBLEM SOLVING (pages 494–497)

▶ You can use information from a random sample of a population to make predictions about the entire population.

▶ Some probability experiments are easier to understand if a simpler experiment is conducted to simulate or imitate the more difficult one.

Describe a simulation you could use to predict the probability.

17. Suppose James was on time for his piano lesson 3 days out of every 5 for the past month. What is the probability that he will be on time for his next ten lessons? **Use a spinner divided into 5 sections; spin the spinner 10 times**

USING DATA Use the data on page 472 to answer Exercise 18.

18. Estimate the amount of rain that will fall over the next 5-year period in Mt. Wai-'ale-'ale based on the greatest amount of rain that fell one year.
approximately 60,000 mm

Review **499**

Follow-Up A coin is tossed and a number cube is rolled. Use the counting principle to find the number of outcomes in the sample space and then use a tree diagram to show them. Then find each probability. **12**
1. P(H, then even number) **1/4**
2. P(T, then a number < 5) **1/3**

Study Skills Tip Remind students that the basic definition of probability,

$$\frac{\text{number of favorable outcomes}}{\text{total number of outcomes}}$$

can also be used for experimental probability.

13 CHAPTER TEST

Introduction The Chapter Test uses a variety of questioning techniques to assess students' mastery of the major objectives of Chapter 13. If you prefer, you may use the Skills Preview (page 471) as an alternative form of the Chapter Test. The items on this test and the Skills Preview correspond in content and level of difficulty.

Alternative Assessment
Toss a bottle cap 20 times.
Answers will vary.
1. Count how many times the cap lands with the top face up and how many times it lands face down.
2. Use the results from your experiment to find the probability the cap lands face down.
3. If you tossed the cap 1,000 times, about how many times do you think it would land face down? Explain.

Find each probability. Use this spinner.

1. $P(A)$ $\frac{1}{6}$
2. $P(E \text{ or } D)$ $\frac{1}{3}$
3. $P(Z)$ 0
4. $P(\text{white})$ $\frac{1}{3}$
5. $P(A, B, C, D, E, \text{ or } F)$ 1

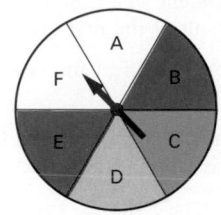

Use a tree diagram to find the number of possible outcomes in each sample space.

6. tossing a nickel and then tossing a number cube
   12: H1, H2, H3, H4, H5, H6, T1, T2, T3, T4, T5, T6
7. spinning a spinner like the one shown at the right and then tossing a quarter
   10: AH, BH, CH, DH, EH, AT, BT, CT, DT, ET

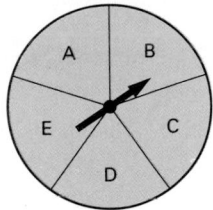

Use the counting principle to find the number of possible outcomes.

8. tossing a coin three times **8**
9. choosing a stereo configuration from 2 receivers, 3 cassette decks, and 5 speakers **30**
10. choosing a dinner consisting of either beef, lamb, fish, or chicken; with either peas or carrots; and with either milk, juice, or iced tea **24**
11. A bag contains 1 red marble, 3 blue marbles, and 1 orange marble. Susan takes one marble from the bag, replaces it, and takes another. Find $P(\text{blue, then orange})$. $\frac{3}{25}$

A number cube is tossed 30 times. The outcomes are recorded in this table.

Number on Cube	1	2	3	4	5	6
Number of Outcomes	2	4	7	3	8	6

12. Find $P(5)$. $\frac{4}{15}$
13. Find $P(\text{even number})$. $\frac{13}{30}$
14. Find $P(2 \text{ or } 6)$. $\frac{1}{3}$

1. To get an average of 85 for five tests, how many total points would you need? **425**

Find each answer.

2. $407.2 - 38.071$ **369.129**

3. $201.7 + 5.654$ **207.354**

4. 6.7×0.44 **2.948**

5. $3.241 \div 0.7$ **4.63**

6. Write two conditional statements using the following two sentences:

You live in Dubuque.
You live in Iowa.
If you live in Dubuque, then you live in Iowa.
If you live in Iowa, then you live in Dubuque.

Add or subtract.

7. $3\frac{3}{8} + 4\frac{5}{6}$ $\mathbf{8\frac{5}{24}}$

8. $7\frac{1}{4} - 2\frac{5}{18}$ $\mathbf{4\frac{35}{36}}$

9. $9 - 3\frac{1}{3}$ $\mathbf{5\frac{2}{3}}$

10. $\frac{7}{9} + \frac{1}{3}$ $\mathbf{1\frac{1}{9}}$

Write each answer in simplest form.

11.   8 yd
 -3 yd 2 ft
  4 yd 1 ft

12.   4 ft 3 in.
 -2 ft 8 in.
  1 ft 7 in.

13.   4 lb 8 oz
 $+2$ lb 10 oz
  7 lb 2 oz

14.   9 yd 2 ft
    $\times 5$
  48 yd 1 ft

Add or subtract.

15. $-65 + 8$ **-57**

16. $-20 - (17)$ **-37**

17. $16 + (-13)$ **3**

18. $-13 - (-18)$ **5**

Solve each equation.

19. $13 = q - 8$ $\mathbf{q = 21}$

20. $7 + b = -2$ $\mathbf{b = -9}$

21. $12 - k = -9$ $\mathbf{k = 21}$

22. $y + 8 = -7$ $\mathbf{y = -15}$

23. Write three equivalent ratios for $6:5$.   **12:10, 24:20, 48:40**

24. What percent of 512 is 384?   **75%**

25. One year's interest on $2,000 is $120. What is the rate of interest?   **6%**

Find the unknown angle measure for each polygon.

26.

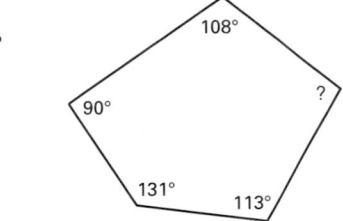

27.

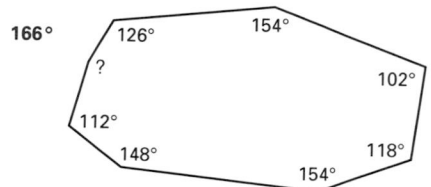

28. Line segment LM is $5\frac{1}{2}$ in. long. NO bisects LM at 0. What is the length of OM? $\mathbf{2\frac{3}{4}}$ **in.**

29. PQ bisects angle RPS. $m\angle RPS = 112°$. What is $m\angle RPQ$?   **56°**

30. Find the area of a circle with a diameter of 9 cm. Round your answer to the nearest tenth.   **63.6 cm²**

Simplify.

31. $\frac{8n + 2}{2}$ $\mathbf{4n + 1}$

32. $\frac{7x^2 + 14y}{7}$ $\mathbf{x^2 + 2y}$

33. Use the counting principle to find the number of possible outcomes for dinner with choices of 4 main courses, 4 vegetables, and 6 desserts. **96 possible outcomes**

13 CUMULATIVE REVIEW

Introduction The purpose of this Cumulative Review is to maintain previously taught skills and concepts and to apply them to the material presented in this chapter. At least one major objective of each chapter is included in the review. Page references direct students back to appropriate sections for additional review and reteaching.

Item Analysis The table below correlates the Cumulative Review items with the chapter and section that are being reviewed.

Section	Items
1–9	1
2–4	2–5
3–3	6
4–6	7–10
5–3	11–14
6–2	15, 17
6–3	16, 18
7–2	19–22
8–3	23
9–3	24
9–9	25
10–5	26–27
10–9	29
10–10	28
11–8	30
12–9	31–32
13–3	33

13 CUMULATIVE TEST

Introduction The Cumulative Test uses a standardized-test format of multiple-choice questions to assess retention of previously learned concepts and test-taking skills. Test results may be used to diagnose students' strengths and weaknesses. Page references direct students back to appropriate sections for additional review and reteaching.

Item Analysis The table below correlates the Cumulative Test items with the chapter and section that are being tested.

Section	Items
1–1	1
2–10	2
3–3	3
4–6	4
5–7	5
6–4	6
7–5	7–8
8–6	9
9–3	10
10–6	11
11–11	12
12–2	13
13–5	14

1. A gasoline company wants to survey its customers. From a list of their credit card holders, they send a questionnaire to every 20th person. What type of sampling are they using?
   A. random   B. clustered
   C. systematic   D. convenience

2. Complete.
   $4(12 + 10) = (\blacksquare \times 12) + (4 \times 10)$
   A. 4   B. 10
   C. 12   D. 88

3. Which is a counterexample for the following statement?
   If x is a number, then x^2 is larger than x.
   A. 2   B. $\frac{1}{2}$
   C. 3   D. none of these

4. Add. $4\frac{2}{3} + 8\frac{5}{6}$
   A. $12\frac{3}{6}$ B. $13\frac{1}{6}$
   C. $13\frac{5}{6}$ D. $13\frac{1}{2}$

5. Find the perimeter of a rectangular figure whose length is 24.3 mi and whose width is 8.1 mi.
   A. 16.2 mi   B. 64.8 mi
   C. 3 mi   D. 196.83 mi

6. Compare.
   $3 \times (-14) \ \bullet \ -13 \times (-3)$
   A. >   B. <
   C. =   D. ≈

7. Solve.
   $22 = 3r + 10$
   A. $r = 12$ B. $r = 15$
   C. $r = 7$ D. $r = 4$

8. Solve. $4(k - 9) = 24$
   A. $k = 15$ B. $k = 11$
   C. $k = 8$ D. $k = -5$

9. The scale of a drawing is 1 cm:3 m. A room in the drawing is 10 cm long. What is the length of the actual room?
   A. 13 m   B. 1 m
   C. 30 cm   D. 30 m

10. What percent of 900 is 270?
    A. $33\frac{1}{3}\%$ B. 25%
    C. 30%   D. $333\frac{1}{3}$

11. Name the figure.
    A. triangular pyramid
    B. pentagonal prism
    C. hexagonal pyramid
    D. triangular prism

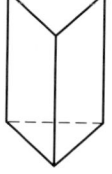

12. Find the surface area.
    A. 40 cm²
    B. 96 cm²
    C. 38 cm²
    D. 76 cm²

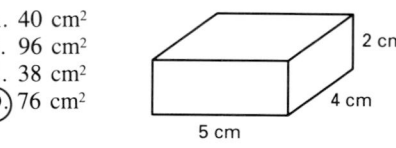

13. Simplify. $(11t^2 + 2) - (3t^2 + 2)$
    A. $8t^2$ B. $8t^4 + 4$
    C. $8t^4$ D. $14t + 4$

14. In a box of 1,000 plugs, 10 are defective. What is the probability of choosing a defective plug?
    A. 1   B. $\frac{1}{10}$
    C. $\frac{1}{100}$ D. $\frac{10}{990}$

Probability

When you flip a coin, there are two possible equally likely **outcomes,** H or T.
An **event** is an outcome or combination of outcomes. The probability of an event
is found using this formula:

$$P(E) = \frac{\text{number of favorable outcomes}}{\text{number of possible outcomes}}$$

The probability is usually written as a fraction in lowest terms.

▶ **Example 1**

Find the probability.
Use the spinner.
a. $P(1)$
b. $P(2 \text{ or } 6)$

Solution

a. $P(1) = \frac{1}{8}$ ← one favorable outcome
  ← total number of outcomes

b. $P(2 \text{ or } 6) = \frac{4}{8}$ ← one 2 and three 6's
  ← total number of outcomes
  $= \frac{1}{2}$

▶ **Example 2**

Find the probability of choosing an M from
a set of cards, with one letter from the word
MATHEMATICS written on each card. You
choose a card without looking.

Solution

$P(M) = \frac{2}{11}$ ← 2 M's
  ← 11 letters in all

EXERCISES

List the outcomes.

1. spinning the spinner in Example 1   **1, 2, 3, 4, 5, 6, 6, 6**

2. drawing a letter in Example 2   **M, A, T, H, E, M, A, T, I, C, S**

3. tossing a number cube   **1, 2, 3, 4, 5, 6**

4. choosing a digit at random from 55,238,175   **5, 5, 2, 3, 8, 1, 7, 5**

Find the probability. Use the spinner in Example 1.

5. $P(6)$ $\frac{3}{8}$ 6. $P(1 \text{ or } 2)$ $\frac{2}{8}$ 7. $P(\text{odd number})$ $\frac{3}{8}$

8. Find the probability that a student chosen at random from a class of
   8 girls and 12 boys is a girl. $\frac{2}{5}$

9. Find the probability of selecting a vowel if a letter is chosen at random
   from the word WINTER. $\frac{1}{3}$

10. A day of the week is chosen at random.

   a. Find the probability it has 6 letters. $\frac{3}{7}$

   b. Find the probability it has fewer than 6 letters. $\frac{4}{7}$

   c. Find the probability it has more than 6 letters. $\frac{0}{7}$

Buffon's Needle Problem

A man who lived more than 200 years ago, Count Buffon (1707–1788)
devised a way to estimate the value of π using probability.

Step 1 : Draw a series of parallel line
   segments 2 in. apart on a large
   piece of paper.

Step 2 : Cut 10 toothpicks to a length of 1 in.

Step 3 : Drop the 10 toothpicks on the ruled
   paper 10 separate times. Each time
   count the number of toothpicks that
   touch a line.

Step 4 : Find the probability of a toothpick
   touching a line. To find this
   probability, divide the total number
   that touch a line by the total number
   dropped.

Step 5 : To find an estimate of the value of π,
   take the reciprocal of the probability
   you found in Step 4. (Remember,
   the reciprocal of a fraction is found
   by interchanging the numerator and
   denominator of the fraction.) Use
   your calculator to find the reciprocal
   to the nearest thousandth. For the
   10 toothpicks above, the probability
   is $\frac{3}{10}$ that a toothpick lands on a line.
   Find the reciprocal of $\frac{3}{10}$ to the
   nearest thousandth.   **3.333**

Repeat the experiment 20 times and average your answers to Step 5.
What value do you get?
   Answers will vary.

Compare this value to 3.1415926, which is an approximate value of π. How close
are you? What is the result if several people pool their results and you average a
greater number of trials?
   Answers will vary; the estimate is closer to the value of π given.

Sample Spaces and Tree Diagrams

The set of all possible outcomes of an experiment is called a **sample
space.** One way to picture a sample space is to make a **tree diagram.**

▶ **Example 1**

Make a tree diagram to show the sample
space if a number cube is rolled and a coin
is tossed.

Solution

There are 6 outcomes for the number cube
(1, 2, 3, 4, 5, 6) and 2 outcomes for the
coin (H, T).

Number cube	Coin	Outcomes
1	H	1,H
	T	1,T
2	H	2,H
	T	2,T
3	H	3,H
	T	3,T
4	H	4,H
	T	4,T
5	H	5,H
	T	5,T
6	H	6,H
	T	6,T

There are twelve possible outcomes.

▶ **Example 2**

You have an equal chance of winning (W) or
losing (L) a game. Make a tree diagram to
show the outcomes when playing three
games. Find $P(\text{winning exactly 1 game})$.

Solution

There are two outcomes each time.

First game	Second game	Third game	Outcomes
W	W	W	W,W,W
		L	W,W,L
	L	W	W,L,W
		L	W,L,L
L	W	W	L,W,W
		L	L,W,L
	L	W	L,L,W
		L	L,L,L

There are eight possible outcomes. Three of
these outcomes involve winning *exactly* one
game.

$P(\text{winning exactly one game}) = \frac{3}{8}$

EXERCISES

Use a tree diagram to find the possible outcomes in the sample space.

1. Roll a number cube and choose a letter from the word AND.

   1,A; 2,A; 3,A; 4,A; 5,A; 6,A; 1,N; 2,N; 3,N; 4,N; 5,N; 6,N; 1,D; 2,D; 3,D; 4,D; 5,D; 6,D

Find each probability if you spin spinner 1 and spinner 2.

2. $P(2 \text{ and A})$ $\frac{1}{40}$

3. $P(3 \text{ and vowel})$ $\frac{1}{20}$

4. $P(\text{even number and consonant})$ $\frac{3}{10}$

5. Find $P(\text{sum of 8})$ if you roll two number cubes $\frac{5}{36}$

6. Find $P(\text{all heads or tails})$ if you toss 6 coins $\frac{1}{32}$

Pascal's Triangle

1. Make a tree diagram to show the possible outcomes if you toss a coin three times.

First Toss	Second Toss	Third Toss	Outcomes
T	T	T	TTT
		F	TTF
	F	T	TFT
		F	TFF
F	T	T	FTT
		F	FTF
	F	T	FFT
		F	FFF

2. Find $P(\text{all heads})$ $\frac{1}{8}$

3. Find $P(\text{exactly 2 heads})$ $\frac{3}{8}$

4. Find $P(\text{exactly 1 head})$ $\frac{3}{8}$

5. Find $P(\text{no heads})$ $\frac{1}{8}$

The triangular pattern of numbers below is part of an arrangement called
Pascal's Triangle:

				Sum of the numbers
1			Row 0	1
1 1			Row 1	2
1 2 1			Row 2	4
1 3 3 1			Row 3	8
1 4 6 4 1			Row 4	16

6. Look for a pattern. Can you complete row 5?   **1   5   10   10   5   1**

7. Compare row 3 of Pascal's triangle to your answers in Questions 2–5. What do
   you discover?
   **The numerators of the probabilities are numbers in the third row of Pascal's
   triangle. The denominator is the sum of the numbers in that row.**

Now consider four tosses of a coin. Use Pascal's triangle to find each probability.

8. $P(\text{all heads})$ $\frac{1}{16}$ 9. $P(\text{exactly 2 heads})$ $\frac{6}{16} = \frac{3}{8}$

10. Find $P(\text{exactly 3 heads in five tosses of a coin})$. $\frac{10}{32} = \frac{5}{16}$

Name _____ Date _____

Using the Counting Principle

You can use the counting principle to find the number of outcomes of an experiment. Just multiply the number of outcomes at each stage of the activity. Remember,

$P(E) = \frac{\text{number of favorable outcomes}}{\text{number of possible outcomes}}$

▶ **Example 1**

Veronica has a choice of four recording artists she likes, each on CD, record, or tape. How many choices does she have?

Solution

There are two stages: choosing an artist and choosing a format.

artist   format
4   ×   3   = 12

Veronica has 12 choices.

▶ **Example 2**

A player tosses a number cube and chooses one of 26 alphabet cards. Using the counting principle, find P(even number; E, F, or G).

Solution

Find the number of possible outcomes.

number cube   alphabet cards
6   ×   26   = 156

Find the number of favorable outcomes.

even number   letters
3   ×   3   = 9

$P(\text{even number; E, F, or G}) = \frac{9}{156} = \frac{3}{52}$

EXERCISES

Use the counting principle to find the number of possible outcomes.

1. making a lunch from 3 choices of soup, 5 choices of sandwich, and 4 choices of a drink
60

2. selecting one of 26 alphabet cards and then tossing a coin twice   **104**

3. tossing a coin 7 times   **2^7 or 128**   4. rolling a number cube 4 times   **6^4 or 1,296**

Use the counting principle to help find each probability.

5. A number cube is rolled 3 times. Find P(all three numbers even)   **$\frac{27}{216}$ or $\frac{1}{8}$**

6. A number cube is rolled and a coin tossed 2 times. Find P(even number and all heads) **$\frac{1}{8}$**

7. A coin is tossed 10 times. Find P(all heads or all tails).   **$\frac{1}{2^9}$ or $\frac{1}{512}$**

8. A card is selected from 26 alphabet cards and a number cube is rolled. Find P(vowel and prime number). Remember: 1 is not a prime number.   **$\frac{5}{52}$**

Name _____ Date _____

What Are Your Chances?

There are 200 residents living in Green Trees, which has a security gate. Each resident has a different code to use to open the gate.

| 1 2 3 4 5 |
| 6 7 8 9 0 |
| WELCOME TO |
| GREEN TREES |

1. How many different 2-digit codes are possible if all digits can be used?   **$10 \times 10 = 100$**

2. How many different 3-digit codes are possible if all digits can be used?   **$10^3 = 1,000$**

3. How many different 4-digit codes are possible if zero cannot be used as the first digit?   **$9 \times 10^3 = 9,000$**

4. How many different 5-digit codes are possible if zero cannot be used as the first digit?   **$9 \times 10^4 = 90,000$**

5. What is the probability of getting a 4-digit code with all digits the same if zero cannot be used as the first digit?   **$\frac{1}{1,000}$**

6. Suppose that, at Green Trees, it is decided to use 4-digit codes. However, no codes starting with zero are allowed, and no codes with the same 4 digits are allowed. The codes 1000, 2000, 3000, etc., are also eliminated. How many codes would there be?   **8,982**

7. Under the conditions given in Exercise 6, what is the probability of someone selecting a code at random and gaining entry?   **$\frac{200}{8,982} = \frac{100}{4,491} \approx 0.022$**

8. Would you recommend using 2-digit codes for Green Trees? Explain.
No; there are not enough codes.

9. Would 3-digit codes have worked for Green Trees? Explain.
No; there is too great a chance of picking a code at random.

10. Why would 10-digit codes make Green Trees more secure? What are the drawbacks of 10-digit codes?
There is not much chance of getting a code at random. Codes would be hard to remember.

11. Suppose Green Trees grows to 1,500 residents. What code system would you recommend? What codes would you eliminate? Design a system.
Answers will vary.

Name _____ Date _____

Independent and Dependent Events

Two events are said to be **independent** if the result of the second event does not depend on the result of the first event. If the result of the first event does affect the result of the second, the two events are said to be **dependent**.

▶ **Example 1**

Eleven cards, each having one of the letters of *MATHEMATICS*, are placed in a bowl. A letter card is selected at random from the bowl. It is replaced and once again, a letter is selected. Find $P(M, \text{then } M)$.

Solution

Since the first letter is replaced, there will be eleven cards in the bowl for both selections. The two events are independent.

$P(M, \text{then } M) = \frac{2}{11} \times \frac{2}{11} = \frac{4}{121}$

▶ **Example 2**

From the bowl with the eleven cards, each having one of the letters of *MATHEMATICS*, a letter card is selected at random. Then a second card is selected at random without the first card being replaced. Find $P(M, \text{then } M)$.

Solution

You know that the probability of selecting an M is $\frac{2}{11}$. Once that M is withdrawn, however, there are only 10 letters left. Only one of those letters is an M. So, the probability of selecting an M as the second letter is $\frac{1}{10}$.

$P(M, \text{then } M) = \frac{2}{11} \times \frac{1}{10} = \frac{2}{110} = \frac{1}{55}$

EXERCISES

A bag contains 8 red marbles, 3 blue marbles, and 5 white marbles. Find each probability if the first marble is replaced before the second one is drawn.

1. P(red, red)   **$\frac{4}{...}$**

2. P(red, blue)   **$\frac{3}{32}$**

3. P(red, green)   **0**

4. P(red, blue or white)   **$\frac{1}{4}$**

Find each probability if the first marble is not replaced before the second one is drawn.

5. P(white, then red)   **$\frac{1}{6}$**

6. P(red, then red)   **$\frac{15}{...}$**

7. P(black, then red)   **0**

8. P(red, then blue or white)   **$\frac{4}{15}$**

Ten cards are numbered from 1 through 10 and placed in a box. Two cards are selected at random, one after another, without returning the first card to the box. Find each probability.

9. P(2, then 3)   **$\frac{1}{90}$**

10. P(3, then odd)   **$\frac{2}{45}$**

11. P(prime, then composite)   **$\frac{8}{45}$**

12. P(even, then prime)   **$\frac{1}{6}$**

Name _____ Date _____

Permutations

An arrangement of objects *in a particular order* is called a **permutation**. To find the number of permutations of any group of objects you can use the counting principle.

▶ **Example 1**

Find the number of permutations if 5 drivers are given a starting position number in order from the inside to the outside of the track.

Solution

Count number of possible choices for each position and then multiply. In a permutation, as each position is filled, the number of drivers available for the next position decreases by 1.

Position 1   Position 2   Position 3   Position 4   Position 5
5   ×   4   ×   3   ×   2   ×   1   = 120

There are 120 possible permutations of drivers in the 5 starting positions.

Suppose that you want to consider the possible arrangements of only part of a group of objects.

▶ **Example 2**

Find the number of permutations for 2 horses out of 6 horses in the first two starting positions.

Solution

Multiply the number of choices for position 1 by the number of choices for position 2.

Position 1   Position 2
6   ×   5   = 30

There are 30 possible permutations for 2 out of 6 horses in the first two starting positions.

EXERCISES

Find the number of permutations possible.

1. 3 runners on a 3-person relay team   **6 permutations**

2. 6 horses in 6 starting positions for a race   **720 permutations**

3. 8 cars in 8 parking stalls in a garage   **40,320 permutations**

Find the number of permutations possible for the part of the group named.

4. One delegate to a convention and a first alternate, chosen from a club of 15 students   **$15 \times 14 = 210$ permutations**

5. 3 out of 10 runners for a 3-person relay team   **$10 \times 9 \times 8 = 720$ permutations**

6. 3 songs out of 8 for the first 3 pieces on a concert program   **$8 \times 7 \times 6 = 336$ permutations**

502B

Name _____ Date _____

Experimental Probability

The probability of an event based on the results of an experiment is called **experimental probability**. For example, if 100 toys are randomly selected from a bin and checked and 2 are found to be defective, the experimental probability of finding a defective toy is $\frac{2}{100} = \frac{1}{50}$.

► **Example**

A spinner divided into 8 equal parts is spun 50 times. The outcomes are shown in the table at the right. Find each experimental probability.

Number on spinner	1	2	3	4	5	6	7	8
Number of outcomes	5	8	6	7	1	9	6	8

a. $P(1)$ b. $P(\text{even number})$

Solution

a. $P(E) = \frac{\text{number of favorable outcomes}}{\text{number of possible outcomes}}$

$P(1) = \frac{5}{50} = \frac{1}{10}$

b. $P(\text{even number}) = \frac{8 + 7 + 9 + 8}{50}$

$= \frac{32}{50} = \frac{16}{25}$

EXERCISES

Rob selected a disk from a bag, recorded the color, and returned the disk to the bag. He repeated the experiment 60 times. Use the results in the table to find each probability.

Outcome	Tally	Total
red	卌 卌 //	12
blue	卌 卌 卌 /	16
green	卌 卌 卌	15
white	卌 卌 卌 //	17

1. $P(\text{red})$ $\frac{5}{}$

2. $P(\text{green})$ $\frac{1}{4}$

3. $P(\text{blue})$ $\frac{4}{15}$

4. $P(\text{white})$ $\frac{17}{60}$

5. $P(\text{red or blue})$ $\frac{7}{15}$

6. $P(\text{green or red})$ $\frac{9}{20}$

7. $P(\text{not red})$ $\frac{4}{5}$

8. $P(\text{not white})$ $\frac{43}{60}$

9. One hundred flashlights were selected at random from a group for sale. Of these, 2 did not work. Find the experimental probability of selecting a flashlight that works. $\frac{49}{50}$

10. On a production line, 14 bottles of carbonated water were found to be over- or underfilled out of 2,000 bottles selected randomly. What is the experimental probability of a bottle not being filled properly? $\frac{7}{1,000}$

Name _____ Date _____

Using Experimental Probability

1. Two number cubes are rolled. If the product of the two numbers is even, you win a point. If the product is odd, you lose a point.

   a. Play the game 50 times and record the results.

Outcome	Tally	Total
Even product		
Odd product		

   b. What is the experimental probability that you will win a point? **Answers will vary.**

   c. What is the experimental probability that you will lose a point? **Answers will vary.**

   d. Does the game appear to be fair? (That is, is there an equal chance you will win or lose a point?) Why? **No**

   e. List the possible outcomes of tossing 2 cubes and finding the products.

   1,2,3,4,5,6; 2,4,6,8,10,12; 3,6,9,12,15,18; 4,8,12,16,20,24; 5,10,15,20,25,30; 6,12,18,24,30,36

   f. Find $P(\text{even})$ $\frac{27}{36} = \frac{3}{4}$ Find $P(\text{odd})$ $\frac{9}{36} = \frac{1}{4}$

   g. Based on your answer to part f, is the game fair? Why?

   No; more even outcomes than odd outcomes are possible.

   h. If you rolled 2 cubes 72 times, how many points would you expect to have at the end, based on the experimental probability? **Answers will vary.**

   i. Based on your answer to part f, how many points would you expect to have after rolling the cubes 72 times? **54 − 18 = 36**

Work with a partner to design and conduct an experiment to find each experimental probability. **Check students' experiments.**

2. that a word in a newspaper has more than four letters

_____

_____

3. that a student chosen at random likes frozen vanilla yogurt

_____

_____

Name _____ Date _____

Problem Solving Skills: Using Samples to Make Predictions

A random sample is a group chosen from a large population in such a way that every member of the population has an equal chance of being chosen. If you use a random sample to find an experimental probability, you can use that probability to make predictions about the population as a whole. To do so, use the following formula:

probability × number in total population = predicted number

► **Example**

Seth and Marla conducted a poll of 50 students selected randomly out of a total of 500 students to see which of the three candidates for president of the student council each student was planning to vote for: Beth Sundfeld, Kim Chung, or Gabriel Melendez. The results of the poll are shown in the table. Predict how many students in the school will vote for Gabriel.

Candidate	Number of Voters Planning to Vote
Beth	17
Kim	15
Gabriel	18

Solution

First, find $P(\text{Gabriel})$. $P(\text{Gabriel}) = \frac{18}{50} = \frac{9}{25}$

Then use the formula to predict the number of all 500 students who will vote for Gabriel.

$P(\text{Gabriel}) \times$ Number of students in all = Predicted number who will vote for Gabriel

$\frac{9}{25} \times 500 = 180$

So, 180 students out of a total of 500 will vote for Gabriel.

EXERCISES

A local radio station asked a random sample of 1,000 out of the 250,000 residents of a city to find out how many people preferred one of four types of music.

Type of music	Number who listen
popular	521
big band	188
classical	207
other	84

1. Find $P(\text{classical})$ $\frac{207}{1,000}$

2. Find $P(\text{popular})$ $\frac{521}{1,000}$

3. Find $P(\text{big band})$ $\frac{47}{250}$

4. Find $P(\text{other})$ $\frac{21}{250}$

If all the people in the city were polled, how many would you expect to select each type of music?

5. Popular **130,250**

6. Big band **47,000**

7. Classical **51,750**

8. Other **21,000**

9. Frank took a random poll of 25 classmates and found that 15 intended to vote for Sarah in the upcoming class election. If 280 students are eligible to vote, about how many votes can Sarah expect? **168 votes**

Name _____ Date _____

Making Predictions

Your class wants to sell caps to raise money for a local charity. The caps come in three colors: red, white, and black. There are 1,800 students in your school. You want to conduct a marketing survey to help decide how many caps to order.

1. You survey 100 students. Of those, 20 indicate they would buy a black cap, 18 indicate they would buy a red cap, and 11 indicate they would buy a white cap. Make a table to show the survey results.

Cap color	No. who would buy each cap
Red	**18**
White	**11**
Black	**20**

2. You can use a formula to predict sales.

$\frac{\text{number of students willing to buy a cap}}{\text{number of students surveyed}} = \frac{\text{expected sales}}{\text{total number of students}}$

   a. Use the formula to predict the number of black caps that will be sold. **360 black caps**

   b. Use the formula to predict the number of red caps that will be sold. **324 red caps**

   c. Use the formula to predict the number of white caps that will be sold. **198 white caps**

3. You do not want any extra caps left unsold. What percent of the predicted number of cap sales do you think you will actually sell? **Answers will vary.**

4. At the percent you selected in Exercise 3, how many caps of each color should you order? **Answers will vary.**

   Black _____ Red _____ White _____

5. Suppose that you are given a bag containing a large number of marbles, the exact number of which is unknown. How could you use random sampling to predict how many marbles in all are in the bag? (Hint: Select as few as 10 marbles from the bag and mark them in some way.)

   Answers will vary. Sample plan: After marking the 10 marbles, return them to the bag. Mix the marbles up thoroughly. Then select randomly a number of marbles, say 20, from the bag, one at a time. Note how many of those 20 are marked. Use this formula, in which n equals the number of marbles in the bag:
   $\frac{10 \text{ marked marbles}}{n} = \frac{\text{number of marked marbles drawn}}{20}$

Problem Solving Strategy: Simulate the Experiment

Some probability problems can be solved by **simulating** the problem, or relating it to a simpler problem that is similar. Types of simulation include tossing a coin, spinning a spinner, and tossing a number cube.

▶ **Example**

1. Read the problem. Ask yourself, What information is given? What must I find?

Estimate the probability that in a family of 3 children, 2 are boys.

Solution

2. Set up a simulation that will fit the problem.

Let H represent a boy and T represent a girl. Toss 3 coins and record the outcomes.

3. Perform the simulation and record the results.

	Boys (H)	Girls (T)
1st toss	3	0
2nd toss	1	2
3rd toss	2	1
4th toss	2	1
5th toss	1	2

4. Solve the problem.

On the first 5 tosses, heads showed on 2 coins 2 times. Based on these results, the probability that 2 are boys in a family of 3 children is
$$P(2 \text{ boys}) = \frac{\text{no. of favorable outcomes}}{\text{no. of possible outcomes}} = \frac{2}{5}$$

5. Check.

A more accurate estimate may be obtained by continuing the simulation for 100 tosses.

EXERCISES

Use a simulation to solve.

1. If you take a *true-false* test and guess at every answer, what is the probability that you will get at least 7 out of 10 correct?
   Answers will vary.

2. If 5 out of every 6 customers entering a store are female, what is the probability that the next 5 customers entering the store will be female?
   Answers will vary.

3. You find that 3 out of 5 people who ride the #34 bus are male. What is the probability that, of the next 10 people who board the bus, 6 will be male?
   Answers will vary.

4. Suppose you find that 8 out of every 10 people leaving a grocery store carry exactly two bags of groceries. What is the probability that the next 5 customers leaving the store, 4 will be carrying two bags of groceries?
   Answers will vary.

Using Random Numbers in Simulations

Random numbers can be used to simulate experiments. A random number table can be generated by a computer, found in a book of random number tables, or created in other ways. One way to create a table of random numbers is to place 10 cards in a box, each containing a digit from 0 through 9. Draw a card, record the digit, and return the card to the box. Mix the cards and repeat the process. Such a table is shown here.

```
5 0 8 4 8 2 8 5 6 1 1 3 5 0 9 9 6 7 9 7 4 3 2 9 2
2 1 5 8 6 5 7 9 3 9 7 9 6 5 9 9 7 8 4 0 7 4 6 2 3
2 4 5 1 7 2 5 8 8 1 4 7 3 3 3 8 5 6 7 5 9 5 1 3 5
5 7 7 3 0 5 3 7 0 5 0 5 9 9 0 9 1 8 5 3 3 2 9 3 5
5 6 3 2 0 3 3 8 9 2 8 2 1 9 6 8 0 0 7 5 4 1 0 1 3
3 6 2 8 8 6 2 9 3 6 9 5 5 6 9 7 8 8 6 5 7 4 5 7 2
8 9 9 8 5 6 8 6 6 3 1 1 1 0 1 1 9 5 4 2 0 2 1 4 0
8 2 5 3 8 0 7 0 2 4 5 7 6 8 8 4 6 0 5 7 6 6 9 5 9
```

The Rockets' basketball coach finds that one of the players on the varsity squad is making 4 out of every 10 baskets attempted. What are the chances that the player will make 4 or more baskets in a row?

Use 150 random digits as the basis for a simulation. Think of the question as follows: Out of 150 attempts, how many times will the player make 4 or more baskets in a row?

1. There are 10 digits from 0 to 9. How many of the digits should you use to represent getting a basket? ___**4**___

2. To do a simulation, from the digits 0 through 9, choose four, each of which represents a basket. (Select 0, 1, 2, 3.) In the first 150 random digits, circle the occurrence of any of these digits. The first few have been circled for you.

```
5 0 8 4 8 2 8 5 6 1 1 3 5 0 9 9 6 7 9 7 4 3 2 9 2
2 1 5 8 6 5 7 9 3 9 7 9 6 5 9 9 7 8 4 0 7 4 6 2 3
2 4 5 1 7 2 5 8 8 1 4 7 3 3 3 8 5 6 7 5 9 5 1 3 5
5 7 7 3 0 5 3 7 0 5 0 5 9 9 0 9 1 8 5 3 3 2 9 3 5
5 6 3 2 0 3 3 8 9 2 8 2 1 9 6 8 0 0 7 5 4 1 0 1 3
3 6 2 8 8 6 2 9 3 6 9 5 5 6 9 7 8 8 6 5 7 4 5 7 2
```

3. What is the longest string of consecutive baskets that seems likely? ___**5**___

4. Out of 150 attempts, how many times will the player get 4 or more baskets in a row? ___**2**___

5. With this simulation, what is the probability that the player will make a basket? ___$\frac{54}{150} = \frac{9}{25}$___

6. Describe a simulation that you could use to determine the longest string of baskets likely for a player making 25% of his or her shots.
   Answers will vary. Sample simulation: Use eight digits, 0–7, and let 7 and 6
   represent baskets (= 25%)

Computer Activity: Probability à la Pascal

When finding the probability for outcomes of coin tosses, there are some interesting patterns that may help you.

Number of coins	Possible Outcomes and (How Many Ways Outcome Can Occur)
1	1H (1) 1T (1)
	2H (1) 1H1T (2) 2T (1)
	3H (1) 2H1T (3) 1H2T (3) 3T (1)
	4H (1) 3H1T (4) 2H2T (6) 1H3T (4) 4T (1)

The numbers that represent how many ways a specific combination can occur can be arranged to form a triangular array of numbers. It is called **Pascal's triangle**, after the French mathematician Blaise Pascal (1623–1662).

row 0	1
row 1	1 1
row 2	1 2 1
row 3	1 3 3 1
row 4	1 4 6 4 1

Pascal's triangle can be used to determine probabilities. For example, to find the probability of getting exactly 2 heads when 3 coins are tossed: Add the numbers in row 3 to get the number of possible outcomes, 8. The second term in row 3 gives the number of favorable outcomes, 3. $P(2H) = \frac{3}{8}$

This program will generate rows of Pascal's triangle. Remember that the row number corresponds to the number of coins tossed. Use the program for the Exercises.

```
10   INPUT "HOW MANY ROWS? ";R: PRINT
20   N = 1: PRINT "ROW 0"; TAB( 20),N
30   FOR M = 1 to R
40   X = M
50   PRINT "ROW ";X; TAB( 20 − X);
60   FOR Y = 1 TO X + 1
70   PRINT N;" ";
80   X = X − 1
90   N = N * (X + 1) / Y
100  NEXT Y
110  PRINT :N = 1
120  NEXT M
```

EXERCISES

1. 5 coins are tossed. Find each of the following probabilities.
   a. $P(4H)$ ___$\frac{5}{32}$___   b. $P(3T)$ ___$\frac{10}{32}$___   c. $P(2T)$ ___$\frac{10}{32}$___   d. $P(5H)$ ___$\frac{1}{32}$___

2. 8 coins are tossed. Find each of the following probabilities.
   a. $P(5H)$ ___$\frac{56}{256}$___   b. $P(6T)$ ___$\frac{28}{256}$___   c. $P(3T)$ ___$\frac{56}{256}$___   d. $P(4H)$ ___$\frac{70}{256}$___

3. State three patterns that you find in the triangle.
   One diagonal is comprised of counting numbers. One diagonal is comprised of
   triangular numbers. The sum of the numbers in row n is 2^n. Answers may vary.

Calculator Activity: Permutations

You can use the counting principle and a calculator to find the number of possible arrangements, or **permutations,** of a group of objects in a given order.

▶ **Example 1**

In how many ways can 4 people line up from left to right in a straight line?

Solution

Any one of the 4 people can go in the first position. That leaves 3 choices for the second position. Once one of these 3 has taken the second position from the left, that leaves 2 possible choices for the third position, and 1 for the last position. The number of possible permutations of 4 people in a given order is

Position 1	Position 2	Position 3	Position 4	Possible Permutations
4 ×	3 ×	2 ×	1 =	24

The pattern of factors above, $4 \times 3 \times 2 \times 1$ is called 4!, or "four factorial." Factorials can be easily found on a scientific calculator. Read your manual to find the technique used on your calculator. One possible way is shown here.

4 [x!]

There are 24 different ways in which the four people can be lined up for the photograph.

▶ **Example 2**

In how many ways can 4 out of 10 people be seated in 4 chairs?

Solution

Any of 10 people can fill the first chair. Nine people are available for the second chair. Only eight are available to fill the third, and seven, to fill the fourth chair. Use your calculator to find this product.

Position 1	Position 2	Position 3	Position 4	Possible Permutations
10 ×	9 ×	8 ×	7 =	5,040

There are 5,040 possible ways in which 10 people could be seated in 4 chairs.

EXERCISES

Use a calculator to find the number of different arrangements that can be made.

1. 10 flags on a flagpole
   3,628,800

2. 6 books on a shelf
   720

3. naming $\triangle ABC$
   6

4. the letters in the word VICTORY
   5040

5. choosing a pitcher and a catcher from 10 players who can play either position
   90

6. choosing a president, vice president, and secretary from a class of 30 students
   24,360

502D

ACHIEVEMENT TEST

Probability
CHAPTER 13 · FORM A

Name _____

Date _____

MATH MATTERS BOOK 1
Chicha Lynch
Eugene Olmstead

SOUTH-WESTERN PUBLISHING CO.

SCORING RECORD	
Possible	Earned
14	

Find each probability. Use this spinner.

1. P(red) $\frac{1}{3}$
2. P(1 or 4) $\frac{1}{3}$
3. P(not green) $\frac{2}{3}$
4. P(8) 0
5. P(red, white, green) 1

Use a tree diagram to find the possible outcomes in each sample space.

6. tossing a quarter and then spinning a spinner such as the one at the right

   8: H1, H2, H3, H4; T1, T2, T3, T4

7. tossing a number cube and tossing a penny

   12: 1H, 2H, 3H, 4H, 5H, 6H; 1T, 2T, 3T, 4T, 5T, 6T

Use the counting principle to find the number of possible outcomes.

8. choosing a sandwich from 4 kinds of bread and 4 sandwich fillings **15**
9. tossing a coin 7 times **14**
10. choosing a pasta lunch from macaroni, spaghetti, or ravioli; tomato sauce, cream sauce, or garlic and oil; a parmesan or a romano cheese topping **28**
11. A bag contains 4 white marbles, 3 red marbles, and 5 blue marbles. Miko takes one marble from the bag, replaces it, and then takes another. Find P(red, then white). $\frac{1}{12}$

Copyright © 1993 by South-Western Publishing Co.
MI01AG

13A-1

Name _____

Date _____

A number cube is tossed 30 times. The outcomes are recorded in this table.

Number on cube	1	2	3	4	5	6
Number of outcomes	4	5	5	4	5	7

12. Find P(4) $\frac{2}{15}$
13. Find P(all even numbers) $\frac{8}{15}$
14. Find P(4 or 5) $\frac{3}{5}$

13A-2

ACHIEVEMENT TEST

Probability
CHAPTER 13 · FORM B

Name _____

Date _____

MATH MATTERS BOOK 1
Chicha Lynch
Eugene Olmstead

SOUTH-WESTERN PUBLISHING CO.

SCORING RECORD	
Possible	Earned
14	

Find each probability. Use this spinner.

1. P(red) $\frac{1}{4}$
2. P(1 or 4) $\frac{1}{4}$
3. P(not blue) $\frac{6}{8}$
4. P(10) 0
5. P(red, blue, green, white) 1

Use a tree diagram to find the possible outcomes in each sample space.

6. tossing a number cube and tossing a nickel

   8: H1, H2, H3, H4; T1, T2, T3, T4

7. tossing a dime and then spinning a spinner such as the one at the right

   12: 1H, 2H, 3H, 4H, 5H, 6H; 1T, 2T, 3T, 4T, 5T, 6T

Use the counting principle to find the number of possible outcomes.

8. choosing a sandwich from 3 kinds of bread and 3 sandwich fillings **9**
9. tossing a coin 10 times **20**
10. choosing a lunch consisting of a sandwich made from either salami, bologna, or boiled ham; either onion soup, chicken noodle soup, or vegetable soup; and either milk, juice, or iced tea **27**
11. A bag contains 2 green marbles, 5 red marbles, and 5 blue marbles. Miko takes one marble from the bag, replaces it, and then takes another. Find P(green, then green). $\frac{1}{36}$

Copyright © 1993 by South-Western Publishing Co.
MI01AG

13B-1

Name _____

Date _____

A number cube is tossed 40 times. The outcomes are recorded in this table.

Number on cube	1	2	3	4	5	6
Number of outcomes	9	8	7	6	6	4

12. Find P(4) $\frac{3}{20}$
13. Find P(all even numbers) $\frac{9}{20}$
14. Find P(5 or 6) $\frac{1}{4}$

13B-2

Teacher's Notes

CHAPTER 14 SKILLS PREVIEW

Use the figure at the right for Exercises 1–2.

1. Give the coordinates of points X and Y.
   X(–3, –2); Y(0, 3)

2. Find the slope of line segment AB. $\frac{4}{5}$

3. From their house, Jane and Bob drove 4 blocks north and 13 blocks east to shop at the supermarket. After that, they drove 5 blocks south and 9 blocks west to deliver a birthday present to Arthur. At that point, how many blocks were they from home?
   5 blocks

4. Graph the points $L(3, -1)$, $M(-2, 0)$, and $N(2, -2)$. **See Additional Answers.**

5. Graph the equation $y = x + 5$. **See Additional Answers.**

6. Complete the table of solutions for the equation $y = x + 6$.

$y = x + 6$

x	y	x, y	
−2	■	−2, ■	4; 4
0	■	0, ■	6; 6
2	■	2, ■	8; 8

7. At a parking lot, there is a fixed fee for using the lot in addition to the hourly rate charged for parking. Find the fixed fee and the hourly rate.

Total Amount Paid for Parking Car	$5.50	$8.00	$13.00
Number of Hours Car Parked	1	2	4

Fee: $3.00; Hourly rate: $2.50

Use the figure at the right for Exercises 8–10.

8. Graph the image of $\triangle DEF$ under a translation 2 units to the right and 1 unit up. **See Additional Answers.**

9. Graph the image of trapezoid $WXYZ$ under a reflection across the x-axis. **See Additional Answers.**

10. Graph the image of $\triangle DEF$ after a 90° turn clockwise about the origin. **See Additional Answers.**

Use the figure at the right for Exercises 11–12.

11. How many lines of symmetry does the figure on the left have? **2**

12. What is the order of rotational symmetry for the figure on the right? **2**

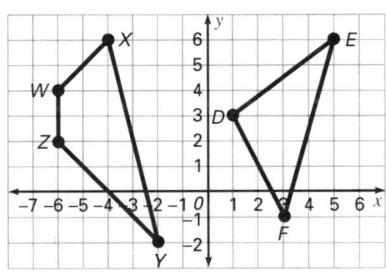

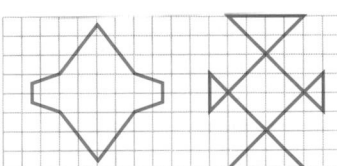

14 SKILLS PREVIEW

Introduction The purpose of this Skills Preview is to assess students' abilities on all the major objectives of Chapter 14. Test results may be used

- to determine those topics which may need only to be reviewed and those topics which need to be more carefully developed;
- for class placement;
- in prescribing for individual differences.

If you prefer, you may use the Skills Preview as an alternative form of the Chapter Test (page 540) to evaluate mastery of chapter objectives. The items on the Skills Preview and the Chapter Test correspond in content and level of difficulty.

Additional Answers
See page 584.

OVERVIEW

This chapter introduces students to the Cartesian coordinate system and deals with graphing points, linear equations, and finding slopes and distances. Students learn the relationship between the preimage and the image of an object under a reflection, a rotation, and a transformation both on and off the coordinate plane.

SPECIAL CONCERNS

Some students may have trouble understanding the correspondence between an equation and a graph, while other students may find it difficult to follow how one variable is related to another. Work individually with these students to construct as much visual support (tables, diagrams) as possible.

VOCABULARY

coordinate plane	order of rotational	rotation	transformation
function	symmetry	rotational	translation
graph of an equation	ordered pairs	symmetry	x-axis
graph of a function	origin	run	x-coordinate
image	preimage	slide	y-axis
linear equation	quadrants	slope	y-coordinate
line of symmetry	reflection	slope-intercept	y-intercept of
model	rise	form	a line

MATERIALS

algebra tiles	graphing calculators
calculators	mirrors
dot graph	rulers
graph paper	tracing paper
	wood blocks

BULLETIN BOARD

Have students use construction paper to cut out congruent block versions of the letters C, E, L, and T. Then have them make tessellation (tiling) designs with these letters using transformations, reflections, and rotations. Display these designs on a bulletin board. Have students tell which move(s) (slide, flip, turn) was used to make each tessellation design.

INTEGRATED UNIT 4

The skills and concepts involved in Chapters 12–14 are included within the special Integrated Unit 4 entitled "Fast Forward." This unit is in the Teacher's Edition beginning on page 545K. Worksheets for this integrated unit appear in the Enrichment Activities booklet, pages 172–174.

TECHNOLOGY CONNECTIONS

- Calculator Worksheets, 281–282
- Computer Worksheets, 283–284
- MicroExam, Apple Version
- MicroExam, IBM Version

- *Graph Wiz, Algebra Xpresser,* William K. Bradford
- *Milo,* Paracomp

TECHNOLOGY NOTES

Computer graphing programs, such as those listed in Technology Connections, allow students to input an expression or equation and have the computer graph it. Such programs enable students to readily generalize about how an equation relates to the form of its graph. They allow students to see the relationships that exist between given equations and their slopes and *y*-intercepts.

GEOMETRY OF POSITION

PLANNING GUIDE

Sections	TEXT PAGES	ASSIGNMENTS		
		BASIC	AVERAGE	ENRICHED
Chapter Opener/Decision Making	504–505			
14–1 The Coordinate Plane	506–509	1–16, 19, 22–23, 27	9–20, 22–25, 27–28	9–21, 22–26, 27–28
14–2 Problem Solving Skills: Using a Grid to Solve Distance	510–511	1–4, 6	1–7	1–7
14–3 Graphing Linear Equations	512–515	1--10, 13, 14	1–10, 11–13, 14	1–10, 11–13, 14
14–4 Working with Slope: Problem Solving Applications	516–519	1–6, 8–9, 14, PSA 1–8	1–7, 8–12, 13–14, PSA 1–8	1–7, 8–12, 13–14, PSA 1–8
14–5 Problem Solving Strategy: Make a Model	520–521	1–5	1–5	1–5
14–6 Translations	522–525	1–8, 9–11, 16	4–8, 9–15, 16–17	4–8, 9–15, 16–19
14–7 Reflections: Problem Solving Applications	526–529	1–10, 11–12, 15, PSA 1–3	7–10, 11–13, 15–16, PSA 1–4	7–10, 11–14, 15–17, PSA 1–4
14–8 Rotations	530–533	1–9, 10–12, 17	4–9, 10–16, 17–20	7–9, 10–16, 17–22
14–9 Working with Symmetry	534–537	1–12, 13–15, 18	7–12, 13–17, 18–19	9–12, 13–17, 18–20
Technology	513, 517, 519, 523	✔	✔	✔

ASSESSMENT

ASSESSMENT				
Skills Preview	503	All	All	All
Chapter Review	538–539	All	All	All
Chapter Test	540	All	All	All
Cumulative Review	541	All	All	All
Cumulative Test	543	All	All	All

ADDITIONAL RESOURCES			
RETEACHING	ENRICHMENT	TECHNOLOGY	TRANSPARENCY
14–1	14–1		TM 59
14–2	14–2		
14–3	14–3		TM 60
14–4	14–4		
14–5	14–5		
14–6	14–6		TM 61
14–7	14–7		TM 62
14–8	14–8	14–8	TM 63, 64
14–9	14–9		

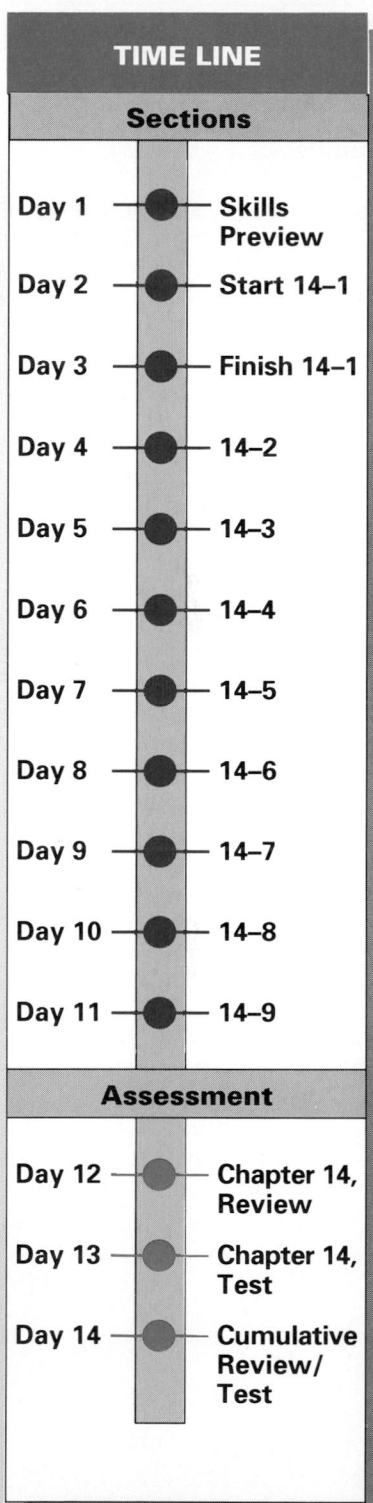

TIME LINE

Sections

Day 1 — Skills Preview
Day 2 — Start 14–1
Day 3 — Finish 14–1
Day 4 — 14–2
Day 5 — 14–3
Day 6 — 14–4
Day 7 — 14–5
Day 8 — 14–6
Day 9 — 14–7
Day 10 — 14–8
Day 11 — 14–9

Assessment

Day 12 — Chapter 14, Review
Day 13 — Chapter 14, Test
Day 14 — Cumulative Review/Test

ASSESSMENT OPTIONS	
Chapter 14, Test Forms A and B	
Chapter 14, Test	Text, 540
Alternative Assessment	TAE, 540
Chapter 14, MicroExam	

GEOMETRY OF POSITION

THEME Patterns

Objective To explore how patterns can be used to represent numbers.

Introduction After reading and discussing the introductory material, have students copy the figures shown on graph paper and count the dots in each figure. Then have them find the differences between the figures. Ask, *What pattern do you observe?* Students should observe that the difference between consecutive figures increases by 1.

Decision Making
Using Data Discuss the chart of numbers and then ask, *What patterns do you observe after analyzing the data?* Elicit from students that the differences between the square numbers are: 3, 5, 7, 9, . . . ; the differences between the hexagonal numbers are 4, 9, 13, 17. After students complete Exercise 3 ask, *How did the data in the chart help you answer the questions?* (The data is organized by the number of dots representing each number.)

Working Together Have students share their drawings. Have students explain what reasoning or systematic methods they used to make each drawing.

If you and a friend who lives on the opposite side of town planned to jog together, you would probably decide on a specific place to meet, such as the corner of Fifth and Main Streets. Because the meeting spot is at the intersection of two streets, you both would know exactly where to go. In this chapter, you will learn how to locate a specific point on a coordinate grid, using ordered pairs of numbers. These numbers are similar to the coordinates you and your friend agreed on: Fifth and Main. You will look for patterns among ordered pairs, and you will learn how to describe these patterns as mathematical functions.

Patterns can be found in the numbers of dots that make up polygons of increasingly larger sizes. These dot patterns represent numbers called **figurate numbers.** A triangular number is one type of figurate number. The first four triangular numbers are represented by these dot patterns:

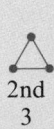

triangular number →	1st	2nd	3rd	4th
number of dots →	1	3	6	10

One way to find the number of dots in a triangular number is to use the dot pattern. Another way is to use algebra. The number of dots in the nth triangular number is represented by the expression $\frac{n(n + 1)}{2}$.

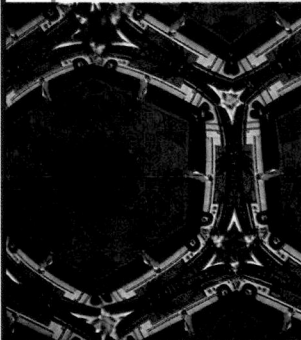

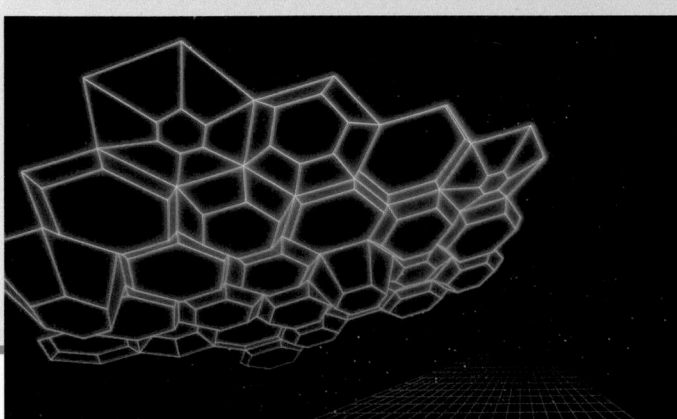

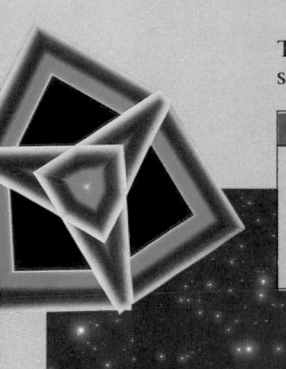

The chart gives the numbers of dots for the first five triangular, square, rectangular, pentagonal, and hexagonal numbers.

	1st	2nd	3rd	4th	5th
triangular	1	3	6	10	15
square	1	4	9	16	25
rectangular	2	6	12	20	30
pentagonal	1	5	12	22	35
hexagonal	1	6	15	28	45

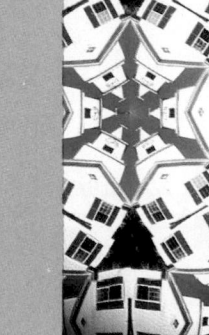

DECISION MAKING

Using Data

1. Use the expression that represents the nth triangular number to find the 6th triangular number. Then draw a dot picture of the 6th triangular number. 21

2. Guess the 7th triangular number. Then use the expression to check your guess. 28

3. Which pentagonal number is represented by the drawing at the right? 3rd

Working Together

Make drawings of the 1st through 10th square numbers and the 1st through 10th pentagonal numbers. Check students' drawings. The 6th through 10th square and pentagonal numbers are as follows. Square: 36, 49, 64, 81, 100; Pentagonal: 51, 70, 92, 117, 145

WARM-UP

Tell if the number is to the left or right of zero on a number line.
1. 7 **right**   2. −9 **left**
3. −23 **left**   4. 1.3 **right**

1 MOTIVATE

Explore Explain to students that the letters and numbers on a map mark off areas in which a city, town, or street can be found. Name some of the cities shown on the map and have students use the letters and numbers to tell the location of each. Repeat this activity to make sure students understand the procedure for locating a place on a map.

2 TEACH

Use the Pages/Skills Development
Have students read this part of the section and then discuss the examples. Draw a coordinate plane on the chalkboard. Then ask for volunteers to come to the chalkboard and label the quadrants, origin, and axes on the coordinate plane.

14-1   The Coordinate Plane

EXPLORE

On a map, a location is specified with a letter followed by a number. For example, on the map at the left, Topeka, Kansas, is located in the box that is both above the letter C and across from the number 2, or region C2.

a. Which city is located in region C4? **Waterloo, Iowa**

b. In which region is Omaha, Nebraska, located? **C3**

c. Between which two regions would you travel if you traveled from Duluth, Minnesota, to the southwest corner of Kansas? **C6 and A1**

d. If you were to travel from St. Louis, Missouri, to Fargo, North Dakota, in which region would your trip begin? In which region would it end? **D2, B6**

SKILLS DEVELOPMENT

The system used to identify locations on a map is much like the mathematical system called a coordinate plane. On a **coordinate plane,** two number lines are drawn perpendicular to each other. The horizontal number line is called the **x-axis.** The vertical number line is called the **y-axis.** The axes separate the plane into four regions called **quadrants.** The point where the axes cross is called the **origin.**

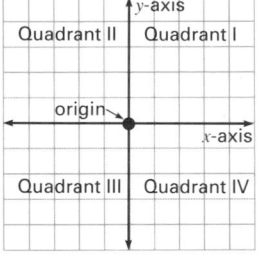
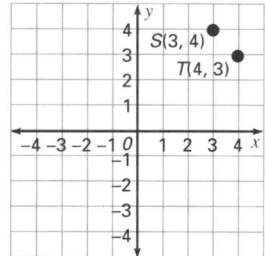

READING MATH

How can a familiar word like *quadruplet* help you remember the meaning of the word *quadrant*?

The prefix *quad-* means four.

In the figure at the right above, point *S* is 3 units to the right of the origin and 4 units up from the origin. So the point is identified as *S*(3, 4), where 3 is the **x-coordinate** of the point and 4 is the **y-coordinate.** The order of the coordinates is important because, for example, point *S*(3, 4) is different from point *T*(4, 3). For this reason, the coordinates for a point, such as (3, 4), are called *ordered pairs.*

506   CHAPTER 14   Geometry of Position

When you read coordinates and graph points in the plane, you need to be aware of directions. The table shows you in which direction to move to locate a point based on the sign of each coordinate in the ordered pair.

	x-coordinate	*y*-coordinate
positive	move *right* →	move *up* ↑
negative	move *left* ←	move *down* ↓

Example 1

Give the coordinates of the point or points.
a. point *D*
b. two points in the third quadrant

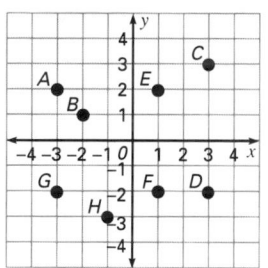

Solution
a. Point *D* is 3 units to the right of the origin and 2 units down from the origin. So the coordinates of *D* are (3, −2).

b. Points *G* and *H* are in the third quadrant. Their coordinates are *G*(−3, −2) and *H*(−1, −3). ◄

Example 2

Graph points *X*(0, 3), *Y*(2, 0) and *Z*(−2, −2).

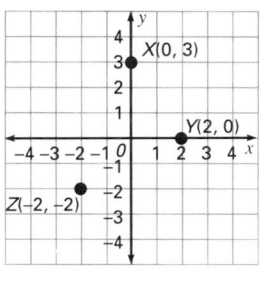

Solution
Point *X* is 0 units to the right or left of the origin and 3 units up.
Point *Y* is 2 units to the right of the origin and 0 units up or down.
Point *Z* is 2 units to the left of the origin and 2 units down from it. ◄

> ✓ **CHECK UNDERSTANDING**
>
> If the *x*-coordinate of a point is 0, what can you tell about the location of the point?
>
> If the *y*-coordinate of a point is 0, what can you tell about the location of the point?

It is located on the *y*-axis.

It is located on the *x*-axis.

Example 3

Sketch the square that has a diagonal whose endpoints are *J*(3, 3) and *L*(−3, −3).

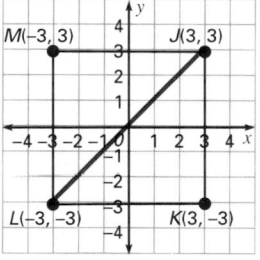

Solution
Begin by graphing *J*(3, 3) and *L*(−3, −3) on a coordinate plane. Point *M* is directly across from *J* and above *L*. So its coordinates, *M*(−3, 3), are at a vertex of the square. Similarly, the coordinates *K*(3, −3) are at another vertex. ◄

BASIC
1–16, 19, 22–23, 27

AVERAGE
9–20, 22–25, 27–28

ENRICHED
9–21, 22–26, 27–28

ADDITIONAL RESOURCES
Reteaching 14–1
Enrichment 14–1
Transparency Master 59

Example 1: Help students understand how to locate a point on a coordinate plane by using a coordinate plane drawn on the chalkboard. Point out that to locate a point, one must begin at the origin. Give an example of each kind of ordered pair: (positive, positive), (negative, negative), (positive, negative), and (negative, positive).

Example 2: Remind students that the *x*-coordinate is always listed first in an ordered pair.

Example 3: Have students recall that a *vertex* is the point of intersection of two sides of a polygon. Point out that a square has four vertices.

Additional Questions/Examples
Use graph paper. Graph each ordered pair on a coordinate plane.

1. *A* (3,3)	**2.** *B* (0,−4)
3. *C* (2,0)	**4.** *D* (−5,3)
5. *E* (3,−3)	**6.** *F* (−4,4)

Name the quadrant(s) in which each point could lie according to the sign of its coordinates.

7. The first coordinate is negative. **II** or **III**

8. The second coordinate is negative. **III** or **IV**

9. Both coordinates are negative. **III**

10. Both coordinates are positive. **I**

Guided Practice/Try These

Guided Practice/Try These Before students complete Exercises 1–8, have them describe the characteristics of points found in each quadrant. Elicit the fact that points whose coordinates are both positive fall in Quadrant I, points with a negative x-coordinate and a positive y-coordinate fall in Quadrant II, points whose coordinates are both negative fall in Quadrant III, and points with a positive x-coordinate and a negative y-coordinate fall in Quadrant IV.

3 SUMMARIZE

Key Questions
1. What do the numbers in an ordered pair represent? **The first number represents the x-coordinate and the second number represents the y-coordinate.**
2. How would you graph point $(-2, -5)$? **Start at $(0, 0)$ and move 2 units to the left along the x-axis, then move 5 units down. Label the point $(-2, -5)$.**

4 PRACTICE

Practice/Solve Problems
For Exercises 1–8, you may wish to have students describe how many units they moved in each direction to identify each ordered pair.

Extend/Solve Problems
For Exercises 22–23, have students recall how to find the area of a square. Elicit the fact that the area of a square is equal to the length of one side squared.

Think Critically/Solve Problems
For Exercise 28 have students interpret the phrase "flip the rectangle over the x-axis" and have them describe how they obtained their results.

508

Find the answer.

1. $\frac{1}{10} + \frac{2}{3}$ $\frac{23}{30}$

2. $3\frac{1}{8} - 1\frac{2}{3}$ $1\frac{11}{24}$

3. $\frac{3}{4} \times \frac{2}{9}$ $\frac{1}{6}$

4. $4^8 \div 4^3$ **1024**

5. What is the square root of 169? **13**

Solve.

6. Find the perimeter of a regular hexagon 6.3 cm on each side. **37.8 cm**

7. Find the fraction that comes next in this pattern:
$\frac{1}{40}, \frac{1}{20}, \frac{1}{10}, \frac{1}{5}$

8. Solve the proportion:
$\frac{14}{n} = \frac{3}{8}$ $37\frac{1}{3}$

TRY THESE

Refer to the figure. Give the coordinates of the point or points.

1. point K $(-5, 0)$
2. point D $(4, -3)$
3. point M $(0, 4)$
4. point H $(-4, -3)$
5. origin $(0, 0)$
6. a point on the x-axis
    $K(-5, 0)$; $E(4, 0)$
7. a point in the fourth quadrant
    $D(4, -3)$
8. two points with the same x-coordinate $C, K,$ and I; B and H; M and J; $L, F, E,$ and D

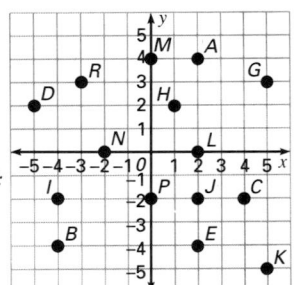

Graph each point on a coordinate plane. **See Additional Answers.**

9. $A(4, 0)$ 10. $B(0, 4)$ 11. $C(0, -2)$

12. $D(3, 5)$ 13. $E(5, 5)$ 14. $F(-4, 2)$

Sketch each figure on a coordinate plane. **See Additional Answers.**

15. triangle with vertices $X(3, 2)$, $Y(5, 6)$, and $Z(2, -4)$

16. square with $C(5, -5)$ and $D(-5, 5)$ as endpoints of a diagonal

EXERCISES

PRACTICE/ SOLVE PROBLEMS

Refer to the figure. Give the coordinates of the point or points.

1. point A $(2, 4)$
2. point B $(-4, -4)$
3. two points on the x-axis
    $N(-2, 0)$; $L(2, 0)$
4. two points with the same y-coordinate M and A; R and G; D and H; N and L; $I, P, J,$ and C; B and E
5. a point in the first quadrant
    $A(2, 4)$; $G(5, 3)$; $H(1, 2)$
6. a point in the third quadrant
    $I(-4, -2)$; $B(-4, -4)$
7. a point whose coordinates are equal
    $B(-4, -4)$
8. a point whose coordinates are opposites $R(-3, 3)$; $J(2, -2)$; $K(5, -5)$

Graph each point on a coordinate plane. **See Additional Answers.**

9. $T(3, 2)$ 10. $V(4, -1)$ 11. $W(0, 4)$ 12. $X(-7, 8)$

13. $Y(-4, 0)$ 14. $Z(3, -11)$ 15. $R(-8, -4)$ 16. $O(0, 0)$

Sketch each figure on a coordinate plane. **See Additional Answers.**

17. rectangle with $S(3, 2)$ and $T(-3, -2)$ as endpoints of a diagonal

18. circle with center $P(3, 2)$ and radius 4 units

19. **USING DATA**   Refer to the table on page 505. Draw the dot pattern for the 2nd through 4th square numbers on a coordinate plane. Give the ordered pair for the corner points of each number. **Answers will vary.**

20. Which pair of points is closer together, $(-2, 7)$ and $(5, 7)$ or $(-1, 3)$ and $(-1, -1)$? **(−1, 3) and (−1, −1)**

21. Find the coordinates of two points other than points on the axes that are the same distance from the origin. **Answers will vary. A possible solution is (3, 2) and (−3, −2).**

Graph each set of points on a coordinate plane. Then, join the points in order, identify the geometric figure, and find its area.
See Additional Answers.

EXTEND/ SOLVE PROBLEMS

22. $O(0, 0)$, $A(0, 5)$, $B(4, 5)$, and $C(4, 0)$

23. $D(-7, 1)$, $E(-3, 1)$, $F(-3, -3)$, and $G(-7, -3)$

24. Graph $L(1, 2)$. Move 3 units to the right and 2 units up to locate point M. What are the coordinates of M? **(4, 4)**

25. Graph $(1, 3)$, $(5, 7)$, $(-1, 1)$, and $(0, 2)$. Write the coordinates of four other points that fit the pattern. **(−2, 0); (2, 4); (3, 5); (4, 6)**

26. All but one of these points suggest a pattern: $A(-7, -4)$, $B(-3, -2)$, $C(3, 1)$, $D(-5, -3)$, $E(1, 0)$, and $F(7, 4)$. Graph these points. Write the coordinates of the one point that does not fit the pattern. Tell why it does not fit the pattern. **All points fit the pattern of lying in a straight line except point $F(7, 4)$.**

27. On $\overleftrightarrow{AC}$, points A and C have coordinates $(6.5, -3)$ and $(-4, 1.6)$, respectively. Find the coordinates of the midpoint of $\overline{AC}$. **(1.25, −0.7)**

THINK CRITICALLY/ SOLVE PROBLEMS

28. Graph the rectangle whose vertices are $A(-2, 4)$, $B(2, 4)$, $C(2, 6)$, and $D(-2, 6)$. Now flip the rectangle over the x-axis so that it is upside down but remains the same distance from the x-axis as before. Give the coordinates of the vertices of the rectangle that results. **$A(-2, -4)$; $B(2, -4)$; $C(2, -6)$; $D(-2, -6)$**

14-1   The Coordinate Plane   **509**

510

14-2 Problem Solving Skills:
USE A GRID TO SOLVE DISTANCE PROBLEMS

A coordinate grid can often be used to find distances.

PROBLEM

John lives 10 blocks east and 4 blocks north of the art museum. Alex lives 6 blocks west and 6 blocks south of the museum. On Saturday, John walked from his house to Alex's; then the boys went to the museum together. These are the routes they took:

To get to Alex's house, John walked 3 blocks west, 8 blocks south, 10 blocks west, 2 blocks south, and 3 blocks west.

From Alex's house, the boys walked north 4 blocks and east 6 blocks. Then Alex suddenly remembered he had left his student pass to the museum at home, so both boys retraced their steps and returned to Alex's house to get the pass. They set out again and reached the museum by walking 2 blocks east, 6 blocks north, and 4 blocks east.

How far did each boy walk between his own home and the museum?

SOLUTION

Use a grid. Let each square of the grid represent one city block. Indicate the directions north, south, east, and west on the grid.

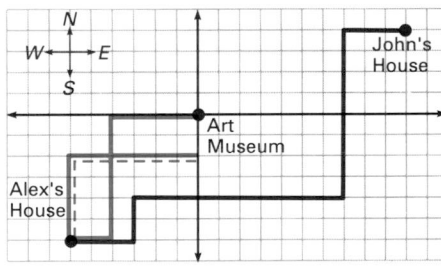

Plot the locations of the museum and each boy's house. Then draw the route John followed to reach Alex's house and the route the boys took together. Then find the total number of blocks each boy walked.

Alex walked 32 blocks.
John walked (26 + 32) blocks, or 58 blocks.

PROBLEMS

1. In a community walk-a-thon, Michele traveled the path shown on the grid at the top of page 511. Each square represents 1 block. What is the total distance Michele walked? **26 blocks**

2. Refer to Problem 1. Assume that the grid lines represent all the sidewalks. If Michele walked only on the sidewalks, what is the shortest distance she could have walked between the start and finish points? **18 blocks**

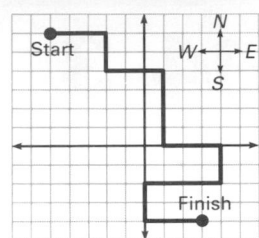

3. From her house, Marie rode her bike 4 blocks east, 4 blocks north, and 5 blocks east to return books to the library. Then she rode 7 blocks west and 3 blocks north to Angela's house. Together the girls biked 3 blocks east to visit Carla. Then Marie rode 5 blocks south and 4 blocks west to the Johnsons' house to baby-sit. How many blocks did Marie ride her bike from her house until she reached the Johnsons' house? How many blocks from home was she while she baby-sat? **35 blocks; 3 blocks**

4. Every Thursday, Kara leaves work and drives 6 mi west, 3 mi south, 5 mi east, 1 mi south, and 3 mi east to get to her exercise class. At about the same time, Kara's husband leaves his work, which is 4 mi due west of Kara's office, and has a friend drive him to baseball practice. They drive 3 mi west, 4 mi south, 5 mi east, 1 mi south, and 2 mi east. After her class, Kara drives the most direct route possible to pick up her husband at practice. How far and in what direction does she drive?
1 mi south, 2 mi west or 2 mi west, 1 mi south

5. One spring, a bear came out of its cave to look for food. After traveling 20 mi due south, 20 mi due west, then 20 mi due north, the bear arrived back at its cave. What was the color of the bear? **white**

PROBLEM SOLVING TIP

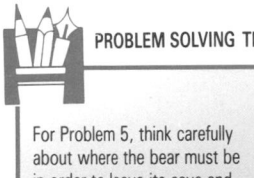

For Problem 5, think carefully about where the bear must be in order to leave its cave and then return to it by going in only three directions.

6. **USING DATA** Use the Data Index on page 546 to find the maximum speeds of animals. Traveling at maximum speed, an African elephant runs north from one spot for 5 min (one twelfth of an hour), then west for 5 min. Starting at the same spot, a Mongolian gazelle runs west for 5 min, then north for 2 min, where it stops at a watering hole. How far is the elephant from the watering hole after 10 min of running? **3 mi**

7. At the same time that the elephant and the gazelle in Problem 6 started toward the watering hole, a game warden set out for the watering hole from a point 5 mi east of the elephant's starting point. Maintaining a speed of 30 mi/h, the warden drove due west for 10 mi and then due north to the watering hole. Did the game warden arrive before or after the gazelle? How many minutes elapsed between the arrival of the warden and the arrival of the gazelle? **after; 17 min**

JUST FOR FUN

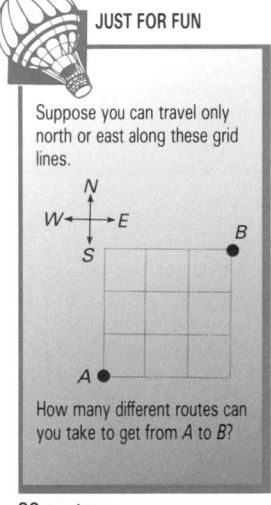

Suppose you can travel only north or east along these grid lines.

How many different routes can you take to get from A to B?

20 routes

ASSIGNMENTS

BASIC
1–4, 6

AVERAGE
1–7

ENRICHED
1–7

ADDITIONAL RESOURCES
Reteaching 14–2
Enrichment 14–2

ative number. From this students should deduce that John's house is at point (10, 4) and Alex''s house is at point (–6, –6).

3 SUMMARIZE

Talk About It Have students discuss how using a grid is helpful in solving distance problems.

4 PRACTICE

An understanding of the steps for solving problems is essential for students to solve problems successfully. For Exercises 3–4 have students use a coordinate grid.

5 FOLLOW-UP

Extra Practice
Robert drove 2 mi east and 5 mi south before he realized that he had forgotten his library card. Robert traced his route and returned home. He set out again for the library. This time he drove 4 mi east, 3 mi south, 6 mi west, and 2 mi north to the point where the library is located. If Robert's house is located at the origin of a coordinate grid, at what point is the library located? **(–2,–1)**

Get Ready graph paper, rulers

5-MINUTE CLINIC

Exercise	Student's Error	Error Diagnosis
Write the ordered pair that represents the location of Mark's house: 3 blocks west 2 blocks north	(3, 2)	• Students forget that west represents a negative direction on a coordinate plane and so they omit the negative sign in the ordered pair.

WARM-UP

Solve each equation.
1. $3x - 5 = 16$ **$x = 7$**
2. $10 + 2x = 32$ **$x = 11$**
3. $6x = 72$ **$x = 12$**
4. $1/2x = 14$ **$x = 28$**

1 MOTIVATE

Explore Elicit from students that the solution to the equation $2x + 5 = 21$ is $x = 8$. Students should conclude that $x = 8$ is the only solution to this equation. Help students understand how x and y are related by the equation $x + y = 14$. Elicit from students that there are many solutions to this equation, since there are two unknown variables.

2 TEACH

Use the Pages/Skills Development
Have students read this part of the section and then discuss the examples. Have them compare the methods used in solving equations with one and two variables.
   Example 1: Have students check the solutions by substituting the values for x and y into the original equation.

512

14-3 Graphing Linear Equations

EXPLORE

Read this equation: $2x + 5 = 21$
What is the solution? Is there another solution? Explain. 8; Only one value (8 for the variable ma the equation true.

Answers will vary. There are an infinite number of solutions.

Now try to solve this equation: $x + y = 14$
What solution did you get?
Can you solve this equation another way? Explain.
How many solutions do you think there are to this equation?

SKILLS DEVELOPMENT

You know how to find the solution of some equations with one variable. Other equations, however, have two variables. Such an equation has an infinite number of solutions. Each of the solutions can be represented by an ordered pair.

To find some of the solutions of an equation with two variables, begin by choosing a value for the first variable, x. Substitute that value into the equation and solve to find the corresponding value of y. Do this for at least three values of x. Make a table to keep track of the ordered pairs that are solutions of the equation.

Example 1

Find three solutions of the equation $y = x - 3$.

TALK IT OVER

Why do you think you are instructed to use at least three different values of x when you graph a linear equation?

Answers will vary. Possible answer: to be certain that the line is really a graph of the equation

Solution

To begin solving $y = x - 3$, choose three values for x. It usually is a good idea to choose a negative number, zero, and a positive number for these values.

Let x equal -2.	Let x equal 0.	Let x equal 3.
$y = x - 3$	$y = x - 3$	$y = x - 3$
$y = -2 - 3$	$y = 0 - 3$	$y = 3 - 3$
$y = -5$	$y = -3$	$y = 0$

Write the solutions in a table.

$y = x - 3$

x	y	(x, y)	
-2	-5	$(-2, -5)$	← When x is -2, y is -5.
0	-3	$(0, -3)$	← When x is 0, y is -3.
3	0	$(3, 0)$	← When x is 3, y is 0. ◄

Since a solution of an equation in two variables is an ordered pair, you can graph it as a point on the coordinate plane. The set of all points whose coordinates are solutions of an equation is called the **graph of the equation.** When these points lie in a straight line, the equation is called a **linear equation.**

512   CHAPTER 14   Geometry of Position

MAKING CONNECTIONS

Introduce the lesson by writing the following word pairs on the board:
   New York / Albany   Iowa / Des Moines
   Alaska / Juneau   Ohio / Columbus
Ask students to identify the relationship between each pair of words. Explain that identifying relationships is a key skill in mathematics, as well.

Example 2

Graph the equation $y = -3x + 2$.

Solution
Find at least three solutions of the equation, using reasonable values for x. The table shows that the values chosen for x are -2, 0, and 2.

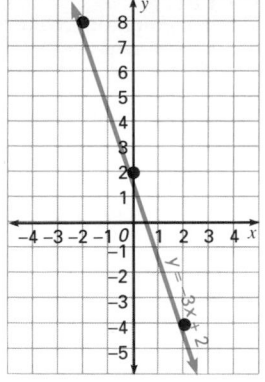

$y = -3x + 2$

x	y	(x, y)
-2	8	$(-2, 8)$
0	2	$(0, 2)$
2	-4	$(2, -4)$

Graph the ordered pairs. Draw a line through the points.

All the points along the line are solutions of the equation $y = -3x + 2$. ◄

When you find solutions of an equation in two variables, the y-value of the ordered pair depends on the x-value that you choose. So we say that *y is a function of x,* and the relationship between y and x is called a **function.**

In everyday life, one quantity often depends on another. Many of these relationships can be represented by mathematical functions. The graph of the equation that represents the function is called the **graph of the function.**

COMPUTER TIP

Some computer software acts as a function grapher and draws graphs quickly and accurately. In most versions of this type of software, when you enter a function and values for the variables, the computer will then graph the function for those values. If you have access to this type of software, you may want to use it to investigate the graphs of functions.

Example 3

Daneesha walks on a treadmill at an average rate of 4 mi/h. The distance she walks is a function of the amount of time she walks. Write an equation that represents this function. Then graph the function.

Solution
Let $x =$ the amount of time and let $y =$ the distance. So $y = 4x$.
Remember: The distance formula is distance = rate × time.

Make a table of solutions.

$y = 4x$

x	y	(x, y)
1	4	$(1, 4)$
2	8	$(2, 8)$
3	12	$(3, 12)$

Use ordered pairs to graph the function.

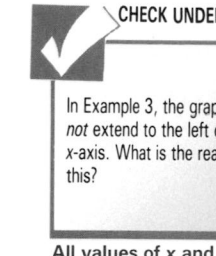

✓ **CHECK UNDERSTANDING**

In Example 3, the graph does *not* extend to the left of the x-axis. What is the reason for this?

◄ All values of x and y are positive.

14-3 Graphing Linear Equations **513**

Example 2: Have students locate another point on the line of the graph of $y = -3x + 2$ and check that the ordered pair is a solution. Students should deduce that every point on the line is a solution of the equation.

Example 3: Have students explain the meaning of the term "function" in this example. Elicit the fact that the distance that Daneesha walks depends on the amount of time she walks.

Additional Questions/Examples
For each equation make a table of three solutions. Then graph the equation. **Solutions may vary.**
1. $y = -4x + 10$
   (2, 2); (0, 10); (–2, 18)
2. $y = x + 2$
   (3,5); (0, 2); (–3, –1)
3. $y = -3x + 2$
   (1, –1); (0, 2); (–1, 5)
4. $y = 5x$
   (0, 0); (1, 5); (–1, –5)
5. $y = -x - 3$
   (0, –3); (1, –4; (–1, –2)

Guided Practice/Try These Have students work in small groups. Point out that all group members must agree on each answer.

AT-RISK STUDENTS

Have students graph each of the following sets of three points on a separate coordinate grid. Then have them use a ruler to determine if all the points in a set fall on the same line.
1. (0, 1); (1, 3); (2, 5) **yes**
2. (0, 0); (1, 2); (2, 3) **no**
3. (0, –5); (–1, –6); (2, –3) **yes**

514

3 SUMMARIZE

Write About Math Have students write in their math journals the procedure(s) for graphing a linear equation.

4 PRACTICE

Practice/Solve Problems For Exercises 1–6, point out that small integers are easiest to work with.

Extend/Solve Problems For these exercises, encourage students to make tables and look at the relationship between the x and y columns. Students should use a guess-and-check-and-revise strategy. They should guess an equation, test the ordered pairs, and revise as needed before writing an equation.

Think Critically/Solve Problems For Exercise 14, help students understand how the coefficient of x indicates whether a line slants upwards or downwards.

5 FOLLOW-UP

Extra Practice For each equation make a table of three solutions. Then graph the equation. **Solutions may vary.**
1. $y = -7x + 12$
(1, 5); (0, 12); (–1, 19)
2. $y = 2x + 3$
(–2, –1); (0, 3); (2, 7)
3. $y = 1/4\, x + 1$
(2, 1 1/2); (0, 1); (–2, 1/2)
4. $y = 3x + 2$
(–3, –7); (0, 2); (3, 11)
5. $y = -1/5x + 1/5$
(0, 1/5); (1, 0); (–1, 2/5)

1.

x	y	(x, y)
–2	–8	(–2, –8)
0	0	(0, 0)
2	8	(2, 8)

2.

x	y	(x, y)
–1	–6	(–1, –6)
0	–1	(0, –1)
1	4	(1, 4)

3.

x	y	(x, y)
–3	–4	(–3, –4)
0	2	(0, 2)
2	6	(2, 6)

TRY THESE

For each equation, make a table of three solutions.
Answers will vary. Sample solutions are given.
1. $y = 4x$ **2.** $y = 5x - 1$ **3.** $y = 2x + 2$

For each equation, complete the table of solutions. Then graph the equation. **See Additional Answers.**

4. $y = 2x - 1$

x	y	(x, y)
–2	–5	(–2, –5)
–1	–3	(–1, –3)
0	–1	(0, –1)
1	■	(1, ■)

5. $y = -3x$

x	y	(x, y)	
–1	■	(–1, ■)	3; 3
0	■	(0, ■)	0; 0
1	■	(1, ■)	–3; –3
2	■	(2, ■)	–6; –6

1; 1

6. $y = x + 4$

x	y	(x, y)	
–2	■	(–2, ■)	2
0	■	(0, ■)	4
2	■	(2, ■)	6
4	■	(4, ■)	8

7. Chicken cutlets cost $3 per pound. This is expressed by the formula $y = 3x$, where y is the cost and x is the number of pounds. Complete the table of solutions for the equation that represents the function. Then graph the function.

x	y	(x, y)	
0	■	(0, ■)	0; 0
1	■	(1, ■)	3; 3
2	■	(2, ■)	6; 6
3	■	(3, ■)	9; 9

EXERCISES

**PRACTICE/
SOLVE PROBLEMS**

For each equation, make a table of three solutions.
Answers will vary. Check students' work.
1. $y = x + 4$ **2.** $y = 3x - 2$ **3.** $y = x - 1$
4. $y = -2x - 4$ **5.** $y = 2x + 4$ **6.** $y = -2x$

For each equation, complete the table of solutions. Then graph the equation. **See Additional Answers.**

7. $y = 4x$

x	y	(x, y)	
–1	■	(–1, ■)	–4; –4
0	■	(0, ■)	0; 0
1	■	(1, ■)	4; 4
2	■	(2, ■)	8; 8

8. $y = -2x + 1$

x	y	(x, y)	
–3	■	(–3, ■)	7; 7
–1	■	(–1, ■)	3; 3
0	■	(0, ■)	1; 1
2	■	(2, ■)	–3; –3

9. $y = \frac{1}{2}x + 5$

x	y	(x, y)	
–4	■	–4, ■	3
–2	■	–2, ■	4
0	■	0, ■	5
2	■	2, ■	6

10. A person's age five years from now (y) is a function of the person's present age (x). Write an equation that represents this function. Then graph the function. $y = x + 5$ **See Additional Answers.**

Write an equation for each graph.

11. $y = 2\frac{1}{2}x$

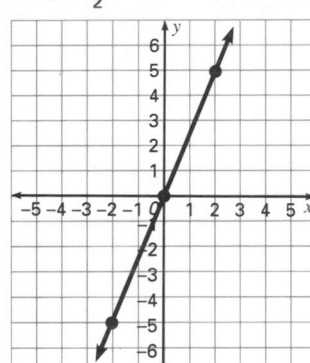

12. $y = -\frac{1}{2}x + 2$

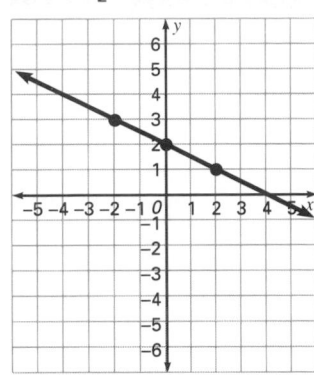

13. On a coordinate plane, graph the ordered pairs $(-2, -1)$ and $\left(1, \frac{1}{2}\right)$. Draw a line through the points. Make a table of these ordered pairs and three others that lie along the line you drew. Write the equation for the line. $y = \frac{1}{2}x$ **See Additional Answers.**

14. Look back at the graphs of the linear equations with which you have been working. Draw some conclusions about which equations are related to the graphs that slant upward and which are related to the graphs that slant downward. **All equations graphed in this lesson are of the form $y = mx + b$. When m is positive, the line slants up to the right. When m is negative, the line slants down to the right.**

**THINK CRITICALLY/
SOLVE PROBLEMS**

14-3   Graphing Linear Equations   **515**

WARM-UP

Simplify.
1. 12/36 **(1/3)**
2. 28/44 **(7/11)**
3. −18/4 **(−9/2)**
4. −3/24 **(−1/8)**

Explore Discuss the Explore activity with students. Make sure students understand the concept of slope before proceeding. If necessary, use additional examples. Ask students to describe the relationship between the vertical and horizontal distances on the ski slope.

Use the Pages/Skills Development Have students read this part of the section and then discuss the examples. Introduce the slope ratio as a means of determining slope. Elicit, for example, what a slope of 20 cm per meter means.

Lead students to understand that a vertical line has no slope and that a horizontal line has a slope of 0.

14-4 Working with Slope

EXPLORE

When did you last climb up or down a slope? Describe the slope.

With your finger, trace the slope of the hill shown at the right. Would a skier coming down the slope be skiing a distance of 326 m? Would the skier be skiing a distance of 621 m? Explain.

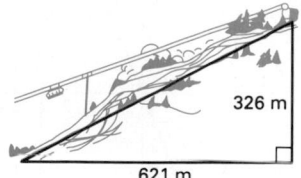

326 m

621 m

Answers will vary.

SKILLS DEVELOPMENT

TALK IT OVER

Name some instances in which someone might need to know the slope of something. Why might it be helpful to use numbers in these instances?

Possible answers include constructing a staircase, finding the pitch of a roof, and building a child's slide.

As you move from one point to another along a straight line, the change in position has two components. There is a vertical change in position, called the **rise,** and a horizontal change, called the **run.** The ratio of the rise to the run is called the **slope** of the line.

$$\text{slope} = \frac{\text{rise}}{\text{run}}$$

You can find the slope of a line graphed on a coordinate plane by counting the units of change between the coordinates of two points on the line.

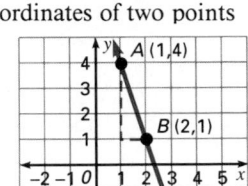

Example 1

Find the slope of the line graphed at the right.

Solution

Choose two points on the line, such as $A(1, 4)$ and $B(2, 1)$. Find the number of units of change in moving from point A to point B.

$$\text{slope} = \frac{\text{rise}}{\text{run}} = \frac{3 \text{ units down}}{1 \text{ unit right}} = \frac{-3}{1} = -3 \quad \blacktriangleleft$$

A move *down* or to the *left* is represented by a negative number.

If you are given the coordinates of two points on a line, you can find the slope of the line by using this formula.

$$\text{slope} = \frac{\text{rise}}{\text{run}} = \frac{\text{difference of } y\text{-coordinates}}{\text{difference of } x\text{-coordinates}}$$

Example 2

Find the slope of the line that passes through the points $M(0, 2)$ and $N(-2, -1)$.

Solution

$$\text{slope} = \frac{\text{rise}}{\text{run}} = \frac{\text{difference in } y\text{-coordinates}}{\text{difference in } x\text{-coordinates}}$$

Be sure to subtract the *y*-coordinates and the *x*-coordinates in the same order.

$$\text{slope} = \frac{2 - (-1)}{0 - (-2)} = \frac{3}{2} \quad \blacktriangleleft$$

TEACHING TIP

To avoid confusion between *rise* and *run,* ask students what they would do if you told them to rise. **stand up** If you told them to run? **move horizontally**

When a linear equation is written with y alone on one side of the equals sign, we say it is written in the form $y = mx + b$. With the equation written in this form, the coefficient of x, which is m, is the slope of the line. The other piece of information that we can read from this equation is that the constant term, b, is the y-intercept of the line. The **y-intercept of a line** is the y-coordinate of the point where it intersects the y-axis.

Since you can use $y = mx + b$ to find the slope and the y-intercept of a line, it is called the **slope-intercept form** of an equation.

Example 3

Write an equation of the line graphed at the right.

Solution

Find the slope by counting units of change from point S to point T.

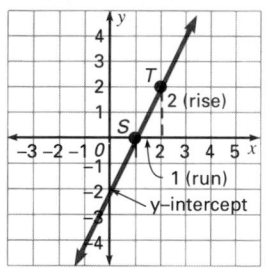

slope $= \dfrac{\text{rise}}{\text{run}} = \dfrac{2}{1} = 2$

The slope of line ST is 2.

The line intersects the y-axis at $(0, -2)$, so the y-intercept is -2.

An equation of the line is $y = 2x - 2$. ◄

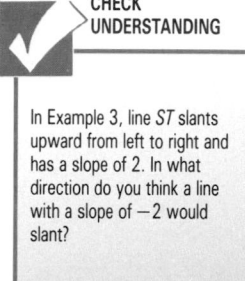

CHECK UNDERSTANDING

In Example 3, line ST slants upward from left to right and has a slope of 2. In what direction do you think a line with a slope of -2 would slant?

downward, from left to right

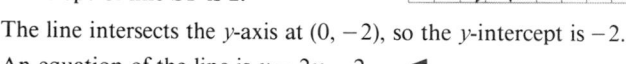

TRY THESE

Find the slope of each line on the coordinate plane at the right.

1. line AB 1

2. line EF $-\dfrac{1}{2}$

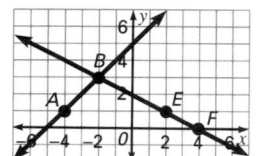

Find the slope of the line that passes through each pair of points.

3. $A(4, 5)$ and $B(3, 7)$ -2

4. $J(1, 6)$ and $K(3, 9)$ $\dfrac{3}{2}$

5. $R(-5, -7)$ and $S(8, -2)$ $\dfrac{5}{13}$

6. $M(-3, 4)$ and $N(-8, 1)$ $\dfrac{3}{5}$

7. $X(6, -5)$ and $Y(9, -3)$ $\dfrac{2}{3}$

8. $P(9, -1)$ and $Q(0, 4)$ $-\dfrac{5}{9}$

Write an equation of each line graphed at the right.

9. line GH $y = 3x - 1$

10. line MN $y = -x + 2$

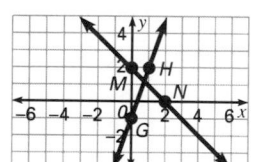

COMPUTER

You may use this program to check your answers to Exercises 3–8 and for other exercises in which you are asked to find the slope.

```
10 INPUT "ENTER THE
   COORDINATES OF THE
   FIRST POINT: ";X1, Y1:
   PRINT
20 INPUT "ENTER THE
   COORDINATES OF THE
   SECOND POINT: ";X2, Y2:
   PRINT
30 PRINT "THE SLOPE IS
   ";Y2 – Y1;"/";X2 – X1
```

14-4   Working with Slope   **517**

Example 1: Call attention to the diagram in the example. Ask students whether the slope of the line is positive or negative and have them justify their responses. Elicit that the slope is negative, since the line slants downward to the right.

Example 2: Help students understand that slope can be found in either of the two following ways:

$$\dfrac{y_2 - y_1}{x_2 - x_1} \text{ or } \dfrac{y_1 - y_2}{x_1 - x_2}$$

Example 3: Help students understand that in order to write an equation of a line they must first determine the slope and then determine the point at which the line crosses the y-axis (the y-intercept).

Additional Questions/Examples

1. Find the slope of the line that passes through points A (3, 8) and B (1, 4). **2**

2. Find the slope of the line that passes through points D (1, –5) and E (0, 3). **–8**

3. Find the slope of the line that passes through points A (1, 8) and B (–3, 2). **3/2**

4. Draw a line with a positive slope. **Lines should slant upward to the right.**

5. Draw a line with a negative slope. **Lines should slant downward to the right.**

Guided Practice/Try These Have students work in small groups. Point out that all group members should come to an agreement on each answer.

3 SUMMARIZE

Key Questions You may wish to have students record their answers in their math journals.
1. In which direction does a line with a negative slope slant? **downward to the right**
2. In which direction does a line with a positive slope slant? **upward to the right**
3. What is the slope of a vertical line? **A vertical line has no slope.**
4. What is the slope of a horizontal line? **0**

4 PRACTICE

Practice/Solve Problems For Exercises 3–4 stress the importance of finding the difference between the coordinates in the same order.

Extend/Solve Problems For Exercise 12 have students name another segment with the same slope. **AC or BD**

Think Critically/Solve Problems For Exercise 13, help students understand that parallel lines have the same slope but different y-intercepts.

Problem Solving Applications Have students work in small groups and take turns positioning blocks, measuring angles, and recording data.

EXERCISES

PRACTICE/ SOLVE PROBLEMS

Find the slope, or pitch, of each roof.

1.

2.

1                    $\frac{4}{7}$

Find the slope of the line that passes through each pair of points.

3. $X(-4, 2)$ and $Z(3, -2)$ $-\frac{4}{7}$ **4.** $M(3, 2)$ and $N(1, 1)$ $\frac{1}{2}$

CHECK UNDERSTANDING

Do you think all lines with the same y-intercept have the same slope?

Do you think all lines with the same slope have the same y-intercept? Explain.

Write an equation of each line.

5. $y = -\frac{1}{2}x + 2$

6. $y = x + 2$

7. The roof of a house rises vertically 3 ft for every 12 ft of horizontal distance. What is the slope, or pitch, of the roof? $\frac{1}{4}$

EXTEND/ SOLVE PROBLEMS

Use the coordinate plane at right. Find the slope of each line segment.

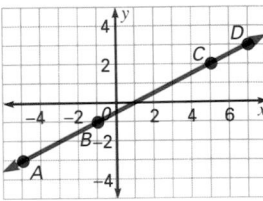

8. $\overline{AB}$ $\frac{1}{2}$ **9.** $\overline{BC}$ $\frac{1}{2}$

10. $\overline{CD}$ $\frac{1}{2}$ **11.** $\overline{AD}$ $\frac{1}{2}$

12. Compare the slopes in Exercises 8–11. What do you notice? **They are all the same.**

THINK CRITICALLY/ SOLVE PROBLEMS

13. What is an equation of the line that has a y-intercept of 2 and is parallel to the line whose equation is $y = \frac{1}{2}x + 1$? $y = \frac{1}{2}x + 2$

14. Write an equation of a line that has a slope of 0.
Answers will vary. Students may give any equation of the form $y = a$, where a is a real number. Samples are $y = 2$ and $y = -5$.

Problem Solving Applications:

INCLINED PLANES

► READ
► PLAN
► SOLVE
► ANSWER
► CHECK

You may already know that the inclined plane is a simple machine that is helpful in moving objects. The effort needed to push an object up an inclined plane to a given point is less than the effort needed to raise the object the vertical distance to that point. For example, in the figure below, it would be easier to push the block of wood to the point at which it appears than to raise the block vertically to that point.

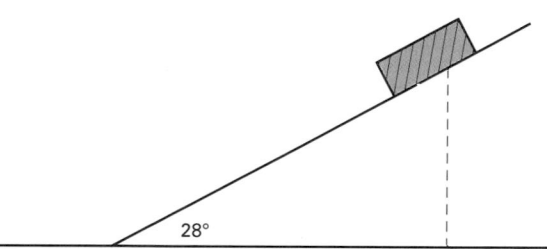

28°

Get a block of wood. Use a smooth plank as an inclined plane.

1. Slant the inclined plane so that the block of wood can be positioned on it without sliding.
2. Position the block on the inclined plane.
3. Slowly incline the plane until the block begins to slide.
4. Record the angle of the plane at the moment when the block begins to slide.
5. Record the slope of the plane when the sliding begins.
6. Repeat steps 1–5 two more times. Record your data in a chart.

	1st trial	2nd trial	3rd trial
angle of plane			
slope of plane			

7. From your chart, how are the values of the angles and slopes related?
8. Repeat steps 1–5 using planes of different materials, such as cardboard or steel, and blocks of different sizes and materials, such as plastic or foam. Perform repeated trials using these materials. How do your data using these materials differ from the data you collected using wooden materials? **Answers will vary.**

14-4   Working with Slope   **519**

CALCULATOR

If you graph the equation $y = 6x + 3$ on a graphing calculator, the graph will show you that the line crosses the x-axis at some point between −1 and 0. The line crosses the y-axis at some point near 3.

If you change the viewing rectangle so that x goes from −1 to 1 and y goes from −1 to 4, you can use the TRACE feature to find the intercepts more accurately. The TRACE feature lets you put the cursor on the intersection points and read the x- and y-values along the bottom of the screen.

With the viewing rectangle described, you see that the y-intercept is about 3 and the x-intercept is about −0.5. Using even more refined viewing rectangles shows that the x-intercept is −0.5 and the y-intercept is 3.

Use a graphing calculator to find the x- and y-intercepts of each of these lines.

1. $y = 2x - 4$ 2; −4
2. $y = x + 3.4$ −3.4; 3.4
3. $y = -2x + 5$ 2.5; 5
4. $y = 3x$ 0; 0

5 FOLLOW-UP

Extra Practice
Find the slope of the line that passes through each pair of points.
1. M (6, −3) and N (1, 2) **−1**
2. A (5, −2) and B (4, 3) **−5**
3. A line has a slope of 1/4. If the rise is 3, what is the run? **12**
4. Find the slope of a hill that has a run of 45 ft and a rise of 9 ft. **1/5**

Extension A line passes through all quadrants except Quadrant III. What do you know about the slope of the line? Explain. **The slope is negative. If the line passes through Quadrants I and II, the y-intercept is positive. If the line passes through I and IV, the x-intercept is positive. If both intercepts are positive, and these coordinates are used to determine the slope, then the slope is negative. The line slants downward to the right.**

Section Quiz Find the slope of the line that passes through each pair of points.
1. S(−3, 5) and T(2, 8) **3/5**
2. F(0, 6) and G(−3, 1) **5/3**
3. Find the slope of the line that passes through the points X(−10, 5) and Y(5, −5). **−2/3**
4. A line has a slope of −4/5 feet. If the rise is 16 feet down, what is the run? **20 feet right**
5. A line has a slope of 6 cm. If the run is 3 cm, what is the rise? **18 cm**

Get Ready graph paper, rulers

CHALLENGE

Have students graph the line with slope 2/3 that goes through point D (−1, −4). **Students should graph the linear equation $y = 2/3x - 3\ 1/3$, or $3y = 2x - 10$.**

SPOTLIGHT

OBJECTIVE
•To solve problems by making a model

MATERIALS NEEDED
graph paper, rulers,

VOCABULARY
model

WARM-UP

Graph each equation.
1. $2x - y = 15$
2. $-y - 4x = 16$
3. $5x + y = 10$
4. $1/3x + y = 4$

1 MOTIVATE

Introduction Emphasize the need to carefully interpret the information given in the problem.

2 TEACH

Use the Pages/Problem Have students read this part of the section and then discuss the problem. Elicit from students that a graph or an equation would be an appropriate model to use to solve this problem.

Use the Pages/Solution Be sure students understand why it is that the *y*-intercept represents the flat fee and the slope represents the commission. Help students understand why the graph cannot extend beyond the first quadrant.

14-5 Problem Solving Strategies:
MAKE A MODEL

► READ
► PLAN
► SOLVE
► ANSWER
► CHECK

Sometimes a problem is easier to solve if you make a model to help you solve it. There are many types of models that you can use, such as equations, sketches, scale drawings, three-dimensional constructions, and computer simulations. When a problem involves a linear function, a model that may be especially helpful is a graph of the function.

PROBLEM

Suppose you supervise employees who make sales calls over the telephone. The employees are paid a flat fee each week for making a minimum of 15 calls. In addition to their flat fee, they receive a fixed commission for each item they sell.

The chart shows the amount of the paycheck and the number of items sold by each of four employees. What is the flat fee and the amount of commission per item that each employee receives?

Number of Items Sold	10	25	30	15
Amount of Paycheck	$300	$675	$800	$425

SOLUTION

Model the situation on a coordinate plane. Label the *x*-axis "Number of Items Sold." Label the *y*-axis "Amount of Paycheck."

When you graph the data from the table as four points on the coordinate plane, the points lie along a straight line. The equation of that line is $y = 25x + 50$. The *y*-intercept, 50, represents the amount each employee receives if 0 items are sold. So the flat fee is $50. The slope of the line, 25, represents the increase in the paycheck for each 1 item sold. So the amount of commission per item is $25.

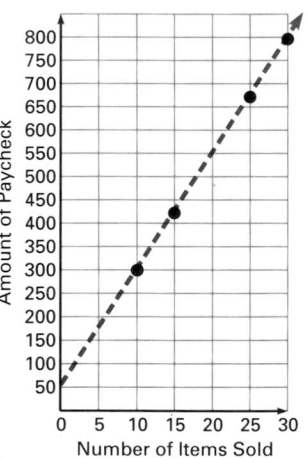

PROBLEMS

Make a model to solve each problem. Use the information in the chart provided with each problem to help you model the situation.

1. Admission to a high school soccer game is charged on a per-vehicle basis. For each vehicle, there is a fixed parking fee. Then an additional amount is charged for each student in the vehicle. What is the parking fee? What is the amount charged per student?
Parking fee = $3.00; per-student charge = $2.00

Amount Charged for Vehicle	$13	$7	$31
Number of Students per Vehicle	5	2	14

2. A telephone company charges a fixed fee for making a service call plus an hourly rate for each hour the service person spends making the call. Find the fixed fee and the hourly rate for each service call. **fee $25, rate $20/h**

Amount Charged for Call	$85	$45	$125
Number of Hours per Call	3	1	5

3. A baby sitter charges a fixed fee for the first hour and thereafter charges an hourly rate depending on the number of children being cared for. Find the fee for the first hour and the hourly rate per child for a four-hour baby-sitting job. **Flat fee = $4.00**
Hourly = $2.50 per child

Amount Charged for Four-Hour Job	$6.50	$11.50	$16.50
Number of Children	1	3	5

4. A health club charges its members a one-time fee when they first join. Thereafter each member is charged annual dues. Find the amount of the one-time fee and the amount of dues each member must pay per year. **One-time fee = $450.00**
Yearly dues = $150.00

Total Amount Paid by Member	$600	$1,050	$1,500
Number of Years of Membership	1	4	7

5. A store charges a fixed amount to ship each package and an added charge per pound. Find the fixed shipping fee and the cost per pound. **Fixed fee = $7.50 Cost per lb = $3.50**

Cost of Shipping Package	$21.50	$32.00	$42.50
Number of Pounds	4	7	10

14-5   Problem Solving Strategies: Make a Model   **521**

CHALLENGE

Choose a model and explain how you would measure exactly 4 quarts of water if you have only a 3-quart measure and a 5-quart measure to use.

3 SUMMARIZE

Talk About It   Have students discuss the advantages of using a model to solve a problem.

4 PRACTICE

An understanding of the steps for solving problems is essential for students to solve problems successfully. For each of Exercises 1–5 you may wish to have students write an equation and model the situation on a coordinate grid.

5 FOLLOW-UP

Extra Practice   A parking garage charges a fixed fee for the first hour and an hourly rate for each additional hour. Find the amount of the fixed fee and the hourly rate for each car.

	Car A	Car B	Car C
Amount charged	$12.50	$17.00	$26.00
Num. of hours	5	8	14

fixed fee: $5.00;
hourly rate: $1.50

Get Ready   graph paper, rulers, algebra tiles

WARM-UP

Add 4 to the x-coordinate. Subtract 2 from the y-coordinate. Write the new ordered pair.
1. (-5, 6) **(-1,4)** 2. (10, -1) **(14,-3)**
3. (0,0) **(4,-2)** 4. (-3,-8) **(1,-10)**

1 MOTIVATE

Explore/Working Together Have students work in pairs to complete the Explore activity. Discuss their results. Elicit from students the fact that the size and shape of the algebra tile did not change, but that the tile merely slid to a new position on the graph.

2 TEACH

Use the Pages/Skills Development Have students read this part of the section and then discuss the examples. Emphasize that when a figure is moved, or slid, without turning, the movement is a translation. The translated figure is identical to the original figure.

14-6 Translations

EXPLORE/ WORKING TOGETHER

Work with a partner. You will need a piece of graph paper, a ruler, and one of the small yellow square tiles from a set of algebra tiles.

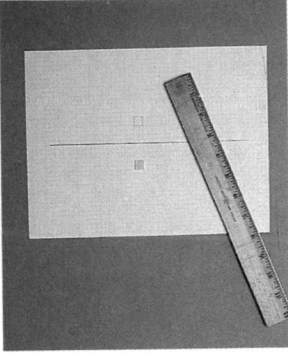

a. Use the ruler to draw a vertical rule down the middle of the graph paper.

b. Place the small yellow tile anywhere on the graph paper to the left of the vertical rule. Position the tile so that it exactly covers four of the squares on the graph paper.

c. You or your partner should hold the tile steady while the other traces it.

d. Now, slowly slide the tile horizontally along the grid lines until it is across the vertical rule and is positioned the same distance from the rule as it was in its starting position.

e. Trace the tile in its new position.

f. Draw an arrow from the first tracing to the second.

How many units did you slide the tile from its original position to its new position?

How did the tile change as you slid it? How did it stay the same?

SKILLS DEVELOPMENT

A **translation,** or *slide,* of a figure produces a new figure that is exactly like the original. As a figure is translated, you imagine all its points sliding along a plane the same distance and in the same direction. So the sides and angles of the new figure are equal in measure to the sides and angles of the original, and each side of the new figure is parallel to the corresponding side of the original.

A move like a translation is called a **transformation** of the figure. The new figure is called the **image** of the original, and the original is called the **preimage** of the new figure.

Example 1

Graph the image of the point $C(-5, 3)$ under a translation 4 units to the right and 3 units down.

TEACHING TIP

Point out to students that the translation image can be traced directly from the preimage once the coordinates of one vertex have been found. Then the coordinates of the other vertices can be checked to make sure the image has been copied correctly.

Solution

Add 4 to the *x*-coordinate and subtract 3 from the *y*-coordinate.

$$C(-5, 3) \rightarrow C'(-5 + 4, 3 - 3)$$
$$\rightarrow C'(-1, 0)$$
↑

Read *C'* as "*C* prime." ◀

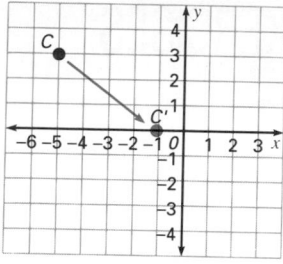

Example 2

Graph the image of $\triangle ABC$ with vertices $A(-3, 3)$, $B(-5, -1)$, and $C(-2, 0)$ under a translation 5 units to the right and 4 units up.

Solution

Add 5 to the *x*-coordinate of each vertex. Add 4 to the *y*-coordinate of each vertex.

$A(-3, 3) \quad \rightarrow A'(-3 + 5, 3 + 4)$
$\qquad\qquad \rightarrow A'(2, 7)$

$B(-5, -1) \rightarrow B'(-5 + 5, -1 + 4)$
$\qquad\qquad \rightarrow B'(0, 3)$

$C(-2, 0) \quad \rightarrow C'(-2 + 5, 0 + 4)$
$\qquad\qquad \rightarrow C'(3, 4)$ ◀

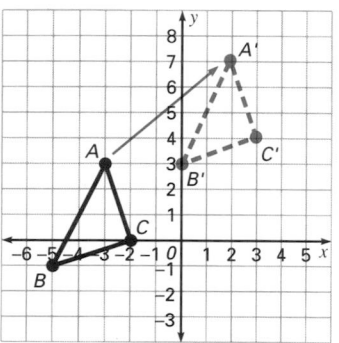

COMPUTER

You can use this program to find the coordinates of the image points in Example 2.

```
10 FOR V = 1 TO 3
20 INPUT "ENTER THE
   COORDINATES OF THE
   VERTEX: ";X, Y: PRINT
30 LET X = X + 5: LET
   Y = Y + 4
40 PRINT "THE IMAGE IS
   (";X;",",";Y;")": PRINT
50 NEXT V
```

You can use the program for a different translation of a triangle by changing the numbers in line 30. For a translation of a different type of polygon, let the V in line 10 go from 1 to the number of vertices.

Try These

1. Graph the image of the point $M(3, -3)$ at the right under a translation 5 units to the left and 6 units down.
$M(3, -3) \rightarrow M'(-2, -9)$

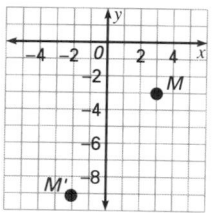

2. Graph the image of $\triangle DEF$ at the right under a translation 0 units to the right or left and 3 units down.
$D(-4, 6) \rightarrow D'(-4, 6 - 3)$
$\qquad\qquad \rightarrow D'(-4, 3)$
$E(1, 7) \quad \rightarrow E'(1, 7 - 3)$
$\qquad\qquad \rightarrow E'(1, 4)$
$F(1, 2) \quad \rightarrow F'(1, 2 - 3)$
$\qquad\qquad \rightarrow F'(1, -1)$

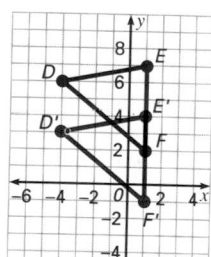

14-6   Translations   **523**

Example 1: Have students identify the preimage, image, and describe the translation of point *C* on the coordinate grid. Elicit from students that the preimage is the original point *C*(–5, 3), the image is the moved point *C'*(–1, 0), and the translation is the move of 4 units to the right and 3 units down.

Example 2: Guide students to summarize the algebraic method for determining translated coordinates. Elicit from students that if a figure is moved *n* units, *n* is added if the translation is to the right or up; *n* is subtracted if the translation is to the left or down. Point out that a move left or right changes the *x*-coordinate and a move up or down changes the *y*-coordinate.

Additional Questions/Examples

1. Draw a 3-unit by 4-unit rectangle on graph paper. Translate the rectangle 5 units to the left and 3 units down.

2. Square *ABCD* has vertices *A*(2, 3), *B*(6, 3), *C*(6, –1), and *D*(2, –1). Its translation image has vertices *A'*(–3, 1), *B'*(1, 1), *C'*(1, –3), and *D'*(–3, –3). Describe the translation. **5 units left, 2 units down**

3. Graph the image of triangle *ABC* with vertices *A*(–3, –1), *B*(–4, –4), and *C*(–2, –3) under a translation 3 units to the left and 5 units up. **A'(–6, 4), B'(–7, 1), C'(–5, 2)**

Guided Practice/Try These Have students work in small groups. Point out that all group members should come to an agreement on each answer.

3 SUMMARIZE

Write About Math Ask students to describe the following procedures in their math journals.
1. finding a translation image by tracing and moving a copy of the preimage
2. finding the coordinates of the translated image algebraically

4 PRACTICE

Practice/Solve Problems For Exercise 4 you may wish to have students describe the translation of the shaded figure that gives figures *A*, *B*, *E*, *F*, *G*, and *H*.

Extend/Solve Problems For Exercise 15 have students work in pairs. Students should each make their own designs and then have their partners graph translations of their designs.

Think Critically/Solve Problems Before students complete Exercises 17–19, give some examples of tessellations so that students can envision what a tessellation is. Include in your examples tile patterns of a floor or ceiling, brick patterns of a wall or sidewalk, and a honeycomb pattern.

524

EXERCISES

PRACTICE/ SOLVE PROBLEMS

1. On a coordinate plane, graph trapezoid *DEFG* with vertices *D*(–4, 3), *E*(6, 3), *F*(4, –3), and *G*(–2, –3). Then graph its image under a translation 4 units left and 0 units up or down. **See Additional Answers.**

Copy each set of figures on a coordinate plane. Then graph the image of each figure under the given translation.

2. 5 units right and 3 units up
Answers will vary. Check students' graphs.

3. 1 unit left and 4 units down
Answers will vary. Check students' graphs.

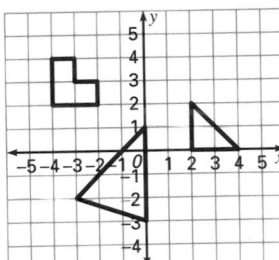 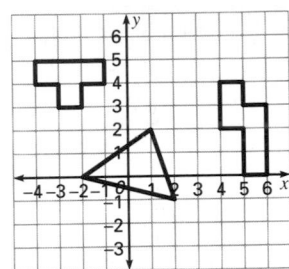

4. Which of the lettered figures are translations of the shaded figure? Give reasons for your answers. **Figures A, B, E, F, G, H are translations of the shaded figure. Reasons will vary.**

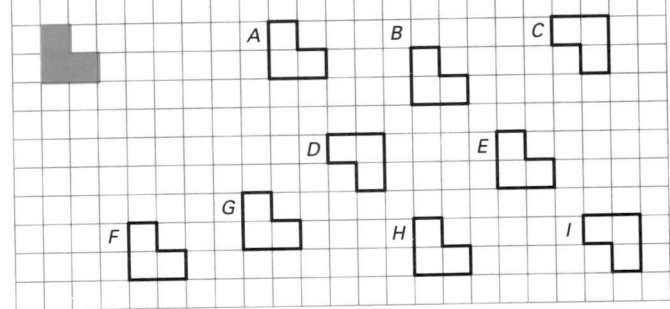

Copy each figure onto graph paper. Then graph the image of each figure under a translation 6 units to the left and 3 units up.
Check students' work.

5.   6.   7.   8.

 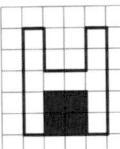 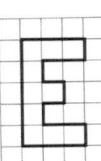

MIXED REVIEW

Find the answer.
1. –8 – (–5) **–3**
2. 0.007 – 0.32 **–0.313**
3. 36 ÷ 1.5 **24**
4. 4,000 × 0.366 **1,464**
5. 0 – (–6) **6**

Solve.
6. Change 345 cm to meters. **3.45 m**
7. Find the volume of a cube that is 8 yd long on each edge. **512 yd³**
8. In a chemistry class of 30 students, 4 students are chosen at random to attend a conference. Does anyone have at least a 50% chance of being chosen? **no**

AT-RISK STUDENTS

Have students work with oaktag shapes. Have them trace their shapes on graph paper, then slide the shapes to another position on the graph paper. Have students trace the shapes in their new positions, then find the translation image for the slide.

Determine the direction and number of units of the translation for each preimage and its image.

9.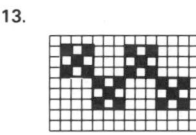

4 right, 3 down

10.

1 right, 3 up

11.

5 right, 2 down

Translations were used to make each design. Decide how each was done. Then copy the design on graph paper. **Check students' work.**

12.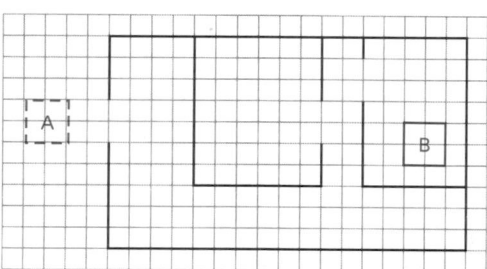

13.

14.

15. Make your own design with translations on graph paper.
Check students' work.

16. Describe the translations needed to move box *B* out of the warehouse to position *A*.
3 up, 4 left, 6 down, 8 left, 4 up, 6 left

A **tessellation** is a pattern in which identical copies of a figure fill a plane so that there are no gaps or overlaps. The computer program is written in Logo. The program will create a tessellation.

17. Describe the tessellation the program will create.
tessellating rectangles

18. Modify the program so that the tessellation is composed of squares. **Answers will vary.**

19. Modify the program so that the tessellation is composed of parallelograms. **Answers will vary.**

```
TO FIGURE
  FD 10 RT 90
  FD 5 RT 90
  FD 10 RT 90
  FD 5 RT 90
END

TO SLIDE
  PU BK 10 BK
END

TO COLUMN
  REPEAT 2 [FIGURE SLIDE]
END

TO TESS
  REPEAT 4 [COLUMN PU
  FD 20 RT 90 FD 5 LT 90 PD]
END
```

Extra Practice
1. Draw a parallelogram on graph paper. Translate the parallelogram 4 units left and 5 units up.

2. A quadrilateral has vertices $W(-5, 5)$, $X(0, 8)$, $Y(4, 3)$, and $Z(3, -3)$. Its translation image has vertices $W'(-2, 3)$, $X'(3, 6)$, $Y'(7, 1)$, and $Z'(6, -5)$. Describe the translation.   **3 units right, 2 units down**

3. Graph the image of square *ABCD* with vertices $A(-2, 1)$, $B(1, 1)$, $C(-2, 4)$, and $D(1, 4)$ under a translation 3 units to the right and 3 units down.
$A'(1, -2)$, $B'(4, -2)$, $C'(1, 1)$, $D'(4, 1)$

Extension Have students work in pairs. Students should each create a design, a preimage, on graph paper and describe a translation. Then have them exchange papers to find the image of each design.

Section Quiz
1. Draw a right triangle. Translate the figure 2 units to the left and four units up.

2. The point $(-8, -5)$ is translated 3 units parallel to the *x*-axis. Write the new coordinates of the point and tell in which direction the point moved.
$(-5, -5)$; right

3. A parallelogram has vertices $A(2, 6)$, $B(5, 6)$, $C(2, 3)$, and $D(-1, 3)$. Its translation image has vertices $A'(-2, 4)$, $B'(1, 4)$, $C'(-2, 1)$, and $D'(-5, 1)$. Describe the translation.   **4 units left, 2 units down**

Get Ready mirror, graph paper, ruler, calculator

Additional Answers
See page 584.

WARM-UP

Write the opposite integer.
1. −7 **7**   2. 18 **−18**   3. −9 **9**
Multiply.
4. $6 \times (-1)$ **−6**
5. $-12 \times (-1)$ **12**

MOTIVATE

Explore Display a mirror in the front of the classroom. Have volunteers look into the mirror and describe what they see. Elicit from students that when you look into a mirror, you are looking at a transformation of yourself.

TEACH

Use the Pages/Skills Development
Have students read this part of the section and then discuss the examples. Help students understand the meaning of the term *line of reflection.* Elicit from students that if the line of reflection is the *y*-axis, points *P* and *P'* have opposite *x*-coordinates and the same *y*-coordinate. Ask when the reflection *Q'* of point *Q* would have the same *x*-coordinate and opposite *y*-coordinates. **When the line of reflection is the x-axis.**

526

14-7   Reflections

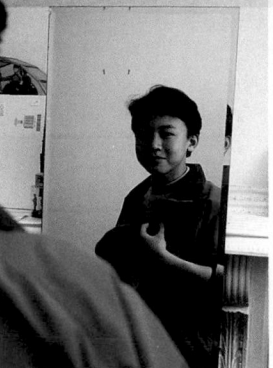

EXPLORE When you think of reflections, you probably think of mirrors.

a. In an ordinary mirror, your image seems to be reversed from right to left. Look into a mirror and raise your right hand. What does your image do? **It raises its left hand.**

b. Hold two mirrors at right angles. Move close enough to the right angle so that you see just one image of yourself. How does this image differ from what you see in just one mirror? **Left and right are not reversed.**

SKILLS DEVELOPMENT You have just learned about one kind of transformation—a translation. Another kind of transformation is a reflection. A **reflection** is a transformation in which a figure is *flipped*, or *reflected*, over a *line of reflection.*

In this diagram, point *P'* is the image of point *P* under a reflection across line *EF*. The two points, *P* and *P'*, are the same distance from $\overleftrightarrow{EF}$, and if line *PP'* were drawn, it would be perpendicular to $\overleftrightarrow{EF}$.

The *x*- and *y*-axes can be used as reflection lines for figures drawn on a coordinate plane.

Example 1

Graph the image of $\triangle PQR$ with vertices $P(-6, 4)$, $Q(-5, -2)$ and $R(-2, -4)$ under a reflection across the *y*-axis.

Solution
Multiply the *x*-coordinate of each vertex by -1.

$P(-6, 4) \rightarrow P'(-6 \times (-1), 4)$
$\rightarrow P'(6, 4)$

$Q(-5, -2) \rightarrow Q'(-5 \times (-1), -2)$
$\rightarrow Q'(5, -2)$

$R(-2, -4) \rightarrow R'(-2 \times (-1), -4)$
$\rightarrow R'(2, -4)$ ◄

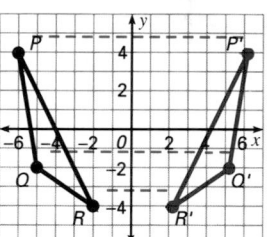

TEACHING TIP

Students will most likely count units in order to determine the position of each reflection image. However, when the original image spans more than one quadrant, students are less likely to make errors if they list the coordinates of the vertices of the reflection image before they draw.

Example 2

Find the image of △STU with vertices S(−2,2), T(3,7), and U(4,2) under a reflection across the x-axis.

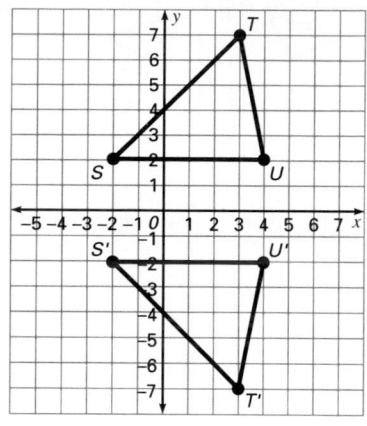

Solution
Multiply the y-coordinate of each vertex by −1.

$S(-2,2) \rightarrow S'(-2, 2 \times (-1))$
$\quad\quad\quad \rightarrow S'(-2,-2)$

$T(3,7) \rightarrow T'(3,7 \times (-1))$
$\quad\quad\quad \rightarrow T'(3,-7)$

$U(4,2) \rightarrow U'(4, 2 \times (-1))$
$\quad\quad\quad \rightarrow U'(4,-2)$

◀

TRY THESE

Give the coordinates of the image of each point under a reflection across the given axis.

1. (4,−5); y-axis (−4, −5)
2. (−6,−1); x-axis (−6, 1)
3. (7,9); x-axis (7, −9)
4. (−3,8); y-axis (3, 8)
5. (−2,0); y-axis (2, 0)
6. (−2,0); x-axis (−2, 0)

EXERCISES

Graph the point on a coordinate plane. Then graph the image of the point under a reflection across the given axis.

**PRACTICE/
SOLVE PROBLEMS**

1. (4, 1); x-axis (4, −1)
2. (−2, 5); x-axis (−2, −5)
3. (7, −2); y-axis (−7, −2)
4. (10, 5); y-axis (−10, 5)
5. (−8, −7); x-axis (−8, 7)
6. (5, 0); y-axis (−5, 0)

ASSIGNMENTS

BASIC
1–10, 11–12, 15, PSA 1–3

AVERAGE
7–10, 11–13, 15–16, PSA 1–4

ENRICHED
7–10, 11–14, 15–17, PSA 1–4

ADDITIONAL RESOURCES
Reteaching 14–7
Enrichment 14–7
Transparency Master 62

Example 1: Draw a coordinate plane on the chalkboard or on acetate for use on an overhead projector. Draw triangle PQR and label each vertex with its coordinates. Have volunteers demonstrate how to draw and label the reflection image, using the y-axis as the reflection line. Discuss students' observations. Have students measure the distance from the preimage and image vertices to the reflection line. Students should notice that the distances are equal for each pair of points.

Example 2: Draw triangle STU on the chalkboard or overhead projector and call on a volunteer to draw and label the reflection image over the x-axis.

Additional Questions/Examples
Graph each point on a coordinate grid. Then find the image of the point under a reflection across the given axis.
1. B (6, −3); y-axis **(−6, −3)**
2. G (−3, −2); x-axis **(−3, 2)**
3. C (1, 0); y-axis **(−1, 0)**
4. E (−5, 8); x-axis **(−5, −8)**

5. Find the image of triangle ABC with vertices A (5, 5), B (5, −1), and C (3, −1) under a reflection across the y-axis. **A'(−5, 5); B'(−5, −1); C'(−3, −1)**

Guided Practice/Try These Have a volunteer complete these exercises on the chalkboard or on an overhead projector while the other students work at their seats. Discuss any problems or questions students might have.

3 SUMMARIZE

Key Questions

1. How is a translation image similar to a reflection? **The images of both are the same size and shape as their preimages.**
2. How is a translation different from a reflection? **A reflection image is flipped, while a translation image is not.**

4 PRACTICE

Practice/Solve Problems For Exercises 1–10 remind students how the x and y-coordinates are affected by the axis used as the reflection line.

Extend/Solve Problems For Exercises 11–12 you may also wish to have students find words, such as TOOT, which can be formed using reflections.

Think Critically/Solve Problems For Exercise 16 have students recall the Pythagorean Theorem formula as $a^2 + b^2 = c^2$.

Problem Solving Applications Have students work in pairs to solve the problems. Suggest that turning their books to see different views of the billiard tables may help them visualize the problems.

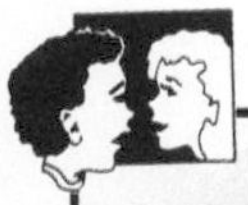

TALK IT OVER

Under what circumstances would the reflection image of a point on the coordinate plane have the same x-coordinate as its preimage?

The preimage and the image of a vertex of a triangle would have the same x-coordinate if it were reflected over the x-axis.

Graph the image of each figure under a reflection across the y-axis.

7.
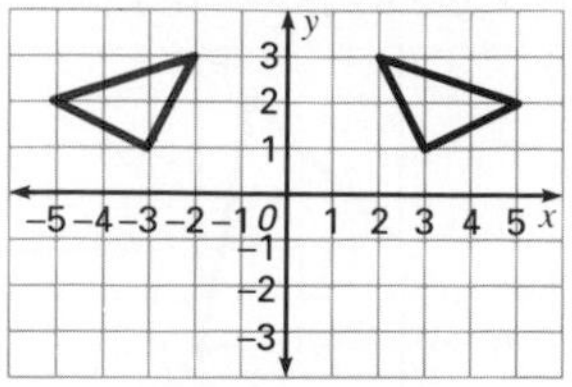

8.
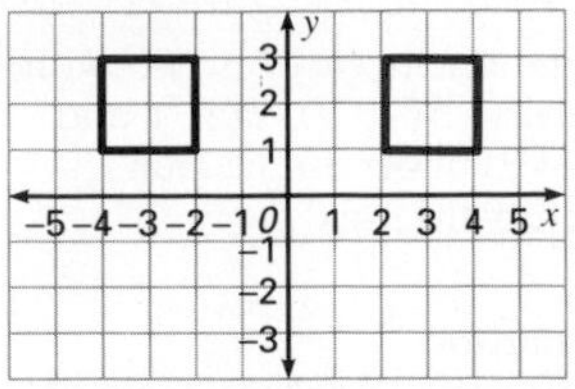

Graph the image of each figure under a reflection across the x-axis.

9.
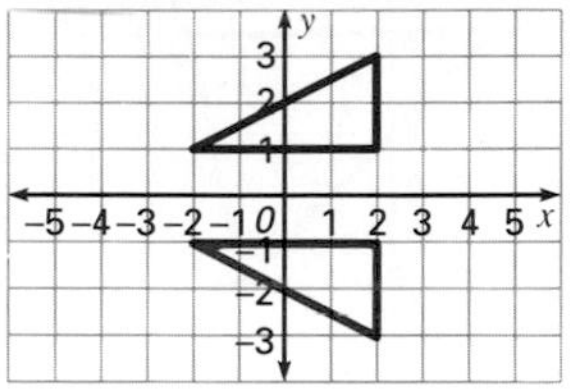

10.
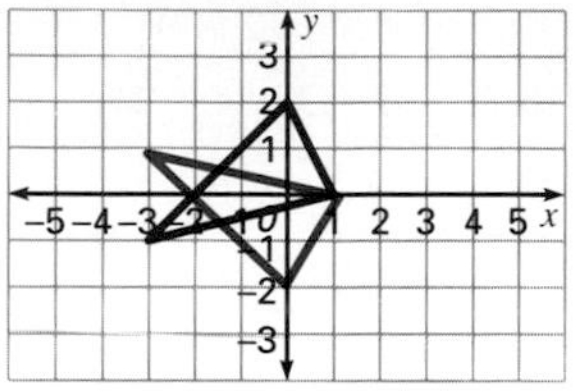

EXTEND/ SOLVE PROBLEMS

11. The letter A is a reflection of itself across the vertical line shown. Which other letters have this property? **H, I, M, O, T, U, V, W, X, Y**

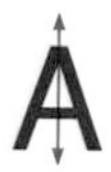

12. The letter H is a reflection of itself in two ways. Which other letters have this property? **I, O, X**

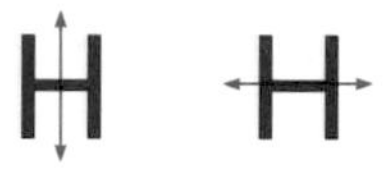

Triangle ABC is reflected in the line GK as shown.

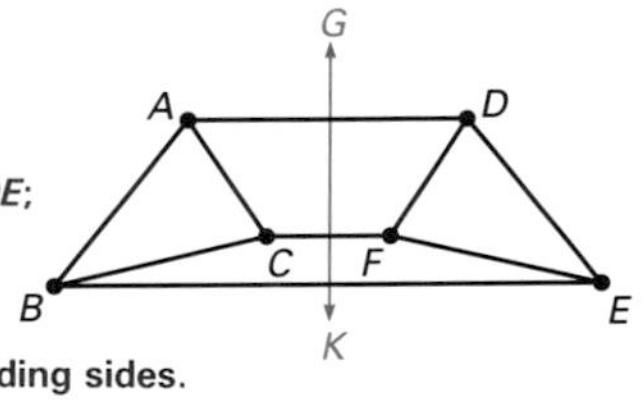

13. Which line segments are equal in length? **AC and DF; AB and DE; BC and EF**
14. Which angles are right angles? **Those formed by GK and the segments connecting corresponding sides.**

THINK CRITICALLY/ SOLVE PROBLEMS

In Exercises 15–17, $\triangle XYZ$ is drawn on a coordinate plane. The vertices are $X(2, 2)$, $Y(2, 5)$, $Z(6, 2)$.

15. Is the order of the vertices of $\triangle XYZ$ clockwise or counterclockwise? Find the reflection image $\triangle X'Y'Z'$ across the x-axis. Is the order of the vertices clockwise or counterclockwise? **clockwise; counterclockwise**
16. Find the lengths of the sides of both triangles XYZ and $X'Y'Z'$. (Hint: You will need to use the Pythagorean Theorem to find the length of one side in each triangle.) **$XY = X'Y' = 3$; $YZ = Y'Z' = 5$; $XZ = X'Z' = 4$**
17. Think of your answers to Exercises 15 and 16. Which properties of a figure change during a reflection? Which stay the same?

17. The order of the vertices changes. The lengths of the sides stay the same.

Problem Solving Applications:

REFLECTIONS IN BILLIARDS

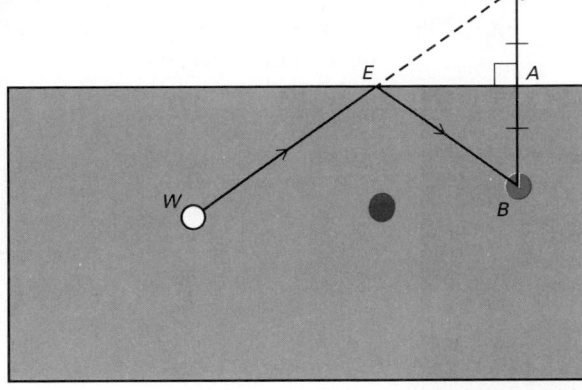

Reflections are important to the game of billiards. Often in this game, you must try to hit the cue ball so that it misses one object ball but gently taps another. To do this, you can deflect your shot off a wall of the table as shown in the diagram. In this case, the object ball, *B*, is blue, and the cue ball, *W*, is white. To miss the red ball, obtain the image of *B* by using the edge of the billiard table as the reflection line.

The distance $AB = AB'$, and $m\angle EAB = 90°$.

To make the shot, aim at the image ball, *B'*.

For each diagram, decide which edge of the billiard table should be the reflection line. Then copy the positions of the balls and draw a diagram so that the white ball will bounce off the reflection line to hit the blue ball.

1.
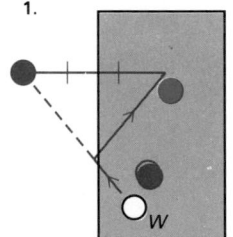

2.
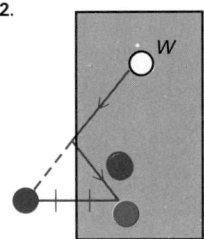

3.
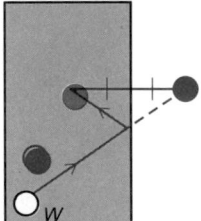

4. A *two-cushion shot* is a shot that uses two different reflecting edges. Copy the figure at the right and show how to use a two-cushion shot to make the white ball hit the blue ball.

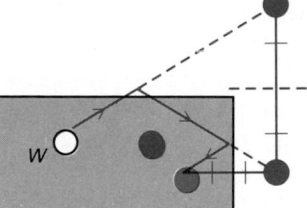

14-7 Reflections **529**

Extra Practice Graph each point on a coordinate plane. Then find the image of the point under a reflection across the given axis.
1. *A* (–4, 1); *x*-axis **(–4, –1)**
2. *B*(10, –2); *x*-axis **(10, 2)**
3. *C*(–6, 6); *y*-axis **(6, 6)**
4. *D*(0, 4); *x*-axis **(0, –4)**
5. Find the image of a triangle with vertices *G* (2, 2), *H* (4, 3), and *I* (3, 6) under a reflection across the *y*-axis.
    G'(–2, 2); H'(–4, 3); I'(–3, 6)
6. Find the image of a triangle with vertices *J* (–4, 3), *K* (–3, 5), and *L* (3, 4) under a reflection across the *x*-axis.
    J'(–4, –3); K'(–3, –5); L'(3, –4)

Extension Have students draw the image of a triangle with vertices *D* (–4, 3), *E* (–2, –1), and *F* (3, 4) under a reflection across the *x*-axis, then a reflecting image across the *y*-axis.

Section Quiz
Write *true* or *false*.
1. Under a reflection, the size and shape of the figure stays the same. **true**
2. Under a reflection, the order of the vertices of a figure stay the same. **false**
3. Under a reflection across the *x*-axis, the *y*-coordinate of a point changes. **true**
4. Find the image of a triangle with vertices *A* (–2, 3), *B* (–1, 2), and *C* (3, –4) under a reflection across the *y*-axis. **A'(2, 3); B'(1, 2); C'(–3, –4)**

Get Ready graph paper or dot paper, tracing paper, ruler

OBJECTIVE
• Identify and draw rotations

MATERIALS NEEDED
graph paper or dot paper, rulers

VOCABULARY
angle of rotation, center of rotation (turn center), clockwise, counterclockwise, rotation

WARM-UP

Find the image of the point reflected across the given axis.
1. B (2, 9); y-axis **(–2, 9)**
2. G (–2, –5); x-axis **(–2, 5)**
3. M (–7, 7); y-axis **(7, 7)**

1 MOTIVATE

Explore On the chalkboard, draw a clockface that is large enough for students to see across the room. Draw hands to indicate 1 o'clock. Ask students what happens to the hands as the time changes from 1 o'clock to 2 o'clock. Elicit that the minute hand turns clockwise in a complete circle during which time the hour hand moves from 1 to 2.

2 TEACH

Use the Pages/Skills Development
Have students read this part of the section and then discuss the examples. Use the clockface drawn on the chalkboard to introduce these terms and concepts: *rotation, center of rotation, angle of rotation, clockwise, counterclockwise.*

14-8 Rotations

EXPLORE The hands of a clock turn, or rotate, around the point at which they are attached in the center of the clockface.

a. What fractional part of a full turn does the minute hand make in 30 minutes? **one half turn**

b. When the minute hand has moved 90° forward, or clockwise, from 3:25, to what number does it point? **to the 8**

c. What degree of rotation of the minute hand will set a clock back by 15 minutes? **a 90° counterclockwise rotation**

SKILLS DEVELOPMENT

CONNECTIONS

A continuous spinning motion in a single direction is called *rotary motion*. To measure rotary motion, try this. Turn a bicycle upside down and mark one tire with chalk. Spin the wheel and count the number of complete turns in one second. From this number and the circumference of the wheel you can compute approximately how far the wheel would roll along the ground in one minute.

See Additional Answers.

You have learned about two kinds of transformations: translations and reflections. A third type of transformation is a rotation.

A **rotation** is a transformation in which a figure is *turned,* or *rotated,* about a point. The movement of the hands of a clock, the turning of the wheels of a bicycle, and moving windshield wipers on a car are just a few examples of a rotation about a point.

To describe a rotation, you need three pieces of information:

► the point about which the figure is rotated (called the *center of rotation,* or *turn center*)
► the amount of turn expressed as a fractional part of a whole turn or in degrees (called the *angle of rotation*)
► the direction of rotation—either *clockwise* or *counterclockwise*

Example 1

Draw the rotation image of the flag when it is turned 90° clockwise about a turn center, *T*.

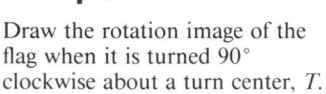

ESL STUDENTS

Use transparencies to help students differentiate between translations, reflections, and rotations. If students work together, they may be able to help one another clarify their thinking.

Solution

Copy the flag onto dot paper. Label point T. Then trace the flag and point T onto a sheet of tracing paper.

Hold your pencil point on the tracing at point T. Turn the tracing one-quarter turn, or $90°$.

Remove the tracing paper and copy the image onto the dot paper. ◄

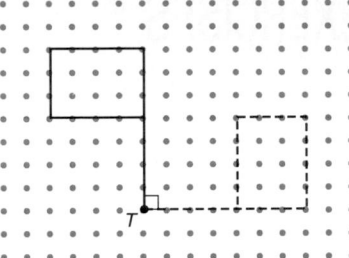

Example 2

Find the image of $\triangle PQR$ with vertices $P(3, 3)$, $Q(2, 1)$, and $R(5, 1)$ after a turn of $180°$ clockwise about the origin.

Solution

Multiply both the x-coordinate and the y-coordinate of each vertex by -1.

$P(3, 3) \rightarrow P'(-1 \times 3, -1 \times 3)$
$ \rightarrow P'(-3, -3)$

$Q(2, 1) \rightarrow Q'(-1 \times 2, -1 \times 1)$
$ \rightarrow Q'(-2, -1)$

$R(5, 1) \rightarrow R'(-1 \times 5, -1 \times 1)$
$ \rightarrow R'(-5, -1)$ ◄

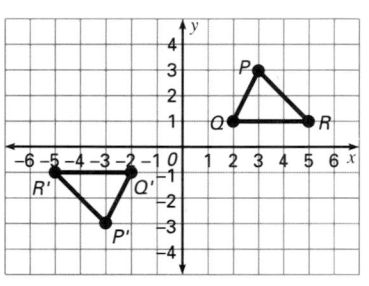

TRY THESE

1. Copy the figure on dot paper. Then draw the image of the figure when it is turned $270°$ clockwise about a turn center, T.

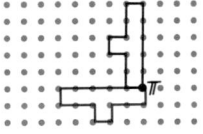

2. Graph the image of $\triangle KLM$ after a turn of $180°$ clockwise about the origin.

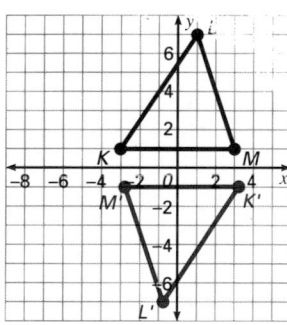

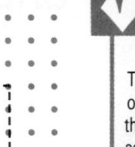
MATH IN THE WORKPLACE

How can an airplane manufacturer design the wing of a jet on a computer? Rosemary Chang does the kind of work that makes this possible. Ms. Chang is an expert on geometric surfaces. Her job is to write math equations that can be used to translate numerical data into computer programs. Armed with one of these programs, the computer designer can create a three-dimensional image of the jet's wing. This image can then be translated, reflected, and rotated on the computer screen, enabling the design to be evaluated from many points of view.

Ms. Chang has a bachelor's degree from New York University and a doctorate in applied mathematics from Brown University.

Example 1: Have students use their tracings to draw a rotation of the flag when it is turned $90°$ counterclockwise about point T.

Example 2: Have students identify the coordinates of the rotation image and compare them with those listed in the solution. Elicit that the coordinates of the rotation image are the opposite of those in the preimage. Point out that this result occurs only for a rotation of $180°$.

Additional Questions/Examples

1. Draw line segment CE 4 units long on graph paper. Place a dot 1 unit to the left of point C and label it D. Draw the rotation image of segment CE after a turn of $90°$ clockwise about point D. What are the coordinates of the rotation image? **Answers will vary.**

2. Find the image of triangle GHI with vertices $G(-2, 1)$, $H(3, 4)$ and $I(4, 0)$ after a rotation of $90°$ counterclockwise about the origin. **$G'(-1, -2)$; $H'(-4, 3)$; $I'(0, 4)$**

3. Find the image of triangle ABC with vertices $A(3, 2)$, $B(0, -1)$, and $C(3, -2)$ after a rotation of $180°$ clockwise about the origin. **$A'(-3, -2)$; $B'(0, 1)$; $C'(-3, 2)$**

MAKING CONNECTIONS

Have students prepare a bulletin board display showing real-life examples of reflections, translations, and rotations.

531

Guided Practice/Try These Guide students through the exercises, showing them how to trace each preimage and rotate it about the turn center until they reach the rotation image.

3 SUMMARIZE

Write About Math Have students describe how to draw a rotation image of a figure in their math journals. You may also wish to have them define a reflection and translation and compare and contrast the transformations.

4 PRACTICE

Practice/Solve Problems Emphasize the fact that many different rotations can produce the same result. For example, a 90°-clockwise rotation about the origin produces the same image as a 270°-counterclockwise rotation. For Exercises 1–6 ask students to name other rotations that would give them the same results.

Extend/Solve Problems For each of Exercises 10–12 have students identify the rotation in terms of degrees.

Think Critically/Solve Problems Exercises 17–19 require students to investigate properties of rotations. Students should conclude that the order of the vertices, side lengths, angle measures, and areas stay the same in a rotation. Only the location and orientation of the figure changes.

5 FOLLOW-UP

Extra Practice
Triangle *ABC* has vertices *A* (2, 4), *B* (5, 1), and *C* (1, 1).

1. Find the rotation image if the triangle is rotated 90° clockwise about the origin.
A'(4, –2); B'(1, –5); C'(1, –1)

EXERCISES

PRACTICE/ SOLVE PROBLEMS

Each drawing shows a line segment, $\overline{AB}$, and its rotation image, $\overline{A'B'}$, about a turn center, *T*. What is the angle of rotation in a clockwise direction?

1. 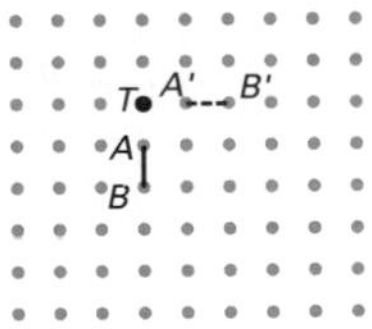

270°

2. 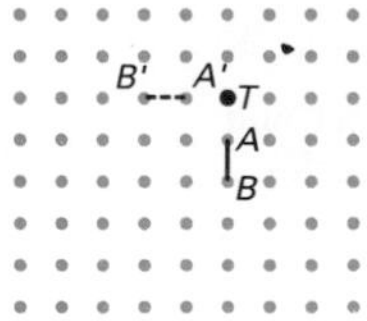

90°

3. 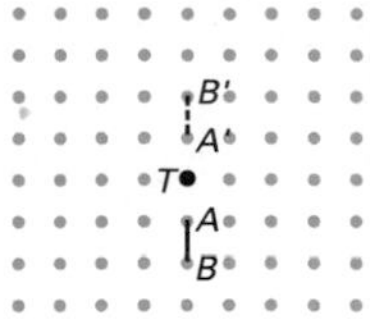

180°

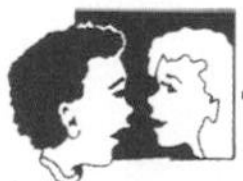

TALK IT OVER

The design is made entirely of straight lines. But parts of the design appear curved. What causes the curves to appear?

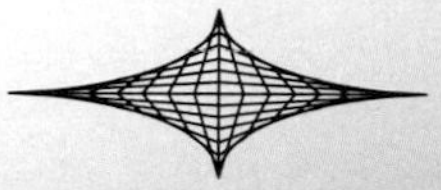

Answers will vary. The endpoints of the line segments form curves.

A figure has been rotated about a turn center, *T*. Describe each angle and direction of rotation.

4. 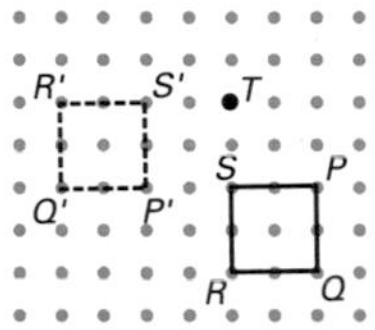

270° counterclockwise or 90° clockwise

5. 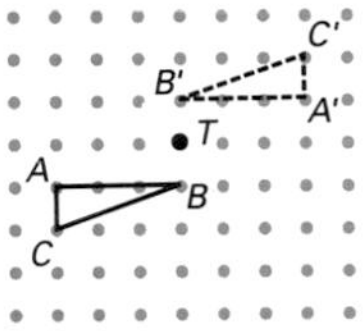

180° counterclockwise or 180° clockwise

6. 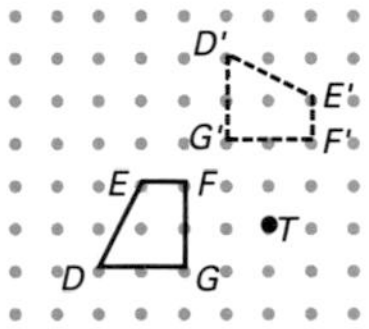

270° counterclockwise or 90° clockwise

Graph the image of each figure after a 90° turn clockwise about the origin. **See Additional Answers.**

7. 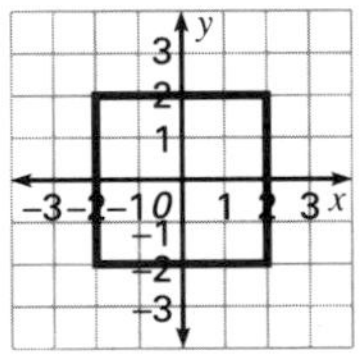

8. 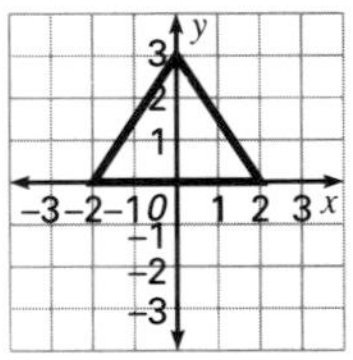

9. 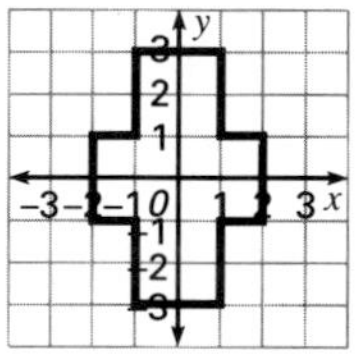

EXTEND/ SOLVE PROBLEMS

PROBLEM SOLVING TIP

Making a tracing of a figure can help you find the rotation image. Be sure to mark the point used as the center of rotation on your tracing.

Copy each figure. Then draw its image after the given rotation about point *T*. **See Additional Answers.**

10. quarter-turn counterclockwise
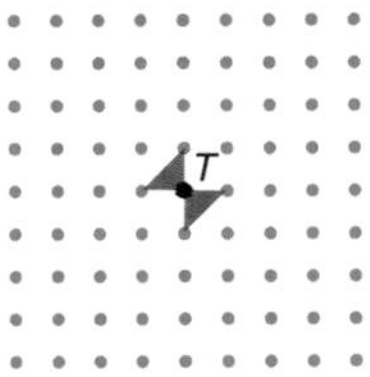

11. half-turn clockwise
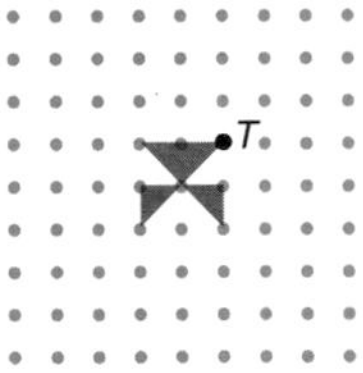

12. complete turn clockwise
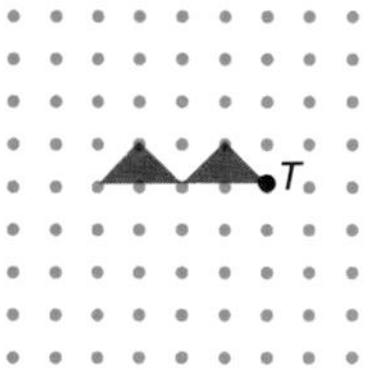

13. quarter-turn
counterclockwise

14. half-turn
clockwise

15. three-quarter-
turn clockwise

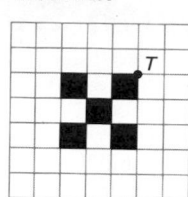

16. Each letter of the message below has been rotated either 90° or 180° clockwise or counterclockwise. The turn centers are shown by •. What is the message? **Hi There**

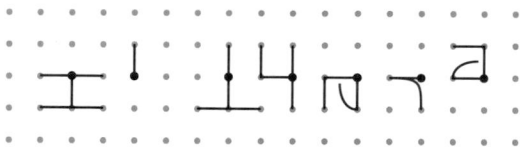

Triangle XYZ has vertices $X(2, 0)$, $Y(5, 0)$, and $Z(2, 4)$.

17. Is the order of the vertices of $\triangle XYZ$ clockwise or counterclockwise? Rotate $\triangle XYZ$ 90° clockwise to obtain the image $\triangle X'Y'Z'$. Are the vertices of $\triangle X'Y'Z'$ in clockwise or counterclockwise order? **counterclockwise; See Additional Answers.**
counterclockwise

18. Find the lengths of the sides of both triangle XYZ and $X'Y'Z'$. (Hint: You will need to use the Pythagorean Theorem to find the length of one side of each triangle.) $XY = X'Y' = 3$; $YZ = Y'Z' = 5$; $ZX = Z'X' = 4$; $\angle X = 90°$; $\angle Y \approx 53°$; $\angle Z \approx 37°$

19. Find the areas of $\triangle XYZ$ and $\triangle X'Y'Z'$. **The area of each is 6 square units.**

20. Generalize from the results of Exercises 17–19. What properties of a figure do you think change during a rotation? What properties do you think stay the same? **Order of vertices, side lengths, angle measures, and areas stay the same. Positions of vertices vary.**

Solve.

21. If a moth sat on the edge of a 12-in. phonograph record rotating at a speed of $33\frac{1}{3}$ revolutions per minute, how far would it travel in 3 min? **3,768 in., or 314 ft**

22. Refer to Exercise 21. How many inches per minute did the moth travel? **1,256 in./min**

THINK CRITICALLY/ SOLVE PROBLEMS

READING MATH

Many mathematical terms have other, related meanings. Give the meaning of the underlined word in each sentence. (You might need to refer to a dictionary.)

1. There was a remarkable <u>transformation</u> in his appearance since the last time I saw him.
2. Her success is a <u>reflection</u> of all her hard work.
3. The meaning of the word was lost in the <u>translation</u> from Spanish to English.
4. There were five players in the baseball team's pitching <u>rotation</u>.

2. Find the rotation image of the triangle rotated 270° counterclockwise about the origin. **A'(4, –2); B'(1, –5); C'(1, –1)**

3. What do you notice about your results? **They are the same.**

Extension
Have students choose a shape and make a design using translations, reflections, and rotations.

Section Quiz
1. Draw a right triangle on graph paper. Draw a turn center and label it R. Draw the rotation image for a 90°-clockwise turn.
2. Triangle MNO has vertices $M(2, –6)$, $N(7, –4)$, $O(4, –2)$. Write the coordinates of the image after a turn of 180° clockwise about the origin. **M'(–2, 6); N'(–7, 4); O'(–4, 2)**
3. Triangle ABC has vertices $A(2, 0)$, $B(5, 0)$, $C(5, 3)$. Triangle $A'B'C'$ has vertices $A'(0, 3)$, $B'(0, 6)$, $C'(–3, 6)$. Graph the two triangles, then tell if triangle $A'B'C'$ is a rotation of triangle ABC 90° counterclockwise about the origin. Explain. **The preimage and the image points are not equidistant from the center of rotation.**

Get Ready graph paper or dot paper, tracing paper, ruler

Additional Answers
See page 584.

14-9 Working with Symmetry

EXPLORE/ WORKING TOGETHER

Look around the classroom. Focus on any small object. Imagine a *vertical* line drawn down the middle of that object. Think about whether such a line would divide the object into two halves that match exactly.

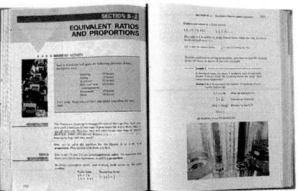

Focus on another object. This time, imagine a *horizontal* line imposed on the object. Do you think such a line would divide the object exactly in half?

Make three lists of things found in the classroom or at home that you think could be divided into halves by a single straight line—those that could be divided by a vertical line, those that could be divided by a horizontal line, and those that could be divided by a line that is neither vertical nor horizontal.

Compare your lists. Could more than one line divide any of the things you listed? Explain.

SKILLS DEVELOPMENT

MATH: WHO, WHERE, WHEN

Many properties of reflections find a particularly striking application in the Japanese art of paper folding. Called *origami* in Japanese, paper folding is used to make such traditional figures as the crane and the tortoise (symbols of good fortune), the carp (symbol of persistency), and the frog (symbol of love). Books with diagrams and instructions for folding these traditional paper decorations were published as early as the first quarter of the eighteenth century.

If you could fold this isosceles triangle along the line *PQ*, one side would fit exactly over the other. So, the triangle is said to have **line symmetry,** and line *PQ* is called a **line of symmetry.**

Example 1

Trace each figure and draw all the lines of symmetry.

a.    b. 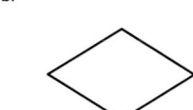   c.

Solution

a. 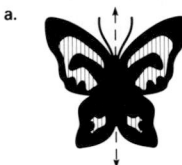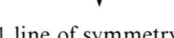   b. 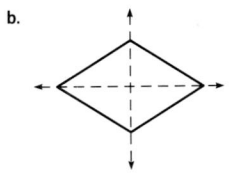   c.

1 line of symmetry   2 lines of symmetry   1 line of symmetry ◄

MAKING CONNECTIONS

To study the significance of mathematics in art, display works of the Dutch artist M.C. Escher (1898–1972). Escher's drawings of interlocking, infinitely repeating forms are based on translations, rotations, and reflections. *Creating Escher-Type Drawings*, Creative Publications, Inc., 1977, is an excellent source of ideas and activities.

The parallelogram does not have line symmetry. However, if you could place the tip of a pencil at point A and turn this parallelogram 180° clockwise, it would fit exactly over its original position. So, the parallelogram is said to have **rotational symmetry.** Its **order of rotational symmetry** is 2, since the parallelogram would fit over its original position 2 times in the process of a complete turn.

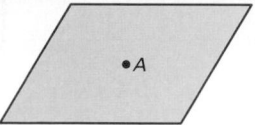

Example 2

Give the order of rotational symmetry for each figure.

a.
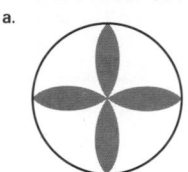

b.

c.

Solution

a.
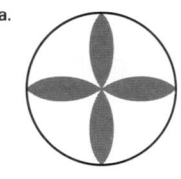

b.

c.

This figure fits over its original position 4 times during a complete turn, so the order of rotational symmetry is 4.

This figure fits over its original position 2 times during a complete turn, so the order of rotational symmetry is 2.

This figure fits over its original position 6 times during a complete turn, so the order of rotational symmetry is 6. ◄

TRY THESE

Trace each figure and draw all the lines of symmetry. If a figure has no lines of symmetry, write *None.*

1.
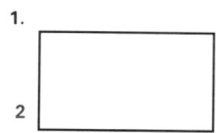

2.
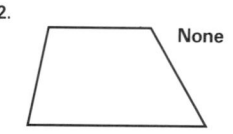
None

3.

Give the order of rotational symmetry for each figure.

4.
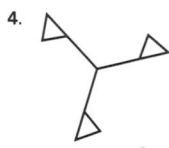

5.
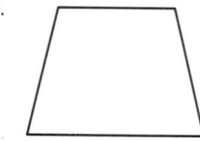

6.
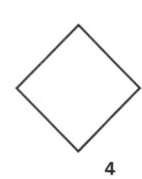

Example 1: Emphasize that each line of symmetry acts as a mirror. Point out that each half of the figure is the mirror image, or reflection, of the other.

Example 2: Explain to students that a figure has rotational symmetry if there is some point about which the figure can be rotated and made to coincide with itself more than once in the process of a complete turn. To test a figure for rotational symmetry, students can trace the figure, then place the tracing directly on top of the original. While holding the center fixed, students should turn the tracing until the tracing and the original again coincide.

Additional Questions/Examples
Draw each figure. How many lines of symmetry does each figure have?
1. a square **4**
2. an equilateral triangle **3**
3. a regular pentagon **5**
4. a semicircle **1**
Give the order of rotational symmetry for each figure.
5. a regular decagon **10**
6. the letter H shown for Exercise 20 in the student text **2**

EXERCISES

Guided Practice/Try These
Have students work together in small groups. Point out that all group members should come to an agreement on each answer.

3 SUMMARIZE

Write About Math Have students define line symmetry and rotational symmetry and describe in their math journals how to find the order of rotational symmetry.

4 PRACTICE

Practice/Solve Problems You may wish to have students use the trace-and-fold technique to complete Exercises 1–6.

Extend/Solve Problems For Exercises 13–15 have students compare drawings.

Think Critically/Solve Problems For Exercise 20 ask students which letters have both vertical and horizontal line symmetry. Elicit the fact that the letters H, I, O, and X have both kinds of symmetry.

5 FOLLOW-UP

Extra Practice
How many lines of symmetry does each figure have?

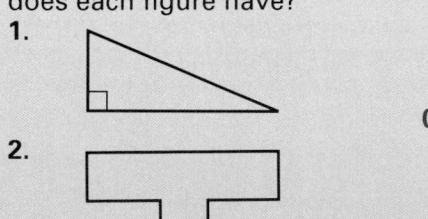

1.   0

2.   1

PRACTICE/ SOLVE PROBLEMS   Tell whether the dashed line is a line of symmetry. (Trace the figure if you need to).

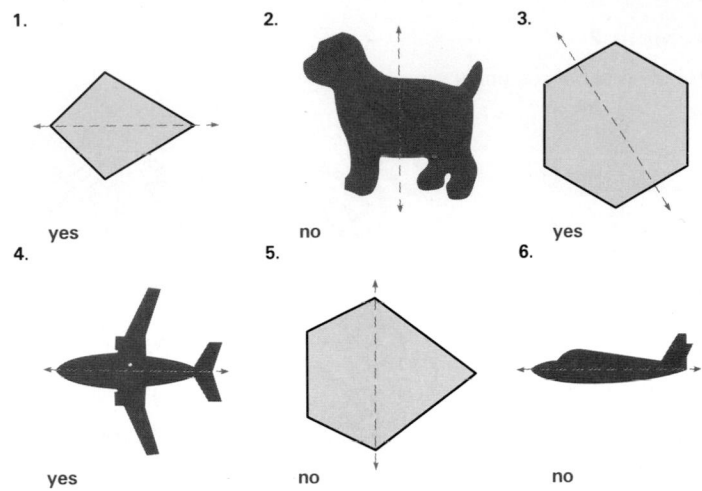

1. yes
2. no
3. yes
4. yes
5. no
6. no

Give the order of rotational symmetry for each figure.

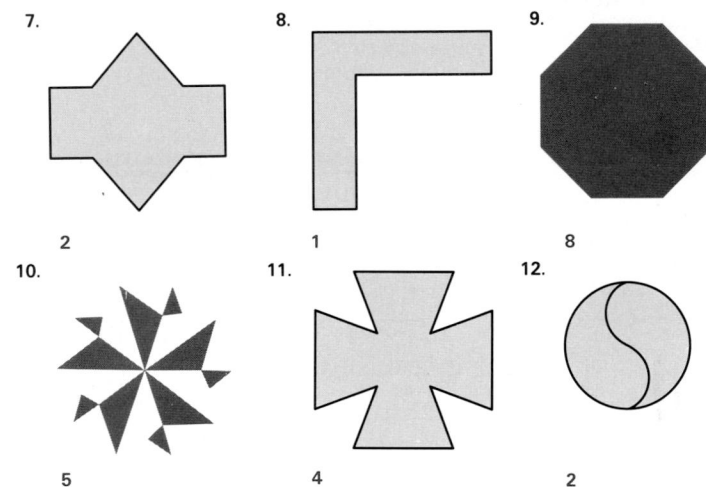

7. 2
8. 1
9. 8
10. 5
11. 4
12. 2

EXTEND/ SOLVE PROBLEMS   Draw a triangle with the given number of lines of symmetry.

13. 1   isosceles triangle
14. 3   equilateral triangle
15. 0   scalene triangle

Trace each drawing on dot paper or graph paper. Then complete the figure to show how it would appear in a mirror held vertically along the line of symmetry.

16.

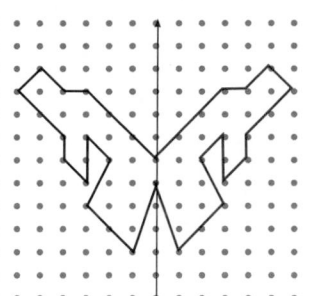

17.

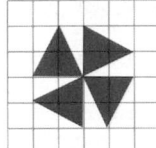

This figure, drawn on graph paper, has rotational symmetry of order 4. Use graph paper to construct a figure having rotational symmetry of the order indicated.
Answers will vary.

18. order 3   19. order 6

20. Some of the letters of the alphabet, when capitalized, have horizontal line symmetry. Others have vertical line symmetry. Still others have both. (Some have no symmetry at all!) Determine which of the capital letters below have horizontal line symmetry. Then, write a sentence using those letters alone. (You may use each letter more than once.) **Students' sentences will vary.**

ABCDEFGHI
JKLMNOPQR
STUVWXYZ

THINK CRITICALLY/ SOLVE PROBLEMS

WRITING ABOUT MATH

Think about which capital letters have rotational symmetry. Write step-by-step directions that someone who is not in your class could use to determine the rotational symmetry of the letters.

Letters H, I, N, O, S, X, and Z have rotational symmetry.

20. Letters with horizontal line symmetry: B, C, D, E, H, I, O, X; Letters with vertical line symmetry: A, H, I, M, O, T, U, V, W, X, Y; Letters F, G, J, L, P, Q, and R have no symmetry.

14-9   Working with Symmetry   **537**

3.

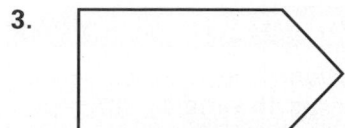

1

Give the order of rotational symmetry for each figure.

4.

4

5.

6

Extension Have students draw a circle on graph paper. Ask them to determine how many lines of symmetry there are in a circle. Elicit from students that a circle has an infinite number of lines of symmetry.

Section Quiz How many lines of symmetry does each figure have?

1.

1

2.

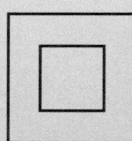

4

3.

4

Find the order of rotational symmetry of each figure.
4. a square   **4**
5. a regular hexagon   **6**
6. a regular octagon   **8**

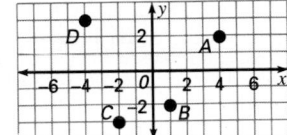

1. The movement of the hands of a clock is an example of a(n) ___?___ about a point. **e**

   a. ordered pair
   b. slope
   c. reflection
   d. translation
   e. rotation
   f. linear

2. A line of ___?___ works something like an ordinary mirror. **c**

3. If the solutions to an equation form a straight line when they are graphed, the equation is ___?___. **f**

4. Because points graphed on a coordinate plane are always identified by first the x-coordinate and then the y-coordinate, (x, y) is called a(n) ___?___. **a**

5. A(n) ___?___ can also be called a slide. **d**

6. For a line segment or a straight line, the ___?___ is the ratio of the rise to the run. **b**

SECTIONS 14–1 AND 14–3   LINEAR EQUATIONS   (pages 506–509, 512–515)

► When graphing an ordered pair (x, y), x is the horizontal distance to the left or right of the origin; y is the vertical distance up or down from the origin. The notation $A(x, y)$ means that point A has x-coordinate x and y-coordinate y.

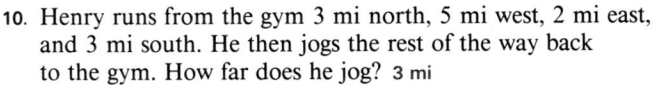

► A **linear equation** in two variables has many solutions, each of which can be represented by an ordered pair. When the points that correspond to these ordered pairs are graphed on a coordinate plane, they lie along a straight line.

7. What are the coordinates of points A, B, C, and D?
   $A(4, 2)$; $B(1, -2)$; $C(-2, -3)$; $D(-4, 3)$

8. On graph paper, graph the points $E(4, -3)$ and $F(-8, -2)$ on a coordinate plane. See Additional Answers.

9. Graph the equation $y = 2x - 3$. **See Additional Answers.**

SECTIONS 14–2 AND 14–4   PROBLEM SOLVING   (pages 510–511, 516–519)

► A grid can often be used to find distances.
► To find the slope of a line, choose any two points on the line. Then find the difference in the y-coordinates and divide by the difference in the x-coordinates.

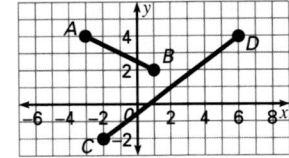

10. Henry runs from the gym 3 mi north, 5 mi west, 2 mi east, and 3 mi south. He then jogs the rest of the way back to the gym. How far does he jog? **3 mi**

11. Find the slope of each line segment. $\overline{AB}$: slope $= -\frac{1}{2}$; $\overline{CD}$: slope $= \frac{3}{4}$

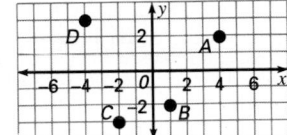

14 CHAPTER REVIEW

Introduction The Chapter Review emphasizes the major concepts, skills, and vocabulary presented in this chapter and can be used for diagnosing students' strengths and weaknesses. Page references direct students back to appropriate sections for additional review and reteaching.

Using Pages 538–539 Allow students to quickly scan the Chapter Review and ask questions about any section they find confusing. For Exercises 1–6 review key vocabulary. For Exercise 11 remind students to be consistent in finding the differences. Tell them to be sure to subtract the y-coordinates and the x-coordinates in the same order. Exercises 12–16 review transformational geometry, both on and off the Cartesian plane. Especially when working with rotations, students may benefit from using a trace-and-turn strategy.

Informal Evaluation Have students explain how they arrived at any incorrect answers. They will probably find their own mistakes and give you some clues as to the nature of their errors. Make sure students understand this material before administering the Chapter Test.

Follow-Up Present the following to the class: Triangle ABC has vertices $A(-4, -1)$, $B(-1, -2)$, $C(-1, -3)$. The triangle is reflected first across the x-axis and then across the y-axis. What are the coordinates of each vertex of the final image?
$A''(4, 1)$; $B''(1, 2)$; $C''(1, 3)$

SECTIONS 14–6 AND 14–7   TRANSLATIONS AND REFLECTIONS   (pages 522–529)

► A **translation** is a transformation in which all the points that make up a figure slide exactly the same distance and in the same direction.
► A **reflection** is a transformation in which a figure is flipped over a line of reflection.

12. Copy the figures onto graph paper. Then graph the image of each figure under a translation 3 units to the right and 3 units up. **Check students' drawings.**

13. Graph the image of each figure under a reflection across the given line.
       See Additional Answers.

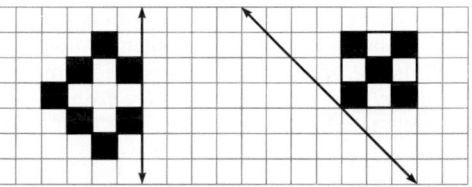

SECTIONS 14–8 AND 14–9   ROTATIONS AND SYMMETRY   (pages 530–537)

► A **rotation** is a transformation in which a figure is turned, or rotated, about a point.
► A figure has **line symmetry** if, when you fold it along a line, one side fits exactly over the other.
► A figure has **rotational symmetry** if, when the figure is turned about a point, the figure fits exactly over its original position at least once before a complete turn has been made.

14. Give the number of lines of symmetry for each figure at the right. **0, 3, 2**

15. Give the order of rotational symmetry for each figure at the right. **2, 3, 2**

16. What is the order of rotational symmetry of the figure shown at the right? **4**

SECTION 14–5   PROBLEM SOLVING STRATEGY: MAKE A MODEL   (pages 520–521)

► Making a model is often helpful in solving problems.
► Types of models include equations, sketches, and graphs.

17. A window washer charged a fixed fee in addition to a charge per window washed. Find the flat fee and the charge per window. **$9.00; $1.50**

Total Amount Charged	$13.50	$16.50	$24.00
Number of Windows Washed	3	5	10

USING DATA   Refer to the table on page 505.

18. Draw the dot pattern for the 5th square number on a coordinate plane. Give the ordered pair for the corner points. **Answers will vary.**

Review   **539**

Study Skills Tip   Encourage students to start to study at least one week before the test by themselves or in cooperative learning groups. Have students organize any notes, handouts, and pretests needed to review for the test. Have them outline the major objectives and vocabulary covered in class and summarize them in a chart. Students may find it helpful to also rewrite class notes to summarize the most important concepts that will be covered on the test.

Additional Answers
See page 584.

14 CHAPTER TEST

Introduction The Chapter Test uses a variety of questioning techniques to assess students' mastery of the major objectives of Chapter 14. If you prefer, you may use the Skills Preview (page 503) as an alternative form of the Chapter Test. The items on this test and the Skills Preview correspond in content and level of difficulty.

Alternative Assessment
Using Manipulatives Provide students with construction-paper shapes (symmetrical and non-symmetrical) and a large coordinate grid. Have them demonstrate their understanding of line symmetry by folding the paper shapes. Have them demonstrate their understanding of translations, reflections, and rotations by moving shapes on the coordinate grid.

Additional Answers
See page 584.

Use the figure at the right for Exercises 1–2.

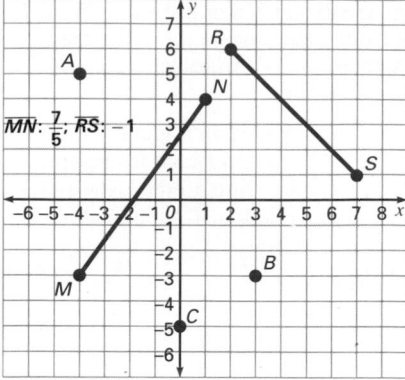

1. Give the coordinates of points A, B, and C.
   $A(-4, 5)$; $B(3, -3)$; $C(0, -5)$
2. Find the slope of line segments MN and RS.
   $MN: \frac{7}{5}$; $RS: -1$

3. From the mall, Paula drove 6 mi east and 2 mi north to her grandmother's house. To go home, she drove 3 mi west and 2 mi south. How far is Paula's house from the mall? In what direction from the mall is her house? **3 mi; east**

4. Graph the points $A(4, -2)$, $B(-1, 5)$, $C(3, 0)$, $D(-1, -1)$, and $E(0, 4)$. **See Additional Answers.**

5. Graph the equation $y = 3x - 2$.
   See Additional Answers.

6. A mail-order store charges handling costs, which consist of a fixed amount for filling an order plus a charge per pound of merchandise shipped. Find the fixed fee and the cost per pound.

Total Handling Costs	$5.50	$7.50	$10.50
Number of Pounds	3	5	8

Fee: $2.50
Cost per pound: $1.00

7. Complete the table of solutions for the equation $y = 2x + 1$.

$y = 2x + 1$

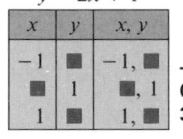

x	y	x, y	
-1	■	$-1,$ ■	**-1; -1**
■	1	■, 1	**0; 0**
1	■	1, ■	**3; 3**

Use the figure at the right for Exercises 8–10.

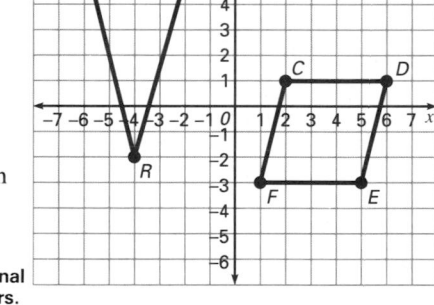

8. Graph the image of parallelogram $CDEF$ under a translation 2 units to the left and 3 units up. **See Additional Answers.**

9. Graph the image of $\triangle PQR$ under a reflection across the x-axis. **See Additional Answers.**

10. Graph the image of $\triangle PQR$ after a 90° turn counterclockwise about the origin. **See Additional Answers.**

Use the figure at the right for Exercises 11–12.

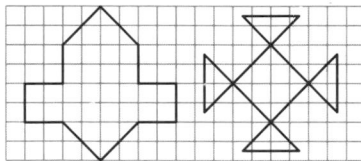

11. How many lines of symmetry does the figure on the left have? **1**

12. What is the order of rotational symmetry for the figure on the right? **4**

1. Members of a cooking class whose last names begin with the letter F were asked to name their favorite dessert. What type of sampling was used? **random**

Use the bar graph for Exercises 2–5.

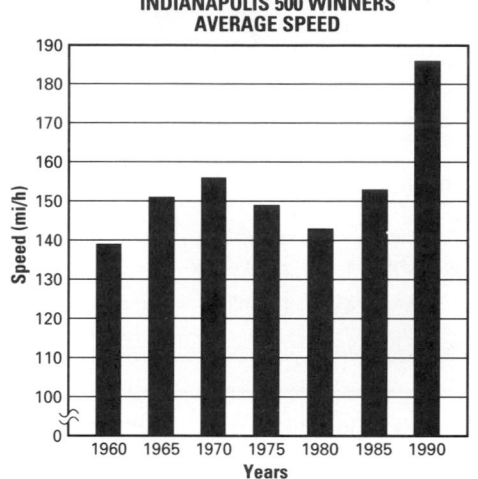

INDIANAPOLIS 500 WINNERS AVERAGE SPEED

2. In 1975, about how fast was the winner's speed? **about 150 mi/h**

3. During which years was the average speed at least 155 mi/h? **1970, 1990**

4. Describe the scale of the speed intervals. **See Additional Answers.**

5. About how much faster did the winner in the 1990 race go than the winner in the 1960 race? **about 50 mph**

A consumer watch group recorded the prices of different brands of apple juice: $1.25 $1.18 $1.29 $1.32 $1.12 $1.20 $1.25

6. Find the mean, median, and mode of the prices. **$1.23, $1.25, $1.25**

7. Which measure is the best indicator of the price of apple juice? Why? **mode; It is the price that occurs most frequently.**

Simplify.

8. $5 \times 4 + (8 - 3)$ **25** 9. $6 \times 3^2 - 2$ **52**
10. $(6 \times 4) \div (2 + 6)$ **3** 11. $3^2 + 5 \times 6 \div 2$ **24**

12. Is the argument *valid* or *invalid*?
    If the number is 15, then the number is divisible by 5.
    The number is divisible by 5.
    Therefore, the number is 15. **invalid**

Multiply or divide.

13. $1\frac{1}{2} \times 2\frac{1}{4}$ **$3\frac{3}{8}$** 14. $1\frac{3}{8} \div 1\frac{1}{4}$ **$1\frac{1}{10}$**
15. $\frac{7}{10} \times \frac{5}{8}$ **$\frac{7}{16}$** 16. $3\frac{1}{2} \div \frac{7}{8}$ **4**

Complete.

17. $0.07 \text{ L} = \blacksquare \text{ mL}$ **70** 18. $0.095 \text{ m} = \blacksquare \text{ mm}$ **95**
19. $3 \text{ km} = \blacksquare \text{ m}$ **3,000** 20. $8,000 \text{ g} = \blacksquare \text{ kg}$ **8**

Divide.

21. $60 \div (-15)$ **−4** 22. $-49 \div (-7)$ **7**
23. $-48 \div 8$ **−6** 24. $80 \div (-5)$ **−16**

Solve each equation.

25. $d + 7 = 2$ **$d = -5$** 26. $3(y - 4) = 27$ **$y = 13$**
27. $-35 = 4x + 1$ **$x = -9$** 28. $b - 19 = -17$ **$b = 2$**

Solve.

29. $\frac{27}{15} = \frac{x}{20}$ **36** 30. $\frac{x}{13} = \frac{3}{2.6}$ **15**
31. $\frac{7}{3} = \frac{35}{x}$ **15** 32. $\frac{10.5}{x} = \frac{5}{10}$ **21**

33. The scale of a drawing is 1 in. : 4 ft. What is the actual length of a room if the length in the scale drawing is $3\frac{1}{2}$ in.? **14 ft**

Cumulative Review **541**

Introduction The purpose of this Cumulative Review is to maintain previously taught skills and concepts and to apply them to the material presented in this chapter. At least one major objective of each chapter is included in the review. Page references direct students back to appropriate sections for additional review and reteaching.

Item Analysis The table below correlates the Cumulative Review items with the chapter and section that are being reviewed.

Section	Items
1–1	1
1–6	2–5
1–9	6–7
2–8	8–11
3–3	12
4–5	13–16
5–2	17–20
6–5	21–24
7–2	25, 28
7–5	26–27
8–4	29–32
8–6	33
9–4	34–35
9–8	36–37
10–3	40–44
10–10	45
11–3	38–39
11–6	46–47
11–8	48
12–2	4, 51
12–4	50
12–8	52
13–1	53–56
14–1	57

Additional Answers
See page 584.

Find the percent of increase or decrease.
Round to the nearest tenth.

34. Original amount: $200
New amount: $158 **21% decrease**

35. Original amount: $35
New amount: $49 **40% increase**

Find the interest and the amount.

36. Principal: $900
Rate: 12%/year
Time: 24 months **I = $216**
A = $1,116

37. Principal: $1,200
Rate: 5.5%/year
Time: 4 years **I = $264**
A = $1,464

Find the area for each figure.

38.

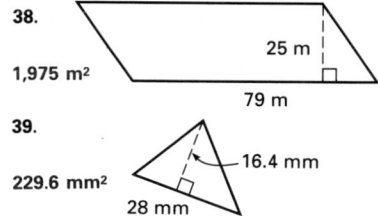

1,975 m²

39.

229.6 mm²

Use the figure below. $\overleftrightarrow{PT} \parallel \overleftrightarrow{UX}$. Find the
measure of each angle.

40. $m\angle 5$ **45°**

41. $m\angle 7$ **135°**

42. $m\angle 9$ **90°**

43. $m\angle 4$ **135°**

44. $m\angle 12$ **90°**

45. Trace $\overleftrightarrow{CD}$. Then construct a
perpendicular to $\overleftrightarrow{CD}$ from point B.
Check students' drawings. • B

C———————————D

Find the unknown length in each triangle.
Round to the nearest tenth.

46.

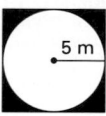

47.

5.7 m

48. Find the area
of the shaded
region to the
nearest tenth.
21.5 cm²

Simplify.

49. $(x^2 + 3x - 7) + (x^2 - 6x - 9)$ **$2x^2 - 3x - 16$**

50. $-3y(4 + 7y - 3)$ **$-3y - 21y^2$**

51. $(2x^2 - 4x + 1) - (x^2 - 2x + 1)$

52. $\frac{25c^6d}{5c^2}$ **$5c^4d$** **$x^2 - 2x$**

Find each probability.

53. P (B) $\frac{1}{3}$

54. P (R or G) $\frac{2}{3}$

55. P (7) **0**

56. P (not R) $\frac{2}{3}$

57. Refer to the figure below. Find four
points that will make a square whose
diagonals lie along the axes. Then give
the coordinates of these four points.
(Call them A, B, C, and D). Is there
only one such square?

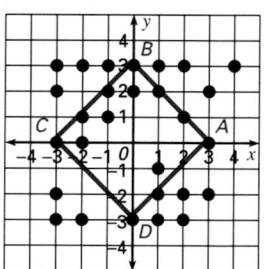

A (3,0); B (0,3); C (−3,0); D (0, −3); **Yes**

Use the bar graph for Exercises 1–4.

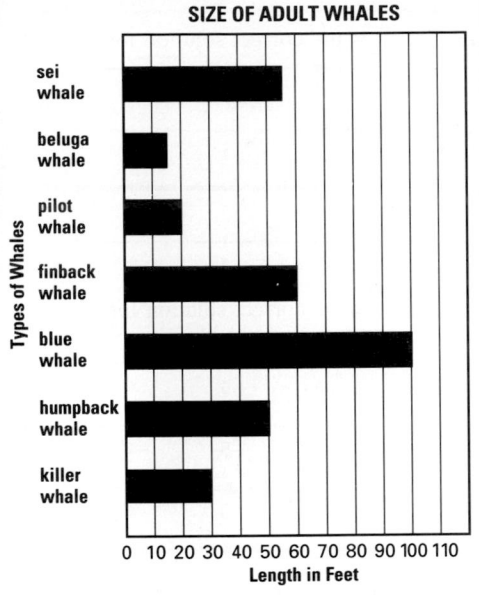

SIZE OF ADULT WHALES

Types of Whales: sei whale, beluga whale, pilot whale, finback whale, blue whale, humpback whale, killer whale

Length in Feet: 0 10 20 30 40 50 60 70 80 90 100 110

1. Which whale measures 20 feet?
   A. killer
   B. sei
   C. beluga
   (D.) pilot

2. How many whales are at least 55 feet long?
   A. 4
   B. 2
   (C.) 3
   D. 1

3. What is the difference in length between the largest whale and the smallest whale?
   (A.) 85 feet
   B. 70 feet
   C. 80 feet
   D. 75 feet

4. About how many times greater in length is the finback whale than the beluga whale?
   A. 2 times
   (B.) 4 times
   C. 3 times
   D. 5 times

5. What is the median for 212, 198, 157, 321, 166, 104, 158, 177?
   A. 186.6
   B. 217
   (C.) 171.5
   D. none of these

6. What is the mean for 31, 25, 39, 29, 27, 30, 33, 28, 28?
   A. 28
   (B.) 30
   C. 11
   D. none of these

7. What is the variable expression for a number divided by 9?
   (A.) $n \div 9$
   B. $9 \div n$
   C. $\frac{9}{n}$
   D. $\frac{1}{9n}$

8. Simplify. $3 \times (4 + 5) - 7$
   A. 6
   B. 2
   C. 189
   (D.) 20

9. Simplify. $7(k - 6)$
   A. $7k - 13$
   B. $k - 42$
   (C.) $7k - 42$
   D. none of these

10. Which is a counterexample for the following statement?
    If a number is divisible by 4, then it is divisible by 12.
    A. 12
    B. 36
    (C.) 32
    D. 60

11. Subtract. $10 - 3\frac{2}{7}$
    (A.) $6\frac{5}{7}$
    B. $7\frac{2}{7}$
    C. $6\frac{2}{7}$
    D. $7\frac{5}{7}$

12. Multiply. $7 \times 2\frac{2}{3}$
    A. $17\frac{1}{3}$
    (B.) $18\frac{2}{3}$
    C. $14\frac{1}{3}$
    D. $\frac{8}{21}$

13. Divide. $1\frac{3}{4} \div 4\frac{3}{8}$
    A. $4\frac{1}{2}$
    B. $2\frac{2}{5}$
    (C.) $\frac{2}{5}$
    D. $2\frac{1}{2}$

Cumulative Test **543**

14 CUMULATIVE TEST

Introduction The Cumulative Test uses a standardized-test format of multiple-choice questions to assess retention of previously learned concepts and test-taking skills. Test results may be used to diagnose students' strengths and weaknesses. Page references direct students back to appropriate sections for additional review and reteaching.

Item Analysis The table below correlates the Cumulative Test items with the chapter and section that are being tested.

Section	Items
1–6	1–4
1–9	5–6
2–2	7
2–8	8
2–10	9
3–3	10
4–5	12–13
4–6	11
5–3	14–16
5–4	17
6–2	18
6–3	19
6–4	20
6–5	21
7–2	22
7–4	23
7–5	24
7–10	25–26
8–2	27
8–3	28
8–4	29
8–6	30
9–3	31
9–8	32
10–3	33
10–5	34
11–7	35
11–9	36
12–2	37
12–6	38–39
12–8	40
13–3	41
13–4	42
13–5	43
14–1	44
14–4	45

14. Add.　2 ft 4 in.
　　　　　　+　　9 in.

A. 2 ft 3 in.　　　(B) 3 ft 1 in.
C. 2 ft 1 in.　　　D. none of these

15. Subtract.　6 gal 2 qt
　　　　　　　　−2 gal 3 qt

A. 3 gal 2 qt　　　B. 3 gal 1 qt
C. 4 gal 1 qt　　　(D) none of these

16. Multiply.　3 ft 5 in.
　　　　　　　×　　3

A. 10 ft 5 in.　　　B. 9 ft 8 in.
(C) 10 ft 3 in.　　　D. 9 ft 5 in.

---

17. What is the perimeter of a rectangular figure whose length is 7.45 dm and whose width is 25 dm?

A. 32.45 dm　　　B. 64.9 dm^2
C. 39.9 dm^2　　　(D) 64.9 dm

---

18. Add. $19 + (-13)$

(A) 6　　　　　　B. 32
C. −6　　　　　D. −32

19. Subtract. $20 - (-5)$

A. −25　　　　(B) 25
C. 15　　　　　D. −15

20. Multiply. -18×3

A. 54　　　　　B. −21
C. −45　　　　(D) −54

21. Divide. $-20 \div (-5)$

(A) 4　　　　　B. 15
C. −4　　　　　D. −25

22. Solve. $-15 = 3 + n$

A. $n = 18$　　　(B) $n = -18$
C. $n = 12$　　　D. $n = -12$

23. Solve. $15b = -60$

A. $b = 5$　　　B. $b = 45$
(C) $b = -4$　　　D. $b = -5$

24. Solve. $12 - 11s = 45$

A. $s = -33$　　　B. $s = 3$
C. $s = 22$　　　(D) $s = -3$

---

25. Which is the correct solution for $x \le 2$?

A.
B.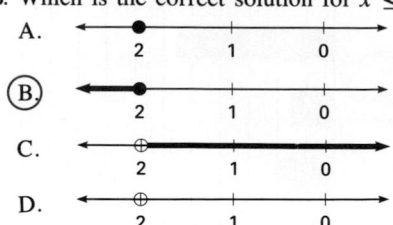
C.
D.

26. Which is the correct solution for $p > -5$?

A.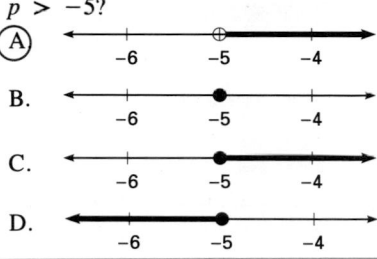
B.
C.
D.

---

27. Which is the unit rate for $\frac{\$108}{13.5 \text{ h}}$?

A. \$7.50/h　　　B. \$9/h
C. \$5/h　　　　(D) none of these

28. Which pair of ratios are equivalent?

A. $\frac{18}{12}, \frac{12}{18}$　　　(B) $\frac{3}{4}, \frac{18}{24}$
C. $\frac{9}{3}, \frac{36}{15}$　　　D. $\frac{4}{9}, \frac{16}{27}$

29. Solve. $\frac{15}{18} = \frac{x}{6}$

A. 90　　(B) 5　　C. 18　　D. 12

---

30. Ellen made a scale drawing of her garden. The width of the garden in her drawing is $5\frac{1}{4}$ in. The actual length is 21 ft. What scale did Ellen use for her drawing?

A. $\frac{1}{2}$ in.:1 ft　　　(B) $\frac{1}{4}$ in.:1 ft
C. $1\frac{1}{2}$ in.:$1\frac{1}{2}$ ft　　　D. none of these

31. The Marricks won 18 of their last 24 games. What percent of the games did the team win?
    A. 133%   B. 55%
    C. 75%   D. 60%

32. What is the amount of interest earned in one year on $900, if the rate of interest is 9.5% per year?
    A. $85.40   B. $85.50
    C. $85.55   D. none of these

33. Which is the alternate interior angle for $\angle 5$? $\overleftrightarrow{EF} \parallel \overleftrightarrow{GH}$
    A. $\angle 1$
    B. $\angle 7$
    C. $\angle 4$
    D. $\angle 8$

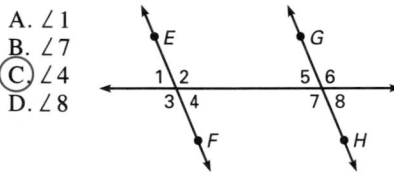

34. Find the sum of the angle measures of the polygon.
    A. 540°
    B. 720°
    C. 1,080°
    D. none of these

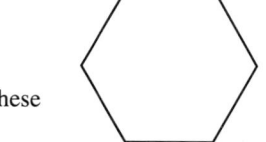

35. Find the volume to the nearest tenth.
    A. 49.6 m³
    B. 12.1 m³
    C. 58.2 m³
    D. 182.9 m³

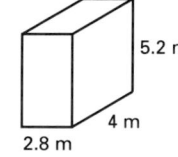

5.2 m
4 m
2.8 m

36. Find the surface area. Round to the nearest tenth.
    A. 108.5 ft²
    B. 301.4 ft²
    C. 150.7 ft²
    D. 125.6 ft²

    4 ft
    8 ft

37. Simplify. $(3c + 3) + (-6c + 8)$
    A. $-3c + 11$ B. $-9c^2 + 11$
    C. $9c^2 + 11$ D. $3c + 11$

38. Find the missing factor.
    $-9x^3 = (3x)(?)$
    A. $3x$ B. $3x^2$ C. $-3x$ D. $-3x^2$

39. Which is the greatest common factor of the monomials? $4y^4z, 4yz, 8y^3z$?
    A. $4y^3z$ B. $4yz$
    C. $4y^2z$ D. yz

40. Simplify. $\frac{-48s^5t^6}{4s^4t^3}$
    A. $12s^9t^9$ B. $-12st^9$
    C. $-12st^3$ D. $12st^9$

41. How many possible outcomes are there for making a sandwich with 3 kinds of bread, 5 fillings, and 3 dressings?
    A. 15   B. 9   C. 45   D. 11

42. A bag contains 3 red pens and 3 green pens. A pen is taken from the bag and replaced. Then a pen is taken from the bag again. Find P (red, then green).
    A. $\frac{1}{4}$ B. $\frac{3}{10}$ C. $\frac{1}{2}$ D. $\frac{1}{9}$

43. In a light bulb factory, 400 bulbs were selected at random. Of these 2 were defective. What is the probability of a bulb being defective?
    A. $\frac{1}{400}$ B. $\frac{1}{100}$ C. $\frac{1}{200}$ D. $\frac{1}{800}$

44. The coordinate pair $(3, -2)$ means
    A. the x coordinate is -2
    B. the y coordinate is 3
    C. both A and B are true
    D. neither A nor B is true

45. Find the slope of the line that passes through point $A(2,4)$ and point $B(-2,-4)$.
    A. 2   B. 4   C. 0   D. 6

Name _____ Date _____

The Coordinate Plane

Two number lines that are perpendicular to one another can be used to locate points on a **coordinate plane**. The horizontal number line is called the **x-axis**. The vertical number line is called the **y-axis**. The point at which the lines intersect is called the **origin**. Its coordinates are (0, 0).

▶ **Example 1** _____

Give the coordinates of point G.

Solution

Find G on the grid. The x-coordinate of G is −4; the y-coordinate is 5. So, the coordinates of point G are (−4, 5).

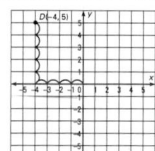

▶ **Example 2** _____

Locate point F(3, −2) on the coordinate plane.

Solution

Point F is 3 units to the right of the origin and 2 units down.

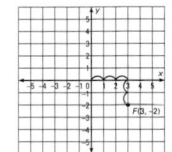

EXERCISES

Refer to the figure. Give the coordinates of the point or points.

1. point A   **(1, 4)**

2. point B   **(−3, −4)**

3. a point in the third quadrant   **S, B, T**

4. a point with a y-coordinate of −4   **B, T**

5. What point has the coordinates (3, 3)?   **C**

6. What point has the coordinates (4, −2)?   **E**

7. What point has the coordinates (−1, −4)?   **T**

Graph each point on the grid at the right.

8. X(2, −3)   9. Y(−3, −2)   10. Z(−1, 1)

Name _____ Date _____

Longitude and Latitude

Any point on the surface of the earth can be located by two coordinates, called longitude and latitude. *Longitude* is described by imaginary north-south lines, called *meridians*, that pass through the poles of the earth. You can think of the earth's surface as a sphere divided into 360 parts by these meridians.

Distances are measured east and west from one line of longitude called the *prime meridian*, which runs from the North Pole to the South Pole through Greenwich, England.

Distances are measured north and south from one line of latitude called the *equator*, which has 0° latitude. Each line of latitude north or south of the equator is a circle that runs around the earth parallel to the equator. These lines of latitude are called *parallels*. Thus, you can identify points on earth by sets of ordered pairs. For example, the coordinates (latitude, longitude) of New York City are approximately, (41°N, 74°W).

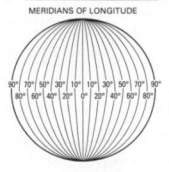

EXERCISES

Use an atlas, map, or globe to answer these questions.

1. Where would you be if you were 90°S of the equator?
   the South Pole

2. Describe the point that would be equivalent to the origin in terms of lines of latitude and longitude.
   the intersections of the equator and the prime meridian

3. How do meridians of longitude and parallels of latitude differ from the grid on a coordinate plane?
   Answers will vary. Possible answers: lines of latitude and longitude are curved, not straight; coordinates on the earth are marked in degrees of arc, not linear units.

4. Between what lines of latitude and longitude does most of your state lie?
   Answers will vary.

5. Try to determine the approximate coordinates of the city or town in which you live.
   Answers will vary.

Name _____ Date _____

Problem Solving Skills: Using a Grid to Solve Distance Problems

Coordinate grids are often used as maps. On a map, *up* represents north, *down* represents south, *left* represents west, and *right* represents east.

▶ **Example** _____

Jenny's paper route is shown on the grid. Each unit on the grid represents one block. The point labeled S is the point at which Jenny starts. F is the point for the last stop on her route. Her home is marked by point H. How many blocks from home is she when she finishes her route?

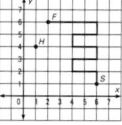

Solution

Start at F and draw a path along the grid lines to H. One route Jenny can take is to go 1 block west and 2 blocks south. Therefore, the distance is 3 blocks.

PROBLEMS

Use the information and the map given in the Example to solve Problems 1 and 2.

1. How many blocks long is Jenny's route?   **17 blocks long**

2. Describe another route Jenny might take to get home.
   2 blocks south, 1 block west

Wayne has a different paper route. He starts at point X and walks 3 blocks north, 2 blocks east, 1 block north, 2 blocks east, 2 blocks north, and 3 blocks east. His home is at point Y on the map shown at the right.

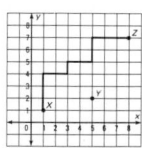

3. Draw his route on the map and label the finishing point of his route, Z.   **Check students' drawings.**

4. How far is the starting point of Wayne's route from his home?   **5 blocks**

5. How many blocks long is Wayne's route?   **13 blocks long**

6. When he finishes his route, how far is Wayne from home?   **8 blocks**

7. Describe a route that Wayne might take to get home from point Z.
   Answers will vary. Sample: 5 blocks south, 3 blocks west

Name _____ Date _____

Circular Grids

Another way to plot locations involves a circular grid.

On this grid, each circle is 1 unit from the next circle. Spokes are drawn from the center outward at 15° intervals. Any point on the grid is located by two coordinates: the distance of the point from the center, as measured along a spoke, and the angle that the spoke makes counterclockwise with a fixed line. In this grid, the fixed line is the line EW.

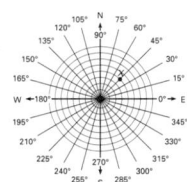

▶ **Example** _____

Find the coordinates of point X on the grid.

Solution

Point X is at the intersection of the fourth circle out from the center and the spoke numbered 45°. The coordinates of X are (4, 45°).

EXERCISES

Find the coordinates of each point on Grid A.

1. A   **(2, 90°)**

2. B   **(4, 120°)**

3. C   **(6, 210°)**

4. D   **(4, 315°)**

Graph each point on Grid B.

5. E(3, 60°)

6. F(4, 150°)

7. G(5, 180°)

8. H(4, 240°)

9. I(6, 300°)

10. J(5, 345°)

11. K(7, 30°)

12. L(8, 225°)

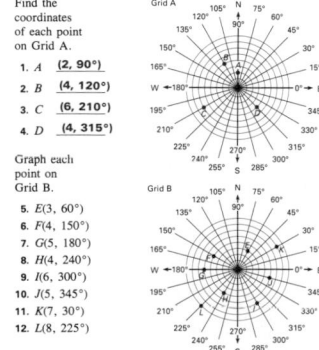

545A

Name _____   Date _____

Graphing Linear Equations

An equation in two variables has an infinite number of solutions. The solution of an equation like $x + y = 7$ is an ordered pair of numbers. For instance, (3, 4), (2, 5), and (−1, 8) are three solutions of $x + y = 7$.

► **Example** _____

Graph the equation $y = -2x + 1$.

Solution

Make a table of three ordered pairs that are solutions of the equation.

x	−2x + 1	y	(x, y)
0	−2(0) + 1	1	(0, 1)
1	−2(1) + 1	−1	(1, −1)
2	−2(2) + 1	−3	(2, −3)

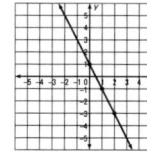

Graph the ordered pairs. Draw a line through the points. All points along the line are solutions of $y = -2x + 1$.

EXERCISES _____

For each equation, complete the table for three solutions. Then graph the equation.

1. $y = 3x$

x	3x	y	(x, y)
0	3(0)	0	(0, 0)
1	3(1)	3	(1, 3)
−1	3(−1)	−3	(−1, −3)

2. $y = -x + 4$

x	−x + 4	y	(x, y)
0	−0 + 4	4	(0, 4)
1	−1 + 4	3	(1, 3)
2	−2 + 4	2	(2, 2)

3. $y = 2x - 5$

x	2x − 5	y	(x, y)
0	2(0) − 5	−5	(0, −5)
1	2(1) − 5	−3	(1, −3)
2	2(2) − 5	−1	(2, −1)

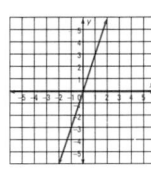 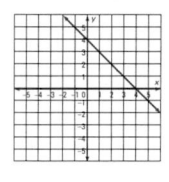 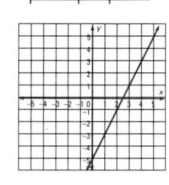

Name _____   Date _____

Functions

A **function** is a set of ordered pairs in which each member of one set of numbers is paired with exactly one member of another set of numbers. A function relates each value of x to a specific value of y. The x-values are called the **domain** of the function. The corresponding y-values are called the **range** of the function.

► **Example** _____

Given the function $y = x + 7$ with domain $D + \{0, 1, 2, 3\}$,
a. Find the range, R, of the function.
b. Write the domain and range as a set of ordered pairs.

Solution

a. Find the corresponding value of y by substituting each member of the domain for x in the rule.

For $x = 0$,   for $x = 1$,   for $x = 2$,   for $x = 3$,
$y = 0 + 7$ $y = 1 + 7$ $y = 2 + 7$ $y = 3 + 7$
$= 7$ $= 8$ $= 9$ $= 10$

The range is $\{7, 8, 9, 10\}$.

b. Match each number in the domain with its corresponding value in the range. The function is

$\{(0, 7), (1, 8), (2, 9), (3, 10)\}$

EXERCISES _____

1. For the function $y = 3x - 1$ and $D = [-3, -2, 0, 1]$, find the range R. Then write the domain and range as ordered pairs.

$R = \{-10, -7, -1, 2\}$; Function: $\{(-3, -10), (-2, -7), (0, -1), (1, 2)\}$

2. For the function $y = 2x + 2$ and $D = [-2, -1, 0, 1]$, find the range R. Then write the domain and range as ordered pairs.

$R = \{-2, 0, 2, 4\}$; Function: $\{(-2, -2), (-1, 0), (0, 2), (1, 4)\}$

3. For the function $y = \frac{2x}{3} + 2$ and $D = [-3, 0, 3, 6]$, find the range R. Then write the domain and range as ordered pairs.

$R = \{0, 2, 4, 6\}$; Function: $\{(-3, 0), (0, 2), (3, 4), (6, 6)\}$

Name _____   Date _____

Working with Slope

The steepness of rise or descent along a line is called the **slope** of the line.

► **Example 1** _____

Find the slope of the line graphed at the right.

Solution

Choose two points on the line, say point A (−2, 3) and point B (1, 1). Find the number of units of change vertically and horizontally in moving from A to B. Subtract the y-coordinate of A from that of B. Subtract the x-coordinate of A from that of B.
$\text{slope} = \frac{\text{rise}}{\text{run}} = \frac{\text{difference of } y\text{-coordinates}}{\text{difference of } x\text{-coordinates}} = \frac{1 - 3}{1 - (-2)} = \frac{-2}{3}$
The slope of the graphed line is $-\frac{2}{3}$.

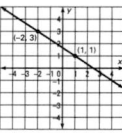

► **Example 2** _____

Write an equation of the line graphed at the right.

Solution

Find the slope of the line from A to B.
$\text{slope} = \frac{2 - 0}{2 - 1} = \frac{2}{1} = 2$
Find the y-intercept. The graph crosses the y-axis at −2.
Substitute the slope (m) and the y-intercept (b) in
$y = mx + b$
$y = 2x - 2$
The equation of the line is $y = 2x - 2$.

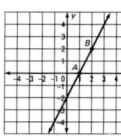

EXERCISES _____

Find the slope of the line that passes through each pair of points.

1. $S(2, 7)$ $T(4, 13)$ **3**   **2.** $V(0, -3)$ $W(2, 8)$ $\frac{11}{2}$ **3.** $X(1, 1)$ $Y(3, -4)$ $-\frac{5}{2}$

Write an equation of each line.

4.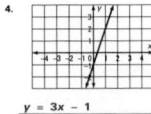

$y = 3x - 1$

5.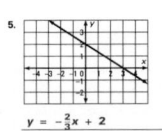

$y = -\frac{2}{3}x + 2$

Name _____   Date _____

Graphing by the Slope-Intercept Method

When a linear equation is in the form $y = mx + b$, or *slope-intercept* form, you can read the slope (m) and the y-intercept (b) directly from the equation. With this information you can graph the equation.

► **Example** _____

Graph the equation $y = \frac{2}{3}x - 1$.

Solution

The y-intercept is −1, which means that the graph crosses the y-axis at −1. The ordered pair is (0, −1).
Use the slope to find another point on the graph. Start at (0, −1). The slope is $\frac{2}{3}$. Go up 2 units and over to the right 3 units. Mark that point. You may want to mark another point, again going up 2 and over to the right 3 units. Draw a straight line through the points you have marked.

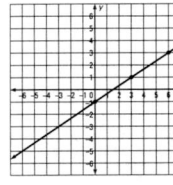

Check: Solve the equation for any value of x.
Let $x = 6$
$y = \frac{2}{3}(6) - 1$
$y = 4 - 1$
$y = 3$

Is the point (6, 3) on the graph? Yes.

EXERCISES _____

Graph each equation using the slope-intercept method. Check your work.

1. $y = \frac{3}{4}x + 3$

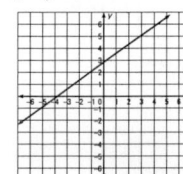

2. $y = -2x - 2$

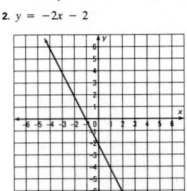

545B

Name _____ Date _____

Problem Solving Strategies: Make a Model

When a problem involves a linear function, a graph of the function can be a useful model.

▶ **Example**

The TV repair service charges a flat fee for making a house call and then an additional charge for each hour of work done. They use this chart. Find the flat fee and the hourly rate.

Repair bill	$20	$30	$40
Hours of work	1	2	3

Solution

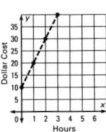

Graph the ordered pairs for the information in the table. Use the three points to draw a line. Find the slope and y-intercept of the line. Then write the equation of the line. The equation of this line is $y = 10x + 10$. For 0 hours, the fee is $10. This is the flat fee. Each hour (x) is multiplied by 10, so the fee per hour is $10.

PROBLEMS

Use the information in the chart. Model the problem on a coordinate plane and solve.

1. A messenger service charges a fixed fee per delivery and then an additional cost per mile. Find the fixed fee and the fee per mile.

Total amount charged for delivery	$7	$9	$13
Number of miles	1	4	4

Fixed fee: $5; fee per mile: $2

2. A parking garage charges a fixed fee per vehicle. In addition, an hourly fee is added. Use the chart to find the fixed fee and the amount charged per hour.

Amount charged for vehicle	$5	$11	$20	$32
Number of hours parked	1	3	6	10

Fixed fee: $2; hourly fee: $3

3. A mail order company charges a fixed per package amount for shipping, in addition to a per pound charge. Find the fixed shipping fee and the cost of shipping per pound.

Total cost of shipping package	$9	$15.50	$25.25	$31.75
Number of pounds	2	4	7	9

Fixed shipping fee: $2.50; cost per pound: $3.25

Name _____ Date _____

Making Comparisons by Graphing

ABC Repair Service charges a $50 fixed fee for a service call plus $5 for each hour of service. Miller Service charges a $40 fixed fee for a service call plus $10 for each hour of service. Which company has the better rate?

Let x = the number of hours required to do a repair job
Let y = the total service charge

Then, for ABC Repair, $y = 5x + 50$
For Miller Service, $y = 10x + 40$

Graph both equations on the same coordinate grid.

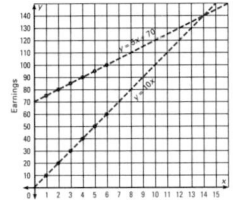

Answer the questions below.

1. For a job of what length is the fee charged by Miller Service less than the fee charged by ABC Repair?
   for a one-hour repair job

2. For a job of what length is the fee charged by Miller Service the same as the fee charged by ABC Repair?
   for a two-hour job

3. For a job of what length is the fee charged by Miller Service more than the fee charged by ABC Repair?
   for a job that takes longer than 2 hours

Mrs. Lopez is offered a job selling computer software. She can choose to get paid $10 per software package sold or $70 a week plus $5 per software package sold.

4. Write and graph two equations to compare the two plans.
   $y = 10x; y = 5x + 70$

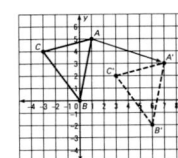

5. Which plan is better if she can sell 20 software packages per week?
   the $10-per-package plan

6. At what point would earnings from the two plans be the same?
   At the point at which 14 packages are sold.

Name _____ Date _____

Translations

A **translation,** or slide, of a figure results in a new figure exactly like the original. The original figure is called the **preimage** of the translated figure. The new figure is called the **image** of the original.

When you translate a figure, try to visualize all the points of the figure moving along a plane the same distance and in the same direction. The sides and angles of the new image are equal in measure to the sides and angles of the preimage. Each side of the new figure is parallel to the corresponding side of the original.

▶ **Example**

Graph the image of △XYZ under the translation 2 units right and 4 units down.

Solution

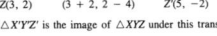

Find the coordinates of each image point of each vertex of the triangle.

Preimage	Translation	Image
$X(-2, -1)$	$(-2 + 2, -1 - 4)$	$X'(0, -5)$
$Y(0, 3)$	$(0 + 2, 3 - 4)$	$Y'(2, -1)$
$Z(3, 2)$	$(3 + 2, 2 - 4)$	$Z'(5, -2)$

△X'Y'Z' is the image of △XYZ under this translation.

EXERCISES

Graph the image of each figure under the given translation.

1. 3 units left, 4 units up

2. 5 units left

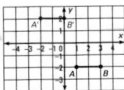

Write the coordinates of the preimage. Then graph the image under the given translation, and write the coordinates of the image.

3. 3 units right, 3 units up

4. 2 units left, 1 unit down

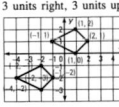 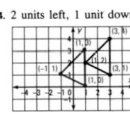

Name _____ Date _____

Mapping a Translation

Given points $A(1, 5)$, $B(0, 0)$, and $C(-3, 4)$, suppose that you want to translate, or slide, △ABC 6 units to the right and 2 units down. The translation can be expressed as a mapping of each point to its image point:

$A(1, 5) \rightarrow A'(7, 3)$ read "$A(1, 5)$ maps to $A'(7, 3)$."
$B(0, 0) \rightarrow B'(6, -2)$
$C(-3, 4) \rightarrow C'(3, 2)$

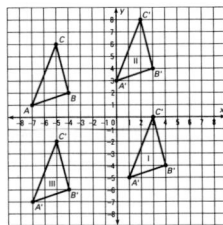

The graph at the right shows the mapping of each point to its image point according to the rule of translation.

You can express the mapping as a rule:
$(x, y) \rightarrow (x + 4, y + 1)$

EXERCISES

Given points $A(-7, 1)$, $B(-4, 2)$, and $C(-5, 6)$, write the coordinates of the image points under each rule. Then graph the image of △ABC under each transformation. Label the images as I, II, and III.

1. I: $(x, y) \rightarrow (x + 8, y - 6)$
   I: $A'(1, -5)$; $B'(4, -4)$; $C'(3, 0)$

2. II: $(x, y) \rightarrow (x + 7, y + 2)$
   II: $A'(0, 3)$; $B'(3, 4)$; $C'(2, 8)$

3. III: $(x, y) \rightarrow (x, y - 8)$
   III: $A'(-7, -7)$; $B'(-4, -6)$; $C'(-5, -2)$

Write a rule for each of the following translations for which B is the image of A.

4. $A(4, 3)$; $B(1, 1)$ $(x, y) \rightarrow (x - 3, y - 2)$ **Answers will vary.**

5. $A(-1, 4)$; $B(-2, 6)$ $(x, y) \rightarrow (x - 1, y + 2)$ **Answers will vary.**

6. $A(-4, -2)$; $B(-1, 5)$ $(x, y) \rightarrow (x + 3, y + 7)$ **Answers will vary.**

545C

Name _____ Date _____

Reflections

A **reflection** is a transformation under which a figure is flipped, or *reflected*, over a line called the **line of reflection**.

▶ **Example**

Graph the image of △ABC with vertices A(2, 3), B(3, 5), and C(4, −1) under a reflection across the y-axis.

Solution

Find the image point for each vertex of the triangle. In a reflection across the y-axis, the y-coordinates remain the same, but the x-coordinates are opposites. Multiply the x-coordinate of each vertex by −1.

Preimage	Image
A(2, 3)	A′(−2, 3)
B(3, 5)	B′(−3, 5)
C(4, −1)	C′(−4, −1)

The reflection of △ABC across the y-axis is △A′B′C′.

EXERCISES

Graph the image of each point under a reflection across the given axis.

1. A(−2, 1); y-axis
2. B(4, 3); x-axis
3. C(−1, −2); y-axis
4. D(3, −4); x-axis

5. Graph the image of △FGH under a reflection across the y-axis.

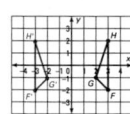

6. Graph the image of △KLM under a reflection across the x-axis. Write the coordinates of the vertices of the image.

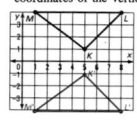

Check students' work.

Name _____ Date _____

Lines of Reflections

Lines other than the x-axis or y-axis can be used as lines of reflection.

▶ **Example**

In the figure at the right, △A′B′C′ is the image of △ABC under a reflection. What is the line of reflection over which △ABC has been reflected?

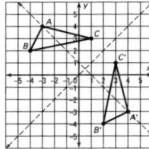

Solution

With a straightedge draw a line segment connecting A and A′ (you may also connect B and B′, or C and C′). Then use a compass to construct the perpendicular bisector of segment AA′. That bisector is the line of reflection.

EXERCISES

For each figure and its image, construct the line of reflection.

1. 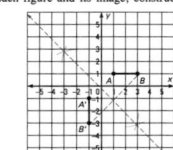   2.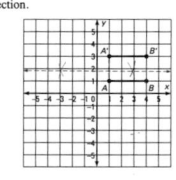

3. 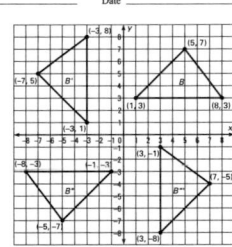   4.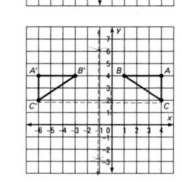

Name _____ Date _____

Rotations

A **rotation** is a transformation that produces an image by turning, or rotating, a figure about a point. A rotation is described in terms of
• the point, or *turn center*, about which the figure is rotated;
• the amount of turn, expressed as a fractional part of a whole turn or by the number of degrees of turn, called the *angle of rotation*;
• the direction of rotation, either clockwise or counterclockwise.

▶ **Example**

Triangle PQR has been rotated about point S. Give its angle of rotation in a clockwise direction.

Solution

Draw a line from S to Q. Draw a line from Q′ to S. Measure the angle between the two lines. The angle of rotation is 90° clockwise.

EXERCISES

Each drawing shows a line segment, LM, and its rotation image L′M′ about turn center T. Give the angle of rotation in a clockwise direction.

1.   2.   3.

   90° clockwise   **180° clockwise**   **270° clockwise**

A figure has been rotated about turn center T. Describe each angle of rotation clockwise and counterclockwise.

4. 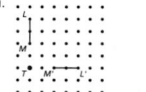   5.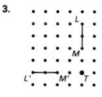

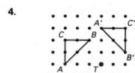
90° clockwise; 270° counterclockwise   **270° clockwise; 90° counterclockwise**

Name _____ Date _____

Transforming Coordinates in a Rotation

How are the coordinates of a figure transformed when the figure is rotated 90°, 180°, or 270°?

Look at the triangles on the grid shown at the right. Triangle B has been rotated about the origin counterclockwise through a 90° turn (△B′), a 180° turn (△B″), and a 270° turn (△B‴). Each turn forms a different rotation image.

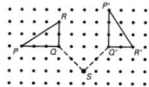

EXERCISES

1. Complete the table by writing the coordinates of each rotation image of △B.

△B	△B′ (90° turn)	△B″ (180° turn)	△B‴ (270° turn)
(1, 3)	(−3, 1)	(−1, −3)	(3, −1)
(8, 3)	(−3, 8)	(−8, −3)	(3, −8)
(5, 7)	(−7, 5)	(−5, −7)	(7, −5)

2. How does a 90° rotation counterclockwise transform each coordinate of △B?
The order of the x- and y-coordinates is reversed, and the sign of the x-coordinate changes.
3. How does a 180° rotation counterclockwise transform each coordinate of △B?
The order of the x- and y-coordinates remains the same, but the sign of each coordinate changes.
4. How does a 270° rotation counterclockwise transform each coordinate of △B?
The order of the coordinates is reversed, and the sign of the y-coordinate changes.

Write the coordinates of the rotation counterclockwise for each point.

5. (2, 3) 90°: **(−3, 2)**   180°: **(−2, −3)**   270°: **(3, −2)**
6. (−2, 4) 90°: **(−4, −2)**   180°: **(2, −4)**   270°: **(4, 2)**
7. (−1, −2) 90°: **(2, −1)**   180°: **(1, 2)**   270°: **(−2, 1)**

Complete the following sentences, based upon your observations.

8. For a rotation of 90° counterclockwise, the point (x, y) becomes point **(−y, x)**
9. For a rotation of 180° counterclockwise, the point (x, y) becomes point **(−x, −y)**
10. For a rotation of 270° counterclockwise, the point (x, y) becomes point **(y, −x)**

Name _____ Date _____

Working with Symmetry

A figure that can be folded along one or more lines so that one half of the figure exactly matches the other half is said to have **line symmetry**.

► **Example 1** _____

Draw all the lines of symmetry.

Solution

If rotated 180°, the figure above will fit exactly over its original position. Such a figure has **rotational symmetry**. The number of times during one full turn that a figure fits exactly over its original position is called the **order of rotational symmetry**.

► **Example 2** _____

Give the order of rotational symmetry for the figure.

Solution

Trace the figure and rotate the tracing over the original figure. The tracing fits over its original position 3 times during a full turn. Its order of rotational symmetry is 3.

EXERCISES _____

Draw all the lines of symmetry in each figure.

1.

2.

3.

Tell whether the figure has rotational symmetry. If so, give the order of rotational symmetry.

4.

yes; 2

5.

yes; 6

6.

no

Reteaching • SECTION 14-9

Name _____ Date _____

Combined Transformations

A new image of a figure can be formed by combining translations, reflections, and rotations.

► **Example** _____

Find an image of line segment XY after a translation of 2 units right, 3 units down, and a reflection across the y-axis.

Solution

First, use the translation to find $X'Y'$. Then reflect $X'Y'$ across the y-axis to find $X''Y''$.

Line segment $X''Y''$ is the result of the translation and the reflection of XY.

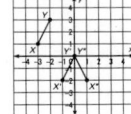

EXERCISES _____

Find the new image of each figure using the given transformations.

1. Translation: 4 units left, 1 unit up
   Reflection across the x-axis

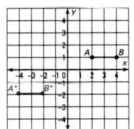

2. Translation: 3 units right
   Rotation 180° about the origin

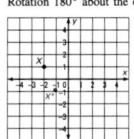

3. Reflection about the y-axis
   Rotation 90° counterclockwise about the origin.

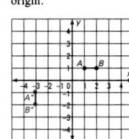

4. Reflection across the x-axis
   Reflection across the y-axis

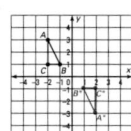

Enrichment • SECTION 14-9

Name _____ Date _____

Calculator Activity: Using a Graphing Calculator

You can use a graphing calculator to graph a function and to trace the graph in order to find coordinates of points on the graph. Be sure that you read the instruction manual carefully before using the calculator.

In general, the following method will allow you to trace along a graph to find ordered pairs of integers that are solutions of an equation of the form $y = mx + b$. (Be aware that if an equation is not in this form, you must rewrite the equation in this form before you can enter it on the y= function screen.)

A word about how the graphs displayed on the screen of a graphing calculator appear: Any graph displayed on the screen will appear somewhat jagged, rather than smooth. The reason for this appearance is that the calculator can plot only a finite number of points; in addition, the display screen itself has only a finite number of dots to represent the infinite number of points in any region of the coordinate plane. Thus, small "gaps" in the line will appear. (It is possible to use the RANGE feature of a graphing calculator to adjust the dot resolution of the graph to some degree.)

► **Example 1** _____

Use a graphing calculator to enter the function $y = x + 3$.

Solution

1. Press Y= . On the screen you will see displayed
   $Y_1 =$
   $Y_2 =$
   $Y_3 =$
   $Y_4 =$

   The calculator allows you to enter up to four different functions. Each function can be stored in memory, and one or more of them can be graphed at a time.

2. To enter the function $y = x + 3$, move the cursor to the space immediately following $Y_1 =$ on the screen. Press the button that enters the domain variable x. On some calculators, this button is labeled X T . After entering x, enter + 3. The first function on the screen now reads $Y_1 = x + 3$. Then press ENTER .

EXERCISES _____

Enter these functions on your graphing calculator. **Check students' calculator screens.**

1. $y = 3x$ 2. $y = x + 4$ 3. $y = 4x - 1$ 4. $y = x - 4$

Technology • SECTION 14-4

Name _____ Date _____

Using a Graphing Calculator (continued)

► **Example 2** _____

Graph the equation $y = x + 3$. Sketch the graph. Then trace the function on the screen to locate at least two ordered pairs that are solutions.

Solution

1. The function $y = x + 3$ has been entered on the screen as the first function on the Y= screen. Press GRAPH . On the screen you will see displayed the graph of $y = x + 3$ on a grid. Sketch the graph as it appears.

2. Press TRACE . The cursor will appear at a position on the line $y = x + 3$. The x- and y-coordinates of that position will be displayed at the lower left and lower right of the screen, respectively.

3. To find integer solutions of the equation,
   a. Use the ► button to move the cursor to the right along the graph until the screen displays integer values for x and for y.
      The first pair of integers displayed is $x = 2$, $y = 5$.
      Move the cursor further to the right. You will see that, as the x-coordinate value increases, the cursor will eventually move off the screen. The calculator will continue to yield values for x and y.
      The next set of values is $x = 6$, $y = 9$.
   b. Use the ◄ button to move the cursor to the left along the graph until you see displayed other integer values for x and for y.
      The first pair of integers displayed is $x = -6$, $y = -3$. Move the cursor further to the right. The next set of values is $x = -10$, $y = -7$.

   So, $(2, 5)$, $(6, 9)$, $(-6, -3)$, and $(-10, -7)$ are ordered pairs of integers that are solutions of the equation $y = x + 3$.

Tracing a graph will not yield all the integer solution sets of the equation of the graph for any given region of the coordinate plane. The limited number of points possible on the graph prevents such high resolution. Thus, the values $x = 0$, $y = 3$ at the y-intercept of the line may appear as decimal approximations, such as $x = .3157895$, $y = 3.3157895$.

EXERCISES _____

Use a graphing calculator to graph. Sketch each graph on graph paper. Then trace each graph and give at least two ordered pairs that are solutions of each equation.
Check students' work.

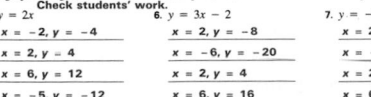

5. $y = 2x$	6. $y = 3x - 2$	7. $y = -3x + 3$
$x = -2, y = -4$	$x = 2, y = -8$	$x = 2, y = 9$
$x = 2, y = 4$	$x = -6, y = -20$	$x = -6, y = 21$
$x = 6, y = 12$	$x = 2, y = 4$	$x = 2, y = -3$
$x = -5, y = -12$	$x = 6, y = 16$	$x = 6, y = -15$

Technology • SECTION 14-4

Name _____ Date _____

Computer Activity: Turtle Tessellations

A **tessellation** is a pattern in which polygons are used to completely cover the plane without gaps or overlaps. A point where the vertices of the polygons meet is a **vertex point** of the tessellation.

Using the computer language, Logo, you can make and investigate tessellations easily.

EXERCISES

1. Enter each procedure. Draw the tessellation that is the result of procedure **e**.

    a. TO TRI
       REPEAT 3[FD 40
       RT 120 WAIT 5]
       END

    b. TO SLIDE
       RT 120 FD 40 LT 90
       END

    c. TO TESS.TRI.TOP
       PU LT 90 FD 40 RT 90 PD
       REPEAT 4[LT 30 TRI SLIDE]
       END

    d. TO TESS.TRI.BOTTOM
       REPEAT 5[LT 30 TRI SLIDE]
       END

    e. TO SUPER.TESS.TRI
       TESS.TRI.TOP
       RT 180
       TESS.TRI.BOTTOM
       END

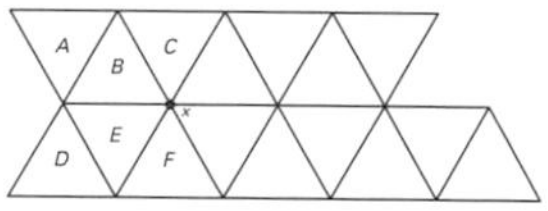

2. On your drawing, label the three triangles at the top left *A*, *B*, and *C*. Label the three triangles directly below them *D*, *E*, and *F*. For each of the following pairs, **Answers** state the transformation(s) that could produce the second triangle from the first. **may vary.**

    a. △*A* to △*D*   reflection or rotation
    b. △*C* to △*A*   translation
    c. △*D* to △*B*   rotation
    d. △*F* to △*A*   reflection + translation

3. Locate the vertex point between △*C* and △*F* and label it *x*. What is the sum of the measures of all the angles at point *x*?   360

Technology • SECTION 14-8   283

Name _____ Date _____

Turtle Tessellations (continued)

4. Enter each procedure. Record the tessellation that is the result of the procedure.

    a. TO HEX
       REPEAT 6[FD 40 RT 60 WAIT 5]
       END

    b. TO MOVE
       LT 90 PU FD 104 RT 90 PD
       END

    c. TO SLIDE.1
       RT 120 PU FD 40
       RT 60 FD 40 RT 180 PD
       END

    d. TO SLIDE.2
       RT 120 PU FD 40
       LT 60 FD 40 LT 60 PD
       END

    e. TO SUPER.TESS.HEX
       MOVE LT 60
       REPEAT 4[HEX SLIDE.1] HEX
       PU LT 120 FD 40 RT 180 PD
       REPEAT 3[HEX SLIDE.2]
       END

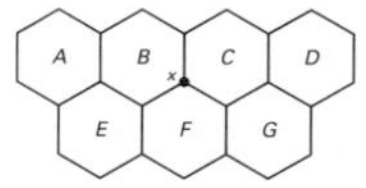

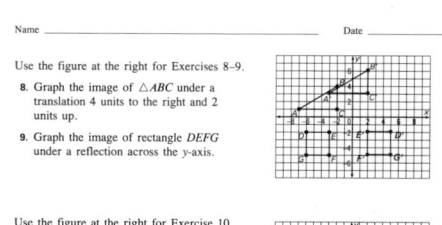

5. Label the hexagons in the top row *A*, *B*, *C*, and *D* and those in the bottom row, *E*, *F*, and *G*. For each of the following pairs, state the transformation(s) that could produce the second hexagon from the first. **Answers may vary.**

    a. *A* to *B*   translation or reflection
    b. *B* to *F*   translation or reflection or rotation
    c. *C* to *A*   translation or double reflection
    d. *D* to *F*   translation + reflection

6. Locate the vertex point between hexagons *B*, *C*, and *F* and label it *x*. What is the sum of all the angles at *x*?   360

7. Based on your answers to Exercises 3 and 6, what do you think must be the sum of the measures of the angles at a vertex point of a tessellation?   360

8. Tessellations formed by regular polygons are called **regular tessellations**. Which regular polygon, other than a hexagon, can form a regular tessellation? Explain.
   Square: because each interior angle measures 90 and 4 × 90 = 360.

9. Refer to Exercise 8. Write a procedure that will create a tessellation for the regular polygon you identified. Draw the tessellation.
   Answers may vary. A sample is given.

    TO SQUARE
    REPEAT 4[FD 40 RT 90 WAIT 5]
    END

    TO SLIDE
    LT 90 BK 40 RT 90
    END

    TO SUPER.TESS.SQ
    REPEAT 3[SQUARE SLIDE]
    LT 90 PU FD 120 RT 90 BK 40 PD
    REPEAT 3[SQUARE SLIDE]
    END

284   Technology • SECTION 14-8

ACHIEVEMENT TEST

Geometry of Position

CHAPTER 14 FORM A

MATH MATTERS BOOK 1
Chicha Lynch
Eugene Olmstead

SOUTH-WESTERN PUBLISHING CO.

Name _____
Date _____

SCORING RECORD	
Possible	Earned
12	

Use the figure at the right for Exercises 1–2.

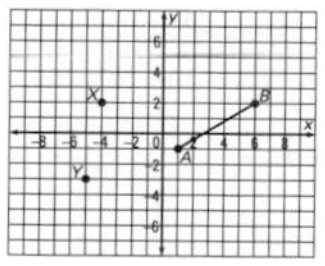

1. Give the coordinates of points *X* and *Y*.
   X(−4, 2); Y(−5, −3)

2. Find the slope of line segment *AB*.
   3/5

3. From their house, Margo and Bill drove 4 blocks east and 7 blocks north to a service station for gasoline. After that, they drove 3 blocks west and 4 blocks south to the supermarket. At that point, how many blocks were they from home?
   4 blocks

Use the grid at the right for Exercises 4–5.

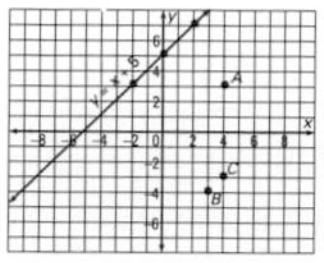

4. Graph the points *A*(4, 3), *B*(3, −4), and *C*(4, −3).

5. Graph the equation *y* = *x* + 5.

6. Complete the table for *y* = *x* + 3.

x	*y*	*x, y*
−3	0	−3, 0
0	3	0, 3
3	6	3, 6

7. At a parking lot, there is a fixed fee for using the lot in addition to an hourly rate charged for parking. Find the fixed fee and the hourly rate.

Total Amount Paid for Parking Car	$4.00	$7.00	$8.50
Number of Hours Car Parked	1	3	4

Fee: $2.50; Hourly rate: $1.50

14A-1

Name _____ Date _____

Use the figure at the right for Exercises 8–9.

8. Graph the image of △*ABC* under a translation 4 units to the right and 2 units up.

9. Graph the image of rectangle *DEFG* under a reflection across the *y*-axis.

Use the figure at the right for Exercise 10.

10. Graph the image of △*HIJ* after a 90° turn clockwise about the origin.

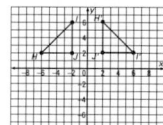

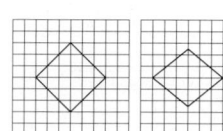

Use the figures at the right for Exercises 11–12.

11. How many lines of symmetry does the figure on the left have?
    4

12. What is the order of rotational symmetry for the figure on the right?
    2

14A-2

ACHIEVEMENT TEST

Geometry of Position

CHAPTER 14   FORM B

MATH MATTERS BOOK 1
Chicha Lynch
Eugene Olmstead

SOUTH-WESTERN PUBLISHING CO.

Name _____

Date _____

SCORING RECORD	
Possible	Earned
12	

Use the figure at the right for Exercises 1–2.

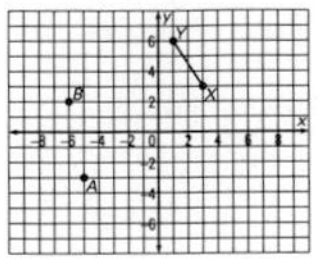

1. Give the coordinates of points A and B.

   A(−5, −3); B(−6, 2)

2. Find the slope of line segment XY.

   $-\frac{3}{2}$

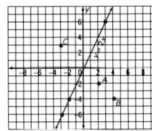

3. From school, Miko and Luanne walked 5 blocks east and 6 blocks north to the library to study. After that, they walked 4 blocks west and 4 blocks south to Miko's house. How many blocks is Miko's house from school?

   3 blocks

Use the figure at the right for Exercises 4–5.

4. Graph the points A(2, −2), B(4, −4), and C(−3, 3).

5. Graph the equation y = 2x.

6. Complete the table for y = 2x − 1.

x	y	x, y
2	3	2, 3
0	−1	0, −1
−1	−3	−1, −3

7. A department store charges a fixed amount for handling, as well as a per-pound fee for merchandise shipped. Find the fixed amount and the cost per pound.

Total Cost for Handling	$4.50	$6.50	$9.50
Number of Pounds	3	5	8

Fee: $1.50; Hourly rate: $1.00

14B-1

Name _____   Date _____

Use the figure at the right for Exercises 8–9.

8. Graph the image of △ABC under a translation 7 units to the right and 1 unit up.

9. Graph the image of square DEFG under a reflection across the x-axis.

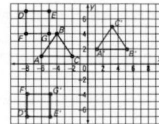

Use the figure at the right for Exercise 10.

10. Graph the image of △HIJ after a 180° turn counterclockwise about the origin.

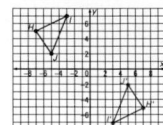

Use the figures at the right for Exercises 11–12.

11. How many lines of symmetry does the figure on the left have?

    2

12. What is the order of rotational symmetry for the figure on the right?

    2

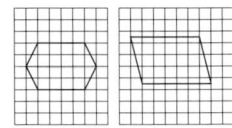

14B-2

ACHIEVEMENT TEST   FINAL

CHAPTERS 1–14   FORM A

MATH MATTERS BOOK 1
Chicha Lynch
Eugene Olmstead

SOUTH-WESTERN PUBLISHING CO.

Name _____

Date _____

SCORING RECORD	
Possible	Earned
70	

To test the quality of a shipment of door locks, 500 locks were chosen randomly from different lots and then tested.

1. What kind of sampling does this represent?

   random sampling

2. What is the advantage of this kind of sampling?

   The sample comes from the whole population.

The pictograph at the right shows the number of students at Plainview High School, by grade, who ride bicycles to school.

3. In which grade are there the fewest bicycle riders?   **grade 12**

4. How many bicycle riders are there in the tenth grade?   **480**

5. How many Plainview High School students in all ride bicycles to school?   **1,400**

Key: represents 80 bicycles

The number of cans of New Aroma coffee beans sold daily during a recent 7-day period was as follows: 29, 41, 45, 57, 72, 89, 45.

6. Find the mean, median, mode, and range of the data.

   mean: 54; median: 45; mode: 45; range: 60

Write in standard form.

7. 8^4 **4,096**

8. 6^5 **7,776**

Write in scientific notation.

9. 7,893   **7.893×10^3**

10. 645   **6.45×10^2**

11. 637,500   **6.375×10^5**

12. 13,603,390   **1.360339×10^7**

Write each answer in exponential form.

13. $6^4 \times 6^7$ **6^{11}**

14. $9^{16} \div 9^{10}$ **9^6**

15. $y^4 \times y^2$ **y^6**

16. $x^8 \div x^3$ **x^5**

FA-1

Name _____   Date _____

Simplify.

17. $3 \times 4 - 36 \div 6$ **6**

18. $3 \times (12 - 9) - 9$ **0**

19. $(2 + 6)^2 \times 3$ **192**

Determine whether the following argument is *valid* or *invalid*.

20. If the number is 21, then the number is divisible by 7. The number is divisible by 7. Therefore, the number is 21.   **invalid**

Find the GCF of each pair of numbers.

21. 9, 15   **3**

22. 10, 35   **5**

23. 48, 72   **24**

24. Write 0.00004275 in scientific notation.   **4.275×10^{-5}**

Complete.

25. 60 in. = **5** ft

26. 4 L = **4,000** mL

Find the circumference. Use 3.14 or $\frac{22}{7}$ for π, as appropriate. Round to the nearest whole number.

27.

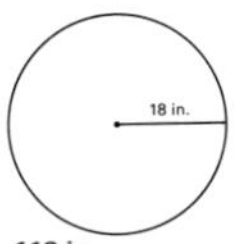

18 in.

113 in.

28.

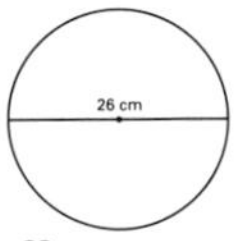

26 cm

82 cm

Divide.

29. $-27 \div (-9)$ **3**

30. $-64 \div 4$ **−16**

31. $-8 \div 8$ **1**

32. $84 \div (-7)$ **−12**

Solve each equation.

33. $a - 4 = 7$ **11**

34. $-63 = 5x + 2$ **13**

35. $4(y - 6) = 32$ **14**

36. $8c - 3 = 5c + 6$ **3**

37. Marc's weekly salary is $162.75 before taxes. He works 21 hours per week. What is his hourly wage?   **$7.75 per hour**

Solve each proportion.

38. $\frac{27}{15} = \frac{x}{20}$ **x = 36**

39. $\frac{42}{3} = \frac{56}{y}$ **y = 4**

40. $\frac{0.9}{4.5} = \frac{2}{x}$ **x = 10**

FA-2

Name _____ Date _____

41. The scale of a drawing is $\frac{1}{2}$ in.:1 ft. Find the drawing length that would be used to represent an actual length of 48 ft.
24 in. or 2 ft

Find the percent of increase or decrease.

42. Original amount: $450
New amount: $549
22% increase

43. Original amount: $128
New amount: $96
25% decrease

44. Original amount: $1,200
New amount: $864
28% decrease

Find the interest and the amount due.

45. Principal: $900
Rate: 7%
Time: 1 year
I = $63; A = $963

46. Principal: $4,500
Rate: 8%
Time: 24 months
I = $720; A = $5,220

47. Principal: $8,500
Rate: 3.5%
Time: 6 months
I = $148.75;
A = $8,648.75

Find the unknown angle measure.

48. 124° 32°
24°

49. 30°
60°

50. 127° 95°
48°

51. Ray BD bisects $\angle ABC$. The measure of $\angle ABC$ is 72°. What is the measure of $\angle DBA$?
36°

Add or subtract.

52. $(7a^2 + 4a - 8) + (2a^2 - a + 5)$ ___ **$9a^2 + 3a - 3$**

53. $(12t^2 + 6t - 16) - (4t^2 - 2t + 18)$ ___ **$8t^2 + 8t + 2$**

Simplify.

54. $(-5m)(2n)$ ___ **$-10mn$**

55. $(9y)(-8z)$ ___ **$-72yz$**

56. $(4t^3)^2$ ___ **$16t^6$**

57. $(-5x^2)(-6x^4)$ ___ **$30x^6$**

58. $\frac{18a^6b^4}{-3a^2b}$ ___ **$-6a^4b^3$**

59. The perimeter of a rectangle is 28 cm. The length is 6 cm greater than the width. Find the area of the rectangle.
w = 4 cm; l = 10 cm; A = 40 cm²

FA-3

Name _____ Date _____

Find each probability.

60. $P(2)$ ___ **$\frac{1}{6}$**

61. P(an even number) ___ **$\frac{1}{2}$**

62. $P(3 \text{ or } 4)$ ___ **$\frac{1}{3}$**

63. P(not a 4 or 5) ___ **$\frac{2}{3}$**

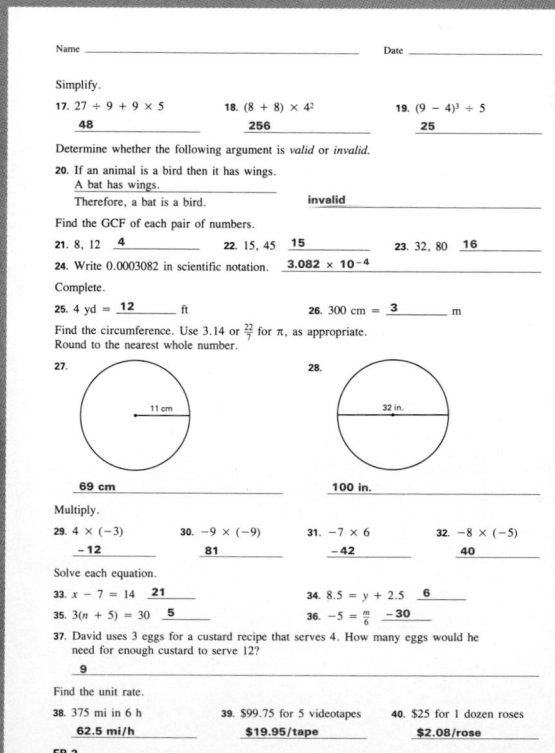

Use a tree diagram to find the number of possible outcomes.

64. tossing a coin four times **16 possible outcomes**

65. tossing a coin and tossing a number cube **12 possible outcomes**

Use the counting principle to find the number of possible outcomes.

66. making a sandwich from white, oatmeal, or rye bread, with beef or ham, and 3 kinds of cheese
18 possible outcomes

Use the figure at the right for Exercises 67 and 68.

67. Give the coordinates of points A, B, and C.
$A(-3, 4)$; $B(1, -3)$; $C(3, 1)$

68. Find the slope of the line that passes through A and C; through B and C.
$-\frac{1}{2}$; 2

Give the order of symmetry of each figure.

69.
2

70.
6

FA-4

ACHIEVEMENT TEST FINAL

Name _____

Date _____

CHAPTERS 1–14 FORM B

MATH MATTERS BOOK 1

Chicha Lynch
Eugene Olmstead

SOUTH-WESTERN PUBLISHING CO.

SCORING RECORD	
Possible	Earned
70	

To test the quality of a day's production of calculators, every twentieth calculator manufactured is tested.

1. What kind of sampling does this represent?
systematic sampling

2. What is a disadvantage of this kind of sampling?
The results could not be applied to calculators manufactured during previous or subsequent runs.

The pictograph at the right shows the number of students at Riverview High School, by grade, who take music lessons at school.

3. In which grade do the fewest students take music lessons?
grade 10

4. How many of the ninth graders take music lessons?
130

5. How many Riverview High School students in all take music lessons at school?
320

Key: represents 20 students taking music lessons

The number of CD's sold daily during a recent 7-day period was as follows: 77, 38, 42, 30, 79, 42, 94.

6. Find the mean, median, mode, and range of the data.
mean: 57; median: 42; mode: 42; range: 64

Write in standard form.

7. 5^5 **3,125**

8. 9^4 **6,561**

Write in scientific notation.

9. 3,268 **3.268×10^3**

10. 818 **8.18×10^2**

11. 289,900 **2.899×10^5**

12. 17,593,017 **1.7593017×10^7**

Write each answer in exponential form.

13. $8^7 \div 8^3$ **8^4**

14. $5^5 \times 5^5$ **5^{10}**

15. $x^8 \div x^2$ **x^6**

16. $y^6 \times y$ **y^7**

FB-1

Name _____ Date _____

Simplify.

17. $27 \div 9 + 9 \times 5$ **48**

18. $(8 + 8) \times 4^2$ **256**

19. $(9 - 4)^3 \div 5$ **25**

Determine whether the following argument is *valid* or *invalid*.

20. If an animal is a bird then it has wings.
A bat has wings.
Therefore, a bat is a bird. **invalid**

Find the GCF of each pair of numbers.

21. 8, 12 **4**

22. 15, 45 **15**

23. 32, 80 **16**

24. Write 0.0003082 in scientific notation. **3.082×10^{-4}**

Complete.

25. 4 yd = **12** ft

26. 300 cm = **3** m

Find the circumference. Use 3.14 or $\frac{22}{7}$ for π, as appropriate. Round to the nearest whole number.

27. 11 cm
69 cm

28. 32 in.
100 in.

Multiply.

29. $4 \times (-3)$ **-12**

30. $-9 \times (-9)$ **81**

31. -7×6 **-42**

32. $-8 \times (-5)$ **40**

Solve each equation.

33. $x - 7 = 14$ **21**

34. $8.5 = y + 2.5$ **6**

35. $3(n + 5) = 30$ **5**

36. $-5 = \frac{m}{6}$ **-30**

37. David uses 3 eggs for a custard recipe that serves 4. How many eggs would he need for enough custard to serve 12?
9

Find the unit rate.

38. 375 mi in 6 h **62.5 mi/h**

39. $99.75 for 5 videotapes **$19.95/tape**

40. $25 for 1 dozen roses **$2.08/rose**

FB-2

Name _____ Date _____

41. A map is drawn to a scale of 1 cm:12 km. What actual distance is represented by 10 cm on the map?
120 km

Find the percent of increase or decrease.

42. Original amount: $350
New amount: $437
25% increase

43. Original amount: $672
New amount: $497
26% decrease

44. Original amount: $22,500
New amount: $18,599
17% decrease

Find the interest and the amount due.

45. Principal: $800
Rate: 5%
Time: 1 year
I = $40; A = $840

46. Principal: $6,200
Rate: 8%
Time: 24 months
I = $992; A = $7,192

47. Principal: $12,500
Rate: 4.5%
Time: 6 months
I = $281.25;
A = $12,781.25

Find the unknown angle measure.

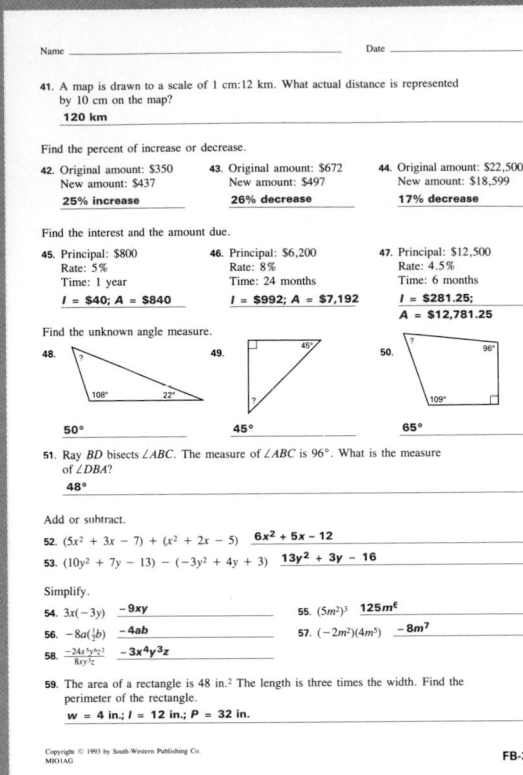

48. **50°**

49. **45°**

50. **65°**

51. Ray BD bisects $\angle ABC$. The measure of $\angle ABC$ is 96°. What is the measure of $\angle DBA$?
48°

Add or subtract.

52. $(5x^2 + 3x - 7) + (x^2 + 2x - 5)$ **$6x^2 + 5x - 12$**

53. $(10y^2 + 7y - 13) - (-3y^2 + 4y + 3)$ **$13y^2 + 3y - 16$**

Simplify.

54. $3x(-3y)$ **$-9xy$**

55. $(5m^2)^3$ **$125m^6$**

56. $-8a(\frac{1}{2}b)$ **$-4ab$**

57. $(-2m^2)(4m^5)$ **$-8m^7$**

58. $\frac{-24x^5y^5z^2}{8xy^2z}$ **$-3x^4y^3z$**

59. The area of a rectangle is 48 in.² The length is three times the width. Find the perimeter of the rectangle.
w = 4 in.; l = 12 in.; P = 32 in.

FB-3

Name _____ Date _____

Find each probability.

60. $P(2)$ $\frac{1}{8}$

61. P(an even number) $\frac{1}{2}$

62. P(3 or 4) $\frac{4}{3}$

63. P(not a 6 or 8) $\frac{3}{4}$

64. Use a tree diagram to find the number of possible outcomes of tossing a penny, a dime, and a nickel.
8 possible outcomes

65. Find the probability of spinning a 5 or a 3 on a spinner divided into six equal sections marked 1–6. $\frac{1}{3}$

66. Use the counting principle to find the number of possible outfits that can be made from 3 pairs of pants and 5 shirts.
15 possible outfits

Use the figure at the right for Exercises 67 and 68.

67. Give the coordinates of points A, B, and C.
$A(-6, -4)$; $B(2, 0)$; $C(5, -3)$

68. Find the slope of the line that passes through A and C; through B and C.
$-\frac{1}{11}$; -1

Give the order of symmetry of each figure.

69.
3

70.
2

FB-4

Teacher's Notes

Fast Forward

Objective: Students will apply mathematical models to real world situations related to the future in technology, the environment, and space exploration. They will find theoretical and sample probabilities; use calculators to determine percent increases and decreases; apply measuring, polling, estimating, and graphing techniques; and plot points and explore reflections, translations, and rotations on a coordinate grid.

UNIT OVERVIEW

In the three lessons in this unit, students will apply polling, probability, percentages, and coordinate geometry while studying future trends. From predicting the extent of use of high-tech communication tools, to monitoring the earth's endangered species, to exploring Mars with robots and computers, students will sample the quickly-approaching 21st century.

LESSON 1   **Technology Within Reach** informs students about telephone technology that is available in some areas now. Students will examine the options for high-tech telephone service and describe the sample space, using both a tree diagram and the counting principle. They will determine theoretical probabilities for some outcomes. Then they will design a poll to sample the percents of a small population who choose certain options for telephone service and use their results to make predictions for the larger population. When appropriate, they will use calculators for their computations.

LESSON 2   **Saving Earth for Tomorrow** involves students in finding data about the environment. They will estimate, using graphs, the percent decrease in certain endangered species and the percent increase due to protective measures. They will measure weight, area, and volume of wasteful packages and determine various ratios. Finally, they will estimate the effect conservation and waste reduction might have in ten, twenty, or fifty years.

LESSON 3   **Mapping Mars** will be a simulation of exciting "virtual reality" in which computers, sensors, robots, and video cameras combine to allow scientists to interact with hostile environments as if they were actually present there. Students will map points of an irregular object on a coordinate grid to simulate what the computer robot might see. Then they will use translations and rotations to help them visualize the object from different perspectives.

TIME   6–15 days

INTRODUCTION AND MOTIVATION

- In groups of 2 to 5, have students brainstorm about what the world might be like in fifty years. Allow time for group discussion. Combine all of the ideas from the groups; then, work as a class to classify the ideas into categories. The categories should include, but not be limited to, *technology, the environment,* and *space exploration.*
- Rearrange students into three groups to consider the topics, *technology, the environment,* and *space exploration.* Ask: What do you know about the current status of this topic? What research has been occurring? What innovations are planned? How do you think mathematics is used in this field? Have students keep a log of these questions and their answers to compare with what they will have learned when the unit is completed.

LESSON 1   Technology Within Reach

OBJECTIVES

Students will:
- describe all possible outcomes (sample space) related to choosing high-tech telephone service using both a tree diagram and the counting principle
- design and implement a poll concerning telephone service
- use their poll results to make predictions for a larger sample
- write a report with recommendations for telephone service
- compute with fractions, decimals, and percents using a calculator

TIME

2 to 6 class periods

MATERIALS

calculators, poster board, markers

INTRODUCING THE LESSON

After forming groups of 3 to 5 students, have the groups think of useful innovations for telephone service. Each group should make a wish list of telephone technology to share with the class.

FACILITATING THE LESSON

1. Hand out **Worksheet 1, Telephone Tech**. Give the groups time to read and discuss the types of service described on the worksheet. In this simulation, the customer must choose **one and only one option** from each box.

    Have students use the worksheet to find the sample space (the set of all possible outcomes) by drawing a tree diagram and by using the counting principle.
2. Have each group design a poll to find out how many people in a small random sample will choose each option. The size of the small sample will vary, relative to the size of the community and the time available for this project. Check group progress as they decide whom to sample and how to conduct the poll. They might adapt the top portion of the worksheet to distribute for their poll.
3. After the results are tallied, have students find the sample probability of each of the 36 outcomes. Each group should identify the most popular options and predict the number of customers in the entire town or city that would choose them. Based on their research, groups will act as "independent consultants" to the telephone company and write a report detailing their recommendations for service.

SUMMARIZING THE RESULTS

Have groups present their reports orally and include graphic displays of their data, if desired. Since groups may have chosen different strategies and/or sample populations for polling, lead a class discussion comparing the results. Did they all find the same options for service to be most popular? If not, why not?

EXTENDING THE LESSON

- Have students use their own ideas for telephone innovations, discussed in the introduction to the lesson, to provide additional choices in their poll.
- Challenge students to create a timeline showing when these innovations might be available to them. Suggest that they contact their local telephone companies, communications firms, communications equipment manufacturers, and state and private research universities or organizations for more information.
- The Caller ID feature being offered by telephone companies is controversial. Students might research why, then take sides on this issue in a debate.

ASSESSING THE LESSON

Check students' worksheets and written reports. Evaluate their use of data, their probability computations, and their understanding and application of mathematical reasoning in making their recommendations.

OBJECTIVES

Students will
- weigh and measure recyclable and nonrecyclable product packaging
- use measures of weight, area and/or volume, both customary and metric
- find the ratio and percentage of waste in packaging for comparable products
- estimate the percent increase and decrease in numbers of endangered species, using graphs and decimal/percent calculations
- estimated effects on the earth's future of various conservation measures

TIME

3 to 7 class periods

MATERIALS

measuring tapes, rulers, balance and other scales, postage scales, world almanacs, information from world wildlife and conservation organizations, newspaper or magazine articles on conservation, wildlife, or recycling

INTRODUCING THE LESSON

1. Have students react to the following statements:
   - Compact disks are sold in packages 10 times the volume of the disk itself.
   - A forest tree absorbs about 13 pounds of carbon dioxide a year—enough to offset the carbon dioxide produced by driving a car 26,000 miles.
   - In South Carolina alone, the alligator population is 16 times greater than its level just a decade ago.
   In the ensuing discussion, have students comment on how they might find out if the facts given in the statements are accurate.
2. Form groups of 3 to 5 students. Have each group obtain one or more packaged items of the same type, find the ratio of product to packaging, and compare the packaging ratios. Have them consider products they buy on a regular basis, so the project will not become costly.

FACILITATING THE LESSON

1. Provide measuring equipment; students may use customary or metric measures, depending on convenience and how the packages are labeled. **Worksheet 2, Be a Waste Watcher** can be used as a guide for calculations and comparisons.
2. Distribute almanacs and other research materials. Challenge each group to find data from which they can perform calculations and write statements like those in the introduction to this lesson. Some issues that groups might study are: percent of decrease or increase in certain endangered or threatened species during a particular time span or percent reductions of rainforest trees.
3. Have students rewrite some of their statements in the form of headlines or captions for photos or drawings that they have created or copied.

SUMMARIZING THE RESULTS

Make a bulletin board to display the headlines, drawings, and captions. Lead a class discussion asking: What did you learn from your research? Which problem facing the earth is most important? Why? What are your predictions for the future?

EXTENDING THE LESSON

- Using their percent increases or decreases, have students predict how one of the environmental issues they studied will change in ten, twenty, or fifty years.
- Challenge students to select one over-packaged product and invent a way to display it attractively without using packaging. How much packaging waste can they save? Have them make or draw a prototype.

ASSESSING THE LESSON

For the packaging experiment, use students' worksheets and essays for evaluation. Throughout the lesson, monitor group participation. Comment on the quality of the group's statements, headlines, and captions. Have students write a short essay remarking on the key points of the class summary discussion.

OBJECTIVES

Students will
- give the coordinates of points depicting an object on a coordinate grid
- graph images of objects under translations, reflections, and rotations
- identify lines of symmetry
- create a design demonstrating one or more kinds of symmetry

TIME

3 to 6 class periods

MATERIALS

graph paper, poster board, markers, rulers, small irregular objects such as stones, publications containing information on virtual reality (Optional: Contact Michael McGreevy, NASA Ames Research Center, Moffett Field, California; Jaron Lanier, VPL Research, Redwood City, California; Henry Fuchs, University of North Carolina at Chapel Hill. Your school media center may be able to obtain videotapes on the subject by contacting Media Magic, Box 507, Nicasio, CA 94946; 415-662-2426.)

INTRODUCING THE LESSON

1. Ask students to think about experiences they have had, perhaps, at amusement theme parks that were designed to mimic reality—that is, to make them see things and feel as if they were actually in the situation being depicted. Students might mention simulated space rides or other such experiences. Inquire whether they have heard of or seen advanced video arcade games in which people interact with the screen by moving their hand within a glove containing sensors. Sum up the discussion by telling students about the future technology under research today called **virtual reality**.

   In one version of virtual reality, a headset is equipped with two tiny video screens, one for each eye. When the wearer turns his or her head, the scene changes accordingly. The headset might also provide appropriate sounds, voices, or music. Some versions include a dataglove having sensors that allows the wearer to manipulate what he or she sees in the virtual reality environment. Pass out copies of one or more articles on the subject if possible.

2. Have students use their creative thinking to think of many unusual ways in which virtual reality technology might be used. Some possibilities are the following:
   - doctors can practice difficult surgical procedures using virtual reality
   - chemists (or students) could interact with the molecular structure of a substance in order to better understand it
   - architects and clients could walk through the designed buildings to discover possible problems
   - scientists could control robots in space as if they were there

3. Tell students that this lesson is based on the idea of using robots to explore Mars. NASA scientists say that virtual reality technology would allow astronauts, either on a base or in orbit around the planet, to explore large areas using cameras mounted on robots. Students will plot irregular objects on a coordinate graph and then make adjustments to show the object in different positions, just as the computer might be doing in a virtual reality experiment. Show students the objects they will use (small, flat, irregular rocks or stones), or ask them to collect some similar objects.

FACILITATING THE LESSON

1. Have students work in pairs. Each partner in the pair should have at least one irregular object with which to work. Using one partner's object at a time, students should:
   - Gently drop the irregular object onto graph paper that has been labeled with x- and y- axes.
   - Trace around the object with a pencil.

- Identify the coordinates of points which might best outline the object. Since the objects are irregular, partners will have to discuss and come to an agreement as to which coordinates best describe the object.

Each person should drop and graph his or her object in two different positions and record the results on **Worksheet 3, Rocking on Mars.**

2. Again using **Worksheet 3,** students should graph each image of their object after a translation 5 units to the right and 3 units down. Then each partner should challenge the other by describing another translation to be completed. There is a space for these coordinates to be recorded on the worksheet.

3. Circulate to check students' work as they continue to follow the directions on the worksheet. Students should use the original sets of coordinates and graph the object after a rotation 90° counterclockwise about the origin. Then students should begin with the translated images specified on the worksheet and graph the new images of the object after a rotation 180° clockwise about the origin.

4. When all groups have completed the graphing activity, have a discussion about how this simulation approximates what the computer screen might be showing when someone is exploring by virtual reality.

5. To complete the entire unit on future prospects for technology, the environment, and space exploration, have students design a poster using reflections and symmetry. Their design may be of any object, animal, or scene related to this unit, but it must have symmetrical elements. Students should be ready to show their poster and describe the line or lines of symmetry.

SUMMARIZING THE RESULTS

Completed packets for this activity should include the worksheet, the graphs as described on the worksheet, and a short written explanation demonstrating an understanding of how these graphing procedures are related to actual computer plotting of the positions of surface features that might be found on Mars. Have students present and describe the symmetry appearing in their posters.

EXTENDING THE LESSON

- Have students choose one of the other uses of virtual reality to investigate further. They should complete a paper describing some of the mathematics relevant to that application.

- You might challenge students to use the graph of their object and one of its translations or rotations as follows: have them choose one particular point, write its coordinates before and after the translation or rotation, draw a line connecting the before and after points, then try to write a linear equation to fit on the line. Interested students may repeat the procedure several times and try to find relationships and make generalizations.

- If supplies and/or help from an artist or art teacher are available, students would enjoy making a T-shirt from a symmetrically designed poster.

- If there is a nearby facility that features advanced video technology, students may wish to plan a group trip. Ask students to write a short paper describing the equipment used, what they experienced, and any suggestions or observations.

- Students may be motivated to become involved in an organization that deals with any of the topics studied in this entire unit.

ASSESSING THE LESSON

Evaluate students' completed packets for accuracy, understanding, imagination, and the degree of challenge the partners attempted for one another. The first or second activity for extending the unit might also be used to assess students' work.

Telephone Tech

Type of Phone (customer chooses one)

> - *Touch-tone phone*
> - *Phone with TV monitor* Allows you and caller, if both have this service, to view one another on TV screen
> - *Phone with computer screen* Allows business transactions by calling up menus on screen
> - *Phone with identification ring* Gives customers two or more other lines with distinctive rings in order to differentiate incoming calls

Incoming options (customer chooses one)

> - *Priority calls* Identifies calls from up to 6 special numbers, that the customer programs in, by special tone
> - *Call blocks* Lets customer screen out calls from up to 6 numbers
> - *Caller ID* Displays the phone number of the caller

Other options (customer chooses one)

> - *Repeat call* Repeatedly dials a busy number while keeping the customer's line open for incoming calls
> - *Select forwarding* Selects up to 5 numbers that will be forwarded while you are at another number
> - *Home intercom* Allows customers to speak to people anywhere in the home where there is a phone in the room

Show all possible outcomes of customer choices below. How many outcomes are there?

Name _____   Date _____

Be a Waste Watcher

Compare a product with and without packaging.

Product (1) _____

	Weight	**Volume**	**Other**
With Packaging			
Without Packaging			

Difference   _____   _____   _____

Ratio   Without:With packaging

Percentage saved $\dfrac{\text{Difference}}{\text{Total}}$

Compare comparable products (different brands of the same product).

Product (2) _____

	Weight	**Volume**	**Other**
With Packaging			
Without Packaging			

Difference   _____   _____   _____

Ratio   Without:With packaging

Percentage saved $\dfrac{\text{Difference}}{\text{Total}}$

Write an essay on the back of this sheet. Include answers to these questions:
- Which brand of product is more environmentally sound? Why?
- Make the same comparisons for different kinds of products.
  Which are the worst offenders? Why?

Name _____ Date _____

Rocking on Mars

Fill in appropriate coordinates.
Then graph each set of data on a coordinate grid.

Original **Position 1**	**Translation 1** **(5 right, 3 down)**	**Translation 2** **(_____)**

Original
Position 2

_____

_____

_____

_____

_____

Rotation 90° **counterclockwise**	**From Original** **Position 1**	**From Original** **Position 2**

Rotation 180° **clockwise**	**From Translation 1** **Position 1**	**From Translation 1** **Position 2**

How to Use the Data File

The *Data File* contains interesting information presented in tables and graphs for your use in solving problems. The data is organized into the following categories: animals, architecture, arts and entertainment, astronomy/space science, earth science, ecology, economics, health and fitness, sports, and United States. You will need to refer to this material often as you work through this book. Whenever you come upon the words "USING DATA" at the beginning of an exercise, refer to the *Data Index* to help you locate the information you will need to complete the exercise. For a quick review of commonly used symbols, formulas, and other information, you may wish to refer to the last part of the *Data File,* Useful Mathematical Data.

DATA INDEX

MAXIMUM SPEEDS OF ANIMALS

The data on this topic are notoriously unreliable because of the many inherent difficulties of timing the movement of most animals—whether running, flying, or swimming—and because of the absence of any standardization of the method of timing, of the distance over which the performance is measured, or of allowance for wind conditions.

The most that can be said is that a specimen of the species below has been timed to have attained as a maximum the speed given.

mi/h		mi/h	
219.5	Spine-tailed Swift	37	Dolphin
180*	Peregrine Falcon	36	Dragonfly
120*	Golden Eagle	35	Flying Fish
96.29	Racing Pigeon	35	Rhinoceros
88	Spurwing Goose	35	Wolf
70†	Cheetah	33	Hawk Head Moth
60	Pronghorn Antelope	32	Giraffe
60	Mongolian Gazelle	32	Guano Bat
57	Quail	30	Blackbird
57	Swordfish	28	Grey Heron
53	Partridge	25	California Sea Lion
45	Red Kangaroo	24	African Elephant
45	English Hare	23	Salmon
40	Red Fox	22.8	Blue Whale
40	Mute Swan	22	Leatherback Turtle
38	Swallow	22	Wren
		20	Monarch Butterfly

*Stooping
†Unable to sustain a speed of over 44 mi/h over 500 yards.

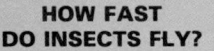

HOW FAST DO INSECTS FLY?

	Wingbeats per second	Flight speed (mi/h)
White butterfly	8–12	4–9
Damselfly	16	2–4
Dragonfly	25–40	16–34
Cockchafer beetle	50	17
Hawkmoth	50–90	11–31
Hoverfly	120	7–9
Bumblebee	130	7
Housefly	200	4
Honeybee	225	4–7
Mosquito	600	0.6–1.2
Midge	1,000	0.6–1.2

LONGEST RECORDED LIFE SPANS OF SOME ANIMALS

Animal	Years	Animal	Years
Marion's tortoise	152	Lobster	50
Deep-sea clam	100	Cow	40
Killer whale	90	Domestic pigeon	35
Blue whale	90	Domestic cat	34
Fin whale	90	Dog (Labrador)	29
Freshwater oyster	80	Budgerigar	28
Cockatoo	70	Sheep	20
Condor	70	Goat	18
Indian elephant	70	Rabbit	18
Ostrich	62	Golden hamster	10
Horse	62	House mouse	6
Chimpanzee	50	Housefly	0.2
Termite	50		

TOP 25
AMERICAN KENNEL CLUB
REGISTRATIONS

Breed	Rank	1989	Rank	1988	Breed	Rank	1989	Rank	1988
Cocker Spaniels	1	111,636	1	108,720	Pomeranians	14	32,109	14	30,516
Labrador Retrievers	2	91,107	2	86,446	Lhasa Apsos	15	28,810	15	30,194
Poodles	3	78,600	3	82,600	Chihuahuas	16	24,917	17	23,487
Golden Retrievers	4	64,269	4	62,950	Pekingese	17	22,986	22	20,134
German Shepherd Dogs	5	58,422	5	57,139	Boxers	18	22,037	20	20,604
Rottweilers	6	51,291	7	42,748	Siberian Huskies	19	21,875	18	21,430
Chow Chows	7	50,150	6	50,781	Doberman Pinschers	20	21,782	16	23,928
Dachshunds	8	44,305	9	41,921	Basset Hounds	21	21,517	19	21,423
Beagles	9	43,314	8	41,983	English Springer				
Miniature Schnauzers	10	42,175	10	41,558	Spaniels	22	20,911	21	20,238
Shetland Sheepdogs	11	39,665	12	38,730	Collies	23	18,227	23	18,931
Yorkshire Terriers	12	39,268	13	36,040	Dalmatians	24	17,488	27	14,109
Shih Tzu	13	38,131	11	38,829	Boston Terriers	25	15,355	24	14,988

GESTATION, LONGEVITY, AND INCUBATION OF ANIMALS

Longevity figures were supplied by Ronald T. Reuther. They refer to animals in captivity; the potential life span of animals is rarely attained in nature. Maximum longevity figures are from the Biology Data Book, 1972. Figures on gestation and incubation are averages based on estimates by leading authorities.

Animal	Gestation (days)	Average longevity (years)	Maximum longevity (yr, mo)	Animal	Gestation (days)	Average longevity (years)	Maximum longevity (yr, mo)
Ass	365	12	35-10	Leopard	98	12	19-4
Baboon	187	20	35-7	Lion	100	15	25-1
Bear: Black	219	18	36-10	Monkey (rhesus)	164	15	—
Grizzly	225	25	—	Moose	240	12	—
Polar	240	20	34-8	Mouse (meadow)	21	3	—
Beaver	122	5	20-6	Mouse (dom. white)	19	3	3-6
Buffalo (American)	278	15	—	Opossum (American)	14-17	1	—
Bactrian camel	406	12	29-5	Pig (domestic)	112	10	27
Cat (domestic)	63	12	28	Puma	90	12	19
Chimpanzee	231	20	44-6	Rabbit (domestic)	31	5	13
Chipmunk	31	6	8	Rhinoceros (black)	450	15	—
Cow	284	15	30	Rhinoceros (white)	—	20	—
Deer (white-tailed)	201	8	17-6	Sea lion (California)	350	12	28
Dog (domestic)	61	12	20	Sheep (domestic)	154	12	20
Elephant (African)	—	35	60	Squirrel (gray)	44	10	—
Elephant (Asian)	645	40	70	Tiger	105	16	26-3
Elk	250	15	26-6	Wolf (maned)	63	5	—
Fox (red)	52	7	14	Zebra (Grant's)	365	15	—
Giraffe	425	10	33-7				
Goat (domestic)	151	8	18				
Gorilla	257	20	39-4		Incubation time (days)		
Guinea pig	68	4	7-6	Chicken 21			
Hippopotamus	238	25	—	Duck .30			
Horse	330	20	46	Goose 30			
Kangaroo	42	7	—	Pigeon 18			
				Turkey 26			

DATA FILE Architecture

THE LONGEST BRIDGE SPANS
(as of 1988)

Bridge	Country	Year Built	Span (meters)	Span (feet)
Akashi-Ohashi*	Japan	1988	1,780	5,840
Humber	England	1981	1,410	4,626
Verrazano Narrows	United States	1964	1,298	4,260
Golden Gate	United States	1937	1,280	4,200
Mackinac	United States	1957	1,158	3,800
Bosphorus	Turkey	1973	1,074	3,524
George Washington	United States	1931	1,067	3,500
Tagus River	Portugal	1966	1,013	3,323
Forth	Scotland	1964	1,006	3,300
Severn	England/Wales	1966	988	3,240

*Scheduled completion information

THE TALLEST BUILDINGS IN THE WORLD
(as of 1990)

Building	City	Year Built	Stories	Height (meters)	Height (feet)
Sears Tower	Chicago	1974	110	443	1,454
World Trade Center, North	New York	1972	110	417	1,368
World Trade Center, South	New York	1973	110	415	1,362
Empire State	New York	1931	102	381	1,250
Bank of China Tower	Hong Kong	1988	72	368	1,209
Amoco	Chicago	1973	80	346	1,136
John Hancock	Chicago	1968	100	344	1,127
Chrysler	New York	1930	77	319	1,046

LAND VEHICULAR TUNNELS IN U.S.
(over 2,000 feet in length)

Name	Location	Length (feet)
E. Johnson Memorial	I-70, Col.	8,959
Eisenhower Memorial	I-70, Col.	8,941
Allegheny (twin)	Penna. Turnpike	6,072
Liberty Tubes	Pittsburgh, Pa.	5,920
Zion Natl. Park	Rte. 9, Utah	5,766
East River Mt. (twin)	Interstate 77, W. Va.–Va.	5,412
Tuscarora (twin)	Penna. Turnpike	5,400
Kittatinny (twin)	Penna. Turnpike	4,660
Blue Mountain (twin)	Penna. Turnpike	4,435
Lehigh	Penna. Turnpike	4,379
Wawona	Yosemite Natl. Park	4,233
Big Walker Mt.	Route I-77, Va.	4,229
Squirrel Hill	Pittsburgh, Pa.	4,225
Fort Pitt	Pittsburgh, Pa.	3,560
Mall Tunnel	Dist. of Columbia	3,400
Caldecott	Oakland, Cal.	3,371
Cody No. 1	U.S. 14, 16, 20, Wyo.	3,202
Kalihi	Honolulu, Ha.	2,780
Ft. Cronkhite	Sausalito, Cal.	2,690
Memorial	W. Va. Tpke. (I-77)	2,669
Cross-Town	178 St., N.Y.C.	2,414
F. D. Roosevelt Dr.	81–89 Sts., N.Y.C.	2,400
Dewey Sq.	Boston, Mass.	2,400
Battery Park	N.Y.C.	2,300
Battery St.	Seattle, Wash.	2,140
Big Oak Flat	Yosemite Natl. Park	2,083

NOTED RECTANGULAR STRUCTURES

Structure	Country	Length (meters)	Width (meters)
Parthenon	Greece	69.5	30.9
Palace of the Governors	Mexico	96	11
Great Pyramid of Cheops	Egypt	230.6	230.6
Step Pyramid of Zosar	Egypt	125	109
Temple of Hathor	Egypt	290	280
Cleopatra's Needle (base)	England*	2.4	2.3
Ziggurat of Ur (base)	Middle East	62	43
Guanyin Pavilion of Dule Monastery	China	20	14
Izumo Shrine	Japan	10.9	10.9
Kibitsu Shrine (main)	Japan	14.5	17.9
Kongorinjo Hondo	Japan	21	20.7
Bakong Temple, Roluos	Cambodia	70	70
Ta Keo Temple	Cambodia	103	122
Wat Kukut Temple, Lampun	Thailand	23	23
Tsukiji Hotel	Japan	67	27

*Gift to England from Egypt

Parthenon

Mosque of Omar.

PENNY SIZES OF NAILS

Penny Size	Length (in centimeters)
2-penny	2.5 cm
3-penny	3.125 cm
4-penny	3.75 cm
6-penny	5 cm
8-penny	6.25 cm
10-penny	7.5 cm
12-penny	8.125 cm
16-penny	8.75 cm

HOUSING UNITS—SUMMARY OF CHARACTERISTICS AND EQUIPMENT, BY TENURE AND REGION: ONE RECENT YEAR

[In thousands of units, except as indicated. Based on the American Housing Survey]

Item	Total housing units	Sea-sonal	Year-round Units							Vacant
			Occupied							
			Total	Owner	Renter	North-east	Mid-west	South	West	
Total units	99,931	3,182	88,425	56,145	32,280	18,729	22,142	30,064	17,490	8,324
Percent distribution	100.0	3.2	88.5	56.2	32.3	21.2	25.0	34.0	19.8	8.3
Units in structure:										
Single family detached	60,607	1,834	55,076	46,703	8,373	9,368	14,958	19,984	10,766	3,697
Single family attached	4,514	64	4,102	2,211	1,890	1,431	814	1,212	645	349
2–4 units	11,655	134	10,217	1,996	8,221	3,324	2,515	2,426	1,952	1,304
5–9 units	5,134	73	4,372	344	4,029	903	967	1,408	1,094	689
10–19 units	4,558	99	3,760	261	3,500	818	766	1,360	817	699
20–49 units	3,530	146	2,913	287	2,627	904	603	651	756	470
50 or more units	3,839	135	3,230	438	2,792	1,540	661	583	446	474
Mobile home or trailer	6,094	698	4,754	3,906	848	440	860	2,440	1,014	642

LONGEST BROADWAY RUNS

Show	Performances
1. A Chorus Line (1975–90)	6,137
2. Oh! Calcutta (1976–89)	5,959
3. 42nd Street (1980–89)	3,486
4. Grease (1972–80)	3,388
5. Cats (1982–)	3,269
6. Fiddler on the Roof (1964–72)	3,242
7. Life with Father (1939–47)	3,224
8. Tobacco Road (1933–41)	3,182
9. Hello, Dolly! (1964–71)	2,844
10. My Fair Lady (1956–62)	2,717
11. Annie (1977–83)	2,377
12. Man of La Mancha (1965–71)	2,328
13. Abie's Irish Rose (1922–27)	2,327
14. Oklahoma! (1943–48)	2,212
15. Pippin (1971–77)	1,944
16. South Pacific (1949–54)	1,925
17. Magic Show (1974–78)	1,920
18. Deathtrap (1978–82)	1,792
19. Gemini (1977–81)	1,788
20. Harvey (1944–49)	1,775
21. Dancin' (1978–82)	1,774
22. La Cage aux Folles (1983–87)	1,761
23. Hair (1968–72)	1,750
24. The Wiz (1975–79)	1,672
25. Born Yesterday (1946–49)	1,642

OPTICAL ILLUSIONS

ALL-TIME TOP TELEVISION PROGRAMS

	Program	Date	Network	Households (000)
1	M*A*S*H	2/28/83	CBS	50,150
2	Dallas	11/21/80	CBS	41,470
3	Roots Pt. VIII	1/30/77	ABC	36,380
4	Super Bowl XVI	1/24/82	CBS	40,020
5	Super Bowl XVII	1/30/83	NBC	40,480
6	Super Bowl XX	1/26/86	NBC	41,490
7	Gone With The Wind— Pt. 1	11/7/76	NBC	33,960
8	Gone With The Wind— Pt. 2	11/8/76	NBC	33,750
9	Super Bowl XII	1/15/78	CBS	34,410
10	Super Bowl XIII	1/21/79	NBC	35,090
11	Bob Hope Christmas Show	1/15/70	NBC	27,260
12	Super Bowl XVIII	1/22/84	CBS	38,800
12	Super Bowl XIX	1/20/85	ABC	39,390
14	Super Bowl XIV	1/20/80	CBS	35,330
15	ABC Theater (The Day After)	11/20/83	ABC	38,550
16	Roots Pt. VI	1/28/77	ABC	32,680
16	The Fugitive	8/29/67	ABC	25,700
18	Super Bowl XXI	1/25/87	CBS	40,030
19	Roots Pt. V	1/27/77	ABC	32,540
20	Ed Sullivan	2/9/64	CBS	23,240
21	Bob Hope Christmas Special	1/14/71	NBC	27,050
22	Roots Pt. III	1/25/77	ABC	31,900
23	Super Bowl XI	1/9/77	NBC	31,610
23	Super Bowl XV	1/25/81	NBC	34,540
25	Super Bowl VI	1/16/72	CBS	27,450

Source: A. C. Nielsen estimates, Jan. 30, 1960 through Apr. 17, 1989, excluding unsponsored or joint network telecasts or programs under 30 minutes long.
Ranked by percent of average audience.

THE RECORDING INDUSTRY ASSOCIATION OF AMERICA
1990 Sales Profile
Released September 1991

PERCENTAGE OF DOLLAR VALUE BY TYPE OF MUSIC

Type of Music	1986	1987	1988	1989	1990
Rock	46.8	47.2	46.2	42.9	37.4
Pop	14.2	12.9	15.2	14.4	13.6
Urban Contemporary	10.1	11.6	13.3	14.0	18.3
Country	9.7	9.5	7.4	6.8	8.8
Classical	5.9	2.6	3.5	4.3	4.1
Jazz	0.4	5.2	4.7	5.7	5.2
Gospel	3.0	3.9	2.5	3.1	2.4
Other	5.6	6.6	6.0	8.0	9.3

PERCENTAGE OF DOLLAR VALUE BY CONFIGURATION

Configuration	1986	1987	1988	1989	1990
Cassette	57.4	59.3	60.4	50.4	48.4
Cassette Singles	0	0	1.0	2.7	2.6
LP's	28.4	18.2	13.9	8.3	4.3
7" Singles	1.5	0.8	3.4	0.5	0.3
12" Singles	2.1	1.5	2.2	1.6	0.7
CD's 3" Singles	0	0	0	0.4	1.1
CD's 5" Full	10.1	19.8	18.6	35.9	42.5

	1986	1987	1988	1989	1990
TOTAL U.S. SALES IN BILLIONS OF DOLLARS— 1986 THROUGH 1990	4.6	5.5	6.2	6.4	7.5

ALL-TIME TOP 25 MOVIES
Source: *Variety*, January, 1990

Title	Total $ Revenue
1. E.T. the Extra-Terrestrial; 1982	228,618,939
2. Star Wars; 1977	193,500,000
3. Return of the Jedi; 1983	168,002,414
4. Batman; 1989	150,500,000
5. The Empire Strikes Back; 1980	141,600,000
6. Ghostbusters; 1984	130,211,324
7. Jaws; 1975	129,549,325
8. Raiders of the Lost Ark; 1981	115,598,000
9. Indiana Jones and the Last Crusade; 1989	115,500,000
10. Indiana Jones and the Temple of Doom; 1984	109,000,000
11. Beverly Hills Cop; 1984	108,000,000
12. Back to the Future; 1985	104,408,738
13. Grease; 1978	96,300,000
14. Tootsie; 1982	96,292,736
15. The Exorcist; 1973	89,000,000
16. The Godfather; 1972	86,275,000
17. Superman; 1978	82,800,000
18. Rain Man; 1989	86,000,000
19. Close Encounters of the Third Kind; 1977/1980	82,750,000
20. Three Men and a Baby; 1987	81,313,000
21. Who Framed Roger Rabbit; 1988	81,244,000
22. Beverly Hills Cop II; 1987	80,857,776
23. The Sound of Music; 1965	79,748,000
24. Gremlins; 1984	79,500,000
25. Top Gun; 1986	79,400,000

Revenue figures are in absolute dollars, reflecting actual amounts received by the distributors (estimated for movies in current release). Ticket price inflation favors recent films, but older films have the advantage of numerous reissues adding to their totals.

MULTIMEDIA AUDIENCES—SUMMARY: 1987

[In percent, except as indicated. As of **spring**. For persons 18 years old and over. Based on sample and subject to sampling error; see source for details]

	Television viewing and coverage	Television prime time viewing and coverage	Cable viewing and coverage	Radio listening and coverage	Newspaper reading and coverage
Total	**91.5**	**77.1**	**45.1**	**86.2**	**85.7**
18–24 years old	88.0	68.5	45.5	93.9	82.6
25–34 years old	90.2	75.8	47.5	93.7	86.2
35–44 years old	90.8	78.3	51.0	91.3	88.2
45–54 years old	91.9	77.6	45.5	87.3	87.4
55–64 years old	94.5	82.9	44.6	79.1	88.1
65 years old and over	94.9	80.8	33.9	66.2	81.7

Source: Mediamark Research Inc., New York, NY, *Multimedia Audiences,* Spring 1988. (Copyright.)

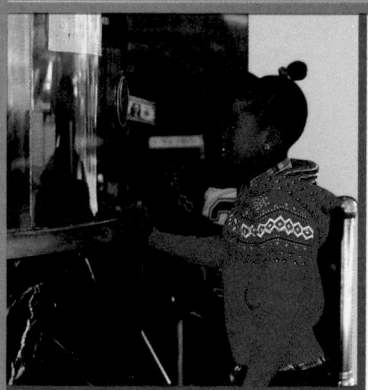

Arts and Entertainment **553**

PLANETARY DIAMETERS AND DISTANCES FROM SUN

Planet	Equatorial Diameter (km)	Average Distance from Sun (km)
Mercury	4,880	58,000,000
Venus	12,100	108,000,000
Earth	12,756	150,000,000
Mars	6,780	228,000,000
Jupiter	142,800	778,000,000
Saturn	120,000	1,427,000,000
Uranus	50,800	2,870,000,000
Neptune	48,600	4,497,000,000
Pluto	2,200	5,900,000,000

PLANETARY DAYS AND YEARS

Planet	Length of Day (in Earth Hours)	Length of Year (in Earth Years; 365.26 days = 1 year)
Mercury	1406.4	0.2409
Venus	5832	0.6152
Earth	23.934	1.00
Mars	24.623	1.88
Jupiter	9.842	11.86
Saturn	10.23	29.46
Uranus	23.0	84.01
Neptune	22.0	164.8
Pluto	153.0	248.4

PERCENTAGE OF MAIN ELEMENTS MAKING UP EARTH'S CRUST

Oxygen	47%	Calcium	4%
Silicon	28%	Magnesium	2%
Aluminum	8%	Sodium	3%
Iron	5%	Potassium	3%

LARGEST MOONS OF THE SOLAR SYSTEM

Moon	Planet	Diameter (mi)
Titan	Saturn	3,500
Triton	Neptune	3,300
Callisto	Jupiter	3,220
Ganymede	Jupiter	3,200
Io	Jupiter	2,310
Moon	Earth	2,160
Europa	Jupiter	1,950

LARGE TELESCOPES
OF THE WORLD

Type	Location	Size of Lens[1]
Refractor	Yerkes Observatory, Wisconsin	102 cm
	Lick Observatory, California	91 cm
	Paris Observatory, Meudon, France	84 cm
Reflector	W. M. Keck Telescope, Mauna Kea, Hawaii	10 m*
	Special Astrophysical Observatory, Zelenchukskaya, USSR	6 m
	Hale Telescope, Mount Palomar, California	5 m
	Cerro Tololo Inter-America Observatory, Chile	4 m
	Kitt Peak National Observatory, Arizona	4 m
	Mount Stromlo Observatory, Australia	3.8 m
	European Southern Observatory, Chile	3.6 m
	Lick Observatory, California	3 m
	McDonald Observatory, Texas	2.7 m
	Hale Observatory, Mt. Wilson, California	2.5 m

[1] for refractor telescopes, size is diameter of objective lens; for reflector telescopes, size is diameter of mirror   *multiple mirrors

ORBITAL PERIODS
OF SELECTED COMETS

Name	Orbital Period (yr)	First Observed
Encke	3.30	A.D. 1786
Giacobini-Zinner	6.42	1900
Biela	6.62	1772
Whipple	7.42	1933
Wolf	8.43	1884
Temple-Tuttle	33.18	1866
Halley	76.03	B.C. 239

SPACE SHUTTLE SYSTEM EXPENDITURES BY NASA
IN CONSTANT (1982) DOLLARS: 1972 TO 1987

[In millions of dollars. For year ending Sept. 30. These data cannot be used alone to estimate the total shuttle cost. Only operating expenditures correspond to an annual cost in the economic sense, and even those estimates require adjustment for the three-year period over which the cost of a shuttle flight is incurred and reimbursements are received.]

Year	Total	D.D.T. and E.[1]	Con-struction	Pro-duction	Opera-tions	Year	Total	D.D.T. and E.[1]	Con-struction	Pro-duction	Opera-tions
1972	40	—	40	—	—	1980	2,751	1,339	42	907	463
1973	58	—	58	—	—	1981	2,724	1,041	13	1,093	578
1974	325	216	109	—	—	1982	2,932	894	20	1,283	735
1975	1,543	1,407	136	—	—	1983	3,014	—	26	1,635	1,354
1976	2,619	2,541	79	—	—	1984	3,127	—	73	690	2,364
1977	2,258	2,089	60	109	—	1985	2,636	—	42	1,376	1,218
1978	2,062	1,898	103	61	—	1986	3,311	—	—	1,504	1,807
1979	2,251	1,713	42	496	—	1987	5,806	—	—	3,841	1,965

— Represents zero.   [1] Design, development, testing, and evaluation.

Source: U.S. Congress, Congressional Budget Office, estimate March 1985 and U.S. National Aeronautics and Space Administration, *NASA Budget Estimates for Fiscal Year 1989.*

DATA FILE Earth Science

HIGHEST AND LOWEST CONTINENTAL ALTITUDES

Source: National Geographic Society, Washington, D.C.

Continent	Highest point	Feet of elevation	Lowest point	Feet below sea level
Asia	Mount Everest, Nepal-Tibet	29,028	Dead Sea, Israel-Jordan	1,312
South America	Mount Aconcagua, Argentina	22,834	Valdes Peninsula, Argentina	131
North America	Mount McKinley, Alaska	20,320	Death Valley, California	282
Africa	Kilimanjaro, Tanzania	19,340	Lake Assai, Djibouti	512
Europe	Mount El'brus, USSR	18,510	Caspian Sea, USSR	92
Antarctica	Vinson Massif	16,864	Unknown	...
Australia	Mount Kosciusko, New South Wales	7,310	Lake Eyre, South Australia	52

From *The World Almanac and Book of Facts*, 1991.

SOME NOTABLE TORNADOES IN U.S. SINCE 1960

Date		Place	Deaths
1960	May 5, 6	SE Oklahoma, Arkansas	30
1965	Apr. 11	Ind., Ill., Oh., Mich., Wis.	271
1966	Mar. 3	Jackson, Miss.	57
1966	Mar. 3	Mississippi, Alabama	61
1967	Apr. 21	Ill., Mich.	33
1968	May 15	Midwest	71
1969	Jan. 23	Mississippi	32
1971	Feb. 21	Mississippi delta	110
1973	May 26–27	South, Midwest (series)	47
1974	Apr. 3–4	Ala., Ga., Tenn., Ky., Oh.	350
1977	Apr. 4	Ala., Miss., Ga.	22
1979	Apr. 10	Tex., Okla.	60
1980	June 3	Grand Island, Neb. (series)	4
1982	Mar. 2–4	South, Midwest (series)	17
1982	May 29	So. Ill.	10
1983	May 18–22	Texas	12
1984	Mar. 28	N. Carolina, S. Carolina	67
1984	Apr. 21–22	Mississippi	15
1984	Apr. 26	Okla. to Minn. (series)	17
1985	May 31	N.Y., Pa., Oh., Ont. (series)	90
1987	May 22	Saragosa, Tex.	29
1990	June 2–3	Midwest, Great Lakes	13

MEASURING EARTHQUAKES

The energy of an earthquake is generally reported using the Richter scale, a system developed by American geologist Charles Richter in 1935, based on measuring the heights of wave measurements on a seismograph.

On the Richter scale, each single-integer increase represents 10 times more ground movement and 30 times more energy released. The change in magnitude between numbers on the scale can be represented by 10^x and 30^x, where x represents the change in the Richter scale measure. Therefore, a 3.0 earthquake has 100 times more ground movement and 900 times more energy released than a 1.0 earthquake.

Richter scale

2.5	Generally not felt, but recorded on seismometers.
3.5	Felt by many people.
4.5	Some local damage may occur.
6.0	A destructive earthquake.
7.0	A major earthquake. About ten occur each year.
8.0 and above	Great earthquakes. These occur once every five to ten years.

From *The Universal Almanac* © 1990.

RECORD HOLDERS FROM THE EARTH SCIENCES

Traditionally, the study of early life by way of its remains—which are called fossils—has been the province of earth scientists, rather than biologists. This is because fossils were the first known way to group rocks of the same age together, since it was assumed that at a given time in the past certain life-forms existed and later became extinct. Earth itself is thought to be about 4.5 billion years old.

By Size

This group of record holders includes a couple of ancient life-forms but concentrates more on those features of Earth that are studied by earth scientists.

Record	Record holder	Size or distance
Longest dinosaur	*Seismosaurus*, who lived about 150 million years ago in what is now New Mexico	100–120 ft long
Largest island	Greenland, also known as Kalaallit Nunnaat	About 840,000 mi²
Largest ocean	Pacific	64,186,300 mi²
Deepest part of ocean	*Challenger* Deep in the Marianas Trench in Pacific Ocean	35,640 ft, or 6.85 mi
Greatest tide	Bay of Fundy between Maine and New Brunswick	47.5 ft between high and low tides
Largest geyser	Steamboat Geyser in Yellowstone Park	Shoots mud and rocks about 1,000 ft in air
Longest glacier	Lambert Glacier in Antarctica (upper section known as Mellor Glacier)	At least 250 mi
Longest known cave	Mammoth Cave, which is connected to Flint Ridge Cave system	Total mapped passageway of over 300 mi
Largest canyon on land	Grand Canyon of the Colorado, in northern Arizona	Over 217 mi long, from 4 to 13 mi wide, as much as 5,300 ft deep
Highest volcano	Cerro Aconagua in the Argentine Andes	22,834 ft high
Most abundant mineral	Magnesium silicate perovskite	About ⅔ of planet, it forms Earth's mantle

By Age

Record	Record holder	Age in years
Oldest rocks	Zircons from Australia	4.4 billion
Oldest fossils	Single-celled algae or bacteria from Australia	3.5 billion
Oldest slime molds (bacteria)	Slime molds—also called slime bacteria—that gathered together to form multicellular bodies for reproduction	2.5 billion
Oldest petroleum	Oil from northern Australia	1.4 billion
Oldest land animal	Millipede? Known only from its burrows	488 million
Oldest fish	*Sacabambasis*, found in Bolivia	470 million
Oldest land plant	Moss or algae	425 million
Oldest insect	Bristletail (relative of modern silverfish)	390 million
Oldest reptile	"Lizzie the Lizard," found in Scotland	340 million
Oldest bird	*Protoavis*, fossils found near Post, Texas	225 million
Oldest dinosaur	Unnamed dinosaur the size of ostrich	225 million

SIZE AND DEPTH OF THE OCEANS

Ocean	Mi²	Greatest Depth-ft
Pacific	63,800,000	36,161
Atlantic	31,800,000	30,249
Indian	28,900,000	24,441
Arctic	5,400,000	17,881

SOME PRINCIPAL RIVERS OF THE WORLD

River	Length (miles)
Amazon	4,000
Arkansas	1,459
Columbia	1,243
Danube	1,776
Ganges	1,560
Indus	1,800
Mackenzie	2,635
Mississippi	2,340
Missouri	2,540
Nile	4,160
Ohio	1,310
Orinoco	1,600
Paraguay	1,584
Red	1,290
Rhine	820
Rio Grande	1,900
St. Lawrence	800
Snake	1,038
Thames	236
Tiber	252
Volga	2,194
Zambezi	1,700

SIZE OF THE CONTINENTS

Continent	Mi²
Asia	17,297,000
Africa	11,708,000
North America	9,406,000
South America	6,883,000
Antarctica	5,405,000
Europe	3,835,000
Australia	2,968,000

Earth Science **557**

FUEL ECONOMY AND CARBON DIOXIDE

The better your car's gas mileage, the less carbon dioxide your car will emit into the environment. As stated earlier, cars and light trucks emit one-fifth of all carbon dioxide in the United States, one of the main causes of the greenhouse effect.

Here are estimates of the amount of carbon dioxide emissions over a single car's lifetime, courtesy of the Energy Conservation Coalition, a project of Environmental Action:

$17\frac{29}{100}$ Tons

60 mi/gal

$25\frac{93}{100}$ Tons

45 mi/gal

$39\frac{29}{100}$ Tons

26 mi/gal

$57\frac{3}{4}$ Tons

18 mi/gal

ACID RAIN TROUBLE AREAS

Source: National Acid Precipitation Assessment Program

Area	No. of lakes over 10 acres	No. acidified	%
Adirondacks			
Southwest Lakes	450	171	38
New England			
Seaboard Lowlands Lakes	848	68	8
Highland Lakes	3,574	71	2
Appalachia			
Forested Lakes	433	43	10
Forested Streams	11,631	1,396	12
Atlantic Coastal Plain			
Northeast Lakes	187	21	11
Pine Barrens Streams	675	378	56
Other Streams	7,452	745	10
Florida			
Northern Highland Lakes	522	329	63
Northern Highland Streams	669	187	28
Eastern Upper Midwest			
Low Silica Lakes	1,254	201	16
High Silica Lakes	1,673	50	3

CONTENTS OF GARBAGE CANS

Our garbage cans are filled with a diverse mix (by weight):

paper and paperboard $\frac{9}{25}$
yard wastes $\frac{19}{100}$
glass $\frac{2}{25}$
metals $\frac{9}{100}$
food $\frac{9}{100}$
plastics $\frac{2}{25}$
wood/fabric $\frac{1}{25}$
rubber and leather $\frac{3}{100}$
textiles $\frac{1}{50}$
other $\frac{4}{250}$
household hazardous waste $\frac{1}{250}$

From: *Nontoxic, Natural, & Earthwise* by Debra Lynn Dood.

QUANTITIES OF HAZARDOUS WASTE BURNED IN U.S.

Incinerator Type	Number of Facilities	Quantity of Hazardous Waste Burned, lbs/yr
Commercial incinerators	17	1.3 billion (1)
Captive/on-site incinerators	154	2.3 billion (1)
Cement kilns	25–30	1.8 billion (2)
Aggregate kilns	6	1.2 billion (2)
Boilers/other furnaces	900 +	1.0 billion (2)
Total	1,100 +	7.6 billion

(1) Highum 1990. Data from review of states' capacity assurance plans.
(2) Holloway 1990. Data from U.S. EPA'S Office of Solid Waste and Emergency Response.

AIR AND WATER POLLUTION ABATEMENT EXPENDITURES IN CONSTANT (1982) DOLLARS: 1975 TO 1986

[In millions of dollars]

Year	Total	AIR						Total[4]	WATER			
		Mobile sources[1]			Stationary sources				Industrial		Public sewer systems[5]	
		Total	Cars	Trucks	Total[2]	Industrial			Facilities	Operations[3]	Facilities	Operations[3]
						Facilities	Operations[3]					
1975	20,768	9,324	7,419	1,904	11,444	6,669	4,348	22,840	4,200	2,950	8,997	3,428
1978	23,774	11.877	8,928	2,949	11,900	5,652	5,711	26,631	4,277	3,934	10,090	4,392
1979	24,462	11,757	8,469	3,288	12,706	5,969	6,108	26,470	4,013	4,222	9,758	4,583
1980	24,744	11.764	8,818	2,946	12,980	5,946	6,304	24,647	3,725	4,081	8,942	4,694
1981	25,850	13,401	10,564	2,837	12,448	5,446	6,299	21,984	3,259	4,180	6,882	4,880
1982	24,961	13,464	10,530	2,934	11,496	5,086	5,675	21,199	3,080	4,022	6,148	5,156
1983	26,367	15,581	12,274	3,307	10,785	4,104	5,990	21,543	2,811	4,509	5,551	5,475
1984	28,591	17,561	13,481	4,081	11,030	4,115	6,260	23,257	2,900	4,795	6,387	5,648
1985	29,524	18,704	14,217	4,488	10,819	3,929	6,342	24,770	2,941	5,042	6,990	5,946
1986, prel.	30,401	19,535	14,830	4,704	10,867	3,881	6,425	26,201	2,852	5,279	7,707	6,567

[1]Excludes expenditures to reduce emissions from sources other than cars and trucks. [2]Includes other expenditures not shown separately. [3]Operation of facilities. [4]Includes nonpoint sources not shown separately. [5]Includes expenditures for private connection to sewer systems, by owners of animal feedlots, and by government enterprises.

Source: U.S. Bureau of Economic Analysis, *Survey of Current Business*, May 1988.

DATA FILE Economics

MONEY AROUND THE WORLD

In the United States, the basic monetary unit is the dollar and the chief fractional unit is the penny. One dollar = 100 pennies. Unless noted otherwise, the basic monetary unit equals 100 chief fractional units for the countries listed below.

Country	Basic Monetary Unit	Chief Fractional Unit	Coin and Paper Denominations
Australia	dollar	cent	100, 50, 20, 10, 5, and 2 dollar notes; 2 and 1 dollar coins; 50, 20, 10, 5, 2, and 1 cent coins
Canada	dollar	cent	1,000, 500, 100, 50, 20, 10, 5, 2, and 1 dollar notes; 1 dollar coin; 25, 10, 5, and 1 cent coins
China	yuan	jiao (10 jiao) fen (1,000 fen)	10, 5, 2, and 1 yuan notes; 5, 2, and 1 jiao notes; 1 yuan coin; 5, 2, and 1 jiao coins; 5, 2, and 1 fen coins
France	franc	centime	500, 200, 100, 50, and 20 franc notes; 10, 5, 2, and 1 franc coins; 50, 20, 10, and 5 centime coins
India	Rupee	paise	1,000, 500, 100, 40, 20, 10, 5, 2, and 1 rupee notes; 2 and 1 rupee coins; 50, 25, and 20 paise coins
Mexico	peso	centavo	50,000, 20,000, 10,000, 5,000, 2,000, 1,000, and 500 peso notes; 500, 200, 100, 50, 20, 10, 5, 2, and 1 peso coins
Netherlands	gulden	cent	1,000, 250, 100, 50, 25, 10, and 5 gulden notes; 5, 2.5, and 1 gulden coins; 25, 10, 5 cents
Sudan	pound	piasters, milliemes (1,000 milliemes)	50, 20, 10, 5, and 1 pound notes; 50 and 25 piaster notes; 50, 10, 5, and 2 piaster coins; 10, 5, 2, and 1 millieme coins

SEIGNIORAGE OF COIN AND SILVER BULLION MADE IN THE UNITED STATES

Seigniorage is the difference between the monetary value of a coin and the cost of making a coin.

Fiscal Year	Dollars
1/1/35–6/30/65 cumulative	2,525,927,763.84
1968	383,141,339.00
1970	274,217,884.01
1972	580,586,683.00
1974	320,706,638.49
1980	662,814,791.48
1983	477,479,387.58
1984	498,371,724.09
1986	692,445,674.57
1988	470,409,480.20
1/1/35–6/30/88 cumulative	13,727,374,271.32

EARLY MONETARY SYSTEMS

Pacific Islands (New Britain, San Cristobel)

10 coconuts = 1 string white whales' teeth
10 strings of white teeth = 1 string of red whales' teeth or 1 dog's tooth
10 strings of red teeth = 50 porpoise teeth
500 porpoise teeth = 1 wife of good quality
1 "marble" (shell) ring = 1 good pig

THE SHRINKING VALUE OF THE DOLLAR

The average retail cost of certain foods in selected years, 1890–1975 (to the nearest cent)

	5 lb flour	1 lb round steak	1 qt milk	10 lb potatoes
1890	$0.15	$0.12	$0.07	$0.16
1910	0.18	0.17	0.08	0.17
1930	0.23	0.43	0.14	0.36
1950	0.49	0.94	0.21	0.46
1970	0.59	1.30	0.33	0.90
1975	0.98	1.89	0.45	0.99

How to Read the Exchange Rates Table →

The first two columns show how many U.S. dollars are needed to equal one unit of another country's currency. For example, on Thursday, one British pound could be exchanged for $1.6045 U.S. The last two columns show the value of one U.S. dollar in another country. For example, on Thursday, $1.00 U.S. had the same value as 0.6232 British pound.

EXCHANGE RATES

Thursday, July 11, 1991

The New York foreign exchange selling rates below apply to trading among banks in amounts of $1 million and more, as quoted at 3 p.m. Eastern time by Bankers Trust Co. and other sources. Retail transactions provide fewer units of foreign currency per dollar.

Country	U.S. $ equiv. Thurs.	U.S. $ equiv. Wed.	Currency per U.S. dollar Thurs.	Currency per U.S. dollar Wed.
Argentina (Austral)	.0001010	.0001010	9902.00	9902.00
Australia (Dollar)	.7667	.7670	1.3043	1.3038
Austria (Schilling)	.07746	.07846	12.91	12.75
Bahrain (Dinar)	2.6525	2.6525	.3770	.3770
Belgium (Franc)	.02648	.02682	37.76	37.28
Brazil (Cruzeiro)	.00320	.00320	312.34	312.34
Britain (Pound)	1.6045	1.6225	.6232	.6163
Canada (Dollar)	.8705	.8716	1.1487	1.1473
Chile (Peso)	.002945	.002944	339.51	339.71
China (Renmimbi)	.186567	.186567	5.3600	5.3600
Colombia (Peso)	.001753	.001751	570.38	571.00
Denmark (Krone)	.1409	.1425	7.0949	7.0163
Ecuador (Sucre)	.000965	.000965	1036.00	1036.00
Finland (Markka)	.22656	.22911	4.4138	4.3647
France (Franc)	.16085	.16260	6.2170	6.1500
Germany (Mark)	.5451	.5516	1.8345	1.8130
Greece (Drachma)	.004995	.005062	200.20	197.55
Hong Kong (Dollar)	.12877	.12882	7.7660	7.7625
India (Rupee)	.04167	.04167	24.00	24.00
Indonesia (Rupiah)	.0005123	.0005123	1952.00	1952.00
Ireland (Punt)	1.4579	1.4765	.6859	.6773
Israel (Shekel)	.4276	.4291	2.3386	2.3302
Italy (Lira)	.0007337	.0007419	1363.01	1347.85
Japan (Yen)	.007211	.007220	138.67	138.50
Jordan (Dinar)	1.4535	1.4535	.6880	.6880
Kuwait (Dinar)	z	z	z	z
Lebanon (Pound)	.001110	.001110	901.00	901.00
Malaysia (Ringgit)	.3588	.3588	2.7870	2.7870
Malta (Lira)	2.9028	2.9028	.3445	.3445
Mexico (Peso)	.0003305	.0003305	3026.00	3026.00
Netherland (Guilder)	.4842	.4918	2.0654	2.0335
New Zealand (Dollar)	.5620	.5627	1.7794	1.7771
Norway (Krone)	.1410	.1410	7.0937	7.0937
Pakistan (Rupee)	.0410	.0410	24.40	24.40
Peru (New Sol)	1.2231	1.2231	.82	.82
Philippines (Peso)	.03724	.03724	26.85	26.85
Portugal (Escudo)	.006389	.006351	156.51	157.46
Saudi Arabia (Riyal)	.26660	.26660	3.7510	3.7510
Singapore (Dollar)	.5690	.5700	1.7575	1.7545
South Africa (Rand)	.3454	.3452	2.8950	2.8968
South Korea (Won)	.0013805	.0013805	724.35	724.35
Spain (Peseta)	.008702	.008787	114.92	113.80
Sweden (Krona)	.1508	.1525	6.6308	6.5576
Switzerland (Franc)	.6285	.6365	1.5910	1.5710
Taiwan (Dollar)	.037397	.037355	26.74	26.77
Thailand (Baht)	.03891	.03891	25.70	25.70
Turkey (Lira)	.0002300	.0002309	4347.01	4330.02
United Arab (Dirham)	.2723	.2723	3.6730	3.6730
Uruguay (New Peso)	.000500	.000500	2000.00	2000.00
Venezuela (Bolivar)	.01803	.01803	55.46	55.46

z—Not quoted.

STATE GENERAL SALES AND USE TAXES, JULY 1988

State	Percent rate	State	Percent rate	State	Percent rate
Alabama	4	Kentucky	5	Ohio	5
Arizona	5	Louisiana	4	Oklahoma	4
Arkansas	4	Maine	5	Pennsylvania	6
California	4.75	Maryland	5	Rhode Island	6
Colorado	3	Massachusetts	5	South Carolina	5
Connecticut	7.5	Michigan	4	South Dakota	4
D.C.	6	Minnesota	6	Tennessee	5.5
Florida	6	Mississippi	6	Texas	6
Georgia	3	Missouri	4.225	Utah	5.09375
Hawaii	4	Nebraska	4	Vermont	4
Idaho	5	Nevada	5.75	Virginia	3.5
Illinois	5	New Jersey	6	Washington	6.5
Indiana	5	New Mexico	4.75	West Virginia	6
Iowa	4	New York	4	Wisconsin	5
Kansas	4	North Carolina	3	Wyoming	3
		North Dakota	5.5		

NOTE: Alaska, Delaware, Montana, New Hampshire, and Oregon have no statewide sales and use taxes.
Source: *Information Please Almanac* questionnaires to the states and Tax Foundation, Inc.

Economics **561**

DATA FILE **Health and Fitness**

VEGETABLES AND RISK OF LUNG CANCER IN WOMEN

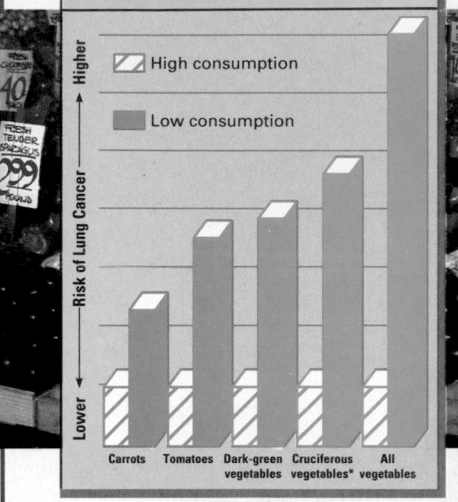

High consumption

Low consumption

Risk of Lung Cancer — Higher / Lower

Carrots, Tomatoes, Dark-green vegetables, Cruciferous vegetables*, All vegetables

VEGETABLES AND RISK OF LUNG CANCER IN MEN

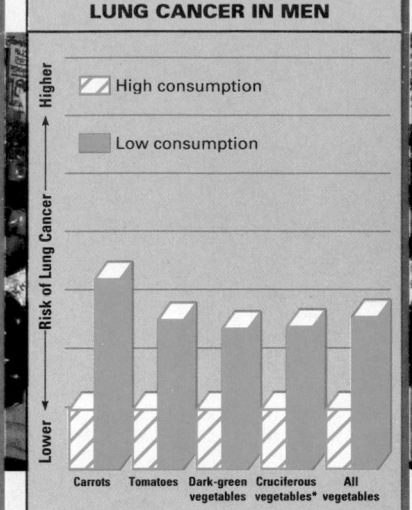

High consumption

Low consumption

Risk of Lung Cancer — Higher / Lower

Carrots, Tomatoes, Dark-green vegetables, Cruciferous vegetables*, All vegetables

*Cruciferous vegetables are vegetables in the cabbage family, such as cabbage, broccoli, cauliflower, and Brussels sprouts.

PATTERNS OF SLEEP

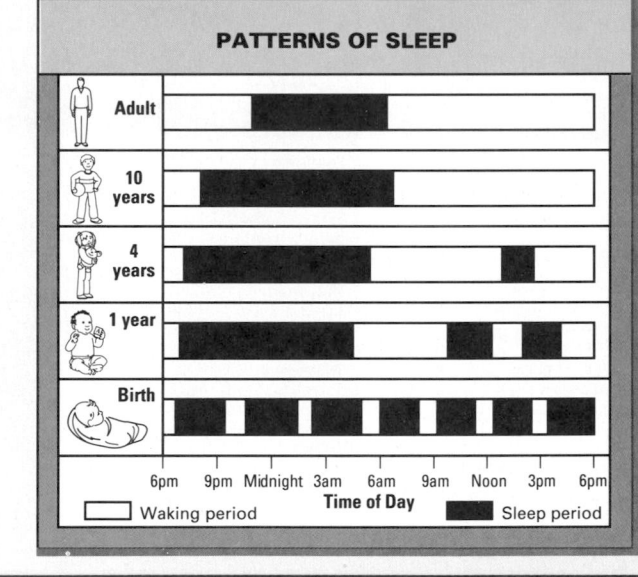

Adult

10 years

4 years

1 year

Birth

6pm 9pm Midnight 3am 6am 9am Noon 3pm 6pm

Time of Day

Waking period   Sleep period

HOW MUCH WALKING IS DONE ON THE JOB?

Job	Miles Walked per Day
Hospital nurse	5.3
Security officer	4.2
City messenger	4.0
Retail salesperson	3.5
Waiter/waitress	3.3
Hotel employee	3.2
Doctor	2.5
Advertising rep.	2.4
Architect	2.3
Secretary	2.2
Newspaper reporter	2.1
Banker	2.0
Accountant	1.8
Teacher	1.7
Lawyer	1.5
Housewife	1.3
Magazine editor	1.3
Radio announcer	1.1
Dentist	0.8

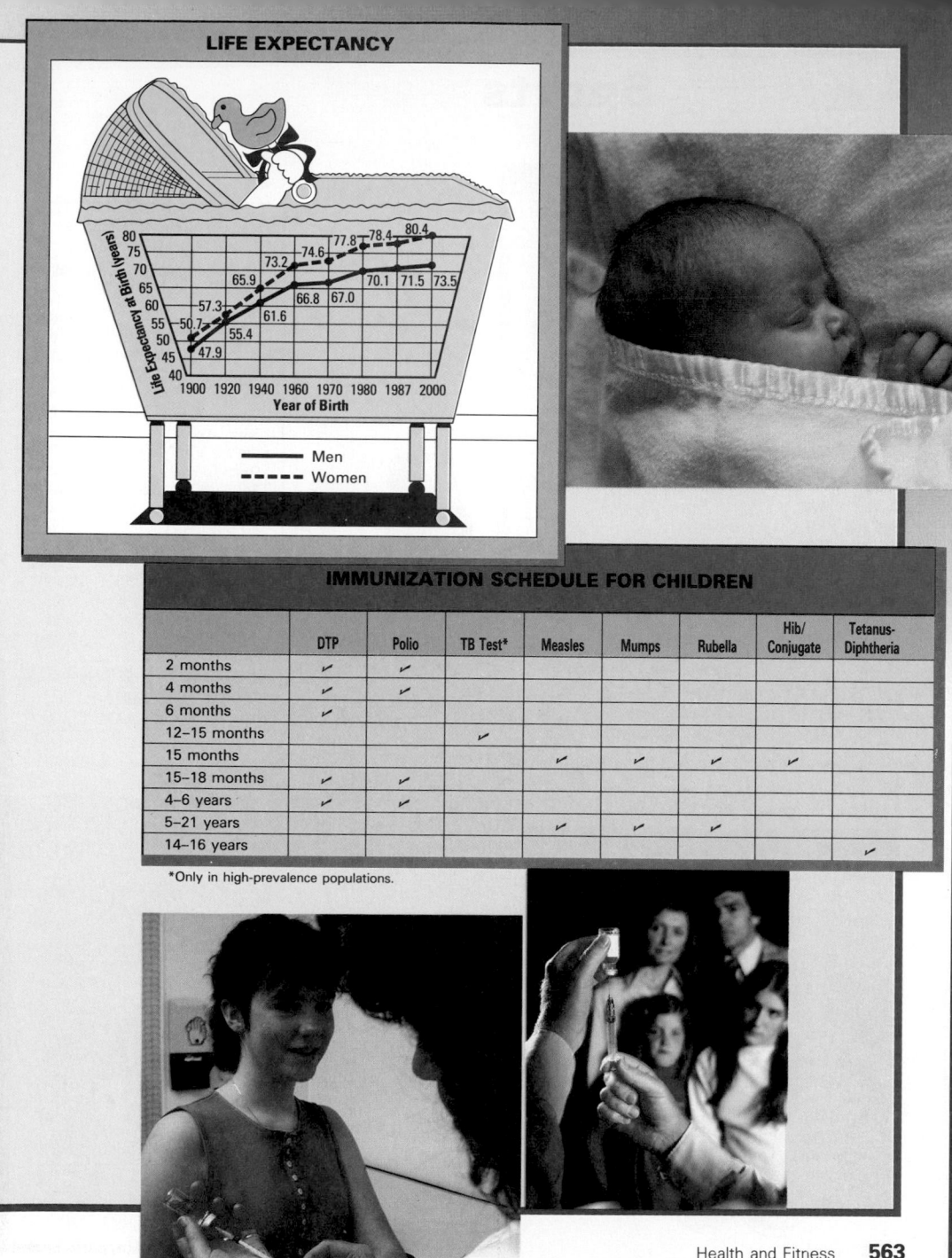

LIFE EXPECTANCY

Life Expectancy at Birth (years)

| Year of Birth | 1900 | 1920 | 1940 | 1960 | 1970 | 1980 | 1987 | 2000 |

Men: 47.9, 55.4, 61.6, 66.8, 67.0, 70.1, 71.5, 73.5
Women: 50.7, 57.3, 65.9, 73.2, 74.6, 77.8, 78.4, 80.4

— Men
--- Women

IMMUNIZATION SCHEDULE FOR CHILDREN

	DTP	Polio	TB Test*	Measles	Mumps	Rubella	Hib/ Conjugate	Tetanus- Diphtheria
2 months	✔	✔						
4 months	✔	✔						
6 months	✔							
12–15 months			✔					
15 months				✔	✔	✔	✔	
15–18 months	✔	✔						
4–6 years	✔	✔						
5–21 years				✔	✔	✔		
14–16 years								✔

*Only in high-prevalence populations.

Health and Fitness **563**

DATA FILE Sports

BASEBALL STADIUMS

Team	Stadium	Seating Capacity
ATLANTA BRAVES	Atlanta—Fulton County Stadium	52,003
BALTIMORE ORIOLES	Memorial Stadium	54,017
BOSTON RED SOX	Fenway Park	34,182
CALIFORNIA ANGELS	Anaheim Stadium	64,593
CHICAGO CUBS	Wrigley Field	39,600
CHICAGO WHITE SOX	Comiskey Park	44,087
CINCINNATI REDS	Riverfront Stadium	52,392
CLEVELAND INDIANS	Cleveland Stadium	74,483
DETROIT TIGERS	Tiger Stadium	52,416
HOUSTON ASTROS	Astrodome	45,000
KANSAS CITY ROYALS	Royals Stadium	40,625
LOS ANGELES DODGERS	Dodger Stadium	56,000
MILWAUKEE BREWERS	Milwaukee County Stadium	53,192
MINNESOTA TWINS	Hubert H. Humphrey Metrodome	55,883
MONTREAL EXPOS	Olympic Stadium	59,149
NEW YORK METS	Shea Stadium	55,300
NEW YORK YANKEES	Yankee Stadium	57,545
OAKLAND A's	Oakland Coliseum	49,219
PHILADELPHIA PHILLIES	Veterans Stadium	64,538
PITTSBURGH PIRATES	Three Rivers Stadium	58,727
SAN DIEGO PADRES	Jack Murphy Stadium	58,433
SAN FRANCISCO GIANTS	Candlestick Park	58,000
SEATTLE MARINERS	Kingdome	58,150
ST. LOUIS CARDINALS	Busch Stadium	54,224
TEXAS RANGERS	Arlington Stadium	43,508
TORONTO BLUE JAYS	Skydome	53,000

LIFETIME EARNED RUN AVERAGE

(minimum 1,500 innings pitched)

Chief Bender	2.46
Three Finger Brown	2.06
Ed Cicotte	2.37
Walter Johnson	2.17
Addie Joss	1.88
Ed Killian	2.38
Sam Leaver	2.47
Christy Matthewson	2.13
Orvie Overall	2.23
Ed Plank	2.34
Nap Rucker	2.42
Ed Ruelbach	2.28
Jim Scott	2.32
Jeff Tesreau	2.43
Rube Waddell	2.16
Ed Walsh	1.82
Doc White	2.38
Joe Wood	2.03

NUMBER OF CALORIES BURNED BY PEOPLE OF DIFFERENT WEIGHTS

Exercise	Calories Burned per Hour		
	110 lb	154 lb	198 lb
Martial arts	620	790	960
Racquetball (2 people)	610	775	945
Basketball (full-court game)	585	750	910
Skiing—cross country (5 mi/h)	550	700	850
downhill	465	595	720
Running—8-min mile	550	700	850
12-min mile	515	655	795
Swimming—crawl, 45 yd/min	540	690	835
crawl, 30 yd/min	330	420	510
Stationary bicycle—15 mi/h	515	655	795
Aerobic dancing—intense	515	655	795
moderate	350	445	540
Walking—5 mi/h	435	555	675
3 mi/h	235	300	365
2 mi/h	145	185	225
Calisthenics—intense	435	555	675
moderate	350	445	540
Scuba diving	355	450	550
Hiking—20-lb pack, 4 mi/h	355	450	550
20-lb pack, 2 mi/h	235	300	365
Tennis—singles, recreational	335	425	520
doubles, recreational	235	300	365
Ice skating	275	350	425
Roller skating	275	350	425

WORLD SPEED SKATING RECORDS
(Ice Skating)

Distance	min:sec	Name and Nationality	Place	Date
MEN				
500 m	36.23*	Nick Thometz (USA)	Medeo, USSR	Mar 26, 1987
	36.45	Jens-Uwe May (E Ger)	Calgary, Canada	Feb 14, 1988
1000 m	1.12.05	Nick Thometz (USA)	Medeo, USSR	Mar 27, 1987
1500 m	1.52.06	Andre Hoffman (E Ger)	Calgary, Canada	Feb 20, 1988
3000 m	3.59.27	Leo Visser (Neth)	Heerenveen, Netherlands	Mar 19, 1987
5000 m	6.47.01	Leo Visser (Neth)	Heerenveen, Netherlands	Feb 14, 1987
10,000 m	13.48.20	Tomas Gustafson (Swe)	Calgary, Canada	Feb 21, 1988
WOMEN				
500 m	39.10	Bonnie Blair (USA)	Calgary, Canada	Feb 22, 1988
1000 m	1.18.11	Karin Kania (*nee* Enke) (GDR)	Calgary, Canada	Dec 5, 1987
1500 m	1.59.30	Karin Kania (GDR)	Medeo, USSR	Mar 22, 1986
3000 m	4.11.94	Yvonne van Gennip (Neth)	Calgary, Canada	Feb 23, 1988
5000 m	6.43.59	Yvonne van Gennip (Neth)	Calgary, Canada	Feb 28, 1988
10,000 m	15.25.25	Yvonne van Gennip (Neth)	Heerenveen, Netherlands	Mar 19, 1988

*Unofficial (represents an average speed of 30.87 mi/h).

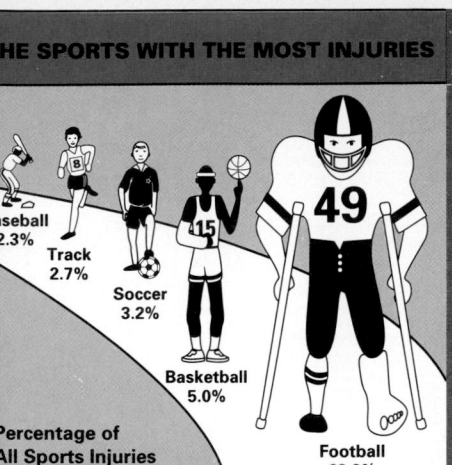

THE SPORTS WITH THE MOST INJURIES

Baseball 2.3%
Track 2.7%
Soccer 3.2%
Basketball 5.0%
Football 63.9%

Percentage of All Sports Injuries

OLYMPIC RECORD TIMES FOR 400-m FREESTYLE SWIMMING (minutes)

YEAR	1924	1928	1932	1936	1948	1952	1956	1960	1964	1968	1972	1976	1980	1984	1988
MALE	5:04.2	5:01.6	4:48.4	4:44.5	4:41.0	4:30.7	4:27.3	4:18.3	4:12.2	4:09.0	4:00.27	3:51.93	3:51.31	3:51.23	3:46.25
FEMALE	6:02.2	5:42.8	5:28.5	5:26.4	5:17.8	5:12.1	4:54.6	4:50.6	4:43.3	4:31.8	4:19.44	4:09.89	4:08.76	4:07.10	4:03.85

SIZES AND WEIGHTS OF BALLS USED IN VARIOUS SPORTS

Type	Diameter (cm)	Average Weight (g)
Baseball	7.6	145
Basketball	24.0	596
Croquet ball	8.6	340
Field hockey ball	7.6	160
Golf ball	4.3	46
Handball	4.8	65
Soccer ball	22.0	425
Softball, large	13.0	279
Softball, small	9.8	187
Table tennis ball	3.7	2
Tennis ball	6.5	57
Volleyball	21.9	256

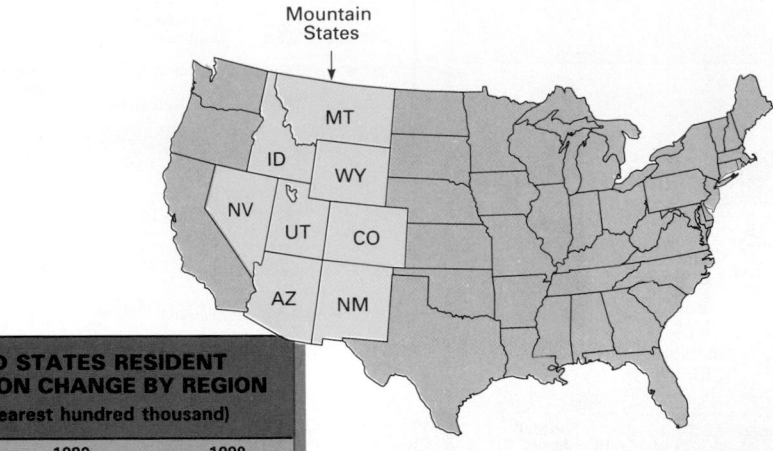

Mountain
States

UNITED STATES RESIDENT POPULATION CHANGE BY REGION
(to the nearest hundred thousand)

	1980	1990
Northeast	49,100,000	50,800,000
Midwest	58,900,000	59,700,000
South	75,300,000	85,400,000
West	43,200,000	52,800,000

UTAH AND THE OTHER MOUNTAIN STATES

In area, Utah ranks 7th among the Mountain states (shaded) and 11th among all 50 states.

SOME ENDANGERED MAMMALS OF THE UNITED STATES

Source: U.S. Fish and Wildlife Service, U.S. Interior Department; as of April 15, 1990

Common name	Scientific name	Range
Ozark big-eared bat	*Plecotus townsendii ingens*	Mo., Okla., Ariz.
Brown or grizzly bear	*Ursus arctos horribilis*	48 conterminous states
Columbian white-tailed deer	*Odocoileus virginianus leucurus*	Wash., Ore.
San Joaquin kit fox	*Vulpes macrotis mutica*	Cal.
Southeastern beach mouse	*Peromyscus polionotus phasma*	Fla.
Ocelot	*Felis pardalis*	Tex., Ariz.
Southern sea otter	*Enhydra lutris hereis*	Wash., Ore., Cal.
Florida panther	*Felis concolor coryi*	La., Ark. east to S.C., Fla.
Utah prairie dog	*Cynomys parvidens*	Ut.
Morro Bay kangaroo rat	*Dipodomys heermanni morroensis*	Cal.
Carolina northern flying squirrel	*Glaucomys sabrinus coloratus*	N.C., Tenn.
Hualapai Mexican vole	*Microtus mexicanus hualpaiensis*	Ariz.
Red wolf	*Canis rufus*	Southeast to central Tex.

FARMS—NUMBER AND ACREAGE, BY STATE: 1980 AND 1988

[1988 data preliminary. Based on 1974 census definition of farms and farmland]

State	Farms (1,000)		Acreage (mil.)		Acreage per farm	
	1980	1988	1980	1988	1980	1988
U.S.	**2,433**	**2,159**	**1,039**	**999**	**427**	**463**
Alabama	59	49	12	11	207	224
Alaska	(z)	1	2	1	3,378	2,123
Arizona	8	8	38	37	5,080	4,506
Arkansas	59	47	17	15	280	319
California	81	78	34	33	417	417
Colorado	27	27	36	34	1,358	1,234
Connecticut	4	4	(z)	(z)	117	119
Delaware	4	3	1	1	186	197
Florida	39	40	13	13	344	325
Georgia	59	49	15	13	254	265
Hawaii	4	4	2	2	458	443
Idaho	24	23	15	14	623	609
Illinois	107	83	29	29	269	345
Indiana	87	72	17	16	193	228
Iowa	119	107	34	34	284	313
Kansas	75	69	48	48	644	694
Kentucky	102	99	15	15	143	146
Louisiana	37	35	10	10	273	271
Maine	8	8	2	2	195	192
Maryland	18	16	3	2	157	147
Massachusetts	6	6	1	1	116	111
Michigan	65	58	11	11	175	193
Minnesota	104	94	30	30	291	319
Mississippi	55	43	15	14	265	314
Missouri	120	113	31	30	261	9
Montana	24	23	62	61	2,601	2,605
Nebraska	65	55	48	47	734	856
Nevada	3	2	9	9	3,100	3,667
New Hampshire	3	3	1	1	160	158
New Jersey	9	7	1	1	109	112
New Mexico	14	14	47	45	3,467	3,333
New York	47	40	9	9	200	213
North Carolina	93	70	12	11	126	150
North Dakota	40	33	42	40	1,043	1,243
Ohio	95	84	16	16	171	186
Oklahoma	72	69	35	33	481	478
Oregon	35	37	18	18	517	488
Pennsylvania	62	56	9	8	145	150
Rhode Island	1	1	(z)	(z)	87	96
South Carolina	34	27	6	5	188	200
South Dakota	39	35	45	44	1,169	1,278
Tennessee	96	94	14	13	142	136
Texas	189	156	138	132	731	846
Utah	14	13	12	11	919	850
Vermont	8	7	2	2	226	223
Virginia	58	49	10	10	169	196
Washington	38	38	16	16	429	421
West Virginia	22	21	4	4	191	176
Wisconsin	93	82	19	18	200	215
Wyoming	9	9	35	35	3,846	4,000

z   represents less than 500 farms or 500,000 acres.

1990 POPULATION AND NUMBER OF REPRESENTATIVES, BY STATE

TOTAL POPULATION 249,632,692

State	Apportionment Population	Number of Representatives Based on the 1990 Census	Change from 1980 Apportionment
U.S. TOTAL*	249,022,783	435	
Alabama	4,062,608	7	–
Alaska	551,947	1	–
Arizona	3,677,985	6	+ 1
Arkansas	2,362,239	4	–
California	29,839,250	52	+ 7
Colorado	3,307,912	6	–
Connecticut	3,295,669	6	–
Delaware	668,696	1	–
Florida	13,003,362	23	+ 4
Georgia	6,508,419	11	+ 1
Hawaii	1,115,274	2	–
Idaho	1,011,986	2	–
Illinois	11,466,682	20	– 2
Indiana	5,564,228	10	–
Iowa	2,787,424	5	– 1
Kansas	2,485,600	4	– 1
Kentucky	3,698,969	6	– 1
Louisiana	4,238,216	7	– 1
Maine	1,233,223	2	–
Maryland	4,798,622	8	–
Massachusetts	6,029,051	10	– 1
Michigan	9,328,784	16	– 2
Minnesota	4,387,029	8	–
Mississippi	2,586,443	5	–
Missouri	5,137,804	9	–
Montana	803,655	1	– 1
Nebraska	1,584,617	3	–
Nevada	1,206,152	2	–
New Hampshire	1,113,915	2	–
New Jersey	7,748,634	13	– 1
New Mexico	1,521,779	3	–
New York	18,044,505	31	– 3
North Carolina	6,657,630	12	+ 1
North Dakota	641,364	1	–
Ohio	10,887,325	19	– 2
Oklahoma	3,157,604	6	–
Oregon	2,853,733	5	–
Pennsylvania	11,924,710	21	– 2
Rhode Island	1,005,984	2	–
South Carolina	3,505,707	6	–
South Dakota	699,999	1	–
Tennessee	4,896,641	9	–
Texas	17,059,805	30	+ 3
Utah	1,727,784	3	–
Vermont	564,964	1	–
Virginia	6,216,568	11	+ 1
Washington	4,887,941	9	+ 1
West Virginia	1,801,625	3	– 1
Wisconsin	4,906,745	9	–
Wyoming	455,975	1	–

*Total population, not including the District of Columbia.

Useful Mathematical Data

SYMBOLS

=	is equal to	%	percent
≠	is not equal to	$0.\overline{3}$	repeating decimal
≅	is congruent to	π	pi: $\approx \frac{22}{7}$ or 3.14
~	is similar to	√	square root
≈	is approximately equal to	|x|	absolute value of x
‖	is parallel to	$\overleftrightarrow{AB}$	line AB
⊥	is perpendicular to	$\overline{AB}$	line segment AB
>	is greater than	$\overrightarrow{AB}$	ray AB
<	is less than	∠ABC	angle ABC
≥	is greater than or equal to	∟	right angle
≤	is less than or equal to	°	degrees
( )	parentheses: "Do this operation first."	A → A'	point A maps onto point A'
3^2 — exponent / base		(1, −2)	coordinates of a point where x = 1 and y = −2

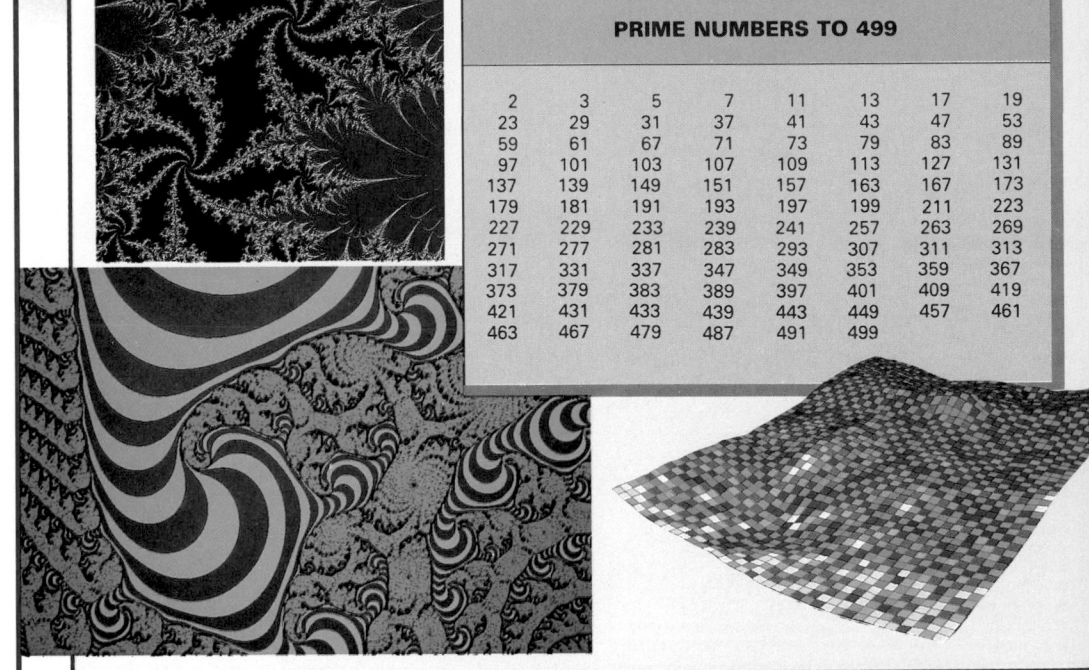

PRIME NUMBERS TO 499

2	3	5	7	11	13	17	19
23	29	31	37	41	43	47	53
59	61	67	71	73	79	83	89
97	101	103	107	109	113	127	131
137	139	149	151	157	163	167	173
179	181	191	193	197	199	211	223
227	229	233	239	241	257	263	269
271	277	281	283	293	307	311	313
317	331	337	347	349	353	359	367
373	379	383	389	397	401	409	419
421	431	433	439	443	449	457	461
463	467	479	487	491	499		

FORMULAS/USEFUL EQUATIONS

Geometric

l = length w = width
b = base h = height
s = side r = radius
d = diameter
B = area of the base of a three-dimensional figure

Perimeter: P

Rectangle: $P = 2 \times l + 2 \times w$
Square: $P = 4 \times s$
Parallelogram: $P = 2 \times (s_1 + s_2)$
    or $P = 2 \times s_1 + 2 \times s_2$
Rhombus: $P = 4 \times s$
Regular Polygon with n sides: $P = n \times s$
Triangle: $P = s_1 + s_2 + s_3$
Circle (Circumference): $C = \pi d$ or $C = 2\pi r$

Area: A

Rectangle: $A = l \times w$
Parallelogram: $A = b \times h$
Trapezoid: $A = \frac{1}{2} \times h \times (b_1 + b_2)$
Triangle: $A = \frac{1}{2} \times b \times h$
Circle: $A = \pi r^2$
Square: $A = s^2$

Surface Area: SA

Rectangular Prism:
    $SA = 2 \times (l \times w + l \times h + w \times h)$
Cube: $SA = 6 \times s^2$
Cylinder: $SA = 2\pi rh + 2\pi r^2$
Sphere: $SA = 4\pi r^2$
Cone: $SA = \pi rs + \pi r^2$

Volume: V

Rectangular Prism:
    $V = l \times w \times h$ or $V = B \times h$
Triangular Prism: $V = B \times h$
Cube: $V = s^3$
Cylinder: $V = \pi r^2 \times h$
Sphere: $V = \frac{4}{3}\pi r^3$
Rectangular Pyramid: $V = \frac{1}{3} \times B \times h$
Cone: $V = \frac{1}{3} \times \pi r^2 \times h$

Pythagorean Theorem

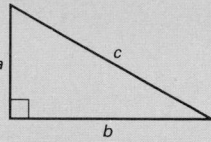

In a right triangle, as shown,
$$a^2 + b^2 = c^2 \text{ or } c = \sqrt{a^2 + b^2}$$

Temperature

C: degrees Celsius
F: degrees Fahrenheit

$C = \frac{5}{9}(F - 32)$

$F = \frac{9}{5}C + 32$

Distance

d: distance
r: rate
t: time

$d = r \times t$
$r = \dfrac{d}{t}$
$t = \dfrac{d}{r}$

Interest

I: interest
p: principal
t: time (in years)
A: total amount
r: rate

$I = p \times r \times t$
$A = p + p \times r \times t$

Laws of Exponents

$a^m \times a^n = a^{m+n}$
$(a^m)^n = a^{m \times n}$
$\dfrac{a^m}{a^n} = a^{m-n}$

MEASUREMENT: METRIC UNITS

Length

1 centimeter (cm) = 10 millimeters (mm)
10 centimeters or
100 millimeters = 1 decimeter (dm)
10 decimeters or
100 centimeters = 1 meter (m)
1,000 meters = 1 kilometer (km)

Area

100 square millimeters (mm^2) = 1 square centimeter (cm^2)
10,000 square centimeters = 1 square meter (m^2)
100 square meters = 1 are (a)
10,000 square meters = 1 hectare (ha)

Volume

1,000 cubic millimeters (mm^3) = 1 cubic centimeter (cm^3)
1,000 cubic centimeters = 1 cubic decimeter (dm^3)
1,000,000 cubic centimeters = 1 cubic meter (m^3)

Capacity

1,000 milliliters = 1 liter (L)
1,000 liters = 1 kiloliter (kL)

Mass

1,000 milligrams (mg) = 1 gram (g)
1,000 grams = 1 kilogram (kg)
1,000 kilograms = 1 metric ton (T)

Temperature

0°C = freezing point of water
37°C = normal body temperature
100°C = boiling point of water

MEASUREMENT: CUSTOMARY UNITS

Length

12 inches (in.) = 1 foot (ft)
3 feet or 36 inches = 1 yard (yd)
1760 yards or 5280 feet = 1 mile (mi)
6,076 feet = 1 nautical mile

Area

144 square inches (in.^{2}) = 1 square foot (ft^2)
9 square feet = 1 square yard (yd^2)
4,840 square yards = 1 acre (A)

Volume

1,728 cubic inches (in.^{3}) = 1 cubic foot (ft^3)
27 cubic feet = 1 cubic yard (yd^3)

Capacity

8 fluid ounces (fl oz) = 1 cup
2 cups = 1 pint (pt)
2 pints = 1 quart (qt)
4 quarts = 1 gallon (gal)

Weight

16 ounces (oz) = 1 pound (lb)
2,000 pounds = 1 ton (T)

Temperature

32°F = freezing point of water
98.6°F = normal body temperature
212°F = boiling point of water

Metric/Customary Comparisons

5 centimeters is about the same length as 2 inches.
1 meter is slightly longer than 1 yard.
5 kilometers is about the same length as 3 miles.

ADDITIONAL ANSWERS

CHAPTER 1

Section 1–1
Exercise **2.** advantage: convenience; disadvantage: these students may not play arcade games.

Section 1–2
Check Understanding

```
5 | 1  3  8  9
6 | 1  4  6  6  7  7  8
7 | 1  3  3  3  6  6  9
8 | 1  2  6  6  6  7
9 | 0  1  3
```

Try These
1.
```
3 | 6  7  2  8  7  2
4 | 9  4  3  6
5 | 2  5  5  6  9  8
6 | 1  2  1  2  3
7 | 0  7  3  2
8 | 2  4
9 | 6
```

2. Answers will vary. One possible answer follows: This set of data indicates that the lowest score for the season was 34 and the highest score was 99. The lowest score, 34, represents an outlier. The scores cluster around 66 and 75. Some gaps are in the mid to high 40s and in the low 80s.

Exercises
2.
```
0 | 8  3  6  9  4  9
1 | 2  7  5  4  1  7
2 | 7  8  3  5  1  9  6
3 | 4  5  6  2  9  8  7  1  4
4 | 1  8  2  2  2  3  1
5 | 0
```

4. Answers will vary. One possible answer follows: This set of data indicates that, out of 50 shots, the least number of baskets sunk was 3 and the greatest number was 50. There are no outliers, clusters, or gaps.

10.
```
1 | 5
2 | 4  4  4  8  8
3 | 3  5  5  6  6  4
4 | 7  4  4
5 | 9  5  5  2  4  2  2
6 | 2  1
```

572

CHAPTER 1   EXPLORING DATA

Section 1-1, pages 6-7

Exercises **1.** advantage: these are the people who play the games; disadvantage: may name only those games available in that arcade **3.** advantage: convenience, some or all of these people may play arcade games; disadvantage: some or all of these people may not play arcade games **5.** The interviewer could ask every tenth person who passes. This should be done on several different days. **7.** Answers will vary. **9.** Answers will vary. **11.** Answers will vary.

Problem Solving Applications **1.** 2 **3.** sports and music videos; mysteries and soap operas; mysteries and news; music videos and comedies **5.** Answers will vary.

Section 1-2, pages 10-11

Exercises **1.** 36 **3.** 2; 1; 0

5.
```
          0 | 8
          1 | 8  7  5  3  3  1
2|5 represents  2 | 5  4  3  2  1  0
2.5 mi.   3 | 5  3  2
          4 | 2  0
          5 | 3
```

7.
```
             1 | 5  5
             2 | 9  7  0  9  3  4  9  9  8  3  9
2|3 represents  3 | 9  2  5  4  8  0  5  0
230 voters.  4 | 6  4  0  2  1  1  9  1
             5 | 8  2  3  3  2  3
             6 | 2  0  2  7  9
```

9. Answers will vary.

Problem Solving Applications

1.

Age at Inauguration		Age at Death
	9	00
	8	3508
	7	3891470128
4912054811	6	785463707034
26514156145504620147877	5	36687
3279689	4	96

3. youngest – T. Roosevelt, 42; oldest – Ronald Reagan, 69 **5.** Answers will vary. **7.** Answers will vary.

Section 1-3, pages 14-15

Exercises **1.** rock **3.** 200 **5.** Answers will vary. **7.** 8 **9.** 60

11. The graph should show the following: Kelly, 6 symbols; Green, $6\frac{1}{2}$ symbols; Currie, 3 symbols; Smith, 4 symbols; Charron, $4\frac{3}{4}$ symbols **13.** Smith **15.** excellent **17.** 170 **19.** Answers will vary. **21.** Answers will vary.

Section 1-4, page 17

Problems **1.** 20 classes **3.** $315 **4.** $937

Section 1-5, pages 18-19

Problems Answers will vary. Samples are given. **1.** 500 or 1,000 (performances) **3.** 400 or 500 (camps) **5.** 0.50 or 1.00 (dollars) **7.** 25 or 30 (millions of yd³)

Section 1-6, pages 22-23

Exercises **1.** male, 15–18 **3.** 800 **5.** 2,200

7.

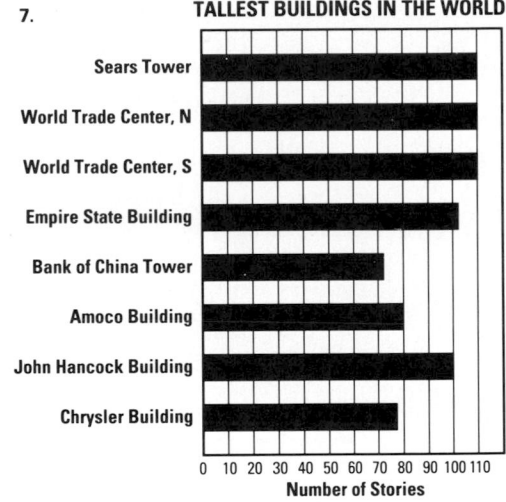

TALLEST BUILDINGS IN THE WORLD

9. Hudson, Mediterranan, Arctic, Black **11.** Answers will vary. **13.** about $2\frac{1}{2}$ times lower **15.** about $14\frac{1}{2}$ h

Section 1-7, pages 24-25

Problems **1.** 1955; 1959 **3.** about 4 times **5.** tobogganing

---

Section 1–3
Talk It Over Since the value of the symbol would double, each category in the pictograph would need only half the number of symbols to represent the data.

Try These **4.** Pictographs will vary. A sample pictograph is given. *(See Fig. 1, p. 588.)*

Exercise **6.** *(See Fig. 2, p. 588.)*

Section 1–5
Problems Answers will vary. Samples are given. **2.** 80 or 100 (mg) **4.** 8 or 10 (hours) **6.** 0.050 (batting average)

572

Section 1–8, pages 28–29

Exercises **1.** $125 **3.** 1985

5.

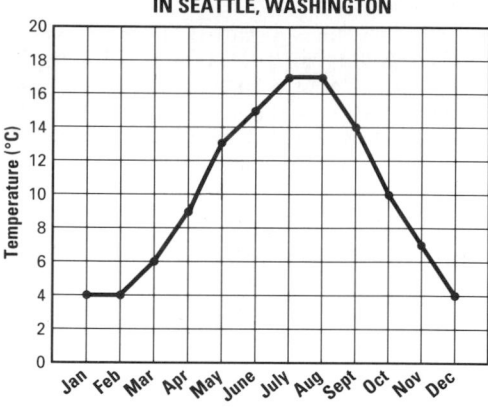

AVERAGE MONTHLY TEMPERATURE
IN SEATTLE, WASHINGTON

7. 30 kg; 45 kg **9.** from birth to 3 mo; 15 kg

11.

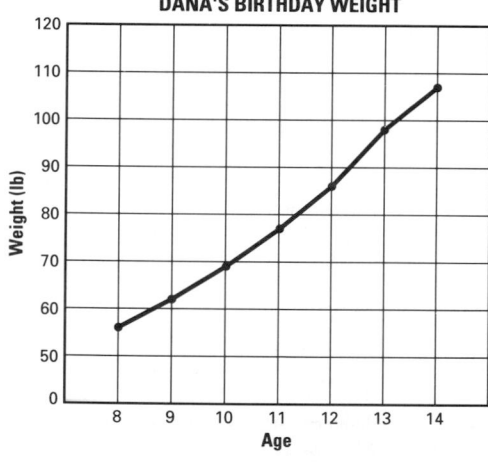

DANA'S BIRTHDAY WEIGHT

13. between her twelfth and thirteenth birthdays
15. about 25 calories **17.** The data could be used in helping to determine the proper amount of exercise needed to gain, lose, or maintain weight. **19.** 79 years

Section 1-9, pages 31–33

Exercises **1.** 9.7; 9.05; 8.7 and 9.2 **3.** mean **5.** mean or median **7.** $69,620.83; $62,250; none **9.** 8.8; 9.1; 7.0 **11.** 191 **13.** 5,730 **15.** 344 **17.** 9,442 **19.** Answers will vary. **21.** Each would be 5 min longer **23.** mean; median **25.** 159 lb **27.** The mean and the median are the same. **29.** no **31.** yes

CHAPTER 2 EXPLORING WHOLE NUMBERS AND DECIMALS

Section 2-1, pages 44–45

Exercises **1.** 30 **3.** 700 **5.** 1,000

	Actual	Estimate
7.	5,972	6,000
9.	8,448	8,000
11.	4,652	5,000
13.	29	30
15.	81	80
17.	61	60

19a. 16,000 people **b.** 2,000 people **21.** 45,000; C **23.** 11,000; A **25.** 9,000; C **27.** $498,000,000 **29.** Answers will vary. Sample: Yes. Rounded to the nearest hundred million, 300,000,000 + 700,000,000 + 700,000,000 = 1,700,000,000. This is close to the answer, so the answer is reasonable.

Estimate:

31. a. $\frac{90 + 80 + 80 + 90 + 80}{5} = 84$; yes
b. median: 85; mode: 75
33. b

Section 2-2, pages 48–49

Exercises **1.** $\frac{n}{2}$ **3.** $6n$ **5.** $n + 7$ **7.** $10n$ **9.** 7 **11.** 3 **13.** 3 **15.** 3 **17.** 27 **19.** 31 **21.** 10 **23.** 4 **25.** 0 **27.** 448 **29.** 3 **31.** 2 **33.** $w + 9$ **35.** $25p$ **37a.** $c \div 8$ or $\frac{c}{8}$ **b.** 33 **39a.** $65 + d$ **b.** 79 **41.** 3 **43.** 11 **45.** 6 **47.** 5 **49.** 25 **51.** 5 **53a.** $3d + 4$ **b.** 79 **55a.** $x + 2$ **b.** $x - 2$ **c.** $x + 1$ **d.** $x - 1$ **e.** $x + x + 1$ **f.** $x + x + 2$ **57.** Answers will vary.

Section 2-3, pages 50–51

Problems **1.** subtract **3.** add; 211 people **5.** multiply; 42 cents **7.** add; 38,321 worms **9.** divide; 8 people

Section 1–5
Mixed Review
1. 85,000; 85,500; 90,000 **2.** 1,000; 700; 0 **3.** 1,035,000; 1,034,500; 1,030,000 **4.** 151,000; 151,200; 150,000 **5.** 57 million; 60 million; 57.3 million **6.** 636 million; 640 million; 636.5 million **7.** 1 million; 0; 0.7 million **8.** 450; 900; 1,800

Section 1–6
Try These **12.** *(See Fig. 3, p. 588.)*
Exercise **8.** *(See Fig. 4, p. 589.)*

Section 1–8
Explore **d.** *(See Fig. 5, p. 589.)*

Exercise **6.** *(See Fig. 6, p. 589.)*

Section 1–9
Who, Where, When Karl mentally paired the numbers to make sums to 101. Then he multiplied $50 \times 101 = 5,050$. *(See Fig. 7, p. 589.)*

Chapter Review
12.
5	3 5 5 5
6	6 8 1 7 9 7 2 7 7 7
7	9 5 9 9 2 2
8	1 2 4 2 7
9	1 0 8

Cumulative Review
6.
0	8 5 9
1	7 9 9
2	4 5 5 1
3	4 7 6 7 1 7
4	3 3 3
5	6 0 6 0
6	2 8
7	4
8	1
13	6

11. *(See Fig. 8, p. 590.)*

CHAPTER 2

Section 2–1
Exercise **30.** No, because an estimate of $10 \times 50 = 500$ seats means that there were many more students than seats.

Section 2–4
Exercise **2.** $120.82

Section 2–8
Exercises **10.** 69; Sample key sequences are given.
First key sequence:
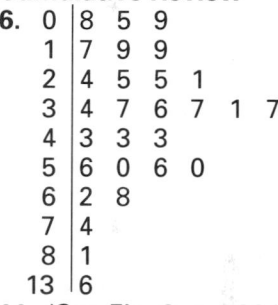
9 × 8 − 6 + 3 =
Second key sequence:
9 × 8 MR − 6 + 3 =

Section 2–8 (continued)

28. Answers will vary. Sample: Some calculators compute using the order of operations, whereas others compute in the order in which the keys are pressed.

First key sequence:

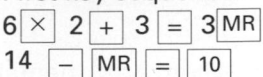

Second key sequence:

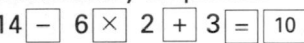

Section 2–9

Try These **13.** 0; Property of Zero for Multiplication **14.** 85; Identity Property of Multiplication **15.** 0.35; Identity Property of Addition **16.** 18; Associative Property **17.** 2.32; Associative and Commutative Properties **18.** 160; Associative and Commutative Properties **19.** 13.2; Associative and Commutative Properties **20.** 29; Associative and Commutative Properties **21.** 81; Associative Property

Section 2–10

Mixed Review **1.** No, because the data are not given over time. **2.** Show the number of cents in each U.S. coin by using a penny as the symbol in a pictograph. **3.** 4.2; 3.785; 2.9 **4.** 7.4; 7.3; 7.3

Exercises **40.** $7 \times 402,000 = 7 \times 400,000 + 2,000) = 2,800,000 + 14,000 = 2,814,000$ gallons

Section 2–11

Problems **12. a.** 512 tickets **b.** 682 tickets

M	T	W	TH	F
2	8	32	128	512

14. 36 laps

1st	2nd	3rd	4th	5th	6th	7th	8th
1	3	6	10	15	21	28	36

574

---

11. divide; 12 cartons **13.** multiply, multiply; 7,200 breaths **15.** 100 coconuts **17.** 50 porpoise teeth

Section 2-4, pages 53–55

Exercises **1.** 187.84 **3.** 1,145.45 **5.** 4,377.9 **7.** 806.646 **9.** $66.65

11. 1890	$0.50
1910	$0.60
1930	$1.16
1950	$2.10
1970	$3.12
1975	$4.31

13. $24.80 **15.** 9,3,6 **17.** 8,6,6,7 **19.** 2,2,5 **21.** $61.51

Problem Solving Applications **1a.** $370.08 **b.** $231.00 **c.** $221.31 **3.** $274.13 **5.** 2,500 pesos **7.** $14.57

Section 2-5, pages 57–59

Exercises **1.** B **3.** B **5.** 26 **7.** 86 **9.** 28.8 **11.** 20.7 **13.** 23.53 **15.** B **17.** B **19.** 0.124 **21.** 0.121 **23.** 0.12 **25.** 1.5 **27.** $6.92 **29.** 326.7 **31.** 1.805 **33.** 3.30 **35.** $49,583.98 **37.** 132.42 **39.** $6,833.36 **41.** True

Section 2-6, pages 62–63

Exercises **1.** 3^3 **3.** 6^4 **5.** 16 **7.** 36 **9.** 6.25984×10^5 **11.** 8.92×10^2 **13.** 63,900 **15.** 300 **17.** No; $9 \neq 8$ **19.** 85 **21.** 89 **23.** 248 **25.** 87 **27.** 62,748,517 **29.** 1,452 **31.** 2^5 **33.** 7^3 **35.** 7 hours

Problem Solving Applications **1.** b **3.** c **5.** 1.39×10^{11}; 139,000,000,000 **7.** 1.0941899×10^{20} **9.** 3.6379788×10^{26} **11.** 2.39245×10^{20}

Section 2-7, pages 66–67

Exercises **1.** 10^{29} **3.** 5^8 **5.** n^6 **7.** m^{12} **9.** 10^2 **11.** 2^3 **13.** b^8 **15.** y^2 **17.** 4^{560} **19.** 5^{63} **21.** a^{30} **23.** n^{49} **25.** 10^5 cm **27.** 4^{150} **29.** 8^{27} **31.** 7^{12} **33.** 12^{25} **35.** 8^{30} **37.** 10^6 **39.** < **41.** = **43.** = **45.** 3.5 shows 10 times more ground movement than 2.5. **47.** 5.5 shows 1,000 times more ground movement than 2.5. **49.** 15 **51.** 4 **53.** $e = 2254$ **55.** $g = 1$ **57.** $i = 30$

Section 2-8, pages 70–71

Exercises **1.** 21 **3.** 81 **5.** 1 **7.** 62

574 Selected Answers

---

9. 28

First key sequence:

$$4 \boxed{\times} 4 \boxed{=} \boxed{\text{MR}}$$
$$2 \boxed{\times} 6 \boxed{+} \boxed{\text{MR}} \boxed{=}$$

Second key sequence:

$$4 \boxed{x^y} 2 \boxed{+} 2 \boxed{\times} 6 \boxed{=}$$

11. $(10 + 3 \times 10) \div 4$, $10 **13.** 1 **15.** 15 **17.** 35 **19.** $25 - (6 \times 4)$ **21.** $5^2 \times (32 - 28) \div 20$ **23.** – **25.** + **27.** ÷ **29.** $4 + (18 + 22 + 10 + 26) + 4 = 23$ miles **31.** $(4 - 3) \times 7 = 7$ **33.** $5 \times (24 - 16) = 40$ **35.** $(16 + 2^2) \div 5 \times 3 = 12$ **37.** $(32 - 6) \div 3 = 1$ **39.** $7 + 3 + -1 \times 1 = 9$ **41.** $(10 + 5 + 5) \div 2 = 10$ **43.** $(30 + 5) \div 5 + 23 = 30$

45. $3 + 5 \times 4 - 1 = 22$
or $3 + (5 \times 4) - 1 = 22$
$(3 + 5) \times 4 - 1 = 31$
$(3 + 5) \times (4 - 1) = 24$
$3 + 5 \times (4 - 1) = 18$

Section 2-9, pages 74–75

Exercises **1.** 8.5 **3.** 963 **5.** 2.58 **7.** 0 **9.** 2 **11.** 6 **13.** 60; Identity Property of Addition **14.** 73; Commutative and Associative Properties **15.** 113; Associative Property **16.** 60; Commutative and Associative Properties **17.** 81; Commutative and Associative Properties **18.** 24; Identity Property of Multiplication **19.** 0; Property of Zero for Multiplication **20.** 55; Identity Property of Multiplication **21.** 820; Commutative and Associative Properties **23.** 40 socks **25.** 720 **27.** 320 **29.** 0 **31.** $21.99 **33.** Explanations may vary. **35.** True. It can be used only to add and multiply. **37.** False. You can use a property as often as necessary to evaluate an expression since the use of the properties does not affect the value of an expression.

Section 2-10, pages 78–79

Exercises **1.** 6 **3.** 8 **5.** $9(7 - 3)$ **7.** $14(12 + 16)$ **9.** 130 **11.** 648 **13.** 594 **15.** $24 + 6b$ **17.** $7d + 7$ **19.** $3f - 6$ **21.** 5,580 **23.** 2,730 **25.** 3,150 **27.** 6,480 **29.** 5,940 **31.** 154,380 **33.** 22,055 **35.** 98,892 **37.** 316,100 **39.** $25 \times 198 = 25(200 - 2) = 5,000 - 50 = 4,950$ packages **41.** Answers will vary. **43.** Yes **45.** True. It is used only with multiplication over addition or subtraction. **47.** True. $ab - ac = ba - ca$. The order

---

Cumulative Review

8. 3|2 represents free throws sunk

```
0 | 8 4 6 3 9 9
1 | 7 5 1 7 2 4
2 | 9 2 3 5 7 8 9 6 7 1
3 | 2 1 5 9 8 4 7 6 4 7 5
4 | 2 8 3 1 1 2
5 | 0
```

of the variables as factors in a product does not affect the value of the product.

Section 2-11, pages 80–81

Problems **1.** a; 0.32, 0.40 **3.** b; 10.765, 11.876 **5.** Subtract 0.2; 0.9, 0.7 **7.** Add 4, subtract 0.1; 8.7; 12.7 **9.** Subtract 1, add 0.20; 7.74, 6.74

11. a. $38 **b.** $258

J	F	M	A	M	J	J	A	S	O	N	D
5	8	11	14	17	20	23	26	29	32	35	38

13. 240 miles

1st	2nd	3rd	4th	5th	6th	7th	8th
9	15	21	27	33	39	45	51

15a. 34, 55, 89 **b.** Answers will vary.

CHAPTER 3   REASONING LOGICALLY

Section 3-1, pages 91–93

Exercises **1.** Lines 1 and m are parallel. **3.** Line segments *DE* and *FG* have the same length. **5.** Viewed from one perspective, the face of an old woman is seen. From another perspective, the side of the head of a young woman is seen. **7.** Answers will vary.

Problem Solving Applications **1–5.** Only 5 is possible. **7.** The connection between the top front cube and the bottom cube at back is impossible.

Section 3-2, pages 96–97

Exercises **1.** 4 **3.** odd **5.** even **7.** The result is always 5. **9.** The digit will always be 9. **11.** If the original number is abc, the final number is abc, abc. **13.** Answers will vary. This conjecture is always true. **15.** yes; no

Section 3-3, pages 100–101

Exercises **1.** If the number is divisible by 9, then the number is divisible by 3. If the number is divisible by 3, then the number is divisible by 9. **3.** valid **5.** valid **7.** If a number is even, then it is divisible by 2. **9.** If a creature is a whale, then it is a mammal. **11.** valid **13.** If a number is a whole number, then it is a rational number.

15. If an animal is a fish, then it is a sea creature.
<u>A shark is a fish.</u>
Therefore, a shark is a sea creature.

17. If a canary is well fed, then it will sing loud.
<u>If a canary sings loud, then it is not melancholy.</u>
Therefore, if a canary is well fed, then it is not melancholy.

Section 3-4, page 103

Problems **1.** 30 students play hockey and basketball. 30 − 12 = 18 **3.** 33 **5.** fall, 8; winter, 7; spring, 9 **7.** 20 **9.** 144

Section 3-5, page 105

Problems **1.** Tillie **3.** Sue married Tom; Betty married Harry; DeeDee married Bill. **5.** Armand is the potter. Belinda is the violinist. Colette is the writer. Dimitri is the painter.

Section 3-6, pages 107–109

Exercises **1.** $2.50 **3.** 20 posts **5.** Fill the 13-gallon container and from it fill the 5-gallon container. You now have 8 gallons left in the 13-gallon container. Pour it into the 11-gallon container for safekeeping. Repeat the process with the 13- and 5-gallon containers to divide up the remaining water. You will end up with 8 gallons of water in each of the 11-, 13-, and 24-gallon containers.

7.

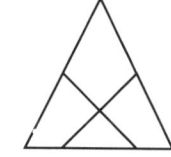 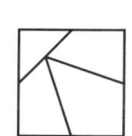

9. 27 cubes **11.** 6 cubes **13.** 8 cubes

15.

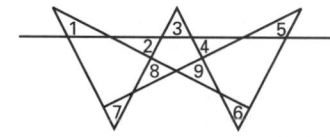

17. He first takes the goat across and returns alone. Next, he takes the cabbage across and returns with the goat. Then he leaves the goat and takes the wolf across and returns alone to get the goat. At no time are the potential enemies alone.

CHAPTER 3

Section 3–2
Computer To find the answers that end in 3, 9, or 7, change line 10 as follows.
For 3: FOR X = 1 TO 13 STEP 4
For 9: FOR X = 2 TO 14 STEP 4
For 7: FOR X = 3 TO 15 STEP 4

Computer Tip To test the conjectures, examine the ones digits of the numbers that result when either of the following programs are run:

```
10  FOR X = 1 TO 5
20  V = INT (4↑X + 0.5)
30  PRINT : PRINT "4↑";X;" = ";V
40  NEXT X
```
                    or
```
10  FOR X = 1 TO 5
20  V = 4↑X
30  PRINT : PRINT "4↑";X;" = ";V
40  NEXT X
```
To check the answers, change line 10 as follows:
For Try These 1: FOR X = 1 TO 6
For Try These 2: FOR X = 1 TO 7

Section 3–3
Exercise **16.** If a number is a multiple of 4, then it is a multiple of 2.
<u>16 is a multiple of 4.</u>
Therefore, 16 is multiple of 2.
18. If a dish is a pudding, then it is nice. If it is nice, then it is not wholesome.
<u>This dish is a pudding.</u>
Therefore, this dish is not wholesome.

Section 3–6
Exercise **6.** First, weigh three marbles against three marbles. Then:
(a) If they do not balance, take the three that weigh more and weigh one against a second. Either one of these will weigh more than the other, or the one not weighed is the heaviest marble; (b) If the marbles do balance, weigh the two originally left out against each other.

Cumulative Review

1.

```
0 | 8 4 3 5 7 9 6 2 1 6 8 4 5 7 6
1 | 2 0 1 4 1 4 0 3 2 5
```

CHAPTER 4

Section 4–4
Explore

$81 = 3 \times 3 \times 3 \times 3 = 3^4$

$27 = 3 \times 3 \times 3 = 3^3$

$9 = 3 \times 3 = 3^2$

$3 = 3 = 3^1$

$1/3 = 1/3^1 = 3^{-1}$

$1/9 = 1/3^2 = 3^{-2}$

$1/27 = 1/3^3 = 3^{-3}$

$1/81 = 1/3^4 = 3^{-4}$

Section 4–5
Problem Solving Applications
2. SAUCE
- 1/2 c tomato sauce
- 1/2 6-oz can tomato paste
- 1/2 T olive oil
- 1 T water
- 1/2 t oregano
- 1/4 t salt
- 1/4 t sugar
- 1/16 t crushed dried hot red chili peppers

4. DOUGH
- 1/2 c warm water
- 2 t dry yeast
- 1 t sugar
- 5 1/2 c all-purpose flour
- 3 t salt
- 1 1/2 c cold water
- 6 T olive oil

6. CHEESE TOPPINGS
- 1 1/2 lb sliced or grated mozzarella cheese
- 2/3 to 1 c grated Parmesan cheese

8. The amount of each ingredient should be multiplied by 2/3.

10. 1/24 t

Cumulative Review
15. If the fruit is a plum, then the fruit has a pit. If the fruit has a pit, then the fruit is a plum.

CHAPTER 4 EXPLORING FRACTIONS

Section 4-1, pages 120–121

Exercises **1.** $2 \times 2 \times 2 \times 13$, or $2^3 \times 13$ **3.** $2 \times 3 \times 5 \times 5$, or $2 \times 3 \times 5^2$ **5.** 2 **7.** 3 **9.** 2 **11.** 8 **13.** 28 **15.** 48 **17.** 28 **19.** 180 **21.** 12th person **23.** 5, 7 **25.** 37, 41, 43, 47, 53, 59, 61 **27.** 2 **29.** $3ab$ **31.** $3a^6b^3$ **33.** 60 **35.** $12a^2b^2$ **37.** $6a^2b^2$ **39.** 2 friends **41.** It is odd. **43.** It is the product of two numbers.

Section 4-2, pages 124–125

Exercises **1.** Samples: $\frac{2}{6}$, $\frac{3}{9}$ **3.** Samples: $\frac{10}{24}$, $\frac{15}{36}$ **5.** Samples: $\frac{6}{9}$, $\frac{2}{3}$ **7.** Samples: $\frac{15}{18}$, $\frac{10}{12}$ **9.** $\frac{1}{2}$ **11.** $\frac{1}{3}$ **13.** $\frac{9}{10}$ **15.** $\frac{2}{3}$ **17.** $<$ **19.** $=$ **21.** $<$ **23.** $\frac{1}{3}$ **25.** $\frac{3}{4}$ **27.** $\frac{1}{3}$, $\frac{2}{3}$, $\frac{5}{6}$ **29.** $\frac{21}{50}$ **31.** $\frac{3}{50}$ **33.** $\frac{51}{100}$ **35.** $\frac{9}{100}$ **37a.** $\frac{1}{4}$ **b.** $\frac{3}{4}$ **39.** $\frac{3}{4}$ **41.** $\frac{3}{5}$ **43.** 1

Section 4-3, pages 128–129

Exercises **1.** $\frac{7}{10}$ **3.** $\frac{233}{250}$ **5.** $\frac{29}{500}$ **7.** $8\frac{1}{20}$ **9.** 0.9 **11.** 0.417 **13.** $0.2\overline{5}$ **15.** 0.375 **17.** 1.4 **19.** 4.875 **21.** $0.8\overline{3}$ **23.** $0.\overline{5}$ **25.** $0.\overline{8}$ **27.** $0.\overline{3}$ **29.** $>$ **31.** $>$ **33.** Christine and her coach are both correct since $\frac{60}{75} = \frac{4}{5}$, 0.8 **35.** 0.25, $\frac{3}{7}$, 0.561, 0.781, $\frac{4}{5}$, $\frac{7}{8}$ **37.** $>$ **39.** Mona **41.** Otis **43.** Answers may vary. Sample: $\frac{3}{4}$, $\frac{6}{8}$, $\frac{9}{12}$, $\frac{75}{100}$ **45.** between 0.8 and 0.832

Section 4-4, pages 132–133

Exercises **1.** $\frac{1}{64}$ **3.** $\frac{1}{9}$ **5.** 8^{-6} **7.** 2^{-2} **9.** 0.0071 **11.** 0.0000091302 **13.** 2.5×10^{-3} **15.** 1.96×10^{-12} **17.** 10^{-15} **19.** $\frac{1}{1,048,576}$ **21.** $\frac{1}{2,562,890,625}$ **23.** 6.2×10^{-5}, 6.27×10^{-5}, 7.3×10^{-4}, $\frac{1}{1000}$ **25.** 8.668×10^{-8}, 10^{-7}, $\frac{1}{10^6}$, 8.886×10^{-6} **27.** 0.001 does not belong. It is equal to $\frac{1}{1000}$, not $\frac{1}{10,000}$. **29.** equal **31.** Picometer. A picometer $= 10^{-12}$, which is greater than 10^{-15}. **33.** 5^{-3} **35.** 2^{-6} or 4^{-3} or 8^{-2} **37.** $\frac{1}{2^2}$ **39.** $6.2193 \times 10,000$ **41a.** neutron **b.** 2×10^{-30} kg **43.** $>$ **45.** $<$

Section 4-5, pages 136–137

Exercises **1.** $\frac{2}{3}$ **3.** 18 **5.** $5\frac{17}{56}$ **7.** $6\frac{3}{10}$ **9.** $1\frac{1}{3}$ **11.** 20 **13.** $1\frac{23}{25}$ **15.** $3\frac{3}{26}$ **17.** 11 hours **19.** $34\frac{29}{50}$ tons **21.** 36 students **23.** 8 **25.** $\frac{1}{4}$ **27–29.** Exercise 29 cannot be true because the product of $\frac{3}{5} \times 24$ is a mixed number, not a whole number.

Problem Solving Applications

1. DOUGH
- $\frac{1}{8}$ c warm water
- $\frac{1}{2}$ t dry yeast
- $\frac{1}{4}$ t sugar
- $1\frac{3}{8}$ c all-purpose flour
- $\frac{3}{4}$ t salt
- $\frac{3}{8}$ c cold water
- $1\frac{1}{2}$ T olive oil

3. CHEESE TOPPINGS
- $\frac{3}{8}$ lb sliced or grated mozzarella cheese
- $\frac{1}{6}$ to $\frac{1}{4}$ c grated Parmesan cheese

5. SAUCE
- 2 c tomato sauce
- 2 6-oz can tomato paste
- 2 T olive oil
- 4 T water
- 2 t oregano
- 1 t salt
- 1 t sugar
- $\frac{1}{4}$ t crushed dried hot red chili peppers

7. 18 pizzas **9.** The amount of each ingredient should be multiplied by $1\frac{1}{2}$. **11.** Unless the recipe is very large, it usually is difficult to divide and still measure accurately. Janet would have a hard time measuring $\frac{1}{24}$ t, so in this case it is probably not a reasonable endeavor.

Section 4-6, pages 139–141

Exercises **1.** $\frac{3}{4}$ **3.** $\frac{1}{5}$ **5.** $\frac{1}{24}$ **7.** $\frac{11}{15}$ **9.** $2\frac{1}{10}$ **11.** $9\frac{1}{5}$ **13.** $8\frac{13}{24}$ **15.** $4\frac{5}{12}$ **17.** $6\frac{7}{8}$ lb **19.** $1\frac{1}{12}$ **21.** $1\frac{7}{24}$ **23.** Answers may vary. Sample answer: Yard wastes, glass, metals, textiles, other, and household hazardous waste total $\frac{40}{100} = \frac{4}{10}$ **25.** 4 h 10 min **27.** 10 h 25 min **29.** 7 **31.** $\frac{1}{8}$ **33.** $\frac{1}{4}$

Problem Solving Applications **1.** eighth note **3.** eighth note **5.** sixteenth note **7.** quarter note **9.** half note **11.** quarter note; quarter note; half note; half note **13.** quarter note; eighth note; quarter note

Section 4-7, page 143

Problems **1.** R **3.** R **5.** NR **7.** R **9.** R **11.** no **13.** no **15.** No. Answers may vary. Sample answers:

6 balloons and 12 favors; 8 balloons and 8 favors
17. yes **19.** about 4 km

Section 4-8, page 145

Problems **1.** 58 lb **3.** 1,350 h **5.** 277.2 mi **7.** $19.82
9. $16.56 **11.** $317.32

CHAPTER 5 MEASUREMENT AND GEOMETRY

Section 5-1, pages 156-157

Exercises **1.** 2 **3.** 7,000 **5.** 1, 1 **7.** 15, 1 **9.** fl oz
11. gal **13.** oz **15.** 8 yd **17.** b **19.** a **21.** 9,700 lb
23. 48 steps **25.** 12 pint jars

Section 5-2, pages 160-161

Exercises **1.** 8,000 **3.** 2,000 **5.** 4.783 **7.** 4 **9.** m
11. L **13.** kg **15.** 7,000 mL **17.** km **19.** cm
21. g **23.** kg **25.** L **27.** 5,013 m **29.** 1,250 mL

31.

Birthday	16	17	18	19	20	21
Grams of Gold	10	20	40	80	160	320
Value of Gold	$200	$400	$800	$1,600	$3,200	$6,400

```
$   200
    400
    800
  1,600
  3,200
+ 6,400
$12,600
```

Section 5-3, pages 164-165

Exercises **1.** 11 lb 5 oz **3.** 5 yd 2 ft **5.** 9 c 4 fl oz
7. 2 yd 2 ft **9.** 3 **11.** 320 **13.** 175 **15.** 650
17. Yes; 2 lb = 32 oz and 3 × 12 oz = 36 oz
19. 4 gal 2 qt **21.** 2 ft 6 in **23.** $4\frac{2}{3}$ ft **25.** $12\frac{1}{2}$ qt
27. $4\frac{3}{4}$ lb **29.** 10 lb 12 oz **31.** 4 oz

Section 5-4, pages 167-169

Exercises **1.** A: 14 ft B: 24 ft C: 28 ft D: 28 ft E: 42 ft
F: 16 ft G: 24 ft H: 50 ft I: 62 ft **3.** 90 yd **5.** 165 ft
7. 468 m **9.** 214 m **11.** 83.4 m **13.** 424 mm or
42.4 cm **15.** $7\frac{2}{3}$ yd **17.** 4.9 m **19.** 12 m
21. 12 m **23.** 17 m, 5.9 m, 5.9 m

Problem Solving Applications **1.** 22.4 m; $57.12 **3.** $163.54
5. $262.44

Section 5-5, pages 170-171

Problems **1a.** $P = a + b + c$ **b.** $P = 208$ ft **3.** 80 ft
5. 66 ft **7.** 40 ft **9.** family room, $143.20; kitchen,
$116.35; dining room, $118.14; bedroom 1, $89.50;
bedroom 2, $71.60; living room, $144.10 **11.** 23 ft,
23 ft **13.** $51\frac{3}{4}$ ft

Section 5-6, pages 174-175

Exercises **1.** 59.7 cm **3.** 42.4 ft **5.** 132 in.
7. 76.6 mm **9.** 30 cm; 94.2 cm **11.** 1,000 in.;
3,140 in **13.** 22.3 m **15.** 31.4 cm **17.** 37.68 cm
19. circumference of circle; perimeter is 222 mm,
circumference is 348.5 mm **21.** 314 ft **23. a.** 126 m
b. 252 m **c.** 2,520 m

Section 5-7, page 177

Problems **1.** 16 thumbtacks **3.** 19 thumbtacks
5. 22 thumbtacks **7.** 24 thumbtacks **9.** 1 photograph **11.** 9 photographs **13.** 25 photographs
15. 49 photos **17.** 33 yd 1 ft

CHAPTER 6 EXPLORING INTEGERS

Section 6-1, pages 188-189

Exercises **1.** −6 and 6 **3.** −2 and 2 **5.** 3 and −3
7. 34 **9.** 17 **11.** > **13.** < **15.** < **17.** 4,278

19. and 21.

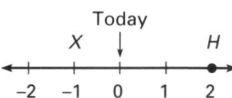

23. −13, −9, 0, 7, 14 **25.** −15, −8, 8, 9, 14
27. False. |7| = |−7|, but 7 > −7. **29.** True.
|6| > |4|, and 6 > 4.

Problem Solving Applications **1.** −10°C to −1°C **3.** +35°C
to +40°C **5.** 100°C **7.** 85°F **9.** 0°C **11.** 198°F
13. 80°F **15.** Less. By comparing the thermometer
scales, one can see that the equivalent temperature, in
degrees Celsius, is about 56°C. **17.** Honolulu, Key
West, (tie between Hartford and Boston), Madison,
(tie between Atlanta and Helena)

CHAPTER 5

Section 5–2
Reading Math **1.** nanosecond—
1 one-billionth of a second
2. megahertz—a unit of frequency equal to one million
cycles per second **3.** micrometer—1 one-millionth of a
meter **4.** gigameter—one
million kilometers **5.** kilowatt—a unit of power equal
to 1,000 watts

Exercise **32.** *(See bottom of
page.)*

Section 5–4
Writing About Math The
perimeter would be 124 m.
Other answers will vary.
Possible responses are garden, playground, or parking
lot that one might want to
enclose with a fence for the
purpose of safety.

Section 5–6
Reading Math **1.** semifinal: next
to last in an elimination tournament **2.** semiprecious: designation for a gemstone having less
value than a precious stone **3.**
semiannual: occurring twice a
year (once every six months) **4.**
semiprivate: designation for a
hospital room accommodating
two patients

CHAPTER 6

Section 6–1
Exercises **20.** "Y" should be
positioned above the 7 point
on the number line. **22.** "Q"
should be positioned above
the −2 point on the number
line. **26.** False. 9 > −13, but
|9| < |−13| **28.** False. −2 > −3,
but |−2| < |−3| **30. a.** True,
3 > 2 > 1; **b.** False, 2 > 1 but
1 < 3; **c.** False, 3 > 2 but
1 < 3; **d.** True, 3 > 1

32.

Birthday	16	17	18	19	20	21	22	23	24	25
Grams of Gold	10	20	40	80	160	320	640	1,280	2,560	5,120
Value of Gold	$300	$600	$1,200	$2,400	$4,800	$9,600	$19,200	$38,400	$76,800	$153,600

Calculator Sequence:

10 [M+] [×] 2 [=] [M+] [×] 2 [=] [M+] [×] 2 [=] [M+] [×] 2 [=] [M+] [×] 2 [=]
[M+] [×] 2 [=] [M+] [×] 2 [=] [M+] [×] 2 [=] [M+] [×] 2 [=] [M+] [×] 2 [=] [M+] [MR] [×] 30 [=]
306,900

Section 6–2

Exercise **20.** $264 + (-69) + (-92) + 21 = 124$

Section 6–3

Try These **1.** 3 **2.** −4 **3.** −3 **4.** 9 **5.** −3 **6.** −9

Exercises **2.** 3 **4.** 21

Section 6–4

Try These Answers may vary. **1.** 20; pattern **2.** −12; chips **3.** −12; commutative property **4.** 21; pattern

Exercise **36.** An even power of a negative integer will have a positive sign. An odd power of a negative integer will have a negative sign.

Section 6–7

Exercises **10.** Intervals chosen for the number line may vary. −2 1/2 will fall between −3 and −2; −1 3/4 will fall between −2 and −1; −1/4 will fall between −1 and 0; 2 1/2 will fall between 2 and 3. **12.** Intervals chosen for the number line may vary. −5/2 will fall between −3 and −2; −1 1/2 will fall between −2 and −1; 1 1/2 will fall between 1 and 2; 7/2 will fall between 3 and 4.

CHAPTER 7

Section 7–4

Exercise **38.** There is no solution, since the product of 0 and any number is 0, and $0 \neq 2$.

Section 7–5

Explore **5.** one x–tile and one 1–tile on one side of the equals sign and five 1–tiles on the other side **6.** one x–tile and four 1–tiles on one side of the equals side and three 1–tiles on the other side **7.** three x–tiles on one side of the equals sign and six 1–tiles on the other side **8.** two x–tiles and two 1–tiles on one side of the equals sign and four 1–tiles on the other side **9.** three x–tiles and one 1–tile on one side of the equals sign and seven 1–tiles on the other side **10.** four x–tiles and three 1–tiles on one side of the equals sign and nine 1–tiles on the other side **11.** one x–tile and four 1–tiles on one side of the equals sign and seven 1–tiles on the other side. Remove four 1–tiles from both sides of the equals sign. The result is one x–tile on one side of the equals sign and three 1–tiles on the other side. x = 3 **12.** three x–tiles on one side of the equals sign and nine 1–tiles on the other side. Separate the tiles on each side of the equals sign onto three equivalent

---

Section 6–2, page 193

Exercises **1.** 10 **3.** −8 **5.** −10 **7.** 0 **9.** 8 **11.** 0 **13.** $49 **15.** > **17.** > **19.** <
21. 264 ⊟ 69 ⊟ 92 ⊞ 21 ⊟ 124
23. 8 and −6 **25.** 32 and −16

Section 6-3, pages 196–197

Exercises **1.** 18 **3.** 6 **5.** 11 **7.** −25 **9.** 64 **11.** −23 **13.** −9 **15.** neg. **17.** pos. **19.** neg. **21.** > **23.** < **25.** > **27.** < **29.** −16°F **31.** − **33.** + **35.** + **37.** −9 **39.** −3 **41.** 16 **43.** −3 **45.** Little Rock: 117°F; Concord: 139°F

Section 6-4, pages 200–201

Exercises **1.** −32 **3.** −64 **5.** −40 **7.** −40 **9.** 15°F higher **11.** $-6 \times (-2) = 12$ **13.** $(-1 + 2) \times 3 = 3$ **15.** < **17.** < **19.** > **21.** 210 **23.** −18 **25.** $5 \times (-15) = -75$, a drop of 75°C **27.** −5 **29.** −9 **31.** −12 **33.** −96, −192, −384, −768 Multiply by 2 **35.** −3 and −4

Section 6-5, pages 203–205

Exercises **1.** −9 **3.** 4 **5.** −5 **7.** 7 **9.** 7 **11.** 2 **13.** −5 **15.** 12 **17.** decreased 2°F **19.** −8 **21.** −21 **23.** 56 **25.** 64 **27.** about 6.7 times deeper **29.** = **31.** > **33.** < **35.** −5 **37.** 3 **39.** decreased 2.5°F **41.** $1.70 **43.** 2 **45.** 50 **47.** −22 **49.** −, × **51.** ×, × **53.** ×, ÷ **55.** No. For example, the answer to 6 ÷ 5 is not an integer. **57.** 4 and −4 **59.** Answers may vary. Sample answer is given. Because a zero quotient times a non-zero divisor equals a dividend of zero, but a non-zero quotient times zero as a divisor will not equal any dividend except zero. So zero as a divisor is meaningless.

Section 6-6, pages 206–207

Problems **1.** −10 **3.** −1 **5.** 18 **7.** 189 **9.** −12 **11.** 8 **13.** −$31 **15.** −540 calories **17.** $40

Section 6-7, pages 210–211

Exercises **1.** Answers may vary. Sample answers: $\frac{7}{1}, \frac{14}{2}$. **3.** Answers may vary. Sample answers: $\frac{10}{3}, \frac{20}{6}$ **5.** B **7.** C

9. Intervals chosen for the number line may vary. $-2\frac{2}{3}$ will fall between −3 and −2; $-\frac{2}{3}$ will fall between −1 and 0; $1\frac{1}{3}$ will fall between 1 and 2; $3\frac{2}{3}$ will fall between 3 and 4.

---

11. Intervals chosen for the number line may vary. $-1\frac{4}{5}$ will fall between −2 and −1; $-\frac{1}{5}$ will fall between −1 and 0; $\frac{2}{5}$ will fall between 0 and 1; $1\frac{2}{5}$ will fall between 1 and 2. **13.** $\frac{7}{20}, \frac{3}{5}, \frac{7}{10}$ **15.** $\frac{7}{12}, \frac{2}{3}, \frac{3}{4}$ **17.** Answers will vary. Possible answers are $\frac{28}{100}$ and $\frac{56}{200}$. **19.** −0.6, $-\frac{18}{30}$ **21.** > **23.** > **25.** = **27.** > **29.** $3\frac{1}{5}$ or 3.2 **31.** $-32\frac{1}{8}$ **33.** $-\frac{3}{4}$ **35.** $-\frac{3}{4}$ **37.** false: $1.5 - 2.5 = -1$ **39.** true **41.** false: $\frac{3}{4} + -\frac{3}{4} = 0$

Section 6-8, pages 212–213

Problems **1.** 36 mi, 81 mi, 182.25 mi; sample strategy: draw a picture **3.** the twelfth week; sample strategy: find a pattern **5.** 28 games; 45 games; 300 games; sample strategy: find a pattern **7.** 13 mi; 6.5 mi; 3.25 mi; sample strategy: find a pattern **9.** $1,200; $4,800; $614,400; sample strategy: find a pattern

CHAPTER 7 EQUATIONS AND INEQUALITIES

Section 7-1, pages 224–225

Exercises **1.** open **3.** true **5.** true **7.** 4 **9.** −8 **11.** −12 **13.** 1 **15.** $8\frac{1}{2}$ **17.** 48 **19.** 2 **21.** 8 **23.** −5 **25.** 4 **27.** 28 points **29.** $\frac{1}{2}$ **31.** 1 **33.** 1.5 **35.** 27 nickels **37.** $72 **39.** 3, −3 **41.** Addie Joss

Section 7-2, pages 228–229

Exercises **1.** add 5 **3.** add 8 **5.** add $\frac{4}{5}$ **7.** 3 **9.** 8 **11.** 4 **13.** $-1\frac{3}{4}$ **15.** 0 **17.** 13 **19.** 9 **21.** −4 **23.** −4.4 **25.** 33.273 **27.** 138,086 **29.** −17°C **31.** Answers may vary. Sample: $x + 3 = 7$ **33.** $x = \pm 3$

Section 7-3, page 231

Problems **1.** c **3.** 58 **5.** $63 **7.** 20 points

Section 7-4, pages 233–235

Exercises **1.** divide by −8 **3.** multiply by 2 **5.** multiply by 16 **7.** 57 **9.** 392 **11.** 33 **13.** 0.6 **15.** $-\frac{4}{3}$ **17.** 2.3 **19.** $8\frac{3}{4}$ **21.** 2.5 **23.** −18 **25.** −0.4 **27.** 21.3 **29.** 108.24 **31.** −3.6162 **33.** 4.7 **35.** $0.85 **37.** There is exactly one solution: $n = 0$ **39.** There are an infinite number of solutions, since the product of 0 and any number is 0 and $0 = 0$. **41.** The equation becomes $0 = 0$.

Problem Solving Applications **1.** $12 + x = 14\frac{1}{2}$; $+2\frac{1}{2}$ **3.** $2x = 300$; 150 shares **5.** $\frac{180}{120} = x$; $-1\frac{1}{2}$

Section 7-5, pages 238–239
Exercises **1.** Add 3 **3.** Subtract 8 **5.** 2 **7.** 24 **9.** −42 **11.** 1.3 **13.** 77 **15.** 7 **17.** $\frac{M}{4} + 6 = 15$ **19.** 10 **21.** −2.6 **23.** 229 **25.** 5 **27.** Answers may vary: Sample: $2f - 5 = 1$

Section 7-6, pages 242–243
Exercises **1.** −8 **3.** 10 **5.** $\frac{5}{4}$ **7.** 30 **9.** 16 **11.** $-\frac{2}{3}$ **13.** $4\frac{11}{14}$ h **15.** 3 **17.** $10\frac{2}{7}$ **19.** $\frac{2}{3}$ **21.** $\frac{1}{2}$ **23.** They are equal; one or the other or both are 0.
Problem Solving Applications **1.** $d = -3$; foal **3.** $p = 1$; squab **5.** $t = 4$; chick

Section 7-7, pages 246–247
Exercises **1.** 24 cm **3.** 11,200 cm³ **5.** 152 chirps per minute **7.** 36 runs **9.** $t = \frac{I}{pr}$ **11.** $i = A - p$ **13.** $C = K - 273$

15. $A = \frac{1}{2}bh$
$2A = bh$
$\frac{2A}{b} = h$

17. $y = mx + b$
$y - b = mx$
$\frac{y - b}{m} = x$

19a. $95 **b.** $105 **c.** $129

Section 7-8, pages 248–249
Problems **1.** $0.22 **3.** 6:45 a.m. **5.** 19 **7.** $9,440

Section 7-9, pages 252–253
Exercises **1.** unlike **3.** like **5.** like **7.** like **9.** like **11.** $5s + 5t$ **13.** $10y - 4$ **15.** $6a$ **17.** $9n + 2x$ **19.** $-6x + 3xy$ **21.** $11s + 9p + 7s + 6k$; $18s + 9p + 6k$ **23.** 7 **25.** $\frac{5}{2}$ **27.** −1 **29.** $4x$ **31.** $18a - 1$ **33.** $700k$ **35.** $15n$ **37.** $10a + 12b + 14c$ **39.** $1,485x - 81y$ **41.** $4(x + y) + 3(2x - y)$; $10x + y$ **43.** −1 **45.** −32 **47a.** $n + 3$ **b.** $4n + 12$ **c.** $4n + 6$ **d.** $2n + 3$ **e.** 3 **f.** The variable n is eliminated from the calculations in the final step, leaving 3 no matter what the value of n.

Section 7-10, pages 256–257
Exercises **9.** yes; no; no; yes **11.** no; no; yes; yes **13.** yes; yes; yes; no **15.** $r > 300$ **17.** $c < 25$ **19.** $x > 0$ (but answers may vary) **21.** $x = 2$ (but answers may vary) **23.** $f \geq 8$ **25.** $p \geq 50,000$ **31.** true **33.** true **35.** false

Section 7-11, pages 260–261
Exercises **1.** $w < 4$ **3.** $x \leq -6$ **5.** $y \leq -3$ **7.** $t > 2$ **9.** $m > -7$ **11.** $x > -2$ **13.** $w > -1$ **15.** $n \geq -2$ **17.** $n \geq -6$ **19.** $y \leq -4$ **21.** $x \geq 12$ **23.** $e \leq 0$ **25.** 15 min **27.** Answers will vary. Sample: Anthony plans to continue going to school until he's twice his present age, then to travel for 4 years. He will still be under 30 years of age. How old is he? **29.** 3, 4, 5, 6, 7, 8 **31.** none

CHAPTER 8 EXPLORING RATIO AND PROPORTION

Section 8-1, pages 272–273
Exercises **1.** 6:9; $\frac{6}{9}$ **3.** 8 to 1; 8:1 **5.** 1 to 3; 1:3 **7.** $\frac{5}{2}$ **9.** $\frac{5}{1}$ **10.** $\frac{16}{3}$ **11.** $\frac{7}{4}$ **13.** Part A: 15; 20; 25 Part B: 6; 12 **15.** $\frac{6}{25}$ **17.** $\frac{13}{100}$ **19.** $\frac{6}{11} > \frac{5}{13}$ **21.** $\frac{4}{9} < \frac{7}{12}$
Problem Solving Applications **1.** Possible answer: The average is a comparison of two numbers. Even though Thomas had fewer at-bats than Brett, the ratio of his at-bats to his hits is about the same as Brett's. **3.** .750

Section 8-2, pages 276–277
Exercises **1.** 50 mi/h **3.** 55 words/min **5.** $5/lb **7.** $65/room **9.** 500 km/day **11.** 55 mi/h **13.** 1,498 km in 7 h **15.** 144 students in 3 buses **17.** $3.96 for 36 oz **19.** 48 mi/gal; about 0.02 gal/mi
Problem Solving Applications **1.** $0.02/sheet **3.** $0.73/lb **5.** the half gallon

Section 8-3, pages 280–281
Exercises **1.** $\frac{12}{14}$; $\frac{18}{21}$; $\frac{24}{28}$ **3.** 15:20; 6:8; 9:12 **5.** 10 to 6; 15 to 9; 20 to 12 **7.** $\frac{16}{18}$; $\frac{24}{27}$; $\frac{32}{36}$ **9.** 6 to 20; 9 to 30; 12:40 **11.** no **13.** no **15.** no **17.** yes **19.** yes; 7 to 10 = 28 to 40 **21.** no; $\frac{6}{9} = \frac{2}{3}$, $\frac{24}{45} = \frac{8}{15}$ **23.** 9 to 10 **25.** 3:4 **27.** 6:17:12 **29.** Possible answer: 6:17:12; 60:170:120

Section 7–5 *Explore* (continued)
groups. The result is that x (one of the three groups on the left) is equal to three (one of the groups on the right). **13.** two x–tiles and five 1–tiles on one side of the equals sign and nine 1–tiles on the other side. Remove five 1–tiles from both sides of the equals sign. The result is two x–tiles on one side of the equals sign and four 1–tiles on the other side. **14.** Divide both sides of the equals sign into two equal groups. The result is one x–tile on one side of the equals sign and two 1–tiles on the other side. **15.** $x = 2$; one x–tile on one side of the equals sign and two 1–tiles on the other side.

Section 7–9
Explore **1.** One group of three x–tiles is being joined to another groups of two x–tiles. **2.** Start with a group of six x–tiles and remove one x–tile.

Computer The result is always 1/2 of the original number entered. Changes for lines 20 and 30 will vary.

Section 7–11
Exercises **1–15.** *(See bottom of page.)*

Computer The hypothesis of line 20 would change to be the conjunction of the two inequalities given in each of Exercises 29–32.

Cumulative Review
7. If the sun has set, the sky will grow dark. If the sky grows dark, the sun has set. **8.** Team A played team Y. Team B played team Z. Team C played team X.

CHAPTER 8

Section 8–1
Exercises **18–21.** To compare ratios, write them as equivalent fractions with the same denominator. Then compare numerators.

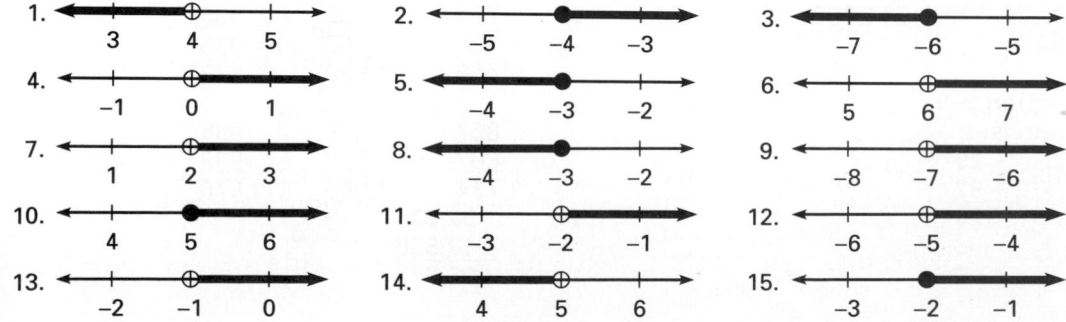

Section 8–4
Computer If the cross-products are not equal, the numbers do not form a proportion.

Calculator
1. 2.4 $\boxed{\times}$ 1.8 $\boxed{\div}$ 3.6 $\boxed{=}$ $x = 1.2$

2. 7 $\boxed{\times}$ 22.5 $\boxed{\div}$ 9 $\boxed{=}$ $m = 17.5$

3. 4.4 $\boxed{\times}$ 2.5 $\boxed{\div}$ 11 $\boxed{=}$ $z = 1$

Section 8–7
Computer Yes, the program would help you see that you could use a calculator to find the factorial of the number.

Cumulative Review
10. Sonia– math; Theo–English; Jay–French; Kim–computer lab
25. $x \geq 8$

26. $e > -5$

27. $t \leq -3$

28. $x < -4$

CHAPTER 9

Section 9–3
Talk It Over Both parts of the whole have been rounded up to the nearest tenth of a percent.

Section 9–7
Exercises 10. $292.22
12. $1,897.42

Section 9–8
Exercises 2. $4.90 $39.90 **4.**
$1,552.96 $6,405.96 **6.** $358.28
$9,315.28 **8.** $1,936.66 $4,137.41

580

Section 8-4, pages 284–285

Exercises 1. yes **3.** yes **5.** no **7.** $n = 8$ **9.** $x = 18$
11. $b = 1$ **13.** $n = 4$ **15.** $y = 16$ **17.** 60 laps
19. $m = 52$ **21.** $m = 35$ **23.** $x = 45$ **25.** $x = 11$
27. $a = 15$ **29.** $y = 1.2$ **31.** 70 yd/min **33.** Answers will vary. Possible answer: $\frac{9}{27} = \frac{35}{105}$ **35.** $w = 60$ ft; $l = 160$ ft

Section 8-5, page 287

Problems 1. 64 statuettes **3.** 104 photographs **5.** 92 students **7.** 45 sculptures

Section 8-6, pages 289–291

Exercises 1. 18 m **3.** 15 in. **5.** 11 cm **7.** 1.4 m
9. 1 m:2.5 mm **11.** 1 in.:10 mi **13.** 1 in.:49 mi
15. 1 in.:5 ft **17.** Answers will vary.

Problem Solving Applications 1. 3.75 mi **3.** 22.5 mi
5. school and post office **7.** 220 km **9.** $1\frac{1}{4}$ in.
11. 4.6 cm **13.** They both are drawings made so actual dimensions can be determined by using a proportion.

Section 8-7, page 293

Problems 1. 17 frames **3.** 6 ways **5.** 8 odd numbers
7. 18 ways

CHAPTER 9 EXPLORING PERCENT

Section 9-1, pages 304–305

Exercises 1. 0.5 **3.** 0.17 **5.** 0.085 **7.** 0.881
9. 0.135 **11.** 0.0575 **13.** $\frac{11}{20}$ **15.** $\frac{1}{5}$ **17.** $\frac{1}{250}$
19. $\frac{7}{1000}$ **21.** $\frac{3}{800}$ **23.** $\frac{13}{80}$ **25.** 20% **27.** 79%
29. 1.5% **31.** 0.5% **33.** $12\frac{1}{2}$% **35.** $16\frac{2}{3}$%
37. 12% **39.** 75% **41.** 87.5% **43.** $41\frac{2}{3}$% **45.** 15%
47. 0.05, 5% **49.** 0.4, 40% **51.** $\frac{3}{4}$% **53.** 45%
55. yes, $66\frac{2}{3}$% **57.** no **59.** $33\frac{1}{3}$%

Section 9-2, pages 308–309

Exercises 1. 69.6 **3.** 18.6 **5.** 33 **7.** 63 **9.** 490
11. 20 **13.** 48 **15.** 27 **17.** 8.5 **19.** 16.6
21. $46.20 **23.** 200 **25.** 40 **27.** 37 or 38 **29.** 845
31. 987 **33.** 162.9 **35.** 0.15 **37.** 0.021
39. $184.47 **41.** $3,000 to $3,200 **43.** 26% of
$7,000 = $1,820 **45.** $5,032 to $5,320. The lowest value is found by subtracting 26% of $6,800 from $6,800; the highest value is found by subtracting 24% of $7,000 from $7,000

580   Selected Answers

Section 9-3, pages 312–313

Exercises 1. 25% **3.** 34% **5.** 10% **7.** 35%
9. 100% **11.** 70% **13.** 70%; 69% **15.** 24,000x

= 1,500; $x = 0.0625 \approx 6.3\%$ 24,000y = 22,500;
$y = 0.9375 \approx 93.8\%$ They save about 6.3% and spend about 93.8%. **17.** It is halved to 3.1%

Problem Solving Applications 1. 56.4%; 60.4%; 55.1%;
59.4%; 55.6% **3.** 48.7%; 57.9%; 53.6%; 59.1%;
56.1%; 53.1%; 51.4% **5.** Esiason, Krieg, Marino, Eason, O'Brien

Section 9-4, pages 316–317

Exercises 1. 10% **3.** 10% **5.** 40% **7.** 25% **9.** 18%
11. 15% **13.** 15% **15.** 30% **17.** 24% increase
19. 5% increase **21.** 120% **23.** Northeast, about
3.5% increase (least); West, about 22.2% increase (greatest)

Section 9-5, pages 320–321

Exercises 1. 156 **3.** 88 **5.** 83 **7.** 25 **9.** 325
11. 600 **13.** 72,000 seats **15.** 100 cm³ **17.** 600
cm³ **19.** 300 cm³ **21.** Nevada **23.** Enter 8.5 for R in Line 20.

Section 9-6, page 323

Problems 1. mental math; $15.25 **3.** calculator;
$16.90 **5.** mental math; 67 **7.** calculator; $382.00 or $382.02, depending on method used

Section 9-7, pages 326–327

Exercises 1. $D = 15.05 $S = 135.45
3. $D = 127.65 $S = 562.34 **5.** $D = 64.44
$S = 446.65 **7.** $102.50 **9.** $68.09 **11.** $17.38
13. the banjo **15.** No. The 35% reduction represents a greater saving. The additional 15% is based on a sale price lower than the original price.

Problem Solving Applications 1. $2,550 **3.** 3% **5.** $5,400
7. $629.55 **9.** $1,375

Section 9-8, pages 330–331

Exercises 1. $10.25 $420.25 **3.** $590.63 $2,165.63
5. $810 $3,810 **7.** $53.08 $558.58 **9.** $348.75
$1,123.75 **11.** $1,313 $11,413 **13.** $2,240
15. $738 **17.** $60.94 **19.** $30,100 **21.** $900
23. $1\frac{1}{2}$ y **25.** 12% **27.** 10% **29.** 10.5%
= $141,440 over 4 years 3% = $137,892.34 over 4 years

---

10. $6,600 $14,600 **12.** $23.72
$93.47

Section 9–9
Exercises
12.   29
          $\underline{+ 92}$
           121

13.

734	765	836
+ 734	or + 765	or + 836
1,468	1,530	1,672

867	928	938
or + 867	or + 928	or +938
1,734	1,856	1,876

14.   92,836
        $\underline{+12,836}$
         105,672

Section 9-9, pages 332–333

Problems **1.** Possible strategy: Work backward; original salary: $237; current salary: $236 **3.** Possible strategy: Use logical reasoning; 9 movies **5.** Possible strategy: Guess and check. There are 12 pentominoes. **7.** Possible strategy: Make a table; 10 days **9.** Possible strategy: Work backward; 960 figurines

11. $\begin{array}{r} 5 \\ \times 5 \\ \hline 25 \end{array}$ $\begin{array}{r} 6 \\ \times 6 \\ \hline 36 \end{array}$

CHAPTER 10 GEOMETRY OF SIZE AND SHAPE

Section 10-1, pages 344–345

Exercises **1.** $\overrightarrow{GH}$ **3.** $\angle PQR$ or $\angle RPQ$ or $\angle Q$ **15.** $\overrightarrow{BA}, \overrightarrow{BC}$ **17.** b **19.** a **21** f **23.** c **25.** false **27.** true **29.** true **31.** 8 vertices in each combination BA **15.** BA, BC **17.** b **19.** a **21.** f **23.** c **25.** false **27.** false **29.** true **31.** 8 vertices in each combination

Section 10-2, pages 348–349

Exercises **1.** 25°, acute **3.** 140°, obtuse **5.** 115°, obtuse **7.** 30°, acute **15.** $\angle LMP$ and $\angle PMN$ **17.** $\angle LMP$ and $\angle PMN$ **19.** acute, 56° **21.** $\angle CAB$, $\angle EAD$, $\angle EAF$, $\angle GAH$, $\angle IAJ$ **23.** $\angle GAD$, $\angle DAB$, $\angle BAI$, $\angle IAG$, $\angle EAC$, $\angle CAJ$, $\angle JAH$, $\angle HAE$ **25.** false **27.** $m\angle C = m\angle B + 90°$

Section 10-3, pages 352–353

Exercises **1.** 30° **3.** 150° **5.** 110° **7.** 70° **9.** 95° **11.** 95° **13.** $m\angle 1 = 22°$; $m\angle 2 = 68°$; $m\angle 3 = 68°$; $m\angle 4 = 90°$ **15.** $m\angle QRN = 60°$; $m\angle PRN = m\angle MRQ = 120°$ **17.** $\overline{AB} || \overline{CD}$ because $\angle BAC$ and $\angle DCE$ are corresponding angles.

Section 10-4, page 357

Exercises **1.** acute **3.** obtuse **5.** isosceles **7.** scalene **9.** 57° **11.** true **13.** true **15.** true

Section 10-5, pages 360–361

Exercises **1.** $3 \times 180° = 540°$ **3.** $5 \times 180° = 900°$ **5.** $7 \times 180° = 1{,}260°$ **7.** 3.74 cm **9.** has no diagonals

13.

number of sides	3	4	5	6	7	8	9
number of diagonals	0	2	5	9	14	20	27

15. true **17.** false **19.** Yes.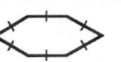

Section 10-6, pages 364–365

Exercises **1.** A polyhedron with two parallel congruent bases that are shaped like triangles. **3.** A right prism having six congruent square faces. **5.** octagonal pyramid **7.** rectangualr prism **9.** hexagonal pyramid **11.** Answers will vary. **13.** Answers will vary.

Polyhedron	Number of Faces (F)	Number of Vertices (V)	Number of Edges (E)	F + V − E
15. rectangular prism	6	8	12	2
17. triangular pyramid	4	4	6	2
19. pentagonal pyramid	6	6	10	2

21. Built in the shape of a pentagon, the Pentagon Building is one of the largest office buildings in the world. **23.** $E = 2t$ **25.** triangular prism

Section 10-7, pages 366–367

Problems **1a.** Answers will vary. **b.** 78° **3a.** Answers will vary. **b.** $1\frac{3}{4}$ in. **5a.** Answers will vary. **b.** 59° **7.** false **9.** true

Section 10-8, pages 368–369

Problems **1**

3. rectangle **5.** square **7.** $\frac{1}{2}$ **9.** 12 ways **11.** 2 quarters + 5 dimes, or 3 quarters + 1 dime + 3 nickels **13.** 6 ft

Section 10-9, page 373

Problem Solving Applications
1. The following are the central angles for the graph: water 144°; fat approximately 122°; and protein approximately 94°.
2. The following are the central angles for the graph: hamburger 117°; kidney beans 63°; tomato sauce approximately 34°; tomato paste approximately 34°; seasoning approximately 14°; and water approximately 99°.

Cumulative Review

8. If Clara can read French, then she can translate this letter. If Clara can translate this letter, then she can read French.

CHAPTER 10

Section 10–1
Computer
```
40 IF S = 4 AND N$ = "QUADRILATERAL"
   THEN GOTO 80
50 IF S = 5 AND N$ =
   "PENTAGON" THEN GOTO 80
60 IF S = 6 AND N$ = "HEXAGON"
   THEN GOTO 80
```

Section 10–5
Computer The derivation of the formula D = (S−2) * 180 is based on the following. The number of overlapping triangular regions into which a polygon can be divided is always less than the number of sides. Since the sum of the measures of the angles of a triangle equals 180°, the sum of the measures of the angles of the polygon equals the number of triangular regions times 180.

Exercise **14.** As the number of sides of the polygons increases by 1, the number of diagonals increases by adding to the number of diagonals of each figure with one less side:

Number of Sides	Number of Diagonals
3	0(+2)
4	2(+3)
5	5(+4)
6	9(+5)
7	14(+6)
8	20(+7)
9	27(+8)
10	35

So, a decagon, a 10-sided figure, has 35 diagonals.

Section 10–6

Explore These eleven patterns can be folded to form a cube.

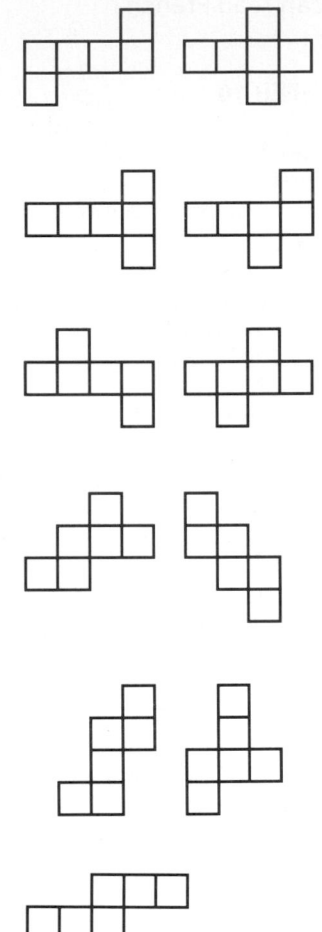

Section 10–9

Problems **2.** The following are the central angles for the graph: sand approximately 130°; water approximately 52°; cement approximately 34°; and gravel approximately 144°. **4.** The following are the central angles for the graph: refrigerators approximately 101°; dishwashers approximately 86°; and color TVs approximately 122°.

Cumulative Review
5. If Rhea helped us study, then we passed the test. If we passed the test, then Rhea helped us study.

Section 10-10, pages 376–377

Section 10-10, pages 376–377

Exercises **1.** Line segment should be 3.5 cm. **3.** Line segment should be 2.5 cm. **5.** The line of intersection should be 90° and cut half way between Point Q and Point R. **7–9.** Line K should intersect CD at 90°.

13. The distance to each vertex is the same. **15.** the distances are the same.

CHAPTER 11 AREA AND VOLUME

Section 11-1, pages 388–389

Exercises **1.** 20 cm² **3.** 56 cm² **5.** perimeter; 22 in.; area, 20 in.² **7.** 50 cm³ **9.** 96 cm³ **11.** 18 in.³ **13.** 14 cm³ **15.** The manufacturer can pack 12 square boxes in each carton. **17.** 29 ft² **19.** 24 cm² **21.** 24 cm² **23.** 30 in.² **25.** 11 m³ **27.** 3 bags; Explanations may vary. Sample: Since the area of the region is 3 ft², 3 bags will be needed. **29.** 5 **31.** 34 unit³

Section 11-2, pages 392–393

Exercises **1.** 360 cm² **3.** 1,280 m² **5.** 4,156 ft² **7.** 497 ft² **9.** 26.5 cm² **11.** 600.25 ft² **13.** $594; area = 72 yd² **15.** 320 in.²
Problem Solving Applications **1.** 468 in.² **3.** 324 in.² **5.** 324 in.² **7.** 444 in.² **9.** 336 in.² **11.** 60 in.²

Section 11-3, pages 396–397

Exercises **1.** 900 cm² **3.** 8.06 m² **5.** 24 cm² **7.** 36 mm² **9.** 60 in.² **11.** 500 cm² **13.** 56 yd² **15.** 250 ft² **17.** 18 cm **19.** For Exercise 17, the picture should be of a parallelogram with an indicated length of 18 cm and an indicated altitude of 10 cm **21.** 72 cm² **23.** 38.605 m² **25.** 25.8 m² **27.** 209 m² **29.** height, length of the base **31.** 32 cm²

Section 11-4, pages 398–399

7. $P = 4 \times s$; 1,700 cm **9.** $A = l \times w$; $6,969.60 **11.** 3.120 ft² **13.** 140 ft **15.** 8,735 ft

Section 11-5, pages 402–403

Exercises **1.** 324 **3.** 441 **5.** 3.24 **7.** 4.41 **9.** $\frac{1}{25}$ **11.** $\frac{64}{81}$ **13.** 16 **15.** −10 **17.** −0.7 **19.** $\frac{2}{9}$ **21.** 3.5 **23.** 9.2 **25.** 6.5 **27.** 2.4 **29.** 12.369 **31.** 31.241 **33.** 10.050 **35.** 28.548 **37.** 4.8 cm **39.** 1,156

---

41. 0.04 **43.** 0.64 **45.** 0.000064 **47.** 8 **49.** 24 **51.** $916.76 or $916.79 **53.** 3.9 s

Section 11-6, pages 406–407

Exercises **1.** 9 **3.** 31.2 **5.** 10 **7.** 45.2 m **9.** 20 cm **11.** 25.5 cm **13.** 30 ft **15.** 3.7 **17.** 5.7 **19.** 10.2 **21.** 23.75 **23.** 2.8 in. **25.** 8.2 m **27.** 14.7 m **29.** 36.2 m **31.** 13 cm

Section 11-7, pages 410–411

Exercises **1.** 24 in.³ **3.** 648 cm³ **5.** 149.5 cm³ **7.** 4,704 in.³, 5,040 in.³ D has the greater volume. **9.** 5.52 cm³ **11.** 1,404 in.³ **13.** 14,861.1 m³ **15.** 66.82¢ or 67¢ **17.** 2 trips **19.** $V = \frac{1}{2} s^3$; 0.864 m³ **21.** 1,890,000 cm³ **23.** 1,890 goldfish **25.** 8,763.3 cm³ or 8.7633 L

Section 11-8, pages 413–415

Exercises **1.** 452.2 cm² **3.** 706.5 ft² **5.** 201.0 ft² **7.** 301.6 in.² **9.** C **11.** B **13.** 379.9 cm² **15.** square **17.** about 1.3 yd² **19.** 78.5 in.² **21.** 201 in.² **23.** 6¢ per in.² **25.** large with two choices, Exercise 24 **27.** 314 in.² **29.** 201.0 ft² **31.** 36.1 in.² **33.** 99.3 yd² **35.** 14,142.0 cm³

Section 11-9, pages 417–419

Exercises **1.** 113 cm³ **3.** 2,703 cm³ **5.** $r = 3$ ft, $V = 226$ ft³ **7.** $d = 13.8$ in., $V = 628$ in.³ **9.** 7,235 in.³ **11.** Answers will vary. **13.** 2,735 cm³ **15.** 401,593 cm³
Problem Solving Applications **1.** 238.5 km³ **3.** 30,711.1 km³ **5.** $A = 426.8$ km³, $B = 375.3$ km³, $A > B$ **7.** 0.3 km³

Section 11-10, pages 422–423

Exercises **1.** 65.3 in.³ **3.** 7,234.6 ft³ **5.** 19.3 ft³ **7.** 2,143.6 in.³ **9.** 1,316.4 ft³ **11.** 3,052.1 m³ **13.** 1,316.4 mm³ **15.** 195.3 cm³ **17.** 2,142.6 m³ **19.** 5,572.5 cm³ **21.** 25.1 m³ **23.** 473.7 m³ **25.** 510.3 yd³ **27.** 237.6 cm³/1,097.5 cm³; 21.6% **29.** The volume is multiplied by 4.

Section 11-11, pages 426–427

Exercises **1.** 222 cm² **3.** 11 m² **5.** 377 ft² **7.** 864 cm² **9.** 312 in.² **11.** 184 in.² **13.** 803.8 cm² **15.** 289.4 cm² **17.** 181.37 cm² **19.** 912 in.²; 984 in.²; 864 in.² **21.** 803.84 cm²

CHAPTER 11
Cumulative Review **1.**

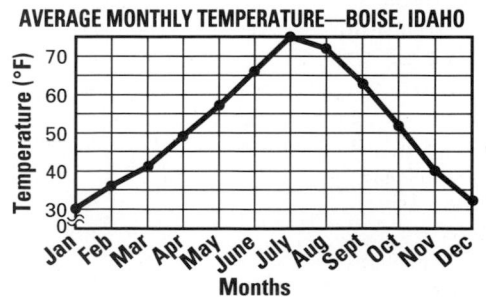

AVERAGE MONTHLY TEMPERATURE—BOISE, IDAHO

Temperature (°F) vs Months

Problems 1. base 10 in.; height 12 in.; 10 **3.** 6 cm
5. base 40 yd; height 120 yd **7.** 12 cm

CHAPTER 12 POLYNOMIALS

Section 12-1, pages 440–441

Exercises 1. $4n^4 + n^3 + 2n^2 + 7n$ **3.** $5z^2 + (-5z) +$
$(-4z) + 4$ **5.** $5n^5 + 2n^3 + (-8n)$ **7.** $3a + (-3)$
9. $-5c + (-2)$ **11.** $12r^2 + r$ **13.** $12s^2 + (-2s)$
15. $-9h^3 + 22h^2 + (-2)$ **17.** $3s^3 + (-2s^2) +$
$(-5s) + 7$ **19.** $5x^2y$ **21.** $3mn$ **23.** $3z^2 + (-11x^2y^2)$
25. $2x^2 + (-x) + 5xy$ **27.** $a^3b + (-2a^2b) + 3ab +$
$4ab^2 + 8ab^3$ **29.** 2 **31.** 1 **33.** false **35.** None of
the other terms are like terms.

Section 12-2, pages 444–445

Exercises 1. $6y - 3$ **3.** $7m - 7$ **5.** $-2a^2 - 1$
7. $2a^2 - 2a - 3$ **9.** $5x$ **11.** $4z - 7$ **13.** $3 + k$
15. $5p^2 - 2p + 14$ **17.** $17m^3 - m^2 - 5m$
19. $3x - y$ **21.** $3a + 3b$ **23.** $13a + b$
25. $-x - 6xy$ **27.** $8r^2t - 2rt^2 - 4rt$ **29.** $3a + b$
31. $-3z + 2y$ **33.** $9f - 11g - 4$ **35.** **35.** $10rt^2$
37. $-a^2bc + 5abc^2 + 2abc - ab^2c$ **39.** xy
41. $5x + y$ **43.** $4x + y$ **45.** They are opposites of
each other. **47.** The sign of each term in the
difference is opposite when the order of polynomials
is reversed in a subtraction.

Section 12-3, pages 448–449

Exercises 1. $6ab$ **3.** $-12km$ **5.** $6kp$ **7.** $7cd$ **9.** $4a^2$
11. $7y^5$ **13.** $-24p^3$ **15.** $-15k^6$ **17.** $9x^2$ **19.** $-8d^9$
21. $81y^3$ **23.** $36c^4$ **25.** $-64b^{12}$ **27.** $6mp$
29. $12a^6b^3$ **31.** $-3p^5r^5$ **33.** $x^{21}y^6$ **35.** $4x^{14}y^{30}$
37. $36p^{10}r^8$ **39.** $24x^2$ **41.** $\frac{27}{64}a^3$ **43.** $9p^3$ **45.** Any
values such that st is negative.

Section 12-4, pages 451–453

Exercises 1. $2x^2 + 6x^3$ **3.** $6a^2 + 2a$ **5.** $2a^2 +$
$10a - 2$ **7.** $-3f^2 + 6f + 24$ **9.** $-r^4 + 9r^2 - 6r$
11. $4x^3 - 8x^2 + 6x$ **13.** $10x^2 - 5x$ **15.** $6x + 2$
17. $-4p + 12$ **19.** $5a - 7b$ **21.** $-x^4 - 2x^3 +$
$2x^2 + 2x$ **23.** $c^4 - 3c^3d - 6c^2d^2$ **25.** $-2r^4 + 6r^3s^2$
$- 8rs^3$ **27.** $c(25t + 24) = 25ct + 24ct$
29. Ganymede **31.** $24m^2 - 12m$ **33.** $x - y$
35. $x - y$ **37.** $w + z$

Section 12-5, pages 454–455

Problems 1. $5n$, $6n$; guess and check **3.** four; $1 + x^6$;

$x + x^5$; $x^2 + x^4$; $x^3 + x^3 = 2x^3$; make an organized
list **5.** 100, 5,625; solve a simpler problem **7.** 44
minutes fast; make an organized list **9.** $24x^2$; draw a
picture **11.** Lums: 6; Mads: 16; Nogs: 14; Ools: 8;
use logical reasoning **13.** 204 squares; find a
pattern.

Section 12-6, pages 457–459

Exercises 1. $4y$ **3.** $5ab$ **5.** $5s^2t$ **7.** d **9.** a **11.** c
13. $9(1 - 2x)$ **15.** $3b(4a - b)$ **17.** $6x(4x - y)$
19. $y(x + xy - y)$ **21.** $7xy(1 - 2xy + 4z)$
23. $-3a^2$ **25.** $4y^2$ **27.** $\frac{1}{3}xyz(1 - y - xz)$
29. $4r(s + u) + 4t(s - u)$ **31.** $x^2(8x - 9)$

Problem Solving Applications 1. $8a$ **3.** $20n + 2$
5. $n(2n + 6) = 2n^2 + 6n$ **7.** SA $= 2\pi r(h + r)$
9. $r^2(\pi - 2)$

Section 12-7, page 461

Problems 1. $n(n + 1) + 3(n + 1) = (n + 3)(n + 1)$
3. $(a + 2)(a - 3)$ **5.** $(r + 2)(r + 6)$
7. $(c + 1)(c^2 + c + 5)$ **9.** Use the chunk $(m + 1)$.
Let $m + 1 = 5$. Then $m = 5 - 1 = 4$ **11.** Since
$(y + 2)^2 = 5^2$ write these two equations:
$y + 2 = 5$ and $y + 2 = -5$. Then $y = 3$ and
$y = -7$.

Section 12-8, pages 464–465

Exercises 1. $5c$ **3.** $4y$ **5.** $-4p$ **7.** y **9.** cd **11.** ab^2
13. $a + 6$ **15.** $3a - 6b$ **17.** $2x^2 - 3x + 1$
19. $3abc - 5c$ **21.** $-6b^8$ **23.** $\frac{x}{y}$ **25.** $\frac{z^2}{3}$ **27.** $-6q^4rs$
29. $8a^2 - 4b + 6ab^2$ **31.** $2z - 3x + 4xy - 6zx^2y^2$
33. $5q$ **35.** $9a$ **37.** $5m^2n^3$ **39.** $90xy^2z^2$
41. $x^2 - 4$

CHAPTER 13 PROBABILITY

Section 13-1, pages 475–477

Exercises 1. $\frac{1}{3}$ **3.** $\frac{1}{2}$ **5.** $\frac{1}{2}$ **7.** $\frac{5}{6}$ **9.** $\frac{1}{6}$ **11.** $\frac{1}{3}$ **13.** $\frac{1}{3}$
15. 0 **17.** $\frac{1}{6}$ **19.** $\frac{1}{2}$ **21.** $\frac{2}{3}$ **23.** 1 to 4 **25.** 3 to 2
27. 4 to 1 **29.** 3 to 2 **31.** $\frac{3}{4}$ **33.** b **35.** e **37.** b

Section 13-2, pages 479–481

Exercises 1. (M,A,T,H,F,U,N) **3.** 8 possible
outcomes **5.** 6 possible outcomes **7.** $\frac{1}{18}$ **9.** $\frac{2}{9}$
11. 36 possible outcomes **13.** 72 possible outcomes
15a. $\frac{1}{4}$ **b.** $\frac{1}{2}$ **c.** $\frac{1}{4}$

CHAPTER 12

Section 12–5
Problems 1. Since the sum of the
monomials has a variable coefficient of n, both monomials must
have variable coefficients of n. By
guess and check, the numerical
coefficients must be 5 and 6, so
the monomials must be $5n$ and $6n$.
4. height $= 25 + 1/2$ height $= 50$ yd;
Guess and check values for the
total height. 50 yd (total height)
5. Note the sum of the first two
odd numbers, the first three odd
numbers, and so on:
$1 + 3 = 4 = 2^2$
$1 + 3 + 5 = 9 = 3^2$
$1 + 3 + 5 + 7 = 16 = 4^2$
$1 + 3 + 5 + 7 + 9 = 25 = 5^2$
The pattern suggests that the sum
of the first n odd numbers is n^2.
So, the sum of the first 10 odd
numbers of $10^2 = 100$, and the
sum of the first 75 odd numbers
is $75^2 = 5,625$. **6.** Work backward,
using inverse operations:
$20x^3 - 5x - (4x^3 + 5x) = 16x^3 - 10x$
$16x^3 - 10x + 10x = 16x^3$
$16x^3 - 7x^3 = 9x^3$
7. Number the half-hour intervals
from 6:00 a.m. to 10:30 a.m. 9.18
minutes fast ÷ 9 half-hour intervals = 2 minutes gained per half
hour. There are 22 half-hour intervals from 6:00 a.m. to 5:00 p.m.
So the watch will be 44 minutes
fast. **11–13.** *(See bottom of page.)*

Cumulative Review
1.

```
4 | 9
5 | 1 3 3 4 7 8
6 | 2 4 4 5 8
7 | 1 2 3 6 9
8 | 2
```

CHAPTER 13

Section 13–2
***Try These* 3.** 16 possible out-
comes: HHHH, HHHT, HHTH,
HHTT, HTHH, HTHT, HTTH, HTTT,
THHH, THHT, THTH, THTT, TTHH,
TTHT, TTTH, TTTT **4.** 12 possible
outcomes: H1, H2, H3, H4, H5, H6,
T1, T2, T3, T4, T5, T6

---

11. Lums must have 6 eyes; if they
had at least 8 eyes, Nogs could not
have more than twice as many.
Nogs cannot have 6 or 8 eyes.
Nogs cannot have 16 eyes,
because 16 is a perfect square, The
only possibility left for Nogs is 14
eyes. Mads must have 16 eyes in
order to have more eyes than
Ools. This leaves 8 eyes as the
only possibility for Ools.

12. $200 = 12n + 8$, so, $n = 16$
There will be $16 + 1 = 17$ benches
on each side for a total of 34.
13. There are *one* 8×8 square,
four 7×7 squares, and *nine* 6×6
squares, and so on. The pattern
shows that the total number of
squares is 204.
$1 + 4 + 9 + 16 + 25 + 36 + 49 + 64 =$
$1^2 + 2^2 + 3^2 + 4^2 + 5^2 + 6^2 + 7^2 + 8^2$
$= 204$

Section 13–2 (continued)

Exercises **3**. 8 possible outcomes: HHH, HHT, HTH, HTT, THH, THT, TTH, TTT **4**. 14 possible outcomes: HM, TM, HA, TA, HT, TT, HH, TH, HF, TF, HU, TU, HN, TN **5**. 6 possible outcomes: H1, H2, H3, T1, T2, T3 **6**. 36 possible outcomes: H11, T11; H12, T12; H13, T13; H21, T21; H22, T22; H23, T23; H31, T31; H32, T32; H33, T33; H41, T41; H42, T42; H43, T43; H51, T51; H52, T52; H53, T53; H61, T61; H62, T62; H63, T63 **11**. 36 possible outcomes: 49AJ, 49JR 49AR, 49RA, 49RJ, 49JA; 43AJ, 43JR, 43AR, 43RA, 43RJ, 43JA; 94AJ, 94JR, 94AR, 94RA, 94RJ, 94JA; 93AJ, 93JR, 93AR, 93RA, 93RJ, 93JA; 39AJ, 39JR, 39AR, 39RA, 39RJ, 39JA **12**. 24 possible outcomes: 71PN, 71NP, 76PN, 76NP, 75PN, 75NP; 16PN, 16NP, 17PN, 17NP, 15PN, 15NP; 61PN, 61NP, 67PN, 67NP, 65PN, 65NP; 51PN, 51NP, 56PN, 56NP, 57PN, 57NP

Problem Solving Applications **1**. There are 24 possible outcomes. In addition to the six listed in the tree diagram on the text page the remaining 18 are: NICG, NIGC, NCIG, NCGI, NGIC, NGCI; CING, CIGN, CNIG, CNGI, CGNI, CGIN; GINC, GICN, GNIC, GNCI, GCIN, GCNI.

CHAPTER 14

Skills Preview

4.

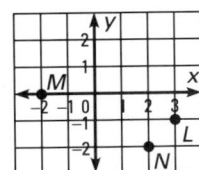

5. The graph is the straight line that includes the points (–4, 1), (–3, 2), and (–2, 3). **8**. The vertices for the translated triangle are *D*'(3, 4), *E*'(7, 7) and *F*'(5, 0). **9**. The vertices for the reflected trapezoid are *W*'(–6, –4), *X*'(–4, –6), *Y*'(–2, 2), and *Z*'(–6, –2).

Section 13-3, pages 483–485

Exercises **1**. 12 **3**. 15 **5**. 8 **7**. $\frac{1}{64}$ **9**. 30 **11**. 3 × 3 × 3 × 3 = 81 **13a**. 3 ways **b**. 2 ways **c**. 1 way **d**. 3 × 2 × 1 = 6; 6 ways

Problem Solving Applications **1**. 24 possible choices **3**. Answers will vary.

Section 13-4, pages 488–489

Exercises **1**. $\frac{1}{30}$ **3**. $\frac{1}{150}$ **5**. $\frac{3}{50}$ **7**. $\frac{4}{75}$ **9**. $\frac{1}{75}$ **11**. $\frac{2}{75}$ **13**. 0 **15**. $\frac{2}{45}$ **17**. $\frac{4}{13}$ **19**. $\frac{24}{91}$ **21**. $\frac{35}{132}$ **23**. $\frac{1}{66}$ **25**. $\frac{1}{6}$ **27**. $\frac{1}{10}$ **29**. pulling out a $20 bill, not replacing it, and then pulling out a $10 bill **31**. pulling out a $10 bill, replacing it, and then pulling out a $5 bill

Section 13-5, pages 492–493

Exercises **1**. $\frac{5}{16}$ **3**. $\frac{1}{4}$ **5**. $\frac{11}{16}$ **7**. $\frac{9}{16}$ **9**. $\frac{1}{6}$ **11**. $\frac{5}{6}$ **13**. $\frac{19}{30}$ **15**. $\frac{1}{10}$ **17**. Answers will vary. **19**. $\frac{31}{60}$ **21**. Answers will vary.

Section 13-6, page 495

Problems **1**. $\frac{4}{25}$ **3**. $\frac{3}{10}$ **5**. 15,000 **7**. 22,500 **9**. 112,500 **11**. 333,750

Section 13-7, page 497

Exercises **1–11**. Answers will vary.

CHAPTER 14 GEOMETRY OF POSITION

Section 14-1, pages 508–509

Exercises **1**. (2,4) **3**. *N*(–2,0); *L*(2,0) **5**. *A*(2,4); *G*(5,3); *H*(1,2) **7**. *B*(–4,–4)

9.
11.
13.
15.

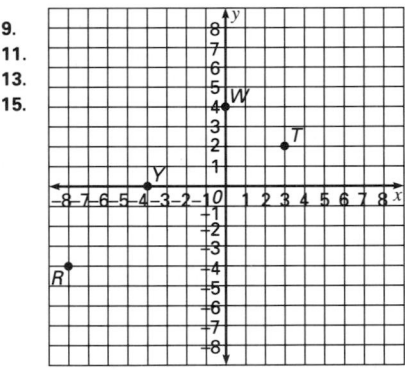

---

10. The vertices for the triangle after a 90°-turn clockwise about the origin are *D*'(3, –1), *E*'(6, –5), and *F*'(–1, –3).

Section 14-1
***Try These* 9–14.**

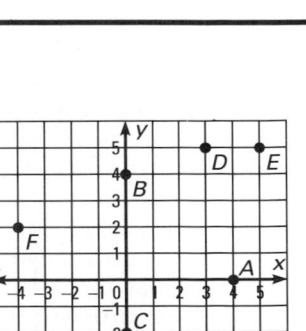

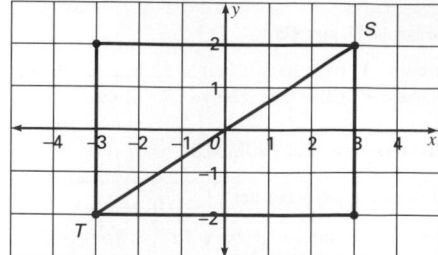

17.

19. Answers will vary. **21**. Answers will vary. A possible solution is (3,2) and (–3, –2).

23. The area is 16 square units.

25. (–2,0); (2,4); (3,5); (4,6) **27**. (1.25, –0.7)

Section 14-2, pages 510–511

Problems **1**. 26 blocks **3**. 35 blocks; 3 blocks **5**. white **7**. after; 17 min

Section 14-3, pages 514–515

Exercises **1**. Answers will vary. **3**. Answers will vary. **5**. Answers will vary. **7**. –4; –4; 0; 0; 4; 4; 8; 8; The straight line that includes points (–1,–4), (1,4), and (2,8) **9**. 3; 3; 4; 4; 5; 5; 6; 6; The straight line that includes points (–4,3), (–2,4), and (2,6) **11**. $y = 2\frac{1}{2}x$ **13**. $y = \frac{1}{2}x$; The straight line that includes points (–2,–1), (–1,–$\frac{1}{2}$), and (1,$\frac{1}{2}$)

Section 14-4, pages 518–519

Exercises **1**. 1 **3**. –$\frac{4}{7}$ **5**. $y = -\frac{1}{2}x + 2$ **7**. $\frac{1}{4}$ **9**. $\frac{1}{2}$ **11**. $\frac{1}{2}$ **13**. $y = \frac{1}{2}x + 2$

Problem Solving Applications **1–8**: Answers will vary.

Section 14-5, page 521

Problems **1**. Parking Charge = $3.00 Ticket Charge = $2.00

Amount charged for vehicle	$13	$7	$31
Number of students per vehicle	5	2	14

3. Flat fee = $4.00 Hourly = $2.50 per child

Amount charged for 4-hour job	$6.50	$11.50	$16.50
Number of children	1	3	5

5. Fixed fee = $7.50 cost per lb = $3.50

Cost of shipping package	$21.50	$32.00	$42.50
Number of pounds	4	7	10

Exercises 1.

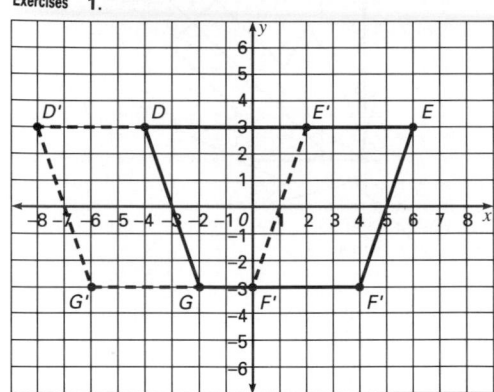

3. The coordinate for the top right corner of the T-shaped figure is $(5, -1)$. For the translated figure, the top right corner would be at the coordinate $(-2, 1)$.

The coordinates for the vertices of the triangle are $(1, 2)$, $(2, -1)$, and $(-2, 0)$. For the translated figure, the coordinates would be at coordinates $(0, -2)$, $(1, -5)$, and $(-3, -4)$.

The coordinate for the top right corner of the remaining figure is $(6, 4)$. For the translated figure, the top right corner would be at coordinate $(5, 0)$.

5. Answers will vary. If for example, the top right corner of the figure was on coordinate $(5, 4)$ the translated figure would have the coordinate $(0, 7)$.

7. Answers will vary. If for example, the top right corner of the figure was on coordinate $(7, -1)$ the translated figure would have the coordinate $(1, 2)$.

9. 4 right, 3 down 11. 5 right, 2 down 13. Various descriptions are possible. For example, for each small square in the design, a corresponding small square is positioned 6 grid squares to the right. 15. Answers will vary. 17. tessellating rectangles. 19. Answers will vary.

Section 14-7, pages 527-529

Exercises 1. $(4, -1)$ 3. $(-7, -2)$ 5. $(-8, 7)$ 7. Coordinates for vertices of reflected image: $(-2, 3)$, $(-5, 2)$, $(-3, 1)$ 9. Coordinates for vertices of reflected image: $(2, -1)$, $(2, -3)$, and $(-2, -1)$
11. $H, I, M, O, T, U, V, W, X, Y$ 13. $\overline{AC}$ and $\overline{DF}$; $\overline{AB}$ and $\overline{DE}$; $\overline{BC}$ and $\overline{EF}$ 15. clockwise; counterclockwise 17. The order of the vertices changes. The lengths of the sides stay the same.

Problem Solving Applications

1.

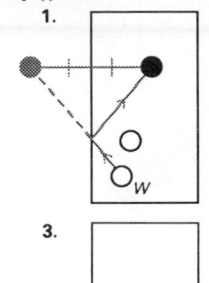

3.

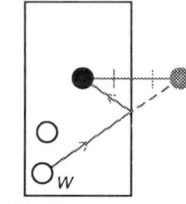

Section 14-8, pages 532-533

Exercises 1. 270° 3. 180° 5. 180° counterclockwise or 180° clockwise 7. The coordinates of the rotated figures are $(2, 2)$, $(-2, 2)$, $(-2, -2)$ and $(2, -2)$.
9. The two top coordinates of the rotated figure are $(1, 2)$ and $(-1, 2)$. The two bottom coordinates of the rotated figure are $(1, -2)$ and $(-1, -2)$.

11.

13.

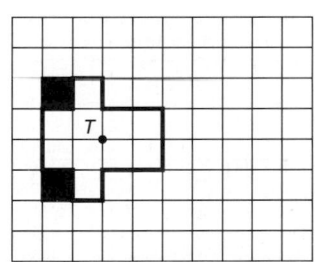

Try These (continued)
15–16. *(See bottom of page.)*

Exercises 9–16.

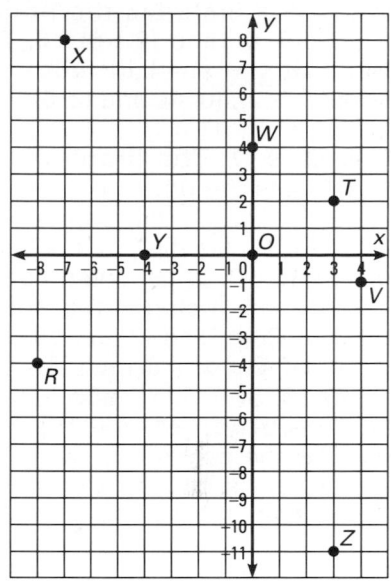

18.

22.

Rectangle *OABC*
Area is 20 square units.

23. Square *DEFG*
Area is 16 square units.

15.

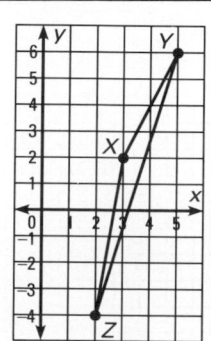

16.

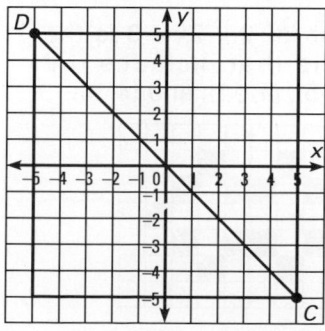

585

Section 14–3

Try These **4.** The graph of $y = 2x - 1$ includes the points $(-2, -5)$, $(-1, -3)$, and $(1, 1)$. **5.** The graph of $y = -3x$ includes the points $(-1, 3)$, $(0, 0)$, and $(2, -6)$. **6.** The graph of $y = x + 4$ includes the points $(-2, 2)$, $(0, 4)$, and $(2, 6)$.

Exercises **7.** The graph of $y = 4x$ includes the points $(-1, -4)$, $(1, 4)$, and $(2, 8)$. **8.** The graph of $y = -2x + 1$ includes the points $(2, -3)$, $(0, 1)$, and $(-1, 3)$. **9.** The graph of $y = 1/2x + 5$ includes the points $(-4, 3)$, $(-2, 4)$, and $(2, 6)$. **10.** The graph of $y = x + 5$ includes the points $(0, 5)$, $(-1, 4)$, and $(-2, 3)$. **13.** The graph of $y = 1/2x - 1$ includes the points $(-7, -4)$, $(2, 1)$, and $(9, 5)$ as well as the points $(-2, -1)$ and $(1, 1/2)$.

Section 14–8

Exercises **8.** The coordinates of the rotated figure are $(0, 2)$, $(3, 0)$, and $(0, -2)$.

10.

12.

14.
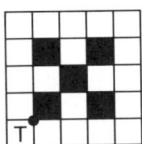

17. The vertices for the rotated triangle are $X'(0, -2)$, $Y'(0, -5)$, and $Z'(4, -2)$.

Chapter Review
8.
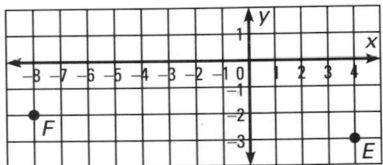

15.
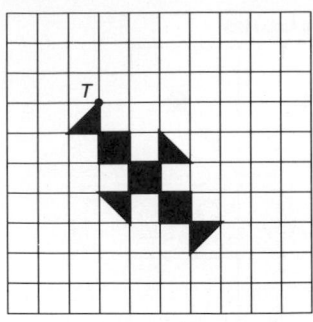

17. counterclockwise; counterclockwise **19.** Both equal 6 square units. **21.** 3,678 in., or 314 ft

Section 14–9, pages 536–537

Exercises **1.** yes **3.** yes **5.** no **7.** 2 **9.** 8 **11.** 4 **13.** isosceles triangle **15.** scalene triangle

---

Prevention's Giant Book of Health Facts, © 1991, Rodale Press, Inc., Emmaus, PA, "Vegetables and Risk of Lung Cancer in Men," "Patterns of Sleep," p. 562; "Life Expectancy," p. 563; "Number of Calories Burned by People of Different Weights," p. 564; "The Sports with the Most Injuries," p. 565.
Reader's Digest Book of Facts, © 1987 The Reader's Digest Association, Inc., Pleasantville, NY, "How Fast Do Insects Fly?" and "Longest Recorded Life Spans of Some Animals," p. 548.
The Recording Industry Association of America, "The Recording Industry Association of America 1990 Sales Profile," p. 553.
The Universal Almanac, © 1990 Andrews and McMeel, Kansas City, MO, "Measuring Earthquakes," p. 556.
U.S. Bureau of the Census, *Statistical Abstract of the United States: 1989* (109th edition) Washington, DC, 1989, "Housing Units — Summary of Characteristics and Equipment, by Tenure and Region: One Recent Year," p. 551; "Multimedia Audiences — Summary 1987," p. 553; "Space Shuttle Expenditures by NASA in Constant Dollars: 1972 to 1987," p. 555; "Air and Water Pollution Abatement Expenditures in Constant Dollars: 1975 to 1986," p. 559; "Farms — Number and Acreage, by State: 1980 and 1988," p. 567.
U.S. Bureau of the Census, "1990 Population and Number of Representatives, by State," p. 567.
The World Almanac and Book of Facts, 1991, © 1990 Pharos Books, "Gestation, Longevity, and Incubation of Animals, " and "Top 25 Kennel Club Registrations," p. 549; "All-Time Top Television Programs," p. 552; "All-Time Top 25 Movies," p. 553; "Highest and Lowest Continental Altitudes," p. 556; "Acid Rain Trouble Areas," p. 558; "Immunization Schedule for Children," p. 563; "Some Endangered Mammals of the United States," p. 566.

586 Selected Answers

17.

19. Answers will vary.

---

9. The graph of $y = 2x - 3$ is the straight line that includes the points $(-1, -5)$, $(1, -1)$, and $(4, 5)$.

13.
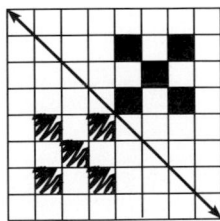

13. (continued)

Glossary

►A

absolute value (p. 186) The distance of any number, *x*, from zero on the number line. Represented by |*x*|.

acute angle (p. 347) An angle measuring between 0° and 90°.

acute triangle (p. 354) A triangle having three acute angles.

addition property of opposites (p. 191) When opposites are added, the sum is always zero.

adjacent angles (p. 347) Have a common vertex and a common side but have no interior points in common.

alternate exterior angles (p. 351) Pairs of angles that are exterior to the lines and on alternate sides of the transversal.

alternate interior angles (p. 351) Pairs of angles that are interior to the lines but on alternate sides of the transversal.

altitude of a triangle (p. 377) Segment from a vertex of a triangle perpendicular to the opposite side or to a line containing that side.

angle (p. 342) A plane figure formed by two rays having a common endpoint.

area (p. 384) The amount of surface covered by a plane figure. Area is measured in square units.

associative property (p. 72) If three or more numbers are added or multiplied, the numbers can be regrouped without changing the result. For example, 4 + (6 + 5) = (4 + 6) + 5; 4 × (6 × 5) = (4 × 6) × 5.

average (p. 30) See *mean*.

axes (p. 506) The perpendicular lines used for reference in a coordinate plane.

►B

back-to-back stem-and-leaf plots (p. 11) See *stem-leaf plot*.

bar graph (p. 20) A means of displaying statistical information in which horizontal or vertical bars are used to compare quantities.

base (of a power) (p. 60) The repeating factor in a power. For example, in 2^3 the base is 2.

binomial (p. 438) A polynomial with two terms.

bisector of an angle (p. 371) A ray that divides the angle into two equal adjacent angles.

►C

Celsius (p. 189) See *metric units*.

chunking (p. 460) The process of collecting several pieces of information and grouping them together as a single piece of information.

circle (p. 172) The set of all points in a plane the same distance from a given point called the center.

circumference (p. 172) The distance around a circle.

closed set (p. 208) If an operation is performed on two numbers in a given set and the result is also a member of the set, then the set is closed, with respect to that operation (closure property).

clusters (p. 9) See *stem-and-leaf plot*.

cluster sampling (p. 4) The members of the population are chosen at random from a part of the population and are then polled in clusters, not individually.

coefficient (p. 438) A number by which a variable or group of variables is multiplied. For example, in $7ab^2$ the coefficient is 7.

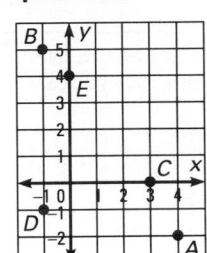

5. The graph of *y* = 3*x* − 2 is the straight line that includes the points (0, −2), (1, 1), and (2, 4).
8. The coordinates for the translated parallelogram are *C*'(0, 4), *D*'(4, 4), *E*'(3, 0), and *F*'(−1, 0).
9. The vertices for the reflected triangle across the *x*-axis are *P*'(−6, −6), *Q*'(−2, −5), and *R*'(−4, 2).
10. The vertices for the triangle after it has gone through a 90°-turn counterclockwise about the origin are *P*'(−6, −6), *Q*'(−5, −2), and *R*'(2, −4).

Cumulative Review
4. Zero to 100, numbered intervals in tens, and halfway marks that fall between numbered intervals in fives.

CHAPTER 1 (Continued)

Figure 1

NUMBER OF STUDENTS IN AFTER-SCHOOL ACTIVITIES	
Writer's Workshop	🧍 🧍 🧍
Math Team	🧍 🧍 🧍 🧍
Crafts Club	🧍 🧍
Softball Team	🧍 🧍
Soccer Team	🧍 🧍 🧍
Runner's Club	🧍 🧍 🧍 🧍 🧍 🧍

Key: 🧍 represents 5 students

Figure 2

ROBERTO'S SCORE FOR THE SEASON	
Game	
1	⬯ ⬯ ⬯
2	⬯ ⬯ ⬯ ⬯
3	⬯ ⬯ ⬯
4	⬯ ⬯
5	⬯ ⬯ ⬯ ⬯
6	⬯
7	⬯ ⬯ ⬯
8	⬯ ⬯ ⬯ ⬯ ⬯ ⬯

Key: ⬯ represents 4 points

Figure 3 *(See bottom of page.)*

---

collinear points (p. 342) Two or more points lying in a straight line.

combining like terms (p. 250) Process using the distributive property to simplify algebraic expressions.

commission (p. 327) An amount of money that is a percent of a sale. The percent is called the *commission rate*.

common factors (p. 119) The factors that are the same for a given set of numbers are the common factors. For example, a common factor of 12 and 18 is 6.

common multiple (p. 119) A number that is a multiple of two different numbers is a common multiple of those two numbers.

commutative property (p. 72) If two numbers are added or multiplied, the operations can be done in any order. For example, $4 \times 5 = 5 \times 4$; $5 + 4 = 4 + 5$.

compass (p. 370) An instrument used to draw circles and arcs and to transfer measurements.

compatible numbers (p. 43) Numbers that are easy to compute mentally.

complementary angles (p. 347) Two angles whose sum measures 90°.

composite numbers (p. 118) A whole number greater than one having more than two factors.

conditional statement (p. 98) One made from two simple sentences. It has an *if* part and a *then* part. The *if* part is called the *hypothesis* and the *then* part is called the *conclusion*.

cone (p. 363) A three-dimensional figure with one circular base and one vertex. The line segment from the vertex perpendicular to the base is the *altitude*. If the endpoint opposite the vertex is the center of the base, the figure is a *right cone*.

conjecture (p. 94) (1) A conclusion reached as part of the process of inductive reasoning.
(p. 366) (2) As a problem solving skill, a guess that seems reasonable based on knowledge available.

constant (p. 438) A monomial that does not contain a variable.

construction (p. 370) A drawing of a geometric figure made by using a compass and a straightedge.

convenience sampling (p. 4) The population is chosen only because it is readily available.

coordinate for a point (p. 506) A number associated with a point on a number line or an ordered pair of numbers associated with a point on a grid.

coordinate plane (p. 506) Two perpendicular number lines forming a grid. The horizontal number line is called the *x*-axis. The vertical number line is called the *y*-axis.

coplanar points (p. 342) Points lying on the same plane.

corresponding angles (p. 351) Angles that are in the same position relative to the transversal and the lines.

counterexample (p. 98) A counterexample shows that a conditional statement is false because it satisfies the hypothesis but not the conclusion.

cross-products (p. 282) For $\frac{a}{b}$ and $\frac{c}{d}$, the cross products are ad and bc. In a proportion cross products are equal.

cube (p. 362) A right prism with six faces that have the same size and shape.

cubed (p. 60) A number multiplied by itself two times. For example, 10^3 is read as "10 cubed."

customary units (p. 154) Units used in the customary system of measurement. Frequently used customary units include the following: length—inch (in.), feet (ft), yard (yd), and mile (mi); capacity—cup (c), pint (pt), quart (qt), and gallon (gal); weight—ounce (oz), pound (lb), and ton (T); and temperature—degrees Fahrenheit.

cylinder (p. 363) A three-dimensional figure having two parallel and congruent circular bases.

Figure 3

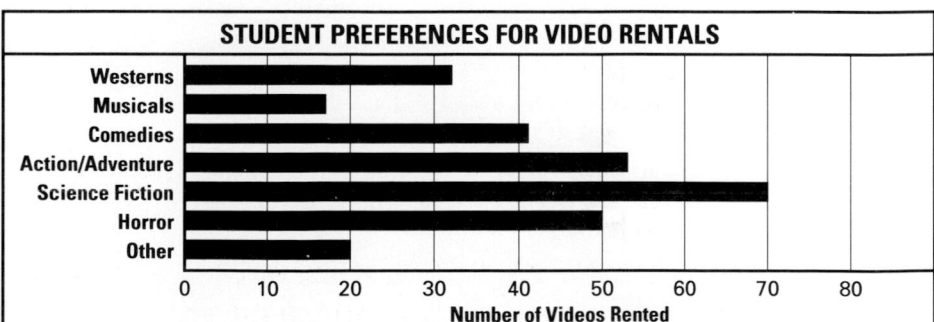

STUDENT PREFERENCES FOR VIDEO RENTALS

►D

data (p. 2) Statistical information most often gathered in numerical form. The word data is the plural form of the Latin word *datum*.

deductive reasoning (p. 99) A process of reasoning in which a conclusion is drawn from a conditional statement and additional information.

degree (p. 189, p. 346) A common unit of measurement for angles or temperature.

degree of a polynomial (p. 446) The greatest degree of any monomial within the polynomial.

denominator (p. 122) In the fraction $\frac{2}{3}$, the number 3 is the denominator. The denominator tells the total number of equal parts into which a whole has been divided.

dependent events (p. 487) Two events such that the outcome of the first affects the outcome of the second. For example, drawing a first card and then a second card without replacing the first.

diagonal (p. 358) A line segment joining two non-adjacent vertices of a polygon.

diameter (p. 172) A line segment passing through the center of a circle and having endpoints on the circle.

discount (p. 324) The difference between the regular selling price of an item and its sale price.

distributive property (p. 76) If one factor in a product is a sum, multiplying each addend by the other factor before adding does not change the product.

►E

endpoint (p. 342) A point at the end of a segment or ray.

equation (p. 222) A mathematical statement that two numbers or expressions are equal.

equiangular (p. 355) Having angles of the same measure.

equilateral triangle (p. 355) A triangle with all three sides having the same length and all angles the same measure.

equivalent fractions (p. 122) Fractions that represent the same amount.

equivalent ratios (p. 278) Two ratios that represent the same comparison. For example, 2:5, 4:10, 6:15 are equivalent ratios.

estimate (p. 42) An approximation of an answer or measurement.

evaluate (p. 46) To find the value of an algebraic expression.

event (p. 474) A set of one or more outcomes.

experimental probability (p. 490) The probability of an event based on the results of an experiment.

exponent (p. 60) A number showing how many times the base is used as a factor. For example, in 2^3 the exponent is 3.

exponential form (p. 60) A number written with a base and an exponent. For example, $2 \times 2 \times 2 \times 2 = 2^4$.

►F

factor (p. 118) Any number multiplied by another number to produce a product.

factoring (p. 118, p. 456) Finding two or more factors of a number or a polynomial.

Fahrenheit (p. 189) See *customary units*.

formula (p. 170) A specific equation giving a rule for relationships between quantities.

frequency table (p. 7) In a frequency table a tally mark is used to record each response. The total number of tally marks for a given response is the frequency of that response.

front-end estimation (p. 53) Estimate with the front-end digits only and then adjust the estimate by approximating the values of the other digits.

function (p. 513) A relation in which each member of the domain is paired with one, and only one, number of the range.

►G

gaps (p. 9) See *stem-and-leaf plot*.

Figure 4 *(See bottom of page.)*

Figure 5

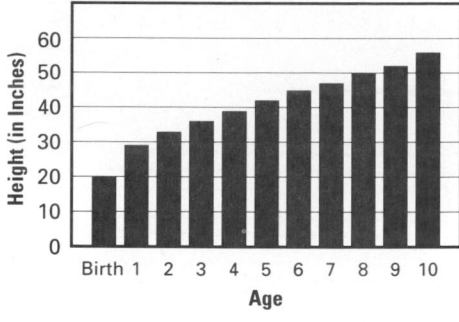

SARAH'S GROWTH GRAPH

Figure 6

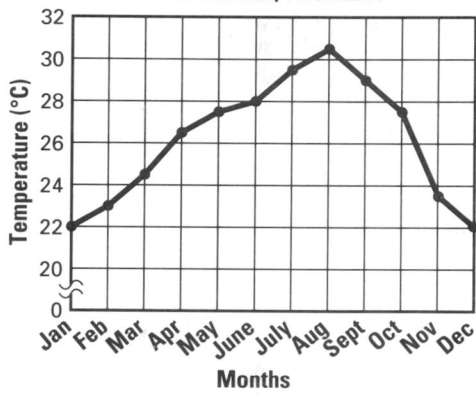

AVERAGE MONTHLY TEMPERATURE IN MIAMI, FLORIDA

Figure 7

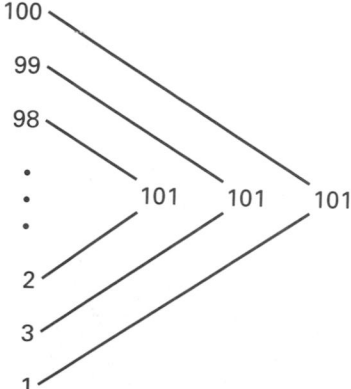

Figure 8 *(See bottom of p. 590.)*

Figure 4

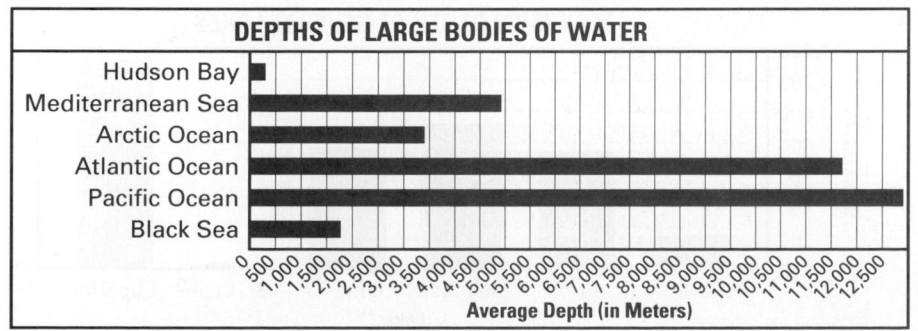

DEPTHS OF LARGE BODIES OF WATER

graph of the equation (p. 512) The set of all points whose coordinates are solutions of an equation.

graph of the function (p. 513) A graph of an equation that represents the function.

greatest common factor (GCF) (p. 119) The greatest integer that is a factor of two or more numbers.

►**H**

hypotenuse (p. 404) The side opposite the right angle in a right triangle.

►**I**

identity property of addition (p. 73) The sum of any number and 1 is that number.

identity property of multiplication (p. 73) The product of any number and 1 is that number.

independent events (p. 486) Events such that the outcome of the first does not affect the outcome of the second. For example, tossing a coin and rolling a number cube.

inductive reasoning (p. 94) Logical reasoning where a conclusion is made based on a set of specific examples. The conclusion is called a *conjecture*.

inequality (p. 254) A mathematical sentence involving one of the symbols $<$, $>$, $\leq$, or $\geq$.

integers (p. 186) The set of whole numbers and their opposites.

interest (p. 328) Amount that is paid for the use of money over a period of time.

intersecting lines (p. 350) Lines that have exactly one point in common.

inverse operations (p. 202, p. 226) Operations that undo each other such as addition and subtraction or multiplication and division.

irrational number (p. 255) Numbers such as π that are non-terminating non-repeating decimals.

isosceles triangle (p. 355) A triangle having at least two sides the same length and at least two angles the same measure.

►**L**

laws of exponents (p. 64) Rules that govern how to perform operations with numbers in exponential form. Examples are: To multiply numbers with the same base, add the exponents. To divide numbers with the same base, subtract the exponents.

least common denominator (LCD) (p. 123) The least common multiple of the denominators of fractions.

least common multiple (LCM) (p. 119) The smallest number that is a multiple of two or more numbers.

legs of a triangle (p. 404) In a right triangle, the two sides that are not the hypotenuse.

like fractions (p. 138) Fractions having the same denominator. For example, $\frac{5}{8}$ and $\frac{1}{8}$.

line (p. 342) A set of points that extends without end in two opposite directions. Two points determine a line.

linear equation (p. 513) An equation that has two variables and many solutions, all of which lie in a straight line when they are graphed on a coordinate grid.

line graph (p. 26) A means of displaying changes in statistical information using points and line segments to represent the data.

line of symmetry (p. 534) A line on which a figure can be folded, so that when one part is reflected over that line it matches the other part exactly.

line plot (p. 7) In a line plot, an X is made to record each response. The X's may be stacked one on top of another until all the data are recorded.

line segment (p. 342) The set of points containing two endpoints and all points between them.

line symmetry (p. 534) A figure has line symmetry if a line drawn through it divides the figure into two matching parts.

lowest terms (p. 123) A fraction in which the numerator and denominator do not have a common factor other than one.

590 Glossary

Figure 8

MAXIMUM DEPTH OF SEVERAL LAKES

Lake

►M

mean (p. 30) The sum of a set of numbers divided by the number of data in that set. Also known as *average.*

measures of central tendency (p. 30) Statistics used to analyze data. These measures are the *mean,* the *median,* and the *mode.*

median (p. 30) In a list of data ordered from least to greatest or greatest to least, the middle number or the *mean* of the middle two numbers.

metric units (p. 158, p. 189) Units used in the metric system of measurement. Frequently used metric units include the following: length—meter (m), millimeter (mm), centimeter (cm), and kilometer (km); liquid capacity—liter (L) and milliliter (mL); mass—kilogram (kg) and gram (g); and temperature—degrees Celsius.

midpoint (p. 374) The midpoint of a line segment is the point that separates it into two line segments of equal length.

mode (p. 30) In a list of data, the number or item occurring most frequently.

monomial (p. 438) An expression that is either a single number, a variable, or the product of a number and one or more variables.

multiple (p. 119) The product of any number and another whole number.

►N

negative integers (p. 186) Integers that are less than zero.

numerator (p. 122) In the fraction $\frac{3}{7}$, the number 3 is the numerator. The numerator tells how many of the equal parts of a whole are being considered.

►O

obtuse angle (p. 347) Any angle measuring between 90° and 180°.

obtuse triangle (p. 354) A triangle having one obtuse angle.

open sentence (p. 222) A sentence that contains one or more variables. It can be true or false, depending upon what values are substituted for the variables. A value of the variable that makes an equation true is called a *solution* of the equation.

opposites (p. 186) Two integers having a sum of zero. For example –27 and 27.

ordered pair (p. 506) Two numbers named in a specific order.

order of operations (p. 68) Rules followed to simplify expressions.

origin (p. 506) The point of intersection of the *x*-axis and *y*-axis in a coordinate plane.

outcomes (p. 474) The possible result of an experiment. For example, tossing a coin.

outliers (p. 9) See *stem-and-leaf plot.*

►P

parallel lines (p. 350) Lines lying on the same plane that do not intersect.

parallelogram (p. 359) A quadrilateral having both pairs of opposite sides parallel.

percent (p. 302) *Per one hundred.* A ratio that compares a number to 100.

perfect square (p. 400) The exact square of any other number or polynomial. For example, 4 is a perfect square.

perimeter (p. 166) The distance around a figure.

permutations (p. 484) An ordered arrangement of a set of objects.

perpendicular bisector (p. 374) The perpendicular bisector of a line segment is the line, ray, or line segment that is perpendicular to a line segment at its midpoint.

perpendicular lines (p. 350) Intersecting lines that form right angles.

pi (π) (p. 172) The ratio of the circumference of a circle to its diameter.

pictograph (p. 12) A graph that uses pictures or symbols to represent data. The *key* identifies the number of data items represented by each symbol.

plane (p. 342) A flat surface that extends without end in all directions.

point (p. 342) A specific location in space.

polygon (p. 343) A closed plane figure that is formed by joining three or more line segments at their endpoints. Each line segment joins exactly two others and is called a *side of the polygon.* Each point where two sides meet is a *vertex.*

polyhedron (p. 362) Plural: polyhedra. A three-dimensional figure in which each surface is a polygon. The surfaces of a polyhedron are called its *faces.* Two faces meet, or intersect, at an *edge.* Three or more edges intersect at a *vertex.*

polynomial (p. 438) The sum of two or more monomials. Each monomial is called a *term* of the polynomial. A polynomial is in *standard form* when its terms are in order from greatest to least powers of one of the variables.

positive integers (p. 186) Integers that are greater than zero.

power (p. 60) A number that can be expressed using an exponent. Read 2^4 as "2 to the fourth power."

power rule (p. 65) To raise an exponential number to a power, multiply the exponents. For example, $(3^2)^4 = 3^{2 \times 4} = 3^8$.

prime factorization (p. 118) A composite number expressed as a product of prime numbers.

prime number (p. 118) A whole number greater than one having exactly two distinct factors, one and the number itself.

principal (p. 328) The amount of money on which interest is paid or the amount of money borrowed at interest.

prism (p. 362) A polyhedron with two *bases* that are parallel and are the same size and shape.

probability (p. 474) The ratio of the number of favorable outcomes to the total number of possible outcomes.

property of zero for multiplication (p. 73) The product of any number and zero is zero.

proportion (p. 282) Two equivalent ratios set up to make a mathematical sentence.

protractor (p. 346) An instrument used to measure angles.

Pythagorean Theorem (p. 404) In a right triangle the sum of the squares of the length of legs is equal to the square of the length of the hypotenuse. $a^2 + b^2 = c^2$

pyramid (p. 363) A three-dimensional figure having three or more triangular faces and a polygonal base.

▶**Q**

quadrant (p. 506) One of the four regions formed by the axes of the coordinate plane.

quadrilateral (p. 359) A polygon having four sides.

▶**R**

radius (p. 172) A line segment having one endpoint at the center of the circle and the other endpoint on the circle.

random sampling (p. 4) Each member of the population is given an equal chance of being selected.

range (p. 30) The difference between the greatest and least number in a set of numerical data.

rate (p. 274) A ratio that compares two different kinds of quantities.

ratio (p. 270) A comparison of two numbers, represented in one of the following ways: 2 to 5, 2 out of 5, 2:5, or $\frac{2}{5}$.

rational number (p. 208) Any number that can be expressed in the form $\frac{a}{b}$, where a is any integer and b is any integer except 0.

ray (p. 342) A part of a line having one endpoint and extending without end in one direction.

real numbers (p. 255) The set of irrational and rational numbers together.

reciprocals (p. 134) Two numbers are reciprocals when their product is 1.

rectangle (p. 359) A parallelogram having four right angles.

reflection (p. 526) A transformation in which a figure is flipped or reflected over a line of reflection.

regular polygon (p. 359) A polygon all of whose sides are equal and all of whose angles are equal.

repeating decimal (p. 127) A decimal in which a digit or group of digits repeats. A bar above the group of repeating decimals is used to express the repeating decimal. For example, $\frac{2}{3} = 0.\overline{6}$.

rhombus (p. 359) A parallelogram with all sides the same length.

right angle (p. 347) An angle having a measure of 90°.

right triangle (p. 354) A triangle having a right angle.

rotation (p. 530) A transformation in which a figure is *turned* or *rotated*.

rotational symmetry (p. 535) A figure has rotational symmetry if when the figure is turned about a point the figure fits exactly over its original position at least once during a complete rotation.

rounding (p. 42) Expressing a number to the nearest ten, hundred, thousand and so on.

►**S**

sale price (p. 324) The purchase price less the discount.

sample space (p. 478) The set of all possible outcomes of an event.

sampling (p. 4) See *cluster sampling, convenience sampling, random sampling,* and *systematic sampling.*

scale drawing (p. 288) A drawing that represents a real object. All lengths in the drawing are proportional to actual lengths in the object. The ratio of the size of the drawing to the size of the actual object is called the *scale* of the drawing.

scalene triangle (p. 355) A triangle with no sides the same length and no angles the same measure.

scientific notation (p. 61) A notation for writing a number as the product of a number between 1 and 10 and a power of 10.

simplify (p. 250) To simplify an expression is to perform as many of the indicated operations as possible.

slide (p. 522) See *translation.*

slope (p. 516) The ratio of the change in y to the change in x, or the ratio of the change in the *rise* (vertical change) to the *run* (horizontal change).

slope-intercept form (p. 517) A linear equation in the form of $y = mx + b$ where m is the slope of the graph of the equation and b is the y-intercept.

solution (p. 222) A replacement set for a variable that makes a mathematical sentence true.

sphere (p. 363) A three-dimensional figure consisting of points that are the same distance from the center.

square (geometric) (p. 359) A parallelogram with four right angles and all sides the same length.

square (numeric) (p. 60, p. 400) The product of a number and itself.

squared (p. 60) A number multiplied by itself. For example, 10^2 is read as "10 squared."

square root (p. 400) One of two equal factors of a number.

statistics (p. 2) The field of mathematics involving the collection, analysis, and presentation of data.

stem-and-leaf plot (p. 8) A means of organizing data in which certain digits are used as *stems,* and the remaining digits are used as *leaves. Outliers* are data values that are much greater than or much less than most of the other values. *Clusters* are isolated groups of values. *Gaps* are large

spaces between values. A *back-to-back stem-and-leaf plot* displays two sets of data simultaneously.

straight angle (p. 347) An angle having a measure of 180°.

supplementary angles (p. 347) Two angles whose sum measures 180°.

surface area (p. 424) The sum of the areas of all the faces of a three-dimensional figure.

survey (p. 4) A means of collecting data for the analysis of some aspect of a group or area.

systematic sampling (p. 4) After a population has been ordered in some way, the members of the population are chosen according to a pattern.

►**T**

terminating decimal (p. 126) A decimal in which the only repeating digit is 0.

terms (p. 250) The parts of an expression separated by addition or subtraction signs. In the expression $2a + 3b + 4a - 5b^2$, the terms $2a$ and $4a$ are *like terms* because they have identical variable parts. The terms $3b$, $4a$, and $5b^2$ are *unlike terms* because they have different variable parts.

time (p. 328) Refers to the period of time during which the principal remains in a bank account. It also refers to the period of time that borrowed money has not been paid back.

transformation (p. 522) A way of moving a geometric figure without changing its size or shape. Prior to the move the figure is called a *preimage*. After the move, the figure is called an *image*.

translation (p. 522) A change in the position of a figure such that all the points in the figure slide exactly the same distance and in the same direction at once.

transversal (p. 351) A line that intersects two or more lines.

trapezoid (p. 359) A quadrilateral having one and only one pair of parallel sides.

tree diagram (p. 478) A diagram that shows all the possible outcomes of an event.

triangle (p. 343) A polygon having three sides.

trinomial (p. 438) A polynomial with three terms.

►**U**

unit price (p. 277) A ratio comparing the price of an item to the unit of its measure.

unit rate (p. 274) A rate that has a denominator of 1 unit.

►**V**

variable expressions (p. 46) Expressions that contain at least one variable.

variables (p. 46) Placeholders in mathematical expressions or sentences.

Venn diagram (p. 102) A means of showing relationships between two or three different classes of things. In a Venn diagram, each class is represented by a circular region inside a rectangle. The rectangle is called the *universal set*.

vertex (p. 342) The common endpoint of two rays that form an angle. The rays are called the *sides* of the angle. Plural form: *vertices*.

vertical angles (p. 350) Angles of the same measure formed by two intersecting lines.

volume (p. 387) The number of cubic units needed to fill a space.

►**X**

***x*-axis** (p. 506) Horizontal number line.

***x*-coordinate** (p. 506) The first number in an ordered pair.

►**Y**

***y*-axis** (p. 506) Vertical number line.

***y*-coordinate** (p. 506) The second number in an ordered pair.

***y*-intercept** (p. 517) The *y*-intercept of a line is the *y*-coordinate of the point where it intersects the *y*-axis.

Index